STUDY GUIDE TO ACCOMPANY
ORGANIC CHEMISTRY

STUDY GUIDE TO ACCOMPANY
ORGANIC CHEMISTRY

Second Edition

FRANCIS A. CAREY
Department of Chemistry
University of Virginia

ROBERT C. ATKINS
Department of Chemistry
James Madison University

McGRAW-HILL, INC.

New York St. Louis San Francisco Auckland Bogotá Caracas
Lisbon London Madrid Mexico Milan Montreal New Delhi
Paris San Juan Singapore Sydney Tokyo Toronto

STUDY GUIDE TO ACCOMPANY ORGANIC CHEMISTRY

4 5 6 7 8 9 0 SEM SEM 9 0 9 8 7 6 5 4 3

ISBN 0-07-009935-9

This book was set in Times Roman.
The editors were Kirk Emry and David A. Damstra;
the production supervisor was Annette Mayeski.
The cover was designed by Rafael Hernandez.
Semline, Inc., was printer and binder.

This book is printed on recycled paper containing a minimum of 50% total recycled fiber with
10% postconsumer de-inked fiber.

CONTENTS

PREFACE

It is our hope that in writing this Study Guide we will be able to assist the student in making the study of organic chemistry a more meaningful and worthwhile experience. To achieve maximum effectiveness, a study guide should be more than just an answer book. What we have written was designed with a larger goal in mind.

The Study Guide contains detailed solutions to all the problems in the text. Learning how to solve a problem is, in our view, more important than merely knowing the correct answer. To that end we have included solutions sufficiently detailed to provide the student with the steps leading to the solution of each problem.

In addition, each chapter includes summaries headed "Important Terms and Concepts" and/or "Important Reactions." These are intended to provide brief overviews of the major points and topics presented in each chapter of the text. At the conclusion of each chapter is a "Self-Test," designed to test the student's mastery of the material.

The completion of this guide was made possible with the help of numerous people who contributed their time and talent. Special thanks go to the reviewers of the first edition: Professors Marye Anne Fox, Raymond C. Fort, William H. Bunnelle, and Roy G. Garvey. Our thanks and appreciation also go to the many users of the first edition who have provided us with helpful suggestions, comments, and corrections. We also wish to acknowledge the assistance and understanding of Randi Rossignol, Kirk Emry, and David Damstra of McGraw-Hill. Last, we thank our wives and families for their understanding of the long hours invested in this work.

Francis A. Carey
Robert C. Atkins

TO THE STUDENT

Before beginning the study of organic chemistry, a few words about "how to do it" are in order. You've probably heard that organic chemistry is difficult; there's no denying that. It need not be overwhelming, though, when approached with the right frame of mind and with sustained effort.

First of all you should realize that organic chemistry tends to "build" on itself. That is, once you have learned a reaction or concept, you will find it being used again and again later on. In this way it is quite different from general chemistry, which tends to be much more compartmentalized. In organic chemistry you will continually find previously learned material cropping up and being used to explain and to help you understand new topics. Often, for example, you will see the preparation of one class of compounds using reactions of other classes of compounds studied earlier in the year.

How to keep track of everything? It might be possible to memorize every bit of information presented to you, but you would still lack a fundamental understanding of the subject. It is far better to *generalize* as much as possible.

You will find that the early chapters of the text will emphasize concepts of *reaction theory*. These will be used, as the various classes of organic molecules are presented, to describe *mechanisms* of organic reactions. A relatively few fundamental mechanisms suffice to describe almost every reaction you will encounter. Once learned and understood, these mechanisms provide a valuable means of categorizing the reactions of organic molecules.

There will be numerous facts to learn in the course of the year, however. For example, chemical reagents necessary to carry out specific reactions must be learned. You might find a study aid known as *flash cards* to be helpful. These take many forms, but one idea is to use 3×5 cards. As an example of how the cards might be used, consider the reduction of alkenes (compounds with carbon-carbon double bonds) to alkanes (compounds containing only carbon-carbon single bonds). The front of the card might look like this:

$$\boxed{\text{Alkenes} \xrightarrow{\;?\;} \text{alkanes}}$$

On the reverse of the card would be found the reagents necessary for this reaction:

$$\boxed{\text{H}_2, \text{ Pt or Pd catalyst}}$$

The card can actually be studied in two ways. You may ask yourself: What reagents will convert alkenes into alkanes? Or, using the back of the card: What chemical reaction is carried out with hydrogen and a platinum or palladium catalyst? This is by no means the only way to use the cards—be creative! Just making up the cards will help you to study.

While study aids such as flash cards will prove helpful, there is only one way to truly master the subject matter in organic chemistry—*work out problems*! The more you work, the more you will learn. Almost certainly the grade you will receive will be a reflection of your ability to solve problems. Don't just think over the problems, either; write them out as if you were handing them in to be graded. Also, be careful of how you use the Study Guide. The solutions contained in this book have been intended to provide explanations to help you understand the problem. Be sure to write out *your* solution to the problem first, and only then look it up to see if what you have done is correct.

One concern frequently expressed by students is that they feel they understand the material but don't do as well as expected on tests. One way to overcome this is to "test" yourself. Each chapter in the Study Guide has a self-test at the end. Work the problems in these tests *without* looking up how to solve them in the text. You'll find it is much harder this way, but it is also a closer approximation to what will be expected of you while taking a test in class.

Success in organic chemistry depends on skills in analytical reasoning. Many of the problems you will be asked to solve require you to proceed through a series of logical steps to the correct answer. Most of the concepts of organic chemistry are fairly simple; stringing them together in a coherent fashion is where the challenge lies. By doing exercises conscientiously you should see a significant increase in your overall reasoning ability. Enhancement of their analytical powers is just one fringe benefit enjoyed by those students who attack the course rather than simply attend it.

Gaining a mastery of organic chemistry is hard work. It is hoped that the hints and suggestions outlined here will be helpful to you and that you will find your efforts rewarded with a knowledge and understanding of an important area of science.

STUDY GUIDE TO ACCOMPANY
ORGANIC CHEMISTRY

CHEMICAL BONDING

IMPORTANT TERMS AND CONCEPTS

Atoms, Electrons, and Orbitals (Sec. 1.1) The electronic organization (configuration) of all atoms may be specified by describing the *occupied orbitals* on the atom. An *orbital* is the region of space in which there is a high probability of finding an electron.

By using *quantum numbers,* each orbital may be completely described. The first shell contains only the $1s$ orbital; the second shell contains the $2s$ orbital and three $2p$ orbitals. The s orbitals are spherical; the p orbitals are orthogonal to each other, and are described as being "dumbbell-shaped." They are designated $2p_x$, $2p_y$, and $2p_z$, respectively.

According to the *Pauli exclusion principle,* no two electrons may have the same set of quantum numbers. Since the first three quantum numbers describe the orbital and the fourth quantum number, known as *spin,* may take only one of two possible values ($+\frac{1}{2}$ or $-\frac{1}{2}$), a maximum of two electrons may exist in the same orbital.

Hund's rule states that when electrons enter orbitals of equal energy (for example, the three $2p$ orbitals), they do so with parallel spin in separate orbitals.

The *valence electrons* of an atom are the outermost electrons, and are the ones most likely to be involved in chemical reactions. The maximum number of electrons in the *valence shell* of any second-row element is 8. Elements that possess eight electrons in their valence shell are said to have a complete *octet* of electrons.

Ionic Bonds (Sec. 1.2) The gain or loss of one or more electrons by a neutral atom results in formation of *ions*. Ions that are positively charged resulting from the loss of one or more electrons are called *cations*. The energy required to completely remove an electron from the highest-energy occupied orbital of an atom is called the *ionization energy*. Ionization is an *endothermic* process, and the ionization energy describes the enthalpy change for the reaction

$$M(g) \longrightarrow M^+(g) + e^-$$

The general trend is for ionization energy to increase going from left to right across the periodic table.

The negative ions formed when an atom gains one or more electrons are called *anions*. The *exothermic* process, that is, the energy liberated when an electron is added to a

1

neutral atom is called the *electron affinity*. Electron affinity is the energy change for the process

$$X(g) + e^- \longrightarrow X^-(g)$$

Ionic bonds form between a metal and a nonmetal, that is, a species of low ionization energy and one of high electron affinity. The transfer of one or more electrons from the metal to the nonmetal gives rise to ions of opposite charge. An *ionic*, or *electrostatic*, *bond* results from the coulombic attraction of the oppositely charged ions for each other.

Covalent Bonds (Sec. 1.3) When nonmetals bond together, the forces which hold the atoms together result from *shared electron pair bonds*, or *covalent bonds*. A discrete unit or collection of atoms held together by covalent bonds is known as a *molecule*.

The most popular method for representing the structure of a molecule is with a *Lewis structure*, or "electron dot" formula. Each valence electron in an atom is represented by a dot, and the dots are paired to form covalent bonds in the molecule. Not all the valence electrons on every element are *shared*, that is, part of a covalent bond. They may also exist as *unshared pairs* in an atom of the molecule. Note in the following examples that each pair of dots may be replaced by a line, representing a *single bond* between two atoms.

For elements in the second row of the periodic table, there must be eight electrons in the outer shell of the atom (dots in a Lewis formula); that is, the *octet* of the atom must be satisfied.

Multiple Bonding (Sec. 1.4) More than one pair of electrons may be shared between two atoms in a molecule. A *double bond* results when two electron pairs (four electrons) are shared; a *triple bond* results from sharing three electron pairs (six electrons). Remember that the octet of a second-row element must not be exceeded when writing Lewis structures. This can become especially important when writing formulas which involve multiple bonds. Some examples include:

Polar Covalent Bonds and Electronegativity (Sec. 1.5) *Electronegativity* is a measure of the tendency of an atom to draw electron density away from another atom to which it is bonded and toward itself. When two atoms of different electronegativity are bonded together, the more electronegative atom will attract electrons, while the less electronegative (or more *electropositive*) atom will donate electrons. The resulting unequal distribution of electron charge, or *dipole*, gives rise to a covalent bond that is *polarized*; such a bond is referred to as a *polar covalent bond*.

Formal Charge (Sec. 1.6) The charge value assigned to an atom in a covalent molecule is the *formal charge* present on that atom. It may be computed as follows:

$$\text{Formal charge} = (\text{group number}) - (\text{electron count})$$

where electron count is the sum of half the number of shared electrons plus the number of unshared electrons.

Certain commonly encountered examples are worth noting:

1. Tetravalent N is positive, e.g., $\overset{+}{N}H_4$, $\overset{+}{N}(CH_3)_4$

2. Trivalent O is positive, e.g., $H—\overset{\overset{+}{\cdot\cdot}}{\underset{\underset{H}{|}}{O}}—H$, $CH_3—\overset{\overset{+}{\cdot\cdot}}{\underset{\underset{CH_3}{|}}{O}}—CH_3$

3. Univalent O is negative, e.g., $H—\overset{\cdot\cdot}{\underset{\cdot\cdot}{O}}:^-$, $CH_3—\overset{\cdot\cdot}{\underset{\cdot\cdot}{O}}:^-$

Note that in all these cases the octet of each atom other than H has been satisfied.

Structural Formulas (Sec. 1.7) The Lewis notation is the most widely used method for representing the structural formulas of organic molecules. *Condensed structural formulas* may be used. For example:

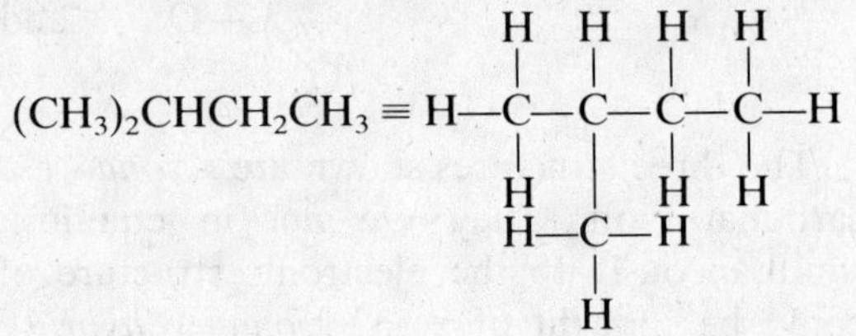

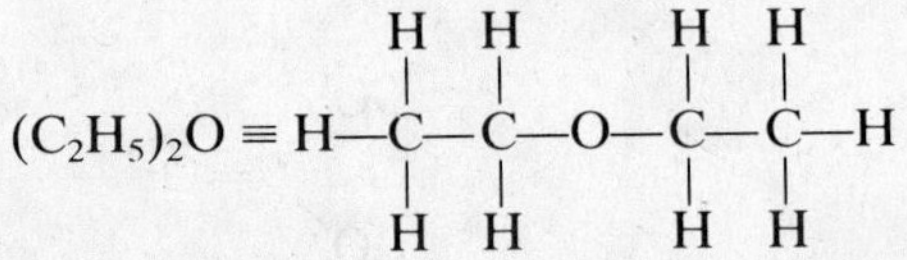

$$(CH_3)_2CHCH_2CH_3 \equiv H-\overset{\overset{\displaystyle H}{|}}{\underset{\underset{\displaystyle H-\overset{\overset{\displaystyle H}{|}}{\underset{\underset{\displaystyle H}{|}}{C}}-H}{|}}{C}}-\overset{\overset{\displaystyle H}{|}}{\underset{\underset{\displaystyle }{|}}{C}}-\overset{\overset{\displaystyle H}{|}}{\underset{\underset{\displaystyle H}{|}}{C}}-\overset{\overset{\displaystyle H}{|}}{\underset{\underset{\displaystyle H}{|}}{C}}-H$$

$$(C_2H_5)_2O \equiv H-\overset{\overset{\displaystyle H}{|}}{\underset{\underset{\displaystyle H}{|}}{C}}-\overset{\overset{\displaystyle H}{|}}{\underset{\underset{\displaystyle H}{|}}{C}}-O-\overset{\overset{\displaystyle H}{|}}{\underset{\underset{\displaystyle H}{|}}{C}}-\overset{\overset{\displaystyle H}{|}}{\underset{\underset{\displaystyle H}{|}}{C}}-H$$

Bond-line, or "stick," formulas are often used to represent complex structures in which the carbons and hydrogens are omitted:

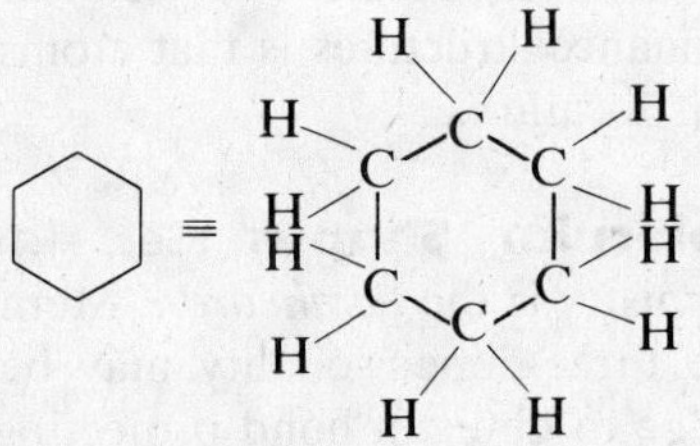

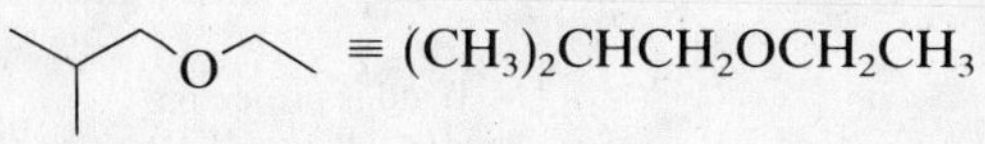

Some simple rules are worth remembering:

Hydrogen:	Forms only one bond
Carbon:	Always has four bonds in a stable molecule
Oxygen:	Has two bonds in a neutral molecule and two unshared electron pairs
Nitrogen:	Has three bonds in a neutral molecule and one unshared electron pair

Constitutional Isomers (Sec. 1.8) Two or more substances which have the same molecular formula but differ from each other in some way are known as *isomers* of each other. *Constitutional isomers,* also known as *structural isomers,* differ in their atomic connections. For example, the following structures each have the formula C_3H_8O, yet they represent different substances:

$$CH_3—CH_2—CH_2—O—H \qquad CH_3—O—CH_2—CH_3 \qquad CH_3—\underset{\underset{O—H}{|}}{CH}—CH_3$$

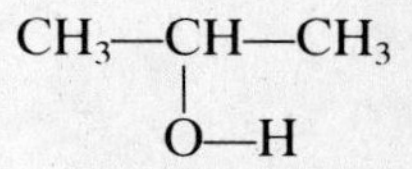

Resonance (Sec. 1.9) For certain chemical substances, the bonding picture depicted by a single Lewis structure does not fully describe the electronic structure of the substance. For example, the three C—O bonds in the carbonate ion, CO_3^{2-}, are equivalent. Yet the following Lewis structure suggests two oxygen atoms connected by single bonds and a third oxygen atom attached by a double bond:

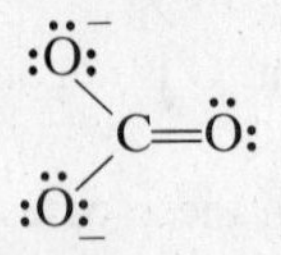

$$\overset{\displaystyle :\overset{..}{\underset{..}{O}}:^-}{\underset{\displaystyle :\overset{..}{\underset{..}{O}}:^-}{\diagdown}}C=\overset{..}{O}:$$

In order to completely describe the bonding of carbonate, two additional Lewis structures are necessary:

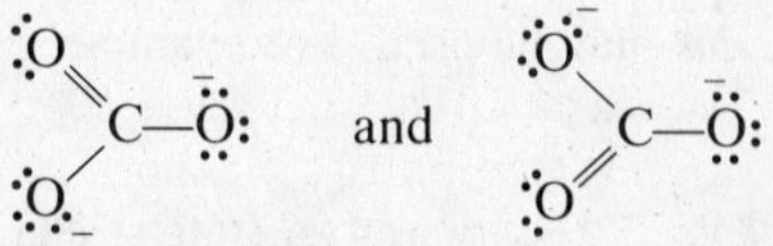

$$\overset{\displaystyle :\overset{..}{O}}{\underset{\displaystyle :\overset{..}{\underset{..}{O}}:^-}{\diagup\diagdown}}C-\overset{..}{\underset{..}{O}}:^- \qquad \text{and} \qquad \overset{\displaystyle :\overset{..}{O}^-}{\underset{\displaystyle :\overset{..}{O}}{\diagup\diagdown}}C-\overset{..}{\underset{..}{O}}:^-$$

The three structures shown are *resonance contributors* to the "true" bonding picture of carbonate ion. They are *not* in equilibrium with each other, but all contribute simultaneously to the electronic structure of the substance. The true bonding picture could be thought of as a *resonance hybrid* to which each of the individual *resonance structures* contributes. A double-headed arrow is used to indicate resonance, as follows:

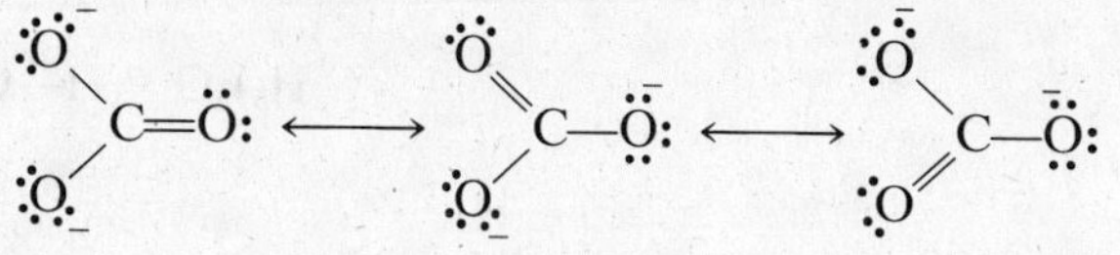

$$\overset{..}{O}^-\diagdown C=\overset{..}{O}: \longleftrightarrow \overset{..}{O}\diagdown C-\overset{..}{O}:^- \longleftrightarrow \overset{..}{O}^-\diagdown C-\overset{..}{O}:^-$$

The smearing of electron density as depicted by the preceding resonance structures is an example of *electron delocalization*. An important rule to remember in writing resonance structures is that atoms may not change position—only electrons may differ in their location.

Molecular Shapes (Sec. 1.10) The most important geometric shape in organic chemistry is the *tetrahedron*. Methane, for example, is described as a *tetrahedral molecule*. The three-dimensionality may be represented in a structural drawing by using a *solid wedge* to depict a bond projecting from the paper toward you, a *dashed wedge* to depict one receding from the paper, and a simple line to depict a bond in the plane of the paper.

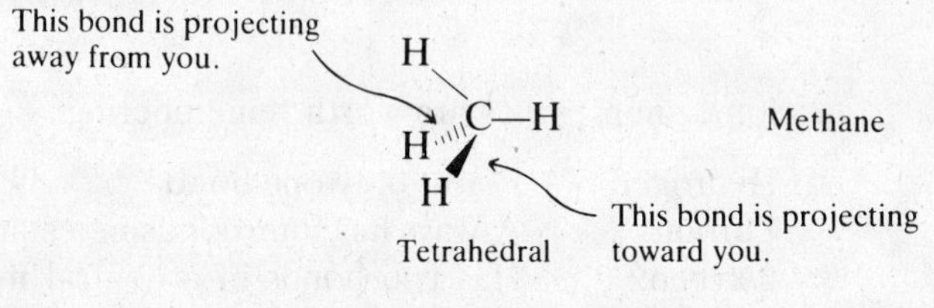

This bond is projecting
away from you.

H
|
H''''C—H   Methane
|
H
                  This bond is projecting
Tetrahedral   toward you.

The *valence-shell electron pair repulsion* (*VSEPR*) *theory* provides a method of explaining molecular shapes. The fundamental idea is that electron pairs, both bonded and nonbonded, will orient themselves as far away from each other as possible around the central atom in a molecule. The geometry of the electron pairs on an atom having two pairs is linear. The geometry of the electron pairs on an atom having three is trigonal planar, and four, tetrahedral. The electron pairs in multiple bonds are treated together.

The *shape* of a molecule is determined by the positions of its atoms. Thus ammonia is pyramidal and water bent, although the electron pairs in both molecules exist in a tetrahedral geometry.

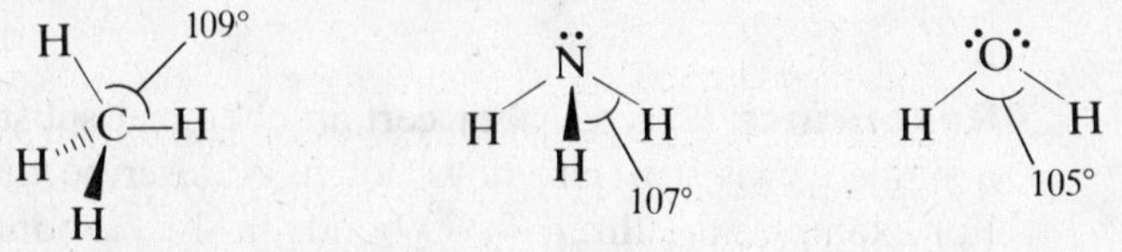

H   109°
\
C—H
H''''/
|
H

H   N̈   H
 \ | /
  |
  H
  107°

H   :Ö: H
 \ | /
 105°

In addition to the tetrahedral shape that was shown for methane, the two shapes most commonly encountered in organic chemistry are *linear* and *trigonal planar*:

$$O=C=O$$

$$\begin{array}{c} H \\ \diagdown \\ C=O \\ \diagup \\ H \end{array}$$

Linear   Trigonal planar

Molecular Dipole Moments (Sec. 1.11) Only for a diatomic molecule (such as HF) can it be stated with certainty that a polar bond will give rise to a polar molecule. *Molecular polarity* is shape-dependent; that is, the dipoles of the individual polar bonds may cancel the effect of each other, depending on the shape of the molecule. For example, CO_2, which is linear, is a nonpolar molecule, although each carbon-oxygen bond is polar. On the other hand, formaldehyde, H_2CO, is polar; there is no cancellation of the dipole of the CO bond.

$$O=C=O$$

$$\begin{array}{c} H \\ \diagdown \\ C=O \\ \diagup \\ H \end{array}$$

Carbon dioxide   Formaldehyde
(nonpolar)   (polar)

Molecular Orbitals (Sec. 1.12) In this chapter the concept of sharing electrons between atoms to form covalent bonds was presented. A more detailed description of these bonds is provided by viewing the orbitals on individual atoms as overlapping and interacting with each other. These regions of interaction are known as *molecular orbitals* (MOs).

Constructive interference, or *in-phase* overlap, of two individual atomic orbitals gives rise to what is called a *bonding molecular orbital*. An electron in a bonding MO is of lower energy than one in the atomic orbitals from which the MO is formed. The result is an *attractive force* between the two atoms.

The subtractive interaction between two atomic orbitals gives rise to an *antibonding molecular orbital*. An electron in an antibonding MO is of higher energy than one in the atomic orbitals from which the antibonding MO arises, resulting in a *repulsion* of the two atoms. When the number of electrons in bonding MOs exceeds the number in antibonding MOs, the net force between the two atoms is attractive, resulting in formation of a stable molecule.

The bonding MOs formed by the overlap of atomic orbitals have the region of highest electron density concentrated between the two nuclei. The electron distribution is symmetric around the internuclear axis. A molecular orbital having this characteristic is known as a σ (*sigma*) *orbital*. Two nuclei bonded by electrons in a σ orbital are said to be connected by a σ *bond*.

Bonding in Methane (Sec. 1.13) While the bonding of an organic compound such as methane may be viewed as a collection of molecular orbitals as described in the previous section, most organic chemists prefer to use a bonding description based on the Lewis model, in which covalent bonds are viewed as localized between pairs of atoms.

The ground-state electronic configuration of carbon ($1s^2$, $2s^2$, $2p^2$) is poorly suited for covalent bonding and for explaining the observed facts regarding carbon compounds. This configuration provides only two orbital vacancies, and these are in p orbitals situated perpendicular to each other. The four carbon-hydrogen bonds of methane are equivalent and are directed toward the corners of a tetrahedron (a *tetrahedral* arrangement).

These problems may be resolved by viewing the ground-state atomic orbitals as mixing together by a process known as *orbital hybridization* to form a new set of orbitals. Each of these new *hybrid orbitals* has characteristics of the ground-state atomic orbitals from which they are derived.

In the case of methane, mixing the $2s$ orbital of the carbon atom with the three $2p$ orbitals yields four new sp^3 *hybrid* orbitals. Each C—H σ bond is formed from the overlap of an sp^3 hybrid orbital with a hydrogen $1s$ orbital, as shown in text Figure 1.18. These orbitals are arranged in a tetrahedral fashion around the carbon atom.

Bonding in Ethane (Sec. 1.14) Bonding in ethane, C_2H_6, may be explained in a manner similar to that used for methane. The hybrid orbitals from each of two carbons may overlap to form an sp^3–sp^3 σ bond. The carbon-carbon single bond framework of larger organic molecules is constructed of orbital overlaps similar to that in ethane.

sp^2 Hybridization: Bonding in Ethylene. (Sec. 1.15) Ethylene is a representative alkene and has the structure

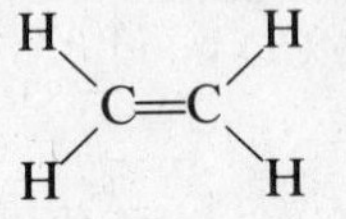

The bonding in ethylene is not explained using the sp^3 orbital hybridization model. As shown in text Figures 1.22 and 1.23, the $2s$ orbital and two $2p$ orbitals mix to yield three equivalent sp^2 hybrid orbitals. One of the $2p$ orbitals remains unhybridized.

The carbon-carbon double bond of an alkene such as ethylene consists of a σ (sigma) component and a π (pi) component. The carbons of the double bond are sp^2 hybridized. The three hybrid orbitals are utilized to form sigma bonds to three atoms, all of which lie in the same plane as the carbon atom. The remaining p orbitals of the double-bonded carbons are parallel to each other, giving rise to a π *orbital*. The result of this arrangement is that the four atoms attached to the double bond are in one plane. Ethylene is a planar molecule:

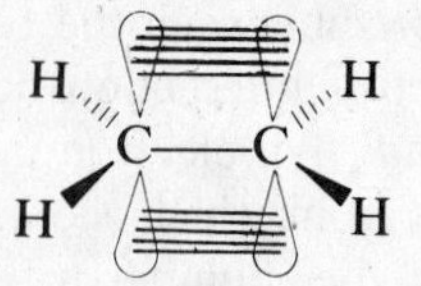

sp Hybridization: Bonding in Acetylene (Sec. 1.16) Acetylene is the simplest alkyne, and has a carbon-carbon triple bond.

The carbons of acetylene are sp hybridized. The $2s$ orbital and one $2p$ orbital of carbon mix to form two equivalent sp hybrid orbitals, as shown in text Figure 1.26. Two p orbitals remain unhybridized in the sp orbital hybridization model. A carbon-carbon triple bond has a σ component resulting from overlap of an sp hybrid orbital from each carbon, and two π components resulting from overlap of the unhybridized p orbitals on each carbon. The bonding of acetylene is shown in text Figure 1.27.

SOLUTIONS TO TEXT PROBLEMS

1.1 Carbon is the element of atomic number 6, and so has a total of six electrons. Two of these electrons are in the $1s$ level. The four electrons in the $2s$ and $2p$ levels (the valence shell) are the valence electrons. Carbon has four valence electrons.

1.2 Electron configurations of elements are derived by applying the following principles:

(a) The number of electrons in a neutral atom is equal to its atomic number Z.

(b) The maximum number of electrons in any orbital is 2.

(c) Electrons are added to orbitals in order of increasing energy, filling the $1s$ orbital before any electrons occupy the $2s$ level. The $2s$ orbital is filled before any of the $2p$ orbitals, and the $3s$ orbital is filled before any of the $3p$ orbitals.

(d) All the $2p$ orbitals are of equal energy, and each is singly occupied before any is doubly occupied. The same holds for the $3p$ orbitals.

With this as background, the electron configuration of the third-row elements is derived as follows:

$$\text{Na } (Z = 11) \qquad 1s^2 2s^2 2p^6 3s^1$$
$$\text{Mg } (Z = 12) \qquad 1s^2 2s^2 2p^6 3s^2$$
$$\text{Al } (Z = 13) \qquad 1s^2 2s^2 2p^6 3s^2 3p_x{}^1$$
$$\text{Si } (Z = 14) \qquad 1s^2 2s^2 2p^6 3s^2 3p_x{}^1 3p_y{}^1$$
$$\text{P } (Z = 15) \qquad 1s^2 2s^2 2p^6 3s^2 3p_x{}^1 3p_y{}^1 3p_z{}^1$$
$$\text{S } (Z = 16) \qquad 1s^2 2s^2 2p^6 3s^2 3p_x{}^2 3p_y{}^1 3p_z{}^1$$
$$\text{Cl } (Z = 17) \qquad 1s^2 2s^2 2p^6 3s^2 3p_x{}^2 2p_y{}^2 3p_z{}^1$$
$$\text{Ar } (Z = 18) \qquad 1s^2 2s^2 2p^6 3s^2 3p_x{}^2 3p_y{}^2 3p_z{}^2$$

1.3 The electron configurations of the designated ions are:

Ion	Z	Number of electrons in ion	Electron configuration of ion
(b) He^+	2	1	$1s^1$
(c) H^-	1	2	$1s^2$
(d) O^-	8	9	$1s^2 2s^2 2p_x{}^2 2p_y{}^2 2p_z{}^1$
(e) F^-	9	10	$1s^2 2s^2 2p^6$
(f) Ca^{2+}	20	18	$1s^2 2s^2 2p^6 3s^2 3p^6$

Those with a noble gas configuration are H^-, F^-, and Ca^{2+}.

1.4 A positively charged ion is formed when an electron is removed from a neutral atom. The equation representing the ionization of carbon and the electron configurations of the neutral atom and the ion are:

$$\underset{1s^2 2s^2 2p_x{}^1 2p_y{}^1}{C} \longrightarrow \underset{1s^2 2s^2 2p_x{}^1}{C^+} + e^-$$

A negatively charged carbon is formed when an electron is added to a carbon atom. The additional electron enters the $2p_z$ *orbital*.

$$\underset{1s^2 2s^2 2p_x{}^1 2p_y{}^1}{C} + e^- \longrightarrow \underset{1s^2 2s^2 2p_x{}^1 p_y{}^1 2p_z{}^1}{C^-}$$

Neither the cation nor the anion has a noble gas electron configuration.

1.5 Hydrogen has one valence electron and fluorine has seven. The covalent bond in hydrogen fluoride arises by sharing the single electron of hydrogen with the unpaired electron of fluorine.

Combine H· and ·F̈: to give the Lewis structure for hydrogen fluoride H:F̈:

1.6 In order to satisfy the Lewis rules, C_2H_6 must have a carbon-carbon bond.

Combine two $\cdot\ddot{C}\cdot$ and six H$\cdot$ to write the Lewis structure of ethane

$$\begin{array}{cc} \text{H} & \text{H} \\ \text{H:}\ddot{\text{C}}\text{:}\ddot{\text{C}}\text{:H} \\ \text{H} & \text{H} \end{array}$$

There are a total of 14 valence electrons distributed as shown. Each carbon is surrounded by eight electrons.

1.7 (*b*) Each carbon contributes four valence electrons and each fluorine contributes seven. Thus, C_2F_4 has 36 valence electrons to be accounted for. The octet rule is satisfied for carbon only if the two carbons are attached by a double bond and there are two fluorines on each carbon. The pattern of connections shown (at left) accounts for 12 electrons. The remaining 24 electrons are divided equally (six each) among the four fluorines. The complete Lewis structure is shown at right.

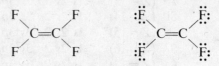

(*c*) Since the problem states that the atoms in C_3H_3N are connected in the order CCCN and all hydrogens are bonded to carbon, the order of attachments can only be as shown (below left) in order to have four bonds to each carbon. Three carbons contribute 12 valence electrons, three hydrogens contribute 3, and nitrogen contributes 5, for a total of 20. The nine bonds indicated in the partial structure account for 18 electrons. Since the octet rule is satisfied for carbon, add the remaining two electrons as an unshared pair on nitrogen (below right).

$$\begin{array}{ccc} \text{H} & & \text{H} \\ & \text{C=C} & \\ \text{H} & & \text{C}\equiv\text{N} \end{array} \qquad \begin{array}{ccc} \text{H} & & \text{H} \\ & \text{C=C} & \\ \text{H} & & \text{C}\equiv\text{N:} \end{array}$$

1.8 The direction of a bond dipole is governed by the electronegativity of the atoms involved. Among the halogens the order of electronegativity is $F > Cl > Br > I$. Therefore fluorine attracts electrons away from chlorine in FCl and chlorine attracts electrons away from iodine in ICl.

$$\overset{\longleftarrow\ +}{\text{F—Cl}} \qquad \overset{+\ \longrightarrow}{\text{I—Cl}}$$

$$\mu = 0.9\,\text{D} \qquad \mu = 0.7\,\text{D}$$

1.9 (*b*) Neither phosphorus nor bromine has a formal charge in PBr_3.

	Valence electrons in neutral atom	Electron count	Formal charge
Phosphorus:	5	$\frac{1}{2}(6) + 2 = 5$	0
Bromine:	7	$\frac{1}{2}(2) + 6 = 7$	0

$$\begin{array}{c} :\ddot{\text{Br}}: \\ | \\ :\ddot{\text{Br}}\text{—P—}\ddot{\text{Br}}: \\ \ddot{} \end{array}$$

(*c*) The formal charges in sulfuric acid are calculated as follows:

	Valence electrons in neutral atom	Electron count	Formal charge
Hydrogen:	1	$\frac{1}{2}(2) = 1$	0
Oxygen (of OH):	6	$\frac{1}{2}(4) + 4 = 6$	0
Oxygen:	6	$\frac{1}{2}(2) + 6 = 7$	-1
Sulfur:	6	$\frac{1}{2}(8) + 0 = 4$	$+2$

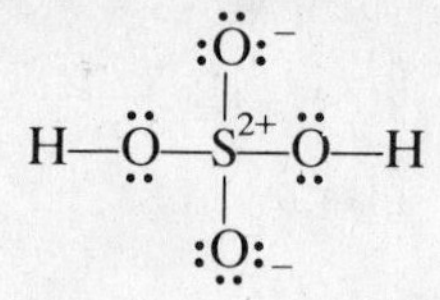

$$H-\overset{\displaystyle :\overset{..}{O}:^-}{\underset{\displaystyle :\overset{..}{O}:_-}{\overset{..}{\underset{..}{O}}-\overset{2+}{S}-\overset{..}{\underset{..}{O}}}-H$$

(*d*) The formal charges in nitrous acid are calculated as follows:

	Valence electrons in neutral atom	Electron count	Formal charge
Hydrogen:	1	$\frac{1}{2}(2) = 1$	0
Oxygen (of OH):	6	$\frac{1}{2}(4) + 4 = 6$	0
Oxygen:	6	$\frac{1}{2}(4) + 4 = 6$	0
Nitrogen:	5	$\frac{1}{2}(6) + 2 = 5$	0

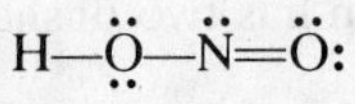

$$H-\overset{..}{\underset{..}{O}}-\overset{..}{N}=\overset{..}{O}:$$

1.10 The electron counts of nitrogen in ammonium ion and boron in borohydride ion are both 4 (one-half of 8 electrons in covalent bonds).

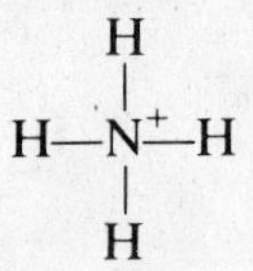

$$\begin{array}{cc} \overset{\displaystyle H}{\underset{\displaystyle H}{H-\overset{+}{N}-H}} & \overset{\displaystyle H}{\underset{\displaystyle H}{H-\overset{-}{B}-H}} \\ \\ \text{Ammonium ion} & \text{Borohydride ion} \end{array}$$

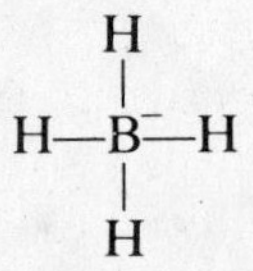

Since a neutral nitrogen has five electrons in its valence shell, an electron count of 4 gives it a formal charge of $+1$. A neutral boron has three valence electrons, and so an electron count of 4 in borohydride ion corresponds to a formal charge of -1.

1.11 (*b*) The compound $(CH_3)_3CH$ has a central carbon to which are attached three CH_3 groups and a hydrogen.

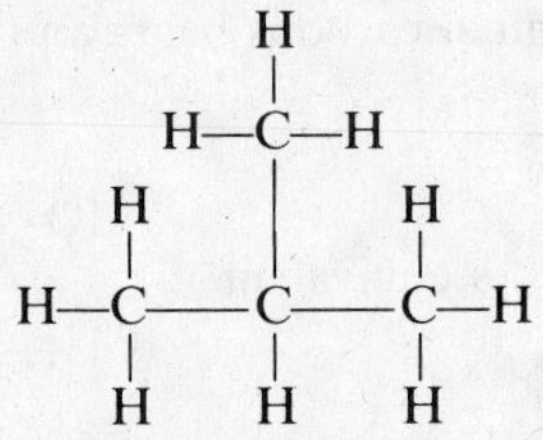

$$\begin{array}{c} H \\ | \\ H-C-H \\ \\ \overset{\displaystyle H}{\underset{\displaystyle H}{H-C}}-\overset{\displaystyle H}{\underset{\displaystyle H}{C}}-\overset{\displaystyle H}{\underset{\displaystyle H}{C-H}} \end{array}$$

Four carbons and 10 hydrogens contribute 26 valence electrons. The structure shown has 13 covalent bonds, and so all the valence electrons are accounted for. The molecule has no unshared electron pairs.

(*c*) The number of valence electrons in $ClCH_2CH_2Cl$ is 26 ($2Cl = 14$; $4H = 4$; $2C = 8$). The structural formula (left) shows 7 covalent bonds accounting for 14 electrons. The remaining 12 electrons are divided equally between the two chlorines as unshared electron pairs. The octet rule is satisfied for both carbon and chlorine (right).

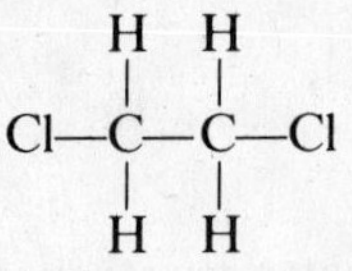

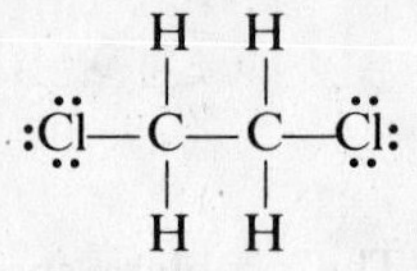

$$\overset{\displaystyle H \; H}{\underset{\displaystyle H \; H}{Cl-C-C-Cl}} \qquad \overset{\displaystyle H \; H}{\underset{\displaystyle H \; H}{:\overset{..}{\underset{..}{Cl}}-C-C-\overset{..}{\underset{..}{Cl}}:}}$$

(*d*) This compound has the same molecular formula as the compound in part (*c*), but a different structure. It, too, has 26 valence electrons, and again only chlorine has unshared pairs.

$$
\begin{array}{ccc}
& H & H \\
& | & | \\
H- & C- & C-\ddot{\underset{..}{C}}l: \\
& | & \\
& H & :\ddot{\underset{..}{C}}l:
\end{array}
$$

(e)   The bonding pattern in $CH_3NHCH_2CH_3$ is shown (left). There are 26 valence electrons, and 24 of them are accounted for by the covalent bonds in the structural formula. The remaining two electrons complete the octet of nitrogen as an unshared pair (right).

$$
\begin{array}{ccccc}
H & & H & H & \\
| & & | & | & \\
H-C-N-C-C-H & & & & \\
| & | & | & | & \\
H & H & H & H &
\end{array}
\qquad
\begin{array}{ccccc}
H & & H & H & \\
| & & | & | & \\
H-C-\ddot{N}-C-C-H & & & & \\
| & | & | & | & \\
H & H & H & H &
\end{array}
$$

(f)   Oxygen has two unshared pairs in $(CH_3)_2CHCH{=}O$.

$$
\begin{array}{c}
H \\
| \\
H-C-H \\
H \quad | \\
| \quad \quad | \\
H-C-\!-C-\!-C{=}\ddot{\underset{..}{O}}: \\
| \quad | \quad | \\
H \quad H \quad H
\end{array}
$$

1.12 (b)   This compound has a four-carbon chain to which are appended two other carbons.

$$\lambda \quad \text{is equivalent to} \quad
\begin{array}{cc}
CH_3 & H \\
| & | \\
CH_3-C-\!\!-C-CH_3 \\
| & | \\
H & CH_3
\end{array}
\quad
\begin{array}{c}\text{which may be}\\ \text{rewritten as}\end{array}
\quad (CH_3)_2CHCH(CH_3)_2$$

(c)   The carbon skeleton is the same as that of the compound in part (b), but one of the terminal carbons bears an OH group in place of one of its hydrogens.

$$ \text{is equivalent to} \quad
\begin{array}{cc}
H \\
| \\
HO-C-H \\
\quad\quad H \\
CH_3-C-\!\!-C-CH_3 \\
| \quad | \\
H \quad CH_3
\end{array}
\quad
\begin{array}{c}\text{which may be}\\ \text{rewritten as}\end{array}
\quad
\begin{array}{c}
CH_2OH \\
| \\
CH_3CHCH(CH_3)_2
\end{array}$$

(d)   The compound is a six-membered ring that bears a $-C(CH_3)_3$ substituent.

is equivalent to

which may be rewritten as

1.13   The problem specifies that nitrogen and both oxygens of carbamic acid are bonded to carbon and one of the carbon-oxygen bonds is a double bond. Since a neutral carbon is associated with four bonds, a neutral nitrogen three (plus one unshared pair), and a neutral oxygen two (plus two unshared pairs), this gives the bonding pattern shown.

:O:
‖
H—N—C—Ö—H   Carbamic acid
|
H

1.14 (b) There are three constitutional isomers of C_3H_8O:

$$CH_3CH_2CH_2OH \qquad CH_3\underset{|}{C}HCH_3 \qquad CH_3CH_2OCH_3$$
(with OH on middle carbon)

(c) Four isomers of $C_4H_{10}O$ have —OH groups:

$$CH_3CH_2CH_2CH_2OH \quad CH_3\underset{OH}{C}HCH_2CH_3 \quad CH_3\underset{CH_3}{C}HCH_2OH \quad CH_3\underset{CH_3}{\overset{CH_3}{C}}OH$$

Three isomers have C—O—C units:

$$CH_3OCH_2CH_2CH_3 \qquad CH_3CH_2OCH_2CH_3 \qquad CH_3O\underset{CH_3}{C}HCH_3$$

1.15 (b) Move electrons from the negatively charged oxygen, as shown by the curved arrows.

The ion shown is bicarbonate ion. The resonance interaction shown is more important than an alternative one involving delocalization of lone-pair electrons in the OH group.

Not equivalent to original structure; not as stable because of charge separation

(c) All three oxygens are equivalent in carbonate ion. Either negatively charged oxygen can serve as the donor atom.

(d) Resonance in borate ion is exactly analogous to that in carbonate.

and

1.16 There are four B—H bonds in BH_4^-. The four electron pairs surround boron in a tetrahedral orientation. The H—B—H angles are 109.5°.

1.17 (*b*) Nitrogen in ammonium ion is surrounded by eight electrons in four covalent bonds. These four bonds are directed toward the corners of a tetrahedron.

Each HNH angle is 109.5°.

(*c*) Double bonds are treated as single bonds when deducing the shape of a molecule. Thus azide ion is linear.

The NNN angle is 180°.

(*d*) Since the double bond in carbonate ion is treated as if it were a single bond, the three sets of electrons are arranged in a trigonal planar arrangement around carbon.

The OCO angle is 120°.

1.18 (*b*) Water is a bent molecule, and so the individual O—H bond moments do not cancel. Water has a dipole moment.

Individual OH bond moments in water

Direction of net dipole moment

(*c*) Methane, CH_4, is perfectly tetrahedral, and so the individual (small) C—H bond moments cancel. Methane has no dipole moment.

(*d*) Methyl chloride has a dipole moment.

Directions of bond moments in CH_3Cl

Direction of molecular dipole moment

(*e*) Oxygen is more electronegative than carbon and attracts electrons from it. The formaldehyde molecule has a dipole moment.

Direction of bond moments in formaldehyde

Direction of molecular dipole moment

(*f*) Nitrogen is more electronegative than carbon. Hydrogen cyanide has a dipole moment.

Direction of bond moments in HCN

Direction of molecular dipole moment

1.19 (*b*) Carbon in formaldehyde ($H_2C=O$) is directly bonded to three other atoms (two hydrogens and one oxygen). It is sp^2 hybridized.

(*c*) Ketene has two carbons in different hybridization states. One is sp^2 hybridized; the other is sp hybridized.

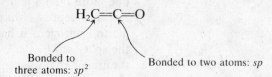

(*d*) One of the carbons in propene is sp^3 hybridized. The doubly bonded carbons are sp^2 hybridized.

$$sp^3 \quad sp^2 \quad sp^2$$
$$CH_3{-}CH{=}CH_2$$

(*e*) The carbons of the CH_3 groups in acetone $[(CH_3)_2C=O]$ are sp^3 hybridized. The $C=O$ carbon is sp^2 hybridized.

(*f*) The carbons in acrylonitrile are hybridized as shown:

$$sp^2 \quad sp^2 \quad sp$$
$$H_2C{=}CH{-}C{\equiv}N$$

1.20 All these species are characterized by the formula $:X{\equiv}Y:$, and each atom has an electron count of 5.

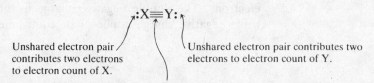

Electron count X = electron count Y = 2 + 3 = 5

(*a*) $:N{\equiv}N:$ A neutral nitrogen atom has five valence electrons. Therefore, each atom is electrically neutral in molecular nitrogen.

(*b*) $:C{\equiv}N:$ Nitrogen, as before, is electrically neutral. A neutral carbon has four valence electrons, and so carbon in this species, with an electron count of 5, has a unit negative charge. The species is cyanide anion; its net charge is −1.

(*c*) $:C{\equiv}C:$ There are two negatively charged carbon atoms in this species. It is a dianion; its net charge is −2.

(*d*) $:N{\equiv}O:$ Here again is a species with a neutral nitrogen atom. Oxygen, with an electron count of 5, has one less electron in its valence shell than a neutral oxygen atom. Oxygen has a formal charge of +1; the net charge is +1.

(*e*) $:C{\equiv}O:$ Carbon has a formal charge of −1; oxygen has a formal charge of +1. Carbon monoxide is a neutral molecule.

1.21 All these species are of the type $:\ddot{Y}{=}X{=}\ddot{Y}:$. Atom X has an electron count of 4, corresponding to half of the eight shared electrons in its four covalent bonds. Each atom Y has an electron count of 6; constituting it are four unshared electrons plus half of the four electrons in the double bond of each Y to X.

(*a*) $:\ddot{O}{=}C{=}\ddot{O}:$ Oxygen, with an electron count of six, and carbon, with an electron count of four, both correspond to the respective neutral atoms in the number of electrons they "own." Carbon dioxide is a neutral molecule, and neither carbon nor oxygen has a formal charge in this Lewis structure.

(b) $:\ddot{N}{=}N{=}\ddot{N}:$ The two terminal nitrogens each have an electron count (6) that is one more than a neutral atom and thus each has a formal charge of -1. The central N has an electron count (4) that is one less than a neutral nitrogen; it has a formal charge of $+1$. The net charge on the species is $(-1 + 1 - 1)$, or -1.

(c) $:\ddot{O}{=}N{=}\ddot{O}:$ As in part (b), the central nitrogen has a formal charge of $+1$. As in part (a), each oxygen is electrically neutral. The net charge is $+1$.

1.22 (a, b)   The problem specifies that ionic bonding is present and that the anion is tetrahedral. The cations are the group I metals Na^+ and Li^+. Both boron and aluminum are group III elements, and thus have a formal charge of -1 in the tetrahedral anions BF_4^- and AlH_4^- respectively.

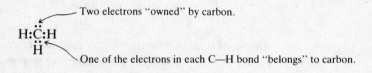

            Sodium tetrafluoroborate   Lithium aluminum hydride

(c, d)   Both of the tetrahedral anions have 32 valence electrons. Sulfur contributes six valence electrons and phosphorus five to the anions. Each oxygen contributes six electrons. The double negative charge in sulfate contributes two more, and the triple negative charge in phosphate contributes three more.

        $2K^+$ Potassium sulfate $3Na^+$ Sodium phosphate

The formal charge on each oxygen in both ions is -1. The formal charge on sulfur in sulfate is $+2$; the charge on phosphorus is $+1$. The net charge of sulfate ion is -2; the net charge of phosphate ion is -3.

1.23 (a)   Each hydrogen has a formal charge of zero, as is always the case when hydrogen is covalently bonded to one substituent. Oxygen has an electron count of 5.

$$H{-}\ddot{O}{-}H \qquad \text{Electron count of oxygen} = 2 + \tfrac{1}{2}(6) = 5$$
$$\underset{H}{|}$$

                                     Unshared   Covalently
                                       pair   bonded electrons

A neutral oxygen atom has six valence electrons; therefore, oxygen in this species has a formal charge of $+1$. The species as a whole has a unit positive charge. It is the hydronium ion, H_3O^+.

(b)   The electron count of carbon is 5; there are two electrons in an unshared pair, and three electrons are counted as carbon's share of the three covalent bonds to hydrogen.

                                     — Two electrons "owned" by carbon.

$$H{:}\overset{..}{\underset{..}{C}}{:}H$$
$$\overset{..}{H}$$

                                     — One of the electrons in each C—H bond "belongs" to carbon.

An electron count of 5 is one more than that for a neutral carbon atom. The formal charge on carbon is -1, as is the net charge on this species.

(c)   This species has one less electron than that of part (b). None of the atoms bears a formal charge. The species is neutral.

$$\text{H}\!-\!\overset{\displaystyle \cdot}{\underset{\displaystyle |}{\text{C}}}\!-\!\text{H} \qquad \text{Electron count of carbon} = 1 + \tfrac{1}{2}(6) = 4$$
$$\underset{\text{H}}{}$$

Unshared electron

Electrons shared in covalent bonds

(d)   The formal charge of carbon in this species is $+1$. Its only electrons are those in its three covalent bonds to hydrogen, and so its electron count is 3. This corresponds to one less electron than in a neutral carbon atom, giving it a unit positive charge.

(e)   In this species the electron count of carbon is 4, or, exactly as in part (c), that of a neutral carbon atom. Its formal charge is zero and the species is neutral.

Two unshared electrons contribute 2 to electron count of carbon.

$$\text{H}\!-\!\overset{\displaystyle \cdot\cdot}{\text{C}}\!-\!\text{H}$$

Half of the four electrons in the two covalent bonds contribute 2 to electron count of carbon.

1.24 Oxygen is surrounded by a complete octet of electrons in each structure but has a different "electron count" in each one because the proportion of shared to unshared pairs is different.

(a) $\text{CH}_3\ddot{\text{O}}:$ (b) $\text{CH}_3\ddot{\text{O}}\text{CH}_3$ (c) $\text{CH}_3\overset{\displaystyle \cdot\cdot}{\text{O}}\text{CH}_3$
$$\underset{\text{CH}_3}{|}$$

| Electron count $=6+\tfrac{2}{2}=7$; formal charge $=-1$ | Electron count $=4+\tfrac{4}{2}=6$; formal charge $=0$ | Electron count $=2+\tfrac{6}{2}=5$; formal charge $=+1$ |

1.25 (a)   Each carbon has four valence electrons, each hydrogen one, and chlorine has seven. Hydrogen and chlorine each can form only one bond, and so the only stable structure must have a carbon-carbon bond. Of the 20 valence electrons, 14 are present in the seven covalent bonds and 6 reside in the three unshared electron pairs of chlorine.

$$\underset{\text{H H}}{\overset{\text{H H}}{\text{H}:\ddot{\text{C}}:\ddot{\text{C}}:\ddot{\text{C}}\text{l}:}} \qquad \text{or} \qquad \underset{\text{H H}}{\overset{\text{H H}}{\text{H}\!-\!\text{C}\!-\!\text{C}\!-\!\ddot{\text{C}}\text{l}:}}$$

(b)   As in part (a) the single chlorine as well as all of the hydrogens must be connected to carbon. There are 18 valence electrons in $\text{C}_2\text{H}_3\text{Cl}$, and the framework of five single bonds accounts for only 10 electrons. Six of the remaining eight are used to complete the octet of chlorine as three unshared pairs, and the last two are used to form a carbon-carbon double bond.

$$\underset{\text{H H}}{\text{H}:\ddot{\text{C}}::\ddot{\text{C}}:\ddot{\text{C}}\text{l}:} \qquad \text{or} \qquad \underset{\text{H}}{\overset{\text{H}}{}}\text{C}\!=\!\text{C}\underset{:\ddot{\text{C}}\text{l}:}{\overset{\text{H}}{}}$$

(c)   All of the atoms except carbon (H, Br, Cl, and F) are monovalent. Therefore they can only be bonded to carbon. The problem states that all three fluorines are bonded to the same carbon, and so one of the carbons is present as a CF_3 group. The other carbon must be present as a CHBrCl group. Connect these groups together to give the structure of halothane.

$$
\begin{array}{l}
:\ddot{F}:H \\
:\ddot{F}:\overset{..}{C}:\overset{..}{C}:\overset{..}{Cl}: \\
:\ddot{F}::\overset{..}{Br}:
\end{array}
\qquad \text{or} \qquad
\begin{array}{c}
F \quad H \\
| \quad | \\
F-C-C-Cl \\
| \quad | \\
F \quad Br
\end{array}
\qquad \text{(Unshared electron pairs omitted for clarity)}
$$

(d) As in part (c) all of the atoms except carbon are monovalent. Since each carbon bears one chlorine, two ClCF$_2$ groups must be bonded together.

$$
\begin{array}{l}
:\ddot{F}::\ddot{F}: \\
:\ddot{Cl}:\overset{..}{C}:\overset{..}{C}:\overset{..}{Cl}: \\
:\ddot{F}::\ddot{F}:
\end{array}
\qquad \text{or} \qquad
\begin{array}{c}
F \quad F \\
| \quad | \\
Cl-C-C-Cl \\
| \quad | \\
F \quad F
\end{array}
\qquad \text{(Unshared electron pairs omitted for clarity)}
$$

1.26 Place hydrogen substituents on the given atoms so that carbon has four bonds, nitrogen three, and oxygen two. Place unshared electron pairs on nitrogen and oxygen so that nitrogen has an electron count of 5 and oxygen has an electron count of 6. These electron counts satisfy the octet rule when nitrogen has three bonds and oxygen two.

$$
\begin{array}{c}
H \\
| \\
(a) \quad H-C-\ddot{N}=\ddot{O}: \\
| \\
H
\end{array}
\qquad\qquad
\begin{array}{c}
(c) \quad H-\ddot{O}-C=\ddot{N}-H \\
| \\
H
\end{array}
$$

$$
\begin{array}{c}
(b) \quad H-C=\ddot{N}-\ddot{O}-H \\
| \\
H
\end{array}
\qquad\qquad
\begin{array}{c}
(d) \quad :\ddot{O}=C-\ddot{N}-H \\
| \quad | \\
H \quad H
\end{array}
$$

1.27 (a) Species A, B, and C have the same molecular formula, the same atomic positions, and the same number of electrons. They differ only in the arrangement of their electrons. Therefore, they are resonance forms of a single compound.

$$
\begin{array}{ccc}
\begin{array}{c}
H \\
\diagdown \\
\quad :\overset{-}{\underset{\diagup}{C}}-\overset{+}{N}\equiv N: \\
H
\end{array}
&
\begin{array}{c}
H \\
\diagdown \\
\quad \underset{\diagup}{C}=\overset{+}{N}=\ddot{N}:^{-} \\
H
\end{array}
&
\begin{array}{c}
H \\
\diagdown \\
\quad {}^{+}\underset{\diagup}{C}-\ddot{N}=\ddot{N}:^{-} \\
H
\end{array} \\
A & B & C
\end{array}
$$

(b) Structure A has a formal charge of -1 on carbon.

(c) Structure C has a formal charge of $+1$ on carbon.

(d) Structures A and B have formal charges of $+1$ on the internal nitrogen.

(e) Structures B and C have a formal charge of -1 on the terminal nitrogen.

(f) All resonance forms of a particular species must have the same net charge. In this case, the net charge on A, B, and C is zero.

(g) Both A and B have the same number of covalent bonds, but the negative charge is on a more electronegative atom in B (nitrogen) than it is in A (carbon). Structure B is more stable than A.

(h) Structure B is more stable than C. Structure B has one more covalent bond, all of its atoms have octets of electrons, and it has a lesser degree of charge separation than C.

1.28 The structures given and their calculated formal charges are:

$$
\begin{array}{cccc}
H-\overset{-1}{\underset{..}{C}}=\overset{+1}{N}=\ddot{O}: &
H-C\overset{+1}{\equiv}\overset{..}{N}-\ddot{O}:^{-1} &
H-C\equiv\overset{+1}{N}=\overset{..}{O}: &
H-C=\overset{+1}{\ddot{N}}-\ddot{O}:^{-1} \\
D & E & F & G
\end{array}
$$

(a) Structure G contains a positively charged carbon.

(b) Structures D and E contain a positively charged nitrogen.

(c) None of the structures contain a positively charged oxygen.
(d) Structure D contains a negatively charged carbon.
(e) None of the structures contain a negatively charged nitrogen.
(f) Structures E and G contain a negatively charged oxygen.
(g) All of the structures are electrically neutral.
(h) Structure E is the most stable. All of the atoms have octets of electrons, and the negative charge resides on the most electronegative element (oxygen).
(i) Structure F is the least stable. Nitrogen has five bonds (10 electrons), and so the octet rule is violated.

1.29 (a) These two structures are resonance forms, since they have the same atomic positions and the same number of electrons.

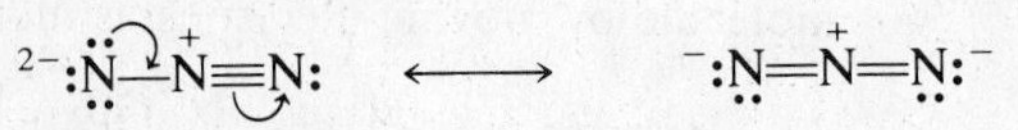

16 valence electrons
(net charge = −1)

16 valence electrons
(net charge = −1)

(b) The two structures have different numbers of electrons and therefore they are not resonance forms.

16 valence electrons
(net charge −1)

14 valence electrons
(net charge +1)

(c) These two structures have different numbers of electrons; they are not resonance forms.

16 valence electrons
(net charge = −1)

20 valence electrons
(net charge = −5)

1.30 Structure J has 10 electrons surrounding nitrogen, but the octet rule limits nitrogen to 8 electrons. Structure J is incorrect.

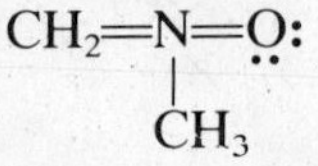

$$CH_2{=}N{=}O:$$ Not a valid Lewis structure
$$|$$
$$CH_3$$

1.31 (a) The terminal nitrogen has only six electrons; therefore, use the unshared pair of the adjacent nitrogen to form another covalent bond.

By moving electrons of the nitrogen lone pair as shown by the arrow

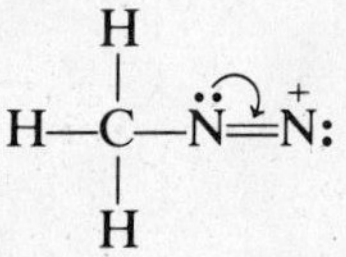

a structure that has octets about both nitrogen atoms is obtained.

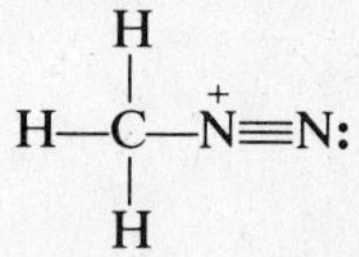

In general, move electrons from sites of high electron density toward sites of low electron density. Notice that the location of formal charge has changed but the net charge on the species remains the same.

(b) The dipolar Lewis structure given can be transformed to one which has no charge separation by moving electron pairs as shown:

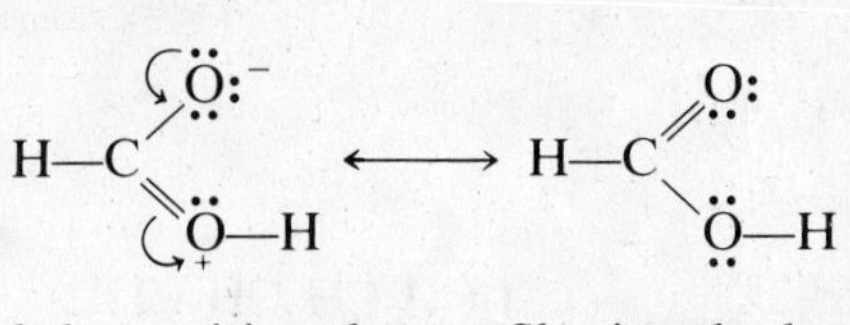

(c) Move electrons toward the positive charge. Sharing the lone pair gives an additional covalent bond and avoids the separation of opposite charges.

$$^+CH_2\frown\ddot{\underline{C}}H_2 \longleftrightarrow CH_2{=}CH_2$$

(d) Octets of electrons at all the carbon atoms can be produced by moving the electrons toward the site of positive charge.

$$^+CH_2\frown CH{=}CH{-}\underline{C}H_2 \longleftrightarrow CH_2{=}CH{-}CH{=}CH_2$$

(e) As in part (d), move the electron pairs toward the carbon atom that has only six electrons.

$$^+CH_2\frown CH{=}CH\frown\ddot{\underline{O}}{:}^- \longleftrightarrow CH_2{=}CH{-}CH{=}\ddot{O}{:}$$

(f) Negative charge can be placed on the most electronegative atom (oxygen) in this molecule by moving electrons as indicated.

(g) Octets of electrons are present around both carbon and oxygen if an oxygen unshared pair is moved toward the positively charged carbon to give an additional covalent bond.

$$H{-}\overset{+}{C}{=}\ddot{O}{:} \longleftrightarrow H{-}C{\equiv}\overset{+}{O}{:}$$

(h) This exercise is similar to part (g); move electrons from oxygen to carbon so as to produce an additional bond and satisfy the octet rule for each atom.

(i) By moving electrons from the site of negative charge toward the positive charge, a structure that has no charge separation is generated.

1.32 (a) In order to generate constitutionally isomeric structures having the molecular formula C_4H_{10}, you need to consider the various ways in which four carbon atoms can be bonded together. These are

$$C{-}C{-}C{-}C \quad \text{and} \quad C{-}C{-}C$$
$$\underset{C}{\overset{|}{\phantom{C{-}C{-}}}}$$

Filling in the appropriate hydrogen substituents gives the correct structures:

$$CH_3CH_2CH_2CH_3 \quad \text{and} \quad CH_3CHCH_3$$
$$\underset{CH_3}{\overset{|}{}}$$

Continue with the remaining parts of the problem using the general approach outlined for part (a).

(b) C_5H_{12}:

$$CH_3CH_2CH_2CH_2CH_3 \qquad CH_3CHCH_2CH_3 \qquad CH_3{-}\overset{\overset{\displaystyle CH_3}{|}}{\underset{\underset{\displaystyle CH_3}{|}}{C}}{-}CH_3$$
$$\underset{CH_3}{\overset{|}{}}$$

(c) $C_2H_4Cl_2$:

$$CH_3CHCl_2 \quad \text{and} \quad ClCH_2CH_2Cl$$

(d) C_4H_9Br:

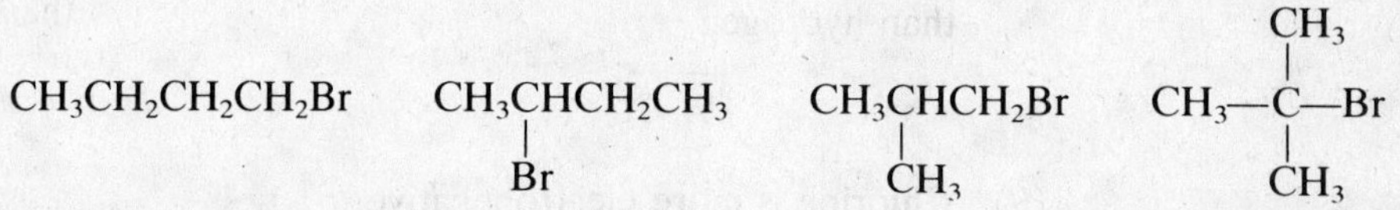

$$CH_3CH_2CH_2CH_2Br \qquad CH_3\underset{\underset{Br}{|}}{CH}CH_2CH_3 \qquad CH_3\underset{\underset{CH_3}{|}}{CH}CH_2Br \qquad CH_3\overset{\overset{CH_3}{|}}{\underset{\underset{CH_3}{|}}{C}}Br$$

(e) C_3H_9N:

$$CH_3CH_2CH_2NH_2 \qquad CH_3CH_2NHCH_3 \qquad CH_3{-}\overset{\overset{CH_3}{|}}{N}\diagdown CH_3 \qquad CH_3\underset{\underset{CH_3}{|}}{CH}NH_2$$

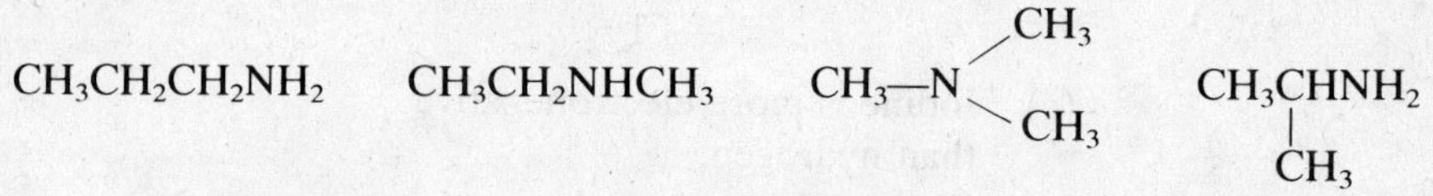

Note that when the three carbons and the nitrogen are arranged in a ring, the molecular formula based on such a structure is C_3H_7N, not C_3H_9N as required.

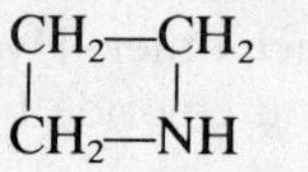

$$\begin{array}{cc} CH_2{-}CH_2 \\ |\qquad\quad | \\ CH_2{-}NH \end{array} \quad \text{(not an isomer)}$$

1.33 (a) All three carbons must be bonded together, and each one has four bonds. Therefore the molecular formula C_3H_8 uniquely corresponds to:

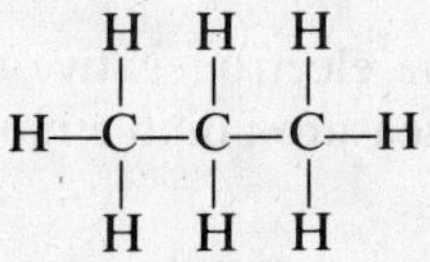

$$\underset{\underset{H}{|}}{\overset{\overset{H}{|}}{H{-}C}}{-}\underset{\underset{H}{|}}{\overset{\overset{H}{|}}{C}}{-}\underset{\underset{H}{|}}{\overset{\overset{H}{|}}{C}}{-}H \qquad (CH_3CH_2CH_3)$$

(b) With two fewer hydrogen atoms than the preceding compound, either C_3H_6 must contain a carbon-carbon double bond or its carbons must be arranged in a ring; thus the following structures are constitutional isomers:

$$CH_2{=}CHCH_3 \qquad \text{and} \qquad \begin{array}{c} CH_2{-}CH_2 \\ \diagdown\;\;\diagup \\ CH_2 \end{array}$$

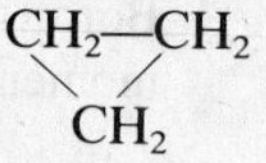

(c) The molecular formula C_3H_4 is satisfied by the structures

$$CH_2{=}C{=}CH_2 \qquad HC{\equiv}CCH_3 \qquad \begin{array}{c} HC{=}CH \\ \diagdown\;\;\diagup \\ CH_2 \end{array}$$

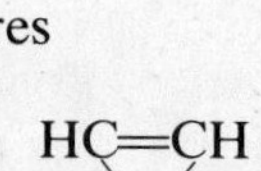

1.34 (a) The only atomic arrangements of C_3H_6O that contain only single bonds must have a ring as part of their structure.

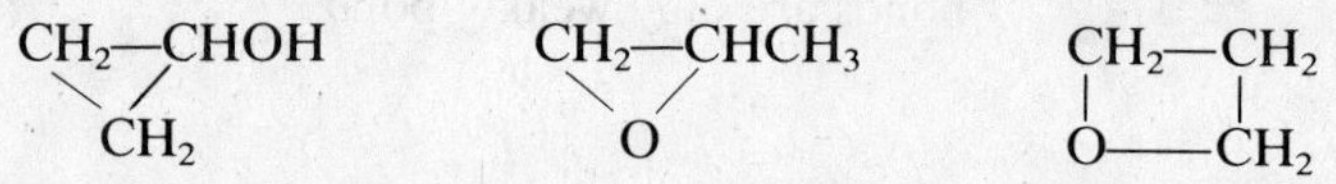

$$\begin{array}{c} CH_2{-}CHOH \\ \diagdown\;\;\diagup \\ CH_2 \end{array} \qquad \begin{array}{c} CH_2{-}CHCH_3 \\ \diagdown\;\;\diagup \\ O \end{array} \qquad \begin{array}{c} CH_2{-}CH_2 \\ |\qquad\quad | \\ O{-\!-}CH_2 \end{array}$$

(b) Structures corresponding to C_3H_6O are possible in noncyclic compounds if they contain a carbon-carbon or carbon-oxygen double bond.

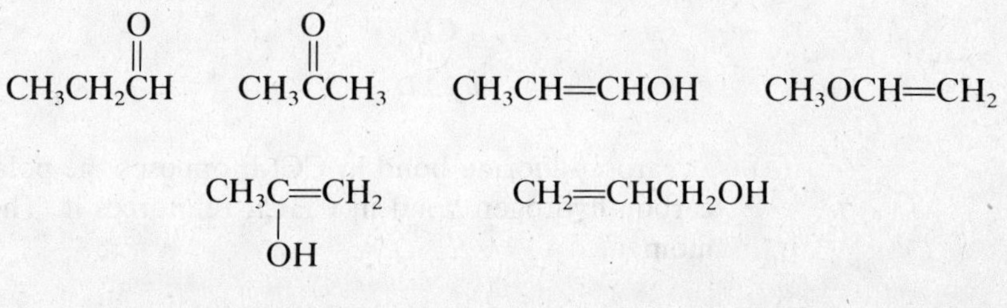

$$\overset{\overset{O}{\|}}{CH_3CH_2CH} \qquad \overset{\overset{O}{\|}}{CH_3CCH_3} \qquad CH_3CH{=}CHOH \qquad CH_3OCH{=}CH_2$$

$$CH_3\underset{\underset{OH}{|}}{C}{=}CH_2 \qquad\qquad CH_2{=}CHCH_2OH$$

1.35 The direction of a bond dipole is governed by the electronegativity of the atoms it connects. In each of the parts to this problem, the more electronegative atom is partially negative and the less electronegative atom is partially positive. Electronegativities of the elements are given in Table 1.2 of the text.

(a) Chlorine is more electronegative than hydrogen.

$$\overset{\longrightarrow}{\text{H}\!-\!\text{Cl}}$$

(b) Chlorine is more electronegative than iodine.

$$\overset{\longrightarrow}{\text{I}\!-\!\text{Cl}}$$

(c) Iodine is more electronegative than hydrogen.

$$\overset{\longrightarrow}{\text{H}\!-\!\text{I}}$$

(d) Oxygen is more electronegative than hydrogen.

(e) Oxygen is more electronegative than either hydrogen or chlorine.

1.36 (a) Sodium chloride is ionic; it has a unit positive charge and a unit negative charge separated from each other. Hydrogen chloride has a polarized bond but is a covalent compound. Sodium chloride has a larger dipole moment. The measured values are as shown.

$$\text{Na}^{+}\,\text{Cl}^{-}\quad \text{is more polar than}\quad \text{H}\!-\!\text{Cl}$$
$$\mu\,9.4\,\text{D} \qquad\qquad\qquad\qquad \mu\,1.1\,\text{D}$$

(b) Fluorine is more electronegative than chlorine, and so its bond to hydrogen is more polar, as the measured dipole moments indicate.

$$\overset{\longrightarrow}{\text{H}\!-\!\text{F}}\quad \text{is more polar than}\quad \overset{\longrightarrow}{\text{H}\!-\!\text{Cl}}$$
$$\mu\,1.7\,\text{D} \qquad\qquad\qquad\qquad \mu\,1.1\,\text{D}$$

(c) Boron trifluoride is planar. Its individual B—F bond dipoles cancel. It has no dipole moment.

$$\overset{\longrightarrow}{\text{H}\!-\!\text{F}}\quad \text{is more polar than}$$

$$\mu\,1.7\,\text{D} \qquad\qquad\qquad\qquad \mu\,0\,\text{D}$$

(d) A carbon-chlorine bond is strongly polar; carbon-hydrogen and carbon-carbon bonds are only weakly polar.

is more polar than

$$\mu\,2.1\,\text{D} \qquad\qquad\qquad\qquad\qquad \mu\,0.1\,\text{D}$$

(e) A carbon-fluorine bond in CCl_3F opposes the polarizing effect of the chlorines. The carbon-hydrogen bond in $CHCl_3$ reinforces it. Therefore $CHCl_3$ has a larger dipole moment.

is more polar than

$$\mu \ 1.0\,D \qquad\qquad\qquad \mu \ 0.5\,D$$

(*f*) Oxygen is more electronegative than nitrogen; its bonds to carbon and hydrogen are more polar than the corresponding bonds formed by nitrogen.

is more polar than

$$\mu \ 1.7\,D \qquad\qquad\qquad \mu \ 1.3\,D$$

(*g*) The Lewis structure for CH_3NO_2 has a formal charge of $+1$ on nitrogen, making it more electron-attracting than the uncharged nitrogen of CH_3NH_2.

is more polar than

$$\mu \ 3.1\,D \qquad\qquad\qquad \mu \ 1.3\,D$$

1.37 (*a*) There are four electron pairs around carbon in $:\bar{C}H_3$; they are arranged in a tetrahedral fashion. The species is trigonal pyramidal.

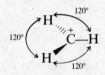

(*b*) Only three electron pairs are present in $\overset{+}{C}H_3$, and so it is trigonal planar.

(*c*) As in part (*b*), there are three electron pairs. When these electron pairs are arranged in a plane, the atoms in $:CH_2$ are not collinear. The atoms of this species are arranged in a bent structure according to VSEPR considerations.

1.38 The structures, written in a form so as to indicate hydrogen substituents and unshared electrons, are as given in the following. Remember a neutral carbon has four bonds, a neutral nitrogen has three bonds plus one unshared pair, and a neutral oxygen has two bonds plus two unshared pairs. Halogen substituents have one bond and three unshared pairs.

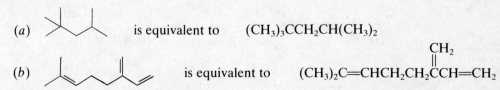

(*a*)   is equivalent to $(CH_3)_3CCH_2CH(CH_3)_2$

(*b*)   is equivalent to $(CH_3)_2C{=}CHCH_2CH_2\overset{\overset{\displaystyle CH_2}{\|}}{C}CH{=}CH_2$

(c)   is equivalent to

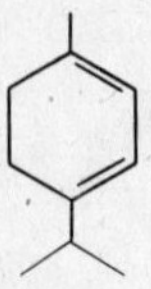

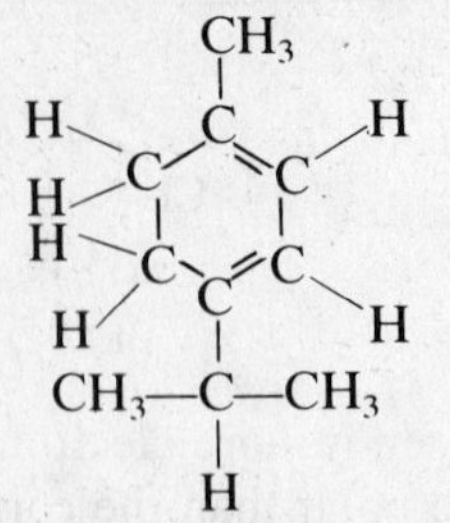

$$\begin{array}{c} CH_3 \\ H-C-C-H \\ H-C=C \\ H-C-C-H \\ H \\ CH_3-C-CH_3 \\ H \end{array}$$

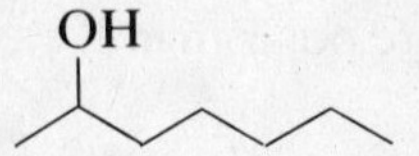

(d)   OH   is equivalent to $\ddot{O}H$ $CH_3CHCH_2CH_2CH_2CH_2CH_3$

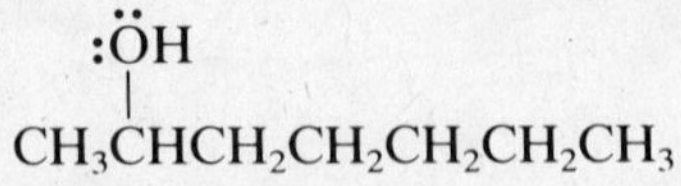

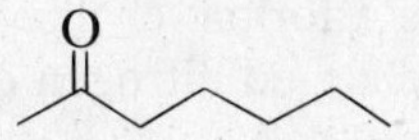

(e)   O   is equivalent to $\ddot{O}:$ $CH_3CCH_2CH_2CH_2CH_2CH_3$

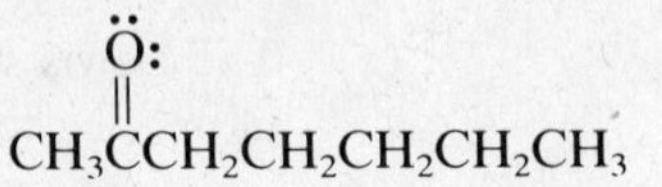

(f)    is equivalent to

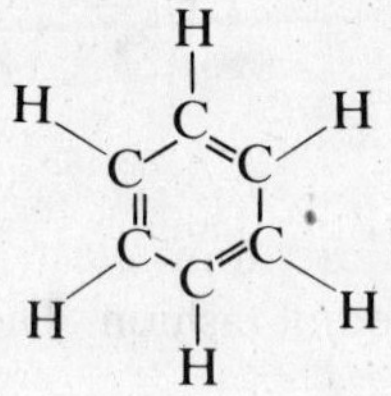

(g) 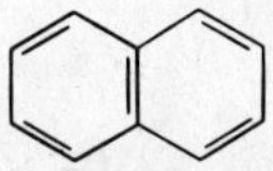   is equivalent to

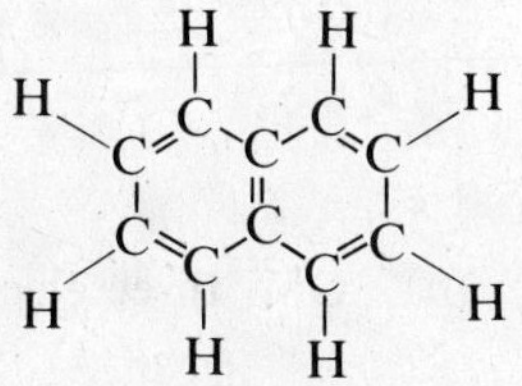

(h) $OCCH_3$ O ... COH O is equivalent to

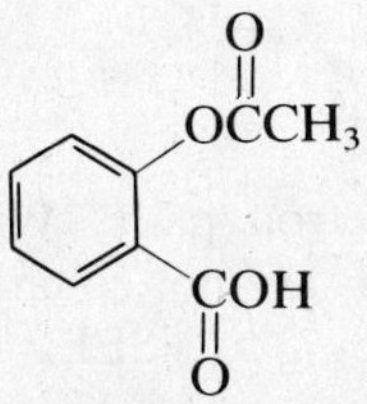

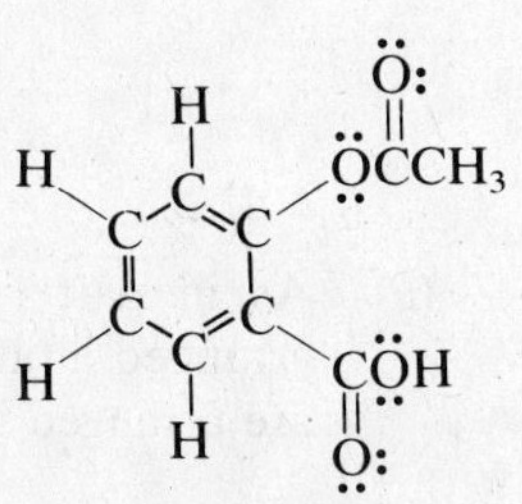

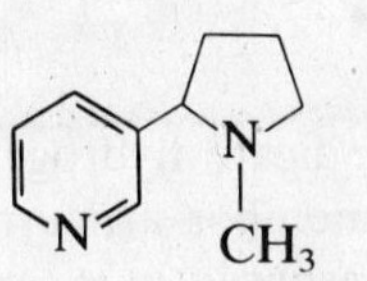

(i)   N CH_3 is equivalent to

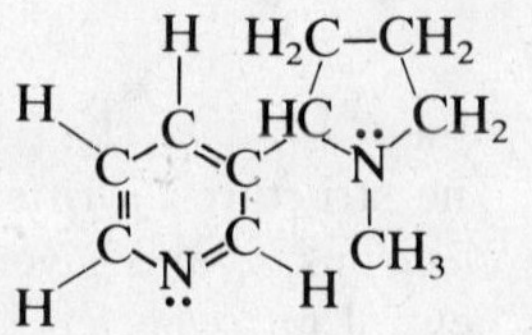

(j)   Br ... CH_3 ... Br   is equivalent to

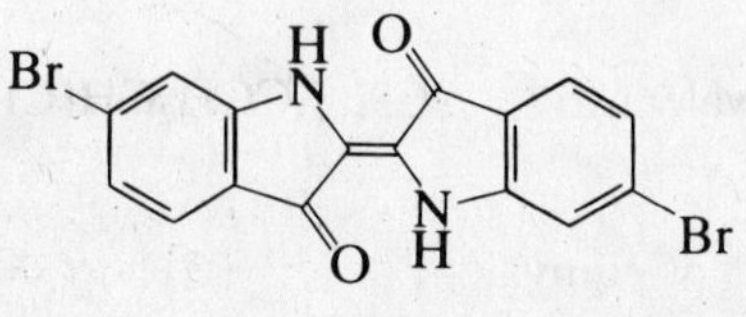

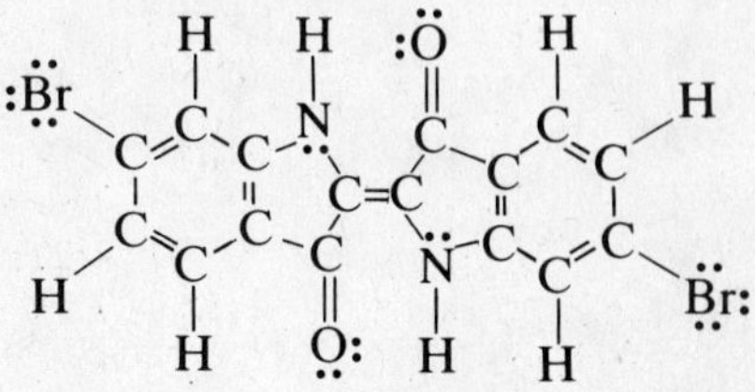

(k) 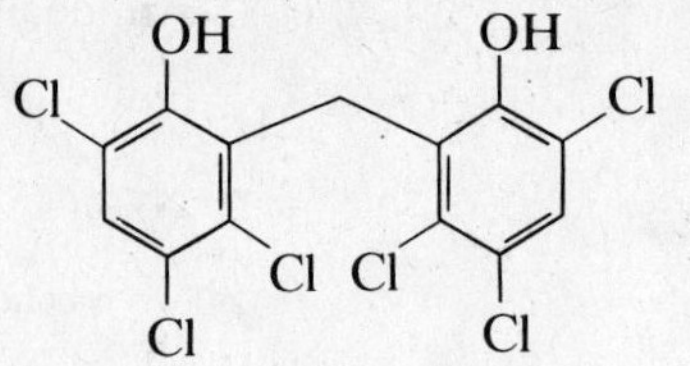   is equivalent to

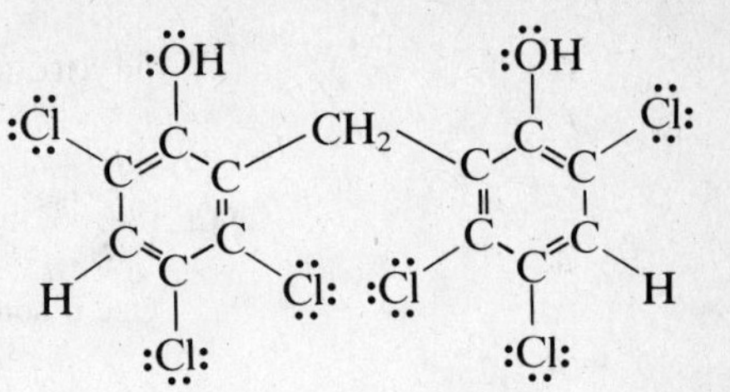

1.39 (a) C_8H_{18} (g) $C_{10}H_8$
(b) $C_{10}H_{16}$ (h) $C_9H_8O_4$
(c) $C_{10}H_{16}$ (i) $C_{10}H_{14}N_2$
(d) $C_7H_{16}O$ (j) $C_{16}H_8Br_2N_2O_2$
(e) $C_7H_{14}O$ (k) $C_{13}H_6Cl_6O_2$
(f) C_6H_6

Isomers are compounds that have different structures but the same molecular formula. Two of these compounds, (b) and (c), have the same molecular formula and are isomers of each other.

1.40 (a) Carbon is sp^3 hybridized when it is directly bonded to four other atoms. Compounds (a) and (d) in Problem 1.38 are the only ones in which *all* of the carbons are sp^3 hybridized.

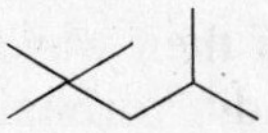

(a) 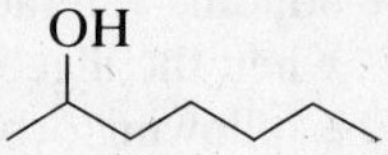   (d)

(b) Carbon is sp^2 hybridized when it is directly bonded to three other atoms. Compounds (f), (g), and (j) in Problem 1.38 have only sp^2 hybridized carbons.

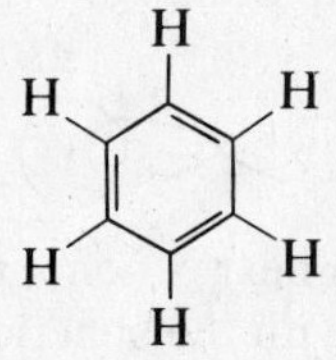

(f) 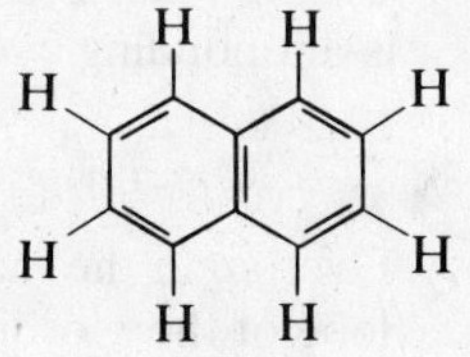   (g)

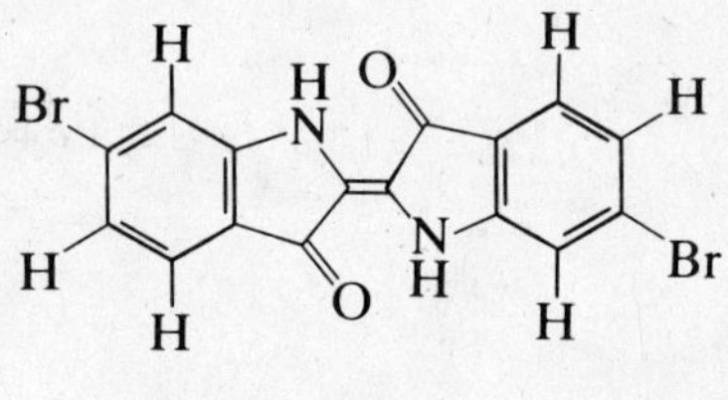

(j)

1.41 The problem specifies that the second-row element is sp^3 hybridized in each of the compounds. Therefore, any unshared electron pairs occupy sp^3 hybridized orbitals and bonded pairs are located in σ orbitals.

(a) Ammonia   (b) Water

sp^3 hybrid orbital

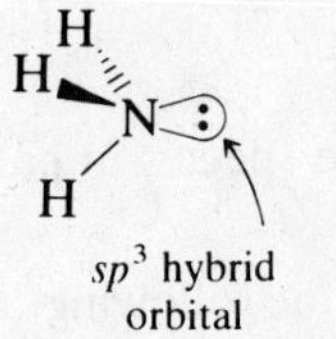

Three σ bonds formed by sp^3–s overlap

sp^3 hybrid orbitals

two sp^3 hybrid orbitals

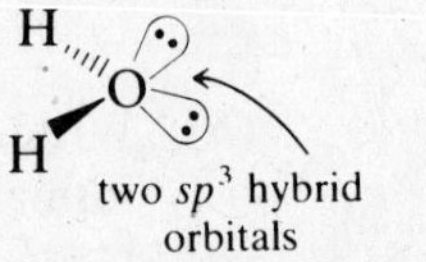

Two σ bonds formed by sp^3–s overlap

(c) Hydrogen fluoride

H—F three sp^3
hybrid orbitals

One σ bond formed
by sp^3–s overlap

(f) Amide anion

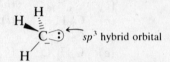

two sp^3
hybrid orbitals

Two σ bonds formed
by sp^3–s overlap

(d) Ammonium ion

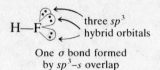

Four σ bonds formed by
sp^3–s overlap

(g) Methyl anion

sp^3 hybrid orbital

Three σ bonds formed
by sp^3–s overlap

(e) Borohydride anion

Four σ bonds formed by
sp^3–s overlap

1.42 A bonding interaction exists when two orbitals overlap "in phase" with each other, that is, when the algebraic signs of their wave functions are the same in the region of overlap. The following orbital is a bonding orbital. It involves overlap of an s orbital with the lobe of a p orbital of the same sign.

   C (bonding)

On the other hand, the overlap of an s orbital with the lobe of a p orbital of opposite sign is antibonding.

   B (antibonding)

Overlap in the manner shown next is nonbonding. Both the positive lobe and the negative lobe of the p orbital overlap with the spherically symmetric s orbital. The bonding overlap between the s orbital and one lobe of the p orbital is exactly canceled by an antibonding interaction between the s orbital and the lobe of opposite sign.

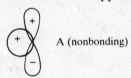   A (nonbonding)

SELF-TEST

PART A

A-1. Write the electronic configuration for each of the following:
   (a) Phosphorus   (b) The stable ion of sulfur

A-2. (a) Determine the formal charge of each atom and the net charge for the following species:

$$:\ddot{N}{=}C{=}\ddot{S}:$$

   (b) Write a second Lewis structure for the species in part (a), and determine the formal charge of each atom.

A-3. Write a correct Lewis structure for each of the following. Be sure to show explicitly any unshared pairs of electrons.
   (a) Methylamine, CH_5N

(*b*) Acetaldehyde, C_2H_4O (the atomic order is CCO; the carbon atoms are connected by a single bond)

A-4. What is the molecular formula of each of the structures shown? Clearly draw any unshared electron pairs that are present.

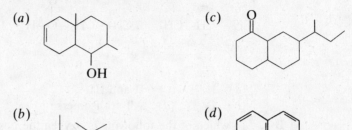

A-5. Account for the fact that all the bonds in SO_3 are the same by drawing the appropriate Lewis structure(s).

A-6. The cyanate ion contains 16 valence electrons, and its three atoms are arranged in the order OCN. Write a Lewis structure for this species and assign a formal charge to each atom. What is the net charge of the ion?

A-7. Using the VSEPR method,
  (*a*) Describe the geometry at each carbon atom and the oxygen atom in the following molecule: $CH_3OCH{=}CHCH_3$.
  (*b*) Deduce the shape of NCl_3 and draw a three-dimensional representation of the molecule. Is NCl_3 polar?

A-8. Consider structures A, B, C, and D:

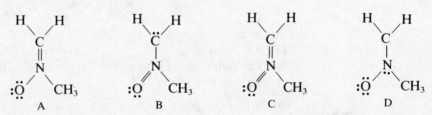

  (*a*) Which structure (or structures) contain a positively charged carbon?
  (*b*) Which structure (or structures) contain a positively charged nitrogen?
  (*c*) Which structure (or structures) contain a positively charged oxygen?
  (*d*) Which structure (or structures) contain a negatively charged carbon?
  (*e*) Which structure (or structures) contain a negatively charged nitrogen?
  (*f*) Which structure (or structures) contain a negatively charged oxygen?
  (*g*) Which structure is the most stable?
  (*h*) Which structure is the least stable?

A-9. Given the following information, write a Lewis structure for urea, CH_4N_2O. The oxygen atom and both nitrogen atoms are bonded to carbon, there is a carbon-oxygen double bond, and none of the atoms bears a formal charge. Be sure to include all unshared electron pairs.

A-10. How many σ and π bonds are present in each of the following?
  (*a*) $CH_3CH{=}CHCH_3$ (*b*) $HC{\equiv}CCH_2CH_3$

A-11. Give the hybridization of each carbon atom in the preceding question.

PART B

B-1. Which one of the following is most likely to have ionic bonds?
  (*a*) HCl (*b*) Na_2O (*c*) N_2O (*d*) NCl_3

B-2. Which one of the following has the highest electron affinity?
(a) Li   (b) B   (c) F   (d) C

B-3. Which of the following is an isomer of compound I?

$$CH_2\!-\!CH\!-\!CH_3 \quad\quad CH_3\!-\!CH_2\!-\!\overset{\overset{\textstyle O}{\|}}{C}\!-\!H \quad\quad CH_3\!-\!\overset{\overset{\textstyle O}{\|}}{C}\!-\!CH_3 \quad\quad CH_3CH\!=\!CH$$

(with O bridging in I, and OH below CH in IV)

 I   II   III   IV

(a) II   (c) II and III
(b) IV   (d) All are isomers

B-4. In which of the following is oxygen the positive end of the dipole?
(a) O—F   (b) O—N   (c) O—S   (d) O—H

B-5. Ionic bonds are likely to form under which of the following sets of conditions?
(a) One atom has a low ionization energy and the other has a high electron affinity.
(b) One atom has a low ionization energy and the other has a low electron affinity.
(c) Both atoms are small.
(d) Each atom has a low ionization energy.

B-6. Which of the following is *not* an electronic configuration for an atom in its ground state?
(a) $1s^2 2s^2 2p_x^2 2p_y^1 2p_z^1$ (c) $1s^2 2s^2 2p_x^2 2p_y^2 2p_z^1$
(b) $1s^2 2s^2 2p_x^2 2p_y^2 2p_z^0$ (d) $1s^2 2s^2 2p_x^2 2p_y^2 2p_z^2$

B-7. The information

$$(Z = 20) \quad\quad 1s^2 2s^2 2p^6 3s^2 3p^6$$

represents
(a) An argon atom   (d) The stable ion of sulfur
(b) The stable ion of calcium (e) None of these
(c) An unstable ion of sulfur

B-8. The formal charge on phosphorus in $(CH_3)_4P$ is
(a) 0   (b) −1   (c) +1   (d) +2

The following two questions refer to the hypothetical compounds:

$$A\!-\!B\!-\!A \quad\quad A\!=\!\overset{..}{B}\!-\!A \quad\quad A\!=\!B\!=\!A \quad\quad A\!-\!\overset{..}{B}\!-\!A$$

(with A below B in IV)

 I   II   III   IV

B-9. Which substance(s) is (are) linear?
(a) I only   (b) I and III   (c) I and II   (d) III only

B-10. Assuming A is more electronegative than B, which substance(s) is (are) polar?
(a) I and III   (b) II only   (c) IV only   (d) II and IV

B-11. The total number of *unshared pairs* of electrons in the molecule O⬡N—H is
(a) 0   (b) 1   (c) 2   (d) 3

B-12. Which of the following contains a triple bond?
(a) SO_2 (b) HCN   (c) C_2H_4 (d) NH_3

B-13. The molecular orbital electronic configuration of Be_2 is $(\sigma_{1s})^2(\sigma_{1s}^*)^2(\sigma_{2s})^2(\sigma_{2s}^*)^2$. One would predict that Be_2 would be held together by
(a) One bond   (b) Two bonds   (c) Four bonds   (d) No bonds

ALKANES

IMPORTANT TERMS AND CONCEPTS

Classes of Hydrocarbons (Sec. 2.1) *Hydrocarbons* are compounds that contain only carbon and hydrogen. The two principal classes of hydrocarbons are *aliphatic hydrocarbons* and *aromatic hydrocarbons*. Aliphatic hydrocarbons are subdivided into three groups:

Alkanes: hydrocarbons in which all the bonds are single bonds
Alkenes: hydrocarbons that contain a carbon-carbon double bond
Alkynes: hydrocarbons that contain a carbon-carbon triple bond

Another name for an aromatic hydrocarbon is an *arene*.

Alkane Nomenclature (Secs. 2.7 to 2.13) Alkanes are hydrocarbons having the general formula C_nH_{2n+2}. For example, the following are all formulas of alkanes:

$$CH_4 \qquad C_5H_{12} \qquad C_{20}H_{42}$$

For C_4H_{10} and higher alkanes it is possible to have *constitutional isomers*, that is, more than one compound having the same formula. In order to distinguish between isomers a method of *systematic nomenclature* has been developed by the International Union of Pure and Applied Chemistry (IUPAC).

According to the IUPAC nomenclature rules, alkanes are named as derivatives of unbranched parent alkanes. Although the unbranched alkane is often referred to as the *straight-chain* isomer, it should be remembered that carbon forms bonds in a tetrahedral framework, and the chain of carbon atoms is anything but "straight." The names of the more common unbranched parent alkanes are listed in text Table 2.4.

Some branched alkanes are often given "common" names. Examples include isobutane and isopentane:

$$
\begin{array}{cc}
\begin{array}{l}
CH_3 \\
\quad\diagdown \\
\qquad CH-CH_3 \\
\quad\diagup \\
CH_3
\end{array}
&
\begin{array}{l}
CH_3 \\
\quad\diagdown \\
\qquad CH-CH_2-CH_3 \\
\quad\diagup \\
CH_3
\end{array}
\\
\text{Isobutane} & \text{Isopentane}
\end{array}
$$

Branched hydrocarbons such as these provide examples of compounds that contain hydrocarbon *substituent groups*. These groups are named according to the number of carbons they contain. The stem of the name is taken from one of the alkane names in Table 2.4 and the suffix *-ane* is replaced by *-yl*. For example:

CH_3-	Methyl group
CH_3CH_2-	Ethyl group
$CH_3CH_2CH_2CH_2CH_2-$	Pentyl group

Note that each of these alkyl groups has one less hydrogen atom than the parent alkane from which its name is derived.

Certain branched alkyl groups are given common names, such as isopropyl and *tert*-butyl. The groups may also be named systematically, numbering from the point of attachment. In the following, the systematic names are shown in parentheses:

$$CH_3-\underset{\underset{\displaystyle CH_3}{|}}{CH}- \qquad\qquad CH_3-\underset{\underset{\displaystyle CH_3}{|}}{\overset{\overset{\displaystyle CH_3}{|}}{C}}-$$

Isopropyl group   *tert*-Butyl group
(1-Methylethyl)   (1,1-Dimethylethyl)

There are an enormous number of isomers possible for larger alkanes (more than (300,000 $C_{20}H_{42}$ isomers, for example); therefore, unique common names for them all would be impossible. The IUPAC rules provide a means of identifying, with a unique name, each and every possible alkane isomer. In fact, as you will see, all organic compounds may be named by following the appropriate IUPAC rules.

Step 1:   Identify the longest *continuous* carbon chain. The stem of the compound name is based on the name of the alkane having this number of carbon atoms (from text Table 2.4).

Step 2:   Number the longest chain from the end that allows the lowest numbers to be assigned to the carbon atoms bearing alkyl group substituents.

Step 3:   Identify the alkyl substituents by name and specify the carbon atoms to which they are attached by the number. If a substituent appears more than once, the appropriate prefix *di-*, *tri-*, *tetra-*, and so on, is used.

Examples of compounds named according to these rules are the following:

$$\underset{\underset{\displaystyle CH_3}{|}}{CH_3CHCH_2CH_2CH_3} \qquad \text{2-Methylpentane}$$

$$(CH_3)_3CCH_2CH(CH_3)_2 \;\equiv\; CH_3-\underset{\underset{\displaystyle CH_3}{|}}{\overset{\overset{\displaystyle CH_3}{|}}{C}}-CH_2-\underset{\underset{\displaystyle H}{|}}{\overset{\overset{\displaystyle CH_3}{|}}{C}}-CH_3$$

2,2,4-Trimethylpentane

$$\underset{\underset{\displaystyle CH_2CH_3}{|}}{CH_3CH_2CHCH_2CHCH_2CH_3} \qquad \text{3-Ethyl-5-methylheptane}$$

Cycloalkane Nomenclature (Sec. 2.14) Alkanes whose carbon backbone forms a ring are commonly encountered in organic chemistry. Such cyclic alkanes, or *cycloalkanes*, have the general formula C_nH_{2n}. The stem names (from text Table 2.4) are preceded by *cyclo*. Therefore cyclopentane, C_5H_{10}, would be represented as:

$$CH_2 \underset{\underset{CH_2-CH_2}{\displaystyle CH_2}}{\overset{\overset{\displaystyle CH_2}{}}{}} CH_2 \qquad \text{or}$$

Cyclopentane

Substituents are named and numbered as described previously. As the following examples demonstrate, a number is not necessary when the ring contains only one substituent. When there are two or more substituents, the first group in the IUPAC name is assigned location 1 on the ring, and the numbering around the ring is done to keep the next number as low as possible.

Methylcyclopentane   1,2-Dimethylcyclopentane   1-Ethyl-3-methylcyclopentane

Physical Properties (Sec. 2.16) Consideration of the relationship between the structure of a substance and its boiling point leads to a discussion of *intermolecular forces*.

Van der Waals attractions constitute the predominant intermolecular force in nonpolar substances. Also known as *induced dipole–induced dipole* attractions, these forces arise from temporary distortion of the electron distribution of molecules in close proximity to each other. The temporary dipoles which result give rise to a weak attraction between the molecules.

The observation that boiling points increase with increasing molecular size, as seen in text Figure 2.3, may be explained by considering the attractive van der Waals forces present in the liquid state. Also, the fact that branched alkanes have lower boiling points than their unbranched isomers is explained by the more compact shape, and hence smaller surface area, of the branched molecules.

Combustion (Sec. 17) Reaction with oxygen is known as *combustion*. A typical example for the combustion of an alkane is:

$$C_5H_{12} + 8O_2 \longrightarrow 5CO_2 + 6H_2O$$

The heat released in the combustion reaction of a substance is known as its *heat of combustion*. By convention the heat of combustion is given by $-\Delta H°$, where

$$\Delta H° = H°_{products} - H°_{reactants}$$

and $H°$ represents the *enthalpy*, or heat content, of a substance in its standard state. Heats of combustion are listed as positive numbers, since combustion reactions are always *exothermic*. Therefore, $\Delta H°$ is always a negative number.

By comparing the heats of combustion, information may be gained regarding the relative energies of isomers. That is, an isomer which is more *stable* will give off less heat on combustion than one which is of higher energy and therefore less stable.

Oxidation-Reduction (Sec. 2.18) *Oxidation* of an organic compound corresponds to an increase in the number of bonds between carbon and oxygen and/or a decrease in the number of carbon-hydrogen bonds. *Reduction* corresponds to an increase in the number of carbon-hydrogen bonds and/or a decrease in the number of carbon-oxygen bonds.

A more general view of oxidation and reduction is that any element *more* electronegative than carbon will have the same effect on oxidation number as oxygen.

Any element *less* electronegative than carbon will have the same effect on oxidation number as hydrogen. This may be summarized as follows:

$$-\overset{|}{\underset{|}{C}}-X \underset{\text{reduction}}{\overset{\text{oxidation}}{\rightleftharpoons}} -\overset{|}{\underset{|}{C}}-Y$$

X is less
electronegative
than carbon

Y is more
electronegative
than carbon

SOLUTIONS TO TEXT PROBLEMS

2.1 Each of the carbons in propane is bonded to four other atoms. All the carbon atoms in propane are sp^3 hybridized.

$$\overset{\displaystyle \text{H} \quad \text{H} \quad \text{H}}{\underset{\displaystyle \text{H} \quad \text{H} \quad \text{H}}{\text{H}-\text{C}-\text{C}-\text{C}-\text{H}}}$$

All 10 of the bonds in propane are σ bonds. The eight C—H σ bonds arise from $C(sp^3)$–$H(1s)$ overlap. The two C—C bonds arise from $C(sp^3)$–$C(sp^3)$ overlap.

2.2 An unbranched alkane (*n*-alkane) of 28 carbons has 26 methylene (CH_2) groups flanked by a methyl (CH_3) group at each end. The condensed formula is $CH_3(CH_2)_{26}CH_3$.

2.3 The alkane represented by the carbon skeleton formula has 11 carbons. The general formula for an alkane is C_nH_{2n+2}, and thus there are 24 hydrogens. The molecular formula is $C_{11}H_{24}$; the condensed structural formula is $CH_3(CH_2)_9CH_3$.

2.4 A second isomer exists having a five-carbon chain with a one-carbon (methyl) branch:

$$\overset{\displaystyle CH_3}{\underset{}{CH_3CH_2CHCH_2CH_3}} \quad \text{or} \quad$$

The remaining two isomers have two methyl branches on a four-carbon chain.

$$\overset{\displaystyle CH_3}{\underset{\displaystyle CH_3}{CH_3CHCHCH_3}} \quad \text{or} \qquad\qquad \overset{\displaystyle CH_3}{\underset{\displaystyle CH_3}{CH_3CH_2CCH_3}} \quad \text{or}$$

2.5 (*b*) Octacosane is not listed in Table 2.4, but its structure can be deduced from its systematic name. The suffix *-cosane* pertains to alkanes that contain 20 to 29 carbons in their longest continuous chain. The prefix *octa-* means "eight." Therefore, octacosane is the unbranched alkane having 28 carbon atoms. It is $CH_3(CH_2)_{26}CH_3$.

(*c*) The alkane has an unbranched chain of 11 carbon atoms and is named *undecane*.

2.6 The ending *-hexadecane* reveals that the longest continuous carbon chain has 16 carbon atoms.

There are four methyl groups (represented by *tetramethyl-*) and they are located at carbons 2, 6, 10, and 14.

2,6,10,14-Tetramethylhexadecane
(phytane)

2.7 (*b*) The structures of the C_5H_{12} isomers and their common names are given in the text. The systematic name of the unbranched isomer is pentane (Table 2.4).

$$CH_3CH_2CH_2CH_2CH_3$$

IUPAC name: **pentane**
common name: *n*-pentane

The isomer $(CH_3)_2CHCH_2CH_3$ has four carbons in the longest continuous chain and so is named as a derivative of butane. Since it has a methyl group at C-2, it is 2-methylbutane.

$$CH_3CHCH_2CH_3$$
$$|$$
$$CH_3$$

IUPAC name: **2-methylbutane**
Common name: isopentane

methyl group at C-2

The isomer $(CH_3)_4C$ has three carbons in its longest continuous chain and so is named as a derivative of propane. There are two methyl groups at C-2, and so it is a 2,2-dimethyl derivative of propane.

$$CH_3$$
$$|$$
$$CH_3CCH_3$$
$$|$$
$$CH_3$$

IUPAC name: **2,2-dimethylpropane**
Common name: neopentane

(*c*) First write out the structure in more detail and identify the longest continuous carbon chain.

$$CH_3 \qquad H$$
$$| \qquad\quad |$$
$$CH_3—C—CH_2—C—CH_3$$
$$| \qquad\quad |$$
$$CH_3 \qquad CH_3$$

There are five carbon atoms in the longest chain and so the compound is named as a derivative of pentane. This five-carbon chain has three methyl substituents attached to it, making it a trimethyl derivative of pentane. We number the chain in the direction that gives the lowest numbers to the substituents at the first point of difference.

$$\overset{1}{CH_3}—\overset{2}{C}—\overset{3}{CH_2}—\overset{4}{C}—\overset{5}{CH_3} \quad not \quad \overset{5}{CH_3}\overset{4}{C}—\overset{3}{CH_2}—\overset{2}{C}—\overset{1}{CH_3}$$

2,2,4-Trimethylpentane   2,4,4-Trimethylpentane

(*d*) The longest continuous chain in $(CH_3)_3CC(CH_3)_3$ contains four carbon atoms.

$$CH_3 \quad CH_3$$
$$| \qquad\quad |$$
$$CH_3—C——C—CH_3$$
$$| \qquad\quad |$$
$$CH_3 \quad CH_3$$

The compound is named as a tetramethyl derivative of butane; it is 2,2,3,3-tetramethylbutane.

2.8 There are three C_5H_{11} alkyl groups with unbranched carbon chains. One is primary, while two are secondary. The IUPAC name of each group is given adjacent to the structure. Remember to number the alkyl groups from the point of attachment.

$$CH_3CH_2CH_2CH_2CH_2—$$ Pentyl group (primary)

$$CH_3CH_2CH_2CHCH_3$$ 1-Methylbutyl group (secondary)
$$|$$

$$CH_3CH_2CHCH_2CH_3$$ 1-Ethylpropyl group (secondary)
$$|$$

There are four alkyl groups derived from $(CH_3)_2CHCH_2CH_3$. Two are primary, one is secondary, and one is tertiary.

$$CH_3CHCH_2CH_2—$$
$$|\atop CH_3$$

3-Methylbutyl group (primary)

$$—CH_2CHCH_2CH_3$$
$$|\atop CH_3$$

2-Methylbutyl group (primary)

$$CH_3CCH_2CH_3$$
$$|\atop CH_3$$

1,1-Dimethylpropyl group (tertiary)

$$CH_3CHCHCH_3$$
$$|\atop CH_3$$

1,2-Dimethylpropyl group (secondary)

2.9 (*b*) The compound is named as a derivative of hexane, because it has six carbons in its longest continuous chain.

$$\overset{6}{C}H_3\overset{5}{C}H_2\overset{4}{C}H\overset{3}{C}H_2\overset{2}{C}H\overset{1}{C}H_3$$
$$\qquad\quad |\qquad\quad |$$
$$\qquad\quad CH_3CH_2\quad CH_3$$

The chain is numbered so as to give the lowest number to the substituent that appears closest to the end of the chain. In this case it is numbered so that the substituents are located at C-2 and C-4 rather than at C-3 and C-5. In alphabetical order the groups are ethyl and methyl; they are listed in alphabetical order in the name. The compound is 4-ethyl-2-methylhexane.

(*c*) The longest continuous chain is shown in the structure; it contains 10 carbon atoms. The structure also shows the numbering scheme that gives the lowest number to the substituent at the first point of difference.

$$\qquad\qquad\qquad CH_3\qquad\quad CH_3$$
$$\qquad\qquad\qquad |\qquad\qquad |$$
$$\overset{10}{C}H_3\overset{9}{C}H_2\overset{8}{C}H\overset{7}{C}H_2\overset{6}{C}H\overset{5}{C}H_2\overset{4}{C}H\overset{}{C}HCH_3$$
$$\qquad\quad |\qquad\qquad\quad |$$
$$\qquad CH_2CH_3\qquad\overset{}{C}H_2\overset{}{C}HCH_3$$
$$\qquad\qquad\qquad\qquad\qquad\quad |$$
$$\qquad\qquad\qquad\qquad\qquad CH_3$$

In alphabetical order, the substituents are ethyl (at C-8), isopropyl at (C-4), and two methyl groups (at C-2 and C-6). The alkane is 8-ethyl-4-isopropyl-2,6-dimethyldecane.

2.10 (*b*) There are 10 carbon atoms in the ring in this cycloalkane. It is cyclodecane.

(*c*) In alphabetical order, an isopropyl group and two methyl groups are substituents on a cyclodecane ring. The numbering pattern is chosen so as to give the lowest number to the substituent at the first point of difference between them. The correct name is 4-isopropyl-1,1-dimethylcyclodecane. Alternatively, the isopropyl group could be named 1-methylethyl, and the compound name would become 1,1-dimethyl-4-(1-methylethyl)cyclodecane.

(d) In this compound there are two cyclopropyl groups attached to the same terminal carbon of a four-carbon chain. The compound is 1,1-dicyclopropylbutane.

(e) There are four carbon atoms in the longest continuous chain in this compound. Cyclopropyl substituents are present at C-1 and C-2. The IUPAC name is 1,2-dicyclopropylbutane.

(f) When two cycloalkyl groups are attached by a single bond, the compound is named as a cycloalkyl-substituted cycloalkane. This compound is cyclohexylcyclohexane.

2.11 The two unbranched-chain alkanes have the greatest surface area, and the highest boiling points. Nonane has the higher boiling point of the two (151°C), followed by octane (126°C). The alkane with the highest degree of branching has the lowest boiling point. Thus 2,2,3,3-tetramethylbutane boils at 106°C and the remaining alkane, 2-methylheptane, boils at 116°C.

2.12 All hydrocarbons burn in air to give carbon dioxide and water. To balance the equation for the combustion of cyclohexane (C_6H_{12}), first balance the carbons and the hydrogens on the right side. Then balance the oxygens on the left side.

| Cyclohexane | Oxygen | Carbon dioxide | Water |

2.13 (b) Icosane (Table 2.4) is $C_{20}H_{42}$. It has four more methylene (CH_2) groups than hexadecane, the last unbranched alkane in Table 2.5. Therefore, its calculated heat of combusion is 4×653 kJ/mol higher.

$$\text{Heat of combustion of icosane} = \frac{\text{heat of combustion}}{\text{of hexadecane}} + 4(653) \text{ kJ/mol}$$

$$= 10{,}701 \text{ kJ/mol} + 2612 \text{ kJ/mol}$$

$$= 13{,}313 \text{ kJ/mol}$$

2.14 Two factors that influence the heats of combustion of alkanes are (1) the number of carbon atoms and (2) the extent of chain branching. Pentane, isopentane, and neopentane are all C_5H_{12}; hexane is C_6H_{14}. Hexane has the largest heat of combustion. Branching leads to a lower heat of combustion; neopentane is the most branched and has the lowest heat of combustion.

Hexane $CH_3(CH_2)_4CH_3$ Heat of combustion 4163 kJ/mol
                                                              (995.0 kcal/mol)

Pentane	$CH_3CH_2CH_2CH_2CH_3$	Heat of combustion 3527 kJ/mol
		(845.3 kcal/mol)
Isopentane	$(CH_3)_2CHCH_2CH_3$	Heat of combustion 3529 kJ/mol
		(843.4 kcal/mol)
Neopentane	$(CH_3)_4C$	Heat of combustion 3514 kJ/mol
		(839.9 kcal/mol)

2.15 (*b*) In the reaction

$$CH_2{=}CH_2 + Br_2 \longrightarrow BrCH_2CH_2Br$$

carbon becomes bonded to an atom (Br) that is more electronegative than itself. Carbon is *oxidized*.

(*c*) In the reaction

$$6CH_2{=}CH_2 + B_2H_6 \longrightarrow 2(CH_3CH_2)_3B$$

one carbon becomes bonded to hydrogen and is, therefore, *reduced*. The other carbon is also reduced, because it becomes bonded to boron, which is less electronegative than carbon.

2.16 It is best to approach problems of this type systematically. Since the problem requires all the isomers of C_7H_{16} to be written, begin with the unbranched isomer heptane.

$CH_3CH_2CH_2CH_2CH_2CH_2CH_3$ Heptane

Two isomers have six carbons in their longest continuous chain. One bears a methyl substituent at C-2, the other a methyl substituent at C-3.

$(CH_3)_2CHCH_2CH_2CH_2CH_3$ 2-Methylhexane

$CH_3CH_2CHCH_2CH_2CH_3$
$\qquad\qquad |$
$\qquad\quad CH_3$ 3-Methylhexane

Now consider all the isomers that have two methyl groups as substituents on a five-carbon continuous chain.

$(CH_3)_3CCH_2CH_2CH_3$ 2,2-Dimethylpentane

$(CH_3CH_2)_2C(CH_3)_2$ 3,3-Dimethylpentane

$(CH_3)_2CHCHCH_2CH_3$
$\qquad\qquad\; |$
$\qquad\qquad CH_3$ 2,3-Dimethylpentane

$(CH_3)_2CHCH_2CH(CH_3)_2$ 2,4-Dimethylpentane

There is one isomer characterized by an ethyl substituent on a five-carbon chain:

$(CH_3CH_2)_3CH$ 3-Ethylpentane

The remaining isomer has three methyl substituents attached to a four-carbon chain.

$(CH_3)_3CCH(CH_3)_2$ 2,2,3-Trimethylbutane

2.17 In the course of doing this problem, you will write and name the 17 alkanes that are constitutional isomers of octane.

(a) The easiest way to attack this part of the exercise is to draw a bond-line depiction of heptane and add a methyl branch to the various positions.

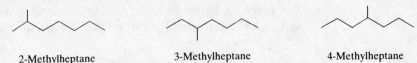

2-Methylheptane   3-Methylheptane   4-Methylheptane

Other structures bearing a continuous chain of seven carbons would be duplicates of these isomers rather than unique isomers. "5-Methylheptane," for example, is an incorrect name for 3-methylheptane, and "6-methylheptane" is an incorrect name for 2-methylheptane.

(b) The major group of isomers named as derivatives of hexane contain two methyl branches on a continuous chain of six carbons.

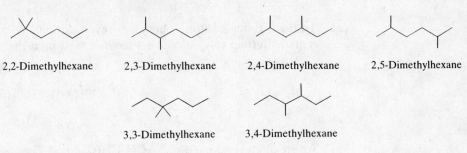

2,2-Dimethylhexane   2,3-Dimethylhexane   2,4-Dimethylhexane   2,5-Dimethylhexane

3,3-Dimethylhexane   3,4-Dimethylhexane

One isomer bears an ethyl substituent:

3-Ethylhexane

(c) Four isomers are trimethyl-substituted derivatives of pentane:

2,2,3-Trimethylpentane   2,3,3-Trimethylpentane   2,2,4-Trimethylpentane   2,3,4-Trimethylpentane

Two bear an ethyl group and a methyl group on a continuous chain of five carbons:

3-Ethyl-2-methylpentane   3-Ethyl-3-methylpentane

(d) Only one isomer is named as a derivative of butane:

2,2,3,3-Tetramethylbutane

2.18 (a) The ending *-hexane* tells us that the longest continuous chain in 3-ethylhexane has six carbons.

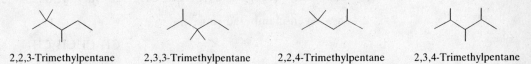

This six-carbon chain bears an ethyl group (CH_3CH_2) at the third carbon.

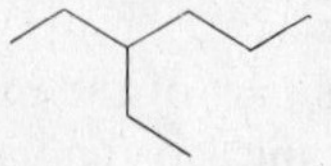

or $(CH_3CH_2)_2CHCH_2CH_2CH_3$

(*b*)   The longest continuous chain contains nine carbon atoms. Begin the problem by writing and numbering the carbon skeleton of nonane.

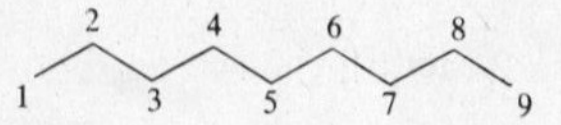

Now add two methyl groups (one to C-2 and the other to C-3) and an isopropyl group (to C-6) to give a structural formula for 2,3-dimethyl-6-isopropylnonane.

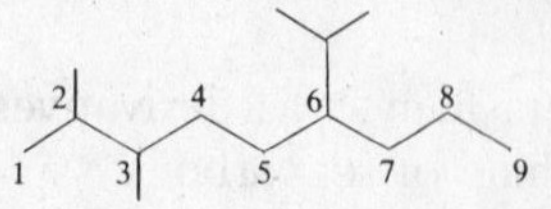

or

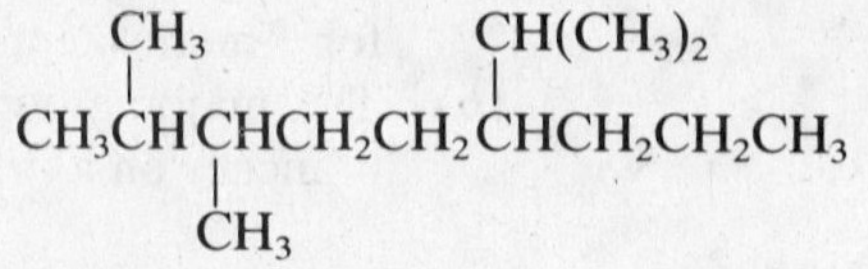

(*c*)   To the carbon skeleton of heptane (seven carbons) add a *tert*-butyl group to C-4 and a methyl group to C-3 to give 4-*tertyl*-butyl-3-methylheptane.

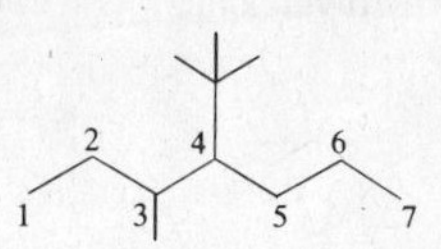

or

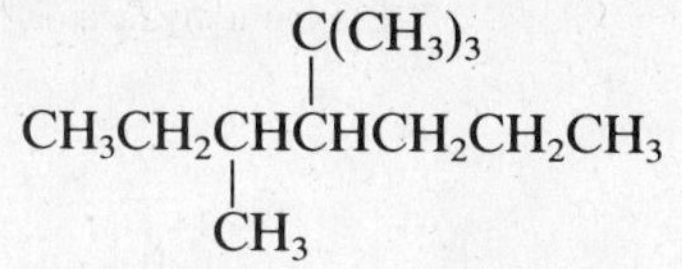

(*d*)   An isobutyl group is —$CH_2CH(CH_3)_2$. The structure of 4-isobutyl-1,1-dimethylcyclohexane is as shown.

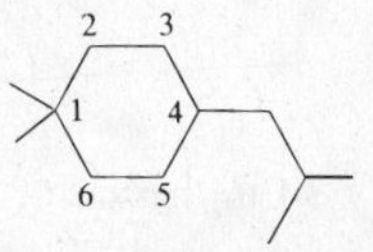

or

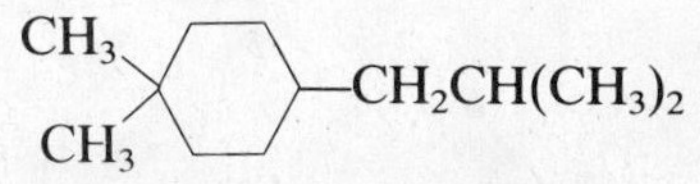

(*e*)   A *sec*-butyl group is $CH_3CHCH_2CH_3$. *sec*-Butylcycloheptane has a *sec*-butyl group on a seven-membered ring.

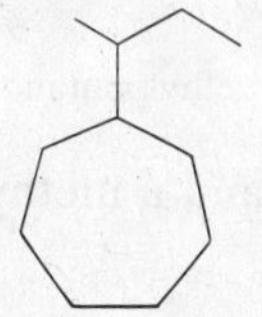

or

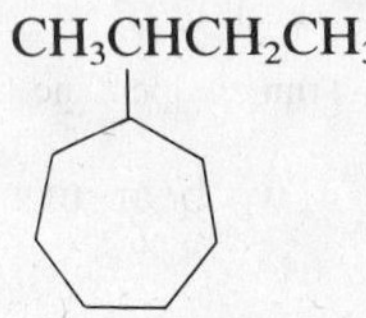

(*f*)   Dicyclopropylmethane has two cyclopropyl groups as substituents on the same carbon.

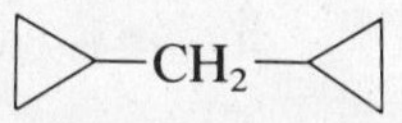

(*g*)   A cyclobutyl group is a substituent on a five-membered ring in cyclobutylcyclopentane.

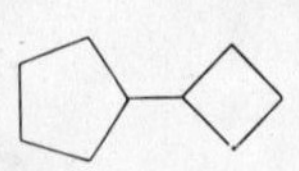

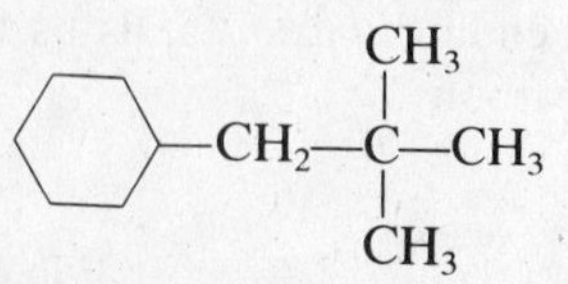

(*h*)   The structure of (2,2-dimethylpropyl)cyclohexane is

(*i*)   The name *pentacosane* contains no numerical locants or suffixes indicating the presence of alkyl groups. Therefore it must be an unbranched alkane. Refer to Table 2.4 in the text to find that the suffix *-cosane* refers to alkanes with 20 to 29 carbons. The prefix *penta-* stands for "five," and so pentacosane must be the unbranched alkane with 25 carbons. Its condensed structural formula is $CH_3(CH_2)_{23}CH_3$.

(*j*)   We need to add a 1-methylpentyl group to C-10 of pentacosane. A 1-methylpentyl group is:

$$\underset{\overset{|}{CH_3}}{-\overset{1}{C}H\overset{2}{C}H_2\overset{3}{C}H_2\overset{4}{C}H_2\overset{5}{C}H_3}$$

It has five carbons in the longest continuous chain counting from the point of attachment, and bears a methyl group at C-1. Therefore, 10-(1-methylpentyl)pentacosane is:

$$\underset{\overset{|}{CH_3CHCH_2CH_2CH_2CH_3}}{CH_3(CH_2)_8CH(CH_2)_{14}CH_3}$$

2.19 (*a*)   This compound is an unbranched alkane with 27 carbons. As noted in parts (*i*) and (*j*) of the preceding problem, inspection of Table 2.4 reveals that unbranched alkanes with 20 to 29 carbons have names ending in *-cosane*. Thus, we add the prefix *hepta-* ("seven") to *-cosane* to name the alkane $CH_3(CH_2)_{25}CH_3$ as *heptacosane*.

(*b*)   The alkane $(CH_3)_2CHCH_2(CH_2)_{14}CH_3$ has 18 carbons in its longest continuous chain. It is named as a derivative of *octadecane*. There is a single substituent, a methyl group at C-2. The compound is *2-methyloctadecane*.

(*c*)   Write the structure out in more detail to reveal that it is *3,3,4-triethylhexane*.

$(CH_3CH_2)_3CCH(CH_2CH_3)_2$ is rewritten as $\underset{\overset{|}{\overset{}{CH_3CH_2}}}{\overset{\overset{CH_3CH_2}{\overset{|}{}}}{\overset{1}{C}H_3\overset{2}{C}H_2\overset{3}{C}\!\!-\!\!-\!\!\overset{4}{C}H\overset{CH_2CH_3}{\overset{|}{}}CH\overset{5}{C}H_2\overset{6}{C}H_3}}$

(*d*)   Each line of a bond-line formula represents a bond between two carbon atoms. Hydrogens are added so that the number of bonds to each carbon atom totals four.

is the same as $\underset{\overset{|}{CH_2CH_3}}{CH_3CH_2CHCH_2C(CH_3)_3}$

The IUPAC name is *4-ethyl-2,2-dimethylhexane*.

(*e*)

is the same as $\underset{\overset{|}{CH_3}\quad\overset{|}{CH_3}}{CH_3CH_2CHCH_2CHCH_2CH_3}$

The IUPAC name is *3,5-dimethylheptane*.

(*f*)

is the same as

The IUPAC name is *1-butyl-1-methylcyclooctane*.

(*g*)   Number the chain in the direction shown to give *3-ethyl-4,5,6-trimethyloctane*. When numbered in the opposite direction, the locants are also 3, 4, 5, and 6. In the case of ties, however, choose the direction that gives the lower number to the substituent that appears first in the name. *Ethyl* precedes *methyl* alphabetically.

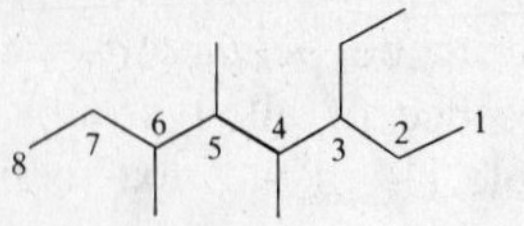

The structural diagram showing carbon numbering 1-8 with branches appears at the top.

2.20 (*a*) The group $CH_3(CH_2)_{10}CH_2-$ is an unbranched alkyl group with 12 carbons. It is a *dodecyl* group. The carbon at the point of attachment is directly attached to only one other carbon. It is a primary alkyl group.

(*b*) The longest continuous chain from the point of attachment is six carbons; it is a hexyl group bearing an ethyl substituent at C-3. The group is a *3-ethylhexyl group*. It is a primary alkyl group.

$$\overset{1}{-CH_2}\overset{2}{CH_2}\overset{3}{CH}\overset{4}{CH_2}\overset{5}{CH_2}\overset{6}{CH_3}$$
$$|$$
$$CH_2CH_3$$

(*c*) By writing the structural formula of this alkyl group in more detail, we see that the longest continuous chain from the point of attachment contains three carbons. It is a *1,1-diethylpropyl* group. Because the carbon at the point of attachment is directly bonded to three other carbons, it is a tertiary alkyl group.

$-C(CH_2CH_3)_3$.   is rewritten as

$$CH_2CH_3$$
$$|$$
$$\overset{1}{-C}-\overset{2}{CH_2}-\overset{3}{CH_3}$$
$$|$$
$$CH_2CH_3$$

(*d*) This group contains four carbons in its longest continuous chain. It is named as a butyl group with a cyclopropyl substituent at C-1. It is a *1-cyclopropylbutyl* group and is a secondary alkyl group.

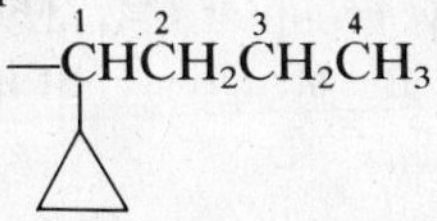

$$\overset{1}{-CH}\overset{2}{CH_2}\overset{3}{CH_2}\overset{4}{CH_3}$$

(*e*, *f*) A two-carbon group that bears a cyclohexyl substituent is a *cyclohexylethyl* group. Number from the point of attachment when assigning a locant to the cyclohexyl group.

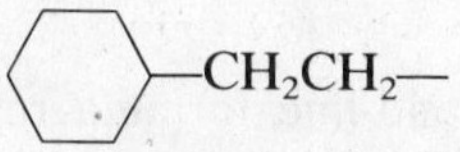

$-CH_2CH_2-$

2-Cyclohexylethyl
(primary)

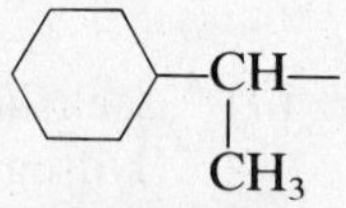

$-CH-$
$|$
CH_3

1-Cyclohexylethyl
(secondary)

2.21 The IUPAC name for pristane reveals that the longest chain contains 15 carbon atoms (as indicated by *-pentadecane*). The chain is substituted with four methyl groups at the positions indicated in the name.

Pristane (2,6,10,14-tetramethylpentadecane)

2.22 All single bonds are σ bonds. Draw a structural formula of each compound to reveal the number of bonds in the molecule.

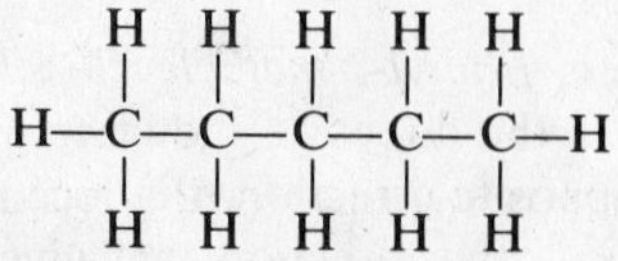

Pentane: 16 σ bonds

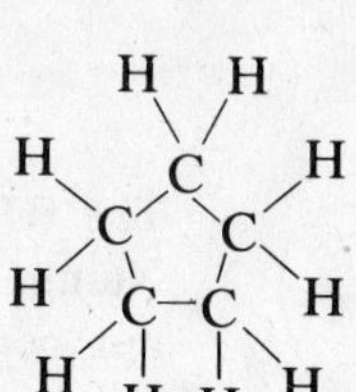

Cyclopentane: 15 σ bonds

2.23 (*a*) An alkane having 100 carbon atoms has $2(100) + 2 = 202$ hydrogens. The molecular formula of hectane is $C_{100}H_{202}$ and the condensed structural formula is $CH_3(CH_2)_{98}CH_3$. The 100 carbon atoms are connected by 99 σ bonds. The total number of σ bonds is 301 (99 C—C bonds + 202 C—H bonds).

(*b*) Unique compounds are formed by methyl substitution at carbons 2 through 50 on the 100-carbon chain (C-51 is identical to C-50, and so on). There are 49 *x*-methylhectanes.

(*c*) Compounds of the type 2,*x*-dimethylhectane can be formed by substitution at carbons 2 through 99. There are 98 of these compounds.

(*d*) Each carbon brings 6 electrons and each hydrogen 1. Therefore 100 carbons contribute 600 electrons and 202 hydrogens contribute 202. The total number of electrons is 802.

2.24 Isomers are different compounds that have the same molecular formula. In all these problems the safest approach is to write a structural formula and then count the number of carbons and hydrogens.

(*a*) Among this group of compounds, only butane and isobutane have the same molecular formula; only these two are isomers.

$$CH_3CH_2CH_2CH_3 \qquad \square \qquad \underset{\substack{|\\CH_3}}{CH_3CHCH_3} \qquad \underset{\substack{|\\CH_3}}{CH_3CHCH_2CH_3}$$

$$\begin{array}{cccc} \text{Butane} & \text{Cyclobutane} & \text{Isobutane} & \text{2-Methylbutane} \\ C_4H_{10} & C_4H_8 & C_4H_{10} & C_5H_{12} \end{array}$$

(*b*) The two compounds that are isomers, i.e., those that have the same molecular formula, are 2,2-dimethylpentane and 2,2,3-trimethylbutane.

$$\underset{\substack{|\\CH_3}}{\overset{\substack{CH_3\\|}}{CH_3CCH_2CH_2CH_3}} \qquad\qquad \underset{\substack{|\\CH_3}}{\overset{\substack{CH_3\ \ CH_3\\|\qquad|}}{CH_3C\!-\!\!-\!\!-\!CHCH_3}}$$

$$\begin{array}{cc} \text{2,2-Dimethylpentane} & \text{2,2,3-Trimethylbutane} \\ C_7H_{16} & C_7H_{16} \end{array}$$

Cyclopentane and neopentane are not isomers of these two compounds, nor are they isomers of each other.

$$\pentagon \qquad\qquad \underset{\substack{|\\CH_3}}{\overset{\substack{CH_3\\|}}{CH_3CCH_3}}$$

$$\begin{array}{cc} \text{Cyclopentane} & \text{Neopentane} \\ C_5H_{10} & C_5H_{12} \end{array}$$

(*c*) The compounds that are isomers are cyclohexane, methylcyclopentane, and 1,1,2-trimethylcyclopropane.

$$\begin{array}{ccc} \hexagon & CH_3\!-\!\pentagon & \triangle\ (CH_3)_2\ CH_3 \end{array}$$

$$\begin{array}{ccc} \text{Cyclohexane} & \text{Methylcyclopentane} & \text{1,1,2-Trimethylcyclopropane} \\ C_6H_{12} & C_6H_{12} & C_6H_{12} \end{array}$$

Hexane, $CH_3CH_2CH_2CH_2CH_2CH_3$, has the molecular formula C_6H_{14}; it is not an isomer of the others.

(*d*) The three that are isomers all have the molecular formula C_5H_{10}.

Ethylcyclopropane
C_5H_{10}

1,1-Dimethylcyclopropane
C_5H_{10}

Cyclopentane
C_5H_{10}

1-Cyclopropylpropane is not an isomer of the others. Its molecular formula is C_6H_{12}.

(e) Only 4-methyltetradecane and pentadecane are isomers. Both have the molecular formula $C_{15}H_{32}$.

$$CH_3(CH_2)_2CH(CH_2)_9CH_3$$
$$\overset{|}{CH_3}$$

4-Methyltetradecane
$C_{15}H_{32}$

$$CH_3(CH_2)_{13}CH_3$$

Pentadecane
$C_{15}H_{32}$

$$CH_3\overset{\overset{CH_3}{|}}{C}H\overset{}{C}H\overset{\overset{CH_3}{|}}{C}H\overset{}{C}H(CH_2)_4CH_3$$
$$\overset{|}{CH_3}\quad\overset{|}{CH_3}$$

2,3,4,5-Tetramethyldecane
$C_{14}H_{30}$

$$CH_3CH_2CH_2CH(CH_2)_5CH_3$$

4-Cyclobutyldecane
$C_{14}H_{28}$

2.25 (a) Ibuprofen is

$$(CH_3)_2CHCH_2—\overset{\overset{CH_3}{|}}{C}H\overset{}{C}\underset{\overset{||}{O}}{O}H$$

(b) Mandelonitrile is

$$H—\overset{\overset{OH}{|}}{C}H—C≡N$$

2.26 Isoamyl acetate is

$$\underset{\underset{Methyl}{\nearrow}}{R}\overset{\overset{O}{||}}{C}O\underset{\underset{3-Methylbutyl}{\nwarrow}}{R'}\quad(ester)$$
which is
$$CH_3\overset{\overset{O}{||}}{C}OCH_2CH_2\overset{\overset{CH_3}{|}}{C}HCH_3$$

2.27 Thiols are characterized by the —SH group. n-Butyl mercaptan is $CH_3CH_2CH_2CH_2SH$.

2.28 (a) Methylene groups are —CH_2—. Therefore $ClCH_2CH_2CH_2CH_2Cl$ is the $C_4H_8Cl_2$ isomer in which all of the carbons belong to methylene groups.
(b) The $C_4H_8Cl_2$ isomer that lacks methylene groups is $(CH_3)_2CHCHCl_2$.

2.29 A quaternary carbon is directly bonded to four other carbons. The compound $C(CH_2Cl)_4$ has one quaternary carbon and four methylene groups.

2.30 Alkanes are characterized by the molecular formula C_nH_{2n+2}. The value of n can be calculated on the basis of the fact that the molecular weight is 240, and the atomic weights of carbon and hydrogen are 12 and 1, respectively.

$$12n + 1(2n + 2) = 240$$
$$14n = 238$$
$$n = 17$$

The molecular formula of the alkane is $C_{17}H_{36}$. Since the problem specifies that the carbon chain is unbranched, the hydrocarbon is heptadecane, $CH_3(CH_2)_{15}CH_3$.

2.31 Since it is an alkane, the sex attractant of the tiger moth has a molecular formula of C_nH_{2n+2}. The number of carbons and hydrogens may be calculated from its molecular weight.

$$12n + 1(2n + 2) = 254$$
$$14n = 252$$
$$n = 18$$

The molecular formula of the alkane is $C_{18}H_{38}$. In the problem it is stated that the sex attractant is a 2-methyl-branched alkane. It is therefore 2-methylheptadecane, $(CH_3)_2CHCH_2(CH_2)_{13}CH_3$.

2.32 When any hydrocarbon is burned in air, the products of combustion are carbon dioxide and water

(a) $CH_3(CH_2)_8CH_3 + \frac{31}{2}O_2 \longrightarrow 10CO_2 + 11H_2O$

      Decane   Oxygen   Carbon   Water
                                  dioxide

(b) $+\ 15O_2 \longrightarrow 10CO_2 + 10H_2O$

    Cyclodecane   Oxygen   Carbon   Water
                                    dioxide

(c) $-CH_3 \quad +\ 15O_2 \longrightarrow 10CO_2 + 10H_2O$

  Methylcyclononane   Oxygen   Carbon   Water
                                    dioxide

(d) $+ \frac{29}{2}O_2 \longrightarrow 10CO_2 + 9H_2O$

 Cyclopentylcyclopentane   Oxygen   Carbon   Water
                                    dioxide

2.33 To determine the quantity of heat evolved per unit mass of material, divide the heat of combustion by the molecular weight.

Methane   Heat of combustion = 890 kJ/mol (212.8 kcal/mol)
               Molecular weight = 16.0 g/mol
               Heat evolved per gram = 55.6 kJ/g (13.2 kcal/g)
Butane   Heat of combustion = 2876 kJ/mol (687.4 kcal/mol)
               Molecular weight = 58.0 g/mol
               Heat evolved per gram = 49.6 kJ/g (11.8 kcal/g)

When equal masses of methane and butane are compared, methane evolves more heat when it is burned.

Equal volumes of gases contain an equal number of moles, so that when equal volumes of methane and butane are compared, the one with the greater heat of combustion in kilocalories per mole gives off more heat. Butane evolves more heat when it is burned than does an equal volume of methane.

2.34 When comparing heats of combustion of alkanes, two factors are of importance:

1. The heats of combustion of alkanes increase as the number of carbon atoms increases.
2. An unbranched alkane has a greater heat of combustion than a branched isomer.

(a) In the group hexane, heptane, and octane, three unbranched alkanes are being compared. Octane (C_8H_{18}) has the most carbons and has the greatest heat of combustion. Hexane (C_6H_{14}) has the fewest carbons and the lowest heat of combustion. The measured values in this group are as follows:

Hexane   Heat of combustion 4163 kJ/mol (995.0 kcal/mol)
Heptane   Heat of combustion 4817 kJ/mol (1151.3 kcal/mol)
Octane   Heat of combustion 5471 kJ/mol (1307.5 kcal/mol)

(b) Isobutane has fewer carbons than either pentane or isopentane and so is the member of the group with the lowest heat of combustion. Isopentane is a 2-methyl-branched isomer of pentane and so has a lower heat of combustion. Pentane has the highest heat of combustion among these compounds.

Isobutane	$(CH_3)_3CH$	Heat of combustion 2868 kJ/mol (685.4 kcal/mol)
Isopentane	$(CH_3)_2CHCH_2CH_3$	Heat of combustion 3529 kJ/mol (843.4 kcal/mol)
Pentane	$CH_3CH_2CH_2CH_2CH_3$	Heat of combustion 3527 kJ/mol 845.3 kcal/mol)

(c) Isopentane and neopentane each have fewer carbons than 2-methylpentane, which therefore has the greatest heat of combustion. Neopentane is more highly branched than isopentane; neopentane has the lowest of combustion.

Neopentane	$(CH_3)_4C$	Heat of combustion 3514 kJ/mol (839.9 kcal/mol)
Isopentane	$(CH_3)_2CHCH_2CH_3$	Heat of combustion 3529 kJ/mol (843.4 kcal/mol)
2-Methylpentane	$(CH_3)_2CHCH_2CH_2CH_3$	Heat of combustion 4157 kJ/mol (993.6 kcal/mol)

(d) Chain branching has a small effect on heat of combustion; the number of carbons has a much larger effect. The alkane with the most carbons in this group is 3,3-dimethylpentane; it has the greatest heat of combustion. Pentane has the fewest carbons in this group and has the smallest heat of combustion.

Pentane	$CH_3CH_2CH_2CH_2CH_3$	Heat of combustion 3527 kJ/mol (845.3 kcal/mol)
3-Methylpentane	$(CH_3CH_2)_2CHCH_3$	Heat of combustion 4159 kJ/mol (994.1 kcal/mol)
3,3-Dimethylpentane	$(CH_3CH_2)_2C(CH_3)_2$	Heat of combustion 4804 kJ/mol (1148.3 kcal/mol)

(e) In this series the heat of combustion increases with increasing number of carbons. Ethylcyclopentane has the lowest heat of combustion; ethylcycloheptane has the greatest.

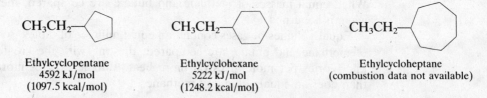

Ethylcyclopentane	Ethylcyclohexane	Ethylcycloheptane
4592 kJ/mol	5222 kJ/mol	(combustion data not available)
(1097.5 kcal/mol)	(1248.2 kcal/mol)	

2.35 (a) The equation for the hydrogenation of ethylene is given by the sum of the following three reactions:

(1) $$H_2(g) + \tfrac{1}{2}O_2(g) \longrightarrow H_2O(l)$$

$$\Delta H° = -286 \text{ kJ } (-68.4 \text{ kcal})$$

(2) $$CH_2{=}CH_2(g) + 3O_2(g) \longrightarrow 2CO_2(g) + 2H_2O(l)$$

$$\Delta H° = -1410 \text{ kJ } (-337.0 \text{ kcal})$$

(3) $$3H_2O(l) + 2CO_2(g) \longrightarrow CH_3CH_3(g) + \tfrac{7}{2}O_2(g)$$
$$\Delta H° = +1560 \text{ kJ } (+372.8 \text{ kcal})$$

Sum: $$CH_2{=}CH_2(g) + H_2(g) \longrightarrow CH_3CH_3(g)$$
$$\Delta H° = -136 \text{ kJ } (-32.6 \text{ kcal})$$

Equations (1) and (2) are the combustion of hydrogen and ethylene, respectively, and $\Delta H°$ values for these reactions are given in the statement of the problem. Equation (3) is the reverse of the combustion of ethane.

(b) Again we need to collect equations of reactions for which the $\Delta H°$ values are known.

(1) $$H_2(g) + \tfrac{1}{2}O_2(g) \longrightarrow H_2O(l)$$
$$\Delta H° = -286 \text{ kJ } (-68.4 \text{ kcal})$$

(2) $$HC{\equiv}CH(g) + \tfrac{5}{2}O_2(g) \longrightarrow 2CO_2(g) + H_2O(l)$$
$$\Delta H° = -1300 \text{ kJ } (-310.7 \text{ kcal})$$

(3) $$2CO_2(g) + 2H_2O(l) \longrightarrow CH_2{=}CH_2(g) + 3O_2(g)$$
$$\Delta H° = +1410 \text{ kJ } (+337.0 \text{ kcal})$$

Sum: $$HC{\equiv}CH(g) + H_2(g) \longrightarrow CH_2{=}CH_2(g)$$
$$\Delta H° = -176 \text{ kJ } (-42.1 \text{ kcal})$$

Equations (1) and (2) are the combustion of hydrogen and acetylene, respectively. Equation (3) is the reverse of the combustion of ethylene, and its value of $\Delta H°$ is the negative of the heat of combustion of ethylene.

The value of $\Delta H°$ for the hydrogenation of acetylene to ethane is equal to the sum of the two reactions just calculated:

$$HC{\equiv}CH(g) + H_2(g) \longrightarrow CH_2{=}CH_2(g)$$
$$\Delta H° = -176 \text{ kJ } (-42.1 \text{ kcal})$$

$$CH_2{=}CH_2(g) + H_2(g) \longrightarrow CH_3CH_3(g)$$
$$\Delta H° = -136 \text{ kJ } (-32.6 \text{ kcal})$$

Sum: $$HC{\equiv}CH(g) + 2H_2(g) \longrightarrow CH_3CH_3(g)$$
$$\Delta H° = -312 \text{ kJ } (-74.7 \text{ kcal})$$

(c) We use the equations for the combustion of ethane, ethylene, and acetylene as shown.

(1) $$2CH_2{=}CH_2(g) + 6O_2(g) \longrightarrow 4CO_2(g) + 4H_2O(l)$$
$$\Delta H° = -2820 \text{ kJ } (-674.0 \text{ kcal})$$

(2) $$2CO_2(g) + H_2O(l) \longrightarrow HC{\equiv}CH(g) + \tfrac{5}{2}O_2(g)$$
$$\Delta H° = +1300 \text{ kJ } (+310.7 \text{ kcal})$$

(3) $$3H_2O(l) + 2CO_2(g) \longrightarrow CH_3CH_3(g) + \tfrac{7}{2}O_2(g)$$
$$\Delta H° = +1560 \text{ kJ } (+372.8 \text{ kcal})$$

Sum: $$2CH_2{=}CH_2(g) \longrightarrow CH_3CH_3(g) + HC{\equiv}CH(g)$$
$$\Delta H° = +40 \text{ kJ } (+9.5 \text{ kcal})$$

The value of $\Delta H°$ for reaction (1) is twice that for the combustion of ethylene because two moles of ethylene are involved.

2.36 (a) The hydrogen content increases in going from $CH_3C{\equiv}CH$ to $CH_3CH{=}CH_2$. The organic compound $CH_3C{\equiv}CH$ is *reduced*.

(b) *Oxidation* occurs because a C—O bond has replaced a C—H bond in going from starting material to product.

(c) There are two carbon-oxygen bonds in the starting material and four carbon-oxygen bonds in the products. *Oxidation* occurs.

$$HO\text{---}CH_2CH_2\text{---}OH \longrightarrow 2H_2C\text{==}O$$

two C—O bonds   four C—O bonds

(d) While the oxidation state of carbon is unchanged in the process

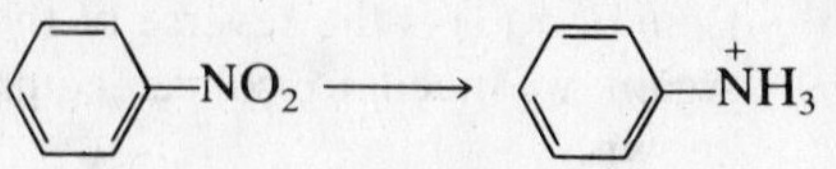

overall, *reduction* of the organic compound has occurred. Its hydrogen content has increased and its oxygen content has decreased.

SELF–TEST

PART A

A-1. Write the structure of each of the four-carbon alkyl groups. Give the common name and an acceptable systematic name for each.

A-2. How many σ bonds are present in each of the following?
(a) Nonane   (b) Cyclononane

A-3. Classify each of the following reactions according to whether the organic substrate is oxidized, reduced, or neither.

(a) $CH_3CH_3 + Br_2 \xrightarrow{\text{light}} CH_3CH_2Br + HBr$

(b) $CH_3CH_2Br + HO^- \longrightarrow CH_3CH_2OH + Br^-$

(c) $CH_3CH_2OH \xrightarrow[\text{heat}]{H_2SO_4} CH_2\text{==}CH_2$

(d) $CH_2\text{==}CH_2 + H_2 \xrightarrow{\text{Pt}} CH_3CH_3$

A-4. (a) Write a structural formula for 3-isopropyl-2,4-dimethylpentane.
(b) How many methyl groups are there in this compound? How many isopropyl groups?

A-5. Give the IUPAC name for the following substance:

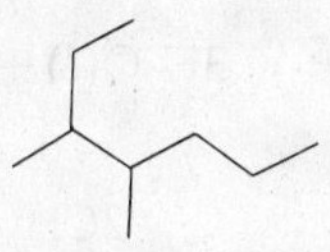

A-6. The compound in the previous question contains _____ primary carbon(s), _____ secondary carbon(s), and _____ tertiary carbon(s).

A-7. Write a balanced chemical equation for the complete combustion of 2,3-dimethylpentane.

A-8. Write structural formulas and give the names of all the constitutional isomers of C_5H_{10} that contain a ring.

A-9. Draw the structure indicated by the name 2,3-dimethyl-3-propylpentane. Is that name the correct one for this substance? If not, write the correct name.

A-10. Draw the structure of the octane isomer which has the highest boiling point.

A-11. Considering the combustion of a particular alkane,

$$CH_3(CH_2)_xCH_3 + yO_2 \longrightarrow zCO_2 + 14H_2O$$

what are the numerical values of x, y, and z? Is the heat evolved in this reaction more or less (per mole) than the heat evolved for the corresponding reaction of tetradecane?

PART B

B-1. Choose the response which best describes the following compounds:

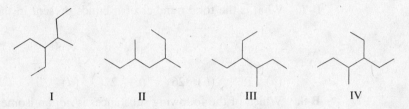

I   II   III   IV

(a) I, III, and IV represent the same compound.
(b) I and III are isomers of II and IV.
(c) I and IV are isomers of II and III.
(d) All the structures represent the same compound.

B-2. Which of the following is a correct name according to the IUPAC rules?
(a) 2-Methylcyclohexane
(b) 3,4-Dimethylpentane
(c) 2-Ethyl-2-methylpentane
(d) 3-Ethyl-2-methylpentane

B-3. Following are the structures of four isomers of hexane. Which of the names given will correctly identify an additional isomer?

$$CH_3CH_2CH_2CH_2CH_2CH_3 \qquad (CH_3)_3CCH_2CH_3$$
$$(CH_3)_2CHCH_2CH_2CH_3 \qquad (CH_3)_2CHCHCH_3$$
$$\qquad\qquad\qquad\qquad\qquad\qquad\qquad | $$
$$\qquad\qquad\qquad\qquad\qquad\qquad\qquad CH_3$$

(a) 2-Methylpentane   (c) 2-Ethylbutane
(b) 2,3-Dimethylbutane   (d) 3-Methylpentane

B-4. Which of the following is cyclohexylcyclohexane?

(a) $CH_2CH_2CH_2CH_2CH_2CH_3$ (c)

(b)   (d)

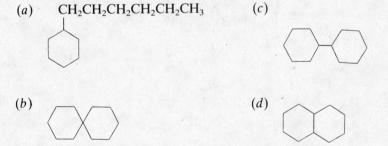

B-5. Which of the following structures is a "3-methylbutyl" group?

(a) $CH_3CH_2CH_2CH_2CH_2—$ (c) $CH_3CH_2CH—$
$$\qquad\qquad\qquad\qquad\qquad\qquad\qquad\qquad\qquad | $$
$$\qquad\qquad\qquad\qquad\qquad\qquad\qquad\qquad\qquad CH_2CH_3$$

(b) $(CH_3)_2CHCH_2CH_2—$ (d) $(CH_3)_3CCH_2—$

B-6. Rank the following substances in decreasing order of heats of combustion (largest → smallest).

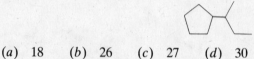

A B C

(a) B > A > C   (c) C > A > B
(b) B > C > A   (d) C > B > A

B-7. What is the total number of σ bonds present in the molecule shown?

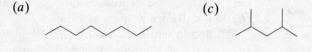

(a) 18   (b) 26   (c) 27   (d) 30

B-8. Which of the following substances is *not* an isomer of 3-ethyl-2-methylpentane?

(a)   (c)

(b)   (d) None of these
                                                              (all are isomers)

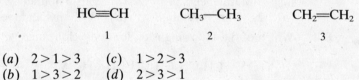

B-9. Which of the following isomers has the weakest intermolecular attractive van der Waals forces in the liquid state?

(a) $(CH_3)_3CCH_2CH_3$ (c) $CH_3CH_2CH_2CH_2CH_2CH_3$
(b) $CH_3CH_2CH_2CH(CH_3)_2$ (d) The attractive van der Waals forces of
                                           all isomers are the same.

B-10. Rank the following in order of decreasing oxidation state of carbon (most to least oxidized).

$HC≡CH$ $CH_3—CH_3$ $CH_2=CH_2$

   1   2   3

(a) 2 > 1 > 3   (c) 1 > 2 > 3
(b) 1 > 3 > 2   (d) 2 > 3 > 1

CONFORMATIONS OF ALKANES AND CYCLOALKANES

CHAPTER

3

IMPORTANT TERMS AND CONCEPTS

Conformations (Sec. 3.1) Structures that differ by rotation around single bonds are said to be different conformations. The study of this subject is known as *conformational analysis*. As shown in text Figure 3.2 for ethane, three types of representations are used to represent conformations; they are *wedge-and-dash* formulas, *sawhorse* drawings, and *Newman projections*.

Ethane Conformations (Secs. 3.1, 3.2) Of the many possible ethane conformations, those having minimum and maximum energy, respectively, are known as *staggered* and *eclipsed*. The angle between adjacent C—H bonds, known as the *torsion angle*, is 60° in the staggered conformation and 0° in the eclipsed conformation.

A *potential energy diagram*, as shown in text Figure 3.4, may be used to depict the energy changes as a molecule of ethane undergoes carbon-carbon bond rotation. As may be seen from the diagram, the staggered conformation is *more stable* than the eclipsed by 12 kJ/mol (2.9 kcal/mol). This difference is due to *torsional strain* in the eclipsed form resulting from repulsion of bonds on adjacent atoms. The difference in energy between the staggered and the eclipsed conformations is termed the *rotational energy barrier* for ethane.

Conformational Analysis of Butane and Higher Alkanes (Secs. 3.3, 3.4) Butane, when viewed looking down the central bond (between C-2 and C-3), has two distinct staggered conformations known as *gauche* and *anti*. In the former the torsion angle between the methyl groups is 60°; in the latter it is 180°. The two eclipsed conformations differ as to whether a methyl group is eclipsed with another methyl or with a hydrogen. The relative energies of the butane conformations may be seen in text Figure 3.7. Note that both the gauche and anti conformations are energy minima, with anti being more stable by 3.2 kJ/mol (0.8 kcal/mol). This energy difference may be accounted for by *van der Waals strain* between the methyl groups; that is, their electrons repel one another.

Higher alkanes are most stable in the all-anti conformation, which is the normal

47

conformation adopted when alkanes crystallize. More efficient packing is possible than would occur with other conformations.

Cycloalkanes (Sec. 3.5) Deviation of the C—C—C bond angles from the ideal tetrahedral value in cyclic hydrocarbons gives rise to destabilization resulting from *angle strain*. This phenomenon is most significant in cyclopropane, which is the only cycloalkane in which all the carbon atoms lie in a plane.

Cyclohexane Conformations (Secs. 3.6 to 3.8) Two of the nonplanar conformations of cyclohexane are named the *chair* and the *boat*. While both conformations have bond angles close to tetrahedral, the chair conformation is more stable than the boat because, in addition to being free of angle strain, it is free of torsional strain as well. A third conformation, the *twist* or *skew boat*, is less stable than the chair but more stable than the boat.

The carbon-hydrogen bonds in cyclohexane may be classified as being of one of two types, *axial* and *equatorial*. As shown in text Figure 3.13, the axial C—H bonds point alternately straight up and down from the ring. The equatorial bonds, on the other hand, alternate slightly up and down around the "equator" of the ring.

By a process of *conformational inversion*, also known as *ring flipping*, the axial and equatorial positions on the ring may be interchanged; that is, a group which is axial becomes equatorial, and vice versa. To draw the ring-flipping process, carbon atoms pointing "up" in the chair cyclohexane are moved "down"; those pointing "down" are moved "up."

Substituted Cyclohexanes (Secs. 3.9, 3.13, 3.14) In a monosubstituted cyclohexane, in which one hydrogen is replaced by a substituent such as a methyl group, the two chair conformations are no longer equivalent. In one the methyl substituent is axial; in the other it is equatorial.

Axial methyl   Equatorial methyl

The equatorial orientation of the methyl group is more stable and is the predominant form present (about 95 percent). Fewer van der Waals repulsions exist when the methyl group is equatorial; the result is that that conformation is more stable.

In general, *any* alkyl substituent is more stable in the equatorial orientation. As the group becomes bulkier, the preference for this orientation increases, being particularly pronounced for a *tert*-butyl group. At equilibrium the amount of axial *tert*-butylcyclohexane is too small to measure.

In disubstituted cyclohexanes the axial and equatorial orientations of each alkyl group must be determined in order to evaluate the relative stability of the two ring conformers, remembering that groups are more stable in the equatorial position. Groups on the same side of the ring are said to be *cis*; those on opposite sides are *trans* to each other. The possible orientations of alkyl disubstituted cyclohexanes (where *a* denotes axial and *e* denotes equatorial) are:

1,2-cis   *a, e* (larger group *e*)
1,2-trans   *e, e* favored over *a, a*
1,3-cis   *e, e* favored over *a, a*

1,3-trans *a, e* (larger group *e*)
1,4-cis *a, e* (larger group *e*)
1,4-trans *e, e* favored over *a, a*

Enthalpy, Free Energy, and Equilibrium Constant (Following Sec. 3.9) An equilibrium constant K is related to the difference in *free energy* ($\Delta G°$) between the initial and final states by

$$\Delta G° = G°_{final} - G°_{initial} = -RT \ln K$$

where T is the absolute temperature and R is the gas constant (8.314 J/mol · K, or 1.99 cal/mol · K).

The difference in free energy is related to enthalpy by the expression

$$\Delta G° = \Delta H° - T \Delta S°$$

where $\Delta S°$ is the difference in *entropy* between reactants and products.

Cyclopropane and Cyclobutane (Sec. 3.10) The planar arrangement of carbon atoms in cyclopropane results in eclipsing of the C—H bonds of each face of the molecule. The resulting torsional strain is in addition to the considerable angle strain from the three-membered ring.

Cyclobutane has less strain than cyclopropane. The nonplanar conformation shown in text Figure 3.18 is preferred, as less torsional strain is present than in the planar conformation.

Polycyclic Ring Systems (Sec. 3.15) When two or more atoms of a cycloalkane are common to more than one ring, the compounds are called *polycyclic*; they may be *bicyclic, tricyclic,* and so on. When a single carbon atom is common to two rings, the substance is called a *spirocyclic* substance.

Bicyclopentane   Spiropentane

The IUPAC name for a bicyclic substance is assigned by identifying the atoms common to more than one ring and naming the number of atoms in each "bridge" connecting these carbons. For example,

Bicyclo[4.3.0]nonane   Bicyclo[2.2.2]octane

where the asterisks denote bridgehead carbons.

Among the most important bicyclic hydrocarbons are *cis-* and *trans*-decalin. These ring systems are frequently found in natural substances such as steroids.

H   H

H   H

cis-Decalin   *trans*-Decalin

Heterocyclic Compounds (Sec. 3.16) Heterocyclic compounds are cyclic substances that contain at least one atom other than carbon in the ring. The most common *heteroatoms* are oxygen and nitrogen, although a number of heterocycles containing sulfur are known. Examples of heterocyclic compounds include tetrahydrofuran and piperidine.

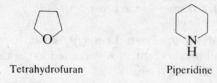

Tetrahydrofuran   Piperidine

SOLUTIONS TO TEXT PROBLEMS

3.1 (*b*) The sawhorse formula contains four carbon atoms in an unbranched chain. The compound is butane, $CH_3CH_2CH_2CH_3$.

(*c*) Rewrite the structure so as to better show its constitution. The compound is $CH_3CH_2CH(CH_3)_2$; it is 2-methylbutane.

(*d*) In this structure, we are sighting down the C-3—C-4 bond of a six-carbon chain. It is $CH_3CH_2CH_2CHCH_2CH_3$, or 3-methylhexane.
CH_3

3.2 All the staggered conformations of propane are equivalent to each other, and all its eclipsed conformations are equivalent to each other. The energy diagram resembles that of ethane in that it is a symmetrical one.

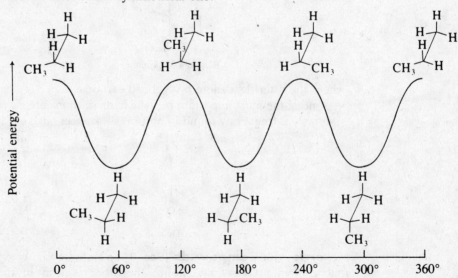

The activation energy for internal rotation in propane is expected to be somewhat higher than that in ethane because its eclipsed conformation incorporates an unfavorable van der Waals interaction between the methyl group and an eclipsed hydrogen. This interaction is, however, less repulsive than the corresponding methyl-methyl interaction of butane, which makes the activation energy for internal rotation less for propane than for butane.

3.3 The formula for the ring angles of a regular polygon is given in the problem as:

$$\frac{(n-2)}{(n)}(180°)$$

where n is the number of sides.

For cyclododecane, $n = 12$ and:

$$\frac{(n-2)}{(n)}(180°) = \frac{10}{12}(180°) = 150°$$

3.4 (*b*) In order to be gauche, substituents X and A must be related by a 60° torsion angle. Therefore, if A is axial as specified in the problem, X must be equatorial.

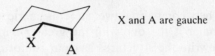

X and A are gauche

(*c*) In order for substituent X at C-1 to be anti to C-3, it must be equatorial.

(*d*) When X is axial at C-1, it is gauche to C-3.

3.5 (*b*) A methyl group is axial when it is "up" at C-1 but is equatorial when it is up at C-4 according to the numbering scheme given in the problem. Since substituents are more stable when they occupy equatorial rather than axial sites, a methyl group that is up at C-1 is less stable than one which is up at C-4.

(*c*) An equatorial substituent at C-3 is "down."

3.6 A *tert*-butyl group is much larger than a methyl group and has a greater preference for the equatorial position. The most stable conformation of 1-*tert*-butyl-1-methylcyclohexane has an axial methyl group and an equatorial *tert*-butyl group.

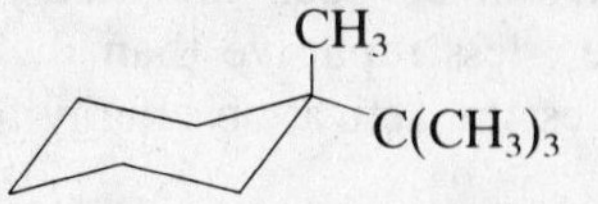

1-*tert*-Butyl-1-methylcyclohexane

3.7 The four constitutional isomers of *cis* and *trans*-1,2-dimethylcyclopropane that do not contain double bonds are

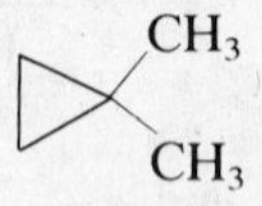

1,1-Dimethylcyclopropane

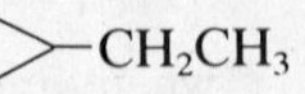

Ethylcyclopropane

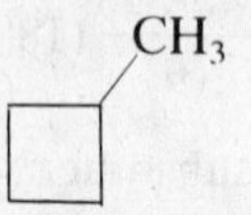

Methylcyclobutane

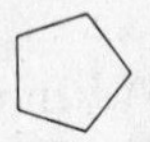

Cyclopentane

3.8 When comparing two stereoisomeric cyclohexane derivatives, the more stable stereo-isomer is the one with the greater number of its substituents in equatorial orientations. Rewrite the structures as chair conformations in order to see which substituents are axial and which are equatorial.

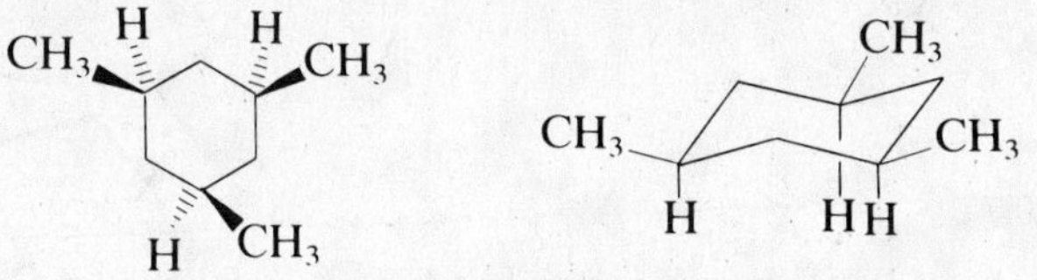

cis-1,3,5-Trimethylcyclohexane

All methyl groups are equatorial in *cis*-1,3,5-trimethylcyclohexane. It is more stable than *trans*-1,3,5-trimethylcyclohexane (shown in the following), which has one axial methyl group in its most stable conformation.

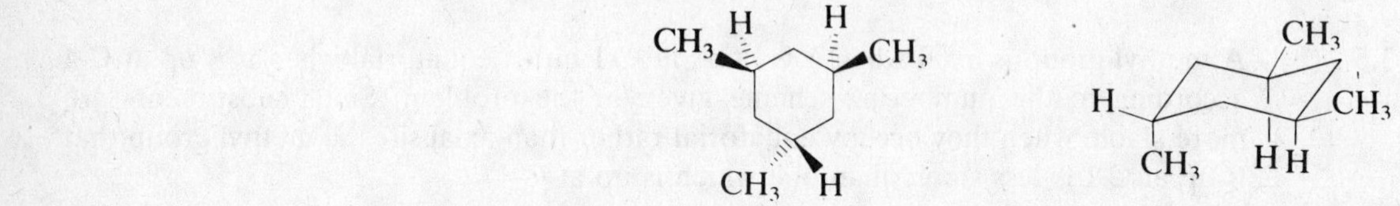

trans-1,3,5-Trimethylcyclohexane

3.9 In each of these problems, a *tert*-butyl group is the larger substituent and will be equatorial in the most stable conformation. Draw a chair conformation of cyclohexane, add an equatorial *tert*-butyl group, and then add the remaining substituent so as to give the required *cis* or *trans* relationship to the *tert*-butyl group.

(*b*) Begin by drawing a chair cyclohexane with an equatorial *tert*-butyl group. In *cis*-1-*tert*-butyl-3-methylcyclohexane the C-3 methyl group is equatorial.

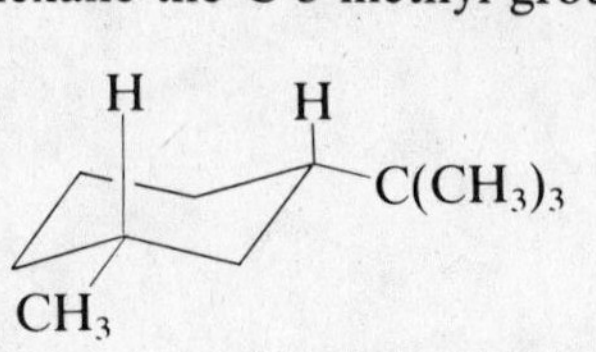

(*c*) In *trans*-1-*tert*-butyl-4-methylcyclohexane both the *tert*-butyl and the C-4 methyl group are equatorial.

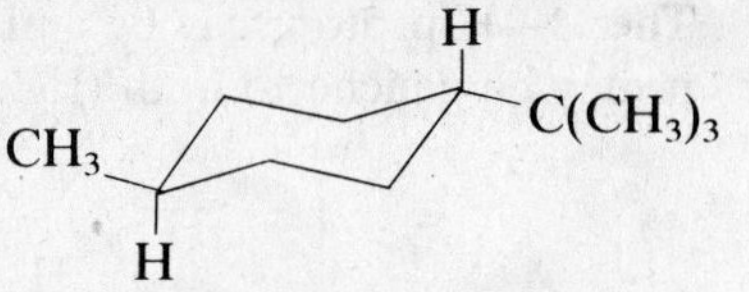

H

CH$_3$ — C(CH$_3$)$_3$

H

(d) Again the *tert*-butyl group is equatorial; however, in *cis*-1-*tert*-butyl-4-methylcyclohexane the methyl group on C-4 is axial.

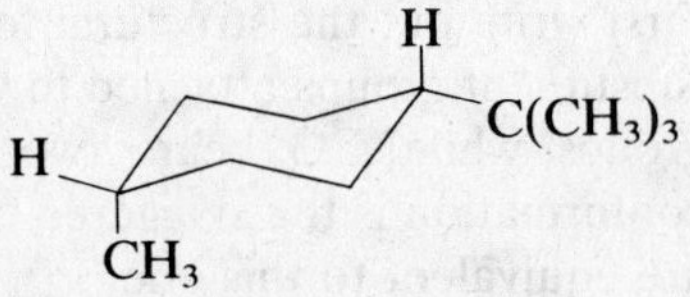

H

H — C(CH$_3$)$_3$

CH$_3$

3.10 (b) This bicyclic compound contains nine carbon atoms. The two carbons that are common to both rings are spanned by a five-carbon bridge and a two-carbon bridge. The 0 in the name bicyclo[5.2.0]nonane tells us that the third bridge has no atoms in it—the carbons that are common to both rings are directly attached to each other.

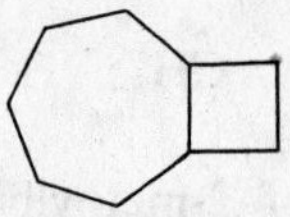

Bicyclo[5.2.0]nonane

(c) The three bridges in bicyclo[3.1.1]heptane contain three carbons, one carbon, and one carbon. The structure can be written in a form which shows the actual shape of the molecule or one that simply emphasizes its constitution.

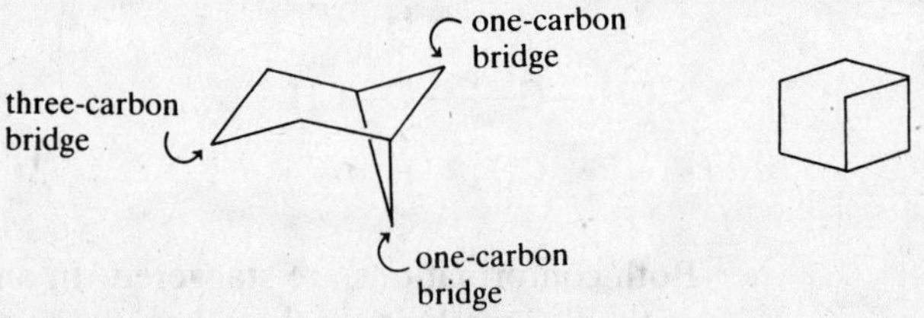

one-carbon bridge

three-carbon bridge

one-carbon bridge

(d) Bicyclo[3.3.0]octane has two five-membered rings that share a common side.

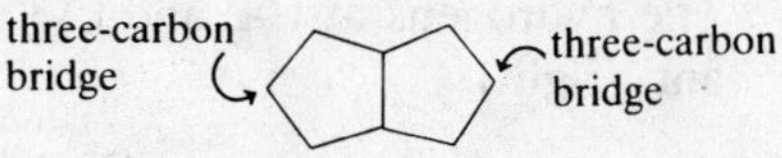

three-carbon bridge

three-carbon bridge

3.11 Since the two conformations are of approximately equal stability when R = H, it is reasonable to expect that the most stable conformation when R = CH$_3$ will have the CH$_3$ group equatorial.

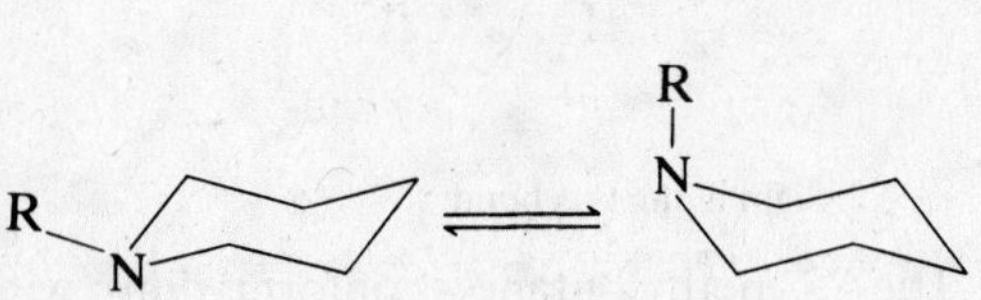

R

R — N ⇌ N

R = H: both conformations similar in energy
R = CH$_3$: more stable conformation has CH$_3$ equatorial

3.12 (a) Recall that a neutral nitrogen atom has three covalent bonds and an unshared electron pair. The three bonds are arranged in a trigonal pyramidal manner around each nitrogen in hydrazine (H$_2$NNH$_2$).

H

H   H

H

H   H

H   H

(*b*)   The O—H proton may be anti to one N—H proton and gauche to the other (left) or it may be gauche to both (right).

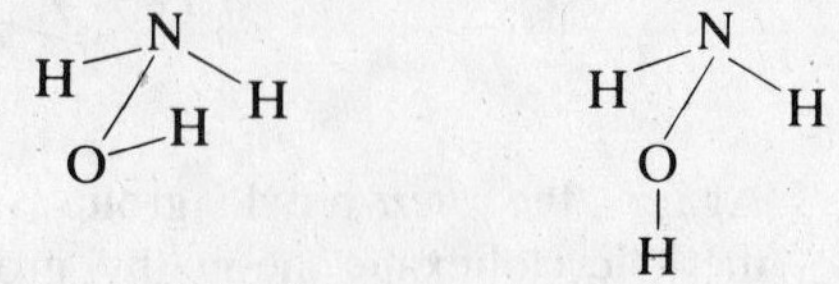

3.13 (*a*)   First write out the structural formula of 2,2-dimethylbutane in order to identify the substituent groups attached to C-2 and C-3. As shown at left, C-2 bears three methyl groups while C-3 bears two hydrogens and a methyl group. The most stable conformation is the staggered one shown at right. All other staggered conformations are equivalent to this one.

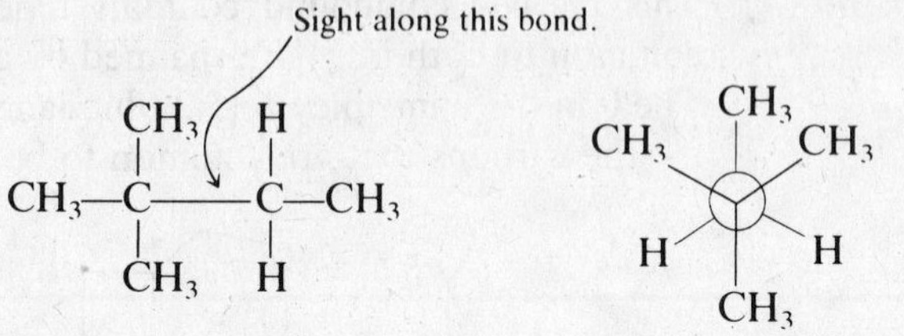

(*b*)   The constitution of 2-methylbutane and its two most stable conformations are shown.

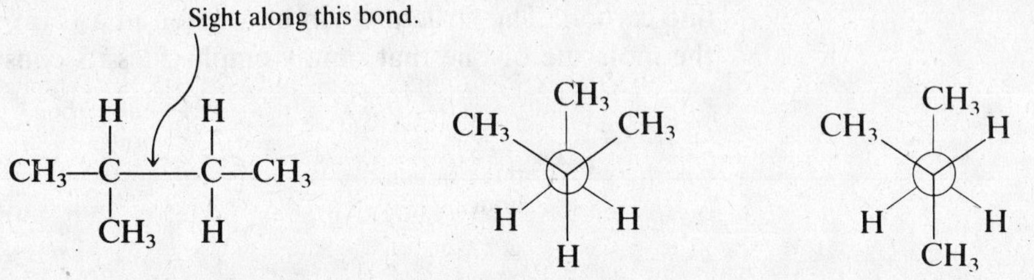

Both conformations are staggered. In one the methyl group at C-3 is gauche to each of the C-2 methyls. In the other, the methyl group at C-3 is gauche to one of the C-2 methyls and anti to the other.

(*c*)   The hydrogens at C-2 and C-3 may be gauche to one another (left) or they may be anti (right).

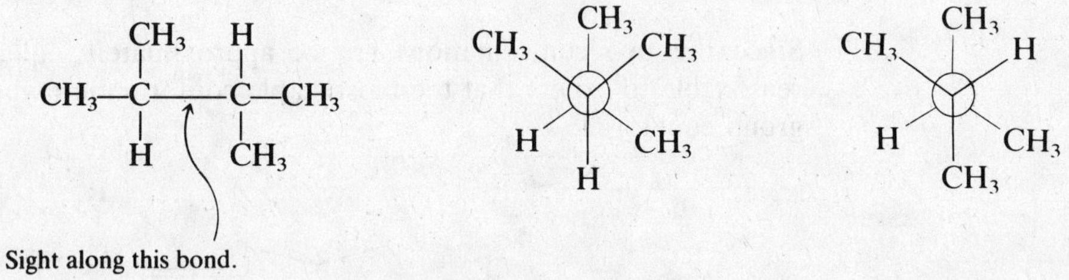

3.14 The 2-methylbutane conformation with one gauche $CH_3 \cdots CH_3$ and one anti $CH_3 \cdots CH_3$ relationship is more stable than the one with two gauche $CH_3 \cdots CH_3$ relationships. The more stable conformation has less van der Waals strain.

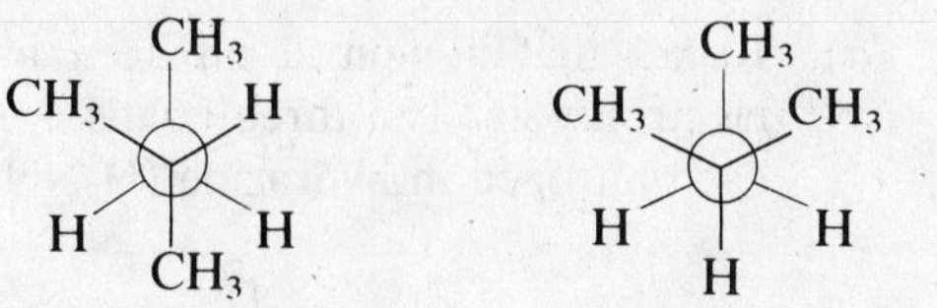

More stable   Less stable

3.15 All the staggered conformations about the C-2—C-3 bond of 2,2-dimethylpropane are equivalent to each other and of equal energy; they represent potential energy minima. All the eclipsed conformations are equivalent and represent potential energy maxima.

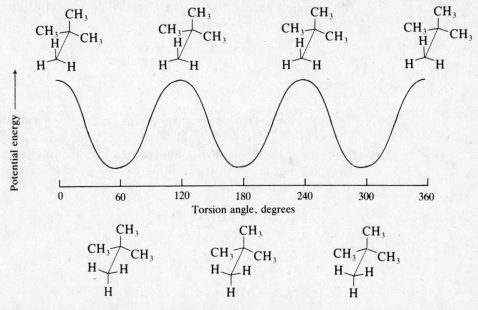

The shape of the potential energy profile for internal rotation in 2,2-dimethylpropane more closely resembles that of ethane than that of butane.

3.16 The potential energy diagram of 2-methylbutane more closely resembles that of butane than that of propane in that the three staggered forms are not all of the same energy. Similarly, not all of the eclipsed forms are of equal energy.

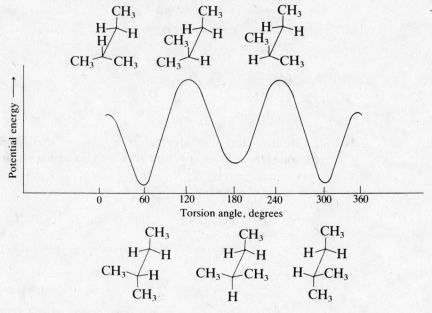

3.17 Van der Waals repulsions between the *tert*-butyl structural units in 2,2,4,4-tetramethylpentane cause the C-2—C-3—C-4 angle to open to 125 to 128°.

$$CH_3-\underset{\underset{CH_3}{|}}{\overset{\overset{CH_3}{|}}{C}}-CH_2-\underset{\underset{CH_3}{|}}{\overset{\overset{CH_3}{|}}{C}}-CH_3 \qquad \text{is equivalent to}$$

This angle is enlarged.

3.18 The structure given is not the most stable conformation, because the bonds of the methyl group are eclipsed with those of the ring carbon to which it is attached. The most stable conformation has the bonds of the methyl group and its attached carbon in a staggered relationship.

Bonds of methyl group eclipsed with those of attached carbon   Bonds of methyl group staggered with those of attached carbon

3.19 Structure A has its methyl group eclipsed with the ring bonds and is less stable than B. The methyl group in structure B has its bonds and those of its attached ring carbon in a staggered relationship.

A (less stable)   B (more stable)

Further, two of the hydrogens of the methyl group of A are uncomfortably close to two axial hydrogens of the ring.

3.20 Conformation D is more stable than C. The methyl groups are rather close together in C and there is a van der Waals repulsive force between them. In D the methyl groups are farther apart and do not repel each other.

Repulsion between cis methyl groups.   Methyl groups remain cis, but are far apart.

C   D

3.21 (a) By rewriting the structures in a form that shows the order of their atomic connections, it is apparent that the two structures are constitutional isomers.

is equivalent to $CH_3C(CH_3)_3$ (2,2-dimethylpropane)

is equivalent to $CH_3CH_2CH(CH_3)_2$ (2-methylbutane)

  (b) Both ball-and-stick models represent alkanes of molecular formula C_6H_{14}. In each one the carbon chain is unbranched. The two models are different conformations of the same compound, $CH_3CH_2CH_2CH_2CH_2CH_3$ (hexane).

  (c) The two compounds have the same constitution; both are $(CH_3)_2CHCH(CH_3)_2$. The Newman projections represent different staggered conformations of the same molecule: in one the hydrogen substituents are anti to each other, whereas in the other they are gauche.

$$H_3C \quad \overset{H}{\underset{H}{\diamond}} \quad CH_3 \qquad H_3C \quad \overset{CH_3}{\underset{CH_3}{\diamond}} \quad H$$

and   are different conformations of 2,3-dimethylbutane

Hydrogens at C-2   Hydrogens at C-2 and
and C-3 are anti.   C-3 are gauche.

(*d*) The compounds differ in the *order* in which the atoms are connected. They are constitutional isomers.

cis-1,2-dimethylcyclopentane   *trans*-1,3-Dimethylcyclopentane

(*e*) Both structures are *cis*-1-ethyl-4-methylcyclohexane (the methyl and ethyl groups are both "up"). In the structure on the left the methyl is axial and the ethyl equatorial. The orientations are opposite to these in the structure on the right. The two structures are ring-flipped forms of one another—different conformations of the same compound.

(*f*) The methyl and the ethyl groups are cis in the first structure but trans in the second. The two compounds are stereoisomers; they have the same constitution but differ in the arrangement of their atoms in space.

cis-1-Ethyl-4-methylcyclohexane
(both alkyl groups are up)

trans-1-Ethyl-4-methylcyclohexane
(ethyl group is down; methyl
group is up)

Do not be deceived because the six-membered rings look like ring-flipped forms. Remember, chair-chair interconversion converts all the equatorial bonds to axial and vice versa. Here the ethyl group is equatorial in both structures.

(*g*) The two structures have the same constitution, but differ in the arrangement of atoms in space. They are stereoisomers. The relationship is more clearly seen when some relevant hydrogens are included.

(*h*) The two structures have the same constitution but differ in the arrangement of their atoms in space; they are stereoisomers. They are not different conformations of the same molecule, because they are not related by rotation about C—C bonds. In the first structure as shown here the methyl group is trans to the darkened bonds, whereas in the second it is cis to these bonds.

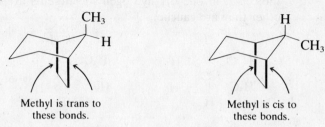

Methyl is trans to
these bonds.

Methyl is cis to
these bonds.

3.22 (*a*) There are three isomers of C_5H_8 that contain two rings and have no alkyl substituents:

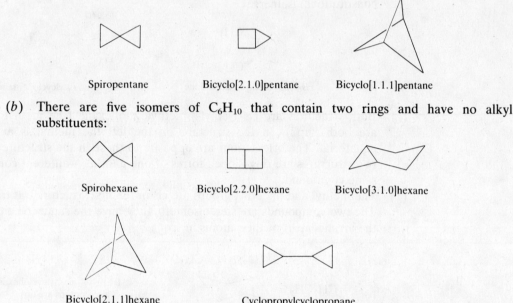

Spiropentane   Bicyclo[2.1.0]pentane   Bicyclo[1.1.1]pentane

(*b*) There are five isomers of C_6H_{10} that contain two rings and have no alkyl substituents:

Spirohexane   Bicyclo[2.2.0]hexane   Bicyclo[3.1.0]hexane

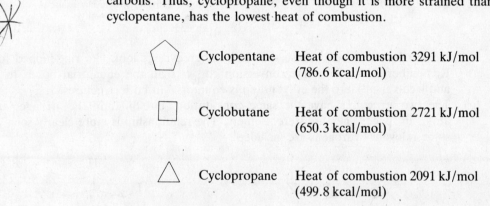

Bicyclo[2.1.1]hexane   Cyclopropylcyclopropane

3.23 (*a*) The heat of combustion is highest for the hydrocarbon with the greatest number of carbons. Thus, cyclopropane, even though it is more strained than cyclobutane or cyclopentane, has the lowest heat of combustion.

Cyclopentane   Heat of combustion 3291 kJ/mol
                 (786.6 kcal/mol)

Cyclobutane   Heat of combustion 2721 kJ/mol
                 (650.3 kcal/mol)

Cyclopropane   Heat of combustion 2091 kJ/mol
                 (499.8 kcal/mol)

(*b*) All the compounds in the group ethylcyclopropane, methylcyclobutane, and cyclopentane are isomers, since they all have the molecular formula C_5H_{10}. Ethylcyclopropane has the most angle strain and so has the highest heat of combustion. Cyclopentane has the least strain and so has the lowest heat of combustion.

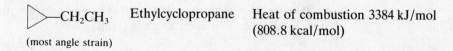

—CH₂CH₃   Ethylcyclopropane   Heat of combustion 3384 kJ/mol
                                   (808.8 kcal/mol)

(most angle strain)

$\diamond$—CH$_3$ Methylcyclobutane   Heat of combustion 3352 kJ/mol
(801.2 kcal/mol)

Cyclopentane   Heat of combustion 3291 kJ/mol
(786.6 kcal/mol)

(least angle strain)

(*c*)   All these compounds have the molecular formula C$_7$H$_{14}$. They are isomers, and so the one with the most strain will have the highest heat of combustion.

CH$_3$ CH$_3$

CH$_3$ CH$_3$

1,1,2,2-Tetramethylcyclopropane (high in angle strain; bonds are eclipsed; van der Waals repulsions between cis methyl groups)

Heat of combustion 4635 kJ/mol (1107.9 kcal/mol)

H   H

CH$_3$ CH$_3$

cis-1,2-Dimethylcyclopentane (low angle strain; some torsional strain; van der Waals repulsions between cis methyl groups)

Heat of combustion 4590 kJ/mol (1097.1 kcal/mol)

CH$_3$

Methylcyclohexane (minimal angle, torsional, and van der Waals strain)

Heat of combustion 4565 kJ/mol (1091.1 kcal/mol)

(*d*)   These hydrocarbons all have different molecular formulas. Their heats of combustion decrease with decreasing number of carbons.

Cyclopropylcyclopropane (C$_6$H$_{10}$)

Heat of combustion 3886 kJ/mol (928.8 kcal/mol)

Spiropentane (C$_5$H$_8$)

Heat of combustion 3296 kJ/mol (787.8 kcal/mol)

Bicyclo[1.1.0]butane (C$_4$H$_6$)

Heat of combustion 2648 kJ/mol (633.0 kcal/mol)

(*e*)   Bicyclo[3.3.0]octane and bicyclo[5.1.0]octane are isomers and their heats of combustion can be compared on the basis of their relative strains. The three-membered ring in bicyclo[5.1.0] octane imparts a significant amount of angle strain to this isomer, making it less stable than bicyclo[3.3.0]octane. The third hydrocarbon, bicyclo[4.3.0]nonane, has a greater number of carbons than either of the others and has the largest heat of combustion.

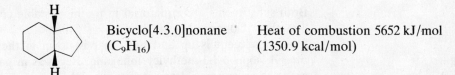

H

H

Bicyclo[4.3.0]nonane (C$_9$H$_{16}$)

Heat of combustion 5652 kJ/mol (1350.9 kcal/mol)

Bicyclo[5.1.0]octane   Heat of combustion 5089 kJ/mol
(C_8H_{14})   (1216.3 kcal/mol)

Bicyclo[3.3.0]octane   Heat of combustion 5016 kJ/mol
(C_8H_{14})   (1198.9 kcal/mol)

3.24 (*a*) The structural formula corresponding to 2,2,5,5-tetramethylhexane is $(CH_3)_3CCH_2CH_2C(CH_3)_3$. The substituents at C-3 are two hydrogens and a *tert*-butyl group. The substituents at C-4 are the same as those at C-3. The most stable conformation has the large *tert*-butyl groups anti to each other.

Anti conformation of
2,2,5,5-tetramethylhexane

(*b*) The zigzag conformation of 2,2,5,5-tetramethylhexane is an alternative way of expressing the same conformation implied in the Newman projection of part (*a*). It is more complete, however, in that it also shows the spatial arrangement of the atoms in the *tert*-butyl substituents.

equivalent to

2,2,5,5-Tetramethylhexane

(*c*) An isopropyl group is more conformationally demanding than a methyl group and so will occupy an equatorial site in the most stable conformation of *cis*-1-isopropyl-3-methylcyclohexane. Draw a chair conformation of cyclohexane and place an isopropyl group in an equatorial position.

$CH(CH_3)_2$

Notice that the equatorial methyl group is down on the carbon atom to which it is attached. Add a methyl group to C-3 so that it is also down.

$CH(CH_3)_2$

CH_3

Both substituents are equatorial in the most stable conformation of *cis*-1-isopropyl-3-methylcyclohexane.

(*d*) One substituent is up and the other is down in the most stable conformation of *trans*-1-isopropyl-3-methylcyclohexane. Begin as in part (*c*) by placing an isopropyl group in an equatorial orientation on a chair conformation of cyclohexane.

In order to be trans to the C-1 isopropyl group, the C-3 methyl group must be up.

The isopropyl group is thus equatorial and the methyl group axial in the most stable conformation.

(e)   In order to be cis to each other, one substituent must be axial and the other equatorial when they are located at positions 1 and 4 on a cyclohexane ring.

Place the larger substituent (the *tert*-butyl group) at the equatorial site and the smaller substituent (the ethyl group) at the axial one.

(f)   First write a chair conformation of cyclohexane, and then add two methyl groups at C-1 and draw in the axial and equatorial bonds at C-3 and C-4. Next, add methyl groups to C-3 and C-4 so that they are cis to each other. There are two different ways that this can be accomplished—either the C-3 and C-4 methyl groups are both up or they are both down.

More stable chair conformation: C-3 methyl group is equatorial; axial C-1 methyl group is not engaged in van der Waals repulsion with C-3 methyl

Less stable chair conformation: C-3 methyl group is axial; axial C-1 and C-3 methyl groups engaged in strong van der Waals repulsion

(g)   Translate the projection formula to a chair conformation.

Check to see if this is the most stable conformation by writing its ring-flipped form.

Less stable conformation: two axial methyl groups

More stable conformation: one axial methyl group

The ring-flipped form, with two equatorial methyl groups and one axial methyl group, is more stable than the originally drawn conformation, with two axial and one equatorial methyl groups.

3.25 Begin by writing each of the compounds in its most stable conformation. Compare them by examining their conformations for sources of strain, particularly van der Waals strain arising from groups located too close together in space.

(*a*) Its axial methyl group makes the cis stereoisomer of 1-isopropyl-2-methylcyclohexane less stable than the trans.

cis-1-Isopropyl-2-methylcyclohexane
(less stable stereoisomer)

trans-1-Isopropyl-2-methylcyclohexane
(more stable stereoisomer)

The axial methyl group in the cis stereoisomer is involved in an unfavorable van der Waals repulsion with the C-4 and C-6 axial hydrogens indicated in the drawing.

(*b*) Both groups are equatorial in the cis stereoisomer of 1-isopropyl-3-methylcyclohexane; cis is more stable than trans in 1,3-disubstituted cyclohexanes.

cis-1-Isopropyl-3-methylcyclohexane
(more stable stereoisomer; both
groups are equatorial)

trans-1-Isopropyl-3-methylcyclohexane
(less stable stereoisomer; methyl group
is axial and involved in van der Waals
repulsion with axial hydrogens at
C-1 and C-5)

(*c*) The more stable stereoisomer of 1,4-disubstituted cyclohexanes is the trans; both alkyl groups are equatorial in *trans*-1-isopropyl-4-methylcyclohexane.

cis-1-Isopropyl-4-methylcyclohexane
(less stable stereoisomer; methyl
group is axial and involved in van
der Waals repulsion with axial
hydrogens at C-2 and C-6)

trans-1-Isopropyl-4-methylcyclohexane
(more stable stereoisomer; both groups
are equatorial)

(*d*) The first stereoisomer of 1,2,4-trimethylcyclohexane is the more stable one. All its methyl groups are equatorial in its most stable conformation. The most stable conformation of the second stereoisomer has one axial and two equatorial methyl groups.

More stable stereoisomer

All methyl groups equatorial in
most stable conformation

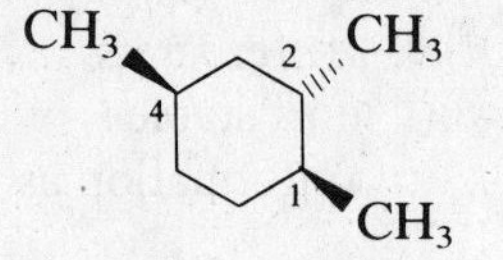

Less stable stereoisomer

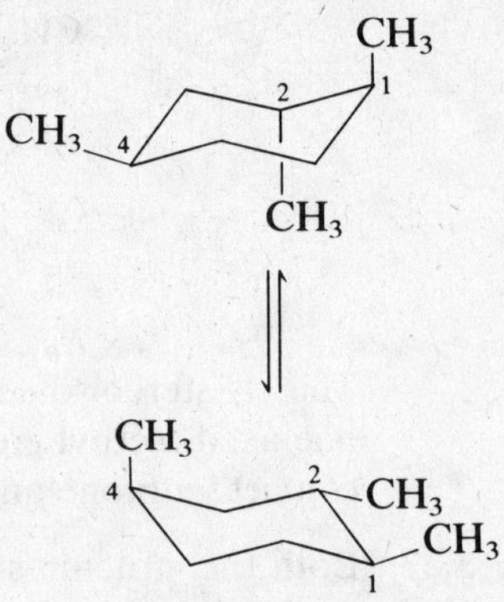

One axial methyl group in most stable conformation

(e) The first stereoisomer of 1,2,4-trimethylcyclohexane is the more stable one here, as it was in part (d). All its methyl groups are equatorial, while one of the methyl groups is axial in the most stable conformation of the second stereoisomer.

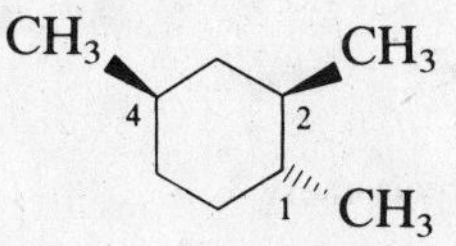

(More stable stereoisomer)

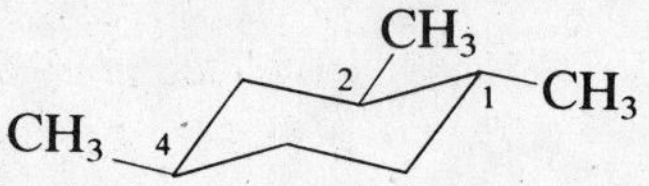

(All methyl groups equatorial in most stable conformation)

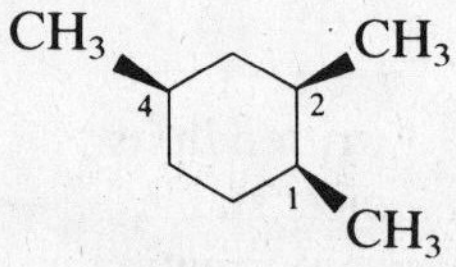

Less stable stereoisomer

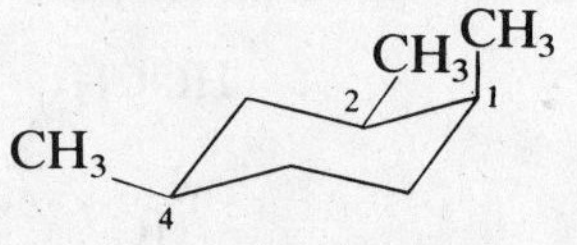

One axial methyl group in most stable conformation

(f) Each stereoisomer of 2,3-dimethylbicyclo[3.2.1]octane has one axial and one equatorial methyl group. The first one, however, has a close contact between its axial methyl group and both methylene groups of the two-carbon bridge. The second stereoisomer has a van der Waals repulsion with only one axial methylene group; it is more stable.

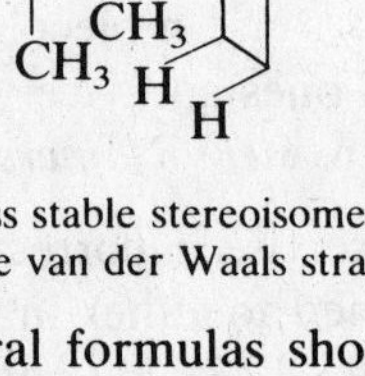

Less stable stereoisomer
(more van der Waals strain)

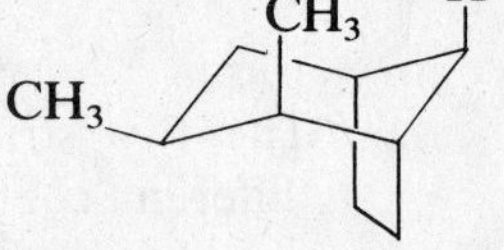

More stable stereoisomer
(less van der Waals strain)

3.26 First write structural formulas showing the relative stereochemistries and the preferred conformations of the two stereoisomers of 1,1,3,5-tetramethylcyclohexane.

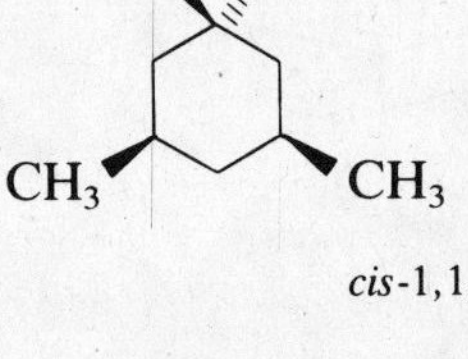

written in its most stable conformation as

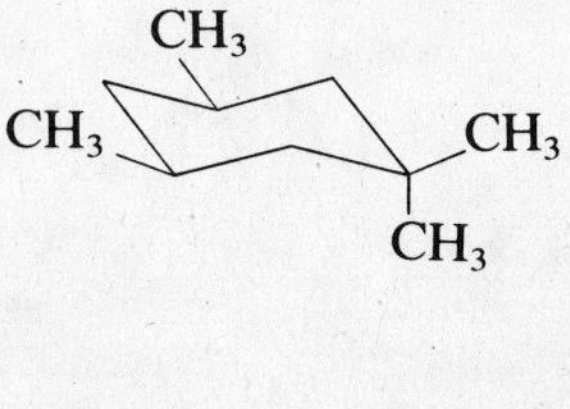

cis-1,1,3,5-Tetramethylcyclohexane

trans-1,1,3,5-Tetramethylcyclohexane

The cis stereoisomer is more stable than the trans. It exists in a conformation with only one axial methyl group, while the trans stereoisomer has two axial methyl groups in close contact with one another. The trans stereoisomer is destabilized by van der Waals strain.

3.27 Both the structures have approximately the same degree of angle strain and of torsional strain. Structure F has more van der Waals strain than E because two pairs of hydrogens (shown here) approach each other at distances that are rather close.

E:
more stable stereoisomer

Van der Waals repulsions
destabilize F

3.28 Conformational representations of the two different forms of glucose are drawn in the usual way. An oxygen atom is present in the six-membered ring, and we are told in the problem that the ring exists in a chair conformation.

written in its
most stable
conformation as

One axial OH substituent

written in its
most stable
conformation as

All substituents equatorial

The two structures are not interconvertible by ring flipping; therefore they are not different conformations of the same molecule. Remember, ring flipping transforms all axial substituents to equatorial ones and vice versa. The two structures differ with respect to only one substituent; they are *stereoisomers* of each other.

3.29 This problem is primarily an exercise in correctly locating equatorial and axial positions in cyclohexane rings that are joined together into a steroid skeleton. Parts (*a*) through (*e*) are concerned with positions 1, 4, 7, 11, and 12 in that order. The following diagram shows the orientation of axial and equatorial bonds at each of those positions.

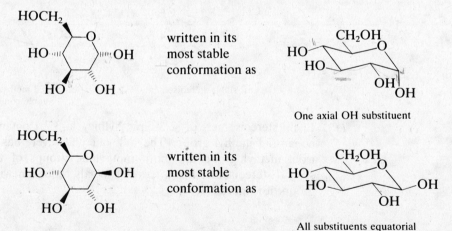

Both methyl groups are up.

(a)　At C-1 the bond that is cis to the methyl groups is equatorial (up).
(b)　At C-4 the bond that is cis to the methyl groups is axial (up).
(c)　At C-7 the bond that is trans to the methyl groups is axial (down).
(d)　At C-11 the bond that is trans to the methyl groups is equatorial (down).
(e)　At C-12 the bond that is cis to the methyl groups is equatorial (up).

3.30　Analyze this problem in exactly the same way as the preceding one by locating the axial and equatorial bonds at each position. It will be seen that the only differences are those at C-1 and C-4.

Both methyl groups are up.

(a)　At C-1 the bond that is cis to the methyl groups is axial (up).
(b)　At C-4 the bond that is cis to the methyl groups is equatorial (up).
(c)　At C-7 the bond that is trans to the methyl groups is axial (down).
(d)　At C-11 the bond that is trans to the methyl groups is equatorial (down).
(e)　At C-12 the bond that is cis to the methyl groups is equatorial (up).

3.31　(a)　The torsion angle between chlorine substituents is 60° in the gauche conformation and 180° in the anti conformation of $ClCH_2CH_2Cl$.

Gauche
(can have a dipole moment)

Anti
(cannot have a dipole moment)

(b)　The anti conformation of $ClCH_2CH_2Cl$ has a center of symmetry; all its individual bond dipole moments cancel and this conformation has no dipole moment. Since $ClCH_2CH_2Cl$ has a dipole moment of 1.12 D, it can exist entirely in the gauche conformation or it can be a mixture of anti and gauche conformations, but it cannot exist entirely in the anti conformation. Statement 1 is false.

3.32　(a)　The planar and nonplanar conformations of *trans*-1,3-dibromocyclobutane are as shown.

Four-membered ring planar
(cannot have a dipole moment)

Four-membered ring nonplanar
(can have a dipole moment)

(b)　The planar conformation of *trans*-1,3-dibromocyclobutane has a center of symmetry and cannot have a dipole moment. The nonplanar conformation can have a dipole moment. Since *trans*-1,3-dibromocyclobutane has a dipole moment of 1.10 D, it may exist entirely in the nonplanar conformation or it may exist as a mixture of the planar and nonplanar forms, but it cannot exist entirely in the planar conformation. Statement 1 is false.

SELF-TEST

PART A

A-1. Draw the gauche conformation of $CH_3CH_2CH_2Cl$ using both a Newman and a sawhorse projection.

A-2. Write Newman projection formulas for
(a) The least stable conformation of butane
(b) Two different staggered conformations of 1,1,2,2-tetrachloroethane.

A-3. Considering the C_5H_{10} isomers which contain a ring:
(a) Which isomer contains both primary and secondary carbons but no tertiary carbons?
(b) Which isomer has the smallest heat of combustion?
(c) Which isomer contains a single methyl group?

A-4. Write the structure of the most stable conformation of the less stable stereoisomer of 1-isopropyl-3-methylcyclohexane.

A-5. Draw the most stable conformation of the following substance:

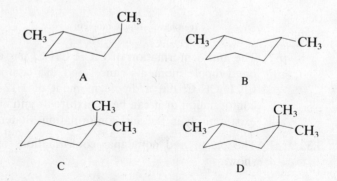

Which substituents are axial and which equatorial?

A-6. Two isomers A and B of a cyclic substance have heats of combustion of 5222 kJ/mol (1248 kcal/mol) and 5217 kJ/mol (1246 kcal/mol), respectively. Which isomer is more stable and which more strained?

A-7. Consider compounds A, B, C, and D.

(a) Which one is a constitutional isomer of two others?
(b) Which two are stereoisomers of one another?
(c) Which one has the highest heat of combustion?
(d) Which one has the stereochemical descriptor *trans* in its name?

A-8. Draw clear depictions of two nonequivalent chair conformations of *cis*-1-isopropyl-4-methylcyclohexane, and indicate which is more stable.

A-9. Which has the lower heat of combustion, *cis*-1-isopropyl-4-methylcyclohexane or *trans*-1-isopropyl-4-methylcyclohexane?

A-10. The high heat of combustion per CH_2 group in cyclopropane is attributed to a combination of _____ strain and _____ strain.

A-11. Draw the structure of *cis*-bicyclo[4.3.0]nonane.

PART B

B-1. Which of the listed terms best describes the relationship between the methyl groups in the chair conformation of the substance shown?

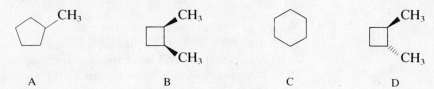

(*a*) eclipsed   (*b*) trans   (*c*) anti   (*d*) gauche

B-2. Rank the following substances in order of decreasing heat of combustion (largest → smallest).

(*a*) A > B > D > C   (*c*) C > D > B > A
(*b*) B > D > A > C   (*d*) A > C > B > D

B-3. Which of the following statements best describes the most stable conformation of *trans*-1,3-dimethylcyclohexane?
(*a*) Both methyl groups are axial.
(*b*) Both methyl groups are equatorial.
(*c*) One methyl group is axial, the other equatorial.
(*d*) The molecule is severely strained and cannot exist.

B-4. Compare the stability of the following two compounds:
  I   *cis*-1-Bromo-3-methylcyclohexane
  II   *trans*-1-Bromo-3-methylcyclohexane
(*a*) I is more stable.
(*b*) II is more stable.
(*c*) I and II are of equal stability.
(*d*) No comparison can be made.

B-5. What, if anything, can be said about the magnitude of the equilibrium constant K for the following process?

(*a*) $K = 1$ (*c*) $K < 1$
(*b*) $K > 1$ (*d*) No estimate of K can be made.

B-6. The two structures shown below are _____ each other.

(*a*) identical with   (*c*) constitutional isomers of
(*b*) conformations of   (*d*) None of these terms applies.

B-7. Which of the following energy versus rotation curves best describe(s) propane?

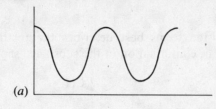

(a)

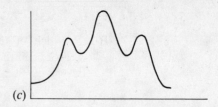

(c)

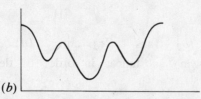

(b)   (d) Both (b) and (c)

B-8. Which of the following statements is *not* true concerning the conformational interconversion of *trans*-1,2-diethylcyclohexane?
(a) An axial group will be changed into the equatorial position.
(b) The energy of repulsions present in the molecule will be changed.
(c) Formation of the cis substance will result.
(d) One chair conformation will be greatly favored over the other.

B-9. The *most stable* conformation of the compound

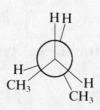

(in which all methyl groups are cis to one another) has:
(a) All methyl groups axial
(b) All methyl groups equatorial
(c) Equatorial methyl groups at C-1 and C-2
(d) Equatorial methyl groups at C-1 and C-4
(e) Equatorial methyl groups at C-2 and C-4

B-10. Which point on the potential energy diagram is represented by the Newman projection shown?

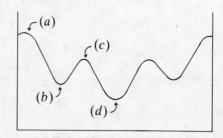

ALCOHOLS AND ALKYL HALIDES

IMPORTANT TERMS AND CONCEPTS

Nomenclature (Secs. 4.1 to 4.3) The functional group of an alkyl halide is a halogen atom (F, Cl, Br, I). Alcohols are a class of organic compounds characterized by the presence of the *hydroxyl (—OH) functional group*. The —OH group of an alcohol is bonded to an sp^3 hybridized carbon.

Alkyl halides and alcohols may be named by either the *radicofunctional* or the *substitutive* nomenclature method. *Ethyl bromide* and *methyl alcohol* are examples of radicofunctional names. The corresponding substitutive names are *bromoethane* and *ethanol*.

$$CH_3CH_2Br \qquad CH_3CH_2OH$$

Radicofunctional:	Ethyl bromide	Ethyl alcohol
Substitutive:	Bromoethane	Ethanol

Substitutive nomenclature is generally preferred.

The substitutive nomenclature of alkyl halides and alcohols is straightforward, using the general rules established in Chapter 2 for alkanes. The steps are:

1. Find the longest chain containing the functional group (—OH or halide).
2. Number the chain, keeping the —OH number as low as possible; halides are numbered as substituents with the same ranking as alkyl groups.
3. Alkyl substituents are named and numbered as usual.
4. For an alcohol, replace the ending −e of the alkane parent name with *-ol*, as shown in the following examples:

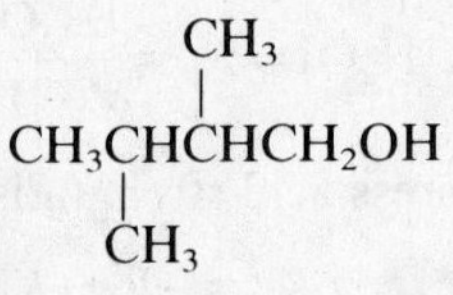

2,3-Dimethyl-1-butanol

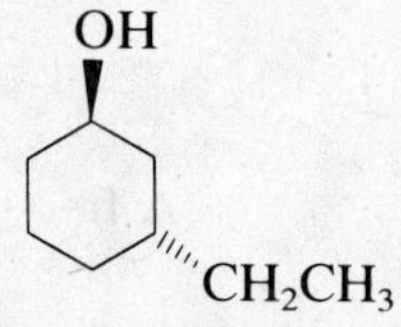

trans-3-Ethylcyclohexanol

5. For an alkyl halide, precede the stem name with the appropriate *halo-* prefix, as follows:

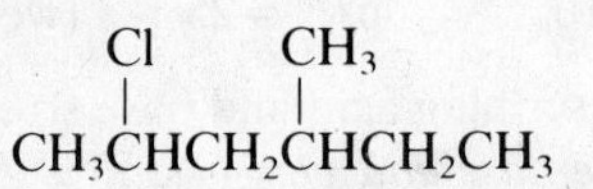

2-Chloro-4-methylhexane

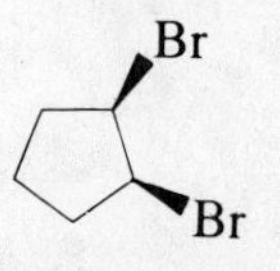

cis-1,2-Dibromocyclopentane

69

Alcohols and alkyl halides may be classified as primary, secondary, or tertiary according to the degree of substitution of the carbon bearing the functional group. Note that —X is often used to represent the halogen in alkyl halides.

$$
\underset{\text{Primary}}{R\!-\!\overset{\displaystyle H}{\underset{\displaystyle H}{C}}\!-\!OH(\!-\!X)} \qquad
\underset{\text{Secondary}}{R\!-\!\overset{\displaystyle H}{\underset{\displaystyle R}{C}}\!-\!OH(\!-\!X)} \qquad
\underset{\text{Tertiary}}{R\!-\!\overset{\displaystyle R}{\underset{\displaystyle R}{C}}\!-\!OH(\!-\!X)}
$$

Keep in mind that it is the carbon that bears the functional group that determines whether the group is described as being primary, secondary, or tertiary. Therefore, 2-methyl-1-propanol, $(CH_3)_2CHCH_2OH$, is a primary alcohol.

Physical Properties (Sec. 4.5) Because of the polar nature of the carbon-halogen bond in alkyl halides, these compounds exhibit *dipole-dipole* attractions. The polar nature of the —OH group of alcohols results in an especially effective form of dipole-dipole association called *hydrogen bonding*. As a result of hydrogen bonding, alcohols exhibit boiling points that are unusually high when compared with other compounds of similar molecular size.

The ability to form hydrogen bonds also affects the solubility of alcohols in water. Low-molecular-weight alcohols tend to be soluble in water. Larger alcohols become more "hydrocarbonlike" and lose their water-solubility.

Acid-Base Properties of Alcohols (Secs. 4.6 to 4.8) Alcohols may act as both *Brønsted bases* and *Brønsted acids*; that is, they may act both as *proton acceptors* (*Brønsted bases*) and as *proton donors* (*Brønsted acids*).

Acceptance of a proton in an acid-base reaction yields the *conjugate acid* of the alcohol:

$$R\!-\!OH + HA \rightleftharpoons R\!-\!\overset{+}{O}H_2 + A^-$$

Donation of a proton results in formation of the *conjugate base* of the alcohol, known as an *alkoxide ion*.

$$R\!-\!OH + OH^- \rightleftharpoons R\!-\!O^- + H_2O$$

The strength of an acid in solution is given by the *equilibrium constant K* for the process:

$$HA + H_2O \rightleftharpoons H_3O^+ + A^-$$

$$K = \frac{[H_3O^+][A^-]}{[HA][H_2O]}$$

Presuming the concentration of water to be a constant, the *acid dissociation constant K_a* may be defined as:

$$K[H_2O] = K_a = \frac{[H_3O^+][A^-]}{[HA]}$$

A frequently used way to express acid strength is pK_a, defined as:

$$pK_a = -\log(K_a)$$

As a result, numbers that are not exponentials and that increase with *decreasing* acid strength are used. For example,

$$K_a = 10^{-16} \qquad pK_a = 16 \qquad \text{(stronger acid)}$$
$$K_a = 10^{-25} \qquad pK_a = 25 \qquad \text{(weaker acid)}$$

An additional means of describing acid and base strength is to note that *the stronger the acid, the weaker the conjugate base and vice versa*.

A potential energy diagram (text Figure 4.3) may be used to illustrate the energy changes that accompany a proton-transfer reaction. Proton-transfer reactions proceed through a single transition state and are examples of *concerted reactions*. Since two molecular species are undergoing change in the transition state, the reaction is said to be *bimolecular*.

Carbocation Intermediates (Secs. 4.11 to 4.13) The *substitution reaction* whereby alcohols react with hydrogen halides to form alkyl halides proceeds through formation of a *carbocation* intermediate. The mechanism of this reaction is outlined in the reaction summary presented later in this chapter.

A carbocation is a *trivalent* intermediate in which the central carbon bears a formal charge of +1. Both the VSEPR and the orbital hybridization bonding model predict a coplanar arrangement of bonds to the positively charged carbon. According to the latter theory, the positively charged carbon adopts an sp^2 hybridization state, as shown in text Figure 4.5.

Substituted carbocations may be classified as *primary, secondary,* or *tertiary,* according to the number of carbons directly attached to the positively charged carbon.

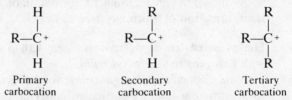

| Primary | Secondary | Tertiary |
| carbocation | carbocation | carbocation |

Alkyl substituents attached to the positively charged carbon stabilize carbocations. The result is the following *carbocation stability series*:

$$\text{Tertiary} > \text{Secondary} > \text{Primary} > CH_3^+$$

Most stable   Least stable

Stabilization of carbocations by alkyl groups is the result of two effects: an *inductive effect,* in which the alkyl groups are electron-releasing, and *delocalization* of electrons from the filled σ orbitals of the alkyl groups into the vacant $2p$ orbital of the carbocation.

Carbocations, having a deficiency of electron density in the carbon valence shell (the octet is not satisfied), are said to be highly *electrophilic*, that is, "electron-seeking" or "electron-loving." An electron-deficient species such as a carbocation is able to react readily with an electron-rich species known as a *nucleophile* (meaning "nucleus-loving"). Such a *nucleophilic substitution reaction* is one example of an important class of organic reactions.

Reaction Mechanisms (Secs. 4.12 to 4.14) The overall rate of a multistep process is determined by the rate of the *slow,* or *rate-determining, step* of the process. In the reaction of secondary and tertiary alcohols with hydrogen halides, formation of the carbocation intermediate is the rate-determining step.

The energy changes which take place as a reaction proceeds to form products may be described by using a *reaction coordinate diagram* (or *potential energy diagram*) as shown in text Figures 4.9 and 4.10.

The rate constant for each step in a chemical process is related to the magnitude of the *activation energy* E_{act} of the step. A reaction step has a larger rate constant k when the activation energy is smaller. The rate-determining step of a multistep process therefore has the largest activation energy.

When a carbocation is formed, as in the reaction of alcohols with hydrogen halides, the energy of the *transition state* is related to the stability of the carbocation. The more stable the carbocation, the lower the activation energy of the reaction step leading to its formation. Therefore the *reactivity series* for the reaction of alcohols with hydrogen halides parallels the series for carbocation stability:

$$R_3COH > R_2CHOH > RCH_2OH > CH_3OH$$

Tertiary, most reactive   Secondary   Primary   Least reactive

Free Radicals (Sec. 4.18) An alkyl group containing a trivalent carbon that possesses an unpaired electron is known as an *alkyl free radical.* Alkyl groups stabilize free radicals in a manner parallel to that described for stabilization of carbocations, resulting in a similar stability series:

$$
\underset{\text{Most stable}}{\overset{\displaystyle R}{\underset{\displaystyle R}{R-\overset{\displaystyle |}{\underset{\displaystyle |}{C}}\cdot}}} >
\overset{\displaystyle R}{\underset{\displaystyle H}{R-\overset{\displaystyle |}{\underset{\displaystyle |}{C}}\cdot}} >
\overset{\displaystyle H}{\underset{\displaystyle H}{R-\overset{\displaystyle |}{\underset{\displaystyle |}{C}}\cdot}} >
\underset{\text{Least stable}}{\overset{\displaystyle H}{\underset{\displaystyle H}{H-\overset{\displaystyle |}{\underset{\displaystyle |}{C}}\cdot}}}
$$

Relative stabilities of free radicals may also be described by using *bond dissociation energies* (BDEs). The BDE is the energy required for the *homolytic* dissociation of a bond; that is, the BDE equals the enthalpy change ($\Delta H°$) for the process

$$ X-Y \longrightarrow X\cdot + Y\cdot $$

As shown in text Figure 4.12, formation of a secondary free radical requires less energy than formation of a primary free radical.

Halogenation of Alkanes (Secs. 4.16 to 4.19) Alkanes undergo *substitution reactions* with halogens to give *alkyl halides.*

Bond dissociation energies (described previously under "Free Radicals") may be used to determine whether a particular halogenation reaction is *exothermic* or *endothermic.* For a reaction to be exothermic, the energy released in bond formation must be greater than the energy absorbed in bond rupture; that is, the enthalpy change $\Delta H°$ must be negative. As an example, BDE data from text Table 4.4 or Appendix B-1 in this study guide may be used to calculate the enthalpy change for chlorination of ethane:

$$ CH_3CH_2-H \ + \ Cl-Cl \ \longrightarrow \ CH_3CH_2-Cl \ + \ H-Cl $$

| 410 kJ/mol | 242 kJ/mol | | 338 kJ/mol | 431 kJ/mol |
| (98 kcal/mol) | (58 kcal/mol) | | (81 kcal/mol) | (103 kcal/mol) |

Bond breaking: 652 kJ/mol   Bond formation: 769 kJ/mol
(156 kcal/mol)   (184 kcal/mol)

$$ \Delta H° = BDE \text{ (bonds broken)} - BDE \text{ (bonds formed)} $$

$$ \Delta H° = (652 - 769) \text{ kJ/mol} = -117 \text{ kJ/mol} \ (-28 \text{ kcal/mol}) $$

The *reaction mechanism* provides a look at the details, on a molecular level, of how reactions proceed; in the case of alkane halogenations, the pathway is a *free-radical chain reaction.* The steps, illustrated for the chlorination of ethane, are:

1. *Initiation,* in which a halogen molecule dissociates to form free radicals:

$$ Cl_2 \xrightarrow[\text{light}]{\text{heat or}} 2Cl\cdot $$

2. *Propagation,* in which halogen radicals are consumed and regenerated while reactants are converted to products:

$$ Cl\cdot + CH_3CH_3 \longrightarrow CH_3\overset{.}{C}H_2 + H-Cl $$
$$ CH_3\overset{.}{C}H_2 + Cl_2 \longrightarrow CH_3CH_2-Cl + Cl\cdot $$

3. *Termination,* in which free radicals are consumed by reacting with each other:

$$ 2Cl\cdot \longrightarrow Cl_2 $$
$$ 2CH_3\overset{.}{C}H_2 \longrightarrow CH_3CH_2CH_2CH_3 $$
$$ CH_3\overset{.}{C}H_2 + Cl\cdot \longrightarrow CH_3CH_2Cl $$

Halogenation of Higher Alkanes (Sec. 4.20) Alkane chlorination has a low selectivity for the replacement of tertiary rather than secondary or primary hydrogens, although a greater selectivity of tertiary substitution is observed for bromination of alkanes. The predominant product formed in halogenation of alkanes is that arising from formation of the *most stable free radical*, according to the stability series noted earlier.

A reaction such as alkane bromination is said to be *regioselective*. This term is used to describe any reaction in which more than one constitutional isomer may be formed but one predominates.

IMPORTANT REACTIONS

Reaction of Alcohols with Metals (Sec. 4.8)

General:

$$2R—OH + 2M \longrightarrow 2R—O^- + 2M^+ + H_2$$

Example:

$$2(CH_3)_2CHOH + 2K \longrightarrow 2(CH_3)_2CHO^-K^+ + H_2$$

PREPARATION OF ALKYL HALIDES

Reaction of Alcohols with Hydrogen Halides (Secs. 4.9, 4.10)

General:

$$R—OH + H—X \longrightarrow R—X + H_2O$$

Examples:

$$\underset{\underset{OH}{|}}{(CH_3)_2CCH_2CH_3} \xrightarrow{HBr} \underset{\underset{Br}{|}}{(CH_3)_2CCH_2CH_3} + H_2O$$

Reaction mechanism:

1. $ROH + HX \underset{}{\overset{fast}{\rightleftharpoons}} R—\overset{+}{O}H_2 + X^-$

2. $R—\overset{+}{O}H_2 \xrightarrow{slow} R^+ + H_2O$

3. $R^+ + X^- \xrightarrow{fast} R—X$

Primary alcohols react by:

$$X^- + R—\overset{+}{O}H_2 \longrightarrow RX + H_2O$$

Reaction of Alcohols with Thionyl Chloride (Sec. 4.15)

General:

$$R—OH + SOCl_2 \xrightarrow{pyridine} R—Cl$$

Example:

$$\underset{\underset{OH}{|}}{CH_3CHCH_2CH_3} + SOCl_2 \xrightarrow{pyridine} \underset{\underset{Cl}{|}}{CH_3CHCHCH_3}$$

Reaction of Alcohols with Phosphorus Tribromide (Sec. 4.15)

General:

$$R{-}OH + PBr_3 \longrightarrow R{-}Br$$

Example:

$$CH_3CH_2CH_2OH \xrightarrow{PBr_3} CH_3CH_2CH_2Br$$

Free-Radical Halogenation of Alkanes (Secs. 4.16 to 4.20)

Examples:

$$CH_3{-}\underset{\underset{CH_3}{|}}{CH}{-}CH_3 + Br_2 \xrightarrow[\text{light}]{\text{heat or}} CH_3{-}\underset{\underset{Br}{|}}{\overset{\overset{CH_3}{|}}{C}}{-}CH_3$$
(major)

$$CH_3CH_2CH_2CH_3 + Cl_2 \xrightarrow[\text{light}]{\text{heat or}} CH_3CH_2CH_2CH_2Cl + CH_3CH_2\underset{\underset{Cl}{|}}{CH}CH_3$$

SOLUTIONS TO TEXT PROBLEMS

4.1 There are four C_4H_9 alkyl groups, and so there are four C_4H_9Cl alkyl chlorides. Each may be named by both the radicofunctional and substitutive methods. The radicofunctional name uses the name of the alkyl group followed by the halide as a second word. The substitutive name modifies the name of the corresponding alkane to show the location of the halogen atom.

	Radicofunctional name	**Substitutive name**
$CH_3CH_2CH_2CH_2Cl$	*n*-Butyl chloride	1-Chlorobutane
$CH_3\underset{\underset{Cl}{|}}{CH}CH_2CH_3$	*sec*-Butyl chloride	2-Chlorobutane
$CH_3\underset{\underset{CH_3}{|}}{CH}CH_2Cl$	Isobutyl chloride	1-Chloro-2-methylpropane
$CH_3\underset{\underset{Cl}{|}}{\overset{\overset{CH_3}{|}}{C}}CH_3$	*tert*-Butyl chloride	2-Chloro-2-methylpropane

4.2 Alcohols may also be named using both the radicofunctional and substitutive methods, as in the previous problem.

	Radicofunctional name	**Substitutive name**
$CH_3CH_2CH_2CH_2OH$	*n*-Butyl alcohol	1-Butanol
$CH_3\underset{\underset{OH}{|}}{CH}CH_2CH_3$	*sec*-Butyl alcohol	2-Butanol

$$CH_3CHCH_2OH$$
$$|$$
$$CH_3$$

Isobutyl alcohol   2-Methyl-1-propanol

$$CH_3$$
$$|$$
$$CH_3CCH_3$$
$$|$$
$$OH$$

tert-Butyl alcohol   2-Methyl-2-propanol

4.3 Alcohols are classified as primary, secondary, or tertiary according to the number of carbon substituents attached to the carbon that bears the hydroxyl group.

$$\begin{array}{c} H \\ | \\ CH_3CH_2CH_2C{-}OH \\ | \\ H \end{array}$$

Primary alcohol
(one alkyl group bonded to —CH₂OH)

$$\begin{array}{c} H \\ | \\ CH_3{-}C{-}CH_2CH_3 \\ | \\ OH \end{array}$$

Secondary alcohol
(two alkyl groups bonded to >CHOH)

$$\begin{array}{c} H \\ | \\ (CH_3)_2CH{-}C{-}OH \\ | \\ H \end{array}$$

Primary alcohol
(one alkyl group bonded to —CH₂OH)

$$\begin{array}{c} CH_3 \\ | \\ CH_3{-}C{-}OH \\ | \\ CH_3 \end{array}$$

Tertiary alcohol
(three alkyl groups bonded to >COH)

4.4 Dipole moment is the product of charge and distance. Although the electron distribution in the carbon-chlorine bond is more polarized than that in the carbon-bromine bond, this effect is counterbalanced by the longer carbon-bromine bond distance.

$$\mu = e \cdot d$$

Dipole moment   Charge Distance

$$CH_3{-}Cl$$

Methyl chloride
(greater value of *e*)
μ 1.9 D

$$CH_3{-}Br$$

Methyl bromide
(greater value of *d*)
μ 1.8 D

4.5 Ammonia is a base and abstracts (accepts) a proton form the acid (proton donor) hydrogen chloride.

$$H_3N{:} \quad + \quad H{-}\overset{..}{\underset{..}{Cl}}{:} \quad \rightleftharpoons \quad {}^+NH_4 \quad + \quad {:}\overset{..}{\underset{..}{Cl}}{:}^-$$

Base   Acid   Conjugate acid   Conjugate base

4.6 An unshared electron pair on oxygen abstracts the proton from hydrogen chloride.

$$(CH_3)_3C \diagdown \quad \quad \quad \quad (CH_3)_3C \diagdown$$
$$\overset{..}{\underset{..}{O}} \quad + \quad H{-}\overset{..}{\underset{..}{Cl}}{:} \quad \rightleftharpoons \quad \overset{..}{\underset{..}{O}}{}^+{-}H + {:}\overset{..}{\underset{..}{Cl}}{:}^-$$
$$| \quad \quad \quad \quad \quad \quad \quad \quad \quad \quad |$$
$$H \quad \quad \quad \quad \quad \quad \quad \quad \quad \quad H$$

Base   Acid   Conjugate acid   Conjugate base

4.7 The equilibrium constant for proton transfer from hydrogen chloride to water is much greater than 1. Alcohols are similar to water in their basicity; thus proton transfer from HCl to *tert*-butyl alcohol will also have an equilibrium constant much greater than 1.

4.8 The proton being transferred is partially bonded to the oxygen of *tert*-butyl alcohol and to chloride at the transition state.

$$(CH_3)_3C \diagdown \underset{\underset{H}{|}}{\overset{\delta+}{O}} \text{---------} H \text{---------} \overset{\delta-}{\ddot{\underset{..}{Cl}}}:$$

4.9 (*b*) Hydrogen chloride converts tertiary alcohols to tertiary alkyl chlorides.

$$(CH_3CH_2)_3COH + \quad HCl \quad \longrightarrow \quad (CH_3CH_2)_3CCl \quad + H_2O$$

3-Ethyl-3-pentanol   Hydrogen chloride   3-Chloro-3-ethylpentane   Water

(*c*) 1-Tetradecanol is a primary alcohol having an unbranched 14-carbon chain. Hydrogen bromide reacts with primary alcohols to give the corresponding primary alkyl bromide.

$$CH_3(CH_2)_{12}CH_2OH + \quad HBr \quad \longrightarrow \quad CH_3(CH_2)_{12}CH_2Br + H_2O$$

1-Tetradecanol   Hydrogen bromide   1-Bromotetradecane   Water

4.10 The order of carbocation stability is tertiary > secondary > primary. There is only one $C_5H_{11}{}^+$ carbocation that is tertiary, and so that is the most stable one.

$$CH_3CH_2\overset{\overset{\textstyle CH_3}{|}}{\underset{\underset{\textstyle CH_3}{|}}{C}}{}^+ \qquad \text{1,1-Dimethylpropyl cation}$$

4.11 1-Butanol is a primary alcohol; 2-butanol is a secondary alcohol. A carbocation intermediate is possible in the reaction of 2-butanol with hydrogen bromide but not in the corresponding reaction of 1-butanol.

The mechanism of the reaction of 1-butanol with hydrogen bromide proceeds by displacement of water by bromide from the protonated form of the alcohol.

Protonation of the alcohol:

$$CH_3CH_2CH_2CH_2\overset{\diagdown}{\underset{\underset{H}{|}}{\ddot{O}}}: \;+\; H{-}\ddot{\underset{..}{Br}}: \longrightarrow CH_3CH_2CH_2CH_2\overset{\overset{\textstyle H}{\diagup}}{\underset{\underset{H}{\diagdown}}{\ddot{O}}}{}^+ \;+\; \ddot{\underset{..}{Br}}:^-$$

1-Butanol   Hydrogen bromide   Butyloxonium ion   Bromide

Displacement of water by bromide:

$$:Br:^- \qquad CH_2{-}\overset{\overset{\textstyle CH_3CH_2CH_2}{|}}{\underset{\underset{H}{\diagdown}}{\ddot{O}}}{}^+\!\!{-}H \longrightarrow CH_3CH_2CH_2CH_2Br \;+\; :\overset{\overset{\textstyle H}{\diagup}}{\underset{\underset{H}{\diagdown}}{O}}:$$

Bromide ion   Butyloxonium ion   1-Bromobutane   Water

The reaction of 2-butanol with hydrogen bromide involves a carbocation intermediate.

Protonation of the alcohol:

$$CH_3CH_2\underset{\underset{\underset{H}{|}}{\overset{|}{\underset{..}{\ddot{O}}:}}}{\overset{|}{C}}HCH_3 + H{-}\ddot{\underset{..}{Br}}: \longrightarrow CH_3CH_2\underset{\underset{\underset{H\;\;\;H}{\diagup\diagdown}}{\overset{|}{\underset{..}{\ddot{O}}{}^+}}}{\overset{|}{C}}HCH_3 + :\ddot{\underset{..}{Br}}:^-$$

2-Butanol   Hydrogen bromide   *sec*-Butyloxonium ion   Bromide ion

Dissociation of the oxonium ion:

$$CH_3CH_2CHCH_3 \longrightarrow CH_3CH_2\overset{+}{C}HCH_3 + :\overset{..}{O}:$$

$$\underset{H \quad H}{:\overset{+}{O}:}$$

$$\underset{H \quad H}{}$$

sec-Butyloxonium ion   *sec*-Butyl cation   Water

Capture of *sec*-butyl cation by bromide

$$:\overset{..}{\underset{..}{Br}}:^- \; + \; \underset{CH_3CH_2}{\overset{+}{C}HCH_3} \longrightarrow CH_3CH_2CHCH_3$$

$$\underset{Br}{|}$$

Bromide ion *sec*-Butyl cation   2-Bromobutane

4.12 The most stable alkyl free radicals are tertiary. The tertiary free radical having the formula C_5H_{11} has the same skeleton as the carbocation in Problem 4.10.

$$CH_3CH_2-\overset{\textstyle CH_3}{\underset{\textstyle CH_3}{C}}\cdot$$

4.13 (*b*) Writing the equations for carbon-carbon bond cleavage in propane and in 2-methylpropane, we see that a primary ethyl radical is produced by a cleavage of propane while a secondary isopropyl radical is produced by cleavage of 2-methylpropane.

$$CH_3CH_2{-}CH_3 \longrightarrow CH_3\dot{C}H_2 \; + \; \cdot CH_3$$

Propane   Ethyl radical   Methyl radical

$$\underset{CH_3CHCH_3}{\overset{\textstyle CH_3}{|}} \longrightarrow CH_3\dot{C}HCH_3 \; + \; \cdot CH_3$$

2-Methylpropane   Isopropyl radical   Methyl radical

A secondary radical is more stable than a primary one, and so carbon-carbon bond cleavage of 2-methylpropane requires less energy than carbon-carbon bond cleavage of propane.

(*c*) Carbon-carbon bond cleavage of 2,2-dimethylpropane gives a tertiary radical.

$$\underset{\underset{\textstyle CH_3}{|}}{\overset{\textstyle CH_3}{\underset{}{\overset{|}{CH_3CCH_3}}}} \longrightarrow CH_3-\overset{\textstyle CH_3}{\underset{\textstyle CH_3}{C}}\cdot \; + \; \cdot CH_3$$

2,2-Dimethylpropane   *tert*-Butyl radical   Methyl radical

As noted in part (*b*), a secondary radical is produced on carbon-carbon bond cleavage of 2-methylpropane. Therefore we expect a lower carbon-carbon bond dissociation energy for 2,2-dimethylpropane than for 2-methylpropane, since a tertiary radical is more stable than a secondary one.

4.14 Writing the structural formula for ethyl chloride reveals that there are two nonequivalent sets of hydrogen atoms, in either of which a hydrogen is capable of being replaced by chlorine.

$$CH_3CH_2Cl \xrightarrow[\substack{\text{light or} \\ \text{heat}}]{Cl_2} CH_3CHCl_2 \ + \ ClCH_2CH_2Cl$$

Ethyl chloride   1,1-Dichloroethane   1,2-Dichloroethane

The two dichlorides are 1,1-dichloroethane and 1,2-dichloroethane.

4.15 Propane has six primary versus two secondary hydrogens. In the chlorination of propane the relative proportions of hydrogen atom removal are given by the product of the statistical distribution and the relative rate per hydrogen. Given that a secondary hydrogen is abstracted 3.9 times as fast as a primary one, we write the expression for the amount of chlorination at the primary relative to that at the secondary position as:

$$\frac{\text{Number of primary hydrogens} \times \text{rate of abstraction of a primary hydrogen}}{\text{Number of secondary hydrogens} \times \text{rate of abstraction of a secondary hydrogen}}$$

$$= \frac{6 \times 1}{2 \times 3.9} = \frac{0.77}{1.00}$$

Thus, the percentage of propyl chloride formed is 0.77/1.77, or 43 percent, and that of isopropyl chloride is 57 percent. (The amounts actually observed are propyl 45 percent, isopropyl 55 percent.)

4.16 (*b*) In contrast with free-radical chlorination, alkane bromination is a highly regioselective process. The major organic product will be the alkyl bromide formed by substitution of a tertiary hydrogen with a bromine.

1-Isopropyl-1-
methylcyclopentane   1-(1-Bromo-1-methylethyl)-
1-methylcyclopentane

(*c*) As in part (*b*), bromination results in substitution of a tertiary hydrogen.

2,2,4-Trimethylpentane   4-Bromo-2,2,4-trimethylpentane

4.17 (*a*) Cyclobutanol has a hydroxyl group attached to a four-membered ring.

—OH   Cyclobutanol

(*b*) *sec*-Butyl alcohol is the radicofunctional name for 2-butanol.

$$\underset{\underset{OH}{|}}{CH_3CHCH_2CH_3} \qquad sec\text{-Butyl alcohol}$$

(*c*) The hydroxyl group is at C-3 of an unbranched seven-carbon chain in 3-heptanol.

$$\underset{\underset{OH}{|}}{CH_3CH_2CHCH_2CH_2CH_2CH_3} \qquad \text{3-Heptanol}$$

(d) 1-Pentadecanol has 15 carbon atoms in an unbranched chain. The hydroxyl group is attached at the end of the chain.

$$CH_3(CH_2)_{13}CH_2OH \qquad \text{1-Pentadecanol}$$

(e) A chlorine at C-2 is on the opposite side of the ring from the C-1 hydroxyl group in *trans*-2-chlorocyclopentanol. Note that it is not necessary to assign a number to the carbon that bears the hydroxyl group; naming the compound as a derivative of cyclopentanol automatically requires the hydroxyl group to be located at C-1.

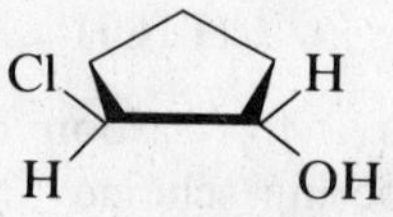

trans-2-Chlorocyclopentanol

(f) The longest continuous chain has six carbon atoms in 4-methyl-2-hexanol. The hydroxyl group is at C-2, the methyl at C-4.

$$CH_3CHCH_2CHCH_2CH_3 \qquad \text{4-Methyl-2-hexanol}$$
$$\underset{OH}{|}\underset{CH_3}{|}$$

(g) This compound is an alcohol in which the longest continuous chain that incorporates the hydroxyl function has eight carbons. It bears chlorine substituents at C-2 and C-6 and methyl and hydroxyl groups at C-4.

$$\overset{\displaystyle CH_3}{\underset{\displaystyle |}{}}$$
$$CH_3CHCH_2CCH_2CHCH_2CH_3 \qquad \text{2,6-Dichloro-4-methyl-4-octanol}$$
$$\underset{Cl}{|}\underset{OH}{|}\underset{Cl}{|}$$

(h) The hydroxyl group is at C-1 in *trans*-4-*tert*-butylcyclohexanol; the *tert*-butyl group is at C-4. The structures of the compound can be represented as shown at the left; the structure at the right depicts it in its most stable conformation.

trans-4-*tert*-Butylcyclohexanol

(i) Each carbon bears a substituent in this compound; it is named as a derivative of 2-propanol, with the halogen substituents cited in alphabetical order.

$$BrCH_2CHCH_2I \qquad \text{1-Bromo-3-iodo-2-propanol}$$
$$\underset{OH}{|}$$

(j) The cyclopropyl group is on the same carbon as the hydroxyl group in 1-cyclopropylethanol.

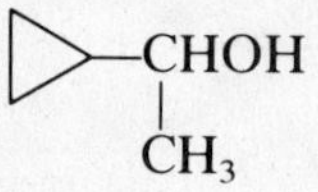

1-Cyclopropylethanol

(k) The cyclopropyl group and the hydroxyl group are on adjacent carbons in 2-cyclopropylethanol.

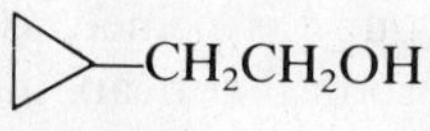

2-Cyclopropylethanol

(l) Chlorotrifluoromethane has a single carbon, which bears a chlorine and three fluorine substituents.

$$F$$
$$F—C—F \qquad \text{Chlorotrifluoromethane}$$
$$Cl$$

4.18 (*a*) This compound has an unbranched nine-carbon chain and so is named as a functionally substituted derivative of nonane. It bears an iodo substituent at C-1; it is 1-iodononane.

$$CH_3(CH_2)_8I \qquad \text{1-Iodononane}$$

(*b*) This compound has a five-carbon chain that bears a methyl substituent and a bromine. The numbering scheme that gives the lower number to the substituent closest to the end of the chain is chosen. Therefore, bromine is at C-1 and methyl is a substituent at C-4.

$$CH_3CHCH_2CH_2CH_2Br \qquad \text{1-Bromo-4-methylpentane}$$
$$CH_3$$

(*c*) This compound has the same carbon skeleton as the compound in part (*b*) but bears a hydroxyl group in place of the bromine and so is named as a derivative of 1-pentanol.

$$CH_3CHCH_2CH_2CH_2OH \qquad \text{4-Methyl-1-pentanol}$$
$$CH_3$$

(*d*) This molecule is a derivative of ethane and bears three chlorines and one bromine. The name 2-bromo-1,1,1-trichloroethane gives a lower number at the first point of difference than 1-bromo-2,2,2-trichloroethane and is the correct substitutive name.

$$Cl_3CCH_2Br \qquad \text{2-Bromo-1,1,1-trichloroethane}$$

(*e*) This compound is a constitutional isomer of the preceding one. Regardless of which carbon the numbering begins at, the substitution pattern is 1,1,2,2. Therefore, alphabetical ranking of the halogens dictates the direction of numbering. Begin with the carbon that bears bromine.

$$Cl_2CHCHBr \qquad \text{1-Bromo-1,2,2-trichloroethane}$$
$$Cl$$

(*f*) This is a trifluoro derivative of ethanol. The direction of numbering is dictated by the hydroxyl group, which is at C-1 in ethanol.

$$CF_3CH_2OH \qquad \text{2,2,2-Trifluoroethanol}$$

(*g*) Italicized prefixes are not considered when ranking substituents alphabetically. Therefore, *tert*-butyl precedes fluoro. Both substituents are on the same side of the ring, and so they are cis.

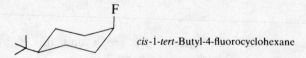

cis-1-*tert*-Butyl-4-fluorocyclohexane

(*h*) Here the compound is named as a derivative of cyclohexanol, and so numbering begins at the carbon that bears the hydroxyl group.

cis-3-*tert*-Butylcyclohexanol

(*i*) This alcohol has its hydroxyl group attached to C-2 of a three-carbon continuous chain; it is named as a derivative of 2-propanol.

2-Cyclopentyl-2-propanol

(*j*) The longest continuous chain that contains the hydroxyl group has six carbons. Number in the direction that assigns the lower number to the hydroxyl group, placing the hydroxyl group at C-2 and the two methyl groups at C-5.

5,5-Dimethyl-2-hexanol

(*k*) The six carbons that form the longest continuous chain have substituents at C-2, C-3, and C-5 when numbering proceeds in the direction that gives the lowest locants to substituents at the first point of difference. The substituents are cited in alphabetical order.

5-Bromo-2,3-dimethylhexane

Had numbering begun in the opposite direction, the locants would be 2,4,5 rather than 2,3,5.

(*l*) Now hydroxyl controls the numbering because the compound is named as an alcohol.

4,5-Dimethyl-2-hexanol

4.19 (*a*) A single carbon bears two chlorines and two fluorines; Cl_2CF_2 in dichlorodifluoromethane. The halogens are cited in alphabetical order; no numerical locants are needed.

(*b*) This compound is named as a derivative of ethane.

$$ClCF_2CF_2Cl \qquad \text{1,2-Dichloro-1,1,2,2-tetrafluoroethane}$$

(*c*) F_2CBrCl is bromochlorodifluoromethane.

(*d*) The correct name for $F_3CCHBrCl$ is 2-bromo-2-chloro-1,1,1-trifluoroethane.

4.20 Primary alcohols are alcohols in which the hydroxyl group is attached to a carbon atom which has one alkyl substituent and two hydrogens. There are four primary alcohols which have the molecular formula $C_5H_{12}O$. The radicofunctional name for each compound is given in parentheses.

$$CH_3CH_2CH_2CH_2CH_2OH$$

1-Pentanol
(Pentyl alcohol)

$$CH_3CH_2\overset{\underset{\displaystyle CH_3}{|}}{CH}CH_2OH$$

2-Methyl-1-butanol
(2-Methylbutyl alcohol)

$$CH_3\overset{\underset{\displaystyle CH_3}{|}}{CH}CH_2CH_2OH$$

3-Methyl-1-butanol
(3-Methylbutyl alcohol)

$$CH_3\overset{\overset{\displaystyle CH_3}{|}}{\underset{\underset{\displaystyle CH_3}{|}}{C}}CH_2OH$$

2,2-Dimethyl-1-propanol
(2,2-Dimethylpropyl alcohol)

Secondary alcohols are alcohols in which the hydroxyl group is attached to a carbon atom which has two alkyl substituents and one hydrogen. There are three secondary alcohols of molecular formula $C_5H_{12}O$:

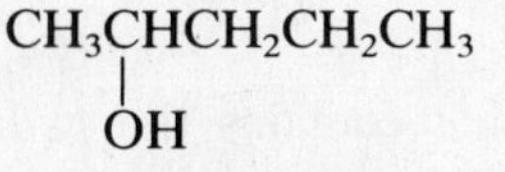

$$CH_3CHCH_2CH_2CH_3 \qquad CH_3CH_2CHCH_2CH_3 \qquad CH_3CHCHCH_3$$
$$|||$$
$$OHOHCH_3$$

with OH above the rightmost structure.

2-Pentanol	3-Pentanol	3-Methyl-2-butanol
(1-Methylbutyl alcohol)	(1-Ethylpropyl alcohol)	(1,2-Dimethylpropyl alcohol)

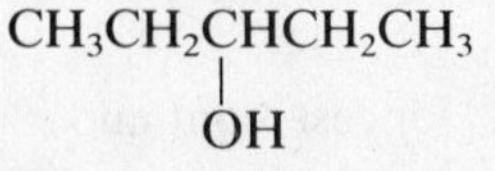

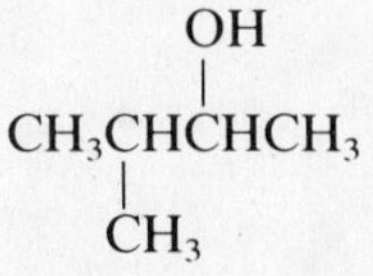

Only 2-methyl-2-butanol is a tertiary alcohol (three alkyl substituents on the hydroxyl-bearing carbon):

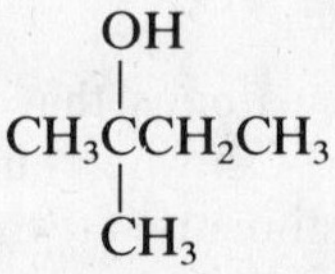

$$OH$$
$$|$$
$$CH_3CCH_2CH_3$$
$$|$$
$$CH_3$$

2-Methyl-2-butanol
(1,1-Dimethylpropyl alcohol)

4.21 The first methylcyclohexanol to be considered is 1-methylcyclohexanol. The preferred chair conformation will have the larger methyl group in an equatorial orientation while the smaller hydroxyl group will be axial.

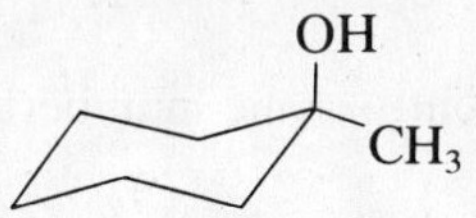

OH ... CH₃   Most stable conformation of 1-methylcyclohexanol

In the other isomers methyl and hydroxyl will be in a 1,2, 1,3, or 1,4 relationship and can be cis or trans in each. We can write the preferred conformation by recognizing that the methyl group will always be equatorial and the hydroxyl either equatorial or axial.

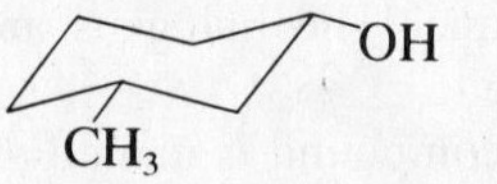

trans-2-Methylcyclohexanol	*cis*-3-Methylcyclohexanol	*trans*-4-Methylcyclohexanol

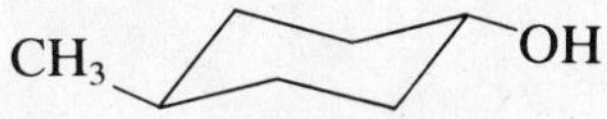

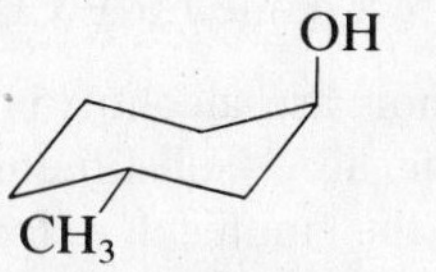

cis-2-Methylcyclohexanol	*trans*-3-Methylcyclohexanol	*cis*-4-Methylcyclohexanol

4.22 The assumption is incorrect for the 3-methylcyclohexanols. *cis*-3-Methylcyclohexanol is more stable than *trans*-3-methylcyclohexanol because the methyl group and the hydroxyl group are both equatorial in the cis isomer while one substituent must be axial in the trans.

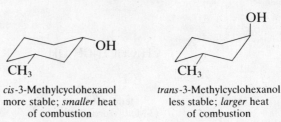

cis-3-Methylcyclohexanol
more stable; *smaller* heat
of combustion

trans-3-Methylcyclohexanol
less stable; *larger* heat
of combustion

4.23 (*a*) The most stable conformation will be the one with all the substituents equatorial.

$$CH_3 \diagdown \diagup OH \diagdown CH(CH_3)_2$$

The hydroxyl group is trans to the isopropyl group and cis to the methyl group.

(*b*) All three substituents need not always be equatorial; instead, one or two of them may be axial. Since neomenthol is the second most stable stereoisomer, we choose the structure with *one* axial substituent. Further, we choose the structure with the smallest substituent (the hydroxyl group) as the axial one. Therefore, neomenthol is shown as follows:

$$CH_3 \diagdown \overset{OH}{\diagup} \diagdown CH(CH_3)_2$$

4.24 In all these reactions the negatively charged atom abstracts a proton from an acid.

(*a*)   HI   +   HO$^-$ $\rightleftharpoons$ I$^-$ +   H$_2$O

Hydrogen iodide: acid   Hydroxide ion: Iodide ion: Water: conjugate acid
(stronger acid, $K_a \sim 10^{10}$)   base   conjugate base   (weaker acid, $K_a \sim 10^{-16}$)

(*b*)

$$CH_3CH_2O^- + \; CH_3\overset{\overset{\displaystyle O}{\|}}{C}OH \; \rightleftharpoons \; CH_3CH_2OH \; + \; CH_3\overset{\overset{\displaystyle O}{\|}}{C}O^-$$

Ethoxide ion: Acetic acid: acid   Ethanol: conjugate acid   Acetate ion:
base   (stronger acid, $K_a \sim 10^{-5}$)   (weaker acid, $K_a \sim 10^{-16}$)   conjugate base

(*c*)   HF   +   H$_2$N$^-$ $\rightleftharpoons$ F$^-$ +   H$_3$N

Hydrogen fluoride: acid   Amide ion: Fluoride ion: Ammonia: conjugate acid
(stronger acid, $K_a \sim 10^{-4}$)   base   conjugate base   (weaker acid, $K_a \sim 10^{-36}$)

(*d*)

$$CH_3\overset{\overset{\displaystyle O}{\|}}{C}O^- + \; HCl \; \rightleftharpoons \; CH_3\overset{\overset{\displaystyle O}{\|}}{C}OH \; + \; Cl^-$$

Acetate ion: Hydrogen chloride: acid   Acetic acid: conjugate acid   Chloride ion:
base   (stronger acid, $K_a \sim 10^7$)   (weaker acid, $K_a \sim 10^{-5}$)   conjugate base

(*e*)   (CH$_3$)$_3$CO$^-$ +   H$_2$O $\rightleftharpoons$ (CH$_3$)$_3$COH   +   HO$^-$

tert-Butoxide ion: Water: acid   *tert*-Butyl alcohol: Hydroxide ion:
base   (stronger acid,   conjugate acid   conjugate base
          $K_a \sim 10^{-16}$)   (weaker acid, $K_a \sim 10^{-18}$)

(*f*)   (CH$_3$)$_2$CHOH   +   H$_2$N$^-$ $\rightleftharpoons$ (CH$_3$)$_2$CHO$^-$ +   H$_3$N

Isopropyl alcohol: Amide ion: Isopropoxide ion: Ammonia: conjugate acid
acid   base   conjugate base   (weaker acid, $K_a \sim 10^{-36}$)
(stronger acid, $K_a \sim 10^{-17}$)

(*g*)   F$^-$ +   H$_2$SO$_4$ $\rightleftharpoons$ HF   +   HSO$_4^-$

Fluoride ion: Sulfuric acid: Hydrogen fluoride: Hydrogen
base   acid   conjugate acid   sulfate ion:
          (stronger acid, $K_a \sim 10^5$)   (weaker acid, $K_a \sim 10^{-4}$)   conjugate base

4.25 *(a)* The proton-transfer transition state can represent the following reaction, or its reverse:

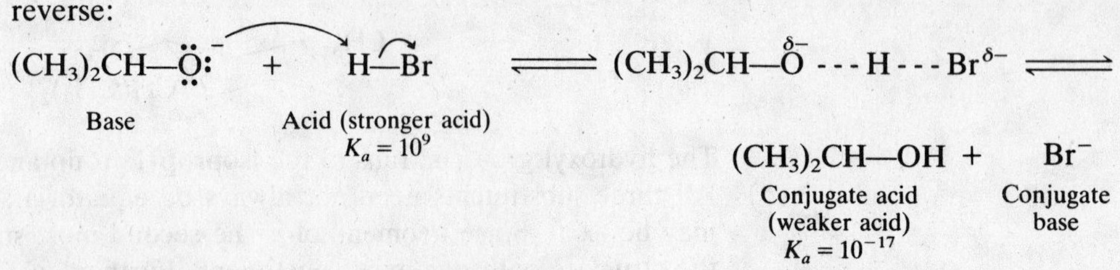

$$(CH_3)_2CH-\overset{..}{\underset{..}{O}}:^- \quad + \quad H-Br \quad \rightleftharpoons \quad (CH_3)_2CH-\overset{\delta-}{O}\cdots H\cdots Br^{\delta-} \rightleftharpoons$$

Base   Acid (stronger acid)
$$K_a = 10^9$$

$$(CH_3)_2CH-OH \quad + \quad Br^-$$

Conjugate acid   Conjugate
(weaker acid)   base
$$K_a = 10^{-17}$$

The equilibrium constant K is greater than 1 for the reaction proceeding from left to right. When the reaction proceeds from left to right, the stronger acid (hydrogen bromide) is on the left and the weaker acid (isopropyl alcohol) is on the right.

(b) Hydroxide is a strong base; methyloxonium ion is a strong acid.

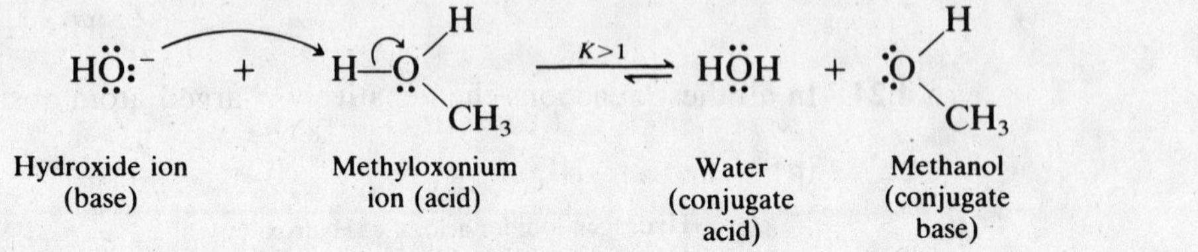

$$H\overset{..}{\underset{..}{O}}:^- \quad + \quad H-\overset{\overset{\displaystyle H}{|}}{\underset{\underset{\displaystyle CH_3}{}}{O}} \quad \overset{K>1}{\rightleftharpoons} \quad H\overset{..}{\underset{..}{O}}H \quad + \quad \overset{\overset{\displaystyle H}{|}}{\underset{\underset{\displaystyle CH_3}{}}{\overset{..}{\underset{..}{O}}}}$$

Hydroxide ion   Methyloxonium   Water   Methanol
(base)   ion (acid)   (conjugate   (conjugate
                                acid)   base)

4.26 *(a)* This problem reviews the relationship between logarithms and exponential numbers. We need to determine K_a, given pK_a. The equation that relates the two is

$$pK_a = -\log_{10} K_a$$

Therefore

$$K_a = 10^{-pK_a}$$
$$= 10^{-3} \times 10^{-0.48} = 10^{-4} \times 10^{0.52}$$
$$= 3.3 \times 10^{-4}$$

(b) As described in part *(a)*, $K_a = 10^{-pK_a}$; therefore K_a for vitamin C is given by the expression:

$$K_a = 10^{-4.17}$$
$$= 10^{-4} \times 10^{-0.17} = 10^{-5} \times 10^{+0.83}$$
$$= 6.7 \times 10^{-5}$$

(c) Similarly, $K_a = 1.8 \times 10^{-4}$ for formic acid (pK_a 3.75).
(d) $K_a = 6.5 \times 10^{-2}$ for oxalic acid (pK_a 1.19).

In ranking the acids in order of decreasing acidity, remember that the larger the equilibrium constant K_a, the stronger the acid; and the lower the pK_a value, the stronger the acid.

Acid	K_a	pK_a
Oxalic (strongest)	6.5×10^{-2}	1.19
Aspirin	3.3×10^{-4}	3.48
Formic acid	1.8×10^{-4}	3.75
Vitamin C (weakest)	6.7×10^{-5}	4.17

4.27 This problem illustrates the reactions of a primary alcohol with the reagents described in the chapter.

(a) $2CH_3CH_2CH_2CH_2OH + 2Li \longrightarrow 2CH_3CH_2CH_2CH_2O^-Li^+ + H_2$

     1-Butanol   Lithium butoxide

(b) $2CH_3CH_2CH_2CH_2OH + 2Na \longrightarrow 2CH_3CH_2CH_2CH_2O^-Na^+ + H_2$

Sodium butoxide

(c) $2CH_3CH_2CH_2CH_2OH + 2K \longrightarrow 2CH_3CH_2CH_2CH_2O^-K^+ + H_2$

Potassium butoxide

(d) $CH_3CH_2CH_2CH_2OH + NaNH_2 \longrightarrow CH_3CH_2CH_2CH_2O^-Na^+ + NH_3$

Sodium butoxide

(e) $CH_3CH_2CH_2CH_2OH \xrightarrow[\text{heat}]{\text{HBr}} CH_3CH_2CH_2CH_2Br$

1-Bromobutane

(f) $CH_3CH_2CH_2CH_2OH \xrightarrow[\text{heat}]{\text{NaBr, H}_2\text{SO}_4} CH_3CH_2CH_2CH_2Br$

1-Bromobutane

(g) $CH_3CH_2CH_2CH_2OH \xrightarrow{\text{PBr}_3} CH_3CH_2CH_2CH_2Br$

1-Bromobutane

(h) $CH_3CH_2CH_2CH_2OH \xrightarrow{\text{SOCl}_2} CH_3CH_2CH_2CH_2Cl$

1-Chlorobutane

4.28   Similar reactions occur with secondary alcohols.

(a)   2 ⬡—OH + 2Li $\longrightarrow$ 2 ⬡—O$^-$Li$^+$ + H$_2$

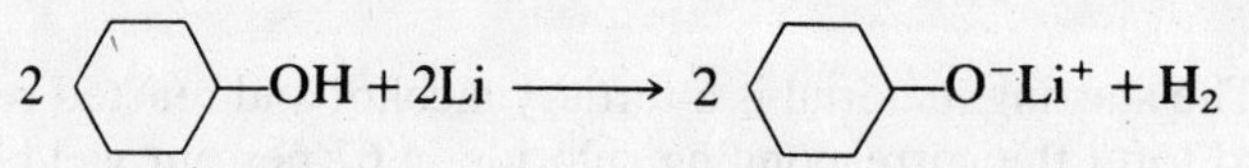

Cyclohexanol   Lithium cyclohexanolate

(b)   2 ⬡—OH + 2Na $\longrightarrow$ 2 ⬡—O$^-$Na$^+$ + H$_2$

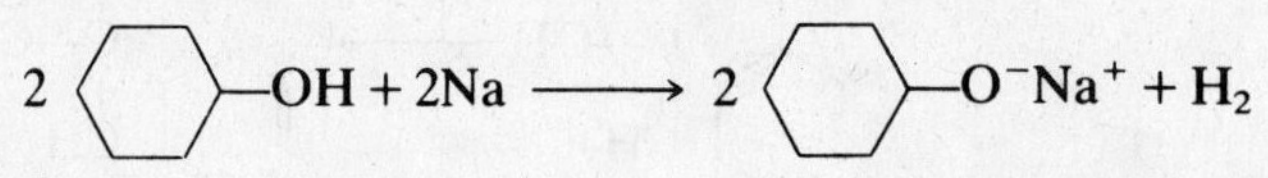

Sodium cyclohexanolate

(c)   2 ⬡—OH + 2K $\longrightarrow$ 2 ⬡—O$^-$K$^+$ + H$_2$

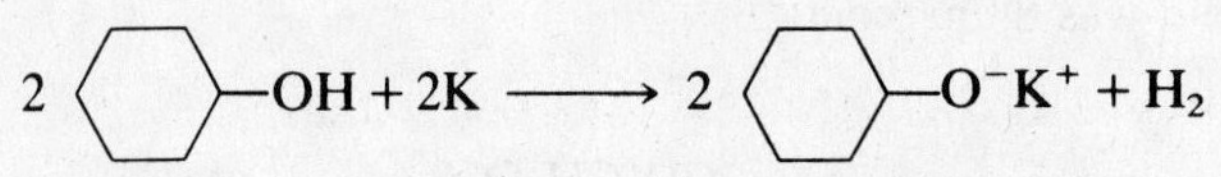

Potassium cyclohexanolate

(d)   ⬡—OH + NaNH$_2$ $\longrightarrow$ ⬡—O$^-$Na$^+$ + NH$_3$

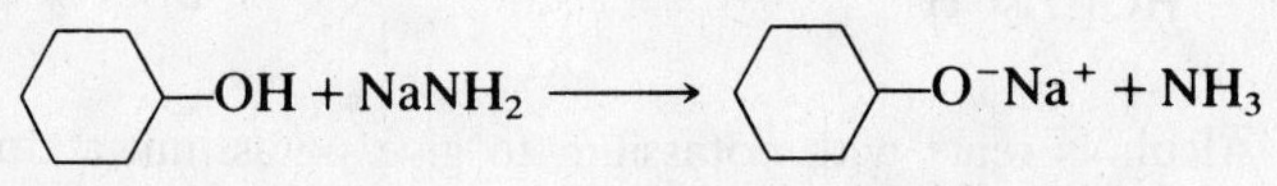

Sodium cyclohexanolate

(e)   ⬡—OH $\xrightarrow[\text{heat}]{\text{HBr}}$ ⬡—Br

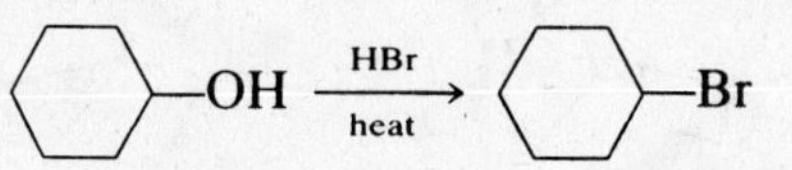

Bromocyclohexane

(f)   ⬡—OH $\xrightarrow[\text{heat}]{\text{NaBr, H}_2\text{SO}_4}$ ⬡—Br

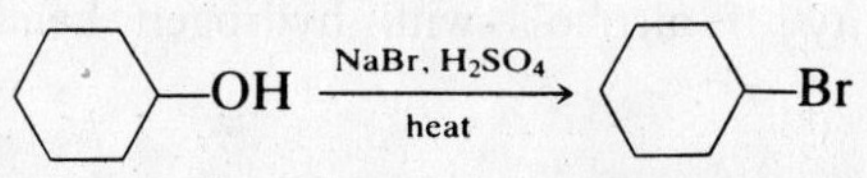

Bromocyclohexane

(*g*) 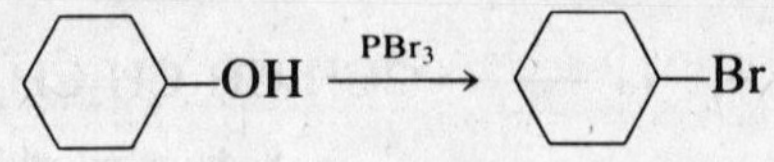
OH
PBr3
Br

Bromocyclohexane

(*h*) 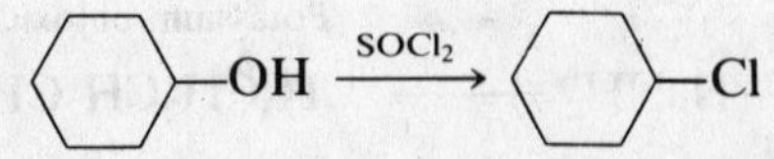
OH
SOCl2
Cl

Chlorocyclohexane

4.29 (*a*) This reaction was used to convert the primary alcohol to the corresponding bromide in 60 percent yield.

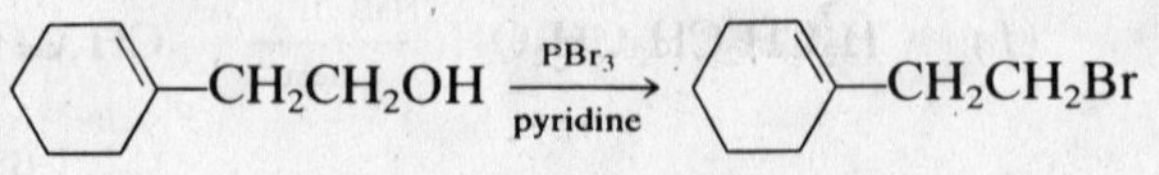
CH2CH2OH
PBr3
pyridine
CH2CH2Br

(*b*) Thionyl chloride treatment of this secondary alcohol gave the chloro derivative in 59 percent yield.

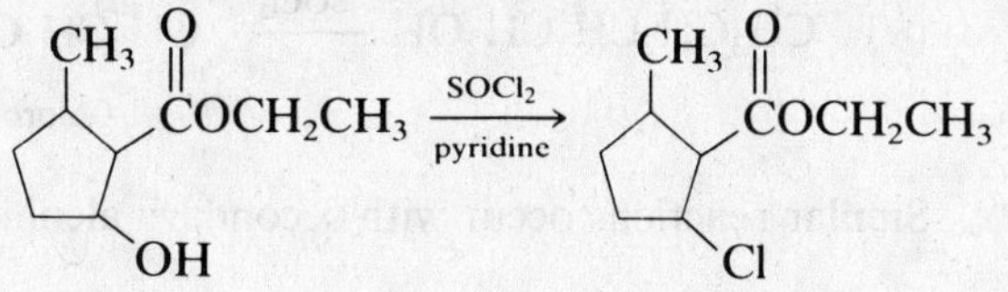
CH3
O
COCH2CH3
OH
SOCl2
pyridine
Cl

(*c*) The starting material is a tertiary alcohol and reacted readily with hydrogen chloride to form the corresponding chloride in 67 percent yield.

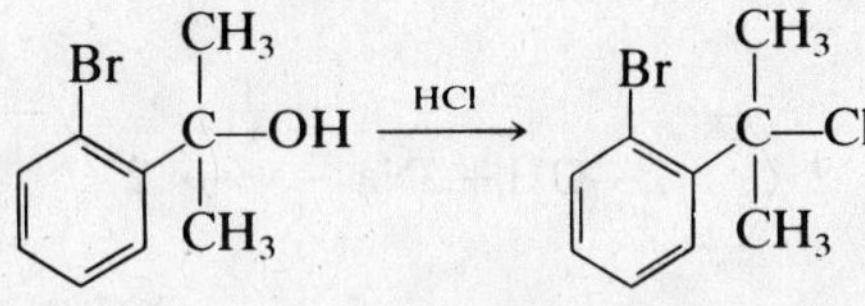
CH3
Br
C—OH
CH3
HCl
C—Cl

(*d*) Both primary alcohol functional groups were converted to primary bromides; the yield was 88 percent.

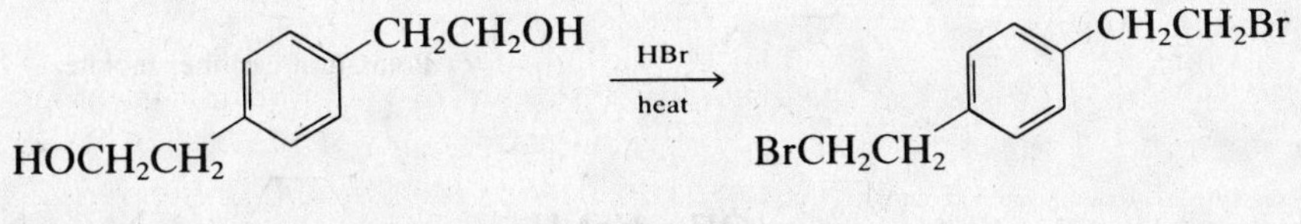
CH2CH2OH
HOCH2CH2
HBr
heat
CH2CH2Br
BrCH2CH2

(*e*) Alcohols react with potassium to give potassium alkoxides, usually in quantitative yield.

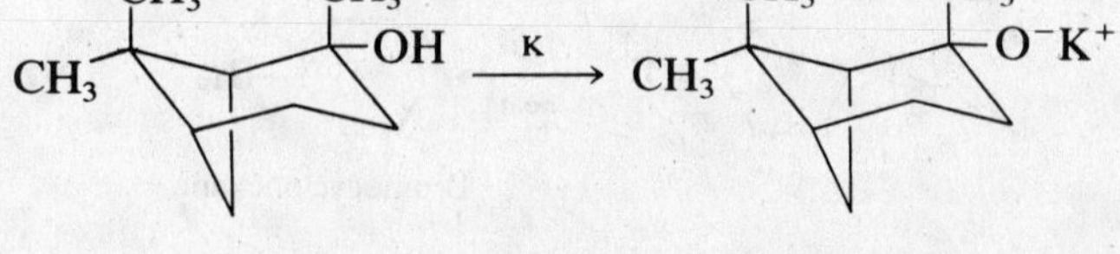
CH3
OH
K
O⁻K⁺

4.30 The order of reactivity of alcohols with hydrogen halides is tertiary > secondary > primary > methyl.

$$ROH + HBr \longrightarrow RBr + H_2O$$

Reactivity of Alcohols with Hydrogen Bromide

Part	More reactive alcohol	Less reactive alcohol
(a)	$CH_3CHCH_2CH_3$ $\quad\vert$ $\quad OH$ 2-Butanol: secondary	$CH_3CH_2CH_2CH_2OH$ 1-Butanol: primary
(b)	$CH_3CHCH_2CH_3$ $\quad\vert$ $\quad OH$ 2-Butanol: secondary	$CH_3CH_2CHCH_2OH$ $\qquad\qquad\vert$ $\qquad\qquad CH_3$ 2-Methyl-1-butanol: primary
(c)	$(CH_3)_2CCH_2CH_3$ $\qquad\vert$ $\qquad OH$ 2-Methyl-2-butanol: tertiary	$CH_3CHCH_2CH_3$ $\quad\vert$ $\quad OH$ 2-Butanol: secondary
(d)	$CH_3CHCH_2CH_3$ $\quad\vert$ $\quad OH$ 2-Butanol	$CH_3CHCH_2CH_3$ $\quad\vert$ $\quad CH_3$ 2-Methylbutane: not an alcohol; does not react with HBr
(e)	$CH_3\quad OH$ 1-Methylcyclopentanol: tertiary	$H\quad OH$ Cyclohexanol: secondary
(f)	$CH_3\quad OH$ 1-Methylcyclopentanol: tertiary	$H\quad OH$ $\qquad\quad H$ $\qquad\quad CH_3$ trans-2-Methylcyclopentanol: secondary
(g)	$CH_3CH_2\quad OH$ 1-Ethylcyclopentanol: tertiary	$—CHCH_3$ $\quad\vert$ $\quad OH$ 1-Cyclopentylethanol: secondary

4.31 (a) The ideal bond angles at the positively charged carbon in carbocations are 120° (sp^2 hybridization). Cyclopropyl cation, because of its geometry, has a C—C—C bond angle that is much smaller than 120°. Cyclopropyl cation is relatively unstable because angle strain at its positively charged carbon precludes sp^2 hybridization.

$$CH_3 \overset{+}{\underset{CH_3}{C}}{-}H \qquad \overset{\pm}{\triangle}H$$

Isopropyl cation   Cyclopropyl cation
(more stable)   (less stable)

(b) Carbocations are stabilized by electron release from substituents. Fluorine is very electronegative and a CF_3 group is far less able to donate electrons to a carbocation than is a CH_3 group. Indeed, the effect of a CF_3 group is to withdraw rather than to donate electrons.

Isopropyl cation
(more stable)

1-(Trifluoromethyl)ethyl cation
(less stable)

4.32 Free radicals and carbocations are both stabilized by substituents that donate electrons to a carbon atom with an unfilled $2p$ orbital. A carbocation is much more sensitive to electron release from alkyl substituents than is a free radical because it has one less electron than a free radical and is positively charged.

4.33 Examine the equations to ascertain which bonds are made and which are broken. Then use the bond dissociation energies in Table 4.4 to calculate $\Delta H°$ for each reaction.

(a) $(CH_3)_2CH—OH$ + $H—F$ $\longrightarrow$ $(CH_3)_2CH—F$ + $H—OH$

385 kJ/mol   569 kJ/mol   439 kJ/mol   498 kJ/mol
(92 kcal/mol)   (136 kcal/mol)   (105 kcal/mol)   (119 kcal/mol)

Bond breaking: 954 kJ/mol (228 kcal/mol)   Bond making: 937 kJ/mol (224 kcal/mol)

   $\Delta H°$ = energy cost of breaking bonds − energy given off in making bonds

       = 954 kJ/mol − 937 kJ/mol (228 kcal/mol − 224 kcal/mol)

       = +17 kJ/mol (+4 kcal/mol)

   The reaction of isopropyl alcohol with hydrogen fluoride is endothermic.

(b) $(CH_3)_2CH—OH$ + $H—Cl$ $\longrightarrow$ $(CH_3)_2CH—Cl$ + $H—OH$

385 kJ/mol   431 kJ/mol   339 kJ/mol   498 kJ/mol
(92 kcal/mol)   (103 kcal/mol)   (81 kcal/mol)   (119 kcal/mol)

Bond breaking: 816 kJ/mol (195 kcal/mol)   Bond making: 837 kJ/mol (200 kcal/mol)

   $\Delta H°$ = energy cost of breaking bonds − energy given off in making bonds

       = 816 kJ/mol − 837 kJ/mol (195 kcal/mol − 200 kcal/mol)

       = −21 kJ/mol (−5 kcal/mol)

   The reaction of isopropyl alcohol with hydrogen chloride is exothermic.

(c) CH_3CHCH_3 + $H—Cl$ $\longrightarrow$ CH_3CHCH_3 + $H—H$
                | |
                H   Cl

395 kJ/mol   431 kJ/mol   339 kJ/mol   435 kJ/mol
(94.5 kcal/mol)   (103 kcal/mol)   (81 kcal/mol)   (104 kcal/mol)

Bond breaking: 826 kJ/mol (197.5 kcal/mol)   Bond making: 774 kJ/mol (185 kcal/mol)

   $\Delta H°$ = energy cost of breaking bonds − energy given off in making bonds

       = 826 kJ/mol − 774 kJ/mol (197.5 kcal/mol − 185 kcal/mol)

       = +52 kJ/mol (+12.5 kcal/mol)

   The reaction of propane with hydrogen chloride is endothermic.

4.34 In the statement of the problem you are told that the starting material is 2,2-dimethyl-propane, that the reaction is one of fluorination, meaning that F_2 is a reactant, and that the product is $(CF_3)_4C$. You need to complete the equation by realizing that HF is also formed in the fluorination of alkanes. Therefore, the balanced equation is:

$$(CH_3)_4C + 12F_2 \longrightarrow (CF_3)_4C + 12HF$$

4.35 The reaction is free-radical chlorination, and substitution occurs at all possible positions that bear a replaceable hydrogen. Write the structure of the starting material and identify the nonequivalent hydrogens

$$
\underset{\substack{| \\ F}}{\overset{\substack{F \\ |}}{Cl-C}}-\underset{\substack{| \\ H}}{\overset{\substack{Cl \\ |}}{C}}-CH_3 \qquad \text{1,2-Dichloro-1,1-difluoropropane}
$$

The problem states that one of the products is 1,2,3-trichloro-1,1-difluoropropane. This compound arises by substitution of one of the methyl hydrogens by chlorine. We are told that the other product is an isomer of 1,2,3-trichloro-1,1-difluoropropane; therefore, it must be formed by replacement of the hydrogen at C-2.

$$
\underset{\substack{| \\ F}}{\overset{\substack{F \\ |}}{Cl-C}}-\underset{\substack{| \\ H}}{\overset{\substack{Cl \\ |}}{C}}-CH_2Cl \qquad\qquad\qquad \underset{\substack{| \\ F}}{\overset{\substack{F \\ |}}{Cl-C}}-\underset{\substack{| \\ Cl}}{\overset{\substack{Cl \\ |}}{C}}-CH_3
$$

1,2,3-Trichloro-1,1-difluoropropane   1,2,2-Trichloro-1,1-difluoropropane

4.36 Free-radical chlorination leads to substitution at each distinct position that bears a hydrogen substituent. This problem essentially requires you to recognize structures that possess various numbers of nonequivalent hydrogens.

(*a*) 2,2-Dimethylpropane is the C_5H_{12} isomer that gives a single monochloride, since all the hydrogens are equivalent.

$$
\underset{\substack{| \\ CH_3}}{\overset{\substack{CH_3 \\ |}}{CH_3C}}CH_3 \quad \xrightarrow[\text{light}]{Cl_2} \quad \underset{\substack{| \\ CH_3}}{\overset{\substack{CH_3 \\ |}}{CH_3C}}CH_2Cl
$$

2,2-Dimethylpropane   1-Chloro-2,2-dimethylpropane

(*b*) The C_5H_{12} isomer that has three nonequivalent sets of hydrogens is pentane. It yields three isomeric monochlorides on free-radical chlorination.

$$
CH_3CH_2CH_2CH_2CH_3 \xrightarrow{Cl_2}
\begin{cases}
\rightarrow ClCH_2CH_2CH_2CH_2CH_3 & \text{1-Chloropentane} \\
\\
\rightarrow \underset{\substack{| \\ Cl}}{CH_3CHCH_2CH_2CH_3} & \text{2-Chloropentane} \\
\\
\rightarrow \underset{\substack{| \\ Cl}}{CH_3CH_2CHCH_2CH_3} & \text{3-Chloropentane}
\end{cases}
$$

Pentane

(*c*) 2-Methylbutane forms four different monochlorides.

$$
\underset{\substack{| \\ CH_3}}{CH_3CHCH_2CH_3} \xrightarrow{Cl_2}
\begin{cases}
\rightarrow \underset{\substack{| \\ CH_3}}{ClCH_2CHCH_2CH_3} & \text{1-Chloro-2-methylbutane} \\
\\
\rightarrow \underset{\substack{| \\ Cl}}{(CH_3)_2CCH_2CH_3} & \text{2-Chloro-2-methylbutane} \\
\\
\rightarrow \underset{\substack{| \\ Cl}}{(CH_3)_2CHCHCH_3} & \text{2-Chloro-3-methylbutane} \\
\\
\rightarrow (CH_3)_2CHCH_2CH_2Cl & \text{1-Chloro-3-methylbutane}
\end{cases}
$$

2-Methylbutane

(*d*) In order that only two dichlorides be formed, the starting alkane must have a structure that is rather symmetrical; that is, one in which most (or all) the hydrogens are equivalent. 2,2-Dimethylpropane satisfies this requirement.

$$CH_3\overset{\underset{CH_3}{|}}{\underset{|}{C}}CH_3 \xrightarrow[\text{light}]{2Cl_2} CH_3\overset{\underset{CH_3}{|}}{\underset{|}{C}}CHCl_2 + ClCH_2\overset{\underset{CH_3}{|}}{\underset{|}{C}}CH_2Cl$$

2,2-Dimethylpropane   1,1-Dichloro-2,2-dimethylpropane   1,3-Dichloro-2,2-dimethylpropane

4.37 (*a*) Heptane has five methylene groups, which on chlorination together contribute 85 percent of the total monochlorinated product.

$$CH_3(CH_2)_5CH_3 \longrightarrow CH_3(CH_2)_5CH_2Cl + (2\text{-chloro} + 3\text{-chloro} + 4\text{-chloro})$$
$$15\% \qquad\qquad 85\%$$

Since the problem specifies that attack at each methylene group is equally probable, the five methylene groups each give rise to 85/5, or 17, percent of the monochloride product.

Since C-2 and C-6 of heptane are equivalent to each other, we calculate that 2-chloroheptane will constitute 34 percent of the monochloride fraction. Similarly, C-3 and C-5 are equivalent, and so there should be 34 percent 3-chloroheptane. The remainder, 17 percent, is 4-chloroheptane.

These predictions are very close to the observed proportions.

	Calculated, %	Observed, %
2-Chloro	34	35
3-Chloro	34	34
4-Chloro	17	16

(*b*) There are a total of 20 methylene hydrogens in dodecane, $CH_3(CH_2)_{10}CH_3$. The 19 percent 2-chlorododecane that is formed arises by substitution of any of the four equivalent methylene hydrogens at C-2 and C-11. Therefore, the total amount of substitution of methylene hydrogens must be:

$$\tfrac{20}{4} \times 19\% = 95\%$$

The remaining 5 percent corresponds to substitution of methyl hydrogens at C-1 and C-12. The proportion of 1-chlorododecane in the monochloride fraction is 5 percent.

4.38 (*a*) Two of the monochlorides derived from chlorination of 2,2,4-trimethylpentane are primary chlorides:

$$ClCH_2\overset{\underset{CH_3}{|}}{\underset{|}{C}}CH_2\overset{\underset{CH_3}{|}}{CH}CH_3 \qquad CH_3\overset{\underset{CH_3}{|}}{\underset{|}{C}}CH_2\overset{\underset{CH_3}{|}}{CH}CH_2Cl$$

1-Chloro-2,2,4-trimethylpentane   1-Chloro-2,4,4-trimethylpentane

The two remaining isomers are a secondary chloride and a tertiary chloride:

$$CH_3\overset{\underset{CH_3}{|}}{\underset{|}{C}}\!-\!\overset{}{CH}\!-\!\overset{\underset{CH_3}{|}}{CH}CH_3 \qquad CH_3\overset{\underset{CH_3}{|}}{\underset{|}{C}}CH_2\overset{\underset{CH_3}{|}}{\underset{|}{C}}CH_3$$

3-Chloro-2,2,4-trimethylpentane   2-Chloro-2,4,4-trimethylpentane

(b) Substitution of any one of the nine hydrogens designated as x in the structural diagram yields 1-chloro-2,2,4-trimethylpentane. Substitution of any one of the six hydrogens designated as y gives 1-chloro-2,4,4-trimethylpentane.

$$
\overset{x}{CH_3} \\
| \\
\overset{x}{CH_3}C\overset{}{CH_2}CH\overset{y}{CH_3} \\
| \quad\; | \\
\underset{x}{CH_3} \;\; \underset{y}{CH_3}
$$

Assuming equal reactivity of a single x hydrogen and a single y hydrogen, the ratio of the two isomers is then expected to be 9:6. Since together the two primary chlorides total 65 percent of the monochloride fraction, there will be 39 percent 1-chloro-2,2,4-trimethylpentane (substitution of x) and 26 percent 1-chloro-2,4,4-trimethylpentane (substitution of y).

4.39 The three monochlorides are shown in the equation

$$CH_3CH_2CH_2CH_2CH_3 \xrightarrow[\text{light}]{Cl_2}$$
Pentane

$$CH_3CH_2CH_2CH_2CH_2Cl \; + \; CH_3CHCH_2CH_2CH_3 \; + \; CH_3CH_2CHCH_2CH_3$$
$$\qquad\qquad\qquad\qquad\qquad\qquad\qquad | \qquad\qquad\qquad\qquad\qquad |$$
$$\qquad\qquad\qquad\qquad\qquad\qquad\qquad Cl \qquad\qquad\qquad\qquad\qquad Cl$$

1-Chloropentane   2-Chloropentane   3-Chloropentane

Pentane has six primary hydrogens (two CH_3 groups) and six secondary hydrogens (three CH_2 groups). Since a single secondary hydrogen is abstracted 3 times faster than a single primary hydrogen and there are equal numbers of secondary and primary hydrogens, the product mixture should contain 3 times as much of the secondary chloride isomers as the primary chloride. The primary chloride 1-chloropentane, therefore, is expected to comprise 25 percent of the product mixture. The secondary chlorides 2-chloropentane and 3-chloropentane are not formed in equal amounts. Rather, 2-chloropentane may be formed by replacement of a hydrogen at C-2 or at C-4, while 3-chloropentane is formed only when a C-3 hydrogen is replaced. Therefore, the amount of 2-chloropentane is 50 percent and that of 3-chloropentane is 25 percent. We predict the major product to be 2-chloropentane, and the predicted proportion of 50 percent corresponds closely to the observed 46 percent.

4.40 A molecular weight of 72 corresponds to a molecular formula of C_5H_{12}. Since only one monochloride is formed, all the hydrogens must be equivalent, which indicates that the compound is 2,2-dimethylpropane.

$$
\overset{CH_3}{\underset{CH_3}{|}}\; CH_3C\!\!-\!\!CH_3 \xrightarrow[\text{light}]{Cl_2} \overset{CH_3}{\underset{CH_3}{|}}\; CH_3CCH_2Cl \xrightarrow[\text{light}]{Cl_2} \overset{CH_3}{\underset{CH_3}{|}}\; ClCH_2CCH_2Cl \;+\; \overset{CH_3}{\underset{CH_3}{|}}\; CH_3CCHCl_2
$$

4.41 The only molecular formula possible for a hydrocarbon that has a molecular weight of 70 is C_5H_{10}. Since only one monochloride is formed, all the hydrogens in this compound must be equivalent. The hydrocarbon is cyclopentane.

$$\text{(pentagon)} \xrightarrow[\text{light}]{Cl_2} \text{(pentagon)}\!\!-\!\!Cl$$

Cyclopentane   Cyclopentyl chloride

4.42 There are two isomers having the molecular formula C_3H_7Cl, 1-chloropropane and 2-chloropropane.

$$CH_3CH_2CH_2Cl \xrightarrow[\text{light}]{Cl_2} ClCH_2CH_2CH_2Cl + CH_3\underset{\underset{Cl}{|}}{C}HCH_2Cl + CH_3CH_2CHCl_2$$

1-Chloropropane

$$CH_3\underset{\underset{Cl}{|}}{C}HCH_3 \xrightarrow[\text{light}]{Cl_2} CH_3\underset{\underset{Cl}{|}}{C}HCH_2Cl + CH_3\underset{\underset{Cl}{|}}{\overset{\overset{Cl}{|}}{C}}CH_3$$

2-Chloropropane

Therefore A is 2-chloropropane and B is 1-chloropropane.

4.43 (a) The monobromination of 2,2-dimethylbutane can give two different primary bromides, C and D:

$$CH_3\overset{\overset{\displaystyle CH_3}{|}}{\underset{\underset{\displaystyle CH_3}{|}}{C}}CH_2CH_3 \xrightarrow[\text{light}]{Br_2} BrCH_2\overset{\overset{\displaystyle CH_3}{|}}{\underset{\underset{\displaystyle CH_3}{|}}{C}}CH_2CH_3 + CH_3\overset{\overset{\displaystyle CH_3}{|}}{\underset{\underset{\displaystyle CH_3}{|}}{C}}CH_2CH_2Br$$

2,2-Dimethylbutane   C   D

1-Bromo-2,2-dimethylbutane   1-Bromo-3,3-dimethylbutane

and a secondary bromide, E:

$$CH_3\overset{\overset{\displaystyle CH_3}{|}}{\underset{\underset{\displaystyle CH_3}{|}}{C}}{\longrightarrow}\underset{\underset{\displaystyle Br}{|}}{C}HCH_3 \qquad \text{3-Bromo-2,2-dimethylbutane}$$

E

(b) If we let x equal the rate of abstraction of a primary hydrogen, then $82x$ is the rate of abstraction of a secondary hydrogen, and c, d, and e, the relative amounts of products C, D, and E, respectively, are given by:

$$c = 9x$$
$$d = 3x$$
$$e = 2(82)x$$

Normalizing to 100 percent we get

$$9x + 3x + 164x = 100$$

$$x = \frac{100}{176} = 0.57$$

and

$$c = 9(0.57) = 5.1\% \text{ 1-bromo-2,2-dimethylbutane}$$
$$d = 3(0.57) = 1.7\% \text{ 1-bromo-3,3-dimethylbutane}$$
$$e = 93.2\% \text{ 3-bromo-2,2-dimethylbutane}$$

4.44 (a) Acid-catalyzed hydrogen-deuterium exchange takes place by a pair of Brønsted acid-base reactions.

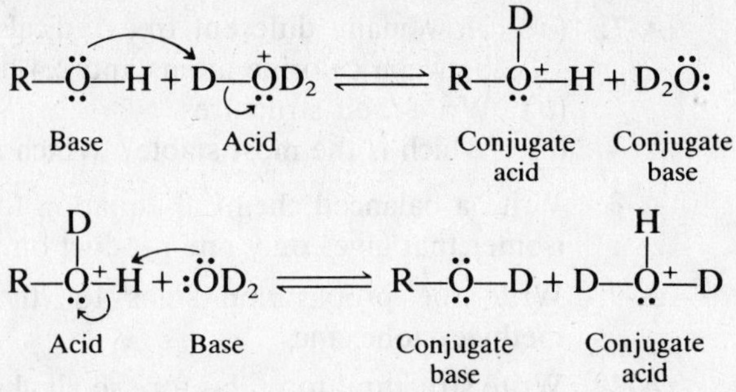

$$R-\overset{..}{\underset{..}{O}}-H + D-\overset{+}{\underset{|}{\overset{D}{O}}}D_2 \rightleftharpoons R-\overset{+}{\underset{|}{O}}-H + D_2\overset{..}{O}:$$

Base   Acid   Conjugate   Conjugate
                                 acid   base

$$R-\overset{+}{\underset{|}{\overset{D}{O}}}H + :\overset{..}{O}D_2 \rightleftharpoons R-\overset{..}{\underset{..}{O}}-D + D-\overset{+}{\underset{|}{\overset{H}{O}}}D$$

Acid   Base   Conjugate   Conjugate
                                 base   acid

(*b*) Base-catalyzed hydrogen-deuterium exchange occurs by a different pair of Brønsted acid-base equilibria.

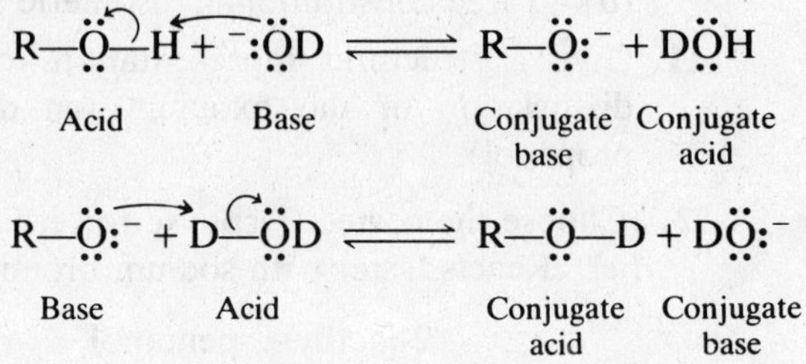

$$R-\overset{..}{\underset{..}{O}}-H + :\overset{..}{O}D \rightleftharpoons R-\overset{..}{\underset{..}{O}}:^- + D\overset{..}{O}H$$

Acid   Base   Conjugate Conjugate
                                 base   acid

$$R-\overset{..}{\underset{..}{O}}:^- + D-\overset{..}{\underset{..}{O}}D \rightleftharpoons R-\overset{..}{\underset{..}{O}}-D + D\overset{..}{\underset{..}{O}}:^-$$

Base   Acid   Conjugate   Conjugate
                                 acid   base

SELF-TEST

PART A

A-1. Give the correct IUPAC name for the following compounds:

(*a*)

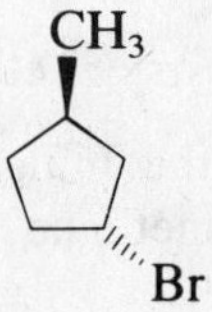

$$CH_3$$

Br

(*b*) $CH_3CH_2\overset{\underset{|}{CH_2OH}}{C}HCH_2\overset{\underset{|}{CH_2CH_3}}{C}HCH_3$

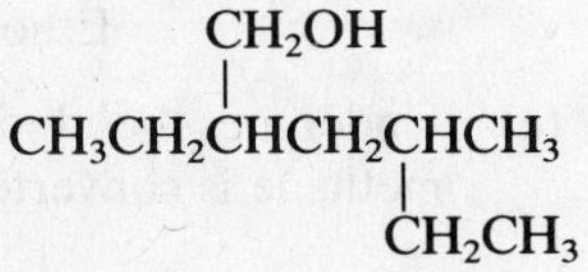

A-2. Draw the structures of the following substances:
(*a*) 2-Chloro-1-iodo-2-methylheptane
(*b*) *cis*-3-Isopropylcyclohexanol

A-3. What are the structures of the conjugate acid and the conjugate base of CH_3OH?

A-4. The acid ionization constant of acetic acid is 1.8×10^{-5}. Calculate the pK_a of this substance.

A-5. Supply the missing component for each of the following reactions:

(*a*) $CH_3CH_2CH_2OH \xrightarrow{SOCl_2} ?$

(*b*) $? \xrightarrow{HBr} CH_3CH_2\overset{\underset{|}{Br}}{C}(CH_3)_2$

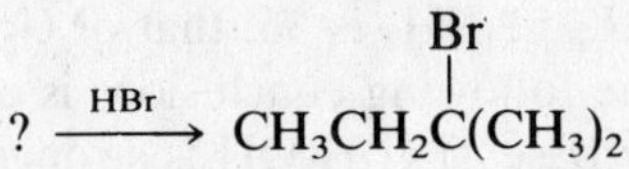

(*c*) ⬡—OH $\xrightarrow{?}$ ⬡—O⁻Na⁺

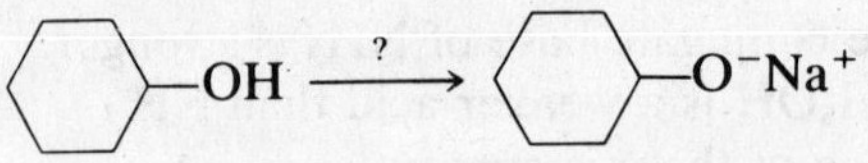

A-6. Write the products of the acid-base reaction that follows, and identify the stronger acid and base and the conjugate of each. Will the equilibrium lie to the left ($K < 1$) or to the right ($K > 1$)? (The pK_a values are given in Problem B-2.)

$$CH_3CH_2O^- + NH_3 \rightleftharpoons$$

A-7. (*a*) How many different free radicals can possibly be produced in the reaction between chlorine atoms and 2,4-dimethylpentane?

(*b*) Write their structures.

(*c*) Which is the most stable? Which is the least stable?

A-8. Write a balanced chemical equation for the reaction of chlorine with the pentane isomer that gives only one product on monochlorination.

A-9. Write the propagation steps for the light-initiated reaction of bromine with methylcyclohexane.

A-10. Write structural formulas for the alkyl chloride of molecular formula C_4H_9Cl which on free-radical chlorination gives:

(*a*) A single dichloride

(*b*) Three constitutionally isomeric dichlorides.

A-11. Give a structural representation for the transition state of the unimolecular dissociation of the oxonium ion derived from *tert*-butyl alcohol (2-methyl-2-propanol).

A-12. (Choose the correct response for each part.) Which species or compound:

(*a*) Reacts faster with sodium bromide and sulfuric acid?

2-methyl-3-pentanol   or   3-methyl-3-pentanol

(*b*) Is a stronger base?

$KOC(CH_3)_3$ or $HOC(CH_3)_3$

(*c*) Reacts more vigorously with cyclohexane?

Fluorine   or   iodine

(*d*) Has an odd number of electrons?

Ethoxide ion   or   ethyl radical

(*e*) Undergoes bond cleavage in the initiation step in the reaction by which methane is converted to chloromethane?

CH_4 or Cl_2

PART B

B-1. Rank the following substances in order of increasing boiling point (lowest→ highest):

$CH_3CH_2CH_2CH_2OH$ $(CH_3)_2CHOCH_3$ $(CH_3)_3COH$

A   B   C

(*a*) A < B < C   (*b*) B < A < C   (*c*) B < C < A   (*d*) C < B < A

B-2. The approximate pK_a of NH_3 is 36; that of C_2H_5OH is 16. From this information, which (if any) of the following conclusions is correct?

(*a*) The conjugate base of C_2H_5OH is stronger.

(*b*) The conjugate base of NH_3 is stronger.

(*c*) C_2H_5OH is a weaker acid than NH_3.

(*d*) None of these is correct.

B-3. When 2-propanol is treated with metallic sodium:

(*a*) Sodium isopropoxide is formed.

(*b*) The sodium is oxidized.

(*c*) The acidic hydrogen is reduced.

(*d*) All these occur.

B-4. The activation energy (E_{act}) of a given reaction is *unrelated* to which of the following parameters?
(*a*) The rate of the slowest step of a multistep reaction
(*b*) The rate of the overall reaction
(*c*) The heat absorbed or given off by the reaction (ΔH)
(*d*) The stability of the transition state

B-5. What is the decreasing stability order (most stable → least stable) of the following carbocations?

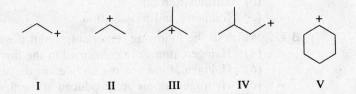

I   II   III   IV   V

(*a*) III > II > I > IV > V
(*b*) I ≈ IV > II ≈ V > III
(*c*) III > II ≈ V > I ≈ IV
(*d*) III > I ≈ IV > II ≈ V

B-6. Which of the following is least able to serve as a nucleophile in a chemical reaction?
(*a*) Br^- (*b*) OH^- (*c*) NH_3 (*d*) CH_3^+

B-7. Thiols are alcohol analogs in which the oxygen has been replaced by sulfur (e.g., CH_3SH). Given the fact that the S—H bond is less polar than the O—H bond, which of the following statements comparing thiols and alcohols is correct?
(*a*) Hydrogen bonding forces are weaker in thiols.
(*b*) Hydrogen bonding forces are stronger in thiols.
(*c*) Hydrogen bonding forces would be the same.
(*d*) No comparison can be made without additional information.

B-8. Rank the *transition states* which occur during the following reaction steps in order of increasing stability (least → most stable):

(A) $CH_3\overset{+}{-}OH_2 \longrightarrow CH_3^+ + H_2O$

(B) $(CH_3)_3C\overset{+}{-}OH_2 \longrightarrow (CH_3)_3C^+ + H_2O$

(C) $(CH_3)_2CH\overset{+}{-}OH_2 \longrightarrow (CH_3)_2CH^+ + H_2O$

(*a*) A < B < C   (*b*) B < C < A   (*c*) A < C < B   (*d*) B < A < C

B-9. Using the data from Appendix B (Table B-1), calculate the heat of reaction $\Delta H°$ for the following:

$$CH_3CH_2\cdot + HBr \longrightarrow CH_3CH_3 + Br\cdot$$

(*a*) +69 kJ/mol (+16.5 kcal/mol)
(*b*) −69 kJ/mol (−16.5 kcal/mol)
(*c*) +44 kJ/mol (+10.5 kcal/mol)
(*d*) −44 kJ/mol (−10.5 kcal/mol)

B-10. An alkane with a molecular formula C_6H_{14} reacts with chlorine in the presence of light and heat. *Four* constitutionally isomeric monochlorides of molecular formula $C_6H_{13}Cl$ are formed (along with a mixture of dichlorides, trichlorides, and so on). What is the most reasonable structure for the starting alkane?
(*a*) $CH_3CH_2CH_2CH_2CH_2CH_3$ (*d*) $(CH_3)_3CCH_2CH_3$
(*b*) $(CH_3)_2CHCH_2CH_2CH_3$ (*e*) $(CH_3)_2CHCH(CH_3)_2$
(*c*) $CH_3CH(CH_2CH_3)_2$

For the remaining four questions, consider the following free-radical reaction:

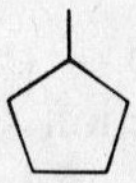

$$\text{cyclopentane} + X_2 \xrightarrow{\text{light}} \text{monohalogenation product}$$

B-11. Light is involved in which of the following reaction steps?
  (a) Initiation only
  (b) Propagation only
  (c) Termination only
  (d) Initiation and propagation

B-12. Which of the following statements about the reaction is *not* true?
  (a) Halogen atoms are consumed in the first propagation step.
  (b) Halogen atoms are regenerated in the second propagation step.
  (c) Hydrogen atoms are produced in the first propagation step.
  (d) Chain termination occurs when two radicals react with each other.

B-13. How many monohalogenation products are possible? (Do not consider stereoisomers.)
  (a) 2   (b) 3   (c) 4   (d) 5

B-14. Which halogen (X_2) will give the best yield of a single monohalogenation product?
  (a) F_2 (b) Cl_2 (c) Br_2 (d) I_2

STRUCTURE AND PREPARATION OF ALKENES. ELIMINATION REACTIONS

IMPORTANT TERMS AND CONCEPTS

Alkene Nomenclature (Sec. 5.1) Alkenes are hydrocarbons which contain a carbon-carbon double bond. The double bond is located in the hydrocarbon chain by numbering in the direction that gives the two double-bonded carbons the lower numbers. In the name the number of the first of these carbons precedes the name stem, the ending of which is *-ene* (replacing the *-ane* of the parent alkane).

$$CH_2{=}CHCH_2CH_2CH_3 \qquad CH_3CH_2CH_2CH{=}CHCH_3$$

1-Pentene   2-Hexene

Branched alkenes are named by using the longest chain *containing the double bond* as the parent name and identifying the substituents as usual.

2,4-Dimethyl-2-pentene   3-Ethyl-1-methylcyclohexene

When a compound contains both a carbon-carbon bond and one or more alkyl groups or halogen substituents, the double bond takes precedence in the systematic name. The hydroxyl group of an alcohol takes precedence over a double bond, however. The longest chain of the molecule is numbered so that the group having precedence has the lower number.

$$\overset{\displaystyle Br}{\underset{\displaystyle CH_3}{CH_3CHCHCH_2CH{=}CH_2}} \qquad\qquad (CH_3)_2C{=}CHCH_2CH_2OH$$

5-Bromo-4-methyl-1-hexene   4-Methyl-3-penten-1-ol
(not 2-bromo-3-methyl-5-hexene)   (not 2-methyl-2-penten-5-ol)

Stereoisomerism (Sec. 5.3) As described in text Sec. 1.15, the carbon-carbon double bond of alkenes consists of a σ (sigma) component and a π (pi) component. The π

overlap of an alkene double bond inhibits the rotation of one carbon atom relative to the other. The rotation about a carbon-carbon bond is imperceptibly slow; in fact, it is presumed not to occur. Thus *stereoisomers* may result which are not interconvertible. For example, there are two stereoisomers of 2-butene, known as *cis*-2-butene and *trans*-2-butene:

cis-2-Butene   trans-2-Butene

When one of the doubly bonded carbon atoms bears two identical substituents, stereoisomerism is not possible. Thus propene and 2-methyl-2-butene are not capable of existing as stereoisomers.

$$CH_2 = CHCH_3 \qquad\qquad (CH_3)_2C = CHCH_3$$

Propene   2-Methyl-2-butene

E−Z System of Nomenclature (Sec. 5.4) Stereoisomeric alkenes may be named by establishing *priorities,* or rankings, within the pair of groups attached to each end of the double bond. The priority ranking is based on the *atomic number* of the groups. If the higher-ranking groups are on the *same* side of the double bond, the alkene is named Z; if they are on opposite sides of the double bond, it is named E. When the atoms attached directly to the double bond are the same, priority rankings are determined outward from the double bond until the first difference is found. For example,

(E)-2-Chloro-2-pentene   (Z)-1-Bromo-2-methyl-2-butene

Alkene Stabilities (Sec. 5.6) Alkyl substitution tends to *stabilize* an alkene. When stereoisomeric alkenes are compared, the trans isomer is generally more stable than the cis (cyclic alkenes would be a notable exception). The following stability series may be written in order of decreasing stability:

$$R_2C = CR_2 > R_2C = CHR > RCH = CHR \text{ (trans)}$$

$$> RCH = CHR \text{ (cis)} > RCH = CH_2 > CH_2 = CH_2$$

An explanation for the increased stability associated with substituted alkenes is the electron-releasing power of an alkyl group when compared with that of a hydrogen substituent on a double bond.

Cyclic Alkenes (Sec. 5.7) Cycloalkenes having fewer than 12 carbons in the ring are more stable in the cis configuration; in fact, a trans double bond is not stable in a ring smaller than 8 carbons. When a ring is larger than 12-membered, trans cycloalkenes are more stable than cis.

Elimination Reactions (Sec. 5.8) Elimination reactions that lead to alkenes are known as β *elimination reactions,* since the reactions proceed by loss of atoms from *adjacent* carbons on the reactant. The reactions discussed in this chapter, *dehydration of alcohols* and *dehydrohalogenation of alkyl halides,* are examples of β elimination reactions.

Zaitsev's Rule (Secs. 5.10, 5.13) Alcohol dehydration is a *regioselective* process; that is although more than one regioisomer can be formed from a single substrate, one is observed to predominate. *Zaitsev's rule* states that the major alkene product formed by an elimination reaction will correspond to *proton loss from the β carbon having the fewest hydrogen substituents.* The rule applies to both dehydration and dehydrohalogenation reactions. In other words, elimination of alcohols and alkyl halides yields the *most highly substituted* alkene as the major product. That is, the major product is the *most stable alkene*. These reactions are also *stereoselective*—when stereoisomeric products are possible, the more stable stereoisomer predominates. For example, dehydration of 3-pentanol yields a $3:1$ ratio of E (or trans) to Z (or cis) isomer of 2-pentene.

$$CH_3CH_2CHCH_2CH_3 \xrightarrow[\text{heat}]{H_2SO_4}$$

$$\underset{\substack{(Z)\\ \text{minor product}}}{\underset{H}{\overset{CH_3}{>}}C=C\underset{H}{\overset{CH_2CH_3}{<}}} \;+\; \underset{\substack{(E)\\ \text{major product}}}{\underset{H}{\overset{CH_3}{>}}C=C\underset{CH_2CH_3}{\overset{H}{<}}}$$

E2 Mechanism (Secs. 5.14, 5.15) The dehydrohalogenation of alkyl halides by strong base is considered to be a *concerted* reaction. That is, the entire reaction proceeds in a single step. Such a β elimination reaction is termed an *E2* (*elimination bimolecular*) *reaction*. The rate of an E2 reaction is described by *second-order kinetics*, the general rate equation being

$$\text{Rate} = k[\text{alkyl halide}][\text{base}]$$

The rate of elimination depends on the nature of the leaving group and *decreases in the order* $I > Br > Cl > F$.

E2 reactions occur by an anti elimination; that is, the β hydrogen and the halide leaving group must be in an anti periplanar arrangement for elimination to proceed.

$$B:^- \quad \overset{H}{\underset{b}{\overset{a}{\diagup}}}\overset{c}{\underset{X}{\diagdown}}d \longrightarrow \overset{a}{\underset{b}{>}}C=C\overset{c}{\underset{d}{<}} + X^- + BH$$

E1 Mechanism for Dehydrohalogenation (Sec. 5.17) An alternative mechanism for elimination is characterized by ionization of the halide leaving group occurring *prior to* loss of the β hydrogen. This *unimolecular ionization* of the substrate gives rise to a carbocation intermediate, in a manner similar to that described for the dehydration of alcohols. Such a reaction exhibits *first-order kinetics* and is known as an *E1* (*elimination unimolecular*) *reaction*.

$$\text{Rate} = k[\text{alkyl halide}]$$

The E1 mechanism is important only with tertiary and some secondary halides and when there is a low concentration of a weak base. In general, however, conditions under which E2 elimination predominates are preferred for the preparation of alkenes from alkyl halides.

IMPORTANT REACTIONS

Dehydration of Alcohols (Secs. 5.9, 5.10)

General:

$$H-\overset{|}{\underset{|}{C}}-\overset{|}{\underset{|}{C}}-OH \xrightarrow{H^+} >C=C< + H_2O$$

Examples:

$$(CH_3)_2\overset{OH}{\underset{|}{C}}CH_2CH_3 \xrightarrow[\text{heat}]{H_2SO_4} (CH_3)_2C=CHCH_3 + CH_2=\overset{CH_3}{\underset{|}{C}}CH_2CH_3$$

$$\text{(major)}$$

(major)

Dehydration Mechanism (Sec. 5.11)

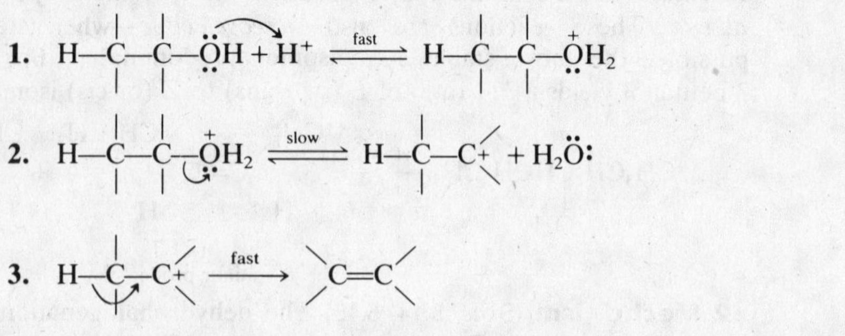

1. $H-\overset{|}{\underset{|}{C}}-\overset{|}{\underset{|}{C}}-\overset{..}{\underset{..}{O}}H + H^+ \underset{fast}{\rightleftharpoons} H-\overset{|}{\underset{|}{C}}-\overset{|}{\underset{|}{C}}-\overset{+}{\underset{..}{O}}H_2$

2. $H-\overset{|}{\underset{|}{C}}-\overset{|}{\underset{+}{C}}-\overset{+}{\underset{..}{O}}H_2 \underset{slow}{\rightleftharpoons} H-\overset{|}{\underset{|}{C}}-C^+ + H_2\overset{..}{O}:$

3. $H-\overset{|}{\underset{|}{C}}-C^+ \overset{fast}{\longrightarrow} C{=}C$

Primary and some secondary alcohols undergo dehydration by a mechanism that avoids carbocation formation.

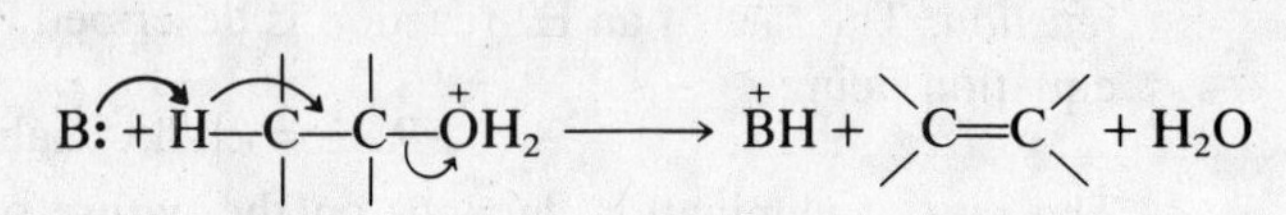

$B: + H-\overset{|}{\underset{|}{C}}-\overset{|}{\underset{|}{C}}-\overset{+}{O}H_2 \longrightarrow BH + C{=}C + H_2O$

Rearrangements (Sec. 5.12)

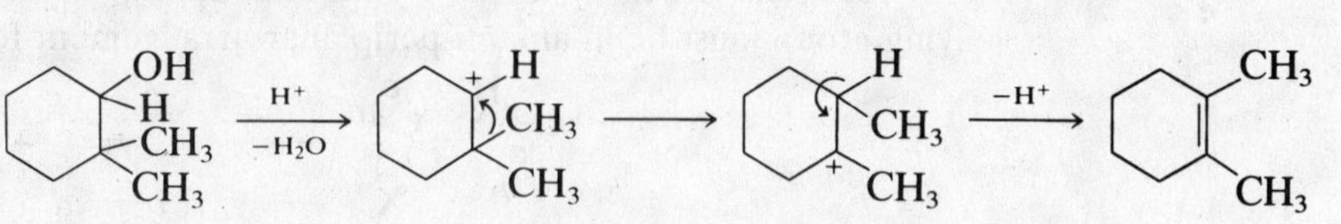

Dehydrohalogenation of Alkyl Halides (Sec. 5.13)

General:

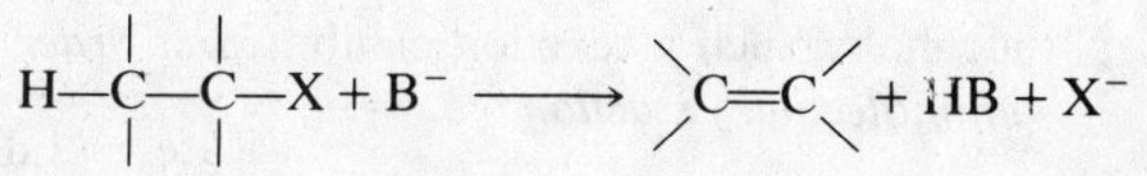

$H-\overset{|}{\underset{|}{C}}-\overset{|}{\underset{|}{C}}-X + B^- \longrightarrow C{=}C + HB + X^-$

Examples:

$(CH_3)_2C-CH(CH_3)_2 \xrightarrow[CH_3CH_2OH]{KOCH_2CH_3} (CH_3)_2C{=}C(CH_3)_2$

with Br above the first carbon.

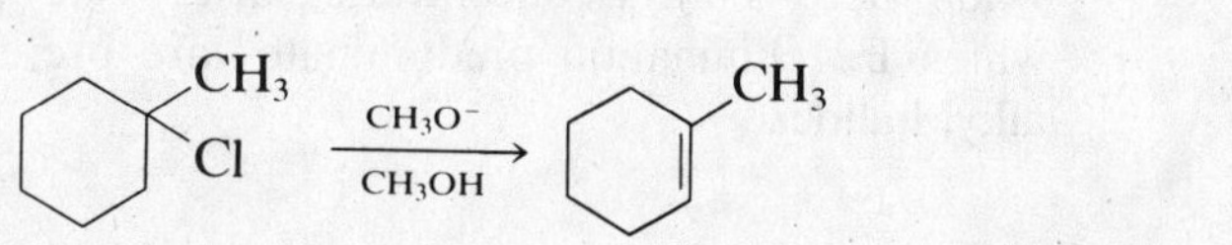

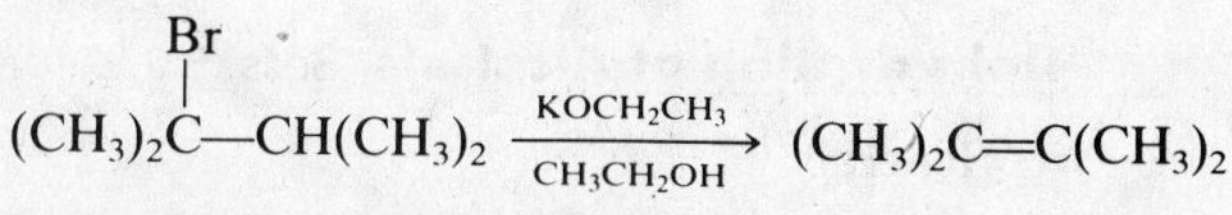

Mechanism of Dehydrohalogenation (Secs. 5.14, 5.15)

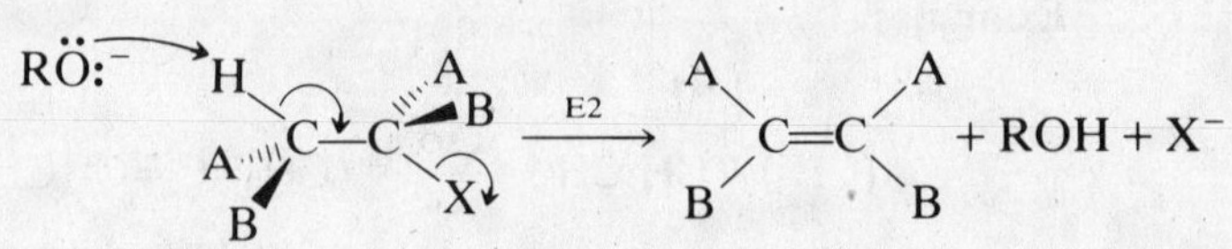

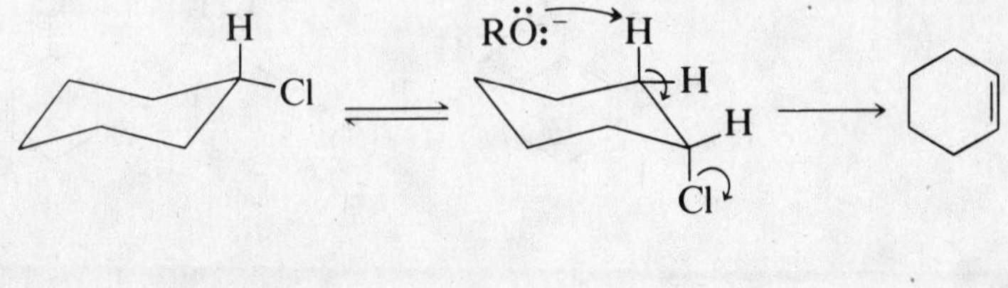

SOLUTIONS TO TEXT PROBLEMS

5.1 (*b*) Writing the structure in more detail, we see that the longest continuous chain contains four carbon atoms.

$$CH_3 \overset{4}{C}H_3 - \overset{3}{\underset{\underset{CH_3}{|}}{\overset{|}{C}}} - \overset{2}{C}H = \overset{1}{C}H_2$$

The double bond is located at the end of the chain, and so the alkene is named as a derivative of 1-butene. Two methyl groups are substituents at C-3. The correct IUPAC name is 3,3-dimethyl-1-butene.

(*c*) Expanding the structural formula reveals the molecule to be a methyl-substituted derivative of hexene.

$$\overset{1}{C}H_3 - \overset{2}{\underset{\underset{CH_3}{|}}{C}} = \overset{3}{C}H\overset{4}{C}H_2\overset{5}{C}H_2\overset{6}{C}H_3 \qquad \text{2-Methyl-2-hexene}$$

(*d*) In compounds containing a double bond and a halogen, the double bond takes precedence in numbering the longest carbon chain.

$$\overset{1}{C}H_2 = \overset{2}{C}H\overset{3}{C}H_2\overset{4}{\underset{\underset{Cl}{|}}{C}}H\overset{5}{C}H_3 \qquad \text{4-Chloro-1-pentene}$$

(*e*) When a hydroxyl group is present in a compound containing a double bond, the hydroxyl takes precedence over the double bond in numbering the longest carbon chain.

$$\overset{5}{C}H_2 = \overset{4}{C}H\overset{3}{C}H_2\overset{2}{\underset{\underset{OH}{|}}{C}}H\overset{1}{C}H_3 \qquad \text{4-Penten-2-ol}$$

5.2 There are three sets of nonequivalent positions on a cyclopentene ring, identified as *a, b,* and *c* on the cyclopentene structure shown:

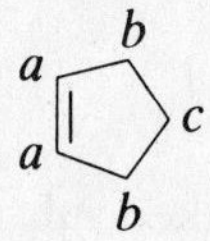

Thus, there are three different monochloro-substituted derivatives of cyclopentene. The carbons that bear the double bond are numbered C-1 and C-2 in each isomer, and the other positions are numbered in sequence in the direction that gives the chlorine-bearing carbon its lower locant.

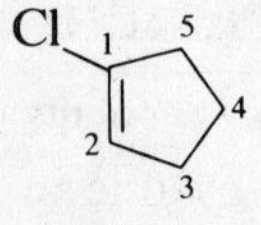

1-Chlorocyclopentene 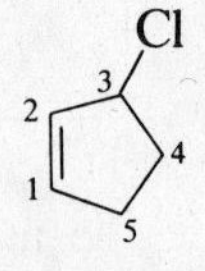   3-Chlorocyclopentene 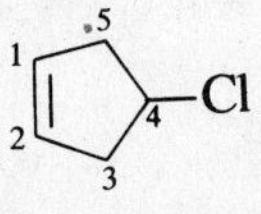   4-Chlorocyclopentene

5.3 (*b*) The alkene is a derivative of 3-hexene regardless of whether the chain is numbered from left to right or from right to left. Number it in the direction that gives the lower number to the substituent.

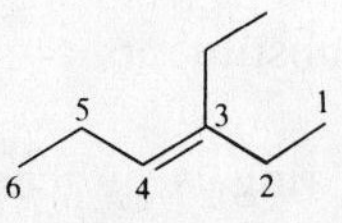

3-Ethyl-3-hexene

(c)   There are only two sp^2 hybridized carbons, the two connected by the double bond. All other carbons (six) are sp^3 hybridized.

(d)   There are three sp^2–sp^3 σ bonds and three sp^3–sp^3 σ bonds.

5.4   Consider first the C_5H_{10} alkenes that have an unbranched carbon chain:

1-Pentene   *cis*-2-Pentene   *trans*-2-Pentene

There are three additional isomers. These have a four-carbon chain with a methyl substituent.

2-Methyl-1-butene   2-Methyl-2-butene   3-Methyl-1-butene

5.5   First, identify the constitution of 9-tricosene. Referring back to Table 2.4 in Section 2.10 of the text, we see that tricosane is the unbranched alkane containing 23 carbon atoms. 9-Tricosene, therefore, contains an unbranched chain of 23 carbons with a double bond between C-9 and C-10. Since the problem specifies that the pheromone has the cis configuration, the first 8 carbons and the last 13 must be on the same side of the C-9—C-10 double bond.

$$CH_3(CH_2)_7 \quad\quad (CH_2)_{12}CH_3$$
$$C{=}C$$
$$H \quad\quad\quad\quad H$$

cis-9-Tricosene

5.6   (b)   One of the carbons bears a methyl group and a hydrogen; methyl is of higher rank than hydrogen. The other carbon bears the groups —CH$_2$CH$_2$F and —CH$_2$CH$_2$CH$_2$CH$_3$. At the first point of difference between these two, fluorine is of higher atomic number than carbon, and so —CH$_2$CH$_2$F is of higher precedence.

higher   CH$_3$ CH$_2$CH$_2$F   higher
              C═C
lower   H   CH$_2$CH$_2$CH$_2$CH$_3$ lower

Higher-ranked substituents are on the same side of the double bond; the alkene has the Z configuration.

(c)   One of the carbons bears a methyl group and a hydrogen; as we have seen, methyl is of higher priority. The other carbon bears —CH$_2$CH$_2$OH and —C(CH$_3$)$_3$. Let us analyze these two groups so that we may determine their order of precedence.

—CH$_2$CH$_2$OH   —C(CH$_3$)$_3$
—C(C,H,H)   —C(C,C,C)

Lower priority   Higher priority

We examine the atoms one by one at the point of attachment before proceeding down the chain. Therefore, —C(CH$_3$)$_3$ outranks —CH$_2$CH$_2$OH.

higher   CH$_3$ CH$_2$CH$_2$OH   lower
              C═C
lower   H   C(CH$_3$)$_3$ higher

Higher-ranked substituents are on opposite sides; the configuration of the alkene is E.

(d)   The cyclopropyl ring is attached to the double bond by a carbon that bears the substituents (C,C,H) and is therefore of higher precedence than an ethyl group —C(C,H,H).

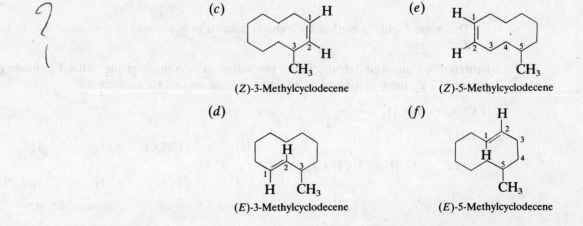

Higher-ranked substituents are on opposite sides; the configuration of the alkene is *E*.

5.7 A trisubstituted alkene has three carbons directly attached to the doubly bonded carbons. There are three trisubstituted C_6H_{12} isomers, two of which are stereoisomers.

2-Methyl-2-pentene   (*E*)-3-Methyl-2-pentene   (*Z*)-3-Methyl-2-pentene

5.8 The most stable C_6H_{12} alkene has a tetrasubstituted double bond:

2,3-Dimethyl-2-butene

5.9 Apply the two general rules for alkene stability to rank these compounds. First, more highly substituted double bonds are more stable than less substituted ones. Second, when two double bonds are similarly constituted, the trans stereoisomer is more stable than the cis. Therefore, the predicted order of decreasing stability is:

2-Methyl-2-butene
(trisubstituted):
most stable

(*E*)-2-Pentene
(disubstituted)

(*Z*)-2-Pentene
(disubstituted)

1-Pentene
(monosubstituted):
least stable

5.10 Double bonds are placed on the parent figure to give the structures shown.

(*c*)

(*Z*)-3-Methylcyclodecene

(*e*)

(*Z*)-5-Methylcyclodecene

(*d*)

(*E*)-3-Methylcyclodecene

(*f*)

(*E*)-5-Methylcyclodecene

5.11 Write out the structure of the alcohol, recognizing that the alkene is formed by loss of a hydrogen and a hydroxyl group from adjacent carbons.

(*b*, *c*) Both 1-propanol and 2-propanol give propene on acid-catalyzed dehydration.

$$CH_3\overset{\beta}{C}H_2\overset{\alpha}{C}H_2OH \xrightarrow[\text{heat}]{H^+} CH_3CH{=}CH_2 \xleftarrow[\text{heat}]{H^+} \overset{\beta}{C}H_3\overset{\alpha}{C}H\overset{\beta}{C}H_3$$
$$\underset{\text{OH}}{|}$$

$$\underset{\text{1-Propanol}}{} \qquad \underset{\text{Propene}}{} \qquad \underset{\text{2-Propanol}}{}$$

(*d*) Carbon-3 has no hydrogen substituents in 2,3,3-trimethyl-2-butanol. Elimination can involve only the hydroxyl group at C-2 and a hydrogen at C-1.

2,3,3-Trimethyl-2-butanol   2,3,3-Trimethyl-1-butene

5.12 (*b*) Elimination can involve loss of a hydrogen from the methyl group or from C-2 of the ring in 1-methylcyclohexanol.

1-Methylcyclohexanol   Methylenecyclohexane   1-Methylcyclohexene
                        (a disubstituted alkene;   (a trisubstituted alkene;
                         minor product)   major product)

According to the Zaitsev rule, the major alkene is the one corresponding to loss of a hydrogen from the alkyl group that has the smaller number of hydrogens. Thus hydrogen is removed from the methylene group in the ring rather than from the methyl group, and 1-methylcyclohexene is formed in greater amounts than methylenecyclohexane.

(*c*) The two alkenes formed are as shown in the equation.

Compound has a   Compound has a
trisubstituted   tetrasubstituted
double bond   double bond;
                      more stable

The more highly substituted alkene is formed in greater amounts, according to the Zaitsev rule.

5.13 2-Pentanol can undergo dehydration in two different directions, giving either 1-pentene or 2-pentene. 2-Pentene is formed as a mixture of the cis and trans isomers.

$$CH_3CHCH_2CH_2CH_3 \xrightarrow[\text{heat}]{H^+}$$
$$\underset{\text{OH}}{|}$$

2-Pentanol

1-Pentene   *cis*-2-Pentene   *trans*-2-Pentene

5.14 (*b*) The site of positive charge in the carbocation is the carbon atom that bears the hydroxyl group in the starting alcohol.

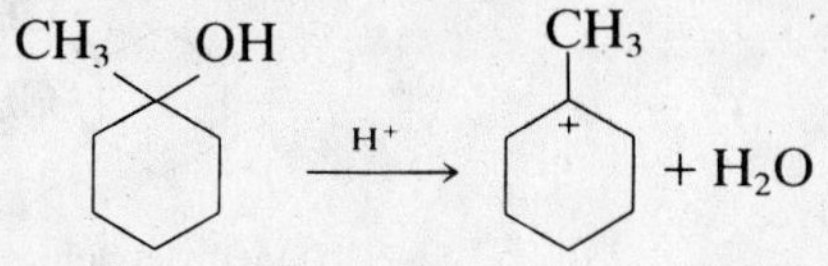

1-Methylcyclohexanol

The curved arrow notation for proton abstraction by water from the methyl group is demonstrated in the following equation:

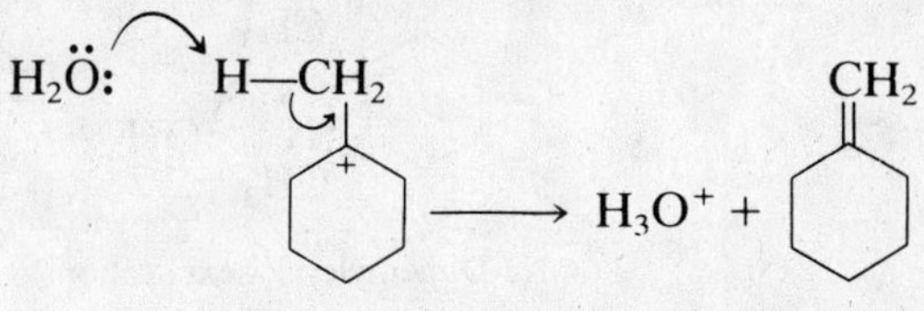

H_3O^+ +

Methylenecyclohexane

Abstraction of a proton from the ring gives 1-methylcyclohexene.

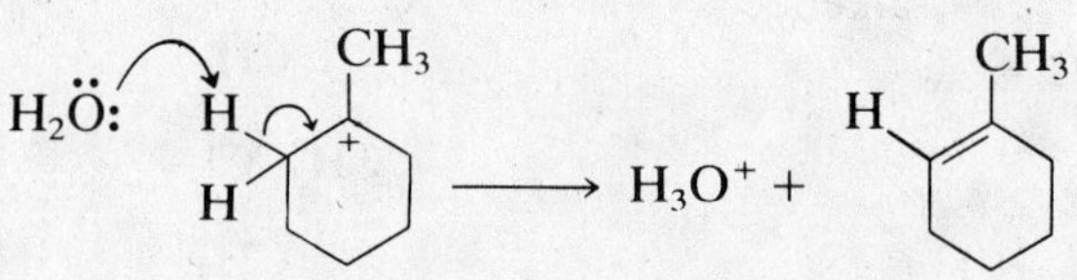

H_3O^+ +

1-Methylcyclohexene

(*c*) Loss of the hydroxyl group under conditions of acid catalysis yields a tertiary carbocation.

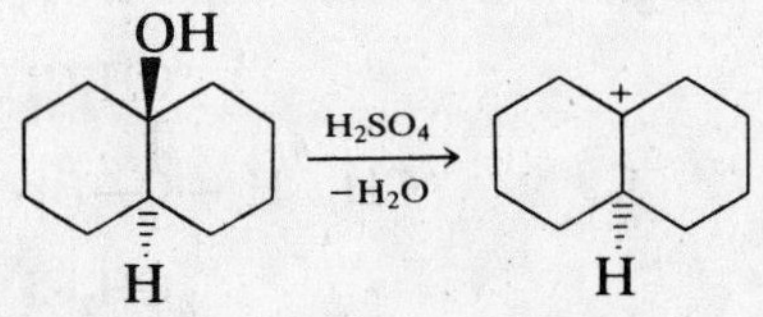

Water may abstract a proton from an adjacent methylene group to give a trisubstituted alkene.

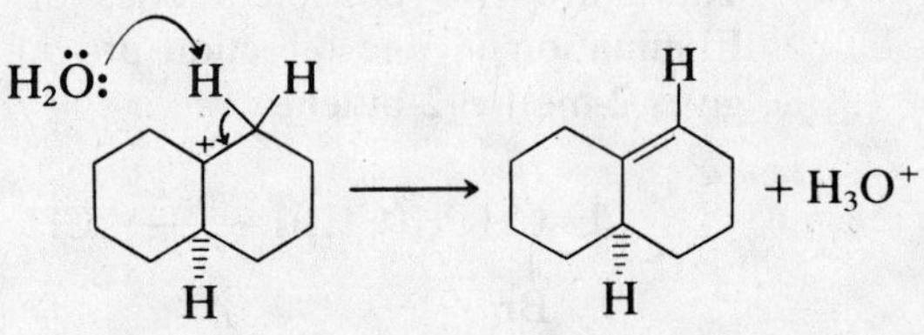

+ H_3O^+

Abstraction of the methine proton affords a tetrasubstituted alkene.

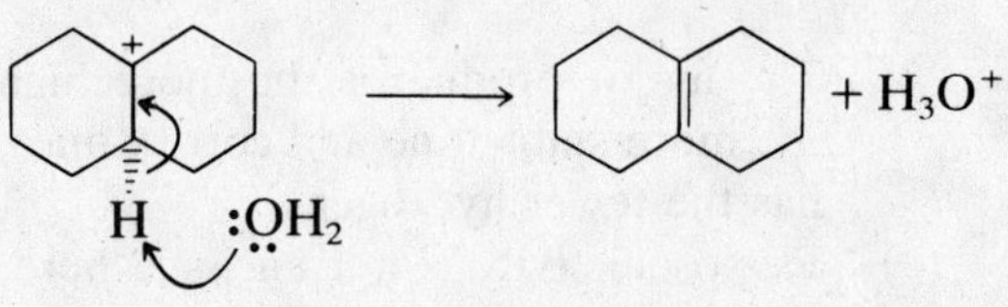

+ H_3O^+

5.15 In writing mechanisms for acid-catalyzed dehydration of alcohols, begin with formation of the carbocation intermediate:

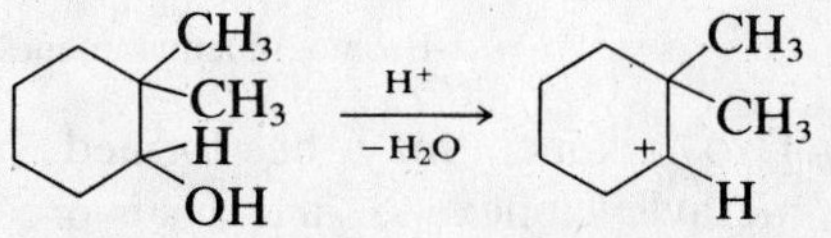

2,2-Dimethylcyclohexanol   2,2-Dimethylcyclohexyl cation

This secondary carbocation can rearrange to a more stable tertiary carbocation by a methyl group shift.

2,2-Dimethylcyclohexyl cation   1,2-Dimethylcyclohexyl cation
(secondary)   (tertiary)

Loss of a proton from the 1,2-dimethylcyclohexyl cation intermediate yields 1,2-dimethylcyclohexene.

1,2-Dimethylcyclohexyl cation   1,2-Dimethylcyclohexene

5.16 (b) All the hydrogens of *tert*-butyl chloride are equivalent. Loss of any of these hydrogens along with the chlorine substituent yields 2-methylpropene as the only alkene.

tert-Butyl chloride   2-Methylpropene

(c) All the β hydrogens of 3-bromo-3-ethylpentane are equivalent, so that β elimination can give only 3-ethyl-2-pentene.

3-Bromo-3-ethylpentane   3-Ethyl-2-pentene

(d) There are two possible modes of β elimination from 2-bromo-3-methylbutane. Elimination in one direction provides 3-methyl-1-butene; elimination in the other gives 2-methyl-2-butene.

$$\overset{\beta}{C}H_3\overset{\beta}{CH}CH(CH_3)_2 \longrightarrow CH_2{=}CHCH(CH_3)_2 + CH_3CH{=}C(CH_3)_2$$
$$\underset{Br}{|}$$

2-Bromo-3-methylbutane   3-Methyl-1-butene   2-Methyl-2-butene
                                (monosubstituted)   (trisubstituted)

The major product is the more highly substituted alkene, 2-methyl-2-butene. It is the more stable one and corresponds to removal of a hydrogen from the carbon that has the fewer hydrogens.

(e) Regioselectivity is not an issue here, because 3-methyl-1-butene is the only alkene that can be formed by β elimination from 1-bromo-3-methylbutane.

$$BrCH_2\overset{\beta}{CH_2}CH(CH_3)_2 \longrightarrow CH_2{=}CHCH(CH_3)_2$$

1-Bromo-3-methylbutene   3-Methyl-1-butene

(f) Two alkenes may be formed here. The more highly substituted one is 1-methylcyclohexene, and this is predicted to be the major product in accordance with Zaitsev's rule.

1-Iodo-1-methylcyclohexane   Methylenecyclohexane   1-Methylcyclohexene
(disubstituted)   (trisubstituted; major product)

5.17 (*b*) The dehydrohalogenation of 2-bromohexane is similar to that of 2-bromobutane described in part (*a*) of this problem. Elimination can occur to give 1-hexene, *cis*-2-hexene, or *trans*-2-hexene.

2-Bromohexane

1-Hexene   *cis*-2-Hexene   *trans*-2-Hexene

The experimental results of an elimination reaction of 2-bromohexane using sodium methoxide as the base in methanol at 99°C are in accord with what we have learned of the regioselectivity and stereoselectivity of these reactions. The major product observed was *trans*-2-hexene (54 percent), with *cis*-2-hexene (18 percent) and 1-hexene (28 percent) as the minor components of the reaction mixture.

(*c*) In 2-iodo-4-methylpentane β elimination can involve a hydrogen at C-1 or at C-3. The alkenes that are capable of being formed are 4-methyl-1-pentene, *trans*-4-methyl-2-pentene, and *cis*-4-methyl-2-pentene.

2-Iodo-4-methylpentane

4-Methyl-1-pentene   *trans*-4-Methyl-2-pentene   *cis*-4-Methyl-2-pentene

When this reaction was performed with use of potassium *tert*-butoxide in dimethyl sulfoxide, *trans*-4-methyl-2-pentene was formed as the major product (60 percent).

(*d*) Elimination can proceed in only one direction in 3-iodo-2,2-dimethylpentane. A mixture of *cis*- and *trans*-4,4-dimethyl-2-pentene is formed.

3-Iodo-2,2-dimethylpentane   *cis*-4,4-Dimethyl-2-pentene   *trans*-4,4-Dimethyl-2-pentene

5.18 The two starting materials are stereoisomers of each other, and so it is reasonable to begin by examining each one in more stereochemical detail. First, write the most stable conformation of each isomer, keeping in mind that isopropyl is the bulkiest of the three substituents and has the greatest preference for an equatorial orientation.

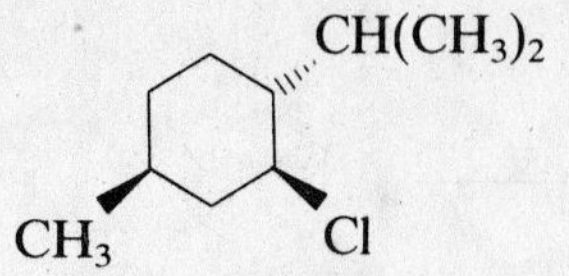

Menthyl chloride

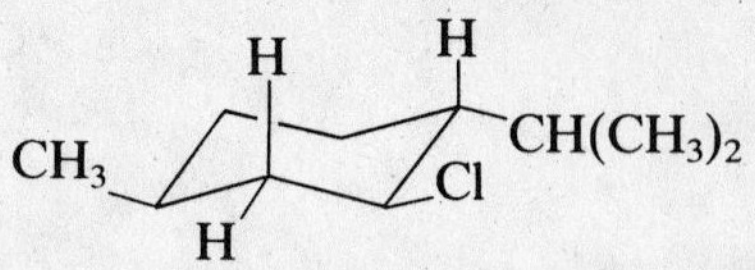

Most stable conformation of menthyl chloride: none of the three β protons is anti to chlorine

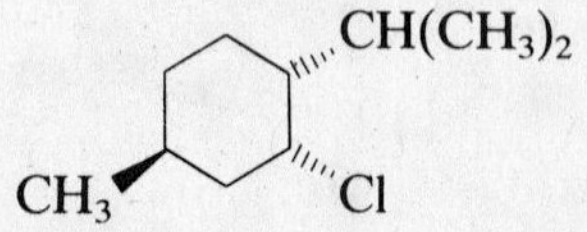

Neomenthyl chloride

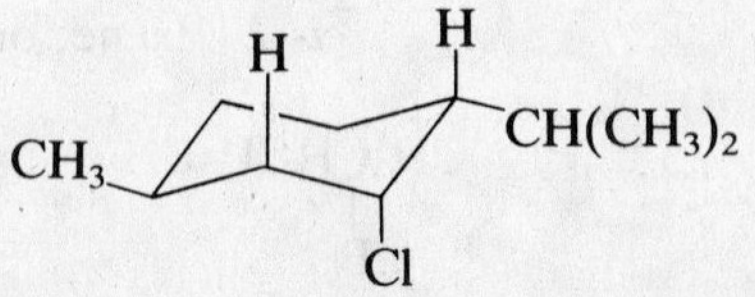

Most stable conformation of neomenthyl chloride: each β carbon has a proton that is anti to chlorine

The anti periplanar relationship of halide and proton can be achieved only when the chlorine is axial; this corresponds to the most stable conformation of neomenthyl chloride. Menthyl chloride, on the other hand, must undergo appreciable distortion of its ring in order to achieve an anti periplanar Cl—C—C—H geometry. Strain increases substantially in going to the transition state for E2 elimination in menthyl chloride but not in neomenthyl chloride. Neomenthyl chloride undergoes E2 elimination at the faster rate.

5.19   (a)   1-Heptene is $CH_2{=}CH(CH_2)_4CH_3$

      (b)   3-Ethyl-2-pentene is $CH_3CH{=}C(CH_2CH_3)_2$

      (c)   *cis*-3-Octene is

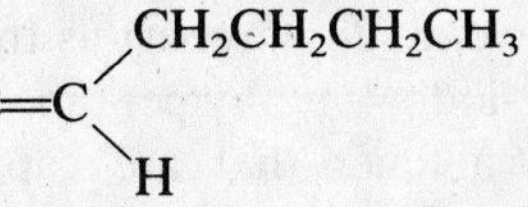

      (d)   *trans*-1,4-Dichloro-2-butene is

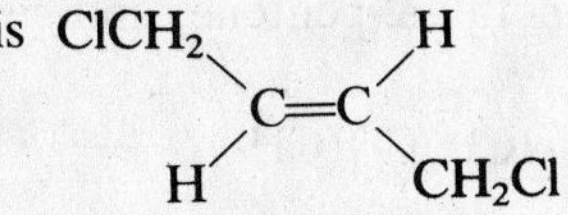

      (e)   (Z)-3-Methyl-2-hexene is

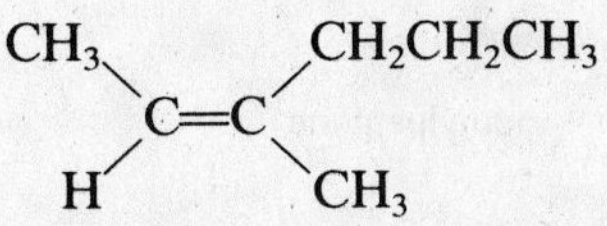

      (f)   (E)-3-Chloro-2-hexene is

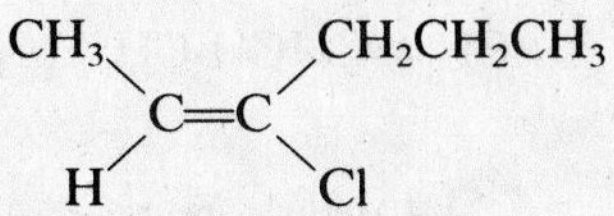

      (g)   1-Bromo-3-methylcyclohexene is

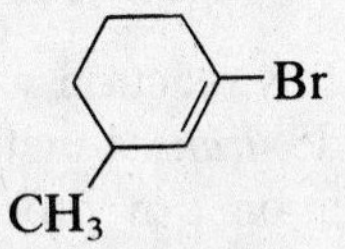

      (h)   1-Bromo-6-methylcyclohexene is

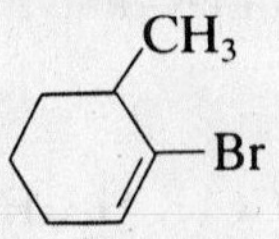

      (i)   4-Methyl-4-penten-2-ol is

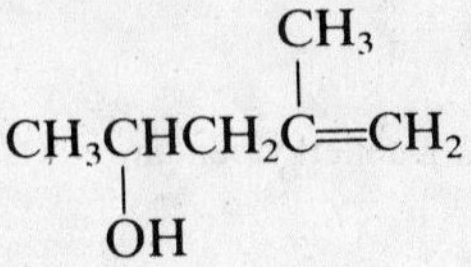

      (j)   Vinylcycloheptane is

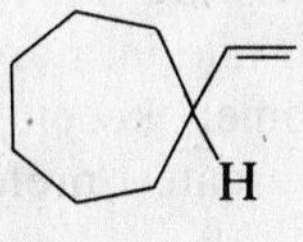

(k)   An allyl group is —CH_2CH=CH_2. 1,1-Diallylcyclopropane is

(l)   An isopropenyl substituent is —$\overset{\overset{\displaystyle CH_3}{|}}{C}$=$CH_2$. *trans*-1-Isopropenyl-3-methylcyclohexane is therefore

5.20   Alkenes with tetrasubstituted double bonds have four alkyl groups attached to the doubly bonded carbons. There is only one alkene of molecular formula C_7H_{14} that has a tetrasubstituted double bond, 2,3-dimethyl-2-pentene.

$$\begin{array}{ccc} CH_3 & & CH_3 \\ & C{=}C & \\ CH_3 & & CH_2CH_3 \end{array}$$ 2,3-Dimethyl-2-pentene

5.21   (a)   The longest chain that includes the double bond in $(CH_3CH_2)_2C$=$CHCH_3$ contains five carbon atoms, and so the parent alkene is a pentene. The numbering scheme which gives the double bond the lowest number is

The compound is 3-ethyl-2-pentene.

(b)   Write out the structure in detail and identify the longest continuous chain that includes the double bond.

$$\overset{1}{C}H_3\overset{2}{C}H_2\underset{CH_3CH_2}{\overset{}{\diagdown}}\overset{3}{C}{=}\overset{4}{C}\underset{CH_2CH_3}{\overset{\overset{5}{C}H_2\overset{6}{C}H_3}{\diagup}}$$

The longest chain contains six carbon atoms, and the double bond is between C-3 and C-4. The compound is named as a derivative of 3-hexene. There are ethyl substituents at C-3 and C-4. The complete name is 3,4-diethyl-3-hexene.

(c)   Write out the structure completely.

The longest carbon chain contains four carbons. Number the chain so as to give the lowest numbers to the doubly bonded carbons, and list the substituents in alphabetical order. This compound is 1,1-dichloro-3,3-dimethyl-1-butene.

(d) The longest chain has five carbon atoms, the double bond is at C-1, and there are two methyl substituents. The compound is 4,4-dimethyl-1-pentene.

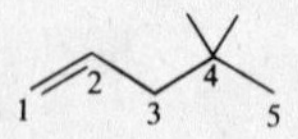

(e) We number this trimethylcyclobutene derivative so as to provide the lowest number for the substituent at the first point of difference. Therefore, we number

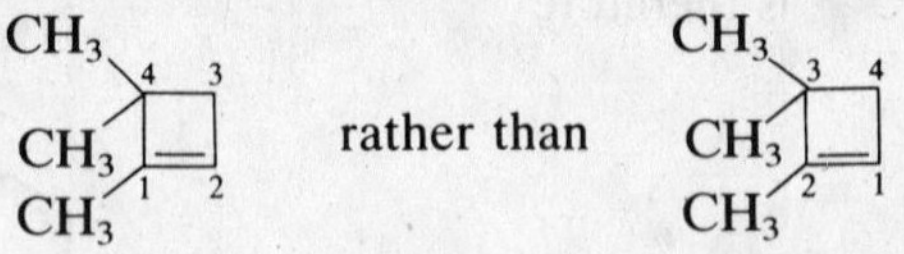

$$CH_3, CH_3, CH_3 \quad \text{rather than} \quad CH_3, CH_3, CH_3$$

The correct IUPAC name is 1,4,4-trimethylcyclobutene, not 2,3,3-trimethyl-cyclobutene.

(f) The cyclohexane ring has a 1,2-cis arrangement of vinyl substituents. The compound is cis-1,2-divinylcyclohexane.

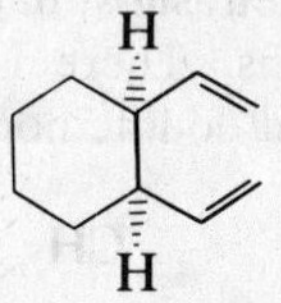

(g) Name this compound as a derivative of cyclohexene. It is 1,2-divinylcyclohexene.

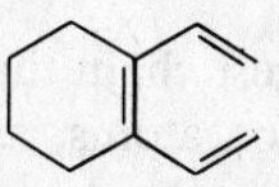

5.22 (a) Go to the end of the name, because this tells you how many carbon atoms are present. In the hydrocarbon name 2,6,10,14-tetramethyl-2-pentadecene, the suffix -2-pentadecene reveals that the longest continuous chain has 15 carbon atoms and that there is a double bond between C-2 and C-3. The rest of the name provides the information that there are four methyl groups and that they are located at C-2, C-6, C-10, and C-14.

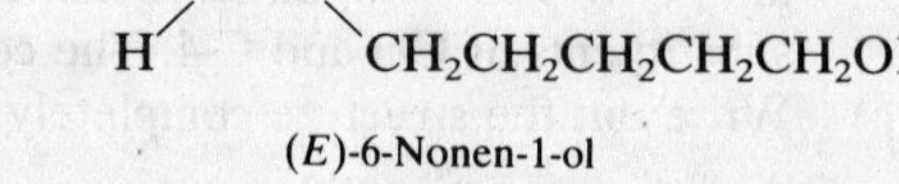

2,6,10,14-Tetramethyl-2-pentadecene

(b) The E configuration means that the higher-priority substituents are on opposite sides of the double bond.

higher CH_3CH_2 H lower
$$C=C$$
lower H $CH_2CH_2CH_2CH_2CH_2OH$ higher

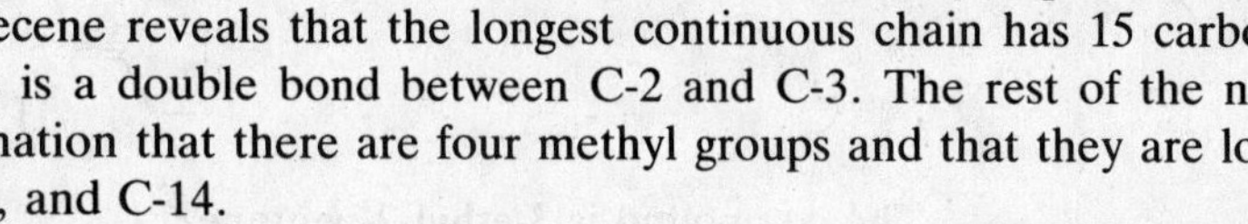

(E)-6-Nonen-1-ol

(c) Geraniol has two double bonds, but only one of them, the one between C-2 and C-3, is capable of stereochemical variation. Of the substituents at C-2, CH_2OH is of higher priority than H. At C-3, CH_2CH_2 outranks CH_3. Higher-priority substituents are on opposite sides of the double bond in the E isomer; hence geraniol has the structure shown.

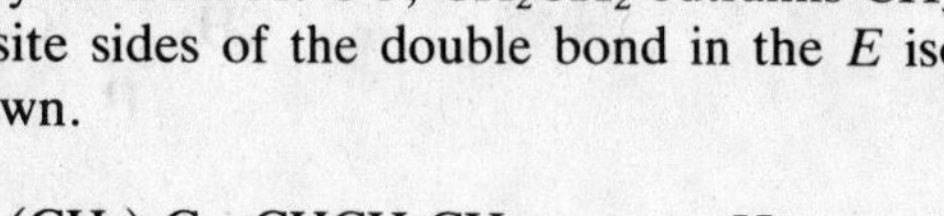

$$(CH_3)_2C=CHCH_2CH_2 \quad H$$
$$C=C$$
$$CH_3 \quad\quad CH_2OH \quad \text{Geraniol}$$

(d) Since nerol is a stereoisomer of geraniol, it has the same constitution and differs from geraniol only in having the Z configuration of the double bond.

$$(CH_3)_2C\text{=}CHCH_2CH_2 \diagdown \underset{CH_3}{\overset{}{C}}\text{=}\underset{H}{\overset{CH_2OH}{C}} \qquad \text{Nerol}$$

(e) Beginning at the 6,7 double bond, we see that the propyl group is of higher priority than the methyl group at C-7. Since the 6,7 double bond is E, the propyl group must be on the opposite side of the higher-priority substituent at C-6, where the CH_2 fragment has a higher priority than hydrogen. Therefore, we write for the stereochemistry of the 6,7 double bond:

$$\underset{\text{lower}}{\overset{\text{higher}}{}} \quad \underset{CH_3}{\overset{\overset{10\quad9\quad8}{CH_3CH_2CH_2}}{\underset{7}{C}}}\text{=}\underset{\underset{5}{CH_2\text{—}}}{\overset{6}{\underset{}{C}}}\underset{\text{higher}}{\overset{H\quad\text{lower}}{}}$$

$$E$$

At C-2, CH_2OH is of higher priority than H; and at C-3, $CH_2CH_2C\text{—}$ is of higher priority than CH_2CH_3. The double-bond configuration at C-2 is Z. Therefore

$$\underset{\text{lower}}{\overset{\text{higher}}{}} \quad \underset{CH_3CH_2}{\overset{\overset{6\ 5\quad4}{\text{—}CCH_2CH_2}}{\underset{3}{C}}}\text{=}\underset{H}{\overset{\overset{1}{CH_2OH}}{\underset{2}{C}}}\underset{\text{lower}}{\overset{\text{higher}}{}}$$

Combining the two partial structures, we obtain for the full structure of the codling moth's sex pheromone

$$\underset{CH_3}{\overset{CH_3CH_2CH_2}{C}}\text{=}\underset{CH_2CH_2}{\overset{H}{C}} \diagdown \underset{CH_3CH_2}{\overset{}{C}}\text{=}\underset{H}{\overset{CH_2OH}{C}}$$

The compound is $(2Z,6E)$-3-ethyl-7-methyl-2,6-decadien-1-ol.

(f) The sex pheromone of the honeybee is (E)-9-oxo-2-decenoic acid, with the structure

$$\underset{\text{lower}}{\overset{\text{higher}}{}} \quad \underset{H}{\overset{\overset{O}{\overset{\|}{CH_3C(CH_2)_4CH_2}}}{C}}\text{=}\underset{CO_2H}{\overset{H}{C}}\underset{\text{higher}}{\overset{\text{lower}}{}}$$

(g) Looking first at the 2,3 double bond of the cecropia moth's growth hormone

$$\underset{CH_3}{\overset{}{\overset{}{\underset{O}{\triangle}}}}\underset{}{\overset{CH_3CH_2}{}} \underset{\underset{H}{6}}{\overset{CH_3CH_2}{\underset{7}{C}}}\text{=}\underset{\underset{H}{2}}{\overset{\overset{CH_3}{3}}{C}}\text{—}CO_2CH_3$$

we find that its configuration is E, since the higher-priority substituents are on opposite sides of the double bond.

$$\underset{\text{higher}}{\overset{\text{lower}}{}} \quad \underset{\overset{5\quad4}{\text{—}CH_2CH_2}}{\overset{CH_3}{\underset{3}{C}}}\text{=}\underset{H}{\overset{\overset{1}{CO_2CH_3}}{\underset{2}{C}}}\underset{\text{lower}}{\overset{\text{higher}}{}}$$

The configuration of the 6,7 double bond is also E.

lower   CH$_3$CH$_2$ ^{5}CH$_2$CH$_2$—   higher

^{7}
C=C^{6}

higher ^{10}C—^{9}CH$_2$^{8}CH$_2$ H   lower

5.23   The alkenes are listed as follows in order of decreasing heat of combustion:

(e)   2,4,4-Trimethyl-2-pentene;   5293 kJ/mol   (1264.9 kcal/mol). Highest heat of combustion because it is C$_8$H$_{16}$ while all others are C$_7$H$_{14}$.

(a)   1-Heptene;   4658 kJ/mol   (1113.4 kcal/mol).   Monosubstituted double bond; therefore least stable C$_7$H$_{14}$ isomer.

(d)   H   H   (Z)-4,4-Dimethyl-2-pentene; 4650 kJ/mol (1111.4 kcal/mol). Disubstituted double bond, but destabilized by van der Waals strain.

(b)   2,4-Dimethyl-1-pentene; 4638 kJ/mol (1108.6 kcal/mol). Disubstituted double bond.

(c)   2,4-Dimethyl-2-pentene; 4632 kJ/mol (1107.1 kcal/mol). Trisubstituted double bond.

5.24   (a)   1-Methylcyclohexene is more stable; it contains a *trisubstituted* double bond while 3-methylcyclohexene has only a disubstituted double bond.

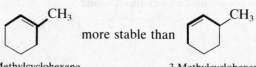

CH$_3$ more stable than   CH$_3$

1-Methylcyclohexene   3-Methylcyclohexene

(b)   Both isopropenyl and allyl are three-carbon alkenyl groups: isopropenyl is CH$_2$=CCH$_3$, allyl is CH$_2$=CHCH$_2$—.

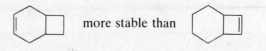

CH$_2$

C

CH$_3$ —CH$_2$CH=CH$_2$

Isopropenylcyclopentane has a disubstituted double bond and so is predicted to be more stable than allylcyclopentane, in which the double bond is monosubstituted.

(c)   A double bond in a six-membered ring is less strained than a double bond in a four-membered ring; therefore bicyclo[4.2.0]oct-3-ene is more stable.

more stable than

Bicyclo[4.2.0]oct-3-ene   Bicyclo[4.2.0]oct-7-ene

(d)   Cis double bonds are more stable than trans double bonds when the ring is smaller than 11-membered. (Z)-Cyclononene has a cis double bond in a 9-membered ring, and is thus more stable than (E)-cyclononene.

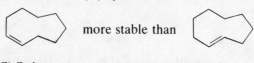

more stable than

(Z)-Cyclononene   (E)-Cyclononene

(*e*) Trans double bonds are more stable than cis when the ring is large. Here the rings are 18-membered, so that (*E*)-cyclooctadecene is more stable than (*Z*)-cyclooctadecene.

more stable than

(*E*)-Cyclooctadecene   (*Z*)-Cyclooctadecene

5.25 (*a*) Carbon atoms that are involved in double bonds are sp^2 hybridized, with ideal bond angles of 120°. Incorporating an sp^2 hybridized carbon into a three-membered ring leads to more angle strain than incorporation of an sp^3 hybridized carbon. 1-Methylcyclopropene has *two* sp^2 hybridized carbons in a three-membered ring and so has substantially more angle strain than methylenecyclopropane.

1-Methylcyclopropene   Methylenecyclopropane

The higher degree of substitution at the double bond in 1-methylcyclopropene is not sufficient to offset the increased angle strain, and so 1-methylcyclopropene is less stable than methylenecyclopropane.

(*b*) 3-Methylcyclopropene has a disubstituted double bond and two sp^2 hybridized carbons in its three-membered ring. It is the least stable of the isomers.

3-Methylcyclopropene

5.26 In all parts of this exercise, write the structure of the alkyl halide in sufficient detail to identify the carbon that bears the halogen and the β carbon atoms that bear at least one hydrogen substituent. These are the carbons that become doubly bonded in the alkene product.

(*a*) 1-Bromohexane can give only 1-hexene under conditions of E2 elimination.

$$\text{BrCH}_2\overset{\beta}{\text{C}}\text{H}_2\text{CH}_2\text{CH}_2\text{CH}_2\text{CH}_3 \xrightarrow[\text{E2}]{\text{base}} \text{CH}_2{=}\text{CHCH}_2\text{CH}_2\text{CH}_2\text{CH}_3$$

1-Bromohexane   1-Hexene (only alkene)

(*b*) 2-Bromohexane can give both 1-hexene and 2-hexene on dehydrobromination. The 2-hexene fraction is a mixture of cis and trans stereoisomers.

$$\overset{\beta}{\text{C}}\text{H}_3\text{CH}\overset{\beta}{\text{C}}\text{H}_2\text{CH}_2\text{CH}_2\text{CH}_3 \xrightarrow[\text{E2}]{\text{base}}$$
$$\underset{\text{Br}}{|}$$

2-Bromohexane

$$\text{CH}_2{=}\text{CHCH}_2\text{CH}_2\text{CH}_2\text{CH}_3 \;+$$

1-Hexene   *cis*-2-Hexene   *trans*-2-Hexene

(c)   Both a cis-trans pair of 2-hexenes and a cis-trans pair of 3-hexenes are capable of being formed from 3-bromohexane.

cis-2-Hexene   *trans*-2-Hexene

cis-3-Hexene   *trans*-3-Hexene

3-Bromohexane

(d)   Dehydrobromination of 2-bromo-2-methylpentane can involve one of the hydrogens of either a methyl group (C-1) or a methylene group (C-3).

2-Bromo-2-methylpentane   2-Methyl-1-pentene   2-Methyl-2-pentene

Neither alkene is capable of existing in stereoisomeric forms, and so these two are the only products of E2 elimination.

(e)   2-Bromo-3-methylpentane can undergo dehydrohalogenation by loss of a proton from either C-1 or C-3. Loss of a proton from C-1 gives 3-methyl-1-pentene.

2-Bromo-3-methylpentane   3-Methyl-1-pentene

Loss of a proton from C-3 gives a mixture of (*E*)- and (*Z*)-3-methyl-2-pentene.

2-Bromo-3-methylpentane   (*E*)-3-Methyl-2-pentene   (*Z*)-3-Methyl-2-pentene

(f)   Three alkenes are possible from 3-bromo-2-methylpentane. Loss of the C-2 proton gives 2-methyl-2-pentene.

3-Bromo-2-methylpentane   2-Methyl-2-pentene

Abstraction of a proton from C-4 can yield either (*E*)- or (*Z*)-4-methyl-2-pentene.

3-Bromo-2-methylpentane   (*E*)-4-Methyl-2-pentene   (*Z*)-4-Methyl-2-pentene

(g) Proton abstraction from the C-3 methyl group of 3-bromo-3-methylpentane yields 2-ethyl-1-butene.

3-Bromo-3-methylpentane   2-Ethyl-1-butene

Stereoisomeric 3-methyl-2-pentenes are formed by proton abstraction from C-2.

3-Bromo-3-methylpentane   (E)-3-Methyl-2-pentene   (Z)-3-Methyl-2-pentene

(h) Only 3,3-dimethyl-1-butene may be formed under conditions of E2 elimination from 2-bromo-3,3-dimethylbutane.

2-Bromo-3,3-dimethylbutane   3,3-Dimethyl-1-butene

5.27 (a) The reaction that takes place with 1-bromo-3,3-dimethylbutane is an E2 elimination involving loss of the bromide at C-1 and abstraction of the proton at C-2 by the strong base potassium *tert*-butoxide, yielding a single alkene.

1-Bromo-3,3-dimethylbutane   3,3-Dimethyl-1-butene

(b) Two alkenes are capable of being formed in this β elimination reaction.

1-Methylcyclopentyl chloride   Methylenecyclopentane   1-Methylcyclopentene

The more highly substituted alkene is 1-methylcyclopentene; it is the major product of this reaction. According to Zaitsev's rule, the major alkene is formed by proton removal from the β carbon that has the fewest hydrogen substituents.

(c) Acid-catalyzed dehydration of 3-methyl-3-pentanol can lead either to 2-ethyl-1-butene or to a mixture of (E)- and (Z)-3-methyl-2-pentene.

$$\underset{\text{3-Methyl-3-pentanol}}{\overset{\beta}{CH_3}\underset{\overset{|}{OH}}{\overset{CH_3}{\underset{|}{\overset{|}{C}}}}\overset{\beta}{CH_2}CH_3} \xrightarrow[80°C]{H_2SO_4}$$

$$\underset{\text{2-Ethyl-1-butene}}{\overset{CH_2}{\underset{CH_3CH_2}{\overset{\|}{C}}CH_2CH_3}} + \underset{(E)\text{-3-Methyl-2-pentene}}{\overset{CH_3}{\underset{H}{C}}=\overset{CH_3}{\underset{CH_2CH_3}{C}}} + \underset{(Z)\text{-3-Methyl-2-pentene}}{\overset{CH_3}{\underset{H}{C}}=\overset{CH_2CH_3}{\underset{CH_3}{C}}}$$

The major product is a mixture of the trisubstituted alkenes, (*E*)- and (*Z*)-3-methyl-2-pentene. Of these two stereoisomers the *E* isomer is slightly more stable and is expected to predominate.

(*d*) Acid-catalyzed dehydration of 2,3-dimethyl-2-butanol can proceed in either of two directions.

$$\underset{\text{2,3-Dimethyl-2-butanol}}{\overset{CH_3\ CH_3}{\underset{OH}{CH_3-C-CHCH_3}}} \xrightarrow[-H_2O]{\overset{H_3PO_4}{120°C}} \underset{\substack{\text{2,3-Dimethyl-1-butene}\\(disubstituted)}}{CH_2=C\overset{CH_3}{\underset{CH(CH_3)_2}{}}} + \underset{\substack{\text{2,3-Dimethyl-2-butene}\\(tetrasubstituted)}}{(CH_3)_2C=C(CH_3)_2}$$

The major alkene is the one with the more highly substituted double bond, 2,3-dimethyl-2-butene. Its formation corresponds to Zaitsev's rule in that a proton is lost from the β carbon that has the fewest hydrogen substituents.

(*e*) Only a single alkene is capable of being formed on E2 elimination from this alkyl iodide. Stereoisomeric alkenes are not possible, and because all the β hydrogens are equivalent, regioisomers cannot be formed either.

$$\underset{\text{3-Iodo-2,4-dimethylpentane}}{\overset{I}{\underset{CH_3\ CH_3}{CH_3CHCHCHCH_3}}} \xrightarrow[\substack{ethanol\\70°C}]{NaOCH_2CH_3} \underset{\text{2,4-Dimethyl-2-pentene}}{(CH_3)_2C=CHCH(CH_3)_2}$$

(*f*) Despite the structural similarity of this alcohol to the alkyl halide in the preceding part of this problem, its dehydration is more complicated. The initially formed carbocation is secondary and can rearrange to a more stable tertiary carbocation by a hydride shift.

$$\underset{\text{2,4-Dimethyl-3-pentanol}}{\overset{OH}{\underset{CH_3\ CH_3}{CH_3CHCHCHCH_3}}} \xrightarrow[-H_2O]{H^+} \underset{\substack{\text{Secondary carbocation}\\(less\ stable)}}{\overset{H}{\underset{CH_3\ CH_3}{CH_3C-\overset{+}{C}HCHCH_3}}} \longrightarrow \underset{\substack{\text{Tertiary carbocation}\\(more\ stable)}}{\overset{+}{\underset{CH_3\ CH_3}{CH_3CCH_2CHCH_3}}}$$

The tertiary carbocation, once formed, can give either 2,4-dimethyl-1-pentene or 2,4-dimethyl-2-pentene by loss of a proton.

$$\underset{CH_3\ CH_3}{\overset{+}{CH_3CCH_2CHCH_3}} \longrightarrow \underset{\substack{\text{2,4-Dimethyl-1-pentene}\\(disubstituted)}}{\underset{CH_3}{CH_2=CCH_2CH(CH_3)_2}} + \underset{\substack{\text{2,4-Dimethyl-2-pentene}\\(trisubstituted)}}{(CH_3)_2C=CHCH(CH_3)_2}$$

The proton is lost from the methylene group in preference to the methyl group. The major alkene is the more highly substituted one, 2,4-dimethyl-2-pentene.

5.28 In all parts of this problem you need to reason backward from an alkene to a bromide of molecular formula $C_7H_{13}Br$ that gives only the desired alkene under E2 elimination conditions. Recall that the carbon-carbon double bond is formed by loss of a proton from one of the doubly bonded carbons and a bromide from the other.

(a) Cycloheptene is the only alkene formed by an E2 elimination reaction of cycloheptyl bromide.

Cycloheptyl bromide   Cycloheptene

(b) (Bromomethyl)cyclohexane is the correct answer. It gives methylenecyclohexane as the *only* alkene under E2 conditions.

(Bromomethyl)cyclohexane   Methylenecyclohexane

1-Bromo-1-methylcyclohexane is not correct. It gives a mixture of 1-methylcyclohexene and methylenecyclohexane on elimination.

1-Bromo-1-methylcyclohexane   Methylenecyclohexane   1-Methylcyclohexene

(c) In order for 4-methylcyclohexene to be the only alkene, the starting alkyl bromide must be 1-bromo-4-methylcyclohexane. Either the cis or the trans isomer may be used, although the cis will react more readily, as the more stable conformation (equatorial methyl) has an axial bromine.

cis- or *trans*-1-bromo-4-methylcyclohexane   4-Methylcyclohexene

1-Bromo-3-methylcyclohexane is incorrect; its dehydrobromination yields a mixture of 3-methylcyclohexene and 4-methylcyclohexene.

1-Bromo-3-methylcyclohexene   3-Methylcyclohexene   4-Methylcyclohexene

(d)   The bromine must be at C-2 in the starting alkyl bromide.

2-Bromo-1,1-dimethylcyclopentane   3,3-Dimethylcyclopentene

If the bromine substituent were at C-3, a mixture of 3,3-dimethyl- and 4,4-dimethylcyclopentene would be formed.

3-Bromo-1,1-dimethylcyclopentane   3,3-Dimethylcyclopentene   4,4-Dimethylcyclopentene

(e)   The alkyl bromide must be primary in order for the desired alkene to be the only product of E2 elimination.

2-Cyclopentylethyl bromide   Vinylcyclopentane

If 1-cyclopentylethyl bromide were used, a mixture of regioisomeric alkenes would be formed, with the desired vinylcyclopentane being the minor component of the mixture.

(f)   Either cis- or trans-1-bromo-3-isopropylcyclobutane would be appropriate here.

cis- or trans-1-bromo-3-isopropylcyclobutane   3-Isopropylcyclobutene

(g)   The desired alkene is the exclusive product formed on E2 elimination from 1-bromo-1-tert-butylcyclopropane.

1-Bromo-1-tert-butylcyclopropane   1-tert-Butylcyclopropene

5.29 (a)   Both 1-bromopropane and 2-bromopropane yield propene as the exclusive product of E2 elimination.

$$CH_3CH_2CH_2Br \quad \text{or} \quad CH_3\overset{\underset{\displaystyle Br}{|}}{C}HCH_3 \xrightarrow{\text{E2, base}} CH_3CH{=}CH_2$$

1-Bromopropane   2-Bromopropane   Propene

(b)   Isobutene is formed on dehydrobromination of either tert-butyl bromide or isobutyl bromide.

$$(CH_3)_3CBr \quad \text{or} \quad (CH_3)_2CHCH_2Br \xrightarrow{\text{E2, base}} (CH_3)_2C{=}CH_2$$

tert-Butyl bromide   Isobutyl bromide   2-Methylpropene

(c)   A tetrabromoalkane is required as the starting material in order to form a tribromoalkene under E2 elimination conditions. Either 1,1,2,2-tetrabromoethane or 1,1,1,2-tetrabromoethane is satisfactory.

$$Br_2CHCHBr_2 \quad\text{or}\quad BrCH_2CBr_3 \xrightarrow{\text{E2, base}} BrCH{=}CBr_2$$

1,1,2,2-Tetrabromoethane   1,1,1,2-Tetrabromoethane   1,1,2-Tribromoethene

(d)   The bromine substituent may be at either C-2 or C-3.

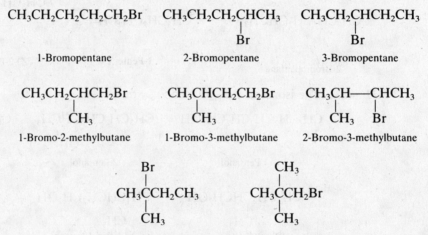

2-Bromo-1,1-dimethylcyclobutane   3-Bromo-1,1-dimethylcyclobutane   3,3-Dimethylcyclobutene

5.30   (a)   The isomeric alkyl bromides having the molecular formula $C_5H_{11}Br$ are:

$$CH_3CH_2CH_2CH_2CH_2Br \qquad CH_3CH_2CH_2\underset{\underset{Br}{|}}{C}HCH_3 \qquad CH_3CH_2\underset{\underset{Br}{|}}{C}HCH_2CH_3$$

1-Bromopentane   2-Bromopentane   3-Bromopentane

$$CH_3CH_2\underset{\underset{CH_3}{|}}{C}HCH_2Br \qquad CH_3\underset{\underset{CH_3}{|}}{C}HCH_2CH_2Br \qquad CH_3\underset{\underset{CH_3}{|}}{C}H{-}\underset{\underset{Br}{|}}{C}HCH_3$$

1-Bromo-2-methylbutane   1-Bromo-3-methylbutane   2-Bromo-3-methylbutane

$$CH_3\overset{\overset{Br}{|}}{\underset{\underset{CH_3}{|}}{C}}CH_2CH_3 \qquad CH_3\overset{\overset{CH_3}{|}}{\underset{\underset{CH_3}{|}}{C}}CH_2Br$$

2-Bromo-2-methylbutane   1-Bromo-2,2-dimethylpropane

(b)   The order of reactivity toward E1 elimination parallels carbocation stability and is tertiary > secondary > primary. The tertiary bromide 2-bromo-2-methylbutane will undergo E1 elimination at the fastest rate.

(c)   1-Bromo-2,2-dimethylpropane has no hydrogens on the β carbon and so cannot form an alkene directly by an E2 process.

$$CH_3\overset{\overset{CH_3}{|}}{\underset{\underset{CH_3}{|}}{C}}{-}CH_2Br \qquad \overset{\curvearrowleft \beta \text{ carbon has no hydrogens}}{}$$

The only available pathway is E1 with rearrangement.

(d)   Only the primary bromides shown can give both regiospecific elimination and no possibility of cis-trans isomers.

$$CH_3CH_2CH_2CH_2CH_2Br \xrightarrow{\text{base, E2}} CH_3CH_2CH_2CH{=}CH_2$$

1-Bromopentane   1-Pentene

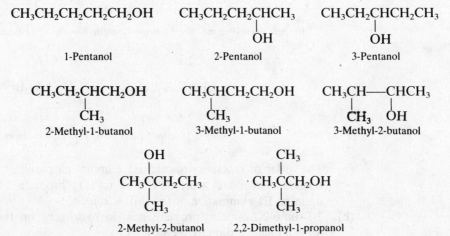

$$CH_3CH_2CHCH_2Br \xrightarrow{\text{base, E2}} CH_3CH_2C{=}CH_2$$
$$\hspace{2.5cm}|\hspace{5.5cm}|$$
$$\hspace{2.5cm}CH_3\hspace{5cm}CH_3$$

1-Bromo-2-methylbutane   2-Methyl-1-butene

$$CH_3CHCH_2CH_2Br \xrightarrow{\text{base, E2}} CH_3CHCH{=}CH_2$$
$$\hspace{1cm}|\hspace{6cm}|$$
$$\hspace{1cm}CH_3\hspace{5.5cm}CH_3$$

1-Bromo-3-methylbutane   3-Methyl-1-butene

(*e*)   Elimination in 3-bromopentane will give (*E*)- and (*Z*)-pentene only.

$$CH_3CH_2CHCH_2CH_3 \xrightarrow{\text{base, E2}}$$
$$\hspace{2cm}|$$
$$\hspace{2cm}Br$$

3-Bromopentane   (*E*)-2-Pentene   (*Z*)-2-Pentene

(*f*)   Three alkenes can be formed from 2-bromopentane.

$$CH_3CH_2CH_2CHCH_3 \xrightarrow[\text{E2}]{\text{base,}} CH_3CH_2CH_2CH{=}CH_2 +$$
$$\hspace{2.2cm}|$$
$$\hspace{2.2cm}Br$$

2-Bromopentane   1-Pentene   (*Z*)-2-Pentene   (*E*)-2-Pentene

5.31 (*a*) The isomeric $C_5H_{12}O$ alcohols are:

$$CH_3CH_2CH_2CH_2CH_2OH \hspace{1cm} CH_3CH_2CH_2CHCH_3 \hspace{1cm} CH_3CH_2CHCH_2CH_3$$
$$\hspace{6cm}|\hspace{4cm}|$$
$$\hspace{6cm}OH\hspace{3.5cm}OH$$

1-Pentanol   2-Pentanol   3-Pentanol

$$CH_3CH_2CHCH_2OH \hspace{1cm} CH_3CHCH_2CH_2OH \hspace{1cm} CH_3CH{-}CHCH_3$$
$$\hspace{1.5cm}|\hspace{3.5cm}|\hspace{4cm}|\hspace{0.6cm}|$$
$$\hspace{1.5cm}CH_3\hspace{3cm}CH_3\hspace{3.5cm}CH_3\hspace{0.3cm}OH$$

2-Methyl-1-butanol   3-Methyl-1-butanol   3-Methyl-2-butanol

$$\hspace{2cm}OH\hspace{4cm}CH_3$$
$$\hspace{2cm}|\hspace{4.5cm}|$$
$$\hspace{1cm}CH_3CCH_2CH_3 \hspace{1.5cm} CH_3CCH_2OH$$
$$\hspace{2cm}|\hspace{4.5cm}|$$
$$\hspace{2cm}CH_3\hspace{4cm}CH_3$$

2-Methyl-2-butanol   2,2-Dimethyl-1-propanol

(*b*)   The order of reactivity in alcohol dehydration parallels carbocation stability, and is tertiary > secondary > primary. The only tertiary alcohol in the group is 2-methyl-2-butanol. It will dehydrate fastest.

(*c*)   The most stable C_5H_{11} carbocation is the tertiary carbocation.

$$CH_3\overset{+}{C}CH_2CH_3 \hspace{1cm} \text{1,1-Dimethylpropyl cation}$$
$$\hspace{1cm}|$$
$$\hspace{1cm}CH_3$$

(*d*)   A proton may be lost from C-1 or C-3:

$$\hspace{1cm}CH_3\hspace{4cm}CH_3\hspace{2.5cm}CH_3$$
$$\hspace{1cm}|\hspace{4.5cm}|\hspace{3cm}|$$
$$CH_3\overset{+}{C}CH_2CH_3 \longrightarrow CH_2{=}CCH_2CH_3 + CH_3C{=}CHCH_3$$

1,1-Dimethylpropyl   2-Methyl-1-butene   2-Methyl-2-butene
cation   (minor alkene)   (major alkene)

(e) In order for the 1,1-dimethylpropyl cation to be formed by a process involving a hydride shift, the starting alcohol must have the same carbon skeleton as the 1,1-dimethylpropyl cation.

$$CH_3\overset{+}{C}CH_2CH_3 \quad \text{has same carbon skeleton as}$$
$$|\atop CH_3$$

$$HOCH_2\underset{CH_3}{\overset{H}{\underset{|}{\overset{|}{C}}}}CH_2CH_3 \quad \text{and} \quad CH_3\underset{CH_3}{\overset{H}{\underset{|}{\overset{|}{C}}}}\!\!-\!\!\underset{OH}{\overset{}{C}}HCH_3$$

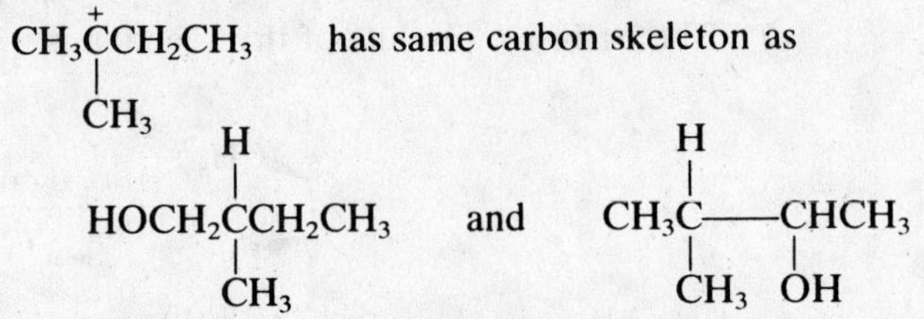

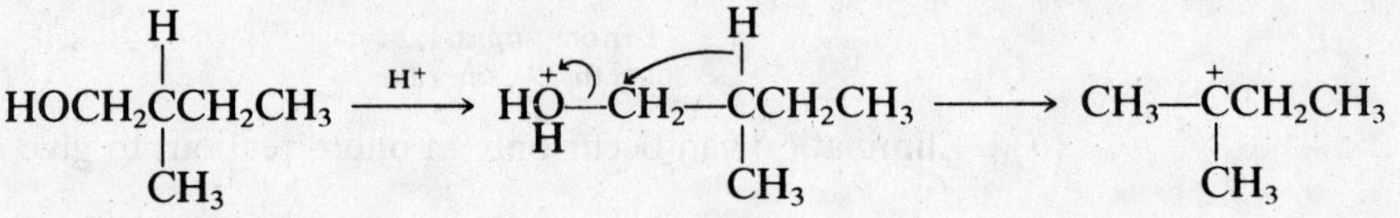

2-Methyl-1-butanol

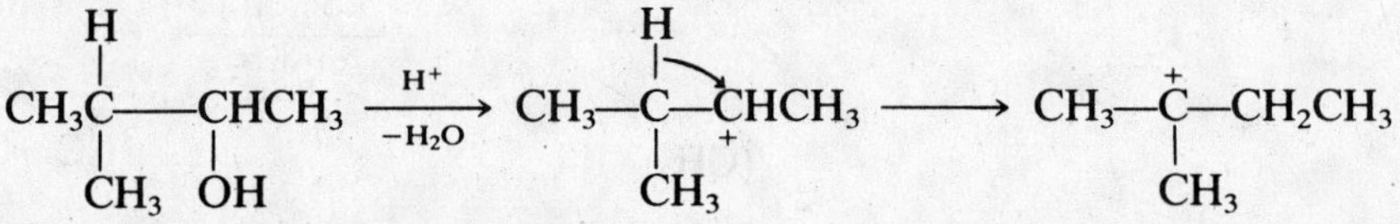

3-Methyl-2-butanol

While the same carbon skeleton is necessary, it alone is not sufficient; the alcohol must also have its hydroxyl group on the carbon atom adjacent to the carbon which bears the migrating hydrogen. Thus, 3-methyl-1-butanol cannot form a tertiary carbocation by a hydride shift. It requires two sequential hydride shifts.

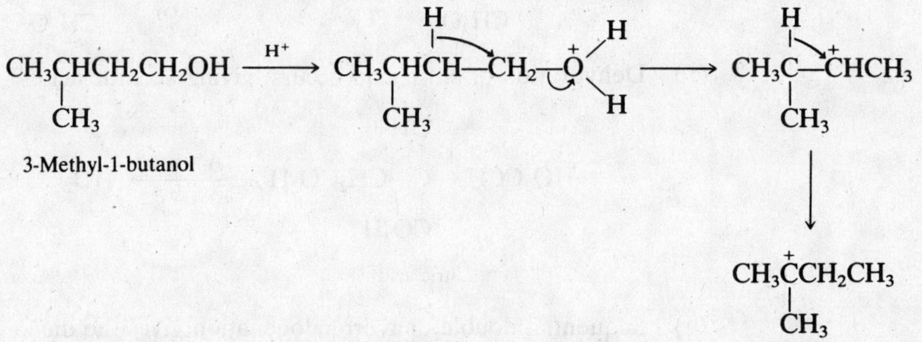

3-Methyl-1-butanol

$$CH_3\overset{+}{C}CH_2CH_3$$
$$|\atop CH_3$$

(f) 2,2-Dimethyl-1-propanol can yield a tertiary carbocation by a process involving a methyl shift.

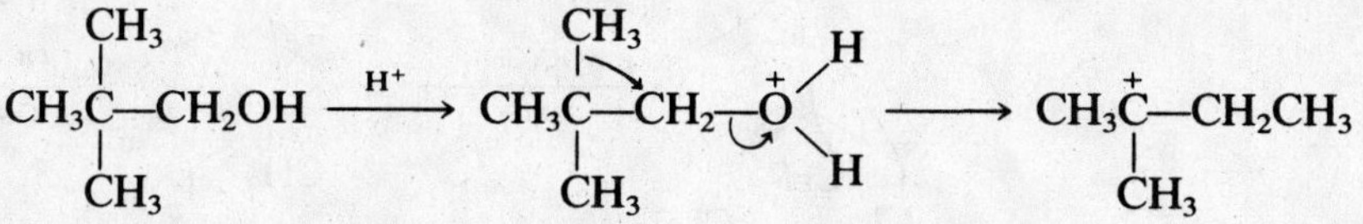

2,2-Dimethyl-1-propanol

5.32 (a) Heating an alcohol in the presence of an acid catalyst ($KHSO_4$) leads to dehydration with formation of an alkene. In this alcohol, elimination can occur in only one direction to give a mixture of cis and trans alkenes.

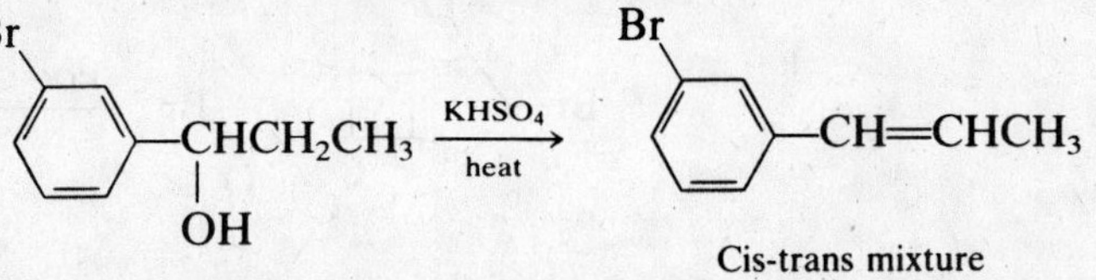

Cis-trans mixture

(b)   Alkyl halides undergo E2 elimination on being heated with potassium *tert*-butoxide.

$$\text{ICH}_2\text{CH(OCH}_2\text{CH}_3)_2 \xrightarrow[\substack{(\text{CH}_3)_3\text{COH} \\ \text{heat}}]{\text{KOC(CH}_3)_3} \text{CH}_2{=}\text{C(OCH}_2\text{CH}_3)_2$$

(c)   The exclusive product of this reaction is 1,2-dimethylcyclohexene.

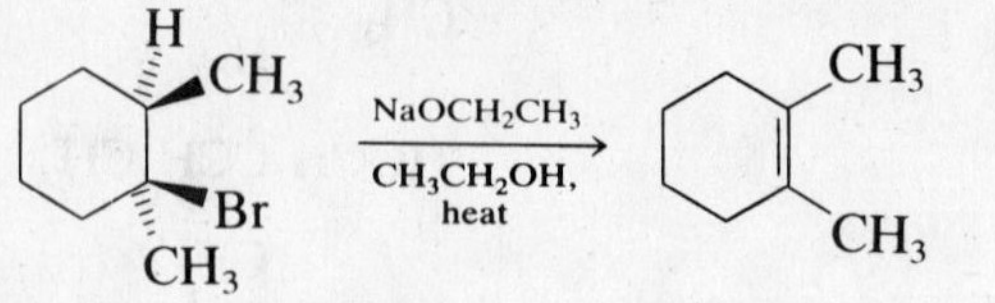

1-Bromo-*trans*-1,2-
dimethylcyclohexane

1,2-Dimethylcyclohexene
(100%)

(d)   Elimination can occur only in one direction, to give the alkene shown.

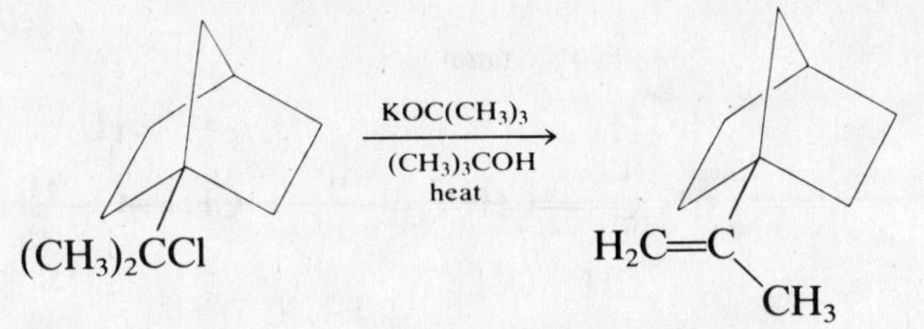

(e)   The reaction is a conventional one of alcohol dehydration and proceeds as written in 76 to 78 percent yield.

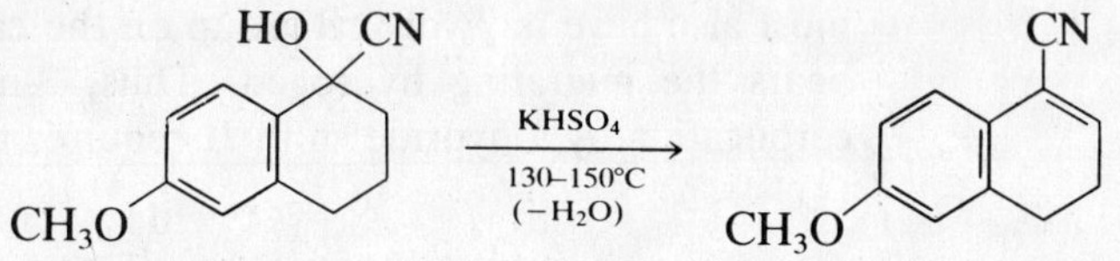

(f)   Dehydration of citric acid occurs, giving aconitic acid.

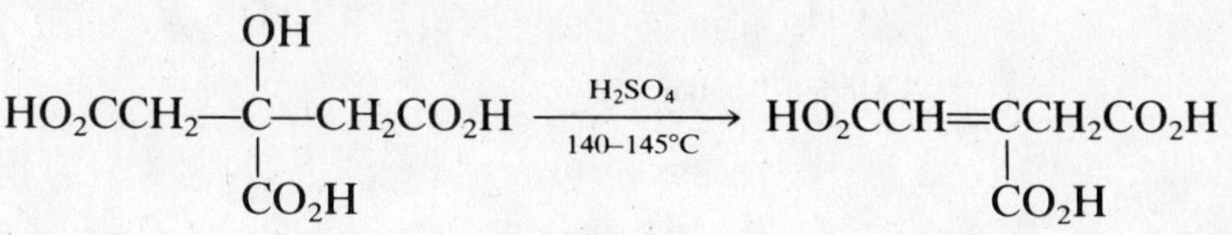

Citric acid

Aconitic acid

(g)   Sequential double dehydrohalogenation gives the diene.

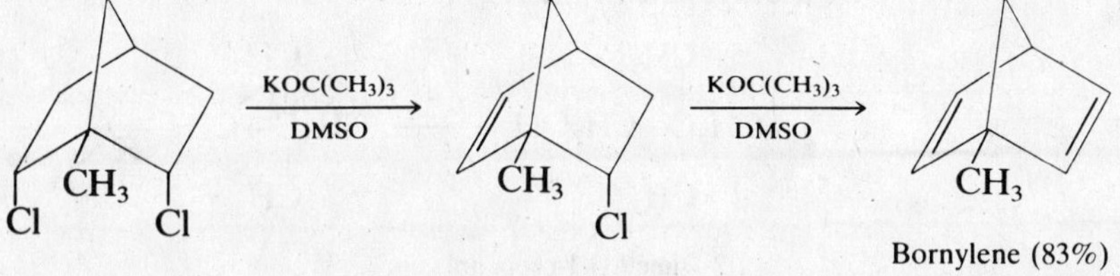

Bornylene (83%)

(h)   This example has been reported in the chemical literature, and in spite of the complexity of the starting material, elimination proceeds in the usual way.

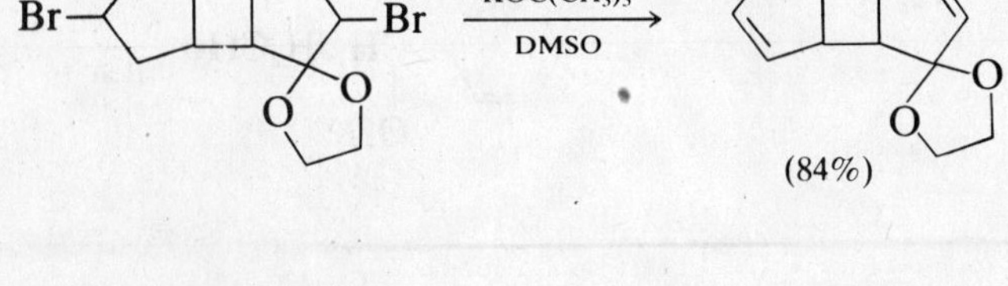

(84%)

(*i*)   Again, we have a fairly complicated substrate, but notice that it is well disposed toward E2 elimination of the axial bromide.

(*j*)   In the most stable conformation of this compound, chlorine occupies an axial site, and so it is ideally situated to undergo an E2 elimination reaction by way of an anti arrangement in the transition state.

4-*tert*-Butyl-
1-methyl-
cyclohexene (95%)

4-*tert*-Butyl-
(methylene)-
cyclohexane (5%)

The minor product is the less highly substituted isomer, in which the double bond is exocyclic to the ring.

5.33   Begin by writing chemical equations for the processes specified in the problem. First consider rearrangement by way of a hydride shift:

Isobutyloxonium ion   Tertiary cation   Water

Rearrangement by way of a methyl group shift is as follows:

Isobutyloxonium ion   Secondary cation   Water

A hydride shift gives a tertiary carbocation; a methyl migration gives a secondary carbocation. It is reasonable to expect that rearrangement will occur so as to produce the more stable of these two carbocations because the transition state has carbocation character at the carbon that bears the migrating group. We predict that rearrangement proceeds by a hydride shift rather than a methyl shift, since the group that remains behind in this process stabilizes the carbocation better.

5.34   Rearrangement proceeds by migration of a hydrogen or an alkyl group from the carbon atom adjacent to the positively charged carbon.

(*a*)   A propyl cation is primary and rearranges to an isopropyl cation, which is secondary, by migration of a hydrogen with its pair of electrons.

Propyl cation
(primary, less stable)

1-Methylethyl cation
(secondary, more stable)

(*b*)   A hydride shift transforms the secondary carbocation to a tertiary one.

1,2-Dimethylpropyl cation   1,1-Dimethylpropyl cation
(secondary, less stable)   (tertiary, more stable)

This hydride shift occurs in preference to methyl migration, which would produce the same secondary carbocation. (Verify this by writing appropriate structural formulas.)

(*c*)   Migration of a methyl group converts this secondary carbocation to a tertiary one.

1,2,2-Trimethylpropyl cation   1,1,2-Trimethylpropyl cation
(secondary, less stable)   (tertiary, more stable)

(*d*)   The group that shifts in this case is the entire ethyl group.

2,2-Diethylbutyl cation   1,1-Diethylbutyl cation
(primary, less stable)   (tertiary, more stable)

(*e*)   Migration of a hydride from the ring carbon that bears the methyl group produces a tertiary carbocation.

2-Methylcyclopentyl cation   1-Methylcyclopentyl cation
(secondary, less stable)   (tertiary, more stable)

5.35 (*a*)   Note that the starting material is an alcohol and that it is treated with an acid. The product is an alkene but its carbon skeleton is different from that of the starting alcohol. The reaction is one of alcohol dehydration accompanied by rearrangement at the carbocation stage. Begin by writing the step in which the alcohol is converted to a carbocation.

The carbocation is tertiary and relatively stable. However, migration of a methyl group from the *tert*-butyl substituent converts it to an isomeric carbocation, which is also tertiary.

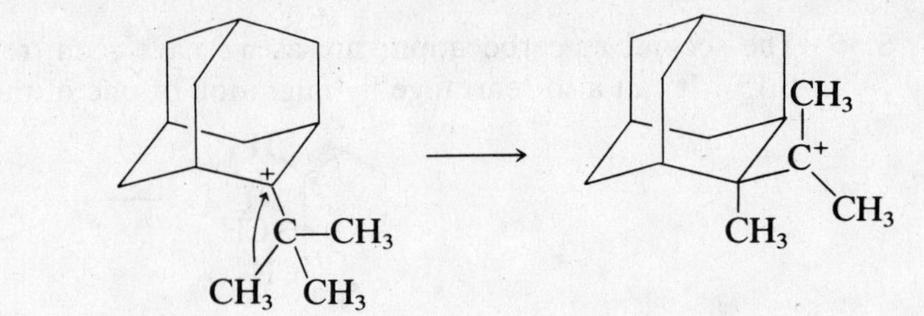

Loss of a proton from this carbocation gives the observed product.

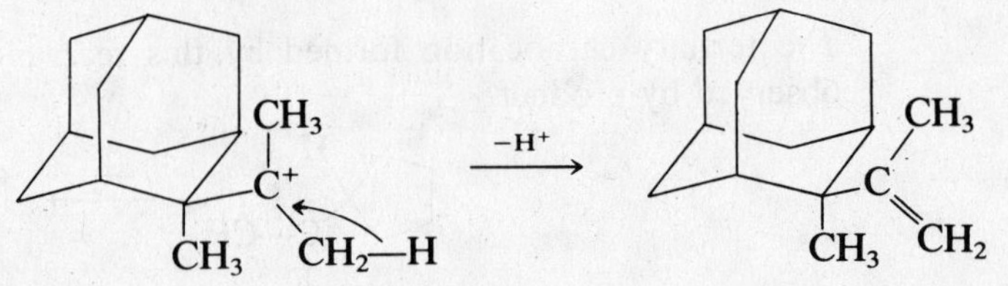

(b) Here also we have an alcohol dehydration reaction accompanied by rearrangement. The first-formed carbocation is secondary.

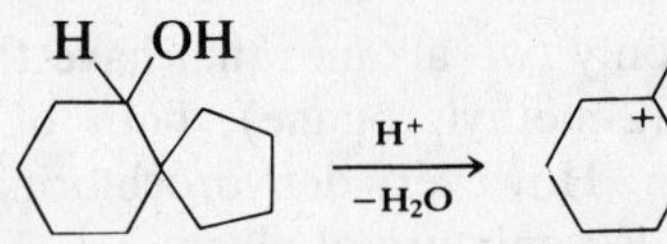

This cation can rearrange to a tertiary carbocation by alkyl group shift.

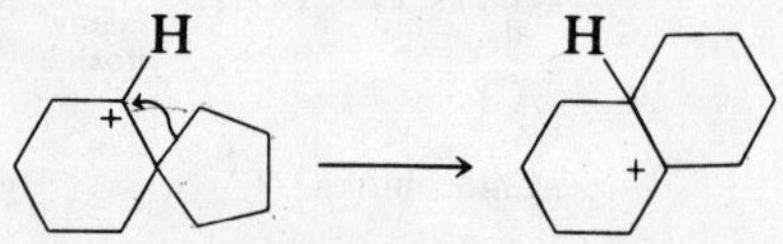

Loss of a proton from the tertiary carbocation gives the observed alkene.

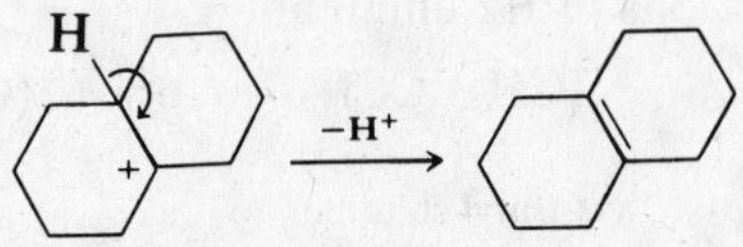

(c) The reaction begins as a normal alcohol dehydration in which the hydroxyl group is protonated by the acid catalyst and then loses water from the oxonium ion to give a carbocation.

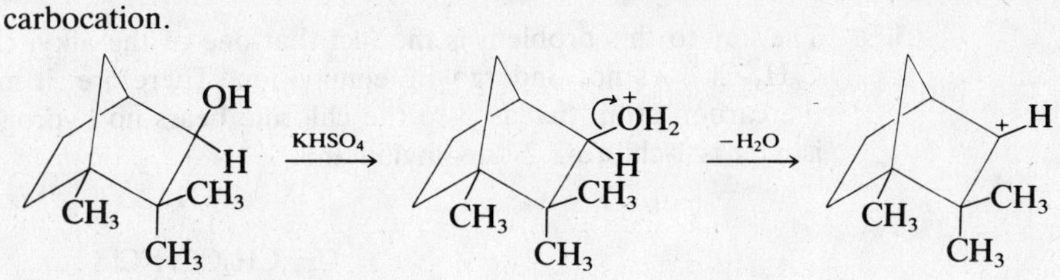

4-Methylcamphenilol

We see that the final product, 1-methylsantene, has a rearranged carbon skeleton corresponding to a methyl shift, and so we consider the rearrangement of the initially formed secondary carbocation to a tertiary ion.

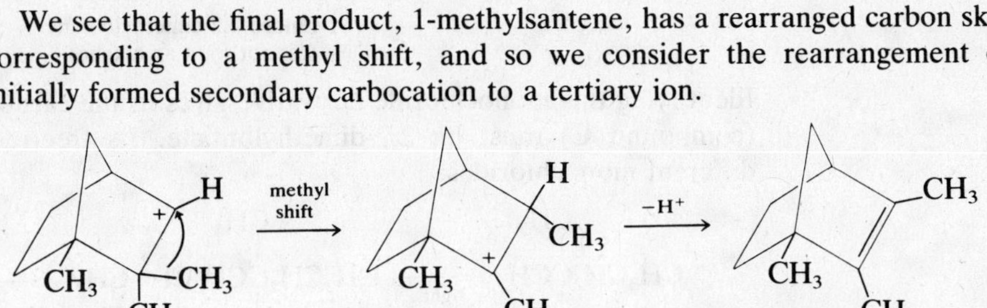

1-Methylsantene

Deprotonation of the tertiary carbocation yields 1-methylsantene.

5.36 The secondary carbocation can, as we have seen, rearrange by a methyl shift (Problem 5.15). It can also rearrange by migration of one of the ring bonds.

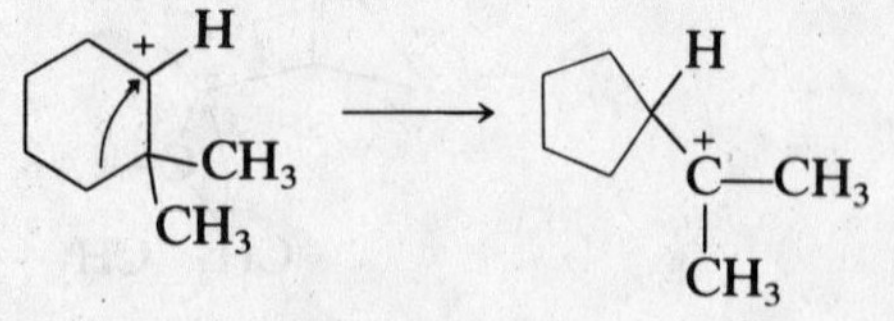

Secondary carbocation   Tertiary carbocation

The tertiary carbocation formed by this rearrangement can lose a proton to give the observed by-product.

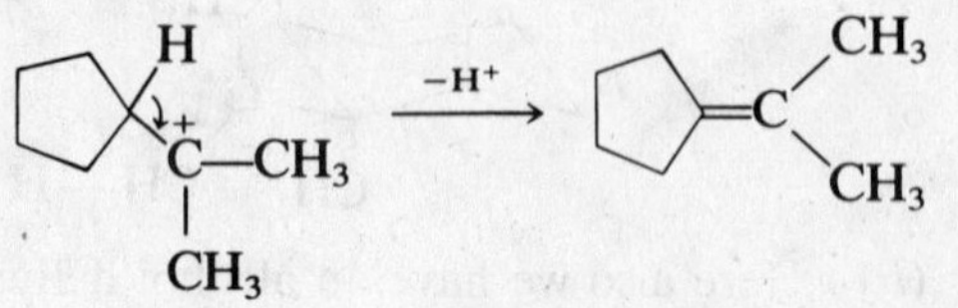

Isopropylidenecyclopentane

5.37 There are only two alkanes that have the molecular formula C_4H_{10}; they are butane and isobutane (2-methylpropane), both of which give two monochlorides on free-radical chlorination. However, dehydrochlorination of one of the monochlorides derived from butane yields a mixture of alkenes.

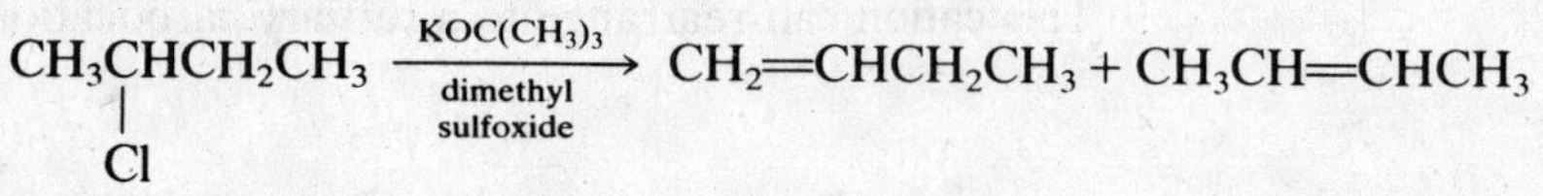

$$CH_3CHCH_2CH_3 \xrightarrow[\substack{dimethyl \\ sulfoxide}]{KOC(CH_3)_3} CH_2{=}CHCH_2CH_3 + CH_3CH{=}CHCH_3$$
$$\underset{Cl}{|}$$

2-Chlorobutane   1-Butene   2-Butene (cis + trans)

Both monochlorides derived from 2-methylpropane yield only 2-methylpropene under conditions of E2 elimination.

$$(CH_3)_3CCl \quad or \quad (CH_3)_2CHCH_2Cl \xrightarrow[\substack{dimethyl \\ sulfoxide}]{KOC(CH_3)_3} (CH_3)_2C{=}CH_2$$

tert-Butyl chloride   Isobutyl chloride   2-Methylpropene

Therefore, compound A is 2-methylpropane, the two alkyl chlorides are tert-butyl chloride and isobutyl chloride, and alkene B is 2-methylpropene.

5.38 The key to this problem is the fact that one of the alkyl chlorides of molecular formula $C_6H_{13}Cl$ does not undergo E2 elimination. Therefore, it must have a structure in which the carbon atom that is β to the chlorine bears no hydrogen substituents. This $C_6H_{13}Cl$ isomer is 1-chloro-2,2-dimethylbutane.

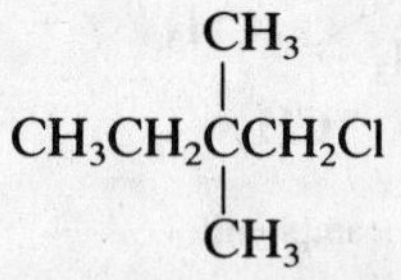

$$\underset{\underset{CH_3}{|}}{\overset{\overset{CH_3}{|}}{CH_3CH_2CCH_2Cl}}$$

1-Chloro-2,2-dimethylbutane
(cannot form an alkene)

Identifying this monochloride derivative gives us the carbon skeleton. The starting alkane (compound C) must be 2,2-dimethylbutane. Its free-radical halogenation gives three different monochlorides:

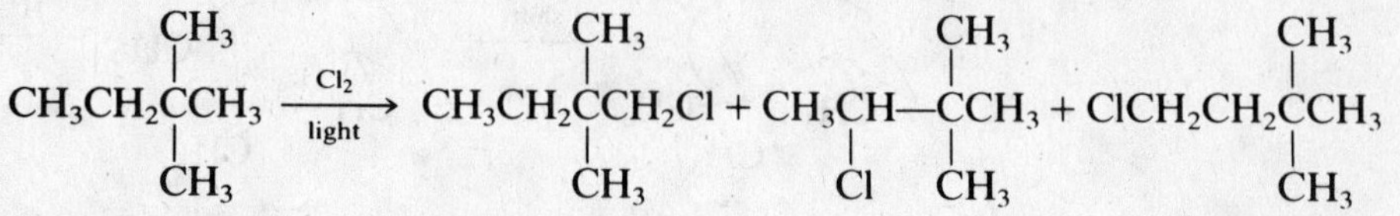

2,2-Dimethylbutane   1-Chloro-2,2-   3-Chloro-2,2-   1-Chloro-3,3-
(compound C)   dimethylbutane   dimethylbutane   dimethylbutane

Both 3-chloro-2,2-dimethylbutane and 1-chloro-3,3-dimethylbutane give only 3,3-dimethyl-1-butene on E2 elimination.

$$\underset{\substack{\text{3-Chloro-2,2-}\\\text{dimethylbutane}}}{\underset{\substack{|\\\text{Cl}}}{\underset{|}{\text{CH}_3\text{CH}}}\text{—}\underset{\substack{|\\\text{CH}_3}}{\overset{\overset{\text{CH}_3}{|}}{\text{C}}}\text{CH}_3} \quad \text{or} \quad \underset{\substack{\text{1-Chloro-3,3-}\\\text{dimethylbutane}}}{\text{ClCH}_2\text{CH}_2\underset{\substack{|\\\text{CH}_3}}{\overset{\overset{\text{CH}_3}{|}}{\text{C}}}\text{CH}_3} \quad \xrightarrow[\text{(CH}_3)_3\text{COH}]{\text{KOC(CH}_3)_3} \quad \underset{\substack{\text{3,3-Dimethyl-1-butene}\\\text{(alkene D)}}}{\text{CH}_2\text{=CH}\underset{\substack{|\\\text{CH}_3}}{\overset{\overset{\text{CH}_3}{|}}{\text{C}}}\text{CH}_3}$$

SELF-TEST

PART A

A-1. Write the correct IUPAC name for each of the following:

    (a) $(CH_3)_3CCH\!=\!C(CH_3)_2$ (b)

A-2. Each of the following is an incorrect name for an alkene. Write the structure and give the correct name for each.
    (a) 2-Ethyl-3-methyl-2-butene   (b) 2,5-Dimethylcyclohexene

A-3. (a) Write the structures of all the pentene (C_5H_{10}) isomers.
    (b) Which isomer is the most stable?
    (c) Which isomers are the least stable?
    (d) Which isomers can exist as a pair of stereoisomers?

A-4. How many carbon atoms are sp^2 hybridized in 2-methyl-2-pentene? How many are sp^3 hybridized? How many σ bonds are of the sp^2–sp^3 type?

A-5. Write the structure, clearly indicating the stereochemistry, of each of the following:
    (a) (Z)-4-Ethyl-3-methyl-3-heptene
    (b) (E)-1,2-Dichloro-3-methyl-2-hexene

A-6. Write structural formulas for two alkenes of molecular formula C_7H_{14} that are stereoisomers of each other and have a trisubstituted double bond. Give systematic names for each.

A-7. Write structural formulas for the reactant or product(s) omitted from each of the following. If more than one product is formed, indicate the major one.

    (a) $(CH_3)_2\underset{\substack{|\\\text{OH}}}{\text{C}}CH_2CH_2CH_3 \xrightarrow[\text{heat}]{\text{H}_2\text{SO}_4}$?

    (b)

         $\xrightarrow[\text{CH}_3\text{OH}]{\text{NaOCH}_3}$?

    (c) ? $\xrightarrow[\text{CH}_3\text{CH}_2\text{OH}]{\text{KOCH}_2\text{CH}_3}$ (only alkene formed)

    (d) $(CH_3)_2\underset{\substack{|\\\text{C(CH}_3)_3}}{\text{COH}} \xrightarrow{\text{H}_3\text{PO}_4}$?

A-8. Write the structure of the $C_6H_{13}Br$ isomer that is not capable of undergoing E2 elimination.

A-9. Outline a mechanism detailing the steps in the following reaction:

$$\text{(structure)} \xrightarrow[\text{heat}]{\text{H}_3\text{PO}_4} \text{(structure)} + \text{other alkenes}$$

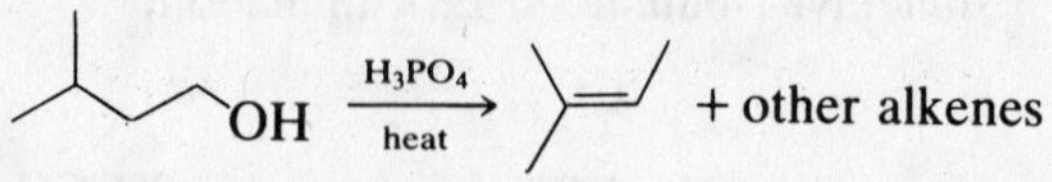

A-10. Using perspective drawings (of chair cyclohexanes), explain formation of the major E2 product from the following reaction:

$$\text{(structure with CH}_3 \text{ and Br)} \xrightarrow{\text{E2}} ?$$

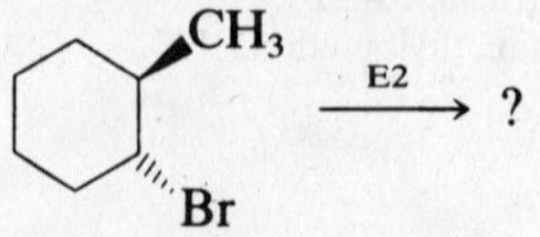

A-11. Compare the relative rate of reaction of *cis*- and *trans*-1-chloro-3-isopropylcyclohexane with sodium methoxide in methanol.

A-12. Provide a synthesis of 1-methylcyclohexene starting with methylcyclohexane.

A-13. Compound A, on reaction with bromine in the presence of light, gave as the major product compound B ($C_9H_{19}Br$). Reaction of B with sodium ethoxide in ethanol gave 3-ethyl-4,4-dimethyl-2-pentene as the only alkene. Identify compounds A and B.

A-14. Draw conformational depictions (using chair cyclohexanes) of the stereoisomers of 1-bromo-3,5-dimethylcyclohexane that undergo E2 elimination at the fastest rate and at the slowest rate.

PART B

B-1. Which (if any) of the following has the word *pentene* in the correct IUPAC name?

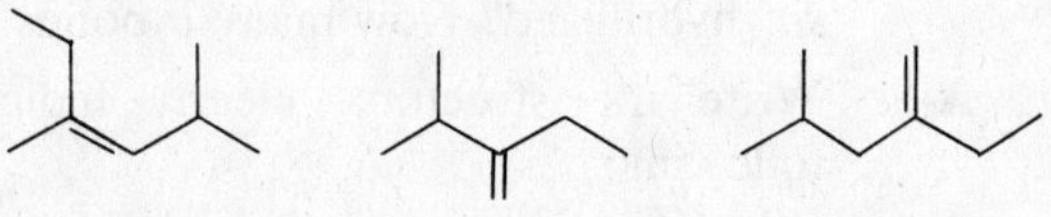

I   II   III

(*a*) I only   (*b*) II only   (*c*) III only   (*d*) I and III

B-2. Rank the following substituent groups in order of decreasing priority according to the Cahn-Ingold-Prelog system:

$$-\text{CH(CH}_3)_2 \qquad -\text{CH}_2\text{Br} \qquad -\text{CH}_2\text{CH}_2\text{Br}$$

A   B   C

(*a*) B > C > A   (*b*) A > C > B   (*c*) C > A > B   (*d*) B > A > C

B-3. The heats of combustion for the four C_6H_{12} isomers shown are (not necessarily in order): 955.3, 953.6, 950.6, and 949.7 (all in kilocalories per mole). Which of these values is most likely the heat of combustion of isomer I?

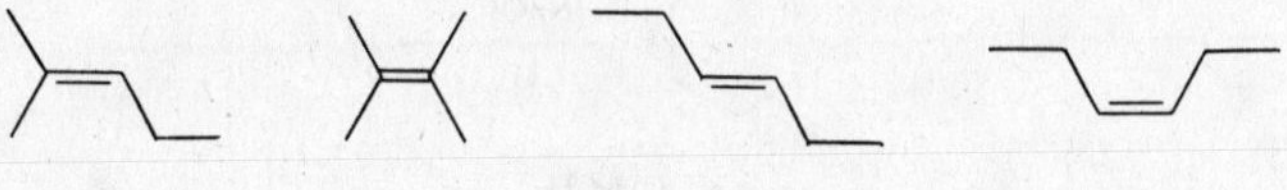

I   II   III   IV

(*a*) 955.3 kcal/mol   (*c*) 950.6 kcal/mol
(*b*) 953.6 kcal/mol   (*d*) 949.7 kcal/mol

B-4. Referring to the structures in the previous question, what can be said about isomers III and IV?
(*a*) III is more stable by 1.7 kcal/mol.
(*b*) IV is more stable by 1.7 kcal/mol.

(c) III is more stable by 3.0 kcal/mol.

(d) III is more stable by 0.9 kcal/mol.

B-5. The correct structure for (Z)-1-chloro-3-methyl-3-hexene is

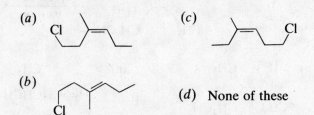

(d) None of these

B-6. For which, if any, of the following do cis and trans stereoisomers exist?

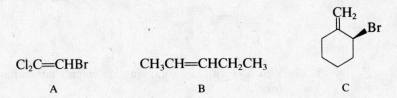

$Cl_2C{=}CHBr$ $CH_3CH{=}CHCH_2CH_3$

A   B   C

(a) C only   (b) B only   (c) A and B   (d) B and C

B-7. Rank the following alcohols in order of decreasing reactivity (fastest→slowest) toward dehydration with 85% H_3PO_4:

$(CH_3)_2CHCH_2CH_2OH$ $(CH_3)_2\overset{\underset{\displaystyle |}{OH}}{C}CH_2CH_3$ $(CH_3)_2CH\overset{\underset{\displaystyle |}{OH}}{C}HCH_3$

A   B   C

(a) B>C>A   (b) A>C>B   (c) B>A>C   (d) A>B>C

B-8. Rank the following in order of increasing reactivity (least→most) toward E2 elimination:

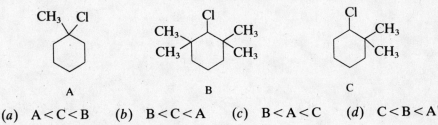

A   B   C

(a) A<C<B   (b) B<C<A   (c) B<A<C   (d) C<B<A

B-9. Consider the following reaction:

$$\overset{\displaystyle \diagup\!\!\diagdown\!\!\diagup}{\underset{\displaystyle OH}{}} \xrightarrow[\text{heat}]{H_3PO_4} \quad + $$

Which response contains all the correct statements about this process and no incorrect ones?

1. Dehydration
2. E2 mechanism
3. Carbon skeleton migration
4. Most stable carbocation forms
5. Single-step reaction

(a) 1, 3   (b) 1, 2, 3   (c) 1, 2, 5   (d) 1, 3, 4

B-10. Select the formula or statement representing the major product(s) of the following reaction:

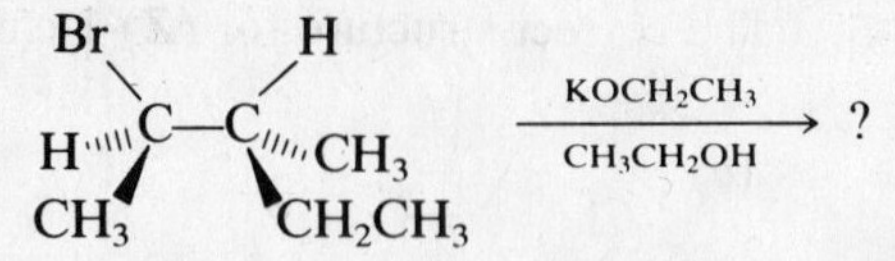

$$\underset{\substack{H\cdots\cdots C \\ CH_3}}{\overset{Br}{\big|}}\!-\!\underset{\substack{C\cdots\cdots CH_3 \\ CH_2CH_3}}{\overset{H}{\big|}} \quad \xrightarrow[\text{CH}_3\text{CH}_2\text{OH}]{\text{KOCH}_2\text{CH}_3} \quad ?$$

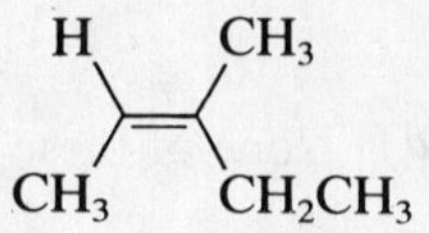

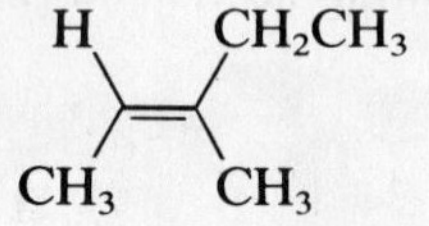

(a)

$$\underset{CH_3}{\overset{H}{\diagup}}C\!=\!C\underset{CH_2CH_3}{\overset{CH_3}{\diagdown}}$$

(b)

$$\underset{CH_3}{\overset{H}{\diagup}}C\!=\!C\underset{CH_3}{\overset{CH_2CH_3}{\diagdown}}$$

(c)

$$CH_2\!=\!CHCH\overset{CH_3}{\underset{|}{}}CH_2CH_3$$

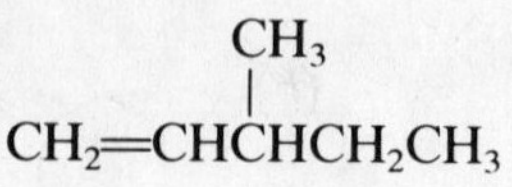

(d) Both (a) and (b) form in approximately equal amounts.

B.11 Which one of the following statements concerning E2 reactions of alkyl halides is *true*?

(a) The rate of an E2 reaction depends only on the concentration of the alkyl halide.

(b) The rate of an E2 reaction depends only on the concentration of the base.

(c) The C—H bond and the C—X (X = halogen) bond are broken in the same step.

(d) Alkyl chlorides generally react faster than alkyl bromides.

REACTIONS OF ALKENES: ADDITION REACTIONS

IMPORTANT TERMS AND CONCEPTS

Heats of Hydrogenation (Sec. 6.2) The heat given off during the catalytic hydrogenation of an alkene (see the section "Important Reactions" for this chapter) may be used to assess the relative stability of alkenes. Less heat will be evolved on hydrogenation of a more stable alkene. The trend of heats of hydrogenation and alkene stability is

$CH_2{=}CH_2$	$RCH{=}CH_2$	$RCH{=}CHR$	$R_2C{=}CHR$	$R_2C{=}CR_2$
Ethylene	Monosubstituted	Disubstituted	Trisubstituted	Tetrasubstituted
Largest heat of hydrogenation; least stable double bond				Smallest heat of hydrogenation; most stable double bond

Electrophilic Addition (Sec. 6.4) Electrophilic addition reactions constitute perhaps the most varied class of alkene reactions. In general, an ionic species E^+Y^- or a polarizable molecule $\overset{\delta+}{E}{-}\overset{\delta-}{Y}$ adds to the double bond.

$$\overset{\delta+}{E}{-}\overset{\delta-}{Y} + \overset{\diagdown}{\diagup}C{=}C\overset{\diagup}{\diagdown} \longrightarrow {-}\overset{|}{\underset{|}{C}}{-}\overset{|}{\underset{|}{C}}{-} \quad \begin{matrix} E & Y \end{matrix}$$

Markovnikov's Rule (Secs. 6.5, 6.6) The addition of a hydrogen halide H—X to an alkene is a regioselective reaction; that is, addition to an unsymmetrical alkene results in predominance of one regioisomer. *Markovnikov's rule* states that the hydrogen atom adds to the carbon having the greater number of hydrogen substituents:

$$H{-}X + R_2C{=}CH_2 \longrightarrow R_2\overset{|}{\underset{}{C}}{-}\overset{|}{\underset{}{CH_2}} \quad \begin{matrix} X & H \end{matrix}$$

Consideration of the mechanism of these reactions provides a theoretical justification

131

of the rule. The more stable carbocation intermediate forms in the process, which results in formation of the more substituted halide product.

$$R_2CH{-}\overset{+}{C}H_2 \quad \text{(Primary carbocation)}$$

$$R_2C{=}CH_2 \quad \xrightarrow{H-X}$$
$$\overset{\curvearrowleft}{H^+}$$

$$\overset{+}{R_2C}{-}CH_3 \xrightarrow{X^-} R_2\overset{X}{\overset{|}{C}}{-}CH_3$$
$$\text{(Tertiary carbocation)}$$

As discussed in Chapter 4, the transition state leading to formation of a tertiary carbocation is more stable than that leading to a secondary carbocation, which in turn is more stable than that leading to a primary. An *alkene reactivity series* may therefore be written which reflects the order of stability of the carbocations formed in the rate-determining step:

$$R_2C{=}CH_2 > RCH{=}CH_2 > CH_2{=}CH_2$$

Le Châtelier's Principle (Sec. 6.9) The statement that a *system at equilibrium adjusts so as to minimize any stress applied to it* is known as *Le Châtelier's principle*. An example is provided by acid-catalyzed hydration of an alkene and the acid-catalyzed dehydration of an alcohol, reactions which are the reverse of one another.

$$R_2C{=}CH_2 + H_2O \underset{}{\overset{H^+}{\rightleftharpoons}} R_2\overset{OH}{\overset{|}{C}}{-}CH_3$$

Addition of water shifts the equilibrium to the right, favoring alcohol formation. Removal of water (or keeping the concentration of water low) shifts the equilibrium to the left, favoring the formation of alkene.

The *principle of microscopic reversibility* is also demonstrated by the same equilibrium. Any intermediates or transition states which appear on the lowest-energy pathway of the forward process must also appear, in the reverse order, on the lowest-energy pathway of the reverse process.

Molecular Formulas (Sec. 6.19) Information regarding the presence of double bonds and rings in a molecule may be gained by consideration of the molecular formula. The *sum of double bonds and rings* (SODAR) is calculated by the following equation, which takes into account the difference in H content between the molecule being considered and the saturated acyclic alkane having the same number of carbon atoms:

SODAR = $\frac{1}{2}$(molecular formula of saturated alkane − molecular formula of compound)

For example, a molecular with the formula C_8H_{12} has a SODAR value of 3; that is, there are a total of three double bonds or rings present.

$$\text{SODAR} = \tfrac{1}{2}(C_8H_{18} - C_8H_{12}) = 3$$

Double bonds and rings may be distinguished from each other on the basis of hydrogenation data. Each double bond will react with one mole of H_2 per mole of compound, thus providing a means of determining the number of rings present.

When an oxygen atom is present in the molecule, it is omitted from the SODAR calculation; a halogen is treated as if it were a hydrogen. For example,

$$C_5H_9Cl \quad \text{SODAR} = \tfrac{1}{2}(C_5H_{12} - C_5H_{10}) = 1$$
$$C_3H_8O \quad \text{SODAR} = \tfrac{1}{2}(C_3H_8 - C_3H_8) = 0$$

IMPORTANT REACTIONS

Catalytic Hydrogenation (Secs. 6.1, 6.3)

General:

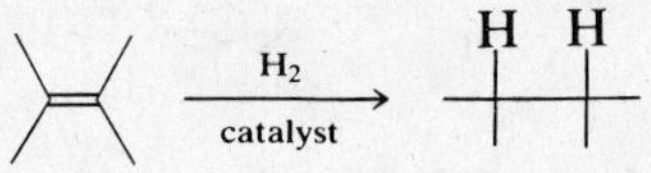

Example:

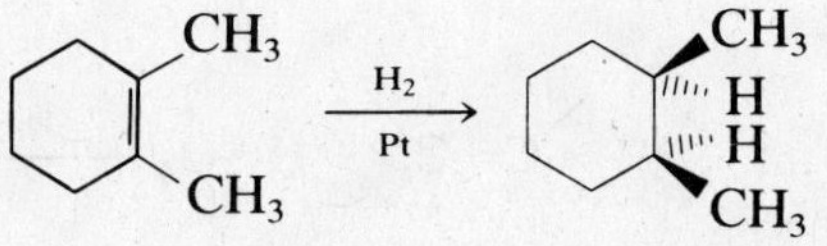

Electrophilic Addition of Hydrogen Halides (Secs. 6.4 to 6.6)

General:

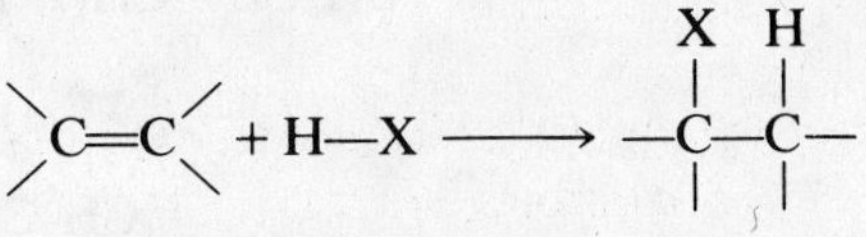

(X = Cl, Br, or I)

Examples:

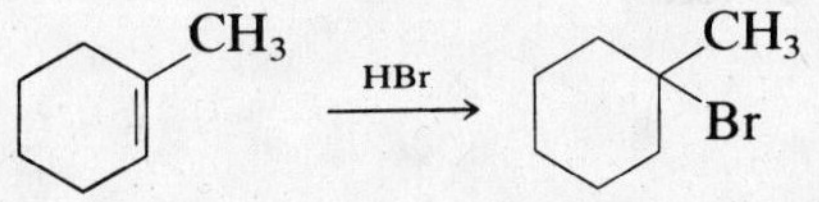

$$(CH_3)_2C = CHCH_3 \xrightarrow{HCl} (CH_3)_2\overset{Cl}{\underset{|}{C}}CH_2CH_3$$

Mechanism:

$$R_2C = CH_2 \underset{\text{slow}}{\overset{H-X}{\rightleftharpoons}} R_2\overset{+}{C} - CH_3 + X^-$$

$$R_2\overset{+}{C} - CH_3 \xrightarrow[\text{fast}]{X^-} R_2\overset{X}{\underset{|}{C}} - CH_3$$

Free-Radical Addition of Hydrogen Bromide (Sec. 6.7)

General:

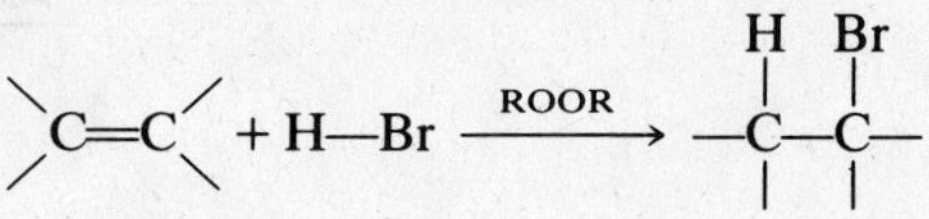

Example:

$$(CH_3)_2C = CHCH_3 + HBr \xrightarrow{ROOR} (CH_3)_2CHCH\overset{Br}{\underset{|}{C}}CH_3$$

Mechanism:

Initiation: $ROOR \longrightarrow 2RO\cdot$

$$RO\cdot \; + \; H-Br \longrightarrow ROH + Br\cdot$$

Propagation: $R_2C\!\!=\!\!CH_2 + Br\cdot \longrightarrow R_2\dot{C}\!\!-\!\!CH_2Br$

$R_2\dot{C}\!\!-\!\!CH_2Br + H\!\!-\!\!Br \longrightarrow R_2\overset{H}{\underset{}{C}}\!\!-\!\!CH_2Br + Br\cdot$

Addition of Sulfuric Acid; Alcohol Formation (Sec. 6.8)

General:

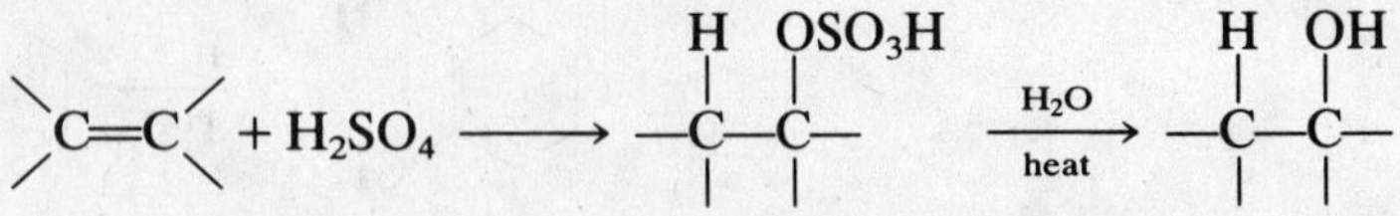

$$\text{C}\!\!=\!\!\text{C} + H_2SO_4 \longrightarrow -\overset{H}{\underset{}{C}}\!-\!\overset{OSO_3H}{\underset{}{C}}- \xrightarrow[\text{heat}]{H_2O} -\overset{H}{\underset{}{C}}\!-\!\overset{OH}{\underset{}{C}}-$$

Example:

$$CH_3CH\!\!=\!\!CH_2 + H_2SO_4 \longrightarrow CH_3\overset{OSO_3H}{\underset{}{C}}HCH_3$$

$$CH_3\overset{OSO_3H}{\underset{}{C}}HCH_3 \xrightarrow[\text{heat}]{H_2O} CH_3\overset{OH}{\underset{}{C}}HCH_3$$

Acid-Catalyzed Hydration (Sec. 6.9)

General:

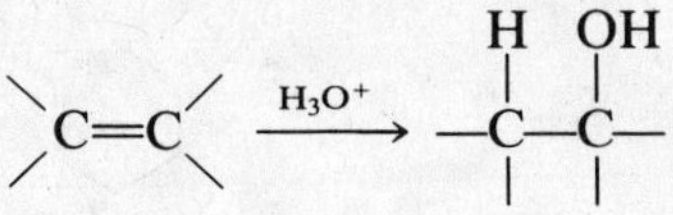

$$\text{C}\!\!=\!\!\text{C} \xrightarrow{H_3O^+} -\overset{H}{\underset{}{C}}\!-\!\overset{OH}{\underset{}{C}}-$$

Examples:

$$(CH_3)_2C\!\!=\!\!CH_2 \xrightarrow{H_3O^+} (CH_3)_3C\!\!-\!\!OH$$

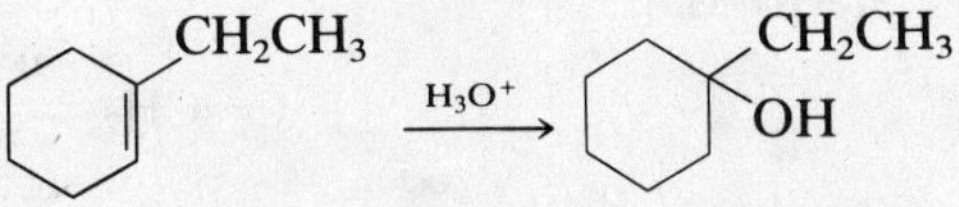

(cyclohexene with CH₂CH₃) $\xrightarrow{H_3O^+}$ (cyclohexane with CH₂CH₃, OH)

Synthesis of Alcohols by Hydroboration-Oxidation (Secs. 6.10 to 6.12)

General:

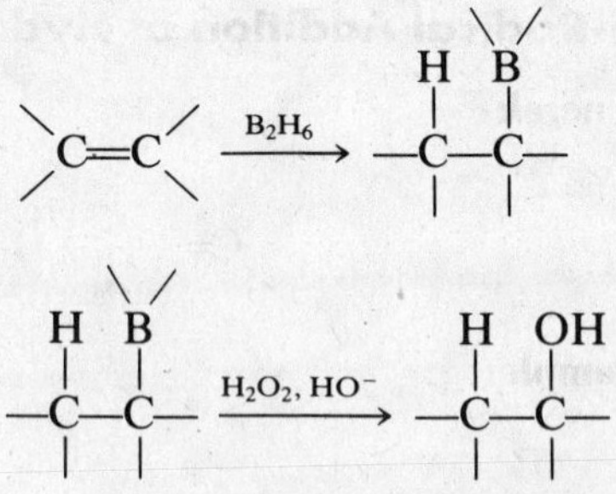

$$\text{C}\!\!=\!\!\text{C} \xrightarrow{B_2H_6} -\overset{H}{\underset{}{C}}\!-\!\overset{B}{\underset{}{C}}-$$

$$-\overset{H}{\underset{}{C}}\!-\!\overset{B}{\underset{}{C}}- \xrightarrow{H_2O_2,\ HO^-} -\overset{H}{\underset{}{C}}\!-\!\overset{OH}{\underset{}{C}}-$$

Examples:

(methylcyclohexene) $\xrightarrow[\text{2. H}_2O_2,\ HO^-]{\text{1. B}_2H_6}$ (product with CH₃, H, OH)

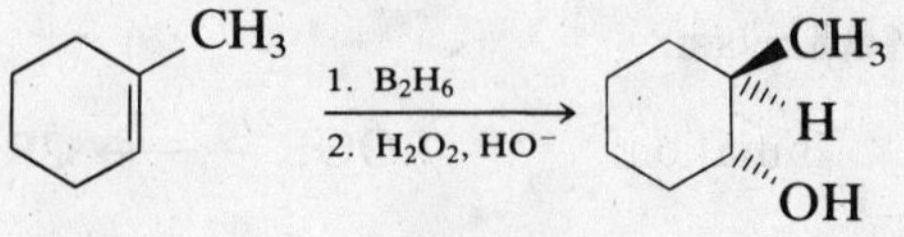

$$(CH_3)_2C\!\!=\!\!CH_2 \xrightarrow[\text{2. H}_2O_2,\ HO^-]{\text{1. B}_2H_6} (CH_3)_2CH\!\!-\!\!CH_2OH$$

Addition of Halogens (Secs. 6.13 to 6.15)

General:

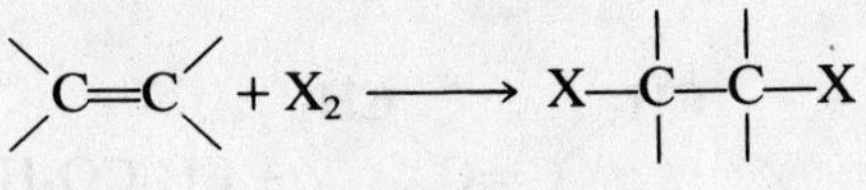

$$\text{C}=\text{C} + X_2 \longrightarrow X-\text{C}-\text{C}-X$$

(X = Cl, Br)

Examples:

$$CH_3CH_2CH{=}CH_2 \xrightarrow{Cl_2} CH_3CH_2\overset{Cl}{\underset{}{C}}HCH_2Cl$$

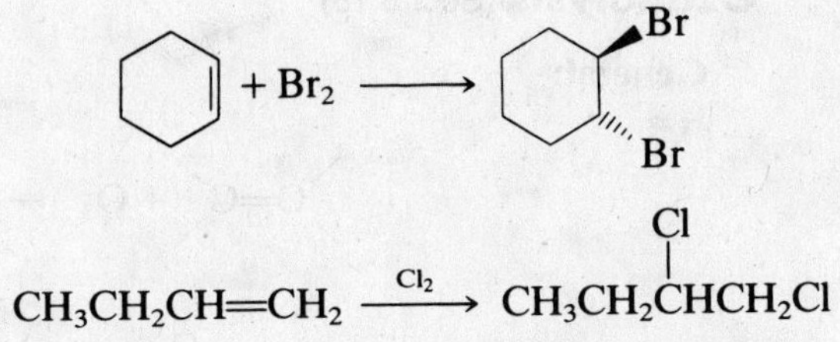

Mechanism:

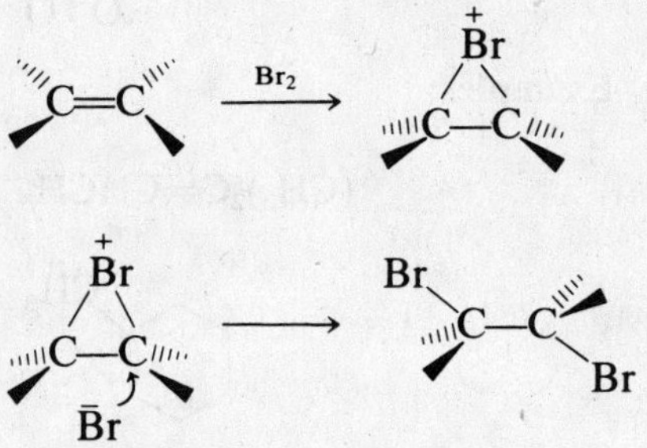

Formation of Vicinal Halohydrins (Sec. 6.16)

General:

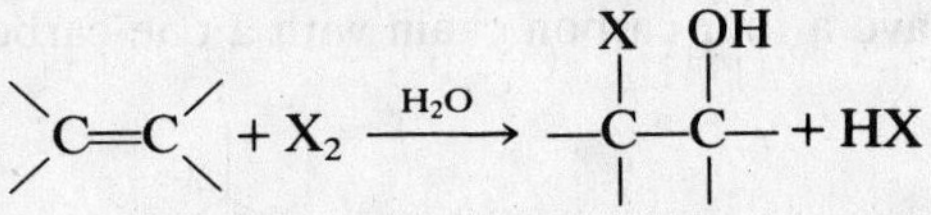

$$\text{C}=\text{C} + X_2 \xrightarrow{H_2O} -\overset{X}{\underset{}{C}}-\overset{OH}{\underset{}{C}}- + HX$$

(X = Cl, Br)

Examples:

$$(CH_3)_2C{=}CH_2 + Cl_2 \xrightarrow{H_2O} (CH_3)_2\overset{OH}{\underset{}{C}}-CH_2Cl + HCl$$

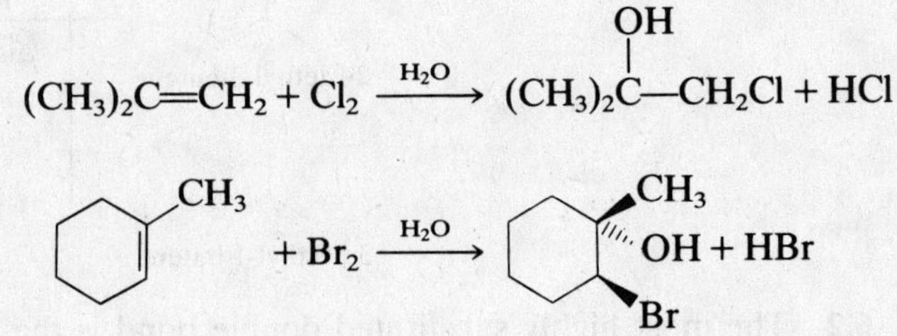

Epoxidation (Sec. 6.17)

General:

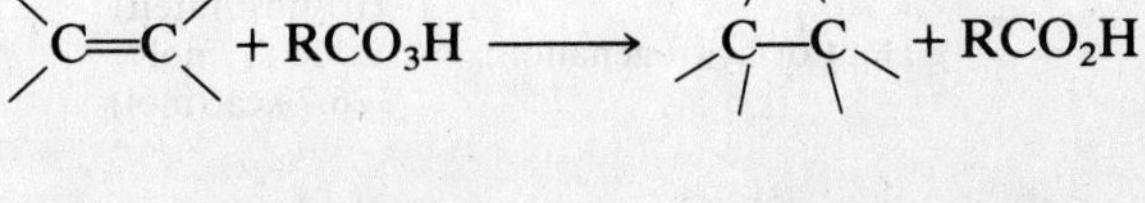

$$\text{C}=\text{C} + RCO_3H \longrightarrow \overset{O}{\text{C}-\text{C}} + RCO_2H$$

Examples:

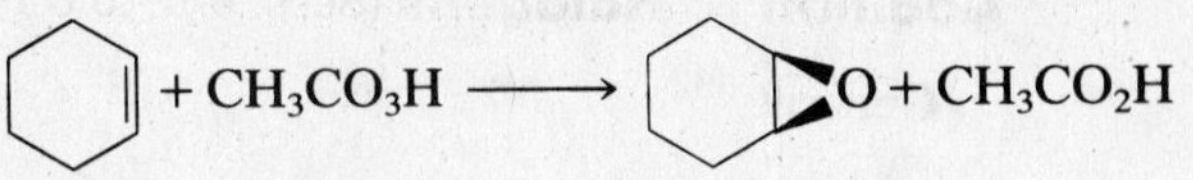

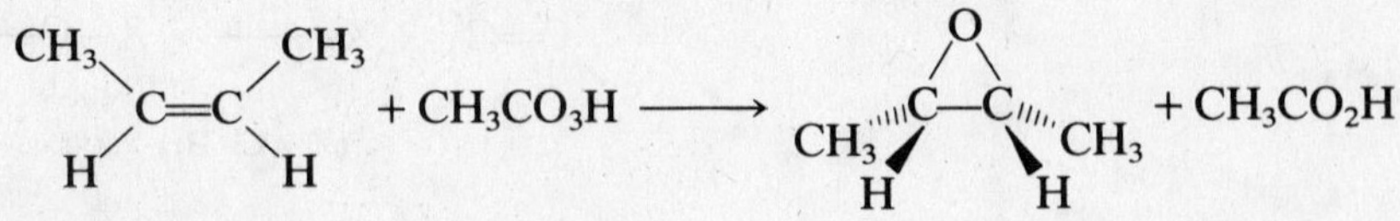

Ozonolysis (Sec. 6.18)

General:

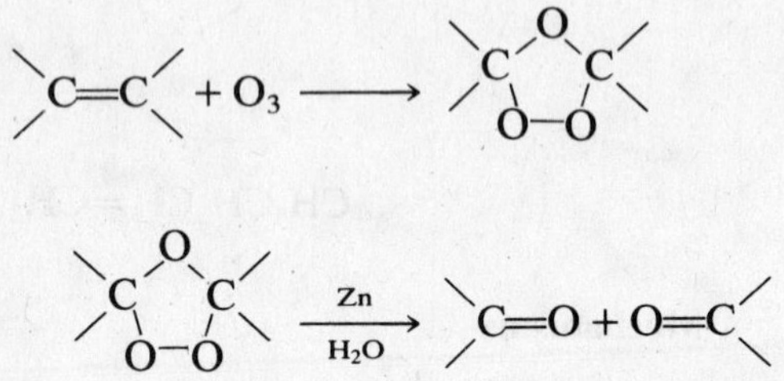

Examples:

$$(CH_3)_2C=CHCH_3 \xrightarrow[\text{2. Zn, H}_2O]{\text{1. O}_3} (CH_3)_2C=O + CH_3\overset{O}{\overset{\|}{C}}H$$

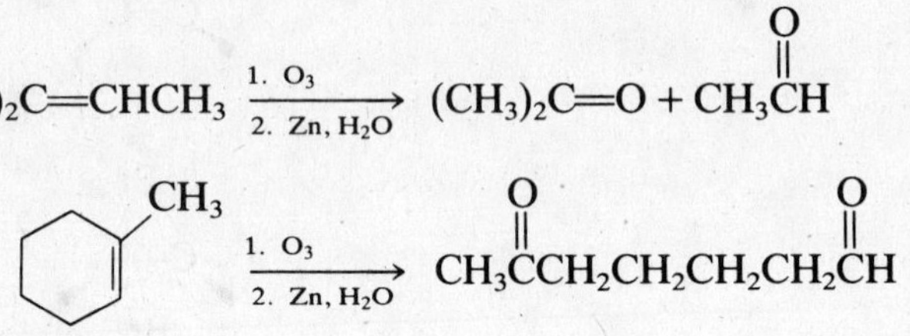

$$\xrightarrow[\text{2. Zn, H}_2O]{\text{1. O}_3} CH_3\overset{O}{\overset{\|}{C}}CH_2CH_2CH_2CH_2\overset{O}{\overset{\|}{C}}H$$

SOLUTIONS TO TEXT PROBLEMS

6.1 Catalytic hydrogenation converts an alkene to an alkane having the same carbon skeleton. Since 2-methylbutane is the product of hydrogenation, all three alkenes must have a four-carbon chain with a one-carbon branch. Therefore, the three alkenes are:

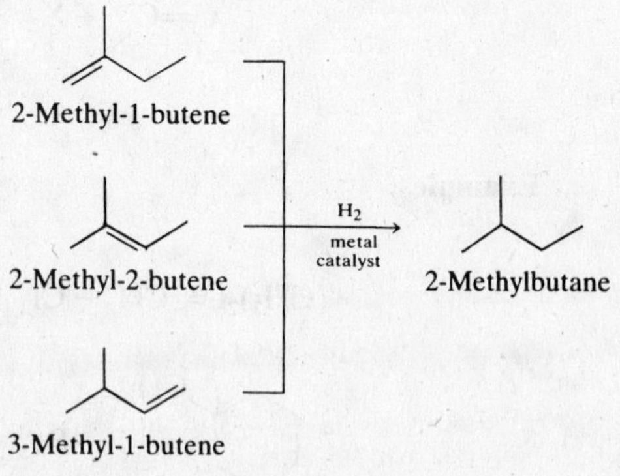

2-Methyl-1-butene

2-Methyl-2-butene

3-Methyl-1-butene

2-Methylbutane

6.2 The most highly substituted double bond is the most stable and has the smallest heat of hydrogenation.

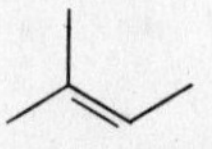

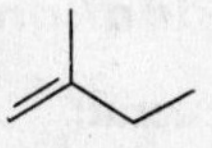

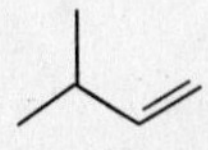

	2-Methyl-2-butene: most stable (trisubstituted)	2-Methyl-1-butene (disubstituted)	3-Methyl-1-butene (monosubstituted)
Heat of hydrogenation:	112 kJ/mol (26.7 kcal/mol)	118 kJ/mol (28.2 kcal/mol)	126 kJ/mol (30.2 kcal/mol)

6.3 (b) Begin by writing out the structure of the starting alkene. Identify the doubly bonded carbon that has the greater number of hydrogen substituents; this is the one to which the proton of hydrogen chloride adds. Chloride adds to the carbon atom of the double bond that has the fewer hydrogen substituents.

2-Methyl-1-butene

HCl

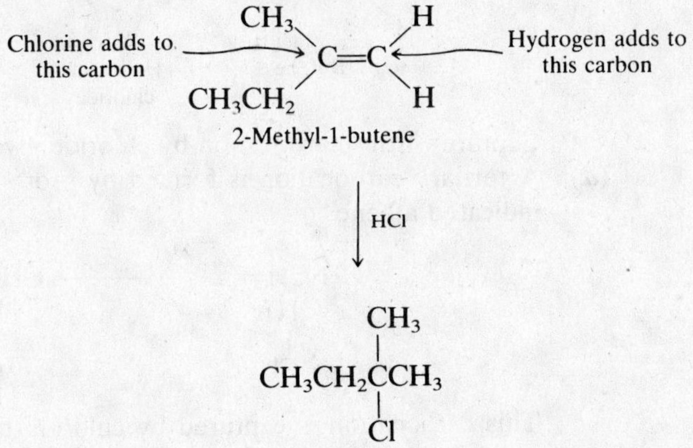

By applying Markovnikov's rule, we see that the major product is 2-chloro-2-methylbutane.

(c) Regioselectivity of addition is not an issue here, because the two carbons of the double bond are equivalent in *cis*-2-butene. Hydrogen chloride adds to *cis*-2-butene to give 2-chlorobutane.

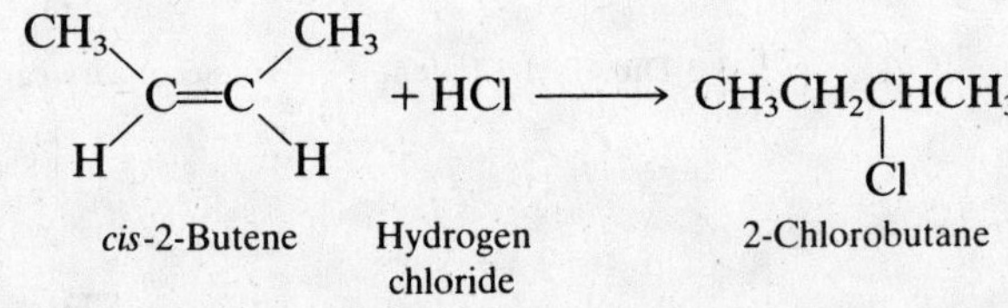

cis-2-Butene   Hydrogen chloride   2-Chlorobutane

(d) One end of the double bond has no hydrogen substituents, while the other end has one. In accordance with Markovnikov's rule, the proton of hydrogen chloride adds to the carbon that already has one hydrogen. The product is 1-chloro-1-ethylcyclohexane.

Ethylidenecyclohexane   Hydrogen chloride   1-Chloro-1-ethylcyclohexane

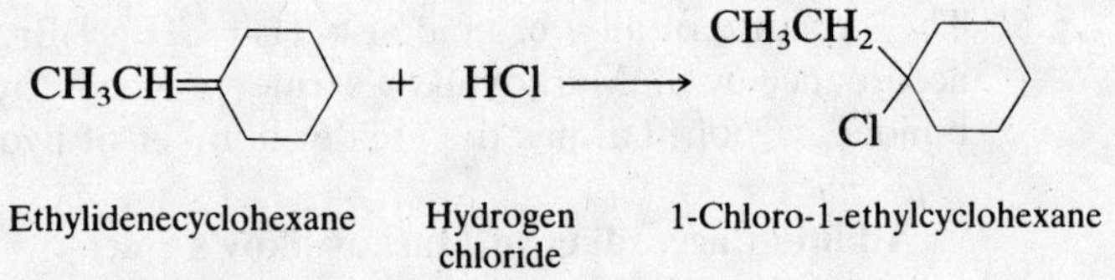

6.4 (b) A proton is transferred to the terminal carbon atom of 2-methyl-1-butene so as to produce a tertiary carbocation.

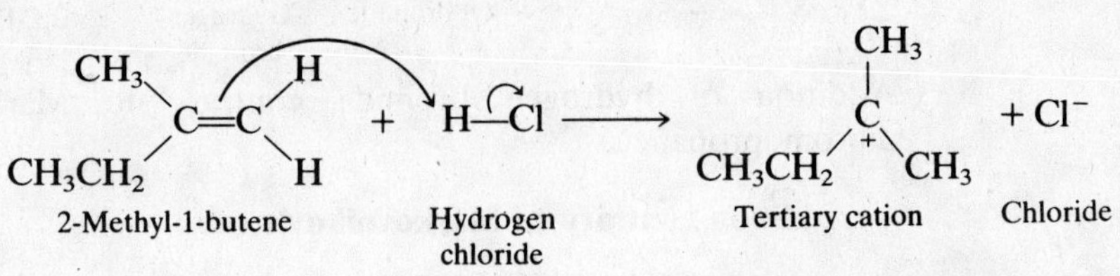

2-Methyl-1-butene   Hydrogen chloride   Tertiary cation   Chloride

This is the carbocation that leads to the observed product, 2-chloro-2-methylbutane.

(c) A secondary carbocation is an intermediate in the reaction of *cis*-2-butene with hydrogen chloride.

$$\underset{\text{cis-2-Butene}}{\text{CH}_3\text{C}=\text{C}\text{CH}_3} + \underset{\substack{\text{Hydrogen}\\\text{chloride}}}{\text{H}-\text{Cl}} \longrightarrow \underset{\text{Secondary cation}}{\overset{+}{\text{C}}-\text{CH}_2\text{CH}_3} + \underset{\text{Chloride}}{\text{Cl}^-}$$

Capture of this carbocation by chloride gives 2-chlorobutane.

(d) A tertiary carbocation is formed by proton transfer from hydrogen chloride to the indicated alkene.

$$\text{CH}_3\text{CH}= \bigcirc \longrightarrow \text{CH}_3\text{CH}_2-\overset{+}{\bigcirc} + \text{Cl}^-$$

$$\underset{\text{Tertiary cation}}{} \qquad \underset{\text{Chloride}}{}$$

This carbocation is captured by chloride to give the observed product, 1-chloro-1-ethylcyclohexane.

6.5 The carbocation formed by protonation of the double bond of 3,3-dimethyl-1-butene is secondary. Methyl migration can occur to give a more stable tertiary carbocation.

$$\underset{\text{3,3-Dimethyl-1-butene}}{\overset{\text{CH}_3}{\underset{\text{CH}_3}{\text{CH}_3\text{C}}}\text{CH}=\text{CH}_2} \xrightarrow{\text{HCl}} \underset{\text{Secondary carbocation}}{\overset{\text{CH}_3}{\underset{\text{CH}_3}{\text{CH}_3\text{C}}}-\overset{+}{\text{C}}\text{HCH}_3} \longrightarrow \underset{\text{Tertiary carbocation}}{\overset{\text{CH}_3}{\underset{\text{CH}_3}{\text{CH}_3\overset{+}{\text{C}}}}-\text{CHCH}_3}$$

$$\downarrow \text{Cl}^- \qquad\qquad\qquad \downarrow \text{Cl}^-$$

$$\underset{\text{3-Chloro-2,2-dimethylbutane}}{\overset{\text{CH}_3}{\underset{\text{CH}_3}{\text{CH}_3\text{C}}}-\underset{\text{Cl}}{\text{CHCH}_3}} \qquad \underset{\text{2-Chloro-2,3-dimethylbutane}}{\overset{\text{Cl}\quad\text{CH}_3}{\underset{\text{CH}_3}{\text{CH}_3\text{C}}-\text{CHCH}_3}}$$

The two chlorides are 3-chloro-2,2-dimethylbutane and 2-chloro-2,3-dimethylbutane.

6.6 The structure of allyl bromide is $\text{CH}_2=\text{CHCH}_2\text{Br}$. Its reaction with hydrogen bromide in accordance with Markovnikov's rule proceeds by addition of a proton to the doubly bonded carbon that has the greater number of hydrogens attached to it.

Addition according to Markovnikov's rule:

$$\underset{\text{Allyl bromide}}{\text{CH}_2=\text{CHCH}_2\text{Br}} + \underset{\substack{\text{Hydrogen}\\\text{bromide}}}{\text{HBr}} \longrightarrow \underset{\text{1,2-Dibromopropane}}{\overset{}{\text{CH}_3\text{CHCH}_2\text{Br}}}$$
$$\underset{}{\overset{}{\quad\quad\quad\quad\quad | \\ \quad\quad\quad\quad\quad \text{Br}}}$$

Addition of hydrogen bromide contrary to Markovnikov's rule leads to 1,3-dibromopropane.

Addition contrary to Markovnikov's rule:

$$\underset{\text{Allyl bromide}}{\text{CH}_2=\text{CHCH}_2\text{Br}} + \underset{\text{Hydrogen bromide}}{\text{HBr}} \longrightarrow \underset{\text{1,3-Dibromopropane}}{\text{BrCH}_2\text{CH}_2\text{CH}_2\text{Br}}$$

6.7 (b) Hydrogen bromide adds to 2-methyl-1-butene in accordance with Markovnikov's rule when peroxides are absent. The product is 2-bromo-2-methylbutane.

$$
\begin{array}{c}
CH_3 \\
\diagdown \\
C = C + HBr \longrightarrow CH_3CH_2CCH_3 \\
\diagup \\
CH_3CH_2 H
\end{array}
$$

2-Methyl-1-butene   Hydrogen bromide   2-Bromo-2-methylbutane

The opposite orientation is observed when peroxides are present. The product is 1-bromo-2-methylbutane.

$$
\begin{array}{c}
CH_3 \\
\diagdown \\
C = C + HBr \xrightarrow{\text{peroxides}} CH_3CH_2CCH_2Br \\
\diagup \\
CH_3CH_2 H
\end{array}
$$

2-Methyl-1-butene   Hydrogen bromide   1-Bromo-2-methylbutane

(c) Both ends of the double bond in *cis*-2-butene are equivalently substituted, so that the same product (2-bromobutane) is formed by hydrogen bromide addition regardless of whether the reaction is carried out in the presence of peroxides or in their absence.

$$
\begin{array}{c}
CH_3 CH_3 \\
\diagdown \diagup \\
C = C + HBr \longrightarrow CH_3CH_2CHCH_3 \\
\diagup \diagdown \\
H H Br
\end{array}
$$

cis-2-Butene   Hydrogen bromide   2-Bromobutane

(d) A tertiary bromide is formed on addition of hydrogen bromide to ethylidenecyclohexane in the absence of peroxides.

$$
CH_3CH = \bigcirc \quad + \quad HBr \longrightarrow \underset{Br}{\overset{CH_3CH_2}{\bigcirc}}
$$

Ethylidenecyclohexane   Hydrogen bromide   1-Bromo-1-ethylcyclohexane

The regioselectivity of addition is reversed in the presence of peroxides, and the product is 1-bromo-1-cyclohexylethane.

$$
CH_3CH = \bigcirc \quad + \quad HBr \xrightarrow{\text{peroxides}} \underset{Br}{CH_3CH} \bigcirc
$$

Ethylidenecyclohexane   Hydrogen bromide   1-Bromo-1-cyclohexylethane

6.8 The presence of hydroxide ion in the second step is incompatible with the medium in which the reaction is carried out. The reaction as shown in step 1

$$1. \ (CH_3)_2C = CH_2 + H_3O^+ \longrightarrow (CH_3)_3C^+ + H_2O$$

is performed in acidic solution. There are, for all practical purposes, no hydroxide ions in aqueous acid, the strongest base present being water itself. It is quite important to pay attention to the species that are actually present in the reaction medium whenever you formulate a reaction mechanism.

6.9 The more stable the carbocation, the faster it is formed. The more reactive alkene gives a tertiary carbocation in the rate-determining step.

Tertiary carbocation

Protonation of ▷—CH=CHCH₃ gives a secondary carbocation.

6.10 Electrophilic addition of hydrogen chloride to 2-methylpropene as outlined in the mechanism of Section 6.6 proceeds through a carbocation intermediate. This mechanism is the reverse of the E1 elimination. The E2 mechanism is concerted—it does not involve an intermediate.

6.11 (*b*) The carbon-carbon double bond is symmetrically substituted in *cis*-2-butene, and so the regioselectivity of hydroboration-oxidation is not an issue. Hydration of the double bond gives 2-butanol.

cis-2-Butene   2-Butanol

(*c*) Hydroboration-oxidation of alkenes is a method that leads to anti-Markovnikov hydration of the double bond.

Methylenecyclobutane   Cyclobutylmethanol

(*d*) Hydroboration-oxidation of cyclopentene gives cyclopentanol.

Cyclopentene   Cyclopentanol

(*e*) When alkenes are converted to alcohols by hydroboration-oxidation, the hydroxyl group is introduced at the less substituted carbon of the double bond.

$$CH_3CH{=}C(CH_2CH_3)_2 \xrightarrow[\text{2. oxidation}]{\text{1. hydroboration}} CH_3CHCH(CH_2CH_3)_2$$
$$\overset{|}{OH}$$

3-Ethyl-2-pentene   3-Ethyl-2-pentanol

(*f*) The less substituted carbon of the double bond in 3-ethyl-1-pentene is at the end of the chain. It is this carbon that bears the hydroxyl group in the product of hydroboration-oxidation.

$$CH_2{=}CHCH(CH_2CH_3)_2 \xrightarrow[\text{2. oxidation}]{\text{1. hydroboration}} HOCH_2CH_2CH(CH_2CH_3)_2$$

3-Ethyl-1-pentene   3-Ethyl-1-pentanol

6.12 The bottom face of the double bond is less hindered than the top face.

Methyl group shields top face.

1. B_2H_6
2. H_2O_2, HO^-

Hydroboration occurs from this direction.

This H comes from B_2H_6.

Syn-addition of H and OH takes place and with a regioselectivity opposite to that of Markovnikov's rule.

6.13 Bromine adds anti to the double bond of 1-bromocyclohexene to give 1,1,2-tribromocyclohexane. The radioactive bromines (^{82}Br) are vicinal and trans to each other.

$$\text{Br} \quad + \; ^{82}Br—^{82}Br \longrightarrow \; ^{82}Br$$

1-Bromocyclohexene   Bromine   1,1,2-Tribromocyclohexane

6.14 Alkyl substituents on the double bond increase the reactivity of the alkene toward addition of bromine.

2-Methyl-2-butene
(trisubstituted double bond; most reactive)

2-Methyl-1-butene
(disubstituted double bond)

3-Methyl-1-butene
(monosubstituted double bond; least reactive)

6.15 (*b*) Bromine adds to the carbon having the greater number of hydrogens, hydroxyl to the carbon having the fewer.

$$(CH_3)_2C{=}CHCH_3 \xrightarrow[H_2O]{Br_2} (CH_3)_2\underset{\underset{HO}{|}}{\overset{\overset{Br}{|}}{C}}CHCH_3$$

2-Methyl-2-butene   3-Bromo-2-methyl-2-butanol

(*c*)
$$(CH_3)_2CHCH{=}CH_2 \xrightarrow[H_2O]{Br_2} (CH_3)_2CH\underset{\underset{OH}{|}}{C}HCH_2Br$$

3-Methyl-1-butene   1-Bromo-3-methyl-2-butanol

(*d*) Anti addition occurs.

$$\xrightarrow[H_2O]{Br_2}$$

1-Methylcyclopentene   *trans*-2-Bromo-1-methylcyclopentanol

6.16 The structure of disparlure is as shown.

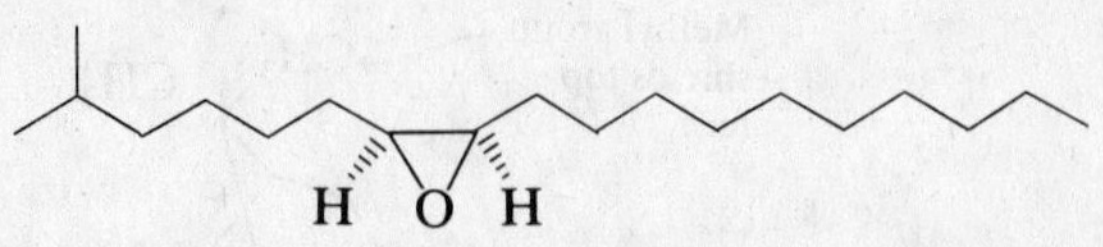

Its longest continuous chain contains 18 carbon atoms, and so it is named as an epoxy derivative of octadecane. Number the chain in the direction that gives the lowest number to the carbons that bear oxygen. Thus, disparlure is *cis*-2-methyl-7,8-epoxyoctadecane.

6.17 Disparlure can be prepared by epoxidation of the corresponding alkene. Epoxidation is stereospecific; cis alkenes yield cis epoxides. Therefore, *cis*-2-methyl-7-octadecene is the alkene chosen to prepare disparlure by epoxidation.

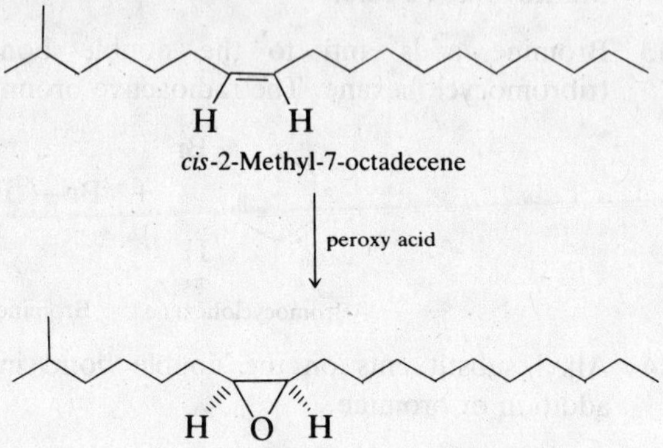

cis-2-Methyl-7-octadecene

peroxy acid

Disparlure

6.18 The products of ozonolysis are formaldehyde and 4,4-dimethyl-2-pentanone.

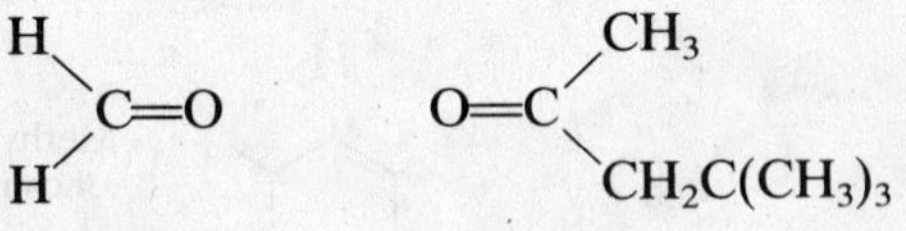

Formaldehyde   4,4-Dimethyl-2-pentanone

The two carbons that were doubly bonded to each other in the alkene become the carbons that are doubly bonded to oxygen in the products of ozonolysis. Therefore, mentally remove the oxygens and connect these two carbons by a double bond to reveal the structure of the starting alkene.

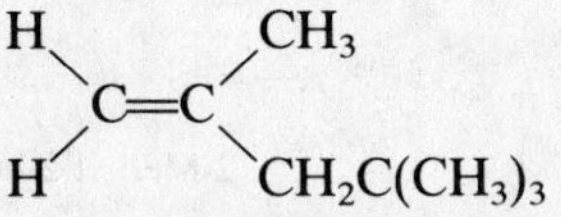

2,4,4-Trimethyl-1-pentene

6.19 (*b*) The sum of double bonds and rings (SODAR) is given by the formula.

$$SODAR = \tfrac{1}{2}(C_nH_{2n+2} - C_nH_x)$$

The compound given contains eight carbons (C_8H_8). Therefore:

$$SODAR = \tfrac{1}{2}(C_8H_{18} - C_8H_8)$$
$$SODAR = 5$$

The problem specifies that it consumes two moles of hydrogen, and so the compound contains two double bonds (or one triple bond). Since the SODAR is equal to 5, there must be three rings.

(*c*) Chlorine substituents are equivalent to hydrogens when calculating the SODAR.

Therefore, consider $C_8H_8Cl_2$ as equivalent to C_8H_{10}. Thus, the SODAR of this compound is 4.

$$SODAR = \tfrac{1}{2}(C_8H_{18} - C_8H_{10})$$

$$SODAR = 4$$

If the compound consumes two moles of hydrogen on catalytic hydrogenation, it must therefore contain two rings.

(d) Oxygen atoms are ignored when calculating the SODAR. Thus, C_8H_8O is treated as if it were C_8H_8.

$$SODAR = \tfrac{1}{2}(C_8H_{18} - C_8H_8)$$

$$SODAR = 5$$

Since the problem specifies that two moles of hydrogen are consumed on catalytic hydrogenation, this compound contains three rings.

(e) Ignoring the oxygen atoms in $C_8H_{10}O_2$, we treat this compound as if it were C_8H_{10}.

$$SODAR = \tfrac{1}{2}(C_8H_{18} - C_8H_{10})$$

$$SODAR = 4$$

If it consumes two moles of hydrogen on catalytic hydrogenation, there must be two rings.

(f) Ignore the oxygen and treat the chlorine as if it were hydrogen. Thus, C_8H_9ClO is treated as if it were C_8H_{10}. Its SODAR is 4, and it contains two rings.

6.20 All that is necessary to convert 2,4,4-trimethyl-1-pentene and 2,4,4-trimethyl-2-pentene to 2,2,4-trimethylpentane is catalytic hydrogenation of the double bond.

2,4,4-Trimethyl-1-pentene   or   2,4,4-Trimethyl-2-pentene

$\big\downarrow$ H_2, Pt

$$(CH_3)_2CHCH_2C(CH_3)_3$$

2,2,4-Trimethylpentane

6.21 From the structural formula of the desired product, we see that it is a *vicinal bromohydrin*. Vicinal bromohydrins are made from alkenes by reaction with bromine in water.

$$BrCH_2C(CH_3)_2 \quad \text{is made from} \quad CH_2{=}C(CH_3)_2$$
$$\mid$$
$$OH$$

Since the starting material given is *tert*-butyl bromide, a practical synthesis is:

$$(CH_3)_3CBr \xrightarrow[\substack{CH_3CH_2OH \\ \text{heat}}]{NaOCH_2CH_3} (CH_3)_2C{=}CH_2 \xrightarrow[H_2O]{Br_2} (CH_3)_2CCH_2Br$$
$$\mid$$
$$OH$$

Na Et   *Ethanol*

tert-Butyl bromide   2-Methylpropene   1-Bromo-2-methyl-2-propanol

6.22 This problem illustrates the reactions of alkenes with various reagents and requires application of the Markovnikov rule in the addition of unsymmetrical electrophiles.

(*a*)   Markovnikov addition of hydrogen chloride to 1-pentene will give 2-chloropentane.

$$CH_2{=}CHCH_2CH_2CH_3 + HCl \longrightarrow CH_3\underset{\underset{Cl}{|}}{C}HCH_2CH_2CH_3$$

1-Pentene   2-Chloropentane

(*b*)   Ionic addition of hydrogen bromide will give 2-bromopentane.

$$CH_2{=}CHCH_2CH_2CH_3 + HBr \longrightarrow CH_3\underset{\underset{Br}{|}}{C}HCH_2CH_2CH_3$$

2-Bromopentane

(*c*)   The presence of peroxides in the reaction medium will cause free-radical addition of hydrogen bromide, and anti-Markovnikov regioselectivity will be observed.

$$CH_2{=}CHCH_2CH_2CH_3 + HBr \xrightarrow{\text{peroxides}} BrCH_2CH_2CH_2CH_2CH_3$$

1-Bromopentane

(*d*)   Hydrogen iodide will add according to Markovnikov's rule.

$$CH_2{=}CHCH_2CH_2CH_3 + HI \longrightarrow CH_3\underset{\underset{I}{|}}{C}HCH_2CH_2CH_3$$

2-Iodopentane

(*e*)   Dilute sulfuric acid will cause hydration of the double bond with regioselectivity in accord with Markovnikov's rule.

$$CH_2{=}CHCH_2CH_2CH_3 + H_2O \xrightarrow{H_2SO_4} CH_3\underset{\underset{OH}{|}}{C}HCH_2CH_2CH_3$$

2-Pentanol

(*f*)   Hydroboration-oxidation of an alkene brings about anti-Markovnikov hydration of the double bond; 1-pentanol will be the product.

$$CH_2{=}CHCH_2CH_2CH_3 \xrightarrow[\text{2. } H_2O_2,\ HO^-]{\text{1. } B_2H_6} HOCH_2CH_2CH_2CH_2CH_3$$

1-Pentanol

(*g*)   Bromine adds across the double bond to give the vicinal dibromide.

$$CH_2{=}CHCH_2CH_2CH_3 + Br_2 \xrightarrow{CCl_4} Br CH_2\underset{\underset{Br}{|}}{C}HCH_2CH_2CH_3$$

1,2-Dibromopentane

(*h*)   Vicinal bromohydrins are formed when bromine in water adds to alkenes. Markovnikov orientation is observed if one considers attack by bromonium ion to be followed by interception of the cation by water as a nucleophile.

$$CH_2{=}CHCH_2CH_2CH_3 + Br_2 \xrightarrow{H_2O} BrCH_2\underset{\underset{OH}{|}}{C}HCH_2CH_2CH_3$$

1-Bromo-2-pentanol

(*i*)   Epoxidation of the alkene occurs on treatment with peroxy acids.

$$CH_2\!\!=\!\!CHCH_2CH_2CH_3 + CH_3CO_2OH \longrightarrow CH_2\!\!-\!\!CHCH_2CH_2CH_3 + CH_3CO_2H$$
$$\underset{O}{\diagdown\diagup}$$

1,2-Epoxypentane   Acetic acid

(*j*)   Ozone reacts with alkenes to give ozonides.

$$CH_2\!\!=\!\!CHCH_2CH_2CH_3 + O_3 \longrightarrow CH_2 \quad CHCH_2CH_2CH_3$$
$$O\!-\!O$$

3-Propyl-1,2,4-trioxolane

(*k*)   When the ozonide in part (*j*) is hydrolyzed in the presence of zinc, formaldehyde and butanal are formed.

$$CH_2 \quad CHCH_2CH_2CH_3 \xrightarrow[Zn]{H_2O} \overset{O}{\underset{\|}{H}}CH + \overset{O}{\underset{\|}{C}}CH_2CH_2CH_3$$
$$O\!-\!O \qquad\qquad\qquad\qquad\quad H$$

Formaldehyde   Butanal

6.23   When we compare the reactions of 2-methyl-2-butene with the analogous reactions of 1-pentene, we find that the reactions proceed in a similar manner.

(*a*) $(CH_3)_2C\!\!=\!\!CHCH_3 + HCl \longrightarrow (CH_3)_2CCH_2CH_3$
$$\underset{Cl}{|}$$

2-Methyl-2-butene   2-Chloro-2-methylbutane

(*b*) $(CH_3)_2C\!\!=\!\!CHCH_3 + HBr \longrightarrow (CH_3)_2CCH_2CH_3$
$$\underset{Br}{|}$$

2-Bromo-2-methylbutane

(*c*) $(CH_3)_2C\!\!=\!\!CHCH_3 + HBr \xrightarrow{\text{peroxides}} (CH_3)_2CHCHCH_3$
$$\underset{Br}{|}$$

2-Bromo-3-methylbutane

(*d*) $(CH_3)_2C\!\!=\!\!CHCH_3 + HI \longrightarrow (CH_3)_2CCH_2CH_3$
$$\underset{I}{|}$$

2-Iodo-2-methylbutane

(*e*) $(CH_3)_2C\!\!=\!\!CHCH_3 + H_2O \xrightarrow{H_2SO_4} (CH_3)_2CCH_2CH_3$
$$\underset{OH}{|}$$

2-Methyl-2-butanol

(*f*) $(CH_3)_2C\!\!=\!\!CHCH_3 \xrightarrow[\text{2. } H_2O_2,\, HO^-]{\text{1. } B_2H_6} (CH_3)_2CHCHCH_3$
$$\underset{OH}{|}$$

3-Methyl-2-butanol

(g) $(CH_3)_2C{=}CHCH_3 + Br_2 \xrightarrow{\text{CCl}_4} (CH_3)_2\overset{\overset{\displaystyle Br}{|}}{C}\underset{\underset{\displaystyle Br}{|}}{C}HCH_3$

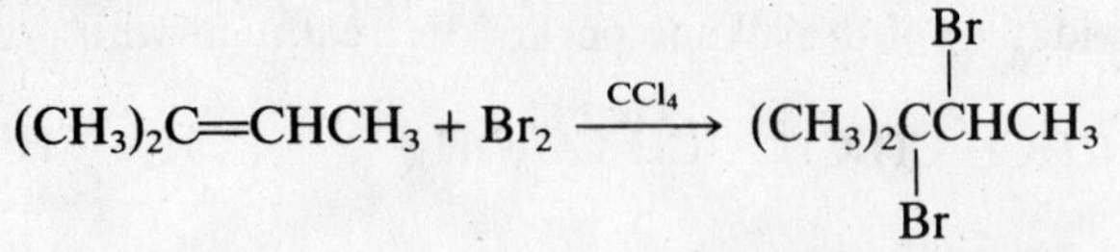

2,3-Dibromo-2-methylbutane

(h) $(CH_3)_2C{=}CHCH_3 + Br_2 \xrightarrow{\text{H}_2\text{O}} (CH_3)_2\overset{\overset{\displaystyle Br}{|}}{C}\underset{\underset{\displaystyle OH}{|}}{C}HCH_3$

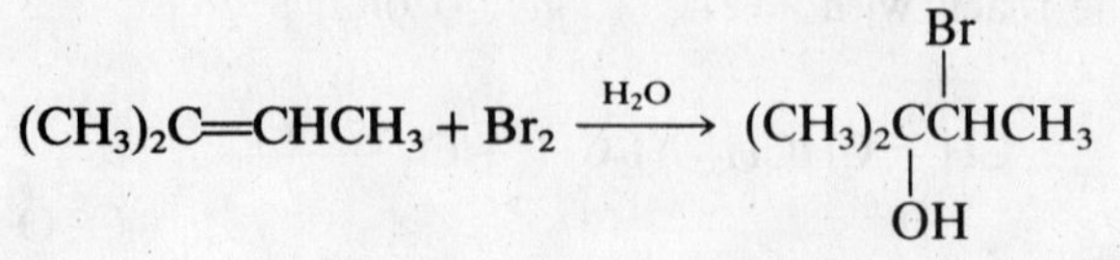

3-Bromo-2-methyl-2-butanol

(i) $(CH_3)_2C{=}CHCH_3 + CH_3CO_2OH \longrightarrow (CH_3)_2\overset{\diagup\!\!\diagdown}{C}\underset{O}{\,}CHCH_3 + CH_3CO_2H$

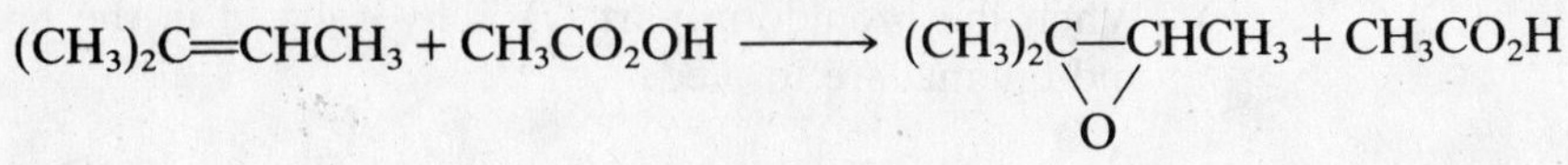

2-Methyl-2,3-epoxybutane

(j) $(CH_3)_2C{=}CHCH_3 + O_3 \longrightarrow$

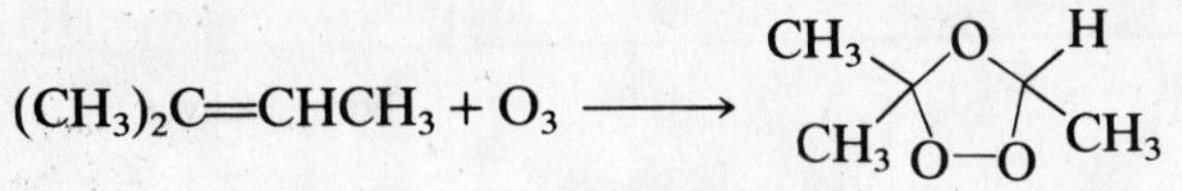

3,3,5-Trimethyl-1,2,4-trioxolane

(k) $\xrightarrow[\text{Zn}]{\text{H}_2\text{O}}$ $CH_3\overset{\overset{\displaystyle O}{\|}}{C}CH_3 + \overset{\overset{\displaystyle O}{\|}}{C}CH_3$

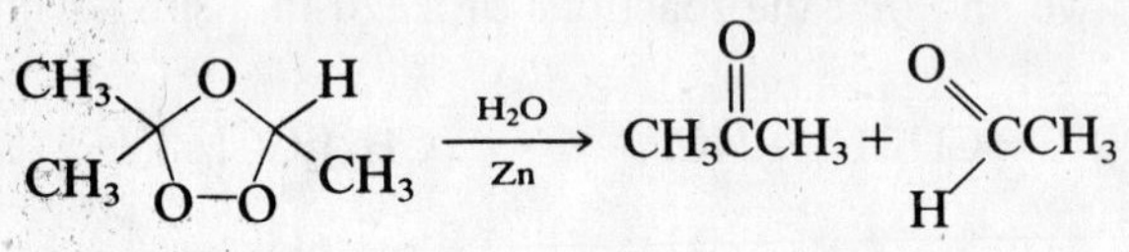

Acetone   Acetaldehyde

6.24   Cycloalkenes undergo the same kinds of reactions as do noncyclic alkenes.

(a)

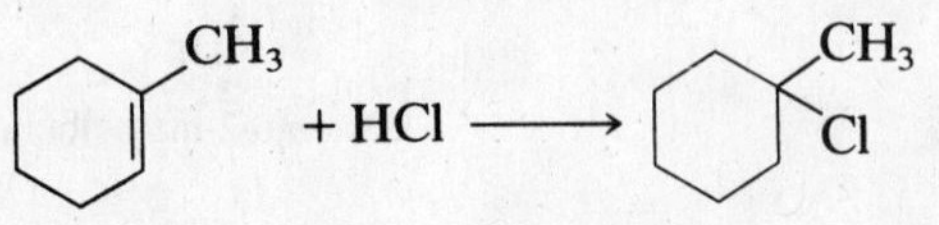

$+ \text{HCl} \longrightarrow$

1-Methylcyclohexene   1-Chloro-1-methylcyclohexane

(b)

$+ \text{HBr} \longrightarrow$

1-Bromo-1-methylcyclohexane

(c)

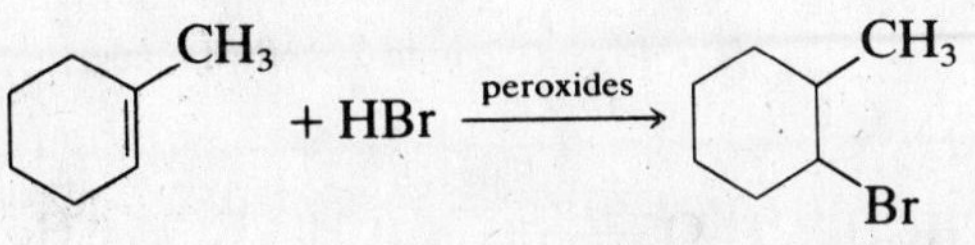

$+ \text{HBr} \xrightarrow{\text{peroxides}}$

1-Bromo-2-methylcyclohexane
(mixture of cis and trans)

(d)

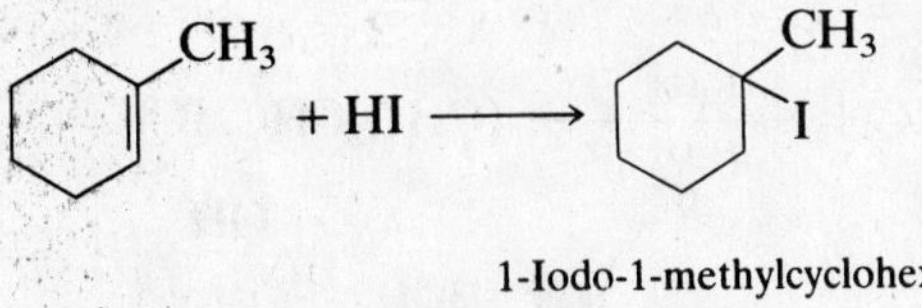

$+ \text{HI} \longrightarrow$

1-Iodo-1-methylcyclohexane

(*e*)

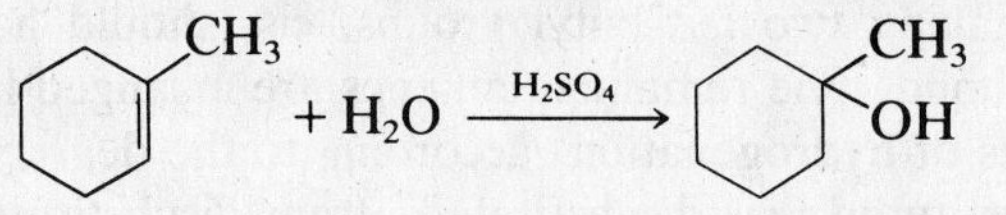
CH3
+ H2O
H2SO4
CH3
OH

1-Methylcyclohexanol

(*f*)

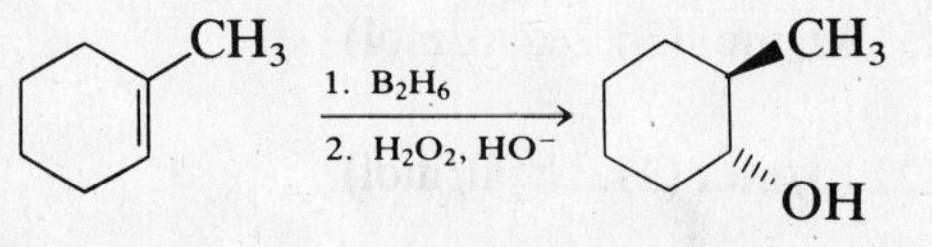
CH3
1. B2H6
2. H2O2, HO−
CH3
OH

trans-2-Methylcyclohexanol

(*g*)

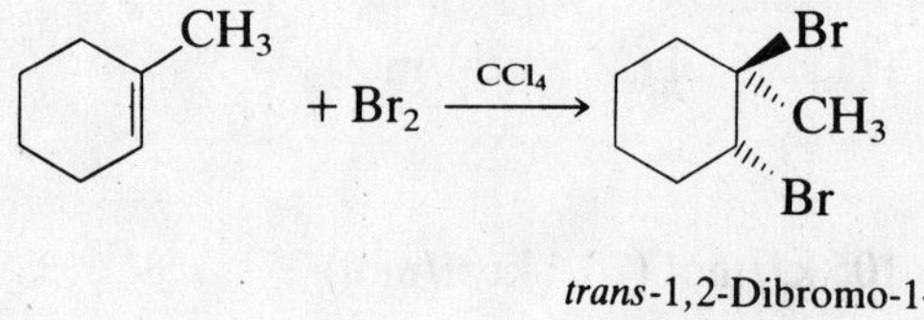
CH3
+ Br2
CCl4
Br
CH3
Br

trans-1,2-Dibromo-1-methylcyclohexane

(*h*)

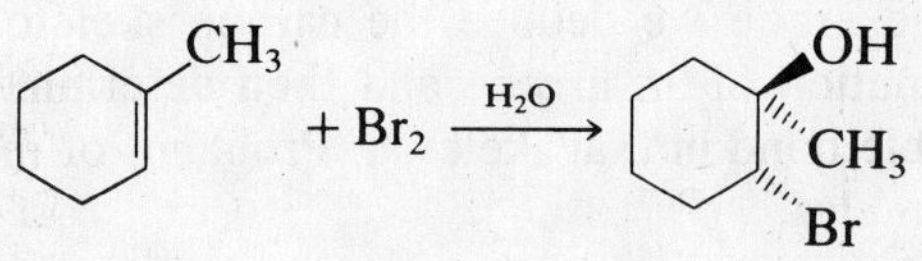
CH3
+ Br2
H2O
OH
CH3
Br

trans-2-Bromo-1-methylcyclohexanol

(*i*)

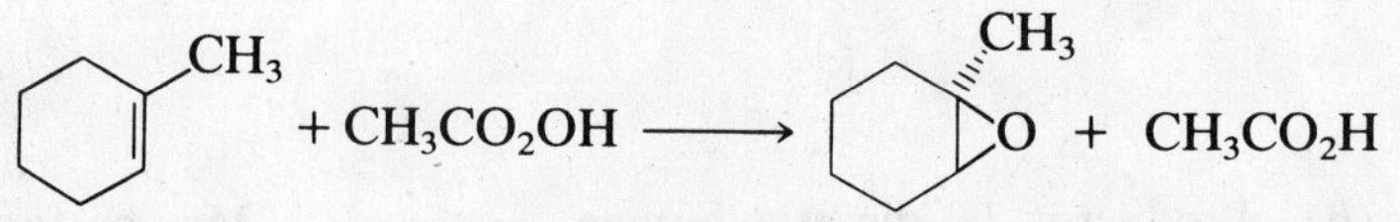
CH3
+ CH3CO2OH
CH3
O
+ CH3CO2H

1,2-Epoxy-1-methylcyclohexane

(*j*)

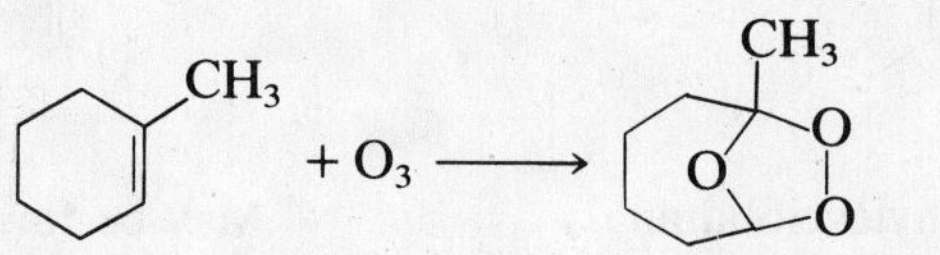
CH3
+ O3
CH3
O
O
O

1-Methyl-7,8,9-trioxabicyclo[4.2.1]nonane

(*k*)

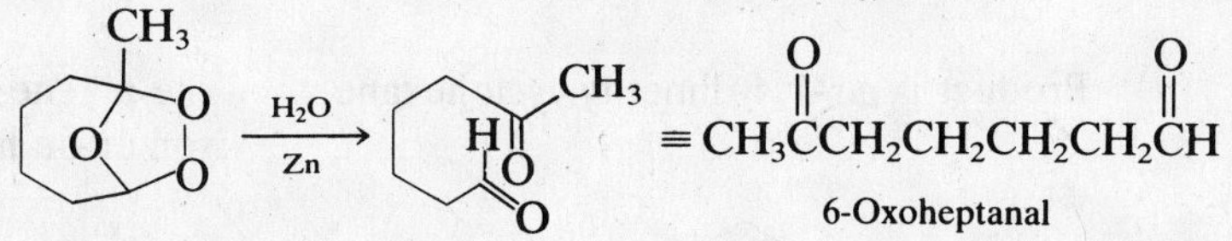
CH3
O
O
O
H2O
Zn
CH3
H
O
O
O
O
≡ CH3CCH2CH2CH2CH2CH

6-Oxoheptanal

6.25 We need first to write out the structures in more detail to evaluate the substitution patterns at the double bonds.

(*a*) 1-Pentene 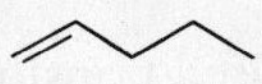 Monosubstituted

(*b*) (*E*)-4,4-Dimethyl-2-pentene 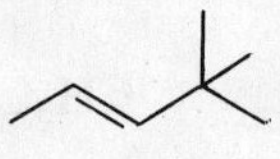 trans-Disubstituted

(*c*) (*Z*)-4-Methyl-2-pentene 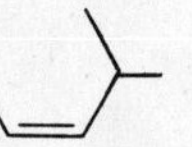 cis-Disubstituted

(*d*) (*Z*)-2,2,5,5-Tetramethyl-3-hexene 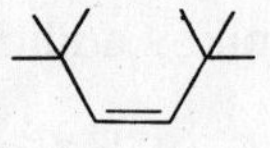 Two *tert*-butyl groups cis

(*e*) 2,4-Dimethyl-2-pentene 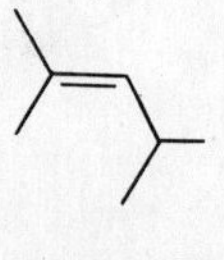 Trisubstituted

Compound *d,* having two *tert*-butyl groups, cis, should have the least stable (highest-energy) double bond. The remaining alkenes are arranged in order of increasing stability (decreasing heats of hydrogenation) according to the degree of substitution of the double bond: monosubstituted, cis-disubstituted, trans-disubstituted, trisubstituted. Therefore the heats of hydrogenation are:

(*d*)   151 kJ/mol (36.2 kcal/mol)

(*a*)   122 kJ/mol (29.3 kcal/mol)

(*c*)   114 kJ/mol (27.3 kcal/mol)

(*b*)   111 kJ/mol (26.5 kcal/mol)

(*e*)   105 kJ/mol (25.1 kcal/mol)

6.26 In all parts of this exercise we deduce the carbon skeleton on the basis of the alkane formed on hydrogenation of an alkene and then determine what carbon atoms may be connected by a double bond in that skeleton. Problems of this type are best done by using carbon skeleton formulas.

(*a*) Product is 2,2,3,4,4-pentamethylpentane.   The only possible alkene precursor is

(*b*) Product is 2,3-dimethylbutane.   May be formed by hydrogenation of

(*c*) Product is methylcyclobutane.   May be formed by hydrogenation of

(*d*) Product is *cis*-1,4-dimethylcyclohexane.   The alkene that gives only the cis isomer on hydrogenation is

6.27 (*a*) The desired transformation is the conversion of an alkene to a vicinal dibromide.

$$CH_3CH{=}C(CH_2CH_3)_2 \xrightarrow[CCl_4]{Br_2} CH_3CHC(CH_2CH_3)_2$$
$$\underset{\displaystyle Br \ \ Br}{\big| \ \ \big|}$$

3-Ethyl-2-pentene   2,3-Dibromo-3-ethylpentane

(*b*) Markovnikov addition of hydrogen chloride is indicated.

$$CH_3CH{=}C(CH_2CH_3)_2 \xrightarrow{HCl} CH_3CH_2C(CH_2CH_3)_2$$
$$\underset{\displaystyle Cl}{\big|}$$

3-Chloro-3-ethylpentane

(c) Free-radical addition of hydrogen bromide will produce the required anti-Markovnikov orientation.

$$CH_3CH=C(CH_2CH_3)_2 \xrightarrow[\text{peroxides}]{HBr} CH_3CHCH(CH_2CH_3)_2$$

with Br substituent below

2-Bromo-3-ethylpentane

(d) Acid-catalyzed hydration will occur in accordance with Markovnikov's rule to yield the desired tertiary alcohol.

$$CH_3CH=C(CH_2CH_3)_2 \xrightarrow[H_2SO_4]{H_2O} CH_3CH_2C(CH_2CH_3)_2$$

with OH substituent below

3-Ethyl-3-pentanol

(e) Hydroboration-oxidation results in hydration of alkenes with a regioselectivity contrary to that of Markovnikov's rule.

$$CH_3CH=C(CH_2CH_3)_2 \xrightarrow[\text{2. } H_2O_2, HO^-]{\text{1. } B_2H_6} CH_3CHCH(CH_2CH_3)_2$$

with OH substituent below

3-Ethyl-2-pentanol

(f) A peroxy acid will convert an alkene to an epoxide.

$$CH_3CH=C(CH_2CH_3)_2 \xrightarrow{CH_3CO_2OH}$$

epoxide structure with CH₃, CH₂CH₃, H, O, CH₂CH₃

3-Ethyl-2,3-epoxypentane

(g) Hydrogenation of alkenes converts them to alkanes.

$$CH_3CH=C(CH_2CH_3)_2 \xrightarrow[\text{Pt}]{H_2} CH_3CH_2CH(CH_2CH_3)_2$$

3-Ethylpentane

6.28 (a) There are four primary alcohols having the molecular formula $C_5H_{12}O$:

$CH_3CH_2CH_2CH_2CH_2OH$

1-Pentanol

$$\underset{CH_3}{CH_3CH_2CHCH_2OH}$$

2-Methyl-1-butanol

$(CH_3)_2CHCH_2CH_2OH$

3-Methyl-1-butanol

$(CH_3)_3CCH_2OH$

2,2-Dimethyl-1-propanol

2,2-Dimethyl-1-propanol cannot be prepared by hydration of an alkene, because no alkene can have this carbon skeleton.

(b) Hydroboration-oxidation of alkenes is the method of choice for converting terminal alkenes to primary alcohols.

$$CH_3CH_2CH_2CH=CH_2 \xrightarrow[\text{2. } H_2O_2, HO^-]{\text{1. } B_2H_6} CH_3CH_2CH_2CH_2CH_2OH$$

1-Pentene   1-Pentanol

$$CH_3CH_2C{=}CH_2 \xrightarrow[\text{2. } H_2O_2,\, HO^-]{\text{1. } B_2H_6} CH_3CH_2CHCH_2OH$$

with CH_3 substituent.

2-Methyl-1-butene   2-Methyl-1-butanol

$$(CH_3)_2CHCH{=}CH_2 \xrightarrow[\text{2. } H_2O_2,\, HO^-]{\text{1. } B_2H_6} (CH_3)_2CHCH_2CH_2OH$$

3-Methyl-1-butene   3-Methyl-1-butanol

(*c*) The only tertiary alcohol is 2-methyl-2-butanol. It can be made by Markovnikov hydration of 2-methyl-1-butene or of 2-methyl-2-butene.

$$CH_2{=}CCH_2CH_3 \xrightarrow{H_2O,\, H_2SO_4} (CH_3)_2CCH_2CH_3$$

with CH_3 and OH substituents.

2-Methyl-1-butene   2-Methyl-2-butanol

$$(CH_3)_2C{=}CHCH_3 \xrightarrow{H_2O,\, H_2SO_4} (CH_3)_2CCH_2CH_3$$

with OH substituent.

2-Methyl-2-butene   2-Methyl-2-butanol

6.29 (*a*) Because the double bond is symmetrically substituted, the same addition product is formed under either ionic or free-radical conditions. Peroxides are absent, and so addition takes place by an ionic mechanism to give 3-bromohexane. (It does not matter whether the starting material is *cis*- or *trans*-3-hexene; both give the same product.)

$$CH_3CH_2CH{=}CHCH_2CH_3 + HBr \xrightarrow{\text{no peroxides}} CH_3CH_2CH_2CHCH_2CH_3$$

with Br substituent.

3-Hexene   Hydrogen bromide   3-Bromohexane (observed yield 76%)

(*b*) In the presence of peroxides hydrogen bromide adds with a regioselectivity opposite to that predicted by Markovnikov's rule. The product is the corresponding primary bromide.

$$(CH_3)_2CHCH_2CH_2CH_2CH{=}CH_2 \xrightarrow[\text{peroxides}]{HBr} (CH_3)_2CHCH_2CH_2CH_2CH_2CH_2Br$$

6-Methyl-1-heptene   1-Bromo-6-methylheptane (observed yield 92%)

(*c*) Hydroboration-oxidation of alkenes leads to hydration of the double bond with a regioselectivity contrary to Markovnikov's rule and without rearrangement of the carbon skeleton.

$$CH_2{=}C(C(CH_3)_3)(C(CH_3)_3) \xrightarrow[\text{2. } H_2O_2,\, HO^-]{\text{1. } B_2H_6} HOCH_2CHC(CH_3)_3$$

with $C(CH_3)_3$ substituent.

2-*tert*-Butyl-3,3-dimethyl-1-butene   2-*tert*-Butyl-3,3-dimethyl-1-butanol (observed yield 65%)

(*d*) Hydroboration-oxidation of alkenes leads to syn hydration of double bonds.

$$\text{1,2-Dimethylcyclohexene} \xrightarrow[\text{2. } H_2O_2, \, HO^-]{\text{1. } B_2H_6} \textit{cis}\text{-1,2-Dimethylcyclohexanol}$$

1,2-Dimethylcyclohexene   *cis*-1,2-Dimethylcyclohexanol
(observed yield 82%)

(*e*) Bromine adds across the double bond of alkenes to give vicinal dibromides.

$$CH_2{=}CCH_2CH_2CH_3 + Br_2 \xrightarrow{CHCl_3} BrCH_2CCH_2CH_2CH_3$$

2-Methyl-1-pentene   1,2-Dibromo-2-methylpentane
(observed yield 60%)

(*f*) In aqueous solution bromine reacts with alkenes to give bromohydrins. Bromine is the electrophile in this reaction and adds to the carbon that has the greater number of hydrogen substituents.

$$(CH_3)_2C{=}CHCH_3 + Br_2 \xrightarrow{H_2O} (CH_3)_2CCHCH_3$$

2-Methyl-2-butene   Bromine   3-Bromo-2-methyl-2-butanol
(observed yield 77%)

(*g*) An aqueous solution of chlorine will react with 1-methylcyclopentene by an anti addition. Chlorine is the electrophile and adds to the less substituted end of the double bond.

$$\text{1-Methylcyclopentene} \xrightarrow[H_2O]{Cl_2} \textit{trans}\text{-2-Chloro-1-methylcyclopentanol}$$

1-Methylcyclopentene   *trans*-2-Chloro-1-
methylcyclopentanol

(*h*) Compounds of the type RCOOH are peroxy acids and react with alkenes to give epoxides.

$$(CH_3)_2C{=}C(CH_3)_2 + CH_3COOH \longrightarrow \text{2,3-Dimethyl-2,3-epoxybutane} + CH_3COH$$

2,3-Dimethyl-2-butene   Peroxyacetic acid   **2,3-Dimethyl-2,3-epoxybutane**   Acetic acid
(observed yield 70–80%)

(*i*) The double bond is cleaved by ozonolysis. Each of the doubly bonded carbons becomes doubly bonded to oxygen in the product.

$$\xrightarrow[\text{2. } H_2O]{\text{1. } O_3}$$

Cyclodecan-1,6-dione
(observed yield 45%)

6.30 (*a*) There is no direct, one-step transformation that moves a hydroxyl group from one carbon to another, and so it is not possible to convert 2-propanol to 1-propanol in a single reaction. Analyze the problem by reasoning backward. 1-Propanol is a primary alcohol. What reactions do we have available for the preparation of primary alcohols? One way is by the hydroboration-oxidation of terminal alkenes.

$$CH_3CH{=}CH_2 \xrightarrow{\text{hydroboration-oxidation}} CH_3CH_2CH_2OH$$

Propene   1-Propanol

The problem now becomes the preparation of propene from 2-propanol. The simplest way is by acid-catalyzed dehydration.

$$\underset{\overset{|}{OH}}{CH_3CHCH_3} \xrightarrow{\text{H}^+,\ \text{heat}} CH_3CH{=}CH_2$$

2-Propanol   Propene

After analyzing the problem in terms of overall strategy, present the synthesis in detail showing the reagents required in each step. Thus, the answer is:

$$\underset{\overset{|}{OH}}{CH_3CHCH_3} \xrightarrow[\text{heat}]{\text{H}_2\text{SO}_4} CH_3CH{=}CH_2 \xrightarrow[\text{2. H}_2\text{O}_2,\ \text{HO}^-]{\text{1. B}_2\text{H}_6} CH_3CH_2CH_2OH$$

2-Propanol   Propene   1-Propanol

(*b*) We analyze this synthetic exercise in a manner similar to the preceding one. There is no direct way to move a bromine from C-2 in 2-bromopropane to C-1 in 1-bromopropane. However, we can prepare 1-bromopropane from propene by free-radical addition of hydrogen bromide in the presence of peroxides.

$$CH_3CH{=}CH_2 + \quad HBr \xrightarrow{\text{peroxides}} CH_3CH_2CH_2Br$$

Propene   Hydrogen bromide   1-Bromopropane

We prepare propene from 2-bromopropane by dehydrohalogenation.

$$\underset{\overset{|}{Br}}{CH_3CHCH_3} \xrightarrow{\text{E2}} CH_3CH{=}CH_2$$

2-Bromopropane   Propene

(KOH in ethanol — handwritten)

Sodium ethoxide in ethanol is a suitable base-solvent system for this conversion. Sodium methoxide in methanol or potassium *tert*-butoxide in *tert*-butyl alcohol could also be used, as could potassium hydroxide in ethanol.

Combining these two transformations gives the complete synthesis.

$$\underset{\overset{|}{Br}}{CH_3CHCH_3} \xrightarrow[\text{CH}_3\text{CH}_2\text{OH, heat}]{\text{NaOCH}_2\text{CH}_3} CH_3CH{=}CH_2 \xrightarrow[\text{peroxides}]{\text{HBr}} CH_3CH_2CH_2Br$$

2-Bromopropane   Propene   1-Bromopropane

(*c*) Planning your strategy in a forward direction can lead to problems when the conversion of 2-bromopropane to 1,2-dibromopropane is considered. There is a temptation to try to simply add the second bromine by free-radical halogenation.

$$\underset{\overset{|}{Br}}{CH_3CHCH_3} \xrightarrow{\text{Br}_2,\ \text{light and heat}} \underset{\overset{|}{Br}}{CH_3CHCH_2Br}$$

2-Bromopropane   1,2-Dibromopropane

This is incorrect! There is no reason to believe that the second bromine will be introduced exclusively at C-1. In fact, the selectivity rules for bromination tell us that 2,2-dibromopropane is the expected major product.

The best approach is to reason backward. 1,2-Dibromopropane is a vicinal dibromide, and we prepare vicinal dibromides by adding elemental bromine to alkenes.

$$CH_3CH{=}CH_2 + Br_2 \longrightarrow CH_3CHCH_2Br$$
$$|$$
$$Br$$

Propene   Bromine   1,2-Dibromopropane

As described in part (*b*), we prepare propene from 2-bromopropane by E2 elimination. Therefore, the correct synthesis is;

$$CH_3CHCH_3 \xrightarrow[\;CH_3CH_2OH\;]{\;NaOCH_2CH_3\;} CH_3CH{=}CH_2 \xrightarrow{\;Br_2\;} CH_3CHCH_2Br$$
$$|\qquad\qquad\qquad\qquad\qquad\qquad\qquad\qquad\qquad\quad |$$
$$Br\qquad\qquad\qquad\qquad\qquad\qquad\qquad\qquad\qquad Br$$

2-Bromopropane   Propene   1,2-Dibromopropane

(*d*) Do not attempt to reason forward and convert 2-propanol to 1-bromo-2-propanol by free radical bromination. Reason backward! The desired compound is a vicinal bromohydrin, and vicinal bromohydrins are prepared by adding bromine to alkenes in aqueous solution. The correct solution is:

$$CH_3CHCH_3 \xrightarrow[heat]{\;H_2SO_4\;} CH_3CH{=}CH_2 \xrightarrow[H_2O]{\;Br_2\;} CH_3CHCH_2Br$$
$$|\qquad\qquad\qquad\qquad\qquad\qquad\qquad\qquad\qquad\qquad |$$
$$OH\qquad\qquad\qquad\qquad\qquad\qquad\qquad\qquad\qquad\qquad OH$$

2-Propanol   Propene   1-Bromo-2-propanol

(*e*) Here we have another problem where reasoning forward can lead to trouble. If we try to conserve the oxygen of 2-propanol so that it becomes the oxygen of 1,2-epoxypropane, we need a reaction in which this oxygen becomes bonded to C-1.

$$CH_3CHCH_3 \longrightarrow CH_3CH{-}CH_2$$
$$|\qquad\qquad\qquad\qquad\qquad\;\; \backslash O /$$
$$OH$$

2-Propanol   1,2-Epoxypropane

No synthetic method for such a single-step transformation exists!

By reasoning backward, recalling that epoxides are made from alkenes by reaction with peroxy acids, we develop a proper synthesis.

$$CH_3CHCH_3 \xrightarrow[heat]{\;H_2SO_4\;} CH_3CH{=}CH_2 \xrightarrow{\;CH_3COOH\;} CH_3CH{-}CH_2$$
$$|\qquad\qquad\qquad\qquad\qquad\qquad\qquad\qquad\qquad\qquad\; \backslash O /$$
$$OH$$

2-Propanol   Propene   1,2-Epoxypropane

(*f*) *tert*-Butyl alcohol and isobutyl alcohol have the same carbon skeleton; all that is required is to move the hydroxyl group from C-1 to C-2. As pointed out in part (*a*) of this problem, we cannot do that directly but we can do it in two efficient steps through a synthesis that involves hydration of an alkene.

$$(CH_3)_2CHCH_2OH \xrightarrow[heat]{\;H_2SO_4\;} (CH_3)_2C{=}CH_2 \xrightarrow{\;H_2O,\;H_2SO_4\;} (CH_3)_3COH$$

Isobutyl alcohol   2-Methylpropene   *tert*-Butyl alcohol

Acid-catalyzed hydration of the alkene gives the desired regioselectivity.

(g) The strategy of this exercise is similar to that of the preceding one. Convert the starting material to an alkene by an elimination reaction, followed by electrophilic addition to the double bond.

$$(CH_3)_2CHCH_2I \xrightarrow[\text{(CH}_3)_3\text{COH, heat}]{\text{KOC(CH}_3)_3} (CH_3)_2C{=}CH_2 \xrightarrow{\text{HI}} (CH_3)_3CI$$

Isobutyl iodide   *E*   2-Methylpropene   *tert*-Butyl iodide

(h) This problem is similar to the one in part (d) in that it requires the preparation of a halohydrin from an alkyl halide. The strategy is the same. Convert the alkyl halide to an alkene, and then form the halohydrin by treatment with the appropriate halogen in aqueous solution.

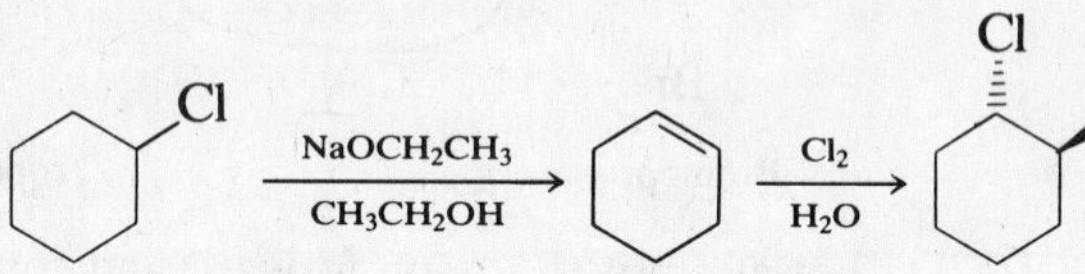

Cyclohexyl chloride   Cyclohexene   *trans*-2-Chlorocyclohexanol

(i) Halogenation of an alkane is required here. Iodination of alkanes, however, is not a feasible reaction, because it is endothermic. We can make alkyl iodides from alcohols or from alkenes by treatment with HI. A reasonable synthesis using reactions that have been presented to this point proceeds as shown:

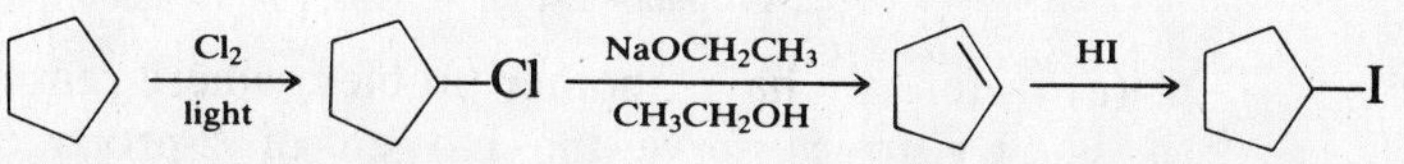

Cyclopentane   Cyclopentyl   Cyclopentene   Cyclopentyl
                    chloride   iodide

(j) Dichlorination of cyclopentane under free-radical conditions is not a realistic approach to the introduction of two chlorines in a trans-1,2 relationship without contamination by isomeric dichlorides. Vicinal dichlorides are prepared by electrophilic addition of chlorine to alkenes. The stereochemistry of addition is anti.

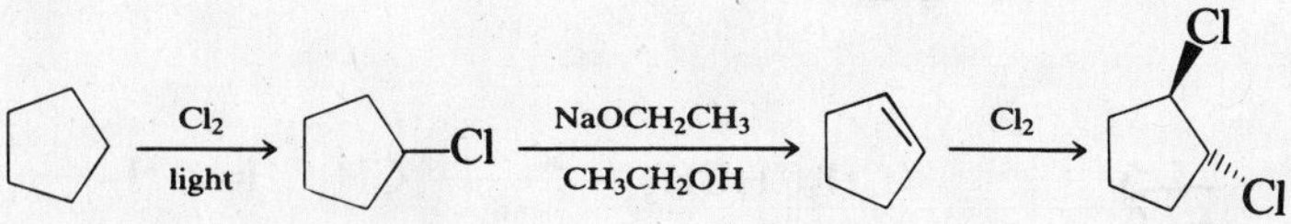

Cyclopentane   Cyclopentyl   Cyclopentene   *trans*-1,2-Dichlorocyclopentane
                    chloride

(k) The desired compound contains all five carbon atoms of cyclopentane but is not cyclic. Two aldehyde functions are present. We know that cleavage of carbon-carbon double bonds by ozonolysis leads to two carbonyl groups, which suggests the synthesis shown in the following equation:

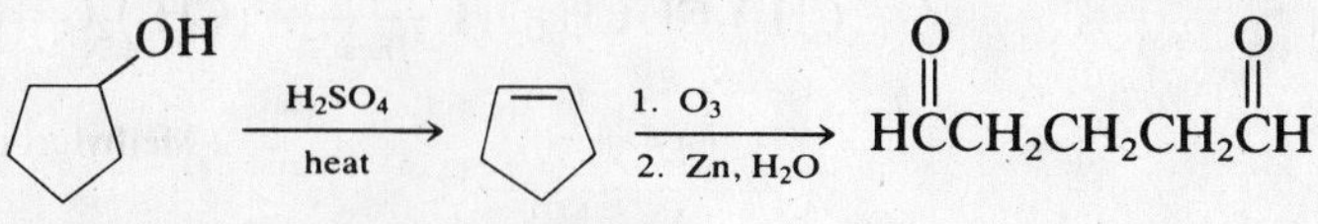

Cyclopentanol   Cyclopentene   Pentanedial

6.31 The two products formed by addition of hydrogen bromide to 1,2-dimethylcyclohexene cannot be regioisomers. Stereoisomers are possible, however.

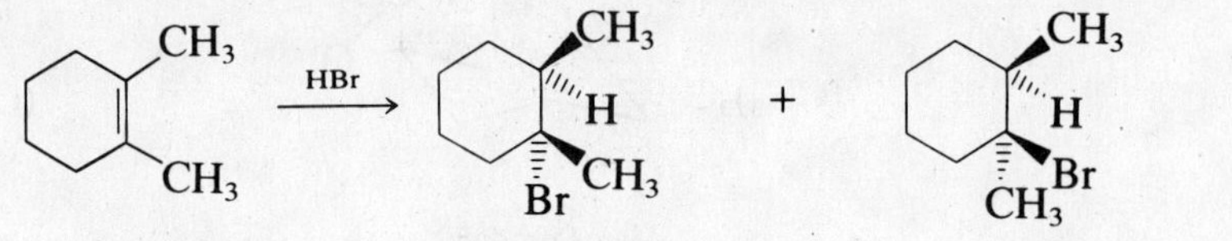

1,2-Dimethylcyclohexene   *cis*-1,2-Dimethylcyclohexyl bromide   *trans*-1,2-Dimethylcyclohexyl bromide

The same two products are formed from 1,6-dimethylcyclohexene because addition of hydrogen bromide follows Markovnikov's rule in the absence of peroxides.

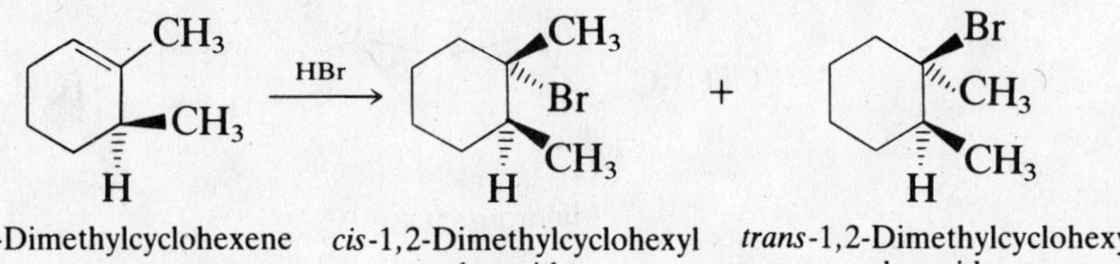

1,6-Dimethylcyclohexene   *cis*-1,2-Dimethylcyclohexyl bromide   *trans*-1,2-Dimethylcyclohexyl bromide

6.32 The reaction is one of electrophilic addition of hydrogen chloride to the double bond. Two regioisomeric modes of addition are indicated:

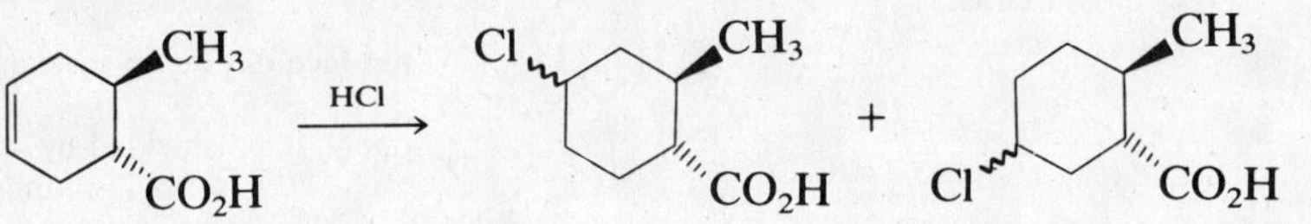

Each of these regioisomeric modes can lead to stereoisomers with the chlorine either cis or trans to the methyl group. The four products are:

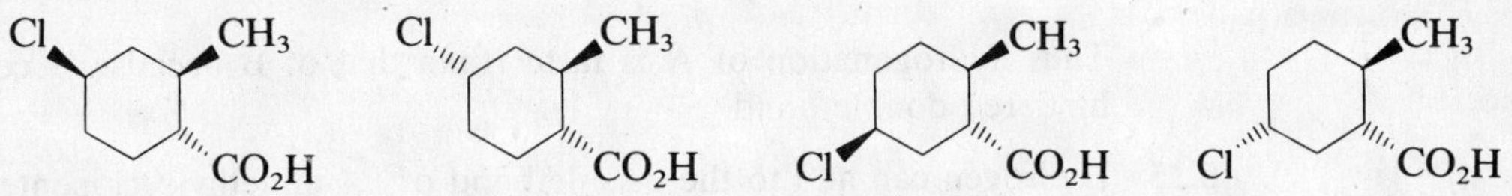

6.33 The problem presents the following experimental observation:

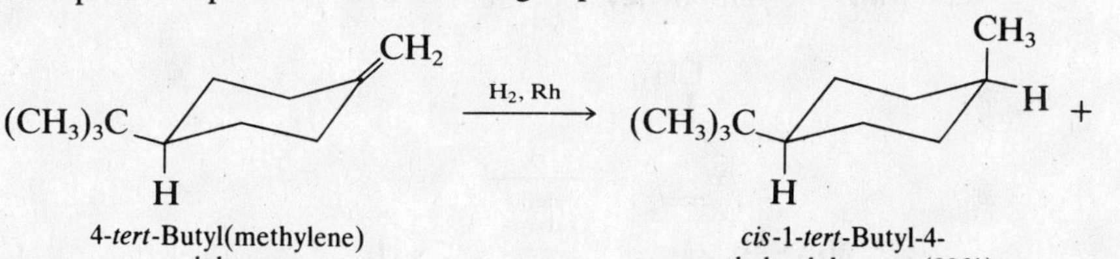

4-*tert*-Butyl(methylene) cyclohexane   *cis*-1-*tert*-Butyl-4-methylcyclohexane (88%)

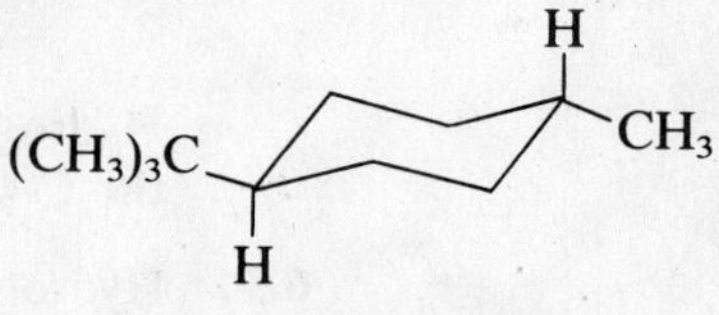

trans-1-*tert*-Butyl-4-methylcyclohexane (12%)

This observation tells us that the less hindered approach to the double bond is from the equatorial direction.

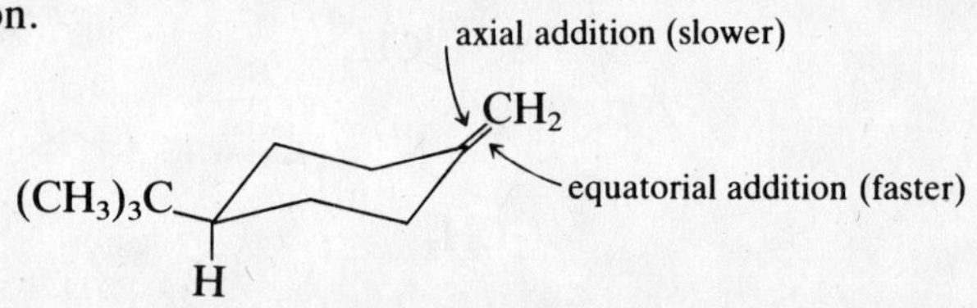

(a) Therefore, epoxidation should give the following products:

Major product   Minor product

The major product is the stereoisomer that corresponds to transfer of oxygen from the equatorial direction.

(b) Hydroboration-oxidation occurs from the equatorial direction.

Major product   Minor product

6.34 The methyl group in compound B shields one face of the double bond from the catalyst surface, so that hydrogen can be transferred only to the bottom face of the double bond. The methyl group in compound A does not interfere with hydrogen transfer to the double bond.

top face of double bond:

open   shielded by methyl group

Compound A   Compound B

Thus hydrogenation of A is faster than that of B because B contains a more sterically hindered double bond.

6.35 Hydrogen can add to the double bond of 1,4-dimethylcyclopentene either from the same side as the C-4 methyl group or from the opposite side. The two possible products are *cis*- and *trans*-1,3-dimethylcyclopentane.

1,4-Dimethylcyclopentene   *cis*-1,3-Dimethylcyclopentane   *trans*-1,3-Dimethylcyclopentane

Hydrogen transfer occurs to the less hindered face of the double bond, that is, trans to the C-4 methyl group. Thus, the major product is *cis*-1,3-dimethylcyclopentane.

6.36 Hydrogen can add to either the top face or the bottom face of the double bond. Syn addition to the double bond requires that the methyl groups in the product be cis.

6.37 3-Carene can in theory undergo hydrogenation to give either *cis*-carane or *trans*-carane.

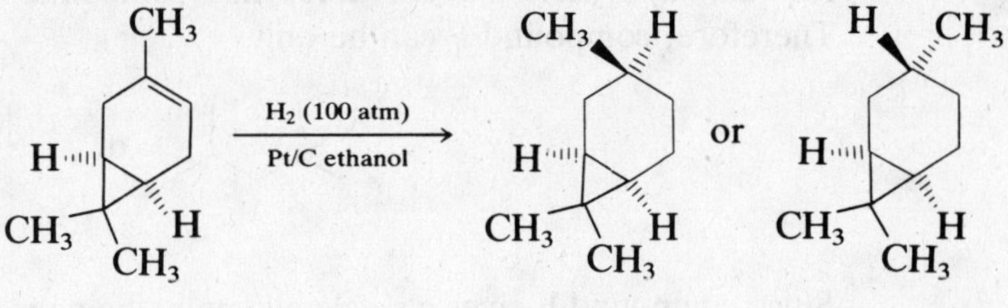

cis-Carane (98%)   trans-Carane

The exclusive product is *cis*-carane, since it corresponds to transfer of hydrogen from the less hindered side.

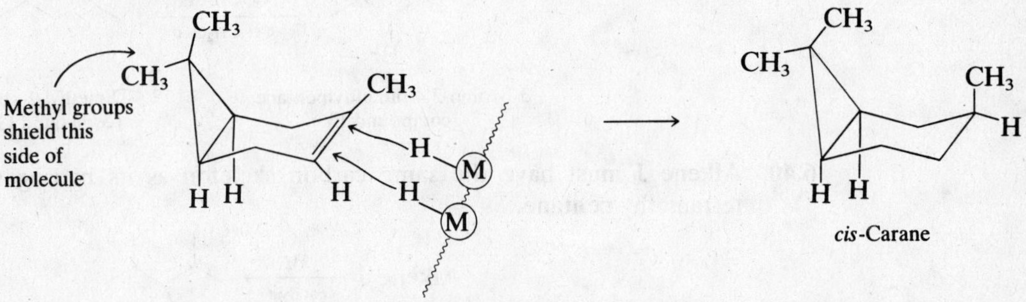

cis-Carane

Stop

6.38 The carbon skeleton is revealed by the hydrogenation experiment. Compounds D and E must have the same carbon skeleton as 3-ethylpentane.

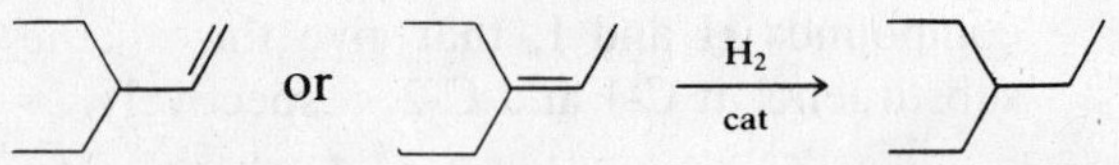

There are three alkyl bromides having this carbon skeleton, namely, 1-bromo-3-ethylpentane, 2-bromo-3-ethylpentane, and 3-bromo-3-ethylpentane. Of these three only 2-bromo-3-ethylpentane will give two alkenes on dehydrobromination.

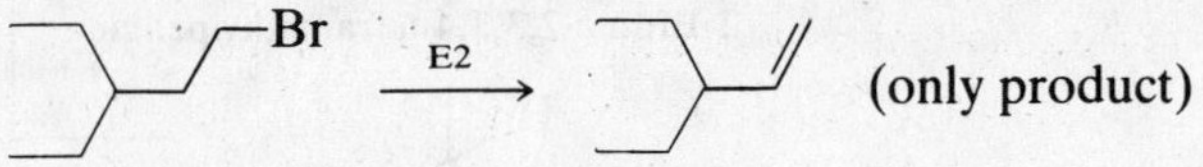

1-Bromo-3-ethylpentane

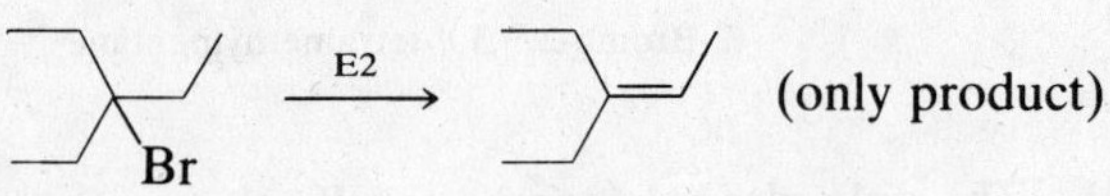

3-Bromo-3-ethylpentane

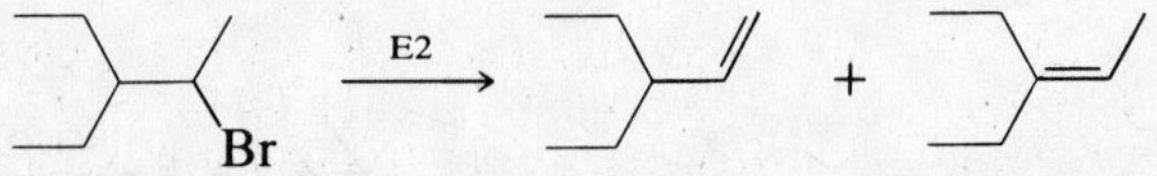

2-Bromo-3-ethylpentane   3-Ethyl-1-pentene   3-Ethyl-2-pentene

Therefore, compound C must be 2-bromo-3-ethylpentane. Dehydrobromination of C will follow the Zaitsev rule, so that the major alkene (compound D) is 3-ethyl-2-pentene and the minor alkene (compound E) is 3-ethyl-1-pentene.

6.39 The information that compound G gives 2,4-dimethylpentane on catalytic hydrogenation establishes its carbon skeleton.

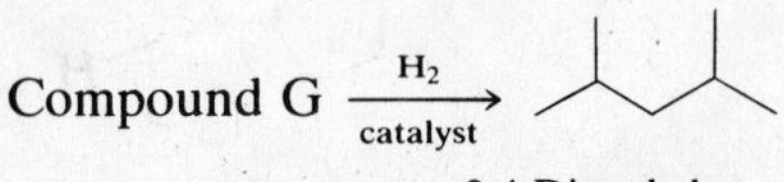

2,4-Dimethylpentane

Compound G is an alkene derived from compound F—an alkyl bromide of molecular formula $C_7H_{15}Br$. We are told that compound F is not a primary alkyl bromide. Therefore, compound F can be only:

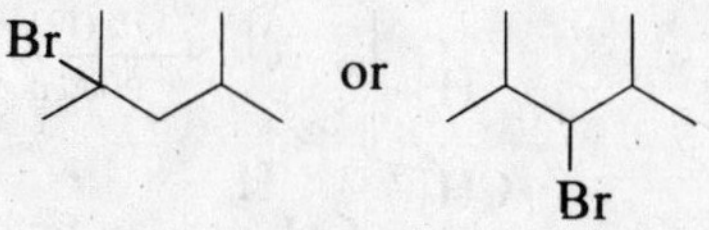

Since compound F gives a single alkene on being treated with sodium ethoxide in ethanol, it can only be 3-bromo-2,4-dimethylpentane, and compound G must be 2,4-dimethyl-2-pentene.

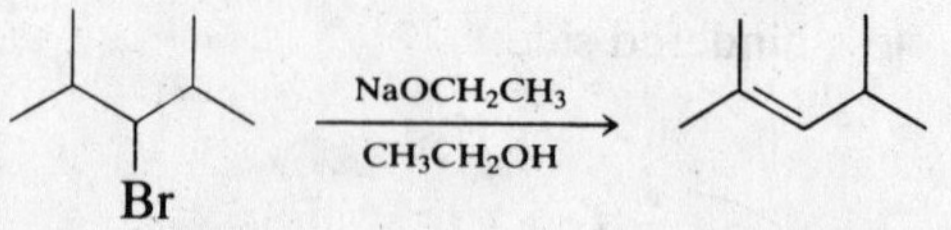

3-Bromo-2,4-dimethylpentane   2,4-Dimethyl-2-pentene
(compound F)   (compound G)

6.40 Alkene J must have the same carbon skeleton as its hydrogenation product, 2,3,3,4-tetramethylpentane.

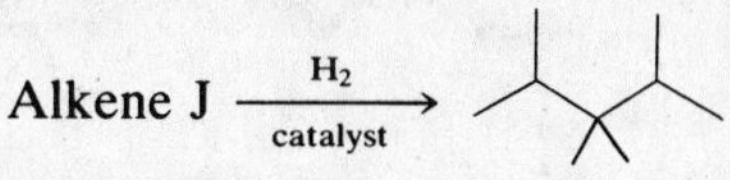

2,3,3,4-Tetramethylpentane

Therefore alkene J can only be 2,3,3,4-tetramethyl-1-pentene. The two alkyl bromides, compounds H and I, that give this alkene on dehydrobromination have their bromine substituents at C-1 and C-2, respectively.

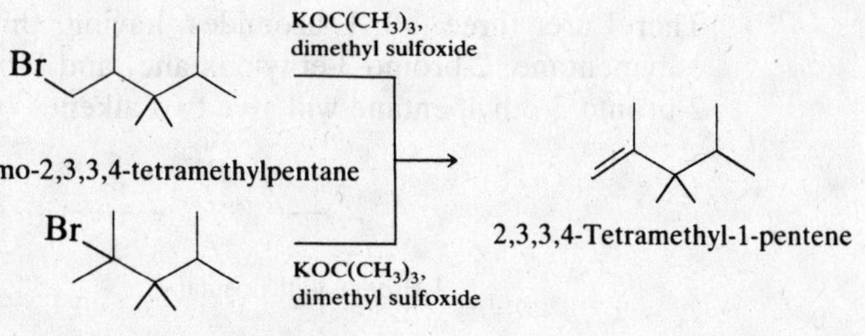

1-Bromo-2,3,3,4-tetramethylpentane

2,3,3,4-Tetramethyl-1-pentene

2-Bromo-2,3,3,4-tetramethylpentane

6.41 The only alcohol (compound K) that can undergo acid-catalyzed dehydration to alkene L without rearrangement is the one shown in the equation

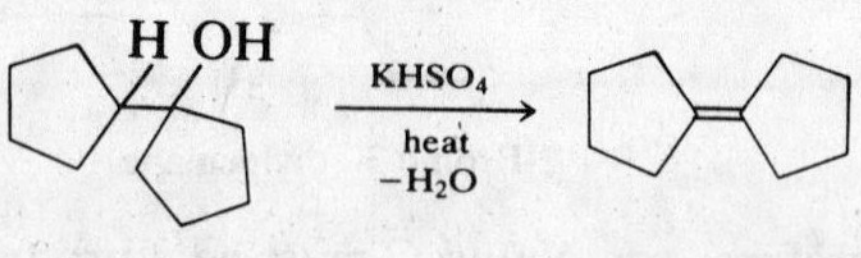

Alcohol K   Alkene L

Dehydration of alcohol K also yields an isomeric alkene under these conditions.

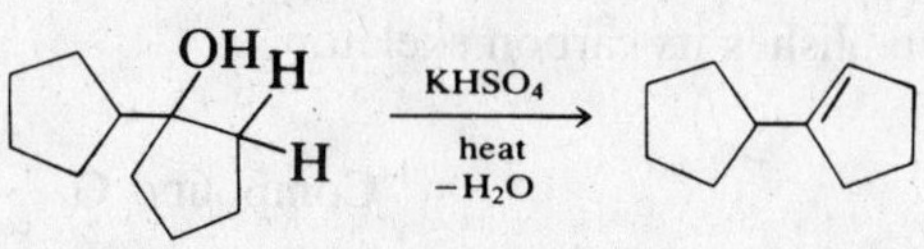

Alcohol K   Alkene M

6.42 Electrophilic addition of hydrogen iodide should occur in accordance with Markovnikov's rule.

$$CH_2=CHC(CH_3)_3 \underset{\text{KOH, n-PrOH}}{\overset{\text{HI}}{\rightleftharpoons}} CH_3\underset{\underset{I}{|}}{C}HC(CH_3)_3$$

3,3-Dimethyl-1-butene   2-Iodo-3,3-dimethylbutane

Treatment of 2-iodo-3,3-dimethylbutane with alcoholic potassium hydroxide should bring about E2 elimination to regenerate the starting alkene. Hence, compound N is 2-iodo-3,3-dimethylbutane.

The carbocation intermediate formed in the addition of hydrogen iodide to the alkene is one which can rearrange by a methyl group migration.

$$CH_2=CHC(CH_3)_3 + H^+ \longrightarrow CH_3-\overset{+}{C}H-\underset{\underset{CH_3}{|}}{\overset{\overset{CH_3}{|}}{C}}CH_3 \longrightarrow CH_3-CH-\underset{\underset{CH_3}{|}}{\overset{\overset{CH_3}{|}}{\overset{+}{C}}}CH_3$$

$$\downarrow I^- \qquad\qquad\qquad \downarrow I^-$$

$$\underset{\underset{I}{|}}{CH_3CH}-C(CH_3)_3 \qquad\qquad (CH_3)_2CH-\underset{\underset{I}{|}}{C}(CH_3)_2$$

Compound N   Compound O
                              (2-iodo-2,3-dimethylbutane)

Therefore, a likely candidate for compound O is the one with a rearranged carbon skeleton, 2-iodo-2,3-dimethylbutane. This is confirmed by the fact that compound O undergoes elimination to give 2,3-dimethyl-2-butene.

$$(CH_3)_2CH-\underset{\underset{I}{|}}{C}(CH_3)_2 \overset{\text{E2}}{\longrightarrow} (CH_3)_2C=C(CH_3)_2$$

Compound O   2,3-Dimethyl-2-butene

6.43 The ozonolysis data are useful in quickly identifying alkenes P and Q.

$$Compound\ P \longrightarrow H\overset{\overset{O}{\|}}{C}H + (CH_3)_3C\overset{\overset{O}{\|}}{C}C(CH_3)_3$$

Therefore, compound P is 2-*tert*-butyl-3,3-dimethyl-1-butene.

$$(CH_3)_3C\overset{\overset{CH_2}{\|}}{C}C(CH_3)_3 \qquad Compound\ P$$

$$Compound\ Q \longrightarrow H\overset{\overset{O}{\|}}{C}H + CH_3\overset{\overset{O}{\|}}{C}-\underset{\underset{CH_3}{|}}{\overset{\overset{CH_3}{|}}{C}}-C(CH_3)_3$$

Therefore, compound Q is 2,3,3,4,4-pentamethyl-1-pentene.

$$CH_3\overset{\overset{H_2C}{\|}}{C}-\underset{\underset{CH_3}{|}}{\overset{\overset{CH_3}{|}}{C}}-C(CH_3)_3 \qquad Compound\ Q$$

Compound Q has a carbon skeleton different from the alcohol which produced it by dehydration. We are therefore led to consider a carbocation rearrangement.

$$(CH_3)_3C-\underset{\underset{OH}{|}}{\overset{\overset{CH_3}{|}}{C}}-C(CH_3)_3 \xrightarrow{H^+} CH_3\overset{\overset{H_3C\ \ CH_3}{|\ \ \ |}}{\underset{\underset{H_3C}{|}}{C}}-\overset{+}{C}C(CH_3)_3 \longrightarrow (CH_3)_3C-\overset{\overset{CH_2}{||}}{C}C(CH_3)_3$$

Compound P

methyl migration

$$CH_3\overset{\overset{H_3C\ \ CH_3}{|\ \ \ |}}{\underset{\underset{CH_3}{|}}{\overset{+}{C}}}-C-C(CH_3)_3 \longrightarrow CH_3\overset{\overset{H_2C\ \ CH_3}{||\ \ \ |}}{\underset{\underset{CH_3}{|}}{C}}-C-C(CH_3)_3$$

Compound Q

6.44 The important clue to deducing the structures of R and S is the ozonolysis product T. Remembering that the two carbonyl carbons of T must have been joined by a double bond in the precursor S, we write

These two carbons must have been connected by a double bond

Compound T

Compound S

The tertiary bromide which gives compound S on dehydrobromination is 1-methylcyclohexyl bromide.

$$\xrightarrow[\text{CH}_3\text{CH}_2\text{OH}]{\text{NaOCH}_2\text{CH}_3}$$

Compound R

Compound S

When tertiary halides are treated with base, they undergo E2 elimination. The regioselectivity of elimination of tertiary halides follows the Zaitsev rule.

6.45 Since santene and 1,3-diacetylcyclopentane (compound U) contain the same number of carbon atoms, the two carbonyl carbons of the diketone must have been connected by a double bond in santene. Therefore, the structure of santene must be

more appropriately represented as

6.46 (*a*) Compound V contains 9 of the 10 carbons and 14 of the 16 hydrogens of sabinene. Ozonolysis has led to the separation of one carbon and two hydrogens from the rest of the molecule. The carbon and the two hydrogens must have been lost as formaldehyde, $H_2C=O$. This H_2C unit was originally doubly bonded to the carbonyl carbon of compound V. Therefore, sabinene must have the structure shown in the equation representing its ozonolysis:

Sabinene   Compound V   Formaldehyde

(*b*) Compound W contains all 10 of the carbons and all 16 of the hydrogens of Δ^3-carene. The two carbonyl carbons of compound W must have been linked by a double bond in Δ^3-carene.

Δ^3-Carene   Compound W

6.47 Since the molecular formula of the sex attractant of the female housefly is $C_{23}H_{46}$, it has a SODAR equal to 1. It takes up one mole of hydrogen on catalytic hydrogenation and so must have one double bond and no rings. The position of the double bond is revealed by the ozonolysis data.

$$C_{23}H_{46} \xrightarrow[\text{2. H}_2\text{O, Zn}]{\text{1. O}_3} CH_3(CH_2)_7\overset{\overset{\displaystyle O}{\|}}{C}H + CH_3(CH_2)_{12}\overset{\overset{\displaystyle O}{\|}}{C}H$$

An unbranched 9-carbon unit and an unbranched 14-carbon unit make up the carbon skeleton, and these two units must be connected by a double bond. Therefore, the housefly sex attractant has the constitution:

$$CH_3(CH_2)_7CH{=\!=}CH(CH_2)_{12}CH_3 \qquad \text{9-Tricosene}$$

The data cited in the problem do not permit the stereochemistry of this natural product to be determined.

6.48 The hydrogenation data tell us that $C_{19}H_{38}$ contains one double bond and has the same carbon skeleton as 2,6,10,14-tetramethylpentadecane. We locate the double bond at C-2 on the basis of the fact that acetone, $(CH_3)_2C{=\!=}O$, is obtained on ozonolysis. The structures of the natural product and the aldehyde produced on its ozonolysis are as follows:

Ozonolysis cleaves molecule here   Aldehyde obtained on ozonolysis

6.49 Since $H\overset{\overset{\displaystyle O}{\|}}{C}CH_2\overset{\overset{\displaystyle O}{\|}}{C}H$ is one of the products of its ozonolysis, the sex attractant of the arctiid moth must contain the unit ${=\!=}CHCH_2CH{=\!=}$. This unit must be bonded to an unbranched

12-carbon unit at one end and an unbranched six-carbon unit at the other in order to give $CH_3(CH_2)_{10}CH{=}O$ and $CH_3(CH_2)_4CH{=}O$ on ozonolysis.

$$CH_3(CH_2)_{10}CH\!\vdots\!CHCH_2CH\!\vdots\!CH(CH_2)_4CH_3$$

Sex attractant of arctiid moth
(dotted lines show positions of cleavage on ozonolysis)

1. O_3
2. H_2O, Zn

$$\underset{\|}{\overset{O}{CH_3(CH_2)_{10}CH}} + \underset{\|}{\overset{O}{HCCH_2CH}} + \underset{\|}{\overset{O}{HC(CH_2)_4CH_3}}$$

The stereochemistry of the double bonds cannot be determined on the basis of the available information.

SELF-TEST

PART A

A-1. How many different alkenes will yield 2,3-dimethylpentane on catalytic hydrogenation? Draw their structures and name them.

A-2. Write structural formulas for the reactant, reagents, or product omitted from each of the following:

(a) $(CH_3)_2C{=}CHCH_3 \xrightarrow{H_2SO_4(dilute)}$?

(b) [cyclohexane ring]$={CH_2} \xrightarrow{?}$ [cyclohexane ring]$-CH_2Br$

(c) ? $\xrightarrow[\text{2. } H_2O, Zn]{\text{1. } O_3}$ [cyclohexanone ring with side chain]$-CH_2CH_2CH_2-\underset{\|}{\overset{}{CH}}{=}O$

(d) $CH_3CH_2\underset{\underset{CH_3}{|}}{C}{=}CH_2 \xrightarrow[H_2O]{Br_2}$?

A-3. Provide a sequence of reaction steps to carry out the following conversions. Write the structure of each intermediate product.

(a) [cyclohexane ring with CH_3 and OH] $\longrightarrow$ [cyclohexane ring with CH_3 and OH]

(b) $CH_3CH_2\underset{\underset{Cl}{|}}{CH}CH(CH_3)_2 \longrightarrow CH_3CH_2\underset{\overset{O}{\diagup\diagdown}}{CH}{-}C(CH_3)_2$

(c) $(CH_3)_3C\underset{\underset{Br}{|}}{C}HCH_3 \longrightarrow (CH_3)_3CCH_2CH_2Br$

A-4. A hydrocarbon A (C_6H_{12}) undergoes reaction with HBr to yield compound B ($C_6H_{13}Br$). Treatment of B with sodium ethoxide in ethanol yields C, an isomer of

A. Reaction of C with ozone followed by treatment with water and zinc gives acetone, $(CH_3)_2C\!=\!O$, as the only organic product. Provide structures for A, B, and C, and outline the reaction pathway.

A-5. Provide a detailed mechanism describing the reaction of 1-butene with HBr in the presence of peroxides.

A-6. Chlorine reacts with an alkene to give the 2,3-dichlorobutane isomer whose structure is shown. What are the structure and name of the alkene? Outline a mechanism for the reaction.

A-7. Write a structural formula, including stereochemistry, for the compound formed from *cis*-3-hexene on treatment with peroxyacetic acid.

A-8. What two alkenes give 2-chloro-2-methylbutane on reaction with hydrogen chloride?

A-9. The physiologically active constituent of marijuana is tetrahydrocannabinol, THC. THC has the molecular formula $C_{21}H_{30}O_2$. Exhaustive catalytic hydrogenation results in uptake of four moles of hydrogen. THC contains _____ rings and _____ double bonds.

PART B

B-1. The heat of hydrogenation of substance I is 119 kJ/mol (28.5 kcal/mol). Which of the following values most likely represents the heat of hydrogenation of compound II?

               I   II

(*a*) 127 kJ/mol (30.3 kcal/mol)   (*c*) 123 kJ/mol (29.5 kcal/mol)
(*b*) 117 kJ/mol (26.7 kcal/mol)   (*d*) 119 kJ/mol (28.5 kcal/mol)

B-2. The product from the reaction of 1-pentene with Cl_2 in H_2O is named:
(*a*) 1-Chloro-2-pentanol   (*c*) 1-Chloro-1-pentanol
(*b*) 2-Chloro-2-pentanol   (*d*) 2-Chloro-1-pentanol

B-3. In the reaction of a reagent such as HBr with an alkene, the first step of the reaction is the _____ to the alkene.
(*a*) Fast addition of an electrophile
(*b*) Fast addition of a nucleophile
(*c*) Slow addition of an electrophile
(*d*) Slow addition of a nucleophile

B-4. The reaction sequence

gives as the major product:

(a)   (c)

(b)   (d)

B-5. Markovnikov's rule "works" because:
(a) The most stable transition state is the one leading to the more substituted carbocation.
(b) The nucleophile adds during the second step of the ionic reaction.
(c) The electrophile adds to the less substituted end of the double bond.
(d) All of these are true.

B-6. Treatment of 2-methyl-2-butene with HBr in the presence of peroxide yields
(a) A primary alkyl bromide
(b) A secondary alkyl bromide
(c) A tertiary alkyl bromide
(d) A vicinal dibromide

B-7. The strongest evidence for the formation of a bridged bromonium ion as an intermediate in the addition of Br_2 to an alkene is:
(a) Markovnikov's rule
(b) Zaitsev's rule
(c) The regioselectivity of the reaction
(d) The stereospecificity of the reaction

B-8. The reaction

$$(CH_3)_2C{=}CH_2 + Br\cdot \longrightarrow (CH_3)_2\dot{C}{-}CH_2Br$$

is an example of a(n) _____ step in a radical chain reaction.
(a) Initiation   (c) Termination
(b) Propagation   (d) Heterolytic cleavage

B-9. A compound having a molecular formula of $C_{20}H_{36}$ is inert to catalytic hydrogenation. Which of the following statements is true?
(a) The substance has at least one double bond and two rings.
(b) The substance is acyclic (i.e., no rings are present).
(c) The substance has two rings.
(d) The substance has three rings.

B-10. A compound having the formula $C_{10}H_{10}Cl_2O$ has a SODAR value of _____. Three moles of hydrogen are consumed on catalytic hydrogenation. The substance has _____ double bonds and _____ rings.

	SODAR	Double bonds	Rings
(a)	6	3	3
(b)	6	6	0
(c)	5	3	2
(d)	5	2	3

STEREOCHEMISTRY

IMPORTANT TERMS AND CONCEPTS

Molecular Chirality (Secs. 7.1 to 7.3) Any molecule which *is not superposable* on its mirror image is said to be *chiral*; a molecule which *is superposable* on its mirror image is *achiral.* The two nonsuperposable mirror images are *stereoisomers* of each other; that is, they differ in the arrangement of their atoms in space. Molecules that are nonsuper-posable mirror images of each other are said to be *enantiomers* of each other. Stereoisomers that are not mirror images are *diastereomers* of each other; for example, (*E*)- and (*Z*)-2-butene are diastereomers.

The most common chiral organic molecules possess a carbon atom having four different groups attached, which is known as the *stereogenic center.* For example, carbon-2 of 2-chlorobutane is the stereogenic center in this molecule.

Recognizing the presence or absence of *elements of symmetry* is important in determining whether a molecule is chiral. Any structure which contains a *plane of symmetry* will be *achiral,* and likewise any structure which contains a *center of symmetry* will also be achiral. Note that only one of these symmetry elements need be present and that the plane is usually easier to recognize when evaluating a structural formula.

Optical Activity (Sec. 7.4) An *optically active* substance is one which *rotates the plane of polarized light* as measured in an instrument known as a *polarimeter.* A substance which does not rotate the plane of polarized light is said to be *optically inactive.*

In order for a substance to exhibit optical activity, *one enantiomer of a chiral substance* must be present in excess over the other. The *observed rotation* is denoted by α. An equal mixture of the two enantiomers of a chiral substance is known as a *racemic mixture* and is optically inactive. The percent enantiomeric excess, or *optical purity,* of a substance is defined as

Optical purity = percent of one enantiomer − percent of other enantiomer

The two enantiomers of a chiral substance differ *only* in the direction of rotation of polarized light. That is, if one enantiomer has an optical rotation of $+\alpha$, that of the other will be $-\alpha$. The positive and negative rotations were formerly termed *dextrorotatory* and *levorotatory* and abbreviated *d* and *l*. Current custom favors use of $(+)$ and $(-)$. A racemic mixture is usually referred to as $(\pm)$.

In order to catalog the optical activity of a substance, the *specific rotation* $[\alpha]$ has been defined as follows:

$$[\alpha] = \frac{100\alpha}{c \cdot l}$$

where α = observed rotation

c = concentration of solution, g/100 mL

l = length of polarimeter tube, decimeters (dm)

Absolute and Relative Configuration (Secs. 7.5, 7.6) The precise arrangement of substituents at a stereogenic center is its *absolute configuration*. The *relative configuration* of a center indicates the spatial relationship when one compound (or stereogenic center) is compared with another. It is important to note that there is no definite relationship between sign of rotation and absolute configuration for an optically active substance.

The absolute configuration of a stereogenic center is best specified by using the *Cahn-Ingold-Prelog,* or *R–S*, system. This method was first encountered in Chapter 5 for describing *E* and *Z* alkenes. The rules are as follows:

1. Identify the substituents attached to the stereogenic center and rank them in order of decreasing precedence.
2. Orient the molecule so that the *lowest*-ranking substituent is pointing away from you. Note that this will often be hydrogen.
3. The three highest-ranking substituents will appear, when viewed as in step 2, pointing out as if they were spokes of a wheel. If the order of decreasing precedence is *clockwise,* the chiral center is *R*. If the order is *counterclockwise* (anticlockwise), the configuration is *S*.

In the following example, the numbers represent group rankings according to the Cahn-Ingold-Prelog rules, where 4 indicates the highest-ranked group and 1 the lowest.

(*R*)-2-Chlorobutane   (*S*)-2-Chlorobutane

From the example, an important relationship may be seen: *the enantiomer of a molecule having the R configuration is S, and vice versa.*

Fischer Projection Formulas (Sec. 7.7) The Fischer projection is a useful abbreviated method for representing stereochemical information. Instead of wedges and dashes, a vertical line represents bonds projecting into the page (away from the reader) and horizontal lines represent bonds projecting above the page (toward the reader).

Note that the stereogenic carbon is not explicitly drawn in a Fischer projection.

Stereochemistry of Chemical Reactions (Sec. 7.9) Many chemical reactions produce a chiral product from an achiral starting material. The chiral product is, however, *optically inactive*, being a *racemic mixture*; that is, equal amounts of the two enantiomers have been produced. As a general rule, optically inactive products result when optically inactive substrates react with optically inactive reagents. This rule holds whether the reactants and reagents are achiral or are racemic mixtures of a chiral substance.

Optically active reactants may or may not lead to formation of optically active products, depending on the nature of the reaction. When an achiral intermediate forms in the reaction, optically inactive products (a racemic mixture) will result. Specific examples of reactions involving chiral reactants will be discussed in the next chapter.

Molecules with Two or More Stereogenic Centers (Secs 7.10 to 7.12) The maximum number of stereoisomers for a given constitution is 2^n where n equals the number of stereogenic centers. Each stereogenic center may be R or S. The number of distinct stereoisomers for a given compound may be less than 2^n, however. If a plane of symmetry is present in the molecule, the mirror images of such a stereoisomer *will* be superposable on and identical to each other. These molecules, which are achiral, may contain two or more stereogenic centers, and are known as *meso forms*. The achiral meso form is a *diastereomer* with respect to each of the chiral stereoisomers, as it is a stereoisomer that is not an enantiomer.

2,3-Dibromobutane, shown as follows, provides an example.

Reactions Which Form Diastereomers (Sec. 7.13) The addition reaction of halogens with alkenes is an example of a *stereospecific reaction*; that is, stereoisomeric reactants yield stereoisomeric products. The reaction of Br_2 with the isomeric 2-butenes yields products that are diastereomers of each other. As shown in text Figure 7.15, the Z, or cis, isomer reacts to give a racemic mixture of enantiomers; the E, or trans, isomer yields the meso form. Epoxidation of alkenes provides another example of a stereospecific reaction which yields diastereomeric products.

Resolution of Enantiomers (Sec. 7.14) The separation of a racemic mixture into its enantiomeric components is known as *resolution*. The racemic substance is allowed to react with a single enantiomer of a chiral reagent, frequently a naturally occurring acid or base. The product is a mixture of *diastereomers,* which may have physical properties (such as solubility) sufficiently different to allow their separation. The appropriate chemical steps then allow the separated enantiomers of the original substance to be recovered in pure enantiomeric form.

Stereogenic Centers Other Than Carbon (Sec. 7.16) The rules employed for identifying stereogenic carbon atoms apply to other atoms as well. In the case of nitrogen, pyramidal inversion is too fast for individual enantiomers to be isolated. The absolute configuration of any stereogenic center may be specified with the Cahn-Ingold-Prelog rules, with the provision that an unshared electron pair is considered to be the lowest-ranking substituent.

SOLUTIONS TO TEXT PROBLEMS

7.1 (c) Carbon-2 is a stereogenic center in 1-bromo-2-methylbutane, as it has four different substituents: H, CH_3, CH_3CH_2, and $BrCH_2$.

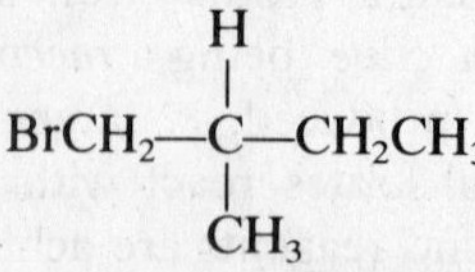

$$BrCH_2 \overset{\overset{\displaystyle H}{|}}{\underset{\underset{\displaystyle CH_3}{|}}{C}} CH_2CH_3$$

(d) There are no stereogenic centers in 2-bromo-2-methylbutane.

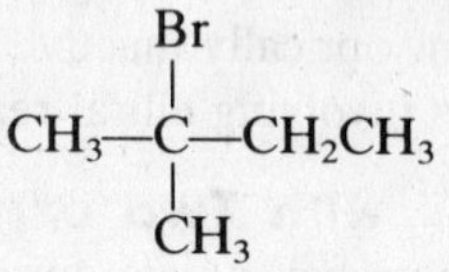

$$CH_3 \overset{\overset{\displaystyle Br}{|}}{\underset{\underset{\displaystyle CH_3}{|}}{C}} CH_2CH_3$$

7.2 (c) Carbon-2 is a stereogenic center in 1,1,2-trimethylcyclobutane.

A stereogenic center; the four substituents to which it is directly bonded [H, CH_3, CH_2, and $C(CH_3)_2$] are all different from one another

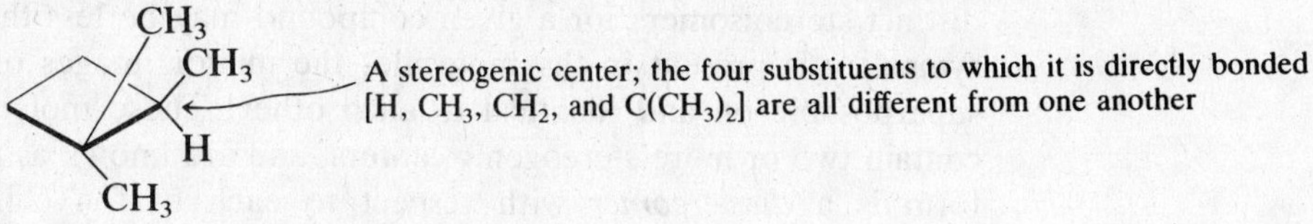

(d) 1,1,3-Trimethylcyclobutane has no stereogenic centers.

Not a stereogenic center; two of its substituents are the same

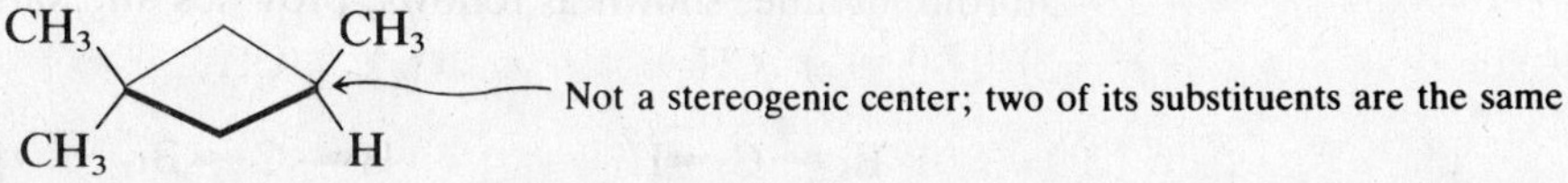

7.3 (b) There are *two* planes of symmetry in (Z)-1,2-dichloroethene, of which one is the plane of the molecule and the second bisects the carbon-carbon bond. There is no center of symmetry. The molecule is achiral.

planes of symmetry

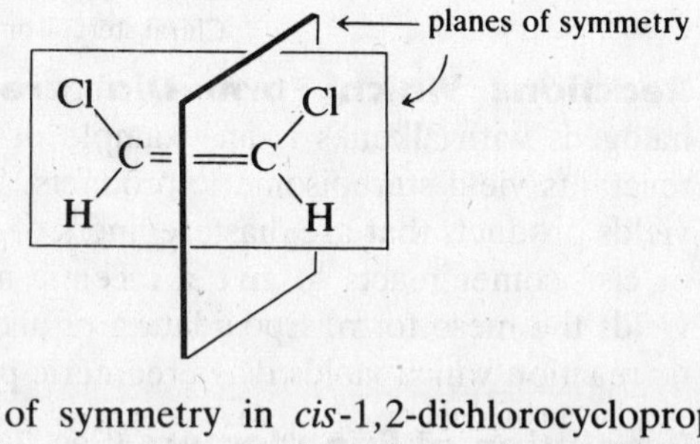

(c) There is a plane of symmetry in *cis*-1,2-dichlorocyclopropane which bisects the C-1—C-2 bond and passes through C-3. The molecule is achiral.

plane of symmetry

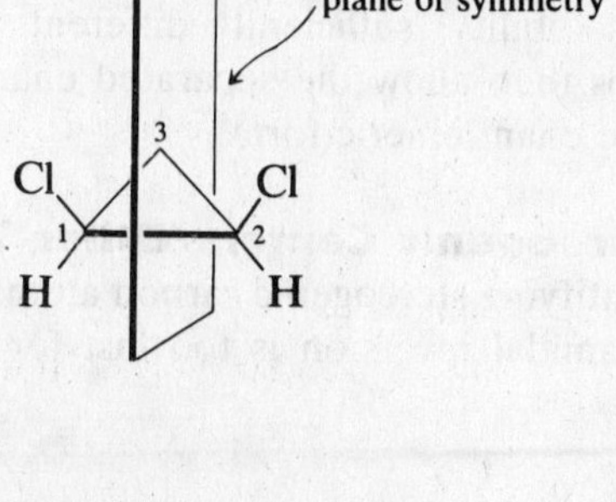

(d) *trans*-1,2-Dichlorocyclopropane has neither a plane of symmetry nor a center of symmetry. Its two mirror images cannot be superposed on one another. The molecule is chiral.

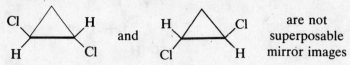

are not superposable mirror images

7.4 The equation relating specific rotation $[\alpha]$ to observed rotation α is

$$[\alpha] = \frac{100\alpha}{cl}$$

The concentration c is expressed in grams per 100 mL and the length l of the polarimeter tube in decimeters. Since the problem specifies the concentration as 0.3 g per 15 mL and the path length as 10 cm, the specific rotation $[\alpha]$ is:

$$[\alpha] = \frac{100(-0.78°)}{100\left(\dfrac{0.3\,\text{g}}{15\,\text{mL}}\right)\left(\dfrac{10\,\text{cm}}{10\,\text{cm/dm}}\right)}$$

$$= -39°$$

7.5 From the previous problem, the specific rotation of natural cholesterol is $[\alpha] = -39°$. The mixture of natural (−)-cholesterol and synthetic (+)-cholesterol specified in this problem has a specific rotation $[\alpha]$ of −13°.

Optical purity = % (−)-cholesterol − % (+)-cholesterol

33.3% = % (−)-cholesterol − [100 − % (−)-cholesterol]

133.3% = 2 [% (−)-cholesterol]

66.7% = % (−)-cholesterol

The mixture is two-thirds natural (−)-cholesterol and one-third synthetic (+)-cholesterol.

7.6 (b) The solution to this problem is exactly analogous to the sample solution given in the text to part (a).

$$
\begin{array}{c}
CH_3 \quad H \\
\diagdown \; \vdots \\
C{-}CH_2F \\
\diagup \\
CH_3CH_2
\end{array}
$$
(+)-1-Fluoro-2-methylbutane

Order of precedence: $CH_2F > CH_3CH_2 > CH_3 > H$

The lowest-ranked substituent (H) at the stereogenic center points away from the reader, and so the molecule is oriented properly as drawn. The three higher-ranked substituents trace a clockwise path from CH_2F to CH_2CH_3 to CH_3.

$$
\begin{array}{c}
CH_3 \qquad CH_2F \\
\diagdown \quad \diagup \\
\overset{\curvearrowright}{C} \\
| \\
CH_2CH_3
\end{array}
$$

The absolute configuration is R; the compound is (R)-1-fluoro-2-methylbutane.

(c) The highest-ranked substituent at the stereogenic center of 1-bromo-2-methylbutane is CH_2Br, and the lowest-ranked substituent is H. Of the remaining two, ethyl outranks methyl.

Order of precedence: $CH_2Br > CH_2CH_3 > CH_3 > H$

The lowest-ranking substituent (H) is directed toward the reader and therefore the molecule needs to be reoriented so that H points in the opposite direction.

$$
\underset{CH_3CH_2}{\overset{H}{\underset{\displaystyle}{\overset{\displaystyle CH_3}{C}}}}{\!-\!CH_2Br} \qquad \circlearrowright \qquad BrCH_2{\!-\!}\underset{CH_2CH_3}{\overset{H}{\overset{\displaystyle CH_3}{C}}}
$$

turn 180°

(+)-1-Bromo-2-methylbutane

The three highest-ranking substituents trace a counterclockwise path when the lowest-ranked substituent is held away from the reader.

$$
\underset{CH_2CH_3}{\overset{BrCH_2 \quad CH_3}{C}}
$$

The absolute configuration is *S*, and thus the compound is (*S*)-1-bromo-2-methylbutane.

(*d*) The highest-ranked substituent at the stereogenic center of 3-buten-2-ol is the hydroxyl group, and the lowest-ranked substituent is H. Of the remaining two, vinyl outranks methyl.

Order of precedence: $HO > CH_2{=}CH > CH_3 > H$

The lowest-ranking substituent (H) is directed away from the reader. We see that the order of decreasing precedence appears in a counterclockwise manner.

$$
\underset{HO}{\overset{CH_3 \quad H}{C}}{\!-\!CH_2{=}CH_2} \qquad\qquad \underset{OH}{\overset{CH_3 \quad CH{=}CH_2}{C}}
$$

(+)-3-Buten-2-ol

The absolute configuration is *S*, and the compound is (*S*)-3-buten-2-ol.

7.7 (*b*) The stereogenic center is the carbon which bears the methyl group. Its substituents are:

$$
{-}CF_2CH_2 > {-}CH_2CF_2 > CH_3 > \qquad H
$$

Highest priority   Lowest priority

When the lowest-priority substituent points away from the reader, the remaining three must appear in descending order of precedence in a counterclockwise fashion in the *S* enantiomer. Therefore, (*S*)-1,1-difluoro-2-methylcyclopropane is

$$
\underset{F \quad H}{\overset{CH_3}{\underset{F}{\triangle}}}
$$

7.8 (*b*) The Fischer projection of (*R*)-(+)-1-fluoro-2-methylbutane is analogous to that of the alcohol in part (*a*). The only difference in the two is that fluorine has replaced hydroxyl as a substituent at C-1.

$$
\underset{CH_3CH_2}{\overset{CH_3 \quad H}{C}}{\!-\!CH_2F} \quad \begin{array}{c}\text{is the}\\\text{same as}\end{array} \quad FCH_2{\blacktriangleright}\underset{CH_2CH_3}{\overset{CH_3}{C}}{\blacktriangleleft}H \quad \begin{array}{c}\text{which becomes the}\\\text{Fischer projection}\end{array} \quad FCH_2{\!-\!}\underset{CH_2CH_3}{\overset{CH_3}{|}}{\!-\!}H
$$

(*c*) The way the structural formula of (*S*)-(+)-1-bromo-2-methylbutane is presented corresponds almost exactly to what is required for a Fischer projection. The methyl group is at the top and the ethyl group at the bottom and both are directed away from you.

(*d*) Here we need to view the molecule from behind the page in order to write the Fischer projection formula of (*S*)-(+)-3-buten-2-ol.

7.9 The reaction which leads to 1,4-dichloro-2-methylbutane from (*S*)-(+)-1-chloro-2-methylbutane is

(*S*)-(+)-1-Chloro-2-methylbutane   (*S*)-1,4-Dichloro-2-methylbutane

The reaction does not involve any of the bonds to the stereogenic center, and so the arrangement of the various groups must be the same as in the starting material. Its configuration is determined to be *S* by the Cahn-Ingold-Prelog system based on the order of substituent priorities:

$$CH_2Cl \quad > CH_2CH_2Cl > CH_3 > \qquad H$$

Highest priority   Lowest priority

Both the starting material and the product are chiral. Since the starting material was a single enantiomer and no bonds were made or broken at its stereogenic center during the reaction, the product is optically active.

7.10 The erythro stereoisomers are characterized by Fischer projections in which analogous substituents, in this case OH and NH_2, are on the same side when the carbon chain is vertical. There are two erythro stereoisomers that are enantiomers of one another:

Erythro   Erythro

Analogous substituents are on opposite sides in the threo isomer:

Threo   Threo

7.11   There are four stereoisomeric forms of 3-amino-2-butanol:

$(2R,3R)$ and its enantiomer $(2S,3S)$
$(2R,3S)$ and its enantiomer $(2S,3R)$

In the text we are told that the $(2R,3R)$ stereoisomer is a liquid. Its enantiomer $(2S,3S)$ has the same physical properties and so must also be a liquid. The text notes that the $(2R,3S)$ stereoisomer is a solid (mp 49°C). Therefore, its enantiomer $(2S,3R)$ must be the other stereoisomer that is a crystalline solid.

7.12   Examine the structural formula of each compound for equivalently substituted stereogenic centers. The only one capable of existing in a meso form is 2,4-dibromopentane.

$$CH_3CHCH_2CHCH_3$$
$$\quad | \qquad\qquad |$$
$$\quad Br \qquad\quad Br$$

2,4-Dibromopentane

Fischer projection of
meso-2,4-dibromopentane

None of the other compounds has equivalently substituted stereogenic centers. No meso forms are possible for:

$$CH_3CHCHCH_2CH_3$$
$$\quad | \;\; |$$
$$\quad Br\; Br$$

2,3-Dibromopentane

$$\qquad\;\; OH$$
$$\qquad\;\; |$$
$$CH_3CHCHCH_2CH_3$$
$$\qquad\qquad |$$
$$\qquad\qquad Br$$

3-Bromo-2-pentanol

$$CH_3CHCH_2CHCH_3$$
$$\quad | \qquad\qquad |$$
$$\quad OH \qquad\quad Br$$

4-Bromo-2-pentanol

7.13   There is a plane of symmetry in the cis stereoisomer of 1,3-dimethylcyclohexane, and so it is an achiral substance—it is a meso form.

Plane of symmetry passes through
C-2 and C-5 and bisects the ring.

The trans stereoisomer is chiral. It is not a meso form.

7.14   A molecule with three stereogenic centers has 2^3, or 8, stereoisomers. The eight combinations of R and S stereogenic centers are:

	Stereogenic center 1 2 3		Stereogenic center 1 2 3
Isomer 1	$R\;R\;R$	Isomer 5	$S\;S\;S$
Isomer 2	$R\;R\;S$	Isomer 6	$S\;S\;R$
Isomer 3	$R\;S\;R$	Isomer 7	$S\;R\;S$
Isomer 4	$S\;R\;R$	Isomer 8	$R\;S\;S$

7.15   2-Hexuloses have three stereogenic centers. They are marked with asterisks in the structural formula.

$$\qquad\qquad\quad O \quad\; OH$$
$$\qquad\qquad\quad \| \quad * \quad\;\; *$$
$$HOCH_2CCHCHCHCH_2OH$$
$$\qquad\qquad\;\; | \quad * \quad |$$
$$\qquad\qquad\;\; OH \quad\;\; OH$$

No meso forms are possible, and so there are a total of 2^3, or 8, stereoisomeric 2-hexuloses.

7.16 Syn addition of bromine to (Z)-2-butene would lead to the meso dibromide.

(Z)-2-Butene   meso-2,3-Dibromobutane

This is not the pathway that is observed, of course. Addition of bromine to an alkene is an anti addition process.

7.17 The tartaric acids incorporate two equivalently substituted stereogenic centers. (+)-Tartaric acid, as noted in the text, is the 2R,3R stereoisomer. There will be two additional stereoisomers, the enantiomeric (−)-tartaric acid (2S,3S) and an optically inactive meso form.

(2S,3S)-Tartaric acid,   meso-Tartaric acid,
mp 170°C, $[\alpha]_D - 12°$ mp 140°C, optically inactive

Interestingly, (−)-tartaric acid is itself a natural product present in the leaves of *bauhinia*, a central African bush, from which it can be extracted with hot water.

7.18 No. Pasteur separated an optically inactive racemic mixture into two optically active enantiomers. A meso form is achiral, is identical to its mirror image, and is incapable of being separated into optically active forms.

7.19 The more soluble salt must have the opposite configuration at the stereogenic center of 1-phenylethylamine, i.e., the S configuration. The malic acid used in the resolution is a single enantiomer, S. Therefore the more soluble salt in this particular case is (S)-1-phenylethylammonium (S)-malate.

7.20 In an earlier exercise (Problem 4.20) the structures of all the isomeric $C_5H_{12}O$ alcohols were presented. Those which lack a stereogenic center and which are *achiral* are:

$$CH_3CH_2CH_2CH_2CH_2OH \qquad CH_3CHCH_2CH_2OH \qquad (CH_3)_3CCH_2OH$$
$$\mid$$
$$CH_3$$

1-Pentanol   3-Methyl-1-butanol   2,2-Dimethyl-1-propanol

$$CH_3$$
$$\mid$$
$$CH_3CH_2CHCH_2CH_3 \qquad CH_3CH_2COH$$
$$\mid \qquad\qquad\qquad \mid$$
$$OH \qquad\qquad\qquad CH_3$$

3-Pentanol   2-Methyl-2-butanol

The chiral isomers are characterized by carbons which bear four different groups. These are:

$$CH_3\overset{*}{C}HCH_2CH_2CH_3 \qquad CH_3\overset{*}{C}HCH(CH_3)_2 \qquad CH_3CH_2\overset{*}{C}HCH_2OH$$
$$\mid \qquad\qquad\qquad \mid \qquad\qquad\qquad\qquad \mid$$
$$OH \qquad\qquad\qquad OH \qquad\qquad\qquad\qquad CH_3$$

2-Pentanol   3-Methyl-2-butanol   2-Methyl-1-butanol

7.21 The isomers of trichlorocyclopropane are:

Enantiomeric forms of 1,1,2-trichlorocyclopropane (both chiral)

cis-1,2,3-Trichlorocyclopropane
(achiral—contains a plane of symmetry)

trans-1,2,3-Trichlorocyclopropane
(achiral—contains a plane of symmetry)

7.22 (*a*) Carbon-2 is a stereogenic center in 3-chloro-1,2-propanediol. Carbon-2 has two equivalent substituents in 2-chloro-1,3-propanediol, where it is not a stereogenic center.

$$ClCH_2\overset{*}{C}HCH_2OH \qquad HOCH_2CHCH_2OH$$
$$\underset{OH}{|} \qquad\qquad\qquad \underset{Cl}{|}$$

3-Chloro-1,2-propanediol, chiral   2-Chloro-1,3-propanediol, achiral

(*b*) The primary bromide is achiral; the secondary bromide contains a stereogenic center and is chiral.

$$CH_3CH{=}CHCH_2Br \qquad CH_3\overset{*}{C}HCH{=}CH_2$$
$$\underset{Br}{|}$$

Achiral   Chiral

(*c*) Both stereoisomers have two equivalently substituted stereogenic centers, and so we must be alert for the possibility of a meso stereoisomer. The structure at the left is chiral. The one at the right has a plane of symmetry, and is the achiral meso stereoisomer.

Chiral   Meso: achiral

(*d*) The structure on the left is chiral; the other is an achiral meso form.

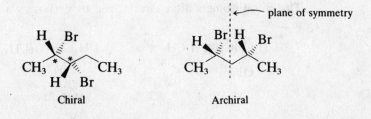

Chiral   Archiral

(*e*)   The first structure is achiral; it has a plane of symmetry.

Plane of symmetry passes through C-1, C-4, and C-7.

The second structure cannot be superposed on its mirror image; it is chiral.

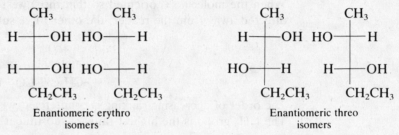

Reference structure   Mirror image   Reoriented mirror image

7.23   There are four stereoisomers of 2,3-pentanediol, represented by the Fischer projections shown. All are chiral.

$$
\begin{array}{cc}
\text{CH}_3 & \text{CH}_3 \\
\text{H}\!\!-\!\!\text{OH} & \text{HO}\!\!-\!\!\text{H} \\
\text{H}\!\!-\!\!\text{OH} & \text{HO}\!\!-\!\!\text{H} \\
\text{CH}_2\text{CH}_3 & \text{CH}_2\text{CH}_3
\end{array}
\qquad
\begin{array}{cc}
\text{CH}_3 & \text{CH}_3 \\
\text{H}\!\!-\!\!\text{OH} & \text{HO}\!\!-\!\!\text{H} \\
\text{HO}\!\!-\!\!\text{H} & \text{H}\!\!-\!\!\text{OH} \\
\text{CH}_2\text{CH}_3 & \text{CH}_2\text{CH}_3
\end{array}
$$

Enantiomeric erythro isomers   Enantiomeric threo isomers

There are three stereoisomers of 2,4-pentanediol. The meso form is achiral; both threo forms are chiral.

$$
\begin{array}{c}
\text{CH}_3 \\
\text{H}\!\!-\!\!\text{OH} \\
\text{H}\!\!-\!\!\text{H} \\
\text{H}\!\!-\!\!\text{OH} \\
\text{CH}_3
\end{array}
\qquad
\begin{array}{cc}
\text{CH}_3 & \text{CH}_3 \\
\text{H}\!\!-\!\!\text{OH} & \text{HO}\!\!-\!\!\text{H} \\
\text{H}\!\!-\!\!\text{H} & \text{H}\!\!-\!\!\text{H} \\
\text{HO}\!\!-\!\!\text{H} & \text{H}\!\!-\!\!\text{OH} \\
\text{CH}_3 & \text{CH}_3
\end{array}
$$

meso-2,4-Pentanediol   Enantiomeric threo isomers

7.24   (*a*)   (−)-2-Octanol has the *R* configuration at C-2. The order of substituent preference is:

$$HO > CH_2CH_2 > CH_3 > H$$

The molecule is oriented so that the lowest-ranking substituent is directed away from the reader and the order of decreasing precedence is clockwise.

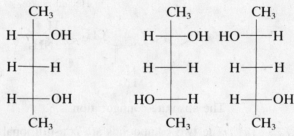

(*b*)   The order of the substituents attached to the stereogenic center is:

$$
\overset{\displaystyle\overset{\text{O}}{\|}}{-\text{C}}-\text{OH} > -\!\!\bigcirc\!\!-\text{CH}_2\text{CH}(\text{CH}_3)_2 > \text{CH}_3 > \text{H}
$$

Highest-ranked   Lowest-ranked

The molecule is oriented so that the lowest-ranked substituent (H) is away from us. When viewed from this perspective the remaining substituents trace a counterclockwise path when proceeding from higher to lower rank. The absolute configuration is S.

$$CH_3 \quad \underset{HO_2C}{\overset{H}{\underset{|}{C}}} - \!\!\!\!\! \bigcirc \!\!\!\!\! -CH_2CH(CH_3)_2 \qquad \text{S-Ibuprofen}$$

(c) In order of decreasing sequence rule preference, the four substituents at the stereogenic center of monosodium L-glutamate are:

$$\overset{+}{N}H_3 > CO_2^- > CH_2 > H$$

$$H_3\overset{+}{N}\!-\!\!\!\underset{CH_2CH_2CO_2^-Na^+}{\overset{CO_2^-}{\underset{|}{\overset{|}{C}}}}\!\!\!-H \qquad \begin{array}{c}\text{is the}\\\text{same}\\\text{as}\end{array} \qquad H_3\overset{+}{N}\!\!\blacktriangleright\!\!\underset{CH_2CH_2CO_2^-Na^+}{\overset{CO_2^-}{\underset{|}{\overset{|}{C}}}}\!\!\blacktriangleleft H$$

When the molecule is oriented so that the lowest-ranking substituent (hydrogen) is directed away from the reader, the other three substituents are arranged as shown.

$$^-O_2C \curvearrowleft \overset{+}{N}H_3$$
$$CH_2CH_2CO_2^-Na^+$$

The order of decreasing ranking is counterclockwise; the absolute configuration is S.

(d) The NH_3 group is the highest-ranking substituent in 5-hydroxytryptophan.

$$\underset{\text{Highest-ranking}}{\overset{+}{N}H_3} > CO_2^- > CH_2 > \underset{\text{Lowest-ranking}}{H}$$

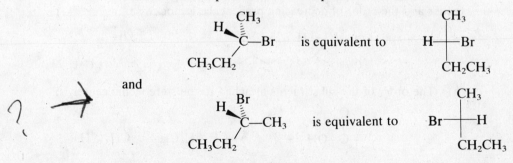

The absolute configuration is S.

7.25 (a) The two compounds are constitutional isomers. Their IUPAC names clearly reflect this difference.

$$\underset{\underset{OH}{|}}{CH_3CHCH_2Br} \qquad \text{and} \qquad \underset{\underset{Br}{|}}{CH_3CHCH_2OH}$$

1-Bromo-2-propanol   2-Bromo-1-propanol

(b) The two structures have the same constitution. Test them for superposability. To do this we need to place them in comparable orientations.

$$\underset{CH_3CH_2}{\overset{CH_3}{\underset{\diagdown}{\overset{H}{\diagdown}}{C}}}\!\!-Br \qquad \text{is equivalent to} \qquad H\!-\!\!\!\underset{CH_2CH_3}{\overset{CH_3}{\underset{|}{\overset{|}{}}}}\!\!\!-Br$$

and

$$\underset{CH_3CH_2}{\overset{Br}{\underset{\diagdown}{\overset{H}{\diagdown}}{C}}}\!\!-CH_3 \qquad \text{is equivalent to} \qquad Br\!-\!\!\!\underset{CH_2CH_3}{\overset{CH_3}{\underset{|}{\overset{|}{}}}}\!\!\!-H$$

The two are nonsuperposable mirror images of each other. They are enantiomers.

To check this conclusion, work out the absolute configuration of each using the Cahn-Ingold-Prelog system.

$$H \overset{CH_3}{\underset{CH_3CH_2}{|}} C-Br \qquad\qquad H \overset{Br}{\underset{CH_3CH_2}{|}} C-CH_3$$

(*S*)-2-Bromobutane   (*R*)-2-Bromobutane

(*c*) Again, place the structures in comparable orientations and examine them for superposability.

$$H \overset{CH_3}{\underset{CH_3CH_2}{|}} C-Br \qquad \text{is equivalent to} \qquad H-\overset{CH_3}{\underset{CH_2CH_3}{|}}-Br$$

and

$$CH_3 \overset{H}{\underset{Br}{|}} C-CH_2CH_3 \qquad \text{is equivalent to} \qquad H-\overset{CH_3}{\underset{CH_2CH_3}{|}}-Br$$

The two structures represent the same compound, since they are superposable. (As a check, notice that both represent the *S* configuration.)

(*d*) If we reorient the first structure,

$$H \overset{CH_3}{\underset{CH_3CH_2}{|}} C-Br \qquad \text{becomes} \qquad CH_3-\overset{CH_2CH_3}{\underset{H}{|}}-Br$$

which is the enantiomer of

$$Br-\overset{CH_2CH_3}{\underset{H}{|}}-CH_3$$

As a check, the first structure is seen to have the *S* configuration, while the second has the *R* configuration.

(*e*) As drawn, the two structures are mirror images of each other; however, they represent an achiral molecule. The two structures are superposable mirror images and are not stereoisomers but identical.

$$\overset{CH_2OH}{\underset{OH}{\overset{|}{\underset{|}{\underset{H-|-H}{H-|-OH}}}}} \quad \text{and} \quad \overset{CH_2OH}{\underset{OH}{\overset{|}{\underset{|}{\underset{H-|-H}{HO-|-H}}}}} \quad \text{are both} \quad HOCH_2CHCH_2OH \atop \qquad\qquad\qquad |\atop \qquad\qquad\qquad OH$$

(*f*) The two structures—one cis, the other trans—are stereoisomers which are not mirror images; they are diastereomers.

trans-1-Chloro-2-methylcyclopropane   cis-1-Chloro-2-methylcyclopropane

(g) Both structures are *cis*-1-chloro-2-methylcyclopropanes. They are related as an object and its nonsuperposable mirror image; they are enantiomers.

(h) The two structures are enantiomers, since they are nonsuperposable mirror images. Checking their absolute configurations reveals one to be *R*, the other *S*. Both have the *E* configuration at the double bond.

(2*R*,3*E*)-3-Penten-2-ol   (2*S*, 3*E*)-3-Penten-2-ol

(i) These two structures are identical; both have the *E* configuration at the double bond and the *R* configuration at the stereogenic center.

Alternatively, we can show their superposability by rotating the second structure 180° about an axis passing through the doubly bonded carbons.

Reference structure   Rotate 180° around this axis   Identical to reference structure

(j) One structure has a cis double bond, the other a trans double bond; therefore, the two are diastereomers.

(2*R*,3*E*)-3-Penten-2-ol   (2*S*,3*Z*)-3-Penten-2-ol

(k) Here it will be helpful to reorient the second structure so that it may be more readily compared with the first.

Reference structure   and   which is equivalent to   Enantiomer of reference structure

The two compounds are enantiomers.

Examining their absolute configurations confirms the enantiomeric nature of the two compounds

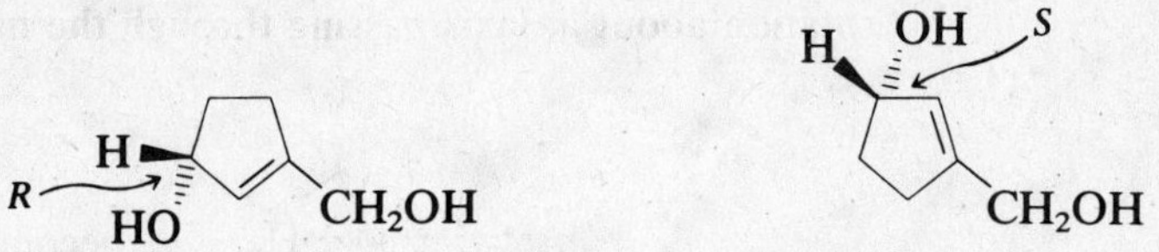

(*l*) These two compounds differ in the order in which their atoms are joined together; they are constitutional isomers.

3-Hydroxymethyl-2-cyclopenten-1-ol   3-Hydroxymethyl-3-cyclopenten-1-ol

(*m*) In order to better compare these two structures, place them both in the same format.

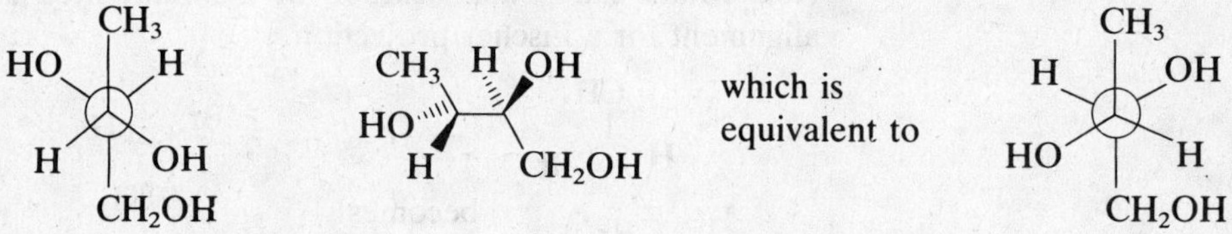

which is equivalent to

The two are enantiomers.

(*n*) Since *cis*-1,3-dimethylcyclopentane has a plane of symmetry, it is achiral and cannot have an enantiomer. The two structures given in the problem are identical.

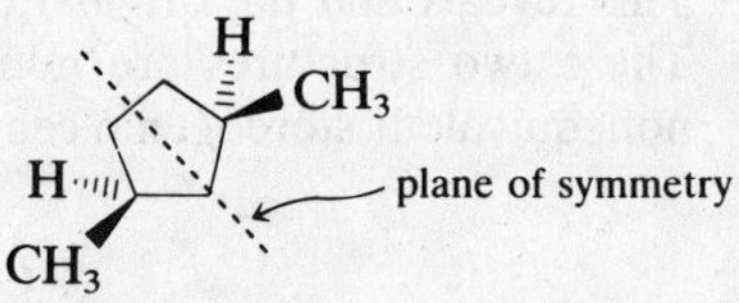

plane of symmetry

(*o*) The structure at the left has a plane of symmetry, and all three of its substituents are cis. The structure at the right is chiral, with one substituent trans to the other two. The compounds are stereoisomers but not enantiomers; therefore they are diastereomers.

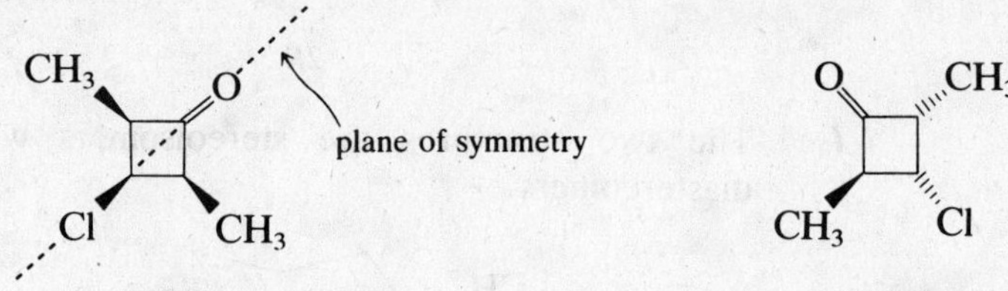

plane of symmetry

Achiral   Chiral

(*p*) These structures are diastereomers, i.e., stereoisomers which are not mirror images. They have the same configuration at C-3 but opposite configurations at C-2.

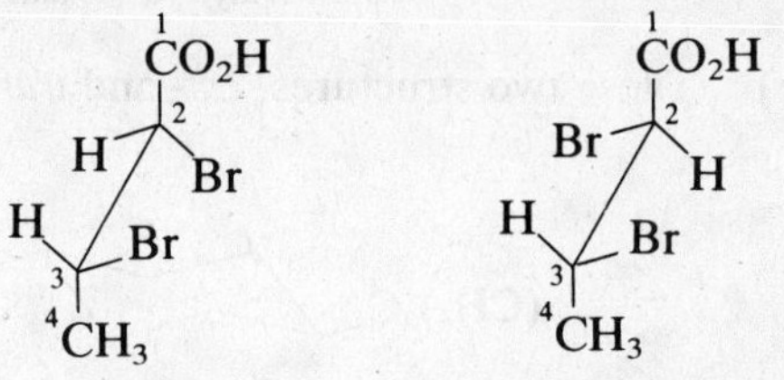

2R,3R   2S,3R

(q) In order to compare these compounds, reorient the first structure so that it may be drawn as a Fischer projection. The first step in the reorientation consists of a 180° rotation about an axis passing through the midpoint of the C-2—C-3 bond.

becomes

Thus

CO_2H

H — Br

H — Br

CH_3

Reference structure

is

CH_3

H — Br

H — Br

CO_2H

Now rotate the "back" carbon of the reoriented structure to give the necessary alignment for a Fischer projection.

CH_3

H — Br

H — Br

CO_2H

becomes

Br — H

CH_3

H — Br

CO_2H

which is the same as

CH_3

Br —|— H

H —|— Br

CO_2H

This reveals that the original two structures in the problem are equivalent.

(r) These two structures are nonsuperposable mirror images of a molecule with two nonequivalent stereogenic centers; they are enantiomers.

$\overset{1}{C}O_2H$

H —$\overset{2}{}$— Br

H —$\overset{3}{}$— Br

$\overset{4}{C}H_3$

2R,3R

$\overset{1}{C}O_2H$

Br —$\overset{2}{}$— H

Br —$\overset{3}{}$— H

$\overset{4}{C}H_3$

2S,3S

(s) The two structures are stereoisomers which are not enantiomers; they are diastereomers.

CH_3 ⟋⟍ OH

cis-3-Methylcyclohexanol

CH_3 ⟋⟍ OH

trans-3-Methylcyclohexanol

(t) These two structures, cis- and trans-4-tert-butylcyclohexyl iodide, are diastereomers.

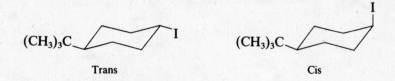

$(CH_3)_3C$ ⟋⟍ I

Trans

$(CH_3)_3C$ ⟋⟍ I

Cis

(u)   The two structures are nonsuperposable mirror images; they are enantiomers.

Reference structure   is equivalent to   Enantiomer of reference structure

(v)   The two structures are identical.

Reference structure   is equivalent to   Identical to reference structure

(w)   As represented, the two structures are mirror images of each other, but because the molecule is achiral (it has a plane of symmetry), the two must be superposable. They represent the same compound.

Reference structure   is equivalent to   Identical to reference structure

The plane of symmetry passes through C-7 and bisects the C-2—C-3 bond and the C-5—C-6 bond.

(x)   The structures are stereoisomers but not enantiomers; they are diastereomers. (Both are achiral and so cannot have enantiomers.)

Achiral   and   Achiral   are stereoisomers but not mirror images

7.26   Write a structural formula for phytol and count the number of structural units capable of stereochemical variation.

$$HOCH_2CH{=}CCH_2CH_2CH_2\overset{*}{C}HCH_2CH_2CH_2\overset{*}{C}HCH_2CH_2CH_2CHCH_3$$

$$\underset{CH_3}{|}\qquad\underset{CH_3}{|}\qquad\underset{CH_3}{|}\qquad\underset{CH_3}{|}$$

3,7,11,15-Tetramethyl-2-hexadecen-1-ol

Phytol has two stereogenic centers (C-7 and C-11) and one double bond. The stereogenic centers may be either R or S, and the double bond may be either E or Z. *Eight* stereoisomers are possible.

	Isomer							
	1	2	3	4	5	6	7	8
Double bond	E	E	E	E	Z	Z	Z	Z
Carbon-7	R	S	R	S	R	S	R	S
Carbon-11	R	S	S	R	R	S	S	R

7.27 (*a*) Muscarine has three stereogenic centers, and so *eight* stereoisomers have this constitution.

(*b*) One-half of the eight stereoisomers (i.e., *four*) have one of the ring substituents trans to the other two.

(*c*) Muscarine is:

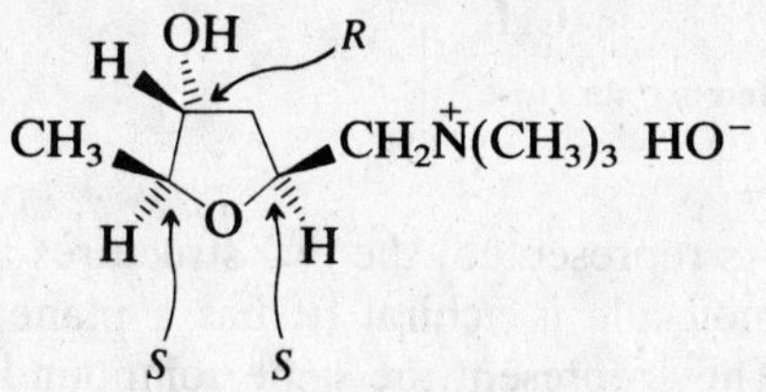

7.28 In order to write a stereochemically accurate representation of ectocarpene, it is best to begin with the configuration of the stereogenic center, which we are told is S.

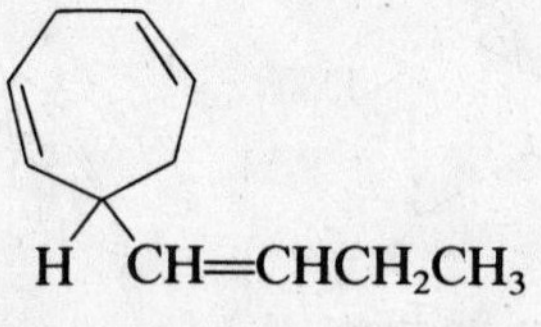

Clearly, hydrogen is the lowest-ranking substituent; among the other three substituents, two are part of the ring and the third is the four-carbon side chain. The priority rankings of these groups are determined by systematically working along the chain.

The substituents

—CH=CHCH$_2$CH=CHCH$_2$ —CH=CHCH$_2$CH$_3$ —CH$_2$CH=CHCH$_2$CH=CH

(Ring)   (Side chain)   (Ring)

are considered as if they were

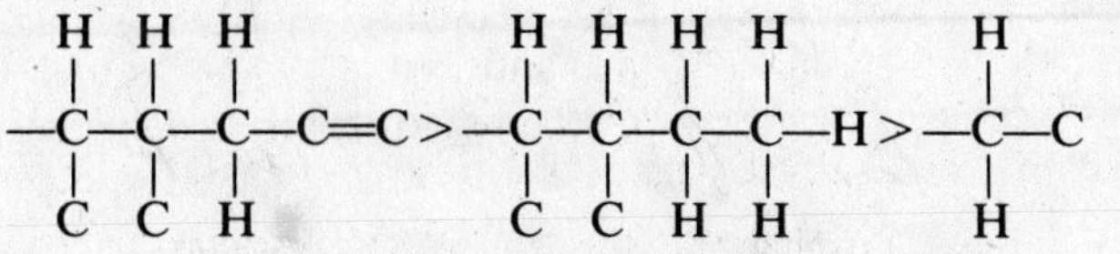

(Ring)   (Side chain)   (Ring)

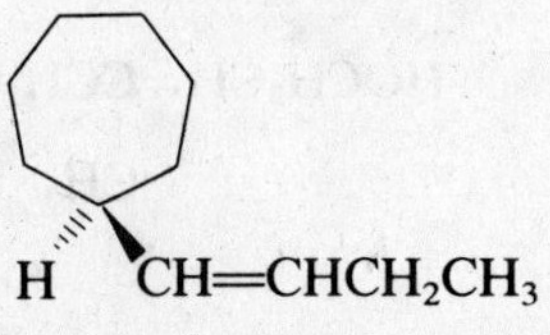

Orienting the molecule with the hydrogen away from the reader we place the double bonds in the ring so that the order of decreasing sequence rule precedence is counterclockwise:

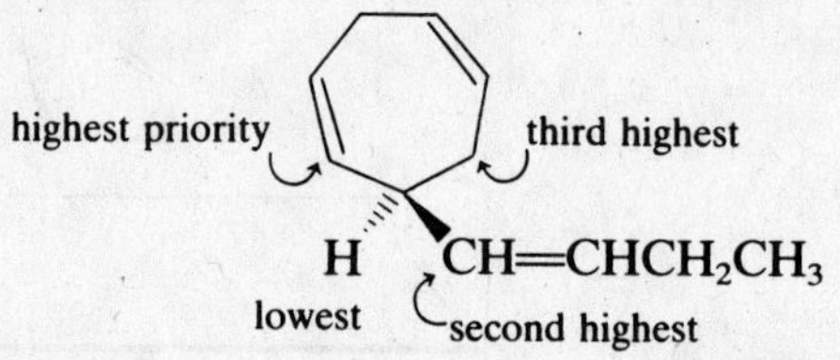

Finally, since all the double bonds are cis, the complete structure becomes:

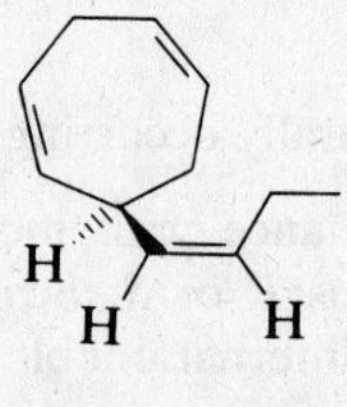

7.29 (*a*) Multifidene has two stereogenic centers and three double bonds. Neither the ring double bond nor the double bond of the vinyl substituent can give rise to stereoisomers, but there is the possibility of *E* and *Z* isomers of the butenyl side chain. Therefore there are eight (2^3) possible stereoisomers. We can rationalize them as:

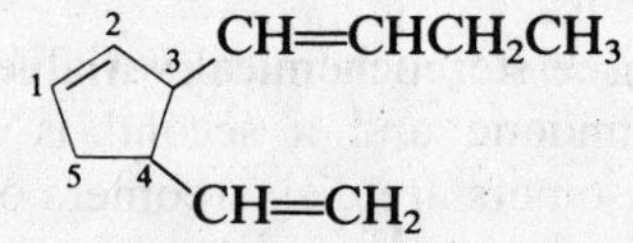

Stereoisomer	C-3	C-4	Butenyl double bond	
1	*R*	*R*	*E*	} enantiomers
2	*S*	*S*	*E*	
3	*R*	*R*	*Z*	} enantiomers
4	*S*	*S*	*Z*	
5	*R*	*S*	*E*	} enantiomers
6	*S*	*R*	*E*	
7	*R*	*S*	*Z*	} enantiomers
8	*S*	*R*	*Z*	

(*b*) Given the information that the alkenyl substituents are cis to each other, the number of stereoisomers is reduced by half. Therefore, four stereoisomers are possible.

(*c*) Knowing that the butenyl group has a *Z* double bond reduces the number of possibilities by half. Two stereoisomers are possible.

(*d*) The two stereoisomers are:

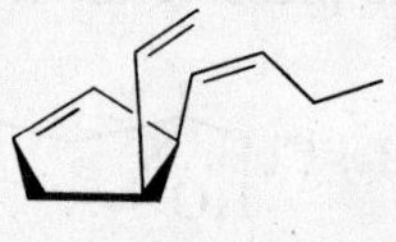

and

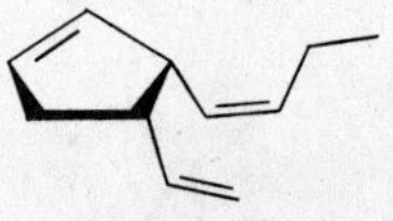

(*e*)   These two stereoisomers are enantiomers. They are nonsuperposable mirror images.

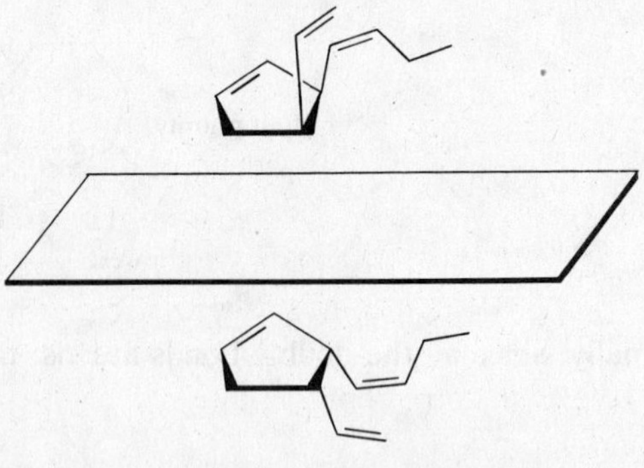

Mirror plane

Naturally occurring (+)-multifidene has the 3*S*,4*S* configuration.

7.30   In a substance with more than one stereogenic center, each center is independently specified as *R* or *S*. Streptimidone has two stereogenic centers and two double bonds. Only the internal double bond is capable of stereoisomerism.

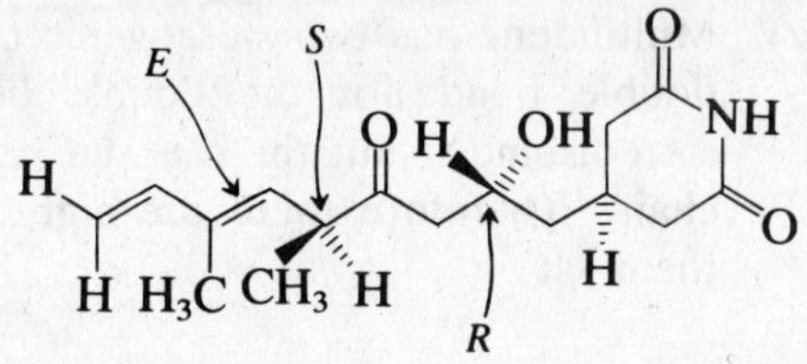

The three stereochemical variables give rise to eight (2^3) stereoisomers, of which one is streptimidone and a second is the enantiomer of streptimidone. The remaining six stereoisomers are diastereomers of streptimidone.

7.31   (*a*)   The first step is to set out the constitution of menthol, which we are told is 2-isopropyl-5-methylcyclohexanol.

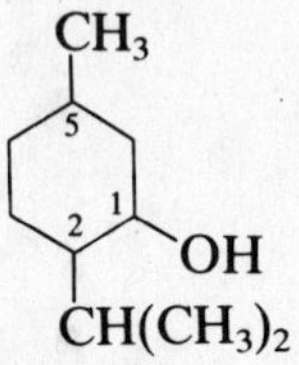

2-Isopropyl-5-methylcyclohexanol

Since the configuration at C-1 is *R* in (−)-menthol, the hydroxyl group must be "up" in our drawing.

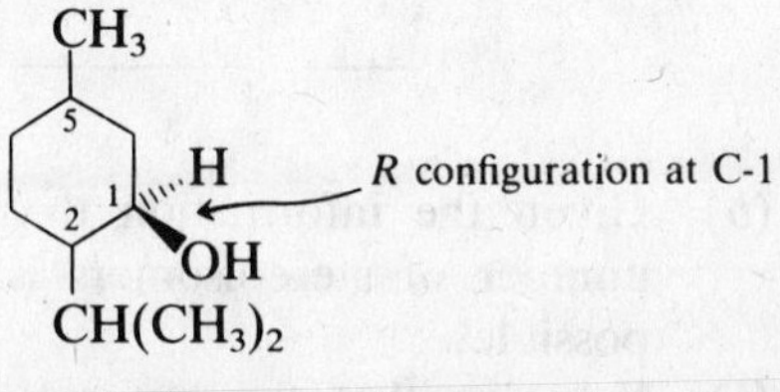

R configuration at C-1

Because menthol is the most stable stereoisomer of this constitution, all three of its substituents must be equatorial. Therefore, we draw the chair form of the preceding structure, which has the hydroxyl group equatorial and up, placing isopropyl and methyl groups so as to preserve the *R* configuration at C-1.

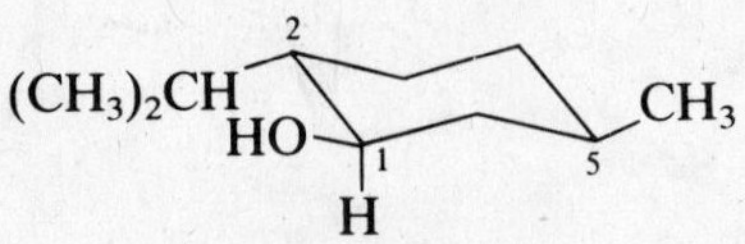

(−)-Menthol

(b)   To transform the structure of (−)-menthol to that of (+)-isomenthol, the configuration at C-5 must remain the same while those at C-1 and C-2 are inverted.

(−)-Menthol   (+)-Isomenthol

(+)-Isomenthol is represented here in its correct configuration, but the conformation with two axial substituents is not the most stable one. The ring-flipped form will be the preferred conformation of (+)-isomenthol:

Most stable conformation of
(+)-isomenthol

7.32   Since the only information available about the compound is its optical activity, examine the two structures for chirality, recalling that only chiral substances can be optically active.

   The structure with the six-membered ring has a plane of symmetry passing through C-1 and C-4. It is achiral and cannot be optically active.

Achiral; $[\alpha]_D$ 0°   Chiral; can be optically active

   The open-chain structure has neither a plane of symmetry nor a center of symmetry; it is not superposable on its mirror image and so is chiral. It can be optically active and is more likely to be the correct choice.

7.33   Compound B has a center of symmetry, is achiral, and thus cannot be optically active.

Compound B: not optically active
(center of symmetry is midpoint of C-16—C-17 bond)

   The diol in the problem is optically active, and so it must be chiral. Compound A is the naturally occurring diol.

7.34   (a)   The equation which relates specific rotation $[\alpha]_D$ to observed rotation α is

$$[\alpha]_D = \frac{100\alpha}{cl}$$

where c is concentration in grams per 100 mL and l is path length in decimeters.

$$[\alpha]_D = \frac{100(-5.20°)}{\left(\frac{2.0\,g}{100\,mL}\right)(2\,dm)}$$

$$[\alpha]_D = -130°$$

(b) The optical purity of the resulting solution is 10/15, or 66.7 percent, since 10 g of optically pure fructose has been mixed with 5 g of racemic fructose. Therefore, the specific rotation will be two-thirds (10/15) of the specific rotation of optically pure fructose:

$$[\alpha]_D = \tfrac{2}{3}(-130°) = -87°$$

7.35 (a) The reaction of 1-butene with hydrogen iodide is one of electrophilic addition. It follows Markovnikov's rule and yields a racemic mixture of (R)- and (S)-2-iodobutane.

$$CH_3CH_2CH{=}CH_2 \xrightarrow{\ HI\ }$$

1-Butene   (R)-2-Iodobutane   (S)-2-Iodobutane

(b) This is a free-radical addition of hydrogen bromide, leading to the formation of equal amounts of (R)- and (S)-2-bromobutane.

(E)-2-Butene   (R)-2-Bromobutane   (S)-2-Bromobutane

(c) Bromine adds anti to carbon-carbon double bonds to give vicinal dibromides.

(E)-2-Pentene   (2R,3S)-2,3-Dibromopentane   (2S,3R)-2,3-Dibromopentane

The two stereoisomers are enantiomers and are formed in equal amounts.

(d) Two enantiomers are formed in equal amounts in this reaction, involving electrophilic addition of bromine to (Z)-2-pentene. These two are diastereomeric with those formed in part (c).

(Z)-2-Pentene   (2R,3R)-2,3-Dibromopentane   (2S,3S)-2,3-Dibromopentane

(e) Epoxidation of 1-butene yields a racemic epoxide mixture.

$$CH_3CH_2CH{=}CH_2 \xrightarrow[\text{CH}_2\text{Cl}_2]{\text{peroxyacetic acid}}$$

(S)-1,2-Epoxybutane   (R)-1,2-Epoxybutane

(f) Two enantiomeric epoxides are formed in equal amounts on epoxidation of (Z)-2-pentene.

$$\xrightarrow[\text{CH}_2\text{Cl}_2]{\text{peroxyacetic acid}}$$

(Z)-2-Pentene   (2S,3R)-2,3-Epoxypentane   (2R,3S)-2,3-Epoxypentane

The reaction is a stereospecific syn addition. The cis alkyl groups in the starting alkene remain cis in the product epoxide.

(g) The starting material is achiral so even though a chiral product is formed, it is a racemic mixture of enantiomers and is optically inactive.

$$\xrightarrow[\text{Pt}]{\text{H}_2}$$

1,5,5-Trimethylcyclopentene   (R)-1,1,2-Trimethylcyclopentane   (S)-1,1,2-Trimethylcyclopentane

(h) Recall that hydroboration-oxidation leads to anti-Markovnikov hydration of the double bond.

$$\xrightarrow[\text{2. H}_2\text{O}_2,\ \text{OH}^-]{\text{1. B}_2\text{H}_6}$$

1,5,5-Trimethylcyclopentene   (1S,2S)-2,3,3-Trimethylcyclopentanol   (1R,2R)-2,3,3-Trimethylcyclopentanol

The product has two stereogenic centers. It is formed as a racemic mixture of enantiomers.

7.36 Hydration of the double bond of aconitic acid can occur in two regiochemically distinct ways:

$$HO_2CCH_2CHCHCO_2H \longleftarrow \quad \longrightarrow HO_2CCH_2CCH_2CO_2H$$

Chiral
(isocitric acid)   Aconitic acid   Achiral
(citric acid)

One of the hydration products lacks a stereogenic carbon. It must be citric acid, the achiral, optically inactive isomer. The other one has two different stereogenic centers and must be isocitric acid, the optically active isomer.

7.37 (a) Structures C and D are chiral. Structure E has a plane of symmetry and is an achiral meso form.

(b) Ozonolysis of the starting material proceeds with the stereochemistry shown. Compound D is the product of the reaction

Compound D

(c) If the methyl groups were cis to one another in the cycloalkene, they would be on the same side of the Fischer projection in the product. Compound E would be formed.

7.38 (a) We need to first convert the Fischer projection into a three-dimensional representation:

The E2 transition state requires bromine and the proton that is lost to be anti to one another.

cis-2-Butene   trans-2-Butene

cis-2-Butene contains deuterium; trans-2-butene does not. 1-Butene contains deuterium.

(*b*)   The starting material in part (*a*) is erythro.

Erythro   Threo

The relative positions of H and D at C-3 are reversed in the threo stereoisomer. Therefore *trans*-2-butene contains deuterium when prepared from threo; *cis*-2-butene does not. 1-Butene again contains deuterium.

7.39   The molecular formula (C_6H_{10}) and the fact that F contains a five-membered ring tell us that it must also contain a double bond. The molecular formula (C_6H_{10}) differs from that of the corresponding alkane (C_6H_{14}) by four hydrogens, and since each double bond or ring causes a decrease of two hydrogens, the sum of rings and double bonds in F must be 2. Therefore, the following possibilities exist for F:

Methylenecyclopentane   1-Methylcyclopentene   3-Methylcyclopentene   4-Methylcyclopentene

Which of these form diastereomeric dibromides on anti addition of bromine?

Compound F must be 3-methylcyclopentene. (However, on the basis of the information available, we cannot tell whether it is the *R* or the *S* enantiomer or is racemic.)

7.40 Dehydration of this tertiary alcohol can yield 2,3-dimethyl-1-pentene or 2,3-dimethyl-2-pentene. Only the terminal alkene in this case is chiral.

2,3-Dimethyl-2-pentanol
(chiral, optically pure)

2,3-Dimethyl-1-pentene
(chiral, optically pure)

2,3-Dimethyl-2-pentene
(achiral, optically inactive)

2,3-Dimethylpentane
(chiral, optically pure)

2,3-Dimethylpentane
(chiral, optically inactive)

The 2,3-dimethyl-1-pentene formed in the dehydration reaction must be optically pure because it arises from optically pure alcohol by a reaction which does not involve any of the bonds to the stereogenic center. When optically pure 2,3-dimethyl-1-pentene is hydrogenated, it must yield optically pure 2,3-dimethylpentane—again, no bonds to the stereogenic center are involved in this step.

The 2,3-dimethyl-2-pentene formed in the dehydration reaction is achiral and must yield racemic 2,3-dimethylpentane on hydrogenation.

Since the alkane is 50 percent optically pure, the alkene fraction must have contained equal amounts of optically pure 2,3-dimethyl-1-pentene and its achiral isomer 2,3-dimethyl-2-pentene.

7.41 The reaction of (R)-3-buten-2-ol with a peroxy acid may be represented as:

(R)-3-Buten-2-ol

Minor stereoisomer

(a) Oxygen may be transferred to the front face or back face of the double bond. Therefore, the major stereoisomer is:

(b) The two epoxides have the same configuration (R) at their secondary alcohol carbons but opposite configurations at the stereogenic center of the epoxide ring; they are diastereomers.

(c) In addition to the two epoxides shown at the beginning of this solution, the corresponding enantiomers will be formed when racemic 3-buten-2-ol is epoxidized. The relative amounts of each of the four will be:

20%

20%

Enantiomeric forms of minor diastereomer, totaling 40%

30%

30%

Enantiomeric forms of major diastereomer, totaling 60%

SELF-TEST

PART A

A-1. For each of the following pairs of drawings, identify the molecules as chiral or achiral and tell whether each pair represents molecules that are enantiomers, diastereomers, or identical.

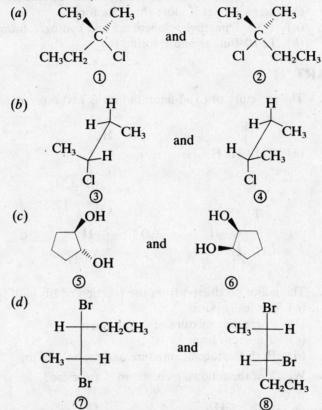

A-2. Specify the configuration of each stereogenic carbon in the preceding problem, using the Cahn-Ingold-Prelog *R-S* system.

A-3. Predict the number of stereoisomers possible for each of the following constitutions. For which of these will meso forms be possible?

(a)

(b) $CH_3CHCHCHCH_3$
              Br Br Cl

A-4. Using the skeletons provided as a guide,

(a)   Draw a perspective view of (2*R*,3*R*)-3-chloro-2-butanol.

(b)   Draw a sawhorse diagram of (*R*)-2-bromobutane.

(c)   Draw Fischer projections of both these compounds.

A-5. (a) The specific rotation of pure (−)-cholesterol is −39°. What is the specific rotation of a sample of cholesterol containing 10 percent (+)-chlolesterol as the only contaminant?

(b) If the rotation of optically pure (R)-2-octanol is −10°, what is the percentage of the S enantiomer in a sample of 2-octanol that has a rotation of −4°?

A-6. Write the organic product(s) expected from each of the following reactions. Show each stereoisomer if more than one forms.

(a) 1,5,5-Trimethylcyclopentene and hydrogen bromide

(b) (E)-2-Butene and chlorine (Cl_2)

PART B

B-1. The structure of (S)-2-fluorobutane is best represented by

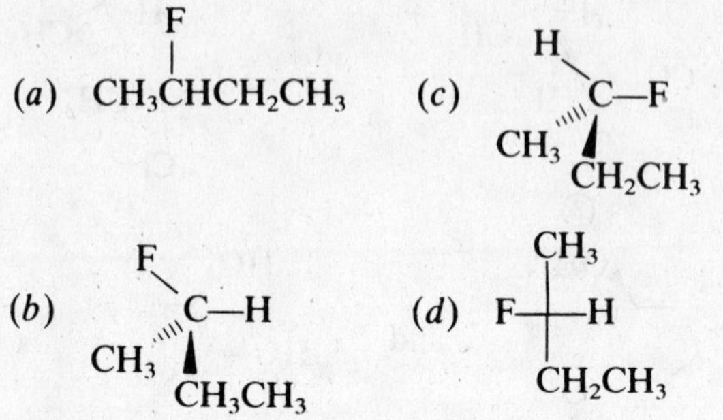

(a) $CH_3CHCH_2CH_3$ with F above

(c) structure with H, C—F, CH_3, CH_2CH_3

(b) structure with F, C—H, CH_3, CH_3CH_3

(d) Fischer projection: CH_3 / F—H / CH_2CH_3

B-2. The major product(s) from the reaction of Br_2 with (Z)-3-hexene is (are)

(a) Optically pure

(b) A racemic mixture of enantiomers

(c) The meso form

(d) Both the racemic mixture and the meso form

B-3. Which of the following compounds are meso?

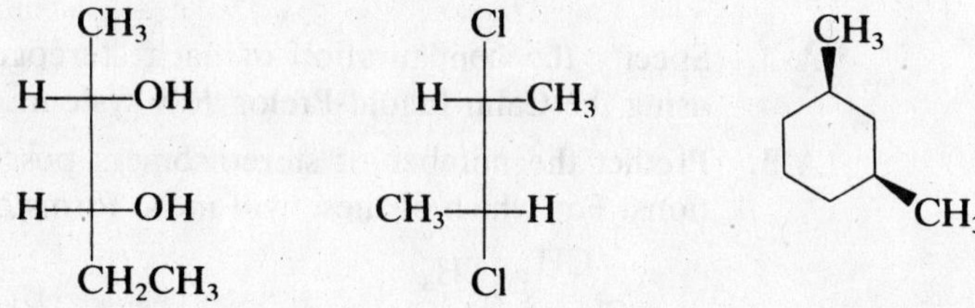

Structure A: Fischer projection with CH_3 / H—OH / H—OH / CH_2CH_3

Structure B: Fischer projection with Cl / H—CH_3 / CH_3—H / Cl

Structure C: cyclohexane with two CH_3 groups

A   B   C

(a) A only   (b) C only   (c) A and B   (d) B and C

B-4. The 2,3-dichloropentane whose structure is shown is

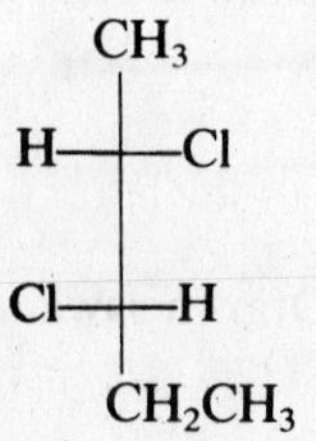

Fischer projection: CH_3 / H—Cl / Cl—H / CH_2CH_3

(a) 2R,3R   (b) 2R,3S   (c) 2S,3R   (d) 2S,3S

B-5. The separation of a 1 to 1 mixture of enantiomers into their pure forms is termed

(a) Racemization   (c) Isomerization

(b) Resolution   (d) Equilibration

B-6. Order the following groups in order of *R-S* ranking (4 is highest):

—CH(CH₃)₂   —CH₂CH₂Br   —CH₂Br   —C(CH₃)₃

A   B   C   D

	4	3	2	1
(*a*)	C	B	D	A
(*b*)	A	D	B	C
(*c*)	C	D	A	B
(*d*)	C	D	B	A

B-7. A meso compound:
   (*a*) Is an achiral molecule which contains stereogenic centers
   (*b*) Contains a plane of symmetry or a center of symmetry
   (*c*) Is optically inactive
   (*d*) Is characterized by all of these

B-8. Which one of the following is chiral?
   (*a*) 1,1-Dibromo-1-chloropropane   (*c*) 1,3-Dibromo-1-chloropropane
   (*b*) 1,1-Dibromo-3-chloropropane   (*d*) 1,3-Dibromo-2-chloropropane

B-9. The reaction sequence

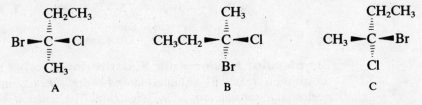

will yield:
   (*a*) A pair of products which are enantiomers
   (*b*) A single product which is optically active
   (*c*) A pair of products which are diastereomers
   (*d*) A pair of products one of which is meso

B-10. Which of the following structures are equivalent (identical)?

CH₂CH₃   CH₃   CH₂CH₃

Br—C◀Cl   CH₃CH₂▶C◀Cl   CH₃▶C◀Br

CH₃   Br   Cl

A   B   C

(*a*) A and B   (*b*) A and C   (*c*) B and C   (*d*) A, B, and C

NUCLEOPHILIC SUBSTITUTION

IMPORTANT TERMS AND CONCEPTS

Nucleophilic Substitution (Secs. 8.1 to 8.3) A nucleophilic substitution reaction may be described in the following general terms:

$$R—X + \quad Y:^- \quad \longrightarrow \quad R—Y + \quad X:^-$$

Substrate   Nucleophile   Product   Leaving group

The *substrate* is normally an alkyl halide in which the halogen is bonded to an sp^3 hybridized carbon. The *nucleophile* is a Lewis base, that is, a species containing an unshared pair of electrons. The *leaving group* must be a weaker base than the nucleophile in order for the substitution reaction to be effective.

$$R—X + \quad Y^- \quad \rightleftharpoons \quad R—Y + \quad X^-$$

Substrate   Stronger   Product   Weaker
            base   base

Bimolecular Nucleophilic Substitution (Secs. 8.4 to 8.7) One class of nucleophilic substitution reactions exhibits *second-order kinetics* and is known as a *bimolecular nucleophilic substitution,* or S_N2, reaction. The reaction, of which a simple example is

$$CH_3Br + OH^- \longrightarrow CH_3OH + Br^-$$

is a single-step process in which *both* the substrate and the nucleophile are involved in the transition state.

An S_N2 reaction proceeds with *inversion of configuration* at the carbon that bears the leaving group. That is, the nucleophile attacks this carbon at the side opposite the bond to the leaving group.

The reactivity of substrates toward bimolecular substitution is controlled by *steric effects*; bulky substrates undergo S_N2 reaction very slowly. The reactivity order may be summarized as follows:

$$CH_3X > RCH_2X > R_2CHX > R_3CX$$

Methyl   Primary   Secondary   Tertiary

Most reactive;   Least reactive;
least crowded   most crowded

Increased substitution on the carbon adjacent to that undergoing reaction will also decrease the reactivity. For example, $(CH_3)_2CHCH_2X$ reacts more slowly than $CH_3CH_2CH_2X$.

Nucleophiles (Sec. 8.8) Nucleophiles are Lewis bases and are often anions. Neutral nucleophiles frequently participate in *solvolysis* reactions, that is, reactions in which the nucleophilic species is the solvent. Examples include water (in which case the reaction is known as *hydrolysis*) and alcohols.

Nucleophilicity is a measure of how reactive a nucleophile is in displacing a leaving group from a substrate. A negatively charged nucleophile is more reactive than its neutral conjugate acid. For example, an alkoxide RO^- is more nucleophilic than the corresponding alcohol ROH. In comparing elements of the same group on the periodic chart in protic solvents, we see that larger anions further down on the chart are less solvated and are better nucleophiles. Therefore I^- is the most nucleophilic halide, and RS^- is more nucleophilic than RO^- in a protic solvent. Protic solvents include water, alcohols, and carboxylic acids.

Unimolecular Nucleophilic Substitution (Secs. 8.9 to 8.12) The class of substitution reactions which exhibit *first-order kinetics*, that is, which have a *unimolecular* rate-determining step, are known as S_N1 reactions. The mechanism of these reactions involves two steps, the first of which is a unimolecular ionization resulting in formation of a carbocation. This step is the *slow*, or rate-determining, step of the reaction. In the second step the carbocation is rapidly captured by the nucleophile.

$$R_3C{-}L \xrightarrow{\text{slow}} R_3C^+L^- \xrightarrow[\text{fast}]{\text{Nu:}} R_3C{-}Nu$$

Reactivity in S_N1 substitution is governed by *electronic effects* and parallels ease of carbocation formation by the substrate.

$$R_3CX > R_2CHX > RCH_2X > CH_3X$$

Tertiary   Secondary   Primary   Methyl

Most reactive; forms   Least reactive; forms
most stable carbocation   least stable carbocation

Although the intermediate carbocation is planar (the charged carbon is sp^2 hybridized), only partial racemization occurs upon S_N1 reaction of an optically active substrate. As shown in text Figure 8.8, a carbocation–halide ion pair forms, allowing the leaving group to shield one side of the carbocation from attack by the nucleophile.

Rearrangement of the carbon skeleton may also accompany an S_N1 reaction. As discussed in Chapter 5, carbocations may rearrange (by migration of a hydrogen or an alkyl group) to form a more stable cation, which then undergoes further reaction. For example, hydrolysis of 2-bromo-3-methylbutane yields mainly 2-methyl-2-butanol.

$$\underset{\underset{Br}{|}}{\underset{\overset{CH_3}{|}}{CH_3CHCHCH_3}} \xrightarrow{H_2O} \underset{\underset{OH}{|}}{\underset{\overset{CH_3}{|}}{CH_3CCH_2CH_3}}$$

Substitution versus Elimination (Sec. 8.14) Two factors are important in determining whether substitution or elimination will predominate in a particular chemical reaction; these are the *structure of the alkyl halide substrate* and the *basicity of the anion*. Primary substrates will tend to favor S_N2 substitution; increasing steric hindrance will cause more E2 elimination to occur.

The principal reaction of *secondary and tertiary alkyl halides* with alkoxide bases is *elimination*. Nucleophiles that are *weaker bases* than alkoxides, for example, CN^-, will favor substitution with secondary as well as primary substrates. The following table summarizes these observations:

	Nucleophile	
Alkyl halide	Strong base*	Weak base
Primary	S_N2	S_N2
Secondary	E2	S_N2, S_N1, or E1, E2
Tertiary	E2	E2, E1, or S_N1

* K_a of conjugate acid less than 10^{-12}.

Sulfonate Esters (Sec. 8.15) Reaction of an alcohol with a sulfonyl chloride yields a sulfonate ester derivative. The *p*-toluenesulfonates, or tosylates, are frequently employed as substrates in nucleophilic substitution reactions.

$$ROH + CH_3\!-\!\!\left\langle\!\!\bigcirc\!\!\right\rangle\!\!-\!SO_2Cl \xrightarrow{\text{pyridine}} CH_3\!-\!\!\left\langle\!\!\bigcirc\!\!\right\rangle\!\!-\!SO_2OR \quad (ROTs)$$

$$ROTs + Y\!:^- \longrightarrow R\!-\!Y + OTs^-$$

IMPORTANT REACTIONS

S_N2 Substitution Reactions

Examples:

$$(CH_3)_2CHCH_2Br \xrightarrow{CN^-} (CH_3)_2CHCH_2CN$$

$$(CH_3)_3CO^- + CH_3Br \longrightarrow (CH_3)_3COCH_3$$

$$(R)\text{-}CH_3\overset{\overset{\displaystyle Br}{|}}{C}HCH_2CH_3 \xrightarrow[\text{acetone}]{NaI} (S)\text{-}CH_3\overset{\overset{\displaystyle I}{|}}{C}HCH_2CH_3$$

S_N1 Substitution Reactions

Examples:

$$(CH_3)_2\overset{\overset{\displaystyle Br}{|}}{C}CH_2CH_3 \xrightarrow{C_2H_5OH} (CH_3)_2\overset{\overset{\displaystyle OCH_2CH_3}{|}}{C}CH_2CH_3 + \text{alkenes}$$

SOLUTIONS TO TEXT PROBLEMS

8.1 Identify the nucleophilic anion in each reactant. The nucleophilic anion replaces bromine as a substituent on carbon.

(*b*) Potassium ethoxide serves as a source of the nucleophilic anion $CH_3CH_2O^-$.

$$CH_3CH_2O^- + CH_3Br \longrightarrow CH_3CH_2OCH_3 + Br^-$$

| Ethoxide ion (nucleophile) | Methyl bromide | Ethyl methyl ether (product) | Bromide ion |

(*c*)

Benzoate ion   Methyl bromide   Methyl benzoate   Bromide ion

(*d*) Lithium azide is a source of the azide ion

$$:\ddot{N}=\overset{+}{N}=\ddot{N}:$$

It reacts with methyl bromide to give methyl azide.

$$:\ddot{N}=\overset{+}{N}=\ddot{N}: + CH_3Br \longrightarrow CH_3\overset{..}{N}=\overset{+}{N}=\ddot{N}: + Br^-$$

| Azide ion (nucleophile) | Methyl bromide | Methyl azide (product) | Bromide ion |

(*e*) The nucleophilic anion in KCN is cyanide ($:\bar{C}\equiv N:$). The carbon atom is negatively charged and is normally the site of nucleophilic reactivity.

$$:N\equiv C:^- + CH_3Br \longrightarrow CH_3C\equiv N: + Br^-$$

| Cyanide ion (nucleophile) | Methyl bromide | Methyl cyanide (product) | Bromide ion |

(*f*) The anion in sodium hydrogen sulfide (NaSH) is ^-SH.

$$HS^- + CH_3Br \longrightarrow CH_3SH + Br^-$$

| Hydrogen sulfide ion | Methyl bromide | Methanethiol | Bromide ion |

8.2 Write out the structure of the starting material. Notice that it contains a primary bromide and a primary chloride. Bromide is a better leaving group than chloride and is the one that is displaced faster by the nucleophilic cyanide ion.

$$ClCH_2CH_2CH_2Br \xrightarrow[\text{ethanol-water}]{\text{NaCN}} ClCH_2CH_2CH_2C\equiv N$$

1-Bromo-3-chloropropane   3-Chloropropyl cyanide

8.3 No, the two-step sequence is not consistent with the observed behavior for the hydrolysis of methyl bromide. The rate-determining step in the two-step sequence shown is the first step, ionization of methyl bromide to give methyl cation.

1. $CH_3Br \xrightarrow{\text{slow}} CH_3^+ + Br^-$

2. $CH_3^+ + HO^- \xrightarrow{\text{fast}} CH_3OH$

In such a sequence the nucleophile would not participate in the reaction until after the rate-determining step is past, and the reaction rate would depend only on the concentration of methyl bromide and be independent of the concentration of hydroxide ion.

$$\text{Rate} = k[\text{CH}_3\text{Br}]$$

The indicated kinetic behavior is first-order, not second-order, as is actually observed for methyl bromide hydrolysis.

8.4 Inversion of configuration occurs at the stereogenic center. When shown in a Fischer projection, this corresponds to replacing the leaving group on the one side by the nucleophile on the opposite side.

$$\underset{\substack{\\ (S)\text{-}(+)\text{-}2\text{-Bromooctane}}}{\overset{\overset{\displaystyle\text{CH}_3}{\big|}}{\text{H} \!-\!\!\!\!\!-\!\! \text{Br}}} \ \underset{\substack{\\ \text{S}_\text{N}2}}{\overset{\text{NaOH}}{\longrightarrow}} \ \underset{\substack{\\ (R)\text{-}(-)\text{-}2\text{-Octanol}}}{\overset{\overset{\displaystyle\text{CH}_3}{\big|}}{\text{HO} \!-\!\!\!\!\!-\!\! \text{H}}}$$

(with substituent $\text{CH}_2(\text{CH}_2)_4\text{CH}_3$ on each)

8.5 The example given in the text illustrates inversion of configuration in the S_N2 hydrolysis of (S)-$(+)$-2-bromooctane, which yields (R)-$(+)$-2-octanol. The hydrolysis of (R)-$(-)$-2-bromooctane exactly mirrors that of its enantiomer and yields (S)-$(+)$-2-octanol.

Hydrolysis of racemic 2-bromooctane gives racemic 2-octanol. Remember, optically inactive reactants must yield optically inactive products.

8.6 Sodium iodide in acetone is a reagent that converts alkyl chlorides and bromides into alkyl iodides by an S_N2 mechanism. Pick the alkyl halide that is most reactive toward S_N2 displacement.

(b) 1-Bromopentane is a primary alkyl halide and so is more reactive than 3-bromopentane, which is secondary.

$$\text{BrCH}_2\text{CH}_2\text{CH}_2\text{CH}_2\text{CH}_3 \qquad\qquad \underset{\overset{\displaystyle|}{\text{Br}}}{\text{CH}_3\text{CH}_2\text{CHCH}_2\text{CH}_3}$$

1-Bromopentane   3-Bromopentane
(primary; more reactive in S_N2)   (secondary; less reactive in S_N2)

The less crowded alkyl halide reacts faster in an S_N2 reaction.

(c) Both halides are secondary, but fluoride is quite a poor leaving group in nucleophilic substitution reactions. Alkyl chlorides are more reactive than alkyl fluorides.

$$\underset{\overset{\displaystyle|}{\text{Cl}}}{\text{CH}_3\text{CHCH}_2\text{CH}_2\text{CH}_3} \qquad\qquad \underset{\overset{\displaystyle|}{\text{F}}}{\text{CH}_3\text{CHCH}_2\text{CH}_2\text{CH}_3}$$

2-Chloropentane   2-Fluoropentane
(more reactive)   (less reactive)

(d) A secondary alkyl bromide reacts faster under S_N2 conditions than a tertiary one.

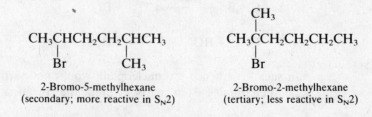

2-Bromo-5-methylhexane   2-Bromo-2-methylhexane
(secondary; more reactive in S_N2)   (tertiary; less reactive in S_N2)

(e) The number of carbons does not matter as much as the degree of substitution at the reaction site. The primary alkyl bromide is more reactive than the secondary.

$$BrCH_2(CH_2)_8CH_3 \qquad\qquad CH_3CHCH_3$$
$$\qquad\qquad\qquad\qquad\qquad\qquad | $$
$$\qquad\qquad\qquad\qquad\qquad\qquad Br$$

1-Bromodecane   2-Bromopropane
(primary; more reactive in S_N2)   (secondary; less reactive in S_N2)

8.7 Nitrite ion has two potentially nucleophilic sites, oxygen and nitrogen.

$$:\ddot{O}{=}\ddot{N}{-}\ddot{O}:^- + R{-}\ddot{I}: \longrightarrow :\ddot{O}{=}\ddot{N}{-}\ddot{O}{-}R + :\ddot{I}:^-$$

Nitrite ion   Alkyl iodide   Alkyl nitrite   Iodide ion

Nitrite ion   Alkyl iodide   Nitroalkane   Iodide ion

Thus, an alkyl iodide can yield either an alkyl nitrite or a nitroalkane depending on whether the oxygen or the nitrogen of nitrite ion attacks carbon. Both do, and the product from 2-iodooctane is a mixture of

$$CH_3CH(CH_2)_5CH_3 \qquad and \qquad CH_3CH(CH_2)_5CH_3$$
$$\qquad | \qquad\qquad\qquad\qquad\qquad\qquad | $$
$$\qquad ONO \qquad\qquad\qquad\qquad\qquad NO_2$$

8.8 Solvolysis of alkyl halides in alcohols yields ethers as the products of reaction.

$$(CH_3)_3CBr + CH_3OH \longrightarrow (CH_3)_3COCH_3 + \quad HBr$$

tert-Butyl   Methanol   tert-Butyl methyl   Hydrogen
bromide   ether   bromide

The reaction proceeds by an S_N1 mechanism.

$$(CH_3)_3C{-}\ddot{B}r: \xrightarrow{\text{slow}} (CH_3)_3C^+ + :\ddot{B}r:^-$$

tert-Butyl   tert-Butyl   Bromide
bromide   cation   ion

tert-Butyl   Methanol   tert-Butyloxonium
cation   ion

tert-Butyloxonium   Bromide   tert-Butyl   Hydrogen
ion   ion   methyl ether   bromide

8.9 The reactivity of an alkyl halide in an S_N1 reaction is dictated by the ease with which it ionizes to form a carbocation. Tertiary alkyl halides are the most reactive, methyl halides the least reactive.

(*b*) Cyclopentyl iodide ionizes to form a secondary carbocation, while the carbocation from 1-methylcyclopentyl iodide is tertiary. The tertiary halide is more reactive.

CH₃ I

H I

1-Methylcyclopentyl iodide
(tertiary; more reactive in S$_N$1)

Cyclopentyl iodide
(secondary; less reactive in S$_N$1)

(*c*) Cyclopentyl bromide ionizes to a secondary carbocation. 1-Bromo-2,2-dimethyl-propane is a primary alkyl halide and is therefore less reactive.

H Br

$(CH_3)_3CCH_2Br$

Cyclopentyl bromide
(secondary; more reactive in S$_N$1)

1-Bromo-2,2-dimethylpropane
(primary; less reactive in S$_N$1)

(*d*) Iodide is a better leaving group than chloride in both S$_N$1 and S$_N$2 reactions.

$(CH_3)_3CI$

$(CH_3)_3CCl$

tert-Butyl iodide
(more reactive)

tert-Butyl chloride
(less reactive)

8.10 The alkyl halide is tertiary and so undergoes hydrolysis by an S$_N$1 mechanism. The carbocation can be captured by water at either face. A mixture of the axial and the equatorial alcohols is formed.

CH₃

CH₃

Br

cis-1,4-Dimethylcyclohexyl bromide

OH

CH₃

CH₃

CH₃

$\xleftarrow{\text{H}_2\text{O}}$

+

CH₃

$\xrightarrow{\text{H}_2\text{O}}$

CH₃

CH₃

OH

trans-1,4-Dimethylcyclohexanol

Carbocation intermediate

cis-1,4-Dimethylcyclohexanol

The same two substitution products are formed from *trans*-1,4-dimethylcyclohexyl bromide because it undergoes hydrolysis via the same carbocation intermediate.

8.11 Write chemical equations illustrating each rearrangement process.

Hydride shift:

$$CH_3{-}\overset{+}{C}{-}CCH_3 \longrightarrow CH_3{-}C{-}\overset{+}{C}{\overset{CH_3}{\underset{CH_3}{}}}$$

Tertiary carbocation

Methyl shift:

$$CH_3{-}\overset{+}{C}{-}C{-}CH_3 \longrightarrow CH_3{-}C{-}\overset{+}{C}{\overset{H}{\underset{CH_3}{}}}$$

Secondary carbocation

Rearrangement by a hydride shift is observed because it converts a secondary carbocation to a more stable tertiary one. A methyl shift gives a secondary carbocation—the same carbocation as the one that existed prior to rearrangement.

8.12 (*b*) Ethyl bromide is a primary alkyl halide and reacts with the potassium salt of cyclohexanol by substitution.

$$CH_3CH_2Br + \bigcirc\!\!-OK \longrightarrow \bigcirc\!\!-OCH_2CH_3$$

Ethyl bromide   Potassium cyclohexanolate   Cyclohexyl ethyl ether

(*c*) No strong base is present in this reaction; the nucleophile is methanol itself, not methoxide. It reacts with *sec*-butyl bromide by substitution, not elimination.

$$CH_3CHCH_2CH_3 \xrightarrow{\ CH_3OH\ } CH_3CHCH_2CH_3$$
$$\quad\ \ |\qquad\qquad\qquad\qquad\quad |$$
$$\quad\ \ Br\qquad\qquad\qquad\qquad\ OCH_3$$

sec-Butyl bromide   *sec*-Butyl methyl ether

(*d*) Secondary alkyl halides react with alkoxide bases by E2 elimination.

$$CH_3CHCH_2CH_3 \xrightarrow[CH_3OH]{NaOCH_3} CH_3CH{=}CHCH_3 + CH_2{=}CHCH_2CH_3$$
$$\quad\ \ |$$
$$\quad\ \ Br$$

sec-Butyl bromide   2-Butene   1-Butene
(major product; mixture of cis and trans)

8.13 Alkyl *p*-toluenesulfonates are prepared from alcohols and *p*-toluenesulfonyl chloride.

$$CH_3(CH_2)_{16}CH_2OH + CH_3{-}\!\!\bigcirc\!\!\overset{\overset{O}{\|}}{\underset{\underset{O}{\|}}{S}}Cl$$

1-Octadecanol   *p*-Toluenesulfonyl chloride

↓ pyridine

$$CH_3(CH_2)_{16}CH_2O\overset{\overset{O}{\|}}{\underset{\underset{O}{\|}}{S}}\!\!\bigcirc\!\!{-}CH_3 + HCl$$

Octadecyl *p*-toluenesulfonate   Hydrogen chloride

8.14 As in part (*a*), identify the nucleophilic anion in each part. The nucleophile replaces the *p*-toluenesulfonate (tosylate) leaving group by an S$_N$2 process. The tosylate group is abbreviated —OTs.

(*b*) $I^- + CH_3(CH_2)_{16}CH_2OTs \longrightarrow CH_3(CH_2)_{16}CH_2I + TsO^-$

Iodide ion   Octadecyl *p*-toluenesulfonate   Octadecyl iodide   *p*-Toluenesulfonate anion

(*c*) $\bar{C}{\equiv}N + CH_3(CH_2)_{16}CH_2OTs \longrightarrow CH_3(CH_2)_{16}CH_2C{\equiv}N + TsO^-$

Cyanide ion   Octadecyl *p*-toluenesulfonate   Octadecyl cyanide   *p*-Toluenesulfonate anion

(d) HS^- + $CH_3(CH_2)_{16}CH_2OTs$ $\longrightarrow$ $CH_3(CH_2)_{16}CH_2SH$ + TsO^-

Hydrogen   Octadecyl   1-Octadecanethiol   p-Toluenesulfonate
sulfide ion   p-toluenesulfonate   anion

(e) $CH_3CH_2CH_2CH_2S^-$ + $CH_3(CH_2)_{16}CH_2OTS$ $\longrightarrow$

Butanethiolate   Octadecyl
     ion   p-toluenesulfonate

$CH_3(CH_2)_{16}CH_2SCH_2CH_2CH_2CH_3$ + TsO^-

Butyl octadecyl   p-Toluenesulfonate
thioether   anion

8.15 The hydrolysis of (S)-(+)-1-methylheptyl p-toluenesulfonate proceeds with inversion of configuration, giving the R enantiomer of 2-octanol.

$$CH_3(CH_2)_5 \overset{\overset{H}{|}}{\underset{\underset{CH_3}{|}}{C}}-OTs \xrightarrow{H_2O} HO-\overset{\overset{H}{|}}{\underset{\underset{CH_3}{|}}{C}}(CH_2)_5CH_3$$

(S)-(+)-1-Methylheptyl   (R)-(−)-2-Octanol
p-toluenesulfonate

In Section 8.15 of the text we are told that optically pure (S)-(+)-1-methylheptyl p-toluenesulfonate is prepared from optically pure (S)-(+)-2-octanol having a specific rotation $[\alpha]_D^{25}$ +9.9°. The conversion of an alcohol to a p-toluenesulfonate proceeds with complete *retention* of configuration. Therefore hydrolysis of this p-toluenesulfonate with *inversion* of configuration yields optically pure (R)-(−)-2-octanol of $[\alpha]_D^{25}$ −9.9°.

8.16 1-Bromopropane is a primary alkyl halide, and so it will undergo predominantly S_N2 displacement regardless of the basicity of the nucleophile.

(a) $CH_3CH_2CH_2Br$ $\xrightarrow[\text{acetone}]{\text{NaI}}$ $CH_3CH_2CH_2I$

1-Bromopropane   1-Iodopropane

(b) $CH_3CH_2CH_2Br$ $\xrightarrow[\text{acetic acid}]{\overset{\overset{O}{\|}}{CH_3CONa}}$ $CH_3CH_2CH_2O\overset{\overset{O}{\|}}{C}CH_3$

Propyl acetate

(c) $CH_3CH_2CH_2Br$ $\xrightarrow[\text{ethanol}]{\text{NaOCH}_2\text{CH}_3}$ $CH_3CH_2CH_2OCH_2CH_3$

Ethyl propyl ether

(d) $CH_3CH_2CH_2Br$ $\xrightarrow[\text{DMSO}]{\text{NaCN}}$ $CH_3CH_2CH_2CN$

Propyl cyanide

(e) $CH_3CH_2CH_2Br$ $\xrightarrow[\text{ethanol-water}]{\text{NaN}_3}$ $CH_3CH_2CH_2N_3$

1-Azidopropane

(f) $CH_3CH_2CH_2Br$ $\xrightarrow[\text{ethanol}]{\text{NaSH}}$ $CH_3CH_2CH_2SH$

1-Propanethiol

(g) $CH_3CH_2CH_2Br$ $\xrightarrow[\text{ethanol}]{\text{NaSCH}_3}$ $CH_3CH_2CH_2SCH_3$

Methyl propyl sulfide

8.17 In the corresponding reactions of 2-bromopropane, the possibility of elimination is more pronounced.

(*a*)   Iodide is weakly basic and very nucleophilic, and so substitution is the principal reaction.

$$CH_3CHCH_3 \xrightarrow[\text{acetone}]{\text{NaI}} CH_3CHCH_3$$
$$\quad | \qquad\qquad\qquad\qquad |$$
$$\quad Br \qquad\qquad\qquad\qquad I$$

2-Bromopropane   2-Iodopropane

(*b*)   Sodium acetate is weakly basic (K_a for acetic acid is 1.8×10^{-5}), and so the product should be isopropyl acetate.

$$CH_3CHCH_3 \xrightarrow[\text{acetic acid}]{\overset{\overset{O}{\|}}{CH_3CONa}} CH_3CHCH_3$$

Isopropyl acetate

(*c*)   Bases as strong as or stronger than hydroxide react with secondary alkyl halides by E2 elimination. Ethoxide is comparable with hydroxide in basicity.

$$CH_3CHCH_3 \xrightarrow{\text{NaOCH}_2\text{CH}_3} CH_3CH{=}CH_2$$
$$\quad | $$
$$\quad Br \qquad\qquad\qquad\qquad \text{Propene}$$

(By way of confirmation, it is known that the elimination-substitution ratio is 87:13 for the reaction of isopropyl bromide with KOH in ethanol-water.)

(*d*)   Sodium cyanide is a weak enough base and a good enough nucleophile that substitution is favored with most secondary alkyl halides.

$$CH_3CHCH_3 \xrightarrow[\text{DMSO}]{\text{NaCN}} CH_3CHCH_3$$
$$\quad | \qquad\qquad\qquad\qquad |$$
$$\quad Br \qquad\qquad\qquad\qquad CN$$

Isopropyl cyanide

(*e*)   Sodium azide is an even weaker base than cyanide and a better nucleophile. Substitution occurs.

$$CH_3CHCH_3 \xrightarrow[\text{ethanol-water}]{\text{NaN}_3} CH_3CHCH_3$$
$$\quad | \qquad\qquad\qquad\qquad |$$
$$\quad Br \qquad\qquad\qquad\qquad N_3$$

2-Azidopropane

(*f*)   Sodium hydrogen sulfide is a relatively weak base and reacts with secondary alkyl halides by substitution.

$$CH_3CHCH_3 \xrightarrow{\text{NaSH}} CH_3CHCH_3$$
$$\quad | \qquad\qquad\qquad\qquad |$$
$$\quad Br \qquad\qquad\qquad\qquad SH$$

2-Propanethiol

(*g*)   Substitution occurs with sodium methanethiolate.

$$CH_2CHCH_3 \xrightarrow{\text{NaSCH}_3} CH_3CHCH_3$$
$$\quad | \qquad\qquad\qquad\qquad |$$
$$\quad Br \qquad\qquad\qquad\qquad SCH_3$$

Isopropyl methyl sulfide

8.18   The alkyl halide is tertiary and reacts with anionic nucleophiles by elimination.

$$CH_3\underset{\underset{CH_3}{|}}{\overset{\overset{Br}{|}}{C}}CH_3 \xrightarrow{\text{anionic nucleophile}} CH_2{=}C(CH_3)_2$$

2-Bromo-2-methylpropane   2-Methylpropene

8.19   (*a*)   The substrate is a primary alkyl bromide and reacts with sodium iodide in acetone to give the corresponding iodide.

$$BrCH_2\overset{\overset{O}{\|}}{C}OCH_2CH_3 \xrightarrow[\text{acetone}]{\text{NaI}} ICH_2\overset{\overset{O}{\|}}{C}OCH_2CH_3$$

Ethyl bromoacetate   Ethyl iodoacetate (89%)

(*b*)   Primary alkyl chlorides react readily with sodium iodide in acetone to yield the corresponding iodides.

$$O_2N{-}\text{⟨benzene⟩}{-}CH_2Cl \xrightarrow[\text{acetone}]{\text{NaI}} O_2N{-}\text{⟨benzene⟩}{-}CH_2I$$

p-Nitrobenzyl chloride   *p*-Nitrobenzyl iodide (100%)

(*c*)   An analogous reaction occurs with sodium acetate to yield an acetate ester.

$$O_2N{-}\text{⟨benzene⟩}{-}CH_2Cl \xrightarrow[\text{acetic acid}]{CH_3\overset{O}{C}ONa} O_2N{-}\text{⟨benzene⟩}{-}CH_2O\overset{\overset{O}{\|}}{C}CH_3$$

p-Nitrobenzyl acetate (78–82%)

(*d*)   The only leaving group in the substrate is bromide. Neither of the carbon-oxygen bonds is susceptible to cleavage by nucleophilic attack.

$$CH_3CH_2OCH_2CH_2Br \xrightarrow[\text{ethanol-water}]{\text{NaCN}} CH_3CH_2OCH_2CH_2CN$$

2-Bromoethyl ethyl ether   2-Cyanoethyl ethyl ether (52–58%)

(*e*)   Hydrolysis of the primary chloride yields the corresponding alcohol.

$$NC{-}\text{⟨benzene⟩}{-}CH_2Cl \xrightarrow{H_2O,\ HO^-} NC{-}\text{⟨benzene⟩}{-}CH_2OH$$

p-Cyanobenzyl chloride   *p*-Cyanobenzyl alcohol (85%)

(*f*)   The substrate is a primary chloride.

$$ClCH_2\overset{\overset{O}{\|}}{C}OC(CH_3)_3 \xrightarrow[\text{acetone-water}]{\text{NaN}_3} N_3CH_2\overset{\overset{O}{\|}}{C}OC(CH_3)_3$$

tert-Butyl chloroacetate   *tert*-Butyl azidoacetate (92%)

(*g*)   Both primary toluenesulfonate leaving group are replaced by cyanide.

cyclohexene with CH_2OTs, CH_2OTs $\xrightarrow[\text{DMSO}]{\text{NaCN}}$ cyclohexene with CH_2CN, CH_2CN

4,5-Bis-(cyanomethyl)cyclohexene

(*h*)   Primary alkyl tosylates yield iodides on treatment with sodium iodide in acetone.

$$\text{TsOCH}_2\text{-(2,2-Dimethyl-1,3-dioxolan-4-yl)} \xrightarrow[\text{acetone}]{\text{NaI}} \text{ICH}_2\text{-(2,2-Dimethyl-4-(iodomethyl)-1,3-dioxolane)}$$

(2,2-Dimethyl-1,3-dioxolan-4-yl)-
methyl *p*-toluenesulfonate

2,2-Dimethyl-4-(iodomethyl)-
1,3-dioxolane (60%)

(*i*)   The nucleophilic sulfur displaces chloride from the —CH₂Cl group.

Benzyl chloride + Sodium benzenethiolate ⟶ Benzyl phenyl sulfide (60%)

(*j, k*)   Nucleophilic substitution takes place with inversion of configuration. The trans chloride yields a cis substitution product.

trans-4-*tert*-Butylcyclohexyl chloride + Sodium benzenethiolate ⟶ *cis*-4-*tert*-Butylcyclohexyl phenyl sulfide

The cis chloride yields a trans substitution product.

cis-4-*tert*-Butylcyclohexyl chloride + Sodium benzenethiolate ⟶ *trans*-4-*tert*-Butylcyclohexyl phenyl sulfide

(*l*)   Cyanide displaces chloride from the —CH₂Cl group.

2-(Chloromethyl)furan $\xrightarrow{\text{NaCN}}$ 2-(Cyanomethyl)furan (90%)

(*m*)   Sulfur displaces bromide from ethyl bromide.

Sodium (2-furyl)-methanethiolate + Ethyl bromide ⟶ Ethyl (2-furyl)methyl sulfide (80%)

(*n*)   The first reaction is one in which a substituted alcohol is converted to a *p*-toluenesulfonate ester. This is followed by an S_N2 displacement with lithium iodide.

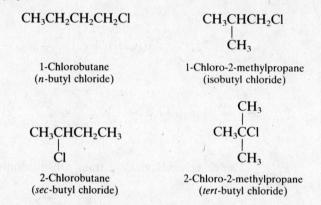

4-(2,3,4-Trimethoxyphenyl)-1-butanol

4-(2,3,4-Trimethoxyphenyl)-1-butyl iodide

8.20   The isomers of C_4H_9Cl are:

$$CH_3CH_2CH_2CH_2Cl$$

1-Chlorobutane
(*n*-butyl chloride)

$$CH_3CHCH_2Cl$$
$$|$$
$$CH_3$$

1-Chloro-2-methylpropane
(isobutyl chloride)

$$CH_3CHCH_2CH_3$$
$$|$$
$$Cl$$

2-Chlorobutane
(*sec*-butyl chloride)

$$CH_3$$
$$|$$
$$CH_3CCl$$
$$|$$
$$CH_3$$

2-Chloro-2-methylpropane
(*tert*-butyl chloride)

The reaction conditions (sodium iodide in acetone) are typical for an S_N2 process.

The order of S_N2 reactivity is primary > secondary > tertiary, and branching of the chain close to the site of substitution hinders reaction. The unbranched primary halide *n*-butyl chloride will be the most reactive and the tertiary halide *tert*-butyl chloride the least. Therefore the order of reactivity will be: 1-chlorobutane > 1-chloro-2-methylpropane > 2-chlorobutane > 2-chloro-2-methylpropane.

8.21   1-Chlorohexane is a primary alkyl halide; 2-chlorohexane and 3-chlorohexane are secondary.

$$CH_3CH_2CH_2CH_2CH_2CH_2Cl$$

$$CH_3CHCH_2CH_2CH_2CH_3$$
$$|$$
$$Cl$$

$$CH_3CH_2CHCH_2CH_2CH_3$$
$$|$$
$$Cl$$

1-Chlorohexane
(primary)

2-Chlorohexane
(secondary)

3-Chlorohexane
(secondary)

Primary and secondary alkyl halides react with potassium iodide in acetone by an S_N2 mechanism, and the rate depends on steric hindrance to attack on the alkyl halide by the nucleophile.

(*a*)   Primary alkyl halides are more reactive than secondary alkyl halides in S_N2 reactions. 1-Chlorohexane is the most reactive isomer.

(*b*)   Branching at the carbon adjacent to the one that bears the leaving group slows down the rate of nucleophilic displacement. In 2-chlorohexane the group adjacent to the point of attack is CH_3 (unbranched). In 3-chlorohexane the group adjacent to the point of attack is CH_2CH_3 (branched). 2-Chlorohexane is more reactive than 3-chlorohexane by a factor of two.

8.22 (*a*) Iodide is a better leaving group than bromide, and so 1-iodobutane should undergo S_N2 attack by cyanide faster than 1-bromobutane.

(*b*) The reaction conditions are typical for an S_N2 process. The methyl branch in 1-chloro-2-methylbutane sterically hinders attack at C-1. The unbranched isomer, 1-chloropentane, reacts faster.

$$\begin{array}{c} CH_3 \\ | \\ CH_3CH_2CHCH_2Cl \end{array} \qquad CH_3CH_2CH_2CH_2CH_2Cl$$

1-Chloro-2-methylbutane is more sterically hindered, therefore less reactive.

1-Chloropentane is less sterically hindered, therefore more reactive.

(*c*) Hexyl chloride is a primary alkyl halide, and cyclohexyl chloride is secondary. Azide ion is a good nucleophile, and so the S_N2 reactivity rules apply; primary is more reactive than secondary.

$$CH_3CH_2CH_2CH_2CH_2CH_2Cl$$

Hexyl chloride is primary, therefore more reactive in S_N2.

Cyclohexyl chloride is secondary, therefore less reactive in S_N2.

(*d*) 1-Bromo-2,2-dimethylpropane is too hindered to react with the weakly nucleophilic ethanol by an S_N2 reaction, and since it is a primary alkyl halide, it is less reactive in S_N1 reactions. *tert*-Butyl bromide will react with ethanol by an S_N1 mechanism at a reasonable rate owing to formation of a tertiary carbocation.

$$(CH_3)_3CBr \qquad (CH_3)_3CCH_2Br$$

tert-Butyl bromide; more reactive in S_N1 solvolysis

1-Bromo-2,2-dimethylpropane; relatively unreactive in nucleophilic substitution reactions

(*e*) Solvolysis of alkyl halides in aqueous formic acid is faster for those which form carbocations readily. The S_N1 reactivity order applies here: secondary > primary.

$$\begin{array}{c} CH_3CHCH_2CH_3 \\ | \\ Br \end{array} \qquad (CH_3)_2CHCH_2Br$$

sec-Butyl bromide; secondary, therefore more reactive in S_N1

Isobutyl bromide; primary, therefore less reactive in S_N1

(*f*) 1-Chlorobutane is a primary alkyl halide and so should react by a direct displacement or S_N2 mechanism. Sodium methoxide is more basic than sodium acetate and is a better nucleophile. Reaction will occur faster with sodium methoxide than with sodium acetate.

(*g*) Azide ion is a very good nucleophile, whereas *p*-toluenesulfonate is a very good leaving group but a very poor nucleophile. In an S_N2 reaction with 1-chlorobutane, sodium azide will react faster than sodium *p*-toluenesulfonate.

8.23 There are only two possible products from free-radical chlorination of the starting alkane:

$$(CH_3)_3CCH_2C(CH_3)_3 \xrightarrow[h\nu]{Cl_2} (CH_3)_3CCH_2C(CH_3)_2 + (CH_3)_3C\overset{6}{C}HC(CH_3)_3$$
$$\qquad\qquad\qquad\qquad\qquad\qquad\qquad\qquad | \qquad\qquad\qquad | $$
$$\qquad\qquad\qquad\qquad\qquad\qquad\qquad\qquad CH_2Cl \qquad\qquad Cl$$

2,2,4,4-Tetramethylpentane

1-Chloro-2,2,4,4-tetramethylpentane (primary)

3-Chloro-2,2,4,4-tetramethylpentane (secondary)

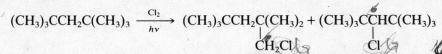

As revealed by their structural formulas, one isomer is a primary alkyl chloride, the other is secondary. The problem states that the major product (compound A) undergoes S_N1 hydrolysis much more slowly than the minor product (compound B). Since secondary halides are much more reactive than primary halides under S_N1 conditions, the major (unreactive) product is the primary alkyl halide 1-chloro-2,2,4,4-tetramethylpentane (compound A) and the minor (reactive) product is the secondary alkyl halide 3-chloro-2,2,4,4-tetramethylpentane (compound B).

8.24 Using the unshared electron pair on its nitrogen, triethylamine acts as a nucleophile in an S_N2 reaction toward ethyl iodide.

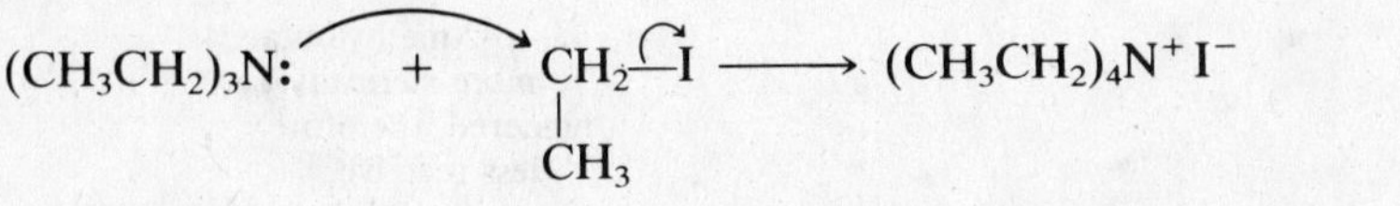

$$(CH_3CH_2)_3N: \quad + \quad CH_2{-}I \longrightarrow (CH_3CH_2)_4N^+ I^-$$
$$\underset{\displaystyle CH_3}{\big|}$$

Triethylamine   Ethyl iodide   Tetraethylammonium iodide

The product of the reaction is a salt and has the structure shown. The properties given in the problem (soluble in polar solvents, high melting point) are typical of those of an ionic compound.

8.25 This reaction has been reported in the chemical literature and proceeds as shown (91% yield):

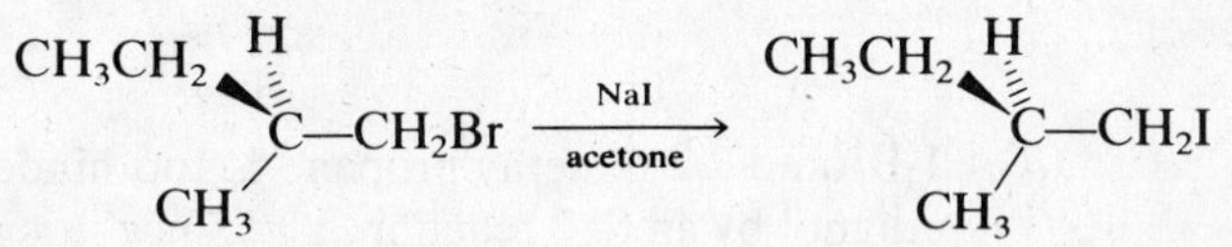

$$
\begin{array}{ccc}
CH_3CH_2 \;\; H & & CH_3CH_2 \;\; H \\
\diagdown \quad | & \xrightarrow[\text{acetone}]{\text{NaI}} & \diagdown \quad | \\
C{-}CH_2Br & & C{-}CH_2I \\
| & & | \\
CH_3 & & CH_3
\end{array}
$$

(S)-1-Bromo-2-methylbutane (S)-1-Iodo-2-methylbutane

Notice that the configuration of the product *is the same* as the configuration of the reactant. This is because the stereogenic center is not involved in the reaction. When we say that S_N2 reactions proceed with inversion of configuration we refer only to the carbon at which substitution takes place, not a stereogenic center elsewhere in the molecule.

8.26 (*a*) The starting material incorporates both a primary chloride and a secondary chloride. The nucleophile (iodide) attacks the less hindered primary position.

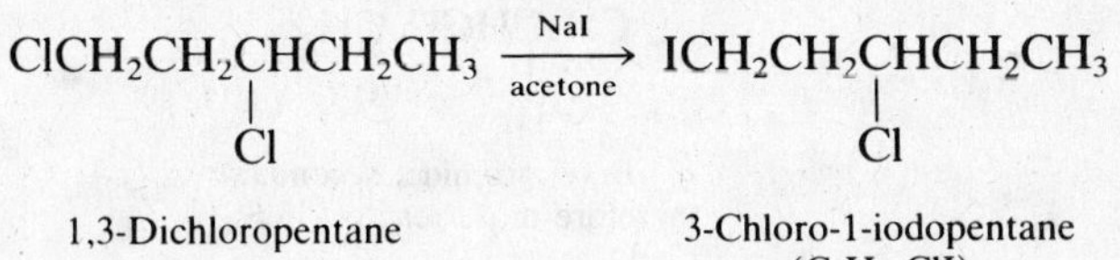

$$
ClCH_2CH_2CHCH_2CH_3 \xrightarrow[\text{acetone}]{\text{NaI}} ICH_2CH_2CHCH_2CH_3
$$
$$
\underset{\displaystyle Cl}{\quad\quad\quad\quad\quad\quad\big|} \qquad\qquad\qquad \underset{\displaystyle Cl}{\quad\quad\quad\quad\quad\quad\big|}
$$

1,3-Dichloropentane   3-Chloro-1-iodopentane
                                   ($C_5H_{10}ClI$)

(*b*) Nucleophilic substitution of the first bromide by sulfur occurs in the usual way.

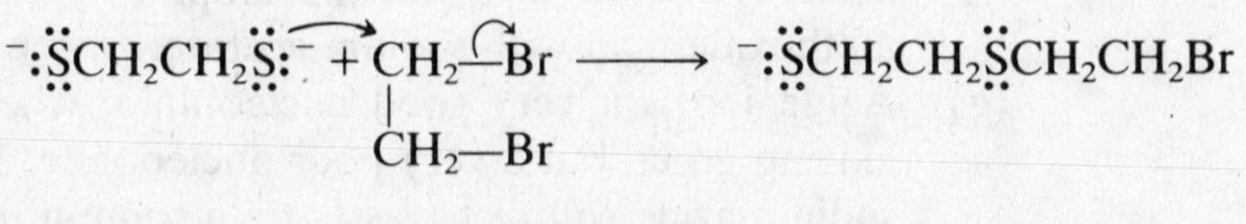

$$^-:\!SCH_2CH_2\ddot{S}:^- + CH_2{-}Br \longrightarrow {}^-:\!SCH_2CH_2\ddot{S}CH_2CH_2Br$$
$$\underset{\displaystyle CH_2{-}Br}{\big|}$$

The product of this step cyclizes by way of an intramolecular nucleophilic substitution.

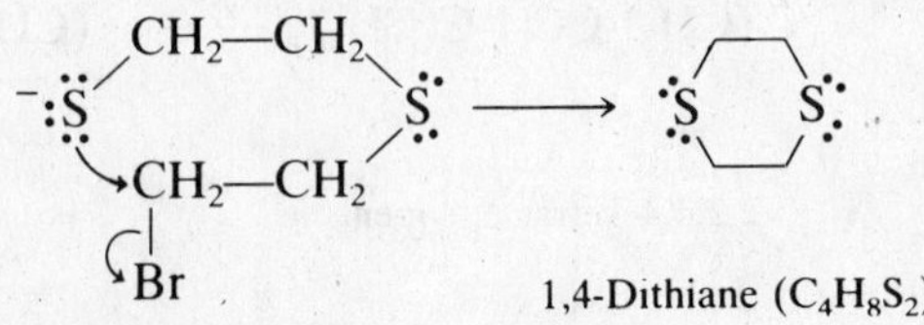

$$
\begin{array}{ccc}
CH_2{-}CH_2 & & \\
\diagup \qquad\qquad & & \\
^-:\!S \qquad\qquad \ddot{S}: & \longrightarrow & :\!\ddot{S} \qquad \ddot{S}: \\
\diagdown \qquad\qquad & & \\
CH_2{-}CH_2 & & \\
| & & \\
Br & & \text{1,4-Dithiane } (C_4H_8S_2)
\end{array}
$$

(c) The nucleophile is a dianion (S^{2-}). Two nucleophilic substitution reactions take place; the second of the two leads to intramolecular cyclization.

$$:\ddot{S}:^{2-} + CH_2{-}Cl \longrightarrow {}^{-}:\ddot{S}CH_2CH_2CH_2CH_2{-}Cl$$
$$CH_2CH_2CH_2Cl$$

Thiolane (C_4H_8S)

8.27 (a) Methyl halides are unhindered and react rapidly by the S_N2 mechanism.

(b) Sodium ethoxide is a good nulceophile and will react with unhindered primary alkyl halides by the S_N2 mechanism.

(c) Cyclohexyl bromide is a secondary halide and will react with a strong base (sodium ethoxide) predominantly by the $E2$ mechanism.

(d) The tertiary halide *tert*-butyl bromide will undergo solvolysis by the S_N1 mechanism.

(e) The presence of the strong base sodium ethoxide will cause the $E2$ mechanism to predominate.

(f) Concerted reactions are those which occur in a single step. The bimolecular mechanisms S_N2 and $E2$ represent concerted processes.

(g) In a stereospecific reaction, stereoisomeric reactants yield products that are stereoisomers of each other. Reactions that occur by the S_N2 and $E2$ mechanisms are stereospecific.

(h) The unimolecular mechanisms S_N1 and $E1$ involve the formation of carbocation intermediates.

(i) Rearrangements are possible when carbocations are intermediates in a reaction. Thus reactions occurring by the S_N1 and $E1$ mechanisms are most likely to have a rearranged carbon skeleton.

(j) Iodide is a better leaving group than bromide, and alkyl iodides will react faster than alkyl bromides by any of the four mechanisms S_N1, S_N2, $E1$, and $E2$.

8.28 (a) Alkyl fluorides cannot be prepared directly from alcohols, but other alkyl halides can be converted to alkyl fluorides by nucleophilic substitution. Therefore, first convert ethanol to ethyl bromide (or chloride or iodide).

$$CH_3CH_2OH + \quad HBr \quad \xrightarrow{heat} \quad CH_3CH_2Br + H_2O$$
Ethanol　Hydrogen bromide　Ethyl bromide　Water

Then use fluoride ion to displace bromide.

$$CH_3CH_2Br \xrightarrow{KF} CH_3CH_2F$$
Ethyl bromide　Ethyl fluoride

Alternatively, ethanol could be converted to its *p*-toluenesulfonate ester and then treated with potassium fluoride. This reaction has been reported in the chemical literature and leads to ethyl fluoride in 86 percent yield.

(b) Cyclopentyl cyanide can be prepared from a cyclopentyl halide by a nucleophilic substitution reaction. The first task, therefore, is to convert cyclopentane to a cyclopentyl halide.

 $+ \quad Cl_2 \xrightarrow{light\ or\ heat}$ [cyclopentyl chloride] $+ \quad HCl$

Cyclopentane　Chlorine　Cyclopentyl chloride　Hydrogen chloride

$$\text{Cyclopentyl-Cl} + \text{NaCN} \xrightarrow[\text{or DMSO}]{\text{ethanol-water}} \text{Cyclopentyl-CN} + \text{NaCl}$$

| Cyclopentyl chloride | Sodium cyanide | Cyclopentyl cyanide | Sodium chloride |

An analogous sequence involving cyclopentyl bromide could be used.

(c) Cyclopentene can serve as a precursor to a cyclopentyl halide.

$$\text{Cyclopentene} + \text{HBr} \longrightarrow \text{Cyclopentyl-Br}$$

Cyclopentene Hydrogen bromide Cyclopentyl bromide

Once cyclopentyl bromide has been prepared, it is converted to cyclopentyl cyanide by nucleophilic substitution, as shown in part (b).

(d) Dehydration of cyclopentanol yields cyclopentene, which can then be converted to cyclopentyl cyanide, as shown in part (c).

$$\text{Cyclopentanol-OH} \xrightarrow[\text{heat}]{\text{H}_2\text{SO}_4} \text{Cyclopentene}$$

Cyclopentanol   Cyclopentene

(e) Two cyano groups are required here, both of which must be introduced in nucleophilic substitution reactions. The substrate in the key reaction is $BrCH_2CH_2Br$.

$$BrCH_2CH_2Br + 2NaCN \xrightarrow[\text{DMSO}]{\text{ethanol-water or}} N\equiv CCH_2CH_2C\equiv N$$

| 1,2-Dibromoethane | Sodium cyanide | 1,2-Dicyanoethane |

1,2-Dibromoethane is prepared from ethylene. Therefore, the overall synthesis from ethyl alcohol is formulated as shown:

$$CH_3CH_2OH \xrightarrow[\text{heat}]{\text{H}_2\text{SO}_4} CH_2{=}CH_2 \xrightarrow{\text{Br}_2} BrCH_2CH_2Br \xrightarrow{\text{NaCN}} NCCH_2CH_2CN$$

| Ethyl alcohol | Ethylene | 1,2-dibromoethane | 1,2-Dicyanoethane |

(f) In this synthesis a primary alkyl chloride must be converted to a primary alkyl iodide. This is precisely the kind of transformation for which sodium iodide in acetone is used.

$$(CH_3)_2CHCH_2Cl + NaI \xrightarrow{\text{acetone}} (CH_3)_2CHCH_2I + NaCl$$

| Isobutyl chloride | Sodium iodide | Isobutyl iodide | Sodium chloride |

(g) First convert *tert*-butyl chloride into an isobutyl halide.

$$(CH_3)_3CCl \xrightarrow{\text{NaOCH}_3} (CH_3)_2C{=}CH_2 \xrightarrow[\text{peroxides}]{\text{HBr}} (CH_3)_2CHCH_2Br$$

| *tert*-Butyl chloride | 2-Methylpropene | Isobutyl bromide |

Treating isobutyl bromide with sodium iodide in acetone converts it to isobutyl iodide.

$$(CH_3)_2CHCH_2Br \xrightarrow[\text{acetone}]{\text{NaI}} (CH_3)_2CHCH_2I$$

| Isobutyl bromide | Isobutyl iodide |

A second approach is by way of isobutyl alcohol.

$$(CH_3)_3CCl \xrightarrow{NaOCH_3} (CH_3)_2C{=}CH_2 \xrightarrow[\text{2. } H_2O_2, HO^-]{\text{1. } B_2H_6} (CH_3)_2CHCH_2OH$$

tert-Butyl   2-Methylpropene   Isobutyl
chloride   alcohol

Isobutyl alcohol is then converted to its *p*-toluenesulfonate ester, which reacts with sodium iodide in acetone in a manner analogous to that of isobutyl bromide.

(*h*) First introduce a leaving group into the molecule by converting isopropyl alcohol to an isopropyl halide. Then convert the resulting isopropyl halide to isopropyl azide by a nucleophilic substitution reaction

$$\underset{\substack{\text{OH} \\ \text{Isopropyl} \\ \text{alcohol}}}{CH_3CHCH_3} \xrightarrow{HBr} \underset{\substack{\text{Br} \\ \text{Isopropyl} \\ \text{bromide}}}{CH_3CHCH_3} \xrightarrow{NaN_3} \underset{\substack{N_3 \\ \text{Isopropyl} \\ \text{azide}}}{CH_3CHCH_3}$$

(*i*) In this synthesis 1-propanol must be first converted to an isopropyl halide.

$$\underset{\text{1-Propanol}}{CH_3CH_2CH_2OH} \xrightarrow[\text{heat}]{H_2SO_4} \underset{\text{Propene}}{CH_3CH{=}CH_2} \xrightarrow{HBr} \underset{\substack{\text{Br} \\ \text{Isopropyl} \\ \textbf{bromide}}}{CH_3CHCH_3}$$

Once an isopropyl halide has been obtained, it can be treated with sodium azide in a nucleophilic substitution reaction.

$$\underset{\substack{\text{Br}}}{CH_3CHCH_3} + NaN_3 \longrightarrow \underset{\substack{N_3}}{CH_3CHCH_3} + NaBr$$

Isopropyl bromide   Sodium azide   Isopropyl azide   Sodium bromide

(*j*) First write out the structure of the starting material and of the product so as to determine their relationship in three dimensions.

(*R*)-*sec*-Butyl alcohol   (*S*)-*sec*-Butyl azide

The hydroxyl group must be replaced by azide with inversion of configuration. First, however, a leaving group must be introduced and it must be introduced in such a way that the configuration at the stereogenic center is not altered. The best way to do this is to convert (*R*)-*sec*-butyl alcohol to its corresponding *p*-toluenesulfonate ester.

(*R*)-*sec*-Butyl alcohol   (*R*)-*sec*-Butyl *p*-toluenesulfonate

Next, convert the *p*-toluenesulfonate to the desired azide by an S_N2 reaction.

(R)-*sec*-Butyl *p*-toluenesulfonate   (S)-*sec*-Butyl azide

(*k*) This problem is carried out in exactly the same way as the preceding one, except that the nucleophile in the second step is HS⁻.

(R)-*sec*-Butyl alcohol   (R)-*sec*-Butyl *p*-toluenesulfonate   (S)-2-Butanethiol

8.29 (*a*) The two possible combinations of alkyl bromide and alkoxide ion which might yield *tert*-butyl methyl ether are:

1. CH_3Br + $(CH_3)_3CO^-$ $\xrightarrow{\text{fast}}$ $(CH_3)_3COCH_3$

   Methyl bromide   *tert*-Butoxide   *tert*-Butyl methyl ether
                              ion

2. CH_3O^- + $(CH_3)_3CBr$ $\xrightarrow{\text{slow}}$ $(CH_3)_3COCH_3$

   Methoxide ion   *tert*-Butyl bromide   *tert*-Butyl methyl ether

We choose the first approach because it is an S_N2 reaction on the unhindered substrate, methyl bromide. The second approach requires an S_N2 reaction on a hindered tertiary alkyl halide, a very poor choice. Indeed, we would expect that the reaction of methoxide ion with *tert*-butyl bromide could not give any ether at all but would proceed entirely by E2 elimination:

$$CH_3O^- + (CH_3)_3CBr \xrightarrow{\text{fast}} CH_3OH + CH_2{=}C(CH_3)_2$$

                                          Methanol   2-Methylpropene

(*b*) Again, the better alternative is to choose the less hindered alkyl halide to permit substitution to predominate over elimination.

Potassium cyclopentoxide Bromomethane   Cyclopentyl methyl ether

An attempt to prepare this compound by the reaction

Chlorocyclopentane Sodium methoxide   Cyclopentyl methyl ether

gave cyclopentyl methyl ether in only 24 percent yield. Cyclopentene was isolated in 31 percent yield.

(*c*) A 2,2-dimethylpropyl halide is too sterically hindered to be a good candidate for this synthesis. The only practical method is:

$$(CH_3)_3CCH_2OK + CH_3CH_2Br \longrightarrow (CH_3)_3CCH_2OCH_2CH_3$$

| Potassium 2,2-dimethylpropoxide | Bromoethane | Ethyl 2,2-dimethylpropyl ether |

8.30 (*a*) The problem states that the reaction type is nucleophilic substitution. Therefore, sodium acetylide is the nucleophile and must be treated with an alkyl halide to give the desired product.

$$Na^+ :\bar{C} \equiv CH + CH_3CH_2Br \longrightarrow CH_3CH_2C \equiv CH + \quad NaBr$$

| Sodium acetylide | Ethyl bromide | 1-Butyne | Sodium bromide |

(*b*) The acidity data given in the problem for acetylene tell us that $HC \equiv CH$ is a very weak acid ($K_a = 10^{-25}$), so that sodium acetylide must be a very strong base—stronger than hydroxide ion. Therefore, *elimination* by the E2 mechanism rather than S_N2 substitution is expected to be the principal (probably the exclusive) reaction observed with secondary and tertiary alkyl halides. The substitution reaction will work well with primary alkyl halides but will likely fail for secondary and tertiary ones. Alkynes such as $(CH_3)_2CHC \equiv CH$ and $(CH_3)_3CC \equiv CH$ could not be prepared by this method.

8.31 (*a*) In order to convert *trans*-2-methylcyclopentanol to *cis*-2-methylcyclopentyl acetate the hydroxyl group must be replaced by acetate with inversion of configuration. Hydroxide is a poor leaving group and so must first be converted to a good leaving group. The best choice is *p*-toluenesulfonate, because this can be prepared by a reaction which alters none of the bonds to the stereogenic center.

trans-2-Methylcyclopentanol   *trans*-2-Methylcyclopentyl *p*-toluenesulfonate

Treatment of the *p*-toluenesulfonate with potassium acetate in acetic acid will proceed with inversion of configuration to give the desired product.

trans-2-Methylcyclopentyl *p*-toluenesulfonate   *cis*-2-Methylcyclopentyl acetate

(*b*) In order to decide on the best sequence of reactions, we must begin by writing structural formulas to determine what kinds of transformations are required.

1-Methylcyclopentanol   *cis*-2-Methylcyclopentyl acetate

We already know from part (*a*) of this problem how to convert *trans*-2-methylcyclopentanol to *cis*-2-methylcyclopentyl acetate, so all that is really necessary is to design a synthesis of *trans*-2-methylcyclopentanol. Therefore:

1-Methylcyclopentanol   1-Methylcyclopentene   *trans*-2-Methylcyclopentanol

Hydroboration-oxidation converts 1-methylcyclopentene to the desired alcohol by anti-Markovnikov syn hydration of the double bond. The resulting alcohol is then converted to its *p*-toluenesulfonate ester and treated with acetate ion as in part (*a*) to give *cis*-2-methylcyclopentyl acetate.

8.32 (*a*) The reaction of an alcohol with a sulfonyl chloride gives a sulfonate ester. The oxygen of the alcohol remains in place and is the atom to which the sulfonyl group becomes attached.

(*S*)-(+)-2-Butanol   (*S*)-*sec*-Butyl methanesulfonate

(*b*) Sulfonate is similar to iodide in its leaving-group behavior. The product in part (*a*) is attacked by $NaSCH_2CH_3$ in an S_N2 reaction. Inversion of configuration occurs at the stereogenic center.

(*S*)-*sec*-Butyl methanesulfonate   (*R*)-(−)-*sec*-Butyl ethyl sulfide

(*c*) In this part of the problem we deduce the stereochemical outcome of the reaction of 2-butanol with PBr_3. We know the absolute configuration of (+)-2-butanol (*S*) from the statement of the problem and the configuration of (−)-*sec*-butyl ethyl sulfide (*R*) from part (*b*). We are told that the sulfide formed from (+)-2-butanol via the bromide has a positive rotation. Therefore it must have the opposite configuration of the product in part (*b*).

(*S*)-(+)-2-Butanol   2-Bromobutane   (*S*)-(+)-*sec*-Butyl ethyl sulfide

Since the reaction of the bromide with $NaSCH_2CH_3$ must proceed with inversion of configuration at the stereogenic center, and since the final product has the same configuration as the starting alcohol, the conversion of the alcohol to the bromide must proceed with inversion of configuration.

(*d*) The conversion of 2-butanol to *sec*-butyl methanesulfonate does not involve any of the bonds to the stereogenic center, and so it must proceed with 100 percent retention of configuration. Assuming that the reaction of the methanesulfonate with $NaSCH_2CH_3$ proceeds with 100 percent inversion of configuration, we conclude that the maximum rotation of *sec*-butyl ethyl sulfide is the value given in the statement of part (*b*), i.e., ±25°. Since the sulfide produced in part (*c*) has a rotation of +23°, it is 92 percent optically pure. It is reasonable to assume that the loss of optical purity occurred in the conversion of the alcohol to the bromide, rather than in the reaction of the bromide with $NaSCH_2CH_3$. Therefore if the bromide is 92 percent optically pure and has a rotation of $-38°$, optically pure bromide has a rotation of 38/0.92, or $\pm 41°$.

8.33 (a) If each act of substitution occurred with retention of configuration, there would be no observable racemization; $k_{rac} = 0$.

(R)-(−)-2-Iodooctane   [(R)-(−)-2-Iodooctane]*
(* indicates radioactive label)

Therefore $k_{rac}/k_{exch} = 0$.

(b) If each act of exchange proceeds with inversion of configuration, (R)-$(−)$-2-iodooctane will be transformed to radioactively labeled (S)-$(+)$-2-iodooctane.

(R)-(−)-2-Iodooctane   [(S)-(+)-2-Iodooctane]*

Thus, if one starts with 100 molecules of (R)-$(−)$-2-iodooctane, the compound will be completely racemized when 50 molecules have become radioactive. Therefore

$$k_{rac}/k_{exch} = 2$$

(c) If radioactivity is incorporated in a stereorandom fashion, then 2-iodooctane will be 50 percent racemized when 50 percent of it has reacted. Therefore,

$$k_{rac}/k_{exch} = 1$$

In fact, Hughes found that the rate of racemization was twice the rate of incorporation of radioactive iodide. This experiment provided strong evidence for the belief that bimolecular nucleophilic substitution proceeds stereospecifically with inversion of configuration.

8.34 The substrate is a tertiary alkyl bromide and can undergo S_N1 substitution and E1 elimination under these reaction conditions. Elimination in either of two directions to give regioisomeric alkenes can also occur.

2-Bromo-2-methylbutane   1,1-Dimethylpropyl acetate   2-Methyl-1-butene   2-Methyl-2-butene

8.35 (a) Tertiary alkyl halides undergo nucleophilic substitution only by way of carbocations: S_N1 is the most likely mechanism for solvolysis of the 2-halo-2-methylbutanes.

2-Halo-2-methylbutanes are tertiary alkyl halides

(b) Tertiary alkyl halides can undergo either E1 or E2 elimination. Since no alkoxide base is present, solvolytic elimination most likely occurs by an E1 mechanism.

(c, d) Iodides react faster than bromides in substitution and elimination reactions irrespective of whether the mechanism is E1, E2, S_N1, or S_N2.

(*e*) Solvolysis in aqueous ethanol can give rise to an alcohol or an ether as product, depending on whether the carbocation is captured by water or ethanol.

$$(CH_3)_2\underset{\underset{X}{|}}{C}CH_2CH_3 \xrightarrow[H_2O]{CH_3CH_2OH} (CH_3)_2\underset{\underset{OH}{|}}{C}CH_2CH_3 + (CH_3)_2\underset{\underset{OCH_2CH_3}{|}}{C}CH_2CH_3$$

2-Methyl-2-butanol Ethyl 1,1-dimethylpropyl ether

(*f*) Elimination can yield either of two isomeric alkenes.

$$(CH_3)_2\underset{\underset{X}{|}}{C}CH_2CH_3 \longrightarrow CH_2{=}\underset{\underset{CH_3}{|}}{C}CH_2CH_3 + (CH_3)_2C{=}CHCH_3$$

2-Methyl-1-butene 2-Methyl-2-butene

Zaitsev's rule predicts that 2-methyl-2-butene should be the major alkene.

(*g*) The product distribution is determined by what happens to the carbocation intermediate. If the carbocation is free of its leaving group, its fate will be the same no matter whether the leaving group is bromide or iodide.

8.36 Solvolysis of 1,2-dimethylpropyl *p*-toluenesulfonate in acetic acid is expected to give one substitution product and two alkenes.

$$(CH_3)_2CH\underset{\underset{OTs}{|}}{C}HCH_3 \xrightarrow{CH_3CO_2H} (CH_3)_2CH\underset{\underset{\underset{\underset{O}{\|}}{OCCH_3}}{|}}{C}HCH_3 + (CH_3)_2C{=}CHCH_3 + (CH_3)_2CHCH{=}CH_2$$

1,2-Dimethylpropyl *p*-toluenesulfonate; 1,2-Dimethylpropyl acetate; 2-Methyl-2-butene; 3-Methyl-1-butene

Since five products are formed, we are led to consider the possibility of carbocation rearrangements in S_N1 and E1 solvolysis.

$$(CH_3)_2CH\overset{+}{C}HCH_3 \longrightarrow (CH_3)_2\overset{+}{C}CH_2CH_3 \longrightarrow$$

1,2-Dimethylpropyl cation 1,1-Dimethylpropyl cation

$$(CH_3)_2\underset{\underset{\underset{\underset{O}{\|}}{OCCH_3}}{|}}{C}CH_2CH_3 + (CH_3)_2C{=}CHCH_3 + CH_2{=}\underset{\underset{CH_3}{|}}{C}CH_2CH_3$$

1,1-Dimethylpropyl acetate 2-Methyl-2-butene 2-Methyl-1-butene

Since 2-methyl-2-butene is a product common to both carbocation intermediates, a total of five different products are accounted for. There are two substitution products:

$$(CH_3)_2CH\underset{\underset{\underset{\underset{O}{\|}}{OCCH_3}}{|}}{C}HCH_3 \qquad (CH_3)_2\underset{\underset{\underset{\underset{O}{\|}}{OCCH_3}}{|}}{C}CH_2CH_3$$

1,2-Dimethylpropyl acetate 1,1-Dimethylpropyl acetate

and three elimination products:

$$(CH_3)_2C{=}CHCH_3 \qquad (CH_3)_2CHCH{=}CH_2 \qquad CH_2{=}\underset{\underset{CH_3}{|}}{C}CH_2CH_3$$

2-Methyl-2-butene 3-Methyl-1-butene 2-Methyl-1-butene

8.37 Solution A contains both acetate ion and methanol as nucleophiles. Acetate is more nucleophilic than methanol, and so the major observed reaction is:

$$CH_3I + CH_3CO^- \xrightarrow{\text{methanol}} CH_3OCCH_3$$

Methyl iodide Acetate Methyl acetate

Solution B prepared by adding potassium methoxide to acetic acid rapidly undergoes an acid-base transfer process:

$$CH_3O^- + CH_3COH \longrightarrow CH_3OH + CH_3CO^-$$

Methoxide Acetic acid Methanol Acetate
(stronger base) (stronger acid) (weaker acid) (weaker base)

Thus the major base present is not methoxide but acetate. Therefore methyl iodide reacts with acetate anion in solution B to give methyl acetate.

8.38 Alkyl chlorides arise by the reaction sequence:

$$RCH_2OH + CH_3\!-\!\!\langle\rangle\!\!-\!SCl + \text{pyridine} \longrightarrow$$

Primary alcohol p-Toluenesulfonyl chloride Pyridine

$$RCH_2OS\!-\!\!\langle\rangle\!\!-\!CH_3 + \text{pyridinium } Cl^-$$

Primary alkyl p-toluenesulfonate Pyridinium chloride

$$RCH_2OS\!-\!\!\langle\rangle\!\!-\!CH_3 + Cl^- \longrightarrow RCH_2Cl + CH_3\!-\!\!\langle\rangle\!\!-\!SO^-$$

Primary alkyl p-toluenesulfonate Primary alkyl chloride

The reaction proceeds to form the alkyl p-toluenesulfonate as expected, but the chloride anion formed in this step subsequently acts as a nucleophile and displaces p-toluenesulfonate from RCH_2OTs.

8.39 Iodide ion is both a better nucleophile than cyanide and a better leaving group than bromide. Therefore, the two reactions shown are together faster than the reaction of cyclopentyl bromide with sodium cyanide alone.

$$\text{Cyclopentyl-Br} \xrightarrow{\text{NaI}} \text{Cyclopentyl-I} \xrightarrow{\text{NaCN}} \text{Cyclopentyl-CN}$$

Cyclopentyl bromide Cyclopentyl iodide Cyclopentyl cyanide

SELF-TEST

PART A

A-1. Write the correct structure of the reactant or product omitted from each of the following. Clearly indicate stereochemistry where it is important.

(a) $CH_3CH_2CH_2CH_2Br \xrightarrow[CH_3CH_2OH]{CH_3CH_2ONa}$?

(b) ? $\xrightarrow{NaCN}$

(c) 1-Chloro-3-methylbutane + sodium azide $\longrightarrow$?

(d) $\xrightarrow[CH_3OH]{CH_3ONa}$? (major)

(e) $+ NaN_3 \longrightarrow$?

(f) $+ NaSCH_3 \longrightarrow$?

A-2. Choose the best pair of reactants to form the following product by an S_N2 reaction:

$$(CH_3)_2CHSCH_2CH_2CH_3$$

A-3. Outline the chemical steps necessary to synthesize (R)-2-iodobutane from (S)-2-butanol as the starting material.

A-4. Hydrolysis of 2-chloro-3,3-dimethylbutane yields 2,3-dimethyl-2-butanol as the major product. Explain this observation, using chemical structures to outline the mechanism of the reaction.

A-5. Identify the class of reaction (e.g., E2) and write the kinetic equation for the solvolysis of *tert*-butyl bromide in methanol.

A-6. Provide a brief explanation why the halogen exchange reaction shown is an acceptable synthetic method:

A-7. Write chemical structures for compounds A through D in the following sequence of reactions:

$$A \xrightarrow{K} B$$

$$C \xrightarrow{HBr, heat} D$$

$$B + D \longrightarrow CH_3CH_2O-\text{⬡}$$

A-8. Write a mechanism describing the solvolysis (S_N1) of 1-bromo-1-methyl-cyclohexane in warm ethanol.

PART B

B-1. The bimolecular substitution reaction

$$CH_3Br + OH^- \longrightarrow CH_3OH + Br^-$$

is represented by the kinetic equation:
(a)   Rate = $k[CH_3Br]^2$
(b)   Rate = $k[CH_3Br][OH^-]$
(c)   Rate = $k[CH_3Br] + k[OH^-]$
(d)   Rate = $k/[CH_3Br][OH^-]$

B-2.   The solvolysis reactions of optically active halides generally proceed with
(a)   Complete racemization
(b)   Partial racemization
(c)   Complete retention of configuration
(d)   Complete inversion of configuration

B-3.   For the reaction

$$\text{(cyclohexyl–Br)} + CH_3CH_2O^-Na^+ \longrightarrow$$

the major product is formed by
(a)   An S_N1 reaction   (c)   An E1 reaction
(b)   An S_N2 reaction   (d)   An E2 reaction

B-4.   Which of the following statements pertaining to an S_N2 reaction are true?
1.   The rate of reaction is independent of the concentration of the nucleophile.
2.   The nucleophile attacks carbon on the side of the molecule opposite the group being displaced.
3.   The reaction proceeds with simultaneous bond formation and bond rupture.
4.   Partial racemization of an optically active substrate results.
(a)   1, 4   (b)   1, 3, 4   (c)   2, 3   (d)   All

B-5.   Ethanol reacts with *tert*-butyl bromide according to the equation shown. If the concentration of ethanol is doubled, by what factor will the rate of the reaction change?

$$(CH_3)_3CBr + CH_3CH_2OH \longrightarrow (CH_3)_3COCH_2CH_3$$

(a)   Remain the same
(b)   Increase by a factor of 2
(c)   Increase by a factor of 4
(d)   Decrease by a factor of 2

B-6.   Which of the following phrases are *not* correctly associated with S_N1 reactions?
1.   Rearrangements possible
2.   Rate affected by solvent polarity
3.   Strength of nucleophile important in determining rate
4.   Reactivity series tertiary > secondary > primary
5.   Proceed with complete inversion of configuration
(a)   3, 5   (b)   5 only   (c)   2, 3, 5   (d)   3 only

B-7.   Rank the following in order of decreasing rate of solvolysis with aqueous ethanol (fastest → slowest):

$$CH_2{=}\underset{\underset{CH_3}{|}}{C}{-}Br \qquad \text{(1-methyl-1-bromocyclohexane)} \qquad \underset{\underset{Br}{|}}{CH_3}CHCH_2CH(CH_3)_2$$

A   B   C

(a)   B > A > C   (b)   A > B > C   (c)   B > C > A   (d)   A > C > B

B-8. Rank the following species in order of decreasing nucleophilicity in a polar protic solvent (most → least nucleophilic):

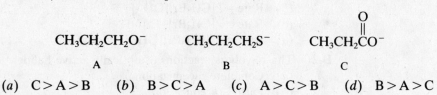

$$CH_3CH_2CH_2O^- \qquad CH_3CH_2CH_2S^- \qquad CH_3CH_2\overset{\overset{\displaystyle O}{\|}}{C}O^-$$

A   B   C

(*a*) C > A > B   (*b*) B > C > A   (*c*) A > C > B   (*d*) B > A > C

B-9. From each of the following pairs select the compound that will react the fastest with sodium iodide in acetone.

2-Chloropropane   or   2-bromopropane
1   2

1-Bromobutane   or   2-bromobutane
3   4

(*a*) 1, 3   (*b*) 1, 4   (*c*) 2, 3   (*d*) 2, 4

B-10. Select the reagent that will yield the greater amount of substitution on reaction with 1-bromobutane.
(*a*) CH_3CH_2OK in dimethyl sulfoxide (DMSO)
(*b*) $(CH_3)_3COK$ in dimethyl sulfoxide (DMSO)
(*c*) Both (*a*) and (*b*) will give comparable amounts of substitution.
(*d*) Neither (*a*) nor (*b*) will give any appreciable amount of substitution.

ALKYNES

IMPORTANT TERMS AND CONCEPTS

Nomenclature (Sec. 9.2) Alkynes are hydrocarbons having a carbon-carbon triple bond and are characterized by the general formula C_nH_{2n-2}. The usual IUPAC rules are followed in naming alkynes, the *-ane* suffix of the parent alkane being replaced with *-yne*. For example,

$$CH_3—C{\equiv}C—\overset{\overset{\displaystyle CH_3}{|}}{C}HCH_2CH_3$$

4-Methyl-2-hexyne

The C—C≡C—C unit of an alkyne is linear; therefore stereoisomerism around the triple bond is not possible.

Structure and Bonding (Sec. 9.3) As described in text Section 1.16, the triple-bond carbons of an alkyne are *sp* hybridized. The hybrid orbitals on each carbon atom are oriented 180° from each other; that is, they are in a linear arrangement. The triple bond itself consists of one σ component from overlap of *sp* hybrid orbitals and two π components at right angles to each other from overlap of the unhybridized *p* orbitals on each carbon atom.

Acidity of Acetylenic C—H Bonds (Sec. 9.6) Terminal alkynes, i.e., those having the triple bond between C-1 and C-2 of the carbon chain, are about 10^{19} times more acidic than terminal alkenes. A typical pK_a for a terminal alkyne is about 26. The greater *s* character of an *sp* orbital (50 percent) compared with an sp^2 orbital (33 percent) explains the greater stability of the acetylide anion. Acetylide anions may be generated by reaction of a terminal acetylene with a suitably strong base, as follows:

$$CH_3CH_2C{\equiv}CH + NaNH_2 \longrightarrow CH_3CH_2C{\equiv}C{:}^-Na^+ + NH_3$$

| Stronger acid | Stronger base | Weaker base | Weaker acid |

221

IMPORTANT REACTIONS

PREPARATION OF ALKYNES

Alkylation of Acetylide Anions (Sec. 9.7)

$$HC \equiv C^- Na^+ + CH_3CH_2CH_2Br \longrightarrow CH_3CH_2CH_2C \equiv CH$$

$$CH_3CH_2C \equiv C^- Na^+ + C_6H_5CH_2Br \longrightarrow C_6H_5CH_2C \equiv CCH_2CH_3$$

Elimination Reactions (Sec. 9.8)

$$(CH_3)_2CHCH_2CHCl_2 \xrightarrow{\text{NaNH}_2} (CH_3)_2CHC \equiv CH$$

$$CH_3CH_2\overset{\overset{\displaystyle Br}{|}}{C}HCH_2Br \xrightarrow{\text{NaNH}_2} CH_3CH_2C \equiv CH$$

REACTIONS OF ALKYNES

Hydrogenation (Sec. 9.10)

$$R-C \equiv C-R \xrightarrow[\text{Pt, Pd, Ni, or Rh}]{\text{H}_2} RCH_2CH_2R$$

$$R-C \equiv C-R \xrightarrow[\substack{\text{Pd/CaCO}_3 \\ \text{(Lindlar catalyst)}}]{\text{H}_2} \underset{H}{\overset{R}{}}C = C\underset{H}{\overset{R}{}} \quad \text{(cis isomer only)}$$

Metal-Ammonia Reduction (Sec. 9.11)

$$R-C \equiv C-R \xrightarrow[\text{NH}_3]{\text{Na}} \underset{H}{\overset{R}{}}C = C\underset{R}{\overset{H}{}} \quad \text{(trans isomer only)}$$

Hydrogen Halide Addition (Sec. 9.12)

$$CH_3CH_2C \equiv CH \xrightarrow{\text{HBr}} CH_3CH_2\overset{\overset{\displaystyle Br}{|}}{C} = CH_2$$

Hydration (Sec. 9.13)

$$R-C \equiv CH \xrightarrow[\text{Hg}^{2+}]{\text{H}_2\text{O, H}^+} R-\overset{\overset{\displaystyle O}{\|}}{C}-CH_3$$

Example:

$$\underset{}{\overset{CH_3}{}}CH_2C \equiv CH \xrightarrow[\text{Hg}^{2+}]{\text{H}_2\text{O, H}^+} \underset{}{\overset{CH_3}{}}CH_2CCH_3 \,(O)$$

Halogen Addition (Sec. 9.14)

$$R-C \equiv C-R \xrightarrow{\text{Br}_2} \underset{Br}{\overset{R}{}}C = C\underset{R}{\overset{Br}{}} \xrightarrow{\text{Br}_2} RCBr_2CBr_2R$$

Ozonolysis (Sec. 9.15)

$$RC \equiv CR' \xrightarrow[\text{2. H}_2\text{O}]{\text{1. O}_3} RCO_2H + R'CO_2H$$

SOLUTIONS TO TEXT PROBLEMS

9.1 A triple bond may connect C-1 and C-2 or C-2 and C-3 in an unbranched chain of five carbons.

$$CH_3CH_2CH_2C \equiv CH \qquad CH_3CH_2C \equiv CCH_3$$

1-Pentyne   2-Pentyne

One of the C_5H_8 isomers has a branched carbon chain.

$$CH_3CHC \equiv CH \qquad \text{3-Methyl-1-butyne}$$
$$|$$
$$CH_3$$

9.2 The methyl group in 1-butyne is attached to an sp^3 hybridized carbon, and so the CH_3—C bond in 1-butyne is longer than in 2-butyne, where the methyl carbon is bonded to an sp hybridized carbon.

$$\overset{sp^3}{CH_3}\text{—}\overset{sp^3}{CH_2}C \equiv CH \qquad \overset{sp^3}{CH_3}\text{—}\overset{sp}{C} \equiv C\text{—}CH_3$$

longer bond   shorter bond

9.3 (b) A proton is transferred from acetylene to ethyl anion.

$$HC \equiv C\text{—}H + {:}CH_2CH_3 \rightleftharpoons HC \equiv \bar{C}{:} + CH_3CH_3$$

Acetylene	Ethyl anion	Acetylide ion	Ethane
(stronger acid)	(stronger base)	(weaker base)	(weaker acid)
$K_a\ 10^{-26}$			$K_a \sim 10^{-62}$
$(pK_a\ 26)$			$(pK_a \sim 62)$

The position of equilibrium lies to the right. Ethyl anion is a very powerful base and deprotonates acetylene quantitatively.

(c) Amide ion is not a strong enough base to remove a proton from ethylene. The equilibrium lies to the left.

$$CH_2{=}CH\text{—}H + {:}\bar{N}H_2 \rightleftharpoons CH_2{=}\ddot{C}\bar{H} + {:}NH_3$$

Ethylene	Amide ion	Vinyl anion	Ammonia
(weaker acid)	(weaker base)	(stronger base)	(stronger acid)
$K_a \sim 10^{-45}$			$K_a\ 10^{-36}$
$(pK_a \sim 45)$			$(pK_a\ 36)$

(d) Alcohols are stronger acids than ammonia; the position of equilibrium lies to the right.

$$CH_3C \equiv CCH_2\ddot{O}\text{—}H + {:}\bar{N}H_2 \rightleftharpoons CH_3C \equiv CCH_2\ddot{O}{:}^- + {:}NH_3$$

2-Butyn-1-ol	Amide ion	2-Butyn-1-olate anion	Ammonia
(stronger acid)	(stronger base)	(weaker base)	(weaker acid)
$K_a \sim 10^{-16}$ to 10^{-20}			$K_a\ 10^{-36}$
$(pK_a \sim 16\text{–}20)$			$(pK_a\ 36)$

9.4 (b) The desired alkyne has a methyl group and a butyl group attached to a —C≡C— unit. Therefore, two alkylations of acetylene are required—one with a methyl halide, the other with a butyl halide.

$$HC \equiv CH \xrightarrow[\text{2. } CH_3Br]{\text{1. } NaNH_2,\ NH_3} CH_3C \equiv CH \xrightarrow[\text{2. } CH_3CH_2CH_2CH_2Br]{\text{1. } NaNH_2,\ NH_3} CH_3C \equiv CCH_2CH_2CH_2CH_3$$

Acetylene   Propyne   2-Heptyne

It does not matter whether the methyl group or the butyl group is introduced first; the order of steps shown in this synthetic scheme may be inverted.

(c)   An ethyl group and a propyl group need to be introduced as substituents on a —C≡C— unit. As in part (b), it does not matter which of the two is introduced first.

$$HC{\equiv}CH \xrightarrow[\text{2. } CH_3CH_2CH_2Br]{\text{1. } NaNH_2, NH_3} CH_3CH_2CH_2C{\equiv}CH \xrightarrow[\text{2. } CH_3CH_2Br]{\text{1. } NaNH_2, NH_3} CH_3CH_2CH_2C{\equiv}CCH_2CH_3$$

Acetylene   1-Pentyne   3-Heptyne

9.5   Both 1-pentyne and 2-pentyne can be prepared by alkylating acetylene. All the alkylation steps involve nucleophilic substitution of a methyl or primary alkyl halide.

$$HC{\equiv}CH \xrightarrow[\text{2. } CH_3CH_2CH_2Br]{\text{1. } NaNH_2, NH_3} CH_3CH_2CH_2C{\equiv}CH$$

Acetylene   1-Pentyne

$$HC{\equiv}CH \xrightarrow[\text{2. } CH_3CH_2Br]{\text{1. } NaNH_2, NH_3} CH_3CH_2C{\equiv}CH \xrightarrow[\text{2. } CH_3Br]{\text{1. } NaNH_2, NH_3} CH_3CH_2C{\equiv}CCH_3$$

Acetylene   1-Butyne   2-Pentyne

The third isomer, 3-methyl-1-butyne, cannot be prepared by alkylation of acetylene, because it requires a secondary alkyl halide as the alkylating agent. The reaction that takes place is elimination, not substitution.

$$HC{\equiv}\bar{C}: + \underset{\underset{Br}{|}}{CH_3CHCH_3} \xrightarrow{E2} HC{\equiv}CH + CH_2{=}CHCH_3$$

Acetylide   Isopropyl   Acetylene   Propene
ion   bromide

9.6   Each of the dibromides shown yields 3,3-dimethyl-1-butyne when subjected to double dehydrohalogenation with strong base.

$$\underset{\underset{Br}{|}}{\overset{\overset{Br}{|}}{(CH_3)_3CCCH_3}} \quad \text{or} \quad (CH_3)_3CCH_2CHBr_2$$

2,2-Dibromo-3,3-   1,1-Dibromo-3,3-
dimethylbutane   dimethylbutane

$$\text{or} \quad \underset{\underset{Br}{|}}{(CH_3)_3CCHCH_2Br} \xrightarrow[\text{2. } H_2O]{\text{1. } 3NaNH_2} (CH_3)_3CC{\equiv}CH$$

1,2-Dibromo-3,3-   3,3-Dimethyl-1-
dimethylbutane   butyne

9.7   (b)   The first task is to convert 1-propanol to propene:

$$CH_3CH_2CH_2OH \xrightarrow[\text{heat}]{H_2SO_4} CH_3CH{=}CH_2$$

1-Propanol   Propene

Once propene is available, it is converted to 1,2-dibromopropane and then to propyne as described in the sample solution for part (a).

(c)   Treat isopropyl bromide with a base to effect dehydrohalogenation.

$$(CH_3)_2CHBr \xrightarrow{NaOCH_2CH_3} CH_3CH{=}CH_2$$

Isopropyl bromide   Propene

Next, convert propene to propyne as in parts (a) and (b).

(*d*) The starting material contains only two carbon atoms, and so an alkylation step is needed at some point. Propyne arises by alkylation of acetylene, and so the last step in the synthesis is:

$$HC\equiv CH \xrightarrow[\text{2. } CH_3Br]{\text{1. } NaNH_2,\ NH_3} CH_3C\equiv CH$$

Acetylene   Propyne

The designated starting material, 1,1-dichloroethane, is a geminal dihalide and can be used to prepare acetylene by a double dehydrohalogenation.

$$CH_3CHCl_2 \xrightarrow[\text{2. } H_2O]{\text{1. } NaNH_2,\ NH_3} HC\equiv CH$$

1,1-Dichloroethane   Acetylene

(*e*) The first task is to convert ethyl alcohol to acetylene. Once acetylene is prepared it can be alkylated with a methyl halide.

$$CH_3CH_2OH \xrightarrow[\text{heat}]{H_2SO_4} CH_2=CH_2 \xrightarrow{Br_2} BrCH_2CH_2Br \xrightarrow[\text{2. } H_2O]{\text{1. } NaNH_2,\ NH_3} HC\equiv CH$$

Ethyl alcohol   Ethylene   1,2-Dibromoethane   Acetylene

$$\xrightarrow[\text{2. } CH_3Br]{\text{1. } NaNH_2,\ NH_3}$$

$$CH_3C\equiv CH$$

Propyne

9.8 The first task is to assemble a carbon chain containing eight carbons. Acetylene has two carbon atoms and can be alkylated via its sodium salt to 1-octyne. Hydrogenation over platinum converts 1-octyne to octane.

$$HC\equiv CH \xrightarrow[NH_3]{NaNH_2} HC\equiv CNa \xrightarrow{BrCH_2(CH_2)_4CH_3} HC\equiv CCH_2(CH_2)_4CH_3 \xrightarrow[Pt]{H_2}$$

Acetylene   Sodium acetylide   1-Octyne

$$CH_3CH_2CH_2(CH_2)_4CH_3$$

Octane

Alternatively, two successive alkylations of acetylene with $CH_3CH_2CH_2Br$ could be carried out to give 4-octyne ($CH_3CH_2CH_2C\equiv CCH_2CH_2CH_3$), which could then be hydrogenated to octane.

9.9 Hydrogenation over Lindlar palladium converts an alkyne to a cis alkene. Therefore, oleic acid has the structure indicated in the following equation:

$$CH_3(CH_2)_7C\equiv C(CH_2)_7CO_2H \xrightarrow[\text{Lindlar Pd}]{H_2} \underset{H \quad\quad H}{\overset{CH_3(CH_2)_7 \quad (CH_2)_7CO_2H}{C=C}}$$

Stearolic acid   Oleic acid

Hydrogenation of alkynes over platinum leads to alkanes.

$$CH_3(CH_2)_7C\equiv C(CH_2)_7CO_2H \xrightarrow[Pt]{H_2} CH_3(CH_2)_{16}CO_2H$$

Stearolic acid   Stearic acid

9.10 Alkynes are converted to trans alkenes on reduction with sodium in liquid ammonia.

$$CH_3(CH_2)_7C\equiv C(CH_2)_7CO_2H \xrightarrow[\text{2. } H_3O^+]{\text{1. Na, NH}_3}$$

CH$_3$(CH$_2$)$_7$ H
   \   /
    C=C
   /   \
  H   (CH$_2$)$_7$CO$_2$H

Stearolic acid   Elaidic acid

9.11 The proper double bond stereochemistry may be achieved by using 2-heptyne as a starting material in the final step. Lithium-ammonia reduction of 2-heptyne gives the trans alkene; hydrogenation over Lindlar palladium gives the cis isomer. Therefore, the first task is the alkylation of propyne to 2-heptyne.

$$CH_3C\equiv CH \xrightarrow[\text{2. } CH_3CH_2CH_2CH_2Br]{\text{1. NaNH}_2, NH_3} CH_3C\equiv CCH_2CH_2CH_2CH_3$$

Propyne   2-Heptyne

Li, NH$_3$

CH$_3$ H
   \   /
    C=C
   /   \
  H   CH$_2$CH$_2$CH$_2$CH$_3$

(*E*)-2-Heptene

H$_2$ / Lindlar Pd

CH$_3$ CH$_2$CH$_2$CH$_2$CH$_3$
   \   /
    C=C
   /   \
  H   H

(*Z*)-2-Heptene

9.12 First convert ethyl bromide to acetylene; then add hydrogen bromide across the carbon-carbon triple bond.

$$CH_3CH_2Br \xrightarrow[\text{DMSO, heat}]{\text{KOC(CH}_3)_3} CH_2{=}CH_2 \xrightarrow{\text{Br}_2} BrCH_2CH_2Br \xrightarrow[\text{2. } H_2O]{\text{1. NaNH}_2, NH_3}$$

Ethyl bromide   Ethylene   1,2-Dibromoethane

$$HC\equiv CH \xrightarrow{\text{HBr}} CH_2{=}CHBr$$

Vinyl bromide

Alternatively, treatment of 1,2-dibromoethane with a base such as KOH or NaOCH$_2$CH$_3$ brings about a single dehydrohalogenation to give vinyl bromide.

$$BrCH_2CH_2Br \xrightarrow{\text{NaOCH}_2CH_3} CH_2{=}CHBr$$

1,2-Dibromoethane   Vinyl bromide

9.13 (*b*) Addition of hydrogen chloride to vinyl chloride gives the geminal dichloride 1,1-dichloroethane.

$$CH_3{=}CHCl \xrightarrow{\text{HCl}} CH_3CHCl_2$$

Vinyl chloride   1,1-Dichloroethane

(*c*) Since 1,1-dichloroethane can be prepared by adding two moles of hydrogen chloride to acetylene, first convert vinyl bromide to acetylene by dehydrohalogenation.

$$CH_2{=}CHBr \xrightarrow[\text{2. } H_2O]{\text{1. NaNH}_2, NH_3} HC\equiv CH \xrightarrow{\text{2HCl}} CH_3CHCl_2$$

Vinyl bromide   Acetylene   1,1-Dichloroethane

(*d*) As in part (*c*), first convert the designated starting material to acetylene.

$$CH_3CHBr_2 \xrightarrow[\text{2. } H_2O]{\text{1. NaNH}_2, NH_3} HC\equiv CH \xrightarrow{\text{2HCl}} CH_3CHCl_2$$

1,1-Dibromoethane   Acetylene   1,1-Dichloroethane

9.14 The enol arises by addition of water across the triple bond.

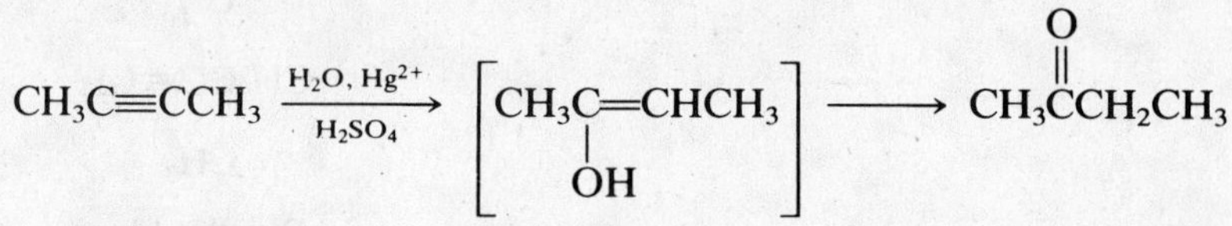

$$CH_3C{\equiv}CCH_3 \xrightarrow[\text{H}_2\text{SO}_4]{\text{H}_2\text{O, Hg}^{2+}} \left[\underset{\underset{\text{OH}}{|}}{CH_3C}{=}CHCH_3 \right] \longrightarrow CH_3\overset{\overset{\text{O}}{\|}}{C}CH_2CH_3$$

2-Butyne   2-Buten-2-ol (enol form)   2-Butanone

The mechanism described in Figure 9.5 is adapted to the case of 2-butyne hydration as shown:

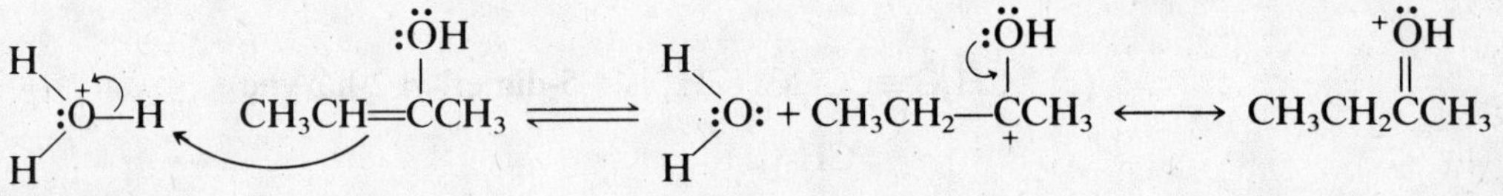

Hydronium   2-Buten-2-ol   Water   Carbocation
ion

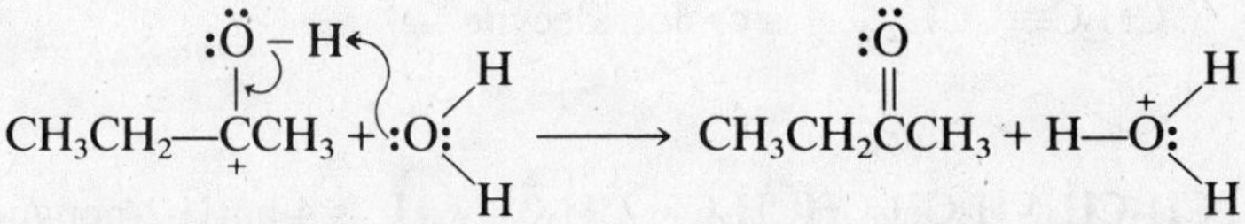

Carbocation   Water   2-Butanone   Hydronium ion

9.15 Hydration of 1-octyne gives 2-octanone according to the equation illustrating the operation of Markovnikov's rule that immediately precedes this problem in the text. Prepare 1-octyne as described in Problem 9.8, and then carry out its hydration in the presence of mercury(II) sulfate and sulfuric acid.

     Hydration of 4-octyne gives 4-octanone. Prepare 4-octyne as described in Problem 9.8.

9.16 Each of the carbons that are part of —CO$_2$H groups was once part of a —C≡C— unit. The two fragments $CH_3(CH_2)_4CO_2H$ and $HO_2CCH_2CH_2CO_2H$ account for only 10 of the original 16 carbons. The full complement of carbons can be accommodated by assuming that two molecules of $CH_3(CH_2)_4CO_2H$ are formed, along with one molecule of $HO_2CCH_2CH_2CO_2H$. The starting alkyne is therefore deduced from the ozonolysis data to be as shown:

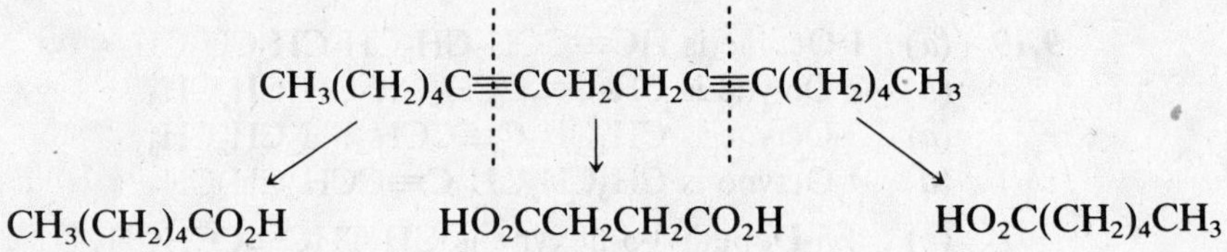

$$CH_3(CH_2)_4C{\equiv}CCH_2CH_2C{\equiv}C(CH_2)_4CH_3$$

$CH_3(CH_2)_4CO_2H$ $HO_2CCH_2CH_2CO_2H$ $HO_2C(CH_2)_4CH_3$

9.17 There are three isomers that have unbranched carbon chains:

$CH_3CH_2CH_2CH_2C{\equiv}CH$ $CH_3CH_2CH_2C{\equiv}CCH_3$ $CH_3CH_2C{\equiv}CCH_2CH_3$

1-Hexyne   2-Hexyne   3-Hexyne

Next consider all the alkynes with a single methyl branch:

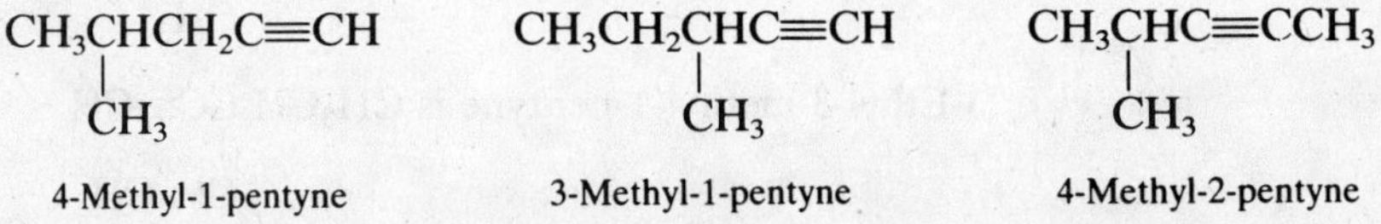

$\underset{\underset{\text{CH}_3}{|}}{CH_3CH}CH_2C{\equiv}CH$ $\underset{\underset{\text{CH}_3}{|}}{CH_3CH_2CH}C{\equiv}CH$ $\underset{\underset{\text{CH}_3}{|}}{CH_3CH}C{\equiv}CCH_3$

4-Methyl-1-pentyne   3-Methyl-1-pentyne   4-Methyl-2-pentyne

There is one isomer with two methyl branches. None is possible with an ethyl branch.

$$
\begin{array}{c}
CH_3 \\
| \\
CH_3CC\!\equiv\!CH \\
| \\
CH_3
\end{array}
$$

3,3-Dimethyl-1-butyne

9.18 (a) $\overset{5}{C}H_3\overset{4}{C}H_2\overset{3}{C}H_2\overset{2}{C}\!\equiv\!\overset{1}{C}H$ is 1-pentyne

(b) $\overset{5}{C}H_3\overset{4}{C}H_2\overset{3}{C}\!\equiv\!\overset{2}{C}\overset{1}{C}H_3$ is 2-pentyne

(c) $\overset{1}{C}H_3\overset{2}{C}\!\equiv\!\overset{3}{C}\overset{4}{C}H\overset{5}{C}H\overset{6}{C}H_3$ is 4,5-dimethyl-2-hexyne

$\qquad\qquad\underset{CH_3\,CH_3}{|\quad\ |}$

(d) ▷—$\overset{5}{C}H_2\overset{4}{C}H_2\overset{3}{C}H_2\overset{2}{C}\!\equiv\!\overset{1}{C}H$ is 5-cyclopropyl-1-pentyne

(e) $\overset{13}{C}H_2\overset{1}{C}\!\equiv\!\overset{2}{C}\overset{3}{C}H_2$ is cyclotridecyne

(f) $CH_3CH_2CH_2CH_2\overset{4}{C}H\overset{5}{C}H_2\overset{6}{C}H_2\overset{7}{C}H_2\overset{8}{C}H_2\overset{9}{C}H_3$ is 4-butyl-2-nonyne

$\qquad\qquad\qquad\underset{\underset{3}{C}\!\equiv\!\underset{2}{C}\underset{1}{C}H_3}{|}$

(Parent chain must contain the triple bond)

(g)
$$
\begin{array}{ccc}
CH_3 & & CH_3 \\
\overset{2}{|} & & \overset{5}{|} \\
\overset{1}{C}H_3\overset{}{C}C\!\equiv\!\overset{}{C}\overset{}{C}\overset{6}{C}H_3 \\
\underset{3}{|} & & \underset{4}{|} \\
CH_3 & & CH_3
\end{array}
$$
is 2,2,5,5-tetramethyl-3-hexyne

(h)
$$
\begin{array}{ccc}
CH_3 & & CH_3 \\
| & & | \\
CH_3C\!-\!\overset{3}{C}H\!-\!\overset{4}{C}\!-\!\overset{5}{C}H_3 \\
| & & | \\
H_3C & \overset{2}{C} & CH_3 \\
& \| & \\
& \overset{1}{C} & \\
& H &
\end{array}
$$
is 3-*tert*-butyl-4,4-dimethyl-1-pentyne

9.19 (a) 1-Octyne is $HC\!\equiv\!CCH_2CH_2CH_2CH_2CH_2CH_3$
(b) 2-Octyne is $CH_3C\!\equiv\!CCH_2CH_2CH_2CH_2CH_3$
(c) 3-Octyne is $CH_3CH_2C\!\equiv\!CCH_2CH_2CH_2CH_3$
(d) 4-Octyne is $CH_3CH_2CH_2C\!\equiv\!CCH_2CH_2CH_3$
(e) 2,5-Dimethyl-3-hexyne is $CH_3CHC\!\equiv\!CCHCH_3$

$\qquad\qquad\qquad\qquad\qquad\qquad\quad\underset{CH_3}{|}\qquad\underset{CH_3}{|}$

(f) 4-Ethyl-1-hexyne is $CH_3CH_2CHCH_2C\!\equiv\!CH$

$\qquad\qquad\qquad\qquad\qquad\qquad\underset{CH_2CH_3}{|}$

(g) Ethynylcyclohexane is ⬡—$C\!\equiv\!CH$

(h) 3-Ethyl-3-methyl-1-pentyne is
$$
\begin{array}{c}
CH_3 \\
| \\
CH_3CH_2CC\!\equiv\!CH \\
| \\
CH_2CH_3
\end{array}
$$

9.20 Ethynylcyclohexane has the molecular formula C_8H_{12}. All the other compounds are C_8H_{14}.

9.21 Only alkynes with the carbon skeletons shown can give 3-ethylhexane on catalytic hydrogenation.

3-Ethyl-1-hexyne   **or**   4-Ethyl-1-hexyne   **or**   4-Ethyl-2-hexyne $\xrightarrow[\text{Pt}]{H_2}$ 3-Ethylhexane

9.22 The carbon skeleton of the unknown acetylenic amino acid must be the same as that of homoleucine. The structure of homoleucine is such that there is only one possible location for a carbon-carbon triple bond in an acetylenic precursor.

$$HC\equiv CCHCH_2CHCO^- \xrightarrow[\text{Pt}]{H_2} CH_3CH_2CHCH_2CHCO^-$$

with CH_3 and $^+NH_3$ substituents (left); CH_3 and $^+NH_3$ substituents (right)

$C_7H_{11}NO_2$ Homoleucine

9.23 (a) $CH_3CH_2CH_2CH_2CH_2CHCl_2 \xrightarrow[\text{2. } H_2O]{\text{1. NaNH}_2, \text{ NH}_3} CH_3CH_2CH_2CH_2C\equiv CH$

1,1-Dichlorohexane   1-Hexyne

(b) $CH_3CH_2CH_2CH_2CH=CH_2 \xrightarrow[\text{CCl}_4]{Br_2} CH_3CH_2CH_2CH_2CHCH_2Br$ with Br substituent

1-Hexene   1,2-Dibromohexane

$\downarrow \begin{array}{l}\text{1. NaNH}_2, \text{ NH}_3 \\ \text{2. } H_2O\end{array}$

$CH_3CH_2CH_2CH_2C\equiv CH$

1-Hexyne

(c) $HC\equiv CH \xrightarrow[\text{NH}_3]{\text{NaNH}_2} HC\equiv C^-Na^+ \xrightarrow{CH_3CH_2CH_2CH_2Br} CH_3CH_2CH_2CH_2C\equiv CH$

Acetylene   1-Hexyne

(d) $CH_3CH_2CH_2CH_2CH_2CH_2I \xrightarrow[\text{DMSO}]{\text{KOC(CH}_3)_3} CH_3CH_2CH_2CH_2CH=CH_2$

1-Iodohexane   1-Hexene

1-Hexene is then converted to 1-hexyne as in part (b).

(e) $\begin{array}{c} CH_3CH_2CH_2CH_2 \\ C=C \\ H Br \end{array} \xrightarrow[\text{2. } H_2O]{\text{1. NaNH}_2, \text{ NH}_3} CH_3CH_2CH_2CH_2C\equiv CH$

(E)-1-Bromo-1-hexene   1-Hexyne

9.24 (a) Working backward from the final product, it can be seen that preparation of 1-butyne will allow the desired carbon skeleton to be constructed.

$CH_3CH_2C\equiv CCH_2CH_3$ prepared from $CH_2CH_2C\equiv C:^- + BrCH_2CH_3$

3-Hexyne

The desired intermediate 1-butyne is available by halogenation followed by dehydrohalogenation of 1-butene.

$$CH_3CH_2CH{=}CH_2 + Br_2 \longrightarrow CH_3CH_2\overset{\overset{\displaystyle Br}{|}}{C}HCH_2Br \xrightarrow[\text{2. } H_2O]{\text{1. } NaNH_2,\ NH_3} CH_3CH_2C{\equiv}CH$$

1-Butene   1-Butyne

Reaction of the anion of 1-butyne with ethyl bromide completes the synthesis.

$$CH_3CH_2C{\equiv}CH \xrightarrow[NH_3]{NaNH_2} CH_3CH_2C{\equiv}C{:}^-\ Na^+ \xrightarrow{CH_3CH_2Br} CH_3CH_2C{\equiv}CCH_2CH_3$$

1-Butyne   3-Hexyne

(b) Dehydrohalogenation of 1,1-dichlorobutane yields 1-butyne. The synthesis is completed as in part (a).

$$CH_3CH_2CH_2CHCl_2 \xrightarrow[\text{2. } H_2O]{\text{1. } NaNH_2,\ NH_3} CH_3CH_2C{\equiv}CH$$

1,1-Dichlorobutane   1-Butyne

(c) Once again dehydrohalogenation with strong base yields 1-butyne, allowing the synthesis to be completed as in part (a).

$$CH_3CH_2CH{=}CHCl \xrightarrow[\text{2. } H_2O]{\text{1. } NaNH_2,\ NH_3} CH_3CH_2C{\equiv}CH$$

1-Chloro-1-butene   1-Butyne

(d) $$HC{\equiv}CH \xrightarrow[NH_3]{NaNH_2} HC{\equiv}C^-\ Na^+ \xrightarrow{CH_3CH_2Br} HC{\equiv}CCH_2CH_3$$

     Acetylene   1-Butyne

1-Butyne is converted to 3-hexyne as in part (a).

9.25 A single dehydrobromination step occurs in the conversion of 1,2-dibromodecane to $C_{10}H_{19}Br$. Bromine may be lost from C-1 to give 2-bromo-1-decene.

$$BrCH_2\overset{\overset{\displaystyle |}{}}{\underset{\underset{\displaystyle Br}{|}}{C}}H(CH_2)_7CH_3 \xrightarrow[\text{ethanol-water}]{KOH} CH_2{=}\overset{}{\underset{\underset{\displaystyle Br}{|}}{C}}(CH_2)_7CH_3$$

         1,2-Dibromodecane   2-Bromo-1-decene

Loss of bromine from C-2 gives (E)- and (Z)-1-bromo-1-decene.

$$BrCH_2\overset{}{\underset{\underset{\displaystyle Br}{|}}{C}}H(CH_2)_7CH_3 \xrightarrow[\text{ethanol-water}]{KOH} \underset{Br}{\overset{H}{}}C{=}C\underset{H}{\overset{(CH_2)_7CH_3}{}} + \underset{H}{\overset{Br}{}}C{=}C\underset{H}{\overset{(CH_2)_7CH_3}{}}$$

     1,2-Dibromodecane   (E)-1-Bromo-1-decene   (Z)-1-Bromo-1-decene

9.26 (a) $$CH_3CH_2CH_2CH_2C{\equiv}CH + 2H_2 \xrightarrow{Pt} CH_3CH_2CH_2CH_2CH_2CH_3$$

           1-Hexyne   Hexane

(b) $$CH_3CH_2CH_2CH_2C{\equiv}CH + H_2 \xrightarrow{\text{Lindlar Pd}} CH_3CH_2CH_2CH_2CH{=}CH_2$$

            1-Hexyne   1-Hexene

(c) $$CH_3CH_2CH_2CH_2C{\equiv}CH \xrightarrow[NH_3]{Li} CH_3CH_2CH_2CH_2CH{=}CH_2$$

           1-Hexyne   1-Hexene

(d) $$CH_3CH_2CH_2CH_2C{\equiv}CH \xrightarrow[NH_3]{NaNH_2} CH_3CH_2CH_2CH_2C{\equiv}C{:}^-\ Na^+$$

           1-Hexyne   Sodium 1-hexynide

(e) $CH_3CH_2CH_2CH_2C{\equiv}C{:}^-Na^+ + CH_3CH_2CH_2CH_2Br \longrightarrow$

Sodium 1-hexynide   1-Bromobutane

$CH_3CH_2CH_2CH_2C{\equiv}CCH_2CH_2CH_2CH_3$

5-Decyne

(f) $CH_3CH_2CH_2CH_2C{\equiv}C{:}^-Na^+ + (CH_3)_3CBr \longrightarrow$

Sodium 1-hexynide   *tert*-Butyl bromide

$CH_3CH_2CH_2CH_2C{\equiv}CH + (CH_3)_2C{=}CH_2$

1-Hexyne   2-Methylpropene

(g) $CH_3CH_2CH_2CH_2C{\equiv}CH \xrightarrow[\text{(1 mol)}]{\text{HCl}} CH_3CH_2CH_2CH_2\underset{\underset{Cl}{|}}{C}{=}CH_2$

1-Hexyne   2-Chloro-1-hexene

(h) $CH_3CH_2CH_2CH_2C{\equiv}CH \xrightarrow[\text{(2 mol)}]{\text{HCl}} CH_3CH_2CH_2CH_2\overset{\overset{Cl}{|}}{\underset{\underset{Cl}{|}}{C}}CH_3$

1-Hexyne   2,2-Dichlorohexane

(i) $CH_3CH_2CH_2CH_2C{\equiv}CH \xrightarrow[\text{(1 mol)}]{\text{Cl}_2}$

$\underset{Cl}{\overset{CH_3CH_2CH_2CH_2}{}}C{=}C\underset{H}{\overset{Cl}{}}$

1-Hexyne   (*E*)-1,2-Dichloro-1-hexene

(j) $CH_3CH_2CH_2CH_2C{\equiv}CH \xrightarrow[\text{(2 mol)}]{\text{Cl}_2} CH_3CH_2CH_2CH_2\overset{\overset{Cl}{|}}{\underset{\underset{Cl}{|}}{C}}CHCl_2$

1-Hexyne   1,1,2,2-Tetrachlorohexane

(k) $CH_3CH_2CH_2CH_2C{\equiv}CH \xrightarrow[\text{HgSO}_4]{\text{H}_2\text{O, H}_2\text{SO}_4} CH_3CH_2CH_2CH_2\overset{\overset{O}{\|}}{C}CH_3$

1-Hexyne   2-Hexanone

(l) $CH_3CH_2CH_2CH_2C{\equiv}CH \xrightarrow[\text{2. H}_2\text{O}]{\text{1. O}_3} CH_3CH_2CH_2CH_2\overset{\overset{O}{\|}}{C}OH + HO\overset{\overset{O}{\|}}{C}OH$

1-Hexyne   Pentanoic acid   Carbonic acid

9.27 (a) $CH_3CH_2C{\equiv}CCH_2CH_3 + 2H_2 \xrightarrow{\text{Pt}} CH_3CH_2CH_2CH_2CH_2CH_3$

3-Hexyne   Hexane

(b) $CH_3CH_2C{\equiv}CCH_2CH_3 + H_2 \xrightarrow[cis]{\text{Lindlar Pd}}$

$\underset{H}{\overset{CH_3CH_2}{}}C{=}C\underset{H}{\overset{CH_2CH_3}{}}$

3-Hexyne   (*Z*)-3-Hexene

(c) $CH_3CH_2C{\equiv}CCH_2CH_3$ $\xrightarrow[\substack{NH_3 \\ anti}]{Li}$

3-Hexyne

$$\underset{H}{\overset{CH_3CH_2}{\diagdown}}C=C\underset{CH_2CH_3}{\overset{H}{\diagup}}$$

(E)-3-Hexene

(d) $CH_3CH_2C{\equiv}CCH_2CH_3$ $\xrightarrow[\text{(1 mol)}]{HCl}$

3-Hexyne

$$\underset{Cl}{\overset{CH_3CH_2}{\diagdown}}C=C\underset{CH_2CH_3}{\overset{H}{\diagup}}$$

(Z)-3-Chloro-3-hexene

(e) $CH_3CH_2C{\equiv}CCH_2CH_3$ $\xrightarrow[\text{(2 mol)}]{HCl}$

3-Hexyne

$$CH_3CH_2\overset{\overset{\displaystyle Cl}{|}}{\underset{\underset{\displaystyle Cl}{|}}{C}}CH_2CH_2CH_3$$

3,3-Dichlorohexane

(f) $CH_3CH_2C{\equiv}CCH_2CH_3$ $\xrightarrow[\text{(1 mol)}]{Cl_2}$

3-Hexyne

$$\underset{Cl}{\overset{CH_3CH_2}{\diagdown}}C=C\underset{CH_2CH_3}{\overset{Cl}{\diagup}}$$

(E)-3,4-Dichloro-3-hexene

(g) $CH_3CH_2C{\equiv}CCH_2CH_3$ $\xrightarrow[\text{(2 mol)}]{Cl_2}$

3-Hexyne

$$CH_3CH_2\overset{\overset{\displaystyle Cl\;\;Cl}{|\;\;\;|}}{\underset{\underset{\displaystyle Cl\;\;Cl}{|\;\;\;|}}{C-C}}CH_2CH_3$$

3,3,4,4-Tetrachlorohexane

(h) $CH_3CH_2C{\equiv}CCH_2CH_3$ $\xrightarrow[\text{HgSO}_4]{H_2O,\,H_2SO_4}$

3-Hexyne

$$CH_3CH_2\overset{\overset{\displaystyle O}{\|}}{C}CH_2CH_2CH_3$$

3-Hexanone

(i) $CH_3CH_2C{\equiv}CCH_2CH_3$ $\xrightarrow[\text{2. H}_2\text{O}]{1.\ O_3}$ $2CH_3CH_2\overset{\overset{\displaystyle O}{\|}}{C}OH$

3-Hexyne   Propanoic acid

9.28 The two carbons of the triple bond are similarly but not identically substituted in $CH_3C{\equiv}CCH_2CH_2CH_2CH_3$. Two regioisomeric enols are formed, each of which gives a different ketone.

$CH_3C{\equiv}CCH_2CH_2CH_2CH_3$ $\xrightarrow[\text{HgSO}_4]{H_2O,\,H_2SO_4}$

2-Heptyne

$$CH_3\overset{\overset{\displaystyle OH}{|}}{C}=CHCH_2CH_2CH_2CH_3 + CH_3CH=\overset{\overset{\displaystyle OH}{|}}{C}CH_2CH_2CH_2CH_3$$

2-Hepten-2-ol   2-Hepten-3-ol

$$CH_3\overset{\overset{\displaystyle O}{\|}}{C}CH_2CH_2CH_2CH_2CH_3$$

2-Heptanone

$$CH_3CH_2\overset{\overset{\displaystyle O}{\|}}{C}CH_2CH_2CH_2CH_3$$

3-Heptanone

9.29 The alkane formed by hydrogenation of (*S*)-3-methyl-1-pentyne is achiral; it cannot be optically active.

(*S*)-3-Methyl-1-pentyne   3-Methylpentane
                                            (does not have a stereogenic
                                            center; optically inactive)

The product of hydrogenation of (*S*)-4-methyl-1-hexyne is optically active because a stereogenic center is present in the starting material and is carried through to the product.

(*S*)-4-Methyl-1-hexyne   (*S*)-3-Methylhexane

Both (*S*)-3-methyl-1-pentyne and (*S*)-4-methyl-1-hexyne yield optically active products when their triple bonds are reduced to double bonds.

9.30 (*a*) The dihaloalkane contains both a primary alkyl chloride and a primary alkyl iodide functional group. Iodide is a better leaving group than chloride and is the one replaced by acetylide.

$$NaC\equiv CH + ClCH_2CH_2CH_2CH_2CH_2CH_2I \longrightarrow ClCH_2CH_2CH_2CH_2CH_2CH_2C\equiv CH$$

Sodium   1-Chloro-6-iodohexane   8-Chloro-1-octyne
acetylide

(*b*) Both vicinal dibromide functions are converted to alkyne units on treatment with excess sodium amide.

1,2,5,6-Tetrabromohexane   1,5-Hexadiyne

(*c*) The starting material is a geminal dichloride. Potassium *tert*-butoxide in dimethyl sulfoxide is a sufficiently strong base to convert it to an alkyne.

1,1-Dichloro-1-   Ethynylcyclopropane
cyclopropylethane

(*d*) Alkyl *p*-toluenesulfonates react similarly to alkyl halides in nucleophilic substitution reactions. The alkynide nucleophile displaces the *p*-toluenesulfonate leaving group from ethyl *p*-toluenesulfonate.

Phenylacetylide ion   Ethyl *p*-toluenesulfonate   1-Phenyl-1-butyne

(e) Both carbons of a —C≡C— unit are converted to carboxyl groups (—CO$_2$H) on ozonolysis.

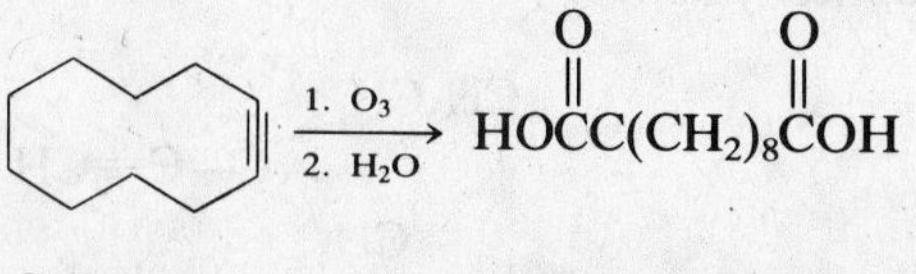

$$\text{Cyclodecyne} \xrightarrow[\text{2. H}_2\text{O}]{\text{1. O}_3} \underset{O}{HOCC(CH_2)_8COH}$$

Cyclodecyne   Decanedioic acid

(f) Ozonolysis cleaves the carbon–carbon triple bond.

$$\xrightarrow[\text{2. H}_2\text{O}]{\text{1. O}_3}$$

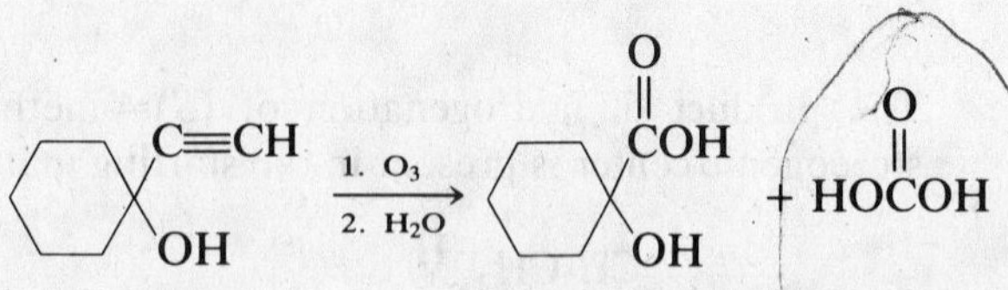

1-Ethynylcyclohexanol   1-Hydroxycyclohexane-   Carbonic
               carboxylic acid   acid

+ HOCOH

(g) Hydration of a terminal carbon–carbon triple bond converts it to a —CCH$_3$ group.

$$CH_3CHCH_2CC\equiv CH \xrightarrow[\text{HgO}]{\text{H}_2\text{O, H}_2\text{SO}_4} CH_3CHCH_2C—CCH_3$$

    CH$_3$ CH$_3$ CH$_3$ CH$_3$

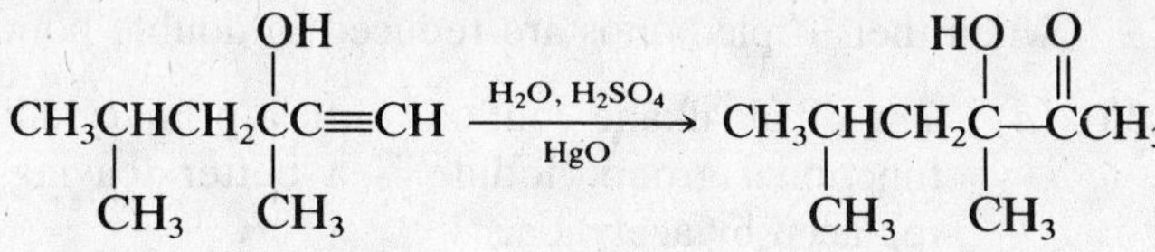

3,5-Dimethyl-1-hexyn-3-ol   3-Hydroxy-3,5-dimethyl-2-hexanone

(h) Sodium-in-ammonia reduction of an alkyne yields a trans alkene. The stereochemistry of a double bond that is already present in the molecule is, of course, not altered during the process.

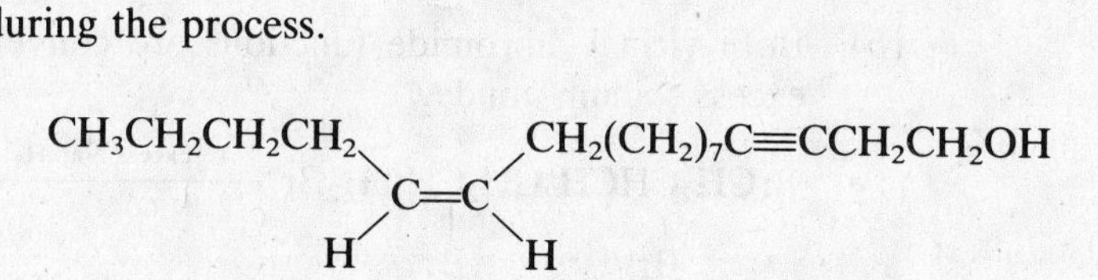

$$\underset{H}{\overset{CH_3CH_2CH_2CH_2}{C}}=\underset{H}{\overset{CH_2(CH_2)_7C\equiv CCH_2CH_2OH}{C}}$$

(Z)-13-Octadecen-3-yn-1-ol

$$\downarrow \begin{array}{l}\text{1. Na, NH}_3 \\ \text{2. H}_2\text{O}\end{array}$$

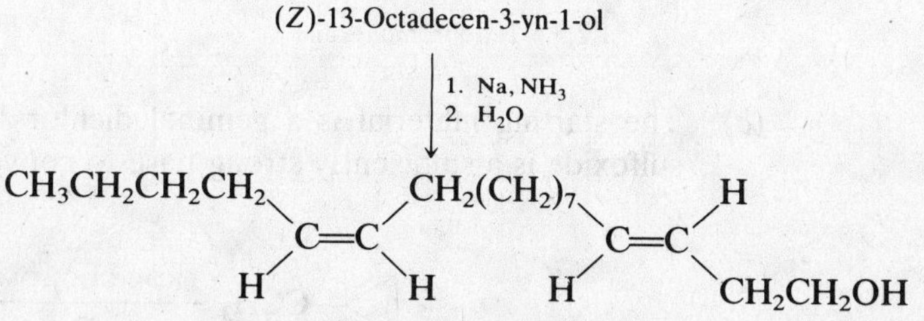

(3E,13Z)-3,13-Octadecadien-1-ol

(i) The primary chloride leaving group is displaced by the alkynide nucleophile.

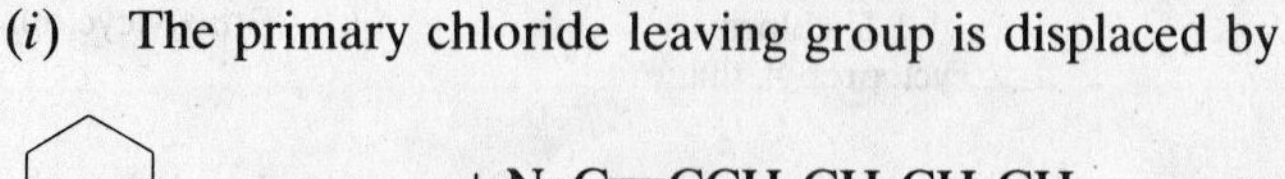

O(CH$_2$)$_8$Cl   + NaC≡CCH$_2$CH$_2$CH$_2$CH$_3$ ⟶

8-Chlorooctyl   Sodium 1-hexynide
tetrahydropyranyl ether

O(CH$_2$)$_8$C≡CCH$_2$CH$_2$CH$_2$CH$_3$

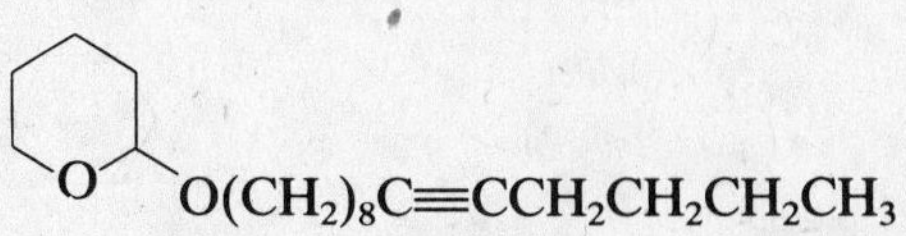

9-Tetradecyn-1-yl tetrahydropyranyl ether

(*j*) Hydrogenation of the triple bond over the Lindlar catalyst converts the compound to a cis alkene.

$$\text{O}\!-\!\text{O(CH}_2)_8\text{C}\!\equiv\!\text{CCH}_2\text{CH}_2\text{CH}_2\text{CH}_3$$

9-Tetradecyn-1-yl tetrahydropyranyl ether

$$\Big\downarrow \begin{array}{c} \text{H}_2 \\ \text{Lindlar Pd} \end{array}$$

$$\text{O}\!-\!\text{O(CH}_2)_8 \quad \underset{\text{H}}{\overset{\text{C}=\text{C}}{\diagup}} \quad \underset{\text{H}}{\overset{\text{CH}_2\text{CH}_2\text{CH}_2\text{CH}_3}{\diagdown}}$$

(*Z*)-9-Tetradecen-1-yl tetrahydropyranyl ether

9.31 Ketones such as 2-heptanone may be readily prepared by hydration of terminal acetylenes. Thus, if we had 1-heptyne, it could be converted to 2-heptanone.

$$\text{HC}\!\equiv\!\text{C(CH}_2)_4\text{CH}_3 \xrightarrow[\text{H}_2\text{SO}_4,\ \text{HgSO}_4]{\text{H}_2\text{O}} \overset{\overset{\text{O}}{\|}}{\text{CH}_3\text{C}}\text{(CH}_2)_4\text{CH}_3$$

1-Heptyne   2-Heptanone

Acetylene, as we have seen in earlier problems, can be converted to 1-heptyne by alkylation.

$$\text{HC}\!\equiv\!\text{CH} \xrightarrow[\text{NH}_3]{\text{NaNH}_2} \text{HC}\!\equiv\!\text{C}^-\text{Na}^+$$

$$\text{HC}\!\equiv\!\text{C}^-\text{Na}^+ + \text{CH}_3\text{CH}_2\text{CH}_2\text{CH}_2\text{CH}_2\text{Br} \longrightarrow \text{HC}\!\equiv\!\text{C(CH}_2)_4\text{CH}_3$$

9.32 Apply the technique of reasoning backward to gain a clue to how to attack this synthesis problem. A reasonable final step is the formation of the *Z* double bond by semi-hydrogenation of an alkyne over Lindlar palladium.

$$\text{CH}_3\text{(CH}_2)_7\text{C}\!\equiv\!\text{C(CH}_2)_{12}\text{CH}_3 \xrightarrow[\text{Lindlar Pd}]{\text{H}_2} \quad \underset{\text{H}}{\overset{\text{CH}_3\text{(CH}_2)_7}{\diagdown}}\overset{\text{C}=\text{C}}{} \underset{\text{H}}{\overset{\text{(CH}_2)_{12}\text{CH}_3}{\diagup}}$$

9-Tricosyne   (*Z*)-9-Tricosene

The necessary alkyne 9-tricosyne can be prepared by a double alkylation of acetylene.

$$\text{HC}\!\equiv\!\text{CH} \xrightarrow[\text{2. }\ \text{CH}_3\text{(CH}_2)_7\text{Br}]{\text{1. }\ \text{NaNH}_2,\ \text{NH}_3} \text{CH}_3\text{(CH}_2)_7\text{C}\!\equiv\!\text{CH} \xrightarrow[\text{2. }\ \text{CH}_3\text{(CH}_2)_{12}\text{Br}]{\text{1. }\ \text{NaNH}_2,\ \text{NH}_3} \text{CH}_3\text{(CH}_2)_7\text{C}\!\equiv\!\text{C(CH}_2)_{12}\text{CH}_3$$

Acetylene   1-Decyne   9-Tricosyne

It does not matter which alkyl group is introduced first.
The alkyl halides are prepared from the corresponding alcohols.

$$\text{CH}_3\text{(CH}_2)_7\text{OH} \xrightarrow[\text{or PBr}_3]{\text{HBr}} \text{CH}_3\text{(CH}_2)_7\text{Br}$$

1-Octanol   1-Bromooctane

$$\text{CH}_3\text{(CH}_2)_{12}\text{OH} \xrightarrow[\text{or PBr}_3]{\text{HBr}} \text{CH}_3\text{(CH}_2)_{12}\text{Br}$$

1-Tridecanol   1-Bromotridecane

9.33 (*a*)   2,2-Dibromopropane is prepared by addition of hydrogen bromide to propyne.

$$CH_3C\equiv CH + 2HBr \longrightarrow CH_3\overset{\displaystyle Br}{\underset{\displaystyle Br}{C}}CH_3$$

| Propyne | Hydrogen bromide | 2,2-Dibromopropane |

The designated starting material, 1,1-dibromopropane, is converted to propyne by a double dehydrohalogenation.

$$CH_3CH_2CHBr_2 \xrightarrow[\text{2. } H_2O]{\text{1. } NaNH_2,\, NH_3} CH_3C\equiv CH$$

| 1,1-Dibromopropane | Propyne |

(*b*)   As in part (*a*), first convert the designated starting material to propyne, and then add hydrogen bromide.

$$CH_3\underset{\displaystyle Br}{CH}CH_2Br \xrightarrow[\text{2. } H_2O]{\text{1. } NaNH_2,\, NH_3} CH_3C\equiv CH \xrightarrow{2HBr} CH_3\overset{\displaystyle Br}{\underset{\displaystyle Br}{C}}CH_3$$

| 1,2-Dibromopropane | Propyne | 2,2-Dibromopropane |

(*c*)   First convert 1-bromopropene to propyne, and then add one mole of hydrogen chloride across the triple bond. Addition occurs in accordance with Markovnikov's rule.

$$BrCH\!=\!CHCH_3 \xrightarrow[\text{2. } H_2O]{\text{1. } NaNH_2,\, NH_3} HC\equiv CCH_3 \xrightarrow{HCl} CH_2\!=\!\underset{\displaystyle Cl}{C}CH_3$$

| 1-Bromopropene | Propyne | 2-Chloropropene |

(*d*)   Instead of trying to introduce two additional chlorines into 1,2-dichloropropane by free-radical substitution (a mixture of products would result), convert the vicinal dichloride to propyne, and then add two moles of Cl_2.

$$CH_3\underset{\displaystyle Cl}{CH}CH_2Cl \xrightarrow[\text{2. } H_2O]{\text{1. } NaNH_2,\, NH_3} CH_3C\equiv CH \xrightarrow{2Cl_2} CH_3\overset{\displaystyle Cl}{\underset{\displaystyle Cl}{C}}CHCl_2$$

| 1,2-Dichloropropane | Propyne | 1,1,2,2-Tetrachloropropane |

(*e*)   The required carbon skeleton can be constructed by alkylating acetylene with ethyl bromide.

$$HC\equiv CH \xrightarrow[NH_3]{NaNH_2} HC\equiv C{:}^- Na^+ \xrightarrow{CH_3CH_2Br} HC\equiv CCH_2CH_3$$

| Acetylene | Sodium acetylide | 1-Butyne |

Addition of hydrogen iodide to 1-butyne gives 2,2-diiodobutane.

$$HC\equiv CCH_2CH_3 + 2HI \longrightarrow CH_3\overset{\displaystyle I}{\underset{\displaystyle I}{C}}CH_2CH_3$$

| 1-Butyne | Hydrogen iodide | 2,2-Diiodobutane |

(*f*) The six-carbon chain is available by alkylation of acetylene with 1-bromobutane.

$$HC\equiv CH \xrightarrow[\text{2. } CH_3CH_2CH_2CH_2Br]{\text{1. } NaNH_2,\ NH_3} HC\equiv CCH_2CH_2CH_2CH_3$$

Acetylene   1-Hexyne

The alkylating agent, 1-bromobutane, is prepared from 1-butene by free-radical (anti-Markovnikov) addition of hydrogen bromide.

$$CH_3CH_2CH\!\!=\!\!CH_2 +\quad HBr \xrightarrow{\text{peroxides}} CH_3CH_2CH_2CH_2Br$$

1-Butene   Hydrogen bromide   1-Bromobutane

Once 1-hexyne is prepared, it can be converted to 1-hexene by semihydrogenation or by sodium-ammonia reduction.

$$CH_3CH_2CH_2CH_2C\equiv CH \xrightarrow[\text{or Na, }NH_3]{H_2,\ \text{Lindlar Pd}} CH_3CH_2CH_2CH_2CH\!\!=\!\!CH_2$$

1-Hexyne   1-Hexene

(*g*) Dialkylation of acetylene with 1-bromobutane, prepared in part (*f*), gives the necessary 10-carbon chain.

$$HC\equiv CH \xrightarrow[\text{2. } CH_3CH_2CH_2CH_2Br]{\text{1. } NaNH_2,\ NH_3} CH_3CH_2CH_2CH_2C\equiv CH$$

Acetylene   1-Hexyne

$$\Bigg\downarrow \begin{array}{l}\text{1. } NaNH_2,\ NH_3 \\ \text{2. } CH_3CH_2CH_2CH_2Br\end{array}$$

$$CH_3CH_2CH_2CH_2C\equiv CCH_2CH_2CH_2CH_3$$

5-Decyne

Hydrogenation of 5-decyne yields decane.

$$CH_3(CH_2)_3C\equiv C(CH_2)_3CH_3 \xrightarrow[Pt]{H_2} CH_3(CH_2)_3CH_2CH_2(CH_2)_3CH_3$$

5-Decyne   Decane

(*h*) A standard method for converting alkenes to alkynes is to add Br_2 and then carry out a double dehydrohalogenation.

Cyclopentadecene   1,2-Dibromocyclopentadecane   Cyclopentadecyne

(*i*) Alkylation of the triple bond gives the required carbon skeleton.

1-Ethynylcyclohexene   1-Propynyl-1-cyclohexene

Semihydrogenation over the Lindlar catalyst converts the carbon-carbon triple bond to a cis double bond.

1-Propynyl-1-cyclohexene   (*Z*)-1-(1-Cyclohexenyl)propene

(*j*) The stereochemistry of *meso*-2,3-dibromobutane is most easily seen with a Fischer projection:

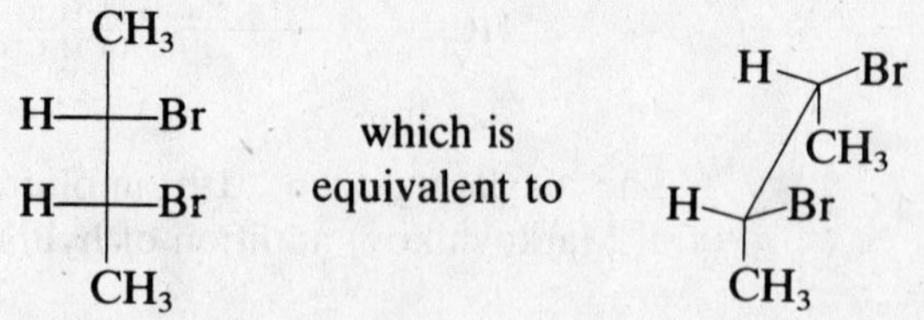

which is equivalent to

Recalling that the addition of Br_2 to alkenes occurs with anti stereochemistry, rotate the sawhorse diagram so that the bromines are anti to one another:

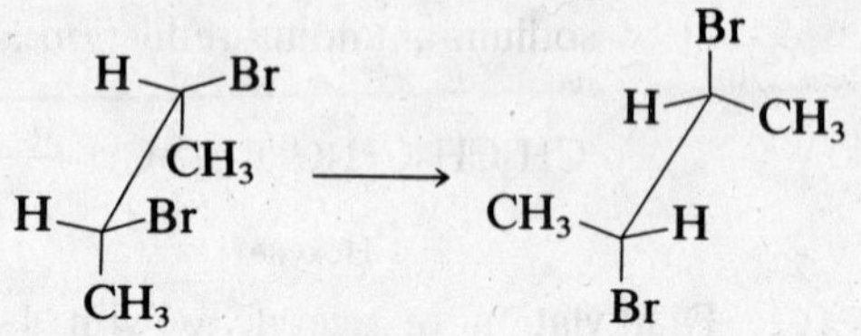

Thus, the starting alkene must be *trans*-2-butene. *trans*-2-Butene is available from 2-butyne by metal-ammonia reduction:

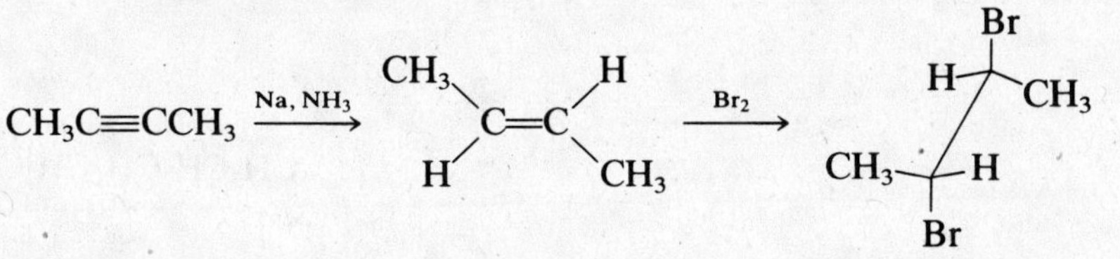

2-Butyne   *trans*-2-Butene   *meso*-2,3-Dibromobutane

9.34 Attack this problem by first planning a synthesis of 4-methyl-2-pentyne from any starting material in a single step. Two different alkyne alkylations suggest themselves:

$$CH_3C{\equiv}CCH(CH_3)_2 \begin{cases} (a) & \text{from } CH_3C{\equiv}C{:}^- \text{ and } BrCH(CH_3)_2 \\ (b) & \text{from } CH_3I \text{ and } {}^-{:}C{\equiv}CCH(CH_3)_2 \end{cases}$$

4-Methyl-2-pentyne

Isopropyl bromide is a secondary alkyl halide and cannot be used to alkylate $CH_3C{\equiv}C{:}^-$ according to reaction (*a*). Therefore, a reasonable last step is the alkylation of $(CH_3)_2CHC{\equiv}CH$ via reaction of its anion with methyl iodide.

The next question that arises from this analysis is the origin of $(CH_3)_2CHC{\equiv}CH$. One of the available starting materials is 1,1-dichloro-3-methylbutane. It can be converted to $(CH_3)_2CHC{\equiv}CH$ by a double dehydrohalogenation. Therefore, the complete synthesis is:

$$(CH_3)_2CHCH_2CHCl_2 \xrightarrow[\text{2. } H_2O]{\text{1. } NaNH_2, NH_3} (CH_3)_2CHC{\equiv}CH \xrightarrow[\text{2. } CH_3I]{\text{1. } NaNH_2} (CH_3)_2CHC{\equiv}CCH_3$$

1,1-Dichloro-3-methylbutane   3-Methyl-1-butyne   4-Methyl-2-pentyne

9.35 The reaction that produces compound A is reasonably straightforward. Compound A is 14-bromo-1-tetradecyne.

$$NaC{\equiv}CH \quad + \quad Br(CH_2)_{12}Br \quad \longrightarrow \quad Br(CH_2)_{12}C{\equiv}CH$$

Sodium acetylide   1,12-Dibromododecane   Compound A ($C_{14}H_{25}Br$)

Treatment of compound A with sodium amide converts it to compound B. Compound B on ozonolysis gives a diacid that retains all the carbon atoms of B. Therefore, compound

B must be a cyclic alkyne, formed by an intramolecular alkylation.

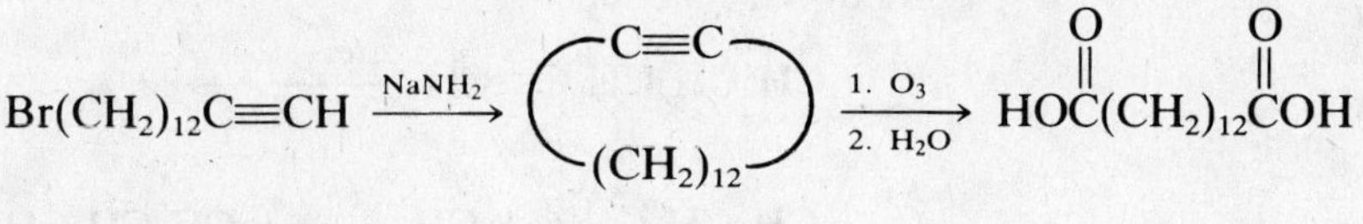

$$Br(CH_2)_{12}C{\equiv}CH \xrightarrow{NaNH_2} \underset{\text{Compound B}}{\left(\overset{C{\equiv}C}{\underset{(CH_2)_{12}}{}}\right)} \xrightarrow[\text{2. H}_2\text{O}]{\text{1. O}_3} HOC(CH_2)_{12}COH$$

Compound A

Compound B is cyclotetradecyne.

Hydrogenation of compound B over Lindlar palladium yields *cis*-cyclotetradecene (compound C).

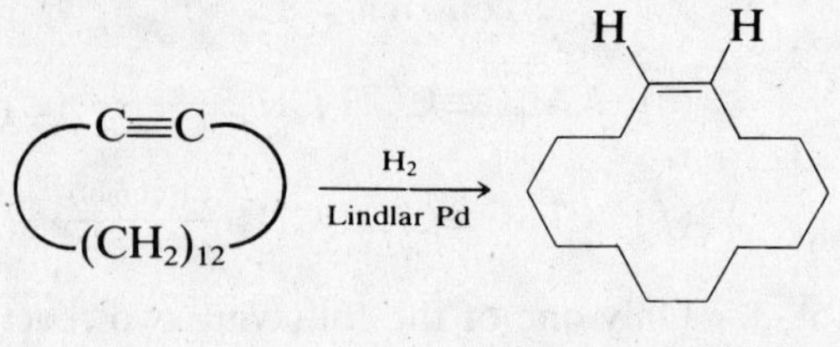

Compound C ($C_{14}H_{26}$)

Hydrogenation over platinum gives cyclotetradecane (compound D).

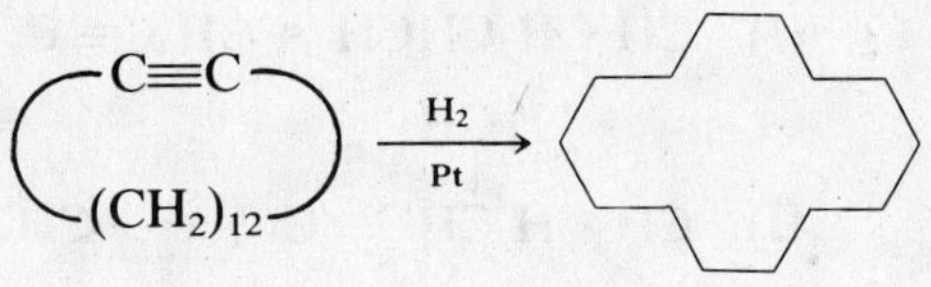

Compound D ($C_{14}H_{28}$)

Sodium-ammonia reduction of compound B yields *trans*-cyclotetradecene.

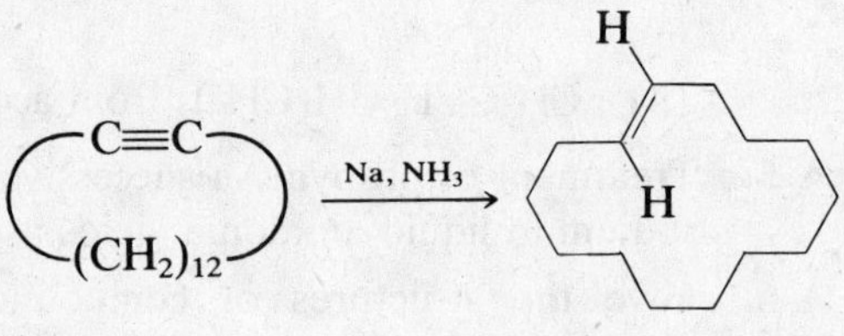

Compound E ($C_{14}H_{26}$)

The cis and trans isomers of cyclotetradecene are both converted to $O{=}CH(CH_2)_{12}CH{=}O$ on ozonolysis, whereas cyclotetradecane does not react with ozone.

SELF-TEST

PART A

A-1. Provide the correct IUPAC names for the following:

(a) $CH_3C{\equiv}CHCH(CH_3)_2$
   |
   CH_3

(b) $CH_3CH_2CH_2CHCHC{\equiv}CH$ with substituents CH_2CH_3 (above) and $CH_2CH_2CH_3$ (below)

A-2. Give the structure of the reactant, reagent, or product omitted from each of the following reactions.

(a) $CH_3CH_2CH_2C\equiv CH \xrightarrow{\text{HCl (1 mol)}}$?

(b) $CH_3CH_2CH_2C\equiv CH \xrightarrow{\text{?}} CH_3CH_2CH_2\overset{\displaystyle O}{\overset{\displaystyle \|}{C}}CH_3$

(c) $CH_3C\equiv CCH_3 \xrightarrow[\text{Lindlar Pd}]{H_2}$?

(d)   ? $\xrightarrow[\text{2. } CH_3CH_2Br]{\text{1. } NaNH_2} (CH_3)_2CHC\equiv CCH_2CH_3$

(e) $CH_3C\equiv CCH_2CH_3 \xrightarrow{\text{?}} (Z)\text{-2-pentene}$

(f) $CH_3C\equiv CCH_2CH_3 \xrightarrow{\text{Cl}_2 \text{ (1 mol)}}$?

A-3. Only one of the following two reactions is effective in the synthesis of 4-methyl-2-hexyne. Which one is that, and why is the other not effective?

(I) $CH_3CH_2\overset{\displaystyle Br}{\overset{\displaystyle |}{C}}HCH_3 + CH_3C\equiv C^-Na^+ \longrightarrow$

(II) $CH_3CH_2\overset{\displaystyle CH_3}{\overset{\displaystyle |}{C}}HC\equiv C^-Na^+ + CH_3I \longrightarrow$

A-4. Outline a series of steps, using any necessary organic and inorganic reagents, for the preparation of:
(a)   3-Hexyne from 1-butene

(b) $CH_3\overset{\displaystyle O}{\overset{\displaystyle \|}{C}}CH_2CH(CH_3)_2$ from acetylene

A-5. Treatment of propyne in successive steps with sodium amide, 1-bromobutane, and sodium in liquid ammonia yields as the final product _____.

A-6. Give the structures of compounds A through D in the following series of equations.

$$A \xrightarrow{\text{NaNH}_2, \text{NH}_3} B$$

$$C \xrightarrow{\text{HBr, heat}} D$$

$$B + D \longrightarrow CH_3CH_2CH_2C\equiv CC(CH_3)_3$$

PART B

B-1. Which of the following statements is (are) true concerning pK_a?
(a)   The larger the pK_a value, the weaker the acid.
(b)   Strong acids have small pK_a values.
(c)   p$K_a = -\log K_a$
(d)   All the statements are correct.

B-2. Referring to the following equilibrium (R = alkyl group)

$$RCH_2CH_3 + RC\equiv C\!:^- \rightleftharpoons RCH_2\overset{\displaystyle ..}{C}H_2^- + RC\equiv C\!-\!H$$

(a) $K << 1$; the equilibrium would lie to the left.
(b) $K >> 1$; the equilibrium would lie to the right.
(c) $K = 1$; equal amounts of all species would be present.
(d)   Not enough information is given; the structure of R must be known.

B-3. Which of the following is an effective way to prepare 1-pentyne?

(a) 1-Pentene $\xrightarrow[\text{2. } \text{NaNH}_2, \text{ heat}]{\text{1. } \text{Cl}_2}$

(b) Acetylene $\xrightarrow[\text{2. } \text{CH}_3\text{CH}_2\text{CH}_2\text{Br}]{\text{1. } \text{NaNH}_2}$

(c) 1,1-Dichloropentane $\xrightarrow[\text{2. } \text{H}_2\text{O}]{\text{1. } \text{NaNH}_2, \text{ NH}_3}$

(d) All of these are effective.

B-4. Which alkyne yields butanoic acid ($CH_3CH_2CH_2CO_2H$) as the only organic product upon treatment with ozone followed by hydrolysis?
(a) 1-Butyne   (b) 4-Octyne   (c) 1-Pentyne   (d) 2-Hexyne

B-5. Which of the following produces a significant amount of acetylide ion on reaction with acetylene?
(a) Conjugate base of CH_3OH (pK_a 16)
(b) Conjugate base of H_2 (pK_a 35)
(c) Conjugate base of H_2O (pK_a 16)
(d) Both (a) and (c).

B-6. The product of the reaction

$$C_6H_5C\equiv CH \xrightarrow[\text{NH}_3]{\text{NaNH}_2}$$

is:
(a) $C_6H_5C\equiv CNH_2$ (c) $C_6H_5CH=CH_2$
(b) $C_6H_5C\equiv CNa$ (d) $C_6H_5CH=CHNH_2$

B-7. Choose the sequence of steps that describes the best synthesis of 1-butene from ethanol.
(a) (1) $NaC\equiv CH$;   (c) (1) HBr, heat; (2) $NaC\equiv CH$;
    (2) H_2, Lindlar Pd   (3) H_2, Lindlar Pd
(b) (1) $NaC\equiv CH$;   (d) (1) HBr, heat; (2) $KOC(CH_3)_3$, DMSO;
    (2) Na, NH_3 (3) $NaC\equiv CH$; (4) H_2, Lindlar Pd

CONJUGATION IN ALKADIENES AND ALLYLIC SYSTEMS

IMPORTANT TERMS AND CONCEPTS

Allylic Carbocations (Sec. 10.2) A carbocation in which the charge is on a carbon atom adjacent to a double bond is said to be *allylic*. Such intermediates are stabilized by delocalization of the electrons of the double bond.

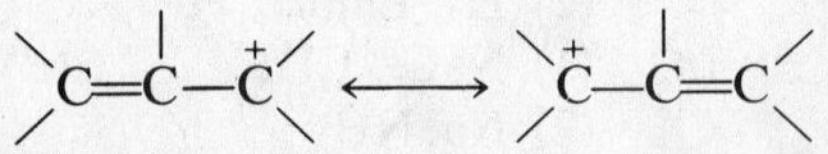

$$\text{C}{=}\text{C}{-}\overset{+}{\text{C}} \longleftrightarrow \overset{+}{\text{C}}{-}\text{C}{=}\text{C}$$

Allylic carbocations are more stable than simple alkyl carbocations.

Allylic Free Radicals (Sec. 10.3) An intermediate in which a carbon bearing an unpaired electron is adjacent to a double bond is known as an allylic free radical. Allylic radicals are stabilized by delocalization of the unpaired electron and π electrons over the three-carbon framework.

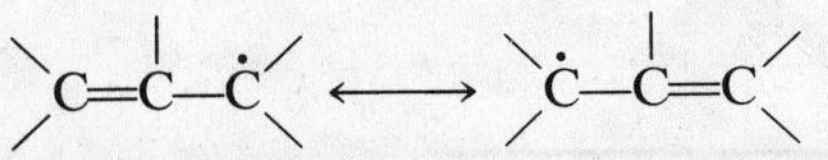

$$\text{C}{=}\text{C}{-}\dot{\text{C}} \longleftrightarrow \dot{\text{C}}{-}\text{C}{=}\text{C}$$

Classes of Dienes (Sec. 10.5) Three classes of dienes may be described according to the relationship between the double bonds:

Isolated: One or more sp^3 carbon atoms separate the double bonds.
Conjugated: The double bonds are joined by one single bond, forming a continuous chain of p orbitals.
Cumulated: The double bonds share a common atom (these dienes are also known as *allenes*).

Diene Stabilities (Secs. 10.6, 10.7) As shown by heats of hydrogenation, a conjugated double bond is 15 kJ/mol (3,5 kcal/mol) more stable than a simple double bond. This stabilization may be explained as arising from *delocalization* of the π electrons. The stabilization is also known as *conjugation energy*. The *s*-trans conformation of a conjugated diene is more stable than the *s*-cis.

$$
\text{s-Trans} \qquad \text{s-Cis}
$$

Direct and Conjugate Addition (Sec. 10.10) Two modes of reaction are possible when a conjugated diene undergoes an addition reaction. Addition across one of the double bonds is known as *direct addition*, or *1,2 addition*; addition to the ends of the diene system is known is *conjugate*, or *1,4 addition*.

$$
\text{H—X} + \quad \text{C=C—C=C} \quad \longrightarrow \quad \underset{\text{X}}{\text{—C—C—C=C}} \; + \; \underset{\text{X}}{\text{—C—C=C—C—}}
$$

Direct
(1,2) addition

Conjugate
(1,4) addition

At low temperature electrophilic addition to a conjugated diene is subject to *kinetic control*, also known as *rate control*. The proportion of products from the reaction is governed by their relative rates of formation, and once formed, the products do not interconvert or equilibrate with one another.

A reaction is subject to *thermodynamic control*, also known as *equilibrium control*, when the product distribution is governed by the relative stabilities of the products. The products are in equilibrium with each other, and so the most stable one predominates regardless of which product is formed fastest.

IMPORTANT REACTIONS

Allylic Halogenation (Sec. 10.4)

General:

$$
\text{C=C—C—H} \xrightarrow[\text{heat}]{N\text{-bromosuccinimide (NBS)}} \text{C=C—C—Br}
$$

Example:

$$
\bigcirc + \text{NBS} \xrightarrow[\text{heat}]{\text{CCl}_4} \bigcirc\text{Br} + \text{succinimide}
$$

Preparation of Dienes (Sec. 10.9)

Example:

$$
\underset{\overset{|}{\text{Br}}}{\text{CH}_2\text{=CHCHCH}_2\text{CH}_3} \xrightarrow[\text{heat}]{\text{KOH}} \text{CH}_2\text{=CHCH=CHCH}_3
$$

Hydrogen Halide Addition to Conjugated Dienes (Sec. 10.10)

Example:

$$
\text{CH}_2\text{=CH—CH=CH}_2 + \text{HCl} \longrightarrow \underset{\overset{|}{\text{Cl}}}{\text{CH}_2\text{=CH—CHCH}_3} + \underset{\overset{|}{\text{Cl}}}{\text{CH}_2\text{—CH=CHCH}_3}
$$

1,2 addition   1,4 addition

Halogen Addition to Conjugated Dienes (Sec. 10.11)

Example:

$$CH_2{=}CH{-}CH{=}CH_2 + Cl_2 \longrightarrow \underset{\text{1,2 addition}}{CH_2{=}CH{-}\underset{|}{\overset{}{CH}}(Cl)CH_2Cl} + \underset{\text{1,4 addition}}{\underset{Cl}{CH_2}{-}CH{=}CH{-}\underset{Cl}{CH_2}}$$

Diels-Alder Reaction (Sec. 10.12)

General:

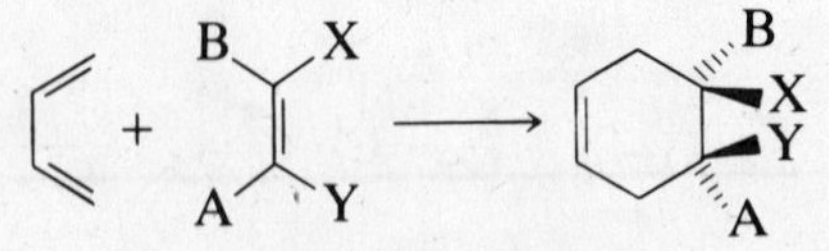

Example:

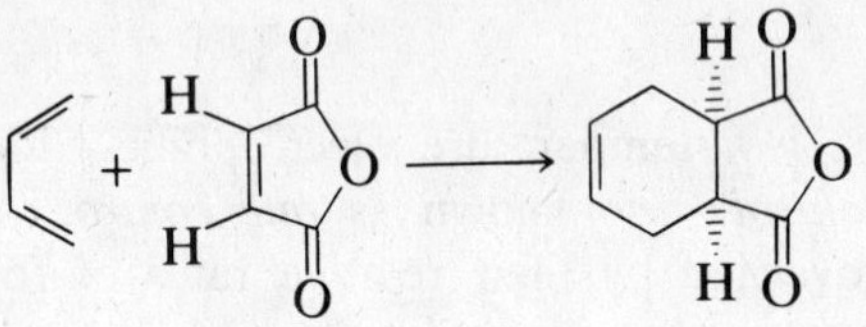

SOLUTIONS TO TEXT PROBLEMS

10.1 As noted in the sample solution to part (*a*), a pair of electrons is moved from the double bond toward the positively charged carbon.

(*b*) $CH_2{=}\underset{CH_3}{C}{-}\overset{+}{C}H_2 \longleftrightarrow \overset{+}{C}H_2{-}\underset{CH_3}{C}{=}CH_2$

(*c*) 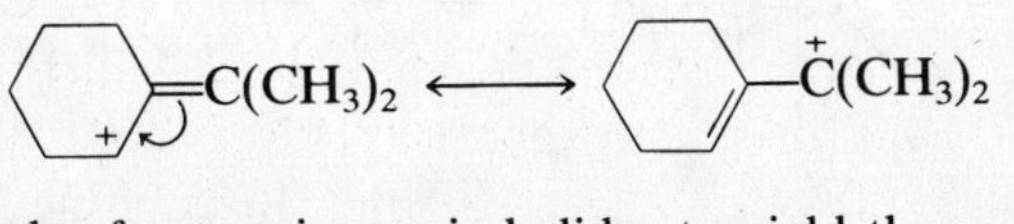

10.2 In order for two isomeric halides to yield the same carbocation on ionization, they must have the same carbon skeleton. They may have their leaving group at a different location, but the carbocations must become equivalent by allylic resonance.

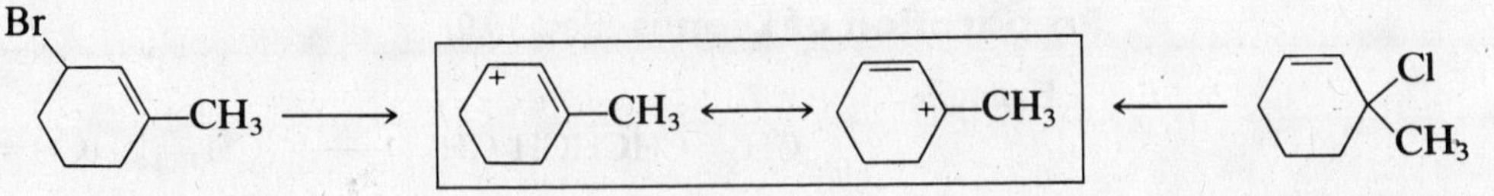

3-Bromo-1-methylcyclohexene 3-Chloro-3-methylcyclohexene

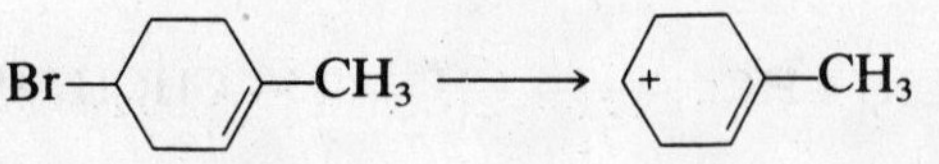 not an allylic carbocation

4-Bromo-1-methylcyclohexene

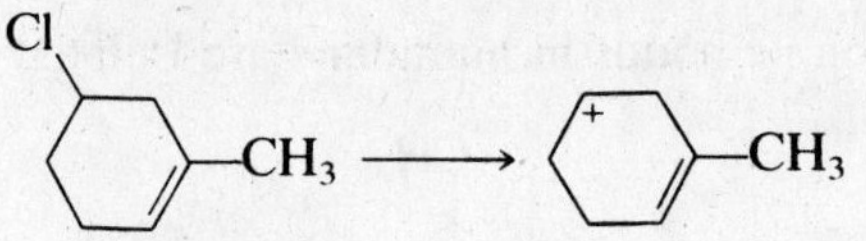

5-Chloro-1-
methylcyclohexene → not an allylic carbocation

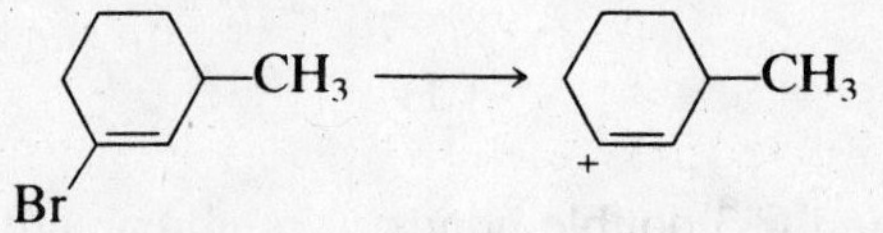

1-Bromo-3-
methylcyclohexene → not an allylic carbocation

10.3 The allylic hydrogens are the ones shown in the structural formulas.

(*b*)   1-Methylcyclohexene

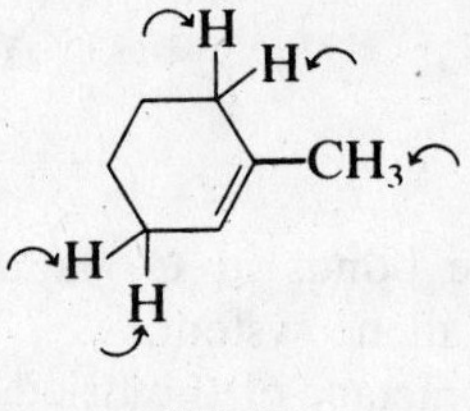

(*c*)   2,3,3-Trimethyl-1-butene

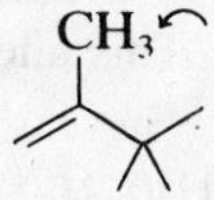

(*d*)   1-Octene

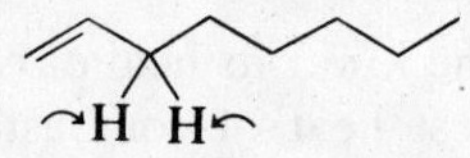

10.4 Write both resonance forms of the allylic radicals produced by hydrogen atom abstraction from the alkene.

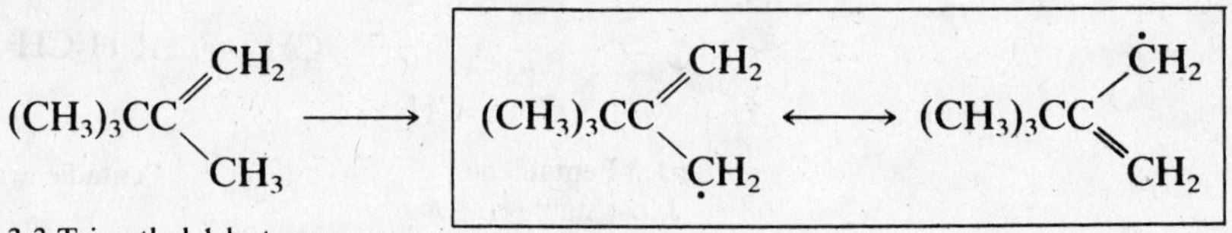

2,3,3-Trimethyl-1-butene

Both resonance forms are equivalent, and so 2,3,3-trimethyl-1-butene gives a single bromide on treatment with *N*-bromosuccinimide (NBS).

$$(CH_3)_3CC{=}CH_2 \xrightarrow{\text{NBS}} (CH_3)_3CC{=}CH_2$$
$$\qquad\quad |\qquad\qquad\qquad\qquad\quad|$$
$$\qquad\quad CH_3 \qquad\qquad\qquad\quad CH_2Br$$

2,3,3-Trimethyl-1-butene   2-(Bromomethyl)-3,3-dimethyl-1-butene

Hydrogen atom abstraction from 1-octene gives a radical in which the unpaired electron is delocalized between two nonequivalent positions.

$$CH_2{=}CHCH_2(CH_2)_4CH_3 \longrightarrow$$
1-Octene

$$CH_2{=}CH\dot{C}H(CH_2)_4CH_3 \longleftrightarrow \dot{C}H_2CH{=}CH(CH_2)_4CH_3$$

Allylic bromination of 1-octene gives a mixture of products

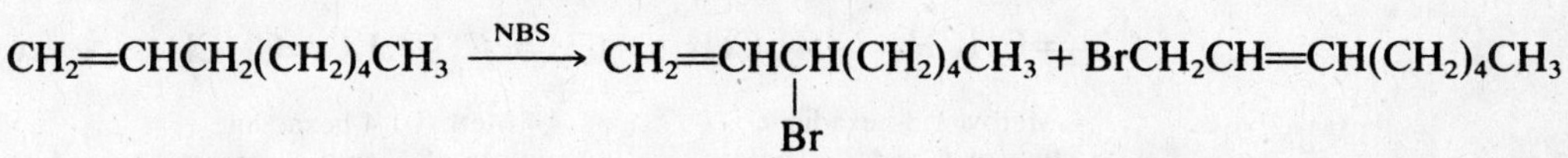

$$CH_2{=}CHCH_2(CH_2)_4CH_3 \xrightarrow{\text{NBS}} CH_2{=}CHCH(CH_2)_4CH_3 + BrCH_2CH{=}CH(CH_2)_4CH_3$$
$$\qquad\qquad\qquad\qquad\qquad\qquad\qquad\qquad\qquad |$$
$$\qquad\qquad\qquad\qquad\qquad\qquad\qquad\qquad\quad Br$$

1-Octene   3-Bromo-1-octene   1-Bromo-2-octene (cis and trans)

10.5 (*b*)   All the double bonds in humulene are isolated from each other

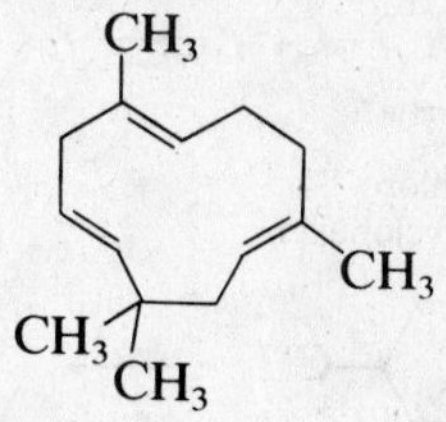

Humulene

(*c*)   The C-1 and C-3 double bonds of cembrene are conjugated to each other.

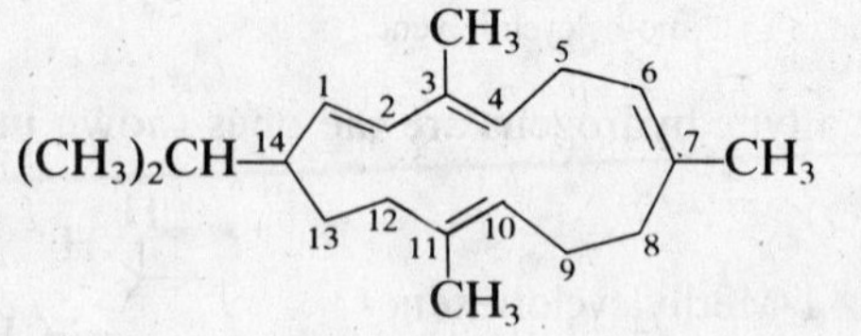

Cembrene

The double bonds at C-6 and C-10 are isolated from each other and from the conjugated diene system.

(*d*)   The sex attractant of the dried-bean beetle has a cumulated diene system involving C-4, C-5, and C-6. This allenic system is conjugated with the C-2 double bond.

$$CH_3(CH_2)_6CH_2\overset{6}{C}H{=}\overset{5}{C}{=}\overset{4}{C}H\overset{3}{C}H{=}\overset{2}{C}H\overset{1}{C}O_2CH_3$$

10.6   The more stable the isomer, the lower its heat of combustion. The conjugated diene is the most stable and has the lowest heat of combustion. The cumulated diene is the least stable and has the highest heat of combustion.

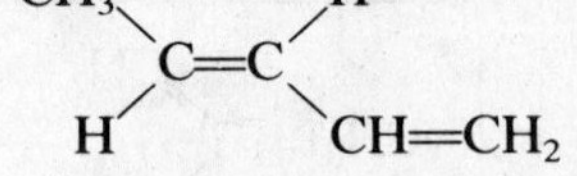

(*E*)-1,3-Pentadiene	1,4-Pentadiene	1,2-Pentadiene
Most stable		Least stable
3186 kJ/mol	3217 kJ/mol	3251 kJ/mol
(761.6 kcal/mol)	(768.9 kcal/mol)	(777.1 kcal/mol)

10.7   Both starting materials undergo β elimination to give a conjugated diene system. There are two minor products, both of which have isolated double bonds.

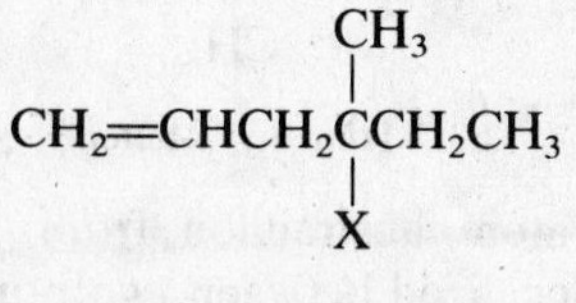

X = OH   3-Methyl-5-hexen-3-ol
X = Br   4-Bromo-4-methyl-1-hexene

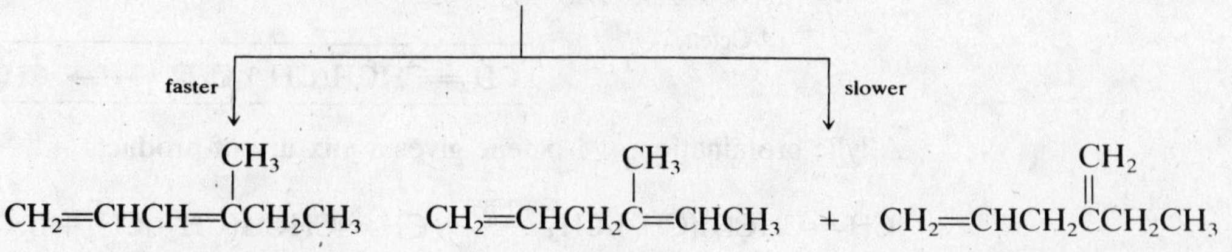

faster   slower

4-Methyl-1,3-hexadiene
(mixture of *E* and *Z* isomers;
major product)

4-Methyl-1,4-hexadiene
(mixture of *E* and *Z* isomers;
minor product)

2-Ethyl-1,4-pentadiene
(minor product)

10.8 The best approach is to work through this reaction mechanistically. Addition of hydrogen halides always proceeds by protonation of one of the terminal carbons of the diene system. Protonation of C-1 gives an allylic cation for which the most stable resonance form is a tertiary carbocation. Protonation of C-4 would give a less stable allylic carbocation for which the most stable resonance form is a secondary carbocation.

CH$_2$=CCH=CH$_2$ $\xrightarrow{\text{HCl}}$ (CH$_3$)$_2$CCH=CH$_2$
 | |
 CH$_3$ Cl

2-Methyl-1,3-butadiene   3-Chloro-3-methyl-1-butene
                                    (major product)

Under kinetically controlled conditions the carbocation is captured at the carbon that bears the greatest share of positive charge, and the product is the tertiary chloride.

10.9 The two double bonds of 2-methyl-1,3-butadiene are not equivalent, and so two different products of direct addition are possible, along with one conjugate addition product.

CH$_2$=CCH=CH$_2$ $\xrightarrow{\text{Br}_2}$ BrCH$_2$CCH=CH$_2$ + CH$_2$=CCHCH$_2$Br
 | | |
 CH$_3$ CH$_3$ CH$_3$

2-Methyl-1,3-   3,4-Dibromo-3-   3,4-Dibromo-2-
butadiene   methyl-1-butene   methyl-1-butene
                   (direct addition)   (direct addition)

\+ BrCH$_2$C=CHCH$_2$Br
         |
         CH$_3$

1,4-Dibromo-2-
methyl-2-butene
(conjugate addition)

10.10 The molecular formula of the product, C$_{10}$H$_9$ClO$_2$, is that of a 1:1 Diels-Alder adduct between 2-chloro-1,3-butadiene and benzoquinone.

2-Chloro-1,3-   Benzoquinone   C$_{10}$H$_9$ClO$_2$
butadiene

10.11 "Unravel" the Diels-Alder adduct as described in the sample solution to part (a).

(b)   is prepared
                     from

Diels-Alder   Diene   Dienophile
adduct   (cyano groups
                                 are cis)

(c) is prepared from

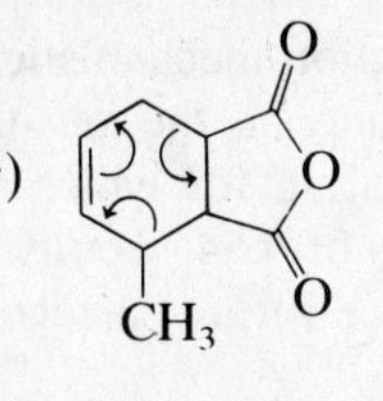

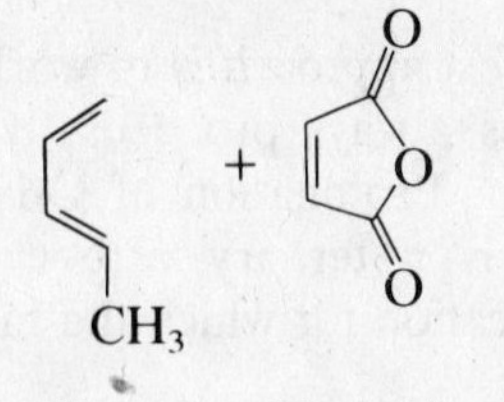

Diene   Dienophile

(d) is prepared from

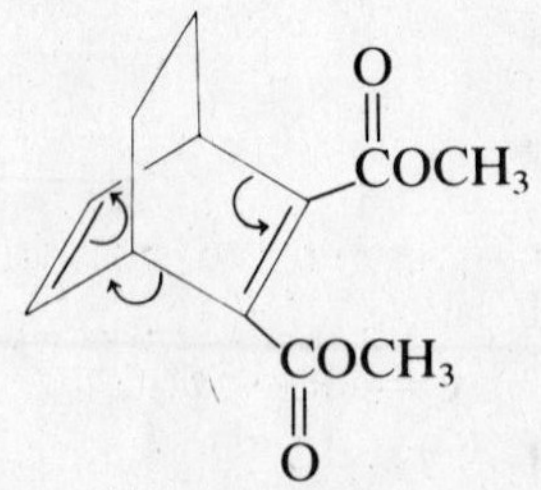

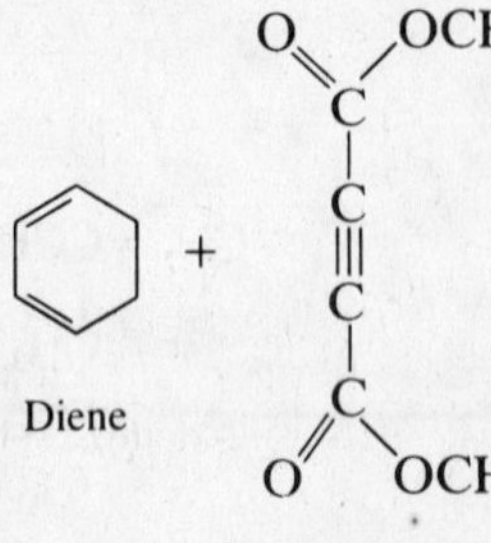

Diene

Dienophile

10.12 In the Diels-Alder reaction of 1,3-cyclopentadiene with dimethyl fumarate described at the top of text page 390, the trans stereochemistry in the dienophile is carried over to a trans arrangement between substituents in the product.

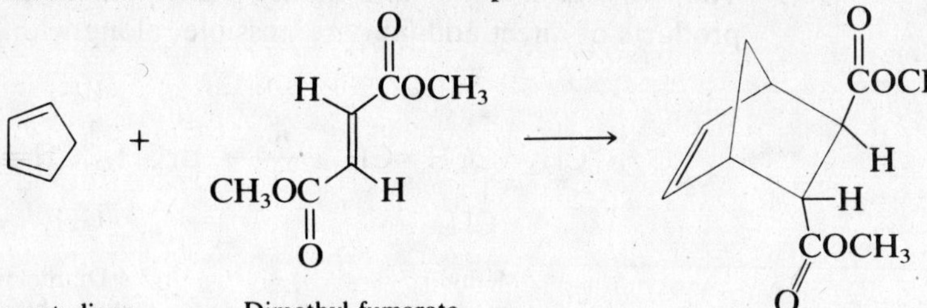

1,3-Cyclopentadiene   Dimethyl fumarate

Dimethyl-bicyclo[2.2.1]hept-2-ene-*trans*-5,6-dicarboxylate

10.13 Take *cis*-3,4-dimethylcyclobutene as an example and consider what happens if the ring-opening step is conrotatory and its reverse (ring closing) is disrotatory.

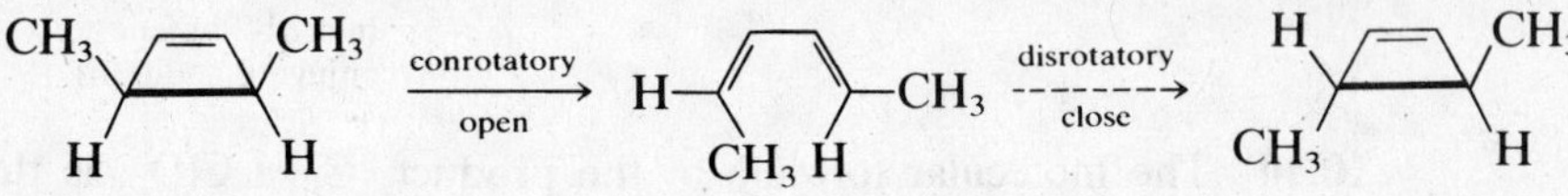

cis-3,4-Dimethylcyclobutene   *cis,trans*-2,4-Hexadiene   *trans*-3,4-Dimethylcyclobutene

If the reaction is reversible, the stereochemistry of the starting materials, and therefore of the products as well, will be scrambled. The reaction should not be stereospecific. Since it is observed to be stereospecific, the forward and reverse steps must both proceed in the same stereochemical sense. Both are conrotatory.

Identical reasoning applies to the 1,3,5-hexatriene to 1,3-cyclohexadiene transformation. Stereochemistry will be preserved only if the ring-opening and ring-closing reactions have the same stereochemical sense. In this case, both are disrotatory.

10.14 One stereoisomer that gives *trans*-7,8-dimethyl-1,3,5-cyclooctatriene upon conrotatory ring closure is *cis,cis,cis,cis*-2,4,6,8-decatetraene.

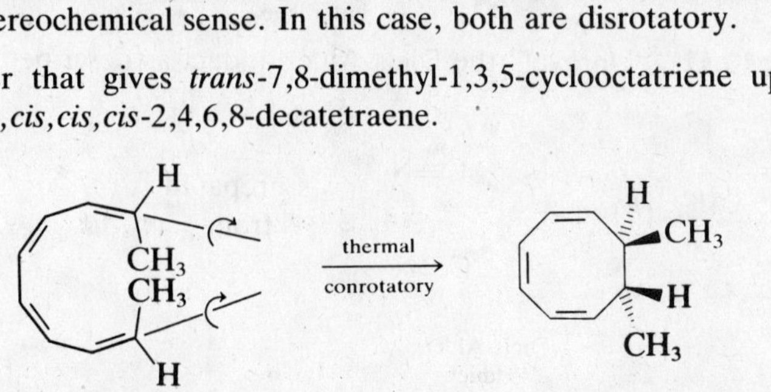

cis,cis,cis,cis-2,4,6,8-Decatetraene   *trans*-7,8-Dimethyl-1,3,5-cyclooctatriene

The second stereoisomer is *trans,cis,cis,trans*-2,4,6,8-decatetraene.

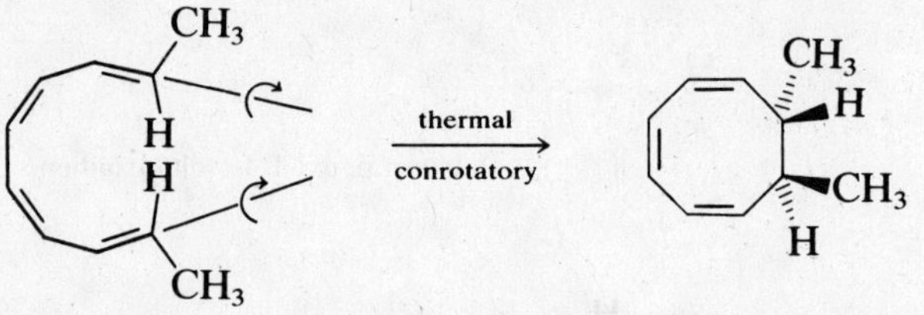

thermal
conrotatory

trans,cis,cis,trans-2,4,6,8-Decatetraene   *trans*-7,8-Dimethyl-1,3,5-cyclooctatriene

10.15 Dienes and trienes are named according to the IUPAC convention by replacing the *-ane* ending of the alkane with *-adiene* or *-atriene* and locating the positions of the double bonds by number. The stereoisomers are identified as *E* or *Z* according to the rules established in Chapter 5.

(*a*) 3,4-Octadiene $CH_3CH_2CH{=}C{=}CHCH_2CH_2CH_3$

(*b*) (*E,E*)-2,4-Octadiene

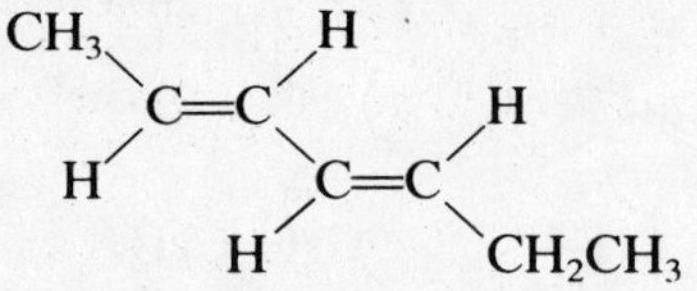

(*c*) (*Z,Z*)-1,3-Cyclooctadiene

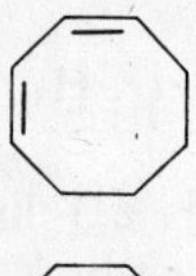

(*d*) (*Z,Z*)-1,4-Cyclooctadiene

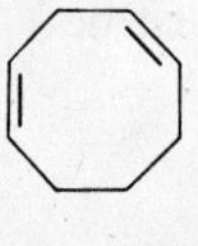

(*e*) (*E,E*)-1,5-Cyclooctadiene

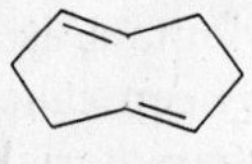

(*f*) (2*E*,4*Z*,6*E*)-2,4,6-Octatriene

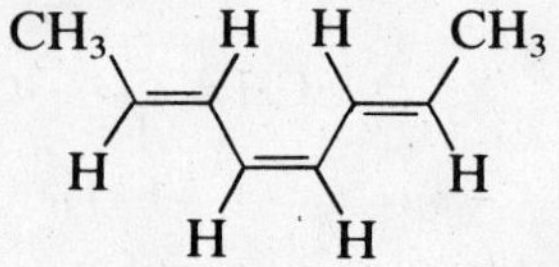

(*g*) 5-Allyl-1,3-cyclopentadiene

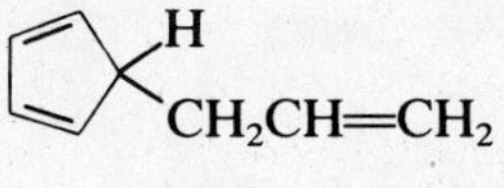

(*h*) *trans*-1,2-Divinylcyclopropane

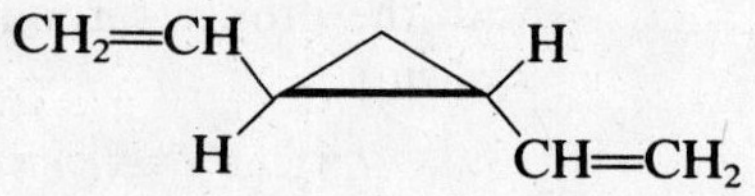

(*i*) 2,4-Dimethyl-1,3-pentadiene $CH_2{=}CCH{=}CCH_3$ with CH_3 CH_3 substituents

10.16 (*a*) $CH_2{=}CH(CH_2)_5CH{=}CH_2$ 1,8-Nonadiene

(*b*) $(CH_3)_2C{=}CC{=}C(CH_3)_2$ with CH_3 substituent   2,3,4,5-Tetramethyl-2,4-hexadiene

(c) $CH_2=CH-CH-CH=CH_2$ 3-Vinyl-1,4-pentadiene
                |
              $CH=CH_2$

(d)   3-Isopropenyl-1,4-cyclohexadiene

(e)   (1Z,3E,5Z)-1,6-Dichloro-1,3,5-hexatriene

(f) $CH_2=C=CHCH=CHCH_3$ 1,2,4-Hexatriene

(g)   (1E,5E,9E)-1,5,9-Cyclododecatriene

(h)   (E)-3-Ethyl-4-methyl-1,3-hexadiene

10.17   (a)   Since the product is 2,3-dimethylbutane we know that the carbon skeleton of the starting material must be

$$C-C-C-C$$
$$\quad\ |\ \ \ |$$
$$\quad\ C\ \ C$$

Since 2,3-dimethylbutane is C_6H_{14} and the starting material is C_6H_{10}, *two* molecules of H_2 must have been taken up and the starting material must have two double bonds. The starting material can only be 2,3-dimethyl-1,3-butadiene.

$$CH_2=C-\!\!\!-C=CH_2 + 2H_2 \xrightarrow{Pt} (CH_3)_2CHCH(CH_3)_2$$
$$\qquad\ |\quad\ |$$
$$\quad\ CH_3\ \ CH_3$$

(b)   Write the carbon skeleton corresponding to 2,2,6,6-tetramethylheptane.

Compounds of molecular formula $C_{11}H_{20}$ have a SODAR of 2. The only compounds with the proper carbon skeleton with a SODAR of 2 are the alkyne and the allene shown.

$(CH_3)_3CC\equiv CCH_2C(CH_3)_3$ $(CH_3)_3CCH=C=CHC(CH_3)_3$

2,2,6,6-Tetramethyl-3-heptyne   2,2,6,6-Tetramethyl-3,4-heptadiene

10.18   The dienes that give 2,4-dimethylpentane on catalytic hydrogenation must have the same carbon skeleton as that alkane.

      or   or $\xrightarrow[Pt]{H_2}$

       (a)   (b)   (c)

   2,4-Dimethyl-   2,4-Dimethyl-   2,4-Dimethyl-   2,4-Dimethylpentane
   1,3-pentadiene   1,4-pentadiene   2,3-pentadiene
   conjugated diene   isolated diene   cumulated diene

10.19 The important piece of information which allows us to complete the structure properly is that the ant repellent is an *allenic* substance. The allenic unit cannot be incorporated into the ring, because the three carbons must be collinear. Therefore, the only possible constitution is

10.20 (*a*) Allylic halogenation of propene with *N*-bromosuccinimide gives allyl bromide.

$$CH_2{=}CHCH_3 \xrightarrow[CCl_4,\ heat]{N\text{-bromosuccinimide}} CH_2{=}CHCH_2Br$$

Propene   Allyl bromide

(*b*) Electrophilic addition of bromine to the double bond of propene gives 1,2-dibromopropane.

$$CH_2{=}CHCH_3 \xrightarrow{Br_2} BrCH_2\underset{\underset{Br}{|}}{C}HCH_3$$

Propene   1,2-Dibromopropane

(*c*) 1,3-Dibromopropane is made from allyl bromide by free-radical addition of hydrogen bromide.

$$CH_2{=}CHCH_2Br \xrightarrow[peroxides]{HBr} BrCH_2CH_2CH_2Br$$

Allyl bromide   1,3-Dibromopropane

(*d*) Addition of hydrogen chloride to allyl bromide proceeds in accordance with Markovnikov's rule.

$$CH_2{=}CHCH_2Br \xrightarrow{HCl} CH_3\underset{\underset{Cl}{|}}{C}HCH_2Br$$

Allyl bromide   1-Bromo-2-chloropropane

(*e*) Addition of bromine to allyl bromide gives 1,2,3-tribromopropane.

$$CH_2{=}CHCH_2Br \xrightarrow{Br_2} BrCH_2\underset{\underset{Br}{|}}{C}HCH_2Br$$

Allyl bromide   1,2,3-Tribromopropane

(*f*) Nucleophilic substitution by hydroxide on allyl bromide gives allyl alcohol.

$$CH_2{=}CHCH_2Br \xrightarrow{NaOH} CH_2{=}CHCH_2OH$$

Allyl bromide   Allyl alcohol

(*g*) Alkylation of sodium acetylide using allyl bromide gives the desired 4-penten-1-yne.

$$CH_2{=}CHCH_2Br \xrightarrow{NaC{\equiv}CH} CH_2{=}CHCH_2C{\equiv}CH$$

Allyl bromide   4-Penten-1-yne

(*h*) Sodium-ammonia reduction of 4-penten-1-yne reduces the triple bond but leaves the double bond intact. Hydrogenation over Lindlar palladium could also be used.

$$CH_2{=}CHCH_2C{\equiv}CH \xrightarrow[\substack{or \\ H_2,\ Lindlar\ Pd}]{Na,\ NH_3} CH_2{=}CHCH_2CH{=}CH_2$$

4-Penten-1-yne   1,4-Pentadiene

10.21 (*a*) The desired allylic alcohol can be prepared by hydrolysis of an allylic halide. Cyclopentene can be converted to an allylic bromide by free-radical bromination with *N*-bromosuccinimide (NBS).

Cyclopentene   3-Bromocyclopentene   2-Cyclopenten-1-ol

(*b*) Reaction of the allylic bromide from part (*a*) with sodium iodide in acetone converts it to the corresponding iodide.

3-Bromocyclopentene   3-Iodocyclopentene

(*c*) Nucleophilic substitution by cyanide converts the allylic bromide to 3-cyanocyclopentene.

3-Bromocyclopentene   3-Cyanocyclopentene

(*d*) Reaction of the allylic bromide with a strong base will yield cyclopentadiene by an E2 elimination.

3-Bromocyclopentene   1,3-Cyclopentadiene

(*e*) Cyclopentadiene formed in part (*d*) is needed in order to form the required Diels-Alder adduct.

1,3-Cyclopentadiene   Dimethyl bicyclo[2.2.1]heptadiene-2,3-dicarboxylate

10.22 The starting material in all cases is 2,3-dimethyl-1,3-butadiene.

(*a*) Hydrogenation of both double bonds will occur to yield 2,3-dimethylbutane.

(*b*) Direct addition of one mole of hydrogen chloride will give the product of Markovnikov addition to one of the double bonds, 3-chloro-2,3-dimethyl-1-butene.

(c)   Conjugate addition will lead to double bond migration and produce 1-chloro-2,3-dimethyl-2-butene.

$$CH_2=\overset{\overset{\displaystyle CH_3}{|}}{C}-\overset{\underset{\displaystyle CH_3}{|}}{C}=CH_2 \xrightarrow{\text{HCl}} (CH_3)_2C=\overset{\overset{\displaystyle CH_3}{|}}{C}CH_2Cl$$

(d)   The direct addition product is 3,4-dibromo-2,3-dimethyl-1-butene.

$$CH_2=\overset{\overset{\displaystyle CH_3}{|}}{C}-\overset{\underset{\displaystyle CH_3}{|}}{C}=CH_2 \xrightarrow{\text{Br}_2} BrCH_2\overset{\overset{\displaystyle CH_3}{|}}{\underset{\underset{\displaystyle Br}{|}}{C}}-\overset{\overset{\displaystyle }{}}{\underset{\underset{\displaystyle CH_3}{|}}{C}}=CH_2$$

(e)   The conjugate addition product will be 1,4-dibromo-2,3-dimethyl-2-butene.

$$CH_2=\overset{\overset{\displaystyle CH_3}{|}}{C}-\overset{\underset{\displaystyle CH_3}{|}}{C}=CH_2 \xrightarrow{\text{Br}_2} BrCH_2\overset{\overset{\displaystyle CH_3}{|}}{C}=\overset{\overset{\displaystyle CH_3}{|}}{C}CH_2Br$$

(f)   Bromination of both double bonds will lead to 1,2,3,4-tetrabromo-2,3-dimethylbutane irrespective of whether the first addition step occurs by direct or conjugate addition.

$$CH_2=\overset{\overset{\displaystyle CH_3}{|}}{C}-\overset{\underset{\displaystyle CH_3}{|}}{C}=CH_2 \xrightarrow{\text{2Br}_2} BrCH_2\overset{\overset{\displaystyle CH_3}{|}}{\underset{\underset{\displaystyle Br}{|}}{C}}-\overset{\overset{\displaystyle CH_3}{|}}{\underset{\underset{\displaystyle Br}{|}}{C}}CH_2Br$$

(g)   The reaction of a diene with maleic anhydride is a Diels-Alder reaction.

10.23   The starting material in all cases is 1,3-cyclohexadiene.

(a)   Cyclohexane will be the product of hydrogenation of 1,3-cyclohexadiene:

(b)   Direct addition will occur according to Markovnikov's rule to give 3-chlorocyclohexene

3-Chlorocyclohexene

(c)   The product of conjugate addition is 3-chlorocyclohexene also. Direct addition and conjugate addition of hydrogen chloride to 1,3-cyclohexadiene gives the same product.

3-Chlorocyclohexene

(d) Bromine can add directly to one of the double bonds to give 3,4-dibromocyclohexene:

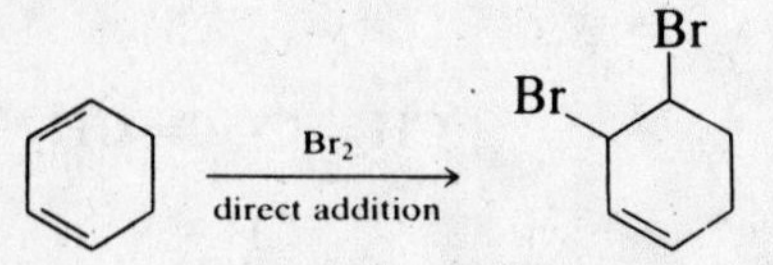

3,4-Dibromocyclohexene

(e) Conjugate addition of bromine will give 3,6-dibromocyclohexene:

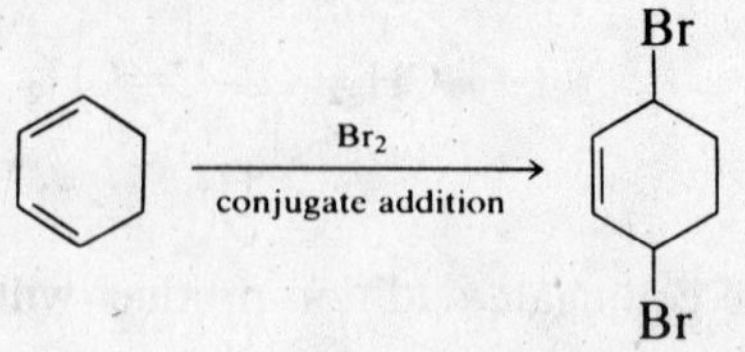

3,6-Dibromocyclohexene

(f) Addition of two moles of bromine will yield 1,2,3,4-tetrabromocyclohexane.

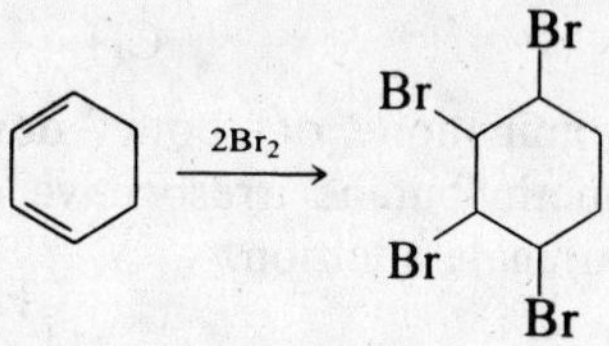

(g) The constitution of the Diels-Alder adduct of 1,3-cyclohexadiene and maleic anhydride will have a bicyclo[2.2.2]octyl carbon skeleton.

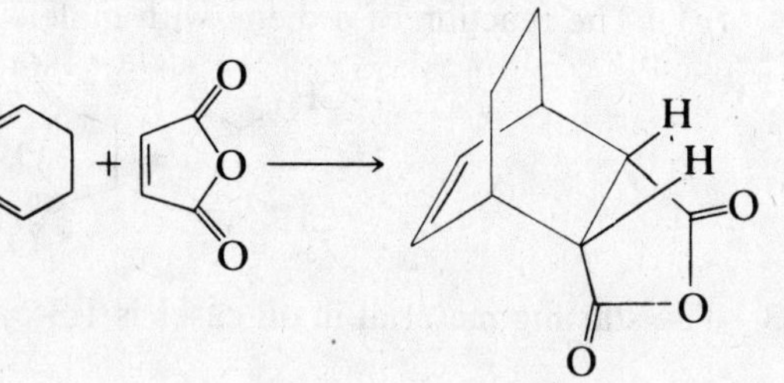

10.24 Compound B must arise by way of a Diels-Alder reaction between compound A and dimethyl acetylenedicarboxylate. Therefore, compound A must have a conjugated diene system.

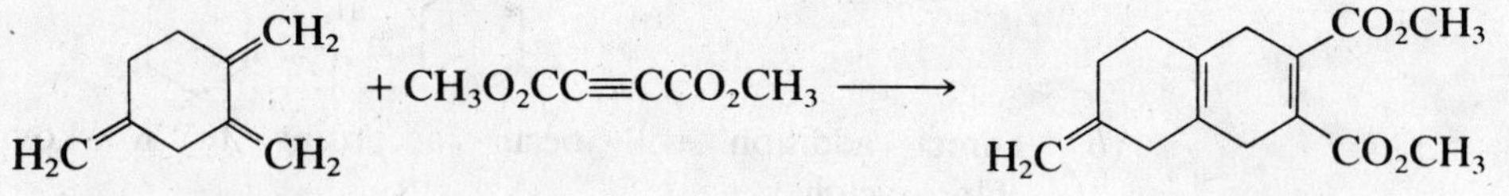

10.25 (a) Solvolysis of $(CH_3)_2C=CHCH_2Cl$ in ethanol proceeds by an S_N1 mechanism and involves a carbocation intermediate.

$$(CH_3)_2C=CHCH_2Cl \quad \text{1-Chloro-3-methyl-2-butene}$$

$$[(CH_3)_2C=CH-\overset{+}{C}H_2 \longleftrightarrow (CH_3)_2\overset{+}{C}-CH=CH_2]$$

This carbocation has some of the character of a tertiary carbocation. It is more stable and is therefore formed faster than allyl cation, $CH_2=CH\overset{+}{C}H_2$.

(b) An allylic carbocation is formed from the alcohol in the presence of an acid catalyst.

$$CH_2=CHCHCH_3 \xrightarrow[H_2O]{H_2SO_4} CH_2=CHCHCH_3 \longrightarrow CH_2=CHCHCH_3$$

3-Buten-2-ol

This carbocation is a delocalized one and can be captured at either end of the allylic system by water acting as a nucleophile.

$$CH_2=CHCHCH_3 \longleftrightarrow \overset{+}{C}H_2CH=CHCH_3$$

$\xrightarrow{H_2O}$

$CH_2=CHCHCH_3 \xrightarrow{-H^+} CH_2=CHCHCH_3$ (3-Buten-2-ol)

$H_2\overset{+}{O}-CH_2CH=CHCH_3 \xrightarrow{-H^+} HOCH_2CH=CHCH_3$ (2-Buten-1-ol)

(c) Hydrogen bromide converts the alcohol to an allylic carbocation. Bromide ion captures this carbocation at either end of the delocalized allylic system.

$$CH_3CH=CHCH_2OH \xrightarrow{HBr} CH_3CH=CHCH_2-\overset{+}{O}H_2 \longrightarrow CH_3CH=CH\overset{+}{C}H_2$$

2-Buten-1-ol

$$CH_3CH=CH\overset{+}{C}H_2 \longleftrightarrow CH_3\overset{+}{C}HCH=CH_2$$

$\xrightarrow{Br^-}$

$CH_3CH=CHCH_2Br$ 1-Bromo-2-butene

$CH_3CHCH=CH_2$ (Br) 3-Bromo-1-butene

(d) The same delocalized carbocation is formed from 3-buten-2-ol as from 2-buten-1-ol.

$$CH_3CHCH=CH_2 \xrightarrow{HBr} CH_3\overset{+}{C}HCH=CH_2 \longleftrightarrow CH_3CH=CH\overset{+}{C}H_2$$

(OH) 3-Buten-2-ol

Since this carbocation is the same as the one formed in part (c), it gives the same mixture of products when it reacts with bromide.

(e) We are told that the major product is 1-bromo-2-butene, not 3-bromo-1-butene.

$BrCH_2CH=CHCH_3$ 1-Bromo-2-butene (major)

$CH_2=CHCHCH_3$ (Br) 3-Bromo-1-butene (minor)

The major product is the more stable one. It is a primary rather than a secondary halide and contains a more substituted double bond. Therefore the reaction is governed by thermodynamic (equilibrium) control.

10.26 Since both products of reaction of hydrogen chloride with vinylacetylene are chloro-substituted dienes, the first step in addition must involve the triple bond. The carbocation produced is an allylic vinyl cation for which two Lewis structures may be written. Capture of this cation gives the products of 1,2 and 1,4 addition. The 1,2 addition product is more stable because of its conjugated system. The observations of the experiment tell us that the 1,4 addition product is formed faster, although we could not have predicted that.

$$HC{\equiv}C{-}CH{=}CH_2 \xrightarrow{\ HCl\ } CH_2{=}\overset{+}{C}{-}CH{=}CH_2 \longleftrightarrow CH_2{=}C{=}CH{-}\overset{+}{C}H_2$$

Vinylacetylene

$$CH_2{=}C{-}CH{=}CH_2 + CH_2{=}C{=}CH{-}CH_2Cl$$
$$|$$
$$Cl$$

2-Chloro-1,3-butadiene   4-Chloro-1,2-butadiene
(1,2 addition)   (1,4 addition)

10.27   (*a*)   The two equilibria are:

For (*E*)-1,3-pentadiene:

s-trans   *s*-cis

For (*Z*)-1,3-pentadiene:

s-trans   *s*-cis

(*b*)   The *s*-cis conformation of (*Z*)-1,3-pentadiene is destabilized by a van der Waals repulsion involving the methyl group.

Methyl-hydrogen replusion   Hydrogen-hydrogen replusion

s-cis conformation of   *s*-cis conformation of
(*Z*)-1,3-pentadiene   (*E*)-1,3-pentadiene

The equilibrium favors the *s*-trans conformation of (*Z*)-1,3-pentadiene more than it does that of the *E* isomer because the *s*-cis conformation of the *Z* isomer has more van der Waals strain.

10.28   Compare the mirror-image forms of each compound for superposability.

(*a*)   2-Methyl-2,3-pentadiene

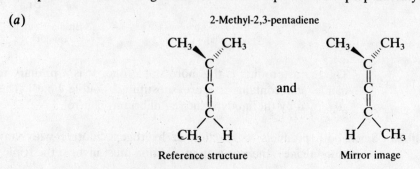

Reference structure   Mirror image

Rotation of the mirror image 180° around an axis passing through the three carbons of the C=C=C unit demonstrates that the reference structure and its mirror image are superposable.

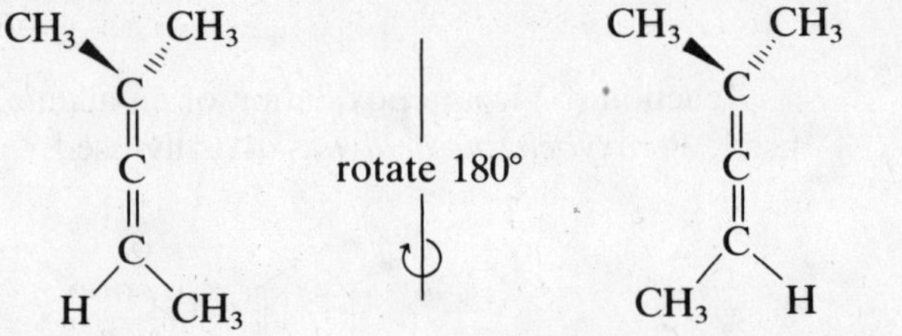

$$CH_3 \quad CH_3$$

rotate 180°

$$CH_3 \quad CH_3$$

Mirror image   Reoriented mirror image

2-Methyl-2,3-pentadiene is an achiral allene.

(b) The stereochemical situation here is the same as in part (a). In part (a) substituent R was a methyl group; here it is an ethyl group. Changing the nature of group R does not affect the superposability of the reference structure and its mirror image. 2-Methyl-2,3-hexadiene is achiral.

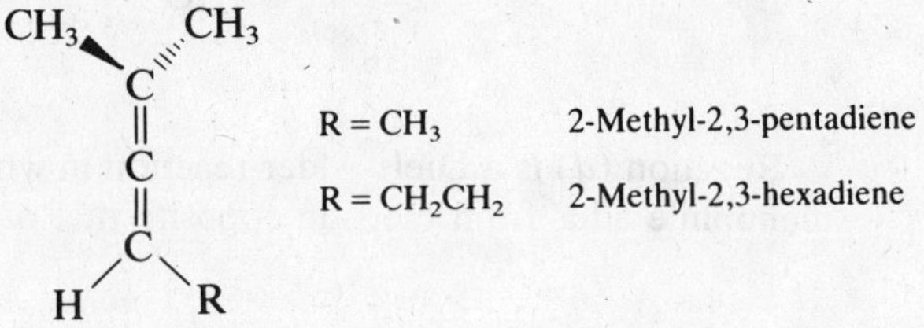

$R = CH_3$ 2-Methyl-2,3-pentadiene

$R = CH_2CH_2$ 2-Methyl-2,3-hexadiene

(c) The two mirror-image forms of 4-methyl-2,3-hexadiene are as shown:

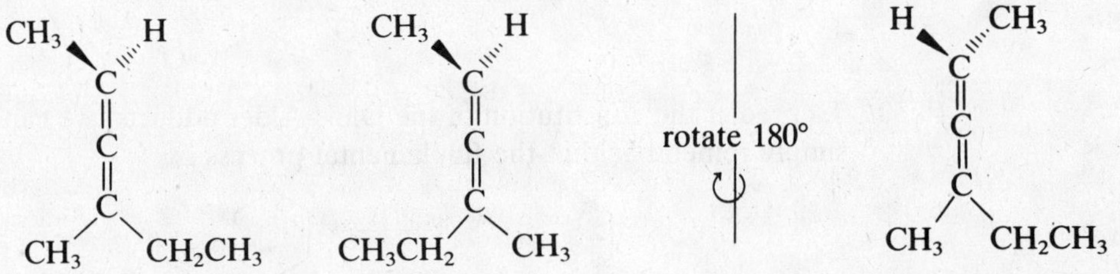

rotate 180°

Reference structure   Mirror image   Reoriented mirror image

The two structures cannot be superposed. 4-Methyl-2,3-hexadiene is chiral. Rotation of either representation 180° around an axis that passes through the three carbons of the C=C=C unit leads to superposition of the groups at the "bottom" carbon but not at the "top."

(d) 2,4-Dimethyl-2,3-pentadiene is achiral. Its two mirror-image forms are superposable.

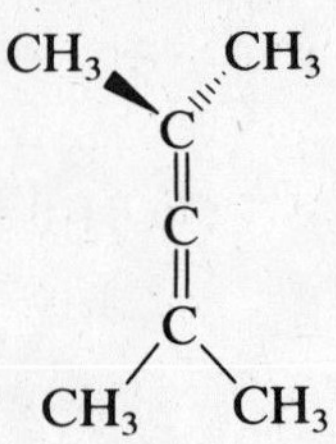

Reference structure 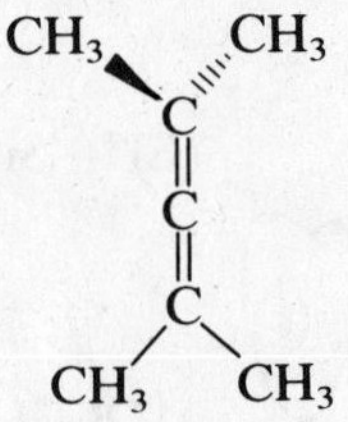   Mirror image

The molecule has two planes of symmetry defined by the three carbons of each CH_3CCH_3 unit.

10.29 Reaction (a) is an electrophilic addition of bromine to an alkene; the appropriate reagent is *bromine in carbon tetrachloride.*

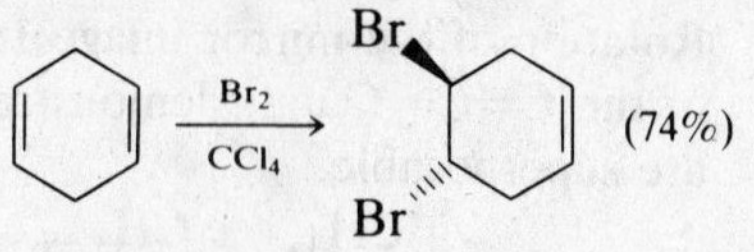

Reaction (b) is an epoxidation of an alkene, for which almost any peroxy acid could be used. *Peroxybenzoic acid* was actually used.

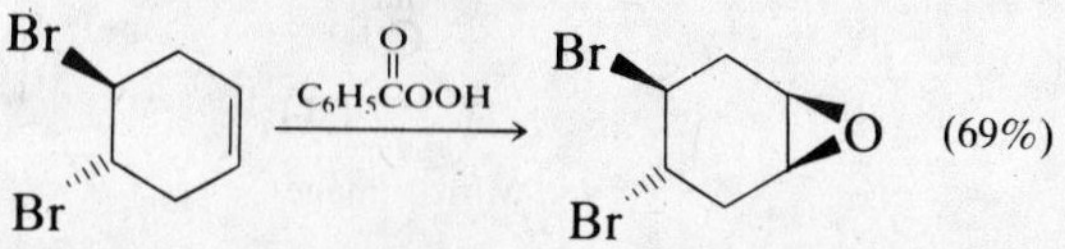

Reaction (c) is an elimination reaction of a vicinal dibromide to give a conjugated diene and requires E2 conditions. *Sodium methoxide in methanol* was used.

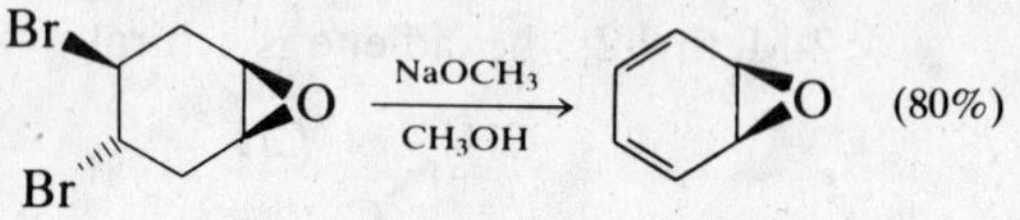

Reaction (d) is a Diels-Alder reaction in which the dienophile is *maleic anhydride*. The dienophile adds from the side opposite that of the epoxide ring.

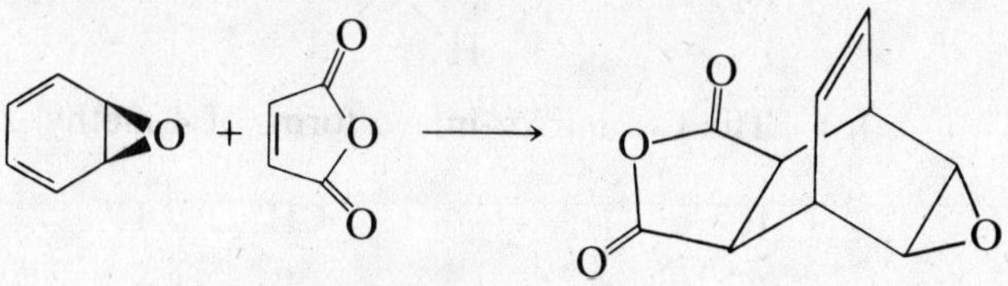

10.30 To predict the constitution of the Diels-Alder adducts, we can ignore the substituents and simply remember that the fundamental process is

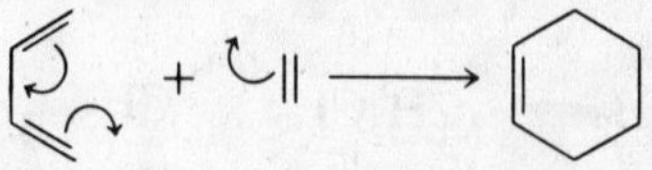

(a)

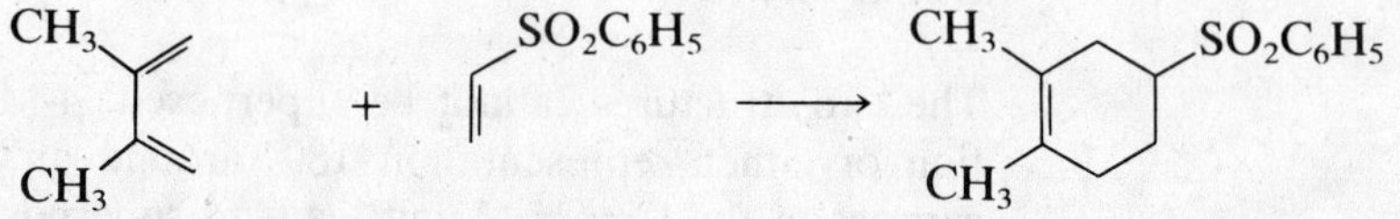

2,3-Dimethyl-1,3-butadiene

(b)

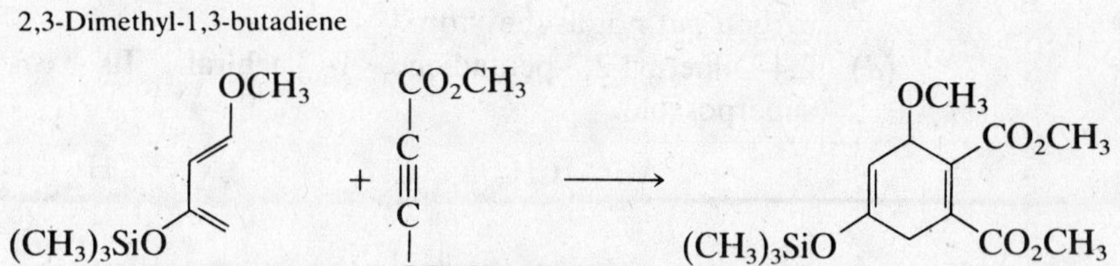

(c)

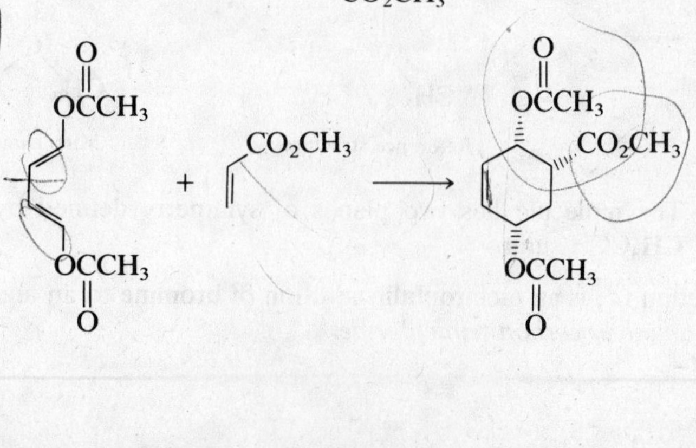

(d)

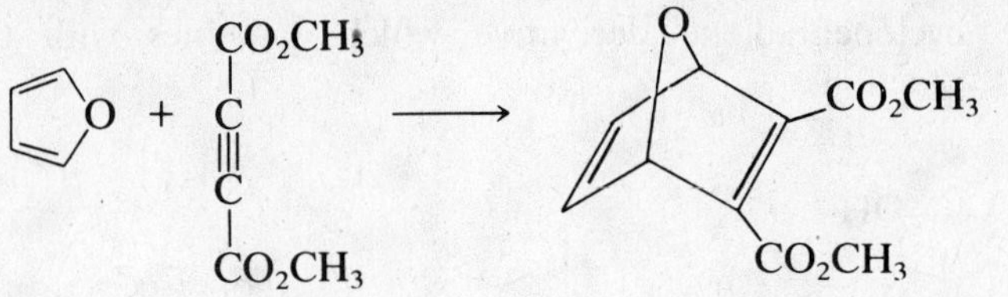

(e)

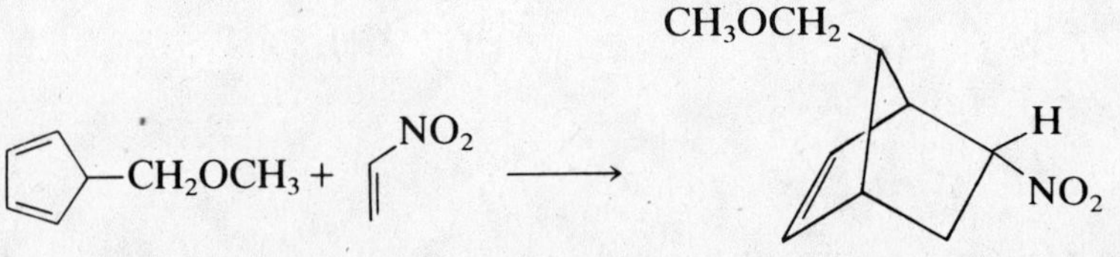

10.31 The carbon skeleton of dicyclopentadiene must be the same as that of its hydrogenation product, and dicyclopentadiene must contain two double bonds, since two moles of hydrogen are consumed in its hydrogenation ($C_{10}H_{12} \longrightarrow C_{10}H_{16}$).

The molecular formula of dicyclopentadiene ($C_{10}H_{12}$) is twice that of 1,3-cyclopentadiene (C_5H_6), and its carbon skeleton suggests that 1,3-cyclopentadiene is undergoing a Diels-Alder reaction with itself. Therefore:

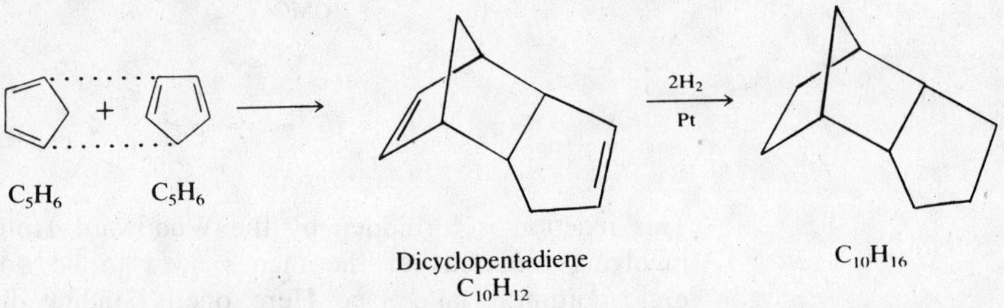

One molecule of 1,3-cyclopentadiene acts as the diene and the other acts as the dienophile in this Diels-Alder reaction.

10.32 The formation of the dibromocyclobutane derivative is nothing more than electrophilic addition to the double bond of the cyclobutene ring. The more unusual product is the dihaloalkene. Heating the cyclobutene causes electrocyclic ring opening to occur and gives a conjugated diene. 1,4 addition of Br_2 to this conjugated diene gives the observed dibromoalkene.

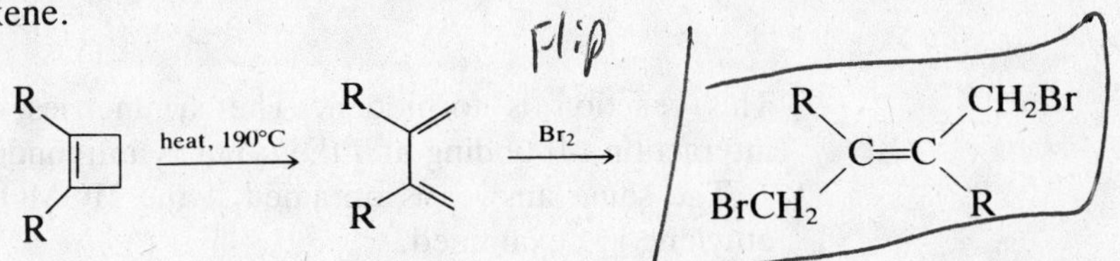

10.33 (a) The cyclobutene ring opens to a substituted 1,3-butadiene on heating. This diene then undergoes a Diels-Alder cycloaddition to maleic anhydride to give the observed product.

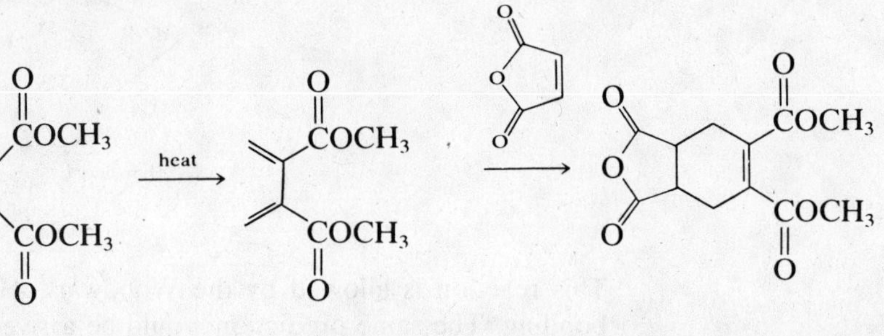

(b) Dehydration of the allylic alcohol under the acidic conditions of the reaction gives a cyclopentadiene derivative which combines with the alkyne in a Diels-Alder reaction.

10.34 (*a*) Analyze the reaction of two butadiene molecules by the Woodward–Hoffmann rules by examining the symmetry properties of the highest occupied molecular orbital (HOMO) of one diene and the lowest unoccupied molecular orbital (LUMO) of the other.

This reaction is forbidden by the Woodward–Hoffmann rules. Both interactions involving the ends of the dienes need to be bonding in order for concerted cycloaddition to take place. Here, one is bonding and the other is antibonding.

(*b*) Since allyl cation is positively charged, examine the process in which electrons "flow" from the HOMO of ethylene to the LUMO of allyl cation.

This reaction is forbidden. The symmetries of the orbitals are such that one interaction is bonding and the other is antibonding.

   The same answer is obtained if the HOMO of allyl cation and the LUMO of ethylene are examined.

(*c*) In this part of the exercise we consider the LUMO of allyl cation and the HOMO of 1,3-butadiene.

This reaction is allowed by the Woodward–Hoffmann rules. Both interactions are bonding. The same prediction would be arrived at if the HOMO of allyl cation and LUMO of 1,3-butadiene were the orbitals considered.

(d)   There is a mismatch between the HOMO of 1,3,5-hexatriene and the LUMO of ethylene. This cycloadditon is *forbidden*.

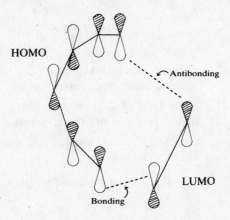

(e)   There is a mismatch between the termini when the HOMO and LUMO of 1,3,5-hexatriene are examined. The cycloaddition is *forbidden*.

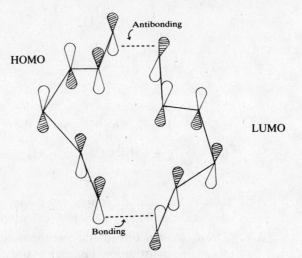

10.35   Since oxygen has two unpaired electrons, it can abstract a hydrogen atom from the allylic position of cyclohexene to give a free-radical intermediate.

The cyclohexenyl radical is resonance-stabilized. It reacts further via the following two propagation steps:

SELF-TEST

PART A

A-1. Give the structures of all the isomeric alkadienes of molecular formula C_5H_8, ignoring stereoisomers. Indicate which are conjugated and which are allenes.

A-2. Provide the IUPAC name for each of the conjugated dienes of the previous problem, *including stereoisomers.*

A-3. Hydrolysis of 3-bromo-3-methylcyclohexene yields two isomeric alochols. Draw their structures and the structure of the intermediate that leads to their formation.

A-4. Give the chemical structure of the reactant, reagent, or product omitted from each of the following:

(a) $CH_3CH{=}CHCH{=}CHCH_3 \xrightarrow{\text{Br}_2}$? (two products)

(b) ? $\xrightarrow{\text{Diels-Alder}}$

(c)

(d)

A-5. One of the isomeric conjugated dienes having the formula C_6H_8 is not able to react with a dienophile in a Diels-Alder reaction. Draw the structure of this compound.

A-6. Draw the structure of the carbocation formed upon ionization of the compound shown. A constitutional isomer of this compound gives the same carbocation; draw its structure.

PART B

B-1. 2,3-Pentadiene, $CH_3CH{=}C{=}CHCH_3$, is
  (a) A planar substance
  (b) An allene
  (c) A conjugated diene
  (d) A substance capable of cis-trans isomerism

B-2. Rank the following carbocations in order of increasing stability (least → most):

$$CH_3\overset{+}{C}HCH_3 \qquad CH_3\overset{+}{C}H{-}CH{=}CHCH_3 \qquad (CH_3)_3C\overset{+}{C}H_2$$

  A   B   C

  (a) A<B<C   (b) B<C<A   (c) C<A<B   (d) B<A<C

B-3. Hydrogenation of cyclohexene releases 120 kJ/mol (28.6 kcal/mol) of heat. Which of the following most likely represents the observed heat of hydrogenation of 1,3-cyclohexadiene?

(*a*) 232 kJ/mol (55.4 kcal/mol)   (*c*) 247 kJ/mol (59.0 kcal/mol)
(*b*) 239 kJ/mol (57.2 kcal/mol)   (*d*) 120 kJ/mol (28.6 kcal/mol)

B-4. Which of the following is *not* a proper resonance form of 1,3-cyclohexadiene?

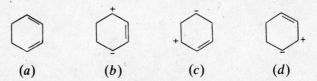

(*a*)   (*b*)   (*c*)   (*d*)

B-5. For the following reactions the major products are shown:

$$CH_2\!=\!CH\!-\!CH\!=\!CH_2 \xrightarrow[0°C]{HBr} CH_2\!=\!CHCHCH_3 \xrightarrow{+25°C} CH_2CH\!=\!CHCH_3$$
$$\qquad\qquad\qquad\qquad\qquad\qquad\quad\underset{Br}{|}\qquad\qquad\qquad\underset{Br}{|}$$

These provide an example of _____ control at low temperature and _____ control
$\qquad\qquad\qquad\qquad\qquad\qquad\qquad 1 \qquad\qquad\qquad\qquad\qquad\qquad\qquad 2$
at higher temperature.

	1	2
(*a*)	kinetic	thermodynamic
(*b*)	thermodynamic	kinetic
(*c*)	kinetic	kinetic
(*d*)	thermodynamic	thermodynamic

B-6. Which of the following C—H bonds would have the smallest bond dissociation energy?

$$(a)\quad CH_3\overset{\overset{H}{|\leftarrow}}{C}HCH_3 \qquad\qquad (c)\quad CH_3CH_2\overset{\overset{H}{|\leftarrow}}{C}H_2$$

$$(b)\quad CH_3CH\!=\!\overset{\overset{H}{|\leftarrow}}{C}H \qquad\qquad (d)\quad CH_2\!=\!CH\!-\!\overset{\overset{H}{|\leftarrow}}{C}HCH_3$$

ARENES AND AROMATICITY

IMPORTANT TERMS AND CONCEPTS

Structure and Bonding of Benzene (Secs. 11.3 to 11.8) Benzene, as the parent *aromatic hydrocarbon*, or *arene*, is best represented as a resonance hybrid of the two Kekulé structures:

Benzene is planar, and all six carbon-carbon bonds are the same length. Each carbon is sp^2 hybridized; the unhybridized p orbitals are all parallel, and their overlap generates a continuous π system encompassing all the carbon atoms of the ring.

Experiments have shown the heat of hydrogenation of benzene to be 152 kJ/mol (36 kcal/mol) *less* than that expected for a hypothetical 1,3,5-cyclohexatriene. This *stabilization energy* is also known as the *resonance energy*, or *delocalization energy*, of benzene.

Nomenclature of Benzene Derivatives (Sec. 11.9) Disubstituted benzene derivatives are named using the prefixes *ortho-* (*o-*), *meta-* (*m-*), and *para-* (*p-*), which refer to 1,2 disubstitution, 1,3 disubstitution, and 1,4 disubstitution, respectively.

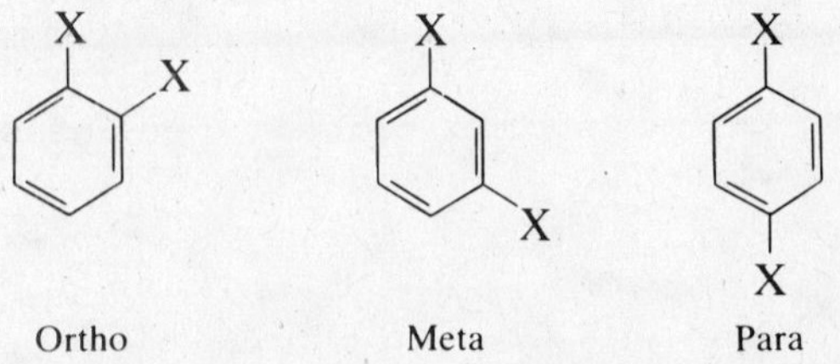

Ortho   Meta   Para

When three or more substituents are present on the ring, the substituents are numbered to specify their position; the *o-*, *m-*, and *p-* prefixes are not used.

Benzylic Conjugation (Secs. 11.14, 11.16) Benzylic cations and radicals are highly stabilized species in which conjugation with the benzene ring results in delocalization of the positive charge in the cation and the unpaired electron in the radical.

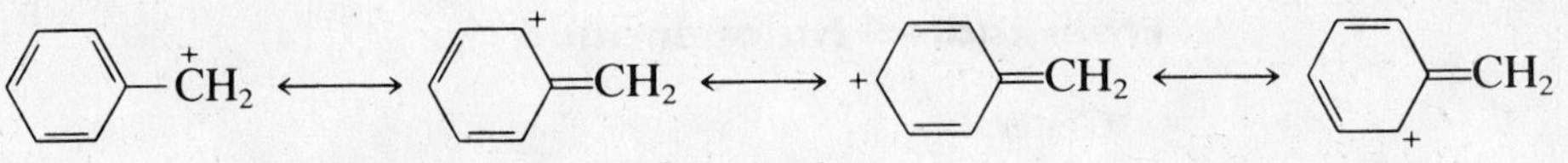

Benzyl cation

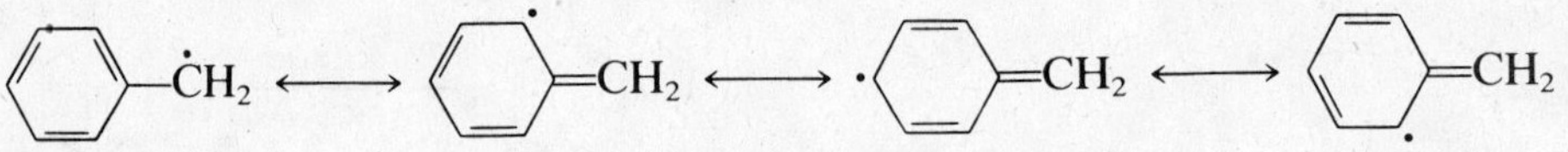

Benzyl radical

Hückel's Rule (Sec. 11.21) Hückel's rule states that among planar monocyclic completely conjugated polyenes only those having $(4n + 2)$ π electrons (where n is an integer) will be aromatic. That is, those molecules containing 2, 6, 10 (and so on) π electrons will exhibit aromatic stabilization. Such molecules are termed *annulenes*; benzene therefore is [6]-annulene.

Aromatic Ions (Sec. 11.22) The preceding statement of Hückel's rule has been extended to planar ions as well. The cycloheptatrienyl cation is aromatic, having six π electrons distributed over a planar cyclic array of seven p orbitals. Likewise, the cyclopentadienide anion contains six π electrons in a five-p-orbital framework. In both cases aromaticity imparts special stability to the ions, making their formation more favorable than would be anticipated otherwise.

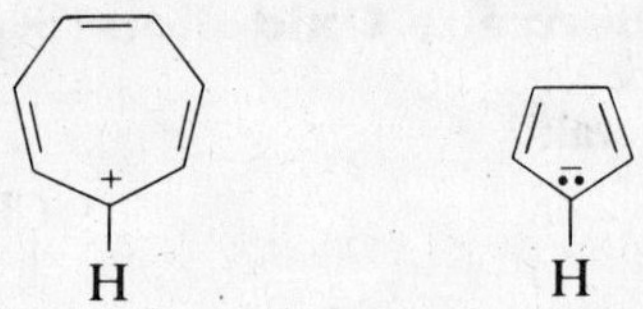

Cycloheptatrienyl cation   Cyclopentadienide anion

Heterocyclic Aromatic Compounds (Secs. 11.23, 11.24) A *heterocycle* is a molecule having an atom other than carbon as part of the ring framework. These *heteroatoms* may contribute an unshared electron pair toward satisfying Hückel's rule. For example, pyrrole and furan are aromatic.

Pyrrole   Furan

IMPORTANT REACTIONS

Birch Reduction (Sec. 11.13)

General:

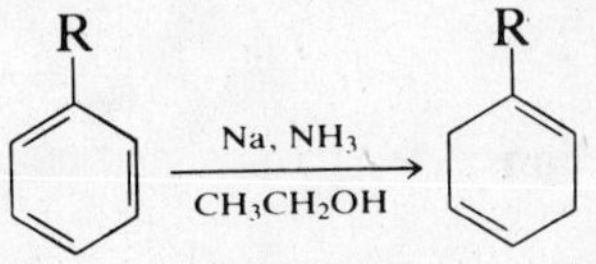

Example:

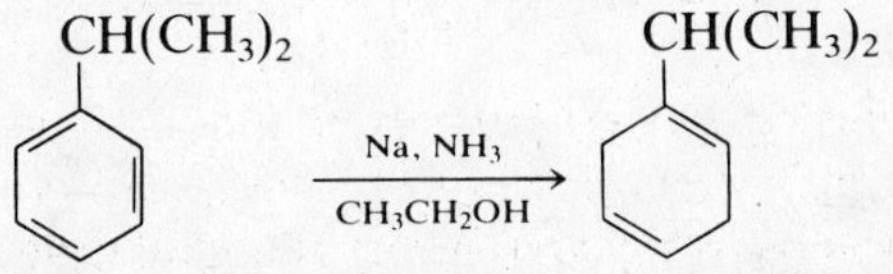

Free-Radical Halogenation (Sec. 11.14)

General:

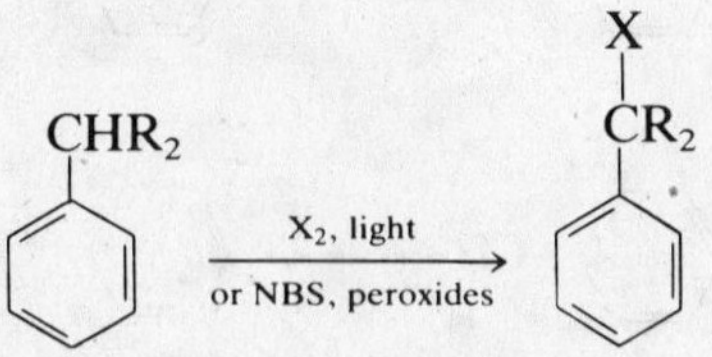

$$X = Cl, Br$$

Examples:

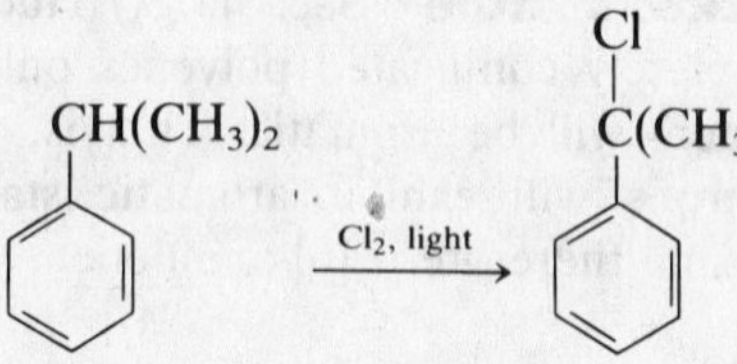

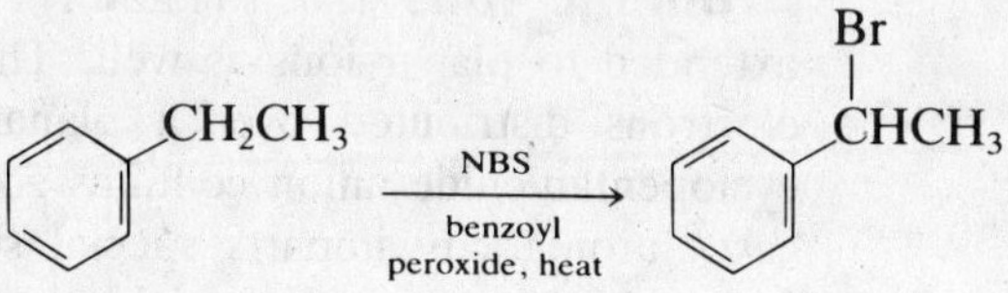

Alkylbenzene Oxidation (Sec. 11.15)

General:

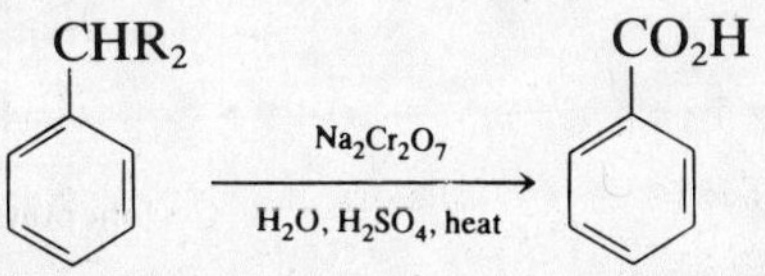

Example:

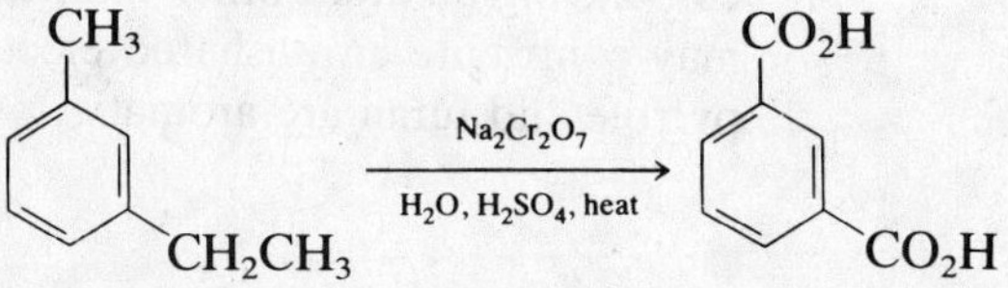

Alkenylbenzenes (Secs. 11.17, 11.18)

Preparation:

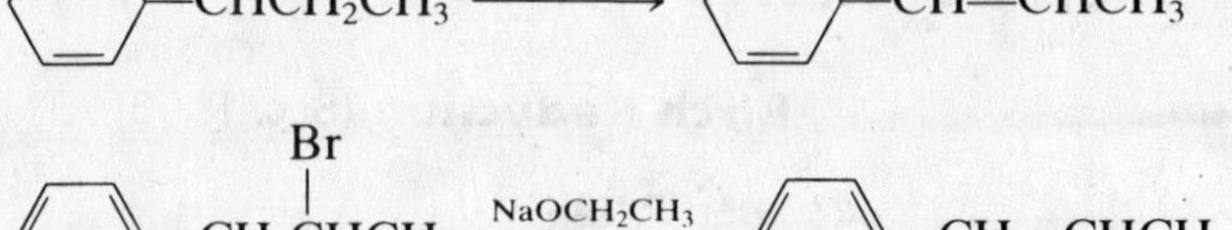

Reactions:

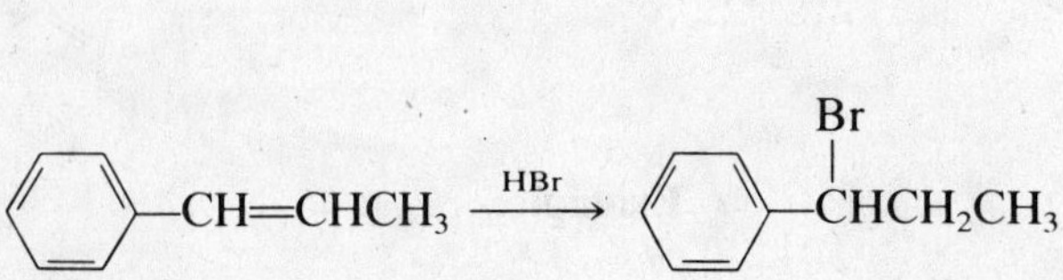

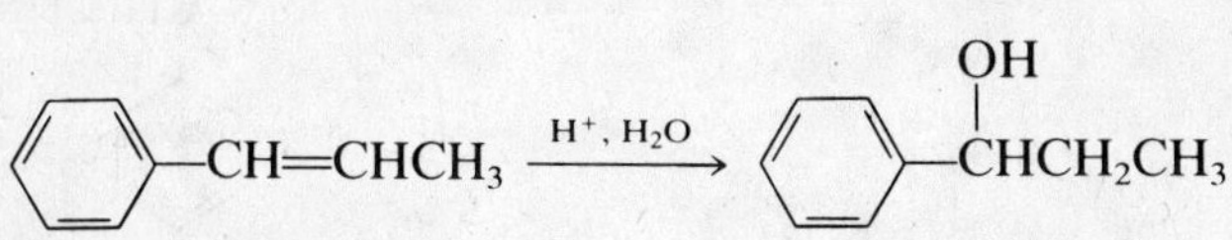

SOLUTIONS TO TEXT PROBLEMS

11.1   Toluene is $C_6H_5CH_3$; it has a methyl group attached to a benzene ring.

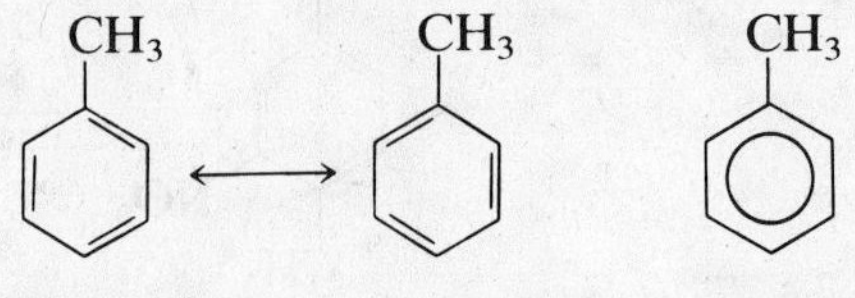

Kekulé forms of toluene   Robinson symbol for toluene

Benzoic acid has a —CO_2H substituent on the benzene ring.

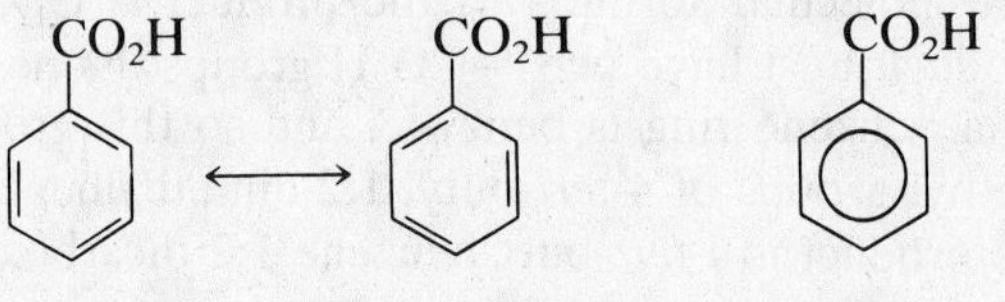

Kekulé forms of benzoic acid   Robinson symbol for benzoic acid

11.2   (*b*)   The parent compound is styrene, C_6H_5—CH=CH_2. The desired compound has a chlorine in the meta position.

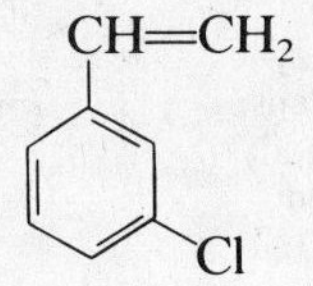

m-Chlorostyrene

(*c*)   The parent compound is aniline, $C_6H_5NH_2$. *p*-Nitroaniline is therefore:

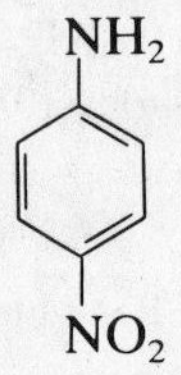

p-Nitroaniline

11.3   The most stable resonance form is the one that has the greatest number of rings that correspond to Kekulé formulations of benzene. For chrysene, electrons are moved in pairs from the structure given to generate a more stable one:

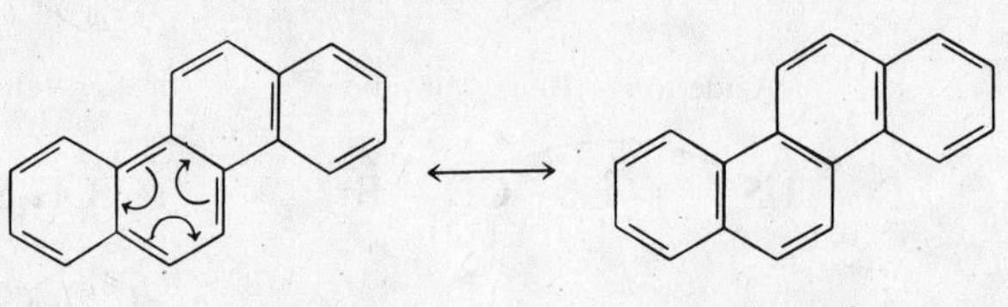

Less stable: two rings have   More stable: four rings have
benzene bonding pattern   benzene bonding pattern

11.4   Birch reductions of monosubstituted arenes yield 1,4-cyclohexadiene derivatives in which the alkyl group is a substituent on the double bond. With *p*-xylene, both methyl groups are double bond substituents in the product.

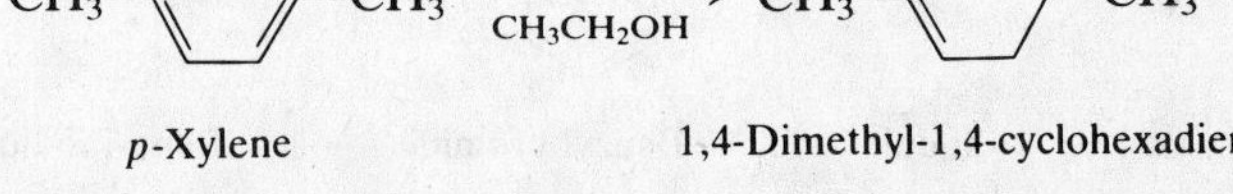

p-Xylene   1,4-Dimethyl-1,4-cyclohexadiene

11.5 (b) Only the benzylic hydrogen is replaced by bromine in the reaction of 4-methyl-3-nitroanisole with *N*-bromosuccinimide.

11.6 The molecular formula of the product is $C_{12}H_{14}O_4$. Since it contains four oxygens, the product must have two —CO_2H groups. None of the hydrogens of a *tert*-butyl substituent on a benzene ring is benzylic, and so this group is inert to oxidation. Only the benzylic methyl groups of 4-*tert*-butyl-1,2-dimethylbenzene are susceptible to oxidation; therefore the product is 4-*tert*-butylbenzene-1,2-dicarboxylic acid.

4-*tert*-Butylbenzene-
1,2-dicarboxylic acid

11.7 Each of these reactions involves nucleophilic substitution of the S_N2 type at the benzylic position of benzyl bromide.

(b) $(CH_3)_3CO^-$ CH_2—Br   → $CH_2OC(CH_3)_3$

tert-Butoxide   Benzyl bromide   Benzyl *tert*-butyl ether
ion

(c) $^-:\ddot{N}{=}\overset{+}{N}{=}\ddot{N}:^-$ CH_2—Br   → CH_2—$\ddot{N}{=}\overset{+}{N}{=}\ddot{N}:^-$

Azide ion   Benzyl bromide   Benzyl azide

(d) HS^- CH_2—Br   → CH_2SH

Hydrogen   Benzyl bromide   Phenylmethanethiol
sulfide ion

(e) I^- CH_2—Br   → CH_2I

Iodide ion   Benzyl bromide   Benzyl iodide

11.8   The dihydronaphthalene in which the double bond is conjugated with the aromatic ring is more stable; thus 1,2-dihydronaphthalene has a lower heat of hydrogenation than 1,4-dihydronaphthalene.

1,2-Dihydronaphthalene
heat of hydrogenation 101 kJ/mol (24.1 kcal/mol)

1,4-Dihydronaphthalene
heat of hydrogenation 113 kJ/mol (27.1 kcal/mol)

11.9   (*b*)   The regioselectivity of alcohol formation is opposite that predicted by Markovnikov's rule on hydroboration-oxidation.

$$\underset{CH_3}{C}=CH_3 \xrightarrow[\text{2. } H_2O_2, \, HO^-]{\text{1. } B_2H_6} \underset{CH_3}{CHCH_2OH}$$

2-Phenylpropene   2-Phenyl-1-propanol (92%)

(*c*)   Bromine adds to alkenes in aqueous solution to give bromohydrins. A water molecule acts as a nucleophile, attacking the bromonium ion at the carbon that can bear most of the positive charge, which in this case is the benzylic carbon.

$$CH=CH_2 \xrightarrow[H_2O]{Br_2} \underset{}{\overset{OH}{CHCH_2Br}}$$

Styrene   2-Bromo-1-phenylethanol (82%)

(*d*)   Peroxy acids convert alkenes to epoxides.

$$CH=CH_2 + \overset{O}{COOH} \longrightarrow CH\!-\!CH_2 + \overset{O}{COH}$$

Styrene   Peroxybenzoic acid   Epoxystyrene   Benzoic acid
(69–75%)

11.10   Styrene contains a benzene ring and will be appreciably stabilized by resonance, which makes it lower in energy than cyclooctatetraene.

$$CH=CH_2$$

structure contains an
aromatic ring

Styrene: heat of
combustion 4393 kJ/mol (1050 kcal/mol)

Cyclooctatetraene (not aromatic):
heat of combustion 4543 kJ/mol (1086 kcal/mol)

11.11   The dimerization of cyclobutadiene is a Diels-Alder reaction in which one molecule of cyclobutadiene acts as a diene and the other as a dienophile.

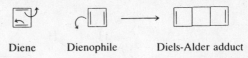

Diene   Dienophile   Diels-Alder adduct

11.12   (*b*)   Since twelve 2*p* orbitals contribute to the cyclic conjugated system of [12]-annulene, there will be 12 π molecular orbitals. These MOs are arranged so that one is of highest energy, one is of lowest energy, and the remaining 10 are found in pairs between the highest- and lowest-energy orbitals. There are 12 π electrons, and so the lowest five orbitals are each doubly occupied, while each of the next two orbitals—orbitals of equal energy—is singly occupied.

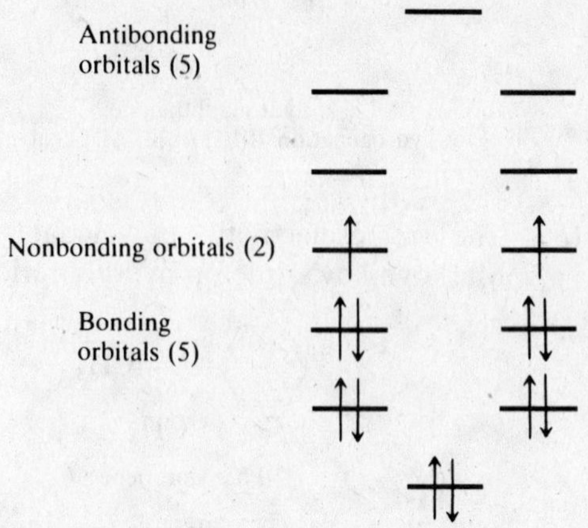

Antibonding
orbitals (5)

Nonbonding orbitals (2)

Bonding
orbitals (5)

11.13 The seven resonance forms for tropylium cation (cycloheptatrienyl cation) may be generated by moving π electrons in pairs toward the positive charge. The resonance forms are simply a succession of allylic carbocations.

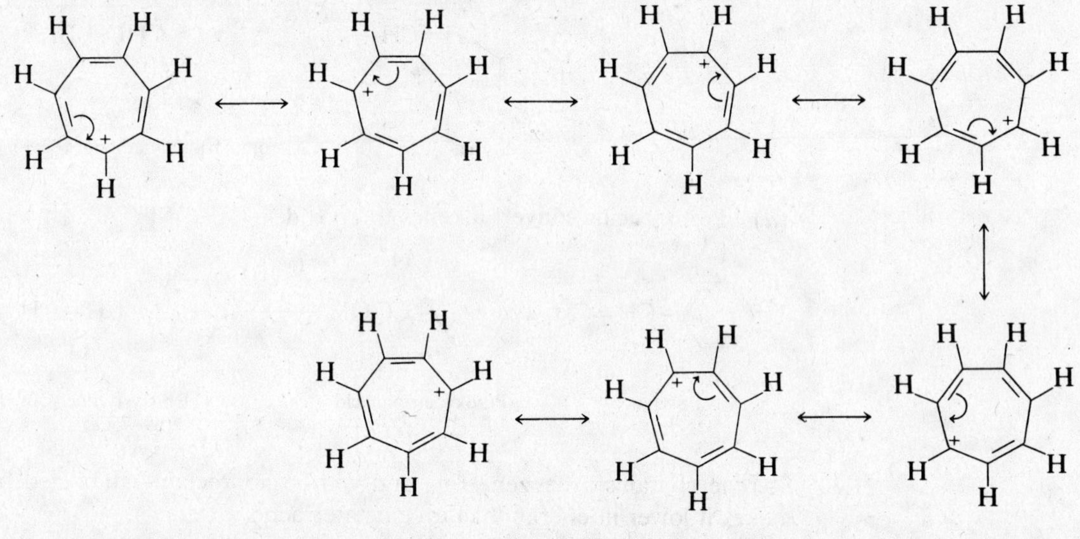

11.14 Resonance structures are generated for cyclopentadienide anion by moving the unshared electron pair from the carbon to which it is attached to a position where it becomes a shared electron pair in a π bond.

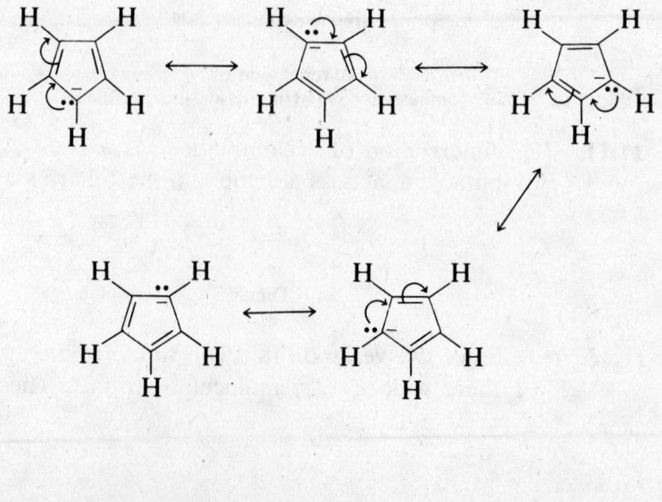

11.15 (b) Cyclononatetraenide anion has 10 π electrons; it is aromatic. The 10 π electrons are most easily seen by writing a Lewis structure for the anion: there are two π electrons for each of four double bonds, and the negatively charged carbon contributes two.

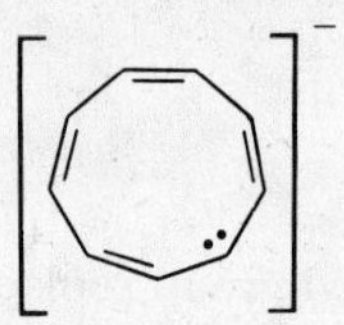

11.16 Indole is more stable than isoindole. While the bonding patterns in both five-membered rings are the same, the six-membered ring in indole has a pattern of bonds identical to benzene and so is highly stabilized. The six-membered ring in isoindole is not of the benzene type.

Six-membered ring corresponds to benzene.

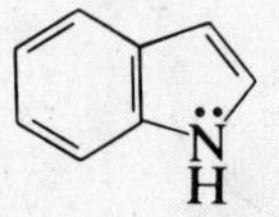

Indole
more stable

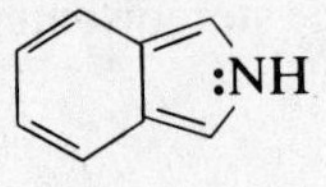

:NH

Six-membered ring does not have same pattern of bonds as benzene.

Isoindole
less stable

11.17 The prefix *benz-* in benzimidazole (structure given in text) signifies that a benzene ring is fused to an imidazole ring. By analogy, benzoxazole has a benzene ring fused to oxazole.

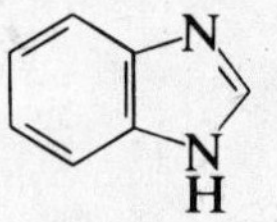

Benzimidazole

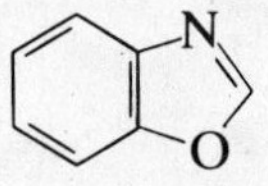

Benzoxazole

Similarly, benzothiazole has a benzene ring fused to thiazole.

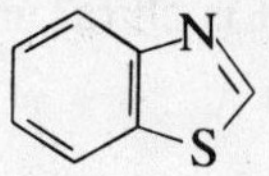

Benzothiazole

11.18 Since the problem requires that the benzene ring be monosubstituted, all that need to be examined are the various isomeric forms of the C_4H_9 substituent.

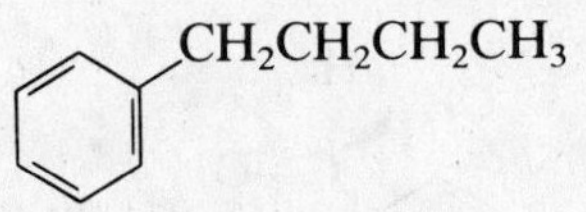

$CH_2CH_2CH_2CH_3$

Butylbenzene
(1-phenylbutane)

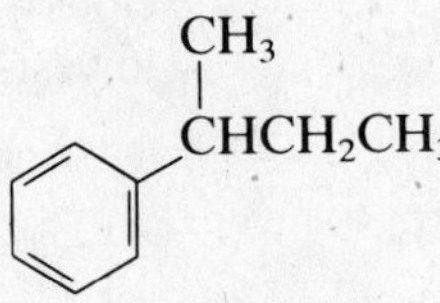

$CHCH_2CH_3$ with CH_3

sec-Butylbenzene
(2-phenylbutane)

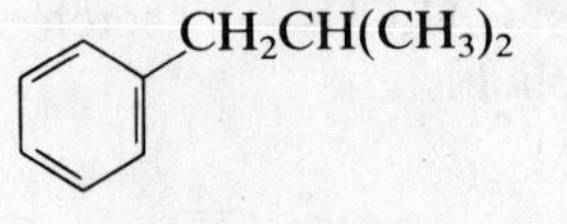

$CH_2CH(CH_3)_2$

Isobutylbenzene
(2-methyl-1-phenylpropane)

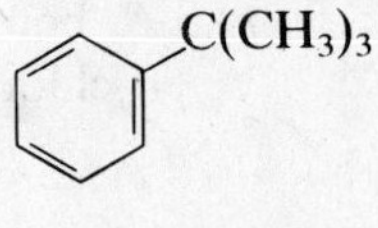

$C(CH_3)_3$

tert-Butylbenzene
(2-methyl-2-phenylpropane)

These are the four constitutional isomers. *sec*-Butylbenzene is chiral and so exists in enantiomeric *R* and *S* forms.

11.19   (*a*)   An allyl substituent is $-CH_2CH=CH_2$.

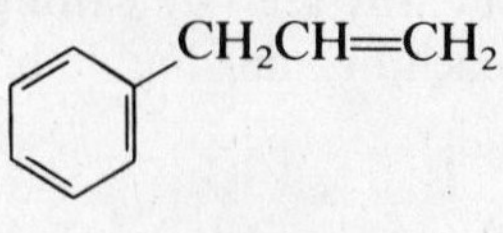

Allylbenzene

(*b*)   A phenyl group replaces one of the hydrogens of acetylene in phenylacetylene.

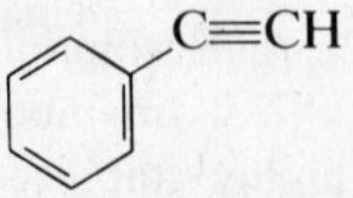

Phenylacetylene
(phenylethyne)

(*c*)   The constitution of 1-phenyl-1-butene is $C_6H_5CH=CHCH_2CH_3$. The *E* stereoisomer is

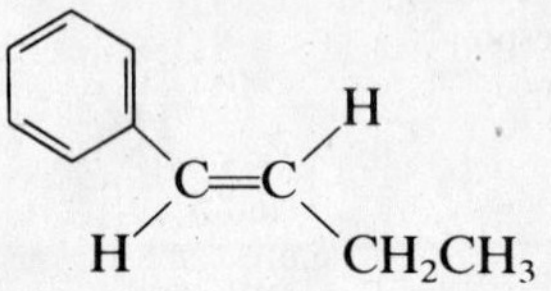

(*E*)-1-Phenyl-1-butene

The two higher-priority substituents, phenyl and ethyl, are on opposite sides of the double bond.

(*d*)   The constitution of 2-phenyl-2-butene is $CH_3C=CHCH_3$. The *Z* stereoisomer is

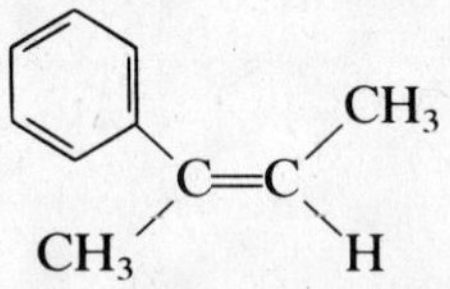

$\overset{\displaystyle |}{\underset{\displaystyle C_6H_5}{}}$

(*Z*)-2-Phenyl-2-butene

The two higher-priority substituents, phenyl and methyl, are on the same side of the double bond.

(*e*)   1-Phenylethanol is chiral and has the constitution $CH_3CHC_6H_5$. Among the substit-

$\overset{\displaystyle |}{\underset{\displaystyle OH}{}}$

uents attached to the stereogenic center, the order of decreasing precedence is

$$HO > C_6H_5 > CH_3 > H$$

In the *R* enantiomer the three highest-priority substituents must appear in a clockwise sense in proceeding from higher priority to next lower priority when the lowest-priority substituent is directed away from the reader.

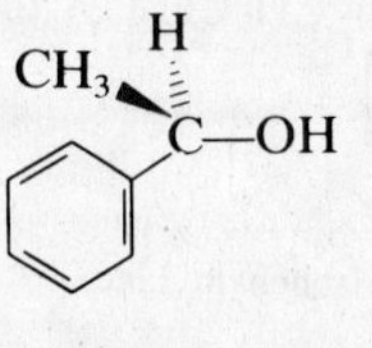

(*R*)-1-Phenylethanol

(*f*)   A benzyl group is $C_6H_5CH_2-$. Therefore, benzyl alcohol is $C_6H_5CH_2OH$ and *o*-chlorobenzyl alcohol is:

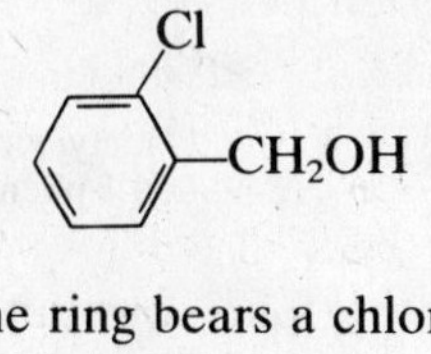

(*g*)   In *p*-chlorophenol the benzene ring bears a chlorine and a hydroxyl substituent in a 1,4 substitution pattern.

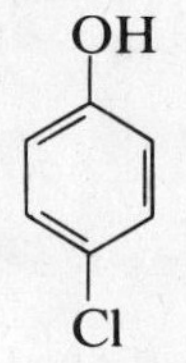

OH

p-Chlorophenol

Cl

(*h*) Benzenecarboxylic acid is an alternative IUPAC name for benzoic acid.

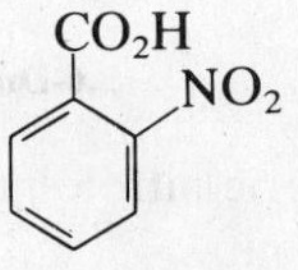

CO_2H

NO_2

2-Nitrobenzenecarboxylic acid

(*i*) Two isopropyl groups are in a 1,4 relationship in *p*-diisopropylbenzene.

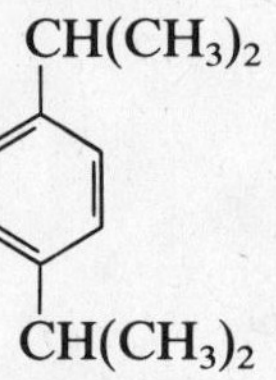

$CH(CH_3)_2$

p-Diisopropylbenzene

$CH(CH_3)_2$

(*j*) Aniline is $C_6H_5NH_2$. Therefore

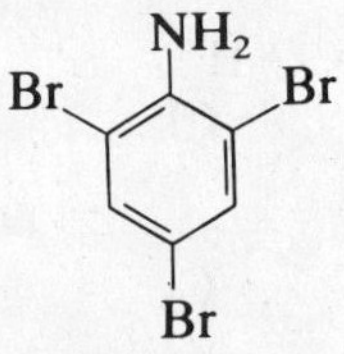

NH_2

Br   Br

2,4,6-Tribromoaniline

Br

(*k*) Acetophenone is acetylbenzene, $C_6H_5CCH_3$. Therefore

CH_3 O

C

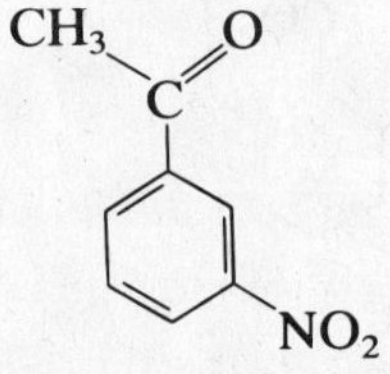

m-Nitroacetophenone

NO_2

(*l*) Styrene is $C_6H_5CH=CH_2$ and numbering of the ring begins at the carbon that bears the side chain.

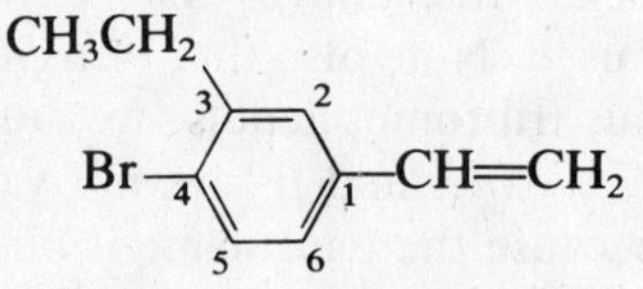

CH_3CH_2

Br—$CH=CH_2$ 4-Bromo-3-ethylstyrene

11.20 (*a*) Anisole is the name for $C_6H_5OCH_3$, and allyl is an acceptable name for the group $CH_2=CHCH_2—$. Number the ring beginning with the carbon that bears the methoxy group.

(*b*) Phenol is the name for C_6H_5OH. The ring is numbered beginning at the carbon that bears the hydroxyl group, and the substituents are listed in alphabetical order.

(*c*) Aniline is the name given to $C_6H_5NH_2$. This compound is named as a dimethyl derivative of aniline. Number the ring sequentially beginning with the carbon that bears the amino group.

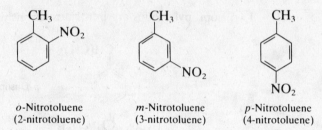

Estragole	Disophenol	*m*-Xylidine
4-Allylanisole	**2,6-Diiodo-4-nitrophenol**	**2,6-Dimethylaniline**

11.21 (*a*) There are three isomeric nitrotoluenes, since the nitro group can be ortho, meta, or para to the methyl group.

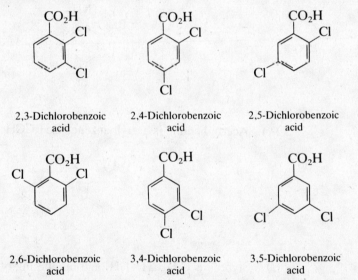

o-Nitrotoluene	*m*-Nitrotoluene	*p*-Nitrotoluene
(2-nitrotoluene)	(3-nitrotoluene)	(4-nitrotoluene)

(*b*) Benzoic acid is $C_6H_5CO_2H$. In the isomeric dichlorobenzoic acids, two of the ring hydrogens of benzoic acid have been replaced by chlorines. The isomeric dichlorobenzoic acids are:

2,3-Dichlorobenzoic	2,4-Dichlorobenzoic	2,5-Dichlorobenzoic
acid	acid	acid

2,6-Dichlorobenzoic	3,4-Dichlorobenzoic	3,5-Dichlorobenzoic
acid	acid	acid

The prefixes *o*-, *m*-, and *p*- may not be used in trisubstituted arenes; numerical prefixes are used. Note also that *benzenecarboxylic* may be used in place of *benzoic*.

(*c*) In the various tribromophenols, we are dealing with tetrasubstitution on a benzene ring. Again, *o*-, *m*-, and *p*- are not valid prefixes. The hydroxyl group is assigned position 1 because the base name is phenol.

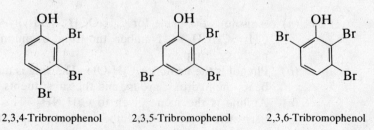

2,3,4-Tribromophenol	2,3,5-Tribromophenol	2,3,6-Tribromophenol

OH ... Br
Br ... Br
2,4,5-Tribromophenol

OH
Br ... Br
Br
2,4,6-Tribromophenol

OH
Br ... Br
Br
3,4,5-Tribromophenol

(*d*) There are only three tetrafluorobenzenes. The two hydrogen substituents may be ortho, meta, or para with respect to each other.

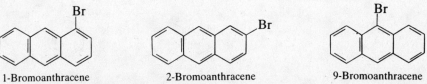

1,2,3,4-Tetrafluorobenzene   **1,2,3,5-Tetrafluorobenzene**   **1,2,4,5-Tetrafluorobenzene**

(*e*) There are only two naphthalenecarboxylic acids.

CO₂H

CO₂H

Naphthalene-1-carboxylic acid   **Naphthalene-2-carboxylic acid**

(*f*) There are three isomeric bromoanthracenes. All other positions are equivalent to one of these.

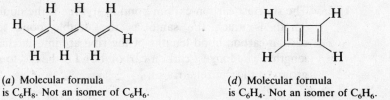

1-Bromoanthracene   **2-Bromoanthracene**   **9-Bromoanthracene**

11.22 Isomers are different compounds that have the same molecular formula. Benzene is C_6H_6. The structure in part (*a*) is C_6H_8, and the structure in part (*d*) is C_6H_4. Neither compound (*a*) nor compound (*d*) is an isomer of benzene.

(*a*) Molecular formula
is C_6H_8. Not an isomer of C_6H_6.

(*d*) Molecular formula
is C_6H_4. Not an isomer of C_6H_6.

The structures in parts (*b*) and (*c*) are C_6H_6 and are isomers of benzene. Structure (*b*) has been given the common name *prismane* (because its structure is that of a prism), and structure (*c*) is called *benzvalene* (because it is a "valence-bond" isomer of benzene).

Prismane

Benzvalene

(*b*) Molecular formula
is C_6H_6; is an isomer of benzene.
All carbons are equivalent.

(*c*) Molecular formula
is C_6H_6; is an isomer of benzene.
Not all carbons are equivalent.

Prismane has six equivalent carbons; benzvalene does not.

11.23   There are three isomeric trimethylbenzenes:

1,2,3-Trimethylbenzene   1,2,4-Trimethylbenzene   1,3,5-Trimethylbenzene

Their relative stabilities are determined by steric effects. Mesitylene (the 1,3,5-trisubstituted isomer) is the most stable because none of its methyl groups are ortho to any other methyl group. Ortho substituents on a benzene ring can, depending on their size, exert repulsive van der Waals interactions upon each other, much as can cis substituents on a carbon-carbon double bond. Because the carbon-carbon bond length in benzene is somewhat longer than in an alkene, these effects are smaller in magnitude, however. The 1,2,4 substitution pattern has one methyl-methyl repulsion between ortho substituents. The least stable isomer is the 1,2,3-trimethyl derivative, because it is the most crowded. The energy differences between isomers are relatively small, heats of combustion being 5198, 5195, and 5193 kJ/mol (1242.4, 1241.6, and 1241.2 kcal/mol) for the 1,2,3, 1,2,4, and 1,3,5 isomers, respectively.

11.24   *p*-Dichlorobenzene has a center of symmetry. Each of its individual bond moments is balanced by an identical bond dipole oriented opposite to it. *p*-Dichlorobenzene has no dipole moment.

o-Dichlorobenzene   *m*-Dichlorobenzene   *p*-Dichlorobenzene
$\mu = 2.27\,D$ $\mu = 1.48\,D$ $\mu = 0\,D$

11.25   The shortest carbon-carbon bond in styrene is the double bond of the vinyl substituent; its length is much the same as the double bond length of any other alkene. The carbon-carbon bond lengths of the ring are intermediate between single and double bond lengths. The longest carbon-carbon bond is the sp^2 to sp^2 single bond connecting the vinyl group to the benzene ring.

134 pm

$-CH=CH_2$

140 pm   147 pm

11.26   Move π electron pairs as shown so that both six-membered rings have an arrangement of bonds that corresponds to benzene.

Less stable   More stable

11.27 (*a*) In the structure shown for naphthalene, one ring but not the other corresponds to a Kekulé form of benzene. We say that one ring is *benzenoid* and the other is not.

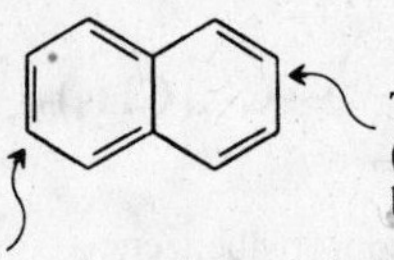

This six-membered ring is not benzenoid (does not correspond to Kekulé form of benzene).

This six-membered ring is benzenoid (corresponds to a Kekulé form of benzene).

By rewriting the benzenoid ring in its alternative Kekulé form, *both* rings become benzenoid.

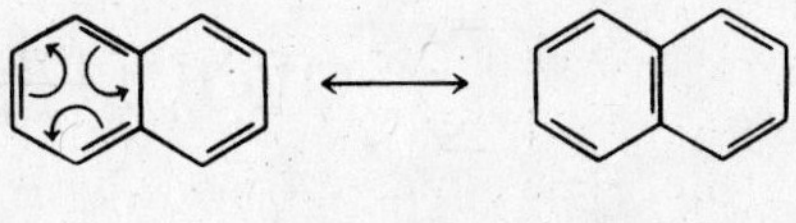

Both rings are benzenoid.

(*b*) Here a cyclobutadiene ring is fused to benzene. By writing the alternative resonance form of cyclobutadiene, the six-membered ring becomes benzenoid.

(*c*) The structure portrayed for phenanthrene contains two terminal benzenoid rings and a nonbenzenoid central ring. All three rings may be represented in benzenoid forms by converting one of the terminal six-membered rings to its alternative Kekulé form as shown:

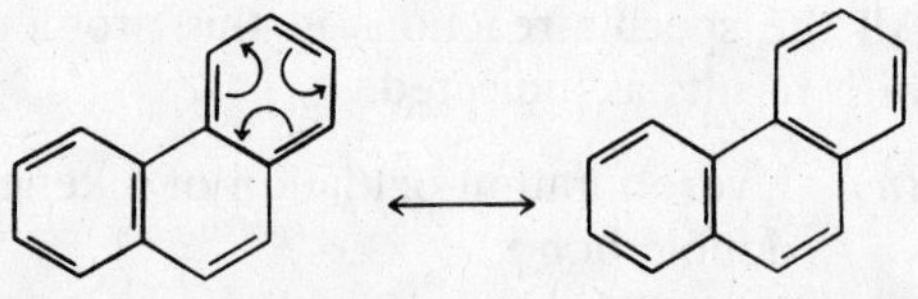

Central ring not benzenoid   All three rings benzenoid

(*d*) Neither of the six-membered rings is benzenoid in the structure shown. By writing the cyclooctatetraene portion of the molecule in its alternative representation, the two six-membered rings become benzenoid.

Six-membered rings are not benzenoid.   Six-membered rings are benzenoid.

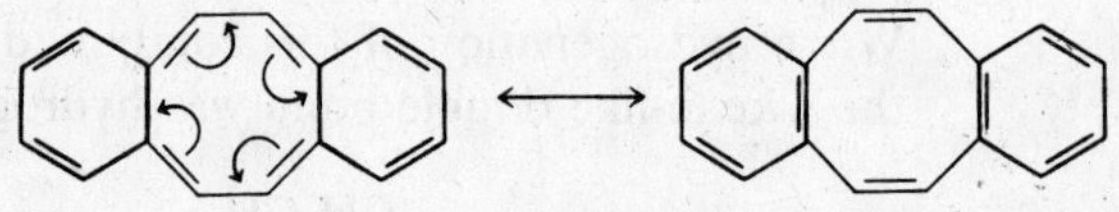

11.28 (*a*) Hydrogenation of isopropylbenzene converts the benzene ring to a cyclohexane unit.

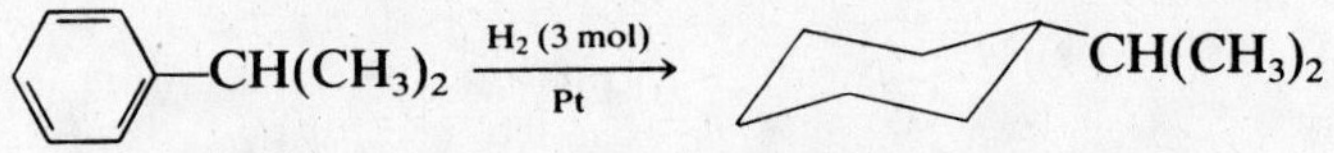

$$\text{—CH(CH}_3)_2 \xrightarrow[\text{Pt}]{\text{H}_2 \text{ (3 mol)}} \text{—CH(CH}_3)_2$$

Isopropylbenzene   Isopropylcyclohexane

(*b*) Sodium and ethanol in liquid ammonia is the combination of reagents that brings about Birch reduction of benzene rings. The 1,4-cyclohexadiene that is formed has its isopropyl group as a substituent on one of the double bonds.

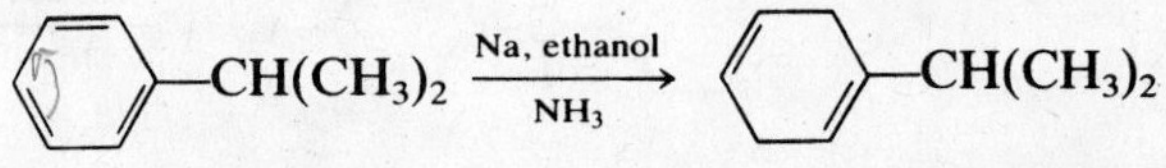

$$\text{—CH(CH}_3)_2 \xrightarrow[\text{NH}_3]{\text{Na, ethanol}} \text{—CH(CH}_3)_2$$

Isopropylbenzene   1-Isopropyl-1,4-cyclohexadiene

(c)   Oxidation of the isopropyl side chain occurs. The benzene ring remains intact.

Isopropylbenzene   Benzoic acid

(d)   N-Bromosuccinimide is a reagent that brings about bromination of arenes by substitution of a benzylic hydrogen.

Isopropylbenzene   2-Bromo-2-phenylpropane

(e)   The tertiary bromide undergoes E2 elimination to give a carbon-carbon double bond.

2-Bromo-2-phenylpropane   2-Phenylpropene

11.29   All the specific reactions in this problem have been reported in the chemical literature with results as indicated.

(a)   Hydroboration-oxidation of alkenes leads to syn anti-Markovnikov hydration of the double bond.

1-Phenylcyclobutene   *trans*-2-Phenyl-1-cyclobutanol (82%)

(b)   The compound contains a substituted benzene ring and an alkene-like double bond. When hydrogenation of this compound was carried out under the usual conditions, the alkene-like double bond was hydrogenated cleanly.

1-Ethylindene   1-Ethylindan (80%)

(c)   Free-radical chlorination will lead to substitution of benzylic hydrogens. The starting material contains four benzylic hydrogens, all of which may eventually be replaced.

(65%)

(*d*) Epoxidation of alkenes is stereospecific.

(*E*)-1,2-Diphenylethene   *trans*-1,2-Diphenylepoxyethane (78–83%)

(*e*) The reaction is one of acid-catalyzed alcohol dehydration.

cis-4-Methyl-1-phenylcyclohexanol   4-Methyl-1-phenylcyclohexene (81%)

(*f*) This reaction illustrates identical reactivity at two equivalent sites in a molecule. Both alcohol functions are tertiary and benzylic and undergo acid-catalyzed dehydration readily.

1,4-Di-(1-hydroxy-1-methylethyl)benzene   1,4-Diisopropenylbenzene (68%)

(*g*) The compound shown is DDT (standing for the nonsystematic name *dichlorodiphenyltrichloroethane*). It undergoes β elimination to form an alkene.

(100%)

(*h*) Alkyl side chains on naphthalene undergo reactions analogous to those of alkyl groups on benzene.

1-Methylnaphthalene   1-(Bromomethyl)naphthalene (46%)

(*i*) Potassium carbonate is a weak base. Hydrolysis of the primary benzylic halide converts it to an alcohol.

p-Cyanobenzyl chloride   *p*-Cyanobenzyl alcohol (85%)

11.30 Only benzylic (or allylic) hydrogens are replaced by *N*-bromosuccinimide. Among the four bromines in 3,4,5-tribromobenzyl bromide, three are substituents on the ring and are not capable of being introduced by benzylic bromination. Therefore, the starting material must have these three bromines already in place.

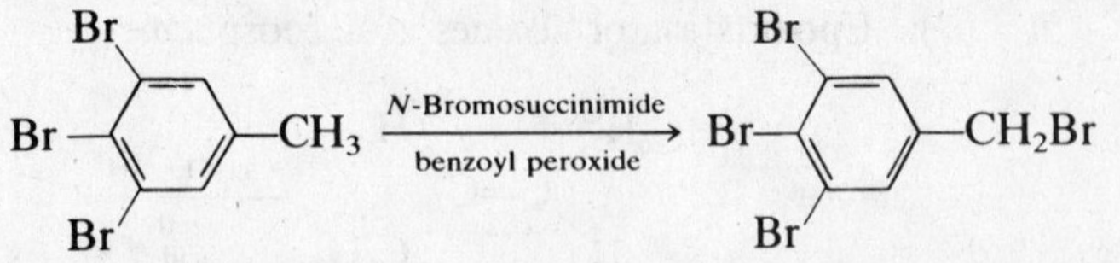

3,4,5-Tribromotoluene
Compound A

3,4,5-Tribromobenzyl bromide

11.31 2,3,5-Trimethoxybenzoic acid has the structure shown. The three methoxy groups occupy the same positions in this oxidation product that they did in compound B. The carboxylic acid function must have arisen by oxidation of the —CH_2CH=$C(CH_3)_2$ side chain. Therefore

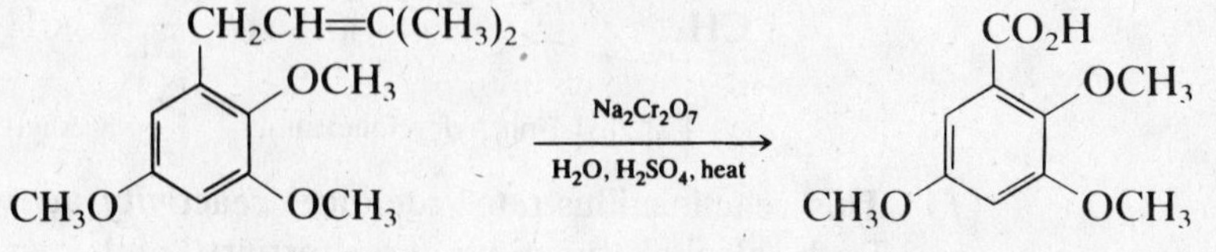

Compound B ($C_{14}H_{20}O_3$)

2,3,5-Trimethoxybenzoic acid

11.32 Hydroboration-oxidation leads to stereospecific syn addition of the elements of water across a carbon-carbon double bond. The regiochemistry of addition is opposite to that predicted by Markovnikov's rule.

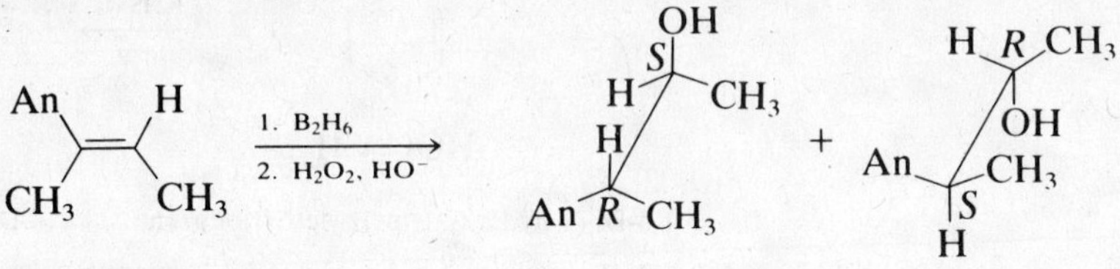

(E)-2-(p-Anisyl)-2-butene

2S,3R

2R,3S

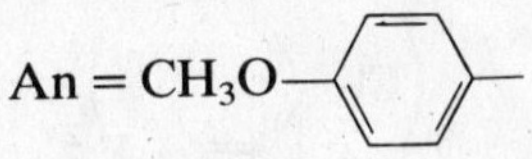

$$An = CH_3O-\text{⟨benzene ring⟩}-$$

Alcohol C is a racemic mixture of the 2S,3R and 2R,3S enantiomers of 3-(p-anisyl)-2-butanol.

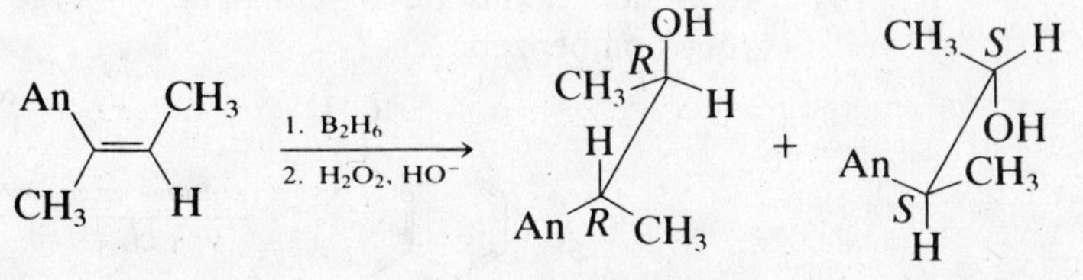

(Z)-2-(p-Anisyl)-2-butene

(2R,3R)

(2S,3S)

Alcohol D is a racemic mixture of the 2R,3R and 2S,3S enantiomers of 3-(p-anisyl)-2-butanol. Alcohols C and D are stereoisomers which are not enantiomers; they are diastereomers.

11.33 Dehydrohalogenation of alkyl halides is stereospecific, requiring an anti arrangement between the hydrogen being lost and the leaving group in the transition state. Therefore (Z)-1,2-diphenylpropene must be formed from the diastereomer shown.

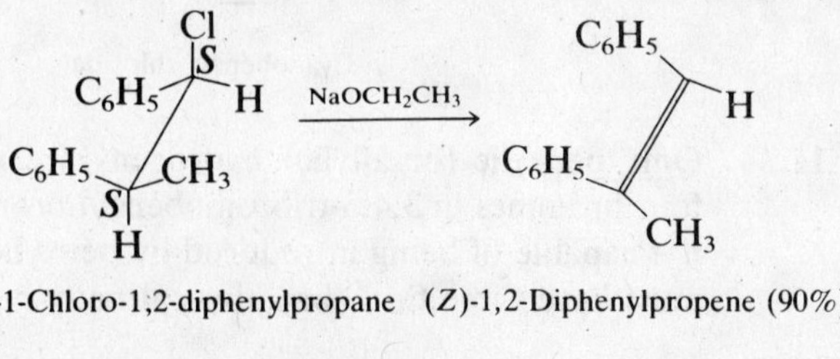

(1S,2S)-1-Chloro-1,2-diphenylpropane   (Z)-1,2-Diphenylpropene (90%)

The mirror-image chloride, 1R,2R, will also give the Z alkene. In fact, the reaction was carried out on a racemic mixture of the 1R,2R and 1S,2S stereoisomers.

The E isomer is formed from either the 1R,2S or the 1S,2R chloride (or from a racemic mixture of the two).

(1R,2S)-1-Chloro-1,2-diphenylpropane   (E)-1,2-Diphenylpropene (87%)

11.34 (a) The conversion of ethylbenzene to 1-phenylethyl bromide is a benzylic bromination. It can be achieved by using either bromine or N-bromosuccinimide (NBS).

$$C_6H_5CH_2CH_3 \xrightarrow[\substack{\text{or} \\ \text{NBS, heat}}]{\text{Br}_2, \text{light}} C_6H_5\underset{\underset{\displaystyle Br}{|}}{C}HCH_3$$

Ethylbenzene   1-Phenylethyl bromide

(b) The conversion of 1-phenylethyl bromide to 1,2-dibromo-1-phenylethane

$$C_6H_5\underset{\underset{\displaystyle Br}{|}}{C}HCH_3 \longrightarrow C_6H_5\underset{\underset{\displaystyle Br}{|}}{C}HCH_2Br$$

cannot be achieved cleanly in a single step. What must be done is to reason backward from the target molecule. Ask yourself, How can I make 1,2-dibromo-1-phenylethane in one step from anything? Vicinal dibromides are customarily prepared by addition of bromine to alkenes. This suggests that 1,2-dibromo-1-phenylethane can be prepared by the reaction

$$C_6H_5CH{=}CH_2 + Br_2 \longrightarrow C_6H_5\underset{\underset{\displaystyle Br}{|}}{C}HCH_2Br$$

Styrene   1,2-Dibromo-1-phenylethane

The necessary alkene, styrene, is available by dehydrohalogenation of the given starting material, 1-phenylethyl bromide.

$$C_6H_5\underset{\underset{\displaystyle Br}{|}}{C}HCH_3 \xrightarrow[\text{CH}_3\text{CH}_2\text{OH}]{\text{NaOCH}_2\text{CH}_3} C_6H_5CH{=}CH_2$$

1-Phenylethyl bromide   Styrene

Thus, by reasoning backward from the target molecule, the synthetic scheme becomes apparent.

$$C_6H_5\underset{\underset{\displaystyle Br}{|}}{C}HCH_3 \xrightarrow[\text{CH}_3\text{CH}_2\text{OH}]{\text{NaOCH}_2\text{CH}_3} C_6H_5CH{=}CH_2 \xrightarrow{\text{Br}_2} C_6H_5\underset{\underset{\displaystyle Br}{|}}{C}HCH_2Br$$

1-Phenylethyl   Styrene   1,2-Dibromo-1-
bromide   phenylethane

(c) The conversion of styrene to phenylacetylene cannot be carried out in a single step. However, as was pointed out in Chapter 9, a standard sequence for converting alkenes to alkynes consists of bromine addition followed by a double dehydrohalogenation in strong base.

$$C_6H_5CH{=}CH_2 \xrightarrow{Br_2} \underset{\underset{\displaystyle Br}{|}}{C_6H_5CHCH_2Br} \xrightarrow[\text{NH}_3]{\text{NaNH}_2} C_6H_5C{\equiv}CH$$

Styrene   1,2-Dibromo-1-   Phenylacetylene
phenylethane

(*d*) The conversion of phenylacetylene to *n*-butylbenzene requires both a hydrogenation step and a carbon-carbon bond formation step. The acetylene function is essential for carbon-carbon bond formation by alkylation. Therefore, the correct sequence is:

$$C_6H_5C{\equiv}CH \xrightarrow[\text{NH}_3]{\text{NaNH}_2} C_6H_5C{\equiv}C{:}^-Na^+$$

Phenylacetylene

$$C_6H_5C{\equiv}C{:}^-Na^+ + CH_3CH_2Br \longrightarrow C_6H_5C{\equiv}CCH_2CH_3$$

$$C_6H_5C{\equiv}CCH_2CH_3 \xrightarrow[\text{Pt}]{\text{H}_2} C_6H_5CH_2CH_2CH_2CH_3$$

n-Butylbenzene

(*e*) The transformation corresponds to alkylation of acetylene, and so the alcohol must first be converted to a species with a good leaving group such as its halide derivative.

$$C_6H_5CH_2CH_2OH \xrightarrow{\text{PBr}_3} C_6H_5CH_2CH_2Br$$

2-Phenylethanol   2-Phenylethyl bromide

$$C_6H_5CH_2CH_2Br + NaC{\equiv}CH \longrightarrow C_6H_5CH_2CH_2C{\equiv}CH$$

2-Phenylethyl   Sodium   4-Phenyl-1-butyne
bromide   acetylide

(*f*) The target compound is a bromohydrin. Bromohydrins are formed by addition of hypobromous acid to alkenes.

$$C_6H_5CH_2CH_2Br \xrightarrow[(\text{CH}_3)_3\text{COH}]{\text{KOC}(\text{CH}_3)_3} C_6H_5CH{=}CH_2 \xrightarrow[\text{H}_2\text{O}]{\text{Br}_2} \underset{\underset{\displaystyle OH}{|}}{C_6H_5CHCH_2Br}$$

2-Phenylethyl bromide   Styrene   2-Bromo-1-phenylethanol

11.35 The stability of free radicals is reflected in their ease of formation. Toluene, which forms a benzyl radical, reacts with bromine 64,000 times faster than does ethane, which forms a primary alkyl radical. Ethylbenzene, which forms a secondary benzylic radical, reacts 1 million times faster than ethane.

Toluene   Primary benzylic
radical

Ethylbenzene   Secondary benzylic
(most reactive)   radical

$$CH_3CH_3 + Br^{\bullet} \longrightarrow CH_3\overset{\bullet}{C}H_2 + HBr$$

Ethane   Primary
(least reactive)   radical

11.36 A good way to develop alternative resonance structures for carbocations is to move electron pairs toward sites of positive charge.

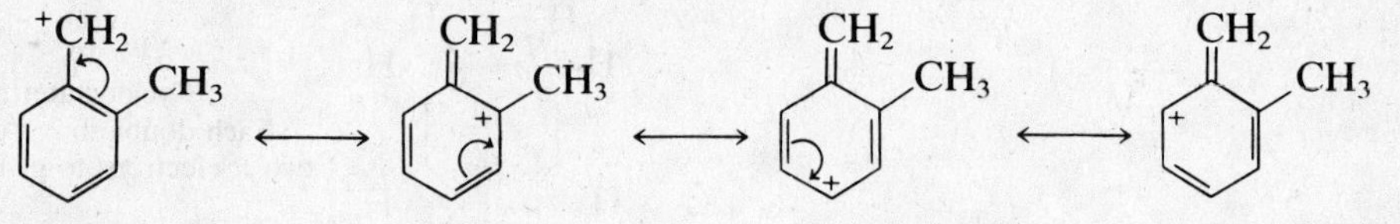

o-Methylbenzyl cation   Tertiary carbocation

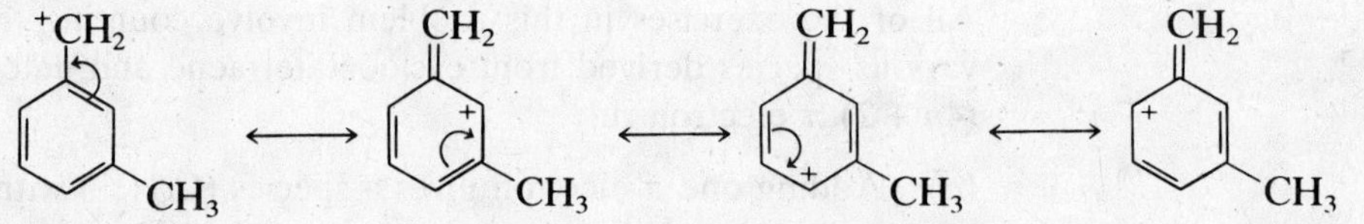

m-Methylbenzyl cation

Only one of the Lewis structures shown is a tertiary carbocation. *o*-Methylbenzyl cation has tertiary carbocation character; *m*-methylbenzyl cation does not.

11.37 The resonance structures for the cyclopentadienide anions formed by loss of a proton from 1-methyl-1,3-cyclopentadiene and 5-methyl-1,3-cyclopentadiene are equivalent.

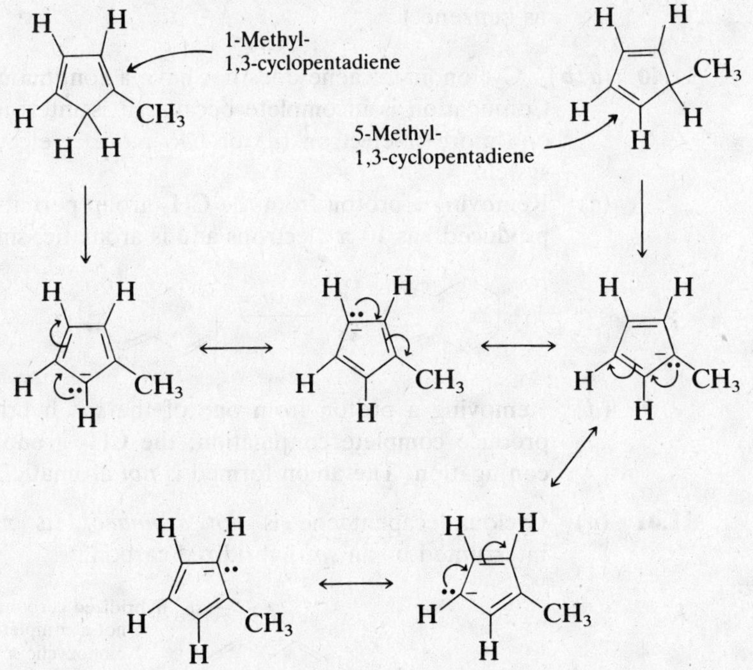

11.38 Cyclooctatetraene is not aromatic. 1,2,3,4-Tetramethylcyclooctatetraene and 1,2,3,8-tetramethylcyclooctatetraene are different compounds.

1,2,3,4-Tetramethylcyclooctatetraene   1,2,3,8-Tetramethylcyclooctatetraene

Leo A. Paquette at Ohio State University synthesized each of these compounds independently of the other and showed them to be stable enough to be stored separately without interconversion.

11.39 Cyclooctatetraene has eight π electrons and thus does not satisfy the $(4n + 2)$ π electron requirement of the Hückel rule.

Cyclooctatetraene.
Each double bond contributes
two π electrons to give a total of 8.

All of the exercises in this problem involve counting the number of π electrons in the various species derived from cyclooctatetraene and determining whether they satisfy the $(4n + 2)$ π electron rule.

(*a*)   Adding one π electron gives a species ($C_8H_8^-$) with nine π electrons. $4n + 2$, where n is a whole number, can never equal 9. Therefore, this species *is not aromatic*.

(*b*)   Adding two π electrons gives a species ($C_8H_8^{2-}$) with 10 π electrons. $4n + 2 = 10$ when $n = 2$. The species $C_8H_8^{2-}$ *is aromatic*.

(*c*)   Removing one π electron gives a species ($C_8H_8^+$) with seven π electrons. $4n + 2$ cannot equal 7. The species $C_8H_8^+$ *is not aromatic*.

(*d*)   Removing two π electrons gives a species ($C_8H_8^{2+}$) with six π electrons. $4n + 2 = 6$ when $n = 1$. The species $C_8H_8^{2+}$ *is aromatic*. (It has the same number of π electrons as benzene.)

11.40   (*a, b*)   Cyclononatetraene does not have a continuous conjugated system of π electrons. Conjugation is incomplete because it is interrupted by a CH_2 group. Thus adding one more π electron (*a*) or two more π electrons (*b*) will *not* give an aromatic system.

(*c*)   Removing a proton from the CH_2 group permits complete conjugation. The species produced has 10 π electrons and is aromatic, since $4n + 2 = 10$ when $n = 2$.

2 π electrons for each double bond
+
2 π electrons for unshared pair
= 10 π electrons

(*d*)   Removing a proton from one of the sp^2 hybridized carbons of the ring does not produce complete conjugation; the CH_2 group remains present to interrupt cyclic conjugation. The anion formed is *not* aromatic.

11.41   (*a*)   Cycloundecapentaene is *not aromatic*. Its π system is not conjugated; it is interrupted by an sp^3 hybridized carbon.

sp^3 hybridized carbon;
not a completely conjugated
monocyclic π system

(*b*)   Cycloundecapentaenyl radical is *not aromatic*. Its π system is completely conjugated and monocyclic but contains 11 π electrons—a number not equal to $(4n + 2)$ where n is an integer.

There are 11 electrons in the conjugated π system.
The five double bonds contribute 10 π electrons;
the odd electron of the radical is the eleventh.

(*c*)   Cycloundecapentaenyl cation is *aromatic*. It includes a completely conjugated π system which contains 10 π electrons (10 equals $4n + 2$ where $n = 2$).

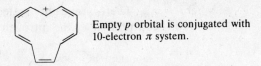

Empty *p* orbital is conjugated with
10-electron π system.

(*d*) Cycloundecapentadienide anion is *not aromatic*. It contains 12 π electrons and thus
does not satisfy the (4*n* + 2) rule.

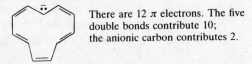

There are 12 π electrons. The five
double bonds contribute 10;
the anionic carbon contributes 2.

11.42 (*a*) The more stable dipolar resonance structure is E, because it has an aromatic
cyclopentadienide anion bonded to an aromatic cyclopropenyl cation. In structure F
neither ring is aromatic.

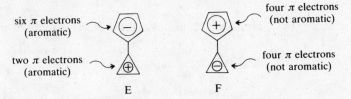

six π electrons
(aromatic)

two π electrons
(aromatic)

four π electrons
(not aromatic)

four π electrons
(not aromatic)

E   F

(*b*) Structure H can be stabilized by resonance involving the dipolar form.

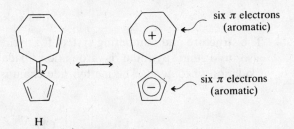

six π electrons
(aromatic)

six π electrons
(aromatic)

H

Comparable stabilization is not possible in structure G, because neither a cyclopropenyl system nor a cycloheptatrienyl system is aromatic in its anionic form. Both are
aromatic as cations.

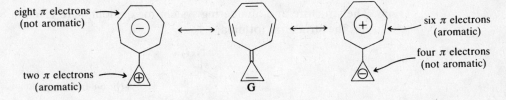

eight π electrons
(not aromatic)

two π electrons
(aromatic)

six π electrons
(aromatic)

four π electrons
(not aromatic)

G

11.43 (*a*) This molecule, called *oxepin*, is *not aromatic*. There are three double bonds, each of
which contributes two π electrons, and an oxygen atom which contributes two π
electrons, to the conjugated system, giving a total of eight π electrons. Only one of
the two unshared pairs on oxygen can contribute to the π system; the other
unshared pair is in an *sp*² hybridized orbital and cannot interact with it.

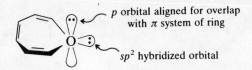

p orbital aligned for overlap
with π system of ring

*sp*² hybridized orbital

(*b*) This compound, called *azonine*, has 10 electrons in a completely conjugated planar
monocyclic π system and therefore satisfies Hückel's rule for (4*n* + 2) π electrons

where $n = 2$. There are eight π electrons from the conjugated tetraene and two electrons contributed by the nitrogen unshared pair.

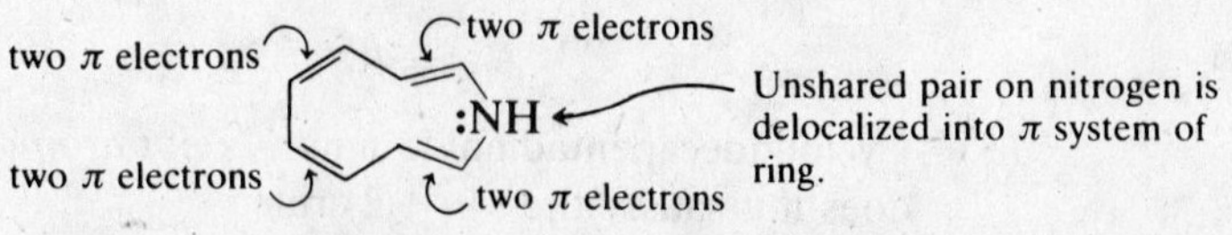

(*c*) Borazole, sometimes called *inorganic benzene*, is *aromatic*. Six π electrons are contributed by the unshared pairs of the three nitrogen atoms. Each boron contributes a *p* orbital to maintain the conjugated system but no electrons.

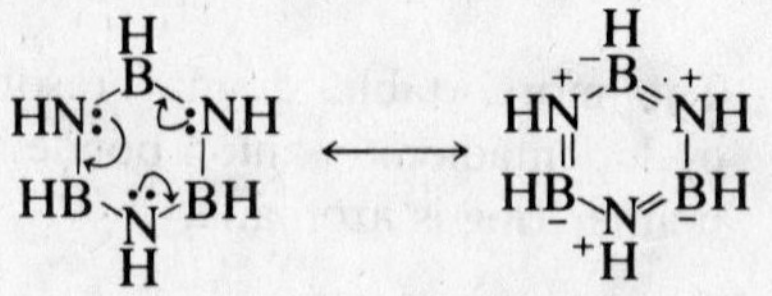

(*d*) This compound has eight π electrons and is *not aromatic*.

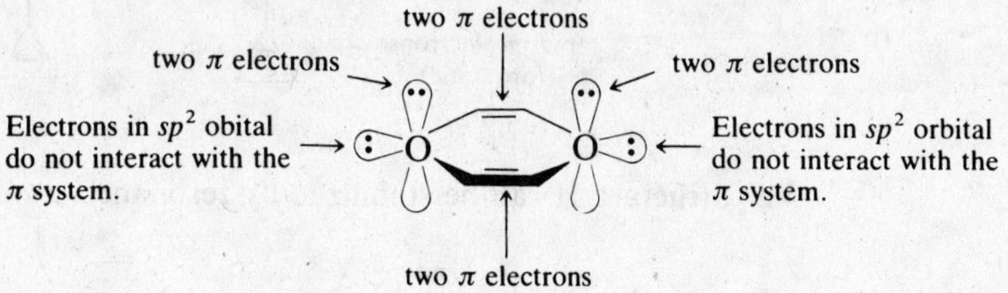

11.44 The structure and numbering system for pyridine are given in Section 11.23, where we are also told that pyridine is aromatic. Oxidation of 3-methylpyridine is analogous to oxidation of toluene. The methyl side chain is oxidized to a carboxylic acid.

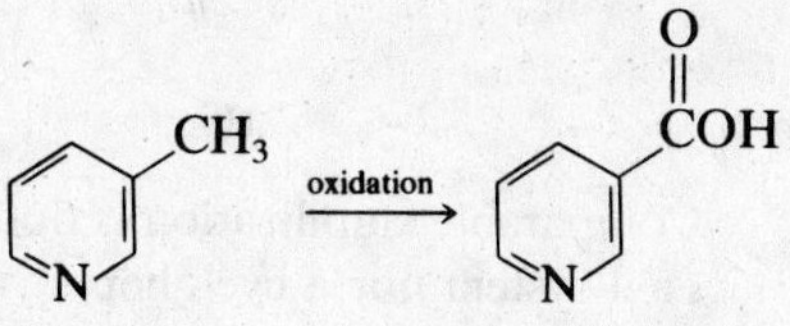

3-Methylpyridine   Niacin

11.45 The structure and numbering system for quinoline are given in Section 11.23. *Nitroxoline* has the structural formula:

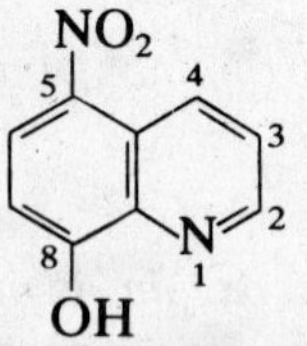

5-Nitro-8-hydroxyquinoline

11.46 We are told that the ring system of *acridine* ($C_{13}H_9N$) is analogous to that of anthracene (that is, tricyclic and linearly fused). Further, the two most stable resonance forms are equivalent to one another. Therefore, the nitrogen atom must be in the central ring, and the structure of acridine is:

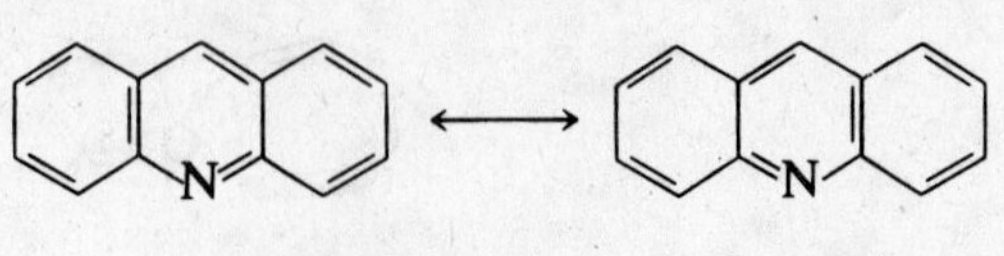

The two resonance forms would not be equivalent if the nitrogen were present in one of the terminal rings. Can you see why?

SELF-TEST

PART A

A-1. Give the IUPAC name of each of the following:

(a) [structure: m-bromotoluene]

(b) $C_6H_5\overset{CH_3}{\underset{Cl}{CHCHCH_3}}$

(c) [structure: 2-chloroacetophenone, with O=CCH_3 and Cl]

(d) [structure: phenol with two NO_2 groups, OH, NO_2, NO_2]

A-2. Draw the structure corresponding to each of the following:
(a) 3,5-Dichlorobenzoic acid
(b) *p*-Nitroanisole
(c) 2,4-Dimethylaniline
(d) *m*-Bromobenzyl chloride

A-3. Write a positive (+) or negative (−) charge at the appropriate position so that each of the following structures contains the proper number of π electrons to permit it to be considered an aromatic ion. For purposes of this problem ignore strain effects that might destabilize the molecule.

(a) [ring structure]   (b) [larger ring structure]

A-4. How many π electrons are counted toward satisfying Hückel's rule in the following substance? It is aromatic?

[ring structure with N and O]

A-5. Azulene, shown as follows, is highly polar. Draw a dipolar resonance structure to explain this fact.

[structure of Azulene]   Azulene

A-6. Give the reactant, reagent, or product omitted from each of the following:

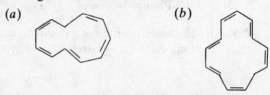

(a) [tetralin structure] $\xrightarrow[\text{peroxides, heat}]{\text{NBS}}$?

(c) [structure CH_2CH_2CH_3 with Cl] $\xrightarrow{?}$ [structure CO_2H with Cl]

(b) ? $\xrightarrow[\text{CH}_3\text{OH}]{\text{NaOCH}_3}$ $C_6H_5CH_2OCH_3$

(d) [dihydronaphthalene structure] $\xrightarrow{\text{CH}_3\text{CO}_2\text{OH}}$?

A-7. Provide two methods for the synthesis of 1-bromo-1-phenylpropane from an aromatic hydrocarbon.

PART B

B-1. The number of possible dichloronitrobenzene isomers is
(*a*) 3   (*b*) 4   (*c*) 6   (*d*) 8

B-2. Which of the following statements is correct concerning the class of reactions to be expected for benzene and cyclooctatetraene?
(*a*) Both substances would undergo addition reactions.
(*b*) Both substances would undergo substitution reactions.
(*c*) Benzene would undergo substitution; cyclooctatetraene would undergo addition.
(*d*) Benzene would undergo addition; cyclooctatetraene would undergo substitution.

B-3. Which, if any, of the following structures represents an aromatic species?

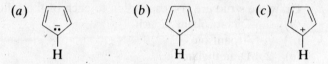

(*a*)   H   (*b*)   H   (*c*)   H

(*d*) None of these is aromatic

B-4. Which of the following compounds has a double bond that is conjugated with the π system of the benzene ring?
(*a*) *p*-Benzyltoluene   (*c*) 3-Phenylcyclohexene
(*b*) 2-Phenyl-1-decene   (*d*) 3-Phenyl-1,4-pentadiene

B-5. Rank the following compounds in order of increasing rate of solvolysis in aqueous acetone (slowest → fastest):

$$(CH_3)_2CHCH_2CH_2Br \qquad\qquad (CH_3)_2CHCHCH_3 \qquad\qquad C_6H_5CHCH(CH_3)_2$$

with Br substituents shown on B and C.

A   B   C

(*a*) A < B < C   (*b*) B < A < C   (*c*) C < B < A   (*d*) A < C < B

B-6. When comparing the hydrogenation of benzene with that of a hypothetical 1,3,5-cyclohexatriene, benzene _____ than the cyclohexatriene.
(*a*) Absorbs 152 kJ/mol (36 kcal/mol) more heat
(*b*) Absorbs 152 kJ/mol (36 kcal/mol) less heat
(*c*) Gives off 152 kJ/mol (36 kcal/mol) more heat
(*d*) Gives off 152 kJ/mol (36 kcal/mol) less heat

B-7. The reaction

$$\text{(bicyclic structure with Br)} \xrightarrow{\text{CH}_3\text{CH}_2\text{O}^- \text{ Na}^+} ?$$

gives as the major elimination product

(*a*)   (*b*)   (*c*) *Equal* amounts of (*a*) and (*b*)

(*d*) Neither (*a*) nor (*b*)

B-8. The molecule having the structure

(a) Does not obey Hückel's rule and is not aromatic
(b) Obeys Hückel's rule and is aromatic
(c) Forms an aromatic cation by loss of one electron
(d) Forms an aromatic dianion by gaining two electrons

B-9. State which of the listed compounds is *not* an effective substrate for the process

$$? \xrightarrow[\text{H}_2\text{O, H}_2\text{SO}_4\text{, heat}]{\text{Na}_2\text{Cr}_2\text{O}_7}$$ benzene with CO$_2$H and CO$_2$H

(a) benzene with CH$_2$CH$_3$ and C(CH$_3$)$_3$

(c) benzene with CH(CH$_3$)$_2$ and CH(CH$_3$)$_2$

(b) tetrahydronaphthalene

(d) benzene with CH$_3$ and CH$_3$

B-10. Which one of the following alcohols undergoes dehydration at the *fastest* rate on being heated with sulfuric acid? (The potential for rearrangement does not affect the rate.)

(a) benzene—CH$_2$CH$_2$CH$_2$CH$_2$OH

(c) benzene—CH$_2$CHCH$_2$CH$_3$ with OH

(b) benzene—CH$_2$CH$_2$CHCH$_3$ with OH

(d) benzene—CHCH$_2$CH$_2$CH$_3$ with OH

REACTIONS OF ARENES. ELECTROPHILIC AROMATIC SUBSTITUTION

IMPORTANT TERMS AND CONCEPTS

General Mechanism of Electrophilic Aromatic Substitution (Sec. 12.2) Electrophilic aromatic substitution is the most widely encountered class of reaction involving the aromatic ring. A general mechanism may be used to describe the various specific reactions: an *electrophile*, i.e., an electron-deficient, often positively charged species, adds to the aromatic ring; the intermediate *cyclohexadienyl cation* thereby formed is resonance-stabilized; and loss of a proton from it generates the aromatic product.

(E = electrophilic species)

(E = NO$_2$, SO$_3$H, halogen, R, $\overset{\text{O}}{\overset{\|}{\text{CR}}}$)

Specific examples of electrophilic aromatic substitution reactions are tabulated on the following page.

Rate and Orientation Effects (Secs. 12.9 to 12.14) Once a substituent is attached to an aromatic ring, its nature affects the rate of reaction (relative to benzene) and the orientation (ortho, meta, or para relative to the group attached to the ring) of further substitution.

Reaction (text section)	Typical reagents	Electrophile	Typical product
Nitration (12.1, 12.3)	HNO_3/H_2SO_4	NO_2^+	$C_6H_5-NO_2$
Sulfonation (12.1, 12.4)	H_2SO_4/heat (SO_3/H_2SO_4)	SO_3	$C_6H_5-SO_3H$
Halogenation (12.1, 12.5)	$Cl_2/FeCl_3$ $Br_2/FeBr_3$	"Cl^+"[a] "Br^+"[b]	C_6H_5-Cl C_6H_5-Br
Friedel-Crafts:			
Alkylation (12.1, 12.6)	$R-Cl/AlCl_3$ ($R-OH$, alkenes/H_2SO_4) *Note*: Rearrangements of R are possible.	R^+	C_6H_5-R
Acylation (12.1, 12.7)	$\overset{\displaystyle O}{\overset{\displaystyle \|}{R}}CCl/AlCl_3$ ($RC\overset{O}{\overset{\|}{}}OC\overset{O}{\overset{\|}{}}R/AlCl_3$)	$R-C\equiv O^+$	$C_6H_5-\overset{O}{\overset{\|}{C}}R$

[a] Electrophile is $Cl_2/FeCl_3$ complex.
[b] Electrophile is $Br_2/FeBr_3$ complex.

Activating groups:

Activating groups are those that increase the rate of substitution. Examples include:

$$-OH, \quad -OR, \quad -O\overset{O}{\overset{\|}{C}}R, \quad -NH_2, \quad -NH\overset{O}{\overset{\|}{C}}CH_3, \quad -R \text{ (alkyl)}, \quad C_6H_5$$

All these groups are *ortho, para directors*. That is, further substitution of the ring will occur in the positions ortho and para to the directing group.

Ortho   Para

Deactivating groups:

Deactivating groups are those that decrease the rate of substitution. Examples include:

$$-CF_3, \quad -\overset{O}{\overset{\|}{C}}R, (R = H, \text{alkyl, aryl, OH, OR}'), \quad -C\equiv N, \quad -NO_2, \quad -SO_3H, -\overset{+}{N}R_3$$

These groups are meta directors, directing further substitution on the ring to the position meta to that occupied by the directing group.

Meta

The *halogens* (F, Cl, Br, I) are an important exception. They are *slightly deactivating* ortho, para directors.

Remember: It is the group *already* attached to the ring (*not* the one being added) which determines rate and orientation effects. For example, nitration of anisole ($C_6H_5OCH_3$) proceeds *faster* than nitration of benzene and gives a mixture of the ortho and para products.

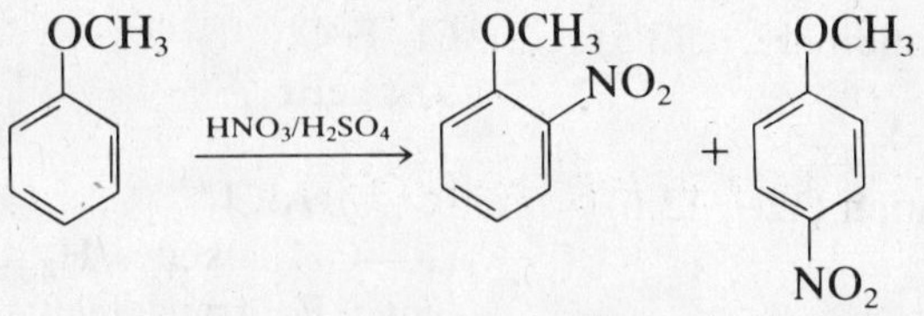

Multiple Substitution (Sec. 12.15) When two or more groups are attached to the benzene ring, orientation is controlled by the *activating groups*. For example,

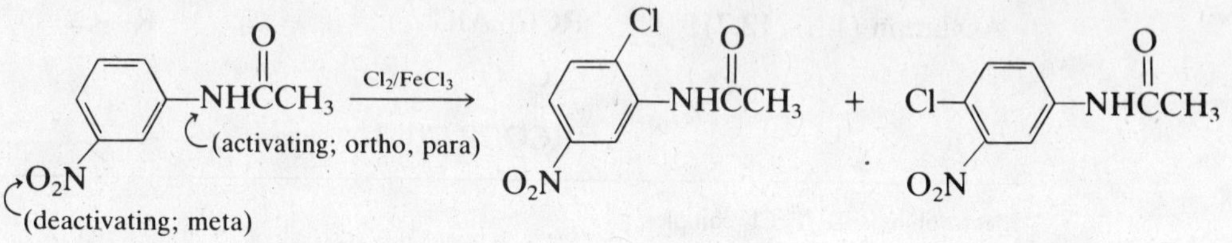

Naphthalene Substitution (Sec. 12.17) Electrophilic aromatic substitution of naphthalene occurs faster at the 1 position. For example,

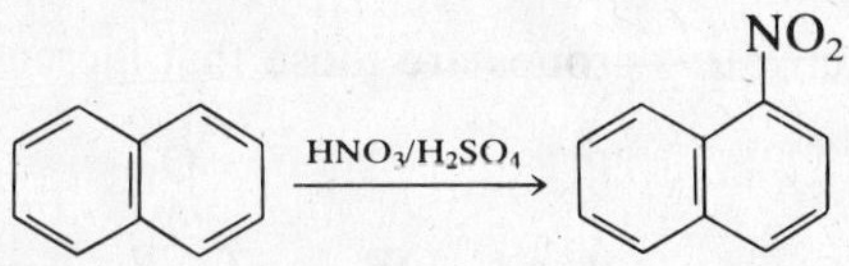

IMPORTANT REACTIONS

Alkylbenzene Synthesis (Secs. 12.7, 12.8) Friedel-Crafts *acylation* followed by *reduction* of the ketone provides a method of synthesis that avoids carbocation rearrangement (which may accompany Friedel-Crafts alkylations).

Friedel-Crafts acylation:

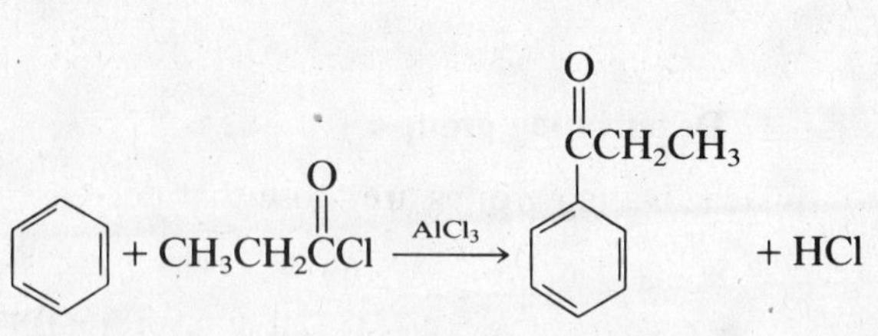

Clemmensen reduction:

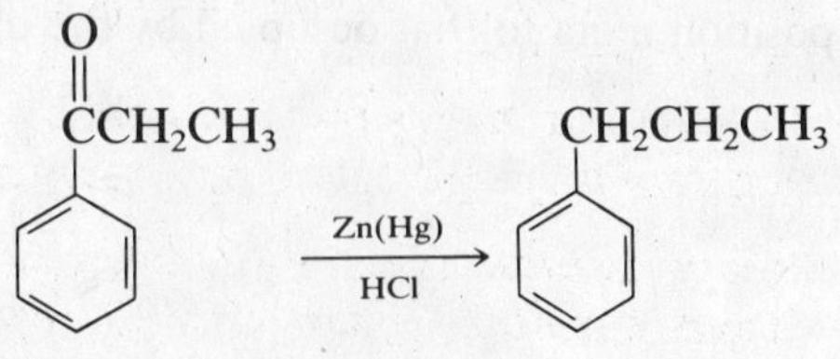

Wolff-Kishner reduction:

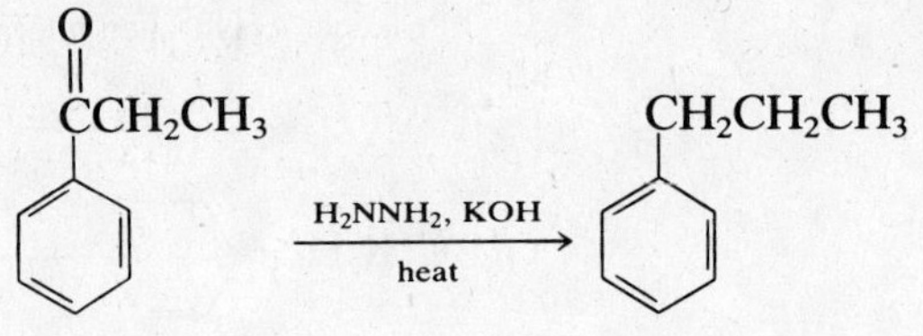

SOLUTIONS TO TEXT PROBLEMS

12.1 To show electron delocalization in a cyclohexadienyl cation intermediate, generate alternative resonance structures by moving pairs of π electrons toward sites of positive charge.

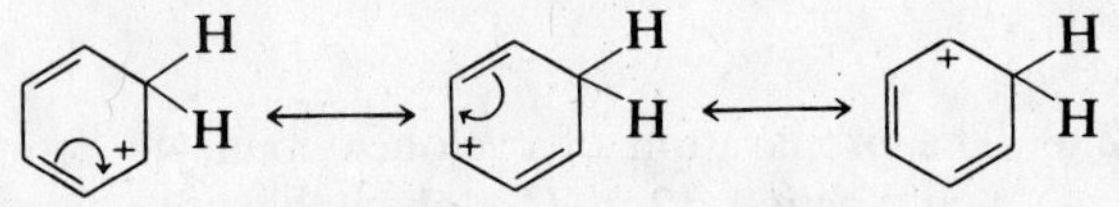

12.2 Electrophilic aromatic substitution leads to replacement of one of the hydrogens directly attached to the ring by the electrophile. All four of the ring hydrogens of *p*-xylene are equivalent, and so it does not matter which one is replaced by the nitro group.

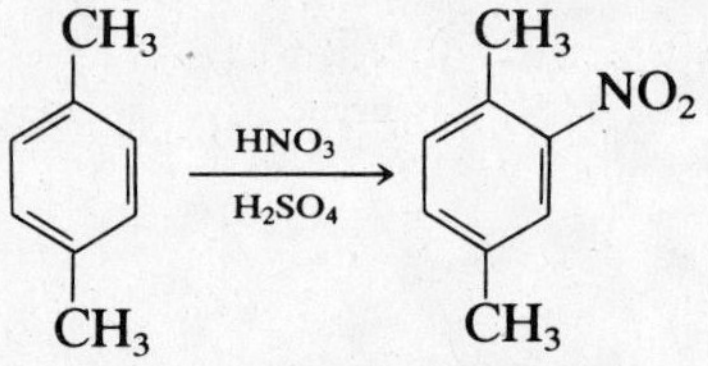

         p-Xylene   1,4-Dimethyl-2-nitrobenzene

12.3 The aromatic ring of 1,2,4,5-tetramethylbenzene has two equivalent hydrogen substituents. Sulfonation of the ring leads to replacement of one of them by —SO$_3$H.

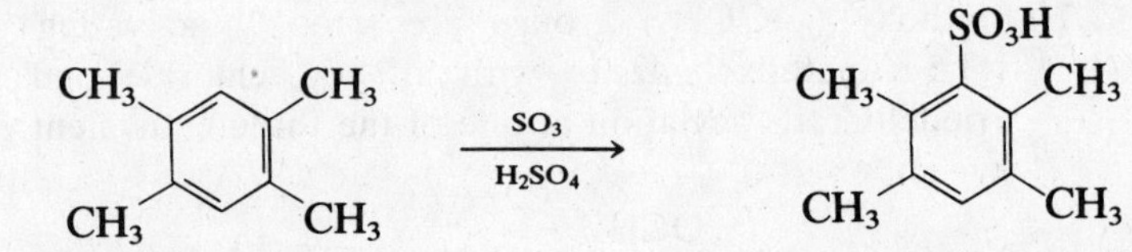

    1,2,4,5-Tetramethylbenzene   2,3,5,6-Tetramethylbenzenesulfonic acid

12.4 The major product is isopropylbenzene.

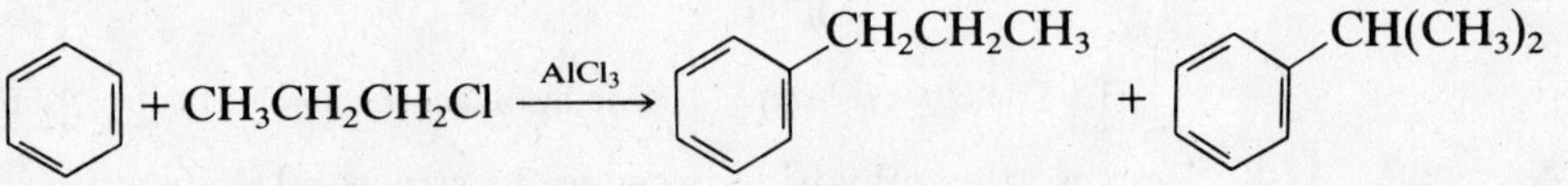

  Benzene   1-Chloropropane   Propylbenzene   Isopropylbenzene
                                     (20% yield)   (40% yield)

Aluminum chloride coordinates with 1-chloropropane to give a Lewis acid–Lewis base complex, which can be attacked by benzene to yield propylbenzene or can undergo an intramolecular hydride shift to produce isopropyl cation. Isopropylbenzene arises by reaction of isopropyl cation with benzene.

12.5 Isopropylbenzene arises from the reaction of isopropyl cation with benzene. Isopropyl cation is formed by protonation of propene with hydrogen fluoride.

$$CH_3CH{=}CH_2 + H{-}F \longrightarrow CH_3\overset{+}{C}HCH_3 + F^-$$

Propene   Hydrogen fluoride   Isopropyl cation   Fluoride ion

Isopropyl   Benzene   Cyclohexadienyl cation
cation   intermediate

Isopropylbenzene

12.6 The preparation of cyclohexylbenzene from cyclohexene and benzene was described in text Section 12.6. Cyclohexylbenzene is converted to 1-phenylcyclohexene by benzylic bromination, followed by dehydrohalogenation.

Benzene   Cyclohexene   Cyclohexylbenzene

N-Bromosuccinimide (NBS),
benzoyl peroxide, heat

1-Phenylcyclohexene   1-Bromo-1-phenylcyclohexane

12.7 Methoxy (—OCH₃) groups are strongly activating and *o,p*-directing. Treatment of 1,3,5-trimethoxybenzene with an acyl chloride and aluminum chloride brings about Friedel-Crafts acylation at one of the three equivalent positions available on the ring.

1,3,5-Trimethoxybenzene   3-Methylbutanoyl chloride   Isobutyl 1,3,5-trimethoxyphenyl ketone

12.8 Because the anhydride is cyclic, its structural units are not incorporated into a ketone and a carboxylic acid as two separate product molecules. Rather, they become part of a four-carbon unit attached to benzene by a ketone carbonyl. The acyl substituent terminates in a carboxylic acid functional group.

Benzene   Succinic anhydride   4-Oxo-4-phenylbutanoic acid

12.9 (b) Neopentyl systems exhibit a pronounced tendency to undergo cationic rearrangements by shift of a methyl group. A Friedel-Crafts alkylation of benzene using 1-chloro-2,2-dimethylpropane would not be a satisfactory method to prepare neopentylbenzene; the best way to prepare this compound is by Friedel-Crafts acylation followed by Clemmensen reduction.

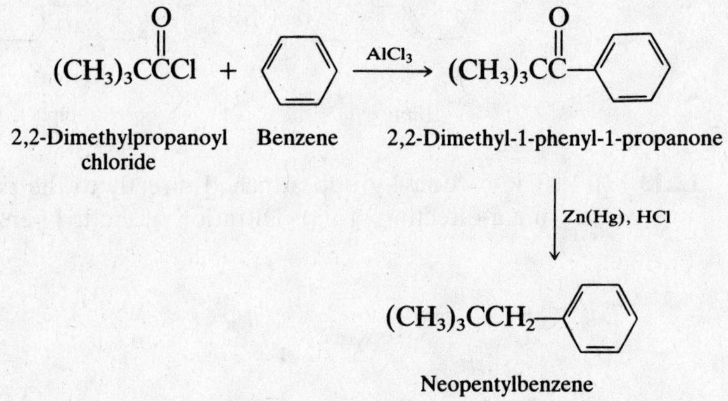

$$(CH_3)_3CCCl \ + \ \bigcirc \ \xrightarrow{\text{AlCl}_3} \ (CH_3)_3CC{-}\bigcirc$$

2,2-Dimethylpropanoyl   Benzene   2,2-Dimethyl-1-phenyl-1-propanone
chloride

$$\Big\downarrow \ \text{Zn(Hg), HCl}$$

$$(CH_3)_3CCH_2{-}\bigcirc$$

Neopentylbenzene

12.10 (b) Partial rate factors for nitration of toluene and *tert*-butylbenzene, relative to a single position of benzene, are as shown:

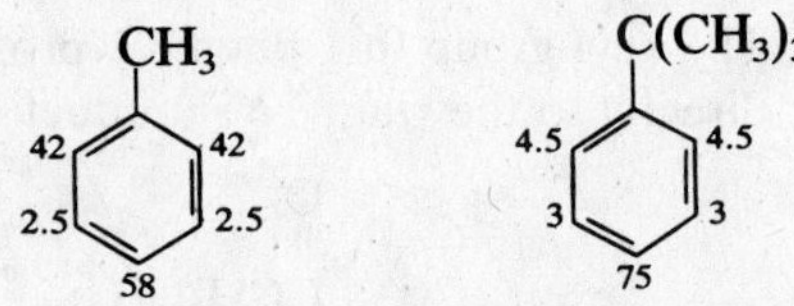

The sum of these partial rate factors is 147 for toluene, 90 for *tert*-butylbenzene. Toluene is 147/90, or 1.7, times more reactive than *tert*-butylbenzene.

(c) The product distribution for nitration of *tert*-butylbenzene is determined from the partial rate factors.

$$\text{Ortho:} \quad \frac{2(4.5)}{90} = 10\%$$

$$\text{Meta:} \quad \frac{2(3)}{90} = 6.7\%$$

$$\text{Para:} \quad \frac{75}{90} = 83.3\%$$

12.11 (b) Attack by bromine at the position meta to the amino group gives a cyclohexadienyl cation intermediate in which delocalization of the nitrogen lone pair cannot participate in dispersal of the positive charge.

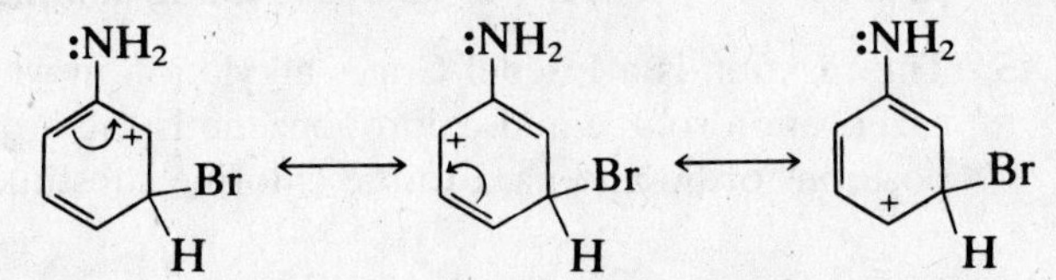

(c) Attack at the position para to the amino group yields a cyclohexadienyl cation intermediate that is stabilized by delocalization of the electron pair of the amino group.

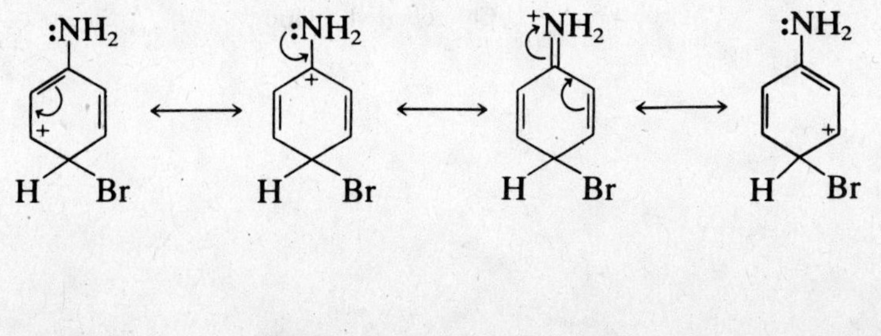

12.12 Electrophilic aromatic substitution in biphenyl is best understood by considering one ring as the functional group and the other as a substituent. An aryl substituent is ortho, para–directing. Nitration of biphenyl gives a mixture of o-nitrobiphenyl and p-nitrobiphenyl.

Biphenyl   o-Nitrobiphenyl   p-Nitrobiphenyl
                                 (37%)   (63%)

12.13 (b) The carbonyl group attached directly to the ring is a signal that the substituent is a meta-directing group. Nitration of methyl benzoate yields methyl m-nitrobenzoate.

Methyl benzoate   Methyl m-nitrobenzoate
                          (isolated in 81–85% yield)

(c) The acyl group in 1-phenyl-1-propanone is meta-directing—the carbonyl is attached directly to the ring. The product is 1-(m-nitrophenyl)-1-propanone.

1-Phenyl-1-propanone   1-(m-Nitrophenyl)-1-propanone
                              (isolated in 60% yield)

12.14 Writing the structures out in more detail reveals that the substituent $-\overset{+}{N}(CH_3)_3$ lacks the unshared electron pair of $-N(CH_3)_2$.

This unshared pair is responsible for the powerful activating effect of an $-N(CH_3)_2$ group. On the other hand, the nitrogen in $-\overset{+}{N}(CH_3)_3$ is positively charged and in that respect resembles the nitrogen of a nitro group. On these bases, we expect the substituent $-\overset{+}{N}(CH_3)_3$ to be deactivating and meta-directing.

12.15 The reaction is a Friedel-Crafts alkylation in which 4-chlorobenzyl chloride serves as the carbocation source and chlorobenzene is the aromatic substrate. Alkylation occurs at the positions ortho and para to the chlorine substituent of chlorobenzene.

Chlorobenzene   4-Chlorobenzyl chloride   1-Chloro-2-(4'-chlorobenzyl)benzene

1-Chloro-4-(4'-chlorobenzyl)benzene

12.16 (b) Halogen substituents are ortho, para–directing, and the disposition in *m*-dichlorobenzene is such that their effects reinforce each other. The major product is 2,4-dichloro-1-nitrobenzene. Substitution at the position between the two chlorines is slow because it is a sterically hindered position.

Positions activated toward
electrophilic aromatic substitution
in *m*-dichlorobenzene

2,4-Dichloro-1-nitrobenzene
(major product of nitration)

(c) Nitro groups are meta-directing. Both nitro groups of *m*-dinitrobenzene direct an incoming substituent to the same position in an electrophilic aromatic substitution reaction. Nitration of *m*-nitrobenzene yields 1,3,5-trinitrobenzene.

Both nitro groups of
m-dinitrobenzene direct
electrophile to same position.

1,3,5-Trinitrobenzene
(principal product of nitration
of *m*-dinitrobenzene)

(d) A methoxy group is ortho, para–directing, and a carbonyl group is meta-directing. The open positions of the ring that are activated by the methoxy group in *p*-methoxyacetophenone are also those that are meta to the carbonyl, and so the directing effects of the two substituents reinforce one another. Nitration of *p*-methoxyacetophenone yields 4-methoxy-3-nitroacetophenone.

Positions ortho to the methoxy
group are meta to the carbonyl.

4-Methoxy-3-nitroacetophenone

(e) The methoxy group of *p*-methylanisole activates the positions that are ortho to it; the methyl activates those ortho to itself. Methoxy is a more powerful activating substituent than methyl, and so nitration occurs ortho to the methoxy group.

Methyl activates C-3 and C-5;
methoxy activates C-2 and C-6.

4-Methyl-2-nitroanisole
(principal product of nitration)

(f) All the substituents in 2,6-dibromoanisole are ortho, para–directing, and their effects are felt at different positions. However, the methoxy group is a far more powerful activating substituent than bromine, and so it controls the regioselectivity of nitration.

Methoxy directs toward C-4;
bromines direct toward C-3 and C-5.

2,6-Dibromo-4-nitroanisole
(principal product of nitration)

12.17 The product that is obtained when benzene is subjected to bromination and nitration depends on the order in which the reactions are carried out. A nitro group is meta-directing, and so if it is introduced prior to the bromination step, *m*-bromonitrobenzene is obtained.

Benzene   Nitrobenzene   *m*-Bromonitrobenzene

Bromine is an ortho, para–directing group. If it is introduced first, nitration of the resulting bromobenzene yields a mixture of *o*-bromonitrobenzene and *p*-bromonitrobenzene.

Benzene   Bromobenzene   *o*-Bromonitrobenzene   *p*-Bromonitrobenzene

12.18 A straightforward approach to the synthesis of *m*-nitrobenzoic acid involves preparation of benzoic acid by oxidation of toluene, followed by nitration. The carboxyl group of benzoic acid is meta-directing. Nitration of toluene prior to oxidation would lead to a mixture of ortho and para products.

Toluene   Benzoic acid   *m*-Nitrobenzoic acid

12.19 The text states that electrophilic aromatic substitution in furan, thiophene, and pyrrole occurs at C-2. The sulfonation of thiophene gives thiophene-2-sulfonic acid.

Thiophene   Thiophene-2-sulfonic acid

12.20 (*a*) Nitration of benzene is the archetypical electrophilic aromatic substitution reaction.

Benzene   Nitrobenzene

(*b*) Nitrobenzene is much less reactive than benzene toward electrophilic aromatic substitution. The nitro group on the ring is a meta director.

Nitrobenzene   *m*-Dinitrobenzene

(*c*) Toluene is more reactive than benzene in electrophilic aromatic substitution. A methyl substituent is an ortho, para director.

Toluene   *o*-Bromotoluene   *p*-Bromotoluene

(*d*) Trifluoromethyl is deactivating and meta-directing.

(Trifluoromethyl)benzene   *m*-Bromo(trifluoromethyl)benzene

(*e*) Anisole is ortho, para–directing, strongly activated toward electrophilic aromatic substitution, and readily sulfonated in sulfuric acid.

Anisole   *o*-Methoxybenzenesulfonic acid   *p*-Methoxybenzenesulfonic acid

Sulfur trioxide, of course, could be added to the sulfuric acid and would facilitate reaction. The para isomer is the predominant product.

(*f*) Acetanilide is quite similar to anisole in its behavior toward electrophilic aromatic substitution.

Acetanilide   *o*-Acetamidobenzenesulfonic acid   *p*-Acetamidobenzenesulfonic acid

(*g*) Bromobenzene is less reactive than benzene. A bromine substituent is ortho, para–directing.

Bromobenzene   o-Bromochlorobenzene   p-Bromochlorobenzene

(h)   Anisole is a reactive substrate toward Friedel-Crafts alkylation and yields a mixture of o- and p-benzylated products when treated with benzyl chloride and aluminum chloride.

Anisole   Benzyl chloride   o-Benzylanisole   p-Benzylanisole

(i)   Benzene will undergo acylation with benzoyl chloride and aluminum chloride.

Benzene   Benzoyl chloride   Benzophenone

(j)   A benzoyl substituent is meta-directing and deactivating.

Benzophenone   m-Nitrobenzophenone

(k)   Clemmensen reduction conditions involve treating a ketone with zinc amalgam and concentrated hydrochloric acid.

Benzophenone   Diphenylmethane

(l)   Wolff-Kishner reduction utilizes hydrazine, a base, and a high-boiling alcohol solvent to reduce ketone functions to methylene groups.

Benzophenone   Diphenylmethane

12.21 (a)   There are three principal resonance forms of the cyclohexadienyl cation intermediate formed by attack of bromine on p-xylene.

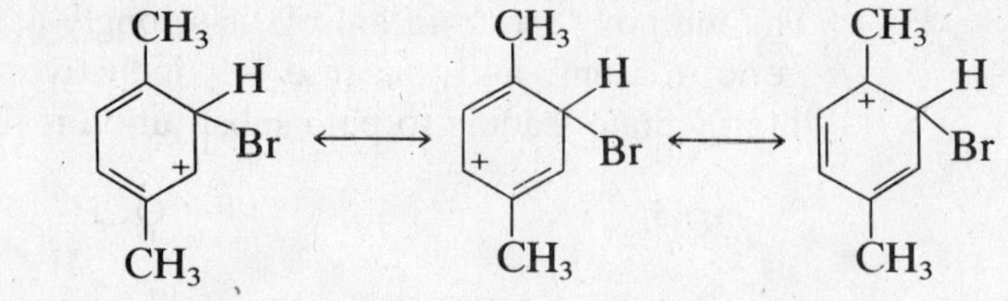

Any one of these resonance forms is a satisfactory answer to the question. Because of its tertiary carbocation character this carbocation is more stable than the corresponding intermediate formed from benzene.

(*b*) Chlorination of *m*-xylene will give predominantly 4-chloro-1,3-dimethylbenzene.

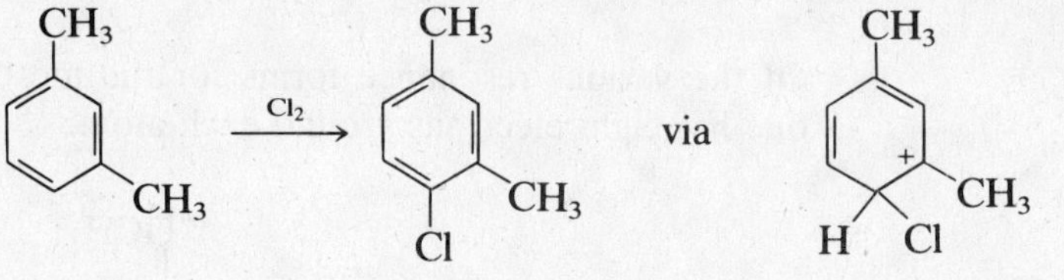

m-Xylene   4-Chloro-1,3-dimethylbenzene   More stable cyclohexadienyl cation

The intermediate shown (or any of its resonance forms) is more stable for steric reasons than

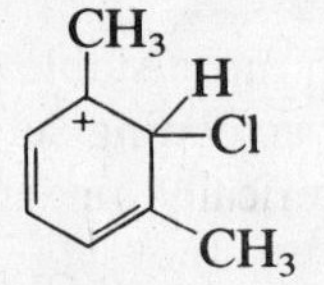

Less stable cyclohexadienyl cation

The cyclohexadienyl cation intermediate leading to 4-chloro-1,3-dimethylbenzene is more stable and is formed faster than the intermediate leading to chlorobenzene because of its tertiary carbocation character.

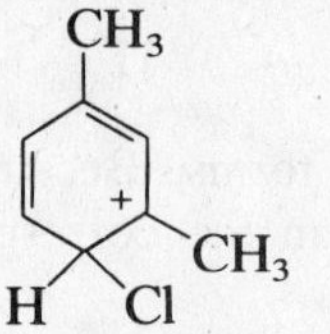

more stable than

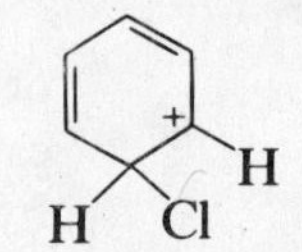

(*c*) The most stable carbocation intermediate formed during nitration of acetophenone is the one corresponding to meta attack.

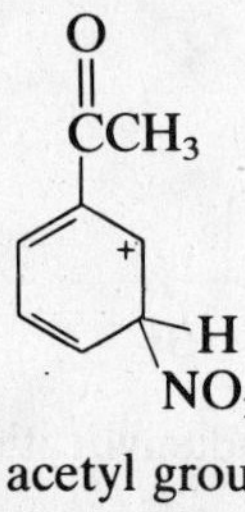

more stable than 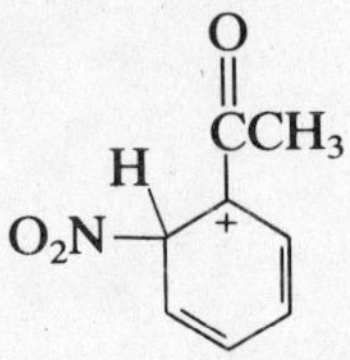   or

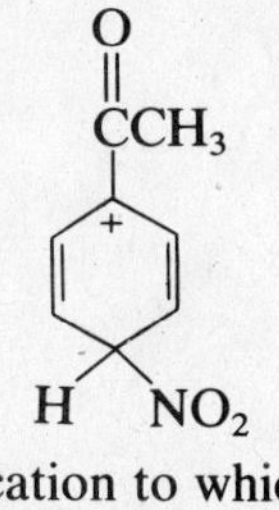

An acetyl group is electron-withdrawing and destabilizes a carbocation to which it is attached. The most stable carbocation intermediate in the nitration of acetophenone is less stable and is formed more slowly than is the corresponding carbocation formed during nitration of benzene.

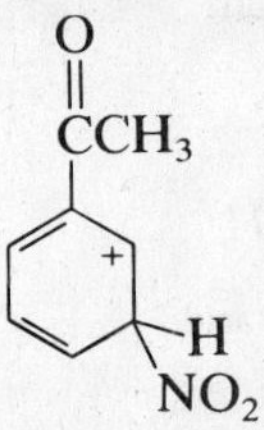

less stable than

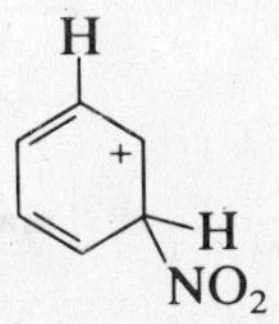

(d) The methoxy group in anisole is strongly activating and ortho, para–directing. For steric reasons and because of inductive electron withdrawal by oxygen the intermediate leading to para substitution is the most stable.

Of the various resonance forms for the most stable intermediate, the most stable one has eight electrons around each atom.

This intermediate is much more stable than the corresponding one from benzene.

(e) An isopropyl group is an activating substituent and is ortho, para–directing. Attack at the ortho position is sterically hindered. The most stable intermediate is

or any of its resonance forms. Because of its tertiary carbocation character this cation is more stable than the corresponding cyclohexadienyl cation intermediate from benzene.

(f) A nitro substituent is deactivating and meta-directing. The most stable cyclohexadienyl cation formed in the bromination of nitrobenzene is:

This ion is less stable than the cyclohexadienyl cation formed during bromination of benzene.

(g) Sulfonation of furan takes place at C-2. The cationic intermediate is more stable than the cyclohexadienyl cation formed from benzene, because it is stabilized by electron release from oxygen.

Furan   Furan-2-sulfonic acid

(h)   Pyridine reacts with electrophiles at C-3. It is less reactive than benzene, and the carbocation intermediate is less stable than the corresponding intermediate formed from benzene.

12.22   (a)   Toluene is more reactive than chlorobenzene in electrophilic aromatic substitution reactions, because a methyl substituent is activating while a halogen substituent is deactivating. Both are ortho, para–directing, however. Nitration of toluene is faster than nitration of chlorobenzene.

Faster:

Toluene   *o*-Nitrotoluene   *p*-Nitrotoluene

Slower:

Chlorobenzene   *o*-Chloronitrobenzene   *p*-Chloronitrobenzene

(b)   A fluorine substituent is not nearly as strongly deactivating as is a trifluoromethyl group. The reaction that takes place is Friedel-Crafts alkylation of fluorobenzene.

o-Benzylfluorobenzene   *p*-Benzylfluorobenzene
(15%)   (85%)

Strongly deactivated aromatic compounds do not undergo Friedel-Crafts reactions.

$+ C_6H_5CH_2Cl \xrightarrow{\text{AlCl}_3}$ no reaction

(c)   A carbonyl group directly bonded to a benzene ring strongly *deactivates* it toward electrophilic aromatic substitution. Methyl benzoate is much less reactive than benzene.

An oxygen substituent directly attached to the ring strongly *activates* it toward electrophilic aromatic substitution. Phenyl acetate is much more reactive than benzene or methyl benzoate.

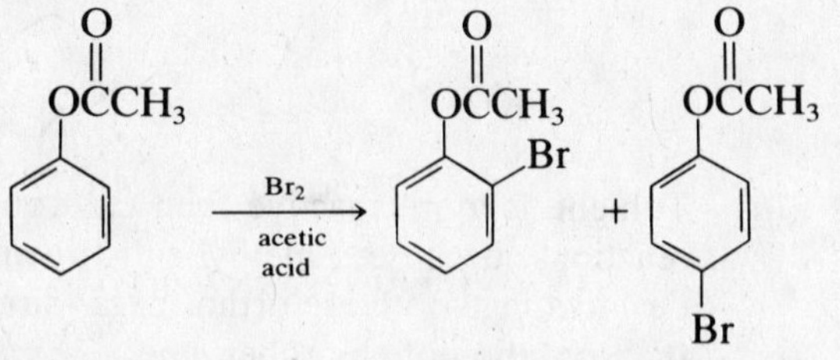

| Phenyl acetate | *o*-Bromophenyl acetate | *p*-Bromophenyl acetate |

Bromination of methyl benzoate requires more vigorous conditions; catalysis by iron(III) bromide is required for bromination of deactivated aromatic rings.

(*d*) Acetanilide is strongly activated toward electrophilic aromatic substitution and reacts faster than nitrobenzene, which is strongly deactivated.

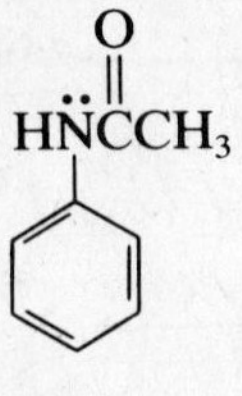

Acetanilide
(lone pair on nitrogen can
stabilize cyclohexadienyl
cation intermediate)

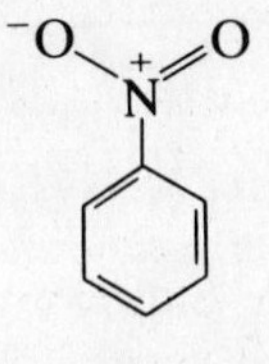

Nitrobenzene
(nitrogen is positively charged
and is electron-withdrawing)

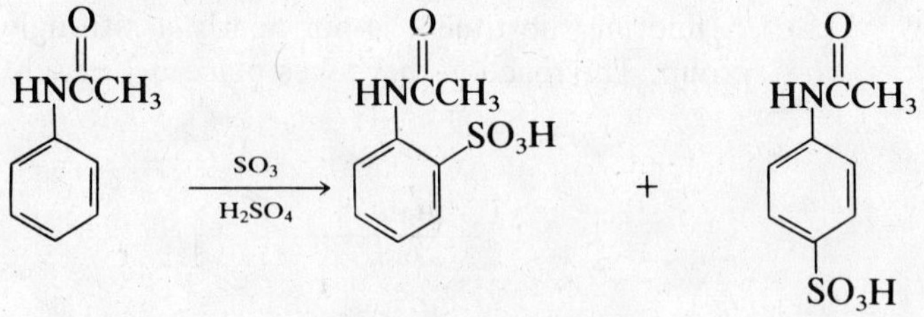

| Acetanilide | *o*-Acetamidobenzenesulfonic acid | *p*-Acetamidobenzenesulfonic acid |

(*e*) Both substrates are of the type

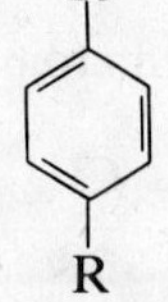

R = alkyl

and are activated toward Friedel-Crafts acylation. Since electronic effects are comparable, we look to differences in steric factors and conclude that reaction will be faster for R = CH_3 than for R = $(CH_3)_3C$—.

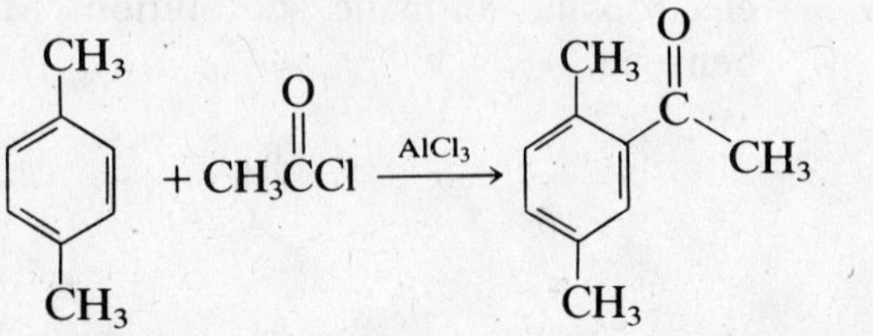

p-Xylene   Acetyl chloride   2,5-Dimethylacetophenone

(f) A phenyl substituent is activating and ortho, para–directing. Biphenyl will undergo chlorination readily.

Biphenyl   o-Chlorobiphenyl   p-Chlorobiphenyl

Each benzene ring of benzophenone is deactivated by the carbonyl group.

Benzophenone

Benzophenone is much less reactive than biphenyl in electrophilic aromatic substitution reactions.

12.23 Reactivity toward electrophilic aromatic substitution increases with increasing number of electron-releasing substituents. Benzene, with no methyl substituents, is the least reactive, followed by toluene, with one methyl group. 1,3,5-Trimethylbenzene, with three methyl substituents, is the most reactive.

Benzene   Toluene   1,3,5-Trimethylbenzene

Relative
reactivity: 1   60 2×10^7

o-Xylene and m-xylene are intermediate in reactivity between toluene and 1,3,5-trimethylbenzene. Of the two, m-xylene is more reactive than o-xylene, because the activating effects of the two methyl groups reinforce each other.

o-Xylene   m-Xylene
(all positions somewhat   (activating effects reinforce
activated)   each other)

Relative
reactivity: 5×10^2 5×10^4

12.24 (a) Chlorine is ortho, para–directing, carboxyl is meta-directing. The positions that are ortho to the chlorine are meta to the carboxyl, so that both substituents direct an incoming electrophile to the same position. Introduction of the second nitro group at the remaining position that is ortho to the chlorine puts it meta to the carboxyl and meta to the first nitro group.

p-Chlorobenzoic acid   4-Chloro-3,5-dinitrobenzoic
acid (90%)

(*b*)   An amino group is one of the strongest activating substituents. The para and both ortho positions are readily substituted in aniline. When aniline is treated with excess bromine, 2,4,6-tribromoaniline is formed in quantitative yield.

Aniline   2,4,6-Tribromoaniline (100%)

(*c*)   The positions ortho and para to the amino group in *o*-aminoacetophenone are the ones most activated toward electrophilic aromatic substitution.

o-Aminoacetophenone   2-Amino-3,5-dibromoacetophenone (65%)

(*d*)   The carboxyl group in benzoic acid is meta-directing, and so nitration gives *m*-nitrobenzoic acid. The second nitration step introduces a nitro group meta to both the carboxyl group and the first nitro group.

Benzoic acid   *m*-Nitrobenzoic acid   3,5-Dinitrobenzoic acid (54–58%)

(*e*)   Both bromine substituents are introduced ortho to the strongly activating hydroxyl group in *p*-nitrophenol.

p-Nitrophenol   2,6-Dibromo-4-nitrophenol (96–98%)

(*f*)   Friedel-Crafts alkylation occurs when biphenyl is treated with *tert*-butyl chloride and iron(III) chloride (a Lewis acid catalyst); the product of monosubstitution is *p-tert*-butylbiphenyl. All the positions of the ring that bears the *tert*-butyl group are sterically hindered, and so the second alkylation step introduces a *tert*-butyl group at the para position of the second ring.

Biphenyl   4,4'-Di-*tert*-butylbiphenyl (70%)

(*g*) Disulfonation of phenol occurs at positions ortho and para to the hydroxyl group. the ortho, para product predominates over the ortho, ortho one.

Phenol   2-Hydroxy-1,5-benzenedisulfonic acid

12.25 When carrying out each of the following syntheses, evaluate how the structure of the product differs from that of benzene or toluene; that is, determine which groups have been substituted on the benzene ring or altered in some way. The sequence of reaction steps when multiple substitution is desired is important; recall that some groups direct ortho, para and others meta.

(*a*) Isopropylbenzene may be prepared by a Friedel-Crafts alkylation of benzene with isopropyl chloride (or bromide, or iodide).

Benzene   Isopropyl chloride   Isopropylbenzene

It would not be appropriate to use propyl chloride and trust that a rearrangement reaction would lead to isopropylbenzene, because a mixture of propylbenzene and isopropylbenzene would be obtained.

Isopropylbenzene may also be prepared by alkylation of benzene with propene in the presence of a proton donor. Liquid hydrogen fluoride is often used in reactions of this type.

Benzene   Propene   Isopropylbenzene

(*b*) Since the isopropyl and sulfonic acid groups are para to each other, the first group introduced on the ring must be the ortho, para director, i.e., the isopropyl group. Therefore we may use the product of part (*a*), isopropylbenzene, in this synthesis. An isopropyl group is a fairly bulky ortho, para director, and so sulfonation of isopropylbenzene gives mainly *p*-isopropylbenzenesulfonic acid.

$$(CH_3)_2CH-\underset{\text{Isopropylbenzene}}{\bigcirc} \xrightarrow[\text{H}_2\text{SO}_4]{\text{SO}_3} (CH_3)_2CH-\underset{\text{p-Isopropylbenzenesulfonic acid}}{\bigcirc}-SO_3H$$

A sulfonic acid group is meta-directing, so that the order of steps must be alkylation followed by sulfonation rather than the reverse.

(*c*)   Free-radical halogenation of isopropylbenzene occurs with high regioselectivity at the benzylic position. *n*-Bromosuccinimide (NBS) is a good reagent to use for benzylic bromination reactions.

$$\underset{\text{Isopropylbenzene}}{\bigcirc}-CH(CH_3)_2 \xrightarrow[\text{or NBS}]{\underset{\text{light}}{\text{Br}_2,}} \underset{\text{2-Bromo-2-phenylpropane}}{\bigcirc}-\overset{\overset{\displaystyle Br}{|}}{\underset{\underset{\displaystyle CH_3}{|}}{C}}-CH_3$$

(*d*)   Toluene is an obvious starting material for the preparation of 4-*tert*-butyl-2-nitrotoluene. Two possibilities, both involving nitration and alkylation of toluene, present themselves; the problem to be addressed is in what order to carry out the two steps. Friedel-Crafts alkylation must precede nitration.

$$\underset{\text{Toluene}}{\overset{\text{CH}_3}{\bigcirc}} \xrightarrow[\text{AlCl}_3]{(\text{CH}_3)_3\text{CCl}} \underset{\text{p-$tert$-Butyltoluene}}{\overset{\text{CH}_3}{\underset{\text{C(CH}_3)_3}{\bigcirc}}} \xrightarrow[\text{H}_2\text{SO}_4]{\text{HNO}_3} \underset{\text{4-$tert$-Butyl-2-nitrotoluene}}{\overset{\text{CH}_3}{\underset{\text{C(CH}_3)_3}{\bigcirc}}\text{NO}_2}$$

Introduction of the nitro group as the first step is not a satisfactory approach, since Friedel-Crafts reactions cannot be carried out on nitro-substituted aromatic compounds.

(*e*)   Two electrophilic aromatic substitution reactions need to be performed, chlorination and Friedel-Crafts acylation. The order in which the reactions are carried out is important; chlorine is an ortho, para director, and the acetyl group is a meta director. Since the groups are meta in the desired compound, introduce the acetyl group first.

$$\underset{\text{Benzene}}{\bigcirc} \xrightarrow[\text{AlCl}_3]{\overset{\overset{\displaystyle O}{\|}}{\text{CH}_3\text{CCl}}} \underset{\text{Acetophenone}}{\bigcirc}-\overset{\overset{\displaystyle O}{\|}}{C}CH_3 \xrightarrow[\text{AlCl}_3]{\text{Cl}_2} \underset{\underset{\text{m-Chloroacetophenone}}{Cl}}{\bigcirc}-\overset{\overset{\displaystyle O}{\|}}{C}CH_3$$

(*f*)   Reverse the order of steps in part (*e*) to prepare *p*-chloroacetophenone.

$$\underset{\text{Benzene}}{\bigcirc} \xrightarrow[\text{FeCl}_3]{\text{Cl}_2} Cl-\underset{\text{Chlorobenzene}}{\bigcirc} \xrightarrow[\text{AlCl}_3]{\overset{\overset{\displaystyle O}{\|}}{\text{CH}_3\text{CCl}}} Cl-\underset{\text{p-Chloroacetophenone}}{\bigcirc}-\overset{\overset{\displaystyle O}{\|}}{C}CH_3$$

Friedel-Crafts reactions can be carried out on halobenzenes but not on arenes that are more strongly deactivated.

(*g*)   Here again the problem involves two successive electrophilic aromatic substitution reactions, in this case using toluene as the initial substrate. The proper sequence is Friedel-Crafts acylation first, followed by bromination of the ring.

Toluene   *p*-Methylacetophenone   3-Bromo-4-methylacetophenone

If the sequence of steps had been reversed, with halogenation preceding acylation, the first intermediate would be *o*-bromotoluene, Friedel-Crafts acylation of which would give a complex mixture of products because both groups are ortho, para–directing. On the other hand, the orienting effects of the two groups in *p*-methylacetophenone reinforce each other, so that its bromination is highly regioselective and in the desired direction.

(*h*) Recalling that alkyl groups attached to the benzene ring by CH_2 may be prepared by reduction of the appropriate ketone, we may reduce 3-bromo-4-methylacetophenone, as prepared in part (*g*), by the Clemmensen or Wolff-Kishner procedure to give 2-bromo-4-ethyltoluene.

3-Bromo-4-methylacetophenone   2-Bromo-4-ethyltoluene

(*i*) This is a relatively straightforward synthetic problem. Bromine is an ortho, para–directing substituent; nitro is meta-directing. Nitrate first, and then brominate to give 1-bromo-3-nitrobenzene.

Benzene   Nitrobenzene   1-Bromo-3-nitrobenzene

(*j*) Take advantage of the ortho, para–directing properties of bromine to prepare 1-bromo-2,4-dinitrobenzene. Brominate first, and then nitrate under conditions that lead to disubstitution. The nitro groups are introduced at positions ortho and para to the bromine and meta to each other.

Benzene   Bromobenzene   1-Bromo-2,4-dinitrobenzene

(*k*) While bromo and nitro substituents are readily introduced by electrophilic aromatic substitution, the only methods we have available so far to prepare carboxylic acids is by oxidation of alkyl side chains. Thus, use toluene as a starting material, planning to convert the methyl group to a carboxyl group by oxidation. Nitrate next; nitro and carboxyl are both meta-directing groups, so that the bromination in the last step occurs with the proper regioselectivity.

| Toluene | Benzoic acid | 3-Nitrobenzoic acid | 3-Bromo-5-nitrobenzoic acid |

If bromination is performed prior to nitration, the bromine substituent will direct an incoming electrophile to positions ortho and para to itself, giving the wrong orientation of substituents in the product.

(*l*) Again toluene is a suitable starting material, with its methyl group serving as the source of the carboxyl substituent. The orientation of the substituents in the final product requires that the methyl group be retained until the final step.

| Toluene | *p*-Nitrotoluene | 2-Bromo-4-nitrotoluene | 2-Bromo-4-nitrobenzoic acid |

Nitration must precede bromination, as in the previous part, in order to prevent formation of an undesired mixture of isomers.

(*m*) Friedel-Crafts alkylation of benzene with benzyl chloride (or benzyl bromide) is a satisfactory route to diphenylmethane.

| Benzene | Benzyl chloride | Diphenylmethane |

Benzyl chloride is prepared by free-radical chlorination of toluene.

| Toluene | Benzyl chloride |

Alternatively, benzene could have been subjected to Friedel-Crafts acylation with benzoyl chloride to give benzophenone. Clemmensen or Wolff-Kishner reduction of benzophenone would then furnish diphenylmethane.

(*n*) 1-Phenyloctane cannot be prepared efficiently by direct alkylation of benzene, because of the probability that rearrangement will occur. Indeed, a mixture of 1-phenyloctane and 2-phenyloctane is formed under the usual Friedel-Crafts conditions, along with 3-phenyloctane.

$$C_6H_6 + CH_3(CH_2)_6CH_2Br \xrightarrow{AlBr_3}$$

Benzene   1-Bromooctane

$$C_6H_5CH_2(CH_2)_6CH_3 + C_6H_5\underset{\underset{CH_3}{|}}{C}H(CH_2)_5CH_3 + C_6H_5\underset{\underset{CH_2CH_3}{|}}{C}H(CH_2)_4CH_3$$

1-Phenyloctane (40%)   2-Phenyloctane (30%)   3-Phenyloctane (30%)

A method which permits the synthesis of 1-phenyloctane free of isomeric compounds is acylation followed by reduction.

$$C_6H_6 + CH_3(CH_2)_6\overset{\overset{\displaystyle O}{\|}}{C}Cl \xrightarrow{\text{AlCl}_3} C_6H_5\overset{\overset{\displaystyle O}{\|}}{C}(CH_2)_6CH_3$$

Benzene   Octanoyl chloride   1-Phenyl-1-octanone

$$\downarrow \begin{smallmatrix} \text{Zn(Hg)} \\ \text{HCl} \end{smallmatrix}$$

$$C_6H_5CH_2(CH_2)_6CH_3$$

1-Phenyloctane

Alternatively, Wolff-Kishner conditions (hydrazine, potassium hydroxide, diethylene glycol) could be used in the reduction step.

(*o*) Direct alkenylation of benzene under Friedel-Crafts reaction conditions does not take place, and so 1-phenyl-1-octene cannot be prepared by the reaction

$$C_6H_6 + ClCH{=}CH(CH_2)_5CH_3 \xrightarrow{\text{AlCl}_3} C_6H_5CH{=}CH(CH_2)_5CH_3$$

Benzene   1-Chloro-1-octene   1-Phenyl-1-octene

No! Reaction effective only with *alkyl* halides, not 1-haloalkenes.

However, having already prepared 1-phenyloctane in part (*n*), we can functionalize the benzylic position by bromination and then carry out a dehydrohalogenation to obtain the target compound.

$$C_6H_5CH_2(CH_2)_6CH_3 \xrightarrow[\text{or NBS}]{\text{Br}_2,\ \text{light}} \underset{\overset{\displaystyle |}{\displaystyle Br}}{C_6H_5CH}(CH_2)_6CH_3 \xrightarrow[\text{CH}_3\text{OH}]{\text{KOCH}_3} C_6H_5CH{=}CH(CH_2)_5CH_3$$

1-Phenyloctane   1-Phenyl-1-bromooctane   1-Phenyl-1-octene

(*p*) 1-Phenyl-1-octyne cannot be prepared in one step from benzene; 1-haloalkynes are not suitable reactants for a Friedel-Crafts process. In Chapter 9, however, we learned that alkynes may be prepared from the corresponding alkene:

$$RC{\equiv}CR \quad \begin{smallmatrix}\text{obtained}\\\text{from}\end{smallmatrix} \quad \underset{\overset{\displaystyle |}{\displaystyle Br}}{RCH}{-}\underset{\overset{\displaystyle |}{\displaystyle Br}}{CHR} \quad \begin{smallmatrix}\text{obtained}\\\text{from}\end{smallmatrix} \quad RCH{=}CHR$$

Using the alkene prepared in part (*o*),

$$C_6H_5CH{=}CH(CH_2)_5CH_3 \xrightarrow{\text{Br}_2} \underset{\overset{\displaystyle |\ \ |}{\displaystyle Br\ Br}}{C_6H_5CHCH}(CH_2)_5CH_3 \xrightarrow[\text{NH}_3]{\text{NaNH}_2}$$

1-Phenyl-1-octene   1,2-Dibromo-1-phenyloctane

$$C_6H_5C{\equiv}C(CH_2)_5CH_3$$

1-Phenyl-1-octyne

(*q*) Nonconjugated cyclohexadienes are prepared by Birch reduction of arenes. Thus the last step in the synthesis of 1,4-di-*tert*-butyl-1,4-cyclohexadiene is the Birch reduction of 1,4-di-*tert*-butylbenzene.

Benzene   *tert*-Butylbenzene   *p*-Di-*tert*-butylbenzene   1,4-Di-*tert*-butyl-1,4-cyclohexadiene

12.26 (*a*) Methoxy is an ortho, para–directing substituent. All that is required to prepare *p*-methoxybenzenesulfonic acid is to sulfonate anisole.

Anisole   *p*-Methoxybenzenesulfonic acid

(*b*) In reactions involving disubstitution of anisole, the better strategy is to introduce the para substituent first. The methoxy group is ortho, para–directing, but para substitution is preferred.

Anisole   *p*-Nitroanisole   2-Bromo-4-nitroanisole

(*c*) Reversing the order of the steps yields 4-bromo-2-nitroanisole.

Anisole   *p*-Bromoanisole   4-Bromo-2-nitroanisole

(*d*) Direct introduction of a vinyl substituent onto an aromatic ring is not a feasible reaction. *p*-Methoxystyrene must be prepared in an indirect way by adding an ethyl side chain and then taking advantage of the reactivity of the benzylic position by bromination (e.g., with *N*-bromosuccinimide) and dehydrohalogenation.

Anisole   *p*-Ethylanisole   1-(*p*-Methoxyphenyl)ethyl bromide   *p*-Methoxystyrene

12.27 (*a*) Methyl is an ortho, para–directing substituent, and toluene yields mainly *o*-nitrotoluene and *p*-nitrotoluene on mononitration. Some *m*-nitrotoluene is also formed.

Toluene   *o*-Nitrotoluene   *m*-Nitrotoluene   *p*-Nitrotoluene

(*b*) There are six isomeric dinitrotoluenes:

2,3-Dinitrotoluene   2,4-Dinitrotoluene   2,5-Dinitrotoluene   2,6-Dinitrotoluene

3,5-Dinitrotoluene   3,4-Dinitrotoluene

The least likely product is 3,5-dinitrotoluene, since neither of its nitro groups is ortho or para to the methyl group.

(*c*) There are six trinitrotoluene isomers:

2,4,6-Trinitrotoluene   2,3,4-Trinitrotoluene   2,3,5-Trinitrotoluene

2,3,6-Trinitrotoluene   3,4,5-Trinitrotoluene   2,4,5-Trinitrotoluene

The most likely major product is 2,4,6-trinitrotoluene, since all the positions activated by the methyl group are substituted. This is, in fact, the compound commonly known as TNT.

12.28 From *o*-xylene:

o-Xylene   Acetyl chloride   3,4-Dimethylacetophenone
                                          (94%)

From *m*-xylene:

m-Xylene   2,4-Dimethylacetophenone
                                          (86%)

From *p*-xylene:

p-Xylene   2,5-Dimethylacetophenone
                                          (99%)

12.29 The ring that bears the nitrogen in benzanilide is activated toward electrophilic aromatic substitution. The ring that bears the C=O is strongly deactivated.

Benzanilide   N-(*o*-Chlorophenyl)benzamide   N-(*p*-Chlorophenyl)benzamide

12.30 (*a*) Nitration of the ring takes place para to the ortho, para–directing chlorine substituent; this position is also meta to the meta-directing carboxyl groups.

2-Chloro-1,3-benzenedicarboxylic   2-Chloro-5-nitro-1,3-benzenedicarboxylic
              acid   acid (86%)

(*b*) Bromination of the ring occurs at the only available position activated by the amino group, a powerful activating substituent and an ortho, para director. This position is meta to the meta-directing trifluoromethyl group and to the meta-directing nitro group.

4-Nitro-2-(trifluoromethyl)aniline   2-Bromo-4-nitro-6-(trifluoromethyl)aniline
(81%)

(*c*) This may be approached as a problem in which there are two aromatic rings. One of them bears two activating substituents and so is more reactive than the other, which bears only one activating substituent. Of the two activating substituents (—OH and C_6H_5—), the hydroxyl substituent is the more powerful and controls the regioselectivity of substitution.

p-Phenylphenol   2-Bromo-4-phenylphenol

(*d*) Both substituents are activating, nitration occurring readily even in the absence of sulfuric acid; both are ortho, para–directing and comparable in activating power. Therefore, the position at which substitution takes place is as follows:

not here; too hindered   not here; too hindered

ortho to isopropyl,
para to *tert*-butyl

1-*tert*-Butyl-3-isopropylbenzene   4-*tert*-Butyl-2-isopropyl-1-nitrobenzene
(78%)

(*e*) Protonation of 1-octene yields a secondary carbocation, which attacks benzene.

Benzene   1-Octene   2-Phenyloctane (84%)

(*f*) The reaction that occurs with arenes and acid anhydrides in the presence of aluminum chloride is Friedel-Crafts acylation. The methoxy group is the more powerful activating substituent, and so acylation occurs para to it.

o-Fluoroanisole   Acetic anhydride   3-Fluoro-4-methoxyacetophenone
                                                                 (70–80%)

(g)   The isopropyl group is ortho, para–directing and the nitro group is meta-directing. In this case their orientation effects reinforce each other. Electrophilic aromatic substitution takes place ortho to isopropyl and meta to nitro.

p-Nitroisopropylbenzene   2,4-Dinitroisopropylbenzene (96%)

(h)   In the presence of an acid catalyst (H_2SO_4), 2-methylpropene is converted to *tert*-butyl cation, which then attacks the aromatic ring ortho to the strongly activating methoxy group.

$$(CH_3)_2C{=}CH_2 + H^+ \longrightarrow (CH_3)_3C^+$$

In this particular example, 2-*tert*-butyl-4-methylanisole was isolated in 98 percent yield.

(i)   There are two things to consider in this problem: (1) In which ring does bromination occur, and (2) what is the orientation of substitution in that ring? All the substituents are activating groups, and so substitution will take place in the ring which bears the greater number of substituents. Orientation is governed by the most powerful activating substituent, the hydroxyl group. Both positions ortho to the hydroxyl group are already substituted, so that bromination takes place para to it. The product shown was isolated from the bromination reaction in 100 percent yield.

3-Benzyl-2,6-dimethylphenol   3-Benzyl-4-bromo-2,6-dimethylphenol
                                                                (100%)

(j)   Wolff-Kishner reduction converts benzophenone to diphenylmethane.

Benzophenone   Diphenylmethane (83%)

(k) Fluorine is an ortho, para–directing substituent. It undergoes Friedel-Crafts alkylation on being treated with benzyl chloride and aluminum chloride to give a mixture of o-fluorodiphenylmethane and p-fluorodiphenylmethane.

Fluorobenzene Benzyl chloride   o-Fluorodiphenylmethane   p-Fluorodiphenylmethane
                                              (15%)   (85%)

(l) The —NHCCH$_3$ substituent is a more powerful activator than the ethyl group. It directs Friedel-Crafts acylation primarily to the position para to itself.

o-Ethylacetanilide   Acetyl chloride   4-Acetamido-3-ethylacetophenone
                                                    (57%)

(m) Clemmensen reduction converts the carbonyl group at a CH$_2$ unit.

2,4,6-Trimethylacetophenone   2-Ethyl-1,3,5-trimethylbenzene
                                            (74%)

(n) Bromination occurs at C-5 on thiophene-3-carboxylic acid. Reaction does not occur at C-2, since substitution at this position would place a carbocation adjacent to the electron-withdrawing carboxyl group.

Thiophene-3-carboxylic acid   5-Bromo-thiophene-3-carboxylic
                                        acid (69%)

12.31 In a Friedel-Crafts acylation reaction an acyl chloride or acid anhydride reacts with an arene to yield an aryl ketone.

$$ArH + RCCl \xrightarrow{AlCl_3} ArCR$$

The ketone carbonyl is bonded directly to the ring. Therefore, in each of these problems you should identify the bond between the aromatic ring and the carbonyl group and realize that it arises as shown in this general reaction.

(a) The compound is derived from benzene and $C_6H_5CH_2CCl$. The observed yield in this reaction is 82 percent.

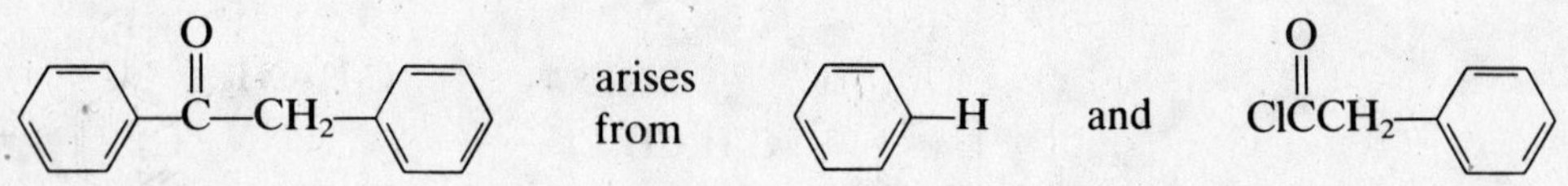

arises from

(b) The presence of the $ArCCH_2CH_2CO_2H$ unit suggests an acylation reaction using succinic anhydride.

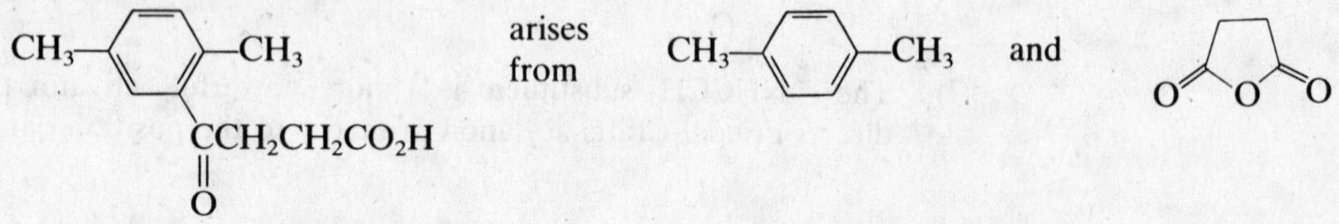

arises from and

In practice, this reaction has been carried out in 55 percent yield.

(c) Two methods seem possible here but only one actually works. The only effective combination is:

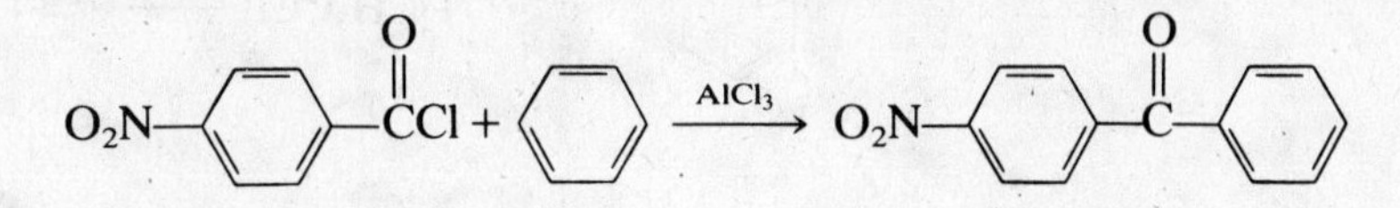

p-Nitrobenzoyl chloride　　　Benzene　　　p-Nitrobenzophenone (87%)

The alternative combination

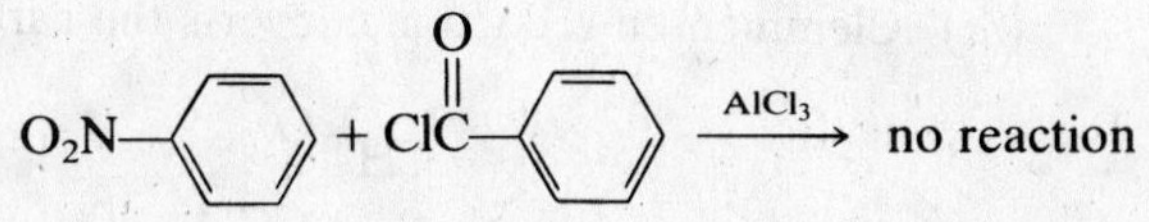

fails because it requires a Friedel-Crafts reaction on a strongly deactivated aromatic ring (nitrobenzene).

(d) Here also two methods seem possible but only one is successful in practice. The valid synthesis is:

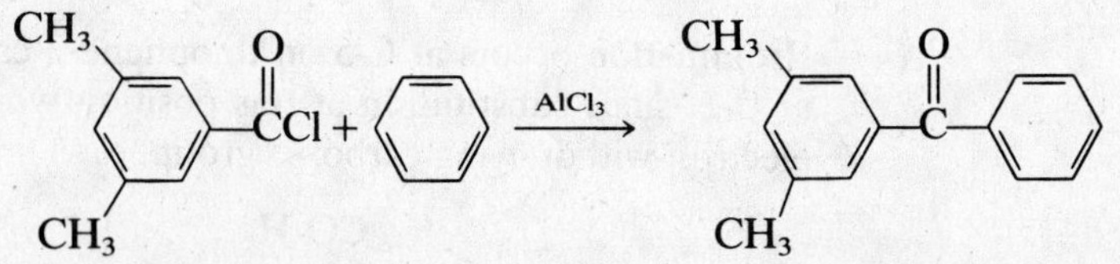

3,5-Dimethylbenzoyl chloride　　　Benzene　　　3,5-Dimethylbenzophenone (89%)

The alternative combination will not give 3,5-dimethylbenzophenone, because of the ortho, para–directing properties of the methyl substituents in m-xylene. The product will be 2,4-dimethylbenzophenone

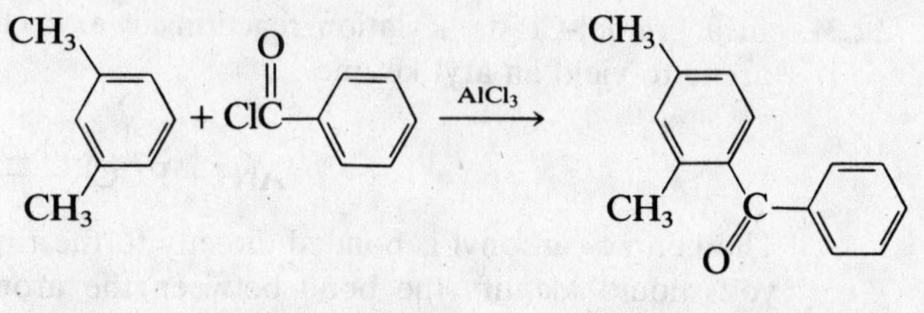

m-Xylene　　　Benzoyl chloride　　　2,4-Dimethylbenzophenone

(e) The combination that follows is not effective, because it involves a Friedel-Crafts reaction on a deactivated aromatic ring.

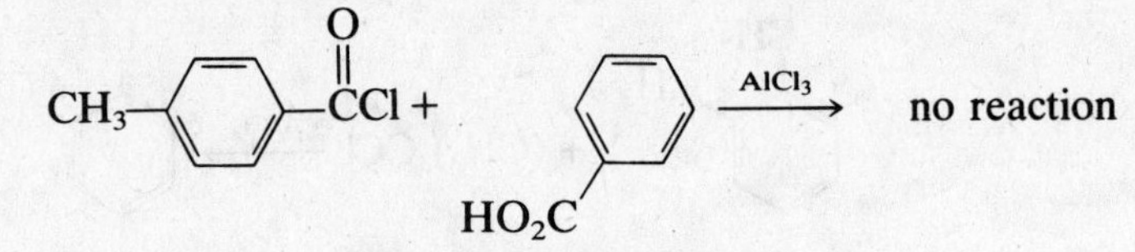

p-Methylbenzoyl chloride   Benzoic acid

Therefore, the following combination, utilizing toluene, seems appropriate:

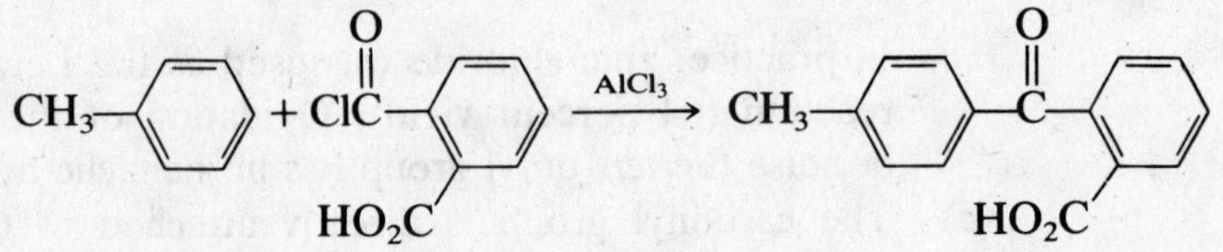

The actual sequence used phthalic anhydride in a reaction analogous to that seen in part (b) which employs succinic anhydride.

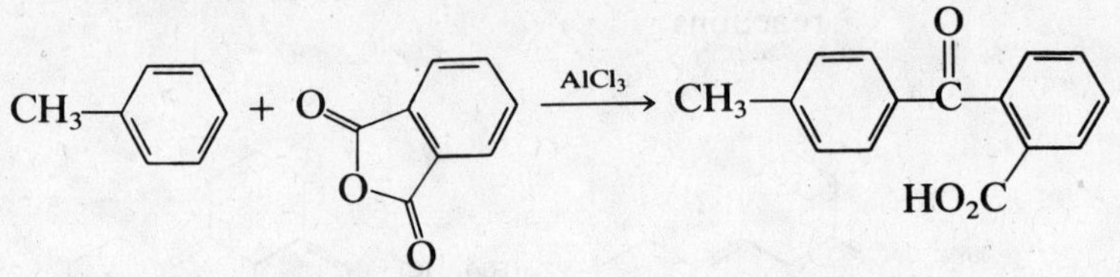

Toluene   Phthalic anhydride   *o*-(4-Methylbenzoyl)benzoic acid (96%)

12.32 (a) The problem to be confronted here is that two meta-directing groups are para to each other in the product. However, by recognizing that the carboxylic acid function can be prepared by oxidation of the isopropyl group

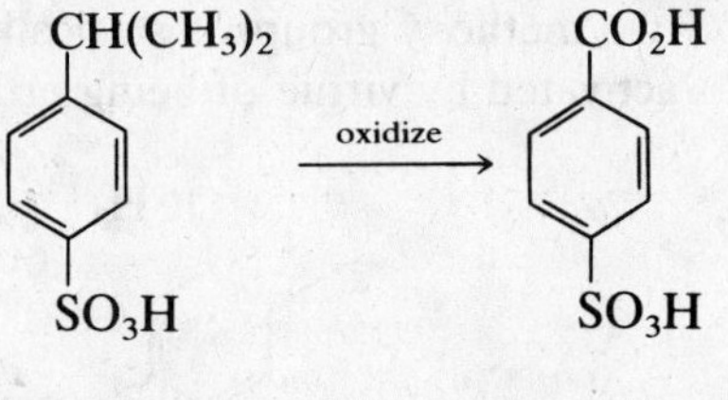

we have a reasonable last step in the synthesis. Happily, the key intermediate has its sulfonic acid group para to the ortho, para–directing isopropyl group, which suggests the following approach:

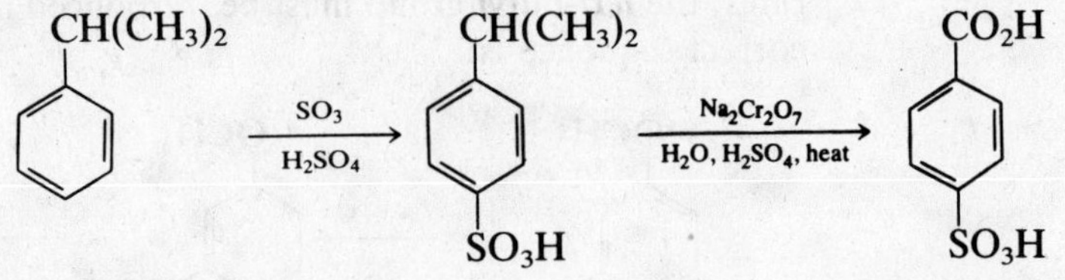

Isopropylbenzene   *p*-Isopropylbenzenesulfonic acid   *p*-Carboxylbenzenesulfonic acid

(b) In this problem two methyl groups must be oxidized to carboxylic acid functions and a *tert*-butyl group must be introduced, most likely by a Friedel-Crafts reaction. Since Friedel-Crafts alkylations cannot be performed on deactivated aromatic rings, oxidation must *follow*, not precede, alkylation. Therefore, the following reaction sequence seems appropriate:

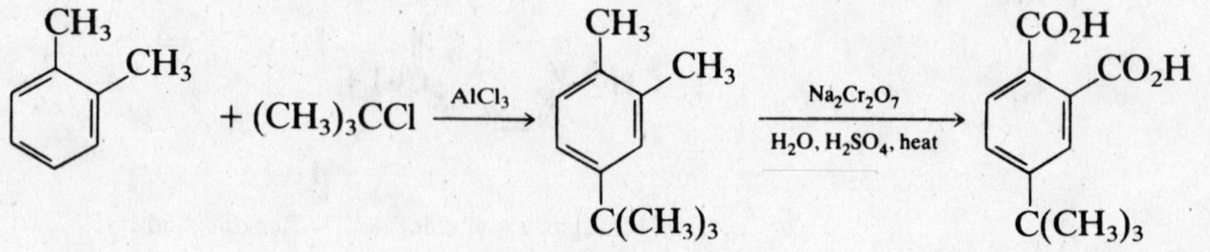

| *o*-Xylene | 4-*tert*-Butyl-1,2-dimethylbenzene | 4-*tert*-Butylbenzene-1,2-dicarboxylic acid |

In practice, zinc chloride was used as the Lewis acid to catalyze the Friedel-Crafts reaction (64 percent yield). Oxidation of the methyl groups occurs preferentially because the *tert*-butyl group has no benzylic hydrogens.

(*c*)   The carbonyl group is directly attached to the naphthalene unit in the starting material. Reduce it in the first step so that a Friedel-Crafts acylation can be accomplished on the napththalene ring. An aromatic ring that bears a strongly electron-withdrawing group such as C=O does not undergo Friedel-Crafts reactions.

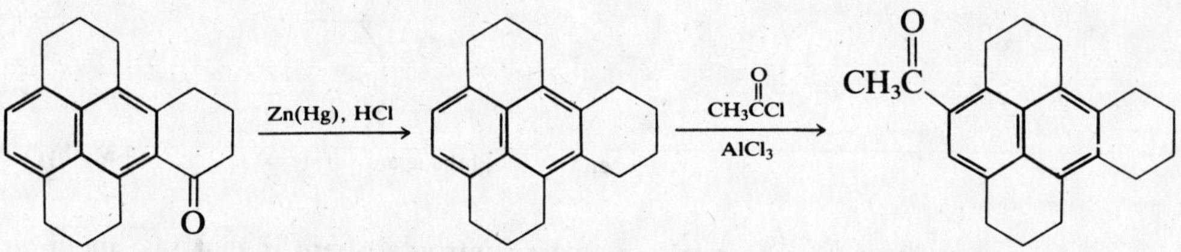

(*d*)   *m*-Dimethoxybenzene is a strongly activated aromatic compound and so will undergo electrophilic aromatic substitution readily. The ring position between the two methoxy groups is sterically hindered and less reactive than the other positions activated by virtue of being ortho or para to a methoxy group.

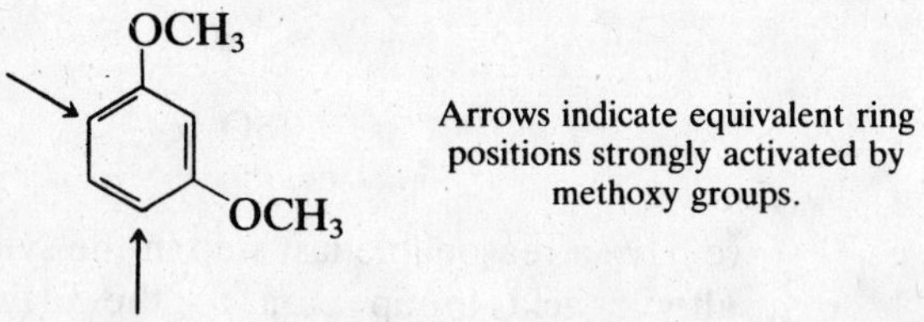

Arrows indicate equivalent ring positions strongly activated by methoxy groups.

Because Friedel-Crafts reactions may not be performed on deactivated aromatic rings, the *tert*-butyl group must be introduced before the nitro group. Therefore, the correct sequence is:

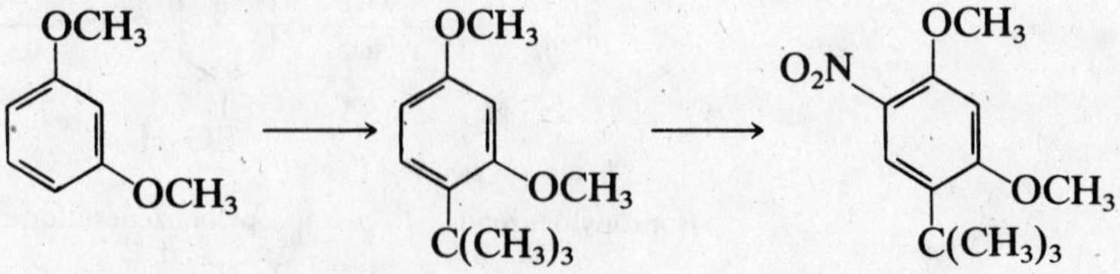

This is essentially the procedure actually followed. Alkylation was effected, however, not with *tert*-butyl chloride and aluminum chloride but with

2-methylpropene and phosphoric acid:

m-Dimethoxybenzene   2-Methylpropene   1-*tert*-Butyl-2,4-dimethoxybenzene
(75%)

Nitration was carried out in the usual way. The orientation of nitration is controlled by the more powerfully activating methoxy group rather than by the weakly activating *tert*-butyl.

1-*tert*-Butyl-2,4-dimethoxy-5-nitrobenzene

12.33 The first step is a Friedel-Crafts acylation reaction. The use of a cyclic anhydride introduces both the acyl and carboxyl groups into the molecule.

Benzene Succinic anhydride   4-Oxo-4-phenylbutanoic acid

The second step is a reduction of the ketone carbonyl to a methylene group. A Clemmensen reduction is normally used for this step.

4-Oxo-4-phenylbutanoic acid   4-Phenylbutanoic acid

The cyclization phase of the process is an intramolecular Friedel-Crafts acylation reaction. It requires conversion of the carboxylic acid to the acyl chloride—thionyl chloride is a suitable reagent—followed by treatment with aluminum chloride.

4-Phenylbutanoic acid   4-Phenylbutanoyl
chloride

12.34 Intramolecular Friedel-Crafts acylation reactions which produce five-membered or six-membered rings occur readily. Cyclization must take place at the position ortho to the reacting side chain.

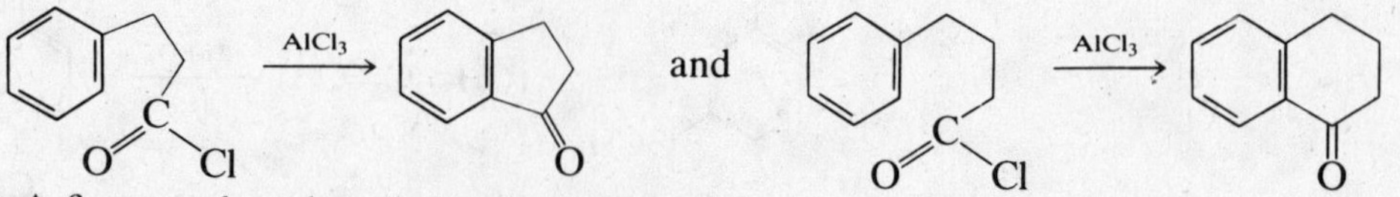

(a) A five-membered cyclic ketone is formed here:

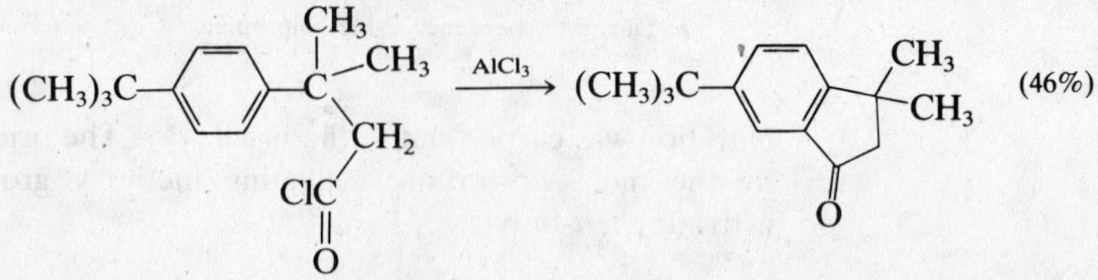

(46%)

(b) This intramolecular Friedel-Crafts acylation takes place to form a six-membered cyclic ketone in excellent yield.

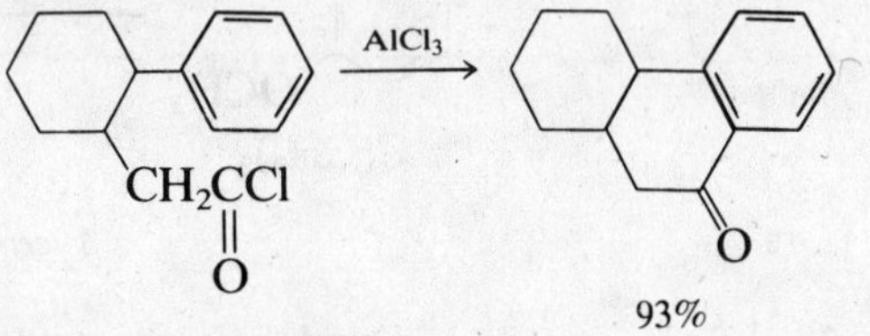

93%

(c) In this case two aromatic rings are available for attack in the acylation reaction. The more reactive ring is the one which bears the two activating methoxy groups, and cyclization occurs on it.

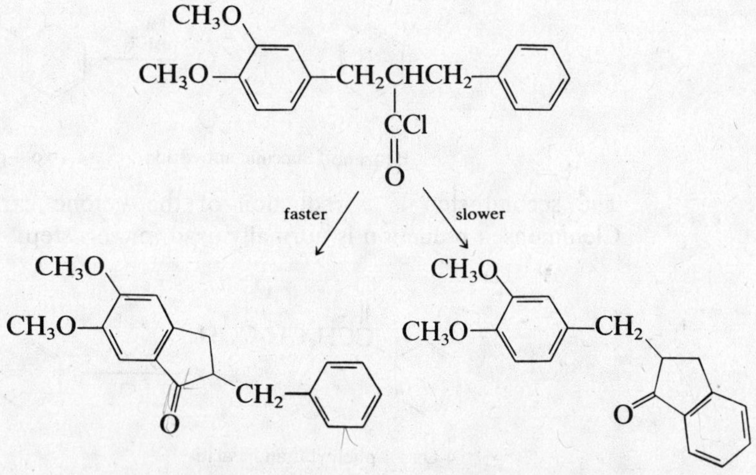

faster   slower

Only product (78% yield)   (not observed)

12.35 (a) To determine the total rate of chlorination of biphenyl relative to that of benzene, we add up the *partial rate factors* for all the positions in each substrate and compare them.

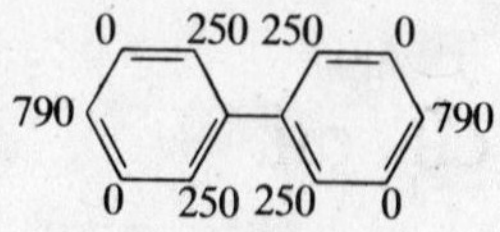

Biphenyl
(sum = 2580)

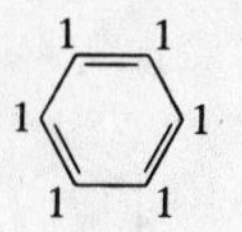

Benzene
(sum = 6)

Relative rate of chlorination: $\dfrac{\text{Biphenyl}}{\text{Benzene}} = \dfrac{2580}{6} = \dfrac{430}{1}$

(*b*) The relative rate of attack at the para compared with the ortho positions is given by the ratio of their partial rate factors.

$$\frac{\text{Para}}{\text{Ortho}} = \frac{1580}{1000}$$

Therefore, 15.8 g of *p*-chlorobiphenyl is formed for every 10 g of *o*-chlorobiphenyl.

12.36 The problem stipulates that the reactivity of various positions in *o*-bromotoluene can be estimated by multiplying the partial rate factors for the corresponding positions in toluene and bromobenzene. Therefore, given the partial rate factors:

the two are multiplied together to give the combined effects of the two substituents at the various ring positions.

The most reactive position is the one which is para to bromine. Therefore, the predicted product is 4-bromo-3-methylacetophenone. Indeed, this is what is observed experimentally.

o-Bromotoluene   Acetyl chloride   4-Bromo-3-methylacetophenone

This was first considered to be "anomalous" behavior on the part of *o*-bromotoluene, but, as can be seen, it is consistent with the individual directing properties of the two substituents.

12.37 The isomerization is triggered by protonation of the aromatic ring, an electrophilic attack by HCl catalyzed by AlCl₃.

2-Isopropyl-1,3,5-trimethylbenzene

The carbocation then rearranges by a methyl shift, and the rearranged cyclohexadienyl cation loses a proton to form the isomeric product

1-Isopropyl-2,4,5-trimethylbenzene

The driving force for rearrangement is relief of the adverse steric interaction between the isopropyl group and one of its adjacent methyl groups. Isomerization is catalytic with respect to H^+, since protonation generates the necessary carbocation intermediate and rearomatization occurs by loss of a proton.

12.38   The relation of compound A to the starting material is

$$C_6H_5(CH_2)_5CCl \longrightarrow A + HCl$$

$$(C_{12}H_{15}ClO) \qquad (C_{12}H_{14}O)$$

The starting acyl chloride has lost the elements of HCl in the formation of A. Because A forms benzene-1,2-dicarboxylic acid on oxidation, it must have two carbon substituents ortho to each other.

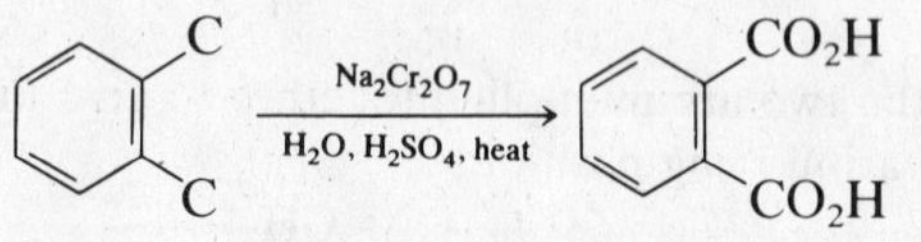

These facts suggest the following process:

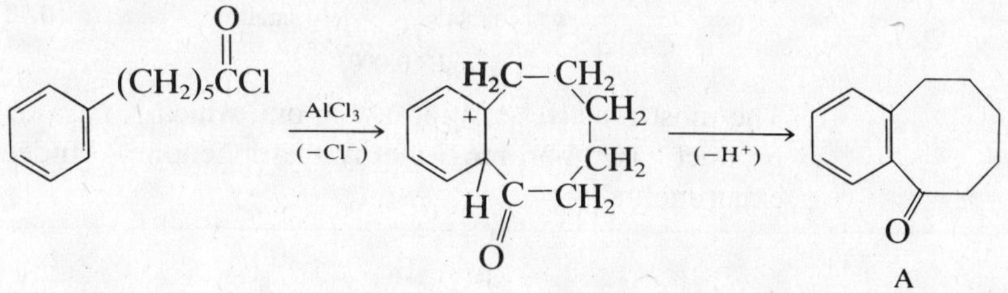

The reaction leading to compound A is an intramolecular Friedel-Crafts acylation. Since cyclization to form an eight-membered ring is difficult, it must be carried out in dilute solution to minimize competition with intermolecular acylation.

12.39   While hexamethylbenzene has no positions available at which ordinary electrophilic aromatic *substitution* might occur, electrophilic *attack* on the ring can still take place to form a cyclohexadienyl cation.

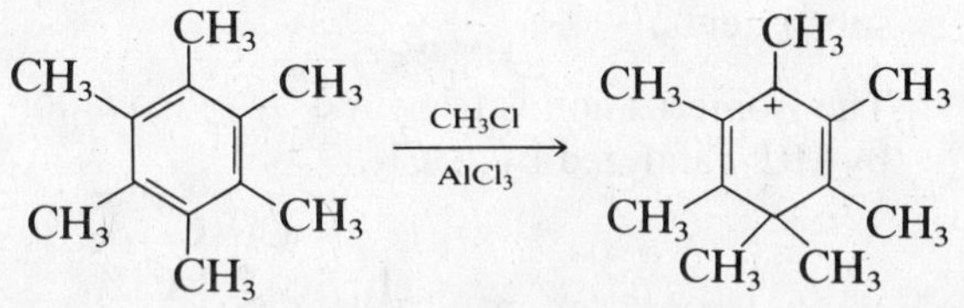

Compound B is the tetrachloroaluminate ($AlCl_4^-$) salt of the carbocation shown. It undergoes deprotonation on being treated with aqueous sodium bicarbonate.

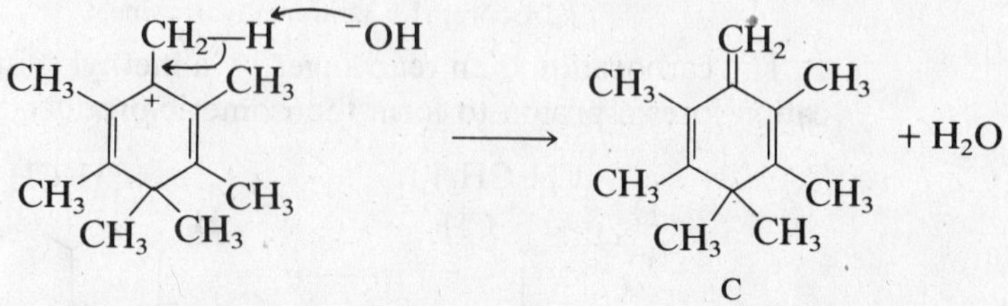

12.40   By examining the structure of the target molecule, compound F, we see that the bond indicated in the following joins two fragments which are related to the given starting materials D and E:

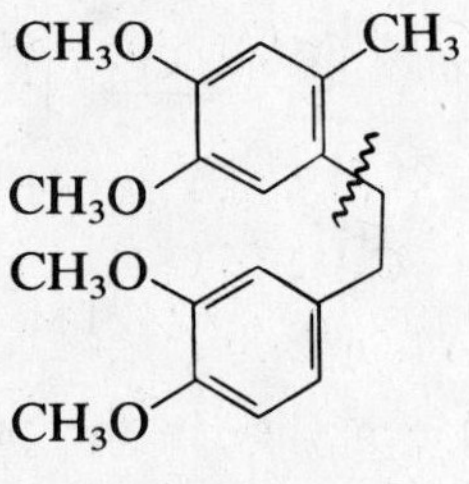

Compound F

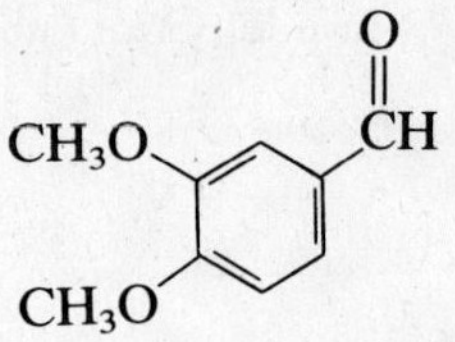

The bond connecting the two fragments can be made by a Friedel-Crafts acylation-reduction sequence using the acyl chloride E.

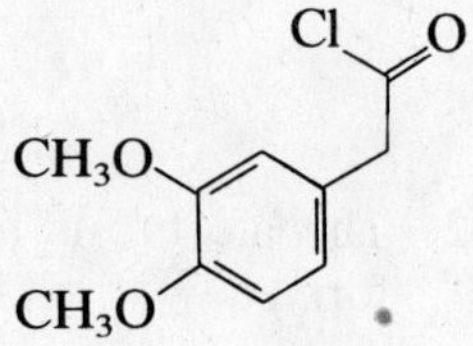

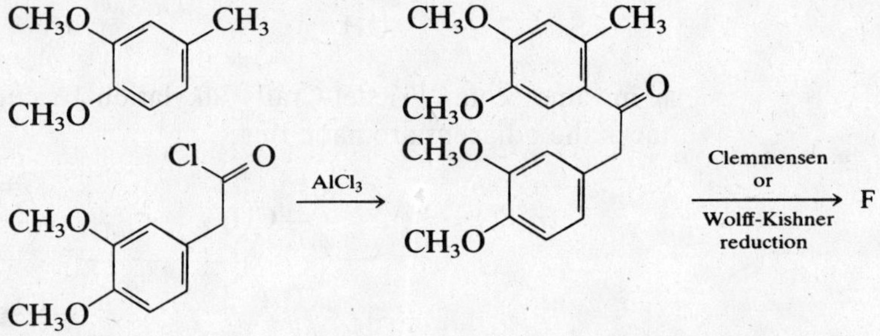

The orientation is right; attack is para to one of the methoxy groups and ortho to the methyl. The substrate for the Friedel-Crafts acylation reaction, 3,4-dimethoxytoluene, is prepared from compound D by a Clemmensen or Wolff-Kishner reduction. Compound D cannot be acylated directly, because it bears a strongly deactivating $-\overset{\underset{\|}{O}}{C}H$ substituent.

12.41 In the presence of aqueous sulfuric acid, the side-chain double bond of styrene undergoes protonation to form a benzylic carbocation:

$$C_6H_5CH{=}CH_2 + H^+ \longrightarrow C_6H_5\overset{+}{C}HCH_3$$

Styrene   1-Phenylethyl cation

This carbocation then reacts with a molecule of styrene in the manner we have seen earlier (Chapter 6) for alkene dimerization:

$$C_6H_5\overset{+}{C}HCH_3 + C_6H_5CH{=}CH_2 \longrightarrow C_6H_5\overset{+}{C}HCH_2\underset{\underset{CH_3}{|}}{C}HC_6H_5$$

The carbocation produced in this step can lose a proton to form 1,3-diphenyl-1-butene

$$C_6H_5\overset{+}{C}HCH_2\underset{\underset{CH_3}{|}}{C}HC_6H_5 \longrightarrow C_6H_5CH{=}CH\underset{\underset{CH_3}{|}}{C}HC_6H_5 + H^+$$

1,3-Diphenyl-1-butene

or it can undergo a cyclization reaction in what amounts to an intramolecular Friedel-Crafts alkylation

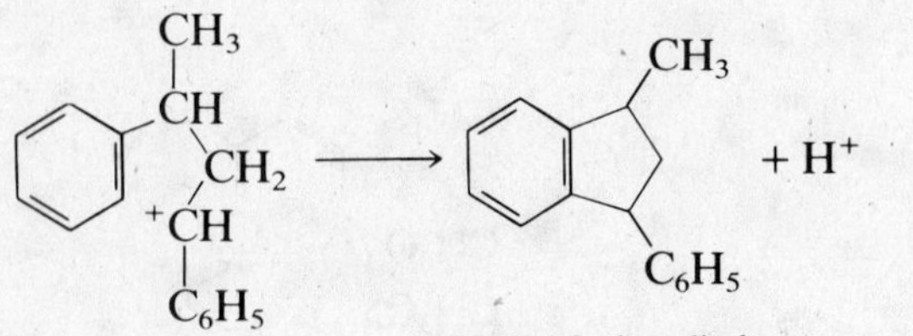

1-Methyl-3-phenylindan

12.42 The alcohol is tertiary and benzylic. In the presence of sulfuric acid a carbocation is formed.

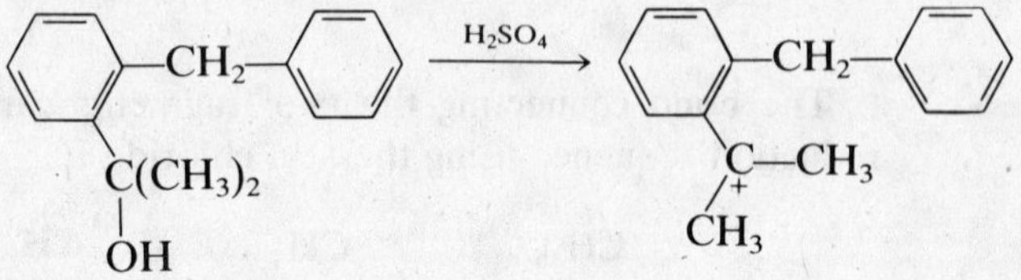

An intramolecular Friedel-Crafts alkylation reaction follows, in which the carbocation attacks the adjacent aromatic ring.

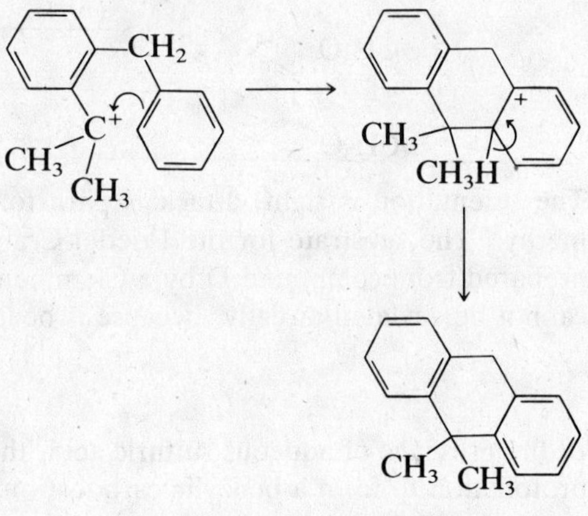

$C_{16}H_{16}$

SELF-TEST

PART A

A-1. Write the three most stable resonance contributors for the cyclohexadienyl cation found in the ortho bromination of toluene.

A-2. Give the major product(s) for each of the following reactions. Indicate whether the reaction proceeds faster or slower than the corresponding reaction of benzene.

(a)

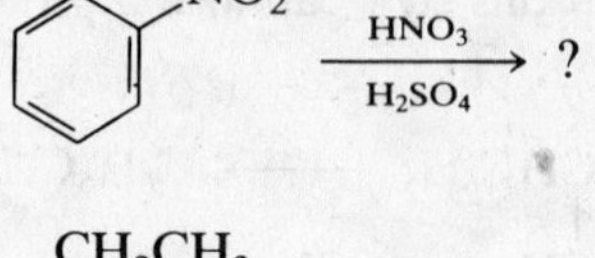

(b)

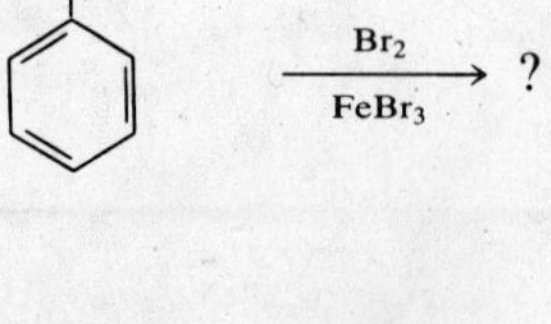

(c)

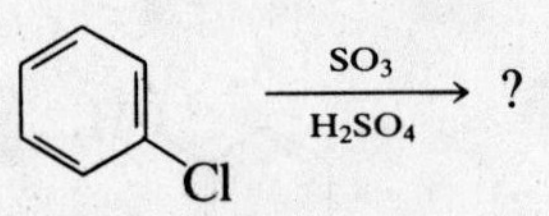

$$\text{C}_6\text{H}_5\text{Cl} \xrightarrow[\text{H}_2\text{SO}_4]{\text{SO}_3} \quad ?$$

A-3. Write the formula of the electrophilic species present in each reaction of the preceding problem.

A-4. Provide the reactant, reagent, or product omitted from each of the following:

(a)

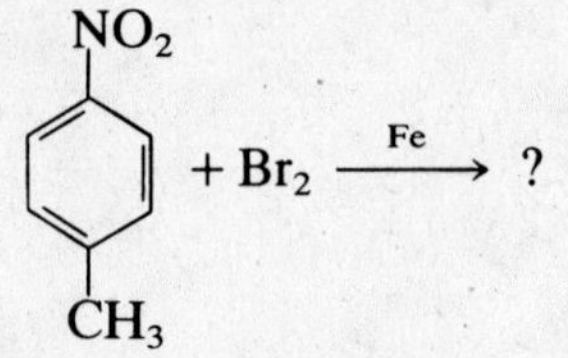

4-nitrotoluene $+ \text{Br}_2 \xrightarrow{\text{Fe}} \quad ?$

(b)

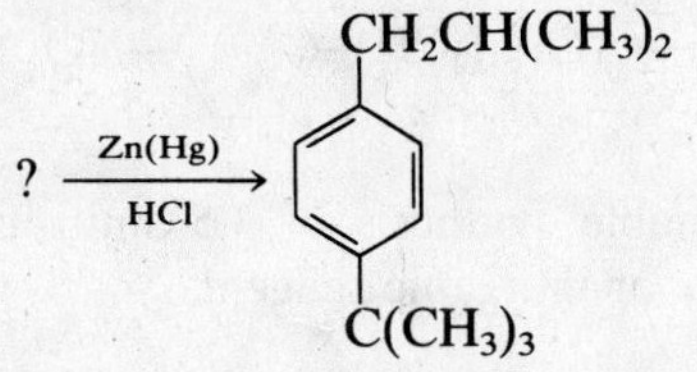

$? \xrightarrow[\text{HCl}]{\text{Zn(Hg)}}$ 1-CH$_2$CH(CH$_3$)$_2$-4-C(CH$_3$)$_3$-benzene

(c)

$\text{C}_6\text{H}_5\text{CH}_2\text{CH}_2\text{CH}_2\text{CHClCH}_3 \xrightarrow{\text{AlCl}_3} \quad ?$

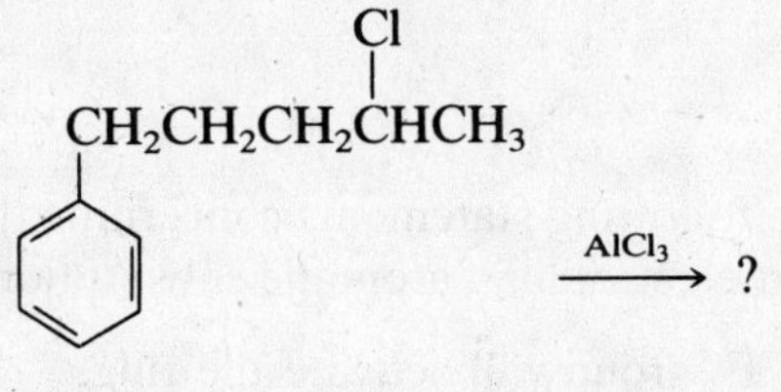

(d)

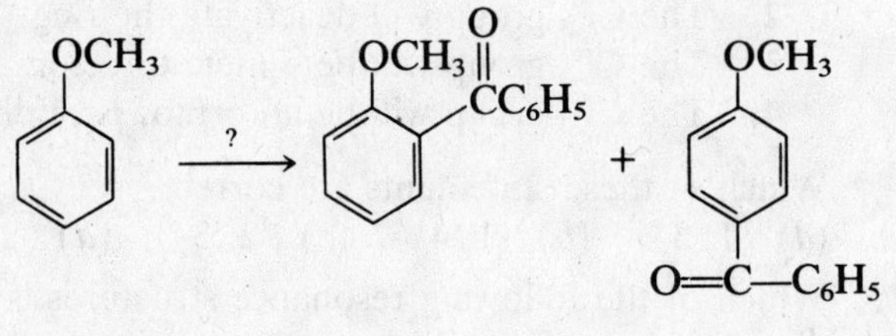

anisole (OCH$_3$) $\longrightarrow$ 2-OCH$_3$-C$_6$H$_4$-CC$_6$H$_5$ (O) $+$ 4-OCH$_3$-C$_6$H$_4$-O=C—C$_6$H$_5$

$? $

(e)

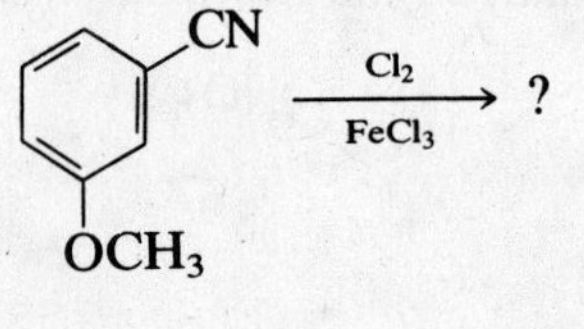

3-OCH$_3$-benzonitrile (CN) $\xrightarrow[\text{FeCl}_3]{\text{Cl}_2} \quad ?$

(f)

pyridine $\xrightarrow[\text{FeCl}_3]{\text{Cl}_2} \quad ?$

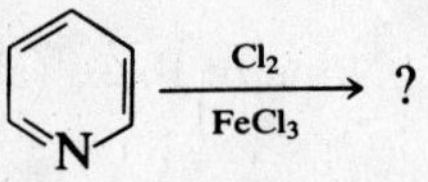

A-5. Provide the necessary reagents for each of the following transformations. More than one step may be necessary.

(a)

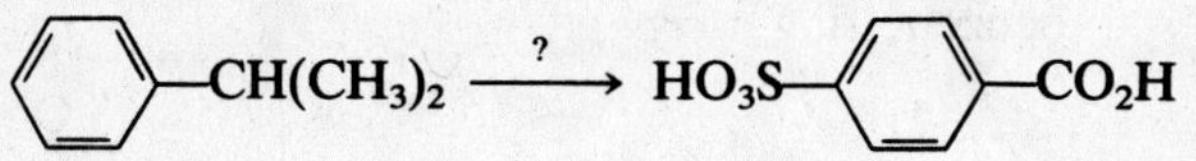

$\text{C}_6\text{H}_5\text{—CH(CH}_3)_2 \xrightarrow{?} \text{HO}_3\text{S—C}_6\text{H}_4\text{—CO}_2\text{H}$

(b)

benzene $\xrightarrow{?}$ 3-Cl-C$_6$H$_4$—CH$_2$CH$_2$C$_6$H$_5$

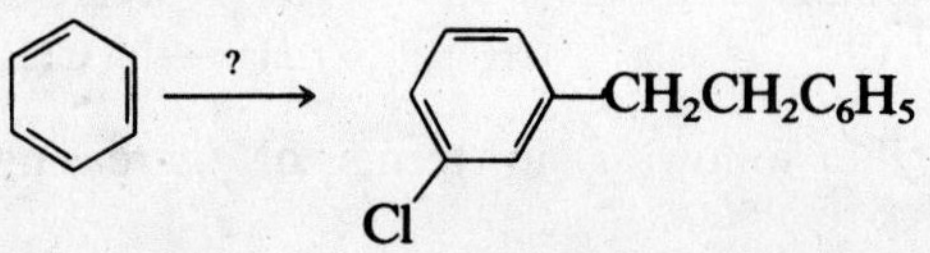

(c)

[structure: benzaldehyde → 4-methylacetophenone]

(d)

[structure: benzene → bromo-substituted cyclohexadiene with SO₃H]

(e)

[structure: benzene → (CH₃)₂CH—C₆H₄—NO₂]

A-6. Outline a reasonable synthesis of 3,5-dinitrobenzoic acid from toluene and any necessary organic or inorganic reagents.

PART B

B-1. Consider the following statements concerning the effect of a trifluoromethyl group, —CF₃, on an electrophilic aromatic substitution.

> **1.** The CF₃ group will activate the ring.
> **2.** The CF₃ group will deactivate the ring.
> **3.** The CF₃ group will be a meta director.
> **4.** The CF₃ group will be an ortho, para director.

Which of these statements are correct?
(a) 1, 3   (b) 1, 4   (c) 2, 3   (d) 2, 4

B-2. Which of the following resonance structures is *not* a contributor to the cyclohexadienyl cation intermediate in the nitration of benzene?

(a)

(c) H   NO₂

(b) H   NO₂

(d) None of these (all are contributors)

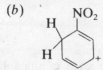

B-3. All the following groups are activating ortho, para directors when attached to a benzene ring *except*

(a) —OCH₃   (b) —NHCCH₃ (with O above the C)

(c) —Cl   (d) —N(CH₃)₂

B-4. Rank the following in terms of increasing reactivity toward nitration with

HNO$_3$, H$_2$SO$_4$ (least → most):

A	B	C

(a)   A < B < C   (c)   C < A < B
(b)   B < A < C   (d)   C < B < A

B-5.   For the reaction

? ⟶

the best reactants are:
(a)   C$_6$H$_5$Br + HNO$_3$, H$_2$SO$_4$ (c)   C$_6$H$_5$Br + H$_2$SO$_4$, heat
(b)   C$_6$H$_5$NO$_2$ + Br$_2$, FeBr$_3$ (d)   C$_6$H$_5$NO$_2$ + HBr

B-6.   For the reaction

? ⟶

the best reactants are

(a)   C$_6$H$_5$Cl + C$_6$H$_5$CCl, AlCl$_3$ (c)   C$_6$H$_5$CH$_2$C$_6$H$_5$ followed by
                                                        oxidation with chromic acid.

(b)   C$_6$H$_5$CC$_6$H$_5$ + Cl$_2$, FeCl$_3$ (d)   None of these yields the
                                                        desired product.

B-7.   The reaction

CH$_3$—⟨ ⟩—⟨ ⟩—Cl $\xrightarrow{\text{HNO}_3/\text{H}_2\text{SO}_4}$

gives as the major product:

(a)   CH$_3$—⟨ ⟩—⟨ ⟩—Cl   (c)   CH$_3$—⟨ ⟩—⟨ ⟩—Cl
                        NO$_2$ NO$_2$

(b)   CH$_3$—⟨ ⟩—⟨ ⟩—Cl   (d)   CH$_3$—⟨ ⟩—⟨ ⟩—Cl
              NO$_2$ NO$_2$

B-8.   Which one of the following compounds undergoes bromination of its aromatic ring
(electrophilic aromatic substitution) at the *fastest* rate?

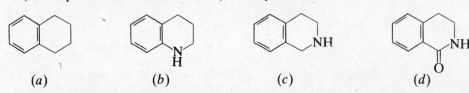

(a)   (b)   (c)   (d)

B-9.   Which one of the following is the *most stable*?

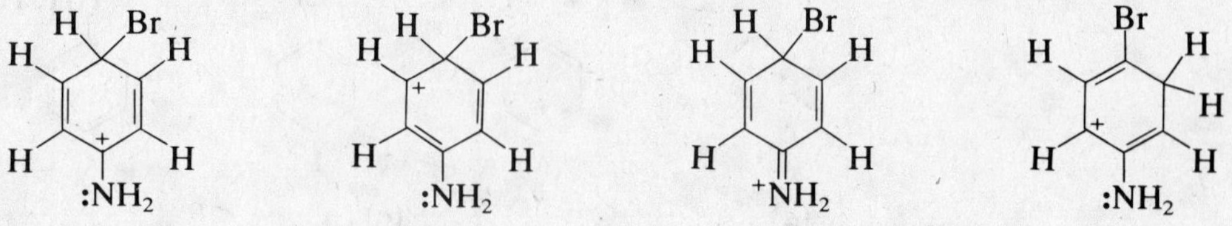

(a)   (b)   (c)   (d)

B-10.   The major product of the reaction

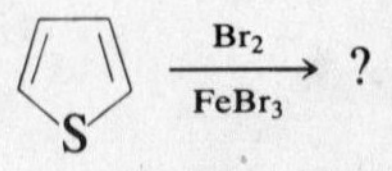

is

(a)   (b)

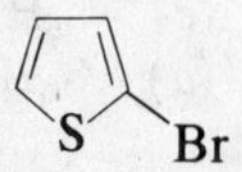

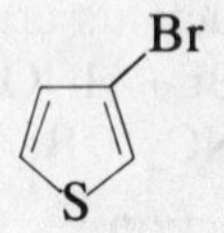

(c)   An *equal* mixture of (a) and (b) would form.
(d)   None of these; substitution would not occur.

SPECTROSCOPY

IMPORTANT TERMS AND CONCEPTS

Electromagnetic Radiation (Secs. 13.1, 13.2) All the important spectroscopic techniques are based on absorption of energy by a molecule. This energy is derived from electromagnetic radiation in nuclear magnetic resonance (nmr) spectroscopy, infrared (ir) spectroscopy, and ultraviolet-visible (uv-vis) spectroscopy. Important relationships for electromagnetic radiation are that *frequency is inversely proportional to wavelength* and that *energy is proportional to frequency*.

Electromagnetic radiation is absorbed by a molecule when the energy of the photons is equal to the difference in energy between two *quantized* energy states. The energy states for each type of spectroscopy are:

nmr: nuclear spin states
ir: vibrational energy states
uv-vis: electronic energy states

Proton Nuclear Magnetic Resonance (^{1}H NMR) Spectroscopy (Secs. 13.3 to 13.6) In the presence of an external magnetic field, a proton may absorb energy in the radio-frequency range to "flip" from one nuclear spin state to the other. When the proton is bound in a molecule, different chemical environments will cause a difference in the precise frequency at which absorption occurs. This difference is known as the *chemical shift*.

Chemical shifts are measured relative to tetramethylsilane (TMS), as a standard, and are expressed in parts per million (ppm).

$$\text{Chemical shift } (\delta) = \frac{\text{position of signal of interest (Hz)} - \text{position of TMS signal (Hz)}}{\text{spectrometer frequency (MHz)}}$$

Interpreting Proton NMR Spectra (Secs. 13.7 to 13.13) Three important aspects of an nmr spectrum must be considered when analyzing an unknown:

1. *The number of signals.* A separate nmr signal is observed for each of the protons in a substance that have nonequivalent chemical environments. All protons which are equivalent to each other have the same chemical shift.

2. *The intensity of the signals as measured by the area under each peak.* The area under each signal, determined electronically by a process known as *integration*, is a measure of the *relative* number of protons giving rise to the signal. For example, ethyl bromide and diethyl ether both give two nmr signals having area ratios of 2 (methylene hydrogens) to 3 (methyl hydrogens).

$$CH_3CH_2Br \qquad\qquad CH_3CH_2OCH_2CH_3$$

Ethyl bromide   Diethyl ether

3. *The multiplicity, or splitting, of each signal.* The number of peaks into which a signal is split, the *multiplicity*, is equal to 1 more than the number of hydrogen atoms vicinal to the one being observed. Common patterns are:

Singlet:	no splitting;	no vicinal H
Doublet:	two peaks;	one vicinal H
Triplet:	three peaks;	two vicinal H's
Quartet:	four peaks;	three vicinal H's

Only those hydrogens that have different chemical shifts split each other's signal. Splitting is usually too small to be detectable if there are more than three single bonds separating the protons.

The following examples illustrate how the nmr spectrum of a compound is interpreted in terms of the preceding three categories:

$$\overset{\overset{\textstyle O}{\|}}{}$$

2-Butanone: $CH_3\!-\!\overset{\overset{\textstyle O}{\|}}{C}\!-\!CH_2\!-\!CH_3$

(a)   (b)   (c)

Signals:	3
Peak areas:	(a) 3H; (b) 2H; (c) 3H
Multiplicity:	(a) singlet; (b) quartet; (c) triplet

1,2-Dimethoxyethane: $CH_3\!-\!O\!-\!CH_2\!-\!CH_2\!-\!O\!-\!CH_3$

(a)   (b)   (b)   (a)

Signals:	2
Peak areas:	(a) 6H; (b) 4H
Multiplicity:	(a) singlet; (b) singlet

Carbon-13 Nuclear Magnetic Resonance Spectroscopy (Secs. 13.16, 13.17) By using instruments known as *Fourier transform* (*FT*) *nmr spectrometers,* the carbon skeleton of a molecule can be deduced by recording the nmr spectrum of the ^{13}C isotope present at natural abundance. When *broadband decoupling* is used, all the peaks in a ^{13}C nmr spectrum appear as singlets. Chemical shifts are measured relative to TMS, as with ^1H nmr.

When *off-resonance decoupling* methods are used, the number of hydrogens attached to each carbon in a molecule may be determined. A methine carbon (CH) appears as a doublet, methylene (CH_2) as a triplet, and methyl (CH_3) as a quartet.

One important distinction of FT nmr is that signal intensities are distorted and peak areas are not as easily obtained as in a ^1H nmr spectrum. However, the chemical-shift range of ^{13}C nmr is so broad that each nonequivalent carbon in a molecule usually gives rise to a separate signal, and the signals overlap to a much smaller extent than is typical for a ^1H nmr spectrum.

Infrared Spectroscopy (Sec. 13.18) The primary usefulness of ir spectroscopy is in identifying the presence of certain functional groups within molecules. An ir spectrum is usually recorded in *wave numbers,* with units of cm^{-1}. Their region ranges from 4000 cm^{-1} (high-energy end) to 625 cm^{-1} (low-energy end).

Absorptions characteristic of particular functional groups are usually found in the region 4000 to 1600 cm^{-1}. In the region 1300 to 625 cm^{-1}, known as the *fingerprint region,* variations of peak patterns are observed for different compounds even if the same functional groups are present.

Ultraviolet-Visible Spectroscopy (Sec. 13.19) Ultraviolet-visible spectroscopy probes the electron distribution in a molecule and is particularly useful when conjugated π electron systems are present. Absorption positions are generally expressed as wavelengths, measured in nanometers (1 nm = 10^{-9} m). The visible region corresponds to 800 to 400 nm and the ultraviolet region to 400 to 200 nm.

Mass Spectrometry (Sec. 13.20) Mass spectrometry examines what happens to a molecule when it is bombarded with high-energy electrons. Ionization of a molecule by *electron impact* gives rise to a positively charged species, known as the *molecular ion,* having the same mass as the neutral molecule. The molecular ion then undergoes *fragmentation* to give various species of lower mass.

A mass spectrometer allows all the ions formed by a particular molecule to be separated according to the mass to charge ratio, m/z, of each ion. The *mass spectrum* obtained is characteristic of the particular compound.

SOLUTIONS TO TEXT PROBLEMS

13.1 The same equation is employed to calculate the chemical shifts as was used in the sample solution to part (*a*).

(*b*) Iodoform (CHI$_3$), 322 Hz at 60 MHz:

$$\delta = \frac{322 \text{ Hz}}{60 \times 10^6 \text{ Hz}} \times 10^6 = 5.37 \text{ ppm}$$

(*c*) Methyl chloride (CH$_3$Cl), 184 Hz at 60 MHz:

$$\delta = \frac{184 \text{ Hz}}{60 \times 10^6 \text{ Hz}} \times 10^6 = 3.07 \text{ ppm}$$

13.2 (*b*) There are four nonequivalent sets of protons bonded to carbon in 1-butanol as well as a fifth distinct type of proton, the one bonded to oxygen. There should be five signals in the ^{1}H nmr spectrum of 1-butanol.

$$\underset{\uparrow \quad \uparrow \quad \uparrow \quad \uparrow \quad \uparrow}{\text{CH}_3\text{CH}_2\text{CH}_2\text{CH}_2\text{OH}} \qquad \begin{array}{l}\text{five different proton environments}\\ \text{in 1-butanol; five signals}\end{array}$$

(*c*) Apply the "proton replacement" test to butane:

$$\underset{\text{Butane}}{\text{CH}_3\text{CH}_2\text{CH}_2\text{CH}_3} \qquad \underset{\text{1-Chlorobutane}}{\text{ClCH}_2\text{CH}_2\text{CH}_2\text{CH}_3} \qquad \underset{\text{2-Chlorobutane}}{\overset{\displaystyle \text{CH}_3\text{CHCH}_3\text{CH}_3}{\underset{\displaystyle \text{Cl}}{|}}}$$

$$\underset{\text{2-Chlorobutane}}{\overset{\displaystyle \text{CH}_3\text{CH}_2\text{CHCH}_3}{\underset{\displaystyle \text{Cl}}{|}}} \qquad \underset{\text{1-Chlorobutane}}{\text{CH}_3\text{CH}_2\text{CH}_2\text{CH}_2\text{Cl}}$$

There are *two* different types of protons in butane; it will exhibit *two* signals in its ^{1}H nmr spectrum.

(*d*) Like butane, 1,4-dibromobutane has two different types of protons. This can be illustrated by using the chlorine atom as a test group.

$$\underset{\text{1,4-Dibromobutane}}{BrCH_2CH_2CH_2CH_2Br} \qquad \underset{\text{1,4-Dibromo-1-chlorobutane}}{\overset{\overset{\displaystyle Cl}{|}}{BrCHCH_2CH_2CH_2Br}} \qquad \underset{\text{1,4-Dibromo-2-chlorobutane}}{\overset{\overset{\displaystyle Cl}{|}}{BrCH_2CHCH_2CH_2Br}}$$

$$\underset{\text{1,4-Dibromo-2-chlorobutane}}{\overset{\overset{\displaystyle Cl}{|}}{BrCH_2CH_2CHCH_2Br}} \qquad \underset{\text{1,4-Dibromo-1-chlorobutane}}{\overset{\overset{\displaystyle Cl}{|}}{BrCH_2CH_2CH_2CHBr}}$$

The ^{1}H nmr spectrum of 1,4-dibromobutane is expected to consists of two signals.

(e) All the carbons in 2,2-dibromobutane are different from each other, and so protons attached to one carbon are not equivalent to the protons attached to any of the other carbons. This compound should have *three* signals in its ^{1}H nmr spectrum.

$$\overset{\overset{\displaystyle Br}{|}}{\underset{\underset{\displaystyle Br}{|}}{CH_3CCH_2CH_3}}$$

There are three nonequivalent sets of protons in 2,2-dibromobutane.

(f) All the protons in 2,2,3,3-tetrabromobutane are equivalent. Its ^{1}H nmr spectrum will consist of one signal.

$$\underset{\underset{\displaystyle Br\ \ Br}{|\ \ \ |}}{\overset{\overset{\displaystyle Br\ \ Br}{|\ \ \ |}}{CH_3C-CCH_3}} \qquad \text{2,2,3,3-Tetrabromobutane}$$

(g) There are *four* nonequivalent sets of protons in 1,1,4-tribromobutane. It will exhibit four signals in its ^{1}H nmr spectrum.

$$\underset{H}{\overset{\overset{\displaystyle Br}{|}}{BrCCH_2CH_2CH_2Br}} \qquad \text{1,1,4-Tribromobutane}$$

(h) The seven protons of 1,1,1-tribromobutane belong to three nonequivalent sets, and hence the ^{1}H nmr spectrum will consists of three signals.

$$Br_3CCH_2CH_2CH_3 \qquad \text{1,1,1-Tribromobutane}$$

13.3 (b) Apply the replacement test to each of the protons of 1,1-dibromoethene:

1,1-Dibromoethene   1,1-Dibromo-2-chloroethene   1,1-Dibromo-2-chloroethene

Replacement of one proton by a test group (Cl) gives exactly the same compound as replacement of the other. The two protons of 1,1-dibromoethene are equivalent, and there is only one signal in the ^{1}H nmr spectrum of this compound.

(c) The replacement test reveals that both protons of *cis*-1,2-dibromoethene are equivalent:

cis-1,2-Dibromoethene   (Z)-1,2-Dibromo-1-chloroethene   (Z)-1,2-Dibromo-1-chloroethene

Because both protons are equivalent, the ^{1}H nmr spectrum of *cis*-1,2-dibromoethene consists of a single signal.

(d) Both protons of *trans*-1,2-dibromoethene are equivalent; each is cis to a bromine substituent.

$$\underset{H}{\overset{Br}{}}C=C\underset{Br}{\overset{H}{}}$$

trans-1,2-Dibromoethene (single signal in the ^{1}H nmr spectrum)

(e) There are *four* nonequivalent sets of protons in allyl bromide.

$$\rightarrow H \qquad H \leftarrow$$
$$C=C$$
$$\rightarrow H \qquad CH_2Br \nearrow$$

Allyl bromide (four signals in the ^{1}H nmr spectrum)

(f) The protons of a single methyl group are equivalent to each other, but all the methyl groups of 2-methyl-2-butene are nonequivalent. The vinyl proton is unique.

$$\underset{CH_3}{\overset{CH_3}{\downarrow}}C=C\underset{H}{\overset{CH_3}{\downarrow}}$$

2-Methyl-2-butene (four signals in the ^{1}H nmr spectrum)

13.4 (b) The three methyl protons of 1,1,1-trichloroethane (Cl_3CCH_3) are equivalent. They have the same chemical shift and do not split each other's signals. The ^{1}H nmr spectrum of Cl_3CCH_3 consists of a single sharp peak.

(c) Separate signals will be seen for the methylene protons and for the methine proton of 1,1,2-trichloroethane.

$$Cl_2C{\overset{|}{\underset{H}{}}}-CH_2Cl \qquad \text{1,1,2-Trichloroethane}$$

triplet   doublet

The methine proton splits the signal for the methylene protons into a doublet. The two methylene protons split the methine proton's signal into a triplet.

(d) Examine the structure of 1,2,2-trichloropropane:

$$ClCH_2\underset{\underset{Cl}{|}}{\overset{\overset{Cl}{|}}{C}}CH_3 \qquad \text{1,2,2-Trichloropropane}$$

The ^{1}H nmr spectrum exhibits a signal for the two equivalent methylene protons and one for the three equivalent methyl protons. Both these signals are sharp singlets. The protons of the methyl group and the methylene group are separated by more than three bonds and do not split each other's signals.

(e) The methine proton of 1,1,1,2-tetrachloropropane splits the signal of the methyl protons into a doublet; its signal is split into a quartet by the three methyl protons.

$$Cl_3C\underset{\underset{H}{|}}{\overset{\overset{Cl}{|}}{C}}-CH_3 \qquad \text{1,1,1,2-Tetrachloropropane}$$

doublet

quartet

13.5 (*b*) The ethyl group appears as a triplet-quartet pattern and the methyl group as a singlet.

$$\text{triplet} \overset{\curvearrowright}{\longrightarrow} \text{CH}_3\text{CH}_2\text{OCH}_3 \overset{\curvearrowleft}{\longleftarrow} \text{singlet; not vicinal to any other protons in molecule}$$

quartet

(*c*) The two ethyl groups of diethyl ether are equivalent to each other. The two methyl groups appear as a single triplet and the two methylene groups as a single quartet.

$$\text{triplet} \overset{\curvearrowright}{\longrightarrow} \text{CH}_3\text{CH}_2\text{OCH}_2\text{CH}_3 \overset{\curvearrowleft}{\longleftarrow} \text{triplet}$$

quartet   quartet

(*d*) The two ethyl groups of *p*-diethylbenzene are equivalent to each other and give rise to a single triplet-quartet pattern.

Three signals:
CH$_3$ triplet;
CH$_2$ quartet;
aromatic H singlet

All four protons of the aromatic ring are equivalent, have the same chemical shift, and split neither each other's signals nor any of the signals of the ethyl group.

(*e*) There are four nonequivalent sets of protons in this compound:

$$\text{ClCH}_2\text{CH}_2\text{OCH}_2\text{CH}_3$$

triplet   triplet   quartet   triplet

Vicinal protons in the ClCH$_2$CH$_2$O group split one another's signals, as do those in the CH$_3$CH$_2$O group.

(*f*) The two methine protons are nonequivalent and each splits the signal of the other into a doublet. The two ethyl groups are equivalent and give a single triplet-quartet pattern.

one triplet for CH$_3$ groups, plus one quartet for CH$_2$ groups

doublet   doublet

13.6 The ^{1}H nmr spectrum of Figure 13.19 is that of Cl$_2$CHCH(OCH$_2$CH$_3$)$_2$. All three compounds contain two equivalent ethyl groups, but only Cl$_2$CHCH(OCH$_2$CH$_3$)$_2$ has its remaining two protons in different environments. The two methine protons split each other's signal into a doublet.

doublet   doublet   singlet   equivalent: therefore a singlet

13.7 Both H$_b$ and H$_c$ in *m*-nitrostyrene appear as doublets of doublets. H$_b$ is coupled to H$_a$ by a coupling constant of 12 Hz and to H$_c$ by a coupling constant of 2 Hz. H$_c$ is coupled to H$_a$ by a coupling constant of 16 Hz and to H$_b$ by a coupling constant of 2 Hz.

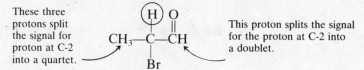

13.8 (*b*) The signal of the proton at C-2 is split into a quartet by the methyl protons, and each line of this quartet is split into a doublet by the aldehyde proton. It appears as a doublet of quartets. (*Note*: It does not matter whether the splitting pattern is described as a doublet of quartets or a quartet of doublets. There is no substantive difference in the two descriptions.)

These three protons split the signal for proton at C-2 into a quartet. ——→

CH_3—$\overset{\overset{\textstyle Br}{|}}{\underset{\overset{\textstyle Br}{|}}{C}}$—CH $\overset{\textstyle O}{\|}$ ⬡H⬡

This proton splits the signal for the proton at C-2 into a doublet.

13.9 (*b*) The two methyl carbons of the isopropyl group are equivalent:

There are four different types of carbons in the aromatic ring and two different types in the isopropyl group. The ^{13}C nmr spectrum of isopropylbenzene contains *six* signals.

(*c*) The methyl substituent at C-2 is different from those at C-1 and C-3:

The four nonequivalent ring carbons and the two different types of methyl carbons give rise to a ^{13}C nmr spectrum that contains *six* signals.

(*d*) The three methyl carbons of 1,2,4-trimethylbenzene are different from one another:

Also, all the ring carbons are different from each other. The nine different carbons give rise to *nine* separate signals.

(*e*) All the methyl carbons of 1,3,5-trimethylbenzene are equivalent.

Because of its high symmetry 1,3,5-trimethylbenzene has only *three* signals in its ^{13}C nmr spectrum.

13.10 The splitting rules for ^{13}C nmr are summarized in the sample solution to part (*a*) given in the text. In all the following solutions, multiplicities are cited as *s* (singlet), *d* (doublet), *t* (triplet), *q* (quartet).

      (*b*) Isopropylbenzene: (*d*) 1,2,4-Trimethylbenzene:

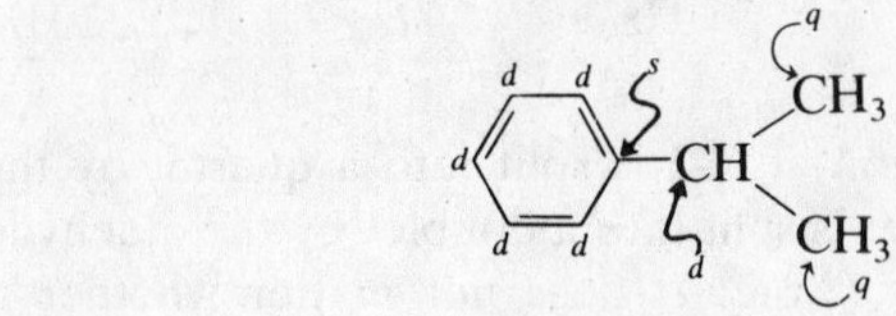

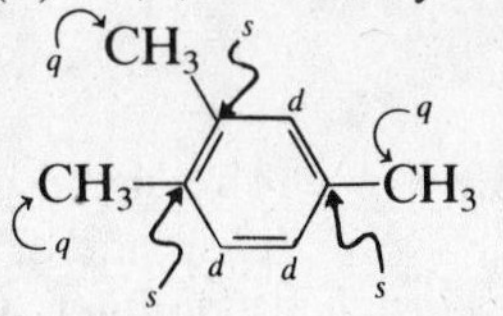

      (*c*) 1,2,3-Trimethylbenzene: 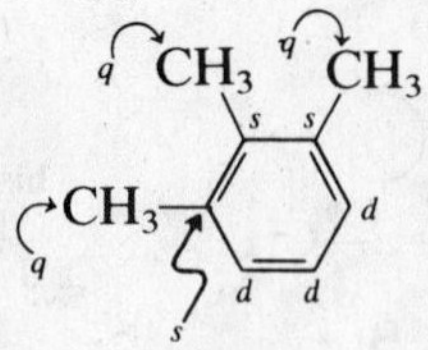   (*e*) 1,3,5-Trimethylbenzene:

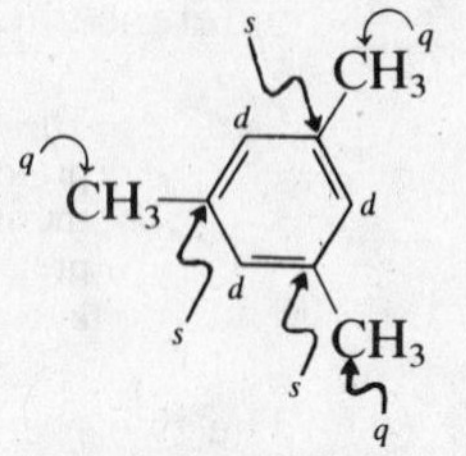

13.11 The ^{13}C nmr spectrum contains three different sp^3 hybridized carbons and six different aromatic carbons. This spectrum is in accord only with 1,2,4-trimethylbenzene.

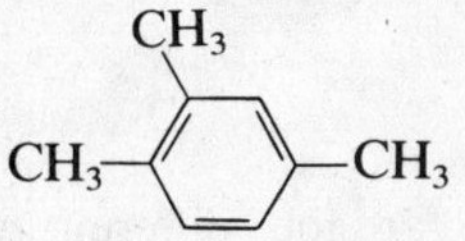

13.12 The infrared spectrum of Figure 13.33 is void of absorption in the 1600 to 1800 cm^{-1} region, and so the unknown compound cannot contain a carbonyl (C=O) group. Therefore, it cannot be acetophenone or benzoic acid.

    There is a broad, intense absorption at 3300 cm^{-1} attributable to a hydroxyl group. While both phenol and benzyl alcohol are possibilities, the peaks at 2800 to 2900 cm^{-1} reveal the presence of hydrogen bonded to sp^3 hybridized carbon. All carbons are sp^2 hybridized in phenol. The infrared spectrum is that of *benzyl alcohol*.

13.13 (*b*) The distribution of molecular-ion peaks in *o*-dichlorobenzene is identical to that in the para isomer. As the sample solution to part (*a*) in the text describes, peaks at m/z 146, 148, and 150 are present for the molecular ion.

      (*c*) The two isotopes of bromine are ^{79}Br and ^{81}Br. When both bromines of *p*-dibromobenzene are ^{79}Br, the molecular ion appears at m/z 234. When one is ^{79}Br and the other is ^{81}Br, m/z for the molecular ion is 236. When both bromines are ^{81}Br, m/z for the molecular ion is 238.

      (*d*) The combinations of ^{35}Cl, ^{37}Cl, ^{79}Br, and ^{81}Br in *p*-bromochlorobenzene and the values of m/z for the corresponding molecular ion are as shown.

$$(^{35}Cl, ^{79}Br) \qquad m/z = 190$$
$$(^{37}Cl, ^{79}Br) \text{ or } (^{35}Cl, ^{81}Br) \qquad m/z = 192$$
$$(^{37}Cl, ^{81}Br) \qquad m/z = 194$$

13.14 The base peak in the mass spectrum of alkylbenzenes corresponds to carbon–carbon bond cleavage at the benzylic carbon.

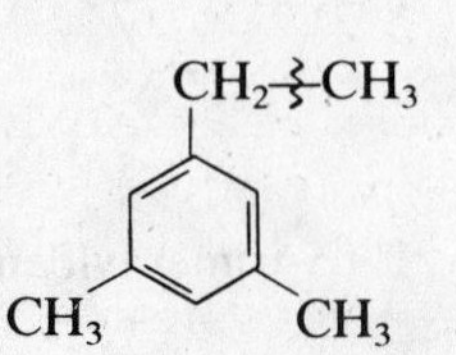

Base peak: $C_9H_{11}^+$
m/z 119

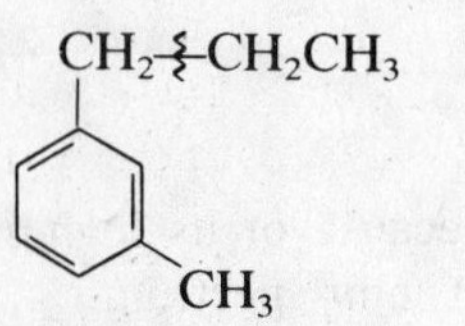

Base peak: $C_8H_9^+$
m/z 105

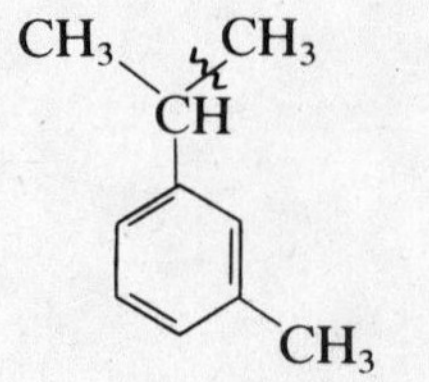

Base peak: $C_9H_{11}^+$
m/z 119

13.15 (*a*) Energy is proportional to frequency and inversely proportional to wavelength. The longer the wavelength, the lower the energy. Microwave photons have a wavelength in the range of 10^{-2} m, which is longer than that of infrared photons (on the order of 10^{-5} m). Thus, microwave radiation is lower in energy than infrared radiation, and the separation between rotational energy levels (measured by microwave) is less than the separation between vibrational energy levels (measured by infrared).

(*b*) Absorption of a photon occurs only when its energy matches the energy difference between two adjacent energy levels in a molecule. Microwave photons have energies that match the differences between the rotational energy levels of water. They are not sufficiently high in energy to excite a water molecule to a higher vibrational or electronic energy state.

13.16 A shift in the uv-vis spectrum of acetone from 279 nm in hexane to 262 nm in water is a shift to shorter wavelength on going from a less polar solvent to a more polar one. This means that the energy difference between the starting electronic state (the *ground* state, *n*) and the excited electronic state (π^*) is *greater* in water than in hexane. Hexane as a solvent does not interact appreciably with either the ground state of acetone or the excited state of acetone. Water is polar and solvates the ground state of acetone, lowering its energy. Since the energy gap between the ground state and the excited state increases, it must mean that the ground state is more solvated than the excited state and therefore more polar than the excited state.

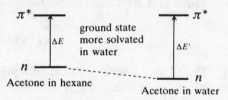

13.17 The dipole moment of carbon dioxide is zero and does not change during the symmetric stretching vibration. The symmetric stretch is not "infrared-active." The antisymmetric stretch generates a dipole moment in carbon dioxide and is infrared-active.

$$\overset{\leftarrow}{O}=C=\vec{O} \qquad\qquad \vec{O}=C=\vec{O}$$

Symmetric stretch: no change in dipole moment   Antisymmetric stretch: dipole moment present as a result of unequal C—O bond distances

13.18 The trans and cis isomers of 1-bromo-4-*tert*-butylcyclohexane can be taken as models to estimate the chemical shift of the proton of the CHBr group when it is axial and equatorial, respectively, in the two chair conformations of bromocyclohexane. An axial proton is more shielded ($\delta = 3.81$ ppm for *trans*-1-bromo-4-*tert*-butylcyclohexane) than an equatorial one ($\delta = 4.62$ ppm for *cis*-1-bromo-4-*tert*-butylcyclohexane).

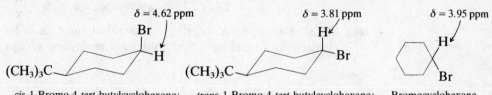

cis-1-Bromo-4-*tert*-butylcyclohexane;   *trans*-1-Bromo-4-*tert*-butylcyclohexane;   Bromocyclohexane
less shielded   more shielded

The difference in chemical shift between these stereoisomers is 0.81 ppm. The corresponding proton in bromocyclohexane is 0.67 ppm more shielded than in the equatorial proton in *cis*-1-bromo-4-*tert*-butylcyclohexane. Therefore the proportion of bromocyclohexane that has an axial hydrogen is 0.67/0.81, or 83 percent. For bromocyclohexane, 83 percent of the molecules have an equatorial bromine, and 17 percent have an axial bromine.

13.19 The two staggered conformations of 1,2-dichloroethane are the anti and the gauche:

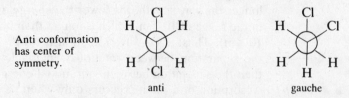

Anti conformation has center of symmetry.

The species present at low temperature (crystalline 1,2-dichloroethane) has a center of symmetry and is therefore the anti conformation. Liquid 1,2-dichloroethane is a mixture of the anti and the gauche conformations.

13.20 Since each compound exhibits only a single peak in its ^1H nmr spectrum, all the hydrogen substituents are equivalent in each one. Structures are assigned on the basis of their molecular formulas and chemical shifts.

(a) This compound has the molecular formula C_8H_{18} and so must be an alkane. The 18 hydrogens are contributed by six equivalent methyl groups.

$$(CH_3)_3CC(CH_3)_3 \qquad \text{2,2,3,3-Tetramethylbutane } (\delta = 0.9 \text{ ppm})$$

(b) A hydrocarbon with the molecular formula C_5H_{10} has a SODAR of 1 and so is either a cycloalkane or an alkene. Since all 10 hydrogens are equivalent, this compound must be cyclopentane.

Cyclopentane ($\delta = 1.5$ ppm)

(c) The chemical shift of the eight equivalent hydrogens in C_8H_8 is $\delta = 5.8$ ppm, which is consistent with protons attached to a carbon–carbon double bond.

1,3,5,7-Cyclooctatetraene ($\delta = 5.8$ ppm)

(d) The compound C_4H_9Br has no rings or double bonds. The nine hydrogens belong to three equivalent methyl groups.

$$(CH_3)_3CBr \qquad \textit{tert}\text{-Butyl bromide } (\delta = 1.8 \text{ ppm})$$

(e) The dichloride has no rings or double bonds (SODAR = 0). The four equivalent hydrogens are present as two —CH_2Cl groups.

$$ClCH_2CH_2Cl \qquad \text{1,2-Dichloroethane } (\delta = 3.7 \text{ ppm})$$

(f) All three hydrogens in $C_2H_3Cl_3$ must be part of the same methyl group in order to be equivalent.

$$CH_3CCl_3 \qquad \text{1,1,1-Trichloroethane } (\delta = 2.7 \text{ ppm})$$

(g) This compound has no rings or double bonds. In order to have eight equivalent hydrogens it must have four equivalent methylene groups.

$$\begin{array}{c} CH_2Cl \\ | \\ ClCH_2CCH_2Cl \\ | \\ CH_2Cl \end{array} \qquad \begin{array}{c} \text{1,3-Dichloro-2,2-di(chloromethyl)propane} \\ (\delta = 3.7 \text{ ppm}) \end{array}$$

(h) A compound with a molecular formula of $C_{12}H_{18}$ has a SODAR of 4. A likely candidate for a compound with 18 equivalent hydrogens is one with 6 equivalent CH_3 groups. Thus, 6 of the 12 carbons belong to CH_3 groups and the other 6 have no hydrogen substituents. The compound is hexamethylbenzene.

CH_3
CH_3 CH_3
CH_3 CH_3
CH_3

A chemical shift of 2.2 ppm is consistent with the fact that all of the protons are benzylic hydrogens.

(*i*) The molecular formula of $C_3H_6Br_2$ tells us that the compound has no double bonds and no rings. All six hydrogens are equivalent, indicating two equivalent methyl groups. The compound is 2,2-dibromopropane, $(CH_3)_2CBr_2$.

13.21 In each of the parts to this problem nonequivalent protons must *not* be bonded to adjacent carbons, because we are told that the two signals in each case are singlets.

(*a*) Each signal corresponds to four protons, and so each must result from two equivalent CH_2 groups. The four CH_2 groups account for four of the carbons of C_6H_8, leaving two carbons that bear no hydrogen substituents. A molecular formula of C_6H_8 corresponds to a SODAR of 3. A compound consistent with these requirements is:

$$CH_2 = \Diamond = CH_2$$

The signal at $\delta = 5.6$ ppm is consistent with that expected for the four vinylic protons. The signal at $\delta = 2.7$ ppm corresponds to that for the allylic protons of the ring.

(*b*) The compound has a molecular formula of $C_5H_{11}Br$ and therefore has no double bonds or rings. A nine-proton singlet at 1.1 ppm indicates three equivalent methyl groups, and a two-proton singlet at 3.3 ppm indicates a CH_2Br group. The correct structure is $(CH_3)_3CCH_2Br$.

(*c*) This compound ($C_6H_{12}O$) has three equivalent CH_3 groups, along with a fourth CH_3 group that is somewhat less shielded. Its molecular formula indicates that it can have either one double bond or one ring. This compound is $(CH_3)_3CCCH_3$, with a $C=O$.

(*d*) A molecular formula of $C_6H_{10}O_2$ corresponds to a SODAR of 2. The signal at 2.2 ppm (6H) is likely due to two equivalent CH_3 groups, and the one at 2.7 ppm (4H) to two equivalent CH_2 groups. The compound is $CH_3CCH_2CH_2CCH_3$, with two $C=O$ groups.

13.22 (*a*) A five-proton signal at $\delta = 7.1$ ppm indicates a monosubstituted aromatic ring. With a SODAR of 4, C_8H_{10} contains this monosubstituted aromatic ring and no other rings or multiple bonds. The triplet-quartet pattern at high field suggests an ethyl group.

$\bigcirc$—CH_2CH_3 Ethylbenzene

quartet   triplet
$\delta = 2.6$ ppm $\delta = 1.2$ ppm
(benzylic)

(*b*) The SODAR of 4 and the five-proton multiplet at $\delta = 7.0$ to 7.5 ppm are accommodated by a monosubstituted aromatic ring. The remaining four carbons and nine hydrogens are most reasonably a *tert*-butyl group, since all nine hydrogens are equivalent.

$$\text{\includegraphics{benzene ring}}\text{—C(CH}_3)_3 \qquad \textit{tert-}\text{Butylbenzene}$$

singlet; $\delta = 1.3$ ppm

(c) Its molecular formula requires that C_6H_{14} be an alkane. The doublet-heptet pattern is consistent with an isopropyl group, and the total number of protons requires that two of these groups be present.

$$(\text{CH}_3)_2\text{CHCH}(\text{CH}_3)_2 \qquad \text{2,3-Dimethylbutane}$$

doublet   heptet
$\delta = 0.8$ ppm $\delta = 1.4$ ppm

Note that the methine (CH) protons do not couple with each other, since they are equivalent.

(d) The molecular formula C_6H_{12} requires the presence of one double bond or ring. A peak at $\delta = 5.1$ ppm is consistent with —C=CH, and so the compound is a noncyclic alkene. The vinyl proton gives a triplet signal, and so the group C=CHCH$_2$ is present. The ^{1}H nmr spectrum shows the presence of the following structural units:

$$\begin{array}{c} \text{H} \quad\text{—5.1 ppm (triplet)} \\ \text{C=C} \\ \text{CH}_2 \quad\text{—2.0 ppm (allylic)} \end{array}$$

$$\text{CH}_2\text{CH}_3 \qquad \text{0.9 ppm (triplet)}$$

$$\begin{array}{c} \text{CH}_3 \quad\text{—1.6 ppm (singlet; allylic)} \\ \text{C=C} \\ \text{CH}_3 \quad\text{—1.7 ppm (singlet; allylic)} \end{array}$$

Putting all these fragments together yields a unique structure:

$$\begin{array}{c} \text{singlet} \quad\quad\quad\quad \text{triplet} \\ \text{CH}_3 \qquad\quad \text{H} \\ \text{C=C} \\ \text{CH}_3 \qquad \text{CH}_2\text{CH}_3 \\ \text{singlet} \qquad\qquad\quad \text{triplet} \\ \text{pentet} \end{array} \qquad \text{2-Methyl-2-pentene}$$

(e) The compound $C_4H_6Cl_4$ contains no double bonds or rings. There are no high-field peaks ($\delta = 0.5$ to 1.5 ppm), and so there are no methyl groups. Therefore at least one chlorine substituent must be at each end of the chain. The most likely structure has the four chlorines divided into two groups of two.

$$\text{Cl}_2\text{CHCH}_2\text{CH}_2\text{CHCl}_2 \qquad \text{1,1,4,4-Tetrachlorobutane}$$

$\delta = 4.6$ ppm $\delta = 3.9$ ppm
(triplet)   (doublet)

(f) The molecular formula $C_4H_6Cl_2$ indicates the presence of one double bond or ring. A signal at $\delta = 5.7$ ppm is consistent with a proton attached to a doubly bonded carbon. The following structural units are present:

$$\begin{array}{c} \text{H} \quad\text{—5.7 ppm (triplet)} \\ \text{C=C} \\ \text{CH}_2 \end{array}$$

$$\text{C=CCH}_3 \qquad \text{—2.2 ppm (allylic)}$$

In order for the methyl group to appear as a singlet and the methylene group to appear as a doublet, the chlorine substituents must be distributed as shown:

singlet
CH_3
$C=CHCH_2Cl$ 1,3-Dichloro-2-butene
Cl

triplet   doublet ($\delta = 4.1$ ppm)

The stereochemistry of the double bond (E or Z) is not revealed by the ^{1}H nmr spectrum.

(g) A molecular formula of C_3H_7ClO is consistent with the absence of rings and multiple bonds. None of the signals is equivalent to three protons, and so no methyl groups are present. There are three methylene groups, all of which are different from each other. Therefore, the compound is:

$ClCH_2CH_2CH_2OH$

$\delta = 3.7$ or 3.8 ppm $\delta = 2.8$ ppm (singlet)
(triplet)

$\delta = 2.0$ ppm $\delta = 3.7$ or 3.8 ppm
(pentet)   (triplet)

(h) The compound has a molecular formula of $C_{14}H_{14}$ and a SODAR of 8. With a 10-proton signal at $\delta = 7.1$ ppm, a logical conclusion is that there are two monosubstituted benzene rings. The other four protons belong to two equivalent methylene groups.

—CH_2CH_2—   1.2-Diphenylethane

$\delta = 2.9$ ppm (singlet; benzylic)

13.23 The compounds of molecular formula C_4H_9Cl are the isomeric chlorides: butyl, isobutyl, *sec*-butyl, and *tert*-butyl chloride.

(a) All nine methyl protons of *tert*-butyl chloride $(CH_3)_3CCl$ are equivalent; its ^{1}H nmr spectrum has only one peak.

(b) A doublet at $\delta = 3.4$ ppm indicates a —CH_2Cl group attached to a carbon that bears a single proton.

$(CH_3)_2CHCH_2Cl$
$\delta = 3.4$ ppm (doublet)

Isobutyl chloride

(c) A triplet at $\delta = 3.5$ ppm means that there is a methylene group attached to the carbon that bears the chlorine.

$CH_3CH_2CH_2CH_2Cl$
$\delta = 3.5$ ppm (triplet)

Butyl chloride

(d) There are two nonequivalent methyl groups in this compound.

$\delta = 1.5$ ppm $CH_3CHCH_2CH_3$ $\delta = 1.0$ ppm
(doublet) | (triplet)
                    Cl

sec-Butyl chloride

13.24 Compounds with the molecular formula C_3H_5Br have either one ring or one double bond.

(a) The two small peaks at $\delta = 5.3$ and 5.5 ppm have chemical shifts consistent with the assumption that each peak is due to a vinyl proton (C=CH). The remaining three protons belong to an allylic methyl group ($\delta = 2.3$ ppm).

The compound cannot be CH₃CH=CHBr, because the methyl signal would be split into a doublet. Isomer A can only be

$$
\begin{array}{c}
CH_3 \\
| \\
CH_2=C \\
| \\
Br
\end{array}
\qquad \text{2-Bromo-1-propene}
$$

(b) Two of the carbons of isomer B have chemical shifts characteristic of sp^2 hybridized carbon. One of these bears two protons ($\delta = 118.8$ ppm, triplet); the other bears one proton ($\delta = 134.2$ ppm, doublet). The remaining carbon is sp^3 hybridized and bears two hydrogens. Isomer B is allyl bromide.

$$
CH_2=CHCH_2Br
$$

$\delta = 118.8$ ppm $\delta = 134.2$ ppm $\delta = 32.6$ ppm
(triplet)   (doublet)   (triplet)

Allyl bromide

(c) All the carbons are sp^3 hybridized in this isomer. Two of the carbons belong to equivalent methylene groups, and the other bears only one hydrogen. Isomer C is cyclopropyl bromide.

$\delta = 12.0$ ppm
(triplet)

$\delta = 16.8$ ppm
(doublet)

H
Br   Cyclopropyl bromide

13.25 All these compounds have the molecular formula $C_4H_{10}O$. They have neither multiple bonds nor rings.

(a) There are two equivalent methyl groups because there is a quartet at $\delta = 18.9$ ppm corresponding to two carbons. There is one carbon that bears a single hydrogen. The least shielded carbon, presumably the one bonded to oxygen, has two hydrogen substituents because its signal is a triplet.

     Putting all the information together reveals this compound to be isobutyl alcohol.

$$
(CH_3)_2CHCH_2OH
$$

$\delta = 18.9$ ppm $\delta = 30.8$ ppm $\delta = 69.4$ ppm
(quartet)   (doublet)   (triplet)

Isobutyl alcohol

(b) There are four distinct peaks in this compound, and so none of the four carbons is equivalent to any of the others. The signal for the least shielded carbon is a doublet, and so the oxygen is attached to a secondary carbon. Only one carbon appears at low field; the compound is an alcohol, not an ether. Therefore:

$\delta = 69.2$ ppm (doublet)

$$
CH_3CHCH_2CH_3 \\
| \\
OH
$$

$\delta = 22.7$ ppm
(quartet)

$\delta = 32.0$ ppm
(triplet)

$\delta = 10.0$ ppm
(quartet)

sec-Butyl alcohol

(c) Signals for three equivalent methyl carbons indicate that this isomer is *tert*-butyl alcohol. This assignment is reinforced by the observation that the least shielded carbon has no hydrogens attached to it.

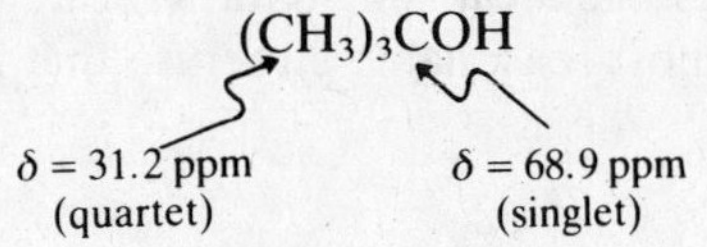

$$(CH_3)_3COH$$

δ = 31.2 ppm   δ = 68.9 ppm
(quartet)   (singlet)

tert-Butyl alcohol

13.26 The molecular formula of C_6H_{14} for each of these isomers requires that all of them be alkanes.

(*a*) This compound contains only methyl and methine carbons.

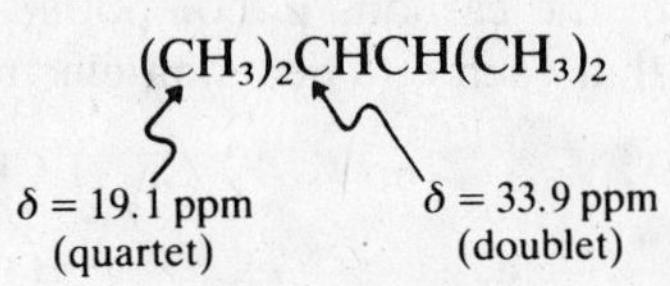

$$(CH_3)_2CHCH(CH_3)_2$$

δ = 19.1 ppm   δ = 33.9 ppm
(quartet)   (doublet)

2,3-Dimethylbutane

(*b*) This isomer has no methine carbons, and there are two different kinds of methylene groups.

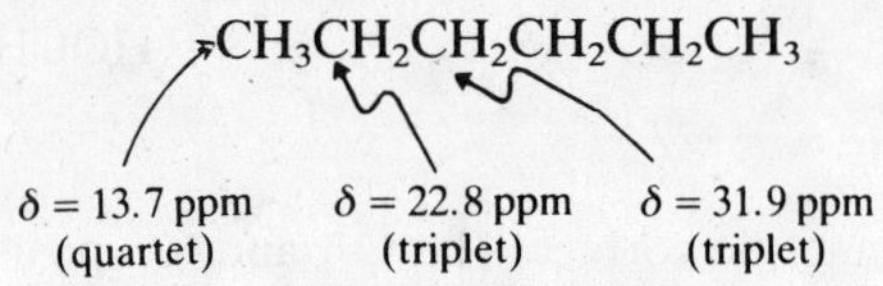

$$CH_3CH_2CH_2CH_2CH_2CH_3$$

δ = 13.7 ppm   δ = 22.8 ppm   δ = 31.9 ppm
(quartet)   (triplet)   (triplet)

Hexane

(*c*) Methyl, methylene, and methine carbons are all present in this isomer. There are two different kinds of methyl groups.

δ = 29.1 ppm   δ = 36.4 ppm (doublet)
(triplet)

$$CH_3CH_2CHCH_2CH_3$$
$$CH_3$$

δ = 11.1 ppm   δ = 18.4 ppm
(quartet)   (quartet)

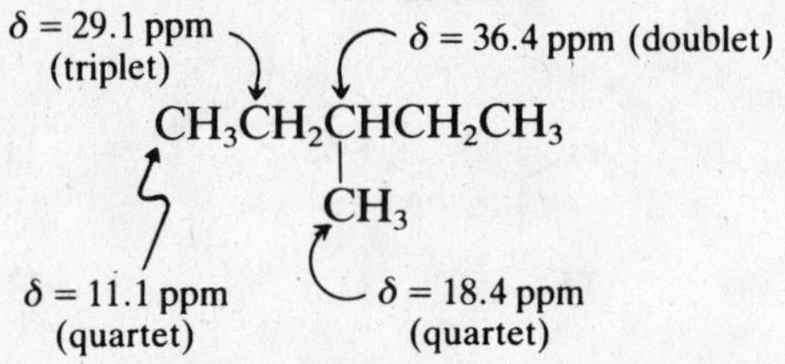

3-Methylpentane

(*d*) This isomer contains a quaternary carbon in addition to a methylene group and two different kinds of methyl groups.

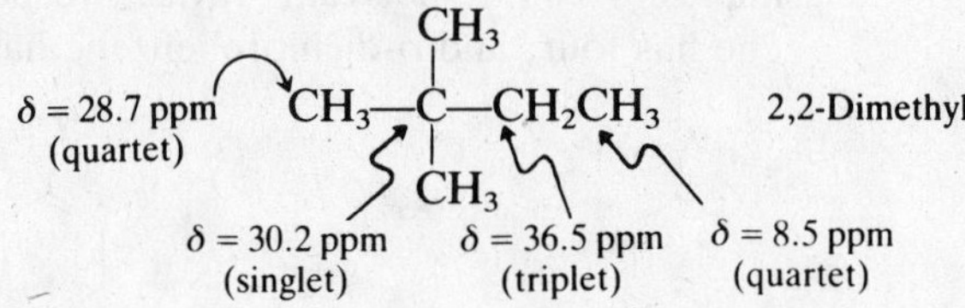

δ = 28.7 ppm
(quartet)

$$CH_3$$
$$CH_3-C-CH_2CH_3$$
$$CH_3$$

2,2-Dimethylbutane

δ = 30.2 ppm   δ = 36.5 ppm   δ = 8.5 ppm
(singlet)   (triplet)   (quartet)

(*e*) This isomer contains two different kinds of methyl groups, two different kinds of methylenes, and a methine carbon.

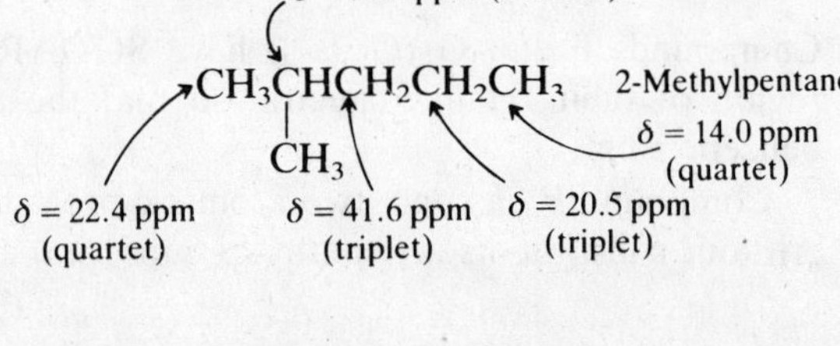

δ = 27.6 ppm (doublet)

$$CH_3CHCH_2CH_2CH_3$$
$$CH_3$$

2-Methylpentane

δ = 14.0 ppm
(quartet)

δ = 22.4 ppm   δ = 41.6 ppm   δ = 20.5 ppm
(quartet)   (triplet)   (triplet)

13.27 The SODAR of compound D (C_4H_6) is 2. It can have two double bonds, two rings, one ring and one double bond, or one triple bond.

The chemical shift data indicate that two carbons are sp^3 hybridized and two are sp^2. The most reasonable structure that is consistent with ^{13}C nmr data is cyclobutene.

$$\delta = 136 \text{ ppm} \qquad \qquad \qquad \text{Cyclobutene}$$
$$\text{(doublet)} \qquad \qquad \delta = 30.2 \text{ ppm}$$
$$\qquad \qquad \qquad \text{(triplet)}$$

Compound D cannot be 1- or 2-methylcyclopropene. Neither of the carbon signals is a quartet, and so compound D cannot contain a methyl group.

13.28 Each of the carbons in compound E gives its ^{13}C nmr signal at relatively low field; it is likely that each one bears an electron-withdrawing substituent. Compound E is:

$$\text{ClCH}_2\text{CHCH}_2\text{OH} \qquad \text{3-Chloro-1,2-propanediol}$$
$$\qquad \quad |$$
$$\qquad \quad \text{OH}$$

$$\delta = 46.8 \text{ ppm} \qquad \delta = 72.0 \text{ ppm} \qquad \delta = 63.5 \text{ ppm}$$
$$\text{(triplet)} \qquad \quad \text{(doublet)} \qquad \quad \text{(triplet)}$$

The regioisomeric compound 2-chloro-1,3-propanediol

$$\text{HOCH}_2\text{CHCH}_2\text{OH}$$
$$\qquad \quad |$$
$$\qquad \quad \text{Cl}$$

cannot be correct. The C-1 and C-3 positions are equivalent; the ^{13}C nmr spectrum of this compound exhibits only two peaks, not three.

13.29 (*a*) All the hydrogens are equivalent in *p*-dichlorobenzene; therefore it has the simplest ^1H nmr spectrum of the three compounds chlorobenzene, *o*-dichlorobenzene, and *p*-dichlorobenzene.

Chlorobenzene
(three different kinds of protons)

o-Dichlorobenzene
(two different kinds of protons)

p-Dichlorobenzene
(all protons are equivalent)

(*b–d*) In addition to giving the simplest ^1H nmr spectrum, *p*-dichlorobenzene gives the simplest ^{13}C nmr spectrum. It has two peaks in its ^{13}C nmr spectrum, chlorobenzene has four, and *o*-dichlorobenzene has three.

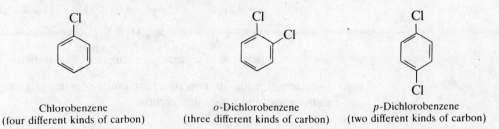

Chlorobenzene
(four different kinds of carbon)

o-Dichlorobenzene
(three different kinds of carbon)

p-Dichlorobenzene
(two different kinds of carbon)

13.30 Compounds F and G ($C_{10}H_{14}$) have SODARs of 4. Both have peaks in the $\delta = 7$ ppm region of their ^1H nmr spectra, so that the SODAR of 4 can be accommodated by a benzene ring.

Compound F has only two aromatic protons, and the ring is therefore tetrasubstituted. All four substituents are methyl groups, and all the methyl groups are equivalent.

$\delta = 2.2$ ppm

$\delta = 6.9$ ppm

1,2,4,5-Tetramethylbenzene
(compound F)

In compound G both aromatic protons are equivalent and there are four methyl groups divided into two pairs.

$\delta = 6.8$ ppm

1,2,3,4-Tetramethylbenzene
(compound G)

$\delta = 2.2$ and 2.3 ppm

13.31 Since compound H has a five-proton signal at $\delta = 7.2$ ppm and a SODAR of 4, we conclude that six of its eight carbons belong to a monosubstituted aromatic ring. The infrared spectrum exhibits absorption at 3300 cm^{-1}, indicating the presence of a hydroxyl group. Compound H is an alcohol. A three-proton doublet at $\delta = 1.4$ ppm, along with a one-proton quartet at $\delta = 4.7$ ppm, signals the presence of a CH$_3$CH unit.

Compound H is 1-phenylethanol.

$\delta = 4.7$ ppm
(quartet)

$\delta = 1.4$ ppm
(doublet)

1-Phenylethanol (compound H)

$\delta = 1.9$ ppm
(singlet)

13.32 The peak at highest m/z in the mass spectrum of compound I is $m/z = 134$; this is likely to correspond to the molecular ion. Among the possible molecular formulas, C$_{10}$H$_{14}$ correlates best with the information from the ^{1}H nmr spectrum. What is evident is that there is a signal due to aromatic protons, as well as a triplet-quartet pattern of an ethyl group. A molecular formula of C$_{10}$H$_{14}$ suggests a benzene ring that bears two ethyl groups. Since the signal for the aryl protons is so sharp, they are probably equivalent. Compound I is p-diethylbenzene.

CH$_3$CH$_2$—   —CH$_2$CH$_3$

p-Diethylbenzene
(compound I)

$\delta = 1.2$ ppm
(triplet)

$\delta = 2.6$ ppm
(quartet)

$\delta = 7.1$ ppm
(singlet)

13.33 The most prominent peak in the infrared spectrum of compound J is at 1725 cm^{-1}, a region characteristic of C=O stretching vibrations.

The ^{1}H nmr spectrum shows only two sets of signals, a triplet at $\delta = 1.1$ ppm and a quartet at $\delta = 2.4$ ppm. Compound J contains a CH$_3$CH$_2$ group as the only source of its protons.

There are three peaks in its ^{13}C nmr spectrum, one of which is at very low field. The signal at $\delta = 211.4$ ppm is in the region characteristic of carbons of C=O groups.

If one assumes that compound J contains only carbon, hydrogen, and one oxygen atom and that the peak at highest m/z in its mass spectrum (m/z 86) corresponds to the molecular ion, then compound J has the molecular formula $C_5H_{10}O$.

All the information points to the conclusion that compound J has the structure shown:

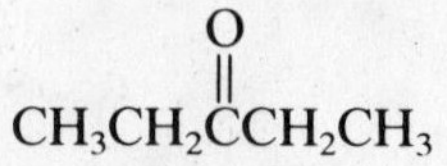

$$\underset{\text{3-Pentanone}}{CH_3CH_2\overset{\overset{\displaystyle O}{\|}}{C}CH_2CH_3}$$

13.34 [18]-Annulene has *two* different kinds of protons; the 12 protons on the outside periphery of the ring are different from the 6 on the inside.

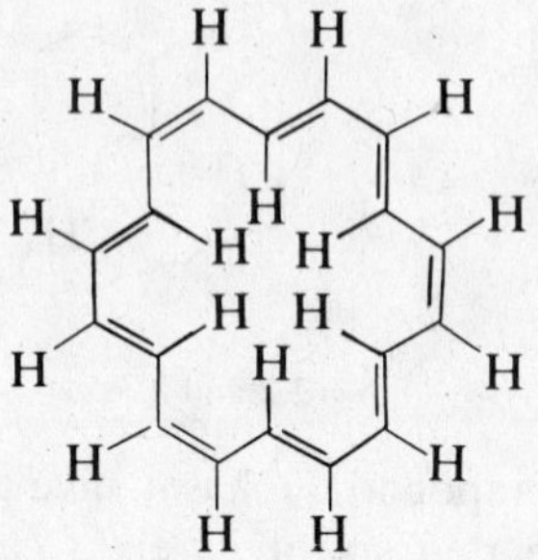

These different environments explain why the ^1H nmr spectrum contains two peaks in a 2:1 ratio. The less intense signal, that for the interior protons, is more shielded than the signal for the outside protons. The reason for this has to do with the magnetic field induced by the circulating π electrons of this aromatic ring, which reinforces the applied field in the region of the outside protons but opposes it in the interior of the ring.

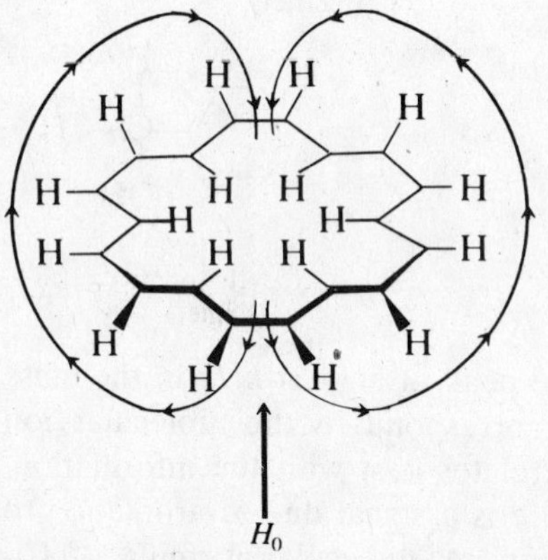

Protons inside the ring are shielded by the induced field to a significant extent—so much so that their signal appears at $\delta = -1.9$ ppm.

13.35 (*a*) The nuclear spin of ^{19}F is $\pm\frac{1}{2}$, i.e., the same as that of a proton. The splitting rules for ^{19}F–^1H couplings are the same as those for ^1H–^1H. Thus the single fluorine atom of CH_3F splits the signal for the protons of the methyl group into a *doublet*.

(*b*) The set of three equivalent protons of CH_3F splits the signal for fluorine into a *quartet*.

(*c*) The proton signal in CH_3F is a doublet centered at $\delta = 4.3$ ppm. The separation between the two halves of this doublet is 45 Hz, which is equivalent to 0.75 ppm at 60 MHz (60 Hz = 1 ppm). Thus one line of the doublet appears at $\delta = (4.3 + 0.375)$ ppm and the other at $\delta = (4.3 - 0.375)$ ppm.

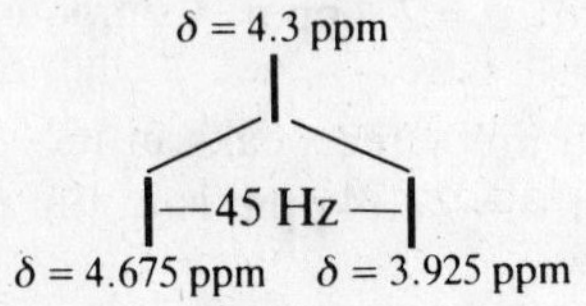

13.36 (a) Both hydrogens are equivalent in 1,1,1,2-tetrafluoroethane, and so there is one signal. This signal is split into a doublet by the fluorine at C-2, and each line of the doublet is split into a quartet by the three fluorines at C-1. The proton signal is split into a doublet of quartets, or a total of eight peaks.

$$|\text{---}45\text{---}|\qquad {}^{2}J_{HF} = 45\ \text{Hz}$$
$$|8|8|8|\qquad |8|8|8|\qquad {}^{3}J_{HF} = 8\ \text{Hz}$$

(b) There are two different kinds of fluorine in 1,1,1,2-tetrafluoroethane, and so there are two distinct signals.

The three equivalent fluorines at C-1 are split into a doublet by vicinal coupling to the fluorine at C-2. Each line of this doublet is further split into a triplet by vicinal coupling to the two protons at C-2. There are six peaks for the fluorines at C-1, a doublet of triplets.

$$|\text{---}15\text{---}|\qquad {}^{3}J_{FF} = 15\ \text{Hz}$$
$$|8|8|\qquad |8|8|\qquad {}^{3}J_{HF} = 8\ \text{Hz}$$

The signal for the C-2 fluorine is split into a triplet by the two geminal hydrogens at C-2. Each line of the triplet is split into a quartet by vicinal coupling to the three equivalent fluorines at C-1. The signal for the C-2 fluorine appears as a 12-line pattern (a triplet of quartets).

(Only 10 lines are actually seen because two of them overlap at each of the positions marked by *x* in the diagram.)

$$|\text{---}45\text{---}|\text{---}45\text{---}|\qquad {}^{2}J_{HF} = 45\ \text{Hz}$$
$$\begin{array}{c} {}_x\qquad {}_x \\ |\ |\ |\ |\ |\ |\ |\ |\ | \\ 15\ 15\ 15\ 15\ 15\ 15\ 15\ 15\ 15 \end{array}\qquad {}^{3}J_{FF} = 15\ \text{Hz}$$

13.37 Since ^{31}P has a spin of $\pm\frac{1}{2}$, it is capable of splitting the ^{1}H nmr signal of protons in the same molecule. The problem stipulates that the methyl protons are coupled through three bonds to phosphorus in trimethyl phosphite.

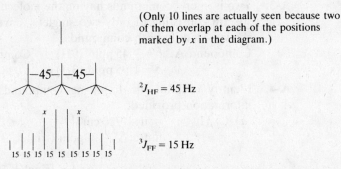

$$CH_3O\text{---}P\overset{\textstyle OCH_3}{\underset{\textstyle OCH_3}{\Big\langle}}$$

(a) The reciprocity of splitting requires that the protons split the ^{31}P signal of phosphorus. There are 9 equivalent protons, and so the ^{31}P signal is split into 10 peaks.

(b) Each peak in the ^{31}P multiplet is separated from the next by a value equal to the $^{1}H-^{31}P$ coupling constant of 12 Hz. There are nine such intervals in a 10-line multiplet, and so the separation is 108 Hz between the highest- and lowest-field peaks in the multiplet.

SELF-TEST

PART A

A-1. Complete the following table relating to ^1H nmr spectra by supplying the missing data for entries 1 through 4.

	Spectrometer frequency	Chemical shift	
		ppm	Hz
(a)	60 MHz	$\overline{1}$	366
(b)	220 MHz	4.35	$\overline{2}$
(c)	$\overline{3}$	3.50	210
(d)	100 MHz	$\overline{4}$ of	TMS

A-2. Indicate the number of signals to be expected and the multiplicity of each in the ^1H nmr spectrum of each of the following substances:

(a) $BrCH_2CH_2CH_2Br$

(b) $CH_3CH_2\overset{\overset{\displaystyle Cl}{|}}{\underset{\underset{\displaystyle Cl}{|}}{C}}CH_2CH_3$

(c) $CH_3OCH_2\overset{\overset{\displaystyle O}{\|}}{C}OCH_3$

A-3. Two isomeric compounds having the molecular formula $C_6H_{12}O_2$ both gave ^1H nmr spectra consisting of only two singlets. Given the chemical shifts and integrations shown, identify both compounds.

Compound A: $\delta = 1.45\,ppm\ (9H)$ Compound B: $\delta = 1.20\,ppm\ (9H)$
    $\delta = 1.95\,ppm\ (3H)$ $\delta = 3.70\,ppm\ (3H)$

A-4. Identify each of the following compounds on the basis of the ir and ^1H nmr information provided:

(a) $C_{10}H_{12}O$:   ir: $1710\,cm^{-1}$
      nmr: $\delta = 1.0\,ppm$ (triplet, 3H)
         $\delta = 2.4\,ppm$ (quartet, 2H)
         $\delta = 3.6\,ppm$ (singlet, 2H)
         $\delta = 7.2\,ppm$ (singlet, 5H)

(b) $C_6H_{14}O_2$:   ir: $3400\,cm^{-1}$
      nmr: $\delta = 1.2\,ppm$ (singlet, 12H)
         $\delta = 2.0\,ppm$ (broad singlet, 2H)

(c) $C_{10}H_{16}O_6$:   ir: $1740\,cm^{-1}$
      nmr: $\delta = 1.3\,ppm$ (triplet, 9H)
         $\delta = 4.2\,ppm$ (quartet, 6H)
         $\delta = 4.4\,ppm$ (singlet, 1H)

(d) C_4H_7NO: ir: $2240\,cm^{-1}$
        $3400\,cm^{-1}$ (broad)
      nmr: $\delta = 1.65\,ppm$ (singlet, 6H)
         $\delta = 3.7\,ppm$ (singlet, 1H)

A-5. Considering the ^{13}C nmr spectrum of the substance shown, give (a) the number of signals expected and (b) the number of peaks each of these signals would be split into in the off-resonance decoupling mode.

$$Cl-\!\!\!\!\bigcirc\!\!\!\!-\overset{\displaystyle O}{\overset{\|}{C}}CH_2CH_3$$

A-6. The ^{13}C nmr of an alkane of molecular formula C_8H_{18} exhibits signals at 22.8 ppm (4C, quartet), 28.5 ppm (2C, doublet), and 37.1 ppm (2C, triplet). Suggest a reasonable structure for this alkane.

PART B

B-1. The ^1H nmr spectrum of acetone consists of a singlet with a chemical shift of 2.07 ppm. What was the spectrometer frequency of the instrument used if this chemical shift equaled 186 Hz?
(*a*) 60 MHz   (*b*) 90 MHz   (*c*) 100 MHz
(*d*) Need more information to determine

The following three problems refer to the ^1H nmr spectrum of $CH_3CH_2OCH_2OCH_2CH_3$.

B-2. How many signals are expected?
(*a*) 12   (*b*) 5   (*c*) 4   (*d*) 3

B-3. The signal farthest downfield (relative to TMS) will be a
(*a*) Singlet   (*c*) Doublet
(*b*) Triplet   (*d*) Quartet

B-4. The signal farthest upfield (closest to TMS) will be a
(*a*) Singlet   (*c*) Doublet
(*b*) Triplet   (*d*) Quartet

B-5. The relationship between magnetic field strength and the energy difference between nuclear spin states is:
(*a*) They are independent of each other.
(*b*) They are directly proportional.
(*c*) They are inversely proportional.
(*d*) The relationship varies from molecule to molecule.

B-6. An infrared spectrum exhibits a broad band in the 3000 to 3500 cm^{-1} region and a strong peak at 1710 cm^{-1}. Which of the following substances best fits the data?

(*a*) $C_6H_5CH_2CH_2OH$ (*c*) $C_6H_5CH_2\overset{\displaystyle O}{\overset{\|}{C}}CH_3$

(*b*) $C_6H_5CH_2\overset{\displaystyle O}{\overset{\|}{C}}OH$ (*d*) $C_6H_5CH_2\overset{\displaystyle O}{\overset{\|}{C}}OCH_3$

B-7. Considering the ^1H nmr spectrum of the following substance, which set of protons appears farthest downfield relative to TMS?

$$a\left\{\!\!\bigcirc\!\!-\underset{b}{OCH_2}\underset{c}{C(CH_3)_3}\right.$$

B-8. Which of the following substances does *not* satisfy a proton nmr spectrum consisting of only two peaks?

(*a*) $CH_3-\underset{\displaystyle CH_3}{\overset{\displaystyle CH_3}{\underset{|}{\overset{|}{C}}}}-OCH_3$ (*b*) $CH_3-\!\!\!\!\bigcirc\!\!\!\!-CH_3$

(*c*) 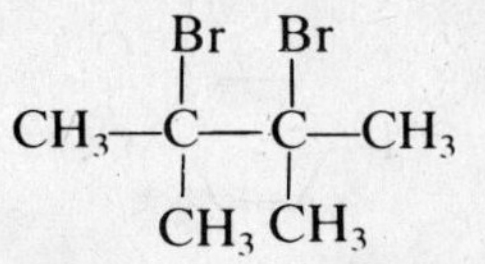

(*d*) None of these (all satisfy the spectrum)

B-9. The multiplicity of the "a" protons in the ^{1}H nmr spectrum of the following substance is:

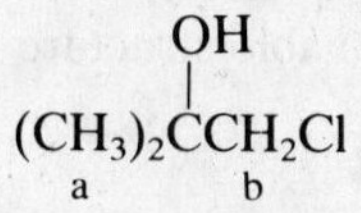

(*a*) Singlet (*b*) Doublet (*c*) Triplet (*d*) Quartet

B-10. How many signals are expected in the ^{13}C nmr spectrum of the following substance?

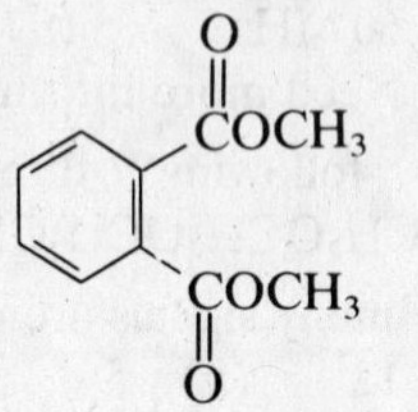

(*a*) 5 (*b*) 6 (*c*) 8 (*d*) 10

ORGANOMETALLIC COMPOUNDS

CHAPTER

14

IMPORTANT TERMS AND CONCEPTS

Bonding and Nomenclature (Secs. 14.1, 14.2) Organic compounds that possess carbon-metal bonds are classified as *organometallic compounds*. Recalling the Pauling scale of electronegativities (from Chapter 1 and Table 14.1), it may be seen that carbon is *more* electronegative than the metallic elements. Therefore an organometallic compound contains a carbon atom that is *negatively polarized*.

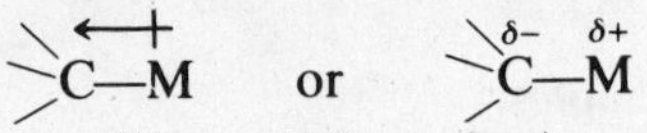

$$\overset{\longleftarrow}{\underset{}{\diagdown}C\text{—}M} \quad \text{or} \quad \overset{\delta-}{\underset{}{\diagdown}C}\text{—}\overset{\delta+}{M}$$

(M = a metallic element)

Covalently bonded organometallic compounds are, as a result of the bonding shown above, said to have *carbanionic character*. It is this property which is responsible for the synthetic utility of organometallic reagents.

Organometallic compounds are named as substituted derivatives of metals, as shown in the following examples:

$(CH_3)_2Hg$ C_6H_5MgBr
Dimethylmercury   Phenylmagnesium bromide

Planning a Synthesis; Retrosynthetic Analysis (Sec. 14.9) The technique of reasoning backward from a target molecule to suitable starting materials in planning a synthesis is known as *retrosynthetic analysis*. Organometallic reagents have as their primary synthetic utility the ability to form new carbon-carbon bonds. Breaking (or disconnecting) carbon-carbon bonds of the target molecule will allow potential precursors to be identified.

The synthesis of 1-phenyl-1-propanol provides an example of the technique. An open arrow is used to represent a retrosynthetic step.

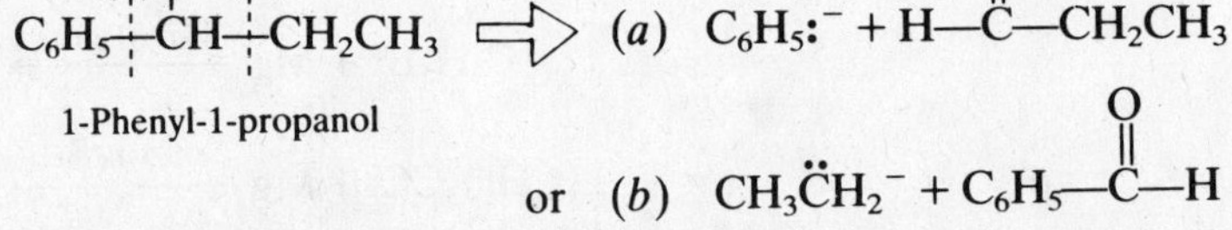

$$\underset{\text{1-Phenyl-1-propanol}}{C_6H_5\overset{(a)}{\underset{|}{\overset{OH}{|}}}CH\overset{(b)}{\underset{|}{\text{—}}}CH_2CH_3} \Longrightarrow (a)\ C_6H_5{:}^- + H\text{—}\overset{O}{\overset{\|}{C}}\text{—}CH_2CH_3$$

$$\text{or} \quad (b)\ CH_3\ddot{C}H_2^- + C_6H_5\text{—}\overset{O}{\overset{\|}{C}}\text{—}H$$

353

Synthetic routes:

(a) $C_6H_5Br + Mg \longrightarrow C_6H_5MgBr$

$$C_6H_5MgBr \xrightarrow[\text{2. } H_3O^+]{\text{1. } \overset{O}{\overset{\|}{HCCH_2CH_3}}} C_6H_5\overset{OH}{\overset{|}{CH}}CH_2CH_3$$

or (b) $CH_3CH_2Br + Mg \longrightarrow CH_3CH_2MgBr$

$$CH_3CH_2MgBr \xrightarrow[\text{2. } H_3O^+]{\text{1. } \overset{O}{\overset{\|}{C_6H_5CH}}} CH_3CH_2\overset{OH}{\overset{|}{CH}}C_6H_5$$

IMPORTANT REACTIONS

ORGANOLITHIUM COMPOUNDS

Preparation (Sec. 14.3)

General:

$$RX + 2Li \longrightarrow RLi + LiX$$
$$(X = \text{halogen})$$

Example:

$$(CH_3)_2CHBr + 2Li \longrightarrow (CH_3)_2CHLi + LiBr$$

Synthesis of Alcohols (Sec. 14.7)

General:

$$RLi + \underset{}{\overset{}{>}}C=O \longrightarrow R-\overset{|}{\underset{|}{C}}-O^-Li^+ \xrightarrow{H_3O^+} R-\overset{|}{\underset{|}{C}}-OH$$

Examples:

$$(CH_3)_2CHLi \xrightarrow[\text{2. } H_3O^+]{\text{1. } \overset{O}{\overset{\|}{CH_3CH}}} CH_3\overset{OH}{\overset{|}{CH}}CH(CH_3)_2$$

$$C_6H_5Li \xrightarrow[\text{2. } H_3O^+]{\text{1. } \overset{O}{\overset{\|}{C_6H_5CCH_3}}} (C_6H_5)_2\overset{OH}{\overset{|}{C}}CH_3$$

ORGANOMAGNESIUM COMPOUNDS (GRIGNARD REAGENTS)

Preparation (Sec. 14.4)

General:

$$RX + Mg \longrightarrow RMgX$$
$$(X = \text{halogen})$$

Examples:

$$C_6H_5Br + Mg \xrightarrow{\text{diethyl ether}} C_6H_5MgBr$$

$$(CH_3)_3CCl + Mg \xrightarrow{\text{diethyl ether}} (CH_3)_3CMgCl$$

Synthesis of Alcohols (Sec. 14.6)

General:

$$RMgX + \underset{}{\overset{}{>}}C{=}O \longrightarrow R{-}\overset{|}{\underset{|}{C}}{-}O^- \,{}^+MgX$$

$$R{-}\overset{|}{\underset{|}{C}}{-}OMgX \xrightarrow{H_3O^+} R{-}\overset{|}{\underset{|}{C}}{-}OH + HOMgX$$

Examples:

$$C_6H_5MgBr \xrightarrow[\text{2. } H_3O^+]{\text{1. } CH_3\overset{O}{\overset{\|}{CH}}} C_6H_5\overset{OH}{\overset{|}{CH}}CH_3$$

$$(CH_3)_3CMgCl \xrightarrow[\text{2. } H_3O^+]{\text{1. } H_2C{=}O} (CH_3)_3CCH_2OH$$

Acid-Base Reactions (Sec. 14.5)

General:

$$R{-}M + R'OH \longrightarrow RH + R'O^-M^+$$

Examples:

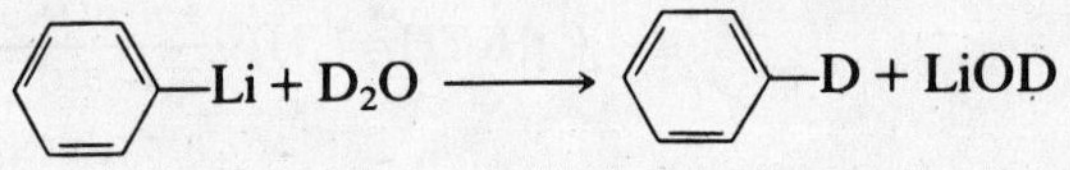

$$\text{⬡}{-}Li + D_2O \longrightarrow \text{⬡}{-}D + LiOD$$

$$(CH_3)_3CMgBr + CH_3CH_2OH \longrightarrow (CH_3)_3CH + BrMgOCH_2CH_3$$

Reaction with Esters; Synthesis of Tertiary Alcohols (Sec. 14.10)

General:

$$2RMgX \xrightarrow[\text{2. } H_3O^+]{\text{1. } R'\overset{O}{\overset{\|}{C}}OR''} R{-}\overset{OH}{\overset{|}{\underset{|}{\underset{R}{C}}}}{-}R' + R''OH$$

Example:

$$2CH_3CH_2MgBr \xrightarrow[\text{2. } H_3O^+]{\text{1. } C_6H_5CO_2CH_3} C_6H_5\overset{OH}{\overset{|}{C}}(CH_2CH_3)_2 + CH_3OH$$

ORGANOCOPPER REAGENTS (Sec. 14.11)

Synthesis of Hydrocarbons:

1. $2CH_3CH_2Li + CuBr \longrightarrow (CH_3CH_2)_2CuLi + LiBr$
2. $(CH_3CH_2)_2CuLi + C_6H_5CH_2I \longrightarrow C_6H_5CH_2CH_2CH_3 + CH_3CH_2Cu + LiI$

ORGANOZINC INTERMEDIATES (Sec. 14.12)

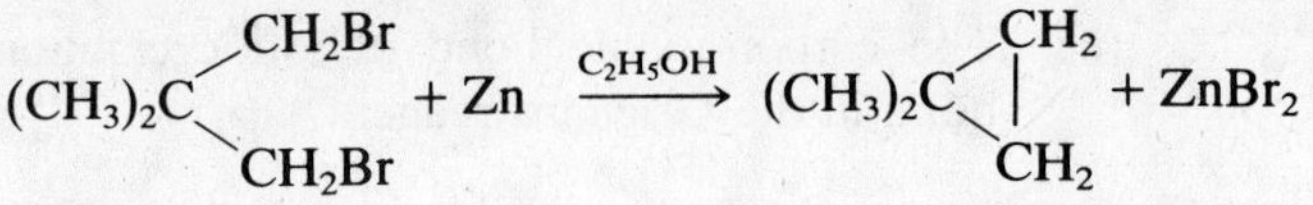

$$(CH_3)_2C\underset{CH_2Br}{\overset{CH_2Br}{<}} + Zn \xrightarrow{C_2H_5OH} (CH_3)_2C\underset{CH_2}{\overset{CH_2}{<}}| + ZnBr_2$$

$$\text{⬡}{=} + CH_2I_2 \xrightarrow{Zn(Cu)} \text{⬡}{◁}$$

ORGANIC DERIVATIVES OF MERCURY (Sec. 14.14)

Synthesis of Alcohols

General:

$$\underset{}{\diagdown}C=C\underset{}{\diagup} \xrightarrow[\text{THF-H}_2\text{O}]{\text{Hg(O}_2\text{CCH}_3)_2} \xrightarrow[\text{HO}^-]{\text{NaBH}_4} \overset{\text{H}}{\underset{}{-\overset{|}{\underset{|}{C}}}}\overset{\text{OH}}{\underset{}{-\overset{|}{\underset{|}{C}}-}}$$

Example:

$$(\text{CH}_3)_2\text{CHCH}=\text{CH}_2 \xrightarrow[\text{2. NaBH}_4,\ \text{HO}^-]{\text{1. Hg(OAc)}_2,\ \text{THF-H}_2\text{O}} (\text{CH}_3)_2\text{CHCHCH}_3 \overset{\text{OH}}{}$$

Synthesis of Ethers

General:

$$\underset{}{\diagdown}C=C\underset{}{\diagup} + \text{ROH} \xrightarrow[\text{2. NaBH}_4,\ \text{HO}^-]{\text{1. Hg(OAc)}_2} \text{H}-\overset{|}{\underset{|}{C}}-\overset{|}{\underset{|}{C}}-\text{OR}$$

Examples:

$$\text{CH}_3\text{CH}=\text{CH}_2 \xrightarrow[\text{2. NaBH}_4,\ \text{HO}^-]{\text{1. Hg(OAc)}_2,\ \text{CH}_3\text{OH}} (\text{CH}_3)_2\text{CHOCH}_3$$

$$\text{C}_6\text{H}_5\text{CH}=\text{CH}_2 \xrightarrow[\text{2. NaBH}_4,\ \text{HO}^-]{\text{1. Hg(OAc)}_2,\ \text{C}_2\text{H}_5\text{OH}} \text{C}_6\text{H}_5\text{CHCH}_3 \overset{\text{OCH}_2\text{CH}_3}{}$$

SOLUTIONS TO TEXT PROBLEMS

14.1 (b) Magnesium bears a cyclohexyl substituent and a chlorine. Chlorine is named in its anionic form. The compound is cyclohexylmagnesium chloride.

(c) This substance is iodomethylzinc iodide.

14.2 (b) The alkyl bromide precursor to *sec*-butyllithium must be *sec*-butyl bromide.

$$\underset{\overset{|}{\text{Br}}}{\text{CH}_3\text{CHCH}_2\text{CH}_3} + 2\text{Li} \longrightarrow \underset{\overset{|}{\text{Li}}}{\text{CH}_3\text{CHCH}_2\text{CH}_3} + \text{LiBr}$$

2-Bromobutane   1-Methylpropyllithium

(*sec*-butyl bromide)   (*sec*-butyllithium)

(c) Benzylsodium is an organometallic compound in which the alkyl group is $\text{C}_6\text{H}_5\text{CH}_2$. Benzyl bromide is the correct precursor.

$$\text{C}_6\text{H}_5-\text{CH}_2\text{Br} + 2\text{Na} \longrightarrow \text{C}_6\text{H}_5-\text{CH}_2\text{Na} + \text{NaBr}$$

Benzyl bromide   Benzylsodium

14.3 (b) Allyl chloride is converted to allylmagnesium chloride on reaction with magnesium.

$$\text{CH}_2=\text{CHCH}_2\text{Cl} \xrightarrow[\text{diethyl ether}]{\text{Mg}} \text{CH}_2=\text{CHCH}_2\text{MgCl}$$

Allyl chloride   Allylmagnesium chloride

(c) The carbon-iodine bond of iodocyclobutane is replaced by a carbon-magnesium bond in the Grignard reagent.

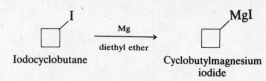

Iodocyclobutane   Cyclobutylmagnesium
iodide

(*d*) Bromine is attached to sp^2 hybridized carbon in 1-bromocyclohexene. The product of its reaction with magnesium has a carbon-magnesium bond in place of the carbon-bromine bond.

1-Bromocyclohexene $\xrightarrow[\text{diethyl ether}]{\text{Mg}}$ 1-Cyclohexenylmagnesium bromide

14.4 (*b*) 1-Hexanol will protonate butyllithium, since its hydroxyl group is a proton donor only slightly less acidic than water. This proton-transfer reaction could be used to prepare lithium 1-hexanolate.

$$CH_3CH_2CH_2CH_2CH_2CH_2OH + CH_3CH_2CH_2CH_2Li \longrightarrow$$

     1-Hexanol   Butyllithium

$$CH_3CH_2CH_2CH_3 + CH_3CH_2CH_2CH_2CH_2CH_2OLi$$

        Butane   Lithium 1-hexanolate

(*c*) The proton donor here is benzenethiol.

$$C_6H_5SH + CH_3CH_2CH_2CH_2Li \longrightarrow CH_3CH_2CH_2CH_3 + C_6H_5SLi$$

Benzenethiol   Butyllithium   Butane   Lithium benzenethiolate

14.5 (*b*) Propylmagnesium bromide reacts with benzaldehyde by addition to the carbonyl group.

(*c*) Tertiary alcohols result from the reaction of Grignard reagents and ketones.

1-Propylcyclohexanol

(*d*) The starting material is a ketone and so reacts with a Grignard reagent to give a tertiary alcohol.

Propylmagnesium bromide + 2-butanone   3-Methyl-3-hexanol

14.6 (*b*) The target alcohol is secondary and so can be prepared by the reaction of a Grignard reagent with an aldehyde. The two retrosynthetic transformations are:

and

$$\text{Ph}-\underset{\underset{\text{OH}}{|}}{\text{CH}}\text{CH}_2\text{CH}_3 \Longrightarrow \text{Ph}-\underset{\overset{\|}{\text{O}}}{\text{CH}} \quad {}^-:\text{CH}_2\text{CH}_3$$

Therefore, two plausible syntheses are:

$$\text{Ph}-\text{MgBr} + \text{CH}_3\text{CH}_2\overset{\overset{\text{O}}{\|}}{\text{CH}} \xrightarrow[\text{2. H}_3\text{O}^+]{\text{1. diethyl}\atop\text{ether}} \text{Ph}-\underset{\underset{\text{OH}}{|}}{\text{CH}}\text{CH}_2\text{CH}_3$$

Phenylmagnesium   Propanal   1-Phenyl-1-propanol
bromide

and

$$\text{CH}_3\text{CH}_2\text{MgBr} + \text{Ph}-\overset{\overset{\text{O}}{\|}}{\text{CH}} \xrightarrow[\text{2. H}_3\text{O}^+]{\text{1. diethyl}\atop\text{ether}} \text{Ph}-\underset{\underset{\text{OH}}{|}}{\text{CH}}\text{CH}_2\text{CH}_3$$

Ethylmagnesium   Benzaldehyde   1-Phenyl-1-propanol
bromide

(c) The target alcohol is tertiary and so is prepared by addition of a Grignard reagent to a ketone.

$$^-:\text{CH}_3 \quad \text{Ph}-\overset{\overset{\text{O}}{\|}}{\text{C}}-\text{CH}_3 \quad \Longleftarrow \quad \text{CH}_3-\underset{\underset{\text{Ph}}{|}}{\overset{\overset{\text{OH}}{|}}{\text{C}}}-\text{CH}_3 \quad \Longrightarrow \quad \text{CH}_3\overset{\overset{\text{O}}{\|}}{\text{C}}\text{CH}_3 \quad \text{Ph}^{\cdot\cdot-}$$

Since two of the substituents on the hydroxyl-bearing carbon are methyl groups, there are only *two*, not three, distinct ways of preparing 2-phenyl-2-propanol by a Grignard reaction:

$$\text{CH}_3\text{MgI} + \text{Ph}-\overset{\overset{\text{O}}{\|}}{\text{C}}\text{CH}_3 \xrightarrow[\text{2. H}_3\text{O}^+]{\text{1. diethyl}\atop\text{ether}} \text{Ph}-\underset{\underset{\text{CH}_3}{|}}{\overset{\overset{\text{OH}}{|}}{\text{C}}}\text{CH}_3$$

Methylmagnesium   Acetophenone   2-Phenyl-2-propanol
iodide

$$\text{Ph}-\text{MgBr} + \text{CH}_3\overset{\overset{\text{O}}{\|}}{\text{C}}\text{CH}_3 \xrightarrow[\text{2. H}_3\text{O}^+]{\text{1. diethyl}\atop\text{ether}} \text{Ph}-\underset{\underset{\text{CH}_3}{|}}{\overset{\overset{\text{OH}}{|}}{\text{C}}}-\text{CH}_3$$

Phenylmagnesium   Acetone   2-Phenyl-2-propanol
bromide

14.7 (b) Begin by drawing the structure of the target molecule, 6-methyl-6-undecanol, to reveal the presence of two pentyl groups. Recall that the two groups that are the same come from the Grignard reagent. This reaction has been described in the chemical literature, and was carried out using pentylmagnesium bromide and ethyl acetate.

$$\text{CH}_3\text{CH}_2\text{CH}_2\text{CH}_2\text{CH}_2-\underset{\underset{\text{CH}_3}{|}}{\overset{\overset{\text{OH}}{|}}{\text{C}}}-\text{CH}_2\text{CH}_2\text{CH}_2\text{CH}_2\text{CH}_3 \qquad \text{6-Methyl-6-undecanol}$$

$$2CH_3CH_2CH_2CH_2CH_2MgBr + CH_3\overset{\overset{\displaystyle O}{\|}}{C}OCH_2CH_3 \xrightarrow[\text{2. } H_3O^+]{\text{1. diethyl ether}}$$

Pentylmagnesium bromide   Ethyl acetate

$$CH_3\overset{\overset{\displaystyle OH}{|}}{C}(CH_2CH_2CH_2CH_2CH_3)_2 + CH_3CH_2OH$$
6-Methyl-6-undecanol   Ethanol

(*c*) The two phenyl substituents arise by addition of a phenyl Grignard reagent to an ester of cyclopropanecarboxylic acid.

$$2C_6H_5MgBr \quad + \quad \triangleright\!-\!\overset{\overset{\displaystyle O}{\|}}{C}OCH_3 \xrightarrow[\text{2. } H_3O^+]{\text{1. diethyl ether}} (C_6H_5)_2\overset{\overset{\displaystyle }{|}}{\underset{\overset{\displaystyle }{OH}}{C}}\!\!\triangleleft + CH_3OH$$

Phenylmagnesium   Methyl   Cyclopropyl-   Methanol
bromide   cyclopropanecarboxylate   diphenylmethanol

(*d*) All the substituents bonded to the $-\overset{|}{\underset{|}{C}}OH$ group in this compound are phenyl groups. Therefore use an ester of benzoic acid and phenylmagnesium bromide.

$$2\,\langle\!\!\bigcirc\!\!\rangle\!-\!MgBr + \langle\!\!\bigcirc\!\!\rangle\!-\!\overset{\overset{\displaystyle O}{\|}}{C}OCH_2CH_3 \xrightarrow[\text{2. } H_3O^+]{\text{1. diethyl ether}} \cdots + CH_3CH_2OH$$

Phenylmagnesium   Ethyl benzoate   Triphenylmethanol   Ethanol
bromide

14.8 (*b*) Of the three methyl groups of 1,3,3-trimethylcyclopentene, only the one connected to the double bond can be attached by way of an organocuprate reagent. Attachment of either of the other methyls would involve a tertiary carbon, a process that does not occur very efficiently.

$$LiCu(CH_3)_2 + \cdots \xrightarrow{\text{diethyl ether}} \cdots$$

Lithium   1-Bromo-3,3-dimethylcyclopentene   1,3,3-Trimethylcyclopentene
dimethylcuprate

14.9 Either the C-1—C-2 bond or the C-2—C-3 bond of methylcyclopropane can be formed in the cyclization reaction.

Mentally Disconnect C-1—C-2 Bond:

$$CH_3 \text{—}\triangle \;\Rightarrow\; \overset{\overset{\displaystyle CH_3}{|}}{\underset{\displaystyle CH_2\!-\!CH_2Br}{CHBr}}$$
1,3-Dibromobutane

Mentally Disconnect C-2—C-3 Bond:

$$\triangle \;\Rightarrow\; BrCH_2\overset{\overset{\displaystyle CH_3}{|}}{\underset{\displaystyle CH_2Br}{CH}}$$
1,3-Dibromo-2-methylpropane

Both 1,3-dibromobutane and 1,3-dibromo-2-methylpropane are suitable starting materials for the preparation of methylcyclopropane by zinc-promoted cyclization.

14.10 (b) Methylenecyclobutane is the appropriate precursor to the spirohexane shown.

Methylenecyclobutane → Spiro[3.2]hexane (22%)

14.11 Syn- addition of dibromocarbene to *cis*-2-butene yields a cyclopropane derivative in which the methyl groups are cis.

cis-2-Butene → *cis*-1,1-Dibromo-2,3-dimethylcyclopropane

Conversely, the methyl groups in the cyclopropane derivative of *trans*-2-butene are trans to one another.

trans-2-Butene → *trans*-1,1-Dibromo-2,3-dimethylcyclopropane

14.12 The two alcohols are 2-pentanol and 3-pentanol.

trans-2-Pentene → $CH_3CHCH_2CH_2CH_3$ + $CH_3CH_2CHCH_2CH_3$ (OH) 2-Pentanol, 3-Pentanol

The substituents at either end of the double bond are so similar that there is little regioselective differentiation between them.

14.13 (b) The tertiary alcohol 1-methylcyclopentanol may be prepared either from methylene-cyclopentane or from 1-methylcyclopentene.

Methylenecyclopentane or 1-Methylcyclopentene → 1-Methylcyclopentanol

(c) Either the cis or the trans isomer of 3-hexene may be used as the starting alkene in order to prepare 3-hexanol by oxymercuration-demercuration.

cis-3-Hexene *trans*-3-Hexene → 3-Hexanol

2-Hexene (either *cis*- or *trans*-$CH_3CH=CHCH_2CH_2CH_3$) is not a satisfactory answer, because there is no reason to believe that addition will be regioselective in the direction that gives the desired 3-hexanol. A mixture of 2- and 3-hexanol will be produced.

14.14 (b) In principle solvomercuration of ethylene could be employed as a synthesis of the desired ethyl 1-methylpentyl ether.

$$CH_2{=}CH_2 \xrightarrow[\text{2. NaBH}_4,\ \text{HO}^-]{\text{1. Hg(OAc)}_2,\ \overset{\displaystyle \overset{OH}{|}}{CH_3CHCH_2CH_2CH_2CH_3}} \underset{\underset{\displaystyle OCH_2CH_3}{|}}{CH_3CHCH_2CH_2CH_2CH_3}$$

Ethylene   Ethyl 1-methylpentyl ether

Since ethylene is a gas, however, it is more convenient to prepare alkyl ethyl ethers by the alternative procedure using ethanol as the solvent in a solvomercuration reaction.

$$CH_2{=}CHCH_2CH_2CH_2CH_3 \xrightarrow[\text{2. NaBH}_4,\ \text{HO}^-]{\text{1. Hg(OAc)}_2,\ CH_3CH_2OH} \underset{\underset{\displaystyle OCH_2CH_3}{|}}{CH_3CHCH_2CH_2CH_2CH_3}$$

1-Hexene   Ethyl 1-methylpentyl ether (98%)

The orientation of addition corresponds to that of Markovnikov's rule. Notice that 2-hexene ($CH_3CH{=}CHCH_2CH_2CH_3$) would not be suitable as a starting material, because it would likely lead to formation of two regioisomeric ethyl ethers.

14.15   (*a*)   Cyclopentyllithium is

It has a carbon-lithium bond. It satisfies the requirement for classification as an organometallic compound.

(*b*)   Ethoxymagnesium chloride does not have a carbon-metal bond. It is not an organometallic compound.

$$CH_3CH_2OMgCl \quad \text{or} \quad CH_3CH_2O^-Mg^{2+}Cl^-$$

(*c*)   2-Phenylethylmagnesium iodide is an example of a Grignard reagent. It is an organometallic compound.

(*d*)   Lithium divinylcuprate has two vinyl groups bonded to copper. It is an organometallic compound.

$$Li^+(CH_2{=}CH{-}\bar{C}u{-}CH{=}CH_2)$$

(*e*)   Mercury(II) acetate is not an organometallic compound. Two acetate groups are ionically bonded to mercury through oxygen, not carbon.

$$Hg^{2+}(^-O\overset{\displaystyle\overset{O}{\|}}{C}CH_3)_2$$

(*f*)   Benzylpotassium is represented as

It has a carbon-potassium bond and thus is an organometallic compound.

(*g*)   Sodium *p*-toluenesulfonate has its sodium atom ionically bonded to oxygen. It is not an organometallic compound.

14.16 (*a*) Grignard reagents such as pentylmagnesium iodide are prepared by reaction of magnesium with the corresponding alkyl halide.

$$\text{CH}_3\text{CH}_2\text{CH}_2\text{CH}_2\text{CH}_2\text{I} + \text{Mg} \xrightarrow{\text{diethyl ether}} \text{CH}_3\text{CH}_2\text{CH}_2\text{CH}_2\text{CH}_2\text{MgI}$$

1-Iodopentane   Pentylmagnesium iodide

(*b*) Acetylenic Grignard reagents are normally prepared by reaction of a terminal alkyne with a readily available Grignard reagent such as an ethylmagnesium halide. The reaction that takes place is an acid-base reaction in which the terminal alkyne acts as a proton donor.

$$\text{CH}_3\text{CH}_2\text{C}\equiv\text{CH} + \text{CH}_3\text{CH}_2\text{MgI} \xrightarrow{\text{diethyl ether}} \text{CH}_3\text{CH}_2\text{C}\equiv\text{CMgI} + \text{CH}_3\text{CH}_3$$

1-Butyne   Ethylmagnesium   1-Butynylmagnesium   Ethane
                       iodide   iodide

(*c*) Alkyllithiums are formed by reaction of lithium with an alkyl halide.

$$\text{CH}_3\text{CH}_2\text{CH}_2\text{CH}_2\text{CH}_2\text{X} + 2\text{Li} \longrightarrow \text{CH}_3\text{CH}_2\text{CH}_2\text{CH}_2\text{CH}_2\text{Li} + \text{LiX}$$

1-Halopentane   Pentyllithium
(X = Cl, Br, or I)

(*d*) Lithium dialkylcuprates arise by the reaction of an alkyllithium with a Cu(I) salt.

$$2\text{CH}_3\text{CH}_2\text{CH}_2\text{CH}_2\text{CH}_2\text{Li} + \quad \text{CuX} \quad \longrightarrow$$

Pentyllithium,   (X = Cl, Br, or I)
from part (c)
                                        $$\text{LiCu}(\text{CH}_2\text{CH}_2\text{CH}_2\text{CH}_2\text{CH}_3)_2 + \text{LiX}$$

                                        Lithium dipentylcuprate

(*e*) The organometallic compound is a β-hydroxyethylmercury(II) acetate. Compounds of this type are formed from alkenes in the oxymercuration stage of the oxymercuration-demercuration sequence. The customary solvent is aqueous tetrahydrofuran.

$$\text{CH}_2{=}\text{CH}_2 + \text{Hg}(\text{OCCH}_3)_2 + \text{H}_2\text{O} \xrightarrow{\text{THF}} \text{HOCH}_2\text{CH}_2\text{HgOCCH}_3 + \text{CH}_3\text{COH}$$

Ethylene   Mercury(II)   Water   2-Hydroxyethylmercury(II)   Acetic
                   acetate   acetate   acid

(*f*) When mercury(II) acetate and an alkene react in an alcohol solvent, the alcohol is incorporated into the product as a β-alkoxy group.

$$\text{CH}_2{=}\text{CH}_2 + \text{Hg}(\text{OCCH}_3)_2 + \text{CH}_3\text{OH} \longrightarrow \text{CH}_3\text{OCH}_2\text{CH}_2\text{HgOCCH}_3 + \text{CH}_3\text{COH}$$

Ethylene   Mercury(II)   Methanol   2-Methoxyethylmercury(II)   Acetic
                   acetate   acetate   acid

14.17 The polarity of a covalent bond increases with an increase in the electronegativity difference between the connected atoms. Carbon has an electronegativity of 2.5 (Table 14.1). Metals are less electronegative than carbon. Therefore, when comparing two metals, the less electronegative one has the more polar bond to carbon.

(*a*) Table 14.1 gives the electronegativity of lithium as 1.0 while that for mercury is 1.9. The carbon-lithium bond in $\text{CH}_3\text{CH}_2\text{Li}$ is more polar than the carbon-mercury bond in $(\text{CH}_3\text{CH}_2)_2\text{Hg}$.

(*b*) The electronegativity of magnesium (1.2) is less than that of zinc (1.6). Therefore $(\text{CH}_3)_2\text{Mg}$ has a more polar carbon-metal bond than $(\text{CH}_3)_2\text{Zn}$.

(*c*) In this part of the problem two Grignard reagents are compared. Magnesium is the metal in both cases. What is different is the hybridization state of carbon. The *sp* hybridized carbon in $\text{HC}\equiv\text{CMgBr}$ is more electronegative than the sp^3 hybridized carbon in $\text{CH}_3\text{CH}_2\text{MgBr}$, so that $\text{HC}\equiv\text{CMgBr}$ has a more polar carbon-magnesium bond.

14.18 (a) $CH_3CH_2CH_2Br + 2Li \xrightarrow{\text{diethyl ether}} CH_3CH_2CH_2Li + LiBr$

      1-Bromopropane   Propyllithium

(b) $CH_3CH_2CH_2Br + Mg \xrightarrow{\text{diethyl ether}} CH_3CH_2CH_2MgBr$

      1-Bromopropane   Propylmagnesium
                                                     bromide

(c) $\underset{\underset{\text{2-Iodopropane}}{|}}{CH_3\overset{}{C}HCH_3} + 2Li \xrightarrow{\text{diethyl ether}} \underset{\underset{\text{Isopropyllithium}}{Li}}{CH_3\overset{}{C}HCH_3} + LiI$

(d) $\underset{\underset{\text{2-Iodopropane}}{I}}{CH_3\overset{}{C}HCH_3} + Mg \xrightarrow{\text{diethyl ether}} \underset{\underset{\underset{\text{iodide}}{\text{Isopropylmagnesium}}}{MgI}}{CH_3\overset{}{C}HCH_3}$

(e) $\underset{\underset{I}{|}}{CH_3\overset{}{C}HCH_3} \xrightarrow[\text{acetic acid}]{Zn} \underset{\text{Propane}}{CH_3CH_2CH_3}$

      2-Iodopropane

✳(f) $2CH_3CH_2CH_2Li + CuI \longrightarrow (CH_3CH_2CH_2)_2CuLi$

      Propyllithium   Lithium dipropylcuprate

(g) $(CH_3CH_2CH_2)_2CuLi + CH_3CH_2CH_2CH_2Br \longrightarrow CH_3CH_2CH_2CH_2CH_2CH_2CH_3$

      Lithium dipropylcuprate   1-Bromobutane   Heptane

(h) $(CH_3CH_2CH_2)_2CuLi +$ (iodobenzene) $\longrightarrow$ (propylbenzene, $CH_2CH_2CH_3$)

      Lithium dipropylcuprate   Iodobenzene   Propylbenzene

(i) $CH_3CH_2CH_2MgBr \xrightarrow[\text{DCl}]{D_2O} CH_3CH_2CH_2D$

      Propylmagnesium   1-Deuteriopropane
             bromide

(j) $\underset{\underset{\underset{\text{iodide}}{\text{Isopropylmagnesium}}}{MgI}}{CH_3\overset{}{C}HCH_3} \xrightarrow[\text{DCl}]{D_2O} \underset{\underset{\underset{\text{2-Deuteriopropane}}{D}}{|}}{CH_3\overset{}{C}HCH_3}$

(k) $CH_3CH_2CH_2Li + H\overset{\overset{O}{\|}}{C}H \xrightarrow[\text{2. } H_3O^+]{\text{1. diethyl ether}} CH_3CH_2CH_2CH_2OH$

      Propyllithium   1-Butanol

(l) $CH_3CH_2CH_2MgBr +$ (benzaldehyde, $C\overset{\overset{O}{\|}}{H}$) $\xrightarrow[\text{2. } H_3O^+]{\text{1. diethyl ether}}$ (1-phenyl-1-butanol, $\underset{\underset{OH}{|}}{C}HCH_2CH_2CH_3$)

      Propylmagnesium   Benzaldehyde   1-Phenyl-1-butanol
            bromide

(m) $\underset{\underset{Li}{|}}{CH_3\overset{}{C}HCH_3} +$ (cycloheptanone) $\xrightarrow[\text{2. } H_3O^+]{\text{1. diethyl ether}}$ (1-isopropylcycloheptanol, $(CH_3)_2CH$ — OH)

      Isopropyllithium   Cycloheptanone   1-Isopropylcycloheptanol

(n) $\underset{\underset{MgI}{|}}{CH_3\overset{}{C}HCH_3} + CH_3\overset{\overset{O}{\|}}{C}CH_2CH_3 \xrightarrow[\text{2. } H_3O^+]{\text{1. diethyl ether}} \underset{\underset{CH_3 \quad CH_3}{|\quad\quad|}}{CH_3\overset{}{C}H-\overset{\overset{OH}{|}}{C}CH_2CH_3}$

      Isopropylmagnesium iodide   2-Butanone   2,3-Dimethyl-3-pentanol

(o) $CH_3CH_2CH_2MgBr + C_6H_5\overset{\overset{O}{\|}}{C}OCH_3 \xrightarrow[\text{2. } H_3O^+]{\text{1. diethyl ether}} C_6H_5\overset{\overset{OH}{|}}{C}(CH_2CH_2CH_3)_2$

Propylmagnesium   Methyl   4-Phenyl-4-heptanol
bromide   benzoate

✳ (p) $CH_2{=}CH(CH_2)_5CH_3 \xrightarrow[\text{diethyl ether}]{\underset{Zn(Cu)}{CH_2I_2}} CH_2{-}CH(CH_2)_5CH_3$ with CH_2 bridge

1-Octene   1-Cyclopropylhexane

✳ (q) (E)-2-Decene $\xrightarrow[\text{diethyl ether}]{\underset{Zn(Cu)}{CH_2I_2}}$ trans-1-Heptyl-2-methylcyclopropane

(r) (Z)-3-Decene $\xrightarrow[\text{diethyl ether}]{\underset{Zn(Cu)}{CH_2I_2}}$ cis-1-Ethyl-2-hexylcyclopropane

✳ (s) $ICH_2\overset{\overset{}{|}}{C}HCH_3 \xrightarrow[\text{ethanol}]{Zn} CH_2{=}CHCH_3$ with I below

1,2-Diiodopropane   Propene

(t) $ICH_2CH_2CH_2I \xrightarrow[\text{ethanol}]{Zn} CH_2{-}CH_2$ with CH_2 bridge

1,3-Diiodopropane   Cyclopropane

✳ (u) $CH_2{=}CHCH_2CH_2CH_3 \xrightarrow[\text{2. } NaBH_4,\ HO^-]{\text{1. } Hg(OCCH_3)_2,\ THF{-}H_2O} CH_3\overset{}{C}HCH_2CH_2CH_3$ with OH

1-Pentene   2-Pentanol

✳ (v) $CH_2{=}CHCH_2CH_2CH_3 \xrightarrow[\text{2. } NaBH_4,\ HO^-]{\text{1. } Hg(OCCH_3)_2,\ CH_3CH_2OH} CH_3\overset{}{C}HCH_2CH_2CH_3$ with OCH_2CH_3

1-Pentene   2-Ethoxypentane

✳ (w) $CH_2{=}CHCH_2CH_2CH_3 \xrightarrow[\text{KOC(CH}_3)_3]{CHBr_3}$ 1,1-Dibromo-2-propylcyclopropane

1-Pentene   1,1-Dibromo-2-propylcyclopropane

14.19   In the solutions to this problem, the Grignard reagent butylmagnesium bromide is used. In each case the use of butyllithium would be equally satisfactory.

(a)   1-Pentanol is a primary alcohol having one more carbon atom than 1-bromobutane. This suggests the reaction of a Grignard reagent with formaldehyde.

$CH_3CH_2CH_2CH_2Br \xrightarrow[\text{ether}]{\underset{\text{diethyl}}{Mg}} CH_3CH_2CH_2CH_2MgBr \xrightarrow[\text{2. } H_3O^+]{\text{1. } H\overset{\overset{O}{\|}}{C}H} CH_3CH_2CH_2CH_2CH_2OH$

1-Bromobutane   Butylmagnesium bromide   1-Pentanol

(b)   2-Hexanol is a secondary alcohol having two more carbon atoms than 1-bromobutane. It may be prepared by reaction of ethanal with butylmagnesium bromide.

$$CH_3CH_2CH_2CH_2Br \xrightarrow[\text{diethyl ether}]{\text{Mg}} CH_3CH_2CH_2CH_2MgBr \xrightarrow[\text{2. H}_3O^+]{\text{1. CH}_3\text{CH}} CH_3CH_2CH_2CH_2CHCH_3$$

1-Bromobutane   Butylmagnesium bromide   2-Hexanol

(c) 1-Phenyl-1-pentanol is a secondary alcohol. This suggests that it can be prepared from butylmagnesium bromide and an aldehyde; benzaldehyde is the appropriate aldehyde.

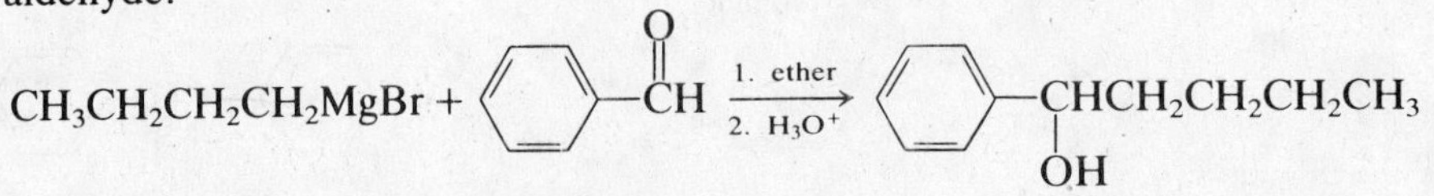

Butylmagnesium bromide   Benzaldehyde   1-Phenyl-1-pentanol

(d) The target molecule 3-methyl-3-heptanol has the structure

$$CH_3CH_2CH_2CH_2-\overset{\overset{\displaystyle CH_3}{|}}{\underset{\underset{\displaystyle OH}{|}}{C}}-CH_2CH_3$$

By retrosynthetically disconnecting the butyl group from the carbon that bears the hydroxyl substituent, we see that the appropriate starting ketone is 2-butanone.

2-Butanone

Therefore

Butylmagnesium bromide   2-Butanone   3-Methyl-3-heptanol

(e) 1-Butylcyclobutanol is a tertiary alcohol. The appropriate ketone is cyclobutanone.

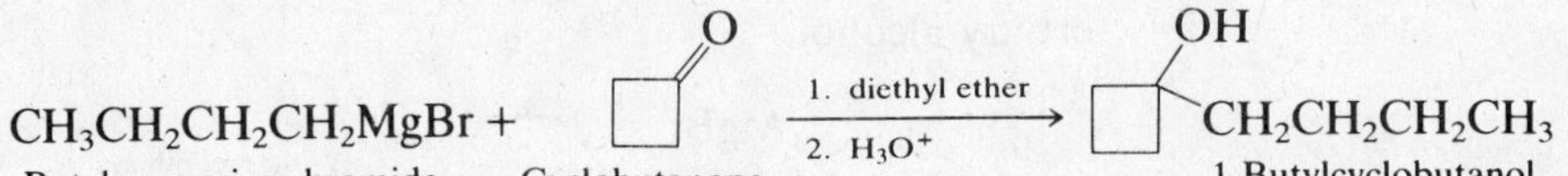

$$CH_3CH_2CH_2CH_2MgBr +$$

Butylmagnesium bromide   Cyclobutanone   1-Butylcyclobutanol

14.20 (a) Conversion of bromobenzene to benzyl alcohol requires formation of the corresponding Grignard reagent and its reaction with formaldehyde.

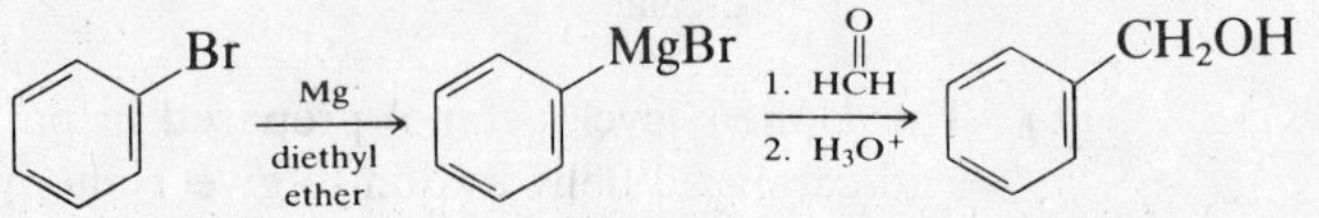

Bromobenzene   Phenylmagnesium bromide   Benzyl alcohol

(b) The product is a secondary alcohol and is formed by reaction of phenylmagnesium bromide with hexanal.

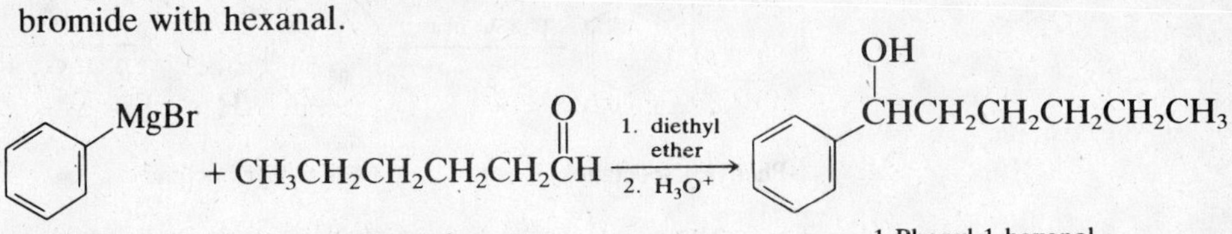

Phenylmagnesium bromide   Hexanal   1-Phenyl-1-hexanol

(*c*) The desired product is a secondary alkyl *bromide*. A reasonable synthesis would be to first prepare the analogous secondary alcohol by reaction of phenylmagnesium bromide with benzaldehyde, followed by a conversion of the alcohol to the bromide.

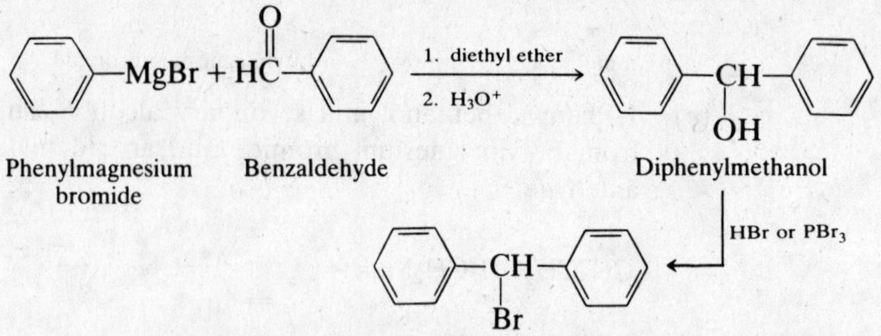

Phenylmagnesium   Benzaldehyde   Diphenylmethanol
bromide

Bromodiphenylmethane

(*d*) The target molecule is a tertiary alcohol, which requires that phenylmagnesium bromide react with a ketone. By mentally disconnecting the phenyl group from the carbon that bears the hydroxyl, we see that the appropriate ketone is 4-heptanone.

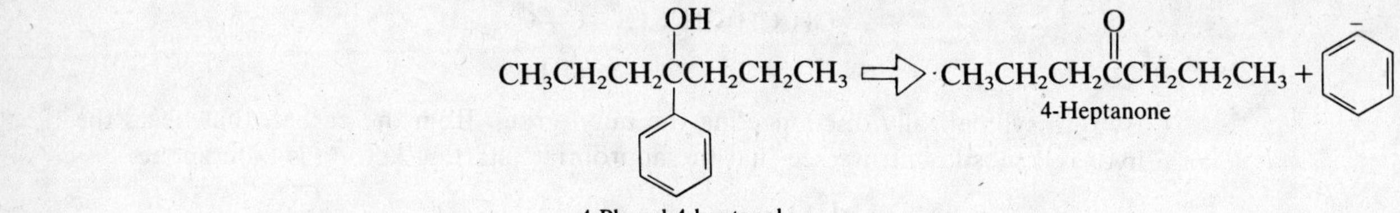

4-Phenyl-4-heptanol

The synthesis is therefore:

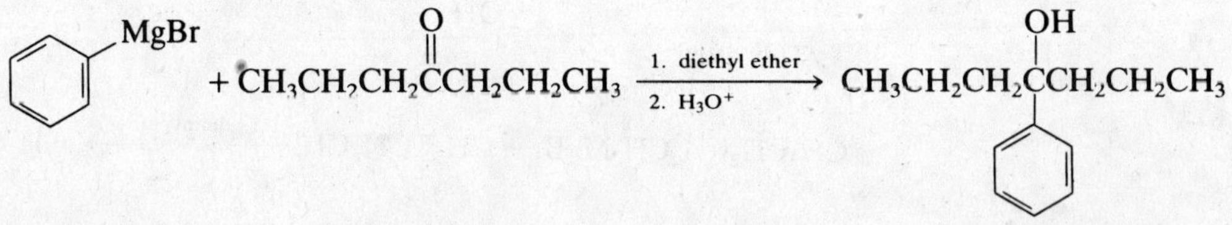

Phenylmagnesium   4-Heptanone   4-Phenyl-4-heptanol
bromide

(*e*) Reaction of phenylmagnesium bromide with cyclooctanone will give the desired tertiary alcohol.

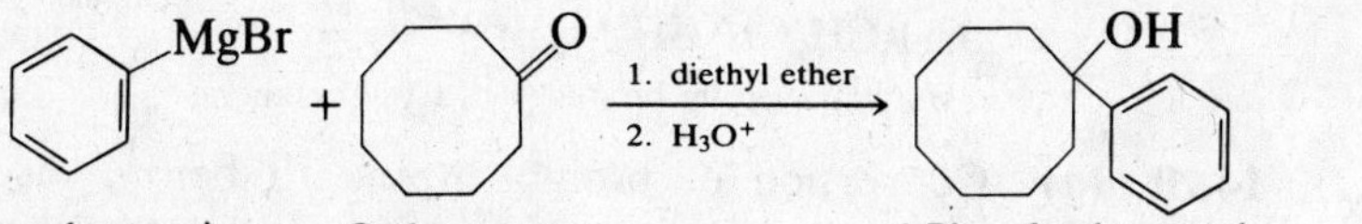

Phenylmagnesium   Cyclooctanone   1-Phenylcyclooctanol
bromide

(*f*) The 1-phenylcyclooctanol prepared in part (*e*) of this problem can be subjected to acid-catalyzed dehydration to give 1-phenylcyclooctene. Hydroboration-oxidation of 1-phenylcyclooctene gives *trans*-2-phenylcyclooctanol.

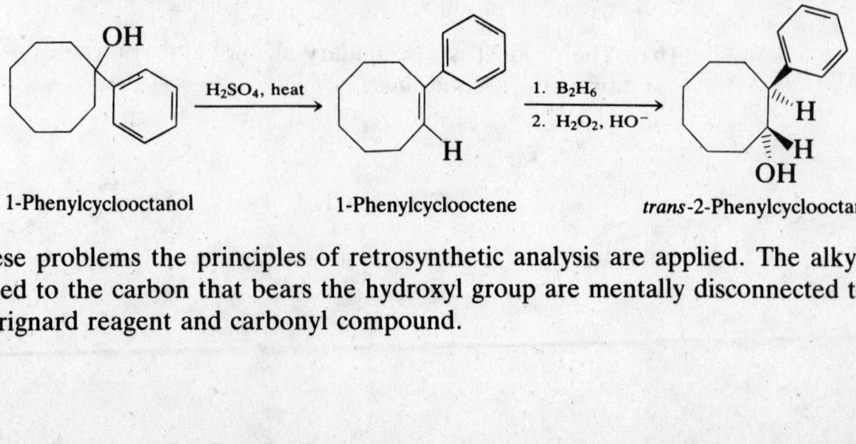

1-Phenylcyclooctanol   1-Phenylcyclooctene   *trans*-2-Phenylcyclooctanol

14.21 In these problems the principles of retrosynthetic analysis are applied. The alkyl groups attached to the carbon that bears the hydroxyl group are mentally disconnected to reveal the Grignard reagent and carbonyl compound.

(*a*)

$$CH_3CH_2CHCH_2CH(CH_3)_2$$
$$|$$
$$OH$$

5-Methyl-3-hexanol

$$\left[\begin{array}{c} CH_3CH_2CH + XMgCH_2CH(CH_3)_2 \\ \parallel \\ O \\ \text{Propanal} \qquad \text{Isobutylmagnesium halide} \end{array} \right]$$

$$\left[\begin{array}{c} CH_3CH_2MgX + HCCH_2CH(CH_3)_2 \\ \parallel \\ O \\ \text{Ethylmagnesium} \qquad \text{3-Methylbutanal} \\ \text{halide} \end{array} \right]$$

(*b*)

$$\triangleright\!-\!\underset{\underset{OH}{|}}{CH}\!-\!\!\!\left\langle\!\!\!\bigcirc\!\!\!\right\rangle\!\!-\!OCH_3$$

1-Cyclopropyl-1-(*p*-anisyl)methanol

$$\left[\begin{array}{cc} \triangleright\!-\!CH & + XMg\!-\!\!\left\langle\!\!\bigcirc\!\!\right\rangle\!\!-\!OCH_3 \\ \parallel & \\ O & \\ \text{Cyclopropanecarbaldehyde} & \text{p-Anisylmagnesium} \\ & \text{halide} \end{array} \right]$$

$$\left[\begin{array}{cc} \triangleright\!-\!MgX & + \; HC\!-\!\!\left\langle\!\!\bigcirc\!\!\right\rangle\!\!-\!OCH_3 \\ & \parallel \\ & O \\ \text{Cyclopropylmagnesium} & \text{p-Anisaldehyde} \\ \text{halide} & \end{array} \right]$$

(*c*)

$$(CH_3)_3CCH_2OH$$
2,2-Dimethyl-1-propanol

$$\begin{array}{ccc} & & O \\ & & \parallel \\ (CH_3)_3CMgX & + & HCH \\ \textit{tert}\text{-Butylmagnesium halide} & & \text{Formaldehyde} \end{array}$$

(*d*)

$$(CH_3)_2C\!\!=\!\!CHCH_2CH_2\underset{\underset{OH}{|}}{CH}CH_3$$

6-Methyl-5-hepten-2-ol

$$\left[\begin{array}{cc} (CH_3)_2C\!\!=\!\!CHCH_2CH_2CH & + \; XMgCH_3 \\ \parallel & \\ O & \\ \text{5-Methyl-4-hexenal} & \text{Methylmagnesium} \\ & \text{halide} \end{array} \right]$$

$$\left[\begin{array}{cc} (CH_3)_2C\!\!=\!\!CHCH_2CH_2MgX & + \; HCCH_3 \\ & \parallel \\ & O \\ \text{4-Methyl-3-hexen-1-ylmagnesium} & \text{Ethanal} \\ \text{halide} & \end{array} \right]$$

(*e*)

4-Ethyl-4-octanol

$$\left[\text{Propylmagnesium halide} + \text{3-Heptanone} \right]$$

$$\left[\text{3-Hexanone} + XMg \; \text{Butylmagnesium halide} \right]$$

$$\left[\text{4-Octanone} + XMg \; \text{Ethylmagnesium halide} \right]$$

14.22 (*a*) Meparfynol is a tertiary alcohol and so can be prepared by addition of a carbanionic species to a ketone. Use the same reasoning which applies to the synthesis of alcohols from Grignard reagents. On mentally disconnecting one of the bonds to the carbon bearing the hydroxyl group

$$CH_3CH_2\overset{\underset{\textstyle CH_3}{|}}{\underset{\textstyle |}{\overset{\textstyle OH}{\overset{|}{C}}}}C\equiv CH \implies CH_3CH_2\overset{O}{\overset{\|}{C}} \quad :C\equiv CH$$

we see that the addition of acetylide ion to 2-butanone will provide the target molecule.

$$HC\equiv CNa + CH_3CH_2\overset{O}{\overset{\|}{C}}CH_3 \xrightarrow[\text{2. } H_3O^+]{\text{1. } NH_3} CH_3CH_2\overset{\underset{\textstyle CH_3}{|}}{\overset{\textstyle OH}{\overset{|}{C}}}C\equiv CH$$

Sodium   2-Butanone   Meparfynol (94%)
acetylide

The alternative, reaction of a Grignard reagent with an alkynyl ketone, is not acceptable in this case. The acidic terminal alkyne C—H would deprotonate, and hence destroy, the anionic Grignard reagent.

(*b*) Diphepanol is a tertiary alcohol and so may be prepared by reaction of a Grignard or organolithium reagent with a ketone. Retrosynthetically, two possibilities seem reasonable:

$$(C_6H_5)_2\overset{\underset{\textstyle OH}{|}}{\overset{\textstyle CH_3}{\overset{|}{C}}}CH-N\bigcirc \implies C_6H_5:^- + C_6H_5\overset{\underset{\textstyle O}{\|}}{\overset{\textstyle CH_3}{\overset{|}{C}}}CH-N\bigcirc$$

and

$$(C_6H_5)_2\overset{\underset{\textstyle OH}{|}}{\overset{\textstyle CH_3}{\overset{|}{C}}}CH-N\bigcirc \implies (C_6H_5)_2C=O + {}^-:\overset{\underset{\textstyle}{}}{\overset{\textstyle CH_3}{\overset{|}{C}}}H-N\bigcirc$$

In principle either strategy is acceptable; in practice the one involving phenylmagnesium bromide is used.

$$C_6H_5MgBr + C_6H_5\overset{\underset{\textstyle O}{\|}}{\overset{\textstyle CH_3}{\overset{|}{C}}}CH-N\bigcirc \xrightarrow[\text{2. } H_3O^+]{\text{1. ether}} (C_6H_5)_2\overset{\underset{\textstyle OH}{|}}{\overset{\textstyle CH_3}{\overset{|}{C}}}CH-N\bigcirc$$

Phenylmagnesium   Diphepanol
bromide

(*c*) A reasonable last step in the synthesis of mestranol is the addition of sodium acetylide to the ketone shown.

Acetylide anion adds to the carbonyl from the less sterically hindered side. The methyl group shields the top face of the carbonyl, and so acetylide adds from the bottom.

14.23 (a) Sodium acetylide adds to ketones to give tertiary alcohols.

Benzophenone   1,1-Diphenyl-2-propyn-1-ol
(50%)

(b) The substrate is a ketone, which reacts with ethyllithium to yield a tertiary alcohol.

2-Adamantanone   2-Ethyl-2-adamantanol (83%)

(c) The first step is conversion of bromocyclopentene to the corresponding Grignard reagent, which then reacts with formaldehyde to give a primary alcohol.

1-Bromocyclopentene   1-Cyclopentenylmagnesium   1-Cyclopentenylmethanol
                       bromide   (53%)

(d) The reaction is one in which an alkene is converted to a cyclopropane through use of the Simmons–Smith reagent, iodomethylzinc iodide.

Allylbenzene   Benzylcyclopropane (64%)

(e) Methylene transfer using the Simmons–Smith reagent is stereospecific. The trans arrangement of substituents in the alkene is carried over to the cyclopropane product.

(E)-1-Phenyl-2-butene   trans-1-Benzyl-2-methylcyclopropane
                            (50%)

(f) Lithium dimethylcuprate transfers a methyl group, which substitutes for iodine on the iodoalkene. Even halogens on sp^2 hybridized carbon are reactive in substitution reactions with lithium dialkylcuprates.

2-Iodo-8-methoxybenzonorbornadiene   8-Methoxy-2-methylbenzonorbornadiene
                                       (73%)

(g) The product of the reaction is the corresponding alkane, formed by reduction of the —CH₂I function to —CH₃. The reaction does not affect any of the bonds to the stereogenic center. By working through the stereochemistry, we see that the absolute configuration of the product is R.

(R)-2-Ethyl-1-iodohexane → (R)-2-Methylheptane

(*h*)   Zinc reacts with vicinal dibromides by dehalogenation. The product is an alkene.

5,6-Dibromocholestanol   Cholesterol (78–80%)

(*i*)   Cyclopropanes result when 1,3-dihalides are treated with zinc.

1,3-Dibromobutane   Methylcyclopropane (70–88%)

14.24   Phenylmagnesium bromide reacts with 4-*tert*-butylcyclohexanone as shown:

4-*tert*-Butylcyclohexanone   4-*tert*-Butyl-1-phenylcyclohexanol

The phenyl substituent can be introduced either cis or trans to the *tert*-butyl group. Therefore the two alcohols are stereoisomers (diastereomers).

Dehydration of either alcohol yields 4-*tert*-butyl-1-phenylcyclohexene.

4-*tert*-Butyl-1-
phenylcyclohexene

14.25   (*a*)   By working through the sequence of reactions that occur when ethyl formate reacts with a Grignard reagent, we can see that this combination leads to *secondary alcohols*:

$$\underset{\substack{\text{Grignard} \\ \text{reagent}}}{RMgX} + \underset{\substack{\text{Ethyl} \\ \text{formate}}}{H\overset{\displaystyle O}{\overset{\|}{C}}OCH_2CH_3} \longrightarrow \underset{\text{Aldehyde}}{R\overset{\displaystyle O}{\overset{\|}{C}}H} + CH_3CH_2OMgX$$

$$\downarrow \begin{array}{l} \text{1. RMgX, diethyl ether} \\ \text{2. } H_3O^+ \end{array}$$

$$\underset{\substack{| \\ \text{OH} \\ \text{Secondary alcohol}}}{RCHR}$$

This is simply because the substituent on the carbonyl carbon of the ester, in this case a hydrogen, is carried through and becomes a substituent on the hydroxyl-bearing carbon of the alcohol.

(*b*) Diethyl carbonate has the potential to react with *three* moles of a Grignard reagent.

$$RMgX + CH_3CH_2O\overset{\displaystyle O}{\overset{\|}{C}}OCH_2CH_3 \longrightarrow R\overset{\displaystyle O}{\overset{\|}{C}}OCH_2CH_3 + CH_3CH_2OMgX$$

Grignard reagent Diethyl carbonate Ester

$$\downarrow RMgX$$

$$R\overset{\displaystyle O}{\overset{\|}{C}}R + CH_3CH_2OMgX$$

Ketone

$$\downarrow RMgX, \text{ then usual workup}$$

$$R\overset{\displaystyle R}{\underset{\displaystyle OH}{\overset{|}{\underset{|}{C}}}}R$$

Tertiary alcohol

The tertiary alcohols that are formed by the reaction of diethyl carbonate with Grignard reagents have three identical R groups attached to the carbon that bears the hydroxyl substituent.

14.26 If we use the 2-bromobutane given, along with the information that the reaction occurs with net inversion of configuration, the stereochemical course of the reaction may be written as:

$$CH_3CH_2\overset{\displaystyle CH_3}{\underset{\displaystyle Br}{\overset{|}{\underset{|}{C}}}}-H \xrightarrow{LiCu(C_6H_5)_2} CH_3\overset{\displaystyle C_6H_5}{\underset{\displaystyle CH_3CH_2 \quad H}{\overset{}{C}}}$$

The phenyl group becomes bonded to carbon from the opposite side of the leaving group.

Applying the Cahn–Ingold–Prelog notational system described in Section 7.6 to the product, the order of decreasing precedence is

$$C_6H_5 > CH_3CH_2 > CH_3 > H$$

Orienting the molecule so that the lowest-priority substituent (H) is away from us, we see that the order of decreasing precedence is clockwise.

$$\underset{CH_3 \qquad CH_2CH_3}{\overset{C_6H_5}{|}}$$

The absolute configuration is *R*.

14.27 The substrates are secondary alkyl *p*-toluenesulfonates, so that elimination is expected to compete with substitution. Compound B is formed in both reactions and has the molecular formula of 4-*tert*-butylcyclohexene. Since the two *p*-toluenesulfonates are diastereomers, it is likely that compounds A and C, especially since they have the same molecular formula, are also diastereomers. Assuming that the substitution reactions proceed with inversion of configuration, we conclude that the products are as shown.

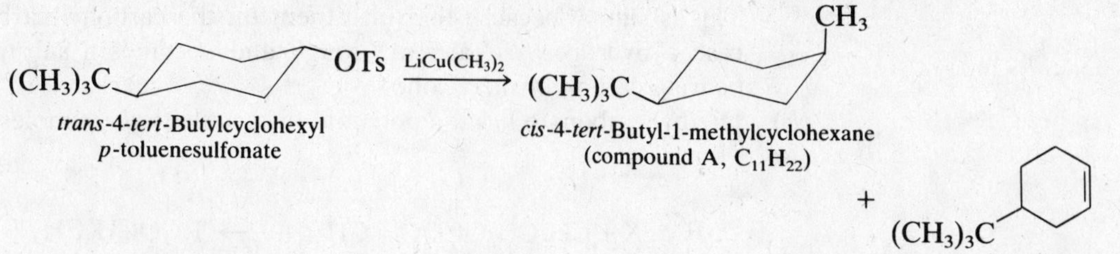

trans-4-tert-Butylcyclohexyl
p-toluenesulfonate

cis-4-tert-Butyl-1-methylcyclohexane
(compound A, $C_{11}H_{22}$)

4-tert-Butylcyclohexene
(compound B, $C_{10}H_{18}$)

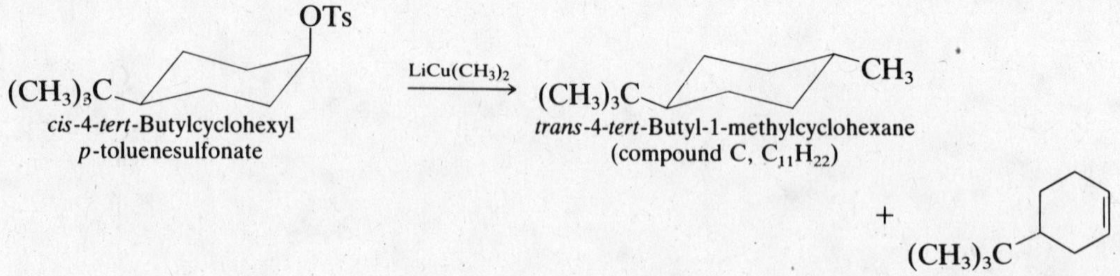

cis-4-tert-Butylcyclohexyl
p-toluenesulfonate

trans-4-tert-Butyl-1-methylcyclohexane
(compound C, $C_{11}H_{22}$)

Compound B

Inversion of configuration is borne out by the fact given in the problem that compound C is more stable than compound A. Both substituents are equatorial in C; the methyl group is axial in A.

14.28 We are told in the statement of the problem that the first step is conversion of the alcohol to the corresponding p-toluenesulfonate. This step is carried out as follows:

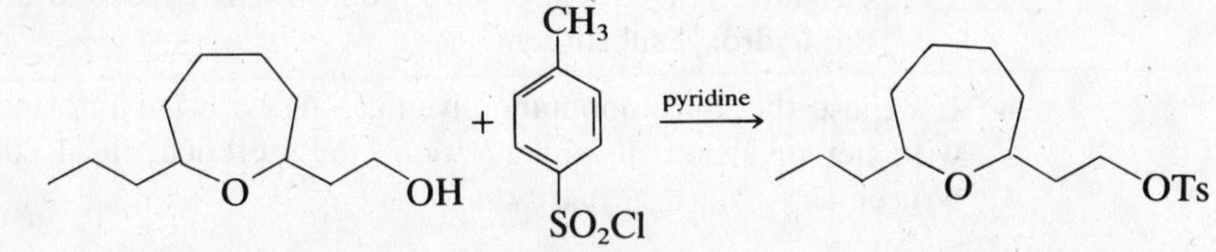

3,8-Epoxy-1-undecanol

p-Toluenesulfonyl
chloride (TsCl)

3,8-Epoxyundecyl
p-toluenesulfonate

Alkyl p-toluenesulfonates react with lithium dialkylcuprates in the same way that alkyl halides do. Treatment of the preceding p-toluenesulfonate with lithium dibutylcuprate gives the desired compound.

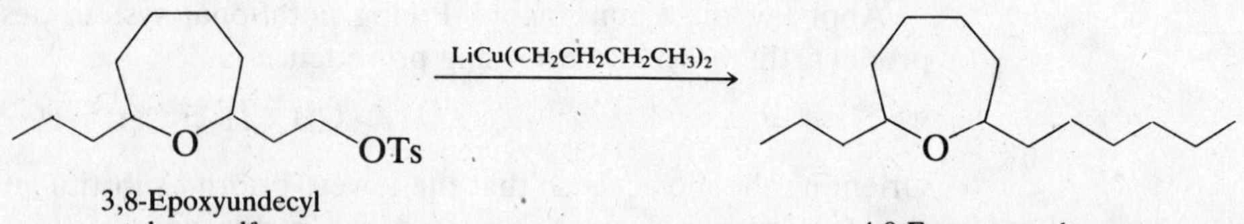

3,8-Epoxyundecyl
p-toluenesulfonate

4,9-Epoxypentadecane

As actually performed, a 91 percent yield of the desired product was obtained in the reaction of the p-toluenesulfonate with lithium dibutylcuprate.

14.29 The regiochemistry of oxymercuration-demercuration follows Markovnikov's rule faithfully. Therefore the products must be the two stereoisomeric tertiary alcohols; one has the hydroxyl group cis to the 3- and 4-methyl groups, the other has the hydroxyl group trans.

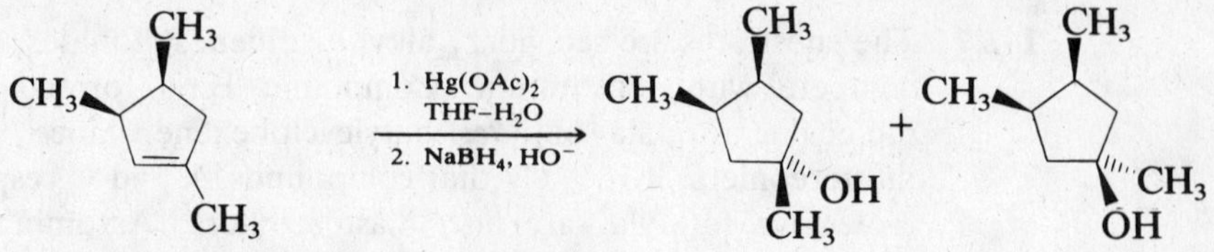

cis-1,3,4-Trimethylcyclopentene

14.30 Vicinal diiodides are converted to alkenes on reaction with zinc. Once formed, the alkene undergoes a Diels-Alder reaction with 1,3-cyclopentadiene to give the observed product.

14.31 (*a*) The desired 1-deuteriobutane can be obtained by reaction of D_2O with butyllithium or butylmagnesium bromide.

$$CH_3CH_2CH_2CH_2Li \; + \; D_2O$$
Butyllithium $\quad$ Deuterium oxide

or

$$CH_3CH_2CH_2CH_2MgBr \; + \; D_2O$$
Butylmagnesium bromide

$$\longrightarrow \; CH_3CH_2CH_2CH_2D$$
1-Deuteriobutane

Preparation of the organometallic compounds requires an alkyl bromide, which is synthesized from the corresponding alcohol.

$$CH_3CH_2CH_2CH_2OH \xrightarrow[\text{or HBr}]{PBr_3} CH_3CH_2CH_2CH_2Br$$
1-Butanol $\qquad\qquad\qquad\qquad\quad$ 1-Bromobutane

$$CH_3CH_2CH_2CH_2Li \xleftarrow[\text{ether}]{Li} CH_3CH_2CH_2CH_2Br \xrightarrow[\text{ether}]{Mg} CH_3CH_2CH_2CH_2MgBr$$
Butyllithium $\qquad\qquad$ 1-Bromobutane $\qquad\qquad$ Butylmagnesium bromide

(*b*) In a sequence identical to that of part (*a*) in design but using 2-butanol as the starting material, 2-deuteriobutane may be prepared.

$$\underset{\underset{\text{2-Butanol}}{OH}}{CH_3CHCH_2CH_3} \xrightarrow{PBr_3} \underset{\underset{\text{2-Bromobutane}}{Br}}{CH_3CHCH_2CH_3} \xrightarrow[\text{ether}]{Mg} \underset{\underset{\textit{sec}\text{-Butylmagnesium bromide}}{MgBr}}{CH_3CHCH_2CH_3}$$

$$\downarrow D_2O$$

$$\underset{\underset{\text{2-Deuteriobutane}}{D}}{CH_3CHCH_2CH_3}$$

An analogous procedure involving *sec*-butyllithium in place of the Grignard reagent can be used.

14.32 A Grignard reagent exists in solution as an equilibrium mixture of an alkylmagnesium halide and a dialkylmagnesium:

$$2RMgX \;\rightleftharpoons\; R_2Mg + MgX_2$$

When dioxane is added to the solution, magnesium halides, being insoluble in dioxane, precipitate. This shifts the position of equilibrium to the right, increasing the concentration of the dialkylmagnesium species at the expense of the alkylmagnesium halide. In effect, the halide is removed from solution by precipitation, so that the alkylmagnesium halide is converted to a dialkylmagnesium.

14.33 All the protons in benzene are equivalent. In diphenylmethane and in triphenylmethane, protons are attached either to the sp^2 hybridized carbons of the ring or to the sp^3 hybridized carbon between the rings. The large difference in acidity between diphenylmethane and benzene suggests that it is not a ring proton that is lost on ionization in diphenylmethane but rather a proton from the methylene group.

$$(C_6H_5)CH_2 \;\rightleftharpoons\; (C_6H_5)_2\ddot{C}H + H^+$$
Diphenylmethane

The anion produced is stabilized by resonance. It is a *benzylic* anion.

Both rings are involved in delocalizing the negative charge. The anion from triphenyl-methane is stabilized by resonance involving all three rings.

Delocalization of the negative charge by resonance is not possible in the anion of benzene. The pair of unshared electrons in phenyl anion is in an sp^2 hybrid orbital that does not interact with the π system.

not delocalized into π system

14.34 The meso form of 2,3-dibromobutane is the stereoisomer shown; syn elimination, that is, loss of both bromine atoms from the same side, will yield (Z)-2-butene.

meso-2,3-Dibromobutane → syn debromination → (Z)-2-Butene

This is at variance with the experimental observation that *meso*-2,3-dibromobutane gives (E)-2-butene. Therefore, a syn elimination may be ruled out. Now let us look at the anti conformation of the meso form:

meso-2,3-Dibromobutane → anti debromination / Zn, ethanol → (E)-2-Butene

Loss of bromines from opposite sides, that is, anti elimination, leads to the observed product, (E)-2-butene.

Similarly, syn debromination of the chiral diastereomer gives the wrong stereoisomeric form of the product:

Chiral diastereomer of 2,3-dibromobutane → syn debromination → (E)-2-Butene (not observed)

Anti debromination of this diastereomer, however, is consistent with the observed stereochemistry.

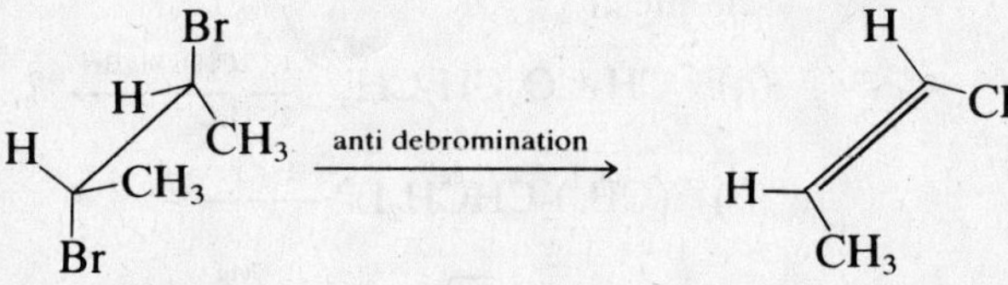

Chiral diastereomer
of 2,3-dibromobutane

(Z)-2-Butene

14.35 (*a*) Compound D, as indicated by its molecular formula, is a dibromide. The selectivity of free-radical bromination is high and favors substitution of tertiary hydrogens. The starting alkane has two tertiary hydrogens that are capable of being replaced by bromine atoms, and so compound D is most reasonably the dibromide shown.

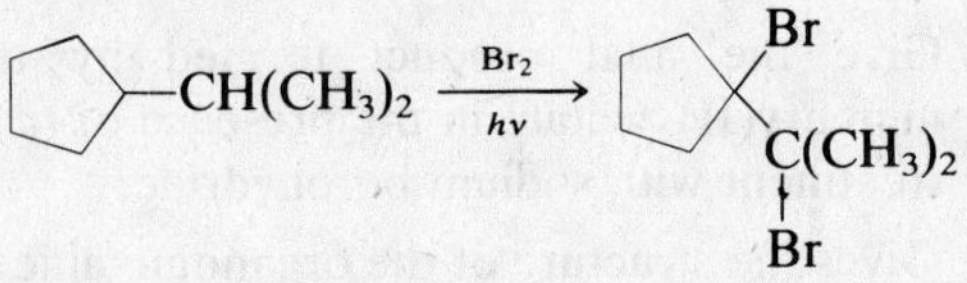

(Compound D)

Compound D is a vicinal dibromide. Vicinal dibromides undergo dehalogenation on being treated with zinc to yield alkenes.

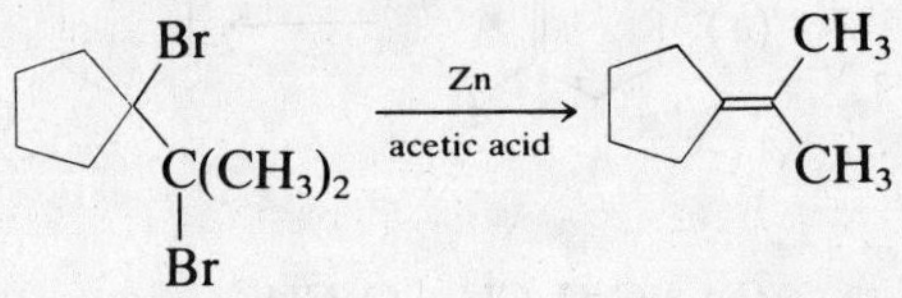

Compound E

(*b*) *N*-Bromosuccinimide reacts with the starting material by benzylic bromination.

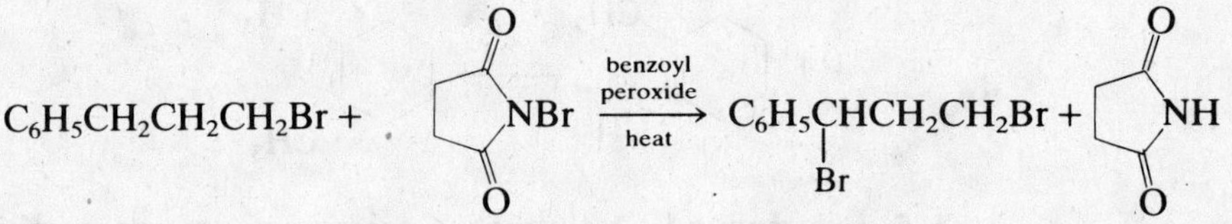

$C_6H_5CH_2CH_2CH_2Br$ +

1-Bromo-3-
phenylpropane

N-Bromosuccinimide

$C_6H_5CHCH_2CH_2Br$ +
  |
  Br

1,3-Dibromo-1-
phenylpropane
(compound F)

Succinimide

Treatment with zinc converts the 1,3-dibromide to a cyclopropane derivative.

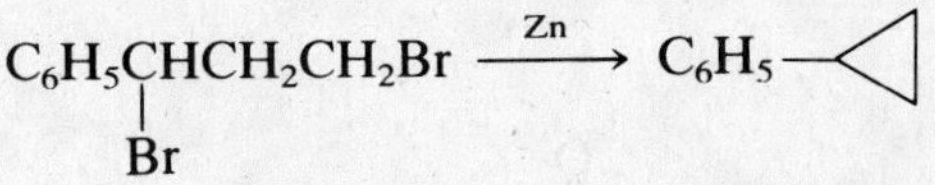

$C_6H_5CHCH_2CH_2Br$
  |
  Br

1,3-Dibromo-1-
phenylpropane

Cyclopropylbenzene
(compound G)

SELF-TEST

PART A

A-1. Give a method for the preparation of each of the following organometallic compounds, using appropriate starting materials:

(*a*) Cyclohexyllithium

(*b*) *tert*-Butylmagnesium bromide

(*c*) Lithium dibenzylcuprate

A-2. Give the structure of the product obtained by each of the following reaction schemes:

(a) $CH_3CO_2CH_2CH_3$ $\xrightarrow[\text{2. H}_3\text{O}^+]{\text{1. 2C}_6\text{H}_5\text{MgBr}}$?

(b) $(CH_3)_2CHCH_2Li$ $\xrightarrow{D_2O}$?

(c) $CH_3-\!\!\!\bigcirc\!\!\!-Br$ $\xrightarrow[\text{3. H}_3\text{O}^+]{\substack{\text{1. Mg} \\ \text{2. H}_2\text{C}=\text{O}}}$?

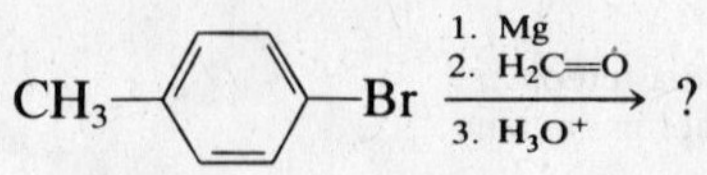

(d) $\xrightarrow[\text{2. H}_3\text{O}^+]{\text{1. CH}_3\text{CH}_2\text{Li}}$?

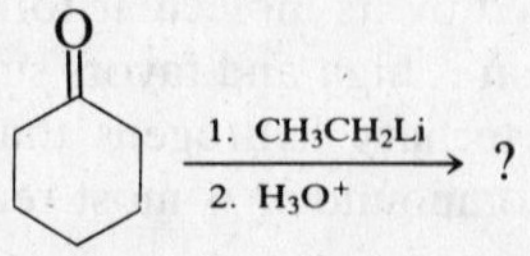

A-3. Give the final product formed by the reaction of 3-methyl-1-butene with mercury(II) acetate in the presence of (a) water and (b) ethanol, both followed by treatment with sodium borohydride.

A-4. Gives the structure of the organometallic reagent necessary to carry out each of the following:

(a)

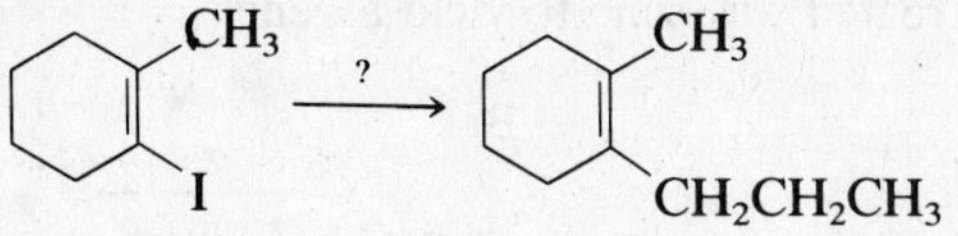

(b) $C_6H_5CH_2CO_2CH_3$ $\xrightarrow[\text{2. H}_3\text{O}^+]{\text{1. ?}}$ $(CH_3)_2CHCCH(CH_3)_2$ with OH above and $CH_2C_6H_5$ below

(c)

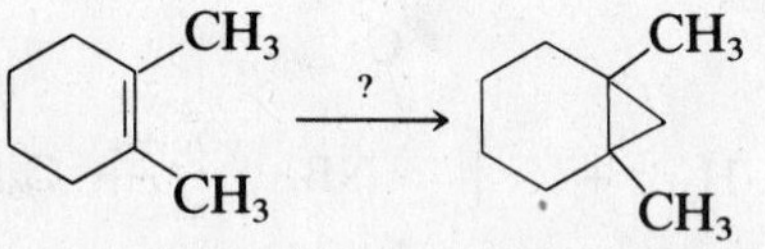

A-5. Compounds A through F are some common organic solvents. Which ones would be suitable for use in the preparation of a Grignard reagent? For those which are not suitable, give a brief reason why.

$$CH_3CH_2CH_2CH_2OCH_2CH_2CH_2CH_3 \qquad CH_3OCH_2CH_2OCH_3 \qquad HOCH_2CH_2OH$$
$$\text{A} \qquad\qquad\qquad \text{B} \qquad\qquad \text{C}$$

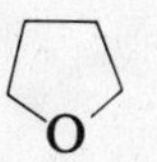

$$CH_3\overset{\text{O}}{\overset{||}{C}}OCH_2CH_3 \qquad \bigcirc\text{(O)} \qquad CH_3\overset{\text{O}}{\overset{||}{C}}OH$$
$$\text{D} \qquad\qquad \text{E} \qquad\qquad \text{F}$$

A-6. Show by a series of chemical equations how you could prepare octane from 1-butanol as the source of all its carbon atoms.

A-7. Synthesis of the following alcohol is possible by three schemes utilizing Grignard reagents. Give the reagents necessary to carry out each of them.

$$\overset{\text{OH}}{\underset{|}{(CH_3)_2CHC(CH_3)_2}}$$

A-8. Using ethylbenzene and any other necessary organic or inorganic reagents, outline a synthesis of 3-phenyl-2-butanol.

PART B

B-1. Which (if any) of the following would *not* be classified as an organometallic substance?

(*a*) Triethylaluminum

(*b*) Ethylmagnesium iodide

(*c*) Potassium *tert*-butoxide

(*d*) None of these (all are organometallic compounds)

B-2. Rank the following species in order of increasing polarity of the carbon-metal bond (least → most polar):

$$CH_3CH_2MgCl \qquad CH_3CH_2Na \qquad (CH_3CH_2)_3Al$$
$$A \qquad\qquad\qquad B \qquad\qquad\qquad C$$

(*a*) C < A < B   (*b*) B < A < C   (*c*) A < C < B   (*d*) B < C < A

B-3. The reaction scheme

$$HOCH_2CH_2CH_2Br \xrightarrow[\substack{2.\ C_6H_5CH=O \\ 3.\ H_3O^+}]{1.\ Mg} HOCH_2CH_2CH_2\overset{\overset{\displaystyle OH}{|}}{C}HC_6H_5$$

(*a*) Is a good synthetic route

(*b*) Is a poor synthetic route

(*c*) May be good or poor, depending on solvent used

(*d*) More information needed to determine

B-4. Arrange the following intermediates in order of decreasing basicity (strongest → weakest):

$$CH_2{=}CHNa \qquad CH_3CH_2Na \qquad CH_3CH_2ONa \qquad HC{\equiv}CNa$$
$$A \qquad\qquad\quad B \qquad\qquad\qquad C \qquad\qquad\qquad D$$

(*a*) B > A > D > C   (*c*) C > D > A > B

(*b*) D > A > B > C   (*d*) C > B > D > A

B-5. Give the major product of the following reaction:

$$(E)\text{-2-pentene} \xrightarrow{CH_2I_2,\ Zn(Cu)} ?$$

(*a*) *cis*-1-Ethyl-2-methylcyclopropane

(*b*) *trans*-1-Ethyl-2-methylcyclopropane

(*c*) 1-Ethyl-1-methylcyclopropane

(*d*) An equimolar mixture of products (*a*) and (*b*)

B-6. Which of the following reagents would be effective for the following reaction sequence?

$$C_6H_5C{\equiv}CH \xrightarrow[\substack{3.\ H_3O^+}]{\substack{1.\ ? \\ 2.\ H_2C=O}} C_6H_5C{\equiv}CCH_2OH$$

(*a*) Sodium ethoxide   (*c*) Butyllithium

(*b*) Mercury(II) acetate   (*d*) Potassium hydroxide

B-7. What is the product of the following reaction?

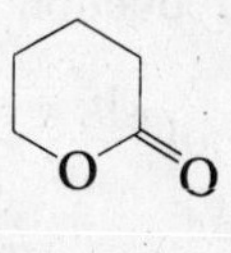

$$\text{(δ-valerolactone)} + 2CH_3MgBr \xrightarrow[\substack{2.\ H_3O^+}]{\substack{1.\ \text{diethyl ether}}}$$

(*a*) $HOCHCH_2CH_2CH_2CHOH$ with CH_3 groups

(*b*) $HOCH_2CH_2CH_2CH_2\overset{\overset{\displaystyle CH_3}{|}}{C}OH$ with CH_3

(*c*) $CH_3OCH_2CH_2CH_2CH_2\overset{\overset{}{}}{C}HCH_3$ with OH

(*d*) $HOCH_2CH_2CH_2CH_2\overset{\overset{}{}}{C}HOCH_3$ with CH_3

ALCOHOLS, DIOLS, AND THIOLS

IMPORTANT TERMS AND CONCEPTS

Oxygen-Containing Functional Groups Several classes of organic compounds contain one or more oxygen atoms. Arranged in order of increasing oxidation states, these groups are:

More reduced:

$$ROH \qquad ROR'$$

Alcohol   Ether

$$\underset{\text{Aldehyde}}{\overset{\overset{\displaystyle O}{\parallel}}{RCH}} \qquad \underset{\text{Ketone}}{\overset{\overset{\displaystyle O}{\parallel}}{RCR'}}$$

More oxidized:

$$\underset{\substack{\text{Carboxylic}\\\text{acid}}}{\overset{\overset{\displaystyle O}{\parallel}}{RCOH}} \qquad \underset{\text{Ester}}{\overset{\overset{\displaystyle O}{\parallel}}{RCOR'}} \qquad \underset{\text{Amide}}{\overset{\overset{\displaystyle O}{\parallel}}{RCNR'_2}}$$

Keeping in mind the order of progression of these groups will aid in determining what types of reaction conditions are needed for their interconversion. For example, an alcohol must be oxidized to yield an aldehyde or a ketone. Likewise, formation of an alcohol from an ester must involve a reduction pathway.

$$\text{Alcohol} \xrightarrow{\text{oxidizing agent}} \text{aldehyde or ketone}$$

$$\text{Ester} \xrightarrow{\text{reducing agent}} \text{alcohol}$$

Preparation of Alcohols and Diols (Secs. 15.2 to 15.5) In previous chapters several methods for the preparation of alcohols have been discussed. These include:

1. Hydration of alkenes
   (a) Acid-catalyzed hydration (Chapter 6)

(*b*)   Hydroboration (Chapter 6)
(*c*)   Oxymercuration-demercuration (Chapter 14)
2.   Hydrolysis of alkyl halides (Chapter 8)
3.   Grignard (and organolithium) reactions with carbonyl derivatives (Chapter 14)

A summary of these methods is given in text Table 15.1. The preparation of alcohols frequently involves the reduction of carbonyl derivatives. The general transformations are summarized in the following.

Reduction reactions:

$$\text{Aldehydes} \longrightarrow \text{primary alcohols}$$
$$\text{Ketones} \longrightarrow \text{secondary alcohols}$$
$$\left.\begin{array}{l}\text{Carboxylic acids} \\ \text{and esters}\end{array}\right\} \longrightarrow \text{primary alcohols}$$

Methods for carrying out these transformations include *catalytic hydrogenation* and reduction with the *metal hydrides* lithium aluminum hydride ($LiAlH_4$) and sodium borohydride ($NaBH_4$). A summary of the functional groups reduced by these reagents follows.

Reagent	Functional groups reduced
H_2, catalyst	Aldehydes, ketones, esters
$LiAlH_4$	Aldehydes, ketones, acids, esters
$NaBH_4$	Aldehydes, ketones (esters reduced slowly)

Specific examples of these reactions are given in the reaction summary which follows this section.

Alcohols may also be prepared by epoxide ring-opening reactions. Diols are prepared by the *hydroxylation* of alkenes.

Reactions of Alcohols (Secs. 15.6 to 15.12) General classes of alcohol reactions may be described which involve the breaking of bonds. The three general possibilities are:

1.   O—H bond breaking results in formation of alkoxides:

$$-\overset{|}{\underset{|}{C}}-O\{H \longrightarrow -\overset{|}{\underset{|}{C}}-O^-$$

2.   C—O bond breaking results in substitution reactions:

$$-\overset{|}{\underset{|}{C}}\{O-H \longrightarrow -\overset{|}{\underset{|}{C}}-X$$

3.   O—H and adjacent C—H bonds are broken in oxidation reactions:

$$-\overset{H}{\underset{|}{C}}-O\{H \longrightarrow\ \diagdown C=O$$

Examples of alcohol reactions are present in the "Important Reactions" section.

Thiols (Secs. 15.13, 15.14) The sulfur analogs of alcohols are *thiols*, RSH. Thiols are named by adding the suffix *-thiol* to the corresponding alkane. The —SH group may be named as a substituent and is called a *mercapto* group.

$$CH_3CH_2SH \qquad\qquad HSCH_2CH_2CH_2OH$$

       Ethanethiol   3-Mercapto-1-propanol

The S—H bond is less polar than the O—H bond, and hydrogen bonding in thiols is weaker than in alcohols. As a result, thiols have lower boiling points than alcohols with the same carbon skeleton. For example, the boiling point of ethanethiol is 36°C while that of ethanol is 78°C.

Spectroscopic Analysis of Alcohols (Sec. 15.15) The infrared spectra of alcohols exhibit characteristic O—H stretching in the 3200 to 3600 cm^{-1} region. In addition, strong absorption due to C—O stretching is observed in the region from 1025 to 1200 cm^{-1}.

The ^{1}H nmr spectra of alcohols exhibit signals due to the hydroxyl proton and the proton of the H—C—O unit when the alcohol is primary or secondary.

$$\overset{\displaystyle |}{\underset{\displaystyle |}{-C}}-O-H \qquad\qquad\qquad H-\overset{\displaystyle |}{\underset{\displaystyle |}{C}}-O$$

           $\delta = 0.5\text{–}5$ ppm $\delta = 3.3\text{–}4.0$ ppm

IMPORTANT REACTIONS

PREPARATION OF ALCOHOLS

Reduction of Aldehydes and Ketones (Sec. 15.2)

General:

$$R-\overset{\displaystyle \overset{O}{\|}}{C}-H \xrightarrow{\text{[H]}} RCH_2OH$$

$$R-\overset{\displaystyle \overset{O}{\|}}{C}-R' \xrightarrow{\text{[H]}} R\overset{\displaystyle \overset{OH}{|}}{C}HR'$$

Examples:

Catalytic hydrogenation:

cyclohexanone $\xrightarrow[\text{Pt}]{\text{H}_2 \text{ (3 atm)}}$ cyclohexanol (H, OH)

Metal hydrides:

$$CH_3CH_2\overset{\displaystyle \overset{O}{\|}}{C}H \xrightarrow[\text{C}_2\text{H}_5\text{OH}]{\text{NaBH}_4} CH_3CH_2CH_2OH$$

$$C_6H_5\overset{\displaystyle \overset{O}{\|}}{C}CH_3 \xrightarrow[\text{2. H}_2\text{O}]{\text{1. LiAlH}_4} C_6H_5\overset{\displaystyle \overset{OH}{|}}{C}HCH_3$$

Reduction of Carboxylic Acids (Sec. 15.3)

General:

$$R\overset{\displaystyle \overset{O}{\|}}{C}OH \xrightarrow[\text{2. H}_2\text{O}]{\text{1. LiAlH}_4} RCH_2OH$$

Example:

$$(CH_3)_3CCO_2H \xrightarrow[\text{2. } H_2O]{\text{1. LiAlH}_4} (CH_3)_3CCH_2OH$$

Reduction of Esters (Sec. 15.3)

General:

$$R\overset{O}{\overset{\|}{C}}OR' \xrightarrow{[H]} RCH_2OH + R'OH$$

Example:

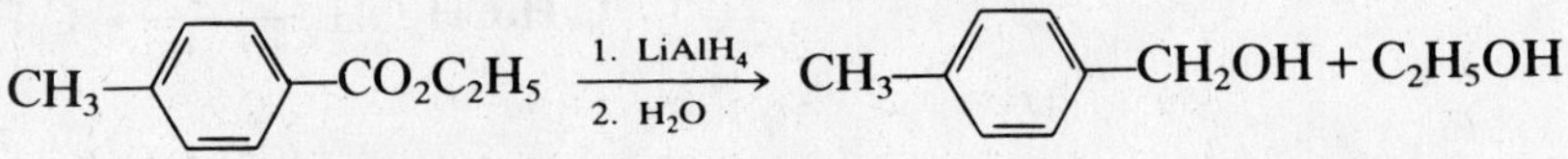

$$CH_3\text{—}\langle\text{—}\rangle\text{—}CO_2C_2H_5 \xrightarrow[\text{2. } H_2O]{\text{1. LiAlH}_4} CH_3\text{—}\langle\text{—}\rangle\text{—}CH_2OH + C_2H_5OH$$

Preparation of Alcohols from Epoxides (Sec. 15.4)

General:

$$RMgX \xrightarrow[\text{2. } H_3O^+]{\text{1. } CH_2\text{—}CH_2 \text{ (epoxide)}} RCH_2CH_2OH$$

Example:

$$C_6H_5MgBr \xrightarrow[\text{2. } H_3O^+]{\text{1. (epoxide)}} C_6H_5CH_2CH_2OH$$

REACTIONS OF ALCOHOLS

Formation of Ethers (Sec. 15.7)

General:

$$2ROH \xrightarrow[\text{heat}]{H^+} ROR + H_2O \quad \text{(R primary)}$$

Example:

$$2CH_3CH_2CH_2CH_2OH \xrightarrow[\text{heat}]{H^+} CH_3CH_2CH_2CH_2OCH_2CH_2CH_2CH_3$$

Esterification (Sec. 15.8)

General:

$$ROH + R'\overset{O}{\overset{\|}{C}}OH \rightleftharpoons R'\overset{O}{\overset{\|}{C}}OR + H_2O$$

(Alcohol or acid in excess; or water removed as formed)

$$ROH + R'\overset{O}{\overset{\|}{C}}\text{—}X \xrightarrow{\text{pyridine}} R'\overset{O}{\overset{\|}{C}}OR$$

(Acyl chloride or anhydride)

Examples:

$$CH_3CH_2OH + C_6H_5CH_2CO_2H \xrightarrow{H^+} C_6H_5CH_2CO_2CH_2CH_3 + H_2O$$

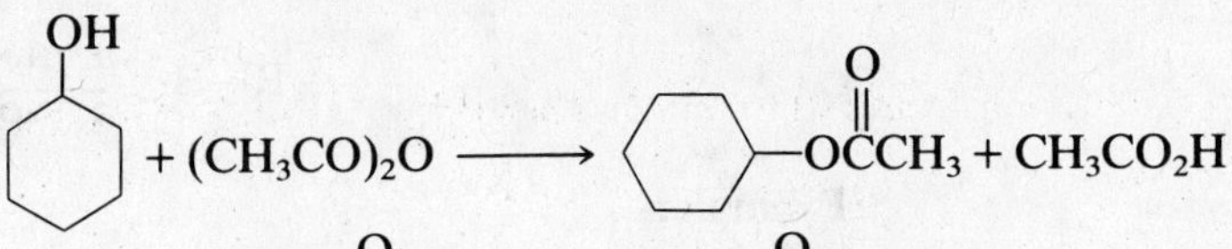

$$\overset{OH}{\underset{}{\bigcirc}} + (CH_3CO)_2O \longrightarrow \bigcirc\text{—}O\overset{O}{\overset{\|}{C}}CH_3 + CH_3CO_2H$$

$$(CH_3)_2CHOH + C_6H_5\overset{O}{\overset{\|}{C}}Cl \xrightarrow{\text{pyridine}} C_6H_5\overset{O}{\overset{\|}{C}}OCH(CH_3)_2$$

Oxidation Reactions (Sec. 15.10)

General:

$$RCH_2OH \xrightarrow{[O]} \underset{\displaystyle \overset{\text{O}}{\|}}{RCH} \xrightarrow{[O]} RCO_2H$$

$$R_2CHOH \xrightarrow{[O]} \underset{\displaystyle \overset{\text{O}}{\|}}{RCR}$$

Examples:

$$C_6H_5CH_2OH \xrightarrow[\text{H}^+,\text{H}_2\text{O}]{\text{K}_2\text{Cr}_2\text{O}_7} C_6H_5CO_2H$$

$$(CH_3)_2CHCH_2OH \xrightarrow{(C_5H_5N)_2CrO_3} (CH_3)_2CH\overset{\displaystyle \overset{\text{O}}{\|}}{C}H$$

DIOLS

Preparation (Sec. 15.5)

General:

$$R_2C{=}CR'_2 \xrightarrow{[O]} \underset{\displaystyle \underset{\text{OH}}{|}}{R_2C}\text{---}\underset{\displaystyle \underset{\text{OH}}{|}}{CR'_2}$$

Example:

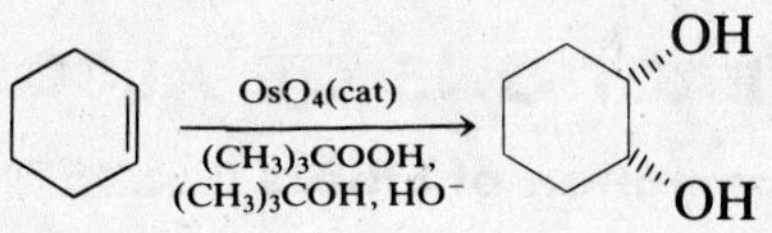

Oxidative Cleavage (Sec. 15.12)

General:

$$R\text{---}\underset{\displaystyle \underset{\text{OH}}{|}}{\overset{\displaystyle \overset{\text{R}}{|}}{C}}\text{---}\underset{\displaystyle \underset{\text{OH}}{|}}{\overset{\displaystyle \overset{\text{R}'}{|}}{C}}\text{---}R' \xrightarrow{\text{HIO}_4} R_2C{=}O + R'_2C{=}O$$

Examples:

$$(CH_3)_2\underset{\displaystyle \underset{\text{OH}}{|}}{C}\text{---}CH_2OH \xrightarrow{\text{HIO}_4} (CH_3)_2C{=}O + H_2C{=}O$$

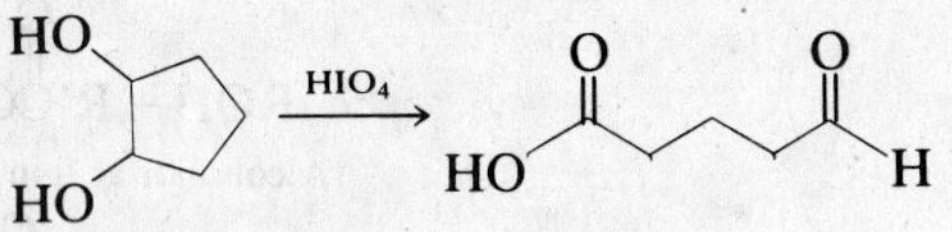

THIOLS

Preparation (Sec. 15.13)

General:

$$R\text{---}Br \xrightarrow[\text{2. NaOH}]{\text{1. (H}_2\text{N})_2\text{C}{=}\text{S}} R\text{---}SH$$

Example:

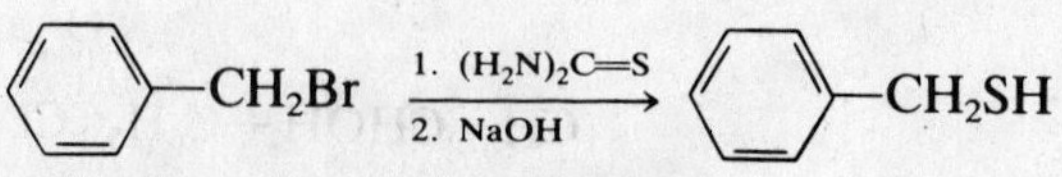

SOLUTIONS TO TEXT PROBLEMS

15.1 The two primary alcohols 1-butanol and 2-methyl-1-propanol can be prepared by hydrogenation of the corresponding aldehydes:

$$CH_3CH_2CH_2\overset{\overset{\textstyle O}{\|}}{C}H \xrightarrow{\text{H}_2, \text{ Ni}} CH_3CH_2CH_2CH_2OH$$
$$\text{Butanal} \qquad\qquad\qquad \text{1-Butanol}$$

$$(CH_3)_2CH\overset{\overset{\textstyle O}{\|}}{C}H \xrightarrow{\text{H}_2, \text{ Ni}} (CH_3)_2CHCH_2OH$$
$$\text{2-Methylpropanal} \qquad\qquad \text{2-Methyl-1-propanol}$$

The secondary alcohol 2-butanol arises by hydrogenation of a ketone.

$$CH_3\overset{\overset{\textstyle O}{\|}}{C}CH_2CH_3 \xrightarrow{\text{H}_2, \text{ Ni}} CH_3\underset{\underset{\textstyle OH}{|}}{C}HCH_2CH_3$$
$$\text{2-Butanone} \qquad\qquad \text{2-Butanol}$$

Tertiary alcohols such as 2-methyl-2-propanol, $(CH_3)_3COH$, cannot be prepared by hydrogenation of a carbonyl compound.

15.2 (*b*) A deuterium atom is transferred from $NaBD_4$ to the carbonyl group of acetone.

$$\underset{\underset{\textstyle CH_3}{|}}{\overset{D-\bar{B}D_3}{CH_3C}}=O \longrightarrow CH_3\underset{\underset{\textstyle CH_3}{|}}{\overset{\overset{\textstyle D}{|}}{C}}-O\bar{B}D_3 \xrightarrow{3(CH_3)_2C=O} \left(CH_3\underset{\underset{\textstyle CH_3}{|}}{\overset{\overset{\textstyle D}{|}}{C}}O-\right)_4\bar{B}$$

On reaction with CH_3OD, deuterium is transferred from the alcohol to the oxygen of $[(CH_3)_2CDO]_4\bar{B}$.

$$CH_3\underset{\underset{\textstyle CH_3}{|}}{\overset{\overset{\textstyle D}{|}}{C}}-O-\bar{B}[OCD(CH_3)_2]_3 \longrightarrow CH_3\underset{\underset{\textstyle CH_3}{|}}{\overset{\overset{\textstyle D}{|}}{C}}OD + \underset{\underset{\textstyle OCH_3}{|}}{\bar{B}[OCD(CH_3)_2]_3}$$
$$\qquad\qquad D-OCH_3$$

Overall:

$$(CH_3)_2C=O \xrightarrow[\text{CH}_3\text{OD}]{\text{NaBD}_4} (CH_3)_2\overset{\overset{\textstyle D}{|}}{C}OD$$
$$\text{Acetone} \qquad\qquad \text{2-Propanol-2-d-}O\text{-d}$$

(*c*) In this case $NaBD_4$ serves as a deuterium donor to carbon while CD_3OH is a proton (not deuterium) donor to oxygen.

$$C_6H_5\overset{\overset{\textstyle O}{\|}}{C}H \xrightarrow[\text{CD}_3\text{OH}]{\text{NaBD}_4} C_6H_5\overset{\overset{\textstyle D}{|}}{C}HOH$$
$$\text{Benzaldehyde} \qquad\qquad \text{Benzyl alcohol-1-d}$$

(*d*) Lithium aluminum deuteride is a deuterium donor to the carbonyl carbon of formaldehyde.

$$\underset{\underset{\textstyle H}{|}}{\overset{D-\bar{A}lD_3}{HC}}=O \longrightarrow \underset{\underset{\textstyle H}{|}}{H\overset{\overset{\textstyle D}{|}}{C}}-O\bar{A}lD_3 \xrightarrow{3HCH} (DCH_2O)_4\bar{A}l$$

On hydrolysis with D_2O, the oxygen-aluminum bond is cleaved and DCH_2OD is formed.

$$\bar{A}l(OCH_2D)_4 \xrightarrow{D_2O} DCH_2OD$$
Methanol-1-d-*O*-d

15.3 The acyl portion of the ester must give a primary alcohol on reduction. The alkyl group bonded to oxygen may be primary, secondary, or tertiary and gives the corresponding alcohol.

$$CH_3CH_2\overset{\overset{\displaystyle O}{\|}}{C}OCH(CH_3)_2 \xrightarrow[\text{2. } H_2O]{\text{1. LiAlH}_4} CH_3CH_2CH_2OH + HOCH(CH_3)_2$$

Isopropyl propanoate   1-Propanol   2-Propanol

15.4 (*b*) Reaction with ethylene oxide results in the addition of a $-CH_2CH_2OH$ unit to the Grignard reagent. Cyclohexylmagnesium bromide (or chloride) is the appropriate reagent.

Cyclohexylmagnesium bromide   Ethylene oxide   2-Cyclohexylethanol

15.5 Lithium aluminum hydride is the appropriate reagent for reducing carboxylic acids or esters to alcohols.

$$HO\overset{\overset{\displaystyle O}{\|}}{C}CH_2\underset{\underset{\displaystyle CH_3}{|}}{C}HCH_2\overset{\overset{\displaystyle O}{\|}}{C}OH \xrightarrow[\text{2. } H_2O]{\text{1. LiAlH}_4} HOCH_2CH_2\underset{\underset{\displaystyle CH_3}{|}}{C}HCH_2CH_2OH$$

3-Methyl-1,5-pentanedioic acid   3-Methyl-1,5-pentanediol

Any alkyl group may be attached to the oxygen of the ester function. In the example shown, it is a methyl group.

$$CH_3O\overset{\overset{\displaystyle O}{\|}}{C}CH_2\underset{\underset{\displaystyle CH_3}{|}}{C}HCH_2\overset{\overset{\displaystyle O}{\|}}{C}OCH_3 \xrightarrow[\text{2. } H_2O]{\text{1. LiAlH}_4} HOCH_2CH_2\underset{\underset{\displaystyle CH_3}{|}}{C}HCH_2CH_2OH + 2CH_3OH$$

Dimethyl 3-methyl-1,5-pentanedioate   3-Methyl-1,5-pentanediol   Methanol

15.6 Hydroxylation of alkenes using osmium tetraoxide is a syn addition of hydroxyl groups to the double bond. *cis*-2-Butene yields the meso diol.

cis-2-Butene   *meso*-2,3-Butanediol

trans-2-Butene yields a racemic mixture of the two enantiomeric forms of the chiral diol.

trans-2-Butene   (2*S*,3*S*)-2,3-Butanediol   (2*R*,3*R*)-2,3-Butanediol

The Fischer projection formulas of the three stereoisomers are:

meso-2,3-Butanediol   (2S,3S)-2,3-Butanediol   (2R,3R)-2,3-Butanediol

15.7 Only the hydroxyl groups on C-1 and C-4 can be involved, since only these two can lead to a five-membered cyclic ether.

1,2,4-Butanetriol   3-Hydroxyoxolane ($C_4H_8O_2$)

Any other combination of hydroxyl groups would lead to a strained three-membered or four-membered ring and is unfavorable under conditions of acid catalysis.

15.8 (b) The relationship of the molecular formula of the ester ($C_{10}H_{10}O_4$) to that of the starting dicarboxylic acid ($C_8H_6O_4$) indicates that the diacid reacted with two moles of methanol to form a diester.

Methanol   1,4-Benzenedicarboxylic acid   Dimethyl 1,4-benzenedicarboxylate

15.9 While neither cis- nor trans-4-tert-butylcyclohexanol is a chiral molecule, the stereochemical course of their reactions with acetic anhydride becomes evident when the relative stereochemistry of the ester function is examined for each case. The cis alcohol yields the cis acetate.

cis-4-tert-Butylcyclohexanol   Acetic anhydride   cis-4-tert-Butylcyclohexyl acetate

The trans alcohol yields the trans acetate.

trans-4-tert-Butylcyclohexanol   Acetic anhydride   trans-4-tert-Butylcyclohexyl acetate

15.10 Glycerol has three hydroxyl groups, each of which is converted to a nitrate ester function in nitroglycerin.

$$CH_2ONO_2$$
$$CHONO_2$$
$$CH_2ONO_2$$

Nitroglycerin

15.11 (b) The substrate is a secondary alcohol and so gives a ketone on oxidation with sodium dichromate. 2-Octanone has been prepared in 92 to 96 percent yield under these reaction conditions.

$$CH_3CH(CH_2)_5CH_3 \xrightarrow[H_2SO_4, H_2O]{Na_2Cr_2O_7} CH_3\overset{O}{\underset{}{C}}(CH_2)_5CH_3$$
$$\quad\quad\;\; OH$$
2-Octanol   2-Octanone

(c) The alcohol is primary, and so oxidation can produce either an aldehyde or a carboxylic acid, depending on the reaction conditions. Here the oxidation is carried out under anhydrous conditions with Collins' reagent. The product, the corresponding aldehyde, has been obtained in 70 to 84 percent yield

$$CH_3CH_2CH_2CH_2CH_2CH_2CH_2OH \xrightarrow[CH_2Cl_2]{(C_5H_5N)_2CrO_3} CH_3CH_2CH_2CH_2CH_2CH_2\overset{O}{\underset{}{C}}H$$
1-Heptanol   Heptanal

15.12 (b) Biological oxidation of CH_3CD_2OH leads to loss of one of the C-1 deuterium atoms to NAD^+. The dihydropyridine ring of the reduced form of the coenzyme will bear a single deuterium substituent.

1,1-Dideuterioethanol NAD^+ 1-Deuterioethanal   NADD

(c) The deuterium atom of CH_3CH_2OD is lost as D^+. The reduced form of the coenzyme contains no deuterium

Ethanol-O-d NAD^+ Ethanal   NADH

15.13 (b) Oxidation of the carbon-oxygen bonds to carbonyl groups accompanies their cleavage.

$$(CH_3)_2CHCH_2CH\!-\!CHCH_2C_6H_5 \xrightarrow{HIO_4} (CH_3)_2CHCH_2\overset{O}{\underset{}{C}}H + H\overset{O}{\underset{}{C}}CH_2C_6H_5$$
$$\quad\quad\quad\quad\quad\; HO\quad OH$$
1-Phenyl-5-methyl-2,3-hexanediol   3-Methylbutanal   2-Phenylethanal

(c) The CH_2OH group is cleaved from the ring as formaldehyde to leave cyclopentanone.

1-(Hydroxymethyl)cyclopentanol   Cyclopentanone   Formaldehyde

15.14 The formation of an alkanethiol by reaction of an alkyl halide or alkyl *p*-toluenesulfonate with thiourea occurs with inversion of configuration in the step in which the carbon-sulfur bond is formed. Thus, the formation of (R)-2-butanethiol requires (S)-sec-butyl p-toluenesulfonate, which then reacts with thiourea by an S_N2 pathway. The p-toluenesulfonate is formed from the corresponding alcohol by a reaction that does not involve any of the bonds to the stereogenic center. Therefore, begin with (S)-2-butanol.

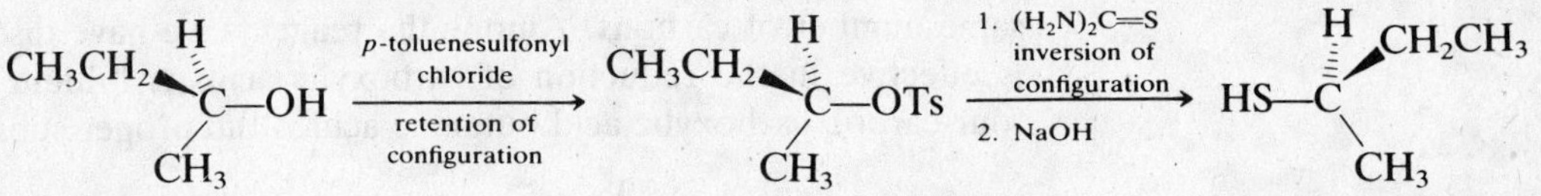

(S)-2-Butanol (S)-sec-Butyl p-toluenesulfonate (R)-2-Butanethiol

15.15 The molecular weight of 2-methyl-2-butanol is 88. A peak in its mass spectrum at m/z 70 corresponds to loss of water from the molecular ion. The peaks at m/z 73 and m/z 59 represent stable cations corresponding to the cleavages shown in the equation:

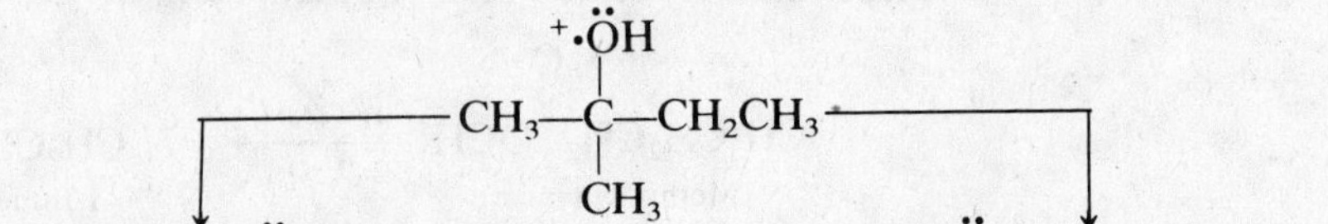

15.16 (a) The appropriate alkene for the preparation of 1-butanol by a hydroboration-oxidation sequence is 1-butene. Remember, hydroboration-oxidation leads to hydration of alkenes with a regioselectivity opposite to that seen in acid-catalyzed hydration and in oxymercuration-demercuration.

$$CH_3CH_2CH{=}CH_2 \xrightarrow[\text{2. } H_2O_2,\ HO^-]{\text{1. } B_2H_6} CH_3CH_2CH_2CH_2OH$$

1-Butene   1-Butanol

(b) 1-Butanol can be prepared by reaction of a Grignard reagent with formaldehyde.

$$CH_3CH_2CH_2CH_2OH \Longrightarrow CH_3CH_2\ddot{C}H_2 + H\overset{O}{\overset{\|}{C}}H$$

An appropriate Grignard reagent is propylmagnesium bromide.

$$CH_3CH_2CH_2Br \xrightarrow[\text{diethyl ether}]{Mg} CH_3CH_2CH_2MgBr$$

1-Bromopropane   Propylmagnesium bromide

$$CH_3CH_2CH_2MgBr + H\overset{O}{\overset{\|}{C}}H \xrightarrow[\text{2. } H_3O^+]{\text{1. diethyl ether}} CH_3CH_2CH_2CH_2OH$$

1-Butanol

(c) Alternatively, 1-butanol may be prepared by the reaction of a Grignard reagent with ethylene oxide.

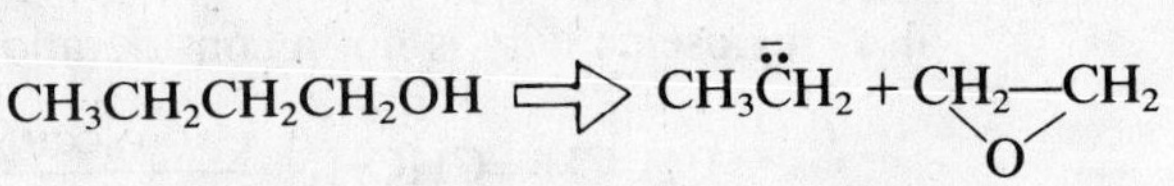

$$CH_3CH_2CH_2CH_2OH \Longrightarrow CH_3\ddot{C}H_2 + CH_2{-}CH_2 \text{ (O)}$$

In this case, ethylmagnesium bromide would be used.

$$CH_3CH_2Br \xrightarrow[\text{diethyl ether}]{Mg} CH_3CH_2MgBr$$

Ethyl bromide   Ethylmagnesium bromide

$$CH_3CH_2MgBr + CH_2\!\!-\!\!CH_2 \xrightarrow[\text{2. } H_3O^+]{\text{1. diethyl ether}} CH_3CH_2CH_2CH_2OH$$
$$\underset{O}{\diagdown\!\!\diagup}$$

Ethylene oxide   1-Butanol

(d) Primary alcohols may be prepared by reduction of the carboxylic acid having the same number of carbons. Among the reagents we have discussed, the only one that is effective in the reduction of carboxylic acids is lithium aluminum hydride. The four-carbon carboxylic acid butanoic acid is the proper substrate.

$$\underset{\text{Butanoic acid}}{CH_3CH_2CH_2\overset{\overset{\displaystyle O}{\|}}{C}OH} \xrightarrow[\text{2. } H_2O]{\text{1. LiAlH}_4\text{, diethyl ether}} \underset{\text{1-Butanol}}{CH_3CH_2CH_2CH_2OH}$$

(e) Reduction of esters can be accomplished using lithium aluminum hydride. The correct methyl ester is methyl butanoate.

$$\underset{\text{Methyl butanoate}}{CH_3CH_2CH_2\overset{\overset{\displaystyle O}{\|}}{C}OCH_3} \xrightarrow[\text{2. } H_2O]{\text{1. LiAlH}_4} \underset{\text{1-Butanol}}{CH_3CH_2CH_2CH_2OH} + \underset{\text{Methanol}}{CH_3OH}$$

(f) 1-Butanol, along with ethanol, is formed when butyl acetate is reduced with lithium aluminum hydride.

$$\underset{\text{Butyl acetate}}{CH_3\overset{\overset{\displaystyle O}{\|}}{C}OCH_2CH_2CH_2CH_3} \xrightarrow[\text{2. } H_2O]{\text{1. LiAlH}_4} \underset{\text{1-Butanol}}{CH_3CH_2CH_2CH_2OH} + \underset{\text{Ethanol}}{CH_3CH_2OH}$$

(g) Since 1-butanol is a primary alcohol having four carbons, butanal must be the aldehyde which is hydrogenated. Suitable catalysts are nickel, palladium, platinum, and ruthenium.

$$\underset{\text{Butanal}}{CH_3CH_2CH_2\overset{\overset{\displaystyle O}{\|}}{C}H} \xrightarrow{H_2,\ Pt} \underset{\text{1-Butanol}}{CH_3CH_2CH_2CH_2OH}$$

(h) Sodium borohydride reduces aldehydes and ketones efficiently. It does not reduce carboxylic acids, and its reaction with esters is too slow to be of synthetic value.

$$\underset{\text{Butanal}}{CH_3CH_2CH_2\overset{\overset{\displaystyle O}{\|}}{C}H} \xrightarrow[\substack{\text{water, ethanol,}\\\text{or methanol}}]{NaBH_4} \underset{\text{1-Butanol}}{CH_3CH_2CH_2CH_2OH}$$

15.17 (a) Both (Z)- and (E)-2-butene yield 2-butanol on hydroboration-oxidation.

$$\underset{\substack{(Z)\text{- or }(E)\text{-2-butene}}}{CH_3CH\!\!=\!\!CHCH_3} \xrightarrow[\text{2. } H_2O_2,\ HO^-]{\text{1. } B_2H_6} \underset{\substack{OH\\\text{2-Butanol}}}{CH_3CHCH_2CH_3}$$

Since the alkene is symmetrically substituted, regioselectivity is not an issue.

(b) Oxymercuration-demercuration of (Z)- and (E)-2-butene yields 2-butanol. Here also, regioselectivity is not a consideration.

$$\underset{\substack{(Z)\text{- or }(E)\text{-2-butene}}}{CH_3CH\!\!=\!\!CHCH_3} \xrightarrow[\text{2. NaBH}_4,\ HO^-]{\text{1. Hg(O}_2CCH_3)_2,\ THF\text{--}H_2O} \underset{\substack{OH\\\text{2-Butanol}}}{CH_3CHCH_2CH_3}$$

1-Butene could also be used, since the hydration follows Markovnikov's rule.

(c) Disconnection of one of the bonds to the carbon that bears the hydroxyl group reveals a feasible route using a Grignard reagent and propanal.

disconnect this bond

$$CH_3 \!\!-\!\! CHCH_2CH_3 \implies \ ^-\!\!:CH_3 + H\overset{\displaystyle O}{\overset{\|}{C}}CH_2CH_3$$
$$\underset{OH}{|}$$
Propanal

The synthetic sequence is:

$$CH_3Br \xrightarrow[\text{diethyl ether}]{Mg} CH_3MgBr \xrightarrow[\text{2. } H_3O^+]{\text{1. } CH_3CH_2\overset{O}{\overset{\|}{C}}H} CH_3\underset{\underset{OH}{|}}{C}HCH_2CH_3$$

Methyl bromide   Methylmagnesium bromide   2-Butanol

(*d*) There is another disconnection related to a synthetic route using a Grignard reagent and acetaldehyde:

$$CH_3CH\!\!-\!\!CH_2CH_3 \implies CH_3\overset{O}{\overset{\|}{C}}H + CH_3\overset{..}{C}H_2$$
$$\underset{OH}{|}$$
disconnect   Acetaldehyde
this bond

$$CH_3CH_2Br \xrightarrow[\text{diethyl ether}]{Mg} CH_3CH_2MgBr \xrightarrow[\text{2. } H_3O^+]{\text{1. } CH_3\overset{O}{\overset{\|}{C}}H} CH_3CH_2\underset{\underset{OH}{|}}{C}HCH_3$$

Ethyl bromide   Ethylmagnesium bromide   2-Butanol

(*e–g*) Since 2-butanol is a secondary alcohol, it can be prepared by reduction of a ketone having the same carbon skeleton, in this case 2-butanone. All three reducing agents indicated in the equations are satisfactory.

$$CH_3\overset{O}{\overset{\|}{C}}CH_2CH_3 \xrightarrow[\substack{\text{(or Pt,} \\ \text{Ni, etc.)}}]{H_2,\ Pd} CH_3\underset{\underset{OH}{|}}{C}HCH_2CH_3$$

2-Butanone   2-Butanol

$$CH_3\overset{O}{\overset{\|}{C}}CH_2CH_3 \xrightarrow[CH_3OH]{NaBH_4} CH_3\underset{\underset{OH}{|}}{C}HCH_2CH_3$$

2-Butanone   2-Butanol

$$CH_3\overset{O}{\overset{\|}{C}}CH_2CH_3 \xrightarrow[\text{2. } H_2O]{\text{1. LiAlH}_4} CH_3\underset{\underset{OH}{|}}{C}HCH_2CH_3$$

2-Butanone   2-Butanol

15.18 (*a*) The desired compound is a tertiary alcohol. The alkene utilized must have the same carbon skeleton as the product. Oxymercuration-demercuration of 2-methylpropene is appropriate.

$$(CH_3)_2C\!\!=\!\!CH_2 \xrightarrow[\text{2. NaBH}_4,\ HO^-]{\text{1. Hg(O}_2\text{CCH}_3)_2,\ \text{THF--H}_2\text{O}} (CH_3)_3COH$$

2-Methylpropene   2-Methyl-2-propanol

(*b*) All the carbon-carbon disconnections are equivalent.

$$
\underset{\substack{\text{CH}_3 \\ | \\ \text{CH}_3-\overset{\displaystyle\text{CH}_3}{\underset{\displaystyle\text{CH}_3}{\text{C}}}-\text{OH}}}{} \implies \text{}^-:\text{CH}_3 + \underset{\text{Acetone}}{\text{CH}_3\overset{\text{O}}{\text{C}}\text{CH}_3}
$$

The synthesis via a Grignard reagent and acetone is:

$$
\underset{\text{Methyl bromide}}{\text{CH}_3\text{Br}} \xrightarrow[\substack{\text{diethyl} \\ \text{ether}}]{\text{Mg}} \underset{\substack{\text{Methylmagnesium} \\ \text{bromide}}}{\text{CH}_3\text{MgBr}} \xrightarrow[\text{2. H}_3\text{O}^+]{\text{1. CH}_3\text{CCH}_3}} \underset{\text{2-Methyl-2-propanol}}{(\text{CH}_3)_3\text{COH}}
$$

(*c*) An alternative route to 2-methyl-2-propanol is addition of a Grignard reagent to an ester.

$$
\underset{\substack{\text{CH}_3 \\ | \\ \text{CH}_3-\text{C}-\text{OH} \\ | \\ \text{CH}_3}}{} \implies 2\text{:CH}_3 + \underset{\text{Acetate ester}}{\text{CH}_3\overset{\text{O}}{\text{COR}}}
$$

Any acetate ester can be used here. Ethyl acetate is satisfactory.

$$
\underset{\text{Methyl bromide}}{\text{CH}_3\text{Br}} \xrightarrow[\substack{\text{diethyl} \\ \text{ether}}]{\text{Mg}} \underset{\substack{\text{Methylmagnesium} \\ \text{bromide}}}{\text{CH}_3\text{MgBr}} \xrightarrow[\text{2. H}_3\text{O}^+]{\text{1. CH}_3\text{COCH}_2\text{CH}_3}} \underset{\text{2-Methyl-2-propanol}}{(\text{CH}_3)_3\text{COH}}
$$

15.19 In order to identify which alcohols might arise by sodium borohydride reduction of carbonyl compounds, write equations for the reduction of all possible aldehydes and ketones that have the molecular formula $C_5H_{10}O$.

$$
\underset{\text{Pentanal}}{\text{CH}_3\text{CH}_2\text{CH}_2\text{CH}_2\overset{\text{O}}{\text{CH}}} \xrightarrow[\text{CH}_3\text{OH}]{\text{NaBH}_4} \underset{\text{1-Pentanol}}{\text{CH}_3\text{CH}_2\text{CH}_2\text{CH}_2\text{CH}_2\text{OH}}
$$

$$
\underset{\text{2-Pentanone}}{\text{CH}_3\text{CH}_2\text{CH}_2\overset{\text{O}}{\text{C}}\text{CH}_3} \xrightarrow[\text{CH}_3\text{OH}]{\text{NaBH}_4} \underset{\substack{\text{2-Pentanol}}}{\text{CH}_3\text{CH}_2\text{CH}_2\underset{\text{OH}}{\text{CHCH}_3}}
$$

$$
\underset{\text{3-Pentanone}}{\text{CH}_3\text{CH}_2\overset{\text{O}}{\text{C}}\text{CH}_2\text{CH}_3} \xrightarrow[\text{CH}_3\text{OH}]{\text{NaBH}_4} \underset{\substack{\text{3-Pentanol}}}{\text{CH}_3\text{CH}_2\underset{\text{OH}}{\text{CHCH}_2\text{CH}_3}}
$$

$$
\underset{\text{2-Methylbutanal}}{\text{CH}_3\text{CH}_2\underset{\text{CH}_3}{\text{CH}}\overset{\text{O}}{\text{CH}}} \xrightarrow[\text{CH}_3\text{OH}]{\text{NaBH}_4} \underset{\substack{\text{2-Methyl-1-butanol}}}{\text{CH}_3\text{CH}_2\underset{\text{CH}_3}{\text{CHCH}_2\text{OH}}}
$$

$$
\underset{\text{3-Methylbutanal}}{(\text{CH}_3)_2\text{CHCH}_2\overset{\text{O}}{\text{CH}}} \xrightarrow[\text{CH}_3\text{OH}]{\text{NaBH}_4} \underset{\substack{\text{3-Methyl-1-butanol}}}{(\text{CH}_3)_2\text{CHCH}_2\text{CH}_2\text{OH}}
$$

$$(CH_3)_2CHCCH_3 \xrightarrow[CH_3OH]{NaBH_4} (CH_3)_2CHCHCH_3$$

O (above C in starting material)

OH (below CH in product)

3-Methyl-2-butanone   3-Methyl-2-butanol

$$(CH_3)_3CCH \xrightarrow[CH_3OH]{NaBH_4} (CH_3)_3CCH_2OH$$

2,2-Dimethylpropanal   2,2-Dimethyl-1-propanol

These are all the *primary* and *secondary alcohols* of molecular formula $C_5H_{12}O$. The tertiary alcohol $(CH_3)_2CCH_2CH_3$ (2-methyl-2-butanol) cannot be prepared by reduction

OH

of a carbonyl compound.

15.20 (*a*) The suggested synthesis

$$CH_3CH_2CH_2CH_3 \xrightarrow[\text{light or heat}]{Br_2} CH_3CH_2CH_2CH_2Br \xrightarrow{KOH} CH_3CH_2CH_2CH_2OH$$

Butane   1-Bromobutane   1-Butanol

is a poor one because bromination of butane yields a mixture of 1-bromobutane and 2-bromobutane, 2-bromobutane being the major product.

$$CH_3CH_2CH_2CH_3 \xrightarrow[\text{heat}]{Br_2 \atop \text{light or}} CH_3CH_2CH_2CH_2Br + CH_3CHCH_2CH_3$$

Br

Butane   1-Bromobutane   2-Bromobutane
(minor product)   (major product)

(*b*) The suggested synthesis

$$(CH_3)_3CH \xrightarrow[\text{heat}]{Br_2 \atop \text{light or}} (CH_3)_3CBr \xrightarrow{KOH} (CH_3)_3COH$$

2-Methylpropane   2-Bromo-2-methylpropane   2-Methyl-2-propanol

will fail because the reaction of 2-bromo-2-methylpropane with potassium hydroxide will proceed by elimination rather than by substitution. The first step in the process, selective bromination of 2-methylpropane to 2-bromo-2-methylpropane, is satisfactory because bromination is selective for substitution of tertiary hydrogens in the presence of secondary and primary ones.

(*c*) Benzyl alcohol, unlike 1-butanol and 2-methyl-2-propanol, can be prepared effectively by this method.

⬡—CH₃ $\xrightarrow[\text{heat}]{Br_2 \atop \text{light or}}$ ⬡—CH₂Br $\xrightarrow{KOH}$ ⬡—CH₂OH

Toluene   Benzyl bromide   Benzyl alcohol

Free-radical bromination of toluene is selective for the benzylic position. Benzyl bromide cannot undergo elimination, and so nucleophilic substitution of bromide by hydroxide will work well.

15.21 Glucose contains five hydroxyl groups and an aldehyde functional group. Its hydrogenation will not affect the hydroxyl groups but will reduce the aldehyde to a primary alcohol.

$$\text{Glucose} \xrightarrow[\text{Ni, 140°C}]{H_2 \text{ (120 atm)}} \text{Sorbitol}$$

Glucose   Sorbitol

15.22 (*a*) 1-Phenylethanol is a secondary alcohol and so can be prepared by the reaction of a Grignard reagent with an aldehyde. One combination is phenylmagnesium bromide and ethanal (acetaldehyde).

$$C_6H_5\underset{\underset{OH}{|}}{C}HCH_3 \implies C_6H_5MgBr \;+\; H\overset{\overset{O}{\|}}{C}CH_3$$

1-Phenylethanol   Phenylmagnesium   Ethanal
                              bromide   (acetaldehyde)

Grignard reagents—phenylmagnesium bromide, in this case—are always prepared by reaction of magnesium metal and the corresponding halide. Starting with bromobenzene, a suitable synthesis is described by the sequence:

$$C_6H_5Br \xrightarrow[\substack{diethyl \\ ether}]{Mg} C_6H_5MgBr \xrightarrow[\text{2. } H_3O^+]{\text{1. } CH_3\overset{\overset{O}{\|}}{C}H} C_6H_5\underset{\underset{OH}{|}}{C}HCH_3$$

Bromobenzene   Phenylmagnesium   1-Phenylethanol
                          bromide

(*b*) An alternative disconnection of 1-phenylethanol reveals a second route using benzaldehyde and a methyl Grignard reagent.

$$C_6H_5\underset{\underset{OH}{|}}{C}HCH_3 \implies C_6H_5\overset{\overset{O}{\|}}{C}H \;+\; CH_3MgI$$

1-Phenylethanol   Benzaldehyde   Methylmagnesium
                                                   iodide

Equations representing this approach are:

$$CH_3I \xrightarrow[\substack{diethyl \\ ether}]{Mg} CH_3MgI \xrightarrow[\text{2. } H_3O^+]{\text{1. } C_6H_5CHO} C_6H_5\underset{\underset{OH}{|}}{C}HCH_3$$

Iodomethane   Methylmagnesium   1-Phenylethanol
                         iodide

(*c*) Aldehydes are, in general, obtainable by oxidation of the corresponding primary alcohol. By recognizing that benzaldehyde can be obtained by Collins' oxidation of benzyl alcohol, we write:

$$C_6H_5CH_2OH \xrightarrow[CH_2Cl_2]{(C_5H_5N)_2CrO_3} C_6H_5\overset{\overset{O}{\|}}{C}H \xrightarrow[\text{2. } H_3O^+]{\substack{\text{1. } CH_3MgI, \\ diethyl \ ether}} C_6H_5\underset{\underset{OH}{|}}{C}HCH_3$$

Benzyl alcohol   Benzaldehyde   1-Phenylethanol

(*d*) Styrene ($C_6H_5CH{=}CH_2$) and the desired product 1-phenylethanol have the same carbon skeleton. The conversion therefore requires hydration of the double bond with a regioselectivity that follows Markovnikov's rule. This is best done by oxymercuration-demercuration.

$$C_6H_5CH{=}CH_2 \xrightarrow[\text{2. } NaBH_4, HO^-]{\text{1. } Hg(O_2CCH_3)_2, THF-H_2O} C_6H_5\underset{\underset{OH}{|}}{C}HCH_3$$

Styrene   1-Phenylethanol

(*e*) The conversion of acetophenone to 1-phenylethanol is a reduction.

$$C_6H_5\overset{\overset{O}{\|}}{C}CH_3 \xrightarrow[\substack{LiAlH_4}]{\substack{H_2, Pt \\ reducing \ agent}} C_6H_5\underset{\underset{OH}{|}}{C}HCH_3$$

Acetophenone   1-Phenylethanol

Any of a number of reducing agents could be used. These include:

1. $NaBH_4$, CH_3OH
2. $LiAlH_4$ in diethyl ether, then H_2O
3. H_2 and a Pt, Pd, Ni, or Ru catalyst

(f) Benzene can be employed as the ultimate starting material in a synthesis of 1-phenylethanol. Friedel-Crafts acylation of benzene gives acetophenone, which can then be reduced as in part (e).

Benzene   Acetyl chloride   Acetophenone

Acetic anhydride (CH_3COCCH_3) can be used in place of acetyl chloride.

15.23 2-Phenylethanol is an ingredient in many perfumes, to which it imparts a roselike fragrance. Numerous methods have been employed for its synthesis.

(a) As a primary alcohol having two more carbon atoms than bromobenzene, it can be formed by reaction of a Grignard reagent, phenylmagnesium bromide, with ethylene oxide.

$$C_6H_5CH_2CH_2OH \Longrightarrow C_6H_5MgBr + CH_2\text{—}CH_2$$

The desired reaction sequence is therefore:

Bromobenzene   Phenylmagnesium   2-Phenylethanol
                bromide

(b) Hydration of sytrene with a regioselectivity contrary to that of Markovnikov's rule is required. This is accomplished readily by hydroboration-oxidation.

$$C_6H_5CH\text{=}CH_2 \xrightarrow[\text{2. } H_2O_2, HO^-]{\text{1. } B_2H_6, \text{ diglyme}} C_6H_5CH_2CH_2OH$$

Styrene   2-Phenylethanol

(c) Phenylacetylene can serve as a precursor to 2-phenylethanol if it is first converted to styrene.

$$C_6H_5C\text{≡}CH \xrightarrow[\text{H}_2, \text{ Lindlar Pd}]{\text{Li, NH}_3 \text{ or}} C_6H_5CH\text{=}CH_2$$

Phenylacetylene   Styrene

Hydration of styrene to 2-phenylethanol as in part (b) completes the synthesis. Alternatively, hydroboration-oxidation of phenylacetylene to yield 2-phenylethanal followed by reduction [see part (d)] also yields 2-phenylethanol.

(d) Reduction of aldehydes yields primary alcohols.

2-Phenylethanal   2-Phenylethanol

Among the reducing agents that could be (and have been) used are:

1. $NaBH_4$, CH_3OH
2. $LiAlH_4$ in diethyl ether, then H_2O
3. H_2 and a Pt, Pd, Ni, or Ru catalyst

(e) Esters are readily reduced to primary alcohols by lithium aluminum hydride.

Ethyl 2-phenylethanoate   2-Phenylethanol

(*f*) The only reagent that is suitable for the direct reduction of carboxylic acids to primary alcohols is lithium aluminum hydride.

$$\underset{\text{2-Phenylethanoic acid}}{C_6H_5CH_2\overset{\displaystyle O}{\overset{\|}{C}}OH} \xrightarrow[\text{2. H}_2\text{O}]{\text{1. LiAlH}_4\text{, diethyl ether}} \underset{\text{2-Phenylethanol}}{C_6H_5CH_2CH_2OH}$$

Alternatively, the carboxylic acid could be esterified with ethanol and the resulting ethyl phenylethanoate reduced.

$$\underset{\substack{\text{2-Phenylethanoic} \\ \text{acid}}}{C_6H_5CH_2\overset{\displaystyle O}{\overset{\|}{C}}OH} + \underset{\text{Ethanol}}{CH_3CH_2OH} \xrightarrow{\text{H}^+} \underset{\text{Ethyl 2-phenylethanoate}}{C_6H_5CH_2\overset{\displaystyle O}{\overset{\|}{C}}OCH_2CH_3} \xrightarrow[\text{in } (e)]{\text{reduce as}} \underset{\text{2-Phenylethanol}}{C_6H_5CH_2CH_2OH}$$

15.24 (*a*) Alkyl fluorides may be prepared from alkyl bromides by heating with sodium or potassium fluoride as described in Sec. 8.2. The mechanism is S_N2. Normally, the alkyl fluoride is the lowest-boiling component of the mixture and is removed by distillation as it is formed. (Alcohols do not give alkyl fluorides on reaction with HF. Nor do they react with sodium fluoride, and so the first step *must* be conversion of 1-butanol to 1-bromobutane.)

$$\underset{\text{1-Butanol}}{CH_3CH_2CH_2CH_2OH} \xrightarrow[\text{or PBr}_3]{\text{HBr, heat}} \underset{\text{1-Bromobutane}}{CH_3CH_2CH_2CH_2Br} \xrightarrow[\substack{\text{heat in} \\ \text{high-boiling} \\ \text{alcohol solvent}}]{\text{NaF}} \underset{\text{1-Fluorobutane}}{CH_3CH_2CH_2CH_2F}$$

(*b*) Thiols are made from alkyl halides by reaction with thiourea, followed by hydrolysis of the isothiouronium salt in base. Again, the first step must be a conversion of the alcohol to an alkyl bromide.

$$\underset{\substack{\text{1-Bromobutane} \\ \text{from part } (a)}}{CH_3CH_2CH_2CH_2Br} \xrightarrow[\text{2. NaOH}]{\text{1. (H}_2\text{N)}_2\text{C=S}} \underset{\text{1-Butanethiol}}{CH_3CH_2CH_2CH_2SH}$$

(*c*) The target molecule 2-hexanol may be mentally disconnected as shown to a four-carbon unit and a two-carbon unit.

$$\underset{\substack{\displaystyle | \\ \displaystyle OH}}{CH_3CH\text{---}CH_2CH_2CH_2CH_3} \Longrightarrow CH_3CH + {}^-{:}CH_2CH_2CH_2CH_3$$

The alternative disconnection to $^-{:}CH_3$ and $H\overset{\displaystyle O}{\overset{\|}{C}}CH_2CH_2CH_2CH_3$ reveals a plausible approach to 2-hexanol but is inconsistent with the requirement of the problem that limits starting materials to four carbons or fewer. The five-carbon aldehyde would have to be prepared first, making for a lengthy overall synthetic scheme.

An appropriate synthesis based on alcohols as starting materials is:

$$\underset{\text{Ethanol}}{CH_3CH_2OH} \xrightarrow[\text{CH}_2\text{Cl}_2]{\text{(C}_5\text{H}_5\text{N)}_2\text{CrO}_3} \underset{\text{Ethanal}}{CH_3\overset{\displaystyle O}{\overset{\|}{C}}H}$$

$$\underset{\text{1-Butanol}}{CH_3CH_2CH_2CH_2OH} \xrightarrow{\text{PBr}_3} \underset{\text{1-Bromobutane}}{CH_3CH_2CH_2CH_2Br} \xrightarrow[\substack{\text{diethyl} \\ \text{ether}}]{\text{Mg}} \underset{\text{Butylmagnesium bromide}}{CH_3CH_2CH_2CH_2MgBr}$$

$$\underset{\text{Butylmagnesium bromide}}{CH_3CH_2CH_2CH_2MgBr} + \underset{\text{Ethanal}}{CH_3\overset{\displaystyle O}{\overset{\|}{C}}H} \xrightarrow[\text{2. H}_3\text{O}^+]{\text{1. diethyl ether}} \underset{\text{2-Hexanol}}{\underset{\substack{\displaystyle | \\ \displaystyle OH}}{CH_3CHCH_2CH_2CH_2CH_3}}$$

(d)   Oxidation of 2-hexanol from part (c) yields 2-hexanone.

$$CH_3CHCH_2CH_2CH_2CH_3 \xrightarrow[\text{H}_2\text{SO}_4,\ \text{water}]{\text{Na}_2\text{Cr}_2\text{O}_7} CH_3\overset{\displaystyle O}{\overset{\|}{C}}CH_2CH_2CH_2CH_3$$

| |
OH

2-Hexanol   2-Hexanone

Collins' reagent, PCC, or PDC can also be used for this transformation.

(e)   In order to obtain hexanoic acid from alcohols having four carbons or fewer, a two-carbon chain extension must be carried out. This suggests reaction of a Grignard reagent with ethylene oxide. The resulting alcohol, 1-hexanol, must then be oxidized to yield the desired carboxylic acid. The retrosynthetic path for this approach is as follows:

$$CH_3(CH_2)_4CO_2H \Longrightarrow CH_3(CH_2)_4CH_2OH \Longrightarrow CH_3CH_2CH_2CH_2MgBr + \overset{\displaystyle O}{CH_2{-}CH_2}$$

The reaction sequence therefore becomes:

$$CH_3CH_2CH_2CH_2Br \xrightarrow[\text{diethyl ether}]{\text{Mg}} CH_3CH_2CH_2CH_2MgBr \xrightarrow[2.\ \text{H}_3\text{O}^+]{1.\ \overset{O}{CH_2{-}CH_2}}$$

1-Bromobutane   Butylmagnesium bromide

$$CH_3CH_2CH_2CH_2CH_2CH_2OH$$

1-Hexanol

$$CH_3(CH_2)_4CH_2OH \xrightarrow[\text{H}_2\text{SO}_4,\ \text{H}_2\text{O}]{\text{K}_2\text{Cr}_2\text{O}_7} CH_3(CH_2)_4CO_2H$$

1-Hexanol   Hexanoic acid

1-Bromobutane can be prepared by reaction of 1-butanol with PBr_3 or with HBr.

$$CH_3CH_2CH_2CH_2OH \xrightarrow[\text{or HBr}]{\text{PBr}_3} CH_3CH_2CH_2CH_2Br$$

1-Butanol   1-Bromobutane

Given the constraints of the problem, we prepare ethylene oxide by the sequence:

$$CH_3CH_2OH \xrightarrow[\text{heat}]{\text{H}_2\text{SO}_4} CH_2{=}CH_2 \xrightarrow{\text{CH}_3\text{COOH}} \overset{\displaystyle CH_2{-}CH_2}{\underset{O}{}}$$

Ethanol   Ethylene

(f)   Fischer esterification of hexanoic acid with ethanol produces ethyl hexanoate.

$$CH_3(CH_2)_4CO_2H + CH_3CH_2OH \xrightarrow{\text{H}^+} CH_3(CH_2)_4\overset{\displaystyle O}{\overset{\|}{C}}OCH_2CH_3$$

Hexanoic acid   Ethanol   Ethyl hexanoate

(g)   Vicinal diols are normally prepared by hydroxylation of alkenes with osmium tetraoxide and tert-butyl hydroperoxide.

$$(CH_3)_2C{=}CH_2 \xrightarrow[\substack{(\text{CH}_3)_3\text{COOH, HO}^- \\ (\text{CH}_3)_3\text{COH}}]{\text{OsO}_4} (CH_3)_2CCH_2OH$$

| |
HO

2-Methylpropene   2-Methyl-1,2-propanediol

The required alkene is available by dehydration of 2-methyl-2-propanol.

$$(CH_3)_3COH \xrightarrow[\text{heat}]{\text{H}_3\text{PO}_4} (CH_3)_2C{=}CH_2$$

2-Methyl-2-propanol   2-Methylpropene

(h) The desired aldehyde can be prepared by oxidation of the corresponding primary alcohol with Collins' reagent, PCC, or PDC.

$$(CH_3)_3CCH_2OH \xrightarrow[CH_2Cl_2]{(C_5H_5N)_2CrO_3} (CH_3)_3CCH$$

2,2-Dimethyl-1-propanol   2,2-Dimethylpropanal

The necessary alcohol is available through reaction of a *tert*-butyl Grignard reagent with formaldehyde, as shown by the disconnection

$$(CH_3)_3CCH_2OH \Longrightarrow (CH_3)_3CMgCl + H_2C=O$$

$$CH_3OH \xrightarrow[CH_2Cl_2]{(C_5H_5N)_2CrO_3} HCH$$

Methanol   Formaldehyde

$$(CH_3)_3COH \xrightarrow{HCl} (CH_3)_3CCl \xrightarrow[diethyl\ ether]{Mg} (CH_3)_3CMgCl$$

2-Methyl-2-propanol   2-Chloro-2-methylpropane   1,1-Dimethylethylmagnesium
(*tert*-butyl   (*tert*-butyl chloride)   chloride (*tert*-butylmagnesium
alcohol)   chloride)

$$(CH_3)_3CMgCl \xrightarrow[2.\ H_3O^+]{1.\ CH_2O,\ diethyl\ ether} (CH_3)_3CCH_2OH$$

1,1-Dimethylethylmagnesium   2,2-Dimethyl-1-propanol
chloride (*tert*-butylmagnesium
chloride)

15.25 (a) The simplest route to this primary chloride from benzene is through the corresponding alcohol. The first step is the two-carbon chain extension used in Problem 15.23a.

Benzene → Bromobenzene → 2-Phenylethanol

$$\text{2-Phenylethanol} \xrightarrow{SOCl_2} \text{1-Chloro-2-phenylethane}$$

The preparation of ethylene oxide is shown in Problem 15.24e.

(b) A Friedel-Crafts acylation is the best approach to the target ketone.

Benzene + (CH_3)_2CHCCl $\xrightarrow{AlCl_3}$ 2-Methyl-1-phenyl-1-propanone

2-Methylpropanoyl chloride

Since carboxylic acid chlorides are prepared from the corresponding acids, we write:

$$(CH_3)_2CHCH_2OH \xrightarrow[H_2SO_4,\ heat]{K_2Cr_2O_7} (CH_3)_2CHCOH \xrightarrow{SOCl_2} (CH_3)_2CHCCl$$

2-Methyl-1-propanol   2-Methylpropanoic acid   2-Methylpropanoyl chloride

(c) Wolff-Kishner or Clemmensen reduction of the ketone just prepared in part (b) affords isobutylbenzene.

$$C_6H_5\overset{\displaystyle O}{\overset{\|}{C}}CH(CH_3)_2 \xrightarrow[\text{or Zn(Hg), HCl}]{\substack{\text{H}_2\text{NNH}_2, \text{ HO}^-, \\ \text{triethylene glycol, heat}}} C_6H_5CH_2CH(CH_3)_2$$

2-Methyl-1-phenyl-1-propanone   Isobutylbenzene

A less direct approach requires three steps:

$$C_6H_5\overset{\displaystyle O}{\overset{\|}{C}}CH(CH_3)_2 \xrightarrow[\text{CH}_3\text{OH}]{\text{NaBH}_4} C_6H_5\underset{\underset{\displaystyle OH}{|}}{C}HCH(CH_3)_2$$

2-Methyl-1-phenyl-1-propanone   2-Methyl-1-phenyl-1-propanol

$$\downarrow \text{H}_2\text{SO}_4, \text{ heat}$$

$$C_6H_5CH_2CH(CH_3)_2 \xleftarrow[\text{Pt}]{\text{H}_2} C_6H_5CH{=}C(CH_3)_2$$

Isobutylbenzene   2-Methyl-1-phenylpropene

15.26 (*a*) Because 1-phenylcyclopentanol is a tertiary alcohol, a likely synthesis would involve reaction of a ketone and a Grignard reagent. Thus, a reasonable last step is treatment of cyclopentanone with phenylmagnesium bromide.

Cyclopentanone   1-Phenylcyclopentanol

Cyclopentanone is prepared by oxidation of cyclopentanol. Any one of a number of oxidizing agents would be suitable. These include PDC, PCC, or $(C_5H_5N)_2CrO_3$ in CH_2Cl_2, or chromic acid (H_2CrO_4) generated from $Na_2Cr_2O_7$ in aqueous sulfuric acid.

Cyclopentanol   Cyclopentanone

(*b*) Acid-catalyzed dehydration of 1-phenylcyclopentanol gives 1-phenylcyclopentene.

1-Phenylcyclopentanol   1-Phenylcyclopentene

(*c*) Hydroboration-oxidation of 1-phenylcyclopentene gives *trans*-2-phenylcyclopentanol. The elements of water (H and OH) are added across the double bond opposite to Markovnikov's rule and syn to one another.

1-Phenylcyclopentene   *trans*-2-Phenylcyclopentanol

(*d*) Oxidation of *trans*-2-phenylcyclopentanol converts this secondary alcohol to the desired ketone. Any of the Cr(VI)–derived oxidizing agents mentioned in part (*a*) for oxidation of cyclopentanol to cyclopentanone is satisfactory.

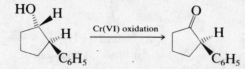

trans-2-Phenylcyclopentanol $\xrightarrow{\text{Cr(VI) oxidation}}$ 2-Phenylcyclopentanone

(*e*) The standard procedure for preparing *cis*-1,2-diols is by hydroxylation of alkenes with osmium tetraoxide.

1-Phenylcyclopentene $\xrightarrow[\text{(CH}_3\text{)}_3\text{COH, HO}^-]{\text{OsO}_4,\ \text{(CH}_3\text{)}_3\text{COOH}}$ 1-Phenyl-*cis*-1,2-cyclopentanediol

(*f*) The desired compound is available either by ozonolysis of 1-phenylcyclopentene:

1-Phenylcyclopentene $\xrightarrow[\text{2. H}_2\text{O, Zn}]{\text{1. O}_3}$ $C_6H_5CCH_2CH_2CH_2CH$

5-Oxo-1-phenyl-1-pentanone

or by periodic acid cleavage of the diol in part (*e*):

1-Phenyl-*cis*-1,2-cyclopentanediol $\xrightarrow{\text{HIO}_4}$ $C_6H_5CCH_2CH_2CH_2CH$

5-Oxo-1-phenyl-1-pentanone

(*g*) Reduction of both carbonyl groups in the product of part (*f*) gives the desired diol.

$C_6H_5CCH_2CH_2CH_2CH$ $\xrightarrow[\substack{\text{or} \\ 1.\ \text{LiAlH}_4,\ \text{diethyl ether} \\ 2.\ \text{H}_2\text{O}}]{\substack{\text{H}_2,\ \text{Pt (or Pd, Ni, Rh)} \\ \text{or} \\ \text{NaBH}_4,\ \text{H}_2\text{O}}}$ $C_6H_5CHCH_2CH_2CH_2CH_2OH$

5-Oxo-1-phenyl-1-pentanone $\quad\quad$ OH

1-Phenyl-1,5-pentanediol

15.27 (*a, b*) Primary alcohols react in two different ways on being heated with acid catalysts: they can condense to form dialkyl ethers or undergo dehydration to yield alkenes. Ether formation is favored at lower temperature, and alkene formation is favored at higher temperature.

$$2CH_3CH_2CH_2OH \xrightarrow[140°C]{\text{H}_2\text{SO}_4} CH_3CH_2CH_2OCH_2CH_2CH_3 + H_2O$$

1-Propanol $\quad\quad\quad$ Dipropyl ether $\quad\quad$ Water

$$CH_3CH_2CH_2OH \xrightarrow[200°C]{\text{H}_2\text{SO}_4} CH_3CH=CH_2 + H_2O$$

1-Propanol $\quad\quad\quad$ Propene $\quad$ Water

(*c*) Nitrate esters are formed by the reaction of alcohols with nitric acid in the presence of a sulfuric acid catalyst.

$$CH_3CH_2CH_2OH + HONO_2 \xrightarrow{\text{H}_2\text{SO}_4\text{(cat)}} CH_3CH_2CH_2ONO_2 + H_2O$$

1-Propanol $\quad$ Nitric acid $\quad\quad\quad$ Propyl nitrate $\quad$ Water

(*d*) Pyridinium chlorochromate (PCC) oxidizes primary alcohols to aldehydes.

$$CH_3CH_2CH_2OH \xrightarrow[\text{CH}_2\text{Cl}_2]{\text{PCC}} CH_3CH_2CH$$

1-Propanol $\quad\quad\quad\quad\quad$ Propanal

(e)   Potassium dichromate in sulfuric acid oxidizes primary alcohols to carboxylic acids.

$$CH_3CH_2CH_2OH \xrightarrow[\text{H}_2\text{SO}_4,\ \text{heat}]{\text{K}_2\text{Cr}_2\text{O}_7,} CH_3CH_2\overset{\displaystyle O}{\overset{\|}{C}}OH$$

1-Propanol   Propanoic acid

(f)   Alcohols react with active metals such as sodium to give alkoxides.

$$2CH_3CH_2CH_2OH + 2Na \longrightarrow 2CH_3CH_2CH_2ONa + H_2$$

1-Propanol   Sodium   Sodium 1-propanolate   Hydrogen

(g)   Amide ion, a strong base, abstracts a proton from 1-propanol to form ammonia and 1-propanolate ion. This is an acid-base reaction.

$$CH_3CH_2CH_2OH + NaNH_2 \longrightarrow CH_3CH_2CH_2ONa + NH_3$$

1-Propanol   Sodium amide   Sodium 1-propanolate   Ammonia

(h)   Acetate ion is a weaker base than 1-propanolate. The equilibrium

$$CH_3CH_2CH_2OH + CH_3\overset{\displaystyle O}{\overset{\|}{C}}O^- \rightleftharpoons CH_3CH_2CH_2O^- + CH_3\overset{\displaystyle O}{\overset{\|}{C}}OH$$

1-Propanol   Acetate   1-Propanolate   Acetic acid

strongly favors starting materials. 1-Propanol does not react with sodium acetate in any significant way.

(i)   With acetic acid and in the presence of an acid catalyst, 1-propanol is converted to its acetate ester.

$$CH_3CH_2CH_2OH + CH_3\overset{\displaystyle O}{\overset{\|}{C}}OH \xrightleftharpoons{\text{HCl}} CH_3\overset{\displaystyle O}{\overset{\|}{C}}OCH_2CH_2CH_3 + H_2O$$

1-Propanol   Acetic acid   Propyl acetate   Water

This is an equilibrium process that slightly favors products.

(j)   Alcohols react with p-toluenesulfonyl chloride to give p-toluenesulfonate esters.

$$CH_3CH_2CH_2OH + CH_3\!-\!\!\langle\ \rangle\!\!-\!SO_2Cl \xrightarrow{\text{pyridine}}$$

1-Propanol   p-Toluenesulfonyl chloride

$$CH_3CH_2CH_2O\overset{\displaystyle O}{\underset{\displaystyle O}{\overset{\|}{\underset{\|}{S}}}}\!-\!\!\langle\ \rangle\!\!-\!CH_3 + HCl$$

Propyl p-toluenesulfonate

(k)   Acyl chlorides convert alcohols to esters.

$$CH_3CH_2CH_2OH + CH_3O\!-\!\!\langle\ \rangle\!\!-\!\overset{\displaystyle O}{\overset{\|}{C}}Cl \xrightarrow{\text{pyridine}}$$

1-Propanol   p-Methoxybenzoyl chloride

$$CH_3CH_2CH_2O\overset{\displaystyle O}{\overset{\|}{C}}\!-\!\!\langle\ \rangle\!\!-\!OCH_3 + HCl$$

Propyl p-methoxybenzoate

(l)   The reagent is benzoic anhydride. Carboxylic acid anhydrides react with alcohols to give esters.

$$CH_3CH_2CH_2OH + C_6H_5\overset{\displaystyle O}{\overset{\|}{C}}O\overset{\displaystyle O}{\overset{\|}{C}}C_6H_5 \xrightarrow{\text{pyridine}} CH_3CH_2CH_2O\overset{\displaystyle O}{\overset{\|}{C}}C_6H_5 + C_6H_5\overset{\displaystyle O}{\overset{\|}{C}}OH$$

1-Propanol   Benzoic anhydride   Propyl benzoate   Benzoic acid

(*m*) The reagent is succinic anhydride, a cyclic anhydride. Esterification occurs, but in this case the resulting ester and carboxylic acid functions remain part of the same molecule.

$$CH_3CH_2CH_2OH + \underset{\text{Succinic anhydride}}{\boxed{O}} \xrightarrow{\text{pyridine}} \underset{\text{Hydrogen propyl succinate}}{CH_3CH_2CH_2O\overset{O}{\overset{||}{C}}CH_2CH_2\overset{O}{\overset{||}{C}}OH}$$

1-Propanol

15.28 (*a, b*) The reaction of 2-propanol with sulfuric acid at elevated temperatures yields propene and diisopropyl ether.

$$\underset{\underset{\text{2-Propanol}}{OH}}{CH_3CHCH_3} \xrightarrow[\text{heat}]{H_2SO_4} \underset{\text{Propene}}{CH_3CH=CH_2} + \underset{\text{Diisopropyl ether}}{(CH_3)_2CHOCH(CH_3)_2}$$

Propene is the major product regardless of whether the temperature is 140°C (*a*) or 200°C (*b*). The fraction of diisopropyl ether formed is greater at the lower temperature.

(*c*) 2-Propanol reacts with nitric acid to form the corresponding alkyl nitrate.

$$\underset{\underset{\text{2-Propanol}}{OH}}{CH_3CHCH_3} + \underset{\text{Nitric acid}}{HONO_2} \xrightarrow{H_2SO_4(\text{cat})} \underset{\underset{\text{Isopropyl nitrate}}{ONO_2}}{CH_3CHCH_3} + \underset{\text{Water}}{H_2O}$$

(*d*) The secondary alcohol 2-propanol is oxidized to the ketone acetone by pyridinium chlorochromate (PCC).

$$\underset{\underset{\text{2-Propanol}}{OH}}{CH_3CHCH_3} \xrightarrow[\text{CH}_2\text{Cl}_2]{PCC} \underset{\text{Acetone}}{CH_3\overset{O}{\overset{||}{C}}CH_3}$$

(*e*) Chromium(VI) reagents oxidize 2-propanol to acetone.

$$\underset{\underset{\text{2-Propanol}}{OH}}{CH_3CHCH_3} \xrightarrow[\text{H}_2\text{SO}_4, \text{H}_2\text{O}]{K_2Cr_2O_7,} \underset{\text{Acetone}}{CH_3\overset{O}{\overset{||}{C}}CH_3}$$

(*f*) Sodium reacts with alcohols to yield the corresponding sodium alkoxide.

$$\underset{\text{2-Propanol}}{(CH_3)_2CHOH} + \underset{\text{Sodium}}{2Na} \longrightarrow \underset{\text{Sodium 2-propanolate}}{(CH_3)_2CHONa} + \underset{\text{Hydrogen}}{H_2}$$

(*g*) The reaction of 2-propanol with sodium amide is an acid-base reaction.

$$\underset{\substack{\text{2-Propanol} \\ \text{(stronger acid)}}}{(CH_3)_2CHOH} + \underset{\substack{\text{Sodium amide} \\ \text{(stronger base)}}}{NaNH_2} \longrightarrow \underset{\substack{\text{Sodium 2-propanolate} \\ \text{(weaker base)}}}{(CH_3)_2CHONa} + \underset{\substack{\text{Ammonia} \\ \text{(weaker acid)}}}{NH_3}$$

(*h*) No net reaction is observed here. An equilibrium is established but it overwhelmingly favors starting materials.

$$\underset{\substack{\text{2-Propanol} \\ \text{(weaker acid)}}}{(CH_3)_2CHOH} + \underset{\substack{\text{Sodium acetate} \\ \text{(weaker base)}}}{NaO\overset{O}{\overset{||}{C}}CH_3} \rightleftharpoons \underset{\substack{\text{Sodium 2-propanolate} \\ \text{(stronger base)}}}{(CH_3)_2CHONa} + \underset{\substack{\text{Acetic acid} \\ \text{(stronger acid)}}}{CH_3\overset{O}{\overset{||}{C}}OH}$$

(*i*) Esterification occurs when 2-propanol is allowed to react with acetic acid in the presence of an acid catalyst.

$$(CH_3)_2CHOH + CH_3\overset{O}{\overset{\|}{C}}OH \rightleftharpoons CH_3\overset{O}{\overset{\|}{C}}OCH(CH_3)_2 + H_2O$$

2-Propanol   Acetic acid   Isopropyl acetate

An equilibrium is established.

(*j*) Secondary alcohols react with *p*-toluenesulfonyl chloride to yield secondary alkyl *p*-toluenesulfonate esters.

2-Propanol   *p*-Toluenesulfonyl chloride   Isopropyl *p*-toluenesulfonate

(*k*) Ester formation is the characteristic reaction between an alcohol and an acyl chloride.

2-Propanol   *p*-Methoxybenzoyl chloride   Isopropyl *p*-methoxybenzoate

(*l*) An acid anhydride converts an alcohol to an ester.

$$(CH_3)_2CHOH + C_6H_5\overset{O}{\overset{\|}{C}}O\overset{O}{\overset{\|}{C}}C_6H_5 \xrightarrow{\text{pyridine}} (CH_3)_2CHO\overset{O}{\overset{\|}{C}}C_6H_5 + C_6H_5\overset{O}{\overset{\|}{C}}OH$$

2-Propanol   Benzoic anhydride   Isopropyl benzoate   Benzoic acid

(*m*) One of the carbonyl groups of succinic anhydride is converted to an ester group, the other to a carboxylic acid group.

2-Propanol   Succinic anhydride   Hydrogen isopropyl succinate

15.29 (*a*) On being heated in the presence of sulfuric acid, tertiary alcohols undergo elimination.

4-Methyl-1-phenylcyclohexanol   4-Methyl-1-phenylcyclohexene (81%)

(*b*) The substrate is a primary alcohol and is oxidized by Cr(VI) in the form of Collins' reagent to an aldehyde. The double bonds are unaffected.

4,6-Heptadien-1-ol   4,6-Heptadienal (87%)

(*c*) The combination of reagents specified converts alkenes to vicinal diols.

2,3-Dimethyl-2-butene   2,3-Dimethyl-2,3-butanediol (72%)

(*d*) Hydroboration-oxidation of the double bond takes place with a regioselectivity that is opposite to Markovnikov's rule. The elements of water are added in a stereospecific syn fashion.

C_6H_5

1-Phenylcyclobutene
    →
1. B_2H_6, diglyme
2. H_2O_2, HO^-

trans-2-Phenylcyclobutanol
(82%)

(*e*) Lithium aluminum hydride reduces carboxylic acids to primary alcohols, but does not reduce carbon-carbon double bonds.

$-CO_2H$

Cyclopentene-4-carboxylic acid
    →
1. LiAlH$_4$, diethyl ether
2. H_2O

$-CH_2OH$

(3-Cyclopentenyl)methanol

(*f*) Chromic acid oxidizes the secondary alcohol to the corresponding ketone but does not affect the triple bond.

$CH_3CHC \equiv C(CH_2)_3CH_3$
|
OH

3-Octyn-2-ol
    →
H_2CrO_4
H_2SO_4, H_2O
acetone

$CH_3 \overset{O}{\overset{\|}{C}}C \equiv C(CH_2)_3CH_3$

3-Octyn-2-one (80%)

(*g*) Lithium aluminum hydride reduces carbonyl groups efficiently but does not normally react with double bonds.

$CH_3\overset{O}{\overset{\|}{C}}CH_2CH=CHCH_2\overset{O}{\overset{\|}{C}}CH_3$

4-Octen-2,7-dione
    →
1. LiAlH$_4$, diethyl ether
2. H_2O

$CH_3CHCH_2CH=CHCH_2CHCH_3$
 | |
 OH OH

4-Octen-2,7-diol (75%)

(*h*) Alcohols react with acyl chlorides to yield esters. The O—H bond is broken in this reaction; the C—O bond of the alcohol remains intact on ester formation.

CH_3 — OH
trans-3-Methylcyclohexanol
 +
3,5-Dinitrobenzoyl chloride (O_2N ... $\overset{O}{\overset{\|}{C}}Cl$)
   pyridine →

trans-3-Methylcyclohexyl
3,5-dinitrobenzoate (74%)

(*i*) Carboxylic acid anhydrides react with alcohols to give esters. Here, too, the spatial orientation of the C—O bond remains intact.

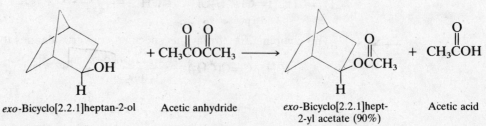

exo-Bicyclo[2.2.1]heptan-2-ol   Acetic anhydride   *exo*-Bicyclo[2.2.1]hept-2-yl acetate (90%)   Acetic acid

(*j*) The substrate is a carboxylic acid and undergoes Fischer esterification with methanol.

4-Chloro-3,5-dinitrobenzoic acid   Methyl 4-chloro-3,5-dinitrobenzoate (96%)

(*k*) Both ester functions are cleaved by reduction with lithium aluminum hydride. The product is a diol.

$$\xrightarrow[\text{2. H}_2\text{O}]{\text{1. LiAlH}_4}$$

+ CH$_3$CH$_2$OH + CH$_3$OH

(96%)

(*l*) Treatment of the diol obtained in part (*k*) with periodic acid brings about its cleavage to two carbonyl compounds.

$$\xrightarrow{\text{HIO}_4}$$

+ HCH

(74%)

15.30 Hydroxylation of alkenes with osmium tetraoxide is a syn addition. A racemic mixture of the 2*R*,3*S* and 2*S*,3*R* stereoisomers is formed from *cis*-2-pentene.

$$\xrightarrow[\text{(CH}_3)_3\text{COH, HO}^-]{\text{OsO}_4,\ \text{(CH}_3)_3\text{COOH}}$$

cis-2-Pentene   2*S*,3*R*-2,3-Pentanediol   2*R*,3*S*-2,3-Pentanediol

trans-2-Pentene gives a racemic mixture of the 2*R*,3*R* and 2*S*,3*S* stereoisomers.

$$\xrightarrow[\text{(CH}_3)_3\text{COH, HO}^-]{\text{OsO}_4,\ \text{(CH}_3)_3\text{COOH}}$$

trans-2-Pentene   2*R*,3*R*-2,3-Pentanediol   2*S*,3*S*-2,3-Pentanediol

15.31 (*a*) The task of converting a ketone to an alkene requires first the reduction of the ketone to an alcohol and then dehydration. In practice the two-step transformation has been carried out in 54 percent yield by treating the ketone with sodium borohydride and then heating the resulting alcohol with *p*-toluenesulfonic acid.

$$\xrightarrow[\text{CH}_3\text{OH}]{\text{NaBH}_4}$$ OH $$\xrightarrow[\text{heat}]{\text{H}^+}$$

Of course, sodium borohydride may be replaced by other suitable reducing agents, and *p*-toluenesulfonic acid is not the only acid that could be used in the dehydration step.

(*b*) This problem and the next one illustrate the value of reasoning backward. The desired product, cyclohexanol, can be prepared cleanly from cyclohexanone.

Once cyclohexanone is recognized to be a key intermediate, the synthetic pathway becomes apparent—what is needed is a method to convert the indicated starting material to cyclohexanone. The reagent ideally suited to this task is periodic acid. Therefore the synthetic sequence to be followed is:

(1-Hydroxycyclohexyl)methanol   Cyclohexanone   Cyclohexanol

(*c*) There is no direct method that allows a second hydroxyl group to be introduced at C-2 of 1-phenylcyclohexanol in a single step. We recognize the product as a vicinal diol and recall that such compounds are available by hydroxylation of alkenes.

This tells us that we must first dehydrate the tertiary alcohol, then hydroxylate the resulting alkene.

1-Phenylcyclohexanol   1-Phenylcyclohexene   1-Phenyl-1,2-cyclohexanediol

The syn stereoselectivity of the hydroxylation step ensures that the product will have its hydroxyl groups cis, as the problem requires.

15.32 Since the target molecule is an eight-carbon secondary alcohol and the problem restricts our choices of starting materials to alcohols of five carbons or fewer, we are led to consider building up the carbon chain by a Grignard reaction.

4-Methyl-3-heptanol

The disconnection shown leads to a three-carbon aldehyde and a five-carbon Grignard reagent. Starting with the corresponding alcohols, the following synthetic scheme seems reasonable.

First, propanal is prepared:

1-Propanol   Propanal

After converting 2-pentanol to its bromo derivative, a solution of the Grignard reagent is prepared:

$$\underset{\substack{|\\ OH}}{CH_3CHCH_2CH_2CH_3} \xrightarrow{PBr_3} \underset{\substack{|\\ Br}}{CH_3CHCH_2CH_2CH_3} \xrightarrow[\substack{diethyl\\ether}]{Mg} \underset{\substack{|\\ MgBr}}{CH_3CHCH_2CH_2CH_3}$$

2-Pentanol 2-Bromopentane 1-Methylbutylmagnesium bromide

Reaction of the Grignard reagent with the aldehyde yields the desired 4-methyl-3-heptanol:

$$\underset{\substack{|\\ MgBr}}{CH_3CHCH_2CH_2CH_3} + \underset{\substack{\\ }}{CH_3CH_2\overset{O}{\overset{\|}{C}}H} \xrightarrow[2.\ H_3O^+]{1.\ diethyl\ ether} \underset{\substack{|\\ HOCHCH_2CH_3}}{CH_3CHCH_2CH_2CH_3}$$

1-Methylbutylmagnesium bromide   Propanal   4-Methyl-3-heptanol

15.33 Our target molecule is void of functionality and so requires us to focus attention on the carbon skeleton. Notice that it can be considered to arise from three ethyl groups.

$$\underset{\substack{CH_3\\|}}{CH_3CH_2\text{--}CH\text{--}CH_2CH_3} \quad \text{3-Methylpentane}$$

Considering the problem retrosynthetically, we can see that a key intermediate having the carbon skeleton of the desired product is 3-methyl-3-pentanol. This becomes apparent from the fact that alkanes may be prepared from alkenes, which in turn are available from alcohols. The desired alcohol may be prepared from reaction of an acetate ester with a Grignard reagent, ethylmagnesium bromide.

$$\underset{\substack{|\\ CH_3}}{CH_3CH_2CHCH_2CH_3} \Rightarrow \underset{\substack{|\\ OH}}{CH_3CH_2\overset{CH_3}{\overset{|}{C}}CH_2CH_3} \Rightarrow CH_3\overset{O}{\overset{\|}{C}}OR + 2CH_3CH_2MgBr$$

The carbon skeleton can be assembled in one step by the reaction of ethylmagnesium bromide and ethyl acetate.

$$2CH_3CH_2MgBr + CH_3\overset{O}{\overset{\|}{C}}OCH_2CH_3 \xrightarrow[2.\ H_3O^+]{1.\ diethyl\ ether} \underset{\substack{|\\ CH_2CH_3}}{CH_3\overset{OH}{\overset{|}{C}}CH_2CH_3}$$

Ethylmagnesium bromide   Ethyl acetate   3-Methyl-3-pentanol

The resulting tertiary alcohol is converted to the desired hydrocarbon by acid-catalyzed dehydration and catalytic hydrogenation of the resulting mixture of alkenes.

$$\underset{\substack{|\\ CH_2CH_3}}{CH_3\overset{OH}{\overset{|}{C}}CH_2CH_3} \xrightarrow{H^+} \underset{\substack{|\\ CH_2CH_3}}{CH_3C{=}CHCH_3} + CH_2{=}C(CH_2CH_3)_2$$

3-Methyl-3-pentanol   3-Methyl-2-pentene (cis + trans)   2-Ethyl-1-butene

$$\xrightarrow{H_2,\ Ni} CH_3CH(CH_2CH_3)_2$$

3-Methylpentane

Since the problem requires that ethanol be the ultimate starting material, we need to show the preparation of the ethylmagnesium bromide and ethyl acetate used in constructing the carbon skeleton.

$$CH_3CH_2OH \xrightarrow{\text{PBr}_3} CH_3CH_2Br \xrightarrow[\text{diethyl ether}]{\text{Mg}} CH_3CH_2MgBr$$

Ethanol   Ethyl bromide   Ethylmagnesium bromide

$$CH_3CH_2OH \xrightarrow[\text{H}_2\text{SO}_4, \text{ heat}]{\text{K}_2\text{Cr}_2\text{O}_7} CH_3\overset{\displaystyle O}{\overset{\|}{C}}OH$$

Ethanol   Acetic acid

$$CH_3\overset{\displaystyle O}{\overset{\|}{C}}OH + CH_3CH_2OH \xrightarrow{\text{H}^+} CH_3\overset{\displaystyle O}{\overset{\|}{C}}OCH_2CH_3$$

Acetic acid   Ethyl acetate

15.34 (*a*) Retrosynthetically, we can see that the cis carbon-carbon double bond is available by hydrogenation of the corresponding alkyne over the Lindlar catalyst.

$$CH_3CH_2CH=CHCH_2CH_2OH \Rightarrow CH_3CH_2C\equiv CCH_2CH_2OH$$

The —CH$_2$CH$_2$OH unit can be appended to an alkynide anion by reaction with ethylene oxide.

$$CH_3CH_2C\equiv CCH_2CH_2OH \Rightarrow CH_3CH_2C\equiv C:^- + \underset{O}{CH_2-CH_2}$$

The alkynide anion is derived from 1-butyne by alkylation of acetylene. This analysis suggests the following synthetic sequence:

$$HC\equiv CH \xrightarrow[\text{2. CH}_3\text{CH}_2\text{Br}]{\substack{\text{1. NaNH}_2\\ \text{NH}_3}} CH_3CH_2C\equiv CH \xrightarrow[\text{2. } \underset{O}{CH_2-CH_2}]{\substack{\text{1. NaNH}_2\\ \text{NH}_3}} CH_3CH_2C\equiv CCH_2CH_2OH$$

Acetylene   1-Butyne   3-Hexyn-1-ol

$$\Big\downarrow \substack{\text{H}_2 \\ \text{Lindlar Pd}}$$

$$\underset{H}{\overset{CH_3CH_2}{}}C=C\underset{H}{\overset{CH_2CH_2OH}{}}$$

cis-3-Hexen-1-ol

(*b*) The compound cited is the aldehyde derived by oxidation of the primary alcohol in part (*a*). Oxidize the alcohol with PDC, PCC, or (C$_5$H$_5$N)$_2$CrO$_3$ in CH$_2$Cl$_2$.

$$\underset{H}{\overset{CH_3CH_2}{}}C=C\underset{H}{\overset{CH_2CH_2OH}{}} \xrightarrow[\substack{\text{(C}_5\text{H}_5\text{N})_2\text{CrO}_3 \\ \text{in CH}_2\text{Cl}_2}]{\text{PDC or PCC or}} \underset{H}{\overset{CH_3CH_2}{}}C=C\underset{H}{\overset{\displaystyle CH_2\overset{O}{\overset{\|}{C}}H}{}}$$

cis-3-Hexen-1-ol   *cis*-3-Hexenal

15.35 Even though we are given the structure of the starting material, it is still better to reason backward from the target molecule rather than forward from the starting material.

The desired product contains a cyano (—CN) group. The only method we have seen so far for introducing such a function into a molecule is by nucleophilic substitution. Therefore, the last step in the synthesis must be:

This step should work very well, since the substrate is a primary benzylic halide, cannot undergo elimination, and is very reactive in S_N2 reactions.

The primary benzylic halide can be prepared from the corresponding alcohol by any of a number of methods.

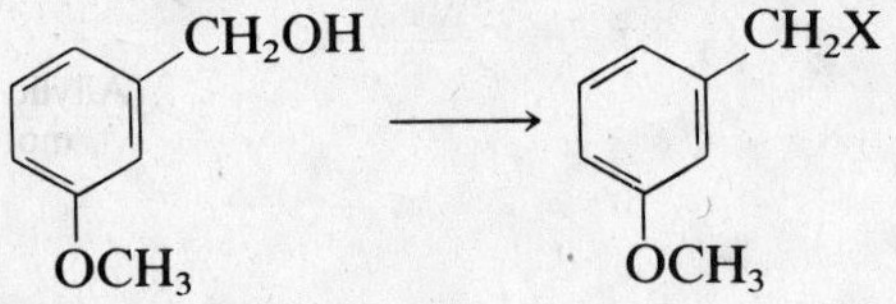
CH₂OH
OCH₃
CH₂X
OCH₃

Suitable reagents include HBr, PBr_3, or $SOCl_2$.

There remains now only to prepare the primary alcohol from the given starting aldehyde, which is accomplished by reduction.

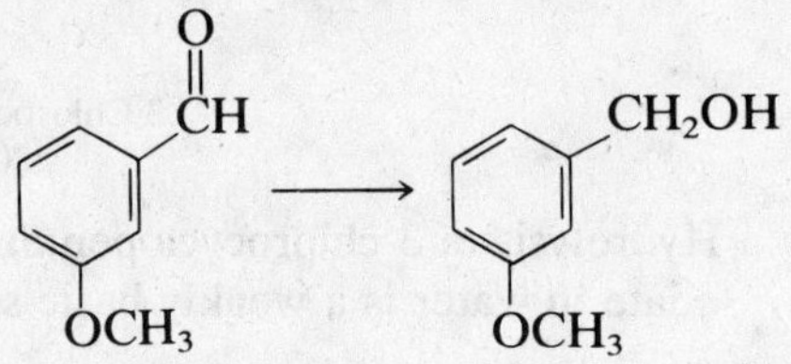
O
CH
OCH₃
CH₂OH
OCH₃

Reduction can be achieved by catalytic hydrogenation, with lithium aluminum hydride, or with sodium borohydride.

The actual sequence of reactions as carried out is as shown:

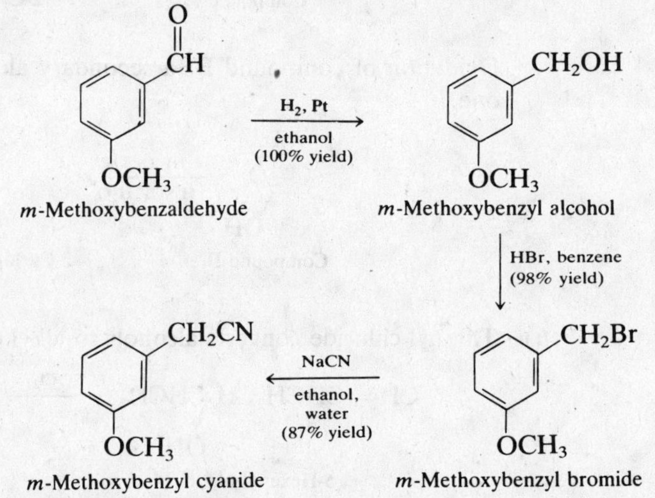
O
CH
OCH₃
m-Methoxybenzaldehyde
H₂, Pt
ethanol
(100% yield)
CH₂OH
OCH₃
m-Methoxybenzyl alcohol
HBr, benzene
(98% yield)
CH₂Br
OCH₃
m-Methoxybenzyl bromide
NaCN
ethanol,
water
(87% yield)
CH₂CN
OCH₃
m-Methoxybenzyl cyanide

Another three-step synthesis, which is reasonable but does not involve an alcohol as an intermediate, is:

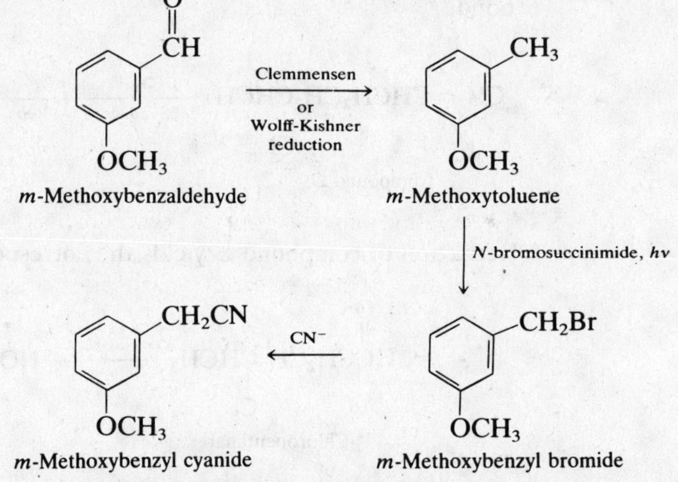
O
CH
OCH₃
m-Methoxybenzaldehyde
Clemmensen
or
Wolff-Kishner
reduction
CH₃
OCH₃
m-Methoxytoluene
N-bromosuccinimide, hν
CH₂Br
OCH₃
m-Methoxybenzyl bromide
CN⁻
CH₂CN
OCH₃
m-Methoxybenzyl cyanide

15.36 (a) Addition of hydrogen chloride to cyclopentadiene takes place by way of the most stable carbocation. In this case it is an allylic carbocation.

(Allylic carbocation; more stable) not (Not allylic; less stable)

3-Chlorocyclopentene (80–90%)
(compound A)

Hydrolysis of 3-chlorocyclopentene gives the corresponding alcohol. Sodium bicarbonate in water is a weakly basic solvolysis medium.

Compound A $\xrightarrow[\text{H}_2\text{O}]{\text{NaHCO}_3}$ 2-Cyclopenten-1-ol (88%)
(compound B)

Oxidation of compound B (a secondary alcohol) gives the ketone 2-cyclopenten-1-one.

Compound B $\xrightarrow[\text{H}_2\text{SO}_4, \text{H}_2\text{O}]{\text{Na}_2\text{Cr}_2\text{O}_7}$ 2-Cyclopenten-1-one (60–68%)
(compound C)

(b) Thionyl chloride converts alcohols to alkyl chlorides.

$$\underset{\substack{\text{OH}\\\text{5-Hexen-2-ol}}}{\text{CH}_2\text{=CHCH}_2\text{CH}_2\text{CHCH}_3} \xrightarrow[\text{pyridine}]{\text{SOCl}_2} \underset{\substack{\text{Cl}\\\text{5-Chloro-1-hexene}\\\text{(compound D)}}}{\text{CH}_2\text{=CHCH}_2\text{CH}_2\text{CHCH}_3}$$

Ozonation, followed by reduction of the ozonide, cleaves the carbon-carbon double bond.

$$\underset{\substack{\text{Cl}\\\text{Compound D}}}{\text{CH}_2\text{=CHCH}_2\text{CH}_2\text{CHCH}_3} \xrightarrow[\text{2. reductive workup}]{\text{1. O}_3} \underset{\substack{\text{Cl}\\\text{4-Chloropentanal}\\\text{(compound E)}}}{\overset{\overset{\text{O}}{\|}}{\text{HCCH}_2\text{CH}_2\text{CHCH}_3}} + \underset{\text{Formaldehyde}}{\overset{\overset{\text{O}}{\|}}{\text{HCH}}}$$

Reduction of compound E yields the corresponding alcohol.

$$\underset{\substack{\text{Cl}\\\text{4-Chloropentanal}}}{\overset{\overset{\text{O}}{\|}}{\text{HCCH}_2\text{CH}_2\text{CHCH}_3}} \xrightarrow{\text{NaBH}_4} \underset{\substack{\text{Cl}\\\text{4-Chloro-1-pentanol}\\\text{(compound F)}}}{\text{HOCH}_2\text{CH}_2\text{CH}_2\text{CHCH}_3}$$

(*c*)   *N*-Bromosuccinimide is a reagent designed to accomplish benzylic bromination.

1-Bromo-2-methylnaphthalene   1-Bromo-2-bromomethylnaphthalene
                                                      (compound G)

Hydrolysis of the benzylic bromide gives the corresponding benzylic alcohol. The bromine that is directly attached to the naphthalene ring does not react under these conditions.

1-Bromo-2-bromomethylnaphthalene   (1-Bromo-2-naphthyl)methanol
                                                      (compound H)

Oxidation of the primary alcohol with PCC gives the aldehyde.

(1-Bromo-2-naphthyl)methanol   1-Bromonaphthalene-2-carboxaldehyde
                                                (compound I)

15.37 The alcohol is tertiary and benzylic and yields a relatively stable carbocation.

2-Phenyl-2-butanol   1-Methyl-1-phenylpropyl cation

The alcohol is chiral, but the carbocation is not. Thus, irrespective of which enantiomer of 2-phenyl-2-butanol is used, the same carbocation is formed. The carbocation reacts with ethanol to give an optically inactive mixture containing equal quantities of enantiomers (racemic).

1-Methyl-1-phenylpropyl cation   Ethanol   2-Ethoxy-2-phenylbutane
                                                       (50% *R*, 50% *S*)

15.38 The difference between the two ethers is that 1-*O*-benzylglycerol contains a vicinal diol function while 2-*O*-benzylglycerol does not. Periodic acid will react with 1-*O*-benzylglycerol but not with 2-*O*-benzylglycerol.

1-*O*-Benzylglycerol   2-Benzyloxyethanal   Formaldehyde

2-*O*-Benzylglycerol

15.39 (a) The presence of aromatic rings in compound J is suggested by the low hydrogen-to-carbon ratio ($C_{14}H_{14}O$) of its molecular formula. The difference of the molecular formula ($C_{14}H_{14}O$) from C_nH_{2n+2} ($C_{14}H_{30}$) tells us that the sum of double bonds plus rings is 8. A SODAR equal to 8 can be accommodated by two benzene rings. That these two benzene rings are monosubstituted is shown by the presence of 10 aromatic protons in the nmr spectrum at $\delta = 7$ to 7.5 ppm. A signal at $\delta = 3.5$ ppm in the ^1H nmr spectrum, equivalent to one proton, can be assigned to the alcohol OH proton.

At this point we have identified three structural units accounting for the oxygen, 12 carbons, and 11 hydrogens:

and   OH

The three remaining hydrogens are part of a methyl group. These methyl protons appear as a singlet in the nmr at $\delta = 1.9$ ppm. The methyl group must be attached to a carbon that bears no hydrogens:

$$CH_3 \!-\! \overset{|}{\underset{|}{C}} \!-\!$$

Putting these structural units together, we arrive at 1,1-diphenylethanol as the correct structure of compound J.

1,1-Diphenylethanol
(compound J)

(b) An aromatic ring in compound K is suggested by the molecular formula, which requires a SODAR of 4. Para substitution is suggested by the four-proton AB signal at $\delta = 6.8$ to 7.5 ppm in the nmr spectrum. The nmr spectrum also contains a one-proton singlet at $\delta = 2.7$ ppm, assignable to the hydroxyl proton.

The hydrogen on the carbon that bears the hydroxyl group appears as a triplet centered at $\delta = 4.4$ ppm, which tells us that it is attached to a methylene group. Since its chemical shift is shifted to lower field than normal, it is probably benzylic. A reasonable conclusion points to the partial structure

There is also a methyl triplet at $\delta = 0.8$ ppm in the nmr, which must arise from a methyl group attached to the methylene unit. There remains only to add the bromine atom to the para position of the ring to complete the structure.

1-(p-Bromophenyl)-1-propanol
(compound K)

15.40 The ratio of carbon to hydrogen in the molecular formula is C_nH_{2n+2} ($C_8H_{18}O_2$), and so compound L has no double bonds or rings. Compound L cannot be a vicinal diol, since it does not react with periodic acid.

The nmr spectrum is rather simple; all peaks are singlets and multiples of 2 with respect to their integrated areas. The 12-proton singlet at $\delta = 1.2$ ppm must correspond to four equivalent methyl groups and the four-proton singlet to two equivalent methylene groups. No nonequivalent protons can be vicinal, since no splitting is observed. The two-proton singlet at $\delta = 2.0$ ppm is due to the hydroxyl protons of the diol.

Compound L must be 2,5-dimethyl-1,2-hexanediol.

$$\begin{array}{cc} CH_3 & CH_3 \\ | & | \\ CH_3CCH_2CH_2CCH_3 \\ | & | \\ OH & OH \end{array}$$

15.41 (a) This compound has only two different types of carbons. One type of carbon comes at low field and, since its ^{13}C peak is a singlet, has no hydrogen substituents. It is most likely a carbon bonded to oxygen and three other carbons. The other peak is a quartet, ascribable to methyl groups. The spectrum leads to the conclusion that this compound is *tert*-butyl alcohol.

$$\begin{array}{c} CH_3 \\ | \\ 31.2\,ppm \longrightarrow CH_3{-}C{-}OH \\ | \\ CH_3 \quad 68.9\,ppm \end{array}$$

(b) There are four different types of carbons in this compound. The only $C_4H_{10}O$ isomers that have four nonequivalent carbons are $CH_3CH_2CH_2CH_2OH$, $CH_3CHCH_2CH_3$, and $CH_3OCH_2CH_2CH_3$. The lowest-field signal, the one at

$\quad\quad\quad |$
$\quad\quad OH$

69.2 ppm from the carbon that bears the oxygen substituent, is a doublet and therefore has only one hydrogen attached to it. The compound is therefore 2-butanol..

$$\begin{array}{c} \text{doublet} \qquad\qquad \text{triplet} \\ \searrow \qquad\qquad \swarrow \\ CH_3CHCH_2CH_3 \\ \text{quartet} \nearrow \quad | \quad \nwarrow \\ OH \qquad \text{quartet} \end{array}$$

(c) This compound has two equivalent methyl groups, as indicated by the quartet of area 2 at 18.9 ppm. Its lowest-field carbon is a triplet, and so the group $-CH_2O$ must be present. The compound must be 2-methyl-1-propanol.

$$\begin{array}{c} 30.8\,ppm\;(d) \\ \nearrow \\ CH_3{-}CH{-}CH_2OH \\ \nearrow \quad | \quad \nwarrow \\ 18.9\,ppm\;(q) \quad CH_3 \quad 69.4\,ppm\;(t) \end{array}$$

15.42 Compound M has only three carbons, none of which can be methyl carbons, because the ^{13}C nmr spectrum contains no quartets. Two of the carbon signals are triplets and so arise from CH_2 groups; the other is a doublet and so corresponds to a CH group. The only structure consistent with the observed data is that of 3-chloro-1,2-propanediol.

$$\begin{array}{c} \text{doublet} \\ \nearrow \\ HOCH_2{-}CH{-}CH_2Cl \\ \text{triplet} \nearrow \quad | \quad \nwarrow \text{triplet} \\ OH \end{array}$$

The structure $HOCH_2CHCH_2OH$ cannot be correct. It would exhibit only two peaks in

$\quad\quad\quad\quad\quad\quad\quad |$
$\quad\quad\quad\quad\quad\quad Cl$

its ^{13}C nmr spectrum, because the two terminal carbons are equivalent to each other.

15.43 The observation of a peak at m/z 31 in the mass spectrum of compound N suggests the presence of a primary alcohol. This fragment is most likely $CH_2\overset{+}{=}OH$. On the basis of this fact and the appearance of four different carbons in the ^{13}C nmr spectrum, compound N is 2-ethyl-1-butanol.

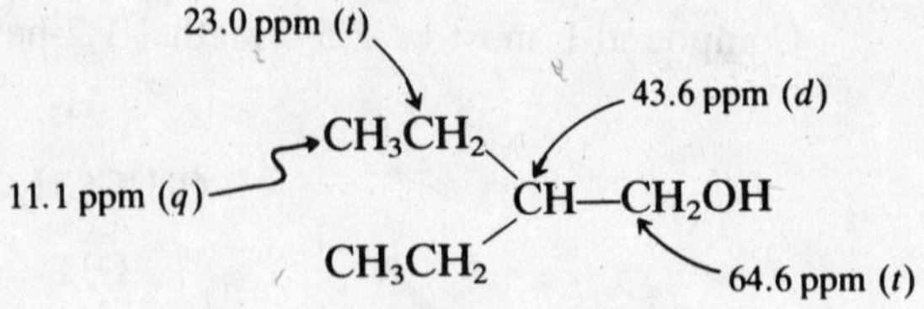

23.0 ppm (t)

43.6 ppm (d)

11.1 ppm (q) $\rightarrow$ CH₃CH₂

CH—CH₂OH

CH₃CH₂

64.6 ppm (t)

SELF-TEST

PART A

A-1. For each of the following reactions give the structure of the missing reactant or reagent.

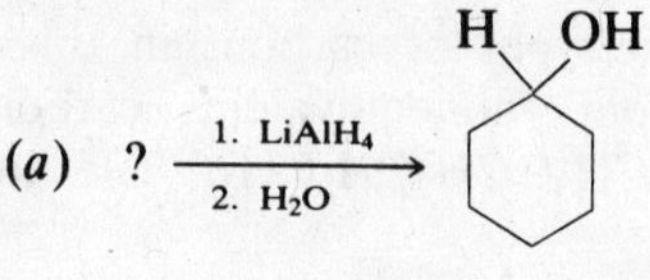

(a) ? $\xrightarrow[\text{2. H}_2\text{O}]{\text{1. LiAlH}_4}$

(b) ? + 2CH₃CH₂MgBr $\xrightarrow[\text{2. H}_3\text{O}^+]{\begin{array}{c}\text{1. diethyl}\\\text{ether}\end{array}}$ C₆H₅C(CH₂CH₃)₂ + CH₃CH₂OH
    (with OH on the central C)

(c) C₆H₅CH₂C(CH₃)=CH₂ $\xrightarrow{?}$ C₆H₅CH₂CH(CH₃)CH₂OH

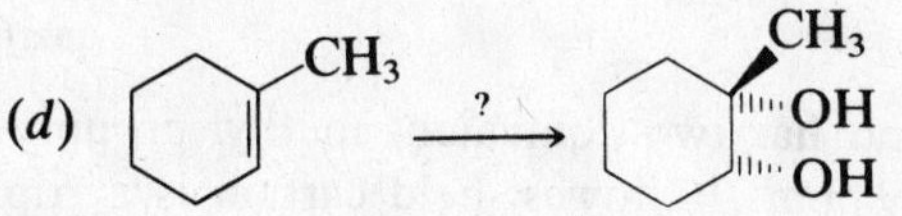

(d) [cyclohexene with CH₃] $\xrightarrow{?}$ [cyclohexane with CH₃, OH, OH]

(e) C₆H₅CH₂Br $\xrightarrow[\text{2. NaOH}]{\text{1. ?}}$ C₆H₅CH₂SH

A-2. For the following reactions of 2-phenylethanol, $C_6H_5CH_2CH_2OH$, give the correct reagent or product(s) omitted from the equation.

(a) C₆H₅CH₂CH₂OH $\xrightarrow[\text{CH}_2\text{Cl}_2]{(\text{C}_5\text{H}_5\text{N})_2\text{CrO}_3}$?

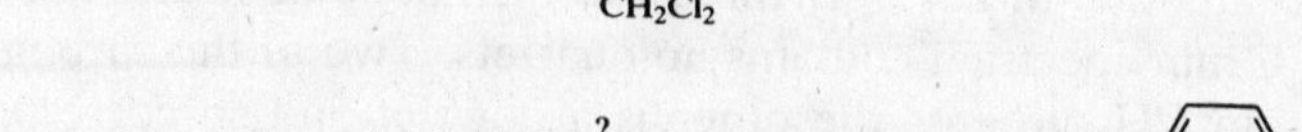

(b) C₆H₅CH₂CH₂OH $\xrightarrow{?}$ CH₃CO₂CH₂CH₂—[phenyl]

(c) C₆H₅CH₂CH₂OH (2 moles) $\xrightarrow[\text{heat}]{\text{H}^+}$ H₂O + ?

(d) C₆H₅CH₂CH₂OH $\xrightarrow{?}$ C₆H₅CH₂CO₂H

A-3. Outline two synthetic schemes for the preparation of 3-methyl-1-butanol using different Grignard reagents.

A-4. Give the structure of the reactant, reagent, or product omitted from each of the following. Show stereochemistry where important.

(a) $\xrightarrow{\text{HIO}_4}$?

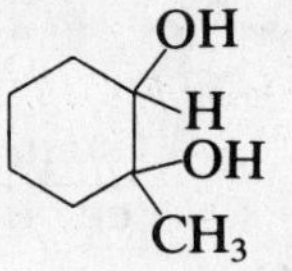

(b) ? (a diol) $\xrightarrow[\text{heat}]{\text{H}^+}$

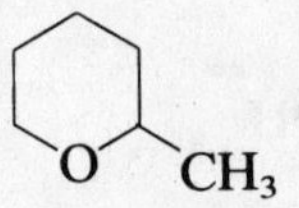

(c) ? $\xrightarrow[\text{(CH}_3)_3\text{COH, HO}^-]{\text{OsO}_4,\ \text{(CH}_3)_3\text{COOH}}$ 2,3-butanediol (chiral diastereomer)

A-5. Give the reagents necessary to carry out each of the following transformations:
  (a) Conversion of benzyl alcohol ($C_6H_5CH_2OH$) to benzaldehyde ($C_6H_5CH{=}O$)
  (b) Conversion of benzyl alcohol to benzoic acid ($C_6H_5CO_2H$)
  (c) Conversion of $CH_2{=}CHCH_2CH_2CO_2H$ to $CH_3CH_2CH_2CH_2CO_2H$
  (d) Conversion of $CH_2{=}CHCH_2CH_2CO_2H$ to $CH_2{=}CHCH_2CH_2CH_2OH$
  (e) Conversion of cyclohexene to *cis*-1,2-cyclohexanediol

A-6. Provide structures for compounds A to C in the following reaction scheme:

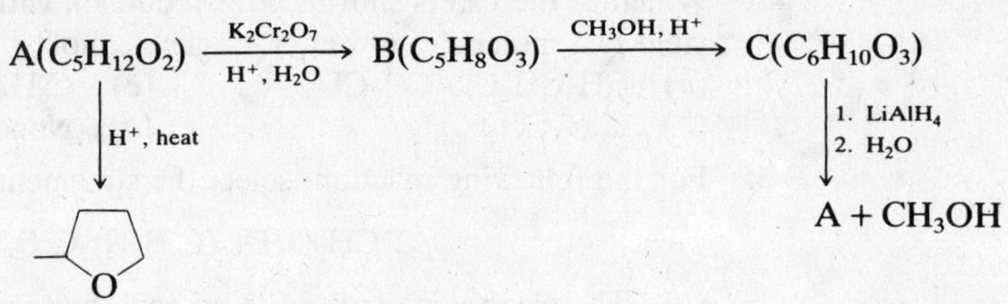

$$A(C_5H_{12}O_2) \xrightarrow[\text{H}^+,\ \text{H}_2\text{O}]{\text{K}_2\text{Cr}_2\text{O}_7} B(C_5H_8O_3) \xrightarrow{\text{CH}_3\text{OH, H}^+} C(C_6H_{10}O_3)$$

A $\downarrow$ H$^+$, heat

C $\downarrow$ 1. LiAlH$_4$ 2. H$_2$O

$A + CH_3OH$

A-7. Using any necessary organic or inorganic reagents, outline a scheme for each of the following conversions.

(a) $(CH_3)_2C{=}CHCH_3 \xrightarrow{\ ?\ }$ $(CH_3)_2CH\overset{\overset{\displaystyle O}{\|}}{C}CH_3$

(b) $\xrightarrow{\ ?\ }$

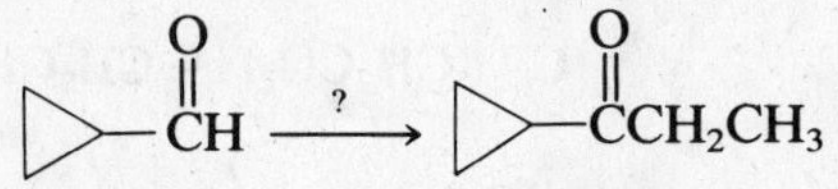

(c) $C_6H_5CH_3 \xrightarrow{\ ?\ } C_6H_5CH_2CH_2CO_2CH_2CH_3$

PART B

B-1. The following chemical change (R = alkyl group) may be best described as a(n)

$$\underset{\text{RCOR}}{\overset{\overset{\displaystyle O}{\|}}{\text{RCOR}}} \longrightarrow \underset{\text{RCR}}{\overset{\overset{\displaystyle O}{\|}}{\text{RCR}}}$$

  (a) Oxidation   (c) Elimination
  (b) Reduction   (d) Addition

B-2. Which of the following would yield a secondary alcohol after the indicated reaction, followed by aqueous hydrolysis if necessary?
  (a) $LiAlH_4$ + a ketone   (c) 2-Butene + mercury(II) acetate, then $NaBH_4$, HO$^-$
  (b) CH_3CH_2MgBr + an aldehyde   (d) All of these

B-3. What is the major product of the following reaction?

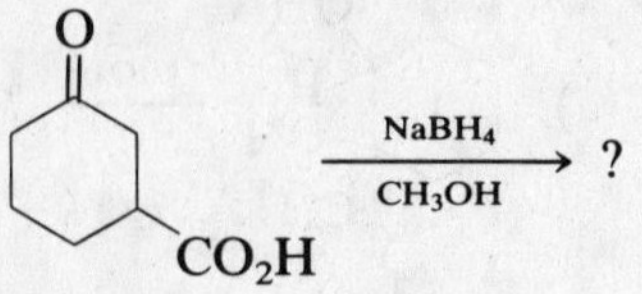

(a) 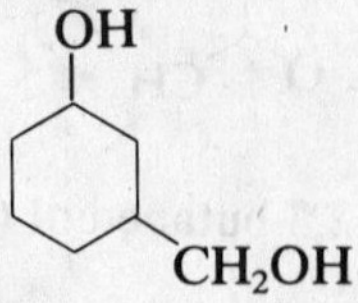 (b) 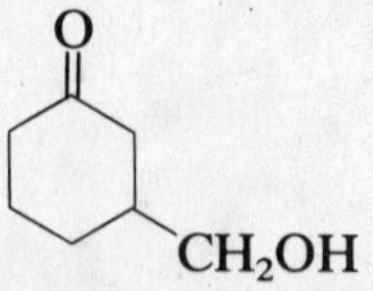

(c) 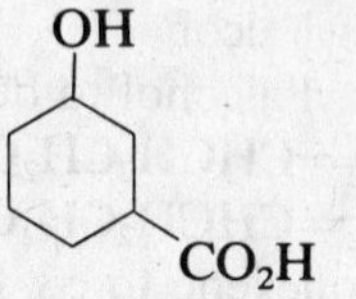 (d) 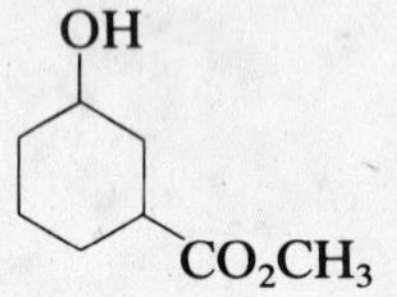

B-4. Which of the esters shown, after reduction with $LiAlH_4$ and aqueous workup, will yield two molecules of only a single alcohol?

(a) $CH_3CH_2CO_2CH_2CH_3$ (c) $C_6H_5CO_2CH_2C_6H_5$
(b) $C_6H_5CO_2C_6H_5$ (d) None of these

B-5. For the following reaction, select the statement which best describes the situation.

$$RCH_2OH + (C_5H_5N)_2CrO_3 \longrightarrow$$

(a) The alcohol is oxidized to an acid, and the Cr(VI) is reduced.
(b) The alcohol is oxidized to an aldehyde, and the Cr(VI) is reduced.
(c) The alcohol is reduced to an aldehyde, and the Cr(III) is oxidized.
(d) The alcohol is oxidized to a ketone, and the Cr(VI) is reduced.

B-6. What is the product from the following esterification?

$$C_6H_5CH_2CO_2H + CH_3CH_2\text{—}^{18}OH \xrightarrow[\text{heat}]{H^+} ?$$

(a) $C_6H_5CH_2\overset{^{18}O}{\overset{\|}{C}}OCH_2CH_3$ (c) $C_6H_5CH_2\overset{^{18}O}{\overset{\|}{C}}\text{—}^{18}OCH_2CH_3$

(b) $C_6H_5CH_2\overset{O}{\overset{\|}{C}}\text{—}^{18}OCH_2CH_3$ (d) $CH_3CH_2\overset{^{18}O}{\overset{\|}{C}}OCH_2C_6H_5$

B-7. The following substance acts as a coenzyme in which of the following biological reactions?

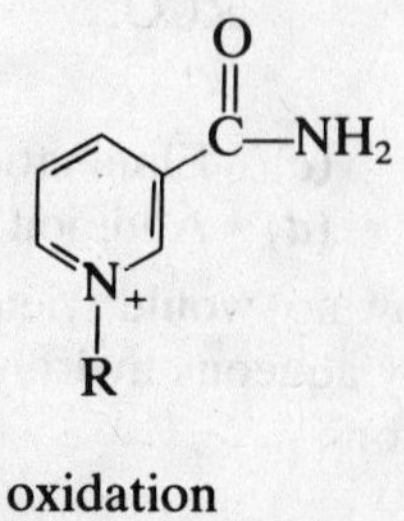

(R = adenine dinucleotide)

(a) Alcohol oxidation (c) Epoxide formation
(b) Ketone reduction (d) None of these

B-8. Which of the following alcohols gives the best yield of dialkyl ether on being heated with a trace of sulfuric acid?

(a) 1-Pentanol   (c) Cyclopentanol
(b) 2-Pentanol   (d) 2-Methyl-2-butanol

B-9. What is the major organic product of the following sequence of reactions?

$$(CH_3)_2CHCH_2OH \xrightarrow{PBr_3} \xrightarrow{Mg} \xrightarrow{\overset{O}{\triangle}} \xrightarrow{H_3O^+} ?$$

(a) $(CH_3)\overset{\overset{\displaystyle OH}{|}}{C}HCHCH_2CH_3$

(c) $(CH_3)_2CHCH_2CH_2OH$

(b) $(CH_3)_2CHCH_2\overset{\overset{\displaystyle OH}{|}}{C}HCH_3$

(d) $(CH_3)_2CHCH_2CH_2CH_2OH$

ETHERS, EPOXIDES, AND SULFIDES

IMPORTANT TERMS AND CONCEPTS

Nomenclature (Sec. 16.1) Ethers are given substitutive names by describing them as alkoxy derivatives of alkanes. Radicofunctional names are derived by specifying the two alkyl groups attached to oxygen, in alphabetical order, and then adding the word *ether*.

$$CH_3CH_2CH_2OCH_2CH_3 \qquad CH_3CH_2CH_2O\!-\!\!\langle \bigcirc \rangle$$

Substitutive name:	1-Ethoxypropane	Propoxybenzene
Radicofunctional name:	Ethyl propyl ether	Phenyl propyl ether

Ethers are described as *symmetrical or unsymmetrical* depending on whether the two groups bonded to oxygen are the same or different.

Epoxides are cyclic ethers whose oxygen atom is part of a three-membered ring.

$$\overset{\textstyle O}{CH_2\!-\!CHCH_2CH_3}$$

1,2-Epoxybutane

The sulfur analogs of alkoxy groups (RS—) are called *alkylthio groups*. Thus sulfides are named by the substitutive method as *alkylthio*alkanes. In radicofunctional names of sulfides, the alkyl groups are followed by the word *sulfide*.

$$CH_3CH_2CH_2SCH_2CH_3 \qquad CH_3CH_2CH_2S\!-\!\!\langle \bigcirc \rangle$$

Substitutive name:	1-Ethylthiopropane	Propylthiobenzene
Radicofunctional name:	Ethyl propyl sulfide	Phenyl propyl sulfide

Structure, Bonding, and Physical Properties (Secs. 16.2, 16.3) The alkyl (or aryl) groups bonded to an ether oxygen are arranged in an approximately tetrahedral fashion. The polar nature of the carbon-oxygen bond and the nonlinear arrangement of groups cause most ethers to be polar molecules.

The boiling points of ethers are similar to those of alkanes of the same chain length and are substantially lower than those of comparably constituted alcohols. Ethers, although polar, are not able to form hydrogen-bonded aggregates as do alcohols. Ethers are, however, able to serve as donors of unshared electron pairs. Therefore their solubility in water tends to parallel that of alcohols.

Crown Ethers (Sec. 16.4) Another property of ethers explained by the presence of unshared electron pairs on oxygen is the ability to coordinate metal ions:

$$R\ddot{O}R + M^+ \rightleftharpoons \overset{+}{R\ddot{O}R} \\ \qquad\qquad\qquad | \\ \qquad\qquad\qquad M$$

A series of *macrocyclic polyethers* has been prepared. These *crown ethers* are capable of complexing metal ions, with resulting drastic alteration of the solubility and chemical reactivity of inorganic salts in nonpolar media. An example is 12-crown-4:

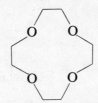

IMPORTANT REACTIONS

PREPARATION OF ETHERS (Secs. 16.5, 16.6)

Williamson Synthesis

General:

$$RO^- + R'X \longrightarrow ROR' + X^-$$

(primary best)

Example:

$$(CH_3)_2CHO^- + CH_3CH_2Br \longrightarrow (CH_3)_2CHOCH_2CH_3$$

REACTIONS OF ETHERS (Secs. 16.7, 16.8)

Acid-Catalyzed Cleavage

General:

$$ROR' + 2HX \xrightarrow{\text{heat}} RX + R'X + H_2O$$

Example:

$$C_6H_5CH_2OCH_2CH_3 \xrightarrow[\text{heat}]{\text{HI}} C_6H_5CH_2I + CH_3CH_2I + H_2O$$

PREPARATION OF EPOXIDES (Secs. 16.9, 16.10)

From Alkenes by Epoxidation (Chapter 6)

General:

$$R_2C{=}CR_2 + R'\overset{\text{O}}{\overset{\|}{C}}OOH \longrightarrow R_2C\overset{\text{O}}{\overset{\diagup \diagdown}{-}}CR_2 + R'CO_2H$$

Example:

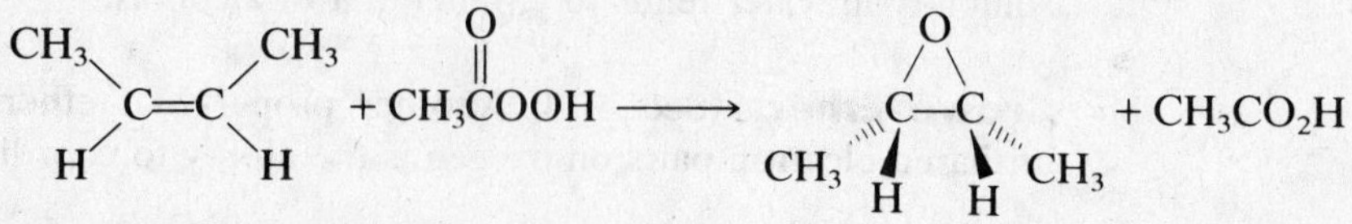

$$CH_3CH=CHCH_3 + CH_3COOH \longrightarrow \underset{\substack{CH_3 \ H \quad H \ CH_3}}{C\text{---}C} + CH_3CO_2H$$

From Vicinal Halohydrins

General:

$$\underset{\substack{| \ X}}{R_2C\text{---}CR_2} \xrightarrow{\text{HO}^-} R_2C\text{---}CR_2$$

with OH on the first carbon.

Example:

$$\underset{\substack{OH}}{CH_3CH_2CHCH_2Br} \xrightarrow{\text{HO}^-} CH_3CH_2CH\text{---}CH_2$$

REACTIONS OF EPOXIDES (Secs. 16.11 to 16.13)

Nucleophilic Ring Opening

General:

$$R_2C\text{---}CR_2 + Y:^- \xrightarrow[\text{(ROH)}]{H_2O} \underset{\substack{OH \qquad | \ Y}}{R_2C\text{---}CR_2}$$

Examples:

$$C_6H_5MgBr + \xrightarrow[\substack{1.\ CH_2\text{---}CH_2 \\ 2.\ H_3O^+}]{} C_6H_5CH_2CH_2OH$$

$$\begin{array}{c}\triangleright O + CH_3CH_2O^- \xrightarrow{CH_3CH_2OH} \end{array} \underset{\substack{OCH_2CH_3 \\ OH}}{}$$

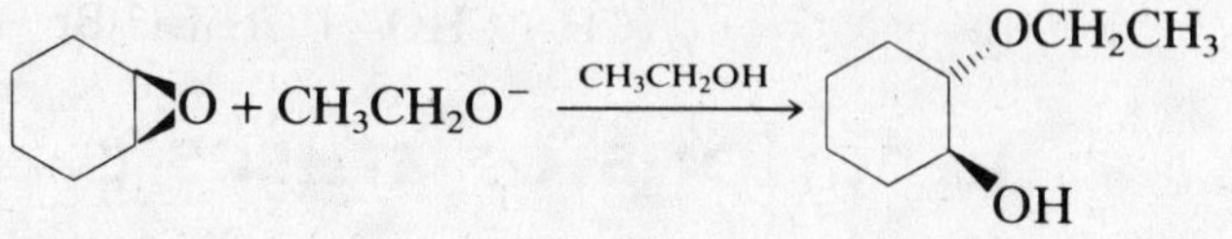

Acid-Catalyzed Ring Opening

General:

$$R_2C\text{---}CR_2 + YH \longrightarrow \underset{\substack{OH \qquad | \ Y}}{R_2C\text{---}CR_2}$$

Example:

$$C_6H_5CH\text{---}CH_2 + HBr \longrightarrow \underset{\substack{Br \\ |}}{C_6H_5CHCH_2OH}$$

PREPARATION OF SULFIDES (Sec. 16.15)

General:

$$RS^- + R'X \longrightarrow RSR' + X^-$$

Example:

$$CH_3CH_2SNa + BrCH_2CH(CH_3)_2 \longrightarrow CH_3CH_2SCH_2CH(CH_3)_2$$

REACTIONS OF SULFIDES

Oxidation (Sec. 16.16)

Examples:

$$(CH_3)_2CHSCH_3 + NaIO_4 \longrightarrow (CH_3)_2CH\overset{O^-}{\underset{}{\overset{|}{\overset{+}{S}}}}CH_3 + NaIO_3$$

$$CH_3CH_2S\text{—} \bigcirc \xrightarrow{H_2O_2} CH_3CH_2\overset{O^-}{\underset{O_-}{\overset{|}{S^{2+}}}}\text{—} \bigcirc$$

Alkylation (Sec. 16.17)

Example:

$$CH_3CH_2SCH_2CH_3 + CH_3I \longrightarrow CH_3CH_2\overset{CH_3}{\underset{+}{\overset{|}{S}}}CH_2CH_3 \quad I^-$$

SOLUTIONS TO TEXT PROBLEMS

16.1 (b) Oxirane is the IUPAC name for ethylene oxide. A chloromethyl group ($ClCH_2$—) is attached to position 2 of the ring in 2-(chloromethyl)oxirane.

$$\overset{3}{CH_2}\text{—}\overset{2}{CH_2} \atop \underset{O}{}$$

Oxirane

$$CH_2\text{—}CHCH_2Cl \atop \underset{O}{}$$

2-(Chloromethyl)oxirane

This compound is more commonly known by its trivial name, *epichlorohydrin*.

(c) Epoxides may be named by adding the prefix *epoxy* to the IUPAC name of a parent compound, specifying by number both atoms to which the oxygen is attached.

$$CH_2\text{=}CHCH_2CH_3$$

1-Butene

$$CH_2\text{=}CHCH\text{—}CH_2 \atop \underset{O}{}$$

3,4-Epoxy-1-butene

16.2 1,2-Epoxybutane and tetrahydrofuran both have the molecular formula C_4H_8O—i.e., they are constitutional isomers—and so it is appropriate to compare their heats of combustion directly. The strained three-membered ring of 1,2-epoxybutane causes it to have more internal energy than tetrahydrofuran, and its combustion is more exothermic.

$$CH_2\text{—}CHCH_2CH_3 \atop \underset{O}{}$$

1,2-Epoxybutane;
heat of combustion 2546 kJ/mol (609.1 kcal/mol)

Tetrahydrofuran;
heat of combustion 2499 kJ/mol (597.8 kcal/mol)

16.3 An ether can function only as a proton acceptor in a hydrogen bond, while an alcohol can be either a proton acceptor or a donor. Therefore the only hydrogen bond possible between an ether and an alcohol is the one shown:

$$\overset{R}{\underset{R}{}}\ddot{O}:\text{----}H\text{—}\overset{\ddot{O}}{\underset{R}{}}$$

16.4 The compound is 1,4-dioxane; it has a six-membered ring and two oxygens separated by CH_2—CH_2 units.

1,4-dioxane ("6-crown-2")

16.5 Protonation of the carbon-carbon double bond leads to the more stable carbocation.

$$(CH_3)_2C{=}CH_2 + H^+ \longrightarrow (CH_3)_2\overset{+}{C}{-}CH_3$$

2-Methylpropene   *tert*-Butyl cation

Methanol acts as a nucleophile to capture *tert*-butyl cation.

Deprotonation of the oxonium ion leads to formation of *tert*-butyl methyl ether.

16.6 (*b*) Methyl propyl ether can be prepared by the reaction of sodium methoxide with a propyl halide:

$$CH_3ONa + CH_3CH_2CH_2Br \longrightarrow CH_3OCH_2CH_2CH_3 + NaBr$$

Sodium methoxide   1-Bromopropane   Methyl propyl   Sodium
             ether   bromide

It can also be made by the reaction of sodium propoxide with a methyl halide:

$$CH_3CH_2CH_2ONa + CH_3Br \longrightarrow CH_3OCH_2CH_2CH_3 + NaBr$$

Sodium propoxide   Bromomethane   Methyl propyl   Sodium
             ether   bromide

(*c*) The two routes to benzyl ethyl ether are:

$$C_6H_5CH_2ONa + CH_3CH_2Br \longrightarrow C_6H_5CH_2OCH_2CH_3 + NaBr$$

Sodium benzyloxide   Bromoethane   Benzyl ethyl ether   Sodium
                    bromide

$$C_6H_5CH_2Br + CH_3CH_2ONa \longrightarrow C_6H_5CH_2OCH_2CH_3 + NaBr$$

Benzyl bromide   Sodium ethoxide   Benzyl ethyl ether   Sodium bromide

16.7 (*b*) A primary carbon and a secondary carbon are attached to the ether oxygen. The secondary carbon can only be derived from the alkoxide, since secondary alkyl halides cannot be used in the preparation of ethers by the Williamson method. The only effective method uses an allyl halide and sodium isopropoxide.

$$(CH_3)_2CHONa + CH_2{=}CHCH_2Br \longrightarrow CH_2{=}CHCH_2OCH(CH_3)_2 + NaBr$$

Sodium isopropoxide   Allyl bromide   Allyl isopropyl ether   Sodium bromide

Elimination will be the major reaction of an isopropyl halide with an alkoxide base.

(*c*) Here the ether is a mixed primary-tertiary one. The best combination is the one that utilizes the primary alkyl halide.

$$(CH_3)_3COK + C_6H_5CH_2Br \longrightarrow (CH_3)_3COCH_2C_6H_5 + KBr$$

Potassium   Benzyl bromide   Benzyl *tert*-butyl   Potassium bromide
tert-butoxide   ether

The reaction between $(CH_3)_3CBr$ and $C_6H_5CH_2O^-$ will be one of elimination, not substitution.

16.8 *(b)*

$$CH_3CH_2CH_2CH_2CH_2OCH_2CH_2CH_2CH_2CH_3 + 15O_2 \longrightarrow 10CO_2 + 11H_2O$$

Dipentyl ethyl · Oxygen · Carbon dioxide · Water

(c)

Tetrahydrofuran · Oxygen · Carbon dioxide · Water

$$\bigcirc + \frac{11}{2}O_2 \longrightarrow 4CO_2 + 4H_2O$$

16.9 *(b)* If benzyl bromide is the only organic product from reaction of a dialkyl ether with hydrogen bromide, then both alkyl groups attached to oxygen must be benzyl.

$$C_6H_5CH_2OCH_2C_6H_5 \xrightarrow[\text{heat}]{\text{HBr}} 2C_6H_5CH_2Br + H_2O$$

Dibenzyl ether · Benzyl bromide · Water

(c) Since *one mole of a dihalide,* rather than two moles of a monohalide, is produced per mole of ether, the ether must be cyclic.

$$\xrightarrow[\text{heat}]{\text{2HBr}} BrCH_2CH_2CH_2CH_2CH_2Br + H_2O$$

Tetrahydropyran · 1,5-Dibromopentane · Water

16.10 Dissociation of the oxonium ion is reasonable, since a tertiary carbocation results.

Di-*tert*-butyloxonium ion · *tert*-Butyl alcohol · *tert*-Butyl cation

Capture of *tert*-butyl cation by chloride then occurs rapidly.

tert-Butyl cation · Chloride · *tert*-Butyl chloride

The *tert*-butyl alcohol generated in the dissociation step goes on to form *tert*-butyl chloride by way of a carbocation intermediate, as described in Chapter 4.

16.11 The cis epoxide is achiral. It is a meso form containing a plane of symmetry. The trans isomer is chiral; its two mirror-image representations are not superposable.

cis-2,3-Epoxybutane (plane of symmetry passes through oxygen and midpoint of carbon-carbon bond)

Nonsuperposable mirror-image forms of *trans*-2,3-epoxybutane

Neither the cis nor the trans epoxide is optically active when formed from the alkene.

The cis epoxide is achiral; it cannot be optically active. The trans epoxide is capable of optical activity but is formed as a racemic mixture, since achiral starting materials are used.

16.12 (b) Azide ion $[:\ddot{N}{=}N{=}\ddot{N}:]^-$ is a good nucleophile, reacting readily with ethylene oxide to yield 2-azidoethanol.

$$\underset{\text{Ethylene oxide}}{\underset{O}{CH_2{-}CH_2}} \xrightarrow[\text{ethanol-water}]{NaN_3} \underset{\text{2-Azidoethanol}}{N_3CH_2CH_2OH}$$

(c) Ethylene oxide is hydrolyzed to ethylene glycol in the presence of aqueous base.

$$\underset{\text{Ethylene oxide}}{\underset{O}{CH_2{-}CH_2}} \xrightarrow[H_2O]{NaOH} \underset{\text{Ethylene glycol}}{HOCH_2CH_2OH}$$

(d) Phenyllithium reacts with ethylene oxide in a manner similar to that of a Grignard reagent.

$$\underset{\text{Ethylene oxide}}{\underset{O}{CH_2{-}CH_2}} \xrightarrow[\text{2. }H_3O^+]{\text{1. }C_6H_5Li,\text{ diethyl ether}} \underset{\text{2-Phenylethanol}}{C_6H_5CH_2CH_2OH}$$

(e) The nucleophilic species here is the acetylenic anion $CH_3CH_2C{\equiv}C:^-$, which attacks a carbon atom of ethylene oxide to give 3-hexyn-1-ol.

$$\underset{\text{Ethylene oxide}}{\underset{O}{CH_2{-}CH_2}} \xrightarrow[NH_3]{NaC{\equiv}CCH_2CH_3} \underset{\text{3-Hexyn-1-ol (48\%)}}{CH_3CH_2C{\equiv}CCH_2CH_2OH}$$

16.13 Nucleophilic attack at C-2 of the starting epoxide will be faster than attack at C-1, because C-1 is more sterically hindered. Compound A, corresponding to attack at C-1, is not as likely as compound B. Compound B not only arises by methoxide ion attack at C-2 but also satisfies the stereochemical requirement that epoxide ring opening take place with inversion of configuration at the site of substitution. Compound B is correct. Compound C, while it is formed by methoxide substitution at the less crowded carbon of the epoxide, is wrong stereochemically. It requires substitution with retention of configuration, a process contrary to the normal mode of epoxide ring opening.

16.14 Acid-catalyzed nuclophilic ring opening proceeds by attack of methanol at the more substituted carbon of the protonated epoxide. Inversion of configuration is observed at the site of attack. The correct product is compound A.

Protonated form of
1-methyl-1,2-epoxycyclopentane

Compound A

16.15 Begin by drawing *meso*-2,3-butanediol, recalling that a meso form has a plane of symmetry.

meso-2,3-Butanediol

Epoxidation followed by acid-catalyzed hydrolysis results in anti addition of hydroxyl groups to the double bond. *trans*-2-Butene is the required starting material.

trans-2-Butene   *trans*-2,3-Epoxybutane   *meso*-2,3-Butanediol

Osmium tetraoxide hydroxylation is a method of achieving syn hydroxylation. The necessary starting material is *cis*-2-butene.

cis-2-Butene   *meso*-2,3-Butanediol

16.16 Dammarenediol can be formed from the same carbocation intermediate that leads to dammaradienol. Capture of the carbocation by water acting as a nucleophile yields dammarenediol.

16.17 Reaction of (*R*)-2-octanol with *p*-toluenesulfonyl chloride yields a *p*-toluenesulfonate ester (tosylate) having the same configuration; the stereogenic center is not involved in this step. Reaction of the tosylate with a nucleophile proceeds by inversion of configuration in an S_N2 process. The product has the *S* configuration.

(*R*)-2-Octanol   *p*-Toluenesulfonyl chloride   (*R*)-1-Methylheptyl tosylate

(*R*)-1-Methylheptyl tosylate   Sodium benzenethiolate   (*S*)-1-Methylheptyl phenyl sulfide

16.18 Phenyl vinyl sulfoxide lacks a plane of symmetry, and is chiral. Phenyl vinyl sulfone is achiral; a plane of symmetry passes through the phenyl and vinyl groups and the central sulfur atom.

Phenyl vinyl sulfoxide   Phenyl vinyl sulfone
(chiral)   (achiral)

16.19 As shown in the text, docecyldimethylsulfonium iodide may be prepared by reaction of dodecyl methyl sulfide with methyl iodide. An alternative method is the reaction of dodecyl iodide with dimethyl sulfide.

$$(CH_3)_2S + CH_3(CH_2)_{10}CH_2I \longrightarrow CH_3(CH_2)_{10}CH_2\overset{+}{S}(CH_3)_2 \quad I^-$$

Dimethyl   Dodecyl iodide   Dodecyldimethylsulfonium
sulfide   iodide

The reaction of a sulfide with an alkyl halide is an S_N2 process. The faster reaction will be the one having the less sterically hindered alkyl halide substrate. The method presented in the text will proceed faster.

16.20 The molecular ion from *sec*-butyl ethyl ether can also fragment by cleavage of a carbon-carbon bond in its ethyl group to give an oxygen-stabilized cation of m/z 87.

$$CH_3CH_2\overset{\cdot\cdot+}{\underset{\cdot\cdot}{O}}-\underset{\underset{CH_3}{|}}{C}HCH_2CH_3 \longrightarrow \cdot CH_3 + CH_2=\overset{+}{\underset{\cdot\cdot}{O}}-\underset{\underset{CH_3}{|}}{C}HCH_2CH_3$$

$$m/z\ 87$$

16.21 All the constitutionally isomeric ethers of molecular formula $C_5H_{12}O$ belong to one of two general groups, $CH_3OC_4H_9$ and $CH_3CH_2OC_3H_7$. Thus, we have

$$CH_3OCH_2CH_2CH_2CH_3 \qquad\qquad CH_3O\underset{\underset{CH_3}{|}}{C}HCH_2CH_3$$

Butyl methyl ether   *sec*-Butyl methyl ether

$$CH_3OCH_2CH(CH_3)_2 \qquad\qquad CH_3OC(CH_3)_3$$

Isobutyl methyl ether   *tert*-Butyl methyl ether

$$CH_3CH_2OCH_2CH_2CH_3 \qquad \text{and} \qquad CH_3CH_2OCH(CH_3)_2$$

Ethyl propyl ether   Ethyl isopropyl ether

These ethers could also have been named as "alkoxyalkanes." Thus, *sec*-butyl methyl ether would become 2-methoxybutane.

16.22 Isoflurane and enflurane are both halogenated derivatives of ethyl methyl ether.

Isoflurane: $F-\underset{\underset{F}{|}}{\overset{\overset{F}{|}}{C}}-\underset{\underset{Cl}{|}}{C}H-O-CHF_2$

1-Chloro-2,2,2-trifluoroethyl difluoromethyl ether

Enflurane: $Cl-\underset{\underset{F}{|}}{\overset{\overset{H}{|}}{C}}-\underset{\underset{F}{|}}{\overset{\overset{F}{|}}{C}}-O-CHF_2$

2-Chloro-1,1,2-trifluoroethyl difluoromethyl ether

16.23 (*a*) The parent compound is cyclopropane. It has a three-membered epoxide function, and thus a reasonable name is epoxycyclopropane. Numbers locating positions of attachment (as in "1,2-epoxycyclopropane") are not necessary, because no other structures (1,3 or 2,3) are possible here.

Epoxycyclopropane

(*b*) The longest continuous carbon chain has seven carbons, and so the compound is named as a derivative of heptane. The epoxy function bridges C-2 and C-4. Therefore

is 2-methyl-2,4-epoxyheptane.

(c)   The oxygen atom bridges the C-1 and C-4 atoms of a cyclohexane ring:

1,4-Epoxycyclohexane

Alternatively this compound may be named 7-oxabicyclo[2.2.1]heptane.

(d)   There are eight carbon atoms continuously linked and bridged by an oxygen. We name the compound as an epoxy derivative of cyclooctane.

1,5-Epoxycyclooctane

Again, an alternative name may be used: 9-oxabicyclo[3.3.1]nonane.

16.24   (a)   There are three methyl-substituted thianes, two of which are chiral:

2-Methylthiane
(chiral)

3-Methylthiane
(chiral)

4-Methylthiane
(achiral)

(b)   The locants in the name indicate the positions of the sulfur atoms in 1,4-dithiane and 1,3,5-trithiane.

1,4-Dithiane

1,3,5-Trithiane

(c)   Disulfides possess two adjacent sulfur atoms. 1,2-Dithiane is a disulfide.

1,2-Dithiane

(d)   Two chair conformations of the sulfoxide derived from thiane are possible; the oxygen atom may be either equatorial or axial.

16.25   Intramolecular hydrogen bonding between the hydroxyl group and the ring oxygens is possible when the hydroxyl group is axial but not when it is equatorial.

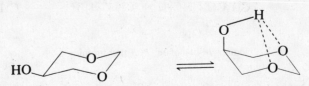

Less stable conformation; no intramolecular
hydrogen bonding

More stable conformation; stabilized
by hydrogen bonding

16.26 The ethers that are to be prepared are:

$$CH_3OCH_2CH_2CH_3 \qquad CH_3OCH(CH_3)_2 \qquad \text{and} \qquad CH_3CH_2OCH_2CH_3$$

Methyl propyl ether   Isopropyl methyl ether   Diethyl ether

First examine the preparation of each ether by the Williamson method. Methyl propyl ether can be prepared in two ways:

$$CH_3ONa \quad + CH_3CH_2CH_2Br \longrightarrow CH_3OCH_2CH_2CH_3$$

Sodium methoxide   1-Bromopropane   Methyl propyl ether

$$CH_3Br \quad + CH_3CH_2CH_2ONa \longrightarrow CH_3OCH_2CH_2CH_3$$

Methyl bromide   Sodium propoxide   Methyl propyl ether

Either combination is satisfactory. The necessary reagents are prepared as shown:

$$CH_3OH \xrightarrow{\text{Na}} CH_3ONa$$

Methanol   Sodium methoxide

$$CH_3CH_2CH_2OH \xrightarrow[\text{(or HBr)}]{\text{PBr}_3} CH_3CH_2CH_2Br$$

1-Propanol   1-Bromopropane

$$CH_3OH \xrightarrow[\text{(or HBr)}]{\text{PBr}_3} CH_3Br$$

Methanol   Methyl bromide

$$CH_3CH_2CH_2OH \xrightarrow{\text{Na}} CH_3CH_2CH_2ONa$$

1-Propanol   Sodium propoxide

Isopropyl methyl ether is best prepared by the reaction

$$CH_3Br \quad + \quad (CH_3)_2CHONa \longrightarrow CH_3OCH(CH_3)_2$$

Methyl bromide   Sodium isopropoxide   Isopropyl methyl ether

The reaction of sodium methoxide with isopropyl bromide will proceed mainly by elimination. Methyl bromide is prepared as shown previously; sodium isopropoxide is prepared by adding sodium to isopropyl alcohol.

Diethyl ether may be prepared as outlined:

$$CH_3CH_2OH \xrightarrow{\text{Na}} CH_3CH_2ONa$$

Ethanol   Sodium ethoxide

$$CH_3CH_2OH \xrightarrow[\text{(or HBr)}]{\text{PBr}_3} CH_3CH_2Br$$

Ethanol   Ethyl bromide

$$CH_3CH_2ONa + CH_3CH_2Br \longrightarrow CH_3CH_2OCH_2CH_3 + NaBr$$

Sodium ethoxide   Ethyl bromide   Diethyl ether   Sodium bromide

16.27 (*a*) This reaction represents a typical Williamson ether synthesis wherein an alkoxide reacts with a primary alkyl halide.

$$CH_3CH_2\underset{\underset{ONa}{|}}{C}HCH_3 + CH_3CH_2Br \longrightarrow CH_3CH_2\underset{\underset{OCH_2CH_3}{|}}{C}HCH_3 + NaBr$$

Sodium 2-butanolate   Ethyl bromide   *sec*-Butyl ethyl ether   Sodium bromide
                                                        (2-ethoxybutane)

(*b*) Secondary alkyl halides react with alkoxide bases by E2 elimination as the major pathway. The Williamson ether synthesis is not a useful reaction with secondary alkyl halides.

$$CH_3CH_2CHCH_3 \ \underset{\underset{ONa}{|}}{} + \ \bigcirc\!\!-Br \longrightarrow CH_3CH_2CHCH_3 + \ \bigcirc + NaBr$$

Sodium 2-butanolate Bromocyclohexane 2-Butanol Cyclohexene Sodium bromide

(c) The starting material is a primary allylic halide, a structural class known to be reactive in nucleophilic substitution reactions of the S_N2 type.

$$CH_3CH{=\!}CHCH_2Cl \xrightarrow{KOC(CH_3)_3} CH_3CH{=\!}CHCH_2OC(CH_3)_3$$

1-Chloro-2-butene 2-Butenyl *tert*-butyl ether (61%)

The reaction produces an ether and is an example of the Williamson ether synthesis.

(d) The potassium alkoxide acts as a nucleophile toward iodoethane to yield an ethyl ether.

(R)-2-Ethoxy-2-phenylbutane

The ether product has the same absolute configuration as the starting alkoxide because no bonds to the stereogenic center are made or broken in the reaction.

(e) Vicinal halohydrins are converted to epoxides on being treated with base.

1-Bromo-2-butanol   1,2-Epoxybutane

(f) The reactants, an alkene plus a peroxy acid, are customary ones for epoxide preparation. The reaction is a stereospecific syn addition of oxygen to the double bond.

(Z)-1-Phenylpropene   Peroxybenzoic acid   cis-2-Methyl-3-phenyloxirane   Benzoic acid

(g) Azide ion is a good nucleophile and attacks the epoxide function. Substitution occurs at carbon with inversion of configuration. The product is *trans*-2-azidocyclohexanol.

1,2-Epoxycyclohexane   *trans*-2-Azidocyclohexanol (61%)

(h) Ammonia is a nucleophile capable of reacting with epoxides. It attacks the less hindered carbon of the epoxide function.

2-(*o*-Bromophenyl)-2-methyloxirane   1-Amino-2-(*o*-bromophenyl)-2-propanol

Aryl halides do not react with nucleophiles under these conditions, and so the bromine substituent on the ring is unaffected.

(*i*) Methoxide ion attacks the less substituted carbon of the epoxide ring with inversion of configuration.

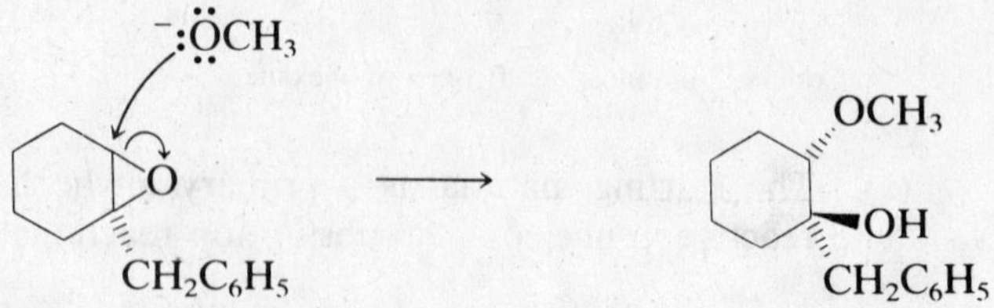

1-Benzyl-1,2-epoxycyclohexane   1-Benzyl-*trans*-2-methoxycyclohexanol (98%)

(*j*) Tosylate esters undergo substitution with nucleophiles such as sodium butane-thiolate.

$$CH_3(CH_2)_{16}CH_2OTs + CH_3CH_2CH_2CH_2SNa \longrightarrow CH_3CH_2CH_2CH_2SCH_2(CH_2)_{16}CH_3$$

Octadecyl tosylate   Sodium butanethiolate   Butyl octadecyl sulfide

(*k*) Nucleophilic substitution proceeds with inversion of configuration.

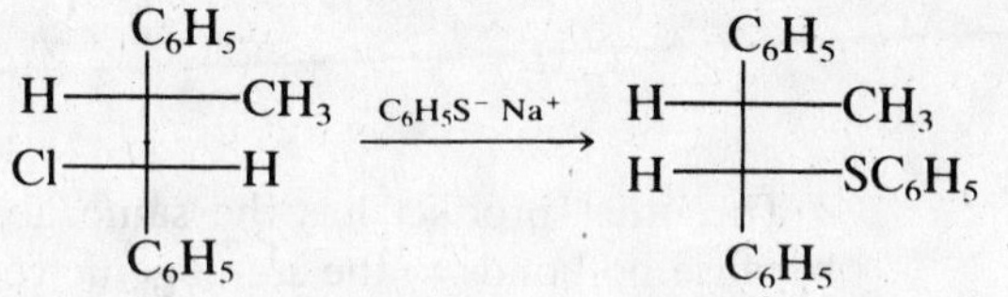

16.28 Oxidation of 4-*tert*-butylthiane yields two sulfoxides that are diastereomers of each other.

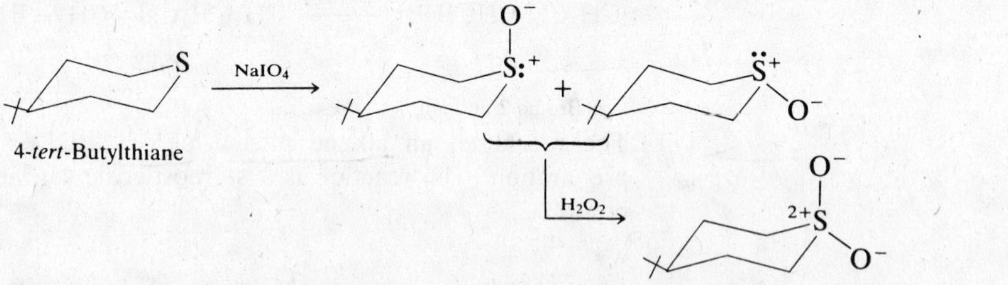

4-*tert*-Butylthiane

Both stereoisomeric sulfoxides yield the same sulfone on further oxidation.

16.29 Protonation of oxygen to form an oxonium ion is followed by loss of water. The resulting carbocation has a plane of symmetry, and is achiral. Capture of the carbocation by methanol yields both enantiomers of 2-methyl-2-phenylbutane. The produce is racemic.

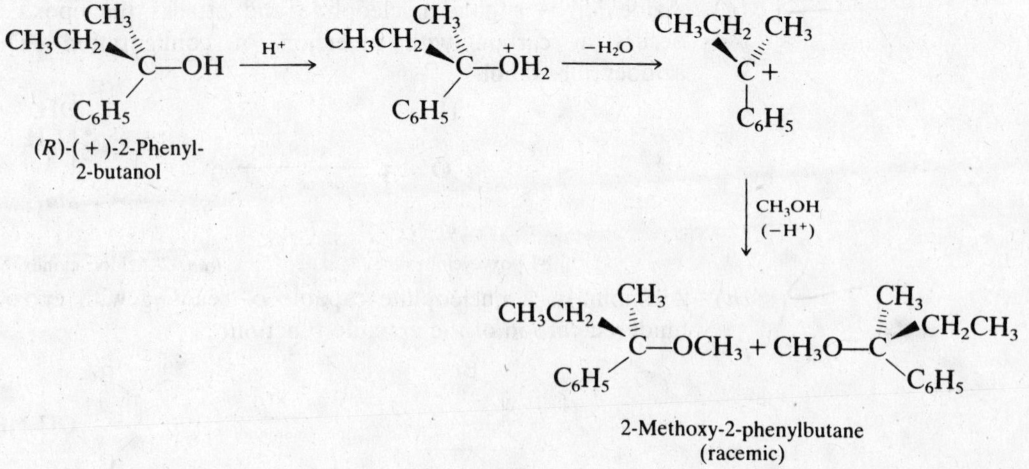

(*R*)-(+)-2-Phenyl-2-butanol

2-Methoxy-2-phenylbutane (racemic)

16.30 The proper approach to this problem is to first write the equations given in full stereochemical detail.

(a)

(R)-1,2-Propanediol

It now becomes clear that the arrangement of groups around the stereogenic center remains unchanged in going from starting materials to products. Therefore, choose conditions such that the nucleophile attacks the CH_2 group of the epoxide rather than the stereogenic center. Base-catalyzed hydrolysis is required; aqueous sodium hydroxide is appropriate.

The nucleophile (hydroxide ion) attacks the less hindered carbon of the epoxide ring.

(b)

(S)-1,2-Propanediol

Inversion of configuration at the stereogenic center is required. Therefore the nucleophile must attack the stereogenic center, and acid-catalyzed hydrolysis should be chosen. Dilute sulfuric acid would be satisfactory.

The nucleophile (a water molecule) attacks that carbon atom of the ring which can better support a positive charge. Carbocation character develops at the transition state and is better supported by the carbon atom that is more highly substituted.

16.31 The key intermediate in the preparation of bis(2-chloroethyl) ether from ethylene is 2-chloroethanol, formed from ethylene by reaction with chlorine in water. Heating 2-chloroethanol in acid gives the desired ether.

$$CH_2{=}CH_2 \xrightarrow{Cl_2,\ H_2O} ClCH_2CH_2OH \xrightarrow{H^+,\ heat} ClCH_2CH_2OCH_2CH_2Cl$$

Ethylene   2-Chloroethanol   Bis(2-chloroethyl) ether

16.32 (a) There is a temptation to try to do this transformation in a single step by using a reducing agent to convert the carbonyl to a methylene group. No reagent is available that reduces esters in this way! The Clemmensen and Wolff-Kishner reduction methods are suitable only for aldehydes and ketones. The best way to approach this problem is by reasoning backward. The desired product is an ether. Ethers can be prepared by the Williamson ether synthesis involving an alkyl halide and an alkoxide ion.

or

Both the alkyl halide and the alkoxide ion are prepared from alcohols. The problem then becomes one of preparing the appropriate alcohol (or alcohols) from the starting ester. This is readily done using lithium aluminum hydride.

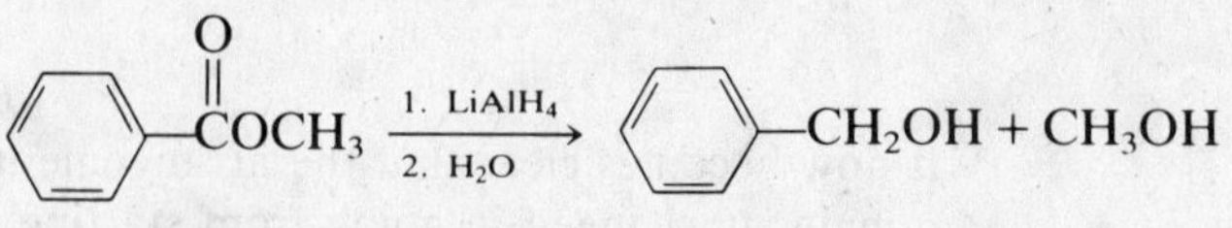

Methyl benzoate → Benzyl alcohol + Methanol

Then

$CH_3OH \xrightarrow{Na} CH_3ONa$

Methanol → Sodium methoxide

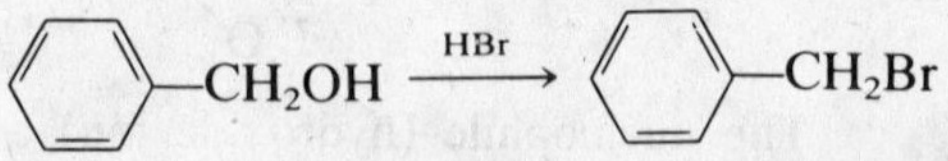

Benzyl alcohol $\xrightarrow{HBr}$ Benzyl bromide

and

Benzyl bromide + $NaOCH_3$ → Benzyl methyl ether + $NaBr$

Sodium methoxide

The following sequence is also appropriate once methanol and benzyl alcohol are obtained by reduction of methyl benzoate:

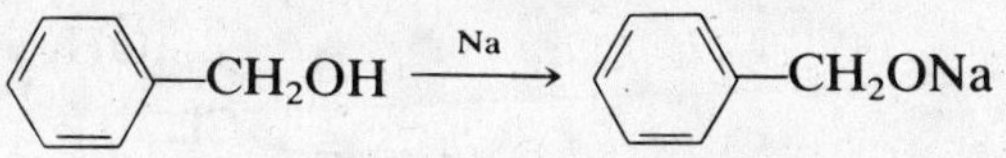

Benzyl alcohol $\xrightarrow{Na}$ Sodium benzyloxide

$CH_3OH \xrightarrow{PBr_3} CH_3Br$

Methanol → Bromomethane

and   Sodium benzyloxide + CH_3Br → Benzyl methyl ether + $NaBr$

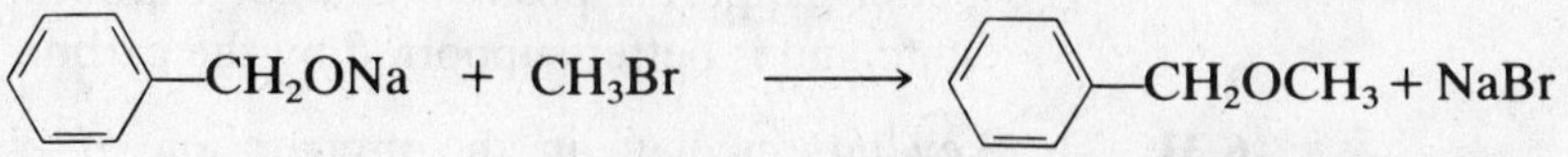

Bromomethane

(b) All the methods that we have so far discussed for the preparation of epoxides are based on alkenes as starting materials. This leads us to consider the partial retrosynthesis shown.

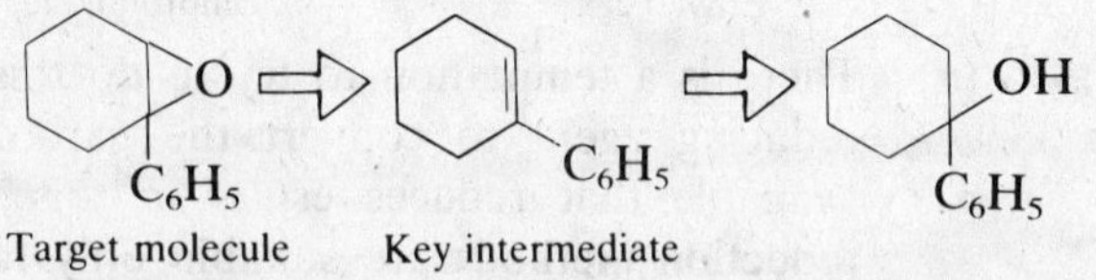

Target molecule   Key intermediate

The key intermediate, 1-phenylcyclohexene, is both a proper precursor to the desired epoxide and readily available from the given starting materials. A reasonable synthesis is:

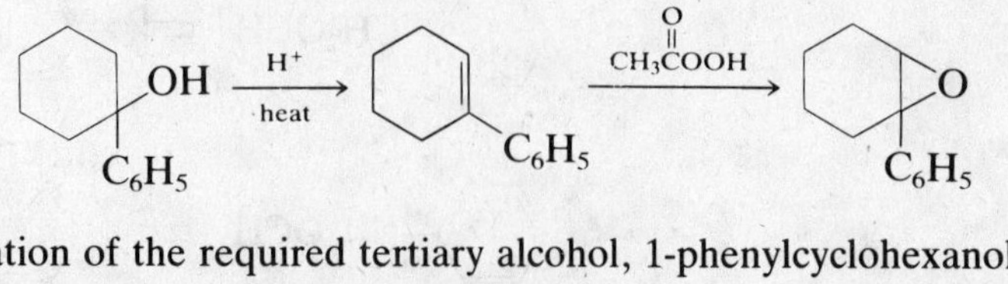

Preparation of the required tertiary alcohol, 1-phenylcyclohexanol, completes the synthesis.

$$\text{Cyclohexanol} \xrightarrow[\text{H}_2\text{SO}_4]{\text{K}_2\text{Cr}_2\text{O}_7} \text{Cyclohexanone}$$

Cyclohexanol   Cyclohexanone

$$\text{C}_6\text{H}_5\text{Br} \xrightarrow[\substack{\text{2. cyclohexanone} \\ \text{3. H}_3\text{O}^+}]{\text{1. Mg, diethyl ether}} \text{1-Phenylcyclohexanol}$$

Bromobenzene   1-Phenylcyclohexanol

(*c*) The necessary carbon skeleton can be assembled through the reaction of a Grignard reagent with 1,2-epoxypropane.

$$\underset{\text{OH}}{\text{C}_6\text{H}_5\text{CH}_2\text{CHCH}_3} \Longrightarrow \text{C}_6\text{H}_5\text{:}^- + \underset{\text{O}}{\text{CH}_2\text{—CHCH}_3}$$

The reaction sequence is therefore:

$$\text{C}_6\text{H}_5\text{MgBr} + \underset{\text{O}}{\text{CH}_2\text{—CHCH}_3} \longrightarrow \underset{\text{OH}}{\text{C}_6\text{H}_5\text{CH}_2\text{CHCH}_3}$$

Phenylmagnesium   1,2-Epoxypropane   1-Phenyl-2-propanol
bromide
(from bromobenzene
and magnesium)

The epoxide required in the first step, 1,2-epoxypropane, is prepared as follows from isopropyl alcohol:

$$\underset{\text{OH}}{\text{CH}_3\text{CHCH}_3} \xrightarrow[\text{heat}]{\text{H}_2\text{SO}_4} \text{CH}_3\text{CH}{=}\text{CH}_2 \xrightarrow{\overset{\text{O}}{\overset{\|}{\text{CH}_3\text{COOH}}}} \underset{\text{O}}{\text{CH}_3\text{CH—CH}_2}$$

2-Propanol   Propene   1,2-Epoxypropane
(isopropyl
alcohol)

(*d*) Since the target molecule is an ether, it ultimately derives from two alcohols:

$$\text{C}_6\text{H}_5\text{CH}_2\text{CH}_2\text{CH}_2\text{OCH}_2\text{CH}_3 \Longrightarrow \text{C}_6\text{H}_5\text{CH}_2\text{CH}_2\text{CH}_2\text{OH} + \text{CH}_3\text{CH}_2\text{OH}$$

Our first task is to assemble 3-phenyl-1-propanol from the designated starting material benzyl alcohol. This requires formation of a primary alcohol with the original carbon chain extended by two carbons. The standard method for this transformation involves reaction of a Grignard reagent with ethylene oxide.

$$\text{C}_6\text{H}_5\text{CH}_2\text{OH} \xrightarrow{\text{PBr}_3} \text{C}_6\text{H}_5\text{CH}_2\text{Br} \xrightarrow[\substack{\text{2. } \underset{\text{O}}{\text{CH}_2\text{—CH}_2} \\ \text{3. H}_3\text{O}^+}]{\substack{\text{1. Mg, diethyl} \\ \text{ether}}} \text{C}_6\text{H}_5\text{CH}_2\text{CH}_2\text{CH}_2\text{OH}$$

Benzyl alcohol   Benzyl bromide   3-Phenyl-1-propanol

Once 3-phenyl-1-propanol has been prepared, its conversion to the corresponding ethyl ether can be accomplished in either of two ways:

$$\text{C}_6\text{H}_5\text{CH}_2\text{CH}_2\text{CH}_2\text{OH} \xrightarrow{\text{PBr}_3} \text{C}_6\text{H}_5\text{CH}_2\text{CH}_2\text{CH}_2\text{Br}$$

3-Phenyl-1-propanol   1-Bromo-3-phenylpropane

$$\downarrow \text{NaOCH}_2\text{CH}_3, \text{ethanol}$$

$$\text{C}_6\text{H}_5\text{CH}_2\text{CH}_2\text{CH}_2\text{OCH}_2\text{CH}_3$$

Ethyl 3-phenylpropyl ether
(1-ethoxy-3-phenylpropane)

or alternatively

$$C_6H_5CH_2CH_2CH_2OH \xrightarrow[\text{2. } CH_3CH_2Br]{\text{1. Na}} C_6H_5CH_2CH_2CH_2OCH_2CH_3$$

3-Phenyl-1-propanol   Ethyl 3-phenylpropyl ether

The reagents in each step are prepared from ethanol:

$$CH_3CH_2OH \xrightarrow{\text{Na}} CH_3CH_2ONa$$

Ethanol   Sodium ethoxide

$$CH_3CH_2OH \xrightarrow{PBr_3} CH_3CH_2Br$$

Ethanol   Ethyl bromide

(*e*)   The target epoxide can be prepared in a single step from the corresponding alkene:

Bicyclo[2.2.2]-oct-2-ene   2,3-Epoxybicyclo[2.2.2]octane

Disconnections show that this alkene is available through a Diels-Alder reaction:

The reaction of 1,3-cyclohexadiene with ethylene gives the desired substance.

1,3-Cyclohexadiene   Ethylene   Bicyclo[2.2.2]-oct-2-ene

1,3-Cyclohexadiene is one of the given starting materials. Ethylene is prepared from ethanol.

$$CH_3CH_2OH \xrightarrow[\text{heat}]{H_2SO_4} CH_2{=}CH_2$$

Ethanol   Ethylene

(*f*)   Retrosynthetic analysis reveals that the desired target molecule may be prepared by reaction of an epoxide with an ethanethiolate ion.

$$C_6H_5\underset{\underset{OH}{|}}{C}HCH_2SCH_2CH_3 \Longrightarrow C_6H_5\overset{O}{\overset{\diagup\diagdown}{CH{-}CH_2}} + {}^-SCH_2CH_3$$

Styrene oxide may be prepared by reaction of styrene with peroxyacetic acid.

$$C_6H_5CH{=}CH_2 + CH_3CO_2OH \longrightarrow C_6H_5\overset{O}{\overset{\diagup\diagdown}{CH{-}CH_2}} + CH_3CO_2H$$

Styrene   Peroxyacetic   Styrene oxide   Acetic acid
                   acid

The necessary thiolate anion is prepared from ethanol by way of the corresponding thiol.

$$CH_3CH_2OH \xrightarrow[\substack{2.\ (H_2N)_2C=S \\ 3.\ NaOH}]{1.\ HBr} CH_3CH_2SH \xrightarrow{3.\ NaOH} CH_3CH_2SNa$$

Ethanol   Ethanethiol   Sodium ethanethiolate

Reaction of styrene oxide with sodium ethanethiolate completes the synthesis.

$$\underset{\text{Styrene oxide}}{C_6H_5CH\!\!-\!\!CH_2} + \underset{\substack{\text{Sodium ethane-}\\\text{thiolate}}}{CH_3CH_2SNa} \xrightarrow{CH_3CH_2OH} \underset{}{C_6H_5\overset{\overset{\displaystyle OH}{|}}{C}HCH_2SCH_2CH_3}$$

16.33 (*a*) A reasonable mechanism is one that parallels the usual one for acid-catalyzed ether formation from alcohols, modified to accommodate these particular starting materials and products. Begin with protonation of one of the oxygen atoms of ethylene glycol.

$$\underset{\text{Ethylene glycol}}{HOCH_2CH_2OH} + H_2SO_4 \rightleftharpoons HOCH_2CH_2\!-\!\overset{+}{\underset{\underset{H}{|}}{\ddot{O}}}\!-\!H + HSO_4^-$$

The protonated alcohol then reacts in the usual way with another molecule of alcohol to give an ether. (This ether is known as *diethylene glycol*.)

$$HOCH_2CH_2\!-\!\overset{+}{\underset{\underset{H}{|}}{\ddot{O}}}\!-\!H \xrightarrow{-H_2O} \underset{\underset{HOCH_2CH_2\!-\!\overset{+}{\ddot{O}}\!-\!H}{}}{HOCH_2CH_2} \xrightarrow{-H^+} HOCH_2CH_2OCH_2CH_2OH$$

$$HOCH_2CH_2\ddot{O}H$$

Diethylene glycol

Diethylene glycol then undergoes intramolecular ether formation to yield 1,4-dioxane.

$$HOCH_2CH_2OCH_2CH_2OH + H_2SO_4 \rightleftharpoons HOCH_2CH_2OCH_2CH_2\!-\!\overset{+}{\underset{\underset{H}{|}}{\ddot{O}}}\!-\!H + HSO_4^-$$

(*b*) The substrate is a primary alkyl halide and reacts with aqueous sodium hydroxide by nucleophilic substitution.

$$\underset{\text{Bis(2-chloroethyl) ether}}{ClCH_2CH_2OCH_2CH_2Cl} + HO^- \xrightarrow{H_2O} HOCH_2CH_2OCH_2CH_2Cl + Cl^-$$

The product of nucleophilic substitution of one of the chloroethyl groups now has an alcohol function and a primary chloride built into the same molecule. It contains the requisite functionality to undergo an intramolecular Williamson reaction.

$$HOCH_2CH_2OCH_2CH_2Cl + HO^- \rightleftharpoons {}^-OCH_2CH_2OCH_2CH_2Cl + H_2O$$

1,4-Dioxane

16.34 (*a*) The first step is a standard Grignard synthesis of a primary alcohol using formaldehyde. Compound A can only be 3-buten-1-ol.

$$CH_2=CHCH_2Br \xrightarrow[\substack{\text{2. } CH_2=O \\ \text{3. } H_3O^+}]{\text{1. Mg}} CH_2=CHCH_2CH_2OH$$

Allyl bromide 3-Buten-1-ol (compound A)

Addition of bromine to the carbon-carbon double bond of 3-buten-1-ol takes place readily to yield the vicinal dibromide.

$$CH_2=CHCH_2CH_2OH \xrightarrow{Br_2} BrCH_2CHCH_2CH_2OH$$
$$| \qquad Br$$

3-Buten-1-ol
3,4-Dibromo-1-butanol (compound B)

When compound B is treated with potassium hydroxide, it loses the elements of HBr to give compound C. Since further treatment of compound C with potassium hydroxide converts it to D by a second dehydrobromination, a reasonable candidate for C is 3-bromotetrahydrofuran.

$$BrCH_2CHCH_2CH_2OH \xrightarrow[25°C]{KOH} \quad \xrightarrow[\text{heat}]{KOH}$$
$$|$$
$$Br$$

3,4-Dibromo-1-butanol (compound B)
3-Bromotetrahydrofuran (compound C)
Compound D

Ring closure occurs by an intramolecular Williamson reaction:

$$BrCH_2CHCH_2CH_2OH \rightleftharpoons$$
$$|$$
$$Br$$
Compound B
Compound C

Dehydrohalogenation of compound C converts it to the final product, D.

The alternative series of events, in which double bond formation proceeds ring closure, is unlikely, because it requires nucleophilic attack by the alkoxide on a vinyl bromide.

$$BrCH_2CHCH_2CH_2OH \xrightarrow{KOH} BrCH=CHCH_2CH_2OH \rightleftharpoons$$
$$|$$
$$Br$$

(Cyclization of this intermediate does not occur.)

(*b*) Lithium aluminum hydride reduces the carboxylic acid to the corresponding primary alcohol, compound E. Treatment of the vicinal chlorohydrin with base results in formation of an epoxide, compound F.

(*S*)-2-Chloro-1-propanol (compound E)

(*R*)-1,2-Epoxypropane (compound F)

As actually carried out, the first step proceeded in 56 to 58 percent yield, the second step in 65 to 70 percent yield.

(c) Treatment of the vicinal chlorohydrin with base results in ring closure to form an epoxide (compound G). Recall that attack occurs on the side opposite that of the carbon-chlorine bond. Compound G undergoes ring opening upon reaction with sodium methanethiolate to give compound H.

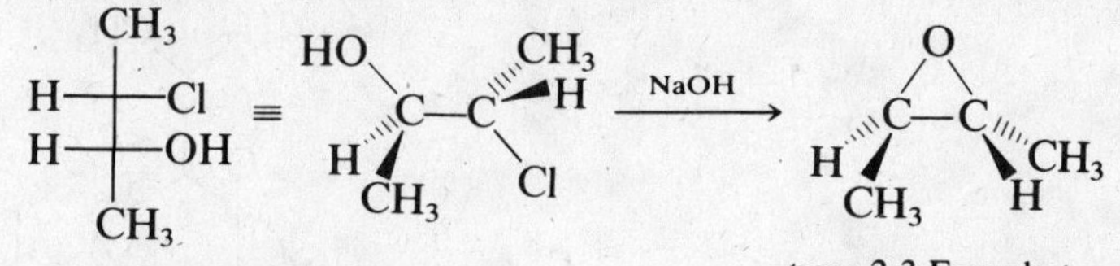

(2R,3S)-3-Chloro-2-butanol

trans-2,3-Epoxybutane
(compound G)

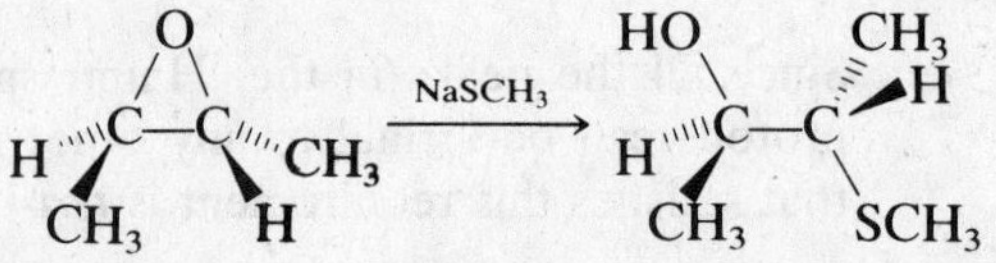

Compound G

Compound H

(d) Since it gives an epoxide on treatment with a peroxy acid, compound I must be an alkene; more specifically, it is 1,2-dimethylcyclopentene.

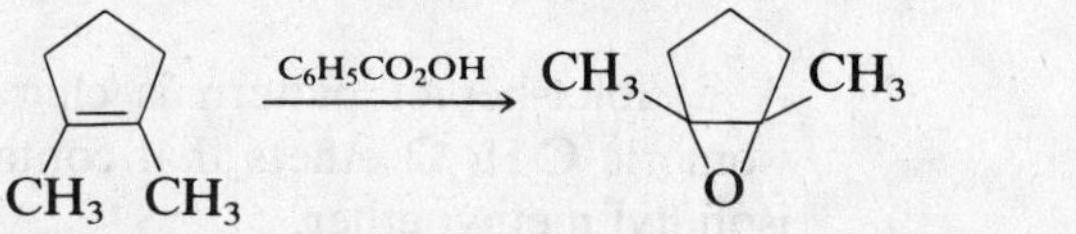

1,2-Dimethylcyclopentene
(compound I)

1,2-Dimethyl-1,2-epoxycyclopentane
(compound K)

Compounds J and L have the same molecular formula, $C_7H_{14}O_2$, but J is a liquid while L is a crystalline solid. Their molecular formulas correspond to the addition of two OH groups to compound I. Osmium tetraoxide brings about syn hydroxylation of an alkene; therefore compound J must be the cis diol.

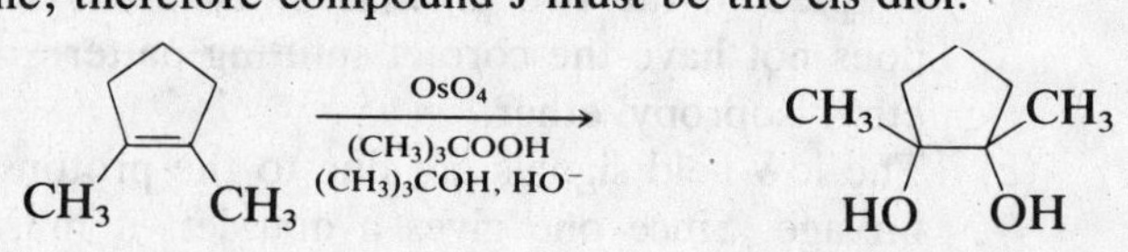

1,2-Dimethylcyclopentene
(compound I)

cis-1,2-Dimethylcyclopentane-1,2-diol
(compound J)

Acid-catalyzed hydrolysis of an epoxide yields a trans diol (compound L):

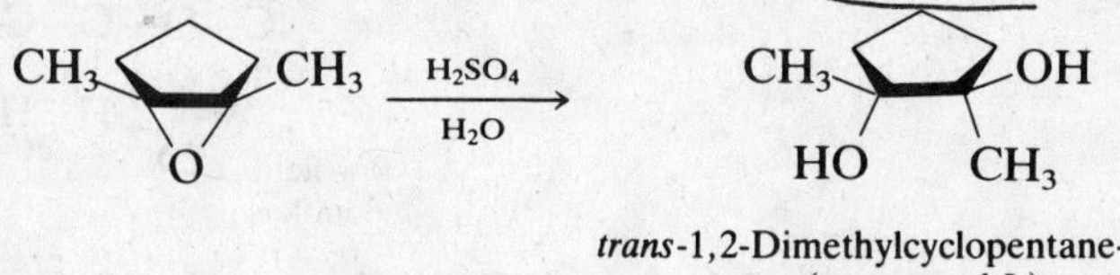

trans-1,2-Dimethylcyclopentane-1,2-diol
(compound L)

16.35 Since cineole contains no double or triple bonds, it must be bicyclic, on the basis of its molecular formula ($C_{10}H_{18}O$, SODAR = 2). When cineole reacts with hydrogen chloride, one of the rings is broken and water is formed.

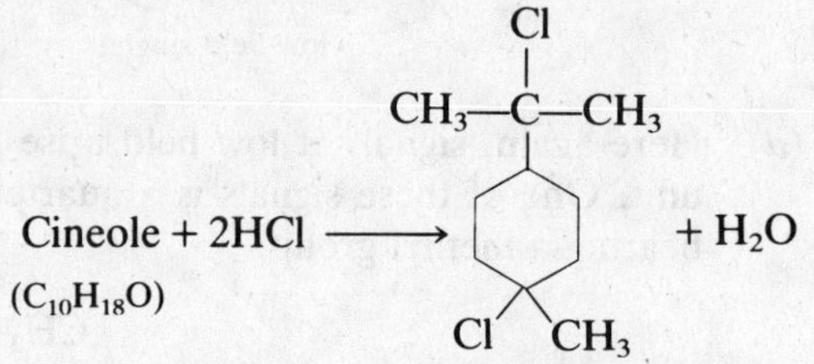

Cineole + 2HCl $\longrightarrow$ + H_2O

($C_{10}H_{18}O$)

($C_{10}H_{18}Cl_2$)

The reaction that takes place is hydrogen halide–promoted ether cleavage. In such a reaction with excess hydrogen halide, the C—O—C unit is cleaved and two carbon-halogen bonds are formed. This suggests that cineole is a cyclic ether because the product contains both newly formed carbon-halogen bonds. A reasonable structure consistent with these facts is:

Cineole

16.36 (a) Since all the peaks in the ^1H nmr spectrum of this ether are singlets, none of the protons can be vicinal to any other nonequivalent proton. The only $C_5H_{12}O$ ether that satisfies this requirement is *tert*-butyl methyl ether.

(b) A doublet-heptet pattern is characteristic of an isopropyl group. There are two isomeric $C_5H_{12}O$ ethers that contain an isopropyl group, ethyl isopropyl ether and isobutyl methyl ether.

$$(CH_3)_2CHOCH_2CH_3 \qquad (CH_3)_2CHCH_2OCH_3$$

Ethyl isopropyl ether   Isobutyl methyl ether

The signal of the methine proton in isobutyl methyl ether will be split into more than a heptet, however, because in addition to being split by two methyl groups, it is coupled to the two protons in the methylene group. Thus, isobutyl methyl ether does not have the correct splitting pattern to be the answer. The correct answer is ethyl isopropyl ether.

(c) The low-field signals are due to the protons on the carbon atoms of the C—O—C linkage. Since one gives a doublet, it must be vicinal to only one other proton. Therefore, we can specify the partial structure:

low-field
doublet

This partial structure contains all the carbon atoms in the molecule. Fill in the remaining valences with hydrogen atoms to reveal isobutyl methyl ether as the correct choice:

$$CH_3—O—CH_2—CH(CH_3)_2$$

low-field singlet   low-field doublet

(d) Here again, signals at low field arise from protons on the carbons of the C—O—C unit. One of these signals is a quartet and so corresponds to a proton on a carbon bearing a methyl group.

quartet

The other carbon of the C—O—C unit has a hydrogen substituent whose signal is split into a triplet. Therefore, this hydrogen must be attached to a carbon that bears a methylene group.

$$CH_3—\underset{\underset{\text{quartet}}{\longrightarrow}\;H}{C}—O—\underset{\underset{H}{\overset{}{C}}\;\underset{\text{triplet}}{\longleftarrow}}{C}—CH_2—$$

These data permit us to complete the structure by adding an additional carbon and the requisite number of hydrogens in such a way that the signals of the protons attached to the carbons of the ether linkage are not split further. The correct structure is ethyl propyl ether.

$$CH_3\underset{\underset{\text{quartet}}{\nearrow}}{CH_2}OCH_2CH_2\underset{\underset{\text{triplet}}{\nwarrow}}{CH_3}$$

16.37 A good way to address this problem is to consider the dibromide derived by treatment of compound M with hydrogen bromide. The presence of an nmr signal equivalent to four protons in the aromatic region at $\delta = 7.3$ ppm indicates that this dibromide contains a disubstituted aromatic ring. The four remaining protons appear as a sharp singlet at $\delta = 4.7$ ppm and are most reasonably contained in two equivalent methylene groups of the type $ArCH_2Br$. Since the dibromide contains all the carbons and hydrogens of the starting material and is derived from it by treatment with hydrogen bromide, it is likely that compound M is a cyclic ether in which a CH_2OCH_2 unit spans two of the carbons of a benzene ring. This can occur only when the positions involved are ortho to each other. Therefore

Compound M

16.38 The molecular formula of compound N ($C_{12}H_{18}O_2$) lacks eight hydrogens as compared with an alkane having the same number of carbon atoms; this indicates a total of four double bonds plus rings (SODAR = 4). A para-substituted aromatic ring is indicated by oxidation of compound N to 1,4-benzenedicarboxylic acid. The benzene ring contributes the equivalent of three double bonds plus one ring, and so both of the side chains are saturated and neither contains a ring. At this point we know the location of 8 of the 12 carbon atoms of compound N.

Turning now to the ^1H nmr spectrum, we see the triplet-quartet pattern characteristic of an ethyl group. Since there are six methyl protons and four methylene protons, as determined by integration of the nmr spectrum, there must be two equivalent ethyl groups in the molecule. Further, because the chemical shift of the methylene protons is at $\delta = 3.4$ ppm, the ethyl group is bonded to oxygen, i.e., there are two $CH_3CH_2O—$ groups in compound N. The two ethoxy groups and the partial structure deduced previously account for all the carbon atoms in compound N. A reasonable candidate structure is:

This structure satisfies all the requirements of the ^{1}H nmr spectrum. All the aromatic ring protons are equivalent and so appear as a sharp singlet, and both benzylic methylene groups are equivalent and appear as a sharp singlet at $\delta = 4.4$ ppm. Their low chemical shift results from their benzylic nature and the fact that they are bonded to oxygen.

16.39 (a) Compound O ($C_6H_{14}O$) has the same number of hydrogen atoms as a six-carbon alkane and so contains neither double bonds nor rings. There are no O—H absorbances in the 3300 cm^{-1} region of its infrared spectrum, and so it is not an alcohol. Completely saturated substances containing oxygen can only be alcohols or ethers; therefore, compound O must be an ether. This conclusion is supported by the observation of a strong C—O stretching vibration at 1100 cm^{-1}.

The ^{1}H nmr spectrum of compound O shows a doublet-heptet pattern typical of an isopropyl group as its only signals. Compound O is diisopropyl ether.

(b) The infrared spectrum of compound P shows C—H absorption above 3000 cm^{-1}, indicating an aromatic ring. A strong band at 700 cm^{-1} points to monosubstitution on that ring. Since there are no O—H peaks in the infrared spectrum, the C—O stretch at 1100 cm^{-1} tells us that P is an ether, not an alcohol.

The ^{1}H nmr spectrum is quite simple, consisting of a 4-proton singlet at $\delta = 4.5$ ppm and a 10-proton singlet at $\delta = 7.3$ ppm. The low-field signal is assigned to two monosubstituted phenyl groups. The signal at $\delta = 4.5$ ppm arises from two CH_2O units. Compound P is dibenzyl ether.

$$C_6H_5CH_2OCH_2C_6H_5$$

16.40 The formula of compound Q ($C_9H_{10}O$) is consistent with a SODAR of 5. Four of the unsaturations are accounted for by a benzene ring, as suggested by the ^{1}H nmr signal at $\delta = 7.0$ ppm. The peak area of this signal (4) indicates that the benzene ring is disubstituted. The two triplets at $\delta = 3.9$ ppm and $\delta = 2.7$ ppm are methylene groups connected to each other (CH_2CH_2). The remaining two-proton singlet at $\delta = 4.7$ ppm is consistent with a methylene group of the $ArCH_2O$— type.

The fact that compound Q undergoes oxidation to form 1,2-benzenedicarboxylic acid indicates that the two substituents on the aromatic ring are ortho.

1,2-Benzenedicarboxylic acid

The structure most consistent with the data given is

SELF-TEST

PART A

A-1. Write the structures of all the isomeric ethers of molecular formula $C_4H_{10}O$ and give the correct name for each.

A-2. Give the structure of the product obtained from each of the following reactions. Show stereochemistry where it is important.

(a) $CH_3CH_2 \overset{H}{\underset{CH_3}{C}}-OH \xrightarrow[\text{2. } CH_3I]{\text{1. Na}} ?$

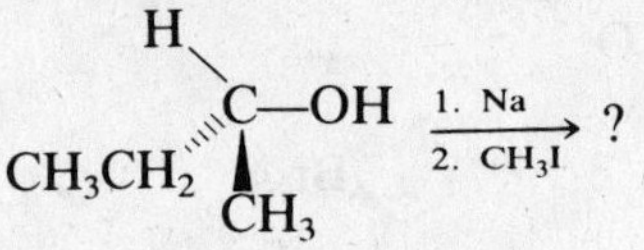

(b) (Z)-2-butene $\xrightarrow[\text{2. } H_3O^+]{\text{1. } CH_3CO_2OH} ?$

(c) $C_6H_5\overset{O}{\overset{\diagup \backslash}{CH}}-\underset{CH_3}{CH_2} \xrightarrow{\text{HI}} ?$

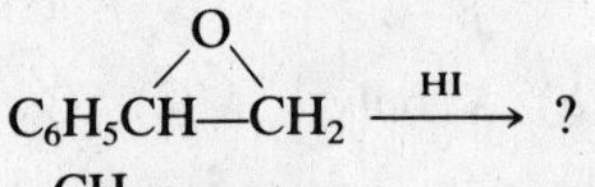

(d) [bicyclic epoxide with two CH₃ groups] $\xrightarrow[CH_3CH_2SH]{CH_3CH_2SNa} ?$

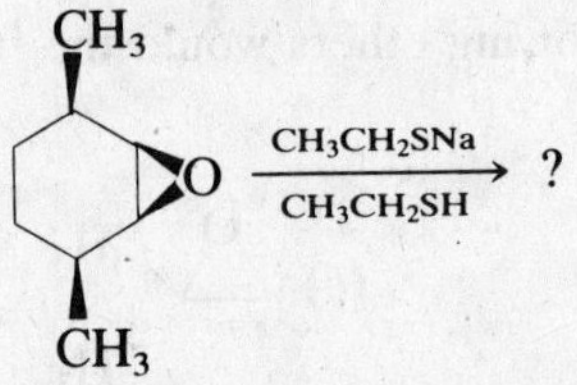

(e) $C_6H_5SNa \xrightarrow{CH_3CH_2Br} ?$

(f) Product of part (e) $\xrightarrow{NaIO_4} ?$

A-3. Outline a scheme for the preparation of cyclohexyl ethyl ether using the Williamson method.

A-4. Outline a synthesis of 2-ethoxyethanol, $CH_3CH_2OCH_2CH_2OH$, using ethanol as the source of all the carbon atoms.

A-5. Provide the reagents necessary to complete each of the following conversions. In each case give the structure of the intermediate product.

(a) [cyclopentane with Br and OH] $\longrightarrow ? \longrightarrow$ [cyclopentane with SCH₃ and OH]

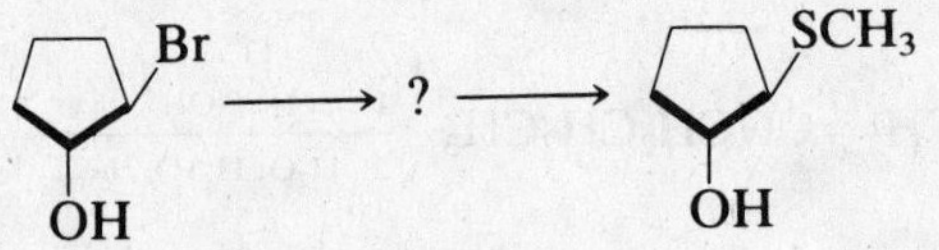

(b) [phenyl C(CH₃)₂OH] $\longrightarrow ? \longrightarrow$ [phenyl C(CH₃) epoxide with CH₂]

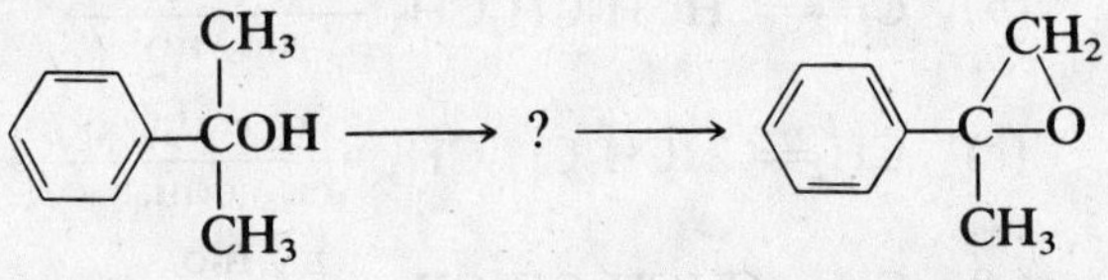

A-6. Provide structures for compounds A and B in the following reaction scheme:

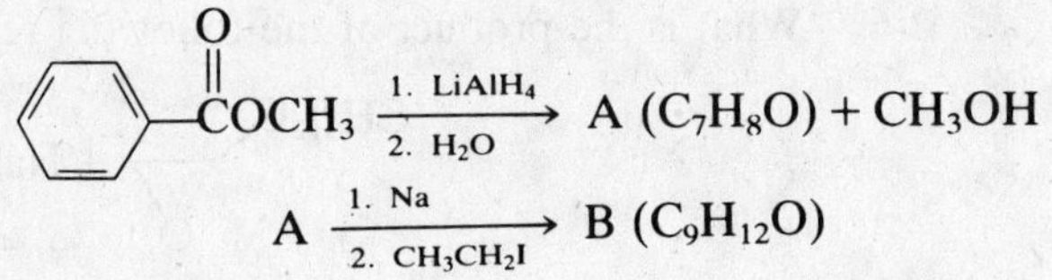

[phenyl COCH₃ with O] $\xrightarrow[\text{2. } H_2O]{\text{1. } LiAlH_4} A\ (C_7H_8O) + CH_3OH$

$A \xrightarrow[\text{2. } CH_3CH_2I]{\text{1. Na}} B\ (C_9H_{12}O)$

PART B

B-1. The most effective pair of reagents for the preparation of *tert*-butyl ethyl ether is:
(a) Potassium *tert*-butoxide and ethyl bromide
(b) Potassium *tert*-butoxide and ethanol
(c) Sodium ethoxide and *tert*-butyl bromide
(d) *tert*-Butyl alcohol and ethyl bromide

B-2. The *best* choice of reactant(s) for the following conversion is

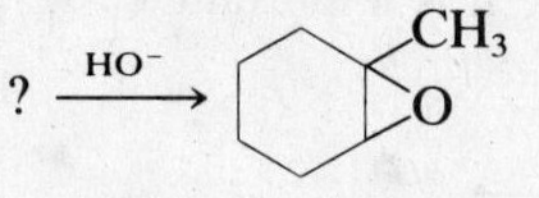

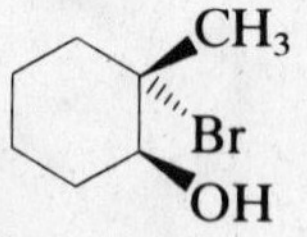

? $\xrightarrow{HO^-}$ [cyclohexene oxide with CH$_3$]

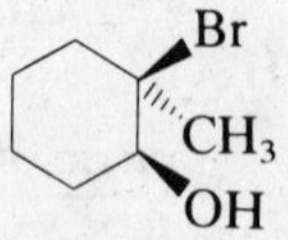

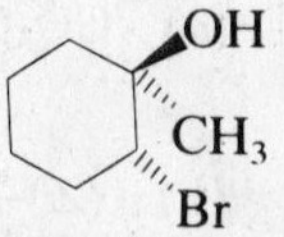

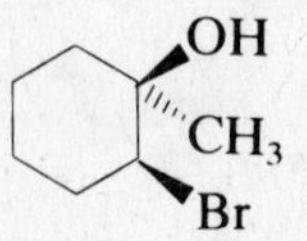

A [cyclohexane with CH$_3$, Br, OH]

B [cyclohexane with Br, CH$_3$, OH]

C [cyclohexane with OH, CH$_3$, Br]

D [cyclohexane with OH, CH$_3$, Br]

(*a*) A and C   (*b*) C only   (*c*) C and D   (*d*) B only

B-3. For which of the following ethers would the ^1H nmr spectrum consist of only two singlets?

(*a*) $CH_3OC(CH_3)_2CH_3$ (with CH$_3$ above and CH$_3$ below central C)

(*c*) [epoxide with C(CH$_3$)$_2$]

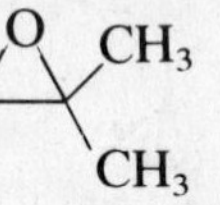

(*b*) $CH_3OCH_2CH_2OCH_3$ (*d*) All of these

B-4. Heating a particular ether with HBr yielded a single organic product. Which of the following conclusions may be reached?
(*a*) The reactant was a methyl ether.
(*b*) The reactant was a symmetrical ether.
(*c*) The reactant was a cyclic ether.
(*d*) Both (*b*) and (*c*) are correct.

B-5. The best method for the preparation of H_2C—$CHCH_2CH_2CH_3$ with epoxide O is

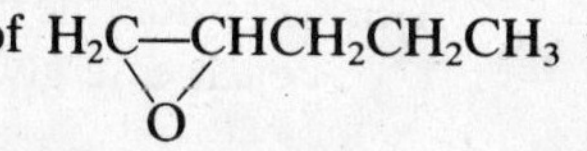

(*a*) $CH_2{=}CHCH_2CH_2CH_3$ $\xrightarrow[\text{2. }H_2O,\ H_2SO_4,\text{ heat}]{\text{1. }CH_3COOH}$

(*b*) $CH_2{=}CHCH_2CH_2CH_3$ $\xrightarrow[\text{2. }H_2O_2,\ HO^-]{\text{1. }B_2H_6,\text{ diglyme}}$

(*c*) $CH_2{=}CHCH_2CH_2CH_3$ $\xrightarrow[\text{(CH}_3)_3COH,\ NaOH]{\text{OsO}_4,\ (CH_3)_3COOH}$

(*d*) $CH_2{=}CHCH_2CH_2CH_3$ $\xrightarrow[\text{2. NaOH}]{\text{1. }Br_2,\ H_2O}$

B-6 What is the product of the following reaction?

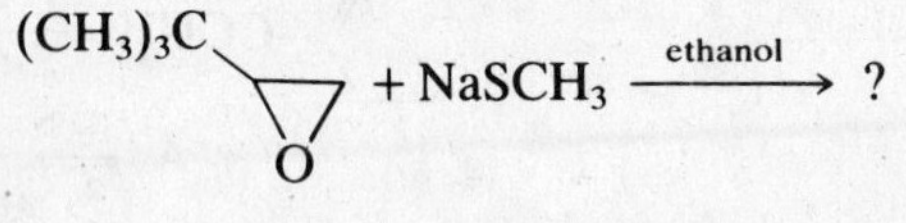

$(CH_3)_3C$—[epoxide with O] $+ NaSCH_3 \xrightarrow{\text{ethanol}}$?

(*a*) $CH_3SCH_2CHC(CH_3)_3$ with OH below

(*c*) $CH_3SCH_2CHC(CH_3)_3$ with OCH$_2$CH$_3$ below

(*b*) $(CH_3)_3CCHCH_2OH$ with SCH$_3$ below

(*d*) $(CH_3)_3CCH_2CHSCH_3$ with OH below

ALDEHYDES AND KETONES. NUCLEOPHILIC ADDITION TO THE CARBONYL GROUP

IMPORTANT TERMS AND CONCEPTS

Nomenclature (Sec. 17.1) In naming both aldehydes and ketones, the base name is derived from the longest chain of carbon atoms containing the carbonyl group. In aldehydes the carbonyl carbon is always C-1 and the *-e* ending of the alkane base name is replaced by *-al*. For example,

$$CH_3CH_2CHCHCH \quad \overset{O}{\overset{\|}{}}$$

2,3-Dimethylpentanal   2-Ethyl-5-methylhexanal

Ketones are named so that the carbonyl position is the lowest possible. In addition, the *-e* ending of the alkane base name is replaced by *-one*. For example,

4,5-Dimethyl-3-hexanone   3-Methylcyclohexanone

The Carbonyl Group (Sec. 17.2) Understanding the structure and bonding of the carbonyl group is fundamental to being able to explain the chemical reactivity of aldehydes and ketones. The carbon utilizes sp^2 hybridized orbitals to bond to the oxygen and two other groups. The unhybridized *p* orbital on carbon forms a π bond to oxygen through overlap with an oxygen $2p$ orbital.

441

The polarity of the carbonyl group is explained by the greater electronegativity of oxygen as compared with carbon. The polarization of the carbonyl group may be represented as

$$\overset{\delta+}{\underset{}{\diagup}}C=\overset{\delta-}{O} \quad \text{or} \quad \underset{}{\diagup}C\overset{+}{=}\overrightarrow{O}$$

The orientation of this dipole explains why *nucleophilic reagents* react with the carbon and *electrophilic reagents* react with the oxygen of the carbonyl group.

Preparation of Aldehydes and Ketones (Sec. 17.5) Several methods have been discussed in previous chapters and are summarized in text Table 17.1. These include:

Ozonolysis of alkenes	Chapter 6
Hydration of alkynes	Chapter 9
Friedel-Crafts acylation	Chapter 12
Oxidation of alcohols	Chapter 15

Methods presented in this chapter for the preparation of aldehydes and ketones are outlined in the reaction summary which follows this section.

Reactions of Aldehydes and Ketones (Secs. 17.6 to 17.17) The principal types of reactivity exhibited by aldehydes and ketones may be divided into two classes: (1) nucleophilic addition to the carbonyl group; and (2) electrophilic substitution at the α carbon by way of enol and enolate intermediates. This latter class of reactions will be discussed in the next chapter.

Nucleophilic addition reactions that are typical of aldehydes and ketones include:

Cyanohydrin formation, Sec. 17.8
Acetal formation, Secs. 17.9, 17.10
Reactions with amines, Secs. 17.11, 17.12
Reactions with ammonia derivatives, Sec. 17.13
The Wittig reaction, Secs. 17.14, 17.15

Examples of these reactions are given in the section which follows.

The hydration reaction (Sec. 17.7), while not a synthetically useful process, provides an example through which a general statement may be made about the mechanisms of nucleophilic addition reactions to carbonyls.

In an acidic medium, protonation of the carbonyl oxygen is the *first* step of the reaction. The protonated species thus formed is more susceptible to attack by a nucleophile, resulting in *acid catalysis* of the reaction. This increased reactivity may be understood by considering the resonance contributions to the cationic species.

Acid-catalyzed hydration:

$$R_2C=O + H_3O^+ \rightleftharpoons R_2C=\overset{+}{O}H + H_2O$$

$$R_2C=\overset{+}{O}H + H_2O \rightleftharpoons R_2C\underset{\overset{|}{{}^+OH_2}}{-OH} \overset{H_2O}{\rightleftharpoons} R_2C\underset{\overset{|}{OH}}{-OH} + H_3O^+$$

In a neutral or basic medium, where protonation of the carbonyl oxygen cannot occur, addition of the nucleophile is the first step of the reaction. This most often occurs when an anionic nucleophile is present.

Base-catalyzed hydration:

$$R_2C=O + HO^- \rightleftharpoons R_2C\underset{\overset{|}{OH}}{-O^-}$$

$$R_2C\underset{\overset{|}{OH}}{-O^-} + H_2O \rightleftharpoons R_2C\underset{\overset{|}{OH}}{-OH} + HO^-$$

Spectroscopic Analysis (Sec. 17.19) Carbonyl groups are easily recognized by infrared spectroscopy, since they give strong absorptions in the region 1710 to 1750 cm^{-1}. In addition, aldehydes exhibit weak absorptions near 2720 to 2820 cm^{-1}.

An aldehyde gives rise to a characteristic peak in the ^1H nmr spectrum in the region $\delta = 9$ to 10 ppm. Protons attached to carbon atoms adjacent to a carbonyl group typically are found in the region $\delta = 2.0$ to 2.4 ppm.

$$\underset{\text{—C—H}}{\overset{\overset{\displaystyle O}{\|}}{}} \quad \delta = 9 \text{ to } 10 \text{ ppm} \qquad\qquad \underset{\text{—C—C—H}}{\overset{\overset{\displaystyle O}{\|}}{}} \quad \delta = 2.0 \text{ to } 2.4 \text{ ppm}$$

IMPORTANT REACTIONS

Cyanohydrin Formation (Sec. 17.8)

General:

$$\underset{\text{RCR'}}{\overset{\overset{\displaystyle O}{\|}}{}} + HC\equiv N \xrightarrow{\text{CN}^-} \underset{\underset{\text{C}\equiv\text{N}}{|}}{\overset{\overset{\displaystyle OH}{|}}{R-C-R'}}$$

Example:

$$\underset{\text{C}_6\text{H}_5\text{CH}}{\overset{\overset{\displaystyle O}{\|}}{}} + HCN \xrightarrow{\text{CN}^-} \underset{\underset{\text{CN}}{|}}{\overset{\overset{\displaystyle OH}{|}}{C_6H_5CH}}$$

Acetal Formation (Secs. 17.9, 17.10)

General:

$$\underset{\text{RCR}}{\overset{\overset{\displaystyle O}{\|}}{}} + 2R'OH \xrightleftharpoons{\text{H}^+} \underset{\underset{\text{OR'}}{|}}{\overset{\overset{\displaystyle OR'}{|}}{R-C-R}} + H_2O$$

Mechanism:

$$R_2C{=}O \rightleftharpoons R_2C{=}\overset{+}{O}H \xrightleftharpoons{\text{R'OH}} \underset{\underset{+}{\overset{|}{HOR'}}}{R_2C{-}OH} \xrightleftharpoons{-\text{H}^+} \underset{\overset{|}{OR'}}{R_2C{-}OH}$$

$$\underset{\overset{|}{OR'}}{R_2C{-}OH} \xrightleftharpoons{\text{H}^+} \underset{\overset{|}{OR'}}{R_2C{-}\overset{+}{O}H_2} \xrightleftharpoons{-\text{H}_2\text{O}} \underset{\overset{|}{OR'}}{R_2C^+} \xrightleftharpoons{\text{R'OH}} \underset{\overset{|}{OR'}}{R_2C{-}\underset{H}{\overset{+}{O}R'}} \xrightleftharpoons{-\text{H}^+} \underset{\overset{|}{OR'}}{R_2C{-}OR'}$$

Examples:

$$\underset{\text{C}_6\text{H}_5\text{CH}}{\overset{\overset{\displaystyle O}{\|}}{}} + 2C_2H_5OH \xrightleftharpoons{\text{H}^+} C_6H_5CH(OC_2H_5)_2 + H_2O$$

+ HOCH$_2$CH$_2$OH $\xrightleftharpoons{\text{H}^+}$ + H$_2$O

Reaction with Primary Amines (Sec. 17.11)

General:

$$R_2C=O + R'NH_2 \longrightarrow R_2C=NR' + H_2O$$

Mechanism:

Nucleophilic addition stage

$$R_2C=O + R'NH_2 \rightleftharpoons R_2\overset{\displaystyle OH}{\underset{\displaystyle |}{C}}-NHR'$$

Elimination stage

$$R_2\overset{\displaystyle OH}{\underset{\displaystyle |}{C}}-NHR' \rightleftharpoons R_2C=NR' + H_2O$$

Examples:

$$C_6H_5\overset{\displaystyle O}{\overset{\displaystyle \|}{C}}CH_3 + CH_3CH_2NH_2 \longrightarrow C_6H_5\overset{\displaystyle N-CH_2CH_3}{\overset{\displaystyle \|}{C}}CH_3$$

$$\text{(cyclohexanone)}=O + (CH_3)_3CNH_2 \longrightarrow \text{(cyclohexylidene)}=NC(CH_3)_3$$

Reaction with Secondary Amines; Formation of Enamines (Sec. 17.12)

General:

$$RCH_2\overset{\displaystyle O}{\overset{\displaystyle \|}{C}}R' + R''_2NH \longrightarrow RCH=\overset{\displaystyle NR''_2}{\underset{\displaystyle |}{C}}R' + H_2O$$

Example:

$$\text{(cyclohexanone)}=O + \text{(pyrrolidine)}\underset{H}{N} \longrightarrow \text{(enamine)}-N\text{(pyrrolidine)} + H_2O$$

Reaction with Ammonia Derivatives (Sec. 17.13)

General:

$$R_2C=O + Z-NH_2 \longrightarrow R_2C=N-Z + H_2O$$

Examples:

Oximes (Z = OH)

$$CH_3CH_2\overset{\displaystyle O}{\overset{\displaystyle \|}{C}}H + NH_2OH \longrightarrow CH_3CH_2CH=NOH + H_2O$$

Hydrazones (Z = NHR')

$$\text{(cyclohexanone)}=O + C_6H_5NHNH_2 \longrightarrow \text{(cyclohexylidene)}=NNHC_6H_5 + H_2O$$

The Wittig Reaction (Secs. 17.14, 17.15)

General:

$$R_2C=O + (C_6H_5)_3\overset{+}{P}-\overset{..}{\overset{..}{C}}R'_2 \longrightarrow R_2C=CR'_2 + (C_6H_5)_3\overset{+}{P}-\bar{O}$$

Mechanism:

$$R-\underset{\underset{O}{\|}}{C}-R \;+\; \underset{\underset{\overset{+}{P}(C_6H_5)_3}{|}}{-:}\overset{R'}{\underset{R'}{C}} \longrightarrow R-\underset{\underset{O}{|}}{\overset{R}{\underset{|}{C}}}-\underset{\underset{P(C_6H_5)_3}{|}}{\overset{R'}{\underset{|}{C}}}-R' \longrightarrow$$

$$\underset{R}{\overset{R}{\diagdown}}C=\underset{R'}{\overset{R'}{\diagup}}C \;+\; (C_6H_5)_3\overset{+}{P}-\bar{O}$$

Examples:

$+ (C_6H_5)_3\overset{+}{P}-\overset{..}{\bar{C}}HCH_3 \longrightarrow$ $=CHCH_3 + (C_6H_5)_3\overset{+}{P}-\bar{O}$

$C_6H_5\overset{\underset{\|}{O}}{C}H + (C_6H_5)_3\overset{+}{P}-\overset{..}{}$ $\longrightarrow C_6H_5CH=$ $+ (C_6H_5)_3\overset{+}{P}-\bar{O}$

Oxidation of Aldehydes (Sec. 17.17)

General:

$$R\overset{\underset{\|}{O}}{C}H \xrightarrow{[O]} RCO_2H$$

Example:

$$CH_3CH_2CH_2\overset{\underset{\|}{O}}{C}H \xrightarrow[H_2SO_4,\ H_2O]{Na_2Cr_2O_7} CH_3CH_2CH_2CO_2H$$

Baeyer-Villiger Oxidation (Sec. 17.18)

General:

$$R\overset{\underset{\|}{O}}{C}R' + R''\overset{\underset{\|}{O}}{C}OOH \longrightarrow R\overset{\underset{\|}{O}}{C}OR' + R''\overset{\underset{\|}{O}}{C}OH$$

Example:

$$(CH_3)_3C\overset{\underset{\|}{O}}{C}CH_3 + CH_3\overset{\underset{\|}{O}}{C}OOH \longrightarrow (CH_3)_3C\overset{\underset{\|}{O}}{C}OCCH_3 + CH_3\overset{\underset{\|}{O}}{C}OH$$

SOLUTIONS TO TEXT PROBLEMS

17.1 (*b*) The longest continuous chain in glutaraldehyde has five carbons and terminates in aldehyde functions at both ends. *Pentanedial* is an acceptable IUPAC name for this compound.

$$\underset{1}{H}\overset{\underset{\|}{O}}{C}\underset{2}{C}H_2\underset{3}{C}H_2\underset{4}{C}H_2\underset{5}{C}H \qquad \text{Pentanedial (glutaraldehyde)}$$

(*c*) The three-carbon parent chain has a double bond between C-2 and C-3 and a phenyl substituent at C-3.

$$\underset{3}{C_6H_5CH}=\underset{2}{CH}\underset{1}{\overset{\underset{\|}{O}}{C}}H \qquad \text{3-Phenyl-2-propenal (cinnamaldehyde)}$$

(d) Vanillin can be named as a derivative of benzaldehyde. Remember to cite the remaining substituents in alphabetical order.

4-Hydroxy-3-methoxybenzaldehyde (vanillin)

17.2 (b) The structure that corresponds to benzyl *tert*-butyl ketone is

The longest carbon chain contains four carbon atoms; the compound is therefore named as a derivative of butane. The substitutive IUPAC name is 3,3-dimethyl-1-phenyl-2-butanone.

(c) First write the structure from the name given. Ethyl isopropyl ketone has an ethyl group and an isopropyl group bonded to a carbonyl group.

$$CH_3CH_2\overset{\displaystyle O}{\overset{\displaystyle \|}{C}}\underset{\underset{\displaystyle CH_3}{|}}{C}HCH_3$$

Ethyl isopropyl ketone may be alternatively named 2-methyl-3-pentanone. Its longest continuous chain has five carbons. The carbonyl carbon is C-3 irrespective of the direction in which the chain is numbered, and so we choose the direction that gives the lower number to the position that bears the methyl group.

(d) Methyl 2,2-dimethylpropyl ketone has a methyl group and a 2,2-dimethylpropyl group bonded to a carbonyl group.

$$CH_3\overset{\displaystyle O}{\overset{\displaystyle \|}{C}}CH_2\underset{\underset{\displaystyle CH_3}{|}}{\overset{\overset{\displaystyle CH_3}{|}}{C}}CH_3$$

The longest continuous chain has five carbons, and the carbonyl carbon is C-2. Thus, methyl 2,2-dimethylpropyl ketone may also be named 4,4-dimethyl-2-pentanone.

(e) The structure correspondng to allyl methyl ketone is

$$CH_3\overset{\displaystyle O}{\overset{\displaystyle \|}{C}}CH_2CH=CH_2$$

Since the carbonyl group is given the lowest possible number in the chain, the substitutive name is 4-penten-2-one.

17.3 No. Lithium aluminum hydride is the only reagent we have discussed that is capable of reducing carboxylic acids (Sec. 15.3).

17.4 The target molecule, 2-butanone, contains four carbon atoms. The problem states that all of the carbon atoms originate in acetic acid, which has two carbon atoms. This suggests the following disconnections:

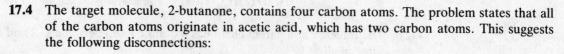

2-Butanone

The necessary aldehyde (acetaldehyde) is prepared from acetic acid by reduction followed by oxidation in an anhydrous medium (Collins oxidation).

$$CH_3CO_2H \xrightarrow[\text{2. } H_2O]{\text{1. } LiAlH_4} CH_3CH_2OH \xrightarrow[CH_2Cl_2]{(C_5H_5N)_2CrO_3} CH_3\overset{\displaystyle O}{\overset{\|}{C}}H$$

Acetic acid   Ethanol   Acetaldehyde

Ethylmagnesium bromide may be obtained from acetic acid by the following sequence:

$$CH_3CH_2OH \xrightarrow[PBr_3]{HBr \text{ or}} CH_3CH_2Br \xrightarrow[\substack{\text{diethyl} \\ \text{ether}}]{Mg} CH_3CH_2MgBr$$

Ethanol   Ethyl bromide   Ethylmagnesium
(Prepared as   bromide
previously)

The preparation of 2-butanone is completed as follows:

$$CH_3\overset{\displaystyle O}{\overset{\|}{C}}H + CH_3CH_2MgBr \xrightarrow[\text{2. } H_3O^+]{\text{1. diethyl ether}} CH_3\overset{\displaystyle OH}{\overset{|}{C}}HCH_2CH_3 \xrightarrow{Cr_2O_7^{2-},\ H^+} CH_3\overset{\displaystyle O}{\overset{\|}{C}}CH_2CH_3$$

Acetaldehyde   Ethylmagnesium   2-Butanol   2-Butanone
                  bromide

17.5 Chloral is trichloroethanal, $CCl_3\overset{\displaystyle O}{\overset{\|}{C}}H$. Chloral hydrate is the addition product of chloral and water.

$$Cl_3C\overset{\displaystyle OH}{\underset{\displaystyle OH}{\overset{|}{\underset{|}{C}}}}H \qquad \text{Chloral hydrate}$$

17.6 Methacrylonitrile is formed by the dehydration of acetone cyanohydrin, and thus has the structure shown.

$$CH_3\overset{\displaystyle OH}{\underset{\displaystyle CN}{\overset{|}{\underset{|}{C}}}}CH_3 \xrightarrow{-H_2O} CH_3\underset{\displaystyle CN}{\overset{|}{C}}=CH_2$$

Acetone cyanohydrin   Methacrylonitrile

17.7 (a) The hemiacetal intermediate corresponds to addition of methanol to the carbonyl group.

$$CH_3\overset{\displaystyle O}{\overset{\|}{C}}H + CH_3OH \underset{H^+}{\rightleftharpoons} CH_3\overset{\displaystyle OCH_3}{\overset{|}{C}}HOH$$

Acetaldehyde   Methanol   Hemiacetal
                              intermediate

(b) This hemiacetal is converted to a carbocation by protonation followed by loss of a water molecule from the protonated form

$$CH_3\overset{\displaystyle OCH_3}{\overset{|}{C}}HOH \underset{H^+}{\rightleftharpoons} CH_3\overset{\displaystyle OCH_3}{\overset{|}{C}}H-\overset{+}{\underset{\displaystyle H}{\overset{H}{\ddot{O}}}} \underset{-H_2O}{\rightleftharpoons} CH_3\overset{+}{C}HOCH_3$$

Hemiacetal   Carbocation
intermediate   intermediate

The two principal forms of this carbocation are:

$$CH_3-\underset{+}{\overset{\displaystyle :\ddot{O}CH_3}{C}}{}_H \longleftrightarrow CH_3-\overset{\displaystyle \overset{+}{\ddot{O}}CH_3}{C}{}_H$$

(c)   The dimethyl acetal of acetaldehyde is $CH_3CH(OCH_3)_2$.

17.8   (a)   Ethanol adds to benzaldehyde to form a hemiacetal.

Benzaldehyde   Ethanol   Hemiacetal intermediate

(b)   Protonation of the hemiacetal triggers its conversion to a carbocation intermediate.

Hemiacetal intermediate   Carbocation intermediate

The two principal resonance forms of this carbocation are:

Other resonance forms involving the aromatic ring are also reasonable, for example:

(c)   The diethyl acetal of benzaldehyde is $C_6H_5CH(OCH_2CH_3)_2$.

17.9   (b)   1,3-Propanediol forms acetals that contain a six-membered 1,3-dioxanc ring.

Benzaldehyde   1,3-Propanediol   2-Phenyl-1,3-dioxane

(c)   The cyclic acetal derived from isobutyl methyl ketone and ethylene glycol bears an isobutyl group and a methyl group at C-2 of a 1,3-dioxolane ring.

Isobutyl methyl ketone   Ethylene glycol   2-Isobutyl-2-methyl-1,3-dioxolane

(d)   Since the starting diol is 2,2-dimethyl-1,3-propanediol, the cyclic acetal is six-membered and bears two methyl substituents at C-5 in addition to isobutyl and methyl groups at C-2.

Isobutyl methyl ketone   2,2-Dimethyl-1,3-propanediol   2-Isobutyl-2,5,5-trimethyl-1,3-dioxane

17.10   The conversion is one of reduction; however, the conditions necessary for this transformation (LiAlH$_4$) would also reduce the ketone carbonyl. The ketone functionality is therefore protected as the cyclic acetal.

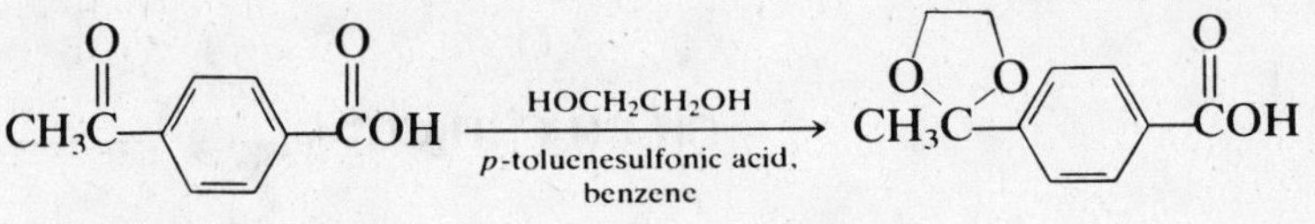

4-Acetylbenzoic acid

Reduction of the carboxylic acid may now be carried out:

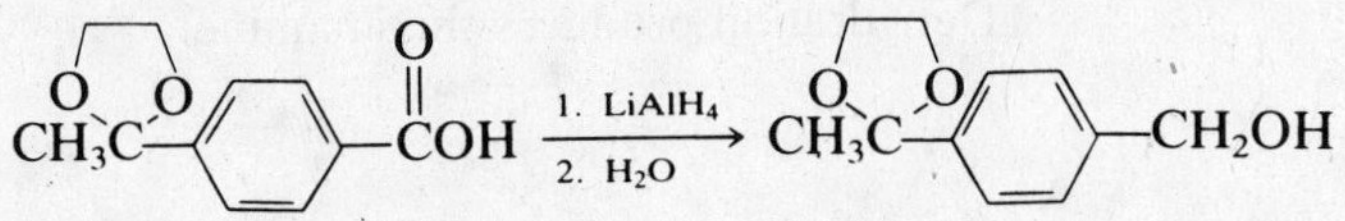

Hydrolysis to remove the protecting group completes the synthesis.

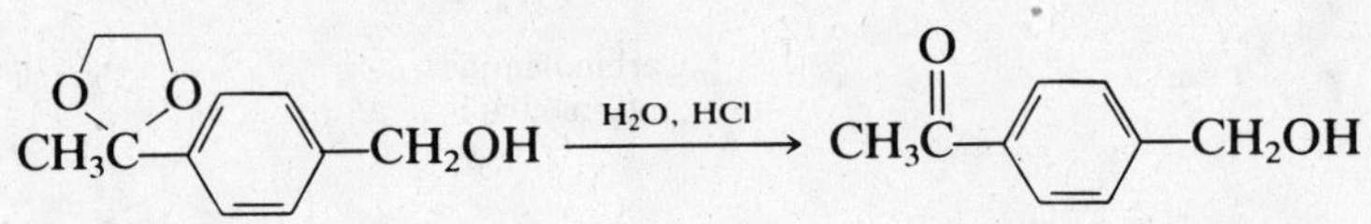

4-Acetylbenzyl alcohol

17.11 (*b*) Nucleophilic addition of butylamine to benzaldehyde gives the carbinolamine.

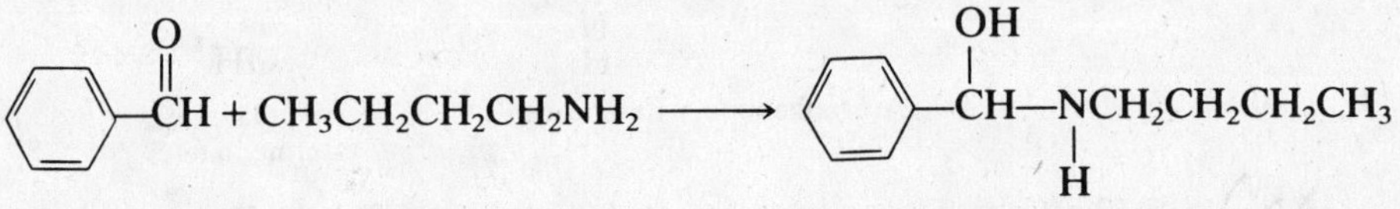

Benzaldehyde   Butylamine   Carbinolamine intermediate

Dehydration of the carbinolamine produces the imine.

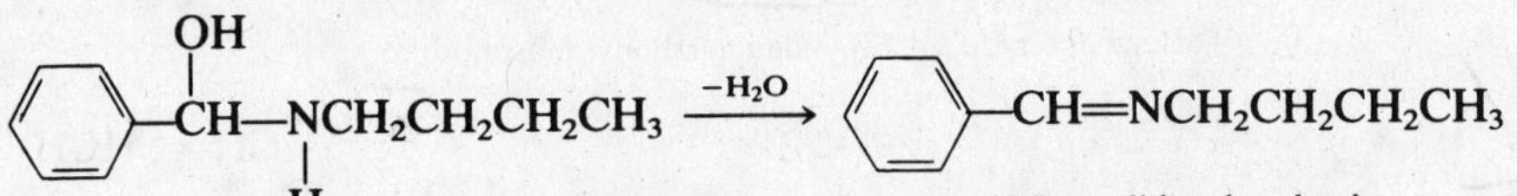

N-Benzylidenebutylamine

(*c*) Cyclohexanone and *tert*-butylamine react according to the equation

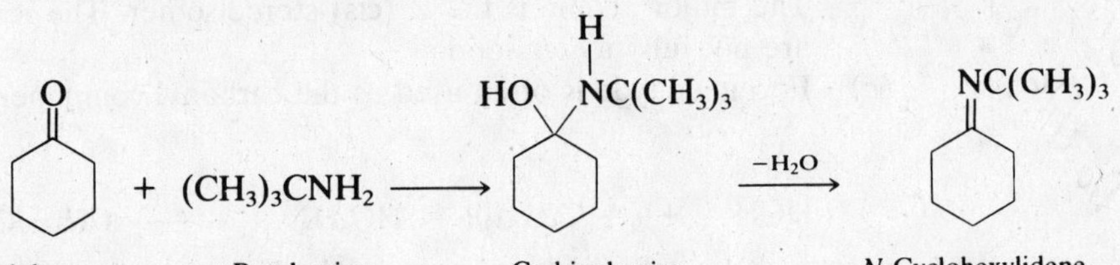

Cyclohexanone   *tert*-Butylamine   Carbinolamine   *N*-Cyclohexylidene-
                                          intermediate   *tert*-butylamine

(*d*)

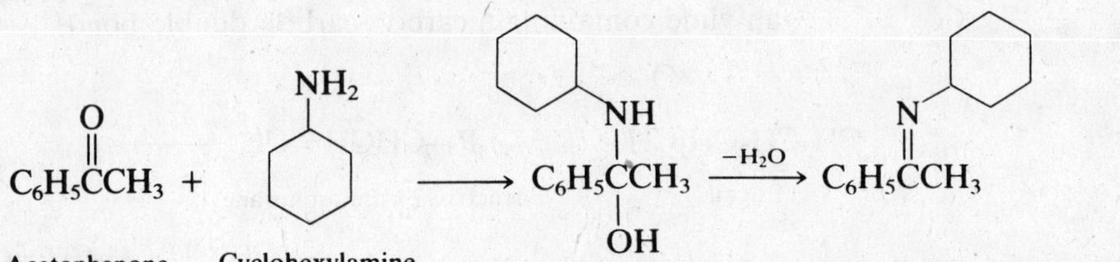

Acetophenone   Cyclohexylamine   Carbinolamine   *N*-(1-Phenylethylidene)-
                                        intermediate   cyclohexylamine

17.12 (b) Pyrrolidine, a secondary amine, adds to 3-pentanone to give a carbinolamine.

$$CH_3CH_2\overset{\overset{O}{\|}}{C}CH_2CH_3 + \underset{H}{\overset{\overset{\displaystyle\bigcirc}{N}}{\ }} \longrightarrow CH_3CH_2\overset{\overset{N}{\ }}{\underset{OH}{C}}CH_2CH_3$$

3-Pentanone　　　Pyrrolidine　　　　Carbinolamine
　　　　　　　　　　　　　　　　　intermediate

Dehydration produces the enamine.

$$CH_3CH_2\overset{\overset{N}{\ }}{\underset{OH}{C}}CH_2CH_3 \longrightarrow CH_3CH{=}\overset{\overset{N}{\ }}{C}CH_2CH_3 + H_2O$$

Carbinolamine　　　　　3-Pyrrolidino-2-pentene
intermediate

(c)

$$C_6H_5\overset{\overset{O}{\|}}{C}CH_3 + \underset{H}{\overset{\overset{\displaystyle\bigcirc}{N}}{\ }} \longrightarrow C_6H_5\overset{\overset{N}{\ }}{\underset{OH}{C}}CH_3 \xrightarrow{-H_2O} C_6H_5\overset{\overset{N}{\ }}{C}{=}CH_2$$

Acetophenone　Piperidine　　　Carbinolamine　　　1-Piperidino-1-phenylethene
　　　　　　　　　　　　　　　intermediate

17.13 (b) The product is a mixture of the E and Z isomers of 1-phenylpropene.

$$\overset{\displaystyle\bigcirc}{}\text{—}\overset{\overset{O}{\|}}{C}H + (C_6H_5)_3\overset{+}{P}\text{—}\overset{..}{C}HCH_3 \longrightarrow$$

Benzaldehyde　　Ethylidenetriphenylphosphorane

$$\overset{\displaystyle\bigcirc}{}\text{—}CH{=}CHCH_3 + (C_6H_5)_3\overset{+}{P}\text{—}O^-$$

1-Phenylpropene　　　　　　Triphenylphosphine oxide
(98% a mixture of E and Z stereoisomers)

The major alkene is the Z (cis) stereoisomer. The reasons for this stereoselectivity are not fully understood.

(c) Formaldehyde is often used as the carbonyl component in Wittig reactions.

$$H\overset{\overset{O}{\|}}{C}H + (C_6H_5)_3\overset{+}{P}\text{—}\overset{..}{C}HC_6H_5 \longrightarrow CH_2{=}CHC_6H_5 + (C_6H_5)_3\overset{+}{P}\text{—}O^-$$

Formaldehyde　Benzylidenetriphenylphosphorane　　　Styrene (75%)　Triphenylphosphine oxide

(d) Here we see an example of the Wittig reaction applied to diene synthesis by use of an ylide containing a carbon-carbon double bond.

$$CH_3CH_2CH_2\overset{\overset{O}{\|}}{C}H + (C_6H_5)_3\overset{+}{P}\text{—}\overset{..}{C}HCH{=}CH_2 \longrightarrow$$

Butanal　　　　Allylidenetriphenylphosphorane

$$CH_3CH_2CH_2CH{=}CHCH{=}CH_2 + (C_6H_5)_3\overset{+}{P}\text{—}O^-$$

1,3-Heptadiene (52%)　　　　Triphenylphosphine oxide

(e) Methylene transfer from methylenetriphenylphosphorane is one of the most commonly used Wittig reactions.

Cyclohexyl methyl ketone   Methylenetriphenylphosphorane

2-Cyclohexylpropene (66%)   Triphenylphosphine oxide

17.14 (b) There are two Wittig reaction routes that lead to 1-pentene. One is represented retrosynthetically by the disconnection

$$CH_3CH_2CH_2CH{=}CH_2 \Rightarrow CH_3CH_2CH_2\overset{O}{\overset{\|}{C}}H + (C_6H_5)_3\overset{+}{P}{-}\overset{..}{\overset{-}{C}}H_2$$

1-Pentene   Butanal   Methylenetriphenylphosphorane

The other route is

$$CH_3CH_2CH_2CH{=}CH_2 \Rightarrow CH_3CH_2CH_2\overset{..}{\overset{-}{C}}H{-}\overset{+}{P}(C_6H_5)_3 + \overset{O}{\overset{\|}{H}}CH$$

1-Pentene   Butylidenetriphenylphosphorane   Formaldehyde

(c) The two retrosyntheses are:

$$C_6H_5CH_2CH{=}C(CH_2CH_3)_2 \Rightarrow C_6H_5CH_2\overset{O}{\overset{\|}{C}}H + (C_6H_5)_3\overset{+}{P}{-}\overset{..}{\overset{-}{C}}(CH_2CH_3)_2$$

3-Ethyl-1-phenyl-2-pentene   2-Phenylethanal   (1-Ethylpropylidene)triphenylphosphorane

and

$$C_6H_5CH_2CH{=}C(CH_2CH_3)_2 \Rightarrow C_6H_5CH_2\overset{..}{\overset{-}{C}}H{-}\overset{+}{P}(C_6H_5)_3 + CH_3CH_2\overset{O}{\overset{\|}{C}}CH_2CH_3$$

3-Ethyl-1-phenyl-2-pentene   2-Phenylethylidenetriphenylphosphorane   3-Pentanone

(d) Cyclopentanone is the carbonyl component in one of the Wittig routes.

Isopropylidenecyclopentane   Cyclopentanone   Isopropylidenetriphenylphosphorane

Acetone is the carbonyl component in the other.

Isopropylidenecyclopentane   Cyclopentylidenetriphenylphosphorane   Acetone

17.15 Ylides are prepared by the reaction of an alkyl halide with triphenylphosphine, followed by treatment with strong base. 2-Bromobutane is the alkyl halide needed in this case.

$$(C_6H_5)_3P + CH_3\overset{Br}{\overset{|}{C}}HCH_2CH_3 \longrightarrow (C_6H_5)_3\overset{+}{P}{-}\overset{|}{C}HCH_2CH_3\ Br^-$$
$$\underset{CH_3}{}$$

Triphenyl phosphine   2-Bromobutane   (1-Methylpropyl)triphenyl-phosphonium bromide

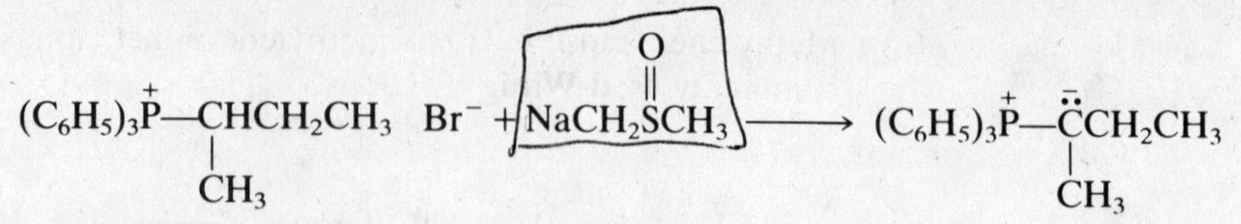

$$(C_6H_5)_3\overset{+}{P}\!-\!CHCH_2CH_3 \;\; Br^- \;+\; NaCH_2SCH_3 \longrightarrow (C_6H_5)_3\overset{+}{P}\!-\!\overset{..}{\overset{-}{C}}CH_2CH_3$$

(1-Methylpropyl)triphenyl-
phosphonium bromide

Sodiomethyl
methyl sulfoxide

Ylide

17.16   The overall reaction is:

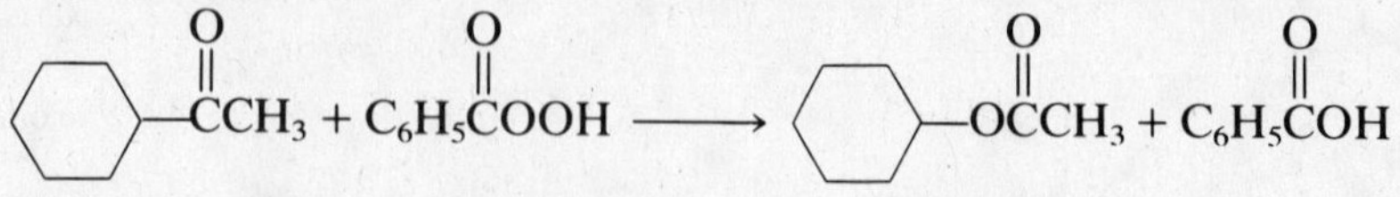

Cyclohexyl methyl
ketone

Peroxybenzoic
acid

Cyclohexyl
acetate

Benzoic
acid

In the first step, the peroxy acid adds to the carbonyl group of the ketone to form a peroxy monoester of a *gem*-diol.

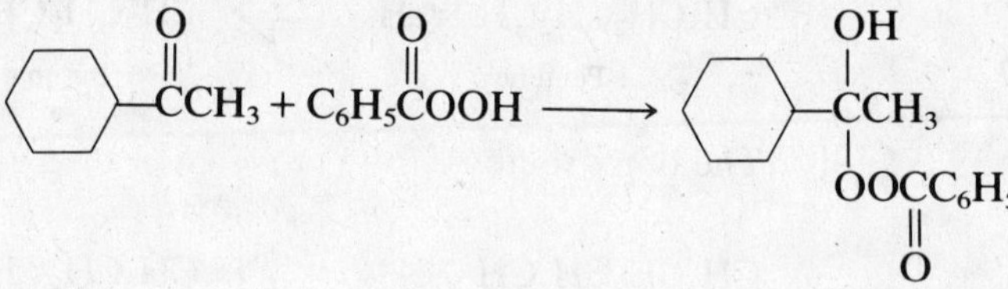

Peroxy monoester

The intermediate then undergoes rearrangement. Alkyl group migration occurs at the same time as cleavage of the O—O bond of the peroxy ester. In general, the more substituted group migrates.

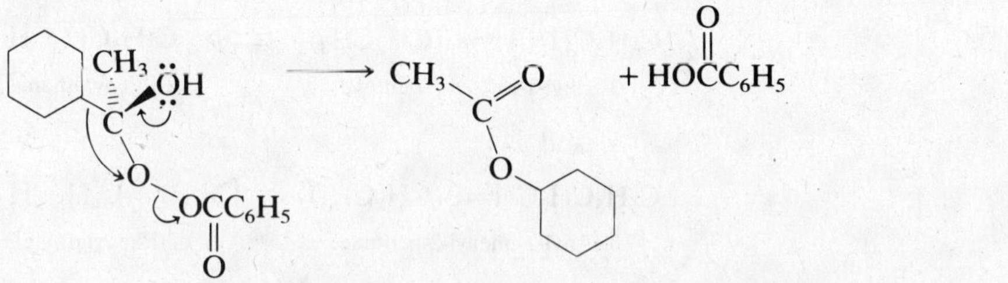

17.17   The formation of a carboxylic acid from Baeyer-Villiger oxidation of an aldehyde requires that hydrogen be the group that migrates.

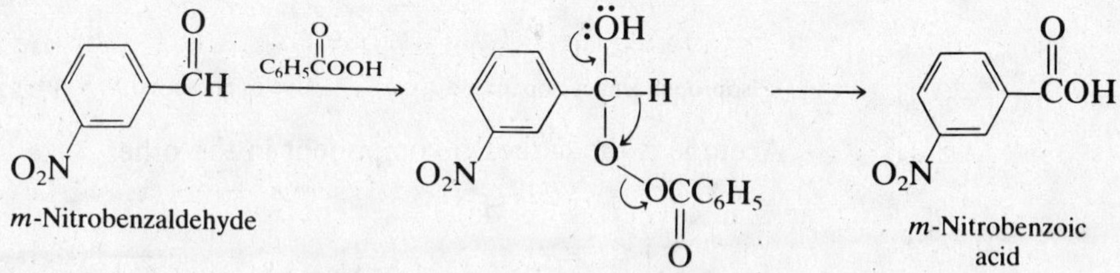

m-Nitrobenzaldehyde

m-Nitrobenzoic
acid

17.18   (*a*)   First consider all the isomeric aldehydes of molecular formula $C_5H_{10}O$.

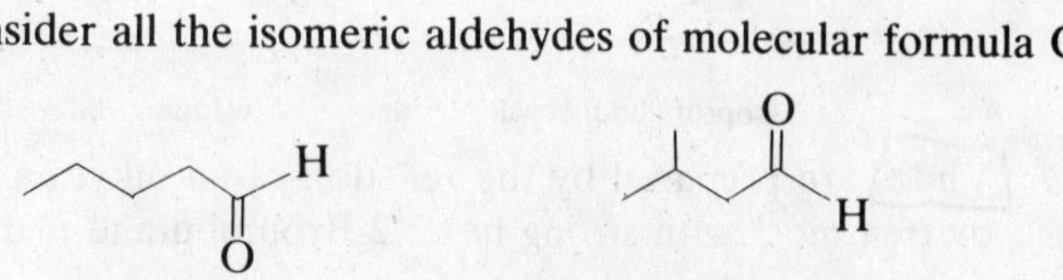

Pentanal

3-Methylbutanal

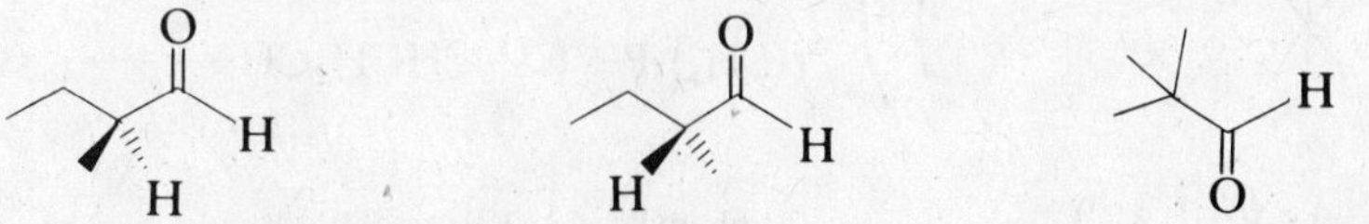

(*S*)-2-Methylbutanal

(*R*)-2-Methylbutanal

2,2-Dimethylpropanal

There are three isomeric ketones:

2-Pentanone 3-Pentanone 3-Methyl-2-butanone

(b) Reduction of an aldehyde to a primary alcohol does not introduce a stereogenic center into the molecule. Therefore the only aldehydes that yield chiral alcohols on reduction are those that already contain a stereogenic center.

(S)-2-Methylbutanal $\xrightarrow[\text{CH}_3\text{OH}]{\text{NaBH}_4}$ (S)-2-Methyl-1-butanol

(R)-2-Methylbutanal $\xrightarrow[\text{CH}_3\text{OH}]{\text{NaBH}_4}$ (R)-2-Methyl-1-butanol

Among the ketones, 2-pentanone and 3-methyl-butanone are reduced to chiral alcohols.

2-Pentanone $\xrightarrow[\text{CH}_3\text{OH}]{\text{NaBH}_4}$ 2-Pentanol (chiral but racemic)

3-Pentanone $\xrightarrow[\text{CH}_3\text{OH}]{\text{NaBH}_4}$ 3-Pentanol (achiral)

3-Methyl-2-butanone $\xrightarrow[\text{CH}_3\text{OH}]{\text{NaBH}_4}$ 3-Methyl-2-butanol (chiral but racemic)

(c) All the aldehydes yield chiral alcohols on reaction with methylmagnesium iodide. Thus

$$\text{C}_4\text{H}_9\overset{\displaystyle O}{\overset{\|}{\text{C}}}\text{H} \xrightarrow[\text{2. H}_3\text{O}^+]{\text{1. CH}_3\text{MgI}} \text{C}_4\text{H}_9\overset{\displaystyle H}{\underset{\displaystyle OH}{\text{C}}}\text{CH}_3$$

A stereogenic center is introduced in each case. None of the ketones yield chiral alcohols.

2-Pentanone $\xrightarrow[\text{2. H}_3\text{O}^+]{\text{1. CH}_3\text{MgI}}$ 2-Methyl-2-pentanol (achiral)

3-Pentanone $\xrightarrow[\text{2. H}_3\text{O}^+]{\text{1. CH}_3\text{MgI}}$ 3-Methyl-3-pentanol (achiral)

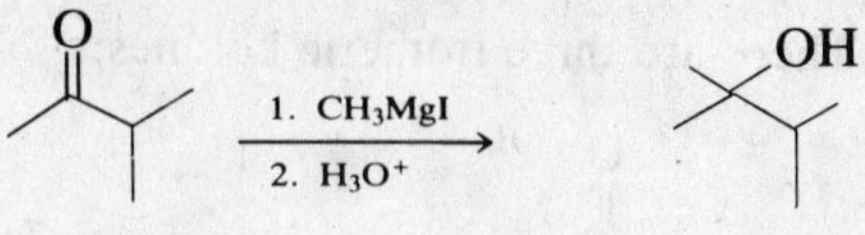

3-Methyl-2-butanone   2,3-Dimethyl-2-butanol (achiral)

17.19 (*a*) Chloral is the trichloro derivative of ethanal (acetaldehyde).

Ethanal   Trichloroethanal
                             (chloral)

(*b*) Pivaldehyde has two methyl groups attached to C-2 of propanal.

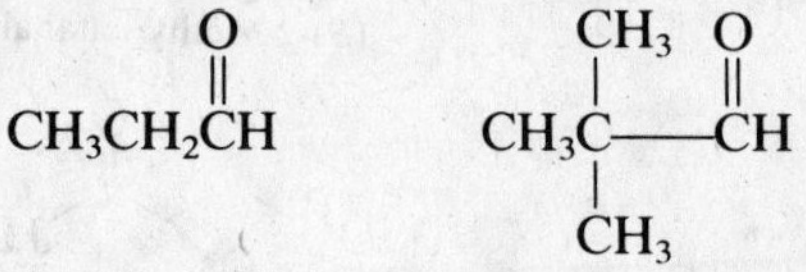

Propanal   2,2-Dimethylpropanal
                        (pivaldehyde)

(*c*) Acrolein has a double bond between C-2 and C-3 of a three-carbon aldehyde.

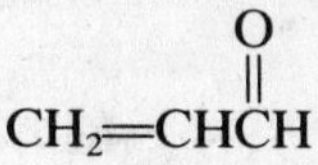

CH_2=CHCH   2-Propenal (acrolein)

(*d*) Crotonaldehyde has a trans double bond between C-2 and C-3 of a four-carbon aldehyde.

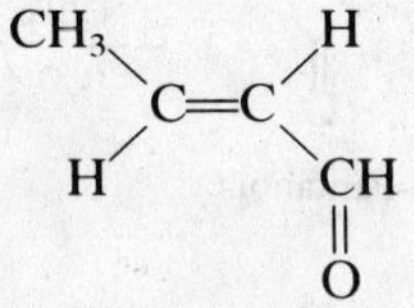

(*E*)-2-Butenal
(crotonaldehyde)

(*e*) Citral has two double bonds, one between C-2 and C-3 and the other between C-6 and C-7. The one at C-2 has the *E* configuration. There are methyl substituents at C-3 and C-7.

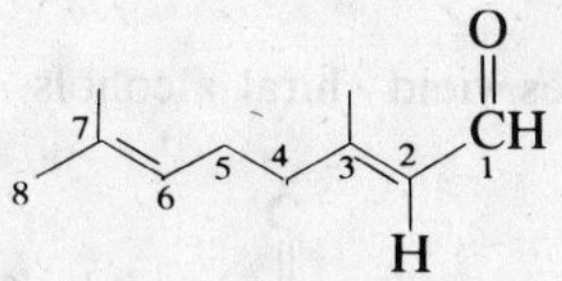

(*E*)-3,7-Dimethyl-2,6-octadienal
(citral)

(*f*) The structure of pinacolone is more easily revealed in its systematic name.

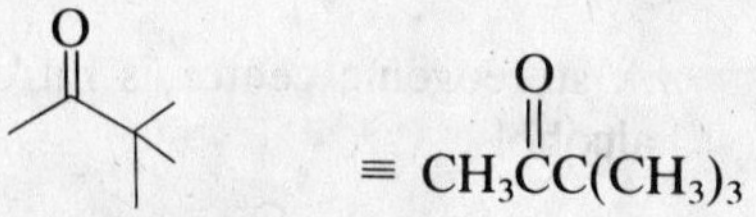

$\equiv$ $CH_3CC(CH_3)_3$

3,3-Dimethyl-2-butanone
(pinacolone)

(*g*) Deoxybenzoin has phenyl groups attached to C-1 and C-2 of a two-carbon chain.

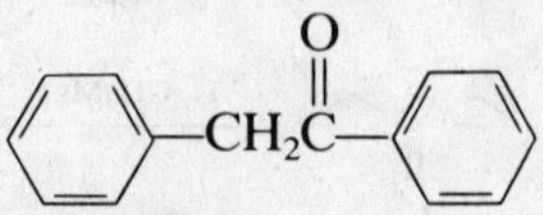

1,2-Diphenylethanone

(*h*)   Diacetone alcohol is

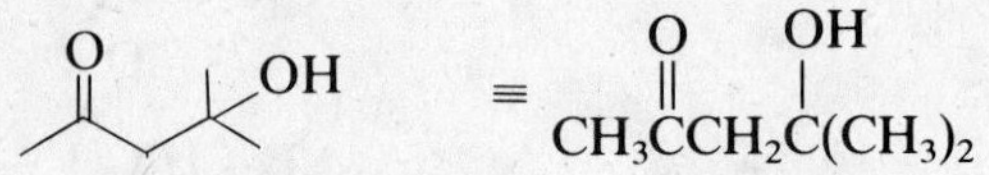

4-Hydroxy-4-methyl-2-pentanone

(*i*)   The systematic name for mesityl oxide tells us that the correct structure is as shown.

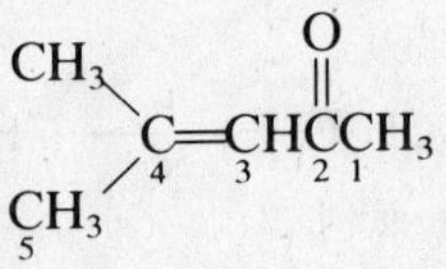

4-Methyl-3-penten-2-one

| methyl substituent at C-4 | five carbons in main chain with double bond at C-3 | C-2 is a ketone carbonyl |

(*j*)   The parent ketone is 2-cyclohexenone.

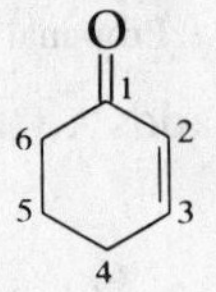

2-Cyclohexenone

Carvone has an isopropenyl group at C-5 and a methyl group at C-2.

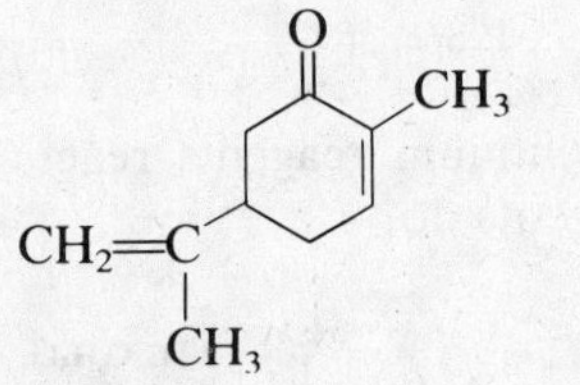

5-Isopropenyl-2-methyl-2-cyclohexenone (carvone)

(*k*)   Biacetyl is 2,3-butanedione. It has a four-carbon chain that incorporates ketone carbonyls as C-2 and C-3.

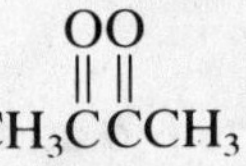

2,3-Butanedione (biacetyl)

(*l*)   Dimedone has two carbonyl groups in a 1,3-relationship on a six-membered ring. There are two methyl substituents at C-5.

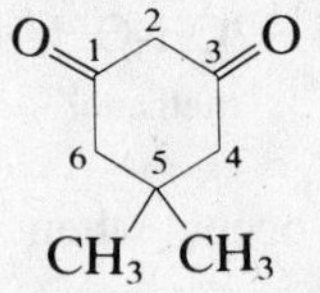

5,5-Dimethyl-1,3-cyclohexanedione (dimedone)

(*m*)   Dypnone is

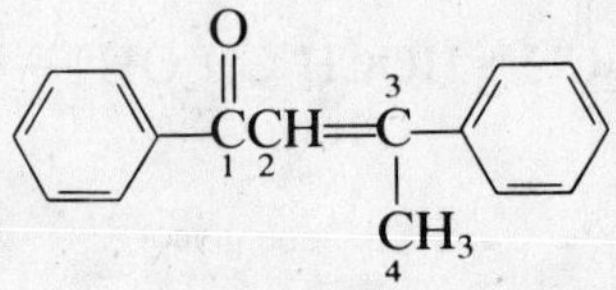

1,3-Diphenyl-2-buten-1-one (dypnone)

17.20   (*a*)   Lithium aluminum hydride reduces aldehydes to primary alcohols.

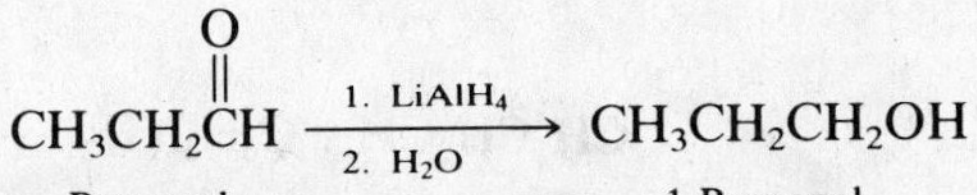

Propanal   1-Propanol

(b) Sodium borohydride reduces aldehydes to primary alcohols.

$$CH_3CH_2\overset{\overset{\displaystyle O}{\|}}{C}H \xrightarrow[CH_3OH]{NaBH_4} CH_3CH_2CH_2OH$$

Propanal   1-Propanol

(c) Aldehydes can be reduced to primary alcohols by catalytic hydrogenation.

$$CH_3CH_2\overset{\overset{\displaystyle O}{\|}}{C}H \xrightarrow[Ni]{H_2} CH_3CH_2CH_2OH$$

Propanal   1-Propanol

(d) Aldehydes react with Grignard reagents to form secondary alcohols.

$$CH_3CH_2\overset{\overset{\displaystyle O}{\|}}{C}H \xrightarrow[\substack{2.\ H_3O^+}]{\substack{1.\ CH_3MgI,\\ diethyl\ ether}} CH_3CH_2\overset{\overset{\displaystyle OH}{|}}{C}HCH_3$$

Propanal   2-Butanol

(e) Sodium acetylide adds to the carbonyl group of propanal to give an acetylenic alcohol.

$$CH_3CH_2\overset{\overset{\displaystyle O}{\|}}{C}H \xrightarrow[\substack{2.\ H_3O^+}]{\substack{1.\ HC{\equiv}CNa,\\ liquid\ ammonia}} CH_3CH_2\overset{\overset{\displaystyle OH}{|}}{C}HC{\equiv}CH$$

Propanal   1-Pentyn-3-ol

(f) Alkyl- or aryllithium reagents react with aldehydes in much the same way that Grignard reagents do.

$$CH_3CH_2\overset{\overset{\displaystyle O}{\|}}{C}H \xrightarrow[\substack{2.\ H_3O^+}]{\substack{1.\ C_6H_5Li,\\ diethyl\ ether}} CH_3CH_2\underset{\underset{\displaystyle OH}{|}}{C}HC_6H_5$$

Propanal   1-Phenyl-1-propanol

(g) Aldehydes are converted to acetals on reaction with alcohols in the presence of an acid catalyst.

$$CH_3CH_2\overset{\overset{\displaystyle O}{\|}}{C}H + 2CH_3OH \xrightarrow{HCl} CH_3CH_2CH(OCH_3)_2$$

Propanal   Methanol   Propanal dimethyl acetal

(h) Cyclic acetal formation occurs when aldehydes react with ethylene glycol.

$$CH_3CH_2\overset{\overset{\displaystyle O}{\|}}{C}H + HOCH_2CH_2OH \xrightarrow[benzene]{\substack{p\text{-toluenesulfonic}\\ acid}}$$

Propanal   Ethylene glycol   2-Ethyl-1,3-dioxolane

(i) Aldehydes react with primary amines to yield imines.

$$CH_3CH_2\overset{\overset{\displaystyle O}{\|}}{C}H + C_6H_5NH_2 \xrightarrow{-H_2O} CH_3CH_2CH{=}NC_6H_5$$

Propanal   Aniline   N-Propylideneaniline

(*j*)   Secondary amines combine with aldehydes to yield enamines.

$$CH_3CH_2\overset{\overset{\displaystyle O}{\|}}{C}H + (CH_3)_2NH \xrightarrow[\text{benzene}]{\overset{p\text{-toluenesulfonic}}{\text{acid}}} CH_3CH{=}\overset{\overset{\displaystyle N(CH_3)_2}{|}}{C}H$$

Propanal   Dimethylamine   1-(Dimethylamino)propene

(*k*)   Oximes are formed on reaction of hydroxylamine with aldehydes.

$$CH_3CH_2\overset{\overset{\displaystyle O}{\|}}{C}H \xrightarrow{H_2NOH} CH_3CH_2CH{=}NOH$$

Propanal   Propanal oxime

(*l*)   Hydrazine reacts with aldehydes to form hydrazones.

$$CH_3CH_2\overset{\overset{\displaystyle O}{\|}}{C}H \xrightarrow{H_2NNH_2} CH_3CH_2CH{=}NNH_2$$

Propanal   Propanal hydrazone

(*m*)   Hydrazone formation is the first step in the Wolf-Kishner reduction (Sec. 12.8).

$$CH_3CH_2CH{=}NNH_2 \xrightarrow[\text{triethylene glycol, heat}]{NaOH} CH_3CH_2CH_3 + N_2$$

Propanal hydrazone   Propane

(*n*)   The reaction of an aldehyde with *p*-nitrophenylhydrazine is analogous to that with hydrazine.

$$CH_3CH_2\overset{\overset{\displaystyle O}{\|}}{C}H + O_2N{-}\langle\bigcirc\rangle{-}NHNH_2 \longrightarrow CH_3CH_2CH{=}NNH{-}\langle\bigcirc\rangle{-}NO_2 + H_2O$$

Propanal   *p*-Nitrophenylhydrazine   Propanal
                                                            p-nitrophenylhydrazone

(*o*)   Semicarbazide converts aldehydes to the corresponding semicarbazone.

$$CH_3CH_2\overset{\overset{\displaystyle O}{\|}}{C}H + H_2NNH\overset{\overset{\displaystyle O}{\|}}{C}NH_2 \longrightarrow CH_3CH_2CH{=}NNH\overset{\overset{\displaystyle O}{\|}}{C}NH_2 + H_2O$$

Propanal   Semicarbazide   Propanal semicarbazone

(*p*)   Phosphorus ylides convert aldehydes to alkenes by a Wittig reaction.

$$CH_3CH_2\overset{\overset{\displaystyle O}{\|}}{C}H + (C_6H_5)_3\overset{+}{P}{-}\overset{..}{C}HCH_3 \longrightarrow CH_3CH_2CH{=}CHCH_3 + (C_6H_5)_3\overset{+}{P}{-}\overset{-}{O}$$

Propanal   Ethylidenetriphenylphosphorane   2-Pentene   Triphenylphosphine
                                                                                    oxide

(*q*)   Acidification of solutions of sodium cyanide generates HCN, which reacts with aldehydes to form cyanohydrins

$$CH_3CH_2\overset{\overset{\displaystyle O}{\|}}{C}H + HCN \longrightarrow CH_3CH_2\overset{\overset{\displaystyle OH}{|}}{C}HCN$$

Propanal   Hydrogen cyanide   Propanal cyanohydrin

(*r*)   Silver oxide oxidizes aldehydes to carboxylic acids with formation of metallic silver.

$$CH_3CH_2\overset{\overset{\displaystyle O}{\|}}{C}H \xrightarrow{Ag_2O} CH_3CH_2CO_2H$$

Propanal   Propanoic acid

(s)   Chromic acid oxidizes aldehydes to carboxylic acids.

$$CH_3CH_2CH \xrightarrow{\ H_2CrO_4\ } CH_3CH_2CO_2H$$

Propanal   Propanoic acid

17.21 (a) Lithium aluminum hydride reduces ketones to secondary alcohols.

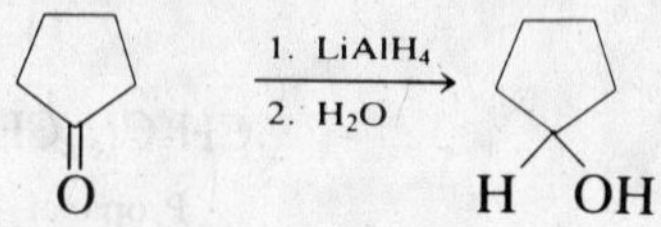

Cyclopentanone → Cyclopentanol

1. LiAlH$_4$
2. H$_2$O

(b) Sodium borohydride converts ketones to secondary alcohols.

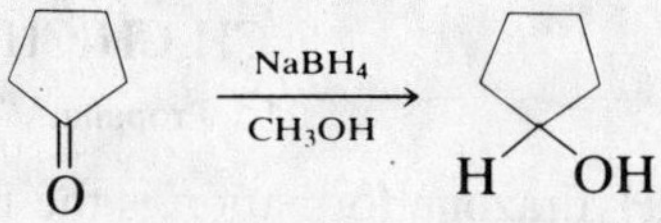

Cyclopentanone → Cyclopentanol

NaBH$_4$ / CH$_3$OH

(c) Catalytic hydrogenation of ketones yields secondary alcohols.

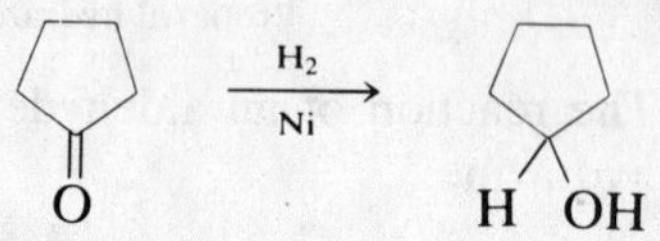

Cyclopentanone → Cyclopentanol

H$_2$ / Ni

(d) Grignard reagents react with ketones to form tertiary alcohols.

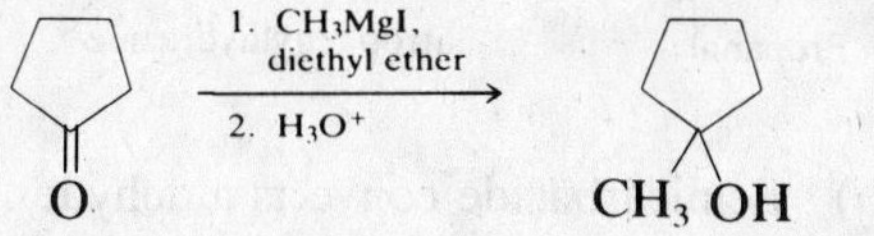

Cyclopentanone → 1-Methylcyclopentanol

1. CH$_3$MgI, diethyl ether
2. H$_3$O$^+$

(e) Addition of sodium acetylide to cyclopentanone yields a tertiary acetylenic alcohol.

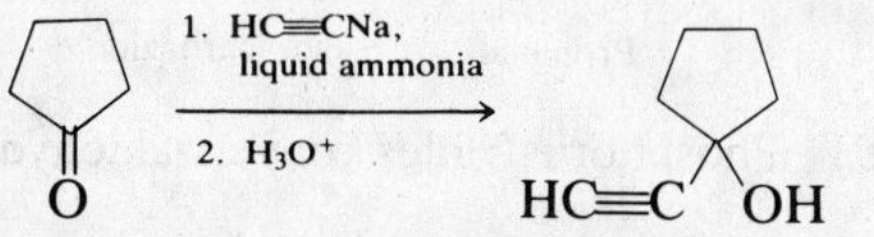

Cyclopentanone → 1-Ethynylcyclopentanol

1. HC≡CNa, liquid ammonia
2. H$_3$O$^+$

(f) Phenyllithium adds to the carbonyl group of cyclopentanone to yield 1-phenylcyclopentanol.

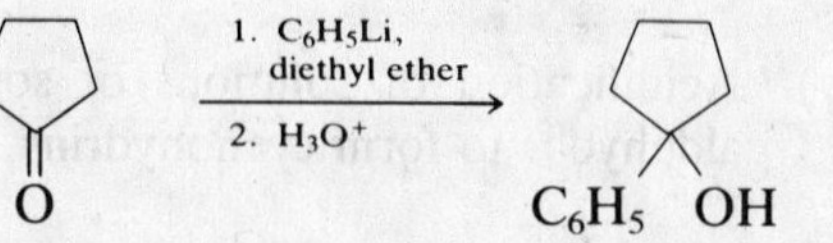

Cyclopentanone → 1-Phenylcyclopentanol

1. C$_6$H$_5$Li, diethyl ether
2. H$_3$O$^+$

(g) The equilibrium constant for acetal formation from ketones is generally unfavorable.

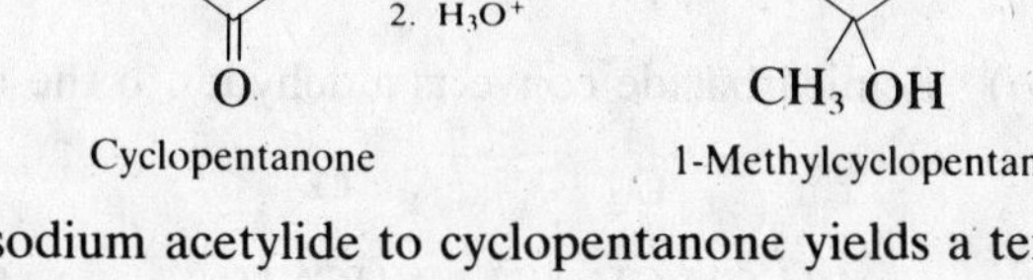

$$\text{Cyclopentanone} + 2CH_3OH \xrightarrow{\ HCl\ } \text{Cyclopentanone dimethyl acetal} \qquad K < 1$$

Cyclopentanone   Methanol

(*h*)   Cyclic acetal formation is favored even for ketones.

Cyclopentanone   Ethylene glycol   1,4-Dioxaspiro[4.4]nonane

(*i*)   Ketones react with primary amines to form imines.

Cyclopentanone   Aniline   *N*-Cyclopentylideneaniline

(*j*)   Dimethylamine reacts with cyclopentanone to yield an enamine.

Cyclopentanone   Dimethylamine   1-(Dimethylamino)cyclopentene

(*k*)   An oxime is formed when cyclopentanone is treated with hydroxylamine.

Cyclopentanone   Cyclopentanone oxime

(*l*)   Hydrazine reacts with cyclopentanone to form a hydrazone.

Cyclopentanone   Cyclopentanone hydrazone

(*m*)   Heating a hydrazone in base with a high-boiling alcohol as solvent converts it to an alkane.

Cyclopentanone hydrazone   Cyclopentane

(*n*)   A *p*-nitrophenylhydrazone is formed.

Cyclopentanone   *p*-Nitrophenylhydrazine   Cyclopentanone
                                                                      p-nitrophenylhydrazone

(*o*)   Cyclopentanone is converted to a semicarbazone on reaction with semicarbazide.

Cyclopentanone   Semicarbazide   Cyclopentanone
                                                    semicarbazone

(*p*) A Wittig reaction takes place, forming ethylidenecyclopentane.

Cyclopentanone   Ethylidenetriphenylphosphorane   Ethylidenecyclopentane   Triphenylphosphine oxide

(*q*) Cyanohydrin formation takes place.

Cyclopentanone   Cyclopentanone cyanohydrin

(*r*) No reaction occurs between silver oxide and cyclopentanone.

(*s*) Cyclopentanone is not oxidized readily with chromic acid. On heating with this reagent, oxidative degradation to a variety of products takes place.

17.22 (*a*) The first step in analyzing this problem is to write the structure of the starting ketone in stereochemical detail.

(*S*)-3-Phenyl-2-butanone   (2*R*,3*S*)-3-Phenyl-2-butanol   (2*S*,3*S*)-3-Phenyl-2-butanol

Reduction of the ketone introduces a new stereogenic center, which may have either the *R* or the *S* configuration; the configuration of the original stereogenic center is unaffected. In practice it is observed that the 2*R*,3*S* diastereomer is formed in greater amounts than the 2*S*,3*S* (ratio 2.5:1 for LiAlH$_4$ reduction).

(*b*) Reduction of the ketone can yield either *cis*- or *trans*-4-*tert*-butylcyclohexanol.

4-*tert*-Butylcyclohexanone

trans-4-*tert*-Butylcyclohexanol   *cis*-4-*tert*-Butylcyclohexanol

It has been observed that the major product obtained on reduction with either lithium aluminum hydride or sodium borohydride is the trans alcohol (trans/cis ~9:1). Although the equatorial alcohol is more stable, the main reason for this stereoselectivity is thought to involve torsional strain in the transition state leading to the axial product.

(*c*) The two diastereomeric alcohols are *cis*- and *trans*-1,2-cyclobutanediol.

2-Hydroxycyclobutanone   *trans*-1,2-Cyclobutanediol   *cis*-1,2-Cyclobutanediol

Both products are formed in approximately equal amounts on reduction of 2-hydroxycyclobutanone with lithium aluminum hydride.

(*d*)   The two reduction products are the exo and endo alcohols.

Bicyclo[2.2.1]heptan-2-one   *exo*-Bicyclo[2.2.1]heptan-2-ol   *endo*-Bicyclo[2.2.1]heptan-2-ol

The major product is observed to be the endo alcohol (endo/exo 9:1) for reduction with $NaBH_4$ or $LiAlH_4$. The stereoselectivity observed in this reaction is due to decreased steric hindrance to attack of the hydride reagent from the exo face of the molecule, giving rise to the endo alcohol.

(*e*)   The hydroxyl group may be on the same side as the double bond or on the opposite side.

Bicyclo[2.2.1]hept-2-en-7-one   *syn*-Bicyclo[2.2.1]hept-2-en-7-ol   *anti*-Bicyclo[2.2.1]hept-2-en-7-ol

The anti alcohol is observed to be formed in greater amounts (85:15) on reduction of the ketone with $LiAlH_4$. Steric factors governing attack of the hydride reagent again explain the major product observed.

17.23   (*a*)   Aldehydes undergo nucleophilic addition faster than ketones; benzaldehyde is reduced by sodium borohydride more rapidly than is acetophenone. The measured relative rates are:

$$k_{rel} = C_6H_5\overset{O}{\overset{\|}{C}}H / C_6H_5\overset{O}{\overset{\|}{C}}CH_3 = 440$$

(*b*)   The carbonyl group of benzaldehyde is stabilized through conjugation with the aromatic ring.

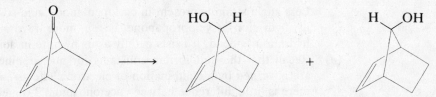

This reduces the electrophilicity of the aldehyde group and makes it less reactive than ethanol toward nucleophilic addition of semicarbazide.

$$k_{rel} = CH_3\overset{O}{\overset{\|}{C}}H / C_6H_5\overset{O}{\overset{\|}{C}}H = 180$$

(*c*)   The presence of an electronegative substituent on the α carbon atom causes a dramatic increase in K_{hydr}. Trichloroethanal (chloral) is almost completely converted to its geminal diol (chloral hydrate) in aqueous solution.

$$Cl_3C\overset{O}{\overset{\|}{C}}H \quad + H_2O \longrightarrow \quad Cl_3C\overset{OH}{\underset{OH}{\overset{|}{\underset{|}{C}}}}H$$

Trichloroethanal   1-Hydroxy-2,2,2-trichloroethanol
(chloral)   (choral hydrate)

Electron-withdrawing groups—and a Cl_3C substituent is strongly electron-withdrawing—destabilize carbonyl groups to which they are attached and make the energy change favoring the products of nucleophilic addition more favorable.

$$K_{rel} = \overset{\overset{O}{\|}}{Cl_3CCH} / \overset{\overset{O}{\|}}{CH_3CH} \cong 20{,}000$$

(d) The hybridization at carbon changes from sp^2 to sp^3 on addition of water to a carbonyl group. Since the preferred bond angles for sp^2 and sp^3 hybridized carbon are 120° and 109.5°, respectively, there is more angle strain in small-ring ketones than in their hydrates. The carbonyl group of cyclopropanone is more strained than the corresponding hydrate. Some angle strain is relieved in the process:

Cyclopropanone   Water   1,1-Cyclopropanediol

Less angle strain is present in cyclopentanone, and so there is little driving force for its hydration. Cyclopropanone has a more favorable equilibrium constant for hydration; indeed, it exists mainly as its hydrate in aqueous solution.

(e) Recall that the equilibrium constants for nucleophilic addition to carbonyl groups are governed by a combination of electronic effects and steric effects. Electronically there is little difference between acetone and 3,3-dimethyl-2-butanone, but sterically there is a significant difference. The cyanohydrin products are more crowded than the starting ketones, and so the bulkier the alkyl groups that are attached to the carbonyl, the more strained and less stable will be the cyanohydrin.

Ketone   Hydrogen cyanide   Cyanohydrin
                                      [less strained for $R = CH_3$
                                      than for $R = C(CH_3)_3$]

$$K_{rel} = \overset{\overset{O}{\|}}{CH_3CCH_3} / \overset{\overset{O}{\|}}{CH_3CC(CH_3)_3} = 40$$

(f) Steric effects influence the rate of nucleophilic addition to these two ketones. Carbon is on its way from tricoordinate to tetracoordinate at the transition state, and alkyl groups are forced closer together than they are in the ketone.

Transition state

The transition state is of lower energy when R is smaller. Acetone (for which R is methyl) is reduced faster than 3,3-dimethyl-2-butanone (where R is *tert*-butyl).

$$k_{rel} = \overset{\overset{O}{\|}}{CH_3CCH_3} / \overset{\overset{O}{\|}}{CH_3CC(CH_3)_3} = 12$$

(g) In this problem we examine the rate of hydrolysis of acetals to the corresponding ketone or aldehyde. The rate-determining step is carbocation formation.

Hybridization at carbon changes from sp^3 to sp^2; crowding at this carbon is relieved as the carbocation is formed. The more crowded acetal ($R = CH_3$) forms a carbocation faster than the less crowded one ($R = H$). Another factor of even greater importance is the extent of stabilization of the carbocation intermediate; the more stable carbocation ($R = CH_3$) is formed faster than the less stable one ($R = H$).

$$k_{rel} = \frac{(CH_3)_2C(OCH_2CH_3)_2}{CH_2(OCH_2CH_3)_2} = 1.8 \times 10^7$$

17.24 (a) The reaction as written is the reverse of cyanohydrin formation, and the principles that govern equilibria in nucleophilic addition to carbonyl groups apply in reverse order to the dissociation of cyanohydrins to aldehydes and ketones. Cyanohydrins of ketones dissociate more at equilibrium than do cyanohydrins of aldehydes. More strain due to crowding is relieved when a ketone cyanohydrin dissociates and a more stabilized carbonyl group is formed. The equilibrium constant K_{diss} is larger for

Acetone cyanohydrin   Acetone   Hydrogen cyanide

than it is for

Propanal cyanohydrin   Propanal   Hydrogen cyanide

(b) Cyanohydrins of ketones have a more favorable equilibrium constant for dissociation than do cyanohydrins of aldehydes. Crowding is relieved to a greater extent when a ketone cyanohydrin dissociates and a more stable carbonyl group is formed. The measured dissociation constants are:

Benzaldehyde cyanohydrin   Benzaldehyde

Acetophenone cyanohydrin   Acetophenone

17.25 (a) Silver oxide is a mild oxidant, which converts aldehydes to carboxylic acids.

Thiophene-3-carbaldehyde   Thiophene-3-carboxylic acid
(95%)

(b) The reaction of an aldehyde with 1,3-propanediol in the presence of *p*-toluenesulfonic acid forms a cyclic acetal.

$$CH_3O\text{-(ring)-}CH_3O\text{-...-CHO} + HOCH_2CH_2CH_2OH \xrightarrow[\text{benzene, heat}]{\substack{p\text{-toluenesulfonic}\\ \text{acid}}} \text{product}$$

2-Bromo-3,4,5-trimethoxy-   1,3-Propanediol   2-(2'-Bromo-3',4',5'-tri-
benzaldehyde   methoxyphenyl)-1,3-dioxane
                                                                                                              (81%)

(c) The reagent CH_3ONH_2 is called *O*-methylhydroxylamine, and reacts with aldehydes in a manner similar to hydroxylamine.

$$\text{(aryl)-CHO} + CH_3ONH_2 \longrightarrow \text{(aryl)-}CH=NOCH_3$$

4-Hydroxy-3-methoxy-   *O*-Methylhydroxylamine   4-Hydroxy-3-methoxy-
benzaldehyde   benzaldehyde *O*-methyloxime

(d) Propanal reacts with 1,1-dimethylhydrazine to yield the corresponding hydrazone.

$$CH_3CH_2\overset{\displaystyle O}{\overset{\displaystyle \|}{C}}H + (CH_3)_2NNH_2 \longrightarrow CH_3CH_2CH=NN(CH_3)_2$$

Propanal   1,1-Dimethylhydrazine   Propanal dimethylhydrazone

(e) Acid-catalyzed hydrolysis of the acetal gives the aldehyde in 87 percent yield.

$$CH_3\text{-(aryl)-}CHCH_2CH_2\text{-(dioxolane)} \xrightarrow[\text{heat}]{H_2O,\ HCl} CH_3\text{-(aryl)-}CHCH_2CH_2CHO$$
with CH_3 substituent

4-(*p*-Methylphenyl)pentanal

(f) Hydrogen cyanide adds to carbonyl groups to form cyanohydrins.

$$C_6H_5\overset{\displaystyle O}{\overset{\displaystyle \|}{C}}CH_3 \xrightarrow[HCl]{NaCN} C_6H_5\underset{\displaystyle CH_3}{\overset{\displaystyle OH}{C}}CN$$

Acetophenone   Acetophenone cyanohydrin

(g) The reagent is a secondary amine known as *morpholine*. Secondary amines react with ketones to give enamines.

$$C_6H_5\overset{\displaystyle O}{\overset{\displaystyle \|}{C}}CH_3 + HN\text{(morpholine)}O \xrightarrow{H^+} C_6H_5\underset{\displaystyle N\text{(morpholine)}}{\overset{\displaystyle OH}{C}}CH_3 \xrightarrow{H^+(-H_2O)} C_6H_5\underset{\displaystyle N\text{(morpholine)}}{C}=CH_2$$

Acetophenone   Morpholine   Carbinolamine   1-Morpholinostyrene
                                                       intermediate   (57–64%)

(h) Migration of the alkyl group in a Baeyer-Villiger oxidation occurs with retention of configuration.

$$CH_3CH_2-\overset{\overset{\displaystyle CH_3}{|}}{\underset{\underset{\displaystyle C_6H_5}{|}}{C}}-\overset{\overset{\displaystyle O}{\|}}{C}CH_3 + C_6H_5\overset{\overset{\displaystyle O}{\|}}{C}OOH \longrightarrow CH_3CH_2-\overset{\overset{\displaystyle CH_3}{|}}{\underset{\underset{\displaystyle C_6H_5}{|}}{C}}-O\overset{\overset{\displaystyle O}{\|}}{C}CH_3$$

(*R*)-3-Methyl-3-phenyl-2-   Peroxybenzoic acid   (*R*)-1-Methyl-1-phenylpropyl
pentanone   acetate

17.26   Wolff-Kishner reduction converts a carbonyl group (C=O) to a methylene group (CH$_2$).

Bicyclo[4.3.0]non-3-en-8-one   N$_2$H$_4$, KOH / HOCH$_2$CH$_2$OH / 130°C →   Bicyclo[4.3.0]non-3-ene
(compound A, 90%)

Treatment of the alkene with *m*-chloroperoxybenzoic acid produces an epoxide, compound B.

Bicyclo[4.3.0]non-3-ene   3,4-Epoxybicyclo[4.3.0]nonane
(compound B, 92%)

Epoxides undergo reduction with lithium aluminum hydride to form alcohols (Sec. 16.12).

3,4-Epoxybicyclo[4.3.0]nonane   1. LiAlH$_4$ / 2. H$_2$O →   Bicyclo[4.3.0]nonan-3-ol
(compound C, 90%)

Chromic acid oxidizes the alcohol to a ketone.

Bicyclo[4.3.0]nonan-3-ol   H$_2$CrO$_4$ →   Bicyclo[4.3.0]nonan-3-one
(compound D, 75%)

17.27   Hydration of formaldehyde by H$_2$^{17}O produces a *gem*-diol in which the labeled and unlabeled hydroxyl groups are equivalent. When this *gem*-diol reverts to formaldehyde, loss of either of the hydroxyl groups is equally likely except for a small isotope effect favoring loss of the common isotope, which leads to eventual replacement of the mass 16 isotope of oxygen by ^{17}O.

$$^{16}O=C\overset{\overset{\displaystyle H}{\diagup}}{\underset{\underset{\displaystyle H}{\diagdown}}{}} + H_2{}^{17}O \rightleftharpoons \overset{\overset{\displaystyle H^{17}O}{\diagdown}}{\underset{\underset{\displaystyle H^{16}O}{\diagup}}{}}CH_2 \rightleftharpoons {}^{17}O=C\overset{\overset{\displaystyle H}{\diagup}}{\underset{\underset{\displaystyle H}{\diagdown}}{}} + H_2{}^{16}O$$

This reaction has been monitored by ^{17}O nmr spectroscopy; ^{17}O gives an nmr signal but ^{16}O does not.

17.28   First write out the chemical equation for the reaction that takes place. Vicinal diols (1,2-diols) react with aldehydes to give cyclic acetals.

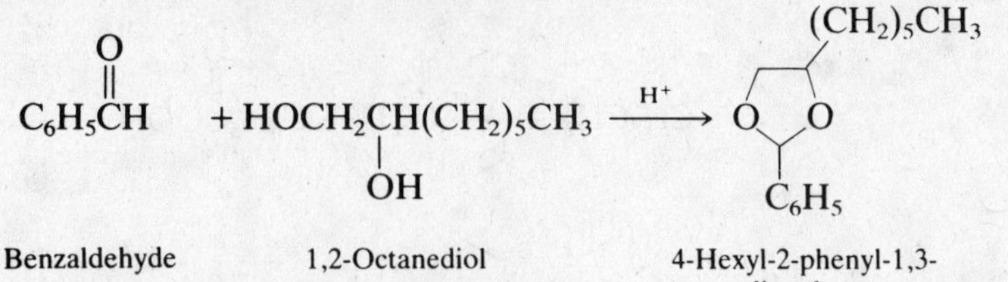

$$C_6H_5CH \xrightarrow{\text{O}} \; + \; HOCH_2CH(CH_2)_5CH_3 \xrightarrow{H^+}$$

Benzaldehyde   1,2-Octanediol   4-Hexyl-2-phenyl-1,3-dioxolane

Notice that the phenyl and hexyl substituents may be either cis or trans to each other. The two products are the cis and trans stereoisomers.

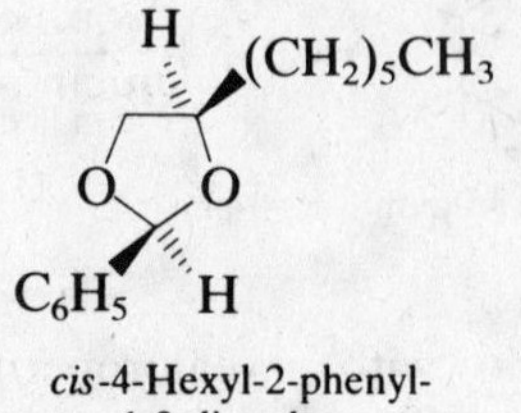

cis-4-Hexyl-2-phenyl-1,3-dioxolane

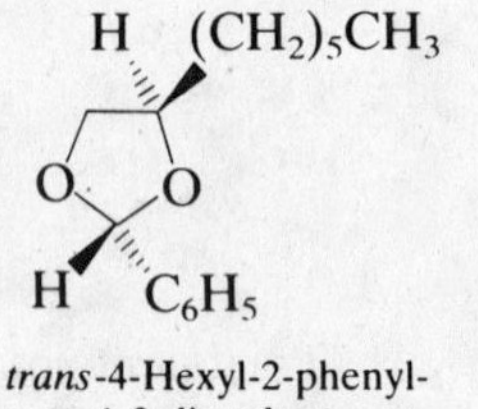

trans-4-Hexyl-2-phenyl-1,3-dioxolane

17.29 Cyclic hemiacetals are formed by intramolecular nucleophilic addition of a hydroxyl group to a carbonyl.

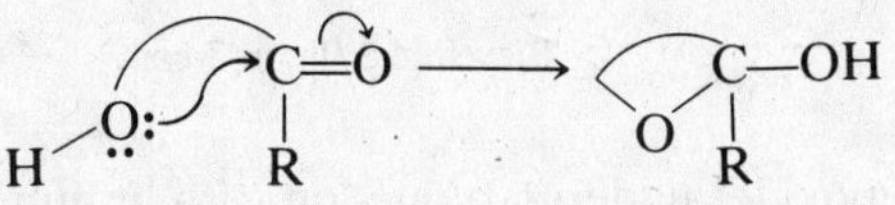

Cyclic hemiacetal

The ring oxygen is derived from the hydroxyl group; the carbonyl oxygen becomes the hydroxyl oxygen of the hemiacetal.

This compound is the cyclic hemiacetal of 5-hydroxypentanal.

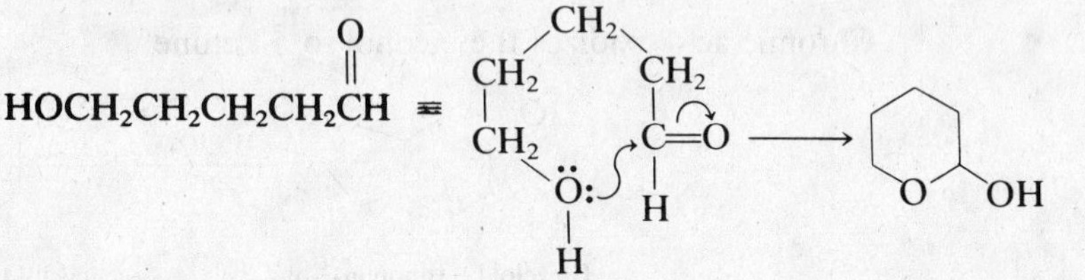

$$HOCH_2CH_2CH_2CH_2CH \equiv$$

Indeed, 5-hydroxypentanal seems to exist entirely as the cyclic hemiacetal. Its infrared spectrum is devoid of absorption in the carbonyl region.

(*b*) The carbon connected to two oxygens is the one that is derived from the carbonyl group. Using retrosynthetic symbolism, disconnect the ring oxygen from this carbon.

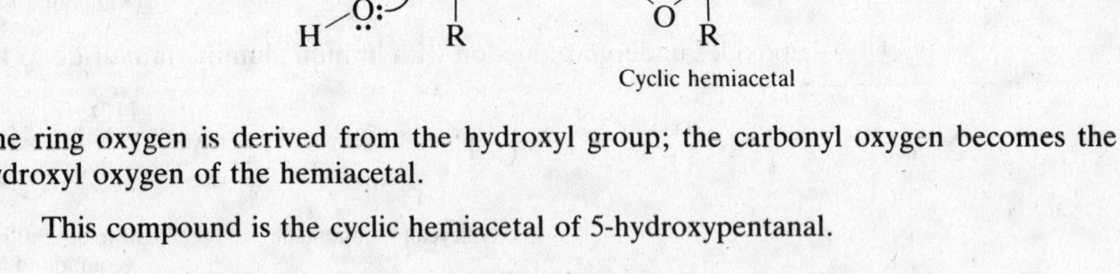

$$\equiv \; HCCH_2CH_2CHCH{=}CHCH{=}CH_2$$

4-Hydroxy-5,7-octadienal

The next two compounds are cyclic acetals. The original carbonyl group is identifiable as the one that bears two oxygen substituents, which originate as hydroxyl oxygens of a diol.

(c)

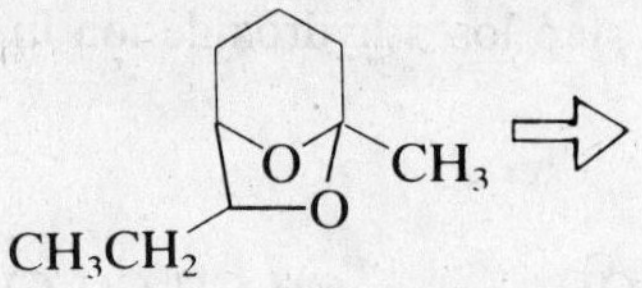

Brevicomin

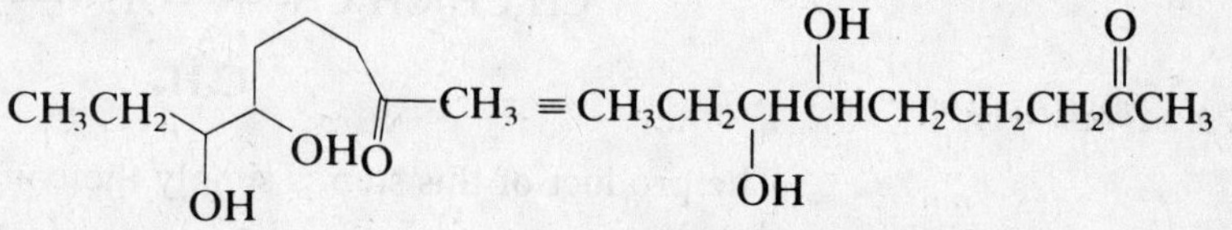

$\equiv CH_3CH_2CHCHCH_2CH_2CH_2CCH_3$

6.7-Dihydroxy-2-nonanone

(d)

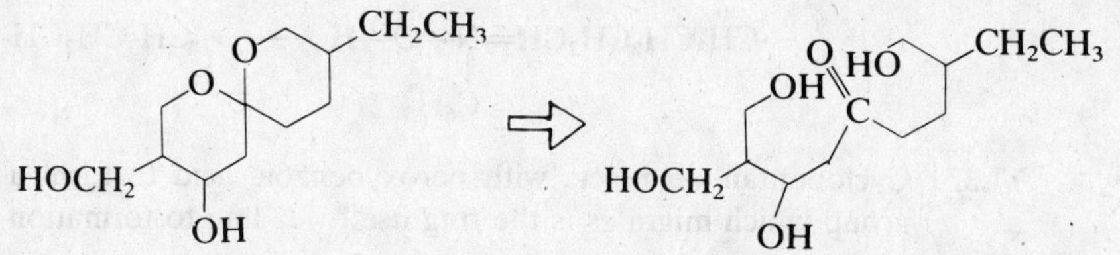

Talaromycin A   2,8-Di(hydroxymethyl)-1,3-dihydroxy-5-decanone

17.30 (a) The Z stereoisomer of $CH_3CH{=}NCH_3$ has its higher-priority substituents on the same side of the double bond,

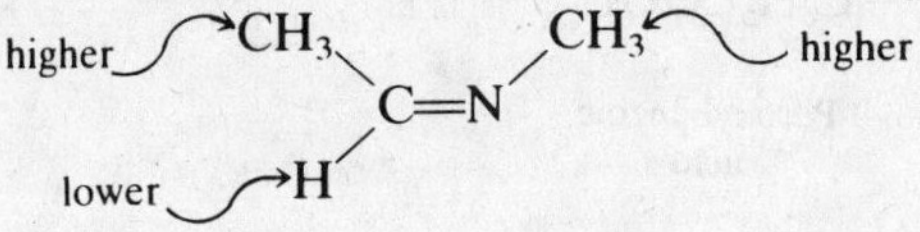

(Z)-N-Ethylidenemethylamine

The lone pair of nitrogen need not be shown. It is considered a "phantom ligand," lower in priority ranking than any other.

(b) Higher-priority groups are on opposite sides of the carbon-nitrogen double bond in the E oxime of acetaldehyde

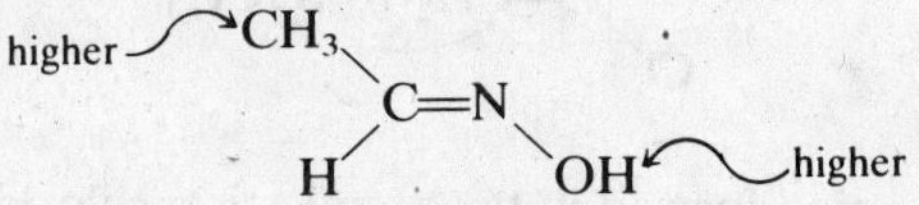

(E)-Acetaldehyde oxime

(c) (Z)-2-Butanone hydrazone is:

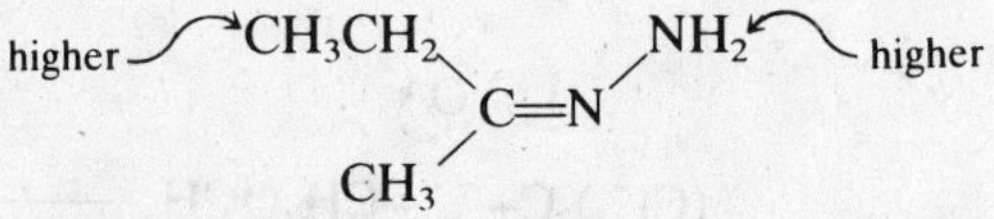

(d) (E)-Acetophenone semicarbazone is:

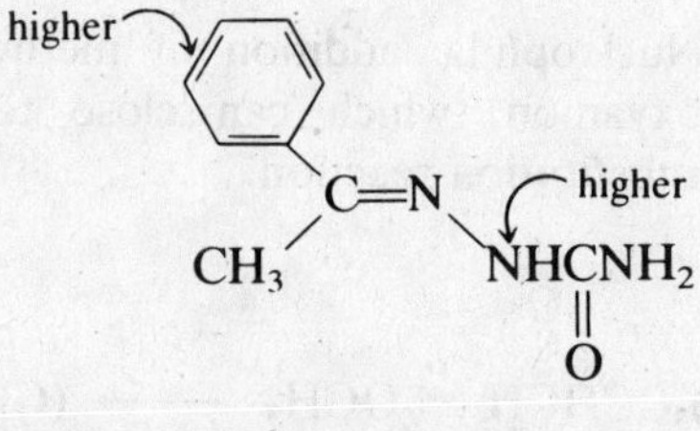

17.31 By analogy to other processes in which derivatives of ammonia react with aldehydes, the first step is nucleophilic addition to the carbonyl group.

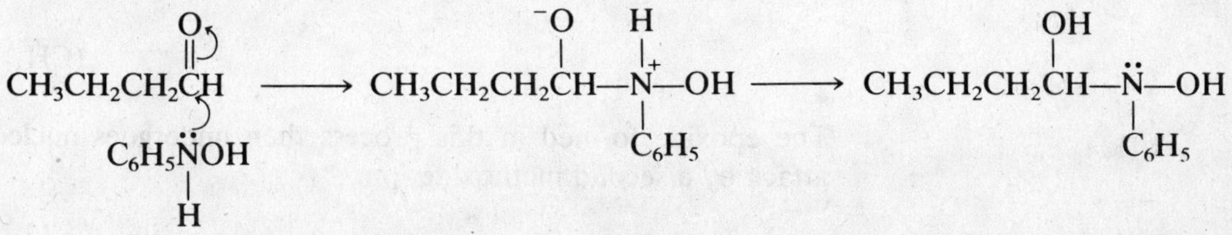

The intermediate formed in this step loses hydroxide ion in a step that is assisted by the nitrogen lone pair.

$$CH_3CH_2CH_2CH\overset{OH}{\underset{C_6H_5}{-}}\ddot{N}-OH \longrightarrow CH_3CH_2CH_2CH=\overset{+}{\underset{C_6H_5}{N}}-OH + {}^-OH$$

The product of this step is simply the conjugate acid of the nitrone.

$$CH_3CH_2CH_2CH=\overset{+}{\underset{C_6H_5}{N}}-O-H \longrightarrow CH_3CH_2CH_2CH=\overset{+}{\underset{C_6H_5}{N}}\overset{O^-}{\diagup} + H^+$$

17.32 Cyclopentanone reacts with peroxybenzoic acid to form a peroxymonoester. The alkyl group which migrates is the ring itself, leading to formation of a six-membered lactone.

$$\text{Cyclopentanone} \quad \begin{array}{c} \text{Peroxybenzoic} \\ \text{acid} \end{array} + C_6H_5COOH \longrightarrow \longrightarrow \begin{array}{c} \text{5-Pentanolide} \end{array} + C_6H_5COH \quad \begin{array}{c} \text{Benzoic} \\ \text{acid} \end{array}$$

17.33 (a) Nucleophilic ring opening of the epoxide occurs by attack of methoxide at the less hindered carbon.

$$(CH_3)_3C\underset{Cl}{\overset{O}{\diagup}}C-CH_2 + {}^-:\ddot{O}CH_3 \longrightarrow (CH_3)_3C-\underset{Cl}{\overset{O^-}{\underset{|}{C}}}-CH_2OCH_3$$

The anion formed in this step loses a chloride ion to form the carbon-oxygen double bond of the product.

$$(CH_3)_3C-\underset{Cl}{\overset{O^-}{\underset{|}{C}}}-CH_2OCH_3 \longrightarrow (CH_3)_3C-\overset{O}{\overset{\|}{C}}-CH_2OCH_3 + Cl^-$$

(b) Nucleophilic addition of methoxide ion to the aldehyde carbonyl generates an oxyanion, which can close to an epoxide by an intramolecular nucleophilic substitution reaction.

$$(CH_3)_3CCH\underset{Cl}{\overset{O}{\overset{\|}{C}}}H + {}^-OCH_3 \longrightarrow (CH_3)_3CCH-\underset{Cl}{\overset{{}^-:\ddot{O}:}{CH}}OCH_3 \longrightarrow$$

$$(CH_3)_3CCH\overset{O}{\diagup}CHOCH_3 + Cl^-$$

The epoxide formed in this process then undergoes nucleophilic ring opening on attack by a second methoxide ion.

$$(CH_3)_3CCH-CHOCH_3 + {}^-:\ddot{O}CH_3 \longrightarrow (CH_3)_3CCH-CHOCH_3 \longrightarrow$$

$$\overset{OH}{\underset{OCH_3}{(CH_3)_3CCHCHOCH_3}}$$

17.34 Amygdalin is a derivative of the cyanohydrin formed from benzaldehyde; thus the structure (without stereochemistry) is

$$\overset{O-R}{\underset{CN}{C_6H_5CH}}$$

The order of decreasing sequence rule precedence is $RO > CN > C_6H_5 > H$. The groups are arranged in a clockwise orientation in order of decreasing precedence in the R enantiomer.

$$C_6H_5-\overset{H}{\underset{CN}{C}}{\cdots}O-R \qquad \begin{array}{l}(R)\text{-Benzaldehyde cyanohydrin}\\ \text{derivative (amygdalin)}\end{array}$$

17.35 (a) The target molecule is the diethyl acetal of acetaldehyde (ethanal).

$$\underset{\substack{\text{Acetaldehyde}\\ \text{diethyl acetal}}}{CH_3CH(OCH_2CH_3)_2} \Longrightarrow \overset{O}{\overset{\|}{CH_3CH}}, CH_3CH_2OH$$

Acetaldehyde may be prepared by oxidation of ethanol.

$$\underset{\text{Ethanol}}{CH_3CH_2OH} \xrightarrow[\text{CH}_2\text{Cl}_2]{(C_5H_5N)_2CrO_3} \underset{\text{Acetaldehyde}}{\overset{O}{\overset{\|}{CH_3CH}}}$$

Reaction with ethanol in the presence of hydrogen chloride yields the desired acetal.

$$\underset{\text{Acetaldehyde}}{\overset{O}{\overset{\|}{CH_3CH}}} + \underset{\text{Ethanol}}{2CH_3CH_2OH} \xrightarrow{\text{HCl}} \underset{\substack{\text{Acetaldehyde}\\ \text{diethyl acetal}}}{CH_3CH(OCH_2CH_3)_2}$$

(b) In this case the target molecule is a cyclic acetal of acetaldehyde.

$$\underset{\text{2-Methyl-1,3-dioxolane}}{\overset{O\quad O}{\underset{H\quad CH_3}{\diagdown\diagup}}} \Longrightarrow \overset{O}{\overset{\|}{CH_3CH}}, HOCH_2CH_2OH$$

Acetaldehyde has been prepared in part (a). Recalling that vicinal diols are available from the hydroxylation of alkenes, 1,2-ethanediol may be prepared by the sequence

$$CH_3CH_2OH \xrightarrow[\text{heat}]{H_2SO_4} CH_2{=}CH_2 \xrightarrow[\text{(CH}_3)_3COH, HO^-]{OsO_4, (CH_3)_3COOH} HOCH_2CH_2OH$$

Ethanol   Ethylene   1,2-Ethanediol

Hydrolysis of ethylene oxide is also reasonable.

$$CH_2{=}CH_2 \xrightarrow{CH_3CO_2OH} CH_2{-}CH_2 \xrightarrow[\text{HO}^-]{H_2O} HOCH_2CH_2OH$$

Ethylene   Ethylene oxide   1,2-Ethanediol

Reaction of acetaldehyde with 1,2-ethanediol yields the cyclic acetal.

$$\underset{\text{Acetaldehyde}}{\overset{\displaystyle O}{\underset{\|}{CH_3CH}}} + \underset{\text{1,2-Ethanediol}}{HOCH_2CH_2OH} \xrightarrow{H^+} \underset{\text{2-Methyl-1,3-dioxolane}}{\overset{\displaystyle O\quad O}{\underset{H\quad CH_3}{\diagdown\!\diagup}}}$$

(c)   The target molecule is, in this case, the cyclic acetal of 1,2-ethanediol and formaldehyde.

$$\underset{\text{1,3-Dioxolane}}{\overset{\displaystyle O\diagdown\!\diagup O}{}} \Longrightarrow \overset{\displaystyle O}{\underset{\|}{HCH}}, HOCH_2CH_2OH$$

The preparation of 1,2-ethanediol was described in part (b). One method of preparing formaldehyde is ozonolysis:

$$CH_3CH_2OH \xrightarrow[\text{heat}]{H_2SO_4} CH_2{=}CH_2 \xrightarrow[\text{2. H}_2O, Zn]{1. O_3} 2\overset{\displaystyle O}{\underset{\|}{HCH}}$$

Ethanol   Ethylene   Formaldehyde

Another method is periodate cleavage of 1,2-ethanediol.

$$HOCH_2CH_2OH \xrightarrow{HIO_4} 2\overset{\displaystyle O}{\underset{\|}{HCH}}$$

1,2-Ethanediol   Formaldehyde

Cyclic acetal formation is then carried out in the usual way.

$$\overset{\displaystyle O}{\underset{\|}{HCH}} + HOCH_2CH_2OH \xrightarrow{H^+} \overset{\displaystyle O\diagdown\!\diagup O}{}$$

Formaldehyde   1,2-Ethanediol   1,3-Dioxolane

(d)   Acetylenic alcohols are best prepared from carbonyl compounds and acetylide anions.

$$\underset{\underset{\text{3-Butyn-2-ol}}{OH}}{CH_3CHC{\equiv}CH} \Longrightarrow \overset{\displaystyle O}{\underset{\|}{CH_3CH}} + {}^-{:}C{\equiv}CH$$

Acetaldehyde is available as in part (a). Alkynes such as acetylene are available from the corresponding alkene by bromination followed by double dehydrobromination. Using ethylene, prepared in part (*b*), the sequence becomes:

$$CH_2\!=\!CH_2 \xrightarrow{Br_2} BrCH_2CH_2Br \xrightarrow[NH_3]{NaNH_2} HC\!\equiv\!CH$$

Ethylene   1,2-Dibromoethane   Acetylene

Then

$$HC\!\equiv\!CH \xrightarrow{NaNH_2} HC\!\equiv\!CNa \xrightarrow[\substack{2.\ H_3O^+}]{\substack{1.\ CH_3CH \\ \ \ \ \ \overset{O}{\|}}} HC\!\equiv\!CCHCH_3$$

$$\overset{\displaystyle |}{OH}$$

Acetylene   Sodium acetylide   3-Butyn-2-ol

(*e*) The target aldehyde may be prepared from the corresponding alcohol.

$$\overset{O}{\overset{\|}{HCCH_2C\!\equiv\!CH}} \implies HOCH_2CH_2C\!\equiv\!CH$$

3-Butynal   3-Butyn-1-ol

The best route to this alcohol is through reaction of an acetylide ion with ethylene oxide.

$$HC\!\equiv\!CNa \ + \ \underset{O}{CH_2\!-\!CH_2} \xrightarrow[\substack{2.\ H_3O^+}]{\substack{1.\ diethyl \\ ether}} HC\!\equiv\!CCH_2CH_2OH$$

Sodium acetylide   Ethylene oxide   3-Butyn-1-ol
[prepared in part (*d*)]   [prepared in part (*b*)]

Oxidation with Collins' reagent is appropriate for the final step.

$$HC\!\equiv\!CCH_2CH_2OH \xrightarrow[CH_2Cl_2]{(C_5H_5N)_2CrO_3} HC\!\equiv\!CCH_2\overset{O}{\overset{\|}{CH}}$$

3-Butyn-1-ol   3-Butynal

(*f*) The target molecule has four carbon atoms, suggesting a route involving reaction of ethyl Grignard with ethylene oxide

$$CH_3CH_2CH_2CH_2OH \implies CH_3\overset{..}{C}H_2 + \underset{O}{CH_2\!-\!CH_2}$$

Ethylmagnesium bromide is prepared in the usual way:

$$CH_3CH_2OH \xrightarrow[PBr_3]{HBr\ or} CH_3CH_2Br \xrightarrow[\substack{diethyl \\ ether}]{Mg} CH_3CH_2MgBr$$

Ethanol   Bromoethane   Ethylmagnesium
                                              bromide

Reaction of the Grignard reagent with ethylene oxide, prepared in part (*b*), completes the synthesis.

$$CH_3CH_2MgBr + \underset{O}{CH_2\!-\!CH_2} \xrightarrow[\substack{2.\ H_3O^+}]{\substack{1.\ diethyl \\ ether}} CH_3CH_2CH_2CH_2OH$$

Ethylmagnesium   Ethylene   1-Butanol
    bromide   oxide

17.36 (*a*) Friedel-Crafts acylation of benzene with benzoyl chloride is a direct route to benzophenone.

$$\text{Benzoyl chloride} + \text{Benzene} \xrightarrow{AlCl_3} \text{Benzophenone}$$

Benzoyl chloride   Benzene   Benzophenone

(b) On analyzing the overall transformation retrosynthetically, it can be seen that the target molecule may be prepared by a Grignard synthesis followed by oxidation of the alcohol formed.

$$C_6H_5\overset{O}{\overset{\|}{C}}C_6H_5 \Rightarrow C_6H_5\overset{OH}{\overset{|}{C}H}C_6H_5 \Rightarrow C_6H_5\overset{O}{\overset{\|}{C}}H + C_6H_5MgBr$$

In the desired synthesis, benzyl alcohol must first be oxidized to benzaldehyde.

$$C_6H_5CH_2OH \xrightarrow[CH_2Cl_2]{(C_5H_5N)_2CrO_3} C_6H_5\overset{O}{\overset{\|}{C}}H$$

Benzyl alcohol   Benzaldehyde

Reaction of benzaldehyde with the Grignard reagent of bromobenzene followed by oxidation of the resulting secondary alcohol gives benzophenone.

$$C_6H_5\overset{O}{\overset{\|}{C}}H + C_6H_5MgBr \xrightarrow[2.\ H_3O^+]{1.\ diethyl\ ether} C_6H_5\overset{}{\underset{OH}{C}}HC_6H_5$$

Benzaldehyde   Phenylmagnesium   Diphenylmethanol
           bromide

$$\downarrow {(C_5H_5N)_2CrO_3,\ CH_2Cl_2}$$

$$C_6H_5\overset{O}{\overset{\|}{C}}C_6H_5$$

Benzophenone

(c) Hydrolysis of bromodiphenylmethane yields the corresponding alcohol, which can be oxidized to benzophenone as in part (b).

$$C_6H_5\underset{Br}{\overset{}{C}}HC_6H_5 \xrightarrow{H_2O} C_6H_5\underset{OH}{\overset{}{C}}HC_6H_5 \xrightarrow{oxidize} C_6H_5\overset{O}{\overset{\|}{C}}C_6H_5$$

Bromodiphenylmethane   Diphenylmethanol   Benzophenone

(d) The starting material is the dimethyl acetal of benzophenone. All that is required is acid-catalyzed hydrolysis.

$$C_6H_5\underset{OCH_3}{\overset{OCH_3}{C}}C_6H_5 + 2H_2O \xrightarrow{H^+} C_6H_5\overset{O}{\overset{\|}{C}}C_6H_5 + 2CH_3OH$$

Dimethoxydiphenylmethane   Water   Benzophenone   Methanol

(e) Oxidative cleavage of the alkene yields benzophenone. Ozonolysis may be used.

$$(C_6H_5)_2C{=}C(C_6H_5)_2 \xrightarrow{O_3,\ then\ H_2O} 2(C_6H_5)_2C{=}O$$

1,1,2,2-Tetraphenylethene   Benzophenone

17.37 The two alcohols given as starting materials contain all the carbon atoms of the desired product.

$$CH_3(CH_2)_8CH=CHCH_2CH=CHCH_2CH\!\!\doteq\!\!CHCH=CH_2 \implies$$

$$CH_3(CH_2)_8CH=CHCH_2CH=CHCH_2CH_2OH \quad \text{and} \quad HOCH_2CH=CH_2$$

3,6-Hexadecadien-1-ol   Allyl alcohol

What is needed is to attach the two groups together so that the two primary alcohol carbons become doubly bonded to each other. This can be accomplished by using a Wittig reaction as the key step.

$$CH_3(CH_2)_8CH=CHCH_2CH=CHCH_2CH_2OH \xrightarrow[CH_2Cl_2]{(C_5H_5N)_2CrO_3}$$

3,6-Hexadecadien-1-ol

$$CH_3(CH_2)_8CH=CHCH_2CH=CHCH_2\overset{\displaystyle O}{\overset{\displaystyle \|}{C}H}$$

3,6-Hexadecadienal

$$CH_2=CHCH_2OH \xrightarrow{PBr_3} CH_2=CHCH_2Br \xrightarrow{(C_6H_5)_3P} (C_6H_5)_3\overset{+}{P}CH_2CH=CH_2$$
$$Br^-$$

Allyl alcohol   Allyl bromide   Allyltriphenylphosphonium bromide

$$\Big\downarrow CH_3CH_2CH_2CH_2Li, THF$$

$$(C_6H_5)_3\overset{+}{P}—\overset{..}{\overset{-}{C}}HCH=CH_2$$

Allylidenetriphenylphosphorane

$$CH_3(CH_2)_8CH=CHCH_2CH=CHCH_2\overset{\displaystyle O}{\overset{\displaystyle \|}{C}H} + (C_6H_5)_3\overset{+}{P}—\overset{..}{\overset{-}{C}}HCH=CH_2 \longrightarrow$$

3,6-Hexadecadienal   Allylidenetriphenylphosphorane

$$CH_3(CH_2)_8CH=CHCH_2CH=CHCH_2CH=CHCH=CH_2$$

1,3,6,9-Nonadecatetraene

Alternatively, allyl alcohol could be oxidized to $CH_2=CHCHO$ for subsequent reaction with the ylide derived from $CH_3(CH_2)_8CH=CHCH_2CH=CHCH_2CH_2OH$ via its bromide and triphenylphosphonium salt.

17.38 Compound E arises by way of an intramolecular Friedel-Crafts acylation:

$C_{21}H_{17}OCl$ $C_{21}H_{16}O$ (compound E)

Since compound F corresponds to a hydrogenation product and its infrared spectrum shows that it is a ketone, its most reasonable constitution is

$C_{21}H_{14}O$ $C_{21}H_{16}O$ (compound F)

Compounds E and F have the same constitution but are isomers; thus they must be stereoisomers. Compound F, arising via hydrogenation, is most likely to be the cis isomer.

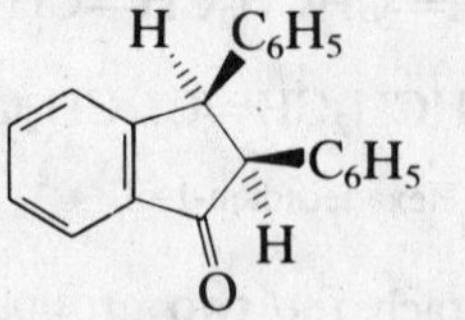

Compound F: syn hydrogenation
yields cis stereoisomer

The acylium ion leading to compound E can cyclize to give either a trans or a cis arrangement of phenyl groups. It yields the more stable trans arrangement because that is the one that has the largest separation of bulky groups in the transition state.

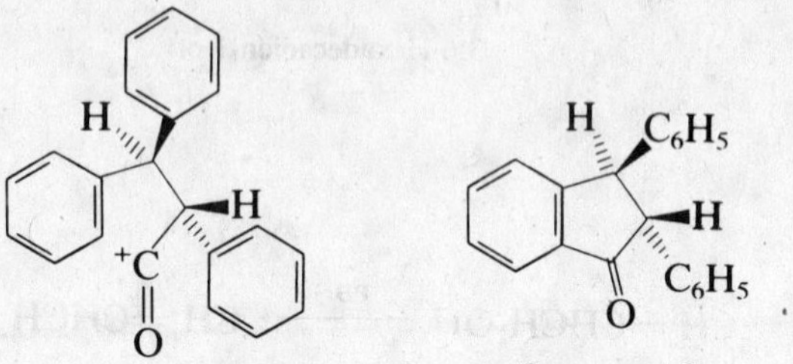

Compound E

17.39 The expected course of the reaction would be hydrolysis of the acetal to the corresponding aldehyde:

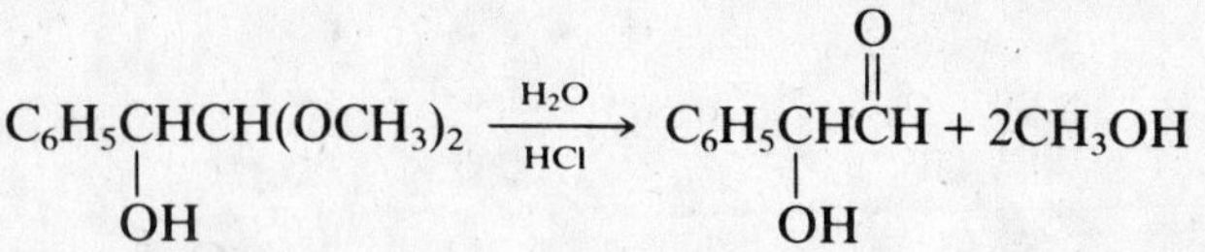

$$C_6H_5CHCH(OCH_3)_2 \xrightarrow[\text{HCl}]{\text{H}_2\text{O}} C_6H_5CHCH + 2CH_3OH$$

Compound G
(mandelaldehyde
dimethyl acetal)

Mandelaldehyde   Methanol

The molecular formula of the observed product (compound H, $C_{16}H_{16}O_4$) is exactly twice that of mandelaldehyde. This suggests that it might be a dimer of mandelaldehyde resulting from hemiacetal formation between the hydroxyl group of one mandelaldehyde molecule and the carbonyl group of another.

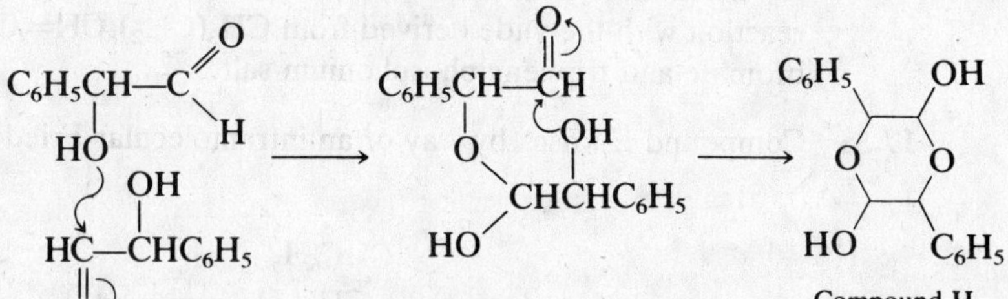

Compound H

Since compound H lacks carbonyl absorption in its infrared spectrum, the cyclic structure is indicated.

17.40 (a) Recalling that alkanes may be prepared by hydrogenation of the appropriate alkene, a synthesis of the desired product becomes apparent. What is needed is to convert —C=O into —C=CH₂; a Wittig reaction is appropriate.

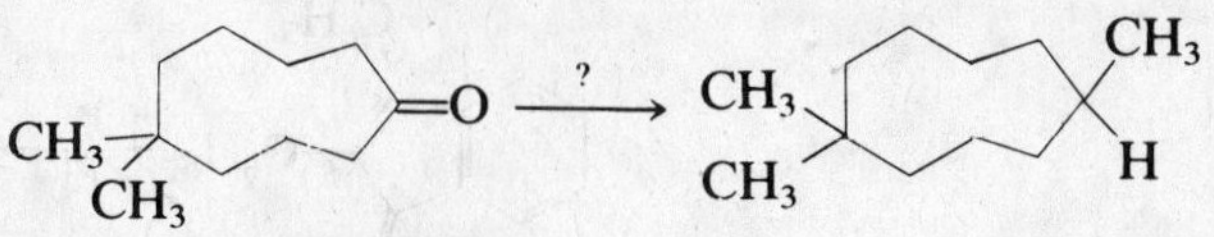

5,5-Dimethylcyclononanone   1,1,5-Trimethylcyclononane

The two-step procedure that was followed used a Wittig reaction to form the carbon-carbon bond, then catalytic hyrdrogenation of the resulting alkene.

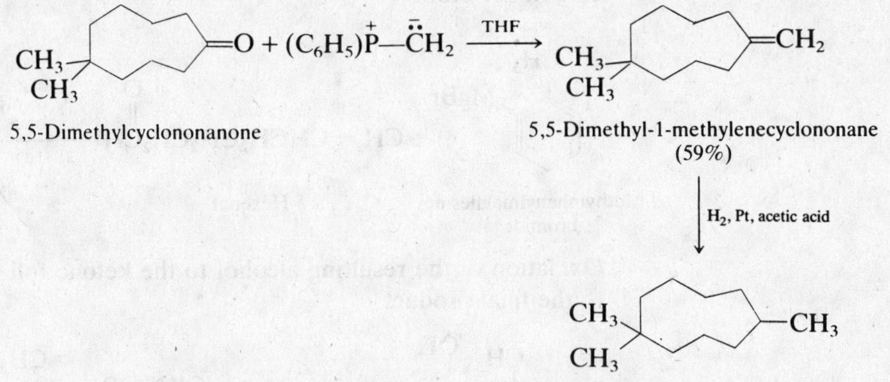

5,5-Dimethylcyclononanone

5,5-Dimethyl-1-methylenecyclononane
(59%)

1,1,5-Trimethylcyclononane (73%)

(b)   In putting together the carbon skeleton of the target molecule, a methyl group has to be added to the original carbonyl carbon.

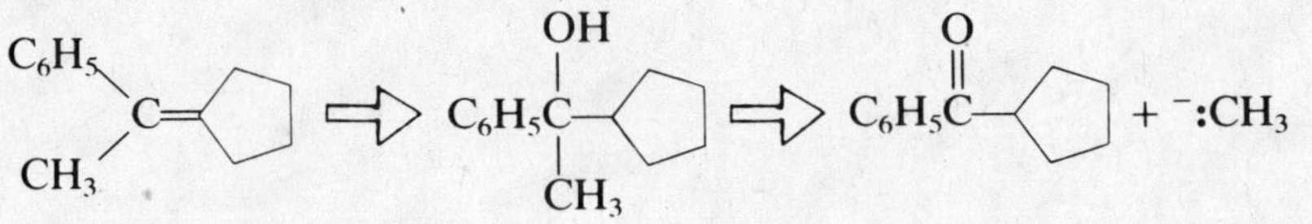

The logical way to do this is by way of a Grignard reagent.

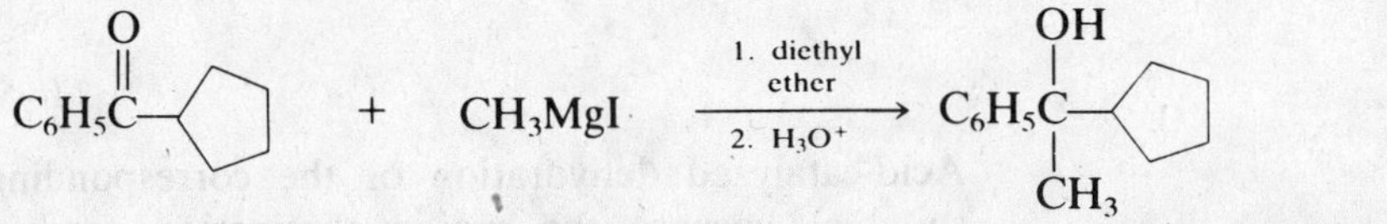

Benzoylcyclopentane

Methylmagnesium
iodide

1-Cyclopentyl-1-phenylethanol

Acid-catalyzed dehydration yields the more highly substituted alkene, the desired product, in accordance with the Zaitsev rule.

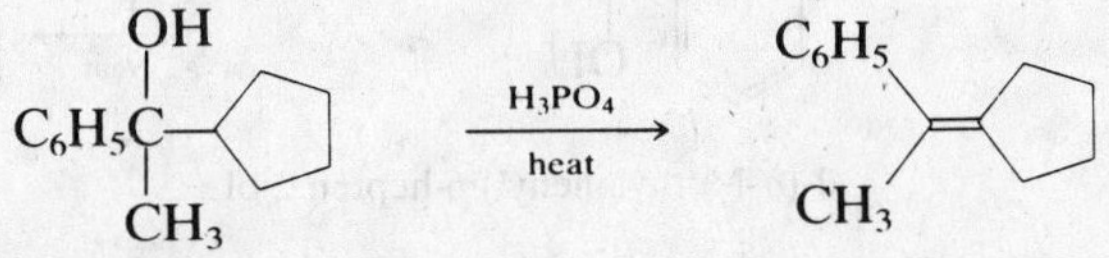

1-Cyclopentyl-1-phenylethanol

(1-Phenylethylidene)cyclopentane

(c)   Analyzing the transformation retrosynthetically, keeping in mind the starting materials stated in the problem, we see that the carbon skeleton may be constructed in a straightforward manner.

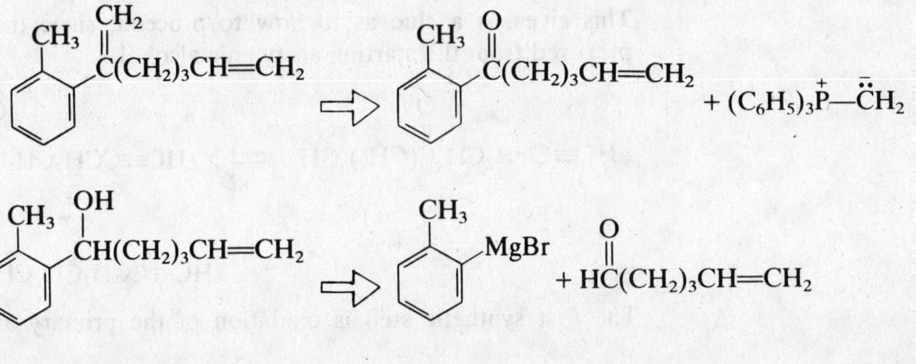

Proceeding with the synthesis in the forward direction, reaction between the Grignard reagent of *o*-bromotoluene and 5-hexenal produces most of the desired carbon skeleton.

$$CH_3\text{-}C_6H_4\text{-}MgBr + CH_2{=}CHCH_2CH_2CH_2CH{=}O \xrightarrow[\text{2. } H_3O^+]{\text{1. diethyl ether}} CH_3\text{-}C_6H_4\text{-}CH(OH)(CH_2)_3CH{=}CH_2$$

o-Methylphenylmagnesium bromide   5-Hexenal   1-(*o*-Methylphenyl)-5-hexen-1-ol

Oxidation of the resulting alcohol to the ketone followed by a Wittig reaction leads to the final product.

$$CH_3\text{-}C_6H_4\text{-}CH(OH)(CH_2)_3CH{=}CH_2 \xrightarrow[CH_2Cl_2]{(C_5H_5N)_2CrO_3} CH_3\text{-}C_6H_4\text{-}C(O)(CH_2)_3CH{=}CH_2$$

1-(*o*-Methylphenyl)-5-hexen-1-ol   1-(*o*-Methylphenyl)-5-hexen-1-one

$$\xrightarrow{(C_6H_5)_3\overset{+}{P}{-}\overset{..}{C}H_2}$$

$$CH_3\text{-}C_6H_4\text{-}C({=}CH_2)(CH_2)_3CH{=}CH_2$$

2-(*o*-Methylphenyl)-1,6-heptadiene

Acid-catalyzed dehydration of the corresponding tertiary alcohol would *not* be suitable, because the major elimination product would have the more highly substituted double bond.

$$CH_3\text{-}C_6H_4\text{-}C(OH)(CH_3)(CH_2)_3CH{=}CH_2 \xrightarrow[\text{heat}]{H^+} CH_3\text{-}C_6H_4\text{-}C(CH_3){=}CHCH_2CH_2CH{=}CH_2$$

2-(*o*-Methylphenyl)-6-hepten-2-ol   6-(*o*-Methylphenyl)-1,5-heptadiene

(*d*) Remember that terminal acetylenes can serve as sources of methyl ketones by hydration.

$$CH_3\overset{O}{\overset{\|}{C}}CH_2CH_2\overset{O}{\overset{\|}{C}}(CH_2)_5CH_3 \Longrightarrow HC{\equiv}CCH_2CH_2\overset{O}{\overset{\|}{C}}(CH_2)_5CH_3$$

This gives us a clue as to how to proceed, since the acetylenic ketone may be prepared from the starting acetylenic alcohol.

$$HC{\equiv}CCH_2CH_2\overset{O}{\overset{\|}{C}}(CH_2)_5CH_3 \Longrightarrow HC{\equiv}CCH_2CH_2\overset{O}{\overset{\|}{C}}H + CH_3(CH_2)_4\overset{..}{C}H_2$$

$$\Downarrow$$

$$HC{\equiv}CCH_2CH_2CH_2OH$$

The first synthetic step is oxidation of the primary alcohol to the aldehyde and construction of the carbon skeleton by a Grignard reaction.

$$HC\equiv CCH_2CH_2CH_2OH \xrightarrow[CH_2Cl_2]{(C_5H_5N)_2CrO_3} HC\equiv CCH_2CH_2\overset{\overset{\displaystyle O}{\|}}{C}H$$

4-Pentyn-1-ol   4-Pentynal

1. $CH_3(CH_2)_5MgBr$
2. H_3O^+

$$HC\equiv CCH_2CH_2\underset{\underset{\displaystyle OH}{|}}{C}H(CH_2)_5CH_3$$

1-Undecyn-5-ol

Oxidation of the secondary alcohol to a ketone and hydration of the terminal triple bond complete the synthesis.

$$HC\equiv CCH_2CH_2\underset{\underset{\displaystyle OH}{|}}{C}H(CH_2)_5CH_3 \xrightarrow[CH_2Cl_2]{(C_5H_5N)_2CrO_3} HC\equiv CCH_2CH_2\overset{\overset{\displaystyle O}{\|}}{C}(CH_2)_5CH_3$$

1-Undecyn-5-ol   1-Undecyn-5-one

H_2O, H_2SO_4, $HgSO_4$

$$CH_3\overset{\overset{\displaystyle O}{\|}}{C}CH_2CH_2\overset{\overset{\displaystyle O}{\|}}{C}(CH_2)_5CH_3$$

2,5-Undecanedione

(e) The desired product is a benzylic ether. In order to prepare it, the aldehyde must first be reduced to the corresponding primary alcohol. Sodium borohydride was used in the preparation described in the literature, but lithium aluminum hydride or catalytic hydrogenation would also be possible. Once the alcohol is prepared, it can be converted to its alkoxide ion and this alkoxide ion treated with methyl iodide.

Alternatively, the alcohol could be treated with hydrogen bromide or with phosphorus tribromide to give the benzylic bromide and the bromide then allowed to react with sodium methoxide.

17.41 Step 1 of the synthesis is formation of a cyclic acetal protecting group; the necessary reagents are ethylene glycol ($HOCH_2CH_2OH$) and p-toluenesulfonic acid, with heating in benzene. In step 2 the ester function is reduced to a primary alcohol. Lithium aluminum hydride ($LiAlH_4$) is the reagent of choice. Oxidation with Collins' reagent, $(C_5H_5N)_2CrO_3$ in CH_2Cl_2, converts the primary alcohol to an aldehyde in step 3. Wolff-Kishner reduction (N_2H_4, KOH, ethylene glycol, heat) converts the aldehyde group to a methyl group in step 4. The synthesis is completed in step 5 by hydrolysis (H_3O^+) of the acetal protecting group.

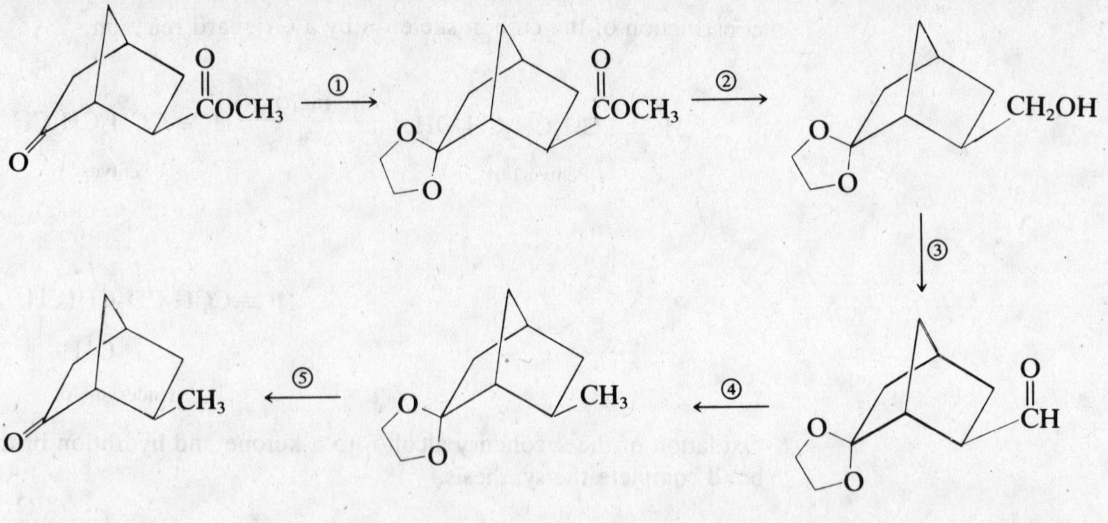

17.42 We need to assess the extent of resonance donation to the carbonyl group by the π electrons of the aromatic rings. Such resonance for the case of benzaldehyde may be written as:

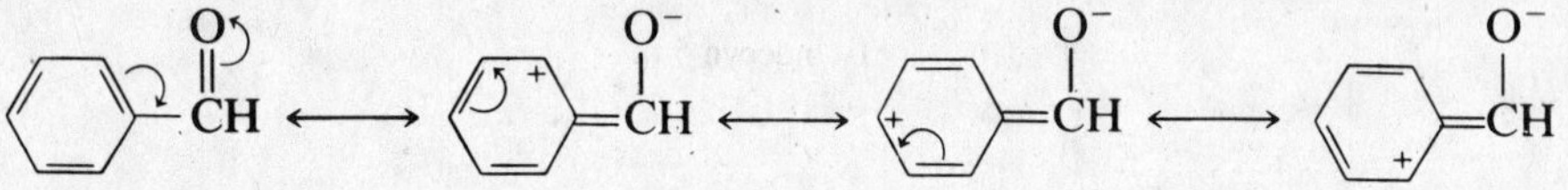

Electron-releasing groups such as methoxy at positions ortho and para to the aldehyde function increase the "single bond character" of the aldehyde by stabilizing the dipolar resonance forms and increasing their contribution to the overall electron distribution in the molecule. Electron-withdrawing groups such as nitro decrease this single bond character. The aldehyde with the lowest carbonyl stretching frequency is 2,4,6-trimethoxybenzaldehyde; the one with the highest is 2,4,6-trinitrobenzaldehyde. The measured values are:

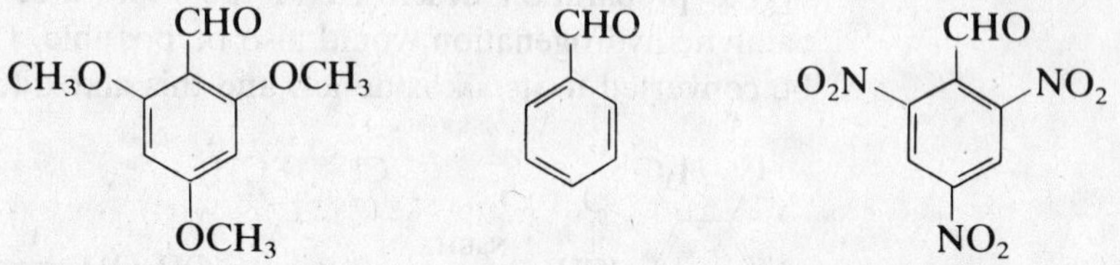

2,4,6-Trimethoxybenzaldehyde
(1665 cm^{-1})

Benzaldehyde
(1700 cm^{-1})

2,4,6-Trinitrobenzaldehyde
(1715 cm^{-1})

17.43 A signal in the ^{1}H nmr spectrum at $\delta = 9.8$ ppm tells us that compound I is an aldehyde rather than a ketone. This aldehyde signal appears as a triplet, and so C-2 must bear two hydrogens. The only C_4H_8O aldehyde that satisfies this requirement is butanal. We can assign all the signals as follows:

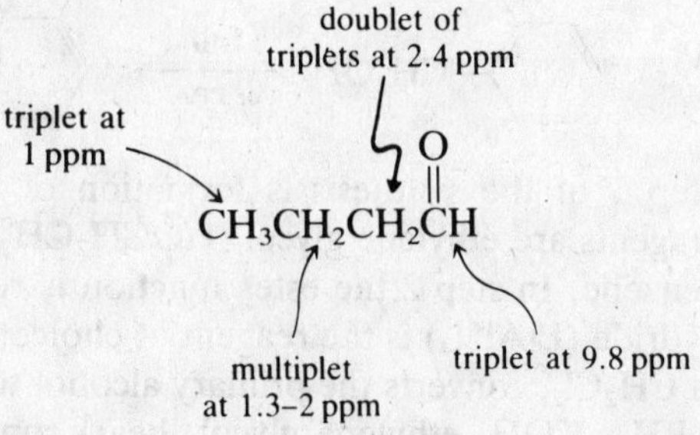

17.44 (*a*) A carbonyl group is evident in the strong infrared absorption at 1710 cm^{-1}. Since all the ^{1}H nmr signals are singlets, there are no nonequivalent hydrogens in a vicinal or "three-bond" relationship. The three-proton signal at $\delta = 2.1$ ppm and the two-proton signal at $\delta = 2.3$ ppm can be understood as arising from a $CH_2\overset{\displaystyle O}{\overset{\|}{C}}CH_3$ unit. The intense nine-proton singlet at $\delta = 1.0$ ppm is due to the three equivalent methyl groups of a $(CH_3)_3C$ unit. Compound J is 4,4-dimethyl-2-pentanone.

$$\overset{\displaystyle O}{\overset{\displaystyle \|}{CH_3C}}CH_2C(CH_3)_3 \qquad \text{Compound J}$$

2.1 ppm   2.3 ppm   1.0 ppm
singlet   singlet   singlet

(b) Compound K has the molecular formula $C_6H_{10}O_2$, which tells us that it has a total of two double bonds and rings ($C_6H_{14} - C_6H_{10} = 4H$). A strong carbonyl band in the infrared at 1710 cm^{-1} tells us that at least one of these elements of unsaturation is a ketone function.

The ^{1}H nmr spectrum has only two peaks, both singlets, at $\delta = 2.2$ and 2.7 ppm. Their intensity ratio (6:4) is consistent with two equivalent methyl groups and two equivalent methylene groups. The chemical shifts are appropriate for $CH_3\overset{\displaystyle O}{\overset{\|}{C}}$ and $CH_2\overset{\displaystyle O}{\overset{\|}{C}}$. The simplicity of the spectrum can be understood if we are dealing with a symmetrical diketone.

The correct structure is:

$$\overset{\displaystyle O}{\overset{\displaystyle \|}{CH_3C}}\underbrace{CH_2CH_2}\overset{\displaystyle O}{\overset{\displaystyle \|}{C}}CH_3 \qquad \text{2,5-Hexanedione (compound K)}$$

Equivalent
methylene groups
do not split
each other.

(c) Compound L has a carbonyl group, as revealed by the strong infrared peak at 1710 cm^{-1}. The lack of an aldehyde signal in its ^{1}H nmr spectrum indicates that compound L must be a ketone. The ketone carbonyl contributes the only element of unsaturation (SODAR = 1) to compound L, based on its molecular formula of $C_6H_{12}O_3$. The infrared peak at 1120 cm^{-1} is in the region expected for C—O stretching. The absence of infrared absorption in the 3300–3600 cm^{-1} region rules out the presence of an alcohol, and so it is reasonable to assume that the two remaining oxygens are part of ether linkages.

A sharp singlet at $\delta = 2.2$ ppm in the ^{1}H nmr spectrum is consistent with a methyl ketone. Two of compound L's six carbons are accounted for by $CH_3\overset{\displaystyle O}{\overset{\|}{C}}$. A six-proton singlet at $\delta = 3.4$ ppm is reasonably assigned to two methoxyl groups. A doublet at $\delta = 2.6$ ppm and a triplet at $\delta = 4.7$ ppm are the only spin-coupled signals in the spectrum. They suggest a $-CH_2-\overset{\displaystyle |}{C}-H$ unit and lead to the conclusion that compound L is the following acetal:

$$\overset{\displaystyle O}{\overset{\displaystyle \|}{CH_3C}}CH_2\overset{\displaystyle OCH_3}{\underset{\displaystyle OCH_3}{\overset{\displaystyle |}{\underset{\displaystyle |}{C}}}}\!\!-\!H$$

Low-field ($\delta = 4.7$ ppm) triplet;
methine carbon bears two electron-
withdrawing oxygen substituents.

17.45 (*a*) The formula for compound M ($C_9H_{10}O$) corresponds to a SODAR of 5. A monosubstituted aromatic ring, accounting for four elements of unsaturation, is indicated by a five-proton signal in the ^1H nmr spectrum at $\delta = 7.2$ ppm. The remaining unsaturation is a carbonyl group, since we are told in the problem that compound M is a ketone. The nmr spectrum reveals a three-proton singlet at $\delta = 2.1$ ppm, typical of a methyl ketone. The remaining signal is a two-proton singlet at $\delta = 3.7$ ppm for a methylene group. The most reasonable structure for this compound is

1-Phenyl-2-propanone

The shift to lower field of the methylene signal is a result of deshielding by both the aromatic ring and the carbonyl group.

(*b*) The molecular formula of compound N (C_9H_9ClO) corresponds to a SODAR of 5. One of the elements of unsaturation is the ketone carbonyl. The other four are accounted for by the aromatic ring, as suggested by the ^1H nmr absorption in the region $\delta = 7.3$ to 8 ppm. That the aromatic ring is para-disubstituted is suggested by the four-proton pair of doublets in the aromatic region of the nmr spectrum, and the presence of the peak in the infrared spectrum at 843 cm^{-1}. The characteristic triplet-quartet pattern of an ethyl group is evident in the nmr spectrum, with the methylene group appearing at $\delta = 3$ ppm. The position of this signal indicates that the methylene group is attached to the electron-withdrawing ketone carbonyl. When put together, this information suggests 1-(*p*-chlorophenyl)-1-propanone as the correct structure.

1-(*p*-Chlorophenyl)-1-propanone
(compound N)

Notice that isomeric structures such as

are ruled out on the basis of nmr splitting patterns and chemical shifts. In each of these incorrectly identified isomers the methyl and methylene signals would appear as singlets.

(*c*) A monosubstituted aromatic ring is indicated in compound O by the five-proton ^1H nmr pattern between $\delta = 7.2$ and 8.2 ppm. The molecular formula of compound O (C_9H_9BrO) corresponds to a SODAR of 5. The aromatic ring accounts for four elements of unsaturation; the remaining unsaturation is the ketone carbonyl.

A doublet-quartet pattern in the nmr spectrum indicates a CH_3—CH group. The chemical shift of the quartet ($\delta = 5.2$ ppm) is at relatively low field and is accommodated by a methine group deshielded by both a bromine and a carbonyl group. The compound is:

2-Bromo-1-phenyl-1-propanone
(compound O)

17.46 With a molecular formula of $C_7H_{14}O$, compound P has a SODAR of 1. Since we are told that it is a ketone, it has no rings or double bonds other than the one belonging to its C=O group. The peak at 211 ppm in the ^{13}C nmr spectrum corresponds to the carbonyl carbon. There are only three signals in the spectrum, and so there are only three types of carbons other than the carbonyl carbon. This suggests that the symmetrical ketone 4-heptanone is compound P.

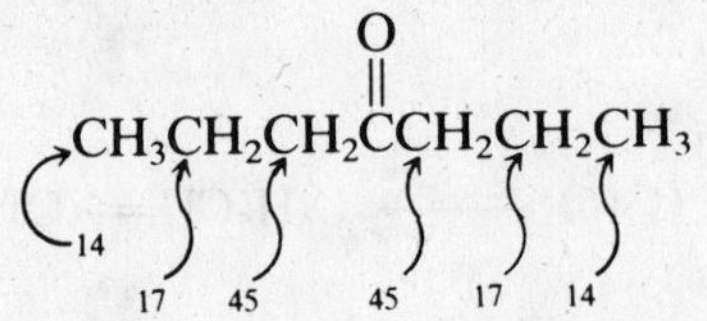

$$CH_3CH_2CH_2CCH_2CH_2CH_3$$

4-Heptanone (compound P)
(all shifts in ppm)

14   17   45   45   17   14

17.47 Compounds Q and R are isomers and have a SODAR of 5. Signals in the region 125 to 140 ppm in their ^{13}C nmr spectra suggest an aromatic ring, and a peak at 200 ppm indicates a carbonyl group. An aromatic ring contributes one ring and three double bonds, and a carbonyl group contributes one double bond, and so the SODAR of 5 is satisfied by a benzene ring and a carbonyl group. The carbonyl group is attached directly to the benzene ring, as evidenced by the presence of a peak at m/z 105 in the mass spectra of compounds Q and R.

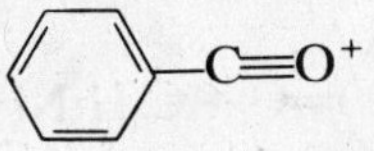

$$\text{Ph}—C{\equiv}O^+ \qquad m/z\ 105$$

Each ^{13}C nmr spectrum shows four aromatic signals, and so the rings are monosubstituted.

Compound Q has three unique carbons apart from its benzoyl group and so must be 1-phenyl-1-butanone. Compound R has only two signals other than those of the benzoyl group and so must be 2-methyl-1-phenyl-1-propanone.

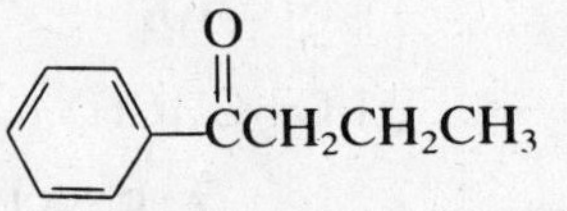

$$\text{Ph}—CCH_2CH_2CH_3 \qquad\qquad \text{Ph}—CCH(CH_3)_2$$

Compound Q 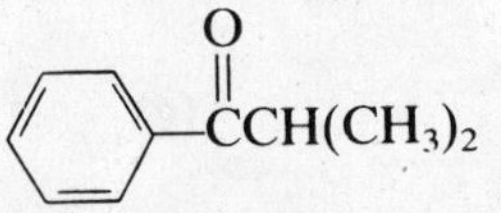   Compound R

SELF-TEST

PART A

A-1. Give the correct IUPAC name for each of the following:

(a) $$CH_3CH_2CHCHCH_2CH$$ with CH_3 substituent, CH_3 substituent, and O (aldehyde)

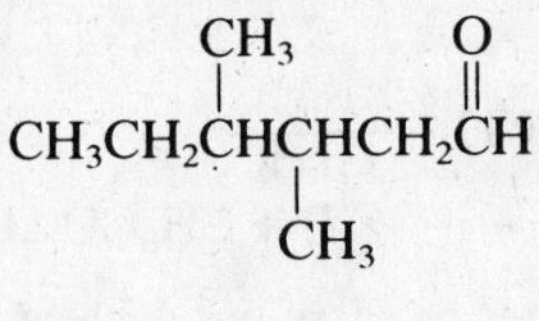

(b) $$(CH_3)_3CCCH_2CH(CH_3)_2$$ with O

(c)   [cyclohexanone ring structure with Br and CH_3 substituents]

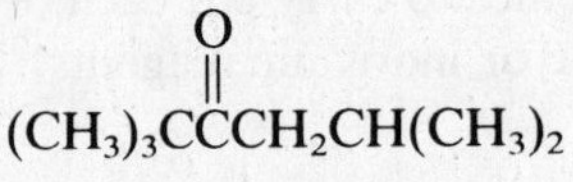

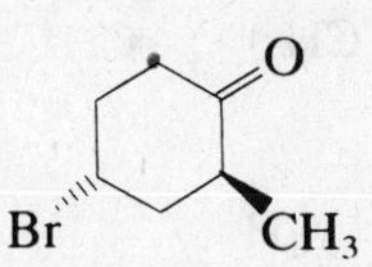

A-2. Write the structural formulas for:
  (a) (E)-3-Hexen-2-one
  (b) 3-Cyclopropyl-2,4-pentanedione
  (c) 3-Ethyl-4-phenylpentanal

A-3. For each of the following reactions supply the structure of the missing reactant, reagent, or product:

(a) $\text{C}_6\text{H}_{10}{=}\text{O} + \text{HCN} \xrightarrow{\text{CN}^-} ?$

(b) $\text{C}_6\text{H}_5\overset{\displaystyle O}{\overset{\|}{\text{C}}}\text{H} + ? \longrightarrow \text{C}_6\text{H}_5\text{CH}{=}\text{NOH}$

(c) $(\text{CH}_3)_2\text{CHCH}\!\!\begin{array}{c}\text{O}\\\text{O}\end{array} \xrightarrow{\text{H}_2\text{O, H}^+} ?$ (two products)

(d) $\text{O} + ? \longrightarrow {=}\text{CHCH}_2\text{CH}_3$

(e) ${=}\text{O} + \text{C}_6\text{H}_5\text{NHNH}_2 \longrightarrow ?$ (with CH_3)

(f) $\text{CH}_3\text{CH}_2\text{CH}_2\overset{\displaystyle O}{\overset{\|}{\text{C}}}\text{H} + 2\text{CH}_3\text{CH}_2\text{OH} \xrightarrow{\text{HCl}} ?$

(g) $? + ? \longrightarrow \text{C}_6\text{H}_5\overset{\displaystyle \text{N(CH}_3)_2}{\overset{|}{\text{C}}}{=}\text{CHCH}_3$

(h) $\text{CH}_3\text{CH}_2\text{CH}_2\text{CH}{=}\text{O} \xrightarrow[\text{2. HCl}]{\text{1. Ag}_2\text{O, NaOH, H}_2\text{O}} ?$

A-4.   Write the structures of the products, compounds A through E, of the reaction steps shown.

    (a) $(\text{C}_6\text{H}_5)_3\text{P} + (\text{CH}_3)_2\text{CHCH}_2\text{Br} \longrightarrow \text{A}$

          $\text{A} + \text{CH}_3\text{CH}_2\text{CH}_2\text{CH}_2\text{Li} \longrightarrow \text{B} + \text{C}_4\text{H}_{10}$

          $\text{B} + \text{benzaldehyde} \longrightarrow \text{C} + (\text{C}_6\text{H}_5)_3\text{PO}$

    (b) $\text{C}_6\text{H}_5\text{CH}_2\overset{\displaystyle \text{OH}}{\overset{|}{\text{C}}}\text{HCH}_2\text{CH}_3 \xrightarrow[\text{CH}_2\text{Cl}_2]{\text{PCC}} \text{D}$

          $\text{D} + \text{CH}_3\text{CO}_2\text{OH} \longrightarrow \text{E} + \text{CH}_3\text{CO}_2\text{H}$

A-5.   Outline reaction schemes to carry out each of the following interconversions, using any necessary organic or inorganic reagents.

    (a) $(\text{CH}_3)_2\text{C}{=}\text{O}$ to $(\text{CH}_3)_2\text{C}\overset{\displaystyle O}{\overset{\triangle}{-}}\text{CHCH}_3$

    (b)   [cyclohexanone with CH_3 and OH]   to   [cyclohexanone with CH_3 and HO, CH_3]

A-6.   What two organic compounds react together (in the presence of an acid catalyst) to give the compound shown, plus a molecule of water?

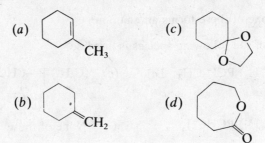

A-7. Draw the structure of the open-chain form of the following cyclic acetal:

A-8. Give the reagents necessary to convert cyclohexanone into each of the following compounds. More than one step may be necessary.

(a)

(c)

(b)

(d)

PART B

B-1. When a nucleophile encounters a ketone, the site of attack is:
(a) The carbon of the carbonyl
(b) The oxygen of the carbonyl
(c) Both the carbon and oxygen, with equal probability
(d) No attack occurs—ketones do not react with nucleophiles.

B-2. What reagent and/or reaction conditions would you choose to bring about the following conversion?

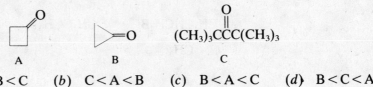

(a) 1. $LiAlH_4$, 2. H_2O (c) H_2O, H_2SO_4, heat
(b) H_2O, NaOH, heat   (d) PCC, CH_2Cl_2

B-3. Rank the following in order of increasing tendency to form a stable hydrate (least → most likely):

A   B $(CH_3)_3CCC(CH_3)_3$

                           C

(a) A < B < C   (b) C < A < B   (c) B < A < C   (d) B < C < A

B-4. The structure

would be best classified as a(n)

(*a*)   Acetal   (*c*)   Hydrate
(*b*)   Hemiacetal   (*d*)   Cyanohydrin

B-5. Which of the following pairs of reactants is most effective in forming an enamine?

(*a*) $CH_3CH_2\overset{\overset{\displaystyle O}{\|}}{C}H + [(CH_3)_2CH]_2NH$

(*b*) $(CH_3)_3C\overset{\overset{\displaystyle O}{\|}}{C}H + (CH_3)_2NH$

(*c*)   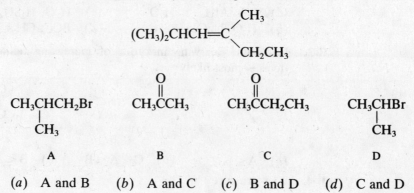 $+ (CH_3)_3CNH_2$

(*d*)   None of these forms an enamine.

B-6. Which of the following species is an ylide?

(*a*) $(C_6H_5)_3\overset{+}{P}CH_2CH_3 \ \ Br^-$ (*c*) $(C_6H_5)_3P\!\!-\!\!CHCH_3$
                                                                                   $\quad\quad\quad\quad\quad\; |\quad\quad\; |$
                                                                                   $\quad\quad\quad\quad\quad O\!\!-\!\!CH_2$

(*b*) $(C_6H_5)_3\overset{+}{P}\overset{..}{C}HCH_3$ (*d*)   None of these

B-7. The species

$$C_6H_5\underset{\underset{\displaystyle NHCH_2CH_3}{|}}{\overset{\overset{\displaystyle OH}{|}}{C}}CH_3$$

is an intermediate in the formation of a(n)

(*a*)   Imine   (*c*)   Oxime
(*b*)   Enamine   (*d*)   Hydrazone

B-8. Which of the following reagents would allow conversion of a ketone to an ester?

(*a*) $NaBH_4$, CH_3OH (*c*)   Acetic anhydride
(*b*) $(C_5H_5N)_2CrO_3$, CH_2Cl_2 (*d*) $C_6H_5CO_2OH$

B-9. Which compounds from the list below would you select as starting materials to prepare the following alkene by the Wittig method?

$$(CH_3)_2CHCH\!=\!C\overset{\displaystyle CH_3}{\underset{\displaystyle CH_2CH_3}{\big\langle}}$$

CH_3CHCH_2Br $CH_3\overset{\overset{\displaystyle O}{\|}}{C}CH_3$ $CH_3\overset{\overset{\displaystyle O}{\|}}{C}CH_2CH_3$ CH_3CHBr
$\quad\; |$ $|$
$\quad\, CH_3$ CH_3

   A   B   C   D

(*a*)   A and B   (*b*)   A and C   (*c*)   B and D   (*d*)   C and D

ENOLS, ENOLATES, AND ENAMINES

IMPORTANT TERMS AND CONCEPTS

Enolization (Secs. 18.4, 18.5) Enols are related to aldehydes or ketones by a proton-transfer equilibrium known as *keto-enol tautomerism*.

$$\underset{\text{Keto form}}{\text{RCH}_2\overset{\displaystyle O}{\overset{\|}{\text{C}}}\text{R}'} \quad \underset{\xrightarrow{\hspace{1cm}}}{\overset{\text{tautomerism}}{\rightleftharpoons}} \quad \underset{\text{Enol form}}{\text{RCH}=\overset{\displaystyle OH}{\overset{|}{\text{C}}}\text{R}'}$$

Only a small amount of enol is present in a simple aldehyde or ketone. In 1,3-dicarbonyl compounds, such as 2,4-pentanedione, the enol isomer is of comparable stability with the keto form and may predominate in the equilibrium.

$$\underset{\text{Keto form (20\%)}}{\text{CH}_3\overset{\displaystyle O}{\overset{\|}{\text{C}}}\text{CH}_2\overset{\displaystyle O}{\overset{\|}{\text{C}}}\text{CH}_3} \quad \rightleftharpoons \quad \underset{\text{Enol form (80\%)}}{\text{CH}_3\overset{\displaystyle OH}{\overset{|}{\text{C}}}=\text{CH}\overset{\displaystyle O}{\overset{\|}{\text{C}}}\text{CH}_3}$$

Enolate Anions (Sec. 18.6) The conjugate base of an aldehyde or a ketone is known as an *enolate*. The K_a's for aldehydes and ketones are in the range of those for alcohols and water, from 10^{-16} to 10^{-20} ($pK_a = 16-20$).

$$\text{RCH}_2\overset{\displaystyle O}{\overset{\|}{\text{C}}}\text{R}' \rightleftharpoons \left[\text{R}\overset{..}{\underset{..}{\text{C}}}\text{H}\overset{\displaystyle :\overset{..}{O}}{\overset{\|}{\text{C}}}\text{R}' \longleftrightarrow \text{RCH}=\overset{\displaystyle :\overset{..}{O}:^-}{\overset{|}{\text{C}}}\text{R}' \right] + \text{H}^+$$

β-Diketones are even more acidic because the negative charge is shared by carbon and both oxygens:

$$\text{CH}_3\overset{\displaystyle O^-}{\overset{|}{\text{C}}}=\text{CH}\overset{\displaystyle O}{\overset{\|}{\text{C}}}\text{CH}_3 \longleftrightarrow \text{CH}_3\overset{\displaystyle O}{\overset{\|}{\text{C}}}\overset{..}{\underset{..}{\text{C}}}\text{H}\overset{\displaystyle O}{\overset{\|}{\text{C}}}\text{CH}_3 \longleftrightarrow \text{CH}_3\overset{\displaystyle O}{\overset{\|}{\text{C}}}\text{CH}=\overset{\displaystyle O^-}{\overset{|}{\text{C}}}\text{CH}_3$$

Conjugation in α,β-Unsaturated Aldehydes and Ketones (Secs. 18.12, 18.13) The electron delocalization resulting from overlap of the carbon-carbon double bond and the carbonyl group in an α,β-unsaturated aldehyde or ketone is reflected in the chemical properties of these compounds.

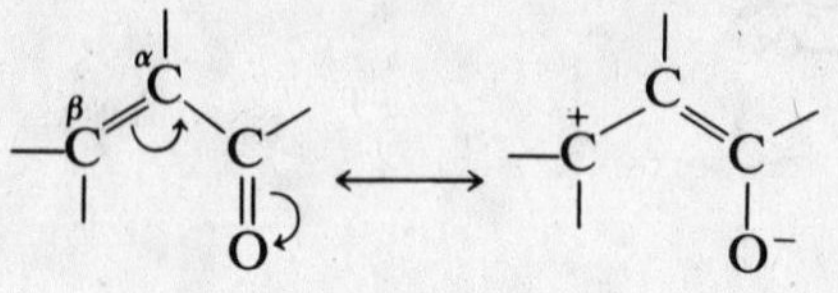

Electrophilic reagents such as bromine or peroxy acids react more slowly with the double bond of an α,β-unsaturated aldehyde or ketone than with simple alkenes. The β carbon is electrophilic and susceptible to attack by nucleophiles. Such an addition is termed *conjugate addition*, or *1,4-addition*.

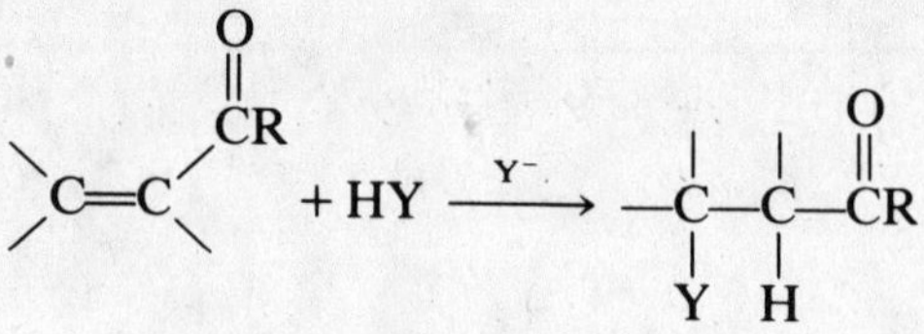

IMPORTANT REACTIONS

The Aldol Condensation (Sec. 18.8)

General:

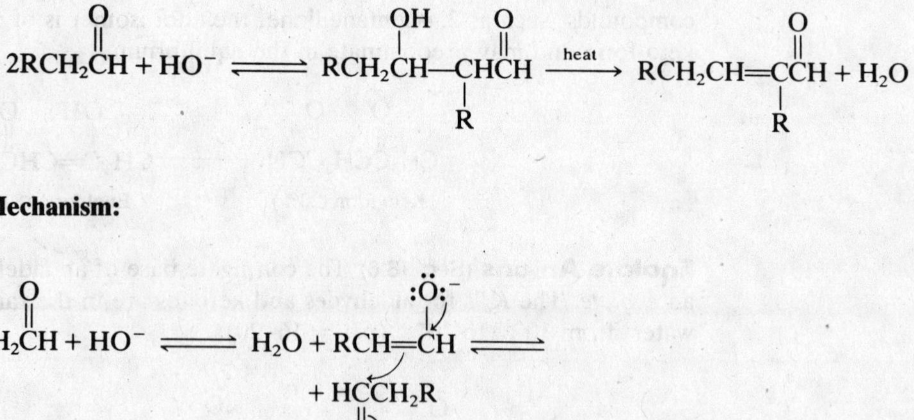

Mechanism:

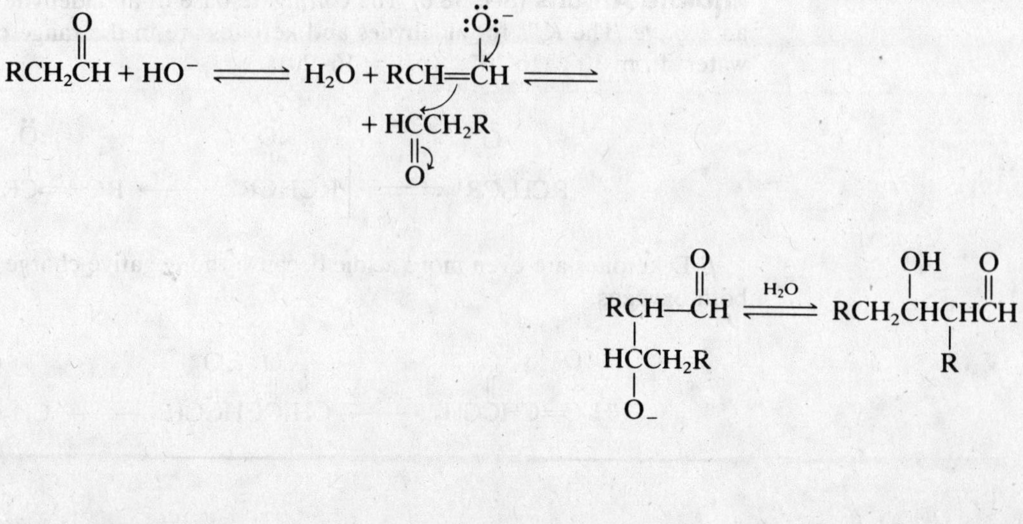

Example:

$$2CH_3CH_2\overset{\overset{\displaystyle O}{\|}}{C}H + HO^- \rightleftharpoons CH_3CH_2\underset{\underset{\displaystyle CH_3}{|}}{\overset{\overset{\displaystyle OH}{|}}{C}H}\overset{\overset{\displaystyle O}{\|}}{C}H \xrightarrow{\text{heat}} CH_3CH_2CH=\underset{\underset{\displaystyle CH_3}{|}}{C}\overset{\overset{\displaystyle O}{\|}}{C}H$$

Mixed (Crossed) Aldol Condensation (Sec. 18.10)

General:

$$RCH_2\overset{\overset{\displaystyle O}{\|}}{C}H + R'\overset{\overset{\displaystyle O}{\|}}{C}H \xrightarrow{HO^-} R'\underset{\underset{\displaystyle R}{|}}{\overset{\overset{\displaystyle OH}{|}}{C}H}CH\overset{\overset{\displaystyle O}{\|}}{C}H \xrightarrow{\text{heat}} R'CH=\underset{\underset{\displaystyle R}{|}}{C}\overset{\overset{\displaystyle O}{\|}}{C}H$$

Example:

$$C_6H_5\overset{\overset{\displaystyle O}{\|}}{C}H + CH_3CH_2\overset{\overset{\displaystyle O}{\|}}{C}H \xrightarrow[\text{heat}]{HO^-} C_6H_5CH=\underset{\underset{\displaystyle CH_3}{|}}{C}\overset{\overset{\displaystyle O}{\|}}{C}H$$

Conjugate Additions (Secs. 18.14, 18.15)

Michael Addition:

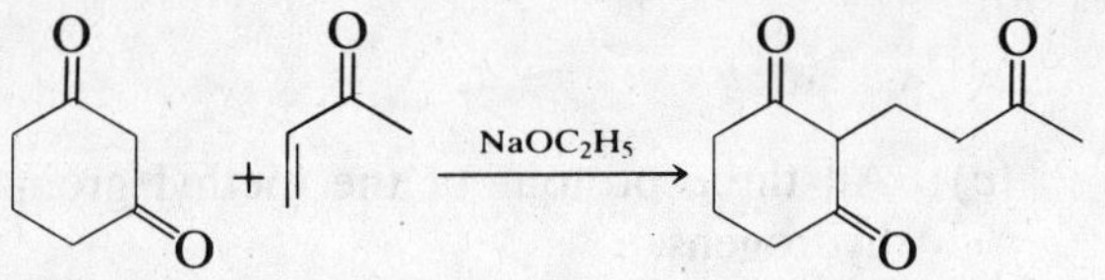

Organocopper Reagents:

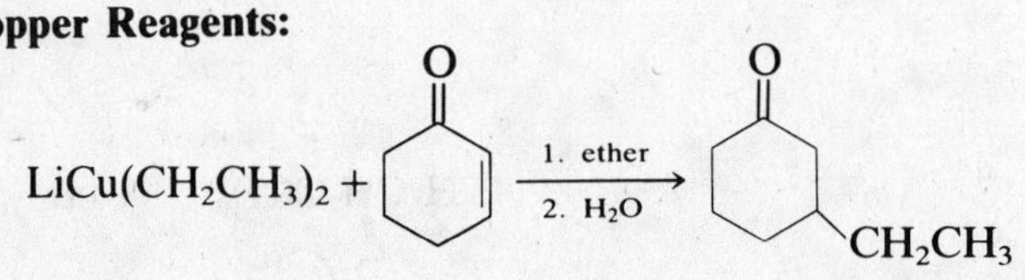

$$LiCu(CH_2CH_3)_2 + $$

Alkylation of Enolate Anions (Sec. 18.16)

General:

$$R_2CH\overset{\overset{\displaystyle O}{\|}}{C}R' \xrightarrow{\text{base}} R_2C=\overset{\overset{\displaystyle O^-}{|}}{C}R' \xrightarrow{R''X} R_2\underset{\underset{\displaystyle R''}{|}}{C}\overset{\overset{\displaystyle O}{\|}}{C}R'$$

Example:

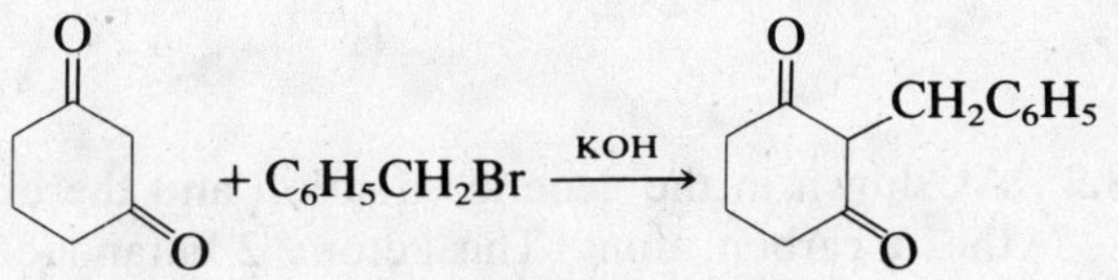

$$+ C_6H_5CH_2Br \xrightarrow{KOH}$$

Alkylation of Enamines (Sec. 18.17)

General:

$$R\overset{\overset{\displaystyle NR'_2}{|}}{C}=CR_2 + R''CH_2X \longrightarrow R\overset{\overset{\displaystyle {}^+NR'_2}{\|}}{C}-\underset{\underset{\displaystyle CH_2R''}{|}}{C}R_2 \xrightarrow{H_2O} R\overset{\overset{\displaystyle O}{\|}}{C}CR_2CH_2R'' + NHR'_2$$

Example:

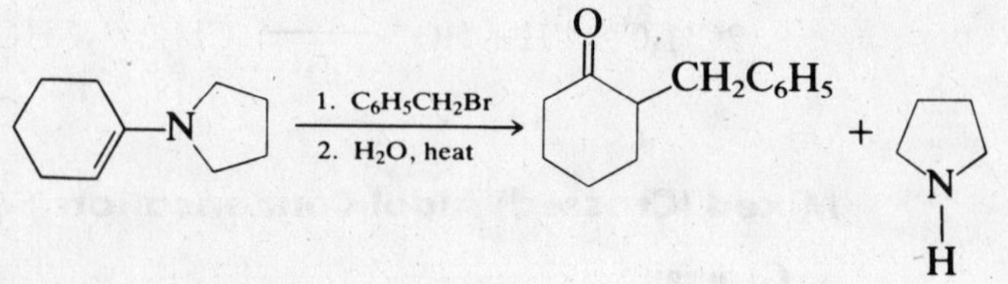

SOLUTIONS TO TEXT PROBLEMS

18.1 (*b*) There are no α hydrogen atoms in 2,2-dimethylpropanal, since the α carbon atom bears three methyl groups.

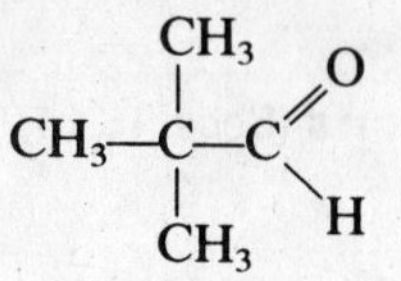

2,2-Dimethylpropanal

(*c*) All three protons of the methyl group as well as the two benzylic protons are α hydrogens.

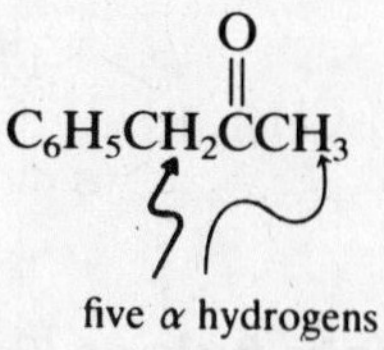

$C_6H_5CH_2CCH_3$ Benzyl methyl ketone

five α hydrogens

(*d*) Cyclohexanone has four equivalent α hydrogens.

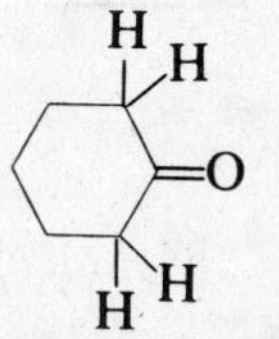

Cyclohexanone (the hydrogens indicated are the α hydrogens)

18.2 As shown in the general equation and the examples, halogen substitution is specific for the α carbon atom. The ketone 2-butanone has two nonequivalent α carbons, and so substitution is possible at both positions. Both 1-chloro-2-butanone and 3-chloro-2-butanone are formed in the reaction.

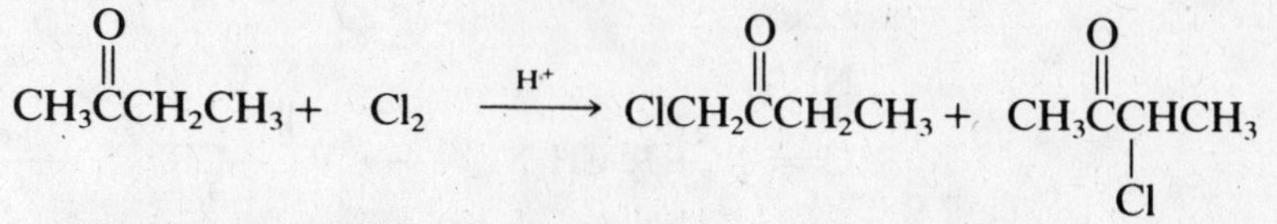

$$CH_3CCH_2CH_3 + Cl_2 \xrightarrow{H^+} ClCH_2CCH_2CH_3 + CH_3CCHCH_3$$
$$\qquad\qquad\qquad\qquad\qquad\qquad\qquad\qquad\qquad\qquad\qquad\qquad Cl$$

2-Butanone   Chlorine   1-Chloro-2-butanone   3-Chloro-2-butanone

18.3 The carbon-carbon double bond of the enol always involves the original carbonyl carbon and the α carbon atom. 2-Butanone can form two different enols, each of which yields a different chloro ketone.

2-Butanone	1-Buten-2-ol (enol)	1-Chloro-2-butanone

2-Butanone	2-Buten-2-ol (enol)	3-Chloro-2-butanone

18.4 Chlorine attacks the enol at its carbon-carbon double bond.

18.5 (*b*) Acetophenone can enolize only in the direction of the methyl group.

Acetophenone　　　　　　　Enol form of acetophenone

(*c*) Enolization of 2-methylcyclohexanone can take place in two different directions.

2-Methylcyclohex-1-enol (enol form)	2-Methylcyclohexanone	6-Methylcyclohex-1-enol (enol form)

18.6 (*b*) Enolization of the central methylene group can involve either of the two carbonyl groups.

Enol form	1-Phenyl-1,3-butanedione	Enol form

18.7 (*b*) Removal of a proton from 1-phenyl-1,3-butanedione occurs on the methylene group between the carbonyls.

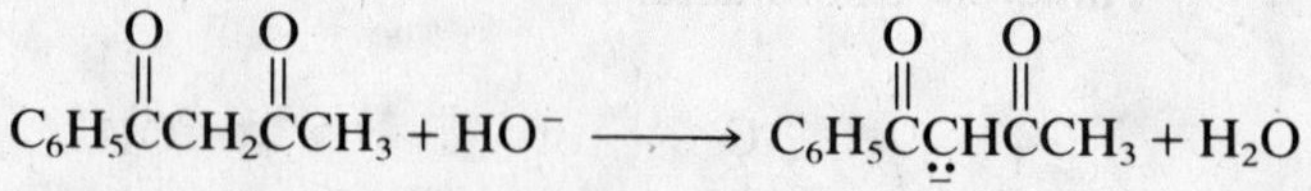

$$C_6H_5\overset{O}{\overset{\|}{C}}CH_2\overset{O}{\overset{\|}{C}}CH_3 + HO^- \longrightarrow C_6H_5\overset{O}{\overset{\|}{C}}\overset{..}{C}HCCH_3 + H_2O$$

The three most stable resonance forms of this anion are:

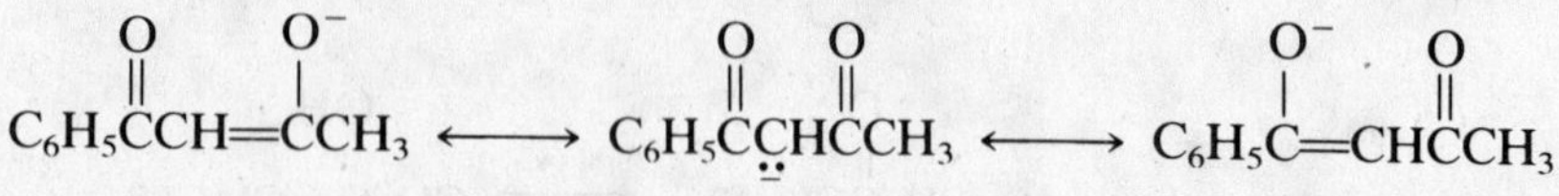

$$C_6H_5\overset{O}{\overset{\|}{C}}CH=\overset{O^-}{\overset{|}{C}}CH_3 \longleftrightarrow C_6H_5\overset{O}{\overset{\|}{C}}\overset{..}{\underset{..}{C}}HCCH_3 \longleftrightarrow C_6H_5\overset{O^-}{\overset{|}{C}}=CHCCH_3$$

(*c*) Deprotonation at C-2 of this β-dicarbonyl compound yields the carbanion shown.

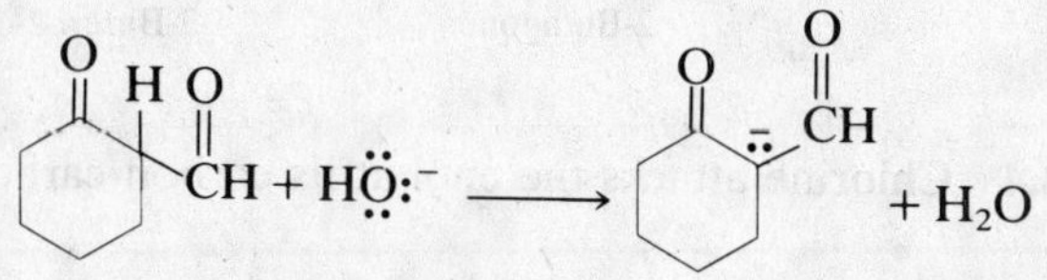

The three most stable resonance forms of the anion are:

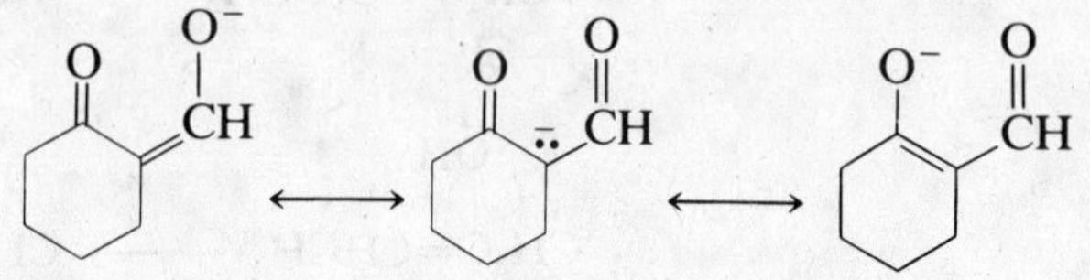

18.8 (*b*) Approaching this problem mechanistically in the same way as part (*a*), write the structure of the enolate ion from 2-methylbutanal.

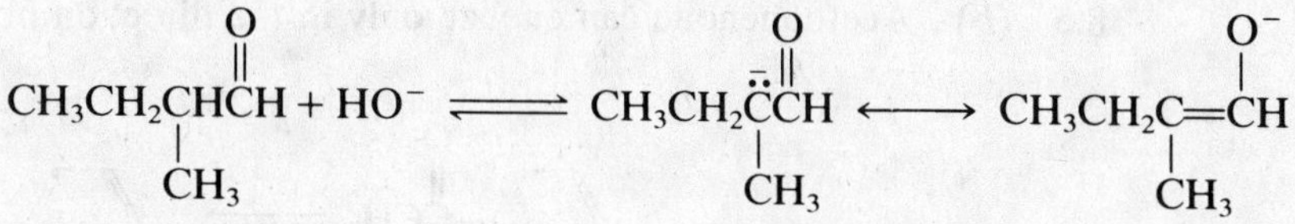

$$CH_3CH_2\underset{CH_3}{\overset{O}{\overset{\|}{\underset{|}{C}H}}}CH + HO^- \rightleftharpoons CH_3CH_2\underset{CH_3}{\overset{O}{\overset{\|}{\underset{|}{\ddot{C}}}}}CH \longleftrightarrow CH_3CH_2\underset{CH_3}{\overset{O^-}{\overset{|}{\underset{|}{C}}}}=CH$$

2-Methylbutanal · Enolate of 2-methylbutanal

This enolate adds to the carbonyl group of the aldehyde.

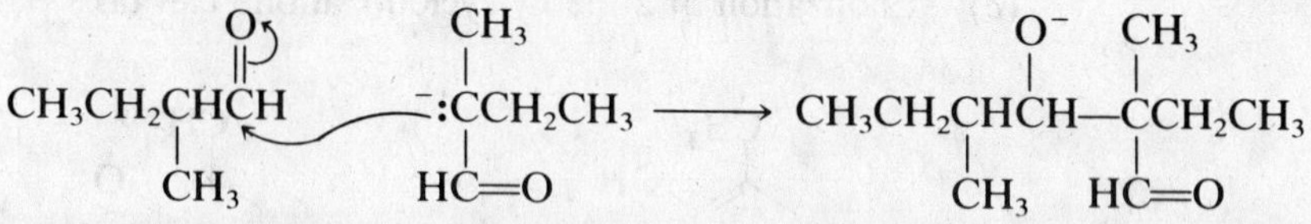

2-Methylbutanal · Enolate of 2-methylbutanal

The alkoxide ion formed in this step accepts a proton from solvent to yield the product of aldol addition.

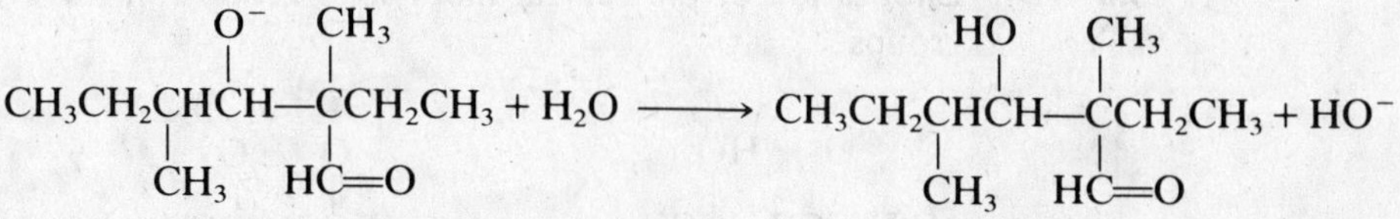

2-Ethyl-2,4-dimethyl-3-
hydroxyhexanal

(c) The aldol addition product of 3-methylbutanal can be identified through the same mechanistic approach.

$$(CH_3)_2CHCH_2CH + HO^- \rightleftharpoons (CH_3)_2CH\ddot{C}HCH + H_2O$$

3-Methylbutanal   Enolate of 3-methylbutanal

$$(CH_3)_2CHCH_2CH + \ :\bar{C}HCH(CH_3)_2 \longrightarrow (CH_3)_2CHCH_2CH—CHCH(CH_3)_2$$

3-Methylbutanal   Enolate of   HC=O
                       3-methylbutanal

$$\downarrow H_2O$$

OH
|
$$(CH_3)_2CHCH_2CH—CHCH(CH_3)_2 + OH^-$$
|
HC=O

3-Hydroxy-2-isopropyl-5-methylhexanal

18.9 Dehydration of the aldol addition product involves loss of a proton from the α carbon atom and hydroxide from the β carbon atom.

$$R_2C—CHCH \longrightarrow R_2C=CHCH + H_2O + HO^-$$

OH   O
| ||
H
|
$$^-:\ddot{O}H$$

(b) The product of aldol addition of 2-methylbutanal has no α hydrogens. It cannot dehydrate to an aldol condensation product.

$$2CH_3CH_2CHCH \xrightleftharpoons{HO^-} CH_3CH_2CHCH—CCH_2CH_3$$

CH_3   CH_3   HC=O

2-Methylbutanal   (no protons on α carbon atom)

(c) Aldol condensation is possible with 3-methylbutanal.

$$2(CH_3)_2CHCH_2CH \xrightleftharpoons{HO^-} (CH_3)_2CHCH_2CHCHCH(CH_3)_2 \xrightarrow{-H_2O}$$

HO
|
HC=O

3-Methylbutanal   Aldol addition
                                    product

$$(CH_3)_2CHCH_2CH=CCH(CH_3)_2$$
|
HC=O

2-Isopropyl-5-methyl-2-hexenal

18.10 The structure of the aldol addition product of acetone was given at the beginning of Section 18.9 in the text. It dehydrates in the direction shown to give the compound called *mesityl oxide*. The systematic name for mesityl oxide is *4-methyl-3-penten-2-one*.

$$(CH_3)_2CCH_2CCH_3 \longrightarrow (CH_3)_2C\!\!=\!\!CHCCH_3 + H_2O$$

4-Hydroxy-4-methyl-2-pentanone

4-Methyl-3-penten-2-one

18.11 Aldol addition involves reaction of the ketone enolate of one cyclohexanone molecule with the carbonyl group of another.

2-(1-Hydroxycyclohexyl)-cyclohexanone
(product of aldol addition)

2-Cyclohexylidenecyclohexanone (78%)
(product of aldol condensation)

18.12 (b) The only enolate that can be formed from *tert*-butyl methyl ketone arises by proton abstraction from the methyl group.

$$(CH_3)_3CCCH_3 + HO^- \rightleftharpoons (CH_3)_3CCCH_2$$

tert-Butyl methyl ketone

Enolate of *tert*-butyl methyl ketone

This enolate adds to the carbonyl group of benzaldehyde to give the mixed aldol addition product, which then dehydrates under the reaction conditions.

$$C_6H_5CH + {}^-\!\!:CH_2CC(CH_3)_3 \longrightarrow C_6H_5CHCH_2CC(CH_3)_3$$

Benzaldehyde   Enolate of *tert*-butyl methyl ketone

$$C_6H_5CH\!\!=\!\!CHCC(CH_3)_3 \xleftarrow{-H_2O} C_6H_5CHCH_2CC(CH_3)_3$$

4,4-Dimethyl-1-phenyl-1-penten-3-one
(product of mixed aldol condensation)

Product of mixed aldol addition

(c) The enolate of cyclohexanone adds to benzaldehyde. Dehydration of the mixed aldol addition product takes place under the reaction conditions to give the mixed aldol condensation product called *benzylidenecyclohexanone*.

Cyclohexanone   Benzaldehyde

Benzylidenecyclohexanone

18.13 The carbon skeleton of 2-ethyl-1-hexanol is the same as that of the aldol condensation product derived from butanal. Hydrogenation of this compound under conditions where both the carbon-carbon double bond and the carbonyl group are reduced gives 2-ethyl-1-hexanol.

$$CH_3CH_2CH_2\overset{\overset{\textstyle O}{\|}}{C}H \xrightarrow[\text{heat}]{\text{NaOH, H}_2\text{O}} CH_3CH_2CH_2CH{=}\overset{\overset{\textstyle O}{\|}}{C}H \xrightarrow{\text{H}_2,\ \text{Ni}}$$
$$\underset{\text{CH}_2\text{CH}_3}{\big|}$$

Butanal   2-Ethyl-2-hexenal

$$CH_3CH_2CH_2CH_2\underset{\underset{\textstyle CH_2CH_3}{\big|}}{C}HCH_2OH$$

2-Ethyl-1-hexanol

18.14 Mesityl oxide is an α,β-unsaturated ketone. Traces of acids or bases can catalyze its isomerization so that some of the less stable β,γ-unsaturated isomer is present.

$$\underset{\text{CH}_3}{\overset{\text{CH}_3}{\big>}}C{=}CH\overset{\overset{\textstyle O}{\|}}{C}CH_3 \rightleftharpoons \underset{\text{CH}_3}{\overset{\text{H}_2\text{C}}{\big>}}C{-}CH_2\overset{\overset{\textstyle O}{\|}}{C}CH_3$$

Mesityl oxide; 4-methyl-3-penten-2-one   4-Methyl-4-penten-2-one
(more stable)   (less stable)

18.15 The relationship between the molecular formula of acrolein (C_3H_4O) and the product ($C_3H_4N_3O$) corresponds to the addition of HN_3 to acrolein. Since propanal ($CH_3CH_2CH{=}O$) does not react under these conditions, it is probable that conjugate addition to the double bond of acrolein is the reaction observed.

$$CH_2{=}CHC\overset{\overset{\textstyle O}{\|}}{H} \xrightarrow[\substack{\text{acetic}\\\text{acid}}]{\text{NaN}_3} N_3CH_2CH_2\overset{\overset{\textstyle O}{\|}}{C}H$$

Acrolein   3-Azidopropanal

18.16 It is the enolate of dibenzyl ketone that adds to methyl vinyl ketone in the conjugate addition step.

$$C_6H_5CH_2\overset{\overset{\textstyle O}{\|}}{C}CH_2C_6H_5 + CH_2{=}CHC\overset{\overset{\textstyle O}{\|}}{C}CH_3 \xrightarrow[\text{CH}_3\text{OH}]{\text{NaOCH}_3} C_6H_5CH_2\overset{\overset{\textstyle O}{\|}}{C}\underset{\underset{\underset{\overset{\textstyle\|}{\textstyle O}}{\textstyle CH_2CH_2CCH_3}}{\big|}}{C}HC_6H_5$$

Dibenzyl ketone   Methyl vinyl ketone   1,3-Diphenyl-2,6-heptanedione

via:

$$C_6H_5CH_2\overset{\overset{\textstyle O}{\|}}{\underset{\underset{\overset{\textstyle\|}{\textstyle O}}{\textstyle CH_2{=}CHCCH_3}}{C}}\overset{..}{C}HC_6H_5 \longrightarrow C_6H_5CH_2\overset{\overset{\textstyle O}{\|}}{\underset{\underset{\overset{\textstyle\|}{\textstyle O}}{\textstyle CH_2{-}\overset{..}{C}HCCH_3}}{C}}HC_6H_5$$

The intramolecular aldol condensation that gives the observed product is:

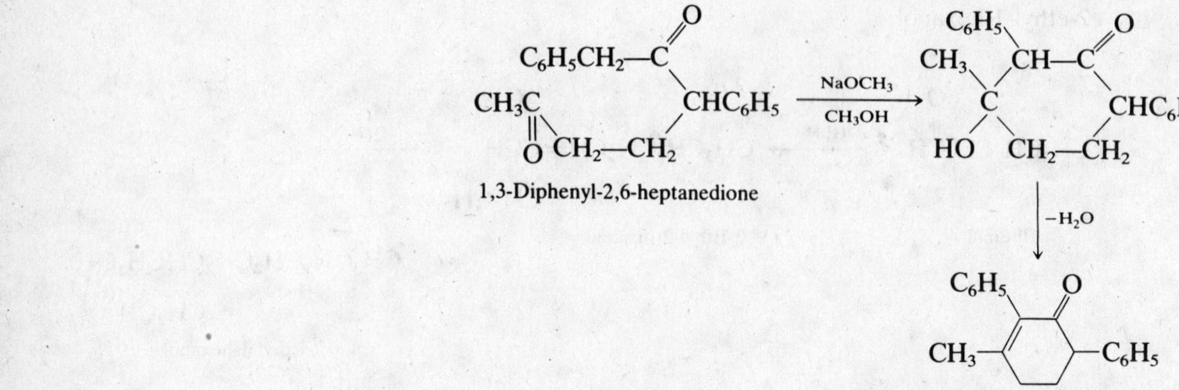

1,3-Diphenyl-2,6-heptanedione

3-Methyl-2,6-diphenyl-2-cyclohexen-1-one

18.17 A second solution to the synthesis of 4-methyl-2-octanone by conjugate addition of a lithium dialkylcuprate reagent to an α,β-unsaturated ketone is revealed by the disconnection shown:

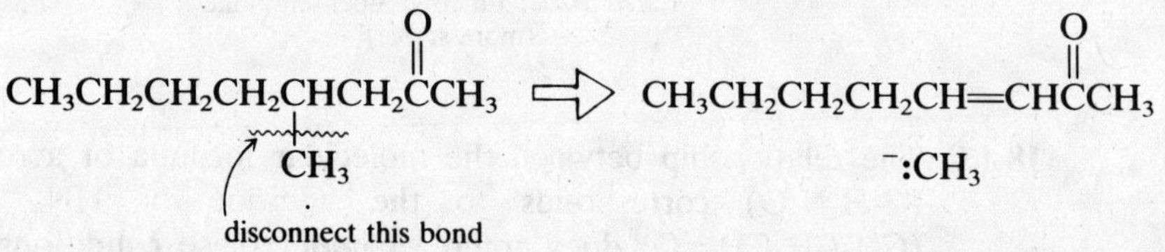

disconnect this bond

According to this disconnection, the methyl group is derived from lithium dimethylcuprate.

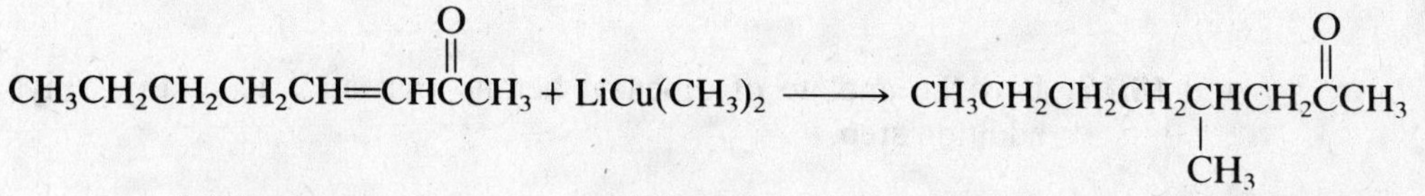

3-Octen-2-one   Lithium   4-Methyl-2-octanone
                     dimethylcuprate

18.18 (b) Examine this problem by the retrosynthetic method. Disconnect a bond to the α carbon atom.

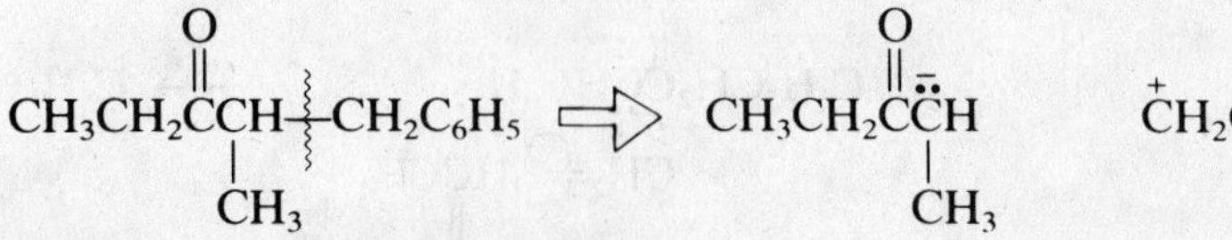

2-Methyl-1-phenyl-3-pentanone

This disconnection reveals the synthetic plan to be followed. A benzyl halide is needed as the alkylating agent, and an enamine of 3-pentanone is the nucleophile.

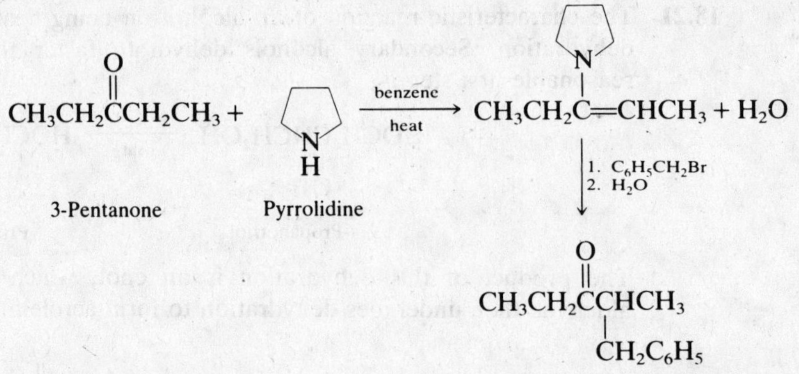

3-Pentanone   Pyrrolidine

2-Methyl-1-phenyl-3-pentanone

18.19 Intramolecular aldol condensation occurs to give an α,β-unsaturated ketone in a six-membered ring.

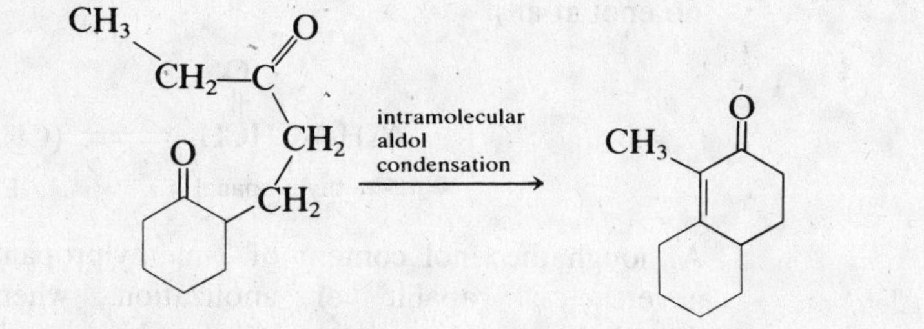

2-(3'-Oxopentyl)cyclohexanone   2-Methyl-3-oxobicyclo[4.4.0]-
1(2)-decene

18.20 (*a*) In addition to the double bond of the carbonyl group, there must be a double bond elsewhere in the molecule in order to satisfy the molecular formula C_4H_6O (the problem states that the compounds are noncyclic). There are a total of five isomers:

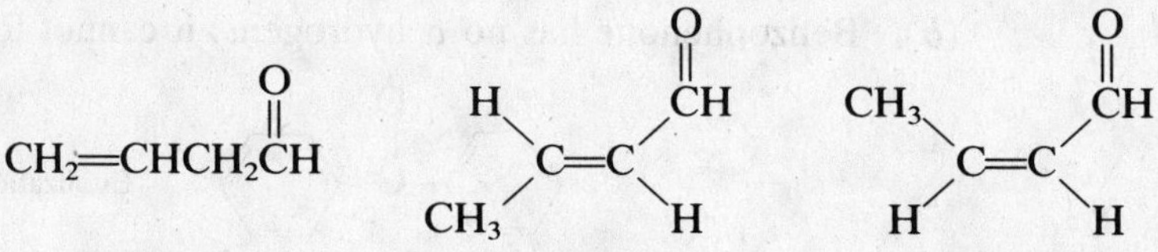

3-Butenal   (*E*)-2-Butenal   (*Z*)-2-Butenal

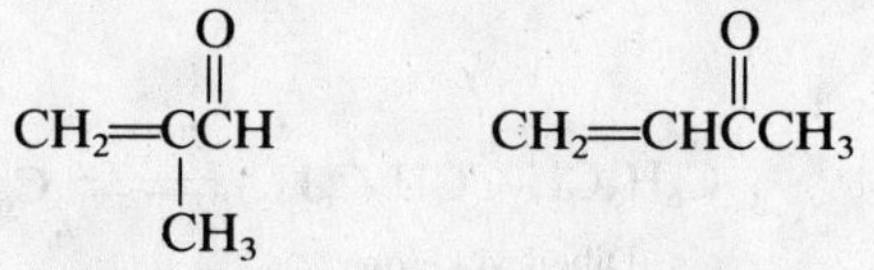

2-Methylpropenal   3-Buten-2-one
(methyl vinyl ketone)

(*b*) The *E* and *Z* isomers of 2-butenal are stereoisomers.
(*c*) None of the C_4H_6O aldehydes and ketones are chiral.
(*d*) The α,β-unsaturated aldehydes are (*E*)- and (*Z*)-CH_3CH=$CHCHO$; and CH_2=$CCHO$.
                     $|$
                     CH_3

There is one α,β-unsaturated ketone in the group: CH_2=$CHCCH_3$.

(e) The E and Z isomers of 2-butenal are formed by the aldol condensation of acetaldehyde.

18.21 The characteristic reaction of an alcohol on being heated with $KHSO_4$ is acid-catalyzed dehydration. Secondary alcohols dehydrate faster than primary alcohols, and so a reasonable first step is:

$$HOCH_2CHCH_2OH \xrightarrow[\text{heat}]{KHSO_4} HOCH_2CH=CHOH$$
$$\quad\quad\quad |$$
$$\quad\quad OH$$

1,2,3-Propanetriol   Propene-1,3-diol

The product of this dehydration is an enol, which rearranges to an aldehyde. The aldehyde then undergoes dehydration to form acrolein.

$$HOCH_2CH=CHOH \longrightarrow HOCH_2CH_2\overset{\overset{\displaystyle O}{\|}}{C}H \xrightarrow[\text{heat}]{KHSO_4} CH_2=CH\overset{\overset{\displaystyle O}{\|}}{C}H$$

Propene-1,3-diol   3-Hydroxypropanal   Acrolein

18.22 (a) 2-Methylpropanal has the greater enol content (since the other compound contains no enol at all).

$$(CH_3)_2CH\overset{\overset{\displaystyle O}{\|}}{C}H \rightleftharpoons (CH_3)_2C=\overset{\overset{\displaystyle OH}{|}}{C}H$$

2-Methylpropanal   Enol form

Although the enol content of 2-methylpropanal is quite small, the compound is nevertheless capable of enolization, whereas the other compound, 2,2-dimethylpropanal, cannot enolize—it has no α hydrogens.

$$CH_3\overset{\overset{\displaystyle CH_3}{|}}{\underset{\underset{\displaystyle CH_3}{|}}{C}}\overset{\overset{\displaystyle O}{\|}}{\underset{\displaystyle H}{C}}$$ (Enolization is impossible.)

(b) Benzophenone has no α hydrogens; it cannot form an enol.

(Enolization is impossible.)

Dibenzyl ketone enolizes slightly to form a small amount of enol.

$$C_6H_5CH_2\overset{\overset{\displaystyle O}{\|}}{C}CH_2C_6H_5 \rightleftharpoons C_6H_5CH=\overset{\overset{\displaystyle OH}{|}}{C}CH_2C_6H_5$$

Dibenzyl ketone   Enol form

(c) Here we are comparing a simple ketone, dibenzyl ketone, with a β-diketone. The β-diketone enolizes to a much greater extent than the simple ketone because its enol form is stabilized by conjugation of the double bond with the remaining carbonyl group and by intramolecular hydrogen bonding.

$$C_6H_5\overset{\overset{\displaystyle O}{\|}}{C}CH_2\overset{\overset{\displaystyle O}{\|}}{C}C_6H_5 \rightleftharpoons$$

1,3-Diphenyl-1,3-propanedione   Enol form

(d) The enol content of cyclohexanone is quite small, whereas the enol form of 2,4-cyclohexadienone is phenol and therefore enolization is essentially complete.

Keto form   Enol form
                  (much more stable)

(e) A small amount of enol is in equilibrium with cyclopentanone.

Cyclopentanone   Enol form

Cyclopentadienone does not form a stable enol. Enolization would lead to a highly strained allene-type compound.

(Not stable; highly strained)

(f) The β-diketone is more extensively enolized.

1,3-Cyclohexanedione   Enol form
                        (double bond conjugated with carbonyl group)

The double bond of the enol form of 1,4-cyclohexanedione is not conjugated with the carbonyl group. Its enol content is expected to be similar to that of cyclohexanone.

1,4-Cyclohexanedione   Enol form
                    (not particularly stable; double bond and
                      carbonyl group not conjugated)

18.23 (a) Chlorination of 3-phenylpropanal under conditions of acid catalysis occurs via the enol form and yields the α-chloro derivative.

$$C_6H_5CH_2CH_2\overset{\overset{O}{\|}}{C}H + Cl_2 \xrightarrow{\text{acetic acid}} C_6H_5CH_2\underset{\underset{Cl}{|}}{C}H\overset{\overset{O}{\|}}{C}H + HCl$$

3-Phenylpropanal   2-Chloro-3-phenylpropanal

(b)   Aldehydes undergo aldol addition on treatment with base.

$$2C_6H_5CH_2CH_2\overset{\displaystyle O}{\overset{\|}{C}}H \xrightarrow[\substack{\text{ethanol,}\\10°C}]{\text{NaOH}} C_6H_5CH_2CH_2\overset{\displaystyle HC=O}{\underset{\displaystyle OH}{\overset{|}{C}H}}\overset{|}{C}HCH_2C_6H_5$$

3-Phenylpropanal   2-Benzyl-3-hydroxy-5-phenylpentanal

(c)   Dehydration of the aldol addition product occurs when the reaction is carried out at elevated temperature.

$$2C_6H_5CH_2CH_2\overset{\displaystyle O}{\overset{\|}{C}}H \xrightarrow[\substack{\text{ethanol,}\\70°C}]{\text{NaOH}} C_6H_5CH_2CH_2CH=\overset{\displaystyle HC=O}{\overset{|}{C}}CH_2C_6H_5$$

3-Phenylpropanal   2-Benzyl-5-phenyl-2-pentenal

(d)   Sodium borohydride reduces the aldehyde function to the corresponding primary alcohol.

$$C_6H_5CH_2CH_2CH=\overset{\displaystyle HC=O}{\overset{|}{C}}CH_2C_6H_5 \xrightarrow[\text{ethanol}]{\text{NaBH}_4} C_6H_5CH_2CH_2CH=\overset{\displaystyle CH_2OH}{\overset{|}{C}}CH_2C_6H_5$$

2-Benzyl-5-phenyl-2-pentenal   2-Benzyl-5-phenyl-2-penten-1-ol

(e)   A characteristic reaction of α,β-unsaturated carbonyl compounds is their tendency to undergo conjugate addition on treatment with weakly basic nucleophiles.

$$C_6H_5CH_2CH_2CH=\overset{\displaystyle HC=O}{\overset{|}{C}}CH_2C_6H_5 \xrightarrow[\text{H}^+]{\text{NaCN}} C_6H_5CH_2CH_2\overset{\displaystyle HC=O}{\underset{\displaystyle CN}{\overset{|}{C}H}}\overset{|}{C}HCH_2C_6H_5$$

2-Benzyl-5-phenyl-2-pentenal   2-Benzyl-3-cyano-5-phenylpentanal

18.24 (a)   Ketones undergo α halogenation by way of their enol form.

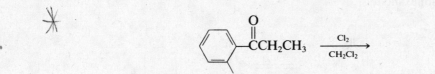

1-(o-Chlorophenyl)-1-propanone   2-Chloro-1-(o-chlorophenyl)-1-propanone

(b)   The combination of $C_6H_5CH_2SH$ and NaOH yields $C_6H_5CH_2S^-$ (as its sodium salt), which is a weakly basic nucleophile and adds to α,β-unsaturated ketones by conjugate addition.

2-Isopropylidene-5-   2-(1-Benzylthio-1-methylethyl)-5-
methylcyclohexanone   methylcyclohexanone (89–90%)

(*c*)   Bromination occurs at the carbon atom that is α to the carbonyl group.

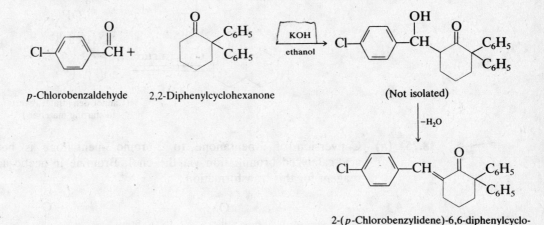

2,2-Diphenylcyclopentanone   2-Bromo-5,5-diphenylcyclopentanone
(76%)

(*d*)   The reaction is a mixed aldol condensation. The enolate of 2,2-diphenylcyclohexanone reacts with *p*-chlorobenzaldehyde. Elimination of the aldol addition product occurs readily to yield the α,β-unsaturated ketone as the isolated product.

p-Chlorobenzaldehyde   2,2-Diphenylcyclohexanone   (Not isolated)

2-(*p*-Chlorobenzylidene)-6,6-diphenylcyclo-
hexanone (84%)

(*e*)   The aldehyde given as the starting material is called *furfural* and is based on a furan unit as an aromatic ring. Furfural cannot form an enolate. It reacts with the enolate of acetone in a manner much as benzaldehyde would.

Furfural   Acetone   (Not isolated)

4-Furyl-3-buten-2-one (60–66%)

(*f*)   Lithium dialkylcuprates transfer an alkyl group to the β carbon atom of α,β-unsaturated ketones.

2,4,4-Trimethyl-2-
cyclohexenone   2,3,4,4-Tetramethylcyclohexanone

A mixture of stereoisomers was obtained in 67 percent yield in this reaction.

(g) There are two nonequivalent α carbon atoms in the starting ketone. While enolate formation is possible at either position, only reaction at the methylene carbon leads to an intermediate that can undergo dehydration.

Observed product
(75% yield)

Reaction at the other α position gives an intermediate that cannot dehydrate.

(Cannot dehydrate; reverts
to starting materials)

18.25 (a) Conversion of 3-pentanone to 2-bromo-3-pentanone is best accomplished by acid-catalyzed bromination via the enol. Bromine in acetic acid is the customary reagent for this transformation.

3-Pentanone   2-Bromo-3-pentanone

(b) Once 2-bromo-3-pentanone has been prepared, its dehydrohalogenation by base converts it to the desired α,β-unsaturated ketone 1-penten-3-one.

2-Bromo-3-pentanone   1-Penten-3-one

Potassium *tert*-butoxide is a good base for bringing about elimination reactions of secondary alkyl halides; suitable solvents include *tert*-butyl alcohol and dimethyl sulfoxide.

(c) Reduction of the carbonyl group of 1-penten-3-one converts it to the desired alcohol.

1-Penten-3-one   1-Penten-3-ol

Catalytic hydrogenation would not be suitable for this reaction, because reduction of the double bond would accompany carbonyl reduction.

(d) Conversion of 3-pentanone to 3-hexanone requires addition of a methyl group to the β carbon atom.

$$CH_3CH_2CH_2\overset{\overset{\displaystyle O}{\|}}{C}CH_2CH_3 \implies {}^-:CH_3 + CH_2{=}CH\overset{\overset{\displaystyle O}{\|}}{C}CH_2CH_3$$

The best way to add an alkyl group to the β carbon of a ketone is via <u>conjugate addition of a dialkylcuprate reagent</u> to an α,β-unsaturated ketone.

$$CH_2{=}CH\overset{\overset{\displaystyle O}{\|}}{C}CH_2CH_3 \quad \xrightarrow[\text{2. }H_2O]{\text{1. }LiCu(CH_3)_2} \quad CH_3CH_2CH_2\overset{\overset{\displaystyle O}{\|}}{C}CH_2CH_3$$

1-Penten-3-one
[prepared as described in part (b)]

3-Hexanone

(e) In this problem a methyl group must be added to the α carbon atom of 3-pentanone. α alkylation of ketones can be achieved by way of the corresponding enamine.

$$CH_3CH_2\overset{\overset{\displaystyle O}{\|}}{C}CH_2CH_3 + \underset{\underset{\displaystyle H}{\overset{\displaystyle |}{N}}}{\bigcirc} \xrightarrow[-H_2O]{\text{benzene}} CH_3CH{=}\overset{\overset{\displaystyle N}{|}}{C}CH_2CH_3 \xrightarrow[\text{2. }H_2O]{\text{1. }CH_3I} CH_3\overset{\overset{\displaystyle O}{|}\atop|}{C}HC\overset{\displaystyle \|}{C}CH_2CH_3$$

3-Pentanone   Pyrrolidine   Pyrrolidine enamine of   2-Methyl-3-pentanone
                                              3-pentanone

(f) The compound to be prepared is the mixed aldol condensation product of 3-pentanone and benzaldehyde.

$$\underset{\underset{\displaystyle C_6H_5CH}{\|}}{CH_3\overset{\overset{\displaystyle O}{\|}}{C}CH_2CH_3} \implies \underset{\underset{\displaystyle C_6H_5CH}{\overset{\displaystyle \|}{O}}}{CH_3\overset{..}{\overset{\displaystyle |}{C}}H\overset{\overset{\displaystyle O}{\|}}{C}CH_2CH_3}$$

2-Methyl-1-phenyl-1-penten-3-one

The desired reaction sequence is:

$$CH_3CH_2\overset{\overset{\displaystyle O}{\|}}{C}CH_2CH_3 \xrightarrow{HO^-} CH_3\overset{..}{C}H\overset{\overset{\displaystyle O}{\|}}{C}CH_2CH_3 \xrightarrow{C_6H_5CH\overset{\displaystyle \|}{}{}}$$

3-Pentanone   Enolate of 3-pentanone

$$CH_3\underset{\underset{\displaystyle C_6H_5CHOH}{\overset{\displaystyle |}{}}}{C}H\overset{\overset{\displaystyle O}{\|}}{C}CH_2CH_3 \xrightarrow{-H_2O} CH_3\underset{\underset{\displaystyle C_6H_5CH}{\overset{\displaystyle \|}{}}}{\overset{\overset{\displaystyle O}{\|}}{C}}CH_2CH_3$$

Aldol addition product
(not isolated; dehydration
occurs under conditions
of its formation)

2-Methyl-1-phenyl-
1-penten-3-one

18.26 (*a*)   The first step is an α halogenation of a ketone. This is customarily accomplished under conditions of acid catalysis.

$$(CH_3)_3CCCH_3 \xrightarrow[H^+]{Br_2} (CH_3)_3CCCH_2Br$$

3,3-Dimethyl-2-butanone   1-Bromo-3,3-dimethyl-2-butanone (58%)

In the second step the carbonyl group of the α-bromo ketone is reduced to a secondary alcohol. As actually carried out, sodium borohydride in water was used to achieve this transformation.

$$(CH_3)_3CCCH_2Br \xrightarrow[H_2O]{NaBH_4} (CH_3)_3CCHCH_2Br$$

1-Bromo-3,3-dimethyl-2-butanone   1-Bromo-3,3-dimethyl-2-butanol (54%)

The third step is conversion of a vicinal bromohydrin to an epoxide in aqueous base.

$$(CH_3)_3CCHCH_2Br \xrightarrow[H_2O]{KOH} (CH_3)_3CC{-}CH_2$$

1-Bromo-3,3-dimethyl-2-butanol   2-*tert*-Butyloxirane (68%)

(*b*)   The overall yield is the product of the yields of the individual steps.

$$\text{Yield} = 100(0.58 \times 0.54 \times 0.68) = 21\%$$

18.27   All these problems begin in the same way, with exchange of all the α protons for deuterium (Sec. 18.7).

Cyclopentanone   Cyclopentanone-2,2,5,5-d_4

Once the tetradeuterated cyclopentanone has been prepared, functional group transformations are employed to convert it to the desired products.

(*a*)   Reduction of the carbonyl group can be achieved by using any of the customary reagents.

Cyclopentanone-2,2,5,5-d_4 Cyclopentanol-2,2,5,5-d_4

(*b*)   Acid-catalyzed dehydration of the alcohol prepared in part (*a*) yields the desired alkene.

Cyclopentanol-2,2,5,5-d_4 Cyclopentene-1,3,3-d_3

(c) Catalytic hydrogenation of the alkene in part (b) yields cyclopentane-1,1,3-d₃.

Cyclopentene-1,3,3-d₃   Hydrogen   Cyclopentane-1,1,3-d₃

(d) Carbonyl reduction of the tetradeuterated ketone under Wolff-Kishner conditions furnishes the desired product.

Cyclopentanone-2,2,5,5-d₄   Cyclopentane-1,1,3,3-d₄

Alternatively, Clemmensen reduction conditions (Zn, HCl) might be used.

18.28 (a) Hydroformylation converts alkenes to aldehydes having one more carbon atom by reaction with carbon monoxide and hydrogen in the presence of a cobalt octacarbonyl catalyst.

$$CH_3CH{=}CH_2 + CO + H_2 \xrightarrow{Co_2(CO)_8} CH_3CH_2CH_2\overset{\displaystyle O}{\overset{\|}{C}}H$$

Propene   Carbon   Hydrogen   Butanal
            monoxide

(b) Aldol condensation of acetaldehyde to 2-butenal, followed by catalytic hydrogenation of the carbon-carbon double bond, gives butanal.

$$2CH_3\overset{\displaystyle O}{\overset{\|}{C}}H \xrightarrow[heat]{NaOH} CH_3CH{=}CH\overset{\displaystyle O}{\overset{\|}{C}}H \xrightarrow[Ni]{H_2} CH_3CH_2CH_2\overset{\displaystyle O}{\overset{\|}{C}}H$$

Acetaldehyde   2-Butenal   Butanal

18.29 (a) The first conversion is the α halogenation of an aldehyde. As described in Sec. 18.2, this particular conversion has been achieved in 80 percent yield simply by treatment with bromine in chloroform.

Cyclohexanecarbaldehyde   1-Bromocyclohexanecarbaldehyde

Dehydrohalogenation of this compound can be accomplished under E2 conditions by treatment with base. Sodium methoxide in methanol would be appropriate, for example, although almost any alkoxide could be employed to dehydrohalogenate this tertiary bromide.

1-Bromocyclohexanecarbaldehyde   Cyclohexene-1-carbaldehyde

As the reaction was actually carried out, the bromide was heated with the weak base N,N-diethylaniline to effect dehydrobromination in 71 percent yield.

(b) Cleavage of vicinal diols to carbonyl compounds can be achieved by using periodic acid (HIO_4) (Sec. 15.12).

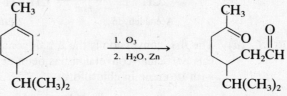

trans-1,2-Cyclohexanediol   1,6-Hexanedial

The conversion of this dialdehyde to cyclopentene-1-carbaldehyde is an intramolecular aldol condensation and is achieved by treatment with potassium hydroxide.

Cyclopentene-1-carbaldehyde

As the reaction was actually carried out, cyclopentene-1-carbaldehyde was obtained in 58 percent yield from *trans*-1,2-cyclohexanediol by this method.

(c) The first transformation requires an oxidative cleavage of a carbon-carbon double bond. Ozonolysis followed by hydrolysis in the presence of zinc is indicated.

4-Isopropyl-1-methylcyclohexene   3-Isopropyl-6-oxoheptanal

Cyclization of the resulting ketoaldehyde is an intramolecular aldol condensation. Base is required.

(*d*)   The first step in this synthesis is the hydration of the alkene function to an alcohol. Notice that this hydration must take place with a regioselectivity opposite to that of Markovnikov's rule and therefore requires a hydroboration-oxidation sequence.

$$(CH_3)_2C{=}CHCH_2CH_2\overset{\displaystyle O}{\overset{\|}{C}}CH_3 \xrightarrow[\text{2. } H_2O_2,\ HO^-]{\text{1. } B_2H_6,\ THF} (CH_3)_2CHCHCH_2CH_2\overset{\displaystyle O}{\overset{\|}{C}}CH_3$$
$$\underset{\text{OH}}{|}$$

6-Methyl-5-hepten-2-one   5-Hydroxy-6-methyl-2-heptanone

Conversion of the secondary alcohol function to a carbonyl group can be achieved with any of a number of oxidizing agents.

$$(CH_3)_2CHCHCH_2CH_2\overset{\displaystyle O}{\overset{\|}{C}}CH_3 \xrightarrow{H_2CrO_4} (CH_3)_2CH\overset{\displaystyle O}{\overset{\|}{C}}CH_2CH_2\overset{\displaystyle O}{\overset{\|}{C}}CH_3$$
$$\underset{\text{OH}}{|}$$

5-Hydroxy-6-methyl-2-heptanone   6-Methyl-2,5-heptanedione

Cyclization of the dione to the final product is a base-catalyzed intramolecular aldol condensation and was accomplished in 71 percent yield by treatment of the dione with a 2 percent solution of sodium hydroxide in aqueous ethanol.

$$(CH_3)_2CH\overset{\displaystyle O}{\overset{\|}{C}}CH_2CH_2\overset{\displaystyle O}{\overset{\|}{C}}CH_3 \xrightarrow{HO^-} (CH_3)_2CH\overset{\displaystyle O}{\overset{\|}{C}}CH_2CH_2\overset{\displaystyle O^-}{\overset{\|}{C}}{=}CH_2$$

18.30   Intramolecular aldol condensations occur best when a five- or six-membered ring is formed. Therefore, carbon-carbon bond formation involves the aldehyde and the methyl group attached to the ketone carbonyl.

2,2-Dimethyl-4-oxopentanal

4,4-Dimethyl-2-cyclopentenone (63%)

18.31 (a) By realizing that the primary alcohol function of the target molecule can be introduced by reduction of an aldehyde, it can be seen that the required carbon skeleton is the same as that of the aldol addition product of 2-methylpropanal.

$$(CH_3)_2CHCHCHCCH_2OH \Rightarrow (CH_3)_2CHCHCHC\overset{O}{\underset{H}{\parallel}} \Rightarrow 2(CH_3)_2CHCH\overset{O}{\parallel}$$

The synthetic sequence is:

$$(CH_3)_2CHCH\overset{O}{\parallel} \xrightarrow[\text{ethanol}]{\text{NaOH}} (CH_3)_2CHCHCHC\overset{O}{\underset{H}{\parallel}} \xrightarrow[\text{CH}_3\text{OH}]{\text{NaBH}_4} (CH_3)_2CHCHCHCCH_2OH$$

2-Methylpropanal   3-Hydroxy-2,2,4-   2,2,4-Trimethyl-1,3-pentanediol
                              trimethylpentanal

The starting aldehyde is prepared by oxidation of 2-methyl-1-propanol.

$$(CH_3)_2CHCH_2OH \xrightarrow[\text{CH}_2\text{Cl}_2]{(C_5H_5N)_2CrO_3} (CH_3)_2CHCH\overset{O}{\parallel}$$

2-Methyl-1-propanol   2-Methylpropanal

(b) Retrosynthetic analysis of the desired product shows that the carbon skeleton can be constructed by a mixed aldol condensation between benzaldehyde and propanal.

$$C_6H_5CH=CCH_2OH \Rightarrow C_6H_5CH=CCH\overset{O}{\parallel} \Rightarrow C_6H_5CH\overset{O}{\parallel} + CH_3CH_2CH\overset{O}{\parallel}$$

The reaction scheme therefore becomes:

$$C_6H_5CH\overset{O}{\parallel} + CH_3CH_2CH\overset{O}{\parallel} \xrightarrow{\text{HO}^-} C_6H_5CH=CCH\overset{O}{\parallel}$$
                                                                    CH_3

Benzaldehyde   Propanal   2-Methyl-3-phenyl-2-propenal

Reduction of the aldehyde to the corresponding primary alcohol gives the desired compound.

$$C_6H_5CH=CCH\overset{O}{\parallel} \xrightarrow[\text{or NaBH}_4, \text{ CH}_3\text{OH}]{\text{LiAlH}_4, \text{ then H}_2\text{O}} C_6H_5CH=CCH_2OH$$
      CH_3 CH_3

2-Methyl-3-phenyl-2-propenal   2-Methyl-3-phenyl-2-propen-1-ol

The starting materials for the mixed aldol condensation, benzaldehyde and propanal, are prepared by oxidation of benzyl alcohol and 1-propanol, respectively.

$$C_6H_5CH_2OH \xrightarrow[\text{CH}_2\text{Cl}_2]{(C_5H_5N)_2CrO_3} C_6H_5CH\overset{O}{\parallel}$$

Benzyl alcohol   Benzaldehyde

$$CH_3CH_2CH_2OH \xrightarrow[\text{CH}_2\text{Cl}_2]{(C_5H_5N)_2CrO_3} CH_3CH_2CH\overset{O}{\parallel}$$

1-Propanol   Propanal

(c) The cyclohexene ring in this case can be assembled by a Diels-Alder reaction.

1,3-Butadiene is one of the given starting materials; the α,β-unsaturated ketone is the mixed aldol condensation product of 4-methylbenzaldehyde and acetophenone.

The complete synthetic sequence is:

4-Methylbenzyl alcohol   4-Methylbenzaldehyde

4-Methylbenzaldehyde   Acetophenone

trans-4-Benzoyl-5-(4-methylphenyl)cyclohexene

α,β-Unsaturated ketones are good dienophiles in Diels-Alder reactions.

18.32   It is the carbon atom flanked by two carbonyl groups that is involved in the enolization of terreic acid.

Terreic acid   Enol A   Enol B

Of these two structures, enol A, with its double bond conjugated to two carbonyl groups, is more stable than enol B, in which the double bond is conjugated to only one carbonyl.

18.33   *(a)*   Recall that aldehydes and ketones are in equilibrium with their *hydrates* in aqueous solution. Thus, the principal substance present when $(C_6H_5)_2CHCH=O$ is dissolved in water is $(C_6H_5)_2CHCH(OH)_2$ (81%).

(*b*) The problem states that the major species present in aqueous base is *not* $(C_6H_5)_2CHCH{=}O$, its enol, or its hydrate. The most reasonable species is the *enolate ion*:

$$(C_6H_5)_2\overset{..}{C}{-}CH{=}\overset{..}{O}: \longleftrightarrow (C_6H_5)_2C{=}CH{-}\overset{..}{\underset{..}{O}}:^-$$

18.34 (*a*) At first glance this transformation seems to be an internal oxidation-reduction reaction. An aldehyde function is reduced to a primary alcohol, while a secondary alcohol is oxidized to a ketone.

$$\underset{\underset{\text{Compound A}}{\overset{\displaystyle |}{\underset{}{OH}}}}{\overset{\displaystyle O}{\underset{}{C_6H_5\overset{}{C}HCH}}} \xrightarrow[\text{H}_2\text{O}]{\text{HO}^-} \underset{\text{Compound B}}{\overset{\displaystyle O}{C_6H_5CCH_2OH}}$$

Once one realizes that enolization can occur, however, a simpler explanation, involving only proton-transfer reactions, emerges:

$$\underset{\underset{\text{Compound A}}{\overset{\displaystyle |}{\underset{}{OH}}}}{\overset{\displaystyle O}{C_6H_5\overset{}{C}HCH}} \rightleftharpoons[\ ]{\text{HO}^-,\ \text{H}_2\text{O}} \underset{\underset{\text{Enol form of compound A}}{\overset{\displaystyle |}{\underset{}{OH}}}}{C_6H_5C{=}CHOH}$$

The enol form of compound A is an enediol; it is at the same time the enol form of compound B. The enediol can revert to compound A or to compound B.

$$\underset{\underset{}{\overset{\displaystyle |}{\underset{}{OH}}}}{C_6H_5C{=}CHOH} \xrightarrow{\text{HO}^-,\ \text{H}_2\text{O}} \begin{cases} \overset{\displaystyle O}{C_6H_5CCH_2OH} \quad \text{(Compound B)} \\[2em] \underset{\overset{\displaystyle |}{OH}}{\overset{\displaystyle O}{C_6H_5CHCH}} \quad \text{(Compound A)} \end{cases}$$

At equilibrium, compound B predominates because it is more stable than A. A ketone carbonyl is more stabilized than an aldehyde, and the carbonyl in B is conjugated with the benzene ring.

(*b*) The isolated product is the double hemiacetal formed between two molecules of compound A.

Compound C

18.35 (*a*) The only stereogenic center in piperitone is adjacent to a carbonyl group. Base-catalyzed enolization causes this carbon to lose its stereochemical integrity.

(−)-Piperitone   Enolate of piperitone   Enol of piperitone

Both the enolate and enol of piperitone are achiral and can revert only to a racemic mixture of piperitones.

(*b*) The enol formed from menthone can revert to either menthone or isomenthone.

Menthone   Enol form   Isomenthone

Only the stereochemistry at the α carbon atom is affected by enolization. The other stereogenic center in menthone (the one bearing the methyl group) is not affected.

18.36 In all parts of this problem the bonding change that takes place is described by the general equation

$$HX-N=Z \rightleftharpoons X=N-ZH$$

(*a*) The compound given is nitrosoethane. Nitrosoalkanes are less stable than their oxime isomers formed by proton transfer.

$$CH_3\overset{\displaystyle |}{\underset{\displaystyle H}{CH}}-N=O \rightleftharpoons CH_3CH=N-OH$$

Nitrosoethane   Acetaldehyde oxime
(less stable)   (more stable)

(*b*) You may recognize this compound as an enamine. It is slightly different, however, from the enamines we discussed earlier in that nitrogen bears a hydrogen substituent. Stable enamines are compounds of the type

where neither R group is hydrogen; both R's must be alkyl or aryl. Enamines that bear a hydrogen substituent are converted to imines in a proton-transfer equilibrium.

$$(CH_3)_2C=CH-\overset{\displaystyle |}{\underset{\displaystyle H}{N}}CH_3 \rightleftharpoons (CH_3)_2CH-CH=NCH_3$$

Enamine (less stable)   Imine (more stable)

(c) The compound given is known as a *nitronic acid*; its more stable tautomeric form is a nitroalkane.

$$CH_3CH=\overset{+}{N}\overset{O^-}{\underset{OH}{}} \rightleftharpoons CH_3CH_2-\overset{+}{N}\overset{O^-}{\underset{O}{}}$$

Nitronic acid   Nitroalkane

(d) The six-membered ring is aromatic in the tautomeric form derived from the compound given.

(e) This compound is called *isourea*. Urea has a carbon-oxygen double bond and is more stable.

$$HN=C\overset{OH}{\underset{NH_2}{}} \rightleftharpoons H_2N-C\overset{O}{\underset{NH_2}{}}$$

Isourea (less stable)   Urea (more stable)

18.37 (a) This reaction is an intramolecular alkylation of a ketone. While alkylation of a ketone with a separate alkyl halide molecule is usually difficult, *intramolecular* alkylation reactions can be carried out effectively. The enolate formed by proton abstraction from the α carbon atom carries out a nucleophilic attack on the carbon that bears the leaving group.

(b) The starting material, known as *citral,* is converted to the two products by a reversal of an aldol condensation. The first step is conjugate addition of hydroxide.

$$(CH_3)_2C=CHCH_2CH_2C=CH-CH \longrightarrow (CH_3)_2C=CHCH_2CH_2C-CH=CH$$

$$\downarrow H_2O$$

$$(CH_3)_2C=CHCH_2CH_2C-CH_2CH$$

The product of this conjugate addition is a β-hydroxy ketone. It undergoes base-catalyzed cleavage to the observed products.

(c) The product is formed by an intramolecular aldol condensation reaction.

(d) In this problem stereochemical isomerization involving a proton attached to the α carbon atom of a ketone takes place. Enolization of the ketone yields an intermediate in which the stereochemical integrity of the α carbon is lost. Reversion to ketone eventually leads to the formation of the more stable stereoisomer at equilibrium.

Less stable ketone;   Enol   More stable ketone;
starting material   preferred at equilibrium

The rate of enolization is increased by heating, or by base catalysis. The cis ring fusion in the product is more stable than the trans because there are not enough atoms in the six-membered ring to span *trans*-1,2 positions in the four-membered ring without excessive strain.

(e) Working backward from the product, we can see that the transformation involves two aldol condensations, one intermolecular and the other intramolecular.

The first reaction is a mixed aldol condensation between the enolate of dibenzyl ketone and one of the carbonyl groups of the dione.

$$C_6H_5CH_2CCH_2C_6H_5 + C_6H_5CCC_6H_5 \longrightarrow C_6H_5CCCH_2C_6H_5$$

Dibenzyl ketone   Benzil

This is followed by an intramolecular aldol condensation.

2,3,4,5-Tetraphenylcyclopentadienone

(f) This is a fairly difficult problem, since it is not obvious at the outset which of the two possible enolates of benzyl ethyl ketone is the one that undergoes conjugate addition to the α,β-unsaturated ketone. A good idea here is to work backward from the final product—in effect, do a retrosynthetic analysis. The first step is to recognize that the enone arises by dehdyration of a β-hydroxy ketone.

Now, mentally disconnect the bond between the α carbon atom and the carbon that bears the hydroxyl group to reveal the intermediate that undergoes intramolecular aldol condensation.

The β-hydroxy ketone is the intermediate formed in the intramolecular aldol addition step, and the diketone that leads to it is the intermediate that is formed in the conjugate addition step. The relationship of the starting materials to the intermediates and product is now more evident.

Intermediate formed in
conjugate addition step

18.38 *(a)* The reduced C=O stretching frequency of α,β-unsaturated ketones is consistent with an enhanced degree of single bond character as compared with simple dialkyl ketones.

Resonance is more important in α,β-unsaturated ketones. Conjugation of the carbonyl group with the carbon-carbon double bond increases opportunities for electron delocalization.

(b) Even more single bond character is indicated in the carbonyl group of cyclopropenone than in that of typical α,β-unsaturated ketones. The dipolar resonance form contributes substantially to the electron distribution because of the aromatic character of the three-membered ring.

equivalent to an oxyanion-substituted cyclopropenyl
cation

(c) The dipolar resonance form is a more important contributor to the electron distribution in diphenylcyclopropenone than in benzophenone.

is more pronounced than

The dipolar resonance form of diphenylcyclopropenone has aromatic character. Its stability leads to increased charge separation and a larger dipole moment.

(d) Decreased electron density at the β carbon atom of an α,β-unsaturated ketone is responsible for its decreased shielding. The decreased electron density arises from the polarization of its π electrons as represented by a significant contribution of the dipolar resonance form.

18.39 Bromination can occur at either of the two α carbon atoms.

3-Methyl-2-butanone   1-Bromo-3-methyl-2-butanone   3-Bromo-3-methyl-2-butanone

The ^{1}H nmr spectrum of the major product, compound D, is consistent with the structure of 1-bromo-3-methyl-2-butanone. The minor product E is identified as 3-bromo-3-methyl-2-butanone on the basis of its nmr spectrum.

Compound D   Compound E

18.40 Three dibromination products are possible from α halogenation of 2-butanone:

1,1-Dibromo-2-butanone   1,3-Dibromo-2-butanone   3,3-Dibromo-2-butanone

The product is *1,3-dibromo-2-butanone*, on the basis of its observed ^{1}H nmr, which showed two signals at low field. One is a two-proton singlet at $\delta = 4.6$ ppm assignable to CH_2Br and the other a one-proton quartet at $\delta = 5.2$ ppm assignable to CHBr.

SELF-TEST

PART A

A-1. Write the correct structure(s) for each of the following:
  (a) The two enol forms of 2-butanone
  (b) The enolate ion derived from reaction of 1,3-cyclohexanedione with sodium methoxide

A-2. Give the correct structures for compounds A through D in the following reaction schemes:

(a) $2C_6H_5CH_2\overset{\displaystyle O}{\overset{\|}{C}}H \xrightarrow[\substack{2.\ heat \\ (-H_2O)}]{1.\ HO^-} A$

(b) $CH_3CH_2CH{=}CH\overset{\displaystyle O}{\overset{\|}{C}}CH_2CH_3 + LiCu(CH_2CH_3)_2 \xrightarrow[2.\ H_2O]{1.\ ether} B$

(c) $C_6H_5\overset{\displaystyle O}{\overset{\|}{C}}CH_3 + $ $\xrightarrow[(-H_2O)]{H^+} C \xrightarrow[2.\ H_2O]{1.\ CH_2=CHCH_2Br} D$

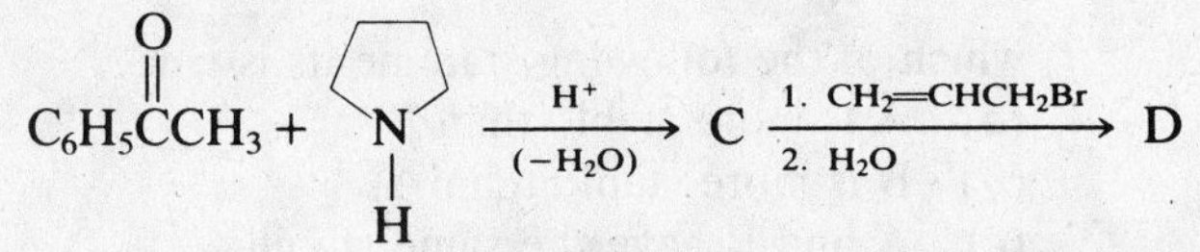

A-3. Write the structures of all the possible aldol addition products which may be obtained by reaction of a mixture of propanal and 2-methylpropanal with base.

$$CH_3CH_2\overset{\displaystyle O}{\overset{\|}{C}}H \qquad CH_3\overset{\displaystyle CH_3}{\overset{|}{C}}H\overset{\displaystyle}{\underset{\displaystyle \underset{\|}{O}}{C}}H$$

Propanal   2-Methylpropanal

A-4. Using any necessary organic or inorganic reagents, outline a synthesis of 1,3-butanediol from ethanol as the only source of carbons.

A-5. Outline a series of reaction steps which will allow the preparation of compound F from 1,3-cyclopentanedione, compound E.

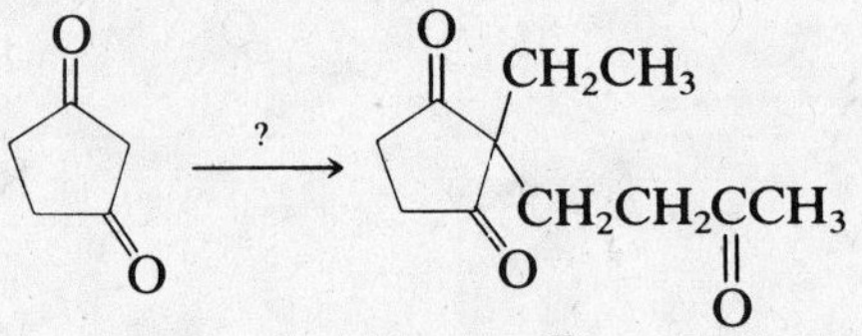

E   F

A-6. Give the structure of the product formed in each of the following reactions:

(a) $CH_3CH_2CH_2CH_2\overset{\displaystyle O}{\overset{\|}{C}}H \xrightarrow[acetic\ acid]{Br_2}$

(b) $2CH_3CH_2CH_2CH_2\overset{\displaystyle O}{\overset{\|}{C}}H \xrightarrow[5°C]{NaOH}$

(c) $\xrightarrow[(-H_2O)]{NaOH,\ heat}$

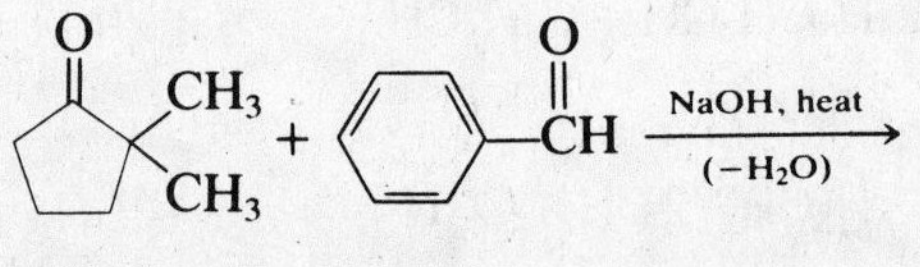

(d) 2 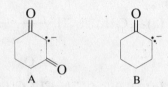 $\xrightarrow[\text{heat}]{\text{Al[OC(CH}_3)_3]_3}$

(e) C₆H₅—CH=CHĊ—C₆H₅ $\xrightarrow[\text{ethanol}]{\text{NaSCH}_3}$

PART B

B-1. When enolate A is compared with enolate B

which of the following statements is true?
(a) A is more stable than B.
(b) B is more stable than A.
(c) A and B have the same stability.
(d) No comparison of stability can be made.

B-2. Which one of the following molecules is most likely to contain deuterium (^{2}H = D) after reaction with NaOD in D$_2$O?

(a) $C_6H_5\overset{O}{\overset{\|}{C}}H$

(c) $C_6H_5\overset{O}{\overset{\|}{C}}C(CH_3)_3$

(b) $C_6H_5CH_2\overset{O}{\overset{\|}{C}}H$

(d) $(CH_3)_3\overset{O}{\overset{\|}{C}}CC(CH_3)_3$

B-3. Which of the following RX compounds is (are) the best alkylating agent(s) in the alkylation reaction shown?

+ RX $\xrightarrow{\text{NaOC}_2\text{H}_5}$

$(CH_3)_3CBr$ $(CH_3)_2CHCH_2OH$ C_6H_5Br $(CH_3)_2CHCH_2Br$
        I   II   II   IV

(a) I and IV   (c) II and IV
(b) IV only   (d) I, II, and IV

B-4. Which of the following pairs of aldehydes gives a single product in a mixed aldol condensation?

(a) $C_6H_5CH_2\overset{O}{\overset{\|}{C}}H + C_6H_5\overset{O}{\overset{\|}{C}}H$

(c) $C_6H_5\overset{O}{\overset{\|}{C}}H + H_2C{=}O$

(b) $C_6H_5\overset{O}{\overset{\|}{C}}H + (CH_3)_3\overset{O}{\overset{\|}{C}}CH$

(d) $CH_3\overset{O}{\overset{\|}{C}}H + (CH_3)_2CH\overset{O}{\overset{\|}{C}}H$

B-5. What is the principal product of the following reaction?

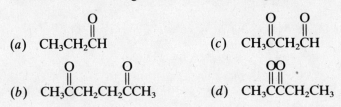

(a) CH=O / CH=O structure

(c) CH=O structure

(b) naphthalene-diol with OH, OH

(d) structure with two O

B-6. Which of the following forms an enol to the greatest extent?

(a) $CH_3CH_2\overset{\displaystyle O}{\overset{\displaystyle \|}{C}}H$

(c) $CH_3\overset{\displaystyle O}{\overset{\displaystyle \|}{C}}CH_2\overset{\displaystyle O}{\overset{\displaystyle \|}{C}}H$

(b) $CH_3\overset{\displaystyle O}{\overset{\displaystyle \|}{C}}CH_2CH_2\overset{\displaystyle O}{\overset{\displaystyle \|}{C}}CH_3$

(d) $CH_3\overset{\displaystyle OO}{\overset{\displaystyle \|\|}{C}}CCH_2CH_3$

CARBOXYLIC ACIDS

IMPORTANT TERMS AND CONCEPTS

Nomenclature (Sec. 19.1) Systematic names of carboxylic acids are derived by locating the longest carbon atom chain containing the *carboxyl group* and replacing the *-e* ending of the corresponding alkane with *-oic acid*. The locants used to identify substituents are always numbered by beginning with the carboxyl carbon. For example,

$$C_6H_5CH_2CH_2CO_2H$$

3-Phenylpropanoic acid

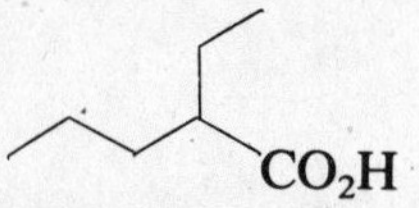

2-Ethylpentanoic acid

As noted in text Table 19.1, certain carboxylic acids are referred to by accepted "common" names. Noteworthy among these are formic and acetic acids (methanoic and ethanoic acid, by the IUPAC rules):

$$HCO_2H \qquad CH_3CO_2H$$

Formic acid · · · · · · · Acetic acid

Structure, Bonding, and Physical Properties (Secs. 19.2, 19.3) The carbon of the carboxyl group is sp^2 hybridized, and it and the three atoms bonded to it lie in the same plane. As a result of strong hydrogen bonding attractions, carboxylic acids exist as cyclic dimers, which often persist in the gas phase. The melting and boiling points of carboxylic acids are much higher than would be expected on the basis of molecular weight.

Acid-Base Properties (Secs. 19.4 to 19.7) Carboxylic acids are the most acidic class of organic compounds containing only C, H, and O atoms. They are, however, considered weak acids, having ionization constants of about 10^{-5}.

$$R\text{---}CO_2H \rightleftharpoons H^+ + R\text{---}CO_2^- \qquad K_a \sim 10^{-5}$$

Carboxylic acid · · · · · · · Carboxylate ion

The acidity of carboxylic acids may be explained by the resonance stabilization of the carboxylate ion. The negative charge is delocalized and shared equally by both oxygens

Electronegative substituents, particularly on the α carbon, increase the acidity of carboxylic acids. For example, the pK_a of chloroacetic acid is 2.9, whereas that of acetic acid is 4.7.

$$ClCH_2CO_2H \quad \text{is more acidic than} \quad CH_3CO_2H$$

Chloroacetic acid   Acetic acid

Benzoic acid is slightly more acidic than acetic acid, owing to the attachment of the carboxyl group to an sp^2 carbon atom. Substituents on the benzene ring affect the acidity of benzoic acid derivatives, the net effect arising as a result of the combination of inductive, field, resonance, and solvation effects.

Esterification (Secs. 19.14, 19.15) A mechanistic model for many of the reactions of carboxylic acid derivatives is provided by the acid-catalyzed esterification of carboxylic acids. A key aspect of the following mechanism is the formation of a *tetrahedral intermediate,* in which the carboxyl carbon has undergone a change from sp^2 hybridization to sp^3.

Tetrahedral intermediate

Hydroxy acids may undergo an intramolecular esterification to form cyclic esters known as *lactones.* For example, 4-hydroxybutanoic acid cyclizes to form a γ-*lactone*:

$$HO\text{—}\bigwedge\text{—}CO_2H \longrightarrow$$

Spectroscopic Properties (Sec. 19.18) Carboxylic acids have characteristic hydroxyl absorptions in the 3500 to 2500 cm^{-1} region of the infrared spectrum. In addition, the carbonyl of the carboxyl group appears as a strong band near 1700 cm^{-1}. The hydroxyl proton of a carboxylic acid is very deshielded relative to most other protons in the nmr spectrum and appears in the region $\delta = 10$ to 12 ppm.

IMPORTANT REACTIONS

SYNTHESIS OF CARBOXYLIC ACIDS

Carboxylation of Grignard Reagents (Sec. 19.11)

General:

$$RMgX + CO_2 \longrightarrow RCO_2MgX \xrightarrow{H_3O^+} RCO_2H$$

Example:

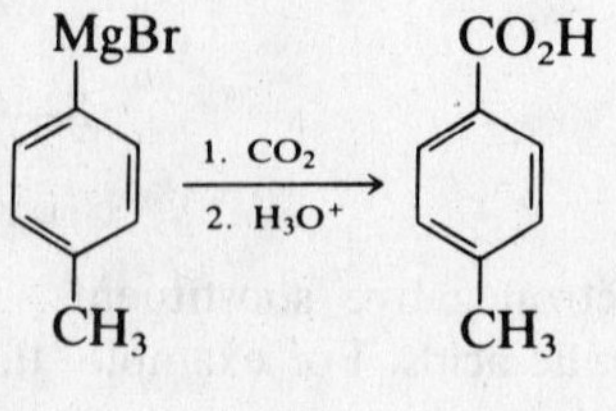

$$\text{(4-methylphenyl-MgBr)} \xrightarrow[\text{2. H}_3\text{O}^+]{\text{1. CO}_2} \text{(4-methylbenzoic acid, CO}_2\text{H)}$$

Preparation and Hydrolysis of Nitriles (Sec. 19.12)

General:

$$\text{R—X} + \text{CN}^- \longrightarrow \text{R—CN} \xrightarrow[\text{heat}]{\text{H}^+, \text{H}_2\text{O}} \text{R—CO}_2\text{H}$$

(primary halides best)

Example:

$$(\text{CH}_3)_2\text{CHCH}_2\text{CH}_2\text{Br} + \text{CN}^- \longrightarrow$$

$$(\text{CH}_3)_2\text{CHCH}_2\text{CH}_2\text{CN} \xrightarrow[\text{heat}]{\text{H}^+, \text{H}_2\text{O}} (\text{CH}_3)_2\text{CHCH}_2\text{CH}_2\text{CO}_2\text{H}$$

REACTIONS OF CARBOXYLIC ACIDS

α-Halogenation; the Hell-Volhard-Zelinsky Reaction (Sec. 19.16)

General:

$$\text{RCH}_2\text{CO}_2\text{H} \xrightarrow{\text{Br}_2, \text{PCl}_3} \overset{\overset{\displaystyle \text{Br}}{|}}{\text{RCHCO}_2\text{H}}$$

Example:

$$\text{CH}_3\text{—C}_6\text{H}_4\text{—CH}_2\text{CO}_2\text{H} \xrightarrow{\text{Br}_2, \text{PCl}_3} \text{CH}_3\text{—C}_6\text{H}_4\text{—}\overset{\overset{\displaystyle \text{Br}}{|}}{\text{CHCO}_2\text{H}}$$

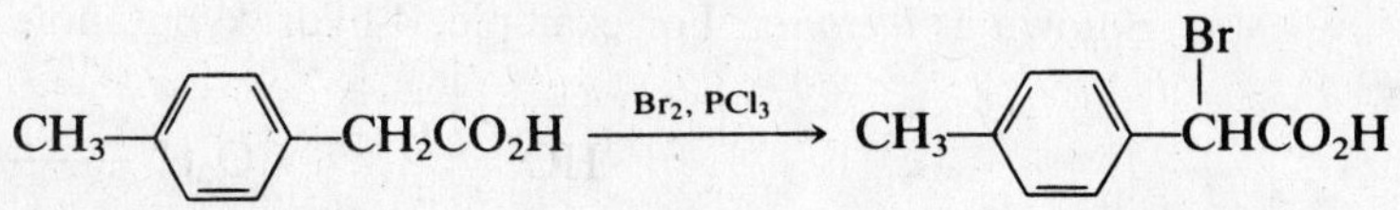

Decarboxylation (Sec. 19.17)

Malonic acid derivatives:

General:

$$\underset{\underset{\displaystyle \text{R} \quad \text{R}'}{}}{\text{HO}_2\text{CCCO}_2\text{H}} \xrightarrow{\text{heat}} \underset{\underset{\displaystyle \text{R}'}{}}{\text{RCHCO}_2\text{H}} + \text{CO}_2$$

Example:

$$\underset{\underset{\displaystyle \text{CH}_3}{}}{\text{CH}_3\text{CH}_2\text{C(CO}_2\text{H)}_2} \xrightarrow{\text{heat}} \overset{\overset{\displaystyle \text{CH}_3}{|}}{\text{CH}_3\text{CH}_2\text{CHCO}_2\text{H}}$$

β-Keto acids:

General:

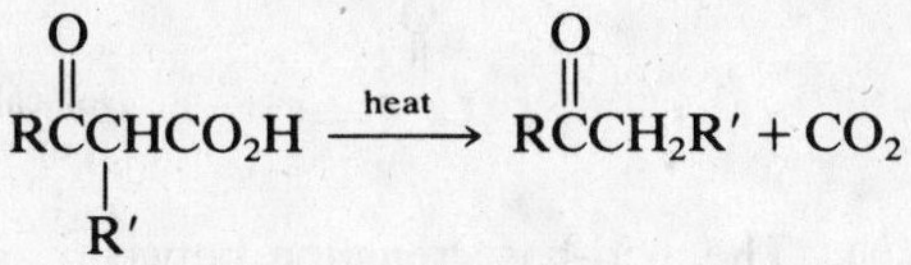

O
O
RCCHCO2H
heat
RCCH2R' + CO2
R'

Example:

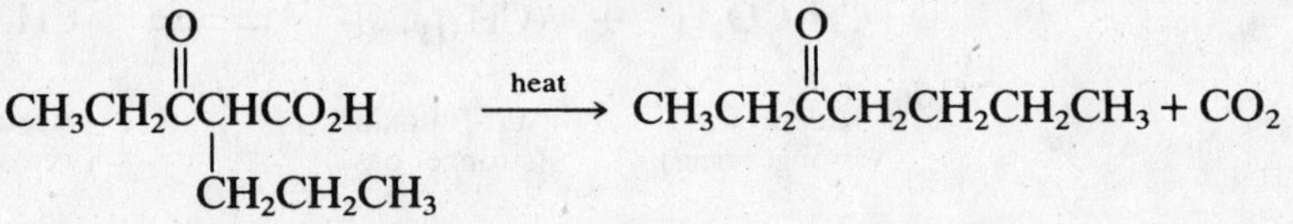

O
O
CH3CH2CCHCO2H
heat
CH3CH2CCH2CH2CH2CH3 + CO2
CH2CH2CH3

SOLUTIONS TO TEXT PROBLEMS

19.1 (*b*) The longest chain in pivalic acid contains three carbons, and therefore it is named as a derivative of propanoic acid. It is *2,2-dimethylpropanoic acid.*

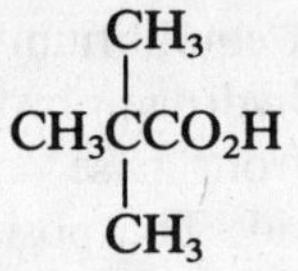

CH3
CH3CCO2H
CH3

2,2-Dimethylpropanoic acid (pivalic acid)

(*c*) The four carbon atoms of crotonic acid form a continuous chain. Since there is a double bond between C-2 and C-3, it is one of the stereoisomers of 2-butenoic acid. The stereochemistry of the double bond is *E*.

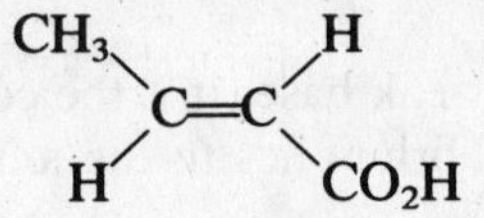

CH3
H
C=C
H
CO2H

(*E*)-2-Butenoic acid (crotonic acid)

(*d*) Oxalic acid is a dicarboxylic acid that contains two carbons. It is *ethanedioic acid.*

HO_2CCO_2H Ethanedioic acid (oxalic acid)

(*e*) The four carbons of maleic acid form a continuous chain with a double bond between C-2 and C-3; it is the *Z* stereoisomer of *butenedioic acid.* No number is necessary to locate the position of the double bond, because the structure permits it to be only at the position shown.

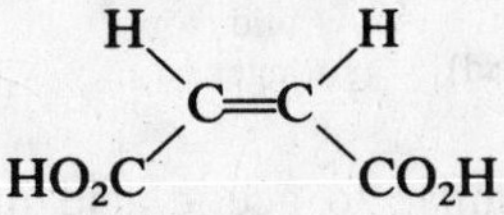

H
H
C=C
HO2C
CO2H

(*Z*)-Butenedioic acid (maleic acid)

(*f*) The systematic name for $C_6H_5CO_2H$ is benzoic acid. Since there is a methyl group at the para position, the compound shown is *p-methylbenzoic acid,* or *4-methylbenzoic acid.*

CH3
CO2H

p-Methylbenzoic acid or 4-methylbenzoic acid (*p*-toluic acid)

19.2 Ionization of peroxy acids such as peroxyacetic acid yields an anion that cannot be stabilized by resonance in the same way that acetate can.

$$CH_3\overset{\overset{\displaystyle O}{\|}}{C}O\text{—}O^-$$

Delocalization of negative charge into carbonyl group is not possible in peroxyacetate ion.

19.3 (b) The acid-base reaction between acetic acid and *tert*-butoxide ion is represented by the equation

$$CH_3CO_2H \ + \ (CH_3)_3CO^- \ \rightleftharpoons \ CH_3CO_2^- \ + \ (CH_3)_3COH$$

Acetic acid	*tert*-Butoxide	Acetate ion	*tert*-Butyl alcohol
(stronger acid)	(stronger base)	(weaker base)	(weaker acid)

Alcohols are weaker acids than carboxylic acids; the equilibrium lies to the right.

(c) Bromide ion is the conjugate base of hydrogen bromide, a strong acid.

$$CH_3CO_2H \ + \ Br^- \ \rightleftharpoons \ CH_3CO_2^- \ + \ HBr$$

Acetic acid	Bromide ion	Acetate ion	Hydrogen bromide
(weaker acid)	(weaker base)	(stronger base)	(stronger acid)

In this case, the position of equilibrium favors the starting materials, because acetic acid is a weaker acid than hydrogen bromide.

(d) Acetylide ion is a rather strong base, and acetylene, with a K_a of 10^{-25}, is a much weaker acid than acetic acid. The position of equilibrium favors the formation of products.

$$CH_3CO_2H \ + \ HC\equiv C:^- \ \rightleftharpoons \ CH_3CO_2^- \ + \ HC\equiv CH$$

Acetic acid	Acetylide ion	Acetate ion	Acetylene
(stronger acid)	(stronger base)	(weaker base)	(weaker acid)

(e) Nitrate ion is a very weak base; it is the conjugate base of the strong acid nitric acid. The position of equilibrium lies to the left.

$$CH_3CO_2H \ + \ NO_3^- \ \rightleftharpoons \ CH_3CO_2^- \ + \ HNO_3$$

Acetic acid	Nitrate ion	Acetate ion	Nitric acid
(weaker acid)	(weaker base)	(stronger base)	(stronger acid)

(f) Amide ion is a very strong base. The position of equilibrium lies to the right.

$$CH_3CO_2H \ + \ H_2N^- \ \rightleftharpoons \ CH_2CO_2^- \ + \ NH_3$$

Acetic acid	Amide ion	Acetate ion	Ammonia
(stronger acid)	(stronger base)	(weaker base)	(weaker acid)

19.4 (b) Propanoic acid is similar to acetic acid in its acidity. A hydroxyl group at C-2 is electron-withdrawing and stabilizes the carboxylate ion of lactic acid by a combination of inductive and field effects.

$$CH_3\underset{\underset{\textstyle OH}{\uparrow|}}{CH}\text{—}C\overset{\displaystyle O}{\underset{\displaystyle O^-}{<}}$$

Hydroxyl group stabilizes negative charge by attracting electrons.

Lactic acid is more acidic than propanoic acid. The measured ionization constants are:

$$CH_3\overset{\underset{\textstyle |}{OH}}{C}HCO_2H \qquad\qquad CH_3CH_2CO_2H$$

Lactic acid
K_a 1.4×10^{-4}
(pK_a 3.8)

Propanoic acid
K_a 1.3×10^{-5}
(pK_a 4.9)

(*c*) Changing the hybridization of C-2 from sp^3 to sp^2 increases its *s* character and makes it more electron-withdrawing; electron-withdrawing groups increase acidity. Acrylic acid, $CH_2{=}CHCO_2H$, is expected to be a stronger acid than 2-methylpropanoic acid.

$$CH_2{=}CHCO_2H \qquad\qquad CH_3\overset{\underset{\textstyle |}{CH_3}}{C}HCO_2H$$

Acrylic acid; C-2 is sp^2 hybridized
K_a 5.5×10^{-5}
(pK_a 4.3)

2-Methylpropanoic acid; C-2 is sp^3 hybridized
K_a 1.3×10^{-5}
(pK_a 4.9)

(*d*) The hybridization at C-2 is the same in both compounds, but a carbonyl group is more strongly electron-withdrawing than a carbon-carbon double bond. Pyruvic acid is a stronger acid than acrylic acid.

$$CH_3\overset{\overset{\textstyle O}{\|}}{C}CO_2H \qquad\qquad CH_2{=}CHCO_2H$$

Pyruvic acid
K_a 5.1×10^{-4}
(pK_a 3.3)

Acrylic acid
K_a 5.5×10^{-5}
(pK_a 4.3)

(*e*) Viewing the two compounds as substituted derivatives of acetic acid, RCH_2CO_2H,

we judge $CH_3\overset{\overset{\textstyle O}{\|}}{\underset{\underset{\textstyle O}{\|}}{S}}$ to be strongly electron-withdrawing and acid-strengthening, whereas

an ethyl group has only a small effect.

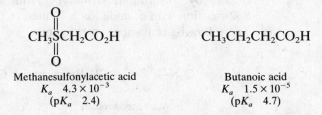

$$CH_3\overset{\overset{\textstyle O}{\|}}{\underset{\underset{\textstyle O}{\|}}{S}}CH_2CO_2H \qquad\qquad CH_3CH_2CH_2CO_2H$$

Methanesulfonylacetic acid
K_a 4.3×10^{-3}
(pK_a 2.4)

Butanoic acid
K_a 1.5×10^{-5}
(pK_a 4.7)

19.5 The compound can only be a carboxylic acid; no other class containing only carbon, hydrogen, and oxygen can be more acidic. A reasonable choice is $HC{\equiv}CCO_2H$; C-2 is *sp* hybridized and therefore rather electron-withdrawing and acid-strengthening. This is borne out by its measured ionization constant K_a, which is 1.4×10^{-2} (pK_a 1.8).

19.6 For carbonic acid, the "true K_1" is given by

$$\text{True } K_1 = \frac{[H^+][HCO_3^-]}{[H_2CO_3]}$$

Since

$$4.3 \times 10^{-7} = \frac{[H^+][HCO_3^-]}{[CO_2]}$$

we have $[H^+][HCO_3^-] = (4.3 \times 10^{-7})[CO_2]$

and   True $K_1 = \dfrac{(4.3 \times 10^{-7})[CO_2]}{[H_2CO_3]}$

$$= \frac{(4.3 \times 10^{-7})(99.7)}{(0.3)}$$

$$= 1.4 \times 10^{-4}$$

Thus, when corrected for the small degree to which carbon dioxide is hydrated, it can be seen that carbonic acid is actually a stronger acid than acetic acid. Carboxylic acids dissolve in sodium bicarbonate solution because the equilibrium leading to carbon dioxide formation is favorable, not because carboxylic acids are stronger acids than carbonic acid.

19.7 (*b*) 2-Chloroethanol has been converted to 3-hydroxypropanoic acid by way of the corresponding nitrile.

$$HOCH_2CH_2Cl \xrightarrow[H_2O]{NaCN} HOCH_2CH_2CN \xrightarrow[heat]{H_3O^+} HOCH_2CH_2CO_2H$$

   2-Chloroethanol   2-Cyanoethanol   3-Hydroxypropanoic acid

The presence of the hydroxyl group in 2-chloroethanol precludes the preparation of a Grignard reagent from this material, and so any attempt at the preparation of 3-hydroxypropanoic acid via the Grignard reagent of 2-chloroethanol is certain to fail.

(*c*) Grignard reagents can be prepared from tertiary halides and react in the expected manner with carbon dioxide. The procedure shown is entirely satisfactory.

$$(CH_3)_3CCl \xrightarrow[\substack{diethyl \\ ether}]{Mg} (CH_3)_3CMgCl \xrightarrow[2.\ H_3O^+]{1.\ CO_2} (CH_3)_3CCO_2H$$

tert-Butyl chloride   *tert*-Butylmagnesium chloride   2,2-Dimethylpropanoic acid
                                                                                                         (61–70%)

Preparation by way of the nitrile will not be feasible. Rather than react with sodium cyanide by substitution, *tert*-butyl chloride will undergo elimination exclusively. The S_N2 reaction with cyanide ion is limited to primary and secondary alkyl halides.

(*d*) The following sequence is satisfactory:

$$O_2N-\!\!\bigcirc\!\!-CH_2Br \xrightarrow[DMSO]{NaCN} O_2N-\!\!\bigcirc\!\!-CH_2CN$$

   p-Nitrobenzyl bromide   *p*-Nitrobenzyl cyanide

$$\Big\downarrow H_3O^+,\ heat$$

$$O_2N-\!\!\bigcirc\!\!-CH_2CO_2H$$

                                   p-Nitrophenylacetic acid

Nitro groups are incompatible with Grignard reagents, and so the alternate route via carboxylation of the Grignard reagent is not suitable.

19.8 Incorporation of ^{18}O into benzoic acid proceeds by a mechanism analogous to that of esterification. The nucleophile that adds to the protonated form of benzoic acid is

^{18}O-enriched water (the ^{18}O atom is represented by the shaded letter Ø in the following equations).

Benzoic acid   Tetrahedral
                                                                                             intermediate

The three hydroxyl groups of the tetrahedral intermediate are equivalent except that one of them is labeled with ^{18}O. Any one of these three hydroxyl groups may be lost in the dehydration step; when the hydroxyl group that is lost is unlabeled, an ^{18}O label is retained in the benzoic acid.

Tetrahedral
intermediate

^{18}O-enriched benzoic acid

19.9 (*b*) The 16-membered ring of 15-pentadecanolide is formed from 15-hydroxypentadecanoic acid.

15-Pentadecanolide   15-Hydroxypentadecanoic acid

(*c*) Vernolepin has two lactone rings, which can be related to two hydroxy acid combinations.

Be sure to keep the relative stereochemistry unchanged. Remember, the carbon-oxygen bond of an alcohol remains intact when the alcohol reacts with a carboxylic acid to give an ester.

19.10 Alkyl chlorides and bromides undergo nucleophilic substitution when treated with sodium iodide in acetone (Sec. 8.2). A reasonable approach is to brominate octadecanoic acid at its α carbon atom, then replace the bromine substituent with iodine by nucleophilic substitution.

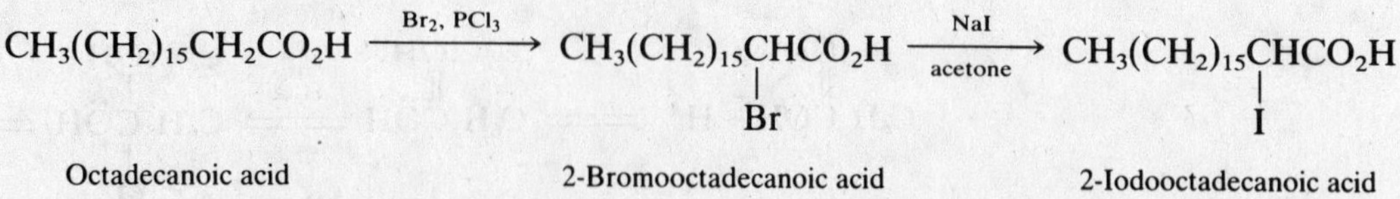
CH3(CH2)15CH2CO2H
Br2, PCl3
CH3(CH2)15CHCO2H
Br
NaI
acetone
CH3(CH2)15CHCO2H
I
Octadecanoic acid
2-Bromooctadecanoic acid
2-Iodooctadecanoic acid

The same procedure could be used by starting with chlorination of octadecanoic acid.

19.11 (*b*) The substrate is recognized as a derivative of malonic acid. It undergoes efficient thermal decarboxylation in the manner shown.

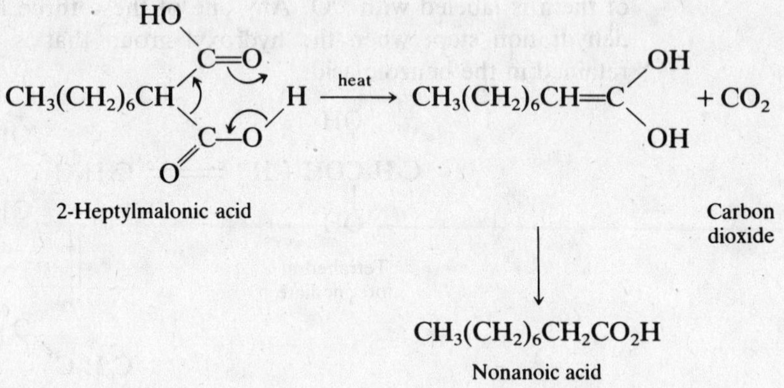
HO
C=O
CH3(CH2)6CH
H
C–O
O
heat
CH3(CH2)6CH=C
OH
OH
+ CO2
2-Heptylmalonic acid
Carbon dioxide
CH3(CH2)6CH2CO2H
Nonanoic acid

(*c*) The phenyl and methyl substituents attached to C-2 of malonic acid play no role in the decarboxylation process.

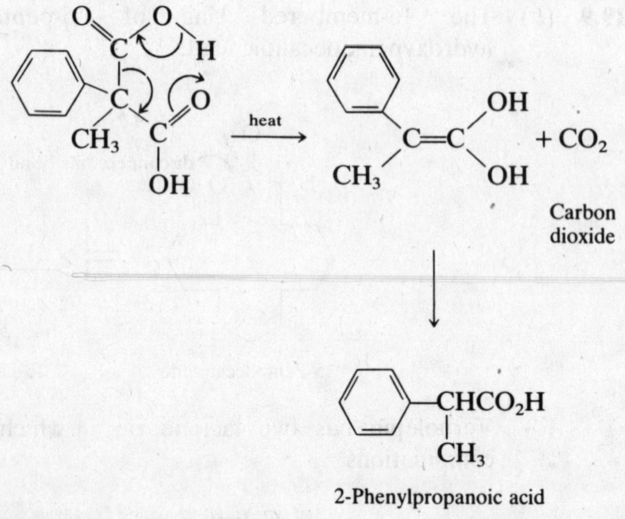
heat
CH3
OH
C=C
OH
OH
+ CO2
Carbon dioxide
CHCO2H
CH3
2-Phenylpropanoic acid

19.12 (*b*) The thermal decarboxylation of β-keto acids resembles that of substituted malonic acids. The structure of 2,2-dimethylacetoacetic acid and the equation representing its decarboxylation were given in the text. The overall process involves the bonding changes shown.

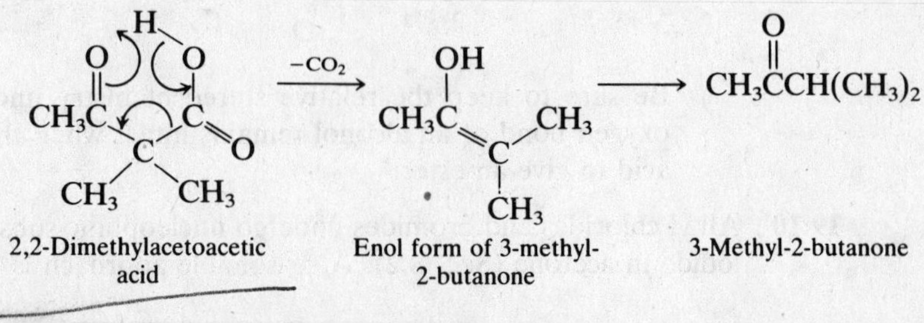
−CO2
CH3C
CH3
CH3
OH
CH3C
CH3
CH3
CH3CCH(CH3)2
2,2-Dimethylacetoacetic acid
Enol form of 3-methyl-2-butanone
3-Methyl-2-butanone

19.13 (*a*) Lactic acid (2-hydroxypropanoic acid) is a three-carbon carboxylic acid that bears a hydroxyl group at C-2.

$$\overset{3}{C}H_3\overset{2}{C}H\overset{1}{C}O_2H \qquad \text{2-Hydroxypropanoic acid}$$
$$|$$
$$OH$$

(*b*) The parent name *ethanoic acid* tells us that the chain that includes the carboxylic acid function contains only two carbons. A hydroxyl group and a phenyl substituent are present at C-2.

$$\text{C}_6\text{H}_5-\text{CHCO}_2\text{H} \qquad \text{2-Hydroxy-2-phenylethanoic acid (mandelic acid)}$$
$$|$$
$$OH$$

(*c*) The parent alkane is *tetradecane,* which has an unbranched chain of 14 carbons. The terminal methyl group is transformed to a carboxyl function in tetradecanoic acid.

$$\overset{O}{\overset{\|}{CH_3(CH_2)_{12}COH}} \qquad \begin{array}{l}\text{Tetradecanoic acid}\\ \text{(myristic acid)}\end{array}$$

(*d*) Undecane is the unbranched alkane with 11 carbon atoms, undecanoic acid is the corresponding carboxylic acid, and *undecenoic acid* is an 11-carbon carboxylic acid that contains a double bond. Since the carbon chain is numbered beginning with the carboxyl group, 10-undecenoic acid has its double bond at the opposite end of the chain from the carboxyl group.

$$CH_2{=}CH(CH_2)_8CO_2H \qquad \begin{array}{l}\text{10-Undecenoic acid}\\ \text{(undecylenic acid)}\end{array}$$

(*e*) Mevalonic acid has a five-carbon chain with hydroxyl groups at C-3 and C-5, along with a methyl group at C-3.

$$\overset{CH_3}{\overset{|}{HOCH_2CH_2CCH_2CO_2H}} \qquad \begin{array}{l}\text{3,5-Dihydroxy-3-methylpentanoic acid}\\ \text{(mevalonic acid)}\end{array}$$
$$|$$
$$OH$$

(*f*) The constitution represented by the systematic name 2-methyl-2-butenoic acid gives rise to two stereoisomers.

$$CH_3CH{=}CCO_2H \qquad \text{2-Methyl-2-butenoic acid}$$
$$|$$
$$CH_3$$

Tiglic acid is the *E* isomer, and the *Z* isomer is known as *angelic acid*. The higher-priority substituents, methyl and carboxyl, are placed on opposite sides of the double bond in tiglic acid and on the same side in angelic acid.

$$\begin{array}{cc} CH_3\diagdown & \diagup CH_3 \\ C{=}C & \\ H\diagup & \diagdown CO_2H \end{array} \qquad\qquad \begin{array}{cc} CH_3\diagdown & \diagup CO_2H \\ C{=}C & \\ H\diagup & \diagdown CH_3 \end{array}$$

(*E*)-2-Methyl-2-butenoic   (*Z*)-2-Methyl-2-butenoic
acid (tiglic acid)   acid (angelic acid)

(g) Butanedioic acid is a four-carbon chain in which both terminal carbons are carboxylic acid groups. Malic acid has a hydroxyl group at C-2.

$$HO_2CCHCH_2CO_2H$$
$$\underset{OH}{|}$$

2-Hydroxybutanedioic acid
(malic acid)

(h) Each of the carbon atoms of propane bears a carboxyl group as a substituent in 1,2,3-propanetricarboxylic acid. In citric acid C-2 also bears a hydroxyl group.

$$\overset{CO_2H}{\underset{OH}{HO_2CCH_2\overset{|}{\underset{|}{C}}CH_2CO_2H}}$$

2-Hydroxy-1,2,3-propanetricarboxylic acid
(citric acid)

(i) There is an aryl substituent at C-2 of propanoic acid in ibuprofen. This aryl substituent is a benzene ring bearing an isobutyl group at the para position.

$$CH_3CHCO_2H$$

2-(*p*-Isobutylphenyl)propanoic acid

$$CH_2CH(CH_3)_2$$

(j) Benzenecarboxylic acid is the systematic name for benzoic acid. *Salicylic acid* is a derivative of benzoic acid bearing a hydroxyl group at the position ortho to the carboxyl.

o-Hydroxybenzenecarboxylic acid (salicylic acid)

19.14 (a) The carboxylic acid contains a linear chain of eight carbon atoms. The parent alkane is *octane,* and so the systematic name of $CH_3(CH_2)_6CO_2H$ is *octanoic acid.*

(b) The compound shown is the potassium salt of octanoic acid. It is *potassium octanoate.*

(c) The presence of a double bond in $CH_2{=}CH(CH_2)_5CO_2H$ is indicated by the ending *-enoic acid.* Numbering of the chain begins with the carboxylic acid, and so the double bond is between C-7 and C-8. The compound is *7-octenoic acid.*

(d) Stereochemistry is systematically described by the *E-Z* notation. Here, the double bond between C-6 and C-7 in octenoic acid has the *Z* configuration; the higher-priority substituents are on the same side.

$$\underset{H}{\overset{CH_3}{\diagdown}}C{=}C\underset{H}{\overset{(CH_2)_4CO_2H}{\diagup}}$$

(*Z*)-6-Octenoic acid

(e) A dicarboxylic acid is named as a *dioic acid.* The carboxyl functions are the terminal carbons of an eight-carbon chain; $HO_2C(CH_2)_6CO_2H$ is *octanedioic acid.* It is not necessary to identify the carboxylic acid locations by number, since they can only be at the ends of the chain when the *-dioic acid* name is used.

(f) Pick the longest continuous chain that includes both carboxyl groups and name the compound as a *-dioic acid.* This chain contains only three carbons and bears a pentyl group as a substituent at C-2. It is not necessary to specify the position of the pentyl group, because it can only be attached to C-2.

$$CH_3(CH_2)_4{-}\overset{2}{\underset{3CO_2H}{\overset{|}{C}H}}\overset{1}{C}O_2H$$

Pentylpropanedioic acid

Malonic acid is an acceptable synonym for propanedioic acid; this compound may also be named *pentylmalonic acid*.

(g) A carboxylic acid function is attached as a substituent on a seven-membered ring. The compound is *cycloheptanecarboxylic acid*.

(h) The aromatic ring is named as a substituent attached to the eight-carbon carboxylic acid. Numbering of the chain begins with the carboxyl group.

6-Phenyloctanoic acid

19.15 (a) The order of decreasing acidity is:

		K_a	pK_a
Acetic acid	CH_3CO_2H	1.8×10^{-5}	4.7
Ethanol	CH_3CH_2OH	10^{-16}	16
Ethane	CH_3CH_3	$\sim 10^{-46}$	~ 46

(b) Here again, the carboxylic acid is the strongest acid and the hydrocarbon the weakest:

		K_a	pK_a
Benzoic acid	$C_6H_5CO_2H$	6.7×10^{-5}	4.2
Benzyl alcohol	$C_6H_5CH_2OH$	$10^{-16}-10^{-18}$	16–18
Benzene	C_6H_6	$\sim 10^{-43}$	~ 43

(c) Propanedioic acid is the stronger acid than propanoic acid, since the electron-withdrawing effect of one carboxyl group enhances the ionization of the other. Propanedial is a 1,3-dicarbonyl compound that yields a stabilized enolate; it is more acidic than 1,3-propanediol.

		K_a	pK_a
Propanedioic acid	$HO_2CCH_2CO_2H$	1.4×10^{-3}	2.9
Propanoic acid	$CH_3CH_2CO_2H$	1.3×10^{-5}	4.9
Propanedial	$OHCCH_2CHO$	$\sim 10^{-9}$	~ 9
1,3-Propanediol	$HOCH_2CH_2CH_2OH$	$\sim 10^{-16}$	~ 16

(d) Trifluoromethanesulfonic acid is by far the strongest acid in the group. It is structurally related to sulfuric acid, but its three fluorine substituents make it much stronger. Fluorine substituents increase the acidity of carboxylic acids and alcohols relative to their nonfluorinated analogs, but not enough to make fluorinated alcohols as acidic as carboxylic acids.

		K_a	pK_a
Trifluoromethanesulfonic acid	CF_3SO_2OH	10^6	-6
Trifluoroacetic acid	CF_3CO_2H	5.9×10^{-1}	0.2
Acetic acid	CH_3CO_2H	1.8×10^{-5}	4.7
2,2,2-Trifluoroethanol	CF_3CH_2OH	4.2×10^{-13}	12.4
Ethanol	CH_3CH_2OH	$\sim 10^{-16}$	~ 16

(e) The order of decreasing acidity is carboxylic acid > β-diketone > ketone > hydrocarbon.

		K_a	pK_a
Cyclopentanecarboxylic acid	(cyclopentane with CO_2H)	1×10^{-5}	5.0
2,4-Pentanedione	$CH_3\overset{O}{\underset{\|}{C}}CH_2\overset{O}{\underset{\|}{C}}CH_3$	10^{-9}	9
Cyclopentanone	(cyclopentanone with O)	10^{-20}	20
Cyclopentene	(cyclopentene)	10^{-45}	45

19.16 (a) A trifluoromethyl group is strongly electron-withdrawing and acid-strengthening. Its ability to attract electrons from the carboxylate ion decreases as its distance down the chain increases. 3,3,3-Trifluoropropanoic acid is a stronger acid than 4,4,4-trifluorobutanoic acid.

$$\boxed{CF_3CH_2CO_2H} \qquad\qquad CF_3CH_2CH_2CO_2H$$

3,3,3-Trifluoropropanoic acid	4,4,4-Trifluorobutanoic acid
$K_a \quad 9.6 \times 10^{-4}$	$K_a \quad 6.9 \times 10^{-5}$
($pK_a \quad 3.0$)	($pK_a \quad 4.2$)

(b) The carbon that bears the carboxyl group in 2-butynoic acid is sp hybridized and is, therefore, more electron-withdrawing than the sp^3 hybridized α carbon of butanoic acid. Therefore, the anion of 2-butynoic acid is stabilized better than the anion of butanoic acid, and 2-butynoic acid is a stronger acid.

$$CH_3C\equiv CCO_2H \qquad CH_3CH_2CH_2CO_2H$$

2-Butynoic acid	Butanoic acid
$K_a \quad 2.5 \times 10^{-3}$	$K_a \quad 1.5 \times 10^{-5}$
($pK_a \quad 2.6$)	($pK_a \quad 4.8$)

(c) Cyclohexanecarboxylic acid is a typical aliphatic carboxylic acid and is expected to be similar to acetic acid in acidity. The greater electronegativity of the sp^2 hybridized carbon attached to the carboxyl group in benzoic acid stabilizes benzoate anion better than the corresponding sp^3 hybridized carbon stabilizes cyclohexane-carboxylate. Benzoic acid is a stronger acid.

Benzoic acid
K_a 6.7×10^{-5}
(pK_a 4.2)

Cyclohexanecarboxylic acid
K_a 1.2×10^{-5}
(pK_a 4.9)

(*d*) Its five fluorine substituents make the pentafluorophenyl group more electron-withdrawing than an unsubstituted phenyl group. Thus, pentafluorobenzoic acid is a stronger acid than benzoic acid.

Pentafluorobenzoic acid
K_a 4.1×10^{-4}
(pK_a 3.4)

Benzoic acid
K_a 6.7×10^{-5}
(pK_a 4.2)

(*e*) The pentafluorophenyl substituent is electron-withdrawing and increases the acidity of a carboxyl group to which it is attached. Its electron-withdrawing effect decreases with distance. Pentafluorobenzoic acid is a stronger acid than *p*-(pentafluorophenyl)benzoic acid.

Pentafluorobenzoic acid
K_a 4.1×10^{-4}
(pK_a 3.4)

p-(Pentafluorophenyl)benzoic acid
(K_a not measured in water; comparable
with benzoic acid in acidity)

(*f*) The oxygen of the ring exercises an acidifying effect on the carboxyl group. This effect is largest when the oxygen is attached directly to the carbon that bears the carboxyl group. Furan-2-carboxylic acid is thus a stronger acid than furan-3-carboxylic acid.

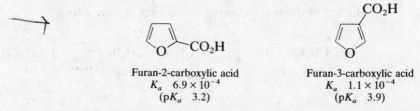

Furan-2-carboxylic acid
K_a 6.9×10^{-4}
(pK_a 3.2)

Furan-3-carboxylic acid
K_a 1.1×10^{-4}
(pK_a 3.9)

(*g*) Furan-2-carboxylic acid has an oxygen attached to the carbon that bears the carboxyl group, while pyrrole-2-carboxylic acid has a nitrogen in that position. Oxygen is more electronegative than nitrogen and so stabilizes the carboxylate anion better. Furan-2-carboxylic acid is a stronger acid than pyrrole-2-carboxylic acid.

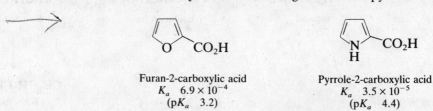

Furan-2-carboxylic acid
K_a 6.9×10^{-4}
(pK_a 3.2)

Pyrrole-2-carboxylic acid
K_a 3.5×10^{-5}
(pK_a 4.4)

19.17 (*a*)   The conversion of 1-butanol to butanoic acid is simply the oxidation of a primary alcohol to a carboxylic acid. Chromic acid is a suitable oxidizing agent.

$$CH_3CH_2CH_2CH_2OH \xrightarrow{H_2CrO_4} CH_3CH_2CH_2CO_2H$$

1-Butanol   Butanoic acid

(*b*)   Aldehydes may be oxidized to carboxylic acids by any of the oxidizing agents that convert primary alcohols to carboxylic acids or by silver oxide, a mild oxidizing agent that oxidizes aldehydes to carboxylic acids.

$$CH_3CH_2CH_2\overset{O}{\overset{\|}{C}}H \xrightarrow[H_2SO_4,\ H_2O]{K_2Cr_2O_7} CH_3CH_2CH_2\overset{O}{\overset{\|}{C}}OH$$

Butanal   Butanoic acid

(*c*)   The starting material has the same number of carbon atoms as does butanoic acid, and so all that is required is a series of functional group transformations. Carboxylic acids may be obtained by oxidation of the corresponding primary alcohol. The alcohol is available from the designated starting material, 1-butene.

$$CH_3CH_2CH_2CO_2H \Longrightarrow CH_3CH_2CH_2CH_2OH \Longrightarrow CH_3CH_2CH{=\!}CH_2$$

Hydroboration-oxidation of 1-butene yields 1-butanol, which can then be oxidized to butanoic acid as in part (*a*).

$$CH_3CH_2CH{=\!}CH_2 \xrightarrow[2.\ H_2O_2,\ HO^-]{1.\ B_2H_6} CH_3CH_2CH_2CH_2OH \xrightarrow{H_2CrO_4} CH_3CH_2CH_2CO_2H$$

1-Butene   1-Butanol   Butanoic acid

(*d*)   Converting 1-propanol to butanoic acid requires the carbon chain to be extended by one atom. Both methods for achieving this conversion, carboxylation of a Grignard reagent and formation and hydrolysis of a nitrile, begin with alkyl halides. Alkyl halides in turn are prepared from alcohols.

$$CH_3CH_2CH_2CO_2H \Longrightarrow CH_3CH_2CH_2MgBr \quad \text{or} \quad CH_3CH_2CH_2CN$$

$$\Downarrow$$

$$CH_3CH_2CH_2OH \Longleftarrow CH_3CH_2CH_2Br$$

Either of the two following procedures is satisfactory:

$$CH_3CH_2CH_2OH \xrightarrow[\text{or HBr}]{PBr_3} CH_3CH_2CH_2Br \xrightarrow[\substack{\text{diethyl}\\ \text{ether}}]{Mg} CH_3CH_2CH_2MgBr$$

1-Propanol   1-Bromopropane

$$\Bigg\downarrow \substack{1.\ CO_2 \\ 2.\ H_3O^+}$$

$$CH_3CH_2CH_2CO_2H$$

Butanoic acid

$$CH_3CH_2CH_2OH \xrightarrow[\text{or HBr}]{PBr_3} CH_3CH_2CH_2Br \xrightarrow[\text{DMSO}]{KCN} CH_3CH_2CH_2CN$$

1-Propanol   1-Bromopropane   Butanenitrile

$$\Bigg\downarrow H_2O,\ HCl,\ heat$$

$$CH_3CH_2CH_2CO_2H$$

Butanoic acid

(e) Dehydration of 2-propanol to propene followed by free-radical addition of hydrogen bromide affords 1-bromopropane.

$$CH_3CHCH_3 \xrightarrow[\text{heat}]{H_2SO_4} CH_3CH=CH_2 \xrightarrow[\text{peroxides}]{HBr} CH_3CH_2CH_2Br$$
$$\underset{OH}{|}$$

2-Propanol   Propene   1-Bromopropane

Once 1-bromopropane has been prepared it is converted to butanoic acid as in (d).

(f) The carbon skeleton of butanoic acid may be assembled by an aldol condensation of acetaldehyde.

$$CH_3CH_2CH_2CO_2H \Rightarrow CH_3CH=CHCH{\overset{O}{\|}} \Rightarrow 2CH_3CH{\overset{O}{\|}}$$

2CH₃CH (O) — KOH, ethanol / heat → CH₃CH=CHCH (O)

Acetaldehyde   2-Butenal

Oxidation of the aldehyde followed by hydrogenation of the double bond yields butanoic acid.

$$CH_3CH=CHCH{\overset{O}{\|}} \xrightarrow{\boxed{Ag_2O}} CH_3CH=CHCO_2H \xrightarrow[\text{Pt}]{H_2} CH_3CH_2CH_2CO_2H$$

2-Butenal   2-Butenoic acid   Butanoic acid

(g) Ethylmalonic acid belongs to the class of substituted malonic acids that undergo ready thermal decarboxylation. Decarboxylation yields butanoic acid.

$$CH_3CH_2CHCO_2H \xrightarrow{\text{heat}} CH_3CH_2CH_2CO_2H + CO_2$$
$$\underset{CO_2H}{|}$$

Ethylmalonic acid   Butanoic acid   Carbon dioxide

19.18 (a) In principle the Friedel-Crafts alkylation of benzene by methyl chloride could be used to prepare ^{14}C-labeled toluene. Once prepared, toluene could be oxidized to benzoic acid.

$$\bigcirc + \overset{*}{C}H_3Cl \xrightarrow{AlCl_3} \bigcirc-\overset{*}{C}H_3$$

Benzene   Methyl chloride   Toluene

1. KMnO₄
2. H⁺
or K₂Cr₂O₇, H₂SO₄, H₂O

$$\bigcirc-\overset{*}{C}O_2H$$

Benzoic acid

In practice, monomethylation of benzene by methyl chloride is difficult to achieve. It would be necessary to perform the alkylation with excess benzene and then separate the toluene from higher-boiling dimethyl, trimethyl, etc., derivatives by distillation.

(b) Formaldehyde can serve as a one-carbon source if it is attacked by the Grignard reagent derived from bromobenzene.

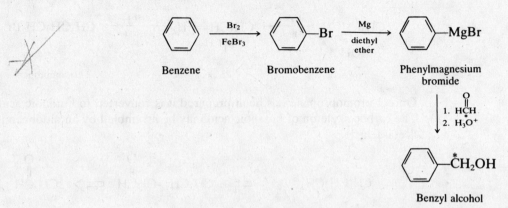

This sequence yields ^{14}C-labeled benzyl alcohol, which can be oxidized to ^{14}C-labeled benzoic acid.

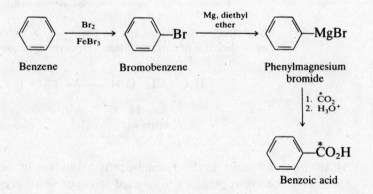

(c) A direct route to ^{14}C-labeled benzoic acid utilizes a Grignard synthesis employing ^{14}C-labeled carbon dioxide.

19.19 (a) An acid-base reaction takes place when pentanoic acid is combined with sodium hydroxide.

$$CH_3CH_2CH_2CH_2CO_2H + NaOH \longrightarrow CH_3CH_2CH_2CH_2CO_2Na + H_2O$$

Pentanoic acid   Sodium   Sodium pentanoate   Water
                      hydroxide

(b) Carboxylic acids react with sodium bicarbonate to give carbonic acid, which dissociates to carbon dioxide and water, so that the actual reaction that takes place is:

$$CH_3CH_2CH_2CH_2CO_2H + NaHCO_3 \longrightarrow$$

Pentanoic acid   Sodium
                      bicarbonate

$$CH_3CH_2CH_2CH_2CO_2Na + CO_2 + H_2O$$

Sodium pentanoate   Carbon   Water
                         dioxide

(c) Thionyl chloride is a reagent that converts carboxylic acids to the corresponding acyl chlorides.

$$CH_3CH_2CH_2CH_2CO_2H + SOCl_2 \longrightarrow CH_3CH_2CH_2CH_2\overset{\overset{\displaystyle O}{\|}}{C}Cl + SO_2 + HCl$$

Pentanoic acid   Thionyl chloride   Pentanoyl chloride   Sulfur dioxide   Hydrogen chloride

(d) Phosphorus tribromide is used to convert carboxylic acids to their acyl bromides.

$$3CH_3CH_2CH_2CH_2CO_2H + PBr_3 \longrightarrow 3CH_3CH_2CH_2CH_2\overset{\overset{\displaystyle O}{\|}}{C}Br + H_3PO_3$$

Pentanoic acid   Phosphorus tribromide   Pentanoyl bromide   Phosphorous acid

(e) Carboxylic acids react with alcohols in the presence of acid catalysts to give esters.

$$CH_3CH_2CH_2CH_2CO_2H + C_6H_5CH_2OH \underset{}{\overset{H_2SO_4}{\rightleftharpoons}} CH_3CH_2CH_2CH_2\overset{\overset{\displaystyle O}{\|}}{C}OCH_2C_6H_5 + H_2O$$

Pentanoic acid   Benzyl alcohol   Benzyl pentanoate   Water

(f) Chlorine is introduced at the α carbon atom of a carboxylic acid. The reaction is catalyzed by a small amount of phosphorus or a phosphorus trihalide and is called the Hell-Volhard-Zelinsky reaction.

$$CH_3CH_2CH_2CH_2CO_2H + Cl_2 \xrightarrow[\text{(catalyst)}]{PBr_3} CH_3CH_2CH_2\underset{\underset{\displaystyle Cl}{|}}{CH}CO_2H + HCl$$

Pentanoic acid   Chlorine   2-Chloropentanoic acid   Hydrogen chloride

The α-halo substituent is derived from the halogen used, not from the phosphorus trihalide.

(g) In the case, bromine is introduced at the α carbon.

$$CH_3CH_2CH_2CH_2CO_2H + Br_2 \xrightarrow[\text{(catalyst)}]{PCl_3} CH_3CH_2CH_2\underset{\underset{\displaystyle Br}{|}}{CH}CO_2H + HBr$$

Pentanoic acid   Bromine   2-Bromopentanoic acid   Hydrogen bromide

(h) α-Halo carboxylic acids are reactive substrates in nucleophilic substitution. Iodide acts as a nucleophile to displace bromide from 2-bromopentanoic acid.

$$CH_3CH_2CH_2\underset{\underset{\displaystyle Br}{|}}{CH}CO_2H + NaI \xrightarrow{\text{acetone}} CH_3CH_2CH_2\underset{\underset{\displaystyle I}{|}}{CH}CO_2H + NaBr$$

2-Bromopentanoic acid   Sodium iodide   2-Iodopentanoic acid   Sodium bromide

(i) Aqueous ammonia converts α-halo acids to α-amino acids.

$$CH_3CH_2CH_2\underset{\underset{\displaystyle Br}{|}}{CH}CO_2H + 2NH_3 \longrightarrow CH_3CH_2CH_2\underset{\underset{\displaystyle NH_2}{|}}{CH}CO_2H + NH_4Br$$

2-Bromopentanoic acid   Ammonia   2-Aminopentanoic acid   Ammonium bromide (*j*) Lithium aluminum hydride is a powerful reducing agent and reduces carboxylic acids to primary alcohols.

$$CH_3CH_2CH_2CH_2CO_2H \xrightarrow[\text{2. H}_2\text{O}]{\text{1. LiAlH}_4} CH_3CH_2CH_2CH_2CH_2OH$$

Pentanoic acid   1-Pentanol

(*k*) Phenylmagnesium bromide acts as a base to abstract the carboxylic acid proton.

$$CH_3CH_2CH_2CH_2CO_2H + \quad C_6H_5MgBr \longrightarrow$$

Pentanoic acid   Phenylmagnesium
bromide

$$CH_3CH_2CH_2CH_2CO_2MgBr + \quad C_6H_6$$

Bromomagnesium pentanoate   Benzene

Grignard reagents are not compatible with carboxylic acids; proton transfer converts the Grignard reagent to the corresponding hydrocarbon.

19.20 (*a*) Conversion of butanoic acid to 1-butanol is a reduction and requires lithium aluminum hydride as the reducing agent.

$$CH_3CH_2CH_2\overset{\displaystyle O}{\overset{\displaystyle \|}{C}}OH \xrightarrow[\text{2. H}_2\text{O}]{\text{1. LiAlH}_4} CH_3CH_2CH_2CH_2OH$$

Butanoic acid   1-Butanol

(*b*) Carboxylic acids cannot be reduced directly to aldehydes. The following two-step procedure may be used:

$$CH_3CH_2CH_2\overset{\displaystyle O}{\overset{\displaystyle \|}{C}}OH \xrightarrow[\text{2. H}_2\text{O}]{\text{1. LiAlH}_4} CH_3CH_2CH_2CH_2OH \xrightarrow[\text{CH}_2\text{Cl}_2]{\text{(C}_5\text{H}_5\text{N)}_2\text{CrO}_3} CH_3CH_2CH_2\overset{\displaystyle O}{\overset{\displaystyle \|}{C}}H$$

Butanoic acid   1-Butanol   Butanal

(*c*) Remember that alkyl halides are usually prepared from alcohols. Therefore, 1-butanol is needed in order to prepare 1-chlorobutane.

$$CH_3CH_2CH_2CH_2OH \xrightarrow{\text{SOCl}_2} CH_3CH_2CH_2CH_2Cl$$

1-Butanol   1-Chlorobutane

Conversion of butanoic acid to 1-butanol requires reduction with lithium aluminum hydride as shown in part (*a*).

(*d*) Carboxylic acids are converted to their corresponding acyl chlorides with thionyl chloride.

$$CH_3CH_2CH_2\overset{\displaystyle O}{\overset{\displaystyle \|}{C}}OH \xrightarrow{\text{SOCl}_2} CH_3CH_2CH_2\overset{\displaystyle O}{\overset{\displaystyle \|}{C}}Cl$$

Butanoic acid   Butanoyl chloride

(*e*) Aromatic ketones are frequently prepared by Friedel-Crafts acylation of the appropriate acyl chloride and benzene. Butanoyl chloride, prepared in part (*d*), can be used to acylate benzene in a Friedel-Crafts reaction.

$$\bigcirc + CH_3CH_2CH_2\overset{\displaystyle O}{\overset{\displaystyle \|}{C}}Cl \xrightarrow{\text{AlCl}_3} \bigcirc\overset{\displaystyle O}{\overset{\displaystyle \|}{C}}CH_2CH_2CH_3$$

Benzene   Butanoyl chloride   Phenyl propyl ketone

(f) The preparation of 4-octanone using compounds derived from butanoic acid may be seen by using disconnections in a retrosynthetic analysis.

$$CH_3CH_2CH_2\overset{\underset{\displaystyle OH}{|}}{CH}CH_2CH_2CH_2CH_3$$

$$CH_3CH_2CH_2\overset{\underset{\displaystyle O}{\|}}{C}CH_2CH_2CH_2CH_3$$

$$CH_3CH_2CH_2\overset{\underset{\displaystyle O}{\|}}{CH} + CH_3CH_2CH_2CH_2MgBr$$

The reaction scheme which may be used is:

$$CH_3CH_2CH_2CH_2Cl \xrightarrow{\text{Mg}} CH_3CH_2CH_2CH_2MgCl$$

1-Chlorobutane [from part (c)]   Butylmagnesium chloride

$$CH_3CH_2CH_2CH_2MgCl + CH_3CH_2CH_2\overset{\underset{\displaystyle O}{\|}}{CH} \xrightarrow[\text{2. } H_3O^+]{\substack{\text{1. diethyl}\\\text{ether}}} CH_3CH_2CH_2\overset{\underset{\displaystyle OH}{|}}{CH}CH_2CH_2CH_2CH_3$$

Butylmagnesium chloride   Butanal [from part (b)]   4-Octanol

$$\xrightarrow{H_2CrO_4}$$

$$CH_3CH_2CH_2\overset{\underset{\displaystyle O}{\|}}{C}CH_2CH_2CH_2CH_3$$

4-Octanone

(g) Carboxylic acids are halogenated at their α carbon atom by the Hell-Volhard-Zelinsky reaction:

$$CH_3CH_2CH_2CO_2H \xrightarrow[\text{Br}_2]{P} CH_3CH_2\overset{\underset{\displaystyle Br}{|}}{CH}CO_2H$$

Butanoic acid   2-Bromobutanoic acid

A catalytic amount of PCl_3 may be used in place of phosphorus in the reaction.

(h) Base-promoted dehydrohalogenation of 2-bromobutanoic acid give 2-butenoic acid.

$$CH_3CH_2\overset{\underset{\displaystyle Br}{|}}{CH}CO_2H \xrightarrow[\text{2. } H^+]{\substack{\text{1. KOC(CH}_3)_3\\\text{DMSO}}} CH_3CH=CHCO_2H$$

2-Bromobutanoic acid   2-Butenoic acid

19.21 (a) The compound to be prepared is *glycine*, an α-amino acid. The amino functional group can be introduced by a nucleophilic substitution reaction on an α-halo acid, which is available by way of the Hell-Volhard-Zelinsky reaction.

$$CH_3CO_2H \xrightarrow[\text{P}]{Cl_2} ClCH_2CO_2H \xrightarrow{NH_3, H_2O} H_2NCH_2CO_2H$$

Acetic acid   Chloroacetic acid   Aminoacetic acid (glycine)

(*b*) Phenoxyacetic acid is used as a fungicide. It can be prepared by a nucleophilic substitution using sodium phenoxide and chloroacetic acid.

$$\underset{\text{Chloroacetic acid}}{ClCH_2CO_2H} \xrightarrow[\text{2. }H^+]{\text{1. }C_6H_5ONa} \underset{\text{Phenoxyacetic acid}}{C_6H_5OCH_2CO_2H} + NaCl$$

(*c*) Cyanide ion is a good nucleophile and will displace chloride from chloroacetic acid.

$$\underset{\substack{\text{Chloroacetic acid}\\ \text{[from part (}a\text{)]}}}{ClCH_2CO_2H} \xrightarrow[Na_2CO_3,\ H_2O]{NaCN} \underset{\text{Sodium cyanoacetate}}{N{\equiv}CCH_2CO_2Na} \xrightarrow{H^+} \underset{\text{Cyanoacetic acid}}{N{\equiv}CCH_2CO_2H}$$

(*d*) Cyanoacetic acid, prepared as in part (*c*), serves as a convenient precursor to malonic acid. Hydrolysis of the nitrile substituent converts it to a carboxyl group.

$$\underset{\text{Cyanoacetic acid}}{N{\equiv}CCH_2CO_2H} \xrightarrow[\text{heat}]{H_2O,\ H^+} \underset{\text{Malonic acid}}{HO_2CCH_2CO_2H}$$

(*e*) Iodoacetic acid is not prepared directly from acetic acid but is derived by nucleophilic substitution of iodide in bromoacetic or chloroacetic acid.

$$\underset{\text{Chloroacetic acid}}{ClCH_2CO_2H} \xrightarrow[\text{acetone}]{NaI} \underset{\text{Iodoacetic acid}}{ICH_2CO_2H}$$

(*f*) Two transformations need to be accomplished, α bromination and esterification. The correct sequence is bromination followed by esterification.

$$\underset{\text{Acetic acid}}{CH_3CO_2H} \xrightarrow[P]{Br_2} \underset{\text{Bromoacetic acid}}{BrCH_2CO_2H} \xrightarrow[H^+]{CH_3CH_2OH} \underset{\text{Ethyl bromoacetate}}{BrCH_2CO_2CH_2CH_3}$$

Reversing the order of steps is not appropriate. It must be the carboxylic acid that is subjected to halogenation, since the Hell-Volhard-Zelinsky reaction is a reaction of carboxylic acids, not esters.

(*g*) The compound shown is an ylide. It can be prepared from ethyl bromoacetate as shown

$$\underset{\text{Ethyl bromoacetate}}{BrCH_2CO_2CH_2CH_3} + \underset{\text{Triphenylphosphine}}{(C_6H_5)_3P} \longrightarrow$$

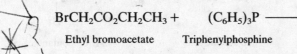

$$(C_6H_5)_3\overset{+}{P}CH_2CO_2CH_2CH_3\ Br^- \xrightarrow{NaOCH_2CH_3} (C_6H_5)_3\overset{+}{P}{-}\overset{..}{\overset{-}{C}}HCO_2CH_2CH_3$$
$$\text{Ylide}$$

The first step is a nucleophilic substitution of bromide by triphenylphosphine. Treatment of the derived triphenylphosphonium salt with base removes the relatively acidic α proton, forming the ylide. (For a review of ylide formation, refer to Sec. 17.14.)

(*h*) Reaction of the ylide formed in part (*g*) with benzaldehyde gives the desired alkene by a Wittig reaction.

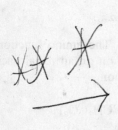

$$\underset{\text{Ylide from part (}g\text{)}}{(C_6H_5)_3\overset{+}{P}{-}\overset{..}{\overset{-}{C}}HCO_2CH_2CH_3} + \underset{\text{Benzaldehyde}}{C_6H_5\overset{\overset{\displaystyle O}{\|}}{C}H} \longrightarrow$$

$$\underset{\text{Ethyl cinnamate}}{C_6H_5CH{=}CHCO_2CH_2CH_3} + \underset{\substack{\text{Triphenylphosphine}\\ \text{oxide}}}{(C_6H_5)_3\overset{+}{P}{-}\overset{-}{O}}$$

19.22 (*a*) Silver oxide is a weak oxidizing agent; it oxidizes aldehydes to the corresponding carboxylic acids.

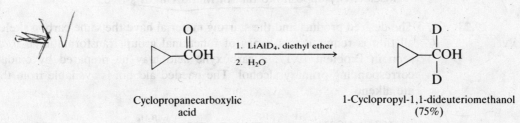

(*E*)-2-Butenal
(*trans*-crotonaldehyde)

(*E*)-2-Butenoic acid
(*trans*-crotonic acid)

(*b*) Carboxylic acids are converted to ethyl esters when they are allowed to stand in ethanol in the presence of an acid catalyst.

(*E*)-2-Methyl-2-butenoic acid   Ethanol   Ethyl (*E*)-2-methyl-2-butenoate   Water
(74–80%)

(*c*) Lithium aluminum hydride, LiAlH$_4$, reduces carboxylic acids to primary alcohols. When LiAlD$_4$ is used, deuterium is transferred to the carbonyl carbon.

Cyclopropanecarboxylic
acid

1-Cyclopropyl-1,1-dideuteriomethanol
(75%)

Notice that deuterium is bonded only to carbon. The hydroxyl proton is derived from water, not from the reducing agent.

(*d*) In the presence of a catalytic amount of phosphorus, bromine reacts with carboxylic acids to yield the corresponding α-bromo derivative.

Cyclohexanecarboxylic
acid

1-Bromocyclohexanecarboxylic
acid (96%)

(*e*) Alkyl fluorides are not readily converted to Grignard reagents, and so it is the bromine substituent that is attacked by magnesium.

m-Bromo(trifluoromethyl)benzene

m-(Trifluoromethyl)benzoic acid

(f) Cyano substituents are hydrolyzed to carboxyl groups in the presence of acid catalysts.

$$\underset{\substack{m\text{-Chlorobenzyl cyanide}}}{\text{CH}_2\text{CN}\ \text{ring-Cl}} \xrightarrow[\text{H}_2\text{SO}_4,\ \text{heat}]{\text{H}_2\text{O, acetic acid}} \underset{\substack{m\text{-Chlorophenylacetic acid}\\(61\%)}}{\text{CH}_2\text{CO}_2\text{H}\ \text{ring-Cl}}$$

(g) The carboxylic acid is converted to its methyl ester under these conditions.

$$\underset{\substack{\text{2-Butynoic acid}}}{\text{CH}_3\text{C}\equiv\text{CCO}_2\text{H}} + \underset{\substack{\text{Methanol}}}{\text{CH}_3\text{OH}} \xrightarrow{\text{H}_2\text{SO}_4} \underset{\substack{\text{Methyl 2-butynoate}}}{\text{CH}_3\text{C}\equiv\text{CCOCH}_3} + \underset{\substack{\text{Water}}}{\text{H}_2\text{O}}$$

(h) The carboxylic acid function plays no part in this reaction; free-radical addition of hydrogen bromide to the carbon-carbon double bond occurs.

$$\underset{\substack{\text{10-Undecenoic acid}}}{\text{CH}_2\!=\!\text{CH}(\text{CH}_2)_8\text{CO}_2\text{H}} \xrightarrow[\substack{\text{benzoyl}\\\text{peroxide}}]{\text{HBr}} \underset{\substack{\text{11-Bromoundecanoic acid (66–70\%)}}}{\text{BrCH}_2\text{CH}_2(\text{CH}_2)_8\text{CO}_2\text{H}}$$

Recall that hydrogen bromide adds to alkenes in the presence of peroxides with a regioselectivity opposite to that of Markovnikov's rule.

19.23 (a) The desired product and the starting material have the same carbon skeleton, and so all that is required is a series of functional group transformations. Recall that, as seen in Problem 19.17, a carboxylic acid may be prepared by oxidation of the corresponding primary alcohol. The needed alcohol is available from the appropriate alkene.

$$\underset{\substack{\textit{tert}\text{-Butyl alcohol}}}{(\text{CH}_3)_3\text{COH}} \xrightarrow[\text{heat}]{\text{H}^+} \underset{\substack{\text{2-Methylpropene}}}{(\text{CH}_3)_2\text{C}\!=\!\text{CH}_2} \xrightarrow[\text{2. H}_2\text{O}_2,\ \text{HO}^-]{\text{1. B}_2\text{H}_6} \underset{\substack{\text{2-Methyl-1-propanol}}}{(\text{CH}_3)_2\text{CHCH}_2\text{OH}}$$

$$\xdownarrow{\substack{\text{K}_2\text{Cr}_2\text{O}_7,\\\text{H}_2\text{SO}_4}}$$

$$\underset{\substack{\text{2-Methylpropanoic acid}}}{(\text{CH}_3)_2\text{CHCO}_2\text{H}}$$

(b) The target molecule contains one more carbon than the starting material, and so a carbon-carbon bond–forming step is indicated. Two approaches are reasonable; one proceeds by way of nitrile formation and hydrolysis, the other by carboxylation of a Grignard reagent. In either case the key intermediate is 1-bromo-2-methylpropane.

$$\underset{\substack{\text{3-Methylbutanoic acid}}}{(\text{CH}_3)_2\text{CHCH}_2\text{CO}_2\text{H}} \implies \underset{\substack{\text{1-Bromo-2-methylpropane}}}{(\text{CH}_3)_2\text{CHCH}_2\text{Br}}$$

The desired bromide may be prepared by free-radical addition of hydrogen bromide to isobutene.

$$\underset{\substack{\textit{tert}\text{-Butyl alcohol}}}{(\text{CH}_3)_3\text{COH}} \xrightarrow[\text{heat}]{\text{H}^+} \underset{\substack{\text{2-Methylpropene}}}{(\text{CH}_3)_2\text{C}\!=\!\text{CH}_2} \xrightarrow[\text{peroxides}]{\text{HBr}} \underset{\substack{\text{1-Bromo-2-methylpropane}}}{(\text{CH}_3)_2\text{CHCH}_2\text{Br}}$$

Another route to the bromide utilizes the alcohol prepared in part (*a*).

$$(CH_3)_2CHCH_2OH \xrightarrow{PBr_3} (CH_3)_2CHCH_2Br$$

2-Methyl-1-propanol   1-Bromo-2-methylpropane

Conversion of the bromide to the desired acid is then carried out as follows:

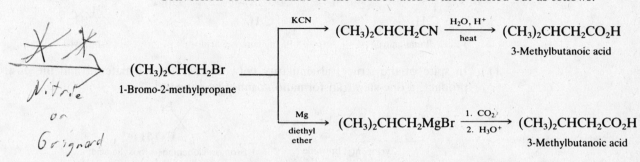

(*Nitrile or Grignard*)

$$(CH_3)_2CHCH_2Br$$
1-Bromo-2-methylpropane

$$\xrightarrow{KCN} (CH_3)_2CHCH_2CN \xrightarrow[heat]{H_2O, \ H^+} (CH_3)_2CHCH_2CO_2H$$

3-Methylbutanoic acid

$$\xrightarrow[diethyl \ ether]{Mg} (CH_3)_2CHCH_2MgBr \xrightarrow[2. \ H_3O^+]{1. \ CO_2} (CH_3)_2CHCH_2CO_2H$$

3-Methylbutanoic acid

(*c*)   Examining the target molecule reveals that it contains two more carbon atoms than the indicated starting material, suggesting use of ethylene oxide in a two-carbon chain-extension process.

$$(CH_3)_3CCH_2CO_2H \Rightarrow (CH_3)_3CCH_2CH_2OH \Rightarrow (CH_3)_3CMgX + CH_2\overset{O}{-}CH_2$$

This suggests the following sequence of steps:

$$(CH_3)_3COH \xrightarrow{HBr} (CH_3)_3CBr \xrightarrow{Mg} (CH_3)_3CMgBr$$

tert-Butyl   2-Bromo-2-   *tert*-Butylmagnesium
alcohol   methylpropane   bromide

$$\downarrow \begin{array}{l} 1. \ CH_2\overset{O}{-}CH_2 \\ 2. \ H_3O^+ \end{array}$$

$$(CH_3)_3CCH_2CO_2H \xleftarrow[H_2O]{K_2Cr_2O_7, \ H^+} (CH_3)_3CCH_2CH_2OH$$

3,3-Dimethylbutanoic   3,3-Dimethyl-1-
acid   butanol

(*d*)   This synthesis requires extending a carbon chain by two carbon atoms. One way to form dicarboxylic acids is by hydrolysis of dinitriles.

$$HO_2C(CH_2)_5CO_2H \Rightarrow NC(CH_2)_5CN \Rightarrow Br(CH_2)_5Br$$

This suggests the following sequence of steps:

$$HO_2C(CH_2)_3CO_2H \xrightarrow[2. \ H_2O]{1. \ LiAlH_4} HOCH_2(CH_2)_3CH_2OH$$

Pentanedioic acid   1,5-Pentanediol

$$\downarrow HBr \ or \ PBr_3$$

$$NCCH_2(CH_2)_3CH_2CN \xleftarrow{KCN} BrCH_2(CH_2)_3CH_2Br$$

1,5-Dicyanopentane   1,5-Dibromopentane

$$\downarrow \begin{array}{c} H_2O, \ H^+ \\ heat \end{array}$$

$$HO_2CCH_2(CH_2)_3CH_2CO_2H \equiv HO_2C(CH_2)_5CO_2H$$

Heptanedioic acid

(e)   The desired alcohol cannot be prepared directly from the nitrile. It is available, however, by lithium aluminum hydride reduction of the carboxylic acid obtained by hydrolysis of the nitrile.

$$CH_3CHCH_2CN \xrightarrow[H^+]{H_2O} CH_3CHCH_2\overset{\displaystyle O}{\overset{\|}{C}}OH \xrightarrow[2.\ H_2O]{1.\ LiAlH_4} CH_3CHCH_2CH_2OH$$

$$\underset{\displaystyle C_6H_5}{} \qquad\qquad \underset{\displaystyle C_6H_5}{} \qquad\qquad \underset{\displaystyle C_6H_5}{}$$

3-Phenylbutanenitrile   3-Phenylbutanoic acid   3-Phenyl-1-butanol

(f)   In spite of the structural similarity between the starting material and the desired product, a one-step transformation cannot be achieved.

Cyclopentyl bromide   1-Bromocyclopentanecarboxylic acid

Instead, recall that α-bromo acids are prepared from carboxylic acids by the Hell-Volhard-Zelinsky reaction:

Cyclopentanecarboxylic acid   1-Bromocyclopentanecarboxylic acid

The problem now simplifies to one of preparing cyclopentanecarboxylic acid from cyclopentyl bromide. Two routes are possible:

Cyclopentyl bromide → Cyclopentyl cyanide → Cyclopentanecarboxylic acid

Cyclopentyl bromide → Cyclopentylmagnesium bromide → Cyclopentanecarboxylic acid

The Grignard route is better; it is a "one-pot" transformation. Converting the secondary bromide to a nitrile will be accompanied by elimination, and the procedure requires two separate operations.

(g)   In this case the halogen substituent is present at the β carbon rather than the α carbon atom of the carboxylic acid. The starting material, a β-chloro unsaturated acid, can lead to the desired carbon skeleton by a Diels-Alder reaction.

1,3-Butadiene   (E)-3-Chloropropenoic acid   trans-2-Chloro-4-cyclohexenecarboxylic acid

The required trans stereochemistry is a consequence of the stereospecificity of the Diels-Alder reaction.

Hydrogenation of the double bond of the Diels-Alder adduct gives the required product.

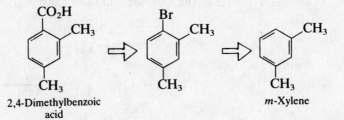

trans-2-Chloro-4-cyclohexenecarboxylic acid

trans-2-Chlorocyclohexanecarboxylic acid

(h) The target molecule is related to the starting material by the retrosynthesis

2,4-Dimethylbenzoic acid

m-Xylene

The necessary bromine substituent can be introduced by electrophilic substitution in the activated aromatic ring of *m*-xylene.

m-Xylene

Br₂
acetic acid
(or Br₂, FeBr₃)

1-Bromo-2,4-dimethylbenzene

The aryl bromide cannot be converted to a carboxylic acid by way of the corresponding cyano compound, because aryl bromides are not reactive toward nucleophilic substitution. The Grignard route is necessary.

1-Bromo-2,4-dimethylbenzene

1. Mg, diethyl ether
2. CO₂
3. H₃O⁺

2,4-Dimethylbenzoic acid

(i) The relationship of the target molecule to the starting material

4-Chloro-3-nitrobenzoic acid

p-Chlorotoluene

requires that there be two synthetic operations, oxidation of the methyl group and nitration of the ring. The orientation of the nitro group requires that nitration of the starting material follow oxidation of the methyl group

$$\text{CH}_3\text{-C}_6\text{H}_4\text{-Cl} \xrightarrow[\text{H}_2\text{O}]{\text{K}_2\text{Cr}_2\text{O}_7,\ \text{H}_2\text{SO}_4} \text{CO}_2\text{H-C}_6\text{H}_4\text{-Cl}$$

p-Chlorotoluene   *p*-Chlorobenzoic acid

Nitration of *p*-chlorobenzoic acid gives the desired product, because the directing effects of the chlorine and the carboxyl groups reinforce each other.

$$\text{CO}_2\text{H-C}_6\text{H}_4\text{-Cl} \xrightarrow[\text{H}_2\text{SO}_4]{\text{HNO}_3} \text{product with NO}_2$$

p-Chlorobenzoic acid   4-Chloro-3-nitrobenzoic acid

(*j*) The desired synthetic route becomes apparent when it is recognized that the *Z* alkene stereoisomer may be obtained from an alkyne which, in turn, is available by carboxylation of the anion derived from the starting material.

$$\underset{\substack{\text{CH}_3\\ \text{H}}}{\text{C}}=\underset{\substack{\text{CO}_2\text{H}\\ \text{H}}}{\text{C}} \ \Longrightarrow\ \text{CH}_3\text{C}\equiv\text{CCO}_2\text{H} \ \Longrightarrow\ \text{CH}_3\text{C}\equiv\text{C:}^- + \text{CO}_2$$

The desired reaction sequence is:

$$\text{CH}_3\text{C}\equiv\text{CH} \xrightarrow[\text{NH}_3]{\text{NaNH}_2} \text{CH}_3\text{C}\equiv\text{CNa} \xrightarrow[\text{2. H}_3\text{O}^+]{\text{1. CO}_2} \text{CH}_3\text{C}\equiv\text{CCO}_2\text{H}$$

Propyne   Propynylsodium   2-Butynoic acid

Hydrogenation of the carbon-carbon triple bond of 2-butynoic acid over the Lindlar catalyst converts this compound to the *Z* isomer of 2-butenoic acid.

$$\text{CH}_3\text{C}\equiv\text{CCO}_2\text{H} \xrightarrow{\text{H}_2,\ \text{Lindlar Pd}} \underset{\substack{\text{H}}}{\overset{\substack{\text{CH}_3}}{\text{C}}}=\underset{\substack{\text{H}}}{\overset{\substack{\text{CO}_2\text{H}}}{\text{C}}}$$

2-Butynoic acid   (*Z*)-2-Butenoic acid

19.24 (*a*) The most stable conformation of formic acid is the one that has both hydrogens anti.

Syn: less stable conformation of formic acid   Anti: more stable conformation of formic acid

A plausible explanation is that the syn conformation is destabilized by lone-pair repulsions:

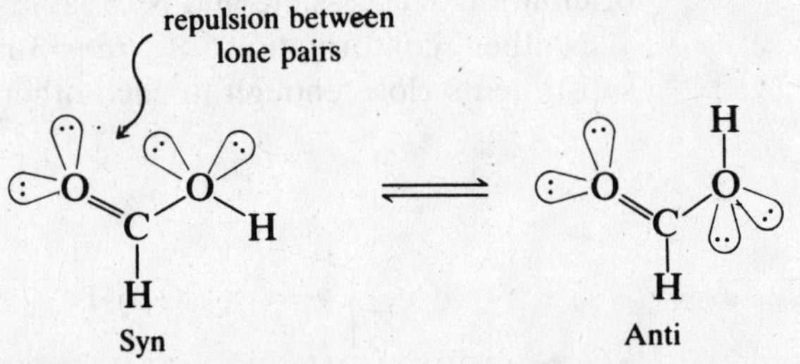

Syn   Anti

(b) A dipole moment of zero can mean that the molecule has a center of symmetry. One structure that satisfies this requirement is characterized by intramolecular hydrogen bonding between the two carboxyl groups and an anti relationship between the two carbonyls.

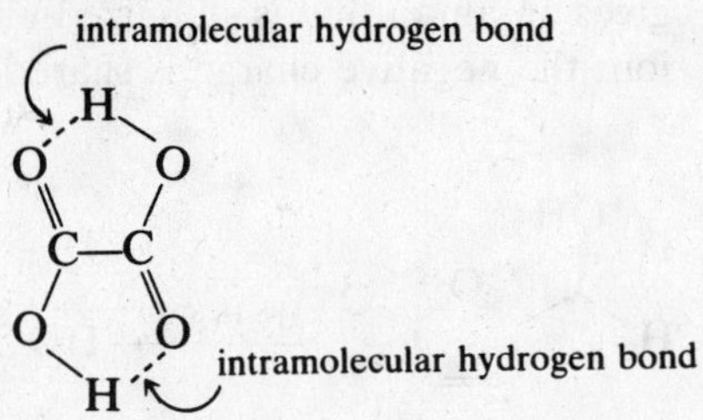

Another possibility is the following structure; it also has a center of symmetry and an anti relationship between the two carbonyls.

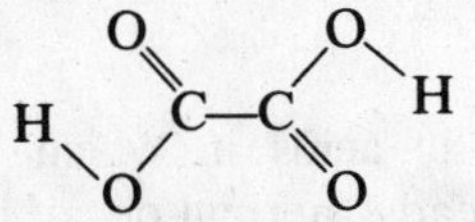

Other centrosymmetric structures can be drawn; these have the two hydrogen atoms out of the plane of the carboxyl groups, however, and are less likely to occur, in view of the known planarity of carboxyl groups. Structures in which the carbonyl groups are syn to each other do not have a center of symmetry.

(c) The anion formed on dissociation of o-hydroxybenzoic acid can be stabilized by an intramolecular hydrogen bond.

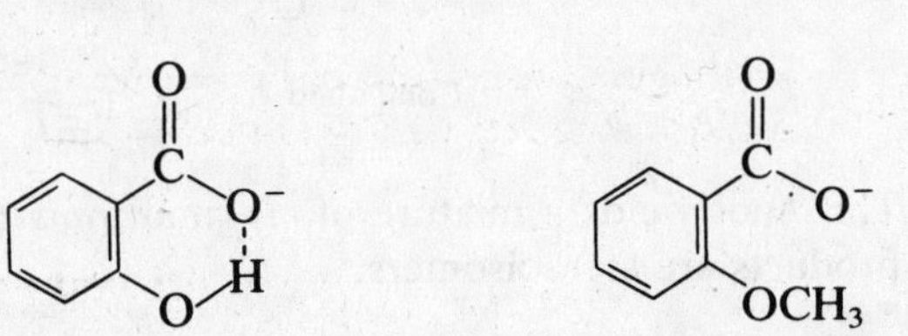

o-Hydroxybenzoate ion
(stabilized by hydrogen
bonding)

o-Methoxybenzoate ion
(hydrogen bonding is not possible)

(d) Lactone formation is possible only when the hydroxyl and carboxyl groups are cis.

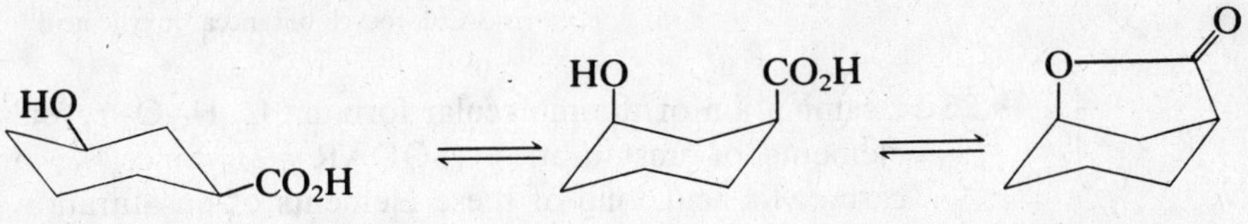

cis-3-Hydroxycyclohexanecarboxylic acid

Lactone

While the most stable conformation of *cis*-3-hydroxycyclohexanecarboxylic acid has both substituents equatorial and is unable to close to a lactone, the diaxial orientation is accessible and is capable of lactone formation.

Neither conformation of *trans*-3-hydroxycyclohexanecarboxylic acid has the substituents close enough to each other to form an unstrained lactone.

trans-3-Hydroxycyclohexanecarboxylic acid: lactone formation impossible

(*e*) Ascorbic acid is relatively acidic because ionization of its enolic hydroxyl at C-3 gives an anion that is stabilized by resonance in much the same way as a carboxylate ion; the negative charge is shared by two oxygens.

acidic proton in
ascorbic acid

19.25 Dicarboxylic acids in which both carboxyl groups are attached to the same carbon undergo ready thermal decarboxylation to produce the enol form of an acid.

Compound A

This enol yields a mixture of *cis*- and *trans*-3-chlorocyclobutanecarboxylic acid. The two products are stereoisomers.

cis-3-Chlorocyclobutanecarboxylic acid   *trans*-3-Chlorocyclobutanecarboxylic acid

19.26 Examination of the molecular formula $C_{14}H_{26}O_2$ reveals that the compound contains two elements of unsaturation (SODAR = 2). Since we are told that the compound is a carboxylic acid, one of these elements of unsaturation must be a carbon-oxygen double bond. The other must be a carbon-carbon double bond, since the compound undergoes cleavage on ozonolysis. Cleavage of the compound serves to locate the position of the double bond.

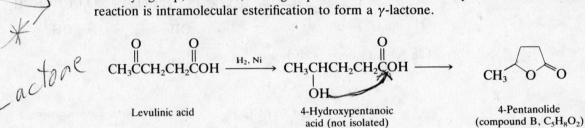

$$CH_3(CH_2)_7CH{=}CH(CH_2)_3CO_2H \xrightarrow[\text{2. } H_2O, Zn]{\text{1. } O_3} CH_3(CH_2)_7\overset{O}{\overset{\|}{C}}H + H\overset{O}{\overset{\|}{C}}(CH_2)_3CO_2H$$

cleavage by ozone occurs here

Nonanal   5-Oxopentanoic acid

The starting acid must be 5-tetradecenoic acid. The stereochemistry of the double bond is not revealed by these experiments.

19.27 Hydrogenation of the starting material is expected to result in reduction of the ketone carbonyl while leaving the carboxyl group unaffected. Since the isolated product lacks a carboxyl group, however, that group must react in some way. The most reasonable reaction is intramolecular esterification to form a γ-lactone.

$$CH_3\overset{O}{\overset{\|}{C}}CH_2CH_2\overset{O}{\overset{\|}{C}}OH \xrightarrow{H_2, Ni} CH_3\underset{OH}{CH}CH_2CH_2\overset{O}{\overset{\|}{C}}OH \longrightarrow$$

Levulinic acid   4-Hydroxypentanoic acid (not isolated)   4-Pentanolide (compound B, $C_5H_8O_2$)

19.28 Compound C is a cyclic acetal and undergoes hydrolysis in aqueous acid to produce acetaldehyde, along with a dihydroxy carboxylic acid.

$$\xrightarrow{H_3O^+}$$

Compound C   3,5-Dihydroxy-3-methylpentanoic acid   Acetaldehyde

The dihydroxy acid that is formed in this step cyclizes to the δ-lactone.

$$\equiv \longrightarrow$$

3,5-Dihydroxy-3-methylpentanoic acid   Mevalonolactone

19.29 Compound D is a δ-lactone. To determine its precursor disconnect the ester linkage to a hydroxy acid.

$$\Rightarrow$$

Compound D

The precursor has the same carbon skeleton as the designated starting material. All that is necessary is to hydrogenate the double bond of the alkynoic acid to the cis alkene. This can be done by using the Lindlar catalyst. Cyclization of the hydroxy acid to the lactone is spontaneous.

$$CH_3CHCH_2C≡CCO_2H \xrightarrow[\text{Lindlar Pd}]{H_2}$$

with OH substituent

5-Hydroxy-2-hexynoic acid

(Not isolated)

Compound D

19.30 Hydration of the double bond can occur in two different directions:

$$\text{maleic-type diacid} \xrightarrow{H_2O} HO_2CCH_2CCH_2CO_2H + HO_2CCHCHCH_2CO_2H$$

(a) The achiral isomer is citric acid.

$$HO_2CCH_2CCH_2CO_2H \qquad \text{Citric acid has no stereogenic centers.}$$

with CO_2H and OH substituents

(b) The other isomer, isocitric acid, has two stereogenic centers. Isocitric acid has the constitution

$$HO_2CC^*HCHCH_2CO_2H \qquad \text{Isocitric acid: stereogenic centers are marked with asterisk.}$$

with CO_2H and OH substituents and asterisks

With two stereogenic centers, there are 2^2, or four, stereoisomers represented by this constitution. The one that is actually formed in this enzyme-catalyzed reaction is the $2R,3S$ isomer.

19.31 Carboxylic acid protons give signals in the range $\delta = 10$ to 12 ppm. A signal in this region suggests the presence of a carboxyl group but tells little about its environment. Thus, in assigning structures to compounds E, F, and G, the data that will be most useful are the chemical shifts of the protons other than the carboxyl protons. Compare the three structures:

$$HCOH \qquad \text{Maleic acid} \qquad HO_2CCH_2CO_2H$$

Formic acid   Maleic acid   Malonic acid

The proton that is diagnostic of structure in formic acid is bonded to a carbonyl group; it is an aldehyde proton. Typical chemical shifts of aldehyde protons are 8 to 10 ppm, and therefore formic acid is compound G.

Compound G $$H-\overset{O}{\overset{\|}{C}}-O-H$$

$\delta = 11.4$ ppm

$\delta = 8.0$ ppm

The critical signal in maleic acid is that of the vinyl protons, which normally is found in the range $\delta = 5$ to 7 ppm. Maleic acid is compound F.

Compound F

$$\underset{HO_2C}{\overset{H}{}}C=C\underset{CO_2H}{\overset{H}{}} \quad \begin{array}{l} \delta = 6.3\ \text{ppm} \\ \delta = 12.4\ \text{ppm} \end{array}$$

Compound E is malonic acid. Here we have a methylene group bearing two carbonyl substituents. These methylene protons are more shielded than the aldehyde proton of formic acid or the vinyl protons of maleic acid.

Compound E $\quad HO_2CCH_2CO_2H \quad\longleftarrow\quad \delta = 12.1\ \text{ppm}$

$$\delta = 3.2\ \text{ppm}$$

19.32 (*a*) The carboxylic acid proton is responsible for the one-proton signal at $\delta = 11.8$ ppm in the ^1H nmr spectrum of this compound. The six-proton singlet at $\delta = 1.3$ ppm reveals the presence of two equivalent methyl groups. The chlorine substituent is bonded to a methylene group, and the peak corresponding to $-CH_2Cl$ appears at $\delta = 3.6$ ppm. The structure consistent with these data is 3-chloro-2,2-dimethylpropanoic acid.

singlet $\delta = 3.6$ ppm $\quad ClCH_2-\overset{\overset{\displaystyle CH_3}{|}}{\underset{\underset{\displaystyle CH_3}{|}}{C}}-C\overset{\displaystyle O}{\underset{\displaystyle O-H}{}} \quad \delta = 11.8$ ppm

singlet $\delta = 1.3$ ppm $\qquad\qquad$ Compound H

(*b*) The formula of compound I ($C_3H_5ClO_2$) has a SODAR of 1—the carboxyl group. Only two structures are possible:

$$ClCH_2CH_2C\overset{\displaystyle O}{\underset{\displaystyle OH}{}} \quad \text{and} \quad CH_3\underset{\underset{\displaystyle Cl}{|}}{CH}C\overset{\displaystyle O}{\underset{\displaystyle OH}{}}$$

3-Chloropropanoic acid $\qquad\qquad$ 2-Chloropropanoic acid

Compound I is determined to be 3-chloropropanoic acid on the basis of its ^1H nmr spectrum, which shows two triplets at $\delta = 2.9$ and $\delta = 3.8$ ppm.

triplet $\delta = 3.8$ ppm $\quad ClCH_2-CH_2-C\overset{\displaystyle O}{\underset{\displaystyle O-H}{}} \quad \delta = 11.7$ ppm

triplet $\delta = 2.9$ ppm $\qquad\qquad$ Compound I

Compound I cannot be 2-chloropropanoic acid, because that compound's ^1H nmr spectrum would show a three-proton doublet for the methyl group and a one-proton quartet for the methine proton.

(*c*) Compound J, with a molecular formula of $C_5H_8O_2$, has two elements of unsaturation (SODAR = 2). Since we are told that it is a carboxylic acid, one of these must be the carbon-oxygen double bond. With a peak in its ^1H nmr spectrum in the vinyl region ($\delta = 5.7$ ppm), a reasonable conclusion is that the second element of unsaturation in compound J is a carbon-carbon double bond. The two peaks at $\delta = 2.0$ and 2.2 ppm are nonequivalent methyl groups and their chemical shift is consistent with their being allylic methyls. Enough information is now available to assign compound J as 3-methyl-2-butenoic acid.

$$\delta = 2.0 \text{ and } 2.2 \text{ ppm} \left\{ \begin{array}{c} CH_3 \\ \\ CH_3 \end{array} \right. C=C \begin{array}{c} H \leftarrow \delta = 5.7 \text{ ppm} \\ \\ CO_2H \leftarrow \delta = 12.1 \text{ ppm} \end{array}$$

Compound J

The slight splitting of the methyl signals and broadening of the vinyl proton signal are a consequence of long-range coupling.

(d) The molecular formula of compound K ($C_{15}H_{14}O_2$) reveals nine elements of unsaturation (SODAR = 9). This is consistent with two monosubstituted aromatic rings (there are 10 aryl hydrogens in the nmr spectrum) and a carboxylic acid group. The two aromatic rings and the carboxyl group account for 13 of the 15 carbon atoms of the molecule. The doublet-triplet pattern in the nmr spectrum suggests the grouping

$$-CH_2\overset{|}{\underset{|}{C}}-H$$

The structure most consistent with the data is 3,3-diphenylpropanoic acid.

$$\delta = 10.6 \text{ ppm} \searrow HO_2CCH_2 \overset{C_6H_5}{\underset{C_6H_5}{\overset{|}{-}C-}}H \leftarrow \text{triplet } \delta = 4.6 \text{ ppm}$$

doublet $\delta = 3.0$ ppm   Compound K

The alternative formulation

$$C_6H_5CH_2 \overset{C_6H_5}{\underset{CO_2H}{\overset{|}{-}C-}}H$$

is not consistent with the given information that compound K is achiral; C-2 is a stereogenic center in the alternative structure.

SELF-TEST

PART A

A-1. Provide an acceptable IUPAC name for each of the following:

(a)
$$\overset{CH_3}{\underset{CH_3}{\overset{|}{\underset{|}{C_6H_5CHCHCH_2CH_2CO_2H}}}}$$

(b) ⬡—CO_2H

(c)
$$\overset{Br}{\underset{CH_2CH_3}{\overset{|}{\underset{|}{CH_3CHCHCO_2H}}}}$$

A-2. Both of the following compounds may be converted into 4-phenylbutanoic acid by one or more reaction steps. Give the reagents and conditions necessary to carry out these conversions.

$$C_6H_5CH_2CH_2CH(CO_2H)_2 \qquad C_6H_5CH_2CH_2CH_2Br$$

(two methods)

A-3. The species whose structure is shown is an intermediate in an esterification reaction. Write the complete, balanced equation for this process.

$$\underset{\underset{\textstyle OH}{|}}{\overset{\overset{\textstyle OH}{|}}{C_6H_5CH_2\overset{}{C}OCH_2CH_3}}$$

A-4. Give the correct structures for compounds A through C in the following reactions.

(a) $(CH_3)_2CHCH_2CH_2CO_2H \xrightarrow{Br_2, P} A \xrightarrow[\text{2. } H^+]{\text{1. NaCN}} B$

(b) $C \xrightarrow{\text{heat}} C_6H_5\overset{\overset{\textstyle O}{\|}}{C}CH(CH_3)_2 + CO_2$

A-5. Give the missing reagent(s) and the missing compound in each of the following:

(a) ⟨benzene⟩—Br $\xrightarrow{?}$? $\xrightarrow[\text{2. } H_3O^+]{\text{1. } CO_2}$ ⟨benzene⟩—CO_2H

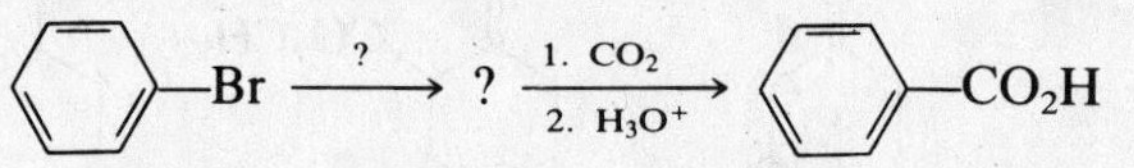

(b) $CH_3CH_2CH_2\overset{\overset{\textstyle O}{\|}}{C}OH \xrightarrow{?}$? $\xrightarrow[\text{CH}_2\text{Cl}_2]{\text{PCC}} CH_3CH_2CH_2\overset{\overset{\textstyle O}{\|}}{C}H$

(c) $CH_3CH_2CH_2\overset{\overset{\textstyle O}{\|}}{C}OH \xrightarrow{?}$? $\xrightarrow{\text{NaSCH}_3} \underset{\underset{\textstyle SCH_3}{|}}{CH_3CH_2\overset{\overset{\textstyle O}{\|}}{C}HCOH}$

(d) $CH_3CH_2\overset{\overset{\textstyle O}{\|}}{C}H \xrightarrow[\text{HCl}]{\text{NaCN}}$? $\xrightarrow{?} \underset{\underset{\textstyle OH}{|}}{CH_3CH_2\overset{\overset{\textstyle O}{\|}}{C}HCOH}$

A-6. Identify the carboxylic acid ($C_4H_7BrO_2$) having the ^1H nmr spectrum consisting of:

$$\delta = 1.1 \text{ ppm, 3H (triplet)}$$

$$\delta = 2.0 \text{ ppm, 2H (pentet)}$$

$$\delta = 4.2 \text{ ppm, 1H (triplet)}$$

$$\delta = 12.1 \text{ ppm, 1H (singlet)}$$

PART B

B-1. Which of the following is a correct IUPAC name for the compound shown?

$$CO_2H$$
$$(CH_3CH_2)_2CCH_2CH(CH_2CH_3)_2$$

(a) 1,1,3-Triethylhexanoic acid
(b) 2,2,4-Triethylhexanoic acid
(c) 3,5-Diethyl-3-heptylcarboxylic acid
(d) 3,5,5-Triethyl-6-hexanoic acid

B-2. Rank the following substances in order of decreasing acid strength (strongest→ weakest):

$$CH_3CH_2CH_2CO_2H \qquad CH_3CH\!=\!CHCO_2H \qquad CH_3CH_2CH_2CH_2OH$$
$$\qquad\quad A \qquad\qquad\qquad\qquad B \qquad\qquad\qquad\qquad C$$
$$CH_3C\!\equiv\!CCO_2H$$
$$D$$

(a) D > B > A > C   (c) C > A > B > D
(b) A > B > D > C   (d) B > D > A > C

B-3. Carboxylic acids exist in the vapor state as:
(a) Monomeric species   (c) Cyclic dimers
(b) Cyclic esters   (d) Cyclic trimers

B-4. Which of the following compounds will undergo decarboxylation on heating?

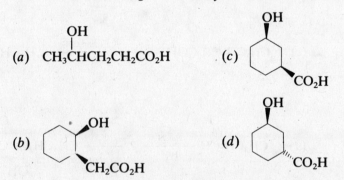

I   II   III   IV

(a) II and III   (c) III only
(b) III and IV   (d) I and IV

B-5. Which of the following is *least* likely to be able to form a lactone?

$$OH$$
(a) $CH_3CHCH_2CH_2CO_2H$ (c)

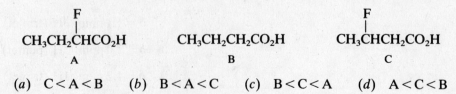

(b)   (d)

B-6. Rank the following in order of increasing acidity (least acidic→ most acidic):

$$F$$
$$CH_3CH_2CHCO_2H \qquad\qquad CH_3CH_2CH_2CO_2H \qquad\qquad CH_3CHCH_2CO_2H$$
$$\qquad A \qquad\qquad\qquad\qquad\qquad B \qquad\qquad\qquad\qquad\qquad C$$

(a) C < A < B   (b) B < A < C   (c) B < C < A   (d) A < C < B

CARBOXYLIC ACID DERIVATIVES. NUCLEOPHILIC ACYL SUBSTITUTION

IMPORTANT TERMS AND CONCEPTS

Nomenclature of Carboxylic Acid Derivatives (Sec. 20.1) Important derivatives of carboxylic acids include:

1. Acyl chlorides:
$$R-\overset{\displaystyle O}{\overset{\|}{C}}-Cl$$

2. Carboxylic acid anhydrides:
$$R-\overset{\displaystyle O}{\overset{\|}{C}}-O-\overset{\displaystyle O}{\overset{\|}{C}}-R'$$

3. Esters:
$$R-\overset{\displaystyle O}{\overset{\|}{C}}-O-R'$$

4. Carboxamides:
$$R-\overset{\displaystyle O}{\overset{\|}{C}}-NH_2, \quad R-\overset{\displaystyle O}{\overset{\|}{C}}-NHR', \quad R-\overset{\displaystyle O}{\overset{\|}{C}}-NR'_2$$

The following examples illustrate the manner in which systematic names are applied to the various derivatives:

$$CH_3CH_2CH_2\overset{\displaystyle O}{\overset{\|}{C}}Cl \qquad \text{Butanoyl chloride}$$

$$CH_3CH_2\overset{\displaystyle O}{\overset{\|}{C}}O\overset{\displaystyle O}{\overset{\|}{C}}CH_2CH_3 \qquad \text{Propanoic anhydride}$$

$$C_6H_5\overset{\displaystyle O}{\overset{\|}{C}}OCH_2CH_3 \qquad \text{Ethyl benzoate}$$

$$CH_3CH_2\overset{\displaystyle O}{\overset{\|}{C}}NHCH_3 \qquad \textit{N}\text{-Methylpropanamide}$$

Certain cyclic derivatives are given special names:

Lactones (cyclic esters):

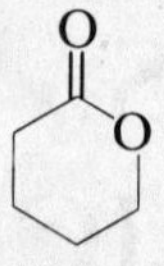

Lactams (cyclic amides, Sec. 20.14):

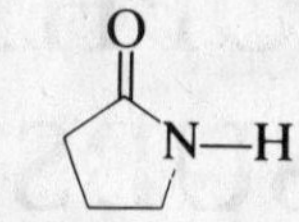

Imides (Sec. 20.15):

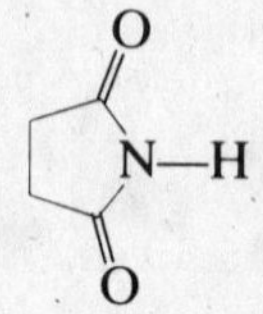

Structure and Reactivity (Sec. 20.2) All the acyl derivatives discussed in this chapter possess one common feature, namely, that the atom attached to the acyl group has an unshared pair of electrons that is capable of interacting with the carbonyl π system. This may be illustrated with resonance structures:

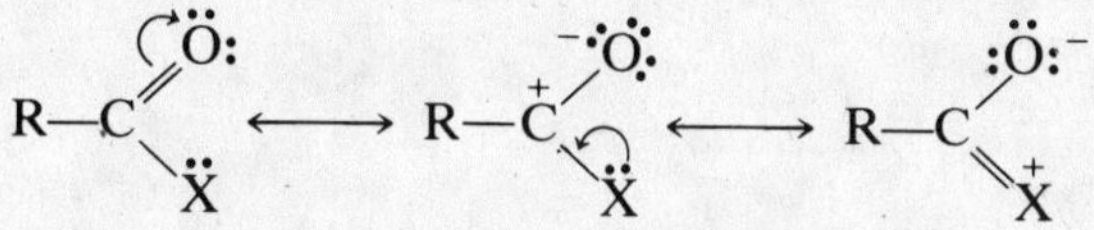

$$R-C \overset{\ddot{O}:}{\underset{\ddot{X}}{}} \longleftrightarrow R-\overset{+}{C} \overset{\ddot{O}:^-}{\underset{X}{}} \longleftrightarrow R-C \overset{:\ddot{O}:^-}{\underset{\overset{+}{X}}{}}$$

The extent of resonance stabilization varies among the derivatives and is a contributing factor in determining the relative reactivity of the groups. The general trend is:

$$\underset{\substack{\text{Most}\\\text{reactive}}}{\overset{O}{\underset{\|}{R C Cl}}} > \overset{O\quad O}{\underset{\|\quad\|}{R C O C R'}} > \overset{O}{\underset{\|}{R C O R'}} > \underset{\substack{\text{Least}\\\text{reactive}}}{\overset{O}{\underset{\|}{R C N R'_2}}}$$

Greater resonance stabilization results in lower susceptibility to attack by nucleophiles at the carbonyl carbon.

General Mechanism

The nucleophilic acyl substitutions discussed in this chapter all involve reaction of a nucleophile with the carbonyl group of an acyl derivative. General mechanisms are as follows:

Acidic medium:

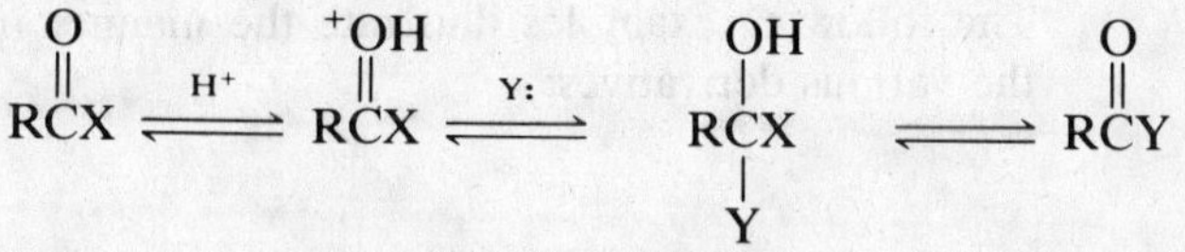

$$\overset{O}{\underset{\|}{R C X}} \overset{H^+}{\rightleftharpoons} \overset{+OH}{\underset{\|}{R C X}} \overset{Y:}{\rightleftharpoons} \underset{\underset{Y}{|}}{\overset{\overset{OH}{|}}{R C X}} \rightleftharpoons \overset{O}{\underset{\|}{R C Y}}$$

Tetrahedral intermediate

Basic medium:

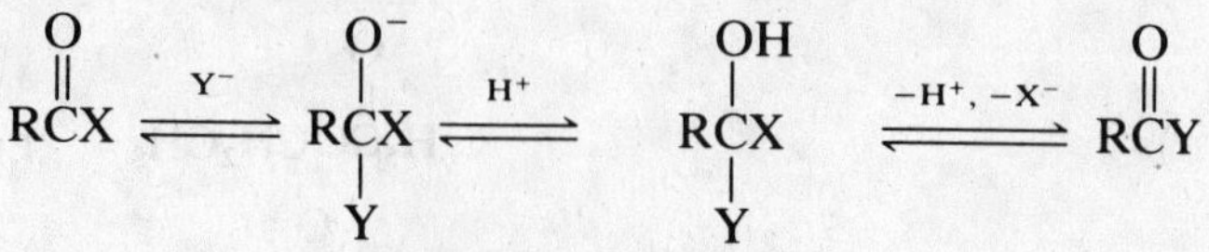

$$\overset{O}{\underset{\|}{R C X}} \overset{Y^-}{\rightleftharpoons} \underset{\underset{Y}{|}}{\overset{\overset{O^-}{|}}{R C X}} \overset{H^+}{\rightleftharpoons} \underset{\underset{Y}{|}}{\overset{\overset{OH}{|}}{R C X}} \overset{-H^+,\ -X^-}{\rightleftharpoons} \overset{O}{\underset{\|}{R C Y}}$$

Tetrahedral intermediate

Spectroscopic Analysis (Sec. 20.21) Infrared spectroscopy may be used to identify carboxylic acid derivatives, since they all exhibit characteristic absorptions in the carbonyl stretching region.

Derivative	Example	$\nu_{C=O}$
Acyl chloride	Acetyl chloride	1822 cm^{-1}
Anhydride	Acetic anhydride	1748 and 1815 cm^{-1}
Ester	Methyl acetate	1736 cm^{-1}
Amide	Acetamide	1694 cm^{-1}

The structures of esters may be assigned using ^{1}H nmr spectroscopy by the chemical shift of protons adjacent to the carbonyl group, as indicated by the following examples:

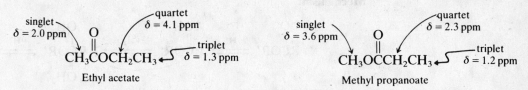

Ethyl acetate	Methyl propanoate

IMPORTANT REACTIONS

ACYL CHLORIDES

Preparation

$$\underset{\text{O}}{\text{R}\overset{\text{O}}{\overset{\|}{\text{C}}}\text{OH}} + \text{SOCl}_2 \longrightarrow \text{R}\overset{\text{O}}{\overset{\|}{\text{C}}}\text{Cl} + \text{SO}_2 + \text{HCl} \quad \text{(Sec. 12.7)}$$

Reactions (Sec. 20.3)

$$\text{ArH} + \text{R}\overset{\text{O}}{\overset{\|}{\text{C}}}\text{Cl} \xrightarrow{\text{AlCl}_3} \text{Ar}\overset{\text{O}}{\overset{\|}{\text{C}}}\text{R} \quad \text{(Friedel-Crafts acylation, Sec. 12.7)}$$

$$\text{R}\overset{\text{O}}{\overset{\|}{\text{C}}}\text{Cl} + \text{H}_2\text{O} \longrightarrow \text{R}\overset{\text{O}}{\overset{\|}{\text{C}}}\text{OH} + \text{HCl} \quad \text{(Sec. 20.3)}$$

CARBOXYLIC ACID ANHYDRIDES

Preparation (Sec. 20.4)

$$\text{R}\overset{\text{O}}{\overset{\|}{\text{C}}}\text{Cl} + \text{R}'\text{CO}_2\text{H} \xrightarrow{\text{pyridine}} \text{R}\overset{\text{O}}{\overset{\|}{\text{C}}}\text{O}\overset{\text{O}}{\overset{\|}{\text{C}}}\text{R}'$$

Reactions (Sec. 20.5)

$$\text{ArH} + \text{R}\overset{\text{O}}{\overset{\|}{\text{C}}}\text{O}\overset{\text{O}}{\overset{\|}{\text{C}}}\text{R} \xrightarrow{\text{AlCl}_3} \text{Ar}\overset{\text{O}}{\overset{\|}{\text{C}}}\text{R} + \text{R}\overset{\text{O}}{\overset{\|}{\text{C}}}\text{OH} \quad \text{(Sec. 12.7)}$$

$$\text{R}\overset{\text{O}}{\overset{\|}{\text{C}}}\text{O}\overset{\text{O}}{\overset{\|}{\text{C}}}\text{R} + \text{R}'\text{OH} \longrightarrow \text{R}\overset{\text{O}}{\overset{\|}{\text{C}}}\text{OR}' + \text{R}\overset{\text{O}}{\overset{\|}{\text{C}}}\text{OH} \quad \text{(Sec. 15.8)}$$

ESTERS

Reactions (Secs. 20.8 to 20.11)

Acid-catalyzed hydrolysis (Sec. 20.9):

General:

$$\underset{RCOR'}{\overset{\overset{\textstyle O}{\|}}{}} + H_2O \overset{H^+}{\rightleftharpoons} \underset{RCOH}{\overset{\overset{\textstyle O}{\|}}{}} + R'OH$$

Example:

$$\underset{C_6H_5COCH(CH_3)_2}{\overset{\overset{\textstyle O}{\|}}{}} + H_2O \overset{H^+}{\rightleftharpoons} \underset{C_6H_5COH}{\overset{\overset{\textstyle O}{\|}}{}} + (CH_3)_2CHOH$$

Mechanism:

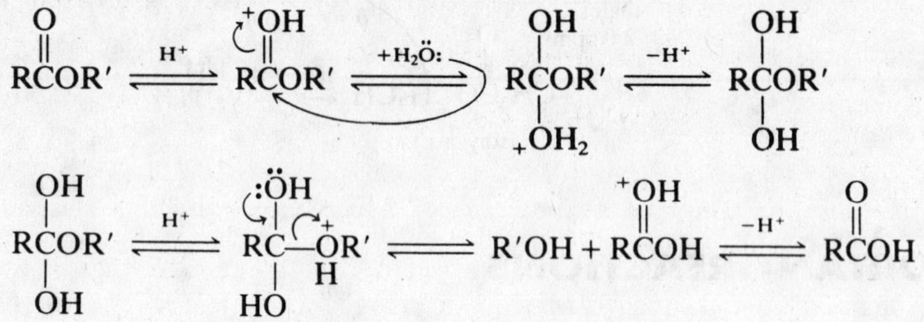

Base-promoted hydrolysis (Sec. 20.10):

General:

$$\underset{RCOR'}{\overset{\overset{\textstyle O}{\|}}{}} + HO^- \overset{H_2O}{\longrightarrow} \underset{RCO^-}{\overset{\overset{\textstyle O}{\|}}{}} + R'OH$$

Example:

$$\underset{(CH_3)_2CHCH_2COCH_3}{\overset{\overset{\textstyle O}{\|}}{}} \overset{HO^-}{\underset{H_2O}{\longrightarrow}} \underset{(CH_3)_2CHCH_2CO^-}{\overset{\overset{\textstyle O}{\|}}{}} + CH_3OH$$

Mechanism:

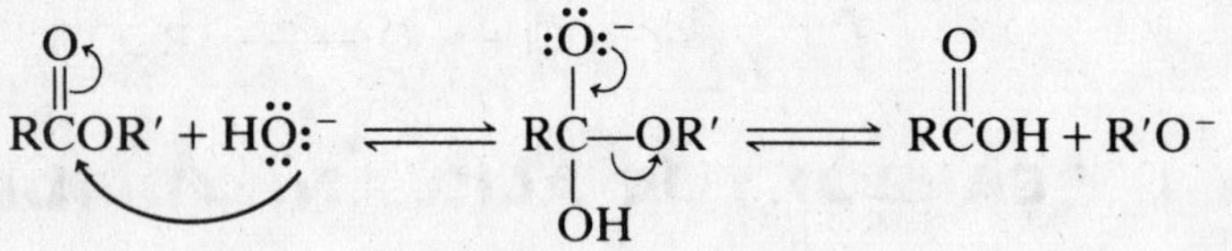

$$R'O^- + H_2O \longrightarrow R'OH + HO^-$$

$$\underset{RCOH}{\overset{\overset{\textstyle O}{\|}}{}} + HO^- \longrightarrow \underset{RCO^-}{\overset{\overset{\textstyle O}{\|}}{}} + H_2O$$

Reaction with ammonia and amines (Sec. 20.11):

$$\underset{RCOR'}{\overset{\overset{\textstyle O}{\|}}{}} + NH_3 \longrightarrow \underset{RCNH_2}{\overset{\overset{\textstyle O}{\|}}{}} + R'OH$$

$$\underset{RCOR'}{\overset{\overset{\textstyle O}{\|}}{}} + HNR''_2 \longrightarrow \underset{RCNR''_2}{\overset{\overset{\textstyle O}{\|}}{}} + R'OH$$

AMIDES

Preparation (Sec. 20.13)

General:

$$2R_2NH + R'\overset{\displaystyle O}{\overset{\|}{C}}Cl \longrightarrow R'\overset{\displaystyle O}{\overset{\|}{C}}NR_2 + R_2\overset{+}{N}H_2 \ Cl^-$$

$$2R_2NH + R'\overset{\displaystyle O}{\overset{\|}{C}}O\overset{\displaystyle O}{\overset{\|}{C}}R' \longrightarrow R'\overset{\displaystyle O}{\overset{\|}{C}}NR_2 + R_2\overset{+}{N}H_2 \ {}^-O_2CR'$$

Examples:

$$2(CH_3CH_2)_2NH + C_6H_5\overset{\displaystyle O}{\overset{\|}{C}}Cl \longrightarrow C_6H_5\overset{\displaystyle O}{\overset{\|}{C}}N(CH_2CH_3)_2$$

$$2CH_3NH_2 + CH_3CH_2\overset{\displaystyle O}{\overset{\|}{C}}O\overset{\displaystyle O}{\overset{\|}{C}}CH_2CH_3 \longrightarrow CH_3CH_2\overset{\displaystyle O}{\overset{\|}{C}}NHCH_3$$

Hydrolysis (Sec. 20.16)

General:

$$R\overset{\displaystyle O}{\overset{\|}{C}}NR'_2 + H_3O^+ \longrightarrow R\overset{\displaystyle O}{\overset{\|}{C}}OH + R'_2\overset{+}{N}H_2$$

$$R\overset{\displaystyle O}{\overset{\|}{C}}NR'_2 + HO^- \longrightarrow R\overset{\displaystyle O}{\overset{\|}{C}}O^- + R'_2NH$$

Example:

$$C_6H_5CH_2\overset{\displaystyle O}{\overset{\|}{C}}NHCH_3 + HO^- \xrightarrow{\ H_2O\ } C_6H_5CH_2\overset{\displaystyle O}{\overset{\|}{C}}O^- + CH_3NH_2$$

Hofmann Rearrangement (Sec. 20.17)

General:

$$R\overset{\displaystyle O}{\overset{\|}{C}}NH_2 + Br_2 \xrightarrow{\ HO^-,\ H_2O\ } RNH_2 + CO_3^{2-} + 2Br^-$$

Example:

$$(CH_3)_3C\overset{\displaystyle O}{\overset{\|}{C}}NH_2 + Br_2 \xrightarrow{\ HO^-,\ H_2O\ } (CH_3)_3CNH_2$$

Mechanism:

$$R-N{=}C{=}O + H_2O \longrightarrow RNH\overset{\displaystyle O}{\overset{\|}{C}}OH \xrightarrow{\ HO^-\ } RNH_2 + CO_3^{2-}$$

NITRILES

Preparation (Sec. 20.18)

$$\underset{\substack{\text{O}\\\|}}{\text{RCNH}_2} \xrightarrow[\text{heat}]{\text{P}_4\text{O}_{10}(-\text{H}_2\text{O})} \text{RC}\equiv\text{N}$$

Hydrolysis (Sec. 20.19)

$$\text{RC}\equiv\text{N} + \text{H}_2\text{O} \xrightarrow{\text{H}^+ \text{ or HO}^-} \underset{\substack{\text{O}\\\|}}{\text{RCOH}} \ (\text{or } \underset{\substack{\text{O}\\\|}}{\text{RCO}^-}) + \overset{+}{\text{NH}}_4 \ (\text{or NH}_3)$$

Reaction with Grignard Reagents (Sec. 20.20)

$$\text{RC}\equiv\text{N} \xrightarrow[\text{2. H}_2\text{O, H}^+, \text{ heat}]{\text{1. R'MgX, diethyl ether}} \underset{\substack{\text{O}\\\|}}{\text{RCR'}}$$

SOLUTIONS TO TEXT PROBLEMS

20.1 (*b*) Carboxylic acid anhydrides bear two acyl groups on oxygen, as in $\underset{\substack{\text{O}\quad\text{O}\\\|\quad\|}}{\text{RCOCR}}$. They are named as derivatives of carboxylic acids.

$$\underset{\substack{|\\\text{C}_6\text{H}_5}}{\underset{\substack{\text{O}\\\|}}{\text{CH}_3\text{CH}_2\text{CHCOH}}}$$

2-Phenylbutanoic acid

$$\underset{\substack{|\qquad\quad|\\\text{C}_6\text{H}_5\qquad\text{C}_6\text{H}_5}}{\underset{\substack{\text{O}\quad\text{O}\\\|\quad\|}}{\text{CH}_3\text{CH}_2\text{CHCOCCHCH}_2\text{CH}_3}}$$

2-Phenylbutanoic anhydride

(*c*) Butyl 2-phenylbutanoate is the butyl ester of 2-phenylbutanoic acid.

$$\underset{\substack{|\\\text{C}_6\text{H}_5}}{\underset{\substack{\text{O}\\\|}}{\text{CH}_3\text{CH}_2\text{CHCOCH}_2\text{CH}_2\text{CH}_2\text{CH}_3}} \qquad \text{Butyl 2-phenylbutanoate}$$

(*d*) In 2-phenylbutyl butanoate the 2-phenylbutyl group is an alkyl group bonded to oxygen of the ester. It is not involved in the acyl group of the molecule.

$$\underset{\substack{|\\\text{C}_6\text{H}_5}}{\underset{\substack{\text{O}\\\|}}{\text{CH}_3\text{CH}_2\text{CH}_2\text{COCH}_2\overset{1\;2\;3\;4}{\text{CHCH}_2\text{CH}_3}}} \qquad \text{2-Phenylbutyl butanoate}$$

(*e*) The ending *-amide* reveals this to be a compound of the type $\underset{\substack{\text{O}\\\|}}{\text{RCNH}_2}$.

$$\underset{\substack{|\\\text{C}_6\text{H}_5}}{\underset{\substack{\text{O}\\\|}}{\text{CH}_3\text{CH}_2\text{CHCNH}_2}} \qquad \text{2-Phenylbutanamide}$$

(*f*) This compound differs from 2-phenylbutanamide in part (*e*) only in that it bears an ethyl substituent on nitrogen.

$$\underset{\substack{|\\C_6H_5}}{CH_3CH_2\overset{\displaystyle O}{\overset{\displaystyle \|}{C}}NHCH_2CH_3}$$

N-Ethyl-2-phenylbutanamide

(*g*) The *-nitrile* ending signifies a compound of the type $RC{\equiv}N$ containing the same number of carbons as the alkane RCH_3.

$$\underset{\substack{|\\C_6H_5}}{CH_3CH_2CHC{\equiv}N}$$

2-Phenylbutanenitrile

20.2 The methyl groups in *N*,*N*-dimethylformamide are nonequivalent; one is cis to oxygen, the other is trans. The two methyl groups have different chemical shifts.

Rotation about the carbon-nitrogen bond is required to average the environments of the two methyl groups, but this rotation is relatively slow in amides as the result of the double bond character imparted to the carbon-nitrogen bond, as shown by these two resonance structures.

20.3 (*b*) Benzoyl chloride reacts with benzoic acid to give benzoic anhydride.

$$\underset{\text{Benzoyl chloride}}{C_6H_5\overset{\displaystyle O}{\overset{\displaystyle \|}{C}}Cl} + \underset{\text{Benzoic acid}}{C_6H_5\overset{\displaystyle O}{\overset{\displaystyle \|}{C}}OH} \longrightarrow \underset{\text{Benzoic anhydride}}{C_6H_5\overset{\displaystyle O}{\overset{\displaystyle \|}{C}}O\overset{\displaystyle O}{\overset{\displaystyle \|}{C}}C_6H_5}$$

(*c*) Acyl chlorides react with alcohols to form esters.

$$\underset{\text{Benzoyl chloride}}{C_6H_5\overset{\displaystyle O}{\overset{\displaystyle \|}{C}}Cl} + \underset{\text{Ethanol}}{CH_3CH_2OH} \longrightarrow \underset{\text{Ethyl benzoate}}{C_6H_5\overset{\displaystyle O}{\overset{\displaystyle \|}{C}}OCH_2CH_3}$$

The organic product is the ethyl ester of benzoic acid, ethyl benzoate.

(*d*) Acyl transfer from benzoyl chloride to the nitrogen of methylamine yields the amide *N*-methylbenzamide.

$$\underset{\text{Benzoyl chloride}}{C_6H_5\overset{\displaystyle O}{\overset{\displaystyle \|}{C}}Cl} + \underset{\text{Methylamine}}{CH_3NH_2} \longrightarrow \underset{\text{\textit{N}-Methylbenzamide}}{C_6H_5\overset{\displaystyle O}{\overset{\displaystyle \|}{C}}NHCH_3}$$

(*e*) In analogy with part (*d*), an amide is formed. In this case the product has two methyl groups on nitrogen.

$$\underset{\text{Benzoyl chloride}}{C_6H_5\overset{\displaystyle O}{\overset{\displaystyle \|}{C}}Cl} + \underset{\text{Dimethylamine}}{(CH_3)_2NH} \longrightarrow \underset{\text{\textit{N},\textit{N}-Dimethylbenzamide}}{C_6H_5\overset{\displaystyle O}{\overset{\displaystyle \|}{C}}N(CH_3)_2}$$

$\longrightarrow$ (f) Acyl chlorides undergo hydrolysis on reaction with water. The product is a carboxylic acid.

$$C_6H_5\overset{\displaystyle O}{\overset{\displaystyle \|}{C}}Cl + H_2O \longrightarrow C_6H_5\overset{\displaystyle O}{\overset{\displaystyle \|}{C}}OH + HCl$$

Benzoyl chloride   Water   Benzoic acid   Hydrogen chloride

20.4 (b) Nucleophilic addition of benzoic acid to benzoyl chloride gives the tetrahedral intermediate shown.

$$C_6H_5\overset{\displaystyle O}{\overset{\displaystyle \|}{C}}Cl + C_6H_5\overset{\displaystyle O}{\overset{\displaystyle \|}{C}}OH \longrightarrow C_6H_5\overset{\displaystyle HO}{\underset{\displaystyle Cl}{C}}\overset{\displaystyle O}{-OC}C_6H_5$$

Benzoyl chloride   Benzoic acid   Tetrahedral intermediate

Dissociation of the tetrahedral intermediate occurs by loss of chloride and of the proton on the oxygen.

$$C_6H_5\overset{\displaystyle H\diagdown}{\underset{\displaystyle Cl}{C}}\overset{\displaystyle O\ O}{OC}C_6H_5 \longrightarrow C_6H_5\overset{\displaystyle O\ O}{\overset{\displaystyle \|\ \|}{C}}OCC_6H_5 + HCl$$

Tetrahedral intermediate   Benzoic anhydride   Hydrogen chloride

(c) Ethanol is the nucleophile that adds to the carbonyl group of benzoyl chloride to form the tetrahedral intermediate.

$$C_6H_5\overset{\displaystyle O}{\overset{\displaystyle \|}{C}}Cl + CH_3CH_2OH \longrightarrow C_6H_5\overset{\displaystyle OH}{\underset{\displaystyle Cl}{C}}OCH_2CH_3$$

Benzoyl chloride   Ethanol   Tetrahedral intermediate

In analogy with parts (a) and (b) of this problem, a proton is lost from the hydroxyl group along with chloride to reform the carbon-oxygen double bond.

$$C_6H_5\overset{\displaystyle H\diagdown}{\underset{\displaystyle Cl}{C}}OCH_2CH_3 \longrightarrow C_6H_5\overset{\displaystyle O}{\overset{\displaystyle \|}{C}}OCH_2CH_3 + HCl$$

Tetrahedral intermediate   Ethyl benzoate   Hydrogen chloride

(d) The tetrahedral intermediate formed from benzoyl chloride and methylamine has a carbon-nitrogen bond.

$$C_6H_5\overset{\displaystyle O}{\overset{\displaystyle \|}{C}}Cl + CH_3NH_2 \longrightarrow C_6H_5\overset{\displaystyle OH}{\underset{\displaystyle Cl}{C}}NHCH_3$$

Benzoyl chloride   Methylamine   Tetrahedral intermediate

Schematically, the dissociation of the tetrahedral intermediate may be shown as:

$$\underset{\substack{\text{Tetrahedral intermediate}}}{\overset{\substack{H\diagdown \\ O \\ | \\ C_6H_5\overset{|}{C}NHCH_3 \\ | \\ Cl}}{}} \longrightarrow \underset{\substack{N\text{-Methylbenzamide}}}{\overset{\substack{O \\ || \\ C_6H_5CNHCH_3}}{}} + \underset{\substack{\text{Hydrogen} \\ \text{chloride}}}{HCl}$$

More realistically, it is a second methylamine molecule that abstracts a proton from oxygen.

$$CH_3\overset{..}{N}H_2$$
$$\underset{\substack{\text{} }}{\overset{\substack{\diagdown H \\ C \\ O \\ | \\ C_6H_5\overset{|}{C}NHCH_3 \\ | \\ Cl}}{}} \longrightarrow \underset{\substack{N\text{-Methylbenzamide}}}{\overset{\substack{O \\ || \\ C_6H_5CNHCH_3}}{}} + \underset{\substack{\text{Methylammonium} \\ \text{chloride}}}{CH_3\overset{+}{N}H_3\ Cl^-}$$

(e) The intermediates in the reaction of benzoyl chloride with dimethylamine are similar to those in part (d). The methyl substituents on nitrogen are not directly involved in the reaction.

$$\underset{\substack{\text{Benzoyl chloride}}}{\overset{\substack{O \\ || \\ C_6H_5CCl}}{}} + \underset{\substack{\text{Dimethylamine}}}{(CH_3)_2NH} \longrightarrow \underset{\substack{\text{Tetrahedral intermediate}}}{\overset{\substack{OH \\ | \\ C_6H_5\overset{|}{C}N(CH_3)_2 \\ | \\ Cl}}{}}$$

Then

$$(CH_3)_2\overset{..}{N}H$$
$$\underset{\substack{}}{\overset{\substack{\diagdown H \\ C \\ O \\ | \\ C_6H_5\overset{|}{C}N(CH_3)_2 \\ | \\ Cl}}{}} \longrightarrow \underset{\substack{N,N\text{-Dimethylbenzamide}}}{\overset{\substack{O \\ || \\ C_6H_5CN(CH_3)_2}}{}} + \underset{\substack{\text{Dimethylammonium} \\ \text{chloride}}}{(CH_3)_2\overset{+}{N}H_2\ Cl^-}$$

(f) Water attacks the carbonyl group of benzoyl chloride to form the tetrahedral intermediate.

$$\underset{\substack{\text{Benzoyl chloride}}}{\overset{\substack{O \\ || \\ C_6H_5CCl}}{}} + \underset{\substack{\text{Water}}}{H_2O} \longrightarrow \underset{\substack{\text{Tetrahedral intermediate}}}{\overset{\substack{OH \\ | \\ C_6H_5\overset{|}{C}Cl \\ | \\ OH}}{}}$$

Dissociation of the tetrahedral intermediate occurs by loss of chloride and the proton on oxygen.

$$\underset{\substack{\text{Tetrahedral intermediate}}}{\overset{\substack{H\diagdown \\ O \\ | \\ C_6H_5\overset{|}{C}\frown Cl \\ | \\ OH}}{}} \longrightarrow \underset{\substack{\text{Benzoic acid}}}{C_6H_5COH} + \underset{\substack{\text{Hydrogen chloride}}}{HCl}$$

20.5 One equivalent of benzoyl chloride reacts rapidly with water to yield benzoic acid.

$$\underset{\text{Benzoyl chloride}}{C_6H_5\overset{O}{\overset{\|}{C}}Cl} + \underset{\text{Water}}{H_2O} \longrightarrow \underset{\text{Benzoic acid}}{C_6H_5\overset{O}{\overset{\|}{C}}OH} + \underset{\substack{\text{Hydrogen} \\ \text{chloride}}}{HCl}$$

The benzoic acid produced in this step reacts with the remaining benzoyl chloride to give benzoic anhydride.

$$\underset{\text{Benzoyl chloride}}{C_6H_5\overset{O}{\overset{\|}{C}}Cl} + \underset{\text{Benzoic acid}}{C_6H_5\overset{O}{\overset{\|}{C}}OH} \longrightarrow \underset{\text{Benzoic anhydride}}{C_6H_5\overset{O}{\overset{\|}{C}}O\overset{O}{\overset{\|}{C}}C_6H_5} + \underset{\substack{\text{Hydrogen} \\ \text{chloride}}}{HCl}$$

20.6 Acetic anhydride serves as a source of acetyl cation.

$$CH_3\overset{O}{\overset{\|}{C}}O{-}\overset{O}{\overset{\|}{C}}CH_3 \longrightarrow \overset{\cdot\cdot}{\underset{\cdot\cdot}{O}}{=}\overset{+}{C}CH_3 \longleftrightarrow {:}\overset{+}{O}{\equiv}CCH_3$$

$$\underset{\text{Acetyl cation}}{}$$

20.7 (b) Acyl transfer from an acid anhydride to ammonia yields an amide.

$$\underset{\text{Acetic anhydride}}{CH_3\overset{O}{\overset{\|}{C}}O\overset{O}{\overset{\|}{C}}CH_3} + \underset{\text{Ammonia}}{2NH_3} \longrightarrow \underset{\text{Acetamide}}{CH_3\overset{O}{\overset{\|}{C}}NH_2} + \underset{\text{Ammonium acetate}}{CH_3\overset{O}{\overset{\|}{C}}O^- \overset{+}{N}H_4}$$

The organic products are acetamide and ammonium acetate.

(c) The reaction of phthalic anhydride with dimethylamine is analogous to that of part (b). The organic products are an amide and the carboxylate salt of an amine.

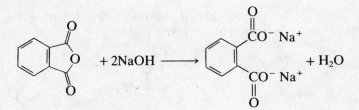

In this case both the amide function and the ammonium carboxylate salt are incorporated into the same molecule.

(d) The disodium salt of phthalic acid is the product of hydrolysis of phthalic acid in excess sodium hydroxide.

20.8 (b) The tetrahedral intermediate is formed by nucleophilic addition of ammonia to one of the carbonyl groups of acetic anhydride.

$$CH_3COCCH_3 \longrightarrow CH_3COCCH_3$$

Tetrahedral
intermediate

Dissociation of the tetrahedral intermediate occurs by loss of acetate as the leaving group.

$$CH_3C{-}OCCH_3 \longrightarrow CH_3CNH_2 + H_4\overset{+}{N}\ {}^-OCCH_3$$

Ammonia + tetrahedral   Acetamide   Ammonium acetate
intermediate

(c) Dimethylamine is the nucleophile; it adds to one of the two equivalent carbonyl groups of phthalic anhydride.

Phthalic anhydride   Dimethylamine   Tetrahedral intermediate

A second molecule of dimethylamine abstracts a proton from the tetrahedral intermediate.

Tetrahedral intermediate + second   Product of reaction
molecule of dimethylamine

(d) Hydroxide acts as a nucleophile to form the tetrahedral intermediate and as a base to facilitate its dissociation.

Hydroxide ion–catalyzed formation of tetrahedral intermediate:

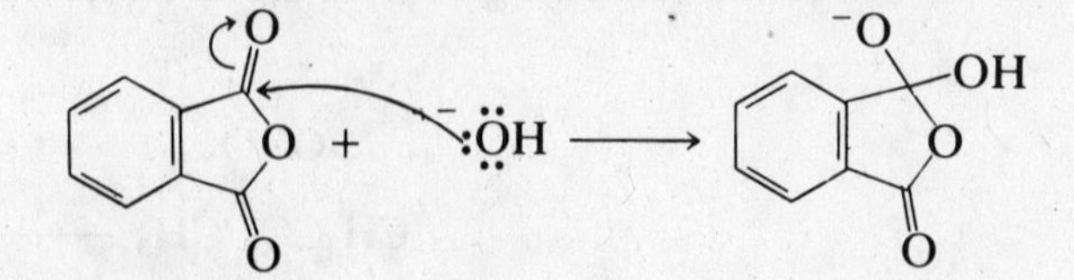

Phthalic anhydride

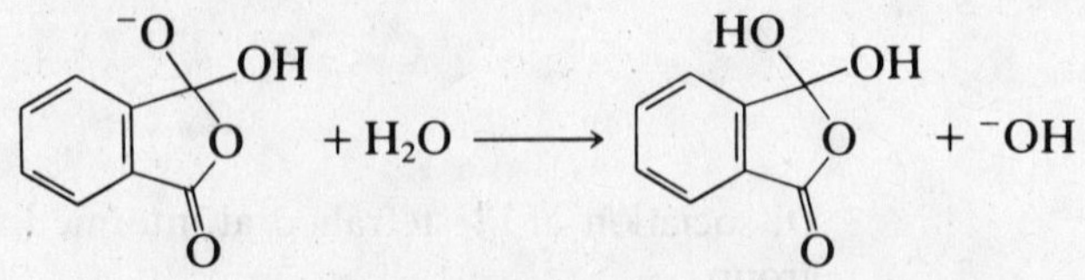

Tetrahedral intermediate

Hydroxide ion–promoted dissociation of tetrahedral intermediate:

$+ H_2O$

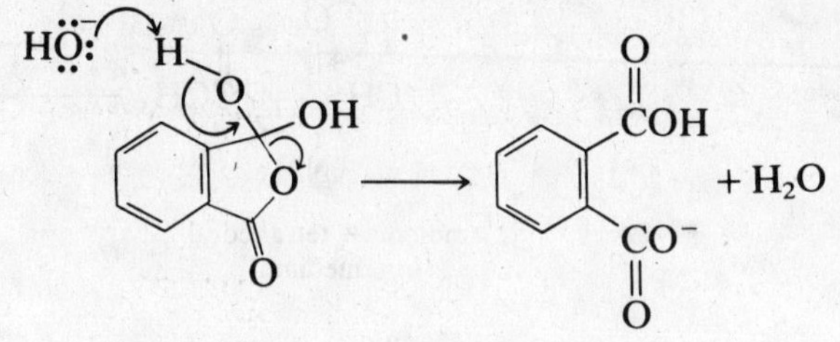

In base, the remaining carboxylic acid group is deprotonated:

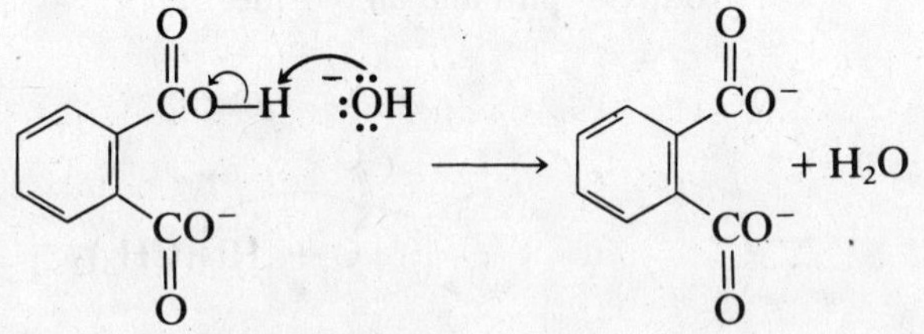

$+ H_2O$

20.9 The starting material contains three acetate ester functions. All these ester groups undergo hydrolysis in aqueous sulfuric acid.

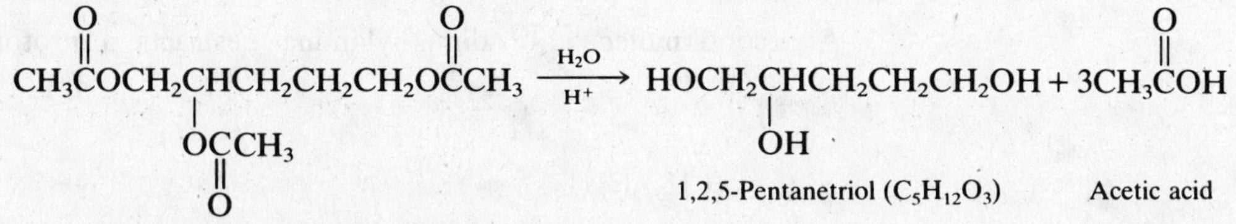

$$CH_3COCH_2CHCH_2CH_2CH_2OCCH_3 \xrightarrow[\text{H}^+]{\text{H}_2\text{O}} HOCH_2CHCH_2CH_2CH_2OH + 3CH_3COH$$

         OCCH$_3$ OH

1,2,5-Pentanetriol ($C_5H_{12}O_3$)   Acetic acid

The product is 1,2,5-pentanetriol. Also formed in the hydrolysis of the starting triacetate are three molecules of acetic acid.

20.10

Step 1: Protonation of the carbonyl oxygen

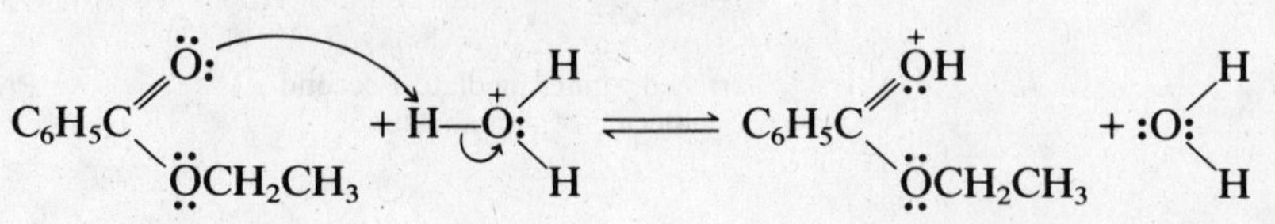

$$C_6H_5C \quad + H\overset{+}{-}\overset{..}{O} \quad \rightleftharpoons \quad C_6H_5C \quad + :O$$

Ethyl benzoate   Hydronium   Protonated form of ester   Water
                     ion

Step 2: Nucleophilic addition of water

$$H_2O + C_6H_5\overset{+}{C}(OH)OCH_2CH_3 \rightleftharpoons C_6H_5C(OH)(OCH_2CH_3)\overset{+}{O}H_2$$

Water   Protonated form of ester   Oxonium ion

Step 3: Deprotonation of oxonium ion to give neutral form of tetrahedral intermediate

$$C_6H_5C(\ddot{O}H)(OCH_2CH_3)\overset{+}{\ddot{O}}H_2 + :\ddot{O}H_2 \rightleftharpoons C_6H_5C(\ddot{O}H)(OCH_2CH_3)HO: + H-\overset{+}{O}H_2$$

       Tetrahedral   Hydronium
       intermediate   ion

Step 4: Protonation of ethoxy oxygen

$$C_6H_5C(\ddot{O}H)(OCH_2CH_3)HO: + H-\overset{+}{O}H_2 \longrightarrow C_6H_5C(\ddot{O}H)(\overset{+}{O}CH_2CH_3)HO:H + :\ddot{O}H_2$$

   Tetrahedral   Hydronium ion   Oxonium ion   Water
   intermediate

Step 5: Dissociation of protonated form of tetrahedral intermediate

This step yields ethyl alcohol and the protonated form of benzoic acid.

$$C_6H_5C(\ddot{O}H)(\overset{+}{O}CH_2CH_3H)(:\ddot{O}H) \rightleftharpoons C_6H_5C(\overset{+}{O}H)(:\ddot{O}H) + H\ddot{O}CH_2CH_3$$

   Oxonium ion   Protonated form of   Ethyl alcohol
                 benzoic acid

Step 6: Deprotonation of protonated form of benzoic acid

$$C_6H_5C(\overset{+}{\ddot{O}}H)(:\ddot{O}H) + :\ddot{O}H_2 \rightleftharpoons C_6H_5C(\ddot{O}:)(:\ddot{O}H) + H-\overset{+}{O}H_2$$

Protonated form   Water   Benzoic acid   Hydronium ion
of benzoic acid

20.11 To determine which oxygen of 4-butanolide becomes labeled with ^{18}O, trace the path of ^{18}O-labeled water ($\varnothing = {}^{18}O$) as it undergoes nucleophilic addition to the carbonyl group to form the tetrahedral intermediate.

$$\text{4-Butanolide} + H_2\varnothing \underset{}{\overset{H^+}{\rightleftharpoons}} \text{Tetrahedral intermediate}$$

4-Butanolide ^{18}O-labeled   Tetrahedral intermediate
          water

The tetrahedral intermediate can revert to unlabeled 4-butanolide by loss of ^{18}O-labeled water. Alternatively it can lose ordinary water to give ^{18}O-labeled lactone.

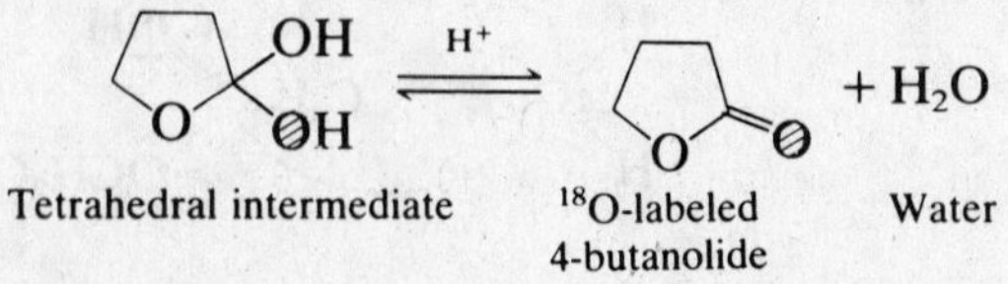

Tetrahedral intermediate ^{18}O-labeled   Water
4-butanolide

The carbonyl oxygen is the one that is isotopically labeled in the ^{18}O-enriched 4-butanolide.

20.12 On the basis of trimyristin's molecular formula $C_{45}H_{86}O_6$ and of the fact that its hydrolysis gives only glycerol and tetradecanoic acid $CH_3(CH_2)_{12}CO_2H$, it must have the structure shown:

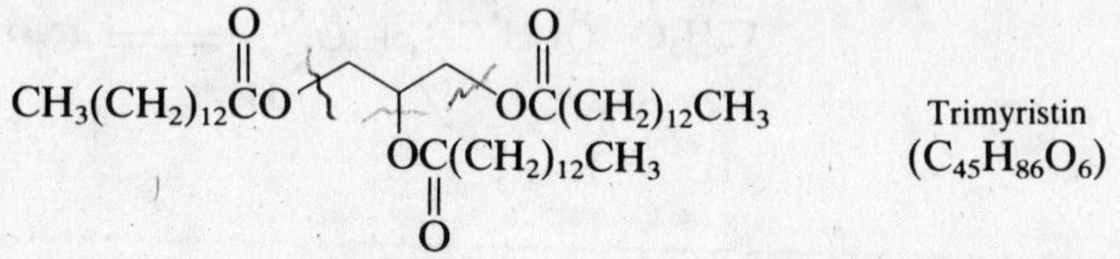

Trimyristin
$(C_{45}H_{86}O_6)$

20.13 Since ester hydrolysis in base proceeds by acyl-oxygen cleavage, the ^{18}O label becomes incorporated in acetate ion ($⊘ = {}^{18}O$).

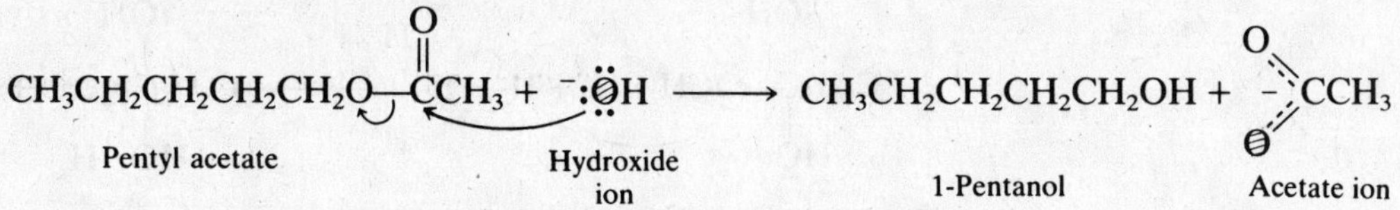

Pentyl acetate   Hydroxide   1-Pentanol   Acetate ion
ion

20.14

Step 1: Nucleophilic addition of hydroxide ion to the carbonyl group

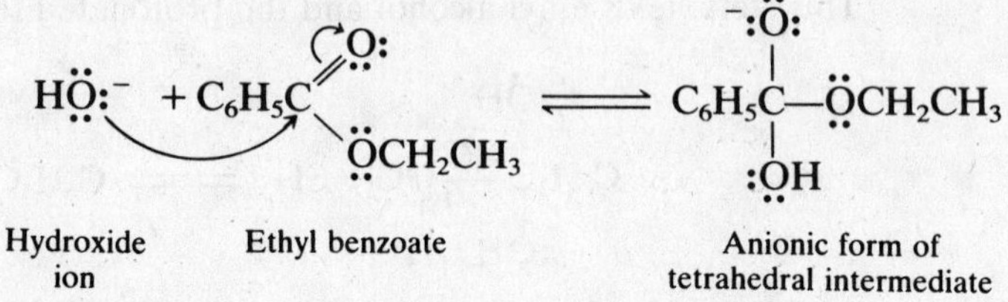

Hydroxide   Ethyl benzoate   Anionic form of
ion   tetrahedral intermediate

Step 2: Proton transfer from water to give neutral form of tetrahedral intermediate

Anionic form of   Water   Tetrahedral   Hydroxide ion
tetrahedral   intermediate
intermediate

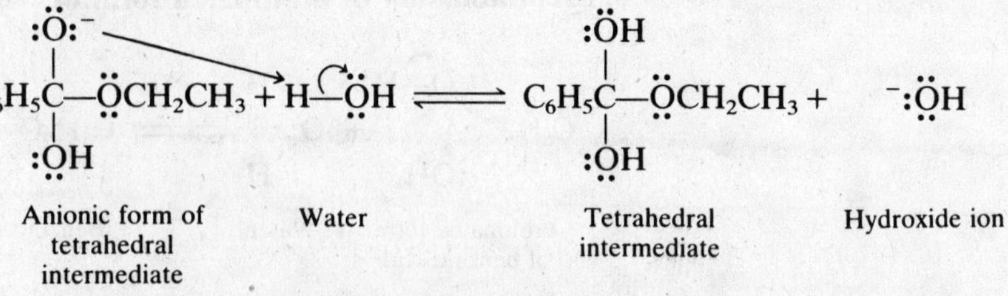

Step 3: Hydroxide ion–promoted dissociation of tetrahedral intermediate

Hydroxide   Tetrahedral   Water   Benzoic acid   Ethoxide
ion   intermediate   ion

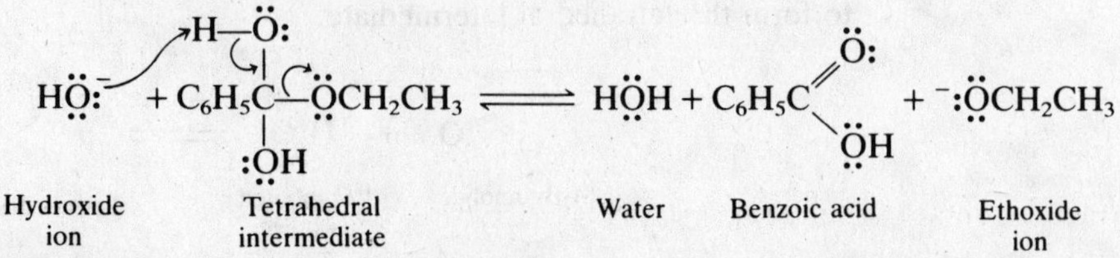

Step 4: Proton abstraction from benzoic acid

$$C_6H_5C \begin{smallmatrix} \ddot{O}: \\ \\ \ddot{O}-H \end{smallmatrix} + \; ^-:\ddot{O}H \longrightarrow C_6H_5C \begin{smallmatrix} \ddot{O}: \\ \\ \ddot{O}:^- \end{smallmatrix} + H\ddot{O}H$$

Benzoic acid   Hydroxide ion   Benzoate ion   Water

20.15 The starting material is a lactone, a cyclic ester. The ester function is converted to an amide by nucleophilic acyl substitution.

$$CH_3\ddot{N}H_2 + \text{[4-Pentanolide]} \longrightarrow CH_3NHCCH_2CH_2CHCH_3 \;(OH)$$

Methylamine   4-Pentanolide   4-Hydroxy-*N*-methylpentanamide

20.16 Methanol is the nucleophile that adds to the carbonyl group of the thioester.

$$CH_3\overset{O}{\overset{\|}{C}}SCH_2CH_2OC_6H_5 + CH_3OH \longrightarrow CH_3\overset{:\ddot{O}H}{\underset{OCH_3}{\overset{|}{C}}}SCH_2CH_2OC_6H_5$$

S-2-Phenoxyethyl ethanethiolate   Methanol   Tetrahedral intermediate

$$CH_3\overset{O}{\overset{\|}{C}}OCH_3 + HSCH_2CH_2OC_6H_5$$

Methyl acetate   2-Phenoxyethanethiol

20.17 (*b*) Acetic anhydride is the anhydride that must be used; it transfers an acetyl group to suitable nucleophiles. The nucleophile in this case is methylamine.

$$CH_3\overset{O}{\overset{\|}{C}}O\overset{O}{\overset{\|}{C}}CH_3 + 2CH_3NH_2 \longrightarrow CH_3\overset{O}{\overset{\|}{C}}NHCH_3 + CH_3\overset{O}{\overset{\|}{C}}O^- \;CH_3\overset{+}{N}H_3$$

Acetic anhydride   Methylamine   *N*-Methylacetamide   Methylammonium acetate

(*c*) The acyl group is $H\overset{O}{\overset{\|}{C}}-$. Since the problem specifies that the acyl transfer agent is a methyl ester, methyl formate is one of the starting materials.

$$H\overset{O}{\overset{\|}{C}}OCH_3 + HN(CH_3)_2 \longrightarrow H\overset{O}{\overset{\|}{C}}N(CH_3)_2 + CH_3OH$$

Methyl formate   Dimethylamine   *N,N*-Dimethylformamide   Methyl alcohol

20.18 Phthalic anhydride reacts with excess ammonia to give the ammonium salt of a compound known as *phthalamic acid*.

$$\text{[Phthalic anhydride]} + 2NH_3 \longrightarrow \text{[Ammonium phthalamate]}$$

Phthalic anhydride   Ammonia   Ammonium phthalamate $(C_8H_{10}N_2O_3)$

Phthalimide is formed when ammonium phthalamate is heated.

Ammonium phthalamate   Phthalimide   Ammonia   Water

20.19

Step 1: Protonation of the carbonyl oxygen

Acetanilide   Hydronium ion   Protonated form   Water
                                              of amide

Step 2: Nucleophilic addition of water

Water   Protonated form   Oxonium ion
          of amide

Step 3: Deprotonation of oxonium ion to give neutral form of tetrahedral intermediate

Oxonium ion   Water   Tetrahedral intermediate   Hydronium ion

Step 4: Protonation of amino group of tetrahedral intermediate

Tetrahedral   Hydronium ion   N-Protonated form of   Water
intermediate   tetrahedral intermediate

Step 5: Dissociation of *N*-protonated form of tetrahedral intermediate

$$CH_3C(\overset{:\ddot{O}H}{\underset{:\ddot{O}H}{}})\overset{H}{\underset{H}{N}}C_6H_5 \rightleftharpoons CH_3C\overset{\overset{+}{\ddot{O}H}}{\underset{\ddot{O}H}{}} + H_2\ddot{N}C_6H_5$$

N-Protonated form of   Protonated form   Aniline
tetrahedral intermediate   of acetic acid

Step 6: Proton-transfer processes

$$\overset{H}{\underset{H}{\overset{+}{\ddot{O}}}}{-}H + H_2\overset{+}{\ddot{N}}C_6H_5 \rightleftharpoons \overset{H}{\underset{H}{\ddot{O}:}} + H_3\overset{+}{N}C_6H_5$$

Hydronium ion   Aniline   Water   Anilinium ion

$$CH_3C\overset{\overset{+}{\ddot{O}}{-}H}{\underset{\ddot{O}H}{}} + :\overset{H}{\underset{H}{\ddot{O}}} \rightleftharpoons CH_3C\overset{\ddot{O}:}{\underset{\ddot{O}H}{}} + H{-}\overset{+}{\underset{H}{\overset{H}{\ddot{O}}}}$$

Protonated form   Water   Acetic acid   Hydronium
of acetic acid   ion

20.20

Step 1: Nucleophilic addition of hydroxide ion to the carbonyl group

$$H\ddot{O}:^- + HC\overset{:\ddot{O}}{\underset{}{}}N(CH_3)_2 \longrightarrow HC\overset{:\ddot{O}:^-}{\underset{:\ddot{O}H}{}}{-}\ddot{N}(CH_3)_2$$

Hydroxide ion   *N*,*N*-Dimethylformamide   Anionic form of tetrahedral
                                     intermediate

Step 2: Proton transfer to give neutral form of tetrahedral intermediate

$$HC\overset{:\ddot{O}:^-}{\underset{:\ddot{O}H}{}}{-}\ddot{N}(CH_3)_2 + H{-}\ddot{O}H \longrightarrow HC\overset{:\ddot{O}H}{\underset{:\ddot{O}H}{}}{-}\ddot{N}(CH_3)_2 + {}^-:\ddot{O}H$$

Anionic form of   Water   Tetrahedral   Hydroxide ion
tetrahedral   intermediate
intermediate

Step 3: Proton transfer from water to nitrogen of tetrahedral intermediate

$$HC\overset{:\ddot{O}H}{\underset{:\ddot{O}H}{}}{-}\ddot{N}(CH_3)_2 + H{-}\ddot{O}H \rightleftharpoons HC\overset{:\ddot{O}H}{\underset{:\ddot{O}H}{}}{-}\overset{+}{N}H(CH_3)_2 + {}^-:\ddot{O}H$$

Tetrahedral   Water   *N*-Protonated form of   Hydroxide
intermediate   tetrahedral intermediate   ion

Step 4: Dissociation of *N*-protonated form of tetrahedral intermediate

| Hydroxide ion | *N*-Protonated form of tetrahedral intermediate | Water | Formic acid | Dimethylamine |

Step 5: Irreversible formation of formate ion

| Formic acid | Hydroxide ion | Formate ion | Water |

20.21   A synthetic scheme becomes apparent when we recognize that a primary amine may be obtained by <u>Hofmann</u> rearrangement of the primary amide having one more carbon in its acyl group. This amide may, in turn, be prepared from the corresponding carboxylic acid.

$$CH_3CH_2CH_2NH_2 \implies CH_3CH_2CH_2\overset{\overset{\displaystyle O}{\|}}{C}NH_2 \implies CH_3CH_2CH_2CO_2H$$

The desired reaction scheme is therefore:

| Butanoic acid | | Butanamide | | Propanamine |

20.22   (*a*)   Ethanenitrile has the same number of carbon atoms as ethyl alcohol. This suggests a reaction scheme proceeding via an amide.

| Ethyl alcohol | Acetamide | Ethanenitrile |

The necessary amide is prepared from ethanol.

| Ethyl alcohol | Acetic acid | Acetamide |

(*b*)   Propanenitrile may be prepared from ethyl alcohol by way of a nucleophilic substitution reaction of the corresponding bromide.

$$CH_3CH_2OH \xrightarrow[\text{or HBr}]{PBr_3} CH_3CH_2Br \xrightarrow{NaCN} CH_3CH_2CN$$

| Ethyl alcohol | Ethyl bromide | Propanenitrile |

20.23 (*a*) The halogen that is attached to the carbonyl group is identified in the name as a separate word following the name of the acyl group.

m-Chlorobenzoyl bromide

(*b*) Trifluoroacetic anhydride is the anhydride of trifluoroacetic acid. Notice that it contains six fluorines.

$$\underset{\text{Trifluoroacetic anhydride}}{CF_3\overset{\displaystyle O}{\overset{\|}{C}}O\overset{\displaystyle O}{\overset{\|}{C}}CF_3}$$

(*c*) This compound is the cyclic anhydride of *cis*-1,2-cyclopropanedicarboxylic acid.

cis-1,2-Cyclopropanedicarboxylic   *cis*-1,2-Cyclopropanedicarboxylic
acid   anhydride

(*d*) Ethyl cycloheptanecarboxylate is the ethyl ester of cycloheptanecarboxylic acid.

$$\overset{\displaystyle O}{\overset{\|}{-C}}OCH_2CH_3 \quad \text{Ethyl cycloheptanecarboxylate}$$

(*e*) 1-Phenylethyl acetate is the ester of 1-phenylethanol and acetic acid.

$$CH_3\overset{\displaystyle O}{\overset{\|}{C}}O\underset{\displaystyle CH_3}{\overset{\displaystyle |}{CH}}\!\!-\!\!\bigcirc \quad \text{1-Phenylethyl acetate}$$

(*f*) 2-Phenylethyl acetate is the ester of 2-phenylethanol and acetic acid.

$$CH_3\overset{\displaystyle O}{\overset{\|}{C}}OCH_2CH_2\!-\!\bigcirc \quad \text{2-Phenylethyl acetate}$$

(*g*) The parent compound in this case is benzamide. *p*-Ethylbenzamide has an ethyl substituent at the ring position para to the carbonyl group.

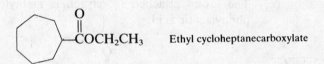

$$CH_3CH_2\!-\!\bigcirc\!-\!\overset{\displaystyle O}{\overset{\|}{C}}NH_2 \quad \text{\textit{p}-Ethylbenzamide}$$

(*h*) The parent compound is benzamide. In *N*-ethylbenzamide the ethyl substituent is bonded to nitrogen.

$$\bigcirc\!-\!\overset{\displaystyle O}{\overset{\|}{C}}NHCH_2CH_3 \quad \text{\textit{N}-Ethylbenzamide}$$

(i)   Nitriles are named by adding the suffix *-nitrile* to the name of the alkane having the same number of carbons. Numbering begins at the nitrile carbon.

$$CH_3CH_2CH_2CH_2CHC{\equiv}N \qquad \text{2-Methylhexanenitrile}$$
$$|$$
$$CH_3$$

20.24 (a)   This compound, with a bromine substituent attached to its carbonyl group, is named as an acyl bromide. It is 3-chlorobutanoyl bromide.

$$\overset{\displaystyle O}{\overset{\displaystyle \|}{CH_3CHCH_2CBr}} \qquad \text{3-Chlorobutanoyl bromide}$$
$$|$$
$$Cl$$

(b)   The group attached to oxygen, in this case *benzyl*, is identified first in the name of the ester. This compound is the benzyl ester of acetic acid.

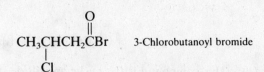

Benzyl acetate

(c)   The group attached to oxygen is methyl; this compound is the methyl ester of phenylacetic acid.

Methyl phenylacetate

(d)   This compound contains the functional group —COC— and thus is an anhydride of a carboxylic acid. We name the acid, in this case 3-chloropropanoic acid, drop the *acid* part of the name, and replace it by *anhydride*.

$$\overset{\displaystyle O\ \ O}{\overset{\displaystyle \|\ \ \|}{ClCH_2CH_2COCCH_2CH_2Cl}} \qquad \text{3-Chloropropanoic anhydride}$$

(e)   This compound is a cyclic anhydride, whose parent acid is 3,3-dimethylpentanedioic acid.

3,3-Dimethylpentanedioic anhydride

(f)   Nitriles are named by adding *-nitrile* to the name of the alkane having the same number of carbons.

$$CH_3CHCH_2CH_2C{\equiv}N$$
$$|$$
$$CH_3$$

4-Methylpentanenitrile

(*g*) This compound is an amide. We name the corresponding acid and then replace the -*oic acid* suffix by -*amide*.

$$CH_3CHCH_2CH_2CNH_2 \quad \text{4-Methylpentanamide}$$
$$\underset{CH_3}{|}$$

(*h*) This compound is the *N*-methyl derivative of 4-methylpentanamide.

$$CH_3CHCH_2CH_2CNHCH_3 \quad \textit{N}\text{-Methyl-4-methylpentanamide}$$
$$\underset{CH_3}{|}$$

(*i*) The amide nitrogen bears two methyl groups. We designate this as an *N,N*-dimethyl amide.

$$CH_3CHCH_2CH_2CN(CH_3)_2 \quad \textit{N,N}\text{-Dimethyl-4-methylpentanamide}$$
$$\underset{CH_3}{|}$$

20.25 (*a*) Acetyl chloride acts as an acyl transfer agent to the aromatic ring of bromobenzene. The reaction is a Friedel-Crafts acylation reaction.

Bromobenzene   Acetyl chloride   *o*-Bromoacetophenone   *p*-Bromoacetophenone

Bromine is an ortho, para–directing substituent.

(*b*) Acyl chlorides react with thiols to give thioesters.

$$CH_3CCl \; + \; CH_3CH_2CH_2CH_2SH \longrightarrow CH_3CSCH_2CH_2CH_2CH_3$$

Acetyl chloride   Butanethiol   *S*-Butyl ethanethioate

(*c*) Sodium propanoate acts as a nucleophile toward propanoyl chloride. The product is propanoic anhydride.

$$CH_3CH_2C\ddot{O}:^- \; + \; CH_3CH_2C-Cl \longrightarrow CH_3CH_2COCCH_2CH_3$$

Propanoate anion   Propanoyl chloride   Propanoic anhydride

(d)   Acyl chlorides convert alcohols to esters.

$$CH_3CH_2CH_2\overset{O}{\overset{\|}{C}}Cl + C_6H_5CH_2OH \longrightarrow CH_3CH_2CH_2\overset{O}{\overset{\|}{C}}OCH_2C_6H_5$$

Butanoyl chloride   Benzyl alcohol   Benzyl butanoate

(e)   Acyl chlorides react with ammonia to yield amides.

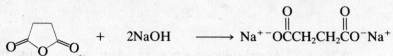

p-Chlorobenzoyl chloride   Ammonia   p-Chlorobenzamide

(f)   The starting material is a cyclic anhydride. Acid anhydrides react with water to yield two carboxylic acid functions; when the anhydride is cyclic, a dicarboxylic acid results.

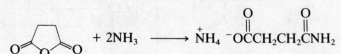

Succinic anhydride   Water   Succinic acid

(g)   In dilute sodium hydroxide the diacid is converted to its disodium salt.

$$\text{(structure)} + 2NaOH \longrightarrow Na^{+-}\overset{O}{\overset{\|}{O}}CCH_2CH_2\overset{O}{\overset{\|}{C}}O^-Na^+$$

Succinic anhydride   Sodium hydroxide   Sodium succinate

(h)   One of the carbonyl groups of the cyclic anhydride is converted to an amide function on reaction with ammonia. The other, the one that would become a carboxylic acid group, is converted to an ammonium carboxylate salt.

$$\text{(structure)} + 2NH_3 \longrightarrow \overset{+}{N}H_4\ ^-\overset{O}{\overset{\|}{O}}CCH_2CH_2\overset{O}{\overset{\|}{C}}NH_2$$

Succinic anhydride   Ammonia   Ammonium succinamate

(i)   Acid anhydrides are used as acylating agents in Friedel-Crafts reactions.

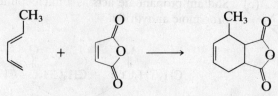

Succinic anhydride   Benzene

3-Benzoylpropanoic acid

(j)   The reactant is maleic anhydride; it is a good dienophile in Diels-Alder reactions.

1,3-Pentadiene   Maleic anhydride   3-Methylcyclohexene-4,5-dicarboxylic anhydride

(k) Acid anhydrides react with alcohols to give an ester and a carboxylic acid.

$$CH_3\overset{O}{\overset{\|}{C}}O\overset{O}{\overset{\|}{C}}CH_3 + CH_3CH_2\underset{\overset{|}{OH}}{CH}CH_2CH_3 \longrightarrow CH_3CH_2\underset{\overset{|}{O}\overset{|}{\underset{\|}{C}}CH_3}{CH}CH_2CH_3 + CH_3CO_2H$$

| Acetic anhydride | 3-Pentanol | 1-Ethylpropyl acetate | Acetic acid |

(l) The starting material is a cyclic ester, a lactone. Esters undergo saponification in aqueous base to give an alcohol and a carboxylate salt.

$$\text{(lactone)} + NaOH \longrightarrow HOCH_2CH_2CH_2\overset{O}{\overset{\|}{C}}O^-Na^+$$

| 4-Butanolide | Sodium hydroxide | Sodium 4-hydroxybutanoate |

(m) Ammonia reacts with esters to give an amide and an alcohol.

$$\text{(lactone)} + NH_3 \longrightarrow HOCH_2CH_2CH_2\overset{O}{\overset{\|}{C}}NH_2$$

| 4-Butanolide | Ammonia | 4-Hydroxybutanamide |

(n) Lithium aluminum hydride reduces esters to two alcohols; the one derived from the acyl group is a primary alcohol.

$$\text{(lactone)} \xrightarrow[\text{2. H}_2\text{O}]{\text{1. LiAlH}_4} HOCH_2CH_2CH_2CH_2OH$$

| 4-Butanolide | 1,4-Butanediol |

(o) Grignard reagents react with esters to give tertiary alcohols.

$$\text{(lactone)} \xrightarrow[\text{2. H}_3\text{O}^+]{\text{1. 2CH}_3\text{MgBr}} HOCH_2CH_2CH_2\underset{\overset{|}{CH_3}}{\overset{\overset{|}{OH}}{C}}CH_3$$

| 4-Butanolide | 4-Methyl-1,4-pentanediol |

(p) In this reaction methylamine acts as a nucleophile toward the carbonyl group of the ester. The product is an amide.

$$CH_3\overset{..}{\overset{..}{N}}H_2 + C_6H_5CH_2\overset{O}{\overset{\|}{C}}OCH_2CH_3 \longrightarrow C_6H_5CH_2\overset{O}{\overset{\|}{C}}NHCH_3 + CH_3CH_2OH$$

| Methylamine | Ethyl phenylacetate | N-Methylphenylacetamide | Ethyl alcohol |

(q) The starting material is a lactam, a cyclic amide. Amides are hydrolyzed in base to amines and carboxylate salts.

$$\text{(lactam, N-CH}_3) + NaOH \longrightarrow CH_3NHCH_2CH_2CH_2\overset{O}{\overset{\|}{C}}O^-Na^+$$

| N-Methylpyrrolidone | Sodium hydroxide | Sodium 4-(methylamino)butanoate |

(r)   In acid solution amides yield carboxylic acids and ammonium salts.

$$CH_3\overset{\overset{\displaystyle H}{|}}{\underset{\underset{\displaystyle H}{|}}{N}}CH_2CH_2CH_2\overset{\overset{\displaystyle O}{\|}}{C}OH$$

N-Methylpyrrolidone   Hydronium ion   4-(Methylammonio)butanoic
                                                              acid

(s)   The starting material is a cyclic imide. Both its amide bonds are cleaved by nucleophilic attack by hydroxide ion.

$$+ \ 2NaOH \longrightarrow Na^+ \ {}^-O\overset{\overset{\displaystyle O}{\|}}{C}CH_2CH_2\overset{\overset{\displaystyle O}{\|}}{C}O^- \ Na^+ \ + \ CH_3NH_2$$

N-Methylsuccinimide   Sodium   Disodium succinate   Methylamine
                                    hydroxide

(t)   In acid the imide undergoes cleavage to give a dicarboxylic acid and the conjugate acid of methylamine.

$$+ \ 2H_2O + \ HCl \longrightarrow HO\overset{\overset{\displaystyle O}{\|}}{C}CH_2CH_2\overset{\overset{\displaystyle O}{\|}}{C}OH + \ CH_3\overset{+}{N}H_3 \ Cl^-$$

N-Methylsuccinimide   Water   Hydrogen   Succinic acid   Methylammonium
                                           chloride   chloride

(u)   Acetanilide is hydrolyzed in acid to acetic acid and the conjugate acid of aniline.

$$C_6H_5NH\overset{\overset{\displaystyle O}{\|}}{C}CH_3 + H_2O + \ HCl \longrightarrow C_6H_5\overset{+}{N}H_3 \ Cl^- + CH_3\overset{\overset{\displaystyle O}{\|}}{C}OH$$

Acetanilide   Water   Hydrogen   Anilinium   Acetic acid
                                       chloride   chloride

(v)   This is another example of amide hydrolysis.

$$C_6H_5\overset{\overset{\displaystyle O}{\|}}{C}NHCH_3 + H_2O + \ H_2SO_4 \longrightarrow C_6H_5\overset{\overset{\displaystyle O}{\|}}{C}OH + \ CH_3\overset{+}{N}H_3 \ HSO_4^-$$

N-Methylbenzamide   Water   Sulfuric acid   Benzoic   Methylammonium hydrogen
                                                                            acid   sulfate

(w)   One way to prepare nitriles is by dehydration of amides.

$$\overset{\overset{\displaystyle O}{\|}}{C}NH_2 \ \xrightarrow{\ P_4O_{10}\ } \ C\equiv N \ + H_2O$$

Cyclopentanecarboxamide   Cyclopentyl cyanide

(x)   Nitriles are hydrolyzed to carboxylic acids in acidic media.

Acid

$$(CH_3)_2CHCH_2C\equiv N \ \xrightarrow[\text{heat}]{\text{HCl, } H_2O} \ (CH_3)_2CHCH_2\overset{\overset{\displaystyle O}{\|}}{C}OH$$

3-Methylbutanenitrile   3-Methylbutanoic acid

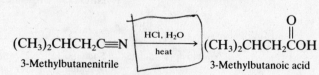

(*y*) Nitriles are hydrolyzed in aqueous base to salts of carboxylic acids.

$$CH_3O-\!\!\!\langle\ \rangle\!\!-\!C\!\equiv\!N \xrightarrow[\text{heat}]{\text{NaOH, H}_2\text{O}} CH_3O-\!\!\!\langle\ \rangle\!\!-\!CO^-Na^+ \ + \ NH_3$$

p-Methoxybenzonitrile   Sodium p-methoxybenzoate   Ammonia

(*z*) Grignard reagents react with nitriles to yield ketones after aqueous acidic workup.

$$CH_3CH_2C\!\equiv\!N \xrightarrow[\text{2. H}_3\text{O}^+]{\text{1. CH}_3\text{MgBr}} CH_3CH_2\overset{\displaystyle O}{\overset{\displaystyle \|}{C}}CH_3$$

Propanenitrile   2-Butanone

(*aa*) Amides undergo the Hofmann rearrangement on reaction with bromine and base. A methyl carbamate is the product isolated when the reaction is carried out in methanol.

$$+ \ Br_2 \xrightarrow[\text{CH}_3\text{OH}]{\text{NaOCH}_3}$$

(*bb*) Saponification of the carbamate in part (*aa*) gives the corresponding amine.

$$\xrightarrow[\text{H}_2\text{O}]{\text{KOH}}$$

20.26 (*a*) Acetyl chloride is prepared by reaction of acetic acid with thionyl chloride. The first task then is to prepare acetic acid by oxidation of ethanol.

$$CH_3CH_2OH \xrightarrow[\text{H}_2\text{O}]{\text{K}_2\text{Cr}_2\text{O}_7, \text{ H}_2\text{SO}_4} CH_3\overset{\displaystyle O}{\overset{\displaystyle \|}{C}}OH \xrightarrow{\text{SOCl}_2} CH_3\overset{\displaystyle O}{\overset{\displaystyle \|}{C}}Cl$$

Ethanol   Acetic acid   Acetyl chloride

(*b*) Acetic acid and acetyl chloride, available from part (*a*), can be combined to form acetic anhydride.

$$CH_3\overset{\displaystyle O}{\overset{\displaystyle \|}{C}}OH + \ CH_3\overset{\displaystyle O}{\overset{\displaystyle \|}{C}}Cl \longrightarrow CH_3\overset{\displaystyle O}{\overset{\displaystyle \|}{C}}O\overset{\displaystyle O}{\overset{\displaystyle \|}{C}}CH_3 + HCl$$

Acetic acid   Acetyl chloride   Acetic anhydride   Hydrogen chloride

(c) Ethanol can be converted to ethyl acetate by reaction with acetic acid, acetyl chloride, or acetic anhydride.

$$CH_3CH_2OH + CH_3\overset{\overset{\displaystyle O}{\|}}{C}OH \xrightarrow{H^+} CH_3\overset{\overset{\displaystyle O}{\|}}{C}OCH_2CH_3 + H_2O$$

Ethanol   Acetic acid   Ethyl acetate   Water

or $$CH_3CH_2OH + CH_3\overset{\overset{\displaystyle O}{\|}}{C}Cl \xrightarrow{pyridine} CH_3\overset{\overset{\displaystyle O}{\|}}{C}OCH_2CH_3$$

Ethanol   Acetyl chloride   Ethyl acetate

or $$CH_3CH_2OH + CH_3\overset{\overset{\displaystyle O}{\|}}{C}O\overset{\overset{\displaystyle O}{\|}}{C}CH_3 \xrightarrow{pyridine} CH_3\overset{\overset{\displaystyle O}{\|}}{C}OCH_2CH_3$$

Ethanol   Acetic anhydride   Ethyl acetate

(d) Ethyl bromoacetate is the ethyl ester of bromoacetic acid; thus the first task is to prepare the acid. We use the acetic acid prepared in part (a), converting it to bromoacetic acid by the Hell-Volhard-Zelinsky reaction.

$$CH_3CO_2H \xrightarrow[P]{Br_2} BrCH_2CO_2H \xrightarrow[H^+]{CH_3CH_2OH} BrCH_2\overset{\overset{\displaystyle O}{\|}}{C}OCH_2CH_3$$

Acetic acid   Bromoacetic acid   Ethyl bromoacetate

Alternatively, bromoacetic acid could be converted to the corresponding acyl chloride, then treated with ethanol. It would be incorrect to try to brominate ethyl acetate; the Hell-Volhard-Zelinsky method requires an acid as starting material, not an ester.

(e) The alcohol $BrCH_2CH_2OH$, needed in order to prepare 2-bromoethyl acetate, is prepared from ethanol by way of ethylene.

$$CH_3CH_2OH \xrightarrow[heat]{H_2SO_4} CH_2{=}CH_2 \xrightarrow[H_2O]{Br_2} BrCH_2CH_2OH$$

Ethanol   Ethylene   2-Bromoethanol

Then $$BrCH_2CH_2OH \xrightarrow[\substack{or \\ \overset{\overset{\displaystyle O}{\|}}{CH_3CCl}}]{\overset{\overset{\displaystyle O\ O}{\| \ \|}}{CH_3COCCH_3}} CH_3\overset{\overset{\displaystyle O}{\|}}{C}OCH_2CH_2Br$$

2-Bromoethanol   2-Bromoethyl acetate

(f) Ethyl cyanoacetate may be prepared from the ethyl bromoacetate obtained in part (d). The bromide may be displaced by cyanide in a nucleophilic substitution reaction.

$$BrCH_2\overset{\overset{\displaystyle O}{\|}}{C}OCH_2CH_3 \xrightarrow[S_N2]{NaCN} N{\equiv}CCH_2\overset{\overset{\displaystyle O}{\|}}{C}OCH_2CH_3$$

Ethyl bromoacetate   Ethyl cyanoacetate

(g) Reaction of the acetyl chloride prepared in part (a) or the acetic anhydride from part (b) with ammonia gives acetamide.

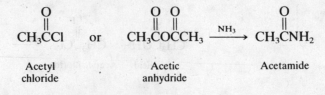

$$CH_3\overset{\overset{\displaystyle O}{\|}}{C}Cl \quad or \quad CH_3\overset{\overset{\displaystyle O}{\|}}{C}O\overset{\overset{\displaystyle O}{\|}}{C}CH_3 \xrightarrow{NH_3} CH_3\overset{\overset{\displaystyle O}{\|}}{C}NH_2$$

Acetyl   Acetic   Acetamide
chloride   anhydride

(*h*)   Methylamine may be prepared from acetamide by a Hofmann rearrangement.

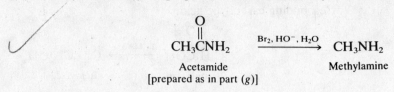

$$CH_3CNH_2 \xrightarrow{Br_2,\ HO^-,\ H_2O} CH_3NH_2$$

Acetamide   Methylamine
[prepared as in part (*g*)]

(*i*)   The desired hydroxy acid is available from hydrolysis of the corresponding cyanohydrin, which may be prepared by reaction of the appropriate aldehyde with cyanide ion.

$$\underset{\underset{\displaystyle OH}{|}}{CH_3CHCOH} \Rightarrow \underset{\underset{\displaystyle OH}{|}}{CH_3CHC\equiv N} \Rightarrow CH_3CH$$

In this synthesis the cyanohydrin is prepared from ethanol by way of acetaldehyde.

$$CH_3CH_2OH \xrightarrow[CH_2Cl_2]{(C_5H_5N)_2CrO_3} CH_3CH \xrightarrow[H^+]{KCN} \underset{\underset{\displaystyle OH}{|}}{CH_3CHC\equiv N}$$

Ethanol   Acetaldehyde   2-Hydroxypropanenitrile

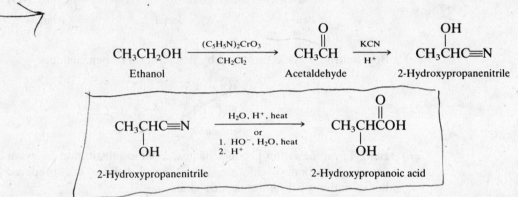

$$\underset{\underset{\displaystyle OH}{|}}{CH_3CHC\equiv N} \xrightarrow[\substack{or \\ 1.\ HO^-,\ H_2O,\ heat \\ 2.\ H^+}]{H_2O,\ H^+,\ heat} \underset{\underset{\displaystyle OH}{|}}{CH_3CHCOH}$$

2-Hydroxypropanenitrile   2-Hydroxypropanoic acid

20.27   (*a*)   Benzoyl chloride is made from benzoic acid. Oxidize toluene to benzoic acid, and then treat with thionyl chloride.

$$C_6H_5CH_3 \xrightarrow[\substack{H_2O \\ or\ 1.\ KMnO_4 \\ 2.\ H^+}]{K_2Cr_2O_7,\ H_2SO_4} C_6H_5COH \xrightarrow{SOCl_2} C_6H_5CCl$$

Toluene   Benzoic acid   Benzoyl chloride

(*b*)   Benzoyl chloride and benzoic acid, both prepared from toluene in part (*a*), react with each other to give benzoic anhydride.

$$C_6H_5COH + C_6H_5CCl \longrightarrow C_6H_5COCC_6H_5$$

Benzoic acid   Benzoyl chloride   Benzoic anhydride

(*c*)   Benzoic acid, benzoyl chloride, and benzoic anhydride have been prepared in parts (*a*) and (*b*) of this problem. Any of them could be converted to benzyl benzoate on reaction with benzyl alcohol. Thus the synthesis of benzyl benzoate requires the preparation of benzyl alcohol from toluene. This is effected by a nucleophilic substitution reaction of benzyl bromide, in turn prepared by halogenation of toluene.

$$C_6H_5CH_3 \xrightarrow[or\ Br_2,\ light]{N\text{-bromosuccinimide (NBS)}} C_6H_5CH_2Br \xrightarrow[HO^-]{H_2O} C_6H_5CH_2OH$$

Toluene   Benzyl bromide   Benzyl alcohol

Alternatively, recall that primary alcohols may be obtained by reduction of the corresponding carboxylic acid.

$$C_6H_5\overset{\overset{\displaystyle O}{\|}}{C}OH \xrightarrow[\text{2. } H_2O]{\text{1. } LiAlH_4} C_6H_5CH_2OH$$

Benzoic acid   Benzyl alcohol

Then

$$C_6H_5\overset{\overset{\displaystyle O}{\|}}{C}Cl + C_6H_5CH_2OH \xrightarrow{\text{pyridine}} C_6H_5\overset{\overset{\displaystyle O}{\|}}{C}OCH_2C_6H_5$$

Benzoyl chloride   Benzyl alcohol   Benzyl benzoate

(*d*) Benzamide is prepared by reaction of ammonia with either benzoyl chloride or benzoic anhydride.

$$C_6H_5\overset{\overset{\displaystyle O}{\|}}{C}Cl \quad \text{or} \quad C_6H_5\overset{\overset{\displaystyle O}{\|}}{C}O\overset{\overset{\displaystyle O}{\|}}{C}C_6H_5 \xrightarrow{NH_3} C_6H_5\overset{\overset{\displaystyle O}{\|}}{C}NH_2$$

Benzoyl   Benzoic   Benzamide
chloride   anhydride

(*e*) Benzonitrile may be prepared by dehydration of benzamide.

$$C_6H_5\overset{\overset{\displaystyle O}{\|}}{C}NH_2 \xrightarrow[\text{heat}]{P_4O_{10}} C_6H_5C{\equiv}N$$

Benzamide   Benzonitrile

(*f*) Benzyl cyanide is the product of nucleophilic substitution by cyanide ion on benzyl bromide or benzyl chloride. The benzyl halides are prepared by free-radical halogenation of the toluene side chain.

$$C_6H_5CH_3 \xrightarrow[\substack{\text{light or}\\\text{heat}}]{Cl_2} C_6H_5CH_2Cl \xrightarrow{NaCN} C_6H_5CH_2C{\equiv}N$$

Toluene   Benzyl chloride   Benzyl cyanide

or

$$C_6H_5CH_3 \xrightarrow[\substack{\text{or } Br_2,\\\text{light}}]{NBS} C_6H_5CH_2Br \xrightarrow{NaCN} C_6H_5CH_2C{\equiv}N$$

Toluene   Benzyl bromide   Benzyl cyanide

(*g*) Hydrolysis of benzyl cyanide yields phenylacetic acid.

$$C_6H_5CH_2C{\equiv}N \xrightarrow[\substack{\text{or}\\\text{1. } NaOH, \text{ heat}\\\text{2. } H^+}]{H_2O, H^+, \text{ heat}} C_6H_5CH_2\overset{\overset{\displaystyle O}{\|}}{C}OH$$

Benzyl cyanide   Phenylacetic acid

Alternatively, the Grignard reagent derived from benzyl bromide may be carboxylated.

$$C_6H_5CH_2Br \xrightarrow[\substack{\text{diethyl}\\\text{ether}}]{Mg} C_6H_5CH_2MgBr \xrightarrow[\text{2. } H_3O^+]{\text{1. } CO_2} C_6H_5CH_2\overset{\overset{\displaystyle O}{\|}}{C}OH$$

Benzyl bromide   Benzylmagnesium   Phenylacetic acid
                        bromide

(*h*) The first goal is to synthesize *p*-nitrobenzoic acid, since this may be readily converted to the desired acyl chloride. First convert toluene to *p*-nitrotoluene; then oxidize. Nitration must precede oxidation of the side chain in order to achieve the desired para orientation.

Toluene → *p*-Nitrotoluene (separate from ortho isomer) → *p*-Nitrobenzoic acid

Treatment of *p*-nitrobenzoic acid with thionyl chloride yields *p*-nitrobenzoyl chloride.

p-Nitrobenzoic acid $\xrightarrow{SOCl_2}$ *p*-Nitrobenzoyl chloride

(i) In order to achieve the correct orientation in *m*-nitrobenzoyl chloride, oxidation of the methyl group must precede nitration.

Toluene → Benzoic acid → *m*-Nitrobenzoic acid

Once *m*-nitrobenzoic acid has been prepared, it may be converted to the corresponding acyl chloride.

m-Nitrobenzoic acid $\xrightarrow{SOCl_2}$ *m*-Nitrobenzoyl chloride

(j) A Hofmann rearrangement of benzamide affords aniline.

Benzamide [prepared as in part *(d)*]   +   Bromine $\xrightarrow[H_2O]{HO^-}$ Aniline

20.28   The problem specifies that $CH_3CH_2\overset{O}{\overset{\|}{C}}OCH_2CH_3$ is to be prepared from ^{18}O-labeled ethyl alcohol ($\varnothing = {}^{18}O$).

$$CH_3CH_2\overset{O}{\overset{\|}{C}}Cl + CH_3CH_2\varnothing H \longrightarrow CH_3CH_2\overset{O}{\overset{\|}{C}}\varnothing CH_2CH_3$$

Propanoyl chloride   Ethyl alcohol   Ethyl propanoate

Thus, we need to prepare ^{18}O-labeled ethyl alcohol from the other designated starting materials, acetaldehyde and ^{18}O-enriched water. First, replace the oxygen of

acetaldehyde with ^{18}O by the hydration-dehydration equilibrium in the presence of ^{18}O-enriched water.

Acetaldehyde	^{18}O-enriched water	Hydrate of acetaldehyde	^{18}O-enriched acetaldehyde	Water

Once ^{18}O-enriched acetaldehyde has been obtained, it can be reduced to ^{18}O-enriched ethanol.

20.29 (*a*) The rate-determining step in base-promoted ester hydrolysis is nucleophilic addition of hydroxide to the carbonyl group. The intermediate formed in this step is negatively charged.

Ethyl acetate	Hydroxide	Rate-determining intermediate

The electron-withdrawing effect of a CF_3 group stabilizes the intermediate formed in the rate-determining step of ethyl trifluoroacetate saponification.

Ethyl trifluoroacetate	Hydroxide	Rate-determining intermediate

Since the intermediate is more stable, it is formed faster than the one from ethyl acetate.

(*b*) Crowding is increased as the transition state for nucleophilic addition to the carbonyl group is approached. The carbonyl carbon undergoes a change in hybridization from sp^2 to sp^3.

Ethyl 2,2-dimethylpropanoate	Hydroxide ion	Rate-determining intermediate; crowded

The *tert*-butyl group of ethyl 2,2-dimethylpropanoate causes more crowding than the methyl group of ethyl acetate; the rate-determining intermediate is less stable and is formed more slowly.

(c) We see here another example of a steric effect of a *tert*-butyl group. The intermediate formed when hydroxide adds to the carbonyl group of *tert*-butyl acetate is more crowded and less stable than the corresponding intermediate formed from methyl acetate.

| *tert*-Butyl acetate   Hydroxide | Rate-determining intermediate; crowded |

Methyl acetate · Rate-determining intermediate; less crowded than intermediate from *tert*-butyl acetate

(d) Here, as in part (a), we have an electron-withdrawing substituent increasing the rate of ester saponification. It does so by stabilizing the negatively charged intermediate formed in the rate-determining step.

more stable than

Rate-determining intermediate   Rate-determining intermediate
from methyl *m*-nitrobenzoate   from methyl benzoate

(e) Addition of hydroxide to 4-butanolide introduces torsional strain in the intermediate because of eclipsed bonds. The corresponding intermediate from 5-butanolide is more stable, because the bonds are staggered in a six-membered ring.

eclipsed bonds   staggered bonds

Less stable; formed   More stable; formed faster
more slowly

(f) Steric crowding increases more when hydroxide adds to the axial carbonyl group.

Cis diastereomer: greater increase in crowding   Trans diastereomer: smaller increase in crowding
when carbon changes from sp^2 to sp^3; formed   when carbon changes from sp^2 to sp^3; formed
more slowly   more rapidly

20.30 Compound A is the *p*-toluenesulfonate ester of *trans*-4-*tert*-butylcyclohexanol. The oxygen atom of the alcohol attacks sulfur of *p*-toluenesulfonyl chloride, and so the reaction proceeds with retention of configuration.

trans-4-*tert*-Butylcyclohexanol   *p*-Toluenesulfonyl chloride

trans-4-*tert*-Butylcyclohexyl *p*-toluenesulfonate (compound A)

The second step is a nucleophilic substitution reaction in which benzoate ion displaces *p*-toluenesulfonate with inversion of configuration.

Benzoate ion   *trans*-4-*tert*-Butylcyclohexyl *p*-toluenesulfonate (compound A)

cis-4-*tert*-Butylcyclohexyl benzoate (compound B)

Saponification of *cis*-4-*tert*-butylcyclohexyl benzoate proceeds with acyl-oxygen cleavage to give *cis*-4-*tert*-butylcyclohexanol.

20.31 Reaction of ethyl trifluoroacetate with ammonia yields the corresponding amide, compound C. Compound C undergoes dehydration on heating with P_4O_{10} to give trifluoroacetonitrile, compound D. Grignard reagents react with nitriles to form ketones. *tert*-Butyl trifluoromethyl ketone is formed from trifluoroacetonitrile by treatment with *tert*-butylmagnesium chloride followed by aqueous hydrolysis.

$$CF_3COCH_2CH_3 \xrightarrow{NH_3} CF_3CNH_2 \xrightarrow[\text{heat}]{P_4O_{10}} CF_3C\equiv N$$

Ethyl trifluoroacetate   Trifluoroacetamide   Trifluoroacetonitrile
                            (compound C)   (compound D)

$$CF_3C\equiv N \; + \; (CH_3)_3CMgCl \xrightarrow[\text{2. } H_3O^+]{\text{1. diethyl ether}} CF_3CC(CH_3)_3$$

Compound D   *tert*-Butylmagnesium   *tert*-Butyl trifluoromethyl
              chloride   ketone

20.32 The first step is acid hydrolysis of an acetal protecting group.

Step 1:

$$\text{Compound E} \xrightarrow[\text{heat}]{\text{H}_2\text{O, H}^+} \underset{\displaystyle \underset{\text{HO} \quad \text{OH}}{\qquad\qquad}}{\text{HOC(CH}_2)_5\text{CH—CH(CH}_2)_7\text{CH}_2\text{OH}}$$

Compound F ($C_{16}H_{32}O_5$)

All three alcohol functions are converted to bromide by reaction with hydrogen bromide in step 2.

Step 2:

$$\text{Compound F} \xrightarrow{\text{HBr}} \underset{\displaystyle \underset{\text{Br} \quad \text{Br}}{\qquad\qquad}}{\text{HOC(CH}_2)_5\text{CH—CH(CH}_2)_7\text{CH}_2\text{Br}}$$

Compound G ($C_{16}H_{29}Br_3O_2$)

Reaction with ethanol in the presence of an acid catalyst converts the carboxylic acid to its ethyl ester in step 3.

Step 3:

$$\text{Compound G} \xrightarrow[\text{H}_2\text{SO}_4]{\text{ethanol}} \underset{\displaystyle \underset{\text{Br} \quad \text{Br}}{\qquad\qquad}}{\text{CH}_3\text{CH}_2\text{OC(CH}_2)_5\text{CH—CH(CH}_2)_7\text{CH}_2\text{Br}}$$

Compound H ($C_{18}H_{33}Br_3O_2$)

Zinc converts vicinal dibromides to alkenes. Of the three bromine substituents in compound H, two of them are vicinal. Step 4 is a dehalogenation reaction.

Step 4:

$$\text{Compound H} \xrightarrow[\text{ethanol}]{\text{Zn}} \text{CH}_3\text{CH}_2\text{OC(CH}_2)_5\text{CH}=\text{CH(CH}_2)_7\text{CH}_2\text{Br}$$

Compound I ($C_{18}H_{33}BrO_2$)

Step 5 is a nucleophilic substitution reaction of the S_N2 type. Acetate ion is the nucleophile and displaces bromide from the primary carbon.

Step 5:

$$\text{Compound I} \xrightarrow[\text{CH}_3\text{CO}_2\text{H}]{\text{NaOCCH}_3} \text{CH}_3\text{CH}_2\text{OC(CH}_2)_5\text{CH}=\text{CH(CH}_2)_7\text{CH}_2\text{OCCH}_3$$

Compound J ($C_{20}H_{36}O_4$)

Step 6 is ester saponification. It yields a 16-carbon chain having a carboxylic acid function at one end and an alcohol at the other.

Step 6:

$$\text{Compound J} \xrightarrow[\text{2. H}^+]{\text{1. KOH, ethanol}} \text{HOC(CH}_2)_5\text{CH}=\text{CH(CH}_2)_7\text{CH}_2\text{OH}$$

Compound K ($C_{16}H_{30}O_3$)

In step 7, compound K cyclizes to ambrettolide on heating.

Step 7:

Compound K   heat ⟶   Ambrettolide

20.33 (*a*) This step requires the oxidation of a primary alcohol to an aldehyde. Collins' reagent would be appropriate.

$$HOCH_2CH=CH(CH_2)_7CO_2CH_3 \xrightarrow[CH_2Cl_2]{(C_5H_5N)_2CrO_3} HCCH=CH(CH_2)_7CO_2CH_3$$

Compound L   Compound M

As reported in the literature, pyridium dichromate in dichloromethane was used to give the desired aldehyde in 84 percent yield.

(*b*) Conversion of —CH to —CH=CH₂ is a typical case in which a Wittig reaction is appropriate.

$$HCCH=CH(CH_2)_7CO_2CH_3 \xrightarrow{(C_6H_5)_3\overset{+}{P}-\overset{-}{C}H_2} CH_2=CHCH=CH(CH_2)_7CO_2CH_3$$

Compound M   Compound N
                                              (observed yield, 53%)

(*c*) Lithium aluminum hydride was used to reduce the ester to a primary alcohol in 81 percent yield.

$$CH_2=CHCH=CH(CH_2)_7CO_2CH_3 \xrightarrow[2.\ H_2O]{1.\ LiAlH_4} CH_2=CHCH=CH(CH_2)_7CH_2OH$$

Compound N   Compound O

(*d*) The desired sex pheromone is the acetate ester of compound O. Compound O was treated with acetic anhydride to give the acetate ester in 99 percent yield.

$$CH_2=CHCH=CH(CH_2)_7CH_2OH \xrightarrow[pyridine]{CH_3\overset{O}{\overset{\|}{C}}O\overset{O}{\overset{\|}{C}}CH_3} CH_2=CHCH=CH(CH_2)_7CH_2O\overset{O}{\overset{\|}{C}}CH_3$$

Compound O   (*E*)-9,11-Dodecadien-1-yl acetate

Acetyl chloride could have been used in this step instead of acetic anhydride.

20.34 (*a*) The reaction given in the problem is between a lactone (cyclic ester) and a difunctional Grignard reagent. Esters usually react with two moles of a Grignard reagent; in this instance both Grignard moieties of the reagent attack the lactone.

The second attack is intramolecular, giving rise to the cyclopentanol ring of the product.

4-Butanolide

1-(3-Hydroxypropyl)-1-cyclopentanol (88%)

(b) An intramolecular acyl transfer process takes place in this reaction. The amine group in the thiolactone starting material replaces sulfur on the acyl group to form a lactam (cyclic amide).

Thiolactone   Tetrahedral   Lactam
     intermediate

20.35 (a) Acyl chlorides react with alcohols to form esters.

p-Methoxybenzoyl   Benzoin
chloride

Benzoin p-methoxybenzoate
(compound P; 95%)

(b) Of the two carbonyl groups in the starting material, the ketone carbonyl is more reactive than the ester. (The ester carbonyl is stabilized by electron release from oxygen.)

$$\underset{\underset{\displaystyle CH_3CCH_2CH_2COCH_2CH_3}{\uparrow}}{\overset{\displaystyle O \qquad\qquad O}{}} \xrightarrow[\text{(1 equiv)}]{CH_3MgI} \underset{\displaystyle CH_3CCH_2CH_2COCH_2CH_3}{\overset{\displaystyle CH_3 \qquad\quad O}{\underset{\displaystyle OMgI}{}}}$$

Compound Q has the molecular formula $C_6H_{10}O_2$. The initial product forms a cyclic ester, with elimination of ethoxide ion.

Compound Q

(c) Only carboxyl groups that are ortho to each other on a benzene ring are capable of forming a cyclic anhydride.

Compound R

20.36 Compound S is an ester but has within it an amine function. Acyl transfer from oxygen to nitrogen converts the ester to a more stable amide.

| Compound S | Tetrahedral | Compound T |
| (Ar = p-nitrophenyl) | intermediate | (Ar = p-nitrophenyl) |

The tetrahedral intermediate is the key intermediate in the reaction.

20.37 (a) The rearrangement in this problem is an acyl transfer from nitrogen to oxygen.

| Compound U | Tetrahedral | Compound V |
| (Ar = p-nitrophenyl) | intermediate | (Ar = p-nitrophenyl) |

This rearrangement takes place in the indicated direction because it is carried out in acid solution. The amino group is protonated in acid and is no longer nucleophilic.

(b) The trans stereoisomer of compound U does not undergo rearrangement because when the oxygen and nitrogen atoms on the five-membered ring are trans, the necessary tetrahedral intermediate would be too strained.

20.38 The ester functions of a polymer such as poly(vinyl acetate) are just like ester functions of simple molecules; they can be cleaved by hydrolysis under either acidic or basic conditions. To prepare poly(vinyl alcohol), therefore, polymerize vinyl· acetate to poly(vinyl acetate), and then cleave the ester groups by hydrolysis.

$$CH_2=CHOCCH_3 \longrightarrow \left(\begin{array}{c} -CH_2CHCH_2CH- \\ | \quad\quad\quad | \\ CH_3CO \quad OCCH_3 \\ \| \quad\quad\quad \| \\ O \quad\quad\quad O \end{array}\right)_n \xrightarrow[\substack{H^+ \text{ or} \\ HO^-}]{H_2O} \left(\begin{array}{c} -CH_2CHCH_2CH- \\ | \quad\quad\quad | \\ HO \quad\quad OH \end{array}\right)_n$$

Vinyl acetate   Poly(vinyl acetate)   Poly(vinyl alcohol)

20.39 First, assume that the peak in the mass spectrum of highest m/z corresponds to the molecular ion. Thus methyl methacrylate is assumed to have a molecular weight of 100. The integrated areas in the ^1H nmr spectrum indicate that there are eight protons in the molecule, which suggests the molecular formula $C_5H_8O_2$. A methyl ester is likely on the basis of the three-proton singlet in the nmr spectrum at $\delta = 3.8$ ppm (OCH_3). Also, there is a prominent peak at m/z 69 in the mass spectrum; this peak corresponds to loss of OCH_3 from the molecular ion.

The ^1H nmr spectrum reveals an allylic methyl group at $\delta = 2.0$ ppm along with two nonequivalent vinyl protons at $\delta = 5.6$ and 6.1 ppm. Two candidate structures present themselves for consideration at this point. The second of these structures is the correct one.

$$CH_3CH=CHCOCH_3 \qquad\qquad CH_2=CCOCH_3$$
$$\quad\quad\quad \| \qquad\qquad\qquad\quad | \quad \|$$
$$\quad\quad\quad O \qquad\qquad\qquad\quad CH_3 \; O$$

Methyl crotonate   Methyl methacrylate

The problem states that methyl methacrylate is made from the cyanohydrin of acetone; therefore it must have the carbon skeleton C—C—C.
$$\qquad\qquad\qquad\qquad |$$
$$\qquad\qquad\qquad\qquad C$$

20.40 Compound W ($C_4H_6O_2$) has a SODAR of 2. With two oxygen atoms and a peak in the infrared at 1760 cm^{-1}, it is likely that one of the elements of unsaturation is the carbon-oxygen double bond of an ester. The ^1H nmr spectrum contains a three-proton singlet at $\delta = 2.1$ ppm, which is consistent with a CH_3C unit. It is likely that compound W
$$\qquad\qquad\qquad\qquad\qquad\qquad\qquad\qquad\qquad\qquad\quad \|$$
$$\qquad\qquad\qquad\qquad\qquad\qquad\qquad\qquad\qquad\qquad\quad O$$
is an acetate ester.

The ^{13}C nmr spectrum reveals that the four carbon atoms of the molecule are contained in one each of the structures CH_3, CH_2, and CH, along with the carbonyl carbon. In addition to the two carbons of the acetate group, the remaining two carbons are the CH_2 and CH carbons of a vinyl group, $CH=CH_2$. Compound W is vinyl acetate.

$$\delta = 20.2 \text{ ppm (quartet)} \quad CH_3C \overset{O}{\underset{OCH=CH_2}{\diagup}} \quad \delta = 96.8 \text{ ppm (triplet)}$$
$$\delta = 167.6 \text{ ppm (singlet)} \qquad\qquad\qquad \delta = 141.8 \text{ ppm (doublet)}$$

Each vinyl proton is coupled to two other vinyl protons; each appears as a doublet of doublets in the ^1H nmr spectrum.

20.41 Compound X contains nitrogen and exhibits a prominent peak in the infrared at 2270 cm^{-1}; it is likely to be a nitrile. Its molecular weight of 69 is consistent with the molecular formula C_4H_7N, and its ^1H nmr spectrum shows the characteristic doublet-heptet pattern of an isopropyl group. Compound X is 2-methylpropanenitrile.

$$CH_3CHC\equiv N \qquad \text{2-Methylpropanenitrile (compound X)}$$
$$\underset{\displaystyle CH_3}{|}$$

20.42 Compound Y has the characteristic triplet-quartet pattern of an ethyl group in its ^1H nmr spectrum. Since these signals correspond to 10 protons, there must be two equivalent ethyl groups in the molecule. The methylene quartet appears at relatively low field ($\delta = 4.1 \text{ ppm}$), which is consistent with ethyl groups bonded to oxygen, as in $-OCH_2CH_3$. There is a peak at 1730 cm^{-1} in the infrared spectrum, suggesting that these ethoxy groups reside in ester functions. The molecular formula $C_8H_{14}O_4$ reveals that if two ester groups are present, there can be no rings or double bonds. The remaining four hydrogens are equivalent in the ^1H nmr spectrum, and so two equivalent CH_2 groups are present. Compound Y is the diethyl ester of succinic acid.

$$\overset{\displaystyle O}{\overset{\displaystyle \|}{}} \qquad \overset{\displaystyle O}{\overset{\displaystyle \|}{}}$$
$$CH_3CH_2OCCH_2CH_2COCH_2CH_3 \qquad \text{Diethyl succinate (compound Y)}$$

20.43 Compound Z contains nitrogen and has a peak in the infrared at 1680 cm^{-1}, suggesting an amide as a possibility. Reinforcing this notion is the appearance of a broad peak, equivalent to two protons, in the ^1H nmr spectrum at $\delta = 6$ to 7 ppm. This is consistent with the $-NH_2$ group of an amide. Since there are only three carbons, the carbon chain cannot be branched and the only point to be determined is the location of the chlorine. The chlorine must be at C-2 because the ^1H nmr shows a three-proton methyl doublet and a one-proton methine quartet. Compound Z is 2-chloropropanamide.

$$\overset{\displaystyle O}{\overset{\displaystyle \diagup\diagup}{CH_3CHC}} \qquad \text{2-Chloropropanamide (compound Z)}$$
$$\underset{\displaystyle Cl}{|} \qquad \diagdown NH_2$$

Compound Z is prepared from propanoic acid as shown.

$$CH_3CH_2CO_2H \xrightarrow[P]{Cl_2} CH_3CHCO_2H \xrightarrow{SOCl_2}$$
$$\underset{\displaystyle Cl}{|}$$

Propanoic acid   2-Chloropropanoic
                           acid

$$CH_3CHC\overset{\displaystyle O}{\diagup\diagup}_{\diagdown Cl} \xrightarrow{NH_3} CH_3CHC\overset{\displaystyle O}{\diagup\diagup}_{\diagdown NH_2}$$
$$\underset{\displaystyle Cl}{|} \qquad\qquad\qquad \underset{\displaystyle Cl}{|}$$

2-Chloropropanoyl   2-Chloropropanamide
chloride

Propanoic acid cannot be converted to its amide and then halogenated at the α carbon, because it is the free acid that must be halogenated by the Hell-Volhard-Zelinsky reaction.

SELF-TEST

PART A

A-1. Give a correct IUPAC name for each of the following acid derivatives:

$$
(a) \quad CH_3CH_2CH_2O\overset{\overset{\textstyle O}{\|}}{C}CH_2CH_2CH_3
$$

$$
(b) \quad C_6H_5\overset{\overset{\textstyle O}{\|}}{C}NHCH_3
$$

$$
(c) \quad (CH_3)_2CHCH_2CH_2\overset{\overset{\textstyle O}{\|}}{C}Cl
$$

A-2. Provide the correct structure of:
  (a) Benzoic anhydride
  (b) *N*-(1-Methylpropyl)acetamide
  (c) Phenyl benzoate

A-3. What is the formula of the reagent(s) needed to carry out each of the following conversions?

$$
(a) \quad C_6H_5CH_2CO_2H \xrightarrow{\;?\;} C_6H_5CH_2\overset{\overset{\textstyle O}{\|}}{C}Cl
$$

$$
(b) \quad (CH_3)_3C\overset{\overset{\textstyle O}{\|}}{C}NH_2 \xrightarrow{\;?\;} (CH_3)_3CNH_2
$$

$$
(c) \quad (CH_3)_2CHCH_2NH_2 \xrightarrow{\;?\;} C_6H_5\overset{\overset{\textstyle O}{\|}}{C}NHCH_2CH(CH_3)_2 + CH_3OH
$$

A-4. Write the structure of the product of each of the following reactions:

  (a)   Cyclohexyl acetate $\xrightarrow[\text{2. H}^+]{\text{1. NaOH, H}_2\text{O}}$? (two products)

  (b)   Cyclopentanol + benzoyl chloride $\xrightarrow{\text{pyridine}}$?

  (c)   [phthalic anhydride structure] O + $CH_3CH_2OH \xrightarrow{\text{H}^+\text{(cat)}}$?

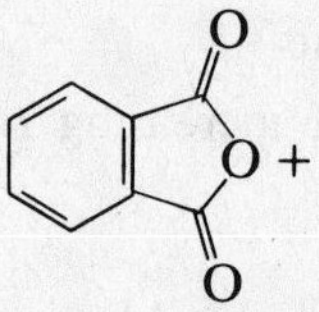

  (d)   Ethyl propanoate + dimethylamine $\longrightarrow$? (two products)

  (e) $CH_3\text{—}\langle\text{benzene ring}\rangle\text{—}\overset{\overset{\textstyle O}{\|}}{C}NHCH_3 \xrightarrow[\text{heat}]{\text{H}_2\text{O, H}_2\text{SO}_4}$? (two products)

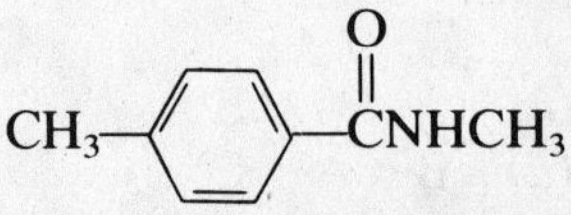

A-5. The following reaction proceeds when the reactant is allowed to stand in pentane. Write the structure of the key intermediate in this process.

$$
\underset{\substack{\text{O}\\\parallel}}{C_6H_5C}OCH_2CH_2NHCH_3 \longrightarrow \underset{\substack{|\\C_6H_5C=O}}{CH_3NCH_2CH_2OH}
$$

A-6. Give the correct structures, clearly showing stereochemistry, of each compound, A through D, in the following sequence of reactions:

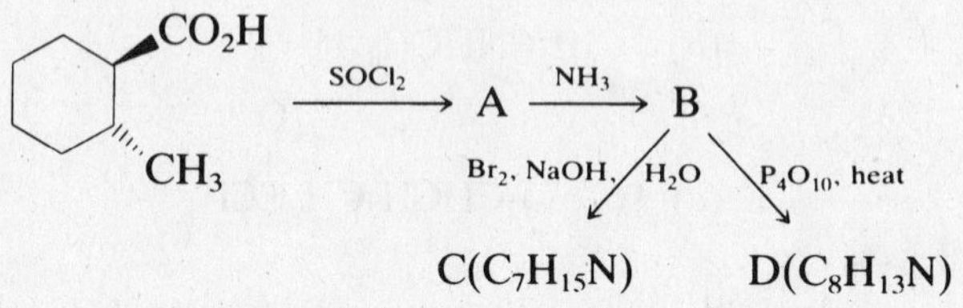

$$
\begin{array}{ccc}
 & \xrightarrow{\text{SOCl}_2} A \xrightarrow{\text{NH}_3} B \\
 & \text{Br}_2, \text{NaOH} \swarrow \text{H}_2\text{O} \quad \searrow \text{P}_4\text{O}_{10}, \text{heat} \\
 & C(C_7H_{15}N) \qquad D(C_8H_{13}N)
\end{array}
$$

(cyclohexane with CO$_2$H and CH$_3$ substituents shown with stereochemistry)

A-7. Write the structure of the neutral form of the tetrahedral intermediate in the:
 (a) Acid-catalyzed hydrolysis of methyl acetate
 (b) Reaction of ammonia with acetic anhydride

A-8. Write the steps necessary to prepare $CH_3\!-\!\!\!\langle\ \rangle\!\!\!-NH_2$ from $CH_3\!-\!\!\!\langle\ \rangle\!\!\!-Br$.

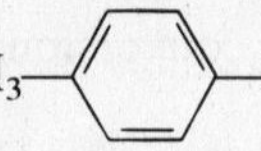

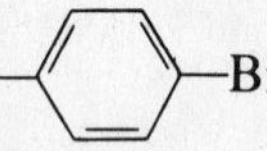

PART B

B-1. The most favorable mode of decomposition of the intermediate species

$$
\underset{\substack{|\\Cl}}{\overset{\substack{OH\\|}}{C_6H_5-C-OH}}
$$

will yield:
 (a) Benzoic acid and HCl
 (b) Benzoyl chloride and H_2O
 (c) Both (a) and (b) with equal likelihood
 (d) Neither (a) nor (b)

B-2. Hydrolysis of propyl butanoate yields:
 (a) Propanoic acid + butanol
 (b) Butanoic acid + butanol
 (c) Butanoic acid + propanol
 (d) Butanoic acid + propylamine

B-3. Rank the following in order of increasing reactivity (least → most) toward acid hydrolysis:

$$
\underset{A}{\overset{\substack{O\\\parallel}}{CH_3COCH_2CH_3}} \qquad \underset{B}{\overset{\substack{O\\\parallel}}{CH_3CCl}} \qquad \underset{C}{\overset{\substack{O\\\parallel}}{CH_3CNHCH_3}}
$$

 (a) A < B < C   (c) A < C < B
 (b) C < A < B   (d) B < A < C

B-4. The structure of *N*-propylacetamide is:

$$(a) \quad CH_3CH_2\overset{\overset{\displaystyle O}{\|}}{C}NHCH_3 \qquad (c) \quad CH_3\overset{\overset{\displaystyle O}{\|}}{C}NHCH_2CH_2CH_3$$

$$(b) \quad CH_3\overset{\overset{\displaystyle O}{\|}}{C}N(CH_2CH_2CH_3)_2 \qquad (d) \quad CH_3CH{=}NCH_2CH_2CH_3$$

B-5. Choose the response which matches the correct functional group classification with the following group of structural formulas.

(*a*) Anhydride	Lactam	Lactone
(*b*) Lactam	Imide	Lactone
(*c*) Imide	Lactone	Anhydride
(*d*) Imide	Lactam	Lactone

B-6. Choose the best sequence of reactions for the transformation given. Semicolons indicate separate reaction steps to be used in the order shown.

$$CH_3{-}\langle\bigcirc\rangle{-}CO_2CH_3 \xrightarrow{\quad ? \quad} CH_3{-}\langle\bigcirc\rangle{-}CH_2\overset{\overset{\displaystyle O}{\|}}{C}NHCH_3$$

(*a*) H_3O^+; $SOCl_2$; CH_3NH_2
(*b*) OH^-/H_2O; PBr_3; Mg; CO_2; H_3O^+; $SOCl_2$; CH_3NH_2
(*c*) $LiAlH_4$; H_2O; HBr; Mg; CO_2; H_3O^+; $SOCl_2$; CH_3NH_2
(*d*) None of these would yield the desired product.

B-7. A key step in the hydrolysis of acetamide in aqueous acid proceeds by attack of:

$$(a) \quad H_3O^+ \text{ on } CH_3\overset{\overset{\displaystyle O}{\|}}{C}NH_2 \qquad (c) \quad HO^- \text{ on } CH_3\overset{\overset{\displaystyle O}{\|}}{C}NH_2$$

$$(b) \quad H_3O^+ \text{ on } CH_3\overset{\overset{\displaystyle {}^+OH}{\|}}{C}NH_2 \qquad (d) \quad H_2O \text{ on } CH_3\overset{\overset{\displaystyle {}^+OH}{\|}}{C}NH_2$$

IMPORTANT REACTIONS

Base-Promoted Ester Condensation. The Claisen Condensation (Sec. 21.1)

General:

$$2RCH_2COR' \xrightarrow[\text{2. } H_3O^+]{\text{1. NaOR'}} RCH_2\overset{O}{\overset{\|}{C}}\overset{\,}{\underset{\underset{R}{|}}{C}}HCOR' + R'OH$$

Example:

$$2CH_3CH_2\overset{O}{\overset{\|}{C}}OCH_2CH_3 \xrightarrow[\text{2. } H_3O^+]{\text{1. NaOCH}_2CH_3} CH_3CH_2\overset{O}{\overset{\|}{C}}\overset{\,}{\underset{\underset{CH_3}{|}}{C}}HCOCH_2CH_3 + CH_3CH_2OH$$

Mechanism:

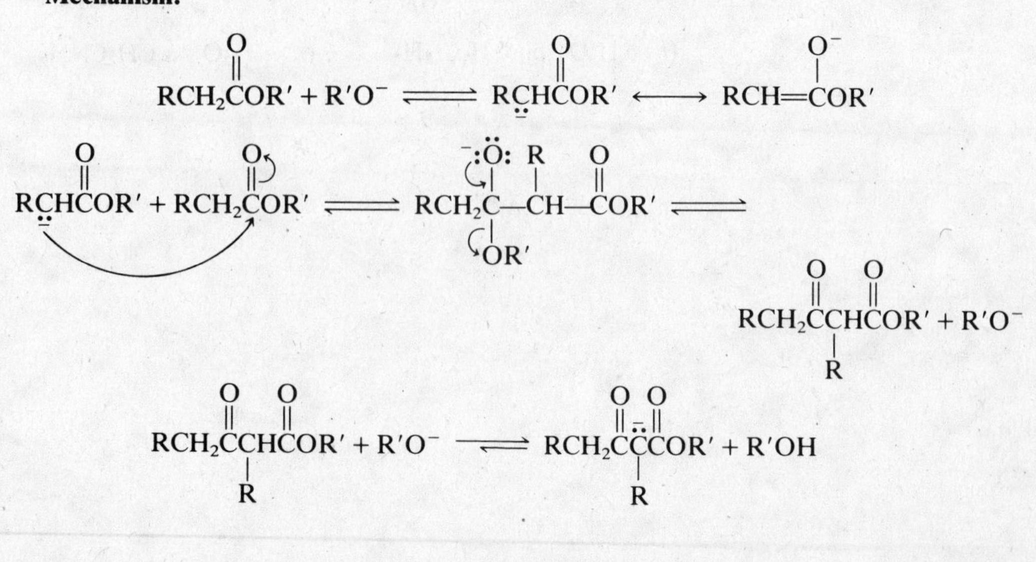

$$RCH_2\overset{O}{\overset{\|}{C}}OR' + R'O^- \rightleftharpoons R\overset{O}{\overset{\|}{C}}HCOR' \longleftrightarrow RCH{=}\overset{O^-}{C}OR'$$

$$R\overset{O}{\overset{\|}{C}}HCOR' + RCH_2\overset{O}{\overset{\|}{C}}OR' \rightleftharpoons RCH_2\overset{\overset{:\overset{..}{O}:}{|}}{\underset{\underset{OR'}{|}}{C}}{-}\overset{R}{\overset{|}{C}}H{-}\overset{O}{\overset{\|}{C}}OR' \rightleftharpoons$$

$$RCH_2\overset{O}{\overset{\|}{C}}\overset{O}{\overset{\|}{C}}\overset{\,}{\underset{\underset{R}{|}}{C}}HCOR' + R'O^-$$

$$RCH_2\overset{O}{\overset{\|}{C}}\overset{O}{\overset{\|}{C}}\overset{\,}{\underset{\underset{R}{|}}{C}}HCOR' + R'O^- \rightleftharpoons RCH_2\overset{O}{\overset{\|}{C}}\overset{O}{\overset{\|}{C}}\overset{\,}{\underset{\underset{R}{|}}{\overset{..}{C}}}COR' + R'OH$$

Intramolecular Claisen Condensation. The Dieckmann Reaction
(Sec. 21.2)

$$R'O\overset{O}{\overset{||}{C}}(CH_2)_n\overset{O}{\overset{||}{C}}OR' \xrightarrow[\text{2. }H_3O^+]{\text{1. NaOR'}} \underset{(CH_2)_{n-1}}{\overset{O}{\overset{||}{C}}}\overset{O}{\overset{||}{CH-COR'}}$$

Example:

$$CH_3CH_2O\overset{O}{\overset{||}{C}}(CH_2)_5\overset{O}{\overset{||}{C}}OCH_2CH_3 \xrightarrow[\text{2. }H_3O^+]{\text{1. NaOCH}_2CH_3}$$

(cyclohexanone with COCH$_2$CH$_3$ substituent)

Mixed Claisen Condensations (Sec. 21.3)

General:

$$R\overset{O}{\overset{||}{C}}OCH_2CH_3 + R'CH_2\overset{O}{\overset{||}{C}}OCH_2CH_3 \xrightarrow[\text{2. }H_3O^+]{\text{1. NaOCH}_2CH_3} R\overset{O}{\overset{||}{C}}\underset{R'}{CH}\overset{O}{\overset{||}{C}}OCH_2CH_3$$

Example:

$$C_6H_5\overset{O}{\overset{||}{C}}OCH_2CH_3 + CH_3\overset{O}{\overset{||}{C}}OCH_2CH_3 \xrightarrow[\text{2. }H_3O^+]{\text{1. NaOCH}_2CH_3} C_6H_5\overset{O}{\overset{||}{C}}CH_2\overset{O}{\overset{||}{C}}OCH_2CH_3$$

Acylation of Ketones (Sec. 21.4)

Examples:

$$H\overset{O}{\overset{||}{C}}OCH_2CH_3 + \text{(cyclohexanone)} \xrightarrow[\text{2. }H_3O^+]{\text{1. NaH}} \text{(2-formylcyclohexanone)}$$

$$CH_3\overset{O}{\overset{||}{C}}CH_2CH_2CH_2\overset{O}{\overset{||}{C}}OCH_2CH_3 \xrightarrow[\text{2. }H_3O^+]{\text{1. NaOCH}_2CH_3} \text{(1,3-cyclohexanedione)}$$

Synthesis of Ketones (Sec. 21.5)

General:

$$R\overset{O}{\overset{||}{C}}\underset{R'}{CH}\overset{O}{\overset{||}{C}}OCH_2CH_3 \xrightarrow[\text{2. }H_3O^+]{\text{1. HO}^-,\,H_2O} R\overset{O}{\overset{||}{C}}\underset{R'}{CH}\overset{O}{\overset{||}{C}}OH \xrightarrow{\text{heat}} R\overset{O}{\overset{||}{C}}CH_2R' + CO_2$$

Example:

$$CH_3CH_2\overset{O}{\overset{||}{C}}\underset{CH_3}{CH}\overset{O}{\overset{||}{C}}OCH_2CH_3 \xrightarrow[\substack{\text{2. }H_3O^+ \\ \text{3. heat}}]{\text{1. HO}^-,\,H_2O} CH_3CH_2\overset{O}{\overset{||}{C}}CH_2CH_3$$

Acetoacetic Ester Synthesis (Sec. 21.6)

General:

$$CH_3\overset{O}{\overset{\|}{C}}CH_2\overset{O}{\overset{\|}{C}}OCH_2CH_3 \xrightarrow[\text{2. R—X}]{\text{1. NaOCH}_2\text{CH}_3} CH_3\overset{O}{\overset{\|}{C}}\underset{\underset{R}{|}}{C}H\overset{O}{\overset{\|}{C}}OCH_2CH_3 \xrightarrow[\substack{\text{2. H}_3\text{O}^+\\ \text{3. heat}}]{\text{1. HO}^-,\,\text{H}_2\text{O}} CH_3\overset{O}{\overset{\|}{C}}CH_2R$$

Example:

$$CH_3\overset{O}{\overset{\|}{C}}CH_2\overset{O}{\overset{\|}{C}}OCH_2CH_3 \xrightarrow[\text{2. (CH}_3)_2\text{CHCH}_2\text{Br}]{\text{1. NaOCH}_2\text{CH}_3} CH_3\overset{O}{\overset{\|}{C}}\underset{\underset{CH_2CH(CH_3)_2}{|}}{C}H\overset{O}{\overset{\|}{C}}OCH_2CH_3 \xrightarrow[\substack{\text{2. H}_3\text{O}^+\\ \text{3. heat}}]{\text{1. HO}^-,\,\text{H}_2\text{O}}$$

$$CH_3\overset{O}{\overset{\|}{C}}CH_2CH_2CH(CH_3)_2$$

Malonic Ester Synthesis (Sec. 21.7)

General:

$$R—X + CH_2(CO_2CH_2CH_3)_2 \xrightarrow[\text{CH}_3\text{CH}_2\text{OH}]{\text{NaOCH}_2\text{CH}_3} RCH(CO_2CH_2CH_3)_2 \xrightarrow[\substack{\text{2. H}_3\text{O}^+\\ \text{3. heat}}]{\text{1. HO}^-,\,\text{H}_2\text{O}} RCH_2CO_2H$$

Example:

$$CH_2{=}CHCH_2Br + CH_2(CO_2CH_2CH_3)_2 \xrightarrow[\text{CH}_3\text{CH}_2\text{OH}]{\text{NaOCH}_2\text{CH}_3}$$

$$CH_2{=}CHCH_2CH(CO_2CH_2CH_3)_2 \xrightarrow[\substack{\text{2. H}_3\text{O}^+\\ \text{3. heat}}]{\text{1. HO}^-,\,\text{H}_2\text{O}} CH_2{=}CHCH_2CH_2CO_2H$$

Michael Reaction (Sec. 21.9)

Example:

$$CH_2(CO_2CH_2CH_3)_2 + CH_2{=}CH\overset{O}{\overset{\|}{C}}CH_3 \xrightarrow[\text{CH}_3\text{CH}_2\text{OH}]{\text{KOH}} CH_3\overset{O}{\overset{\|}{C}}CH_2CH_2CH(CO_2CH_2CH_3)_2$$

1,4-Addition product

Knoevenagel Condensation (Sec. 21.10)

General:

$$R\overset{O}{\overset{\|}{C}}R' + CH_2(CO_2CH_2CH_3)_2 \xrightarrow{\text{piperidine}} \underset{R'}{\overset{R}{>}}C{=}C(CO_2CH_2CH_3)_2 + H_2O$$

Example:

$$CH_3CH_2\overset{O}{\overset{\|}{C}}H + CH_2(CO_2CH_2CH_3)_2 \xrightarrow{\text{piperidine}} CH_3CH_2CH{=}C(CO_2CH_2CH_3)_2$$

Reformatskii Reaction (Sec. 21.11)

General:

$$R\overset{O}{\overset{\|}{C}}R' + R''\underset{\underset{Br}{|}}{C}H\overset{O}{\overset{\|}{C}}OCH_2CH_3 \xrightarrow[\text{2. H}_3\text{O}^+]{\text{1. Zn}} R\underset{\underset{R'}{|}}{\overset{\overset{OH}{|}}{C}}\underset{\underset{R''}{|}}{C}H\overset{O}{\overset{\|}{C}}OCH_2CH_3$$

Example:

$$\underset{\text{CH}_3\text{CH}_2\overset{\displaystyle O}{\overset{\|}{\text{C}}}\text{H}}{} + \underset{\text{BrCH}_2\overset{\displaystyle O}{\overset{\|}{\text{C}}}\text{OCH}_2\text{CH}_3}{} \xrightarrow[\text{2. H}_3\text{O}^+]{\text{1. Zn}} \text{CH}_3\text{CH}_2\overset{\displaystyle\text{OH}}{\underset{}{\text{CHCH}_2\overset{\displaystyle O}{\overset{\|}{\text{C}}}\text{OCH}_2\text{CH}_3}}$$

SOLUTIONS TO TEXT PROBLEMS

21.1 Ethyl benzoate cannot undergo the Claisen condensation, because it has no protons on its α carbon atom and so cannot form an enolate. Ethyl pentanoate and ethyl phenylacetate can undergo the Claisen condensation.

$$2\text{CH}_3\text{CH}_2\text{CH}_2\text{CH}_2\overset{\displaystyle O}{\overset{\|}{\text{C}}}\text{OCH}_2\text{CH}_3 \xrightarrow[\text{2. H}_3\text{O}^+]{\text{1. NaOCH}_2\text{CH}_3} \text{CH}_3\text{CH}_2\text{CH}_2\text{CH}_2\overset{\displaystyle O}{\overset{\|}{\text{C}}}\overset{}{\underset{\displaystyle\text{CH}_2\text{CH}_2\text{CH}_3}{\text{CH}}}\overset{\displaystyle O}{\overset{\|}{\text{C}}}\text{OCH}_2\text{CH}_3$$

Ethyl pentanoate   Ethyl 3-oxo-2-propylheptanoate

$$2\text{C}_6\text{H}_5\text{CH}_2\overset{\displaystyle O}{\overset{\|}{\text{C}}}\text{OCH}_2\text{CH}_3 \xrightarrow[\text{2. H}_3\text{O}^+]{\text{1. NaOCH}_2\text{CH}_3} \text{C}_6\text{H}_5\text{CH}_2\overset{\displaystyle O}{\overset{\|}{\text{C}}}\overset{}{\underset{\displaystyle\text{C}_6\text{H}_5}{\text{CH}}}\overset{\displaystyle O}{\overset{\|}{\text{C}}}\text{OCH}_2\text{CH}_3$$

Ethyl phenylacetate   Ethyl 3-oxo-2,4-diphenylbutanoate

21.2 (*b*) The enolate formed by proton abstraction from the α carbon atom of diethyl 4-methylheptanedioate cyclizes to form a six-membered β-keto ester.

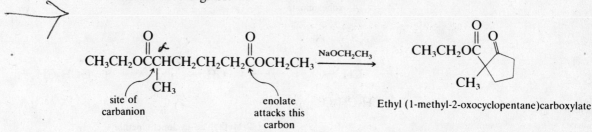

Ethyl (5-methyl-2-oxocyclohexane)carboxylate

(*c*) The two α carbons of this diester are not equivalent. Cyclization by attack of the enolate at C-2 gives:

This β-keto ester cannot form a stable enolate by deprotonation. It is present in only small amounts at equilibrium. The major product is formed by way of the other enolate.

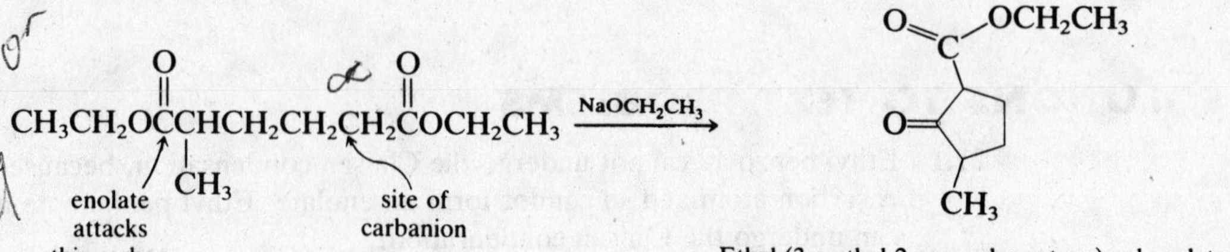

Major Product

$$CH_3CH_2OCCHCH_2CH_2CH_2COCH_2CH_3 \xrightarrow{\text{NaOCH}_2\text{CH}_3}$$

enolate attacks this carbon

site of carbanion

Ethyl (3-methyl-2-oxocyclopentane)carboxylate

This β-keto ester is converted to a stable enolate on deprotonation, causing the equilibrium to shift in its favor.

21.3 (*b*) Both carbonyl groups of diethyl oxalate are equivalent. The enolate of ethyl phenylacetate attacks one of them.

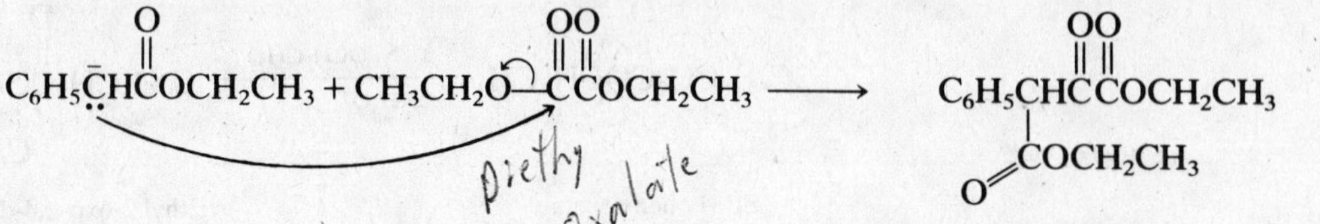

$$C_6H_5\overset{..}{C}HCOCH_2CH_3 + CH_3CH_2O-CCOCH_2CH_3 \longrightarrow C_6H_5CHCCOCH_2CH_3$$

Diethyl oxalate

COCH_2CH_3

Diethyl 2-oxo-3-phenylbutanedioate

(*c*) The enolate of ethyl phenylacetate attacks the carbonyl group of ethyl formate.

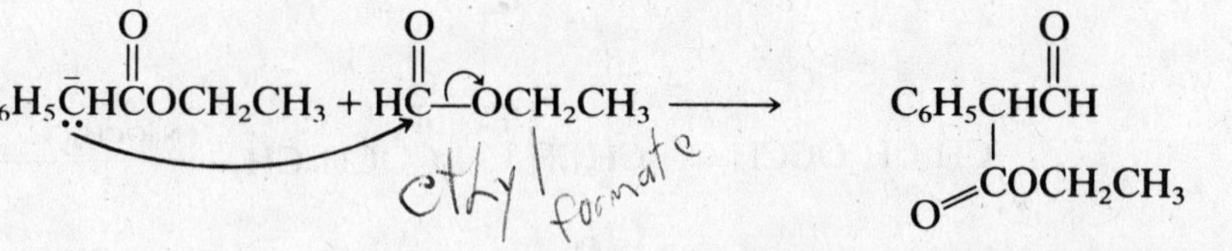

$$C_6H_5\overset{..}{C}HCOCH_2CH_3 + HC-OCH_2CH_3 \longrightarrow C_6H_5CHCH$$

Ethyl formate

COCH_2CH_3

Ethyl 3-oxo-2-phenylpropanoate

21.4 In order for a five-membered ring to be formed, C-5 must be the carbanionic site that attacks the ester carbonyl.

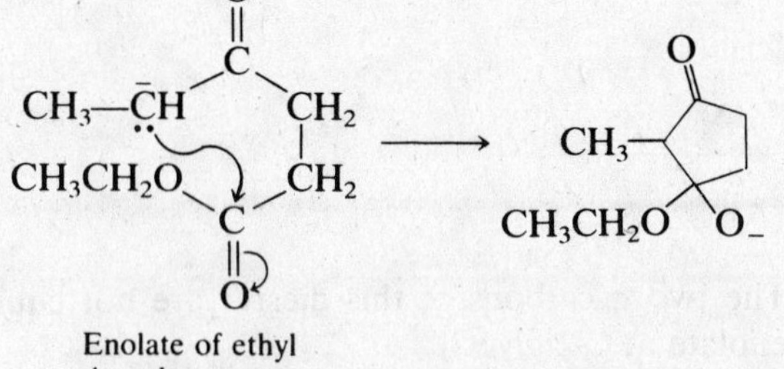

$$CH_3-\overset{..}{C}H$$

CH_3CH_2O

Enolate of ethyl
4-oxohexanoate

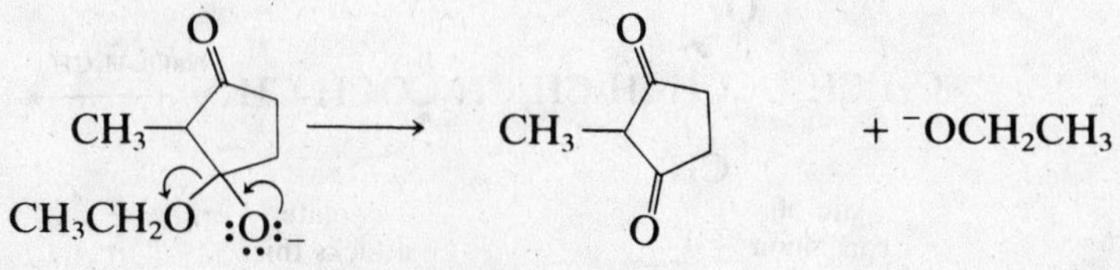

CH_3

CH_3CH_2O O_

CH_3

CH_3CH_2O :O:_

CH_3

O

+ ⁻OCH_2CH_3

2-Methyl-1,3-cyclopentanedione

21.5 The desired ketone, cyclopentanone, is derived from the corresponding β-keto ester. This key intermediate is obtained from a Dieckmann cyclization of the starting material, diethyl hexanedioate.

First treat the diester with sodium ethoxide to effect the Dieckmann cyclization.

$$CH_3CH_2OCCH_2CH_2CH_2CH_2COCH_2CH_3 \xrightarrow[\text{2. } H_3O^+]{\text{1. } NaOCH_2CH_3}$$

Ethyl (2-oxocyclopentane)carboxylate

Next convert the β-keto ester to the desired product by saponification and decarboxylation.

$$\xrightarrow[\substack{\text{2. } H^+ \\ \text{3. heat}}]{\text{1. } HO^-, H_2O}$$

Ethyl
(2-oxocyclopentane)carboxylate   Cyclopentanone

21.6 (*b*) Write a structural formula for the desired product; then disconnect a bond to the α carbon atom.

$$\text{C}_6\text{H}_5-CH_2 \!\mid\! CH_2CCH_3 \Rightarrow \text{C}_6\text{H}_5-\underset{X}{CH_2} + {}:CH_2CCH_3$$

disconnect   Required   Derived from
here   alkyl halide   ethyl
                                  acetoacetate

Therefore

$$\text{C}_6\text{H}_5-CH_2Br + CH_3CCH_2COCH_2CH_3 \xrightarrow[\substack{\text{2. } HO^-, H_2O \\ \text{3. } H^+ \\ \text{4. heat}}]{\text{1. } NaOCH_2CH_3} \text{C}_6\text{H}_5-CH_2CH_2CCH_3$$

Benzyl bromide   Ethyl acetoacetate   4-Phenyl-2-butanone

(*c*) The disconnection approach to retrosynthetic analysis reveals that the preparation of 5-hexen-2-one by the acetoacetic ester synthesis requires an allylic halide.

$$CH_2=CHCH_2 \!\mid\! CH_2CCH_3 \Rightarrow CH_2=CHCH_2 + \underset{X}{:CH_2CCH_3}$$

disconnect   Required alkyl   Derived from ethyl
here   halide   acetoacetate

$$CH_2=CHCH_2Br + CH_3CCH_2COCH_2CH_3 \xrightarrow[\substack{\text{2. } HO^-, H_2O \\ \text{3. } H^+ \\ \text{4. heat}}]{\text{1. } NaOCH_2CH_3} CH_2=CHCH_2CH_2CCH_3$$

Allyl bromide   Ethyl acetoacetate   5-Hexen-2-one

21.7 (*b*) Nonanoic acid has a $CH_3(CH_2)_5CH_2-$ unit attached to the $\overset{\overset{O}{\parallel}}{CH_2COH}$ synthon.

$$CH_3(CH_2)_5CH_2\overset{|}{+}CH_2\overset{\overset{O}{\parallel}}{COH} \implies CH_3(CH_2)_5CH_2X + \quad :\overset{\overset{O}{\parallel}}{CH_2COH}$$

disconnect here   Required alkyl halide   Derived from diethyl malonate

Therefore the anion of diethyl malonate is alkylated with a 1-haloheptane.

$$CH_3(CH_2)_5CH_2Br + CH_2(COOCH_2CH_3)_2 \xrightarrow[\text{ethanol}]{NaOCH_2CH_3} CH_3(CH_2)_5CH_2CH(COOCH_2CH_3)_2$$

1-Bromoheptane   Diethyl malonate   Diethyl heptylmalonate

$$\downarrow \begin{array}{l} 1.\ HO^-,\ H_2O \\ 2.\ H^+ \\ 3.\ heat \end{array}$$

$$CH_3(CH_2)_5CH_2CH_2CO_2H$$

Nonanoic acid

(*c*) Disconnection of the target molecule adjacent to the α carbon reveals the alkyl halide needed to react with the enolate derived from diethyl malonate.

$$\underset{\underset{CH_3}{|}}{CH_3CH_2CHCH_2}\overset{|}{+}CH_2\overset{\overset{O}{\parallel}}{COH} \implies \underset{\underset{CH_3}{|}}{CH_3CH_2CHCH_2}X + \quad :\overset{\overset{O}{\parallel}}{CH_2COH}$$

Required alkyl halide   Derived from diethyl malonate

The necessary alkyl halide in this synthesis is 1-bromo-2-methylbutane.

$$\underset{\underset{CH_3}{|}}{CH_3CH_2CHCH_2Br} + CH_2(COOCH_2CH_3)_2 \xrightarrow[\text{ethanol}]{NaOCH_2CH_3} \underset{\underset{CH_3}{|}}{CH_3CH_2CHCH_2CH}(COOCH_2CH_3)_2$$

1-Bromo-2-methylbutane   Diethyl malonate   Diethyl 2-(2-methylbutyl)malonate

$$\downarrow \begin{array}{l} 1.\ HO^-,\ H_2O \\ 2.\ H^+ \\ 3.\ heat \end{array}$$

$$\underset{\underset{CH_3}{|}}{CH_3CH_2CHCH_2CH_2}\overset{\overset{O}{\parallel}}{COH}$$

4-Methylhexanoic acid

(*d*) Once again disconnection reveals the necessary halide, which is treated with diethyl malonate.

$$C_6H_5CH_2\overset{|}{+}CH_2\overset{\overset{O}{\parallel}}{COH} \implies C_6H_5CH_2X + \quad :\overset{\overset{O}{\parallel}}{CH_2COH}$$

Required halide   Derived from diethyl malonate

Alkylation of diethyl malonate with benzyl bromide is the first step in the preparation of 3-phenylpropanoic acid.

$$C_6H_5CH_2Br + CH_2(COOCH_2CH_3)_2 \xrightarrow[\text{ethanol}]{\text{NaOCH}_2\text{CH}_3} C_6H_5CH_2CH(COOCH_2CH_3)_2$$

Benzyl bromide   Diethyl malonate   Diethyl benzylmalonate

1. HO⁻, H₂O
2. H⁺
3. heat

$$\underset{\text{3-Phenylpropanoic acid}}{C_6H_5CH_2CH_2\overset{\displaystyle O}{\overset{\|}{C}}OH}$$

21.8 Retrosynthetic analysis of the formation of 3-methyl-2-butanone is carried out in the same way as for other ketones.

$$\underset{\substack{\text{3-Methyl-2-butanone} \\ \text{(two disconnections} \\ \text{as shown)}}}{CH_3\overset{\displaystyle O}{\overset{\|}{C}}CH \vdots CH_3} \qquad \Longrightarrow \qquad \underset{\substack{\text{Derived from} \\ \text{ethyl acetoacetate}}}{CH_3\overset{\displaystyle O}{\overset{\|}{C}}\ddot{C}H} \quad + 2CH_3X$$

$$\overset{\displaystyle |}{\underset{\displaystyle CH_3}{}}$$

The two alkylation steps are carried out sequentially.

$$\underset{\substack{\text{Ethyl} \\ \text{acetoacetate}}}{CH_3\overset{\displaystyle O}{\overset{\|}{C}}CH_2\overset{\displaystyle O}{\overset{\|}{C}}OCH_2CH_3} \xrightarrow[\text{CH}_3\text{Br}]{\text{NaOCH}_2\text{CH}_3} \underset{\substack{\text{Ethyl} \\ \text{2-methyl-3-oxobutanoate}}}{CH_3\overset{\displaystyle O}{\overset{\|}{C}}\underset{\displaystyle CH_3}{\overset{\displaystyle |}{C}}H\overset{\displaystyle O}{\overset{\|}{C}}OCH_2CH_3} \xrightarrow[\text{CH}_3\text{Br}]{\text{NaOCH}_2\text{CH}_3}$$

$$\underset{\substack{\text{Ethyl} \\ \text{2,2-dimethyl-3-oxobutanoate}}}{CH_3\overset{\displaystyle O}{\overset{\|}{C}}\underset{\displaystyle CH_3 \quad CH_3}{\overset{\displaystyle | \quad |}{C}}COCH_2CH_3}$$

1. HO⁻, H₂O
2. H⁺
3. heat

$$\underset{\text{3-Methyl-2-butanone}}{CH_3\overset{\displaystyle O}{\overset{\|}{C}}CH(CH_3)_2}$$

21.9   Alkylation of ethyl acetoacetate with 1,4-dibromobutane gives a product that can cyclize to a five-membered ring.

$$CH_3CCH_2COCH_2CH_3 + BrCH_2CH_2CH_2CH_2Br \xrightarrow[\text{ethanol}]{NaOCH_2CH_3} BrCH_2CH_2CH_2CH_2CHCCH_3$$

Ethyl acetoacetate   1,4-Dibromobutane

Ethyl 1-acetylcyclopentanecarboxylate

Saponification and decarboxylation give cyclopentyl methyl ketone.

Ethyl 1-acetylcyclopentanecarboxylate   Cyclopentyl methyl ketone

21.10   The last step in the synthesis of pentobarbital is the reaction of the appropriately substituted derivative of diethyl malonate with urea.

Diethyl 2-ethyl-2-(1-methylbutyl)malonate   Urea   Pentobarbital

The dialkyl derivative of diethyl malonate is made in the usual way. It does not matter whether the ethyl group or the 1-methylbutyl group is introduced first.

Diethyl malonate   Diethyl 2-ethyl-2-(1-methylbutyl)malonate

21.11 The carbonyl oxygen at C-2 of pentobarbital is replaced by sulfur in Pentothal (thiopental).

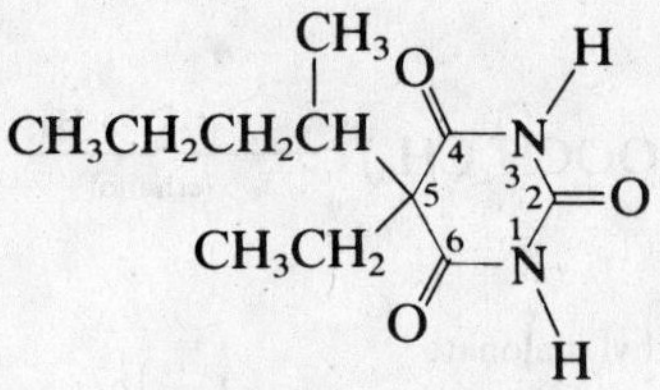

Pentobarbital; prepared
from urea, $(H_2N)_2C=O$

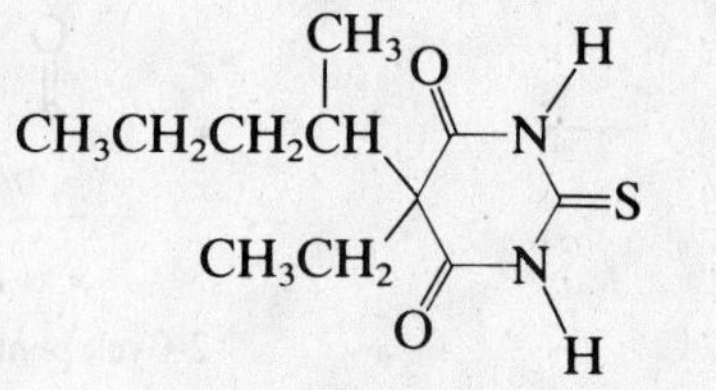

Pentothal; prepared
from thiourea, $(H_2N)_2C=S$

The sodium salt of Pentothal is formed by removal of a proton from one of the N—H groups by sodium hydroxide.

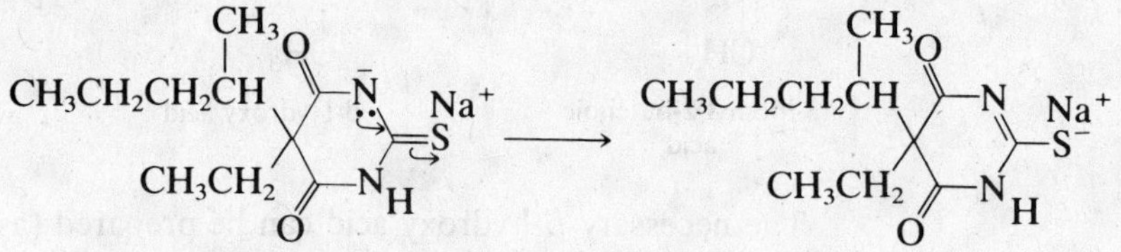

Pentothal sodium

21.12 The synthesis of phenobarbital requires diethyl phenylmalonate as the starting material.

$$C_6H_5CH(COOCH_2CH_3)_2 \xrightarrow[CH_3CH_2Br]{NaOCH_2CH_3} C_6H_5C(COOCH_2CH_3)_2 \xrightarrow{H_2NCNH_2}$$
$$\underset{CH_2CH_3}{|}$$

Diethyl
phenylmalonate

Diethyl
2-ethyl-2-phenylmalonate

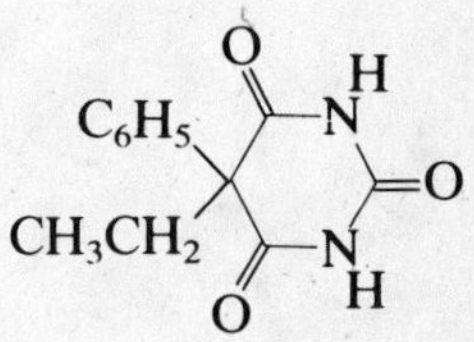

Phenobarbital

Diethyl phenylmalonate is prepared by a mixed Claisen condensation between ethyl phenylacetate and diethyl carbonate.

$$C_6H_5CH_2COCH_2CH_3 + CH_3CH_2OCOCH_2CH_3 \xrightarrow{NaOCH_2CH_3} C_6H_5CH(COOCH_2CH_3)_2$$

Ethyl phenylacetate   Diethyl carbonate   Diethyl phenylmalonate

21.13 (b) The reaction of diethyl malonate with ethyl crotonate is analogous to its reaction with acrolein; addition occurs at the β carbon atom of ethyl crotonate.

$$(CH_3CH_2OOC)_2CH_2 + CH_3CH=CHCOCH_2CH_3 \xrightarrow[ethanol]{NaOCH_2CH_3}$$

Diethyl malonate   Ethyl crotonate

$$(CH_3CH_2OOC)_2CHCHCH_2COCH_2CH_3$$
$$\underset{CH_3}{|}$$

Triethyl 2-methyl-1,1,3-propanetricarboxylate

Michael

Additions

(*c*) The reaction of diethyl malonate with 2-cyclopentenone is an example of a Michael addition to an α,β-unsaturated ketone.

$$\text{2-Cyclopentenone} + CH_2(COOCH_2CH_3)_2 \xrightarrow[\text{ethanol}]{NaOCH_2CH_3} \text{Diethyl (3-oxocyclopentyl)malonate}$$

2-Cyclopentenone	Diethyl malonate	Diethyl (3-oxocyclopentyl)malonate

21.14 (*b*) Analyze 3-methyl-2-butenoic acid retrosynthetically to reveal the source of its structural units.

$$CH_3C{=}CHCOH \Longrightarrow CH_3C{-}CH_2COH \Longrightarrow CH_3C{=}O + {:}\bar{C}H_2C{=}O$$

3-Methyl-2-butenoic acid · β-Hydroxy acid

The necessary β-hydroxy acid can be prepared (as its ethyl ester) by a Reformatskii reaction:

$$CH_3CCH_3 + BrCH_2COCH_2CH_3 \xrightarrow[\text{2. } H_3O^+]{\text{1. Zn}} CH_3C{-}CH_2COCH_2CH_3$$

Acetone · Ethyl bromoacetate · Ethyl 3-hydroxy-3-methylbutanoate

Dehydration of the β-hydroxy ester followed by ester hydrolysis gives the desired product.

$$(CH_3)_2CCH_2COCH_2CH_3 \xrightarrow[\text{heat}]{H_2SO_4} (CH_3)_2C{=}CHCOCH_2CH_3 \xrightarrow[\text{2. } H^+]{\text{1. } HO^-, H_2O}$$

Ethyl 3-hydroxy-3-methylbutanoate · Ethyl 3-methyl-2-butenoate

$$(CH_3)_2C{=}CHCOH$$

3-Methyl-2-butenoic acid

(*c*) Lithium aluminum hydride reduction of the β-hydroxy ester prepared in part (*b*) provides the desired diol.

$$(CH_3)_2CCH_2COCH_2CH_3 \xrightarrow[\text{2. } H_2O]{\text{1. } LiAlH_4} (CH_3)_2CCH_2CH_2OH$$

Ethyl 3-hydroxy-3-methylbutanoate (prepared by Reformatskii reaction between acetone and ethyl bromoacetate) · 3-Methyl-1,3-butanediol

(d) Disconnection of the product shows that the carbon skeleton of 2,3-dimethyl-1,3-butanediol may be derived from a Reformatskii reaction between acetone and ethyl 2-bromopropanoate.

$$\underset{\underset{CH_3}{|}}{\overset{\overset{OH\ \ CH_3}{|\ \ \ \ |}}{CH_3C-}}\!\!\!-\!\!CHCH_2OH \implies \underset{}{\overset{\overset{O}{||}}{CH_3CCH_3}} + \underset{\underset{CH_3}{|}}{\overset{\overset{O}{||}}{BrCHCOCH_2CH_3}}$$

The β-hydroxy ester resulting from the Reformatskii reaction gives the desired diol upon reduction with lithium aluminum hydride.

$$\overset{\overset{O}{||}}{CH_3CCH_3} + \underset{\underset{Br}{|}}{\overset{\overset{O}{||}}{CH_3CHCOCH_2CH_3}} \xrightarrow[\text{2. } H_3O^+]{\text{1. Zn}} \underset{\underset{CH_3\ \ CH_3}{|\ \ \ \ \ |}}{\overset{\overset{OH\ \ \ \ \ \ O}{|\ \ \ \ \ \ \ \ ||}}{CH_3C-\!\!\!-\!\!CHCOCH_2CH_3}} \xrightarrow[\text{2. } H_2O]{\text{1. LiAlH}_4}$$

Acetone	Ethyl 2-bromopropanoate	Ethyl 2-hydroxy-2,3-dimethylbutanoate

$$\underset{\underset{CH_3\ \ CH_3}{|\ \ \ \ \ |}}{\overset{\overset{OH}{|}}{CH_3C-\!\!\!-\!\!CHCH_2OH}}$$

2,3-Dimethyl-1,3-butanediol

21.15 (b) The α carbon atom of the ester bears a phenyl substituent and a methyl group. Only the methyl group can be attached to the α carbon by a nucleophilic substitution reaction. Therefore generate the enolate of methyl phenylacetate with lithium diisopropylamide (LDA) in tetrahydrofuran (THF) and then alkylate with methyl iodide.

$$C_6H_5CH_2CO_2CH_3 \xrightarrow[\text{THF}]{\text{LDA}} \underset{\underset{OCH_3}{\diagdown}}{\overset{\overset{OLi}{\diagup}}{C_6H_5CH=C}} \xrightarrow{\text{CH}_3\text{I}} \underset{\underset{CH_3}{|}}{C_6H_5CHCO_2CH_3}$$

Methyl phenylacetate	Enolate of methyl phenylacetate	Methyl 2-phenylpropanoate

(c) The desired product corresponds to an aldol addition product.

$$\underset{}{\overset{\overset{O\ \ \ \ \ \ \overset{OH}{|}}{||\ \ \ \ \ CHC_6H_5}}{}} \implies \underset{}{\overset{O}{}} + \overset{\overset{O}{||}}{HCC_6H_5}$$

Therefore convert cyclohexanone to its enolate and then treat with benzaldehyde.

$$\underset{\text{Cyclohexanone}}{\overset{O}{}} \xrightarrow[\substack{\text{2. } C_6H_5CHO \\ \text{3. } H_3O^+}]{\text{1. LDA, THF}} \underset{\text{1-(2-Oxocyclohexyl)-1-phenylmethanol}}{\overset{\overset{O\ \ \ \ \ \ \ \overset{OH}{|}}{\ \ \ \ \ \ \ \ CHC_6H_5}}{}}$$

(d) This product corresponds to the addition of the enolate of *tert*-butyl acetate to cyclohexanone.

Generate the enolate of *tert*-butyl acetate with lithium diisopropylamide; then add cyclohexanone.

$$CH_3CO_2C(CH_3)_3 \xrightarrow[\substack{\text{2. cyclohexanone} \\ \text{3. } H_3O^+}]{\text{1. LDA, THF}}$$

tert-Butyl
acetate

tert-Butyl
(1-hydroxycyclohexyl)acetate

21.16 In order to undergo a Claisen condensation, an ester must have at least two protons on the α carbon:

$$2RCH_2\overset{O}{\overset{\|}{C}}OCH_2CH_3 + NaOCH_2CH_3 \longrightarrow$$

$$\left[RCH_2\overset{O}{\overset{\|}{C}}\overset{O}{\overset{\|}{\overset{\cdot\cdot}{C}}}\underset{R}{\underset{|}{C}}OCH_2CH_3 \right] Na^+ + 2CH_3CH_2OH$$

The equilibrium constant for condensation is unfavorable unless the β-keto ester can be deprotonated to form a stable anion.

(a) Among the esters given, ethyl pentanoate and ethyl 3-methylbutanoate undergo the Claisen condensation

$$CH_3CH_2CH_2CH_2\overset{O}{\overset{\|}{C}}OCH_2CH_3 \xrightarrow[\text{2. } H^+]{\text{1. NaOCH}_2CH_3} CH_3CH_2CH_2CH_2\overset{O}{\overset{\|}{C}}\overset{O}{\underset{\underset{CH_2CH_2CH_3}{|}}{\overset{\|}{C}}H}OCH_2CH_3$$

Ethyl pentanoate

Ethyl 3-oxo-2-propylheptanoate

$$(CH_3)_2CHCH_2\overset{O}{\overset{\|}{C}}OCH_2CH_3 \xrightarrow[\text{2. } H^+]{\text{1. NaOCH}_2CH_3} (CH_3)_2CHCH_2\overset{O}{\overset{\|}{C}}\overset{O}{\underset{\underset{CH(CH_3)_2}{|}}{\overset{\|}{C}}H}OCH_2CH_3$$

Ethyl 3-methylbutanoate

Ethyl 2-isopropyl-5-methyl-
3-oxohexanoate

(b) The Claisen condensation product of ethyl 2-methylbutanoate cannot be deprotonated; the equilibrium constant for its formation is less than 1.

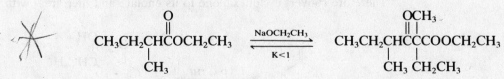

Ethyl 2-methylbutanoate

No protons on α carbon atom;
cannot form stabilized enolate
by deprotonation

(c) Ethyl 2,2-dimethylpropanoate has no protons on its α carbon; it cannot form the ester enolate required in the first step of the Claisen condensation.

$$\underset{\underset{CH_3}{|}}{\overset{\overset{CH_3}{|}}{CH_3CCOOCH_2CH_3}} + \ ^-OCH_2CH_3 \longrightarrow \text{no reaction}$$

Ethyl 2,2-dimethylpropanoate

21.17 (a) The Claisen condensation of ethyl phenylacetate is given by the equation

$$\underset{}{\overset{\overset{O}{\|}}{C_6H_5CH_2COCH_2CH_3}} \xrightarrow[\text{2. } H^+]{\text{1. } NaOCH_2CH_3} \underset{\underset{C_6H_5}{|}}{\overset{\overset{O\quad O}{\|\ \ \|}}{C_6H_5CH_2CCHCOCH_2CH_3}}$$

Ethyl phenylacetate   Ethyl 3-oxo-2,4-diphenylbutanoate

(b) Saponification and decarboxylation of this β-keto ester gives dibenzyl ketone.

$$\underset{\underset{C_6H_5}{|}}{\overset{\overset{O\quad O}{\|\ \ \|}}{C_6H_5CH_2CCHCOCH_2CH_3}} \xrightarrow[\substack{\text{2. } H^+ \\ \text{3. heat}}]{\text{1. } OH^-,\ H_2O} \underset{}{\overset{\overset{O}{\|}}{C_6H_5CH_2CCH_2C_6H_5}}$$

Ethyl 3-oxo-2,4-diphenylbutanoate   Dibenzyl ketone

(c) This process illustrates the alkylation of a β-keto ester with subsequent saponification and decarboxylation.

$$\underset{\underset{C_6H_5}{|}}{\overset{\overset{O\quad O}{\|\ \ \|}}{C_6H_5CH_2CCHCOCH_2CH_3}} \xrightarrow[CH_2=CHCH_2Br]{NaOCH_2CH_3} \underset{\underset{C_6H_5}{|}}{\overset{\overset{OCH_2CH=CH_2}{\|\ |}}{C_6H_5CH_2CCCOOCH_2CH_3}}$$

Ethyl 3-oxo-2,4-diphenylbutanoate

$$\Big\downarrow \substack{\text{1 } HO^-,\ H_2O \\ \text{2. } H^+ \\ \text{3. heat}}$$

$$\underset{\underset{C_6H_5}{|}}{\overset{\overset{O}{\|}}{C_6H_5CH_2CCHCH_2CH=CH_2}}$$

1,3-Diphenyl-5-hexen-2-one

(d) The enolate ion of ethyl phenylacetate attacks the carbonyl carbon of ethyl benzoate.

$$\overset{\overset{O}{\|}}{C_6H_5C}-OCH_2CH_3$$

$$\overset{\underset{O}{\|}}{C_6H_5\ddot{C}HCOCH_2CH_3} \longrightarrow \underset{\underset{C_6H_5}{|}}{\overset{\overset{O\quad O}{\|\ \ \|}}{C_6H_5CCHCOCH_2CH_3}}$$

Ethyl 2,3-diphenyl-3-oxopropanoate

(*e*) Saponification and decarboxylation yield benzyl phenyl ketone.

$$\underset{\substack{\text{Ethyl} \\ \text{3-Oxo-2,3-diphenylpropanoate}}}{C_6H_5\overset{O}{\overset{||}{C}}\underset{\underset{C_6H_5}{|}}{CH}\overset{O}{\overset{||}{C}}OCH_2CH_3} \xrightarrow[\substack{2.\ H^+ \\ 3.\ \text{heat}}]{1.\ HO^-,\ H_2O} \underset{\substack{\text{Benzyl phenyl ketone}}}{C_6H_5\overset{O}{\overset{||}{C}}CH_2C_6H_5}$$

(*f*) This sequence is analogous to that of part (*c*).

$$\underset{\substack{C_6H_5}}{C_6H_5\overset{O}{\overset{||}{C}}\underset{\underset{C_6H_5}{|}}{CH}\overset{O}{\overset{||}{C}}OCH_2CH_3} \xrightarrow[CH_2=CHCH_2Br]{NaOCH_2CH_3} C_6H_5\overset{\overset{OCH_2CH=CH_2}{|}}{\underset{\underset{C_6H_5}{|}}{C}}COOCH_2CH_3$$

$$\Big\downarrow \begin{array}{l} 1.\ HO^-,\ H_2O \\ 2.\ H^+ \\ 3.\ \text{heat} \end{array}$$

$$\underset{\substack{\text{1,2-Diphenyl-4-penten-1-one}}}{C_6H_5\overset{O}{\overset{||}{C}}\underset{\underset{C_6H_5}{|}}{CH}CH_2CH=CH_2}$$

21.18 (*a*) The Dieckmann reaction is the intramolecular version of the Claisen condensation. It employs a diester as starting material.

$$\underset{\substack{\text{Diethyl heptanedioate}}}{CH_3CH_2O\overset{O}{\overset{||}{C}}(CH_2)_5\overset{O}{\overset{||}{C}}OCH_2CH_3} \xrightarrow[2.\ H^+]{1.\ NaOCH_2CH_3} \underset{\substack{\text{Ethyl 2-oxo-1-} \\ \text{cyclohexanecarboxylate}}}{\text{[cyclohexanone ring with COCH}_2\text{CH}_3\text{]}}$$

(*b*) Acylation of cyclohexanone with diethyl carbonate yields the same β-keto ester formed in part (*a*).

$$\underset{\substack{\text{Cyclohexanone}}}{\text{[cyclohexanone]}} + \underset{\substack{\text{Diethyl carbonate}}}{CH_3CH_2O\overset{O}{\overset{||}{C}}OCH_2CH_3} \xrightarrow[2.\ H^+]{1.\ NaOCH_2CH_3} \underset{\substack{\text{Ethyl} \\ \text{2-oxo-1-cyclohexanecarboxylate}}}{\text{[cyclohexanone ring with COCH}_2\text{CH}_3\text{]}}$$

(*c*) The two most stable enol forms are those that involve the proton on the carbon flanked by the two carbonyl groups.

(d) Deprotonation of the β-keto ester involves the acidic proton at the carbon flanked by the two carbonyl groups 3   *most stable*

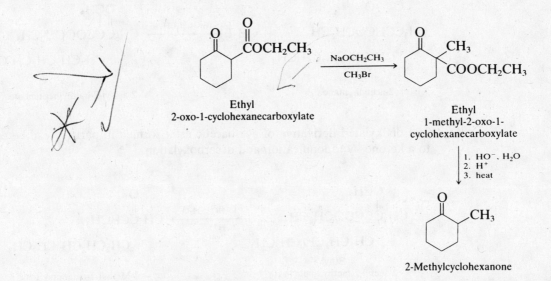

(e) The methyl group is introduced by alkylation of the β-keto ester. Saponification and decarboxylation complete the synthesis.

Ethyl
2-oxo-1-cyclohexanecarboxylate

$\xrightarrow[\text{CH}_3\text{Br}]{\text{NaOCH}_2\text{CH}_3}$

Ethyl
1-methyl-2-oxo-1-
cyclohexanecarboxylate

1. HO⁻, H₂O
2. H⁺
3. heat

2-Methylcyclohexanone

(f) The enolate ion of the β-keto ester [see part (d)] undergoes Michael addition to the carbon-carbon double bond of acrolein.

Ethyl
2-oxo-1-cyclohexanecarboxylate

+ CH_2=CHCH

Acrolein

$\xrightarrow[\text{CH}_3\text{CH}_2\text{OH}]{\text{NaOCH}_2\text{CH}_3}$

Michael adduct

This reaction has been reported in the chemical literature and proceeds in 65 to 75 percent yield.

21.19 (a) Ethyl acetoacetate is converted to its enolate ion with sodium ethoxide; this anion then acts as a nucleophile toward 1-bromopentane.

$$CH_3CCH_2COCH_2CH_3 + CH_3CH_2CH_2CH_2CH_2Br \xrightarrow{\text{NaOCH}_2\text{CH}_3} CH_3CCHCOCH_2CH_3$$
$$\qquad\qquad\qquad\qquad\qquad\qquad\qquad\qquad\qquad\qquad\qquad\qquad\qquad\qquad CH_2CH_2CH_2CH_2CH_3$$

Ethyl acetoacetate   1-Bromopentane   Ethyl 2-acetylheptanoate (*b*) Saponification and decarboxylation of the product in part (*a*) yields 2-octanone.

$$\underset{\substack{\text{Ethyl}\\ \text{2-acetylheptanoate}}}{\underset{\underset{(CH_2)_4CH_3}{|}}{CH_3\overset{O}{\overset{||}{C}}CH\overset{O}{\overset{||}{C}}OCH_2CH_3}} \xrightarrow[\substack{2.\ H^+ \\ 3.\ heat}]{1.\ HO^-,\ H_2O} \underset{\text{2-Octanone}}{CH_3\overset{O}{\overset{||}{C}}CH_2CH_2CH_2CH_2CH_2CH_3}$$

(*c*) The product derived from the reaction in part (*a*) can be alkylated again:

$$\underset{\substack{\text{Ethyl}\\ \text{2-acetylheptanoate}}}{\underset{\underset{CH_2CH_2CH_2CH_2CH_3}{|}}{CH_3\overset{O}{\overset{||}{C}}CH\overset{O}{\overset{||}{C}}OCH_2CH_3}} + CH_3I \xrightarrow{NaOCH_2CH_3} \underset{\substack{\text{Ethyl}\\ \text{2-acetyl-2-methylheptanoate}}}{\underset{\underset{CH_2CH_2CH_2CH_2CH_3}{|}}{CH_3\overset{OCH_3}{\overset{||}{C}}CCOOCH_2CH_3}}$$

(*d*) The dialkylated derivative of acetoacetic ester formed in part (*c*) can be converted to a ketone by saponification and decarboxylation.

$$\underset{\substack{\text{Ethyl}\\ \text{2-acetyl-2-methylheptanoate}}}{\underset{\underset{CH_2CH_2CH_2CH_2CH_3}{|}}{CH_3\overset{OCH_3}{\overset{||}{C}}CCOOCH_2CH_3}} \xrightarrow[\substack{2.\ H^+ \\ 3.\ heat}]{1.\ HO^-,\ H_2O} \underset{\substack{\text{3-Methyl-2-octanone}}}{\underset{\underset{CH_2CH_2CH_2CH_2CH_3}{|}}{CH_3\overset{O}{\overset{||}{C}}CHCH_3}}$$

(*e*) The anion of ethyl acetoacetate acts as a nucleophile toward 1-bromo-3-chloropropane. Bromide is a better leaving group than chloride and is displaced preferentially.

$$\underset{\text{Ethyl acetoacetate}}{CH_3\overset{O}{\overset{||}{C}}CH_2\overset{O}{\overset{||}{C}}OCH_2CH_3} + \underset{\substack{\text{1-Bromo-3-}\\\text{chloropropane}}}{BrCH_2CH_2CH_2Cl} \xrightarrow{NaOCH_2CH_3} \underset{\substack{\text{Ethyl}\\ \text{2-acetyl-5-chloropentanoate}}}{\underset{\underset{CH_2CH_2CH_2Cl}{|}}{CH_3\overset{O}{\overset{||}{C}}CH\overset{O}{\overset{||}{C}}OCH_2CH_3}}$$

(*f*) Treatment of the product of part (*e*) with sodium ethoxide gives an enolate ion that cyclizes by intramolecular nucleophilic substitution of chloride.

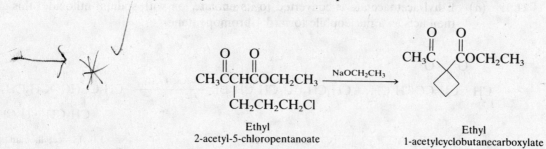

$$\underset{\substack{\text{Ethyl}\\ \text{2-acetyl-5-chloropentanoate}}}{\underset{\underset{CH_2CH_2CH_2Cl}{|}}{CH_3\overset{O}{\overset{||}{C}}CH\overset{O}{\overset{||}{C}}OCH_2CH_3}} \xrightarrow{NaOCH_2CH_3} \underset{\substack{\text{Ethyl}\\ \text{1-acetylcyclobutanecarboxylate}}}{CH_3\overset{O}{\overset{||}{C}}\quad\overset{O}{\overset{||}{C}}OCH_2CH_3}$$

(g) Cyclobutyl methyl ketone is formed by saponification and decarboxylation of the product in part (f).

$$\underset{\substack{\text{Ethyl}\\\text{1-acetylcyclobutanecarboxylate}}}{CH_3\overset{O}{\overset{\|}{C}}\;\overset{O}{\overset{\|}{C}}OCH_2CH_3} \xrightarrow[\substack{3.\ \text{heat}}]{\substack{1.\ HO^-,\ H_2O\\2.\ H^+}} \underset{\text{Cyclobutyl methyl ketone}}{CH_3\overset{O}{\overset{\|}{C}}}$$

(h) Ethyl acetoacetate undergoes Michael addition to phenyl vinyl ketone in the presence of base.

$$\underset{\substack{\text{Ethyl}\\\text{acetoacetate}}}{CH_3\overset{O}{\overset{\|}{C}}CH_2\overset{O}{\overset{\|}{C}}OCH_2CH_3} + \underset{\text{Phenyl vinyl ketone}}{CH_2{=}CH\overset{O}{\overset{\|}{C}}C_6H_5} \xrightarrow[\text{ethanol}]{NaOCH_2CH_3} \underset{\substack{\text{Ethyl}\\\text{2-acetyl-5-oxo-5-}\\\text{phenylpentanoate}}}{CH_3\overset{O}{\overset{\|}{C}}\underset{\underset{\underset{O}{\overset{\|}{C}}C_6H_5}{\overset{|}{CH_2CH_2}}}{CH}\overset{O}{\overset{\|}{C}}OCH_2CH_3}$$

(i) A diketone results from saponification and decarboxylation of the Michael adduct.

$$\underset{\substack{\text{Ethyl 2-acetyl-5-oxo-5-}\\\text{phenylpentanoate}}}{CH_3\overset{O}{\overset{\|}{C}}\underset{\underset{\underset{O}{\overset{\|}{C}}C_6H_5}{\overset{|}{CH_2CH_2}}}{CH}\overset{O}{\overset{\|}{C}}OCH_2CH_3} \xrightarrow[\substack{2.\ H^+\\3.\ \text{heat}}]{1.\ HO^-,\ H_2O} \underset{\text{1-Phenyl-1,5-hexanedione}}{CH_3\overset{O}{\overset{\|}{C}}CH_2CH_2CH_2\overset{O}{\overset{\|}{C}}C_6H_5}$$

(j) Ethyl acetoacetate reacts with ketones in the presence of piperidine to give alkenes by way of the Knoevenagel condensation.

$$\underset{\text{Ethyl acetoacetate}}{CH_3\overset{O}{\overset{\|}{C}}CH_2\overset{O}{\overset{\|}{C}}OCH_2CH_3} + \underset{\text{3-Pentanone}}{CH_3CH_2\overset{O}{\overset{\|}{C}}CH_2CH_3} \xrightarrow{\text{piperidine}} \underset{\text{Ethyl 2-acetyl-3-ethyl-2-pentenoate}}{(CH_3CH_2)_2C{=}C\underset{\underset{O}{\overset{\|}{C}}OCH_2CH_3}{\overset{\overset{O}{\overset{\|}{C}}CH_3}{}}}$$

21.20 Diethyl malonate reacts with the reagents given in the preceding problem in a manner analogous to that of ethyl acetoacetate.

(a) $\underset{\text{Diethyl malonate}}{CH_2(COOCH_2CH_3)_2} + \underset{\text{1-Bromopentane}}{CH_3CH_2CH_2CH_2CH_2Br} \xrightarrow{NaOCH_2CH_3}$

$$\underset{\substack{\text{Diethyl hexane-1,1-dicarboxylate}\\\text{(diethyl pentylmalonate)}}}{CH_3CH_2CH_2CH_2CH_2CH(COOCH_2CH_3)_2}$$

(b) $CH_3CH_2CH_2CH_2CH_2CH(COOCH_2CH_3)_2$ $\xrightarrow[\begin{array}{l}\text{2. } H^+\\\text{3. heat}\end{array}]{\text{1. } HO^-, H_2O}$

Diethyl 1,1-hexanedicarboxylate

$$CH_3CH_2CH_2CH_2CH_2CH_2\overset{\displaystyle O}{\overset{\|}{C}}OH$$

Heptanoic acid

(c) $CH_3CH_2CH_2CH_2CH_2CH(COOCH_2CH_3)_2$ $\xrightarrow[NaOCH_2CH_3]{CH_3I}$

Diethyl 1,1-hexanedicarboxylate

$$CH_3CH_2CH_2CH_2CH_2\underset{\underset{\displaystyle CH_3}{|}}{C}(COOCH_2CH_3)_2$$

Diethyl 2,2-heptanedicarboxylate

(d) $CH_3CH_2CH_2CH_2CH_2\underset{\underset{\displaystyle CH_3}{|}}{C}(COOCH_2CH_3)_2$ $\xrightarrow[\begin{array}{l}\text{2. } H^+\\\text{3. heat}\end{array}]{\text{1. } HO^-, H_2O}$ $CH_3CH_2CH_2CH_2CH_2\underset{\underset{\displaystyle CH_3}{|}}{C}H\overset{\displaystyle O}{\overset{\|}{C}}OH$

Diethyl 2,2-heptanedicarboxylate   2-Methylheptanoic acid

(e) $CH_2(COOCH_2CH_3)_2$ + $BrCH_2CH_2CH_2Cl$ $\xrightarrow{NaOCH_2CH_3}$

Diethyl malonate   1-Bromo-3-chloropropane

$$ClCH_2CH_2CH_2CH(COOCH_2CH_3)_2$$

Diethyl 4-chloro-1,1-butanedicarboxylate

(f) $ClCH_2CH_2CH_2CH(COOCH_2CH_3)_2$ $\xrightarrow{NaOCH_2CH_3}$

Diethyl 4-chloro-1,1-butanedicarboxylate   Diethyl
cyclobutane-1,1-dicarboxylate

(g) $\xrightarrow[\begin{array}{l}\text{2. } H^+\\\text{3. heat}\end{array}]{\text{1. } HO^-, H_2O}$

Diethyl   Cyclobutanecarboxylic
cyclobutane-1,1-dicarboxylate   acid

(h) $CH_2(COOCH_2CH_3)_2$ + $C_6H_5\overset{\displaystyle O}{\overset{\|}{C}}CH{=}CH_2$ $\xrightarrow[CH_3CH_2OH]{NaOCH_2CH_3}$

Diethyl malonate   Phenyl vinyl ketone

$$C_6H_5\overset{\displaystyle O}{\overset{\|}{C}}CH_2CH_2CH(COOCH_2CH_3)_2$$

Diethyl 4-oxo-4-phenylbutanedicarboxylate

(i)　$C_6H_5CCH_2CH_2CH(COOCH_2CH_3)_2$ $\xrightarrow[\substack{2.\ H^+ \\ 3.\ heat}]{1.\ HO^-,\ H_2O}$ $C_6H_5CCH_2CH_2CH_2COH$

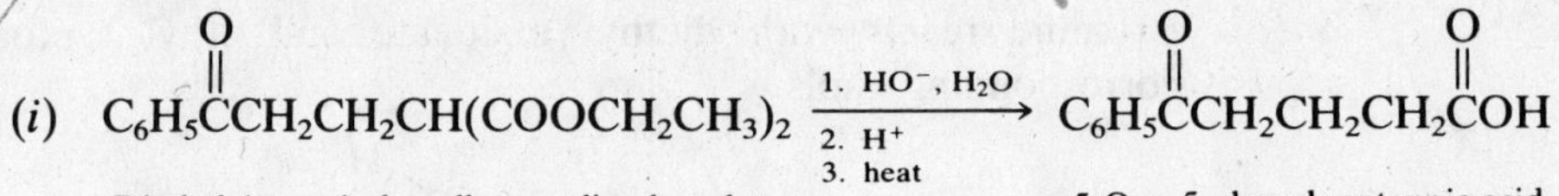

　　　　Diethyl 4-oxo-4-phenylbutanedicarboxylate　　　　　5-Oxo-5-phenylpentanoic acid

(j)　$CH_2(COOCH_2CH_3)_2 + CH_3CH_2CCH_2CH_3$ $\xrightarrow{\text{piperidine}}$

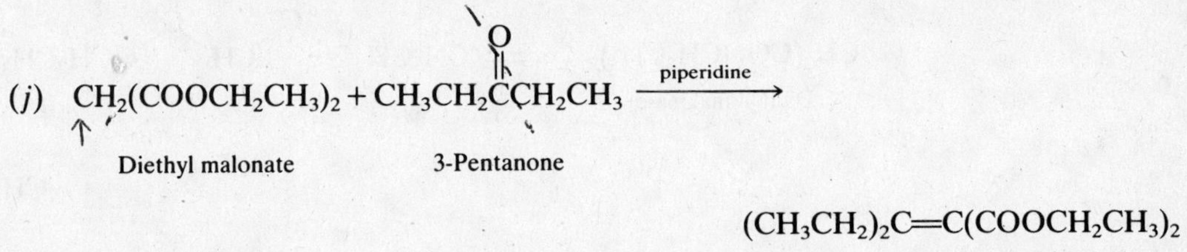

　　　　Diethyl malonate　　　　　3-Pentanone

　　　　　　　　　　　　　　　　$(CH_3CH_2)_2C{=}C(COOCH_2CH_3)_2$

　　　　　　　　　　　　　　Diethyl 2-ethyl-1-butene-1,1-dicarboxylate

21.21　(*a*)　Both carbonyl groups of diethyl malonate are equivalent, and so enolization can occur in either direction.

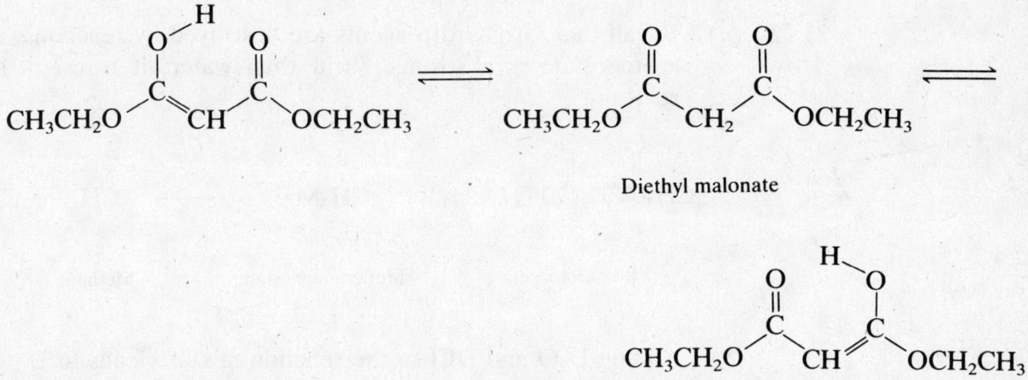

Diethyl malonate

(*b*)　Ethyl acetoacetate can give three constitutionally isomeric enols:

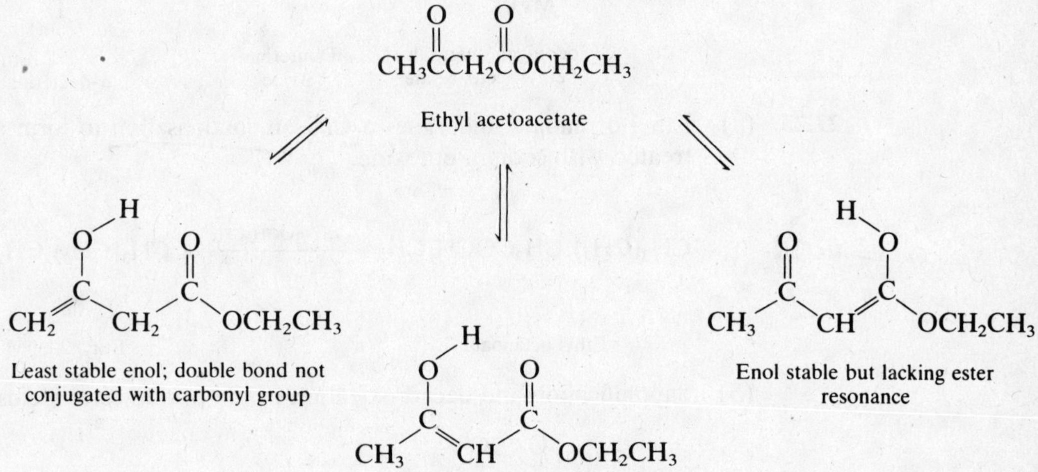

$CH_3CCH_2COCH_2CH_3$

Ethyl acetoacetate

Least stable enol; double bond not conjugated with carbonyl group

Enol stable but lacking ester resonance

Most stable enol; double bond conjugated with carbonyl group; ester carbonyl stabilized by resonance

(c) Bromine reacts with diethyl malonate and ethyl acetoacetate by way of the corresponding enols:

$$CH_2(COOCH_2CH_3)_2 \rightleftharpoons CH_3CH_2O-\overset{OH}{\underset{}{C}}=CH-\overset{O}{\underset{}{C}}-OCH_2CH_3 \xrightarrow{Br_2}$$

Diethyl malonate

$$CH_3CH_2O\overset{O}{\underset{}{C}}\overset{}{\underset{Br}{C}}H\overset{O}{\underset{}{C}}OCH_2CH_3$$

Diethyl bromomalonate

$$CH_3\overset{O}{\underset{}{C}}CH_2\overset{O}{\underset{}{C}}OCH_2CH_3 \rightleftharpoons CH_3\overset{OH}{\underset{}{C}}=CH\overset{O}{\underset{}{C}}OCH_2CH_3 \xrightarrow{Br_2} CH_3\overset{O}{\underset{}{C}}\overset{}{\underset{Br}{C}}H\overset{O}{\underset{}{C}}OCH_2CH_3$$

Ethyl acetoacetate   Ethyl α-bromoacetoacetate

21.22 (a) Recall that Grignard reagents are destroyed by reaction with proton donors. Ethyl acetoacetate is a stronger acid than water; it transfers a proton to a Grignard reagent.

$$CH_3\overset{O}{\underset{}{C}}CH_2\overset{O}{\underset{}{C}}OCH_2CH_3 + CH_3MgI \longrightarrow CH_4 + CH_3\overset{O}{\underset{}{C}}\overset{}{\underset{MgI}{C}}H\overset{O}{\underset{}{C}}OCH_2CH_3$$

Ethyl acetoacetate   Methylmagnesium   Methane   Iodomagnesium salt of
                     iodide   ethyl acetoacetate

(b) Adding D_2O and DCl to the reaction mixture leads to D^+ transfer to the α carbon atom of ethyl acetoacetate.

$$CH_3\overset{O}{\underset{}{C}}\overset{}{\underset{MgI}{C}}H\overset{O}{\underset{}{C}}OCH_2CH_3 + D_2O \xrightarrow{DCl} CH_3\overset{O}{\underset{}{C}}\overset{}{\underset{D}{C}}H\overset{O}{\underset{}{C}}OCH_2CH_3$$

Iodomagnesium salt   Deuterium   Ethyl
of ethyl acetoacetate   oxide   α-deuterioacetoacetate

21.23 (a) Ethyl octanoate undergoes a Claisen condensation to form a β-keto ester on being treated with sodium ethoxide.

$$CH_3(CH_2)_5CH_2\overset{O}{\underset{}{C}}OCH_2CH_3 \xrightarrow[\text{2. } H^+]{\text{1. } NaOCH_2CH_3} CH_3(CH_2)_5CH_2\overset{O}{\underset{}{C}}\overset{O}{\underset{(CH_2)_5CH_3}{C}}H\overset{O}{\underset{}{C}}OCH_2CH_3$$

Ethyl octanoate   Ethyl 2-hexyl-3-oxodecanoate

(b) Saponification and decarboxylation of the β-keto ester yields a ketone.

$$CH_3(CH_2)_5CH_2\overset{O}{\underset{}{C}}\overset{O}{\underset{(CH_2)_5CH_3}{C}}H\overset{O}{\underset{}{C}}OCH_2CH_3 \xrightarrow[\substack{\text{2. } H^+ \\ \text{3. heat}}]{\text{1. } NaOH, H_2O} CH_3(CH_2)_5CH_2\overset{O}{\underset{}{C}}CH_2(CH_2)_5CH_3$$

Ethyl 2-hexyl-3-oxodecanoate   8-Pentadecanone

(c). On treatment with base, ethyl acetoacetate is converted to its enolate, which reacts as a nucleophile toward 1-bromobutane.

$$\underset{\text{Ethyl acetoacetate}}{CH_3\overset{O}{\overset{||}{C}}CH_2\overset{O}{\overset{||}{C}}OCH_2CH_3} + \underset{\text{1-Bromobutane}}{CH_3CH_2CH_2CH_2Br} \xrightarrow[\text{ethanol}]{NaOCH_2CH_3} \underset{\text{Ethyl 2-acetylhexanoate}}{CH_3\overset{O}{\overset{||}{C}}\overset{}{\underset{\underset{CH_2CH_2CH_2CH_3}{|}}{CH}}\overset{O}{\overset{||}{C}}OCH_2CH_3}$$

(d) Alkylation of ethyl acetoacetate, followed by saponification and decarboxylation, gives a ketone. The two steps constitute the acetoacetic ester synthesis.

$$\underset{\text{Ethyl 2-acetylhexanoate}}{CH_3\overset{O}{\overset{||}{C}}\overset{}{\underset{\underset{CH_2CH_2CH_2CH_3}{|}}{CH}}\overset{O}{\overset{||}{C}}OCH_2CH_3} \xrightarrow[\substack{2.\ H^+ \\ 3.\ heat}]{1.\ NaOH,\ H_2O} \underset{\text{2-Heptanone}}{CH_3\overset{O}{\overset{||}{C}}CH_2CH_2CH_2CH_2CH_3}$$

(e) An alkylated derivative of ethyl acetoacetate is capable of being alkylated a second time.

$$\underset{\text{Ethyl 2-acetylhexanoate}}{CH_3\overset{O}{\overset{||}{C}}\overset{}{\underset{\underset{CH_2CH_2CH_2CH_3}{|}}{CH}}\overset{O}{\overset{||}{C}}OCH_2CH_3} + \underset{\text{1-Iodobutane}}{CH_3CH_2CH_2CH_2I} \xrightarrow[\text{ethanol}]{NaOCH_2CH_3} \underset{\text{Ethyl 2-acetyl-2-butylhexanoate}}{CH_3\overset{O}{\overset{||}{C}}\overset{}{\underset{\underset{COOCH_2CH_3}{|}}{C}}(CH_2CH_2CH_2CH_3)_2}$$

(f) The dialkylated derivative of acetoacetic ester formed in part (e) is converted to a ketone by saponification and decarboxylation.

$$\underset{\text{Ethyl 2-acetyl-2-butylhexanoate}}{CH_3\overset{O}{\overset{||}{C}}\overset{}{\underset{\underset{COOCH_2CH_3}{|}}{C}}(CH_2CH_2CH_2CH_3)_2} \xrightarrow[\substack{2.\ H^+ \\ 3.\ heat}]{1.\ NaOH} \underset{\text{3-Butyl-2-heptanone}}{CH_3\overset{O}{\overset{||}{C}}CH(CH_2CH_2CH_2CH_3)_2}$$

(g) The enolate of acetophenone attacks the carbonyl group of diethyl carbonate.

$$\underset{\text{Acetophenone}}{C_6H_5\overset{O}{\overset{||}{C}}CH_3} + \underset{\text{Diethyl carbonate}}{CH_3CH_2O\overset{O}{\overset{||}{C}}OCH_2CH_3} \xrightarrow[2.\ H^+]{1.\ NaOCH_2CH_3} \underset{\text{3-Oxo-3-phenylpropanoate}}{C_6H_5\overset{O}{\overset{||}{C}}CH_2\overset{O}{\overset{||}{C}}OCH_2CH_3}$$

(h) Diethyl oxalate acts as an acylating agent toward the enolate of acetone.

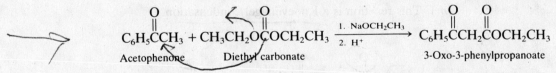

$$\underset{\text{Acetone}}{CH_3\overset{O}{\overset{||}{C}}CH_3} + \underset{\text{Diethyl oxalate}}{CH_3CH_2O\overset{O\ O}{\overset{||\ ||}{C\ C}}OCH_2CH_3} \xrightarrow[2.\ H^+]{1.\ NaOCH_2CH_3} \underset{\text{Ethyl 2,4-dioxopentanoate}}{CH_3\overset{O}{\overset{||}{C}}CH_2\overset{O\ O}{\overset{||\ ||}{C\ C}}OCH_2CH_3}$$

(*i*) The first stage of the malonic ester synthesis is the alkylation of diethyl malonate with an alkyl halide.

$$CH_2(COOCH_2CH_3)_2 + BrCH_2\underset{\underset{CH_3}{|}}{C}HCH_2CH_3 \xrightarrow[\text{ethanol}]{NaOCH_2CH_3}$$

Diethyl malonate   1-Bromo-2-methylbutane

$$CH_3CH_2\underset{\underset{CH_3}{|}}{C}HCH_2CH(COOCH_2CH_3)_2$$

Diethyl 3-methylpentane-1,1-dicarboxylate

(*j*) Alkylation of diethyl malonate is followed by saponification and decarboxylation to give a carboxylic acid.

$$CH_3CH_2\underset{\underset{CH_3}{|}}{C}HCH_2CH(COOCH_2CH_3)_2 \xrightarrow[\substack{\text{2. H}^+ \\ \text{3. heat}}]{\text{1. NaOH, H}_2\text{O}} CH_3CH_2\underset{\underset{CH_3}{|}}{C}HCH_2CH_2\overset{\overset{O}{\|}}{C}OH$$

Diethyl 3-methylpentane-1,1-dicarboxylate

4-Methylhexanoic acid
(57% yield from 1-bromo-2-methylbutane)

(*k*) The anion of diethyl malonate undergoes <u>Michael addition</u> to 6-methyl-2-cyclohexenone.

$$CH_2(COOCH_2CH_3)_2 + \quad\text{(6-methyl-2-cyclohexenone)} \xrightarrow[\text{ethanol}]{NaOCH_2CH_3} \quad\text{(product)}$$

Diethyl malonate   6-methyl-2-cyclohexenone   Diethyl 2-(4-methyl-3-oxocyclohexyl)malonate
(isolated yield, 50%)

(*l*) Acid hydrolysis converts the diester in part (*k*) to a malonic acid derivative, which then undergoes decarboxylation.

$$\text{(diester)} \xrightarrow[\text{heat}]{\text{H}_2\text{O, HCl}} \text{(product)}$$

Diethyl 2-(4-methyl-3-oxocyclohexyl)malonate   (4-Methyl-3-oxocyclohexyl)acetic acid
(isolated yield, 80%)

(*m*) This reaction is a <u>Knoevenagel condensation</u>.

2,3-Dimethoxybenzaldehyde   Ethyl acetoacetate   Ethyl
2-acetyl-3-(2,3-dimethoxyphenyl)propenoate
(isolated yield, 64–72%)

(*n*) Lithium diisopropylamide (LDA) is used to convert esters quantitatively to their enolate ions. In this reaction the enolate of *tert*-butyl acetate adds to benzaldehyde.

$$\underset{\substack{\text{\textit{tert}-Butyl acetate}}}{CH_3\overset{O}{\overset{\|}{C}}OC(CH_3)_3} \xrightarrow{LDA} \underset{\substack{\text{Lithium enolate of}\\ \text{\textit{tert}-butyl acetate}}}{CH_2=C\overset{OLi}{\underset{OC(CH_3)_3}{}}} \xrightarrow[\text{2. H}^+]{\text{1. C}_6\text{H}_5\overset{O}{\overset{\|}{C}}H} \underset{\substack{\text{\textit{tert}-Butyl 3-hydroxy-3-phenylpropanoate}}}{C_6H_5\underset{OH}{\overset{}{C}}HCH_2\overset{O}{\overset{\|}{C}}OC(CH_3)_3}$$

21.24 (*a*) Both ester functions in this molecule are β to a ketone carbonyl. Hydrolysis is followed by decarboxylation.

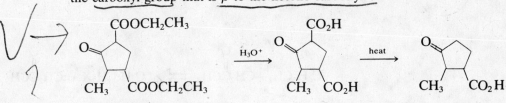

Diethyl 3-ethylcyclopentanone-2,5-dicarboxylate $\xrightarrow[\text{heat}]{\text{H}_2\text{O, H}_2\text{SO}_4}$ 3-Ethylcyclopentanone ($C_7H_{12}O$)

(*b*) Examine each carbon that is α to an ester function to see if it can lead to a five-, six-, or seven-membered cyclic β-keto ester by a Dieckmann cyclization.

Cyclization to a five-membered ring possible, but β-keto ester cannot be deprotonated to give a stable anion.

Cyclization not likely; resulting ring is four-membered and highly strained.

Cyclization gives a five-membered ring; β-keto ester deprotonated under reaction conditions; this is the observed product ($C_{12}H_{18}O_5$).

(*c*) Both ester functions undergo hydrolysis in acid, but decarboxylation occurs only at the carboxyl group that is β to the ketone carbonyl.

Diethyl 2-methylcyclopentanone-3,5-dicarboxylate $\xrightarrow{\text{H}_3\text{O}^+}$ $\xrightarrow{\text{heat}}$ 2-Methylcyclopentanone-3-carboxylic acid ($C_7H_{10}O_3$)

(d) A Dieckmann cyclization occurs, giving a five-membered ring fused to the original three-membered ring.

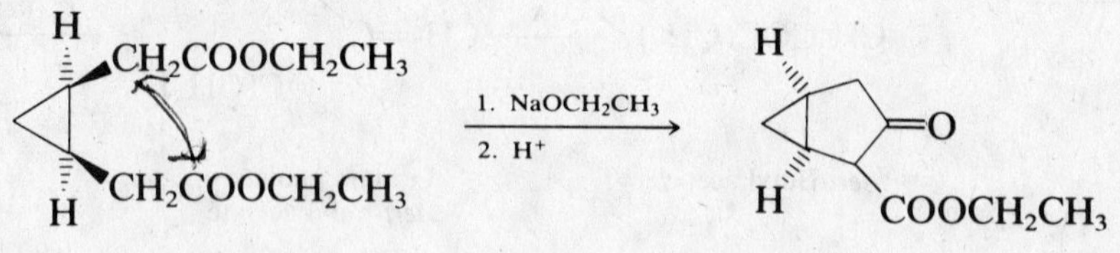

Diethyl *cis*-1,2-cyclopropanediacetate

1. NaOCH₂CH₃
2. H⁺

Ethyl bicyclo[3.1.0]hexan-3-one-2-carboxylate ($C_9H_{12}O_3$, 79%)

(e) Saponification and decarboxylation convert the β-keto ester to a ketone.

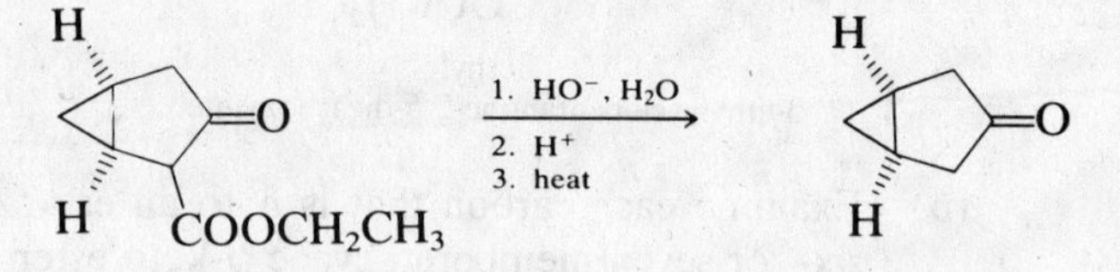

Ethyl bicyclo[3.1.0]hexan-3-one-2-carboxylate

1. HO⁻, H_2O
2. H⁺
3. heat

Bicyclo[3.1.0]hexan-3-one (C_6H_8O, 43%)

(f) The reactants are functionalized in such a way that two Knoevenagel reactions can take place.

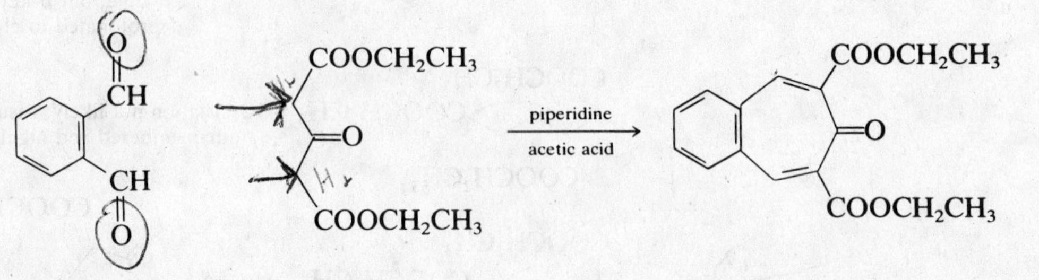

Benzene-1,2-dicarbaldehyde

Diethyl 3-oxopentanedioate

piperidine
acetic acid

Diethyl 4,5-benzocycloheptadienone-2,7-dicarboxylate ($C_{17}H_{16}O_5$, 63% yield)

The two Knoevenagel condensations occur sequentially; the first is intermolecular, the second intramolecular.

21.25 (a) First write out the structure of 4-phenyl-2-butanone and identify the synthon that is derived from ethyl acetoacetate.

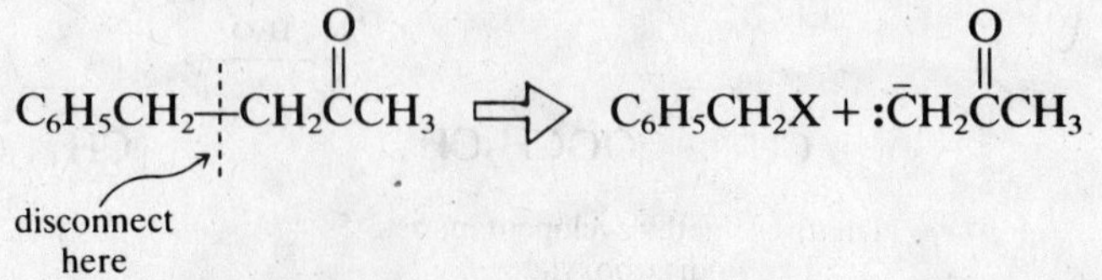

disconnect
here

Therefore carry out the acetoacetic ester synthesis using a benzyl halide as the alkylating agent.

$$C_6H_5CH_2OH \xrightarrow[\text{or PBr}_3]{\text{HBr}} C_6H_5CH_2Br$$

Benzyl alcohol   Benzyl bromide

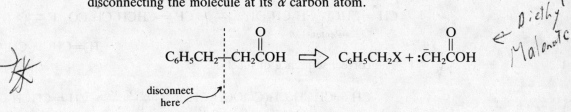

$$CH_3CCH_2COCH_2CH_3 + C_6H_5CH_2Br \xrightarrow[\text{ethanol}]{\text{NaOCH}_2\text{CH}_3} CH_3CCHCOCH_2CH_3$$
$$CH_2C_6H_5$$

Ethyl acetoacetate   Benzyl bromide   Ethyl 2-benzyl-3-oxobutanoate

1. HO⁻, H₂O
2. H⁺
3. heat

$$CH_3CCH_2CH_2C_6H_5$$

4-Phenyl-2-butanone

(*b*) Identify the synthon in 3-phenylpropanoic acid that is derived from malonic ester by disconnecting the molecule at its α carbon atom.

$$C_6H_5CH_2{-}CH_2COH \Rightarrow C_6H_5CH_2X + {:}\bar{C}H_2COH$$

← Diethyl Malonate

disconnect
here

Here, as in part (*a*), a benzyl halide is the required alkylating agent.

$$CH_2(COOCH_2CH_3)_2 + C_6H_5CH_2Br \xrightarrow[\text{ethanol}]{\text{NaOCH}_2\text{CH}_3} C_6H_5CH_2CH(COOCH_2CH_3)_2$$

Diethyl malonate   Benzyl bromide   Diethyl benzylmalonate

1. HO⁻, H₂O
2. H⁺
3. heat

$$C_6H_5CH_2CH_2COOH$$

3-Phenylpropanoic acid

(*c*) The target molecule can be prepared by a Knoevenagel condensation of benzaldehyde and diethyl malonate.

$$C_6H_5CH{=}C(CO_2CH_2CH_3)_2 \Rightarrow C_6H_5CH + CH_2(CO_2CH_2CH_3)_2$$

First benzaldehyde must be prepared by oxidation of benzyl alcohol.

$$C_6H_5CH_2OH \xrightarrow{(C_5H_5N)_2CrO_3, \ CH_2Cl_2} C_6H_5CH$$

Benzyl alcohol   Benzaldehyde

$$C_6H_5CH + CH_2(COOCH_2CH_3)_2 \xrightarrow[\text{acetic acid}]{\text{piperidine}} C_6H_5CH{=}C(COOCH_2CH_3)_2$$

Benzaldehyde   Diethyl malonate   Diethyl benzylidenemalonate

(d) In this synthesis the desired 1,3-diol function can be derived by reduction of a malonic ester derivative. First propene must be converted to an allyl halide for use as an alkylating agent.

$$CH_2\!\!=\!\!CHCH_2CH(CH_2OH)_2 \implies CH_2\!\!=\!\!CHCH_2X + {}^-\!:\!CH(COOCH_2CH_3)_2$$

$$CH_2\!\!=\!\!CHCH_3 \xrightarrow[heat]{Cl_2} CH_2\!\!=\!\!CHCH_2Cl$$

Propene   Allyl chloride

$$CH_2(COOCH_2CH_3)_2 + CH_2\!\!=\!\!CHCH_2Cl \xrightarrow[ethanol]{NaOCH_2CH_3} CH_2\!\!=\!\!CHCH_2CH(COOCH_2CH_3)_2$$

Diethyl malonate   Allyl chloride   Diethyl 2-allylmalonate

$$\Big\downarrow \begin{array}{l} 1.\ LiAlH_4 \\ 2.\ H_2O \end{array}$$

$$CH_2\!\!=\!\!CHCH_2CH(CH_2OH)_2$$

2-Allyl-1,3-propanediol

(e) The desired primary alcohol may be prepared by reduction of the corresponding carboxylic acid, which in turn is available from the malonic ester synthesis using allyl chloride, including saponification and decarboxylation of the diester [prepared in part (d)].

$$CH_2\!\!=\!\!CHCH_2CH_2CH_2OH \implies CH_2\!\!=\!\!CHCH_2CH_2CO_2H \implies$$

4-Penten-1-ol

$$CH_2\!\!=\!\!CHCH_2CH(CO_2CH_2CH_3)_2$$

The correct sequence of reactions is:

$$CH_2\!\!=\!\!CHCH_2CH(COOCH_2CH_3)_2 \xrightarrow[\substack{2.\ H^+ \\ 3.\ heat}]{1.\ HO^-,\ H_2O} CH_2\!\!=\!\!CHCH_2CH_2COOH$$

Diethyl 2-allylmalonate [prepared as in part (d)]   4-Pentenoic acid

$$\Big\downarrow \begin{array}{l} 1.\ LiAlH_4 \\ 2.\ H_2O \end{array}$$

$$CH_2\!\!=\!\!CHCH_2CH_2CH_2OH$$

4-Penten-1-ol

(f) The desired product is an alcohol. It can be prepared by reduction of a ketone, which in turn can be prepared by the acetoacetic ester synthesis.

$$\underset{\quad}{CH_2\!\!=\!\!CHCH_2CH_2\overset{\displaystyle OH}{\overset{|}{C}H}CH_3} \implies CH_2\!\!=\!\!CHCH_2CH_2\overset{\displaystyle O}{\overset{\|}{C}}CH_3 \implies$$

$$CH_2\!\!=\!\!CHCH_2X + {}^-\!:\!CH_2\overset{\displaystyle O}{\overset{\|}{C}}CH_3$$

Therefore

$$CH_3\overset{\displaystyle O}{\overset{\|}{C}}CH_2\overset{\displaystyle O}{\overset{\|}{C}}OCH_2CH_3 + CH_2\!\!=\!\!CHCH_2Cl \xrightarrow[ethanol]{NaOCH_2CH_3} CH_3\overset{\displaystyle O}{\overset{\|}{C}}\overset{\displaystyle}{\underset{\underset{\displaystyle CH_2CH\!\!=\!\!CH_2}{|}}{C}}H\overset{\displaystyle O}{\overset{\|}{C}}OCH_2CH_3$$

Ethyl acetoacetate   Allyl chloride

$$\Big\downarrow \begin{array}{l} 1.\ HO^-,\ H_2O \\ 2.\ H^+ \\ 3.\ heat \end{array}$$

$$\underset{\underset{\displaystyle OH}{|}}{CH_3\overset{\displaystyle}{C}HCH_2CH_2CH\!\!=\!\!CH_2} \xleftarrow[CH_3OH]{NaBH_4} CH_3\overset{\displaystyle O}{\overset{\|}{C}}CH_2CH_2CH\!\!=\!\!CH_2$$

5-Hexen-2-ol

(g) Cyclopropanecarboxylic acid may be prepared by a malonic ester synthesis, as retrosynthetic analysis shows.

The desired reaction sequence is:

$$CH_2(COOCH_2CH_3)_2 + BrCH_2CH_2Br \xrightarrow[\text{ethanol}]{NaOCH_2CH_3}$$

Diethyl malonate   1,2-Dibromoethane

COOCH$_2$CH$_3$
COOCH$_2$CH$_3$

1. HO$^-$, H$_2$O
2. H$^+$
3. heat

Cyclopropanecarboxylic acid

(h) Treatment of the diester formed in part (g) with ammonia gives a diamide.

Diethyl cyclopropane-1,1-dicarboxylate $\xrightarrow{NH_3}$ Cyclopropane-1,1-dicarboxamide
[prepared as in part (g)]

(i) We need to extend the carbon chain of the starting material by *four* carbons. One way to accomplish this is by way of a malonic ester synthesis at each end of the chain.

$$HOC(CH_2)_6COH \xrightarrow[\text{2. H}_2\text{O}]{\text{1. LiAlH}_4} HOCH_2(CH_2)_6CH_2OH \xrightarrow[\text{or PBr}_3]{HBr} BrCH_2(CH_2)_6CH_2Br$$

Octanedioic acid   1,8-Dibromooctane

$$2CH_2(COOCH_2CH_3)_2 + Br(CH_2)_8Br \xrightarrow{NaOCH_2CH_3}$$

Diethyl malonate   1,8-Dibromooctane

$$(CH_3CH_2OOC)_2CH(CH_2)_8CH(COOCH_2CH_3)_2$$

1. HO$^-$, H$_2$O
2. H$^+$
3. heat

$$HOCCH_2(CH_2)_8CH_2COH$$

Dodecanedioic acid

21.26 The problem states that diphenadione is prepared from 1,1-diphenylacetone and dimethyl 1,2-benzenedicarboxylate. Therefore, disconnect the molecule in a way that reveals the two reactants.

Diphenadione   and $CH_3CCH(C_6H_5)_2$

Thus all that is required is to treat dimethyl 1,2-benzenedicarboxylate and 1,1-diphenylacetone with base. Two successive acylations of a ketone enolate occur; the first is intermolecular, the second intramolecular.

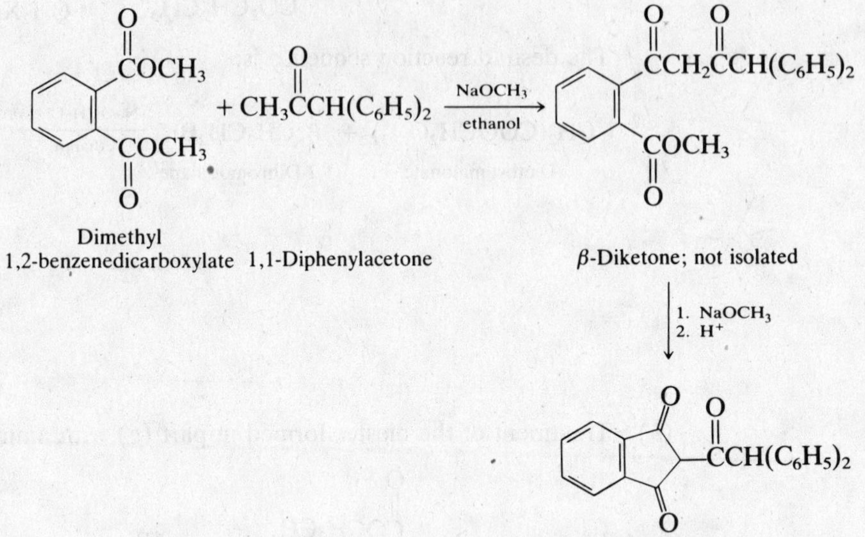

Dimethyl
1,2-benzenedicarboxylate 1,1-Diphenylacetone β-Diketone; not isolated

1. NaOCH$_3$
2. H$^+$

Diphenadione

21.27 Esters react with amines to give amides. Each nitrogen of 1,2-diphenylhydrazine reacts with a separate ester function of diethyl butylmalonate.

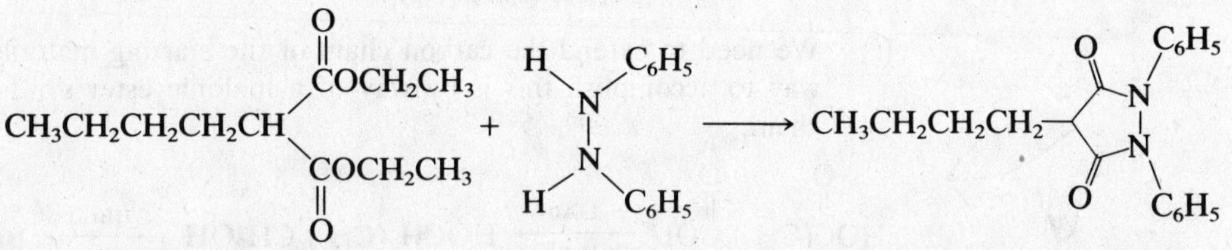

Diethyl butylmalonate 1,2-Diphenylhydrazine Phenylbutazone (C$_{19}$H$_{20}$N$_2$O$_2$)

21.28 The sodium salt of ethyl acetoacetate reacts with chloroacetone in a manner analogous to that of its reaction with alkyl halides.

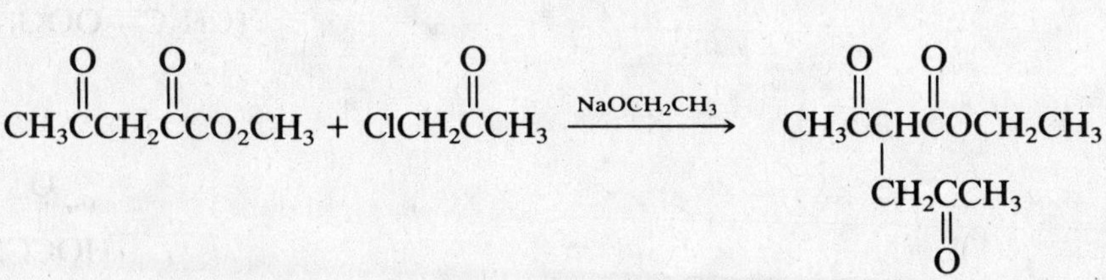

$$CH_3CCH_2CCO_2CH_3 + ClCH_2CCH_3 \xrightarrow{\text{NaOCH}_2\text{CH}_3} CH_3CCHCOCH_2CH_3$$

Ethyl acetoacetate Chloroacetone Ethyl 2-acetyl-4-oxopentanoate
(compound A, C$_9$H$_{14}$O$_4$)

One possible mechanism for the acid-catalyzed conversion of compound A to compound B begins with the protonation of one of the ketone carbonyl groups.

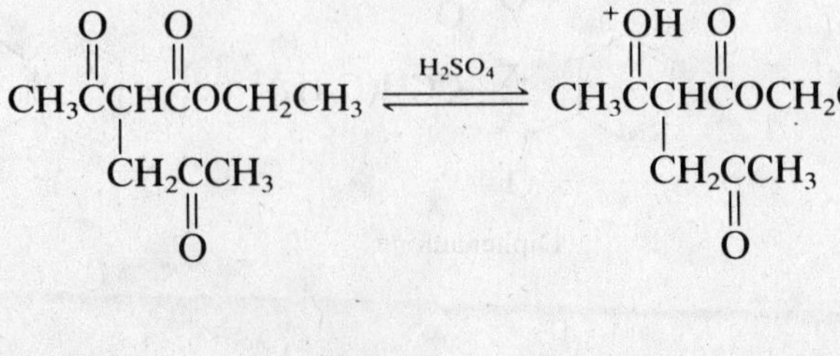

This is followed by intramolecular nucleophilic attack by the oxygen of the other ketone carbonyl. Subsequent loss of a proton and dehydration complete the process.

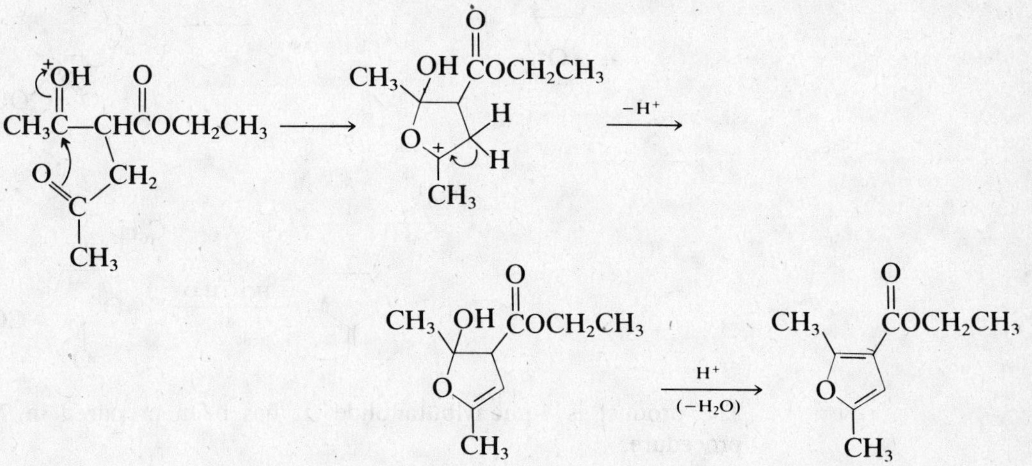

The same product (compound B) is formed if the initial protonation step involves the other ketone function.

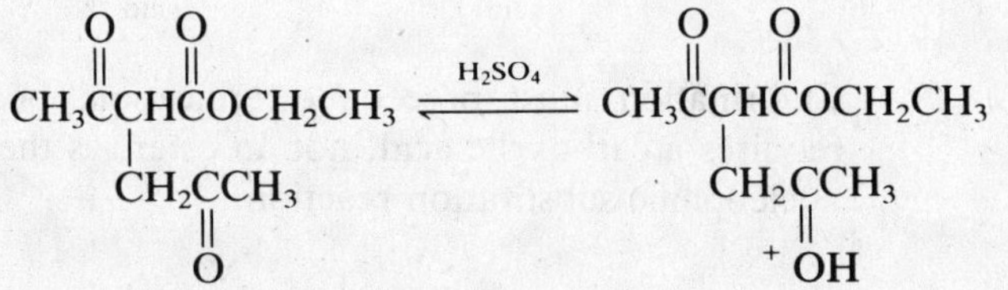

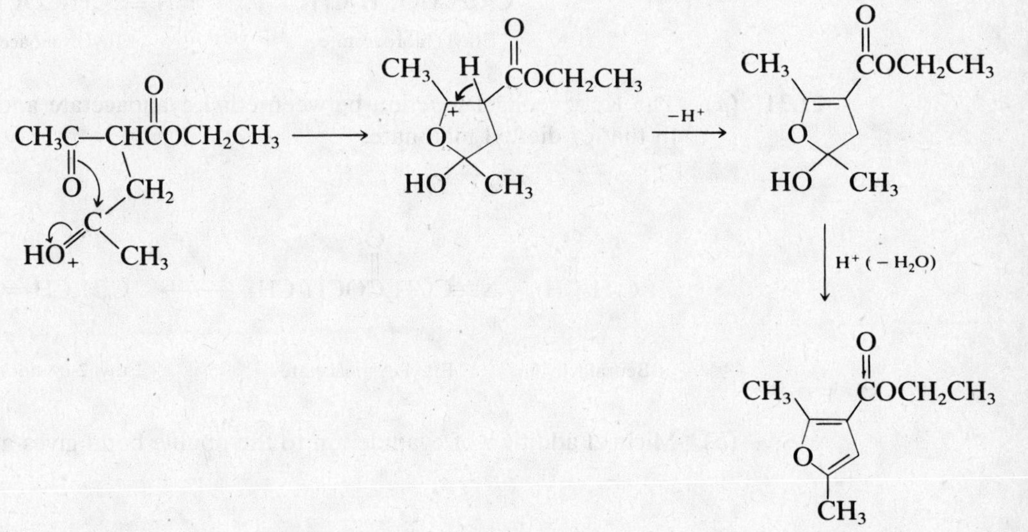

We cannot choose between these two very similar pathways; indeed, *both* may be occurring. Other mechanisms are also possible, including some that involve enol forms of compound A.

21.29 Styrene oxide will be attacked by the anion of diethyl malonate at its less hindered ring position.

The product is 4-phenylbutanolide. It has been prepared in 72 percent yield by this procedure.

21.30 The first task is to convert acetic acid to ethyl chloroacetate.

$$
\underset{\text{Acetic acid}}{CH_3COH} \xrightarrow[P]{Cl_2} \underset{\text{Chloroacetic acid}}{ClCH_2COH} \xrightarrow[H^+]{CH_3CH_2OH} \underset{\text{Ethyl chloroacetate}}{ClCH_2COCH_2CH_3}
$$

Chlorination must precede esterification, because the Hell-Volhard-Zelinsky reaction requires a carboxylic acid, not an ester, as the starting material. The remaining step is a nucleophilic substitution reaction.

$$
\underset{\text{Ethyl chloroacetate}}{ClCH_2COCH_2CH_3} \xrightarrow{NaCN} \underset{\text{Ethyl cyanoacetate}}{N\equiv CCH_2COCH_2CH_3}
$$

21.31 (a) The Knoevenagel reaction between ethyl cyanoacetate and benzaldehyde is similar to that of diethyl malonate.

(b) Michael addition of cyanide ion to the double bond gives a dinitrile.

(*c*) Acid hydrolysis converts both cyano groups and the ester to carboxylic acid functions. The triacid then undergoes decarboxylation.

$$
\begin{array}{c}
\underset{\substack{| \\ \text{C}\equiv\text{N}}}{\text{N}\equiv\text{C}} \quad \text{O} \\
\text{C}_6\text{H}_5\text{CHCHCOCH}_2\text{CH}_3
\end{array}
\xrightarrow{\text{H}_3\text{O}^+}
\left[\begin{array}{c}
\text{HOOC} \\
\text{C}_6\text{H}_5\text{CHCHCOOH} \\
\text{COOH}
\end{array}\right]
\xrightarrow[-\text{CO}_2]{\text{heat}}
\begin{array}{c}
\text{O} \quad\quad \text{O} \\
\text{HOCCHCH}_2\text{COH} \\
\text{C}_6\text{H}_5
\end{array}
$$

Ethyl
2,3-dicyano-3-phenylpropanoate   (Not isolated)   2-Phenylsuccinic acid

The diacid is 2-phenylsuccinic acid. It is converted to a cyclic imide on being heated with methylamine.

2-Phenylsuccinic acid   Methylamine   Phensuximide

21.32 From the hint given in the problem, it can be seen that synthesis of 2-methyl-2-propyl-1,3-propanediol is required. This diol is obtained by a sequence involving dialkylation of diethyl malonate.

Begin the synthesis by dialkylation of diethyl malonate.

$$
\text{CH}_2(\text{COOCH}_2\text{CH}_3)_2 \xrightarrow[\substack{\text{2. CH}_3\text{Br, NaOCH}_2\text{CH}_3}]{\substack{\text{1. CH}_3\text{CH}_2\text{CH}_2\text{Br,} \\ \text{NaOCH}_2\text{CH}_3}}
\begin{array}{c}
\text{CH}_3\text{CH}_2\text{CH}_2\quad\quad\text{COOCH}_2\text{CH}_3 \\
\text{C} \\
\text{CH}_3\quad\quad\text{COOCH}_2\text{CH}_3
\end{array}
$$

Diethyl malonate   Diethyl 2-methyl-2-propylmalonate

Convert the ester functions to primary alcohols by reduction.

$$
\begin{array}{c}
\text{CH}_3\text{CH}_2\text{CH}_2\quad\quad\text{COOCH}_2\text{CH}_3 \\
\text{C} \\
\text{CH}_3\quad\quad\text{COOCH}_2\text{CH}_3
\end{array}
\xrightarrow[\text{2. H}_2\text{O}]{\text{1. LiAlH}_4}
\begin{array}{c}
\text{CH}_3\text{CH}_2\text{CH}_2\quad\quad\text{CH}_2\text{OH} \\
\text{C} \\
\text{CH}_3\quad\quad\text{CH}_2\text{OH}
\end{array}
$$

Diethyl 2-methyl-2-propylmalonate   2-Methyl-2-propyl-1,3-propanediol

Conversion of the primary alcohol groups to carbamate esters completes the synthesis.

2-Methyl-2-propyl-1,3-propanediol   Meprobamate

21.33 The compound given in the problem contains three functionalities that can undergo acid-catalyzed hydrolysis: an acetal and two equivalent ester groups. Hydrolysis yields 3-oxo-1,1-cyclobutanedicarboxylic acid and two moles each of methanol and 2-propanol. The hydrolysis product is a malonic acid derivative which decarboxylates on heating. The final product of the reaction is 3-oxocyclobutanecarboxylic acid ($C_5H_6O_3$).

Diisopropyl 3,3-dimethoxycyclobutane-   3-Oxo-1,1-cyclobutanedicarboxylic   Methanol   2-Propanol
1,1-dicarboxylate   acid

3-Oxocyclobutanecarboxylic   Carbon
acid   dioxide

SELF-TEST

PART A

A-1. Give the structure of the reactant, reagent, or product omitted from each of the following:

(*a*) $CH_3CH_2CH_2COCH_2CH_3 \xrightarrow[\text{2. } H_3O^+]{\text{1. } NaOCH_2CH_3} \; ?$

(*b*) $HCOCH_2CH_3 + ? \xrightarrow[\text{2. } H_3O^+]{\text{1. } NaOCH_2CH_3} C_6H_5CHCOCH_2CH_3$

(*c*)

     $\xrightarrow[\substack{\text{2. } H_3O^+ \\ \text{3. heat}}]{\text{1. } HO^-, H_2O} \; ?$ (two isomeric products; $C_5H_7ClO_2$)

(d) $(CH_3CH_2OOC)_2CH_2 + CH_2{=}CHCOCH_2CH_3 \xrightarrow[\text{ethanol}]{\text{NaOCH}_2\text{CH}_3}$?

(e) $CH_3CCH_2COCH_2CH_3 \xrightarrow[\text{2. C}_6\text{H}_5\text{CH}_2\text{Br}]{\text{1. NaOCH}_2\text{CH}_3}$?

(f) Product of part (e) $\xrightarrow{?}$ $C_6H_5CH_2CH_2CCH_3$

(g) $CH_3CCHCH_2CO_2H \xrightarrow{\text{heat}} CO_2 + ?$
      $|$
      CO_2H

A-2. Provide the correct structures of compounds A through E in the following reaction sequences:

(a) A $\xrightarrow[\text{2. H}_3\text{O}^+]{\text{1. NaOCH}_2\text{CH}_3}$ [cyclopentanone-$COCH_2CH_3$] $\xrightarrow[\text{2. CH}_3\text{CH}_2\text{I}]{\text{1. NaOCH}_2\text{CH}_3}$ B $\xrightarrow[\substack{\text{2. H}_3\text{O}^+ \\ \text{3. heat}}]{\text{1. HO}^-,\ \text{H}_2\text{O}}$ C

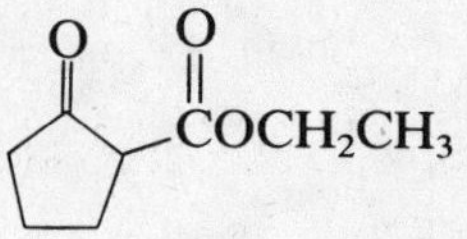

(b) $CH_3CH_2CH_2COCH_2CH_3 \xrightarrow[\text{2. H}_3\text{O}^+]{\text{1. NaOCH}_2\text{CH}_3}$ D $\xrightarrow[\substack{\text{2. H}_3\text{O}^+ \\ \text{3. heat}}]{\text{1. HO}^-,\ \text{H}_2\text{O}}$ E + CO_2

A-3. Give a series of steps that will enable preparation of each of the following compounds from the starting material(s) given and any other necessary reagents:

(a) $CH_3CCH_2CH_2COH$ from ethyl acetoacetate

(b) $C_6H_5CCH_2CH_2CH_2CCH_3$ from $C_6H_5CCH_3$ and diethyl carbonate

A-4. Write a stepwise mechanism for the reaction of ethyl propanoate with sodium ethoxide in ethanol.

PART B

B-1. Which of the following compounds is the strongest acid?
(a) $HCO_2CH_2CH_3$
(b) $CH_3CH_2O_2CCH_2CO_2CH_2CH_3$
(c) $CH_3CH_2O_2CCH_2CH_2CO_2CH_2CH_3$
(d) $CH_3CO_2CH_2CH_3$

B-2. Which of the following will yield a ketone and carbon dioxide following saponification, acidification, and heating?

(a) $CH_3CH_2CHCH_2CCH_3$
      $|$
      $O{=}COCH_2CH_3$

(c) $CH_3CH_2CHCCH_3$
      $|$
      $O{=}CCH_2CH_3$

(b) $CH_3CH_2CHCOCH_2CH_3$
      $|$
      $O{=}COCH_2CH_3$

(d) $CH_3CH_2CHCCH_3$
      $|$
      $O{=}COCH_2CH_3$

B-3. Which of the following keto esters is *not* likely to have been prepared by a Claisen condensation?

(a) $CH_3CH_2CCHCOCH_2CH_3$
      (with two $C=O$ groups above, and CH_3 below the central carbon)

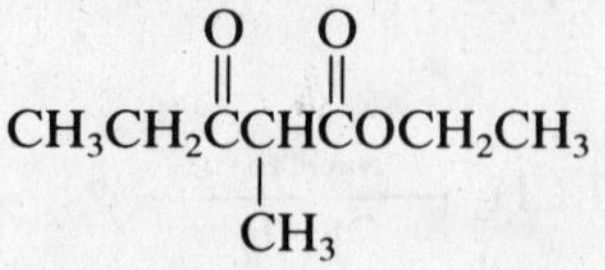

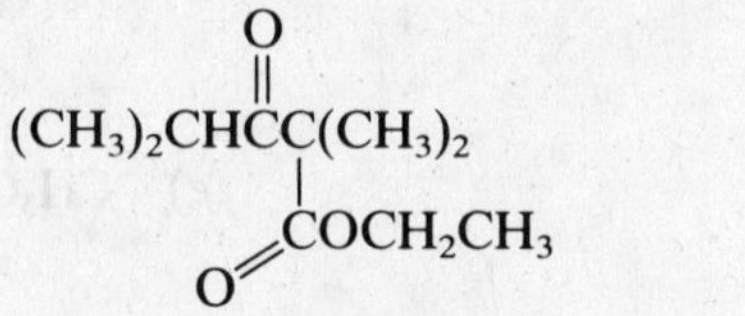

(c) $(CH_3)_2CHCC(CH_3)_2$
      (with $C=O$ above, and $COCH_2CH_3$ / O below)

(b) $C_6H_5CCHCOCH_2CH_3$
      (with two $C=O$ groups above, and CH_3 below the central carbon)

(d) $(CH_3)_2CHCH_2CCHCOCH_2CH_3$
      (with two $C=O$ groups above, and $CH(CH_3)_2$ below the central carbon)

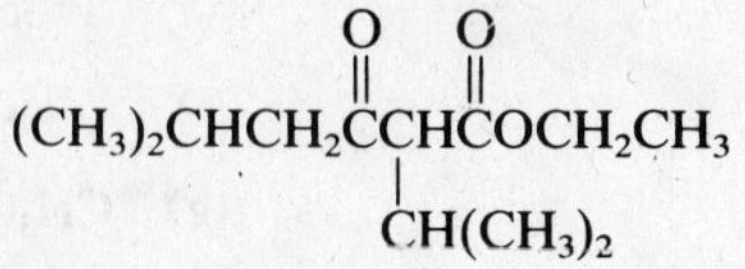

B-4. Dieckmann cyclization of $CH_3CH_2OC(CH_2)_5COCH_2CH_3$ (with two $C=O$ groups) will yield:

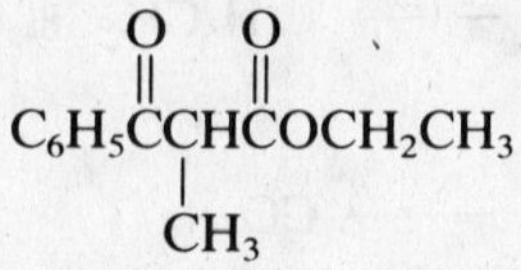

(a) cyclohexanone with a $COCH_2CH_3$ substituent

(c) cyclopentane with two $COCH_2CH_3$ substituents

(b) cyclohexanone with a $COCH_2CH_3$ substituent

(d) cyclohexane with a $COCH_2CH_3$ substituent

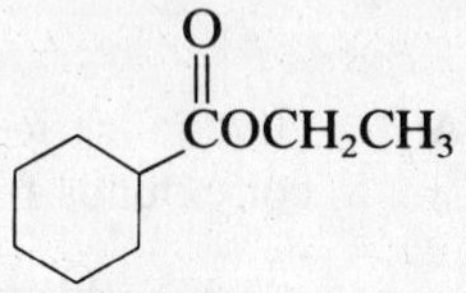

B-5. What is the final product of this sequence?

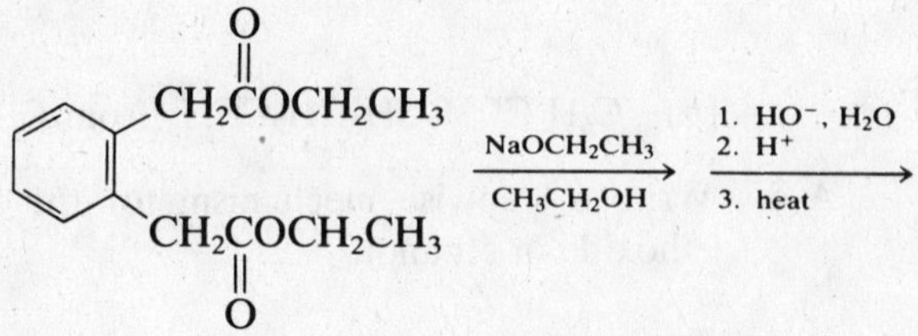

benzene ring with $CH_2COCH_2CH_3$ ($C=O$) and $CH_2COCH_2CH_3$ ($C=O$) substituents

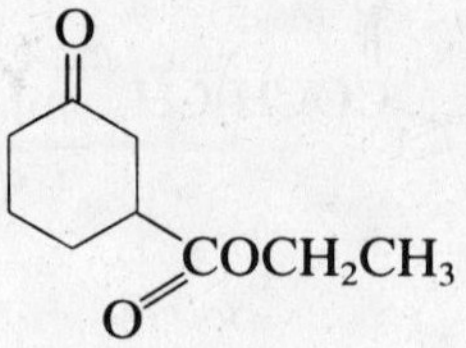

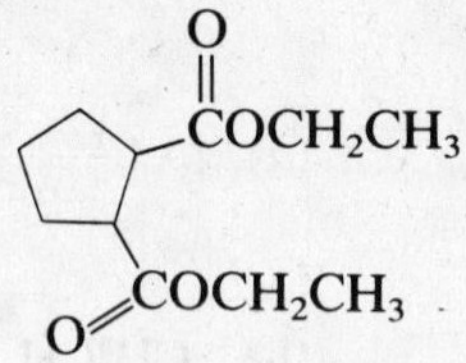

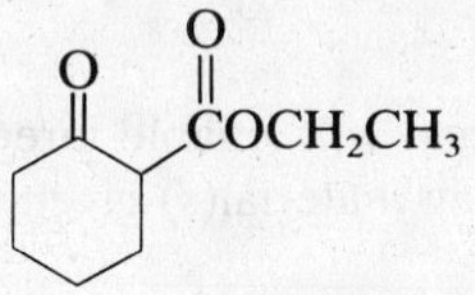

$\xrightarrow[\text{CH}_3\text{CH}_2\text{OH}]{\text{NaOCH}_2\text{CH}_3}$ $\xrightarrow{\begin{array}{l}1.\ \text{HO}^-,\ \text{H}_2\text{O}\\2.\ \text{H}^+\\3.\ \text{heat}\end{array}}$

(a) indanone structure

(c) indanone structure

(b) tetralone structure

(d) benzene with CH ($C=O$) and $CH=CH_2$ substituents

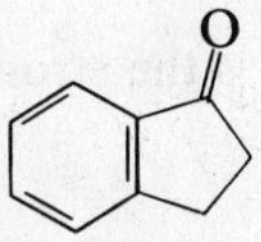

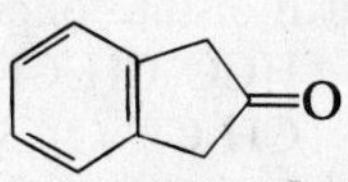

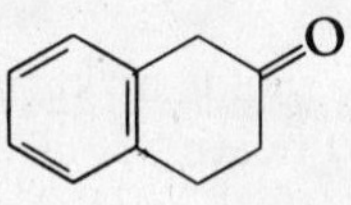

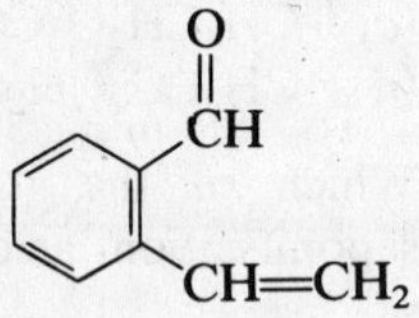

AMINES

IMPORTANT TERMS AND CONCEPTS

Nomenclature (Sec. 22.1) Amines are organic derivatives of ammonia. *Alkylamines* have their nitrogen attached to sp^3 hybridized carbon; *arylamines* have their nitrogen attached to sp^2 hybridized carbon. Amines are classified as being either primary, secondary, or tertiary, depending on the number of carbon substituents attached to the nitrogen atom.

$$R—\overset{..}{N}—H \qquad R—\overset{..}{N}—R \qquad R—\overset{..}{N}—R$$
$$\underset{\text{Primary amine}}{\overset{|}{H}} \qquad \underset{\text{Secondary amine}}{\overset{|}{H}} \qquad \underset{\text{Tertiary amine}}{\overset{|}{R}}$$

Primary amines may be named either by adding *-amine* to the alkyl group name or by replacing the *-e* ending of the longest carbon chain with *-amine*. For example,

Pentylamine, or 1-pentanamine

1-Methylbutylamine, or 2-pentanamine

Secondary and tertiary amines are named as *N*-substituted derivatives of primary amines. For example,

N-Methylcyclohexylamine, or *N*-methylcyclohexanamine

N-Methyl-*N*-propyl-*sec*-butylamine, or *N*-methyl-*N*-propyl-2-butanamine

Arylamines are named as derivatives of aniline, $C_6H_5NH_2$. The ring is numbered beginning with the amino nitrogen, and substituents are arranged in alphabetical order. The prefix *N*- is used to specify substituents attached to the amino nitrogen.

3-Bromo-*N*-methylaniline

Structure and Bonding (Sec. 22.2) Alkylamines have a pyramidal arrangement of the three substituents attached to the nitrogen atom, the fourth valence position being occupied by an unshared pair of electrons. It is this pair of electrons that enables alkylamines to act as bases or nucleophiles. The nitrogen atom is sp^3 hybridized and undergoes rapid pyramidal inversion:

The carbon-nitrogen bond of an arylamine such as aniline is shorter than that of an alkylamine such as cyclohexylamine.

Aniline   Cyclohexylamine

Two factors account for this. First, bonds to sp^2 hybridized carbon atoms are shorter than those to sp^3 hybridized carbon. Second, the carbon-nitrogen bond in an arylamine also has partial double bond character due to resonance delocalization of the lone-pair electrons into the aromatic system.

This phenomenon also accounts for the fact that the bond angles in aniline are greater than tetrahedral; the nitrogen atom adopts an orbital hybridization that is between sp^3 and sp^2.

Physical Properties (Sec. 22.3) Hydrogen bonding interactions in amines are weaker than those in alcohols; therefore primary and secondary amines have lower boiling points than alcohols of comparable molecular weight. Tertiary amine molecules lack the ability to form hydrogen bonds among themselves, although they may act as hydrogen bond acceptors with molecules of other substances.

Basicity (Secs. 22.4, 22.5) Amines act as bases by donating their unshared electron pairs to suitable acceptors. They react with acids by proton transfer to form *ammonium salts*.

$$R_3N: + HX \longrightarrow R_3\overset{+}{N}H \ X^-$$

The base strength of an amine is described by the equation

$$R_3N + H_2O \rightleftharpoons R_3\overset{+}{N}H + OH^-$$

The basicity constant K_b is related to the equilibrium constant for this equation by:

$$K_b = K[H_2O] = \frac{[R_3\overset{+}{N}H][OH^-]}{[R_3N]}$$

In aqueous solution the basicity trend of alkylamines may be summarized as:

$$NH_3 < RNH_2 \sim R_3N < R_2NH$$

Least   Most
basic   basic

The *gas-phase* basicity trend, however, is somewhat different:

$$NH_3 < RNH_2 < R_2NH < R_3N$$

Least   Most
basic   basic

The difference between the trends is explained by the presence of hydrogen bonds between NH protons and water molecules, which cannot be formed in the gas phase.

Arylamines are *less* basic than alkylamines by a factor of about 10^5. The lone-pair electrons of an arylamine such as aniline are delocalized through the π system of the aromatic ring; such delocalization is not possible in the corresponding ammonium ion. The resulting stabilization of the amine, compared with the ammonium ion, shifts the acid-base equilibrium to the left, lowering the basicity of an arylamine. Such delocalization is not possible in alkylamines, which lack an adjacent π system.

$$\ddot{N}H_2\text{-C}_6H_5 + H_2O \rightleftharpoons {}^+NH_3\text{-C}_6H_5 + HO^- \qquad K_b = 3.8 \times 10^{-10}$$

$$\ddot{N}H_2\text{-C}_6H_{11} + H_2O \rightleftharpoons {}^+NH_3\text{-C}_6H_{11} + HO^- \qquad K_b = 4.4 \times 10^{-4}$$

Substituents on the aromatic ring affect the basicity of arylamines. In general, electron-donating groups result in an increase in basicity, while electron-withdrawing groups tend to decrease it. Electron-withdrawing groups that are conjugated with the amine nitrogen, such as a para nitro group, exert a considerable effect on the basicity of the amine; thus *p*-nitroaniline is almost 4000 times less basic than aniline.

$$K_b = 1.0 \times 10^{-13}$$

p-Nitroaniline

Spectroscopic Analysis (Sec. 22.20) The infrared spectra of primary and secondary alkyl- and arylamines exhibit characteristic N—H stretching vibrations in the range 3000 to 3500 cm^{-1}. Two bands are observed for primary amines (RNH$_2$), while secondary amines (R$_2$NH) exhibit only one. The ^{1}H nmr chemical shifts of N—H protons, like those of hydroxyl protons, are variable and depend on the solvent, concentration, and temperature.

IMPORTANT REACTIONS

PREPARATION OF AMINES

The Gabriel Synthesis (Sec. 22.9)

General:

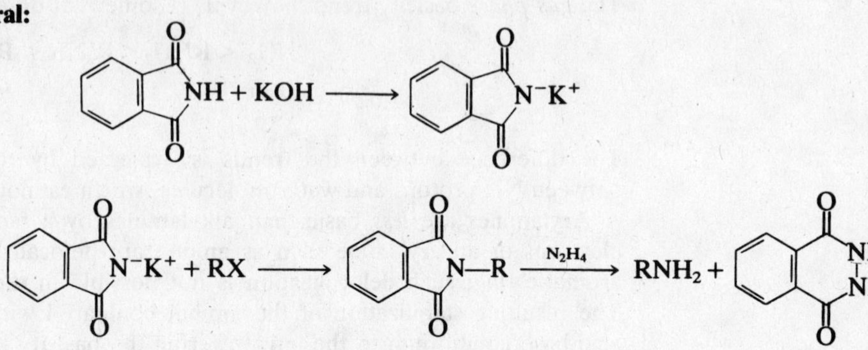

$$\text{(phthalimide)}\ \text{NH} + \text{KOH} \longrightarrow \text{N}^-\text{K}^+$$

$$\text{N}^-\text{K}^+ + \text{RX} \longrightarrow \text{N}-\text{R} \xrightarrow{\text{N}_2\text{H}_4} \text{RNH}_2 + \text{(phthalhydrazide)}$$

Example:

$$\text{NH} \xrightarrow[\substack{\text{2. (CH}_3)_2\text{CHCH}_2\text{Cl} \\ \text{3. N}_2\text{H}_4}]{\text{1. KOH}} (\text{CH}_3)_2\text{CHCH}_2\text{NH}_2$$

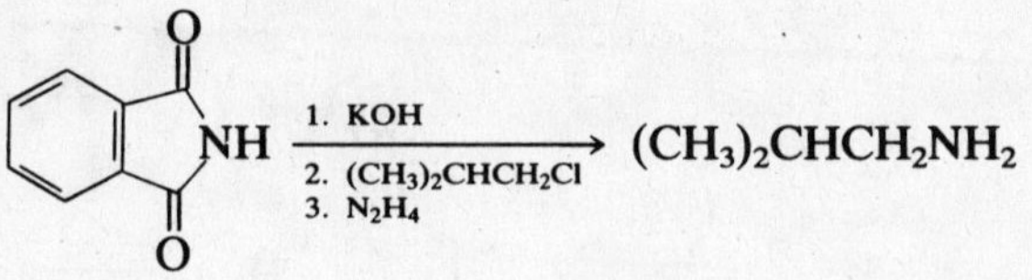

Reduction (Sec. 22.10)

General:

Azides

$$\text{RN}_3 \xrightarrow[\text{2. H}_2\text{O}]{\text{1. LiAlH}_4} \text{RNH}_2$$

Nitriles

$$\text{RC}\equiv\text{N} \xrightarrow[\text{2. H}_2\text{O}]{\text{1. LiAlH}_4} \text{RCH}_2\text{NH}_2$$

Amides

$$\overset{\displaystyle O}{\overset{\|}{\text{RCNR}'_2}} \xrightarrow[\text{2. H}_2\text{O}]{\text{1. LiAlH}_4} \text{RCH}_2\text{NR}'_2$$

Nitro compounds

$$\text{ArNO}_2 \xrightarrow[\text{2. NaOH}]{\text{1. Sn, HCl}} \text{ArNH}_2$$

Examples:

$$\text{CH}_3\text{CH}_2\text{Br} \xrightarrow{\text{NaN}_3} \text{CH}_3\text{CH}_2\text{N}_3 \xrightarrow[\text{2. H}_2\text{O}]{\text{1. LiAlH}_4} \text{CH}_3\text{CH}_2\text{NH}_2$$

$$\text{C}_6\text{H}_5\text{CH}_2\text{Cl} \xrightarrow{\text{NaCN}} \text{C}_6\text{H}_5\text{CH}_2\text{CN} \xrightarrow[\text{2. H}_2\text{O}]{\text{1. LiAlH}_4} \text{C}_6\text{H}_5\text{CH}_2\text{CH}_2\text{NH}_2$$

$$C_6H_5CH_2CO_2H \xrightarrow{SOCl_2} C_6H_5CH_2\overset{\displaystyle O}{\overset{\|}{C}}Cl \xrightarrow{(CH_3CH_2)_2NH} C_6H_5CH_2\overset{\displaystyle O}{\overset{\|}{C}}N(CH_2CH_3)_2$$

$$\xrightarrow[\substack{1.\ LiAlH_4 \\ 2.\ H_2O}]{} C_6H_5CH_2CH_2N(CH_2CH_3)_2$$

$$(CH_3)_3C\!-\!\!\bigcirc\!\!- \xrightarrow[H_2SO_4]{HNO_3} (CH_3)_3C\!-\!\!\bigcirc\!\!-NO_2 \xrightarrow[\substack{1.\ Sn,\ HCl \\ 2.\ NaOH}]{} (CH_3)_3C\!-\!\!\bigcirc\!\!-NH_2$$

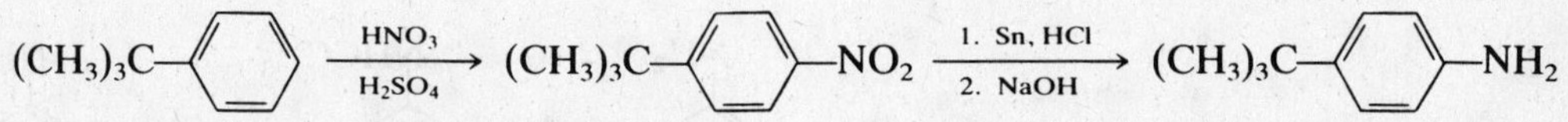

Reductive Amination (Sec. 22.11)

General:

$$R_2C{=}O + R'NH_2 \xrightarrow{[H]} R_2CHNHR'$$

Examples:

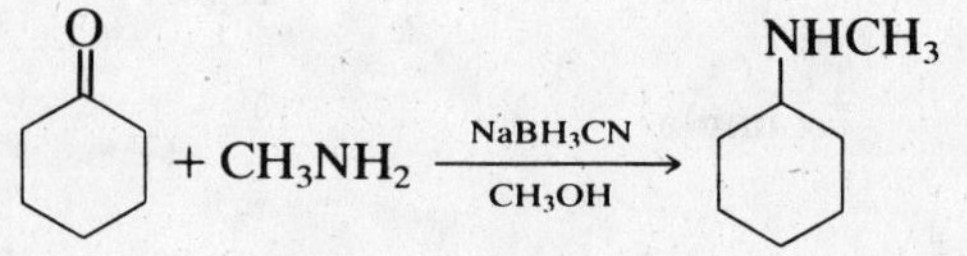

$$\bigcirc\!\!{=}O + CH_3NH_2 \xrightarrow[CH_3OH]{NaBH_3CN} \overset{NHCH_3}{\bigcirc}$$

$$C_6H_5CH_2\overset{\displaystyle O}{\overset{\|}{C}}H + NH_3 \xrightarrow{H_2,\ Ni} C_6H_5CH_2CH_2NH_2$$

REACTIONS OF AMINES

Reaction with Alkyl Halides (Sec. 22.13)

General:

$$RNH_2 \xrightarrow{R'X} RNHR' \xrightarrow{R'X} RNR'_2 \xrightarrow{R'X} R\overset{+}{N}R'_3\ X^-$$

Example:

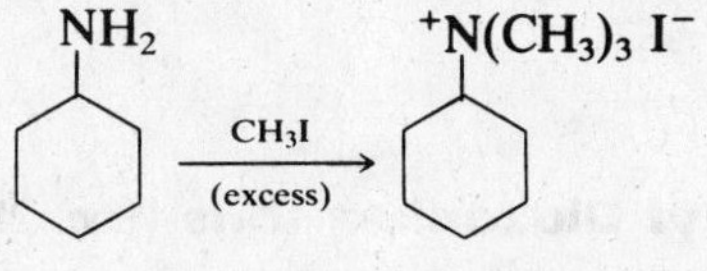

$$\overset{NH_2}{\bigcirc} \xrightarrow[(excess)]{CH_3I} \overset{+N(CH_3)_3\ I^-}{\bigcirc}$$

The Hofmann Elimination (Sec. 22.14)

General:

$$RCH_2CH_2\overset{+}{N}R_3\ I^- \xrightarrow[H_2O]{Ag_2O} RCH_2CH_2\overset{+}{N}R_3\ OH^- \xrightarrow{heat} RCH{=}CH_2 + R'_3N + H_2O$$

Example:

$$(CH_3)_2CHCH_2\overset{\overset{\displaystyle CH_3}{\underset{\displaystyle |}{+}}}{\underset{\underset{\displaystyle CH_3}{|}}{N}}CH_2CH_3\ I^- \xrightarrow[\substack{1.\ Ag_2O,\ H_2O \\ 2.\ heat}]{} (CH_3)_2CHCH_2N(CH_3)_2 + CH_2{=}CH_2$$

Electrophilic Aromatic Substitution of Arylamines (Sec. 22.15) The amine group is often protected by acetylation prior to carrying out electrophilic aromatic substitution reactions.

General:

$$ArNH_2 \xrightarrow{\underset{}{\overset{\overset{\displaystyle O}{\|}}{CH_3CCl}}} ArNHCCH_3 \xrightarrow{E^+} \text{ortho, para substitution}$$

Example:

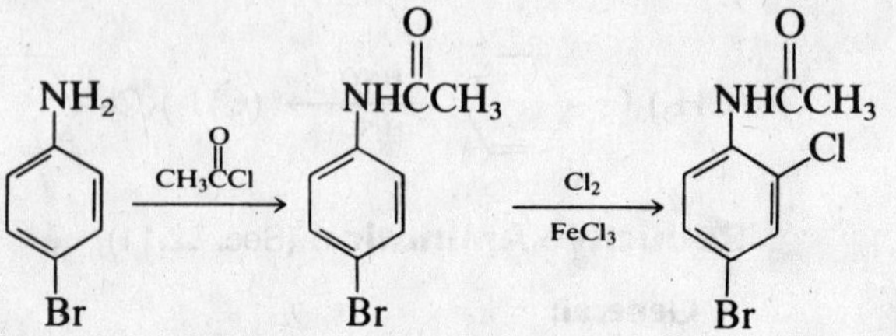

Nitrosation of Alkylamines (Sec. 22.16)

General:

$$R_2NH \xrightarrow[\underset{H_2O}{}]{NaNO_2,\ HCl} R_2N{-}N{=}O$$

Example:

$$C_6H_5CH_2NHCH_3 \xrightarrow[\underset{H_2O}{}]{NaNO_2,\ HCl} C_6H_5CH_2\underset{CH_3}{\overset{\overset{\displaystyle N{=}O}{|}}{N}}$$

Nitrosation of Arylamines (Sec. 22.17) Primary aromatic amines yield aryl diazonium salts on nitrosation.

General:

$$ArNH_2 \xrightarrow[\underset{H_2O}{}]{NaNO_2,\ HCl} ArN_2^+Cl^-$$

Example:

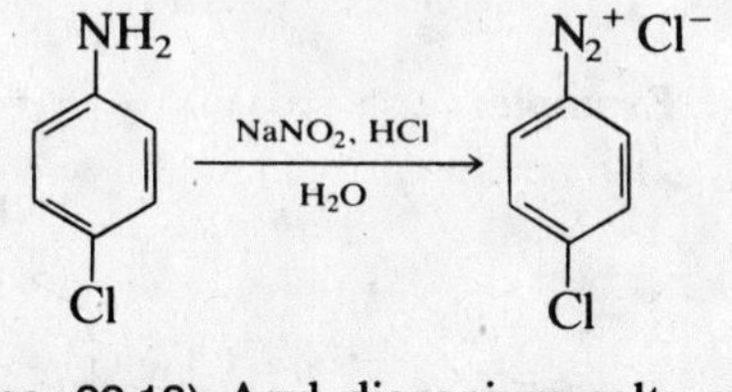

Aryl Diazonium Ions (Sec. 22.18) Aryl diazonium salts are useful intermediates in the synthesis of a variety of compounds.

General:

Phenols

$$Ar{-}N_2^+ \xrightarrow[\underset{heat}{}]{H_2O} Ar{-}OH$$

Halides

$$Ar{-}N_2^+ \xrightarrow{CuX} Ar{-}X \quad (X = Cl,\ Br)$$

$$Ar{-}N_2^+ \xrightarrow{I^-} Ar{-}I$$

Cyanides

$$Ar-N_2^+ \xrightarrow{\text{CuCN}} Ar-CN$$

Deamination

$$Ar-N_2^+ \xrightarrow[\text{or CH}_3\text{CH}_2\text{OH}]{\text{H}_3\text{PO}_2} Ar-H$$

Examples:

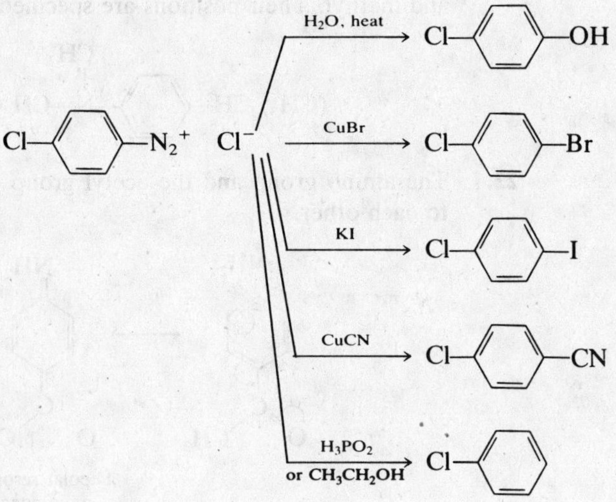

Cl—⟨benzene⟩—N₂⁺ Cl⁻

- $\xrightarrow{\text{H}_2\text{O, heat}}$ Cl—⟨benzene⟩—OH
- $\xrightarrow{\text{CuBr}}$ Cl—⟨benzene⟩—Br
- $\xrightarrow{\text{KI}}$ Cl—⟨benzene⟩—I
- $\xrightarrow{\text{CuCN}}$ Cl—⟨benzene⟩—CN
- $\xrightarrow[\text{or CH}_3\text{CH}_2\text{OH}]{\text{H}_3\text{PO}_2}$ Cl—⟨benzene⟩

Azo Coupling (Sec. 22.19)

General:

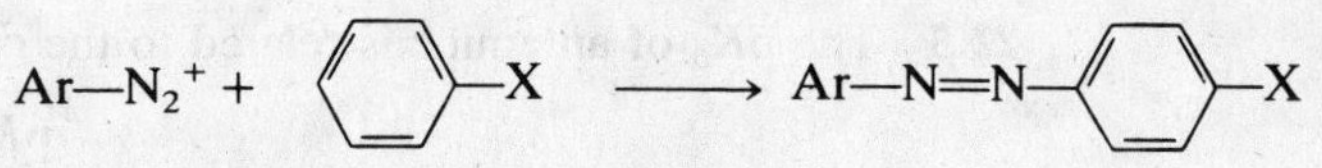

$$Ar-N_2^+ + \text{⟨benzene⟩}-X \longrightarrow Ar-N=N-\text{⟨benzene⟩}-X$$

$$(X = -OH, -NR_2)$$

Example:

⟨benzene⟩—OH + ⟨benzene⟩—N₂⁺ Cl⁻ (with CH₃) $\longrightarrow$ CH₃—⟨benzene⟩—N=N—⟨benzene⟩—OH

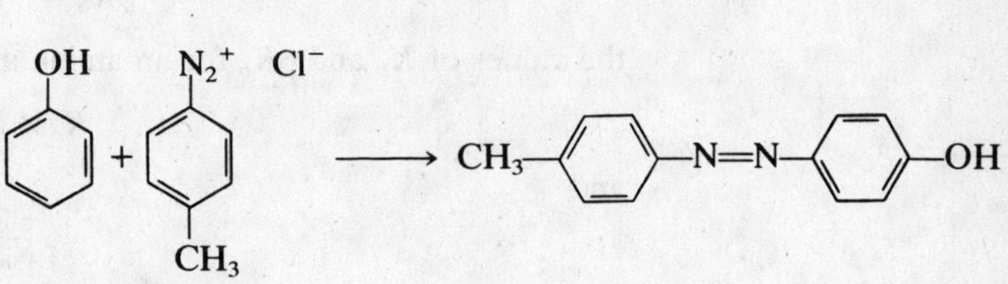

SOLUTIONS TO TEXT PROBLEMS

22.1 (*b*) The amino and phenyl groups are both attached to C-1 of an ethyl group.

$$\underset{\underset{\text{NH}_2}{|}}{\text{C}_6\text{H}_5\text{CHCH}_3} \quad \begin{array}{l}\text{1-Phenylethylamine, or}\\ \text{1-phenylethanamine}\end{array}$$

(*c*) $CH_2{=}CHCH_2NH_2$ Allylamine, or
2-propen-1-amine

22.2   *N,N*-Dimethylcycloheptylamine may also be named as a dimethyl derivative of cycloheptanamine:

     —N(CH$_3$)$_2$ *N,N*-Dimethylcycloheptanamine

22.3   Three substituents are attached to the nitrogen atom; the amine is tertiary. In alphabetical order, the substituents present on the aniline nucleus are ethyl, isopropyl, and methyl. Their positions are specified as *N*-ethyl, 4-isopropyl, and *N*-methyl.

(CH$_3$)$_2$CH— —N—CH$_2$CH$_3$ *N*-Ethyl-4-isopropyl-*N*-methylaniline

22.4   The amino group and the acetyl group of *p*-aminoacetophenone are directly conjugated to each other.

| Dipolar resonance form of *p*-aminoacetophenone | | Dipolar resonance form of aniline |

The dipolar resonance form has its negative charge on the electronegative oxygen and is more stable than the dipolar resonance form of aniline itself. Thus, there is more double bond character in the carbon-nitrogen bond of *p*-aminoacetophenone than in that of aniline, and *p*-aminoacetophenone has a shorter carbon-nitrogen bond.

22.5   The pK_b of an amine is related to the equilibrium constant K_b by

$$pK_b = -\log K_b$$

The pK_b of quinine is therefore

$$pK_b = -\log (1 \times 10^{-6}) = 6$$

the values of K_b and pK_b for an amine and K_a and pK_a of its conjugate acid are given by

$$K_a \times K_b = 1 \times 10^{-14}$$

and

$$pK_a + pK_b = 14$$

Therefore the values of K_a and pK_a for the conjugate acid of quinine are

$$K_a = \frac{10^{-14}}{K_b} = \frac{1 \times 10^{-14}}{1 \times 10^{-6}} = 1 \times 10^{-8}$$

and

$$pK_a = 14 - pK_b = 14 - 6 = 8$$

22.6   The Henderson–Hasselbalch equation described in Section 19.4 can be applied to bases such as amines, as well as carboxylic acids. The ratio [CH$_3$NH$_3^+$]/[CH$_3$NH$_2$] is given by

$$\frac{[CH_3NH_3^+]}{[CH_3NH_2]} = \frac{[H^+]}{K_a}$$

The ionization constant of methylammonium ion is given in the text as 2×10^{-11}. At pH $= 7$ the hydrogen ion concentration is 1×10^{-7}. Therefore

$$\frac{[CH_3NH_3^+]}{[CH_3NH_2]} = \frac{1 \times 10^{-7}}{2 \times 10^{-11}} = 5 \times 10^3$$

22.7 (*b*) As was seen in part (*a*), a 2-fluoro substituent weakens the basicity of an amine. Successive fluorine substitution reduces the basicity even further. Thus, 2,2-difluoroethylamine is a relatively weak base, but it is a stronger base than its trifluoro analog.

$$F_2CHCH_2NH_2 \qquad\qquad F_3CCH_2NH_2$$

2,2-Difluoroethylamine,   2,2,2-Trifluoroethylamine,
stronger base: K_b 1.3×10^{-7} weaker base: K_b 4×10^{-9}
(pK_b 6.9)   (pK_b 8.4)

22.8 Nitrogen is attached directly to the aromatic ring in tetrahydroquinoline, which is thus an arylamine, and the nitrogen lone pair is delocalized into the π system of the aromatic ring. It is less basic than tetrahydroisoquinoline, in which the nitrogen is insulated from the ring by an sp^3 hybridized carbon.

Tetrahydroisoquinoline (an alkylamine):   Tetrahydroquinoline (an arylamine):
more basic, K_b 2.5×10^{-5} less basic, K_b 1.0×10^{-9}
(pK_b 4.6)   (pK_b 9.0)

22.9 (*b*) An acetyl group attached directly to nitrogen as in acetanilide delocalizes the nitrogen lone pair into the carbonyl group. Amides are weaker bases than amines.

(*c*) An acetyl group in a position para to an amine function is conjugated to it and delocalizes the nitrogen lone pair.

22.10 The reaction that leads to allylamine is nucleophilic substitution by ammonia on allyl chloride.

$$CH_2{=}CHCH_2Cl + 2NH_3 \longrightarrow CH_2{=}CHCH_2NH_2 + NH_4Cl$$

Allyl chloride   Ammonia   Allylamine   Ammonium
                                                     chloride

Allyl chloride is prepared by free-radical chlorination of propene.

$$CH_2{=}CHCH_3 + Cl_2 \xrightarrow{400°C} CH_2{=}CHCH_2Cl + HCl$$

Propene   Chlorine   Allyl chloride   Hydrogen
                                                        chloride

22.11 (*b*) Isobutylamine is $(CH_3)_2CHCH_2NH_2$. It is a primary amine and can be prepared from a primary alkyl halide by the Gabriel synthesis.

$(CH_3)_2CHCH_2Br$ +

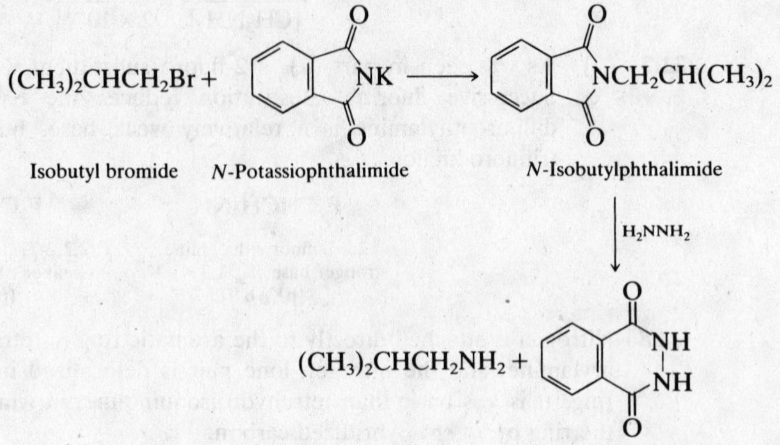

Isobutyl bromide   *N*-Potassiophthalimide   *N*-Isobutylphthalimide

H_2NNH_2

$(CH_3)_2CHCH_2NH_2$ +

Isobutylamine   Phthalhydrazide

(*c*) While *tert*-butylamine $(CH_3)_3CNH_2$ is a primary amine, it cannot be prepared by the Gabriel method, because it would require an S_N2 reaction on a tertiary alkyl halide in the first step. Elimination occurs instead.

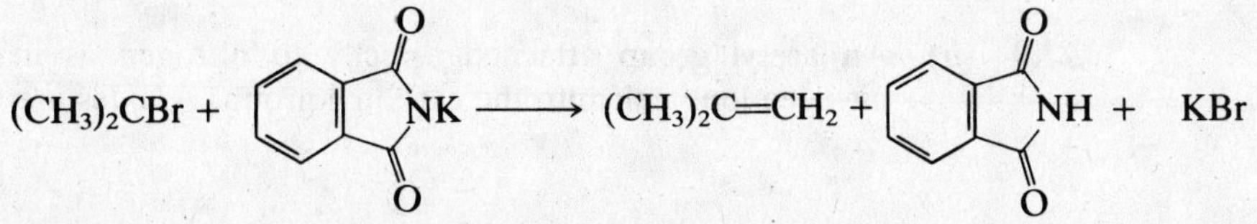

$(CH_3)_2CBr$ + $(CH_3)_2C{=}CH_2$ + NH + KBr

tert-Butyl   *N*-Potassiophthalimide   2-Methylpropene   Phthalimide   Potassium
bromide   bromide

(*d*) The preparation of 2-phenylethylamine by the Gabriel synthesis has been described in the chemical literature.

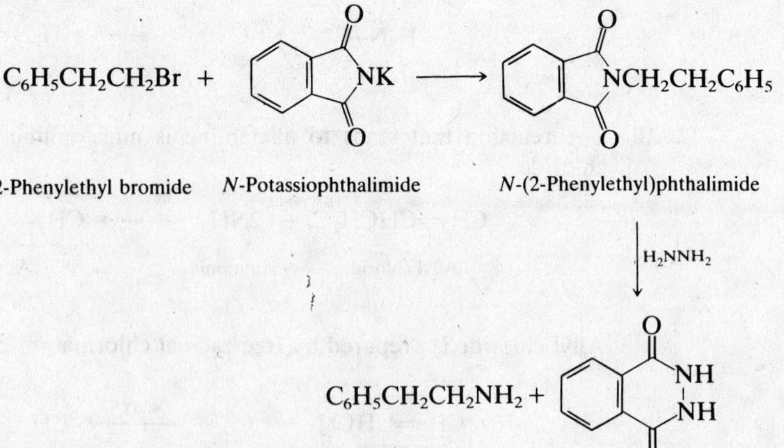

$C_6H_5CH_2CH_2Br$ +

2-Phenylethyl bromide   *N*-Potassiophthalimide   *N*-(2-Phenylethyl)phthalimide

H_2NNH_2

$C_6H_5CH_2CH_2NH_2$ +

2-Phenylethylamine   Phthalhydrazide

(e) The Gabriel synthesis leads to primary amines; *N*-methylbenzylamine is a secondary amine and cannot be prepared by this method.

N-Methylbenzylamine
(two carbon substituents on nitrogen;
a secondary amine)

(f) Aniline cannot be prepared by the Gabriel method. Aryl halides do not undergo nucleophilic substitution under these conditions.

Bromobenzene *N*-Potassiophthalimide

22.12 For each part of this problem, keep in mind that aromatic amines are derived by reduction of the corresponding aromatic nitro compound. Each synthesis should be approached from the standpoint of how best to prepare the necessary nitroaromatic compound.

$$Ar\text{—}NH_2 \implies Ar\text{—}NO_2 \implies Ar\text{—}H$$

(Ar = substituted aromatic ring)

(b) The para isomer of isopropylaniline may be prepared by a procedure analogous to that used for its ortho isomer in part (a).

Benzene   Isopropylbenzene   *o*-Isopropylnitrobenzene   *p*-Isopropylnitrobenzene

After separating the ortho, para mixture by distillation, the nitro group of *p*-isopropylnitrobenzene is reduced to yield the desired *p*-isopropylaniline.

(c) The target compound is the reduction product of 1-isopropyl-2,4-dinitrobenzene.

1-Isopropyl-2,4-dinitrobenzene   4-Isopropyl-1,3-benzenediamine

This reduction is carried out in the same way as reduction of an arene that contains only a single nitro group. In this case hydrogenation over a nickel catalyst gave the desired product in 90 percent yield.

The starting dinitro compound is prepared by nitration of isopropylbenzene.

Isopropylbenzene   1-Isopropyl-2,4-dinitrobenzene
(43%)

(*d*) The conversion of *p*-chloronitrobenzene to *p*-chloroaniline was cited as an example in the text to illustrate reduction of aromatic nitro compounds to arylamines. *p*-Chloronitrobenzene is prepared by nitration of chlorobenzene.

Benzene   Chlorobenzene   *o*-Chloronitrobenzene   *p*-Chloronitrobenzene

The para isomer accounts for 69 percent of the product in this reaction (30 percent is ortho, 1 percent meta). Separation of *p*-chloronitrobenzene and its reduction completes the synthesis.

p-Chloronitrobenzene   *p*-Chloroaniline

Chlorination of nitrobenzene would not be a suitable route to the required intermediate, because it would produce mainly *m*-chloronitrobenzene.

(*e*) The synthesis of *m*-aminoacetophenone may be carried out by the scheme shown:

Benzene   Acetophenone   *m*-Nitroacetophenone   *m*-Aminoacetophenone

The acetyl group is attached to the ring by Friedel-Crafts acylation. It is a meta director, and its nitration gives the proper orientation of substituents. The order of the first two steps cannot be reversed, because Friedel-Crafts acylation of nitrobenzene is not possible (Section 12.16). Once prepared, *m*-nitroacetophenone can be reduced to *m*-nitroaniline by any of a number of reagents. Indeed, all three reducing combinations described in the text have been employed for this transformation.

	Reducing agent	Yield, %
m-Nitroacetophenone	H$_2$, Pt	94
↓	Fe, HCl	84
m-Aminoacetophenone	Sn, HCl	82

22.13 (*b*) Dibenzylamine is a secondary amine and can be prepared by reductive amination of benzaldehyde with benzylamine.

$$\underset{\text{Benzaldehyde}}{C_6H_5\overset{\overset{O}{\|}}{C}H} + \underset{\text{Benzylamine}}{C_6H_5CH_2NH_2} \xrightarrow{\text{H}_2, \text{ Ni}} \underset{\text{Dibenzylamine}}{C_6H_5CH_2NHCH_2C_6H_5}$$

(*c*) *N*,*N*-Dimethylbenzylamine is a tertiary amine. Its preparation from benzaldehyde requires dimethylamine, a secondary amine.

$$\underset{\text{Benzaldehyde}}{C_6H_5\overset{\overset{O}{\|}}{C}H} + \underset{\text{Dimethylamine}}{(CH_3)_2NH} \xrightarrow{\text{H}_2, \text{ Ni}} \underset{\textit{N,N}\text{-Dimethylbenzylamine}}{C_6H_5CH_2N(CH_3)_2}$$

(*d*) The preparation of *N*-butylpiperidine by reductive amination is described in the text in Section 22.11. An analogous procedure is used to prepare *N*-benzylpiperidine.

Benzaldehyde   Piperidine   *N*-Benzylpiperidine

22.14 (*b*) First identify the available β hydrogens. Elimination must involve a proton from the carbon atom adjacent to the one that bears the nitrogen.

It is a proton from one of the methyl groups, rather than one from the more sterically hindered methylene, that is lost on elimination.

(1,1,3,3-Tetramethylbutyl)trimethylammonium
hydroxide

$$\underset{\substack{\text{2,4,4-Trimethyl-1-pentene (only alkene} \\ \text{formed, 70\% isolated yield)}}}{(CH_3)_3CCH_2\overset{\overset{CH_3}{|}}{C}\!=\!CH_2} + \underset{\text{Trimethylamine}}{(CH_3)_3N\!:}$$

(*c*)   The base may abstract a proton from either of two β carbons. Deprotonation of the β methyl carbon yields ethylene.

N-Ethyl-*N*,*N*-dimethylbutylammonium hydroxide

$$CH_2{=}CH_2 + (CH_3)_2\ddot{N}CH_2CH_2CH_2CH_3$$

Ethylene   *N*,*N*-Dimethylbutylamine

Deprotonation of the β methylene carbon yields 1-butene.

N-Ethyl-*N*,*N*-dimethylbutylammonium   *N*,*N*-Dimethylethylamine   1-Butene
hydroxide

The preferred order of proton removal in Hofmann elimination reactions is $\beta\,CH_3 > \beta\,CH_2 > \beta\,CH$. Ethylene is the major alkene formed, the observed ratio of ethylene to 1-butene being 98:2.

22.15   (*b*)   The pattern of substituents in 2,4-dinitroaniline suggests that they can be introduced by dinitration. Since nitration of aniline itself is accompanied by oxidation, the amino group must be protected by conversion to its *N*-acetyl derivative.

Aniline   Acetanilide   2,4-Dinitroacetanilide

Hydrolysis of the amide bond in 2,4-dinitroacetanilide furnishes the desired 2,4-dinitroaniline.

2,4-Dinitroacetanilide   2,4-Dinitroaniline

(*c*)   Retrosynthetically, *p*-aminoacetanilide may be derived from *p*-nitroacetanilide.

p-Aminoacetanilide   *p*-Nitroacetanilide

This suggests the sequence

Aniline   Acetanilide   p-Nitroacetanilide (separate from ortho isomer)

reduce | Fe, HCl or Sn, HCl or H_2, Pt

p-Aminoacetanilide

22.16 The principal resonance forms of *N*-nitrosodimethylamine are:

All atoms (except hydrogen) have octets of electrons in each of these structures. Other resonance forms are less stable because they do not have a full complement of electrons around each atom.

22.17 Deamination of 1,1-dimethylpropylamine gives products that result from 1,1-dimethylpropyl cation. Since 2,2-dimethylpropylamine gives the same products, it is likely that 1,1-dimethylpropyl cation is formed from 2,2-dimethylpropylamine by way of its diazonium ion. A carbocation rearrangement reaction is indicated.

2,2-Dimethylpropylamine $\xrightarrow{HONO}$ 2,2-Dimethylpropyldiazonium ion $\xrightarrow{-N_2}$ 1,1-Dimethylpropyl cation

Once formed, 1,1-dimethylpropyl cation loses a proton to form an alkene or is captured by water to give an alcohol.

1,1-Dimethylpropyl cation $\xrightarrow{-H^+}$ $CH_2=CCH_2CH_3$ 2-Methyl-1-butene   + $(CH_3)_2C=CHCH_3$ 2-Methyl-2-butene

$\xrightarrow{H_2O}$ $(CH_3)_2CCH_2CH_3$ | OH

2-Methyl-2-butanol

22.18 Phenols may be prepared by diazotization of the corresponding aniline derivative. The problem simplifies itself, therefore, to the preparation of *m*-bromoaniline. Recognizing

that arylamines are ultimately derived from nitroarenes, we derive the retrosynthetic sequence of intermediates:

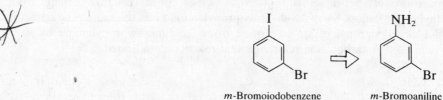

m-Bromophenol   *m*-Bromoaniline   *m*-Bromonitrobenzene   Nitrobenzene

The desired reaction sequence is straightforward, using reactions that have been discussed previously in the text.

22.19 The key to this problem is to recognize that the iodine substituent in *m*-bromoiodobenzene is derived from an arylamine by diazotization.

m-Bromoiodobenzene   *m*-Bromoaniline

The preparation of *m*-bromoaniline from benzene has been described in Problem 22.18. All that remains is to write the equation for its conversion to *m*-bromoiodobenzene.

m-Bromoaniline   1. NaNO$_2$, HCl, H$_2$O   2. KI   *m*-Bromoiodobenzene

22.20 Since the final step in the preparation of ethyl *m*-fluorophenyl ketone is shown in the text example immediately preceding this problem, all that is necessary is to describe the preparation of *m*-aminophenyl ethyl ketone.

Ethyl *m*-fluorophenyl ketone   *m*-Aminophenyl ethyl ketone   Ethyl *m*-nitrophenyl ketone

Recalling that arylamines are normally prepared by reduction of nitroarenes, we see that ethyl *m*-nitrophenyl ketone is a pivotal synthetic intermediate. It is prepared by nitration of ethyl phenyl ketone, which is analogous to that of acetophenone, shown in Section 12.16. The preparation of ethyl phenyl ketone by Friedel-Crafts acylation of benzene is shown in Section 12.7.

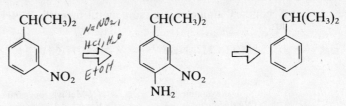

Ethyl *m*-nitrophenyl   Ethyl phenyl ketone
ketone

Reversing the order of introduction of the nitro and acyl groups is incorrect. It is possible to nitrate ethyl phenyl ketone but not possible to carry out a Friedel-Crafts acylation on nitrobenzene, owing to the strong deactivating influence of the nitro group.

22.21 Direct nitration of the prescribed starting material cumene (isopropylbenzene) is not suitable, because isopropyl is an ortho, para–directing substituent and will give the target molecule *m*-nitrocumene as only a minor component of the nitration product. However, the conversion of 4-isopropyl-2-nitroaniline to *m*-isopropylnitrobenzene, which was used to illustrate reductive deamination of arylamines, establishes the last step in the synthesis.

m-Nitrocumene   4-Isopropyl-2-nitroaniline   Cumene

Our task simplifies itself to the preparation of 4-isopropyl-2-nitroaniline from cumene. The following procedure is a straightforward extension of the reactions and principles developed in this chapter.

Cumene   *p*-Nitrocumene   *p*-Isopropylaniline

p-Isopropylacetanilide

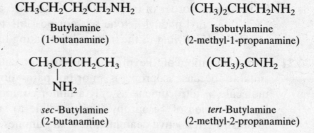

p-Isopropylacetanilide   4-Isopropyl-2-nitroacetanilide   4-Isopropyl-2-nitroaniline

Reductive deamination of 4-isopropyl-2-nitroaniline by diazotization in the presence of ethanol or hypophosphorous acid yields *m*-nitrocumene and completes the synthesis.

22.22 Amines may be primary, secondary, or tertiary. The $C_4H_{11}N$ primary amines, compounds of the type $C_4H_9NH_2$, and their systematic names are:

$$CH_3CH_2CH_2CH_2NH_2 \qquad\qquad (CH_3)_2CHCH_2NH_2$$

Butylamine   Isobutylamine
(1-butanamine)   (2-methyl-1-propanamine)

$$\underset{\underset{NH_2}{|}}{CH_3CHCH_2CH_3} \qquad\qquad (CH_3)_3CNH_2$$

sec-Butylamine   *tert*-Butylamine
(2-butanamine)   (2-methyl-2-propanamine)

Secondary amines have the general formula R_2NH. Those of molecular formula $C_4H_{11}N$ are:

$$(CH_3CH_2)_2NH \qquad \underset{\underset{H}{|}}{CH_3NCH_2CH_2CH_3} \qquad \underset{\underset{H}{|}}{CH_3NCH(CH_3)_2}$$

Diethylamine   *N*-Methylpropylamine   *N*-Methylisopropylamine
(*N*-ethylethanamine)   (*N*-methyl-1-propanamine)   (*N*-methyl-2-propanamine)

There is only one tertiary amine (R_3N) of molecular formula $C_4H_{11}N$:

$$(CH_3)_2NCH_2CH_3 \qquad \begin{array}{l}\text{\textit{N,N}-Dimethylethylamine}\\ \text{(\textit{N,N}-dimethylethanamine)}\end{array}$$

22.23   (a)   The name 2-ethyl-1-butanamine designates a four-carbon chain terminating in an amino group and bearing an ethyl group at C-2.

$$\underset{\underset{CH_2CH_3}{|}}{CH_3CH_2CHCH_2NH_2} \qquad \text{2-Ethyl-1-butanamine}$$

(b)   The prefix *N*- in *N*-ethyl-1-butanamine identifies the ethyl group as a substituent on nitrogen in a secondary amine.

$$\underset{\underset{H}{|}}{CH_3CH_2CH_2CH_2NCH_2CH_3} \qquad \text{\textit{N}-Ethyl-1-butanamine}$$

(c)   Dibenzylamine is a secondary amine. It bears two benzyl groups on nitrogen.

$$\underset{\underset{H}{|}}{C_6H_5CH_2NCH_2C_6H_5} \qquad \text{Dibenzylamine}$$

(d)   Tribenzylamine is a tertiary amine.

$$(C_6H_5CH_2)_3N \qquad \text{Tribenzylamine}$$

(e) Tetraethylammonium hydroxide is a quaternary ammonium salt.

$$(CH_3CH_2)_4\overset{+}{N}\ HO^-$$ Tetraethylammonium hydroxide

(f) This compound is a secondary amine; it bears an allyl substituent on the nitrogen of cyclohexylamine.

N-Allylcyclohexylamine

(g) Piperidine is a cyclic secondary amine that contains nitrogen in a six-membered ring. *N*-Allylpiperidine is a tertiary amine.

$NCH_2CH{=}CH_2$ *N*-Allylpiperidine

(h) The compound is the benzyl ester of 2-aminopropanoic acid.

$$\underset{\underset{NH_2}{|}}{CH_3CHC}\overset{\overset{O}{\|}}{}OCH_2C_6H_5$$ Benzyl 2-aminopropanoate

(i) The parent compound is cyclohexanone. The substituent $(CH_3)_2N$— group is attached to C-4.

4-(*N,N*-Dimethylamino)cyclohexanone

(j) The suffix *-diamine* reveals the presence of two amino groups, one at either end of a three-carbon chain that bears two methyl groups at C-2.

$$H_2NCH_2\underset{\underset{CH_3}{|}}{\overset{\overset{CH_3}{|}}{C}}CH_2NH_2$$ 2,2-Dimethyl-1,3-propanediamine

22.24 (a) A phenyl group and an amino group are trans to each other on a three-membered ring in this compound.

trans-2-Phenylcyclopropylamine (tranylcypromine)

(b) This compound is a tertiary amine. It bears a benzyl group, a methyl group, and a 2-propynyl group on nitrogen.

N-Benzyl-*N*-methyl-2-propynylamine (pargyline)

(c) The amino group is at C-2 of a three-carbon chain that bears a phenyl substituent at its terminus.

$$C_6H_5CH_2\underset{\underset{NH_2}{|}}{CH}CH_3$$ 1-Phenyl-2-propanamine (amphetamine)

(d) Phenylephrine is named systematically as an ethanol derivative.

1-(*m*-Hydroxyphenyl)-2-(methylamino)ethanol (phenylephrine)

22.25 (a) There are five isomers of C_7H_9N that contain a benzene ring:

$$C_6H_5CH_2NH_2 \qquad C_6H_5NHCH_3$$

Benzylamine   *N*-Methylaniline

o-Methylaniline   *m*-Methylaniline   *p*-Methylaniline

(b) Benzylamine is the strongest base, since its amine group is bonded to an sp^3 hybridized carbon. Benzylamine is a typical alkylamine, with a K_b of 2×10^{-5}. All the other isomers are arylamines, with K_b values in the 10^{-10} range.

(c) The formation of *N*-nitrosoamines on reaction with sodium nitrite and hydrochloric acid is a characteristic reaction of secondary amines. The only C_7H_9N isomer in this problem that is a secondary amine is *N*-methylaniline.

(d) Ring nitrosation is a characteristic reaction of tertiary arylamines.

Tertiary arylamine   *p*-Nitroso-*N*,*N*-dialkylaniline

None of the C_7H_9N isomers in this problem are tertiary amines; hence none will undergo ring nitrosation.

22.26 (a) Basicity decreases in proceeding across a row in the periodic table. The increased nuclear charge as one progresses from carbon to nitrogen to oxygen to fluorine causes the electrons to be bound more strongly to the atom and thus less readily shared.

$$H_3\bar{C}: \; > H_2\bar{N}: > H\ddot{O}:^- > \; :\ddot{F}:^-$$

Strongest base   Weakest base

K_a of conjugate acid 10^{-60} 10^{-36} 10^{-16} 3.5×10^{-4}

(b) The strongest base in this group is amide ion, H_2N^-, while the weakest base is water, H_2O. Ammonia is a weaker base than hydroxide ion; the equilibrium lies to the left.

$$:NH_3 \; + \; H_2O \; \rightleftharpoons \; \overset{+}{N}H_4 \; + \; OH^-$$

Weaker base   Weaker acid   Stronger acid   Stronger base

The correct order is:

$$H_2\ddot{N}: \; > H\ddot{O}:^- > :NH_3 > \; H_2\ddot{O}:$$

Strongest   Weakest
base   base

(*c*) These anions can be ranked according to their basicity by considering the respective acidities of their conjugate acids.

Base	Conjugate acid	K_a of conjugate acid
H_2N^-	H_3N	10^{-36}
HO^-	H_2O	10^{-16}
$^-:C\equiv N:$	$HC\equiv N:$	7.2×10^{-10}
$^-O{-}\overset{\overset{\displaystyle O}{\|}}{\underset{\underset{\displaystyle O^-}{+}}{N}}$	$HO\overset{\overset{\displaystyle O}{\|}}{\underset{\underset{\displaystyle O^-}{+}}{N}}$	2.5×10^{1}

The order of basicities is the opposite of the order of acidities of their conjugate acids.

$$H_2N^- \; > HO^- > :\bar{C}\equiv N: > \; NO_3^-$$

Strongest   Weakest
base   base

(*d*) A carbonyl group attached to nitrogen stabilizes its negative charge. The strongest base is the anion that has no carbonyl groups on nitrogen; the weakest base is phthalimide anion, which has two carbonyl groups.

Strongest base   Weakest base

22.27 (*a*) An alkyl substituent on nitrogen is electron-releasing and base-strengthening; thus methylamine is a stronger base than ammonia. An aryl substituent is electron-withdrawing and base-weakening, and so aniline is a weaker base than ammonia.

$$CH_3NH_2 \; > \; NH_3 \; > \; C_6H_5NH_2$$

Methylamine,   Ammonia: Aniline,
strongest base: weakest base:
K_b 4.4×10^{-4} K_b 1.8×10^{-5} K_b 3.8×10^{-10}
pK_b 3.4 pK_b 4.7 pK_b 9.4

(*b*) An acetyl group is an electron-withdrawing and base-weakening substituent, especially when bonded directly to nitrogen. Amides are weaker bases than amines, and thus acetanilide is a weaker base than aniline. Alkyl groups are electron-releasing; *N*-methylaniline is a slightly stronger base than aniline.

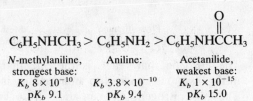

$$C_6H_5NHCH_3 > C_6H_5NH_2 > C_6H_5NH\overset{\overset{\displaystyle O}{\|}}{C}CH_3$$

N-methylaniline,   Aniline: Acetanilide,
strongest base: weakest base:
K_b 8×10^{-10} K_b 3.8×10^{-10} K_b 1×10^{-15}
pK_b 9.1 pK_b 9.4 pK_b 15.0

(c) Chlorine substituents are slightly electron-withdrawing and methyl groups are slightly electron-releasing. Therefore, 2,4-dimethylaniline is a stronger base than 2,4-dichloroaniline. Nitro groups are strongly electron-withdrawing, their base-weakening effect being especially pronounced when a nitro group is ortho or para to an amino group because the two groups are then directly conjugated.

2,4-Dimethylaniline,
strongest base:
K_b 8×10^{-10}
pK_b 9.1

2,4-Dichloroaniline:
K_b 1×10^{-12}
pK_b 12.0

2,4-Dinitroaniline,
weakest base:
K_b 3×10^{-19}
pK_b 18.5

(d) Nitro groups are more electron-withdrawing than chlorine, and the base-weakening effect of a nitro substituent is greater when it is ortho or para to an amino group than when it is meta to it.

3,4-Dichloroaniline,
strongest base:
$K_b \cong 10^{-11}$
$pK_b \cong 11$

4-Chloro-3-nitroaniline:
K_b 8×10^{-13}
pK_b 12.1

4-Chloro-2-nitroaniline,
weakest base:
K_b 1×10^{-15}
pK_b 15.0

(e) According to the principle applied in part (a) (alkyl groups increase basicity, aryl groups decrease it), the order of decreasing basicity is as shown:

$$(CH_3)_2NH > C_6H_5NHCH_3 > (C_6H_5)_2NH$$

Dimethylamine,
strongest base:
K_b 5.1×10^{-4}
pK_b 3.3

N-Methylaniline:
K_b 8×10^{-10}
pK_b 9.1

Diphenylamine,
weakest base:
K_b 6×10^{-14}
pK_b 13.2

22.28 The most nucleophilic of the three nitrogen atoms of physostigmine will be the one that reacts with methyl iodide. Nitrogen (a) is the most nucleophilic.

Physostigmine

Methyl
iodide

"Physostigmine methiodide"

The nitrogen that reacts is the one that is a tertiary alkylamine. Of the other two nitrogens, (b) is attached to an aromatic ring and is much less basic and less nucleophilic. The third nitrogen, (c), is an amide nitrogen; amides are less nucleophilic than amines.

22.29 (*a*) Looking at the problem retrosynthetically, it can be seen that a variety of procedures are available for preparing ethylamine from ethanol. The methods by which a primary amine may be prepared include:

$$CH_3CH_2NH_2 \Rightarrow \begin{cases} \text{phthalimide-N-CH}_2\text{CH}_3 & \text{Gabriel synthesis} \\ CH_3CH_2N_3 & \text{Reduction of an azide} \\ CH_3\overset{\displaystyle O}{\overset{\|}{C}}H & \text{Reductive amination} \\ CH_3\overset{\displaystyle O}{\overset{\|}{C}}NH_2 & \text{Reduction of an amide} \end{cases}$$

Two of these methods, the Gabriel synthesis and the preparation and reduction of the corresponding azide, begin with ethyl bromide:

$$CH_3CH_2OH \xrightarrow[\text{or HBr}]{PBr_3} CH_3CH_2Br$$

Ethanol → Ethyl bromide

$$CH_3CH_2Br + \text{(}N^-K^+\text{ phthalimide)} \longrightarrow \text{(}NCH_2CH_3\text{ phthalimide)} \xrightarrow{H_2NNH_2} CH_3CH_2NH_2$$

Ethyl bromide — *N*-Potassiophthalimide — *N*-Ethylphthalimide — Ethylamine

$$CH_3CH_2Br \xrightarrow{NaN_3} CH_3CH_2N_3 \xrightarrow[2.\ H_2O]{1.\ LiAlH_4} CH_3CH_2NH_2$$

Ethyl bromide — Ethyl azide — Ethylamine

To use reductive amination, we must begin with oxidation of ethanol to acetaldehyde.

$$CH_3CH_2OH \xrightarrow[CH_2Cl_2]{(C_5H_5N)_2CrO_3 \text{ or PCC or PDC}} CH_3\overset{\displaystyle O}{\overset{\|}{C}}H$$

Ethanol — Acetaldehyde

$$CH_3\overset{\displaystyle O}{\overset{\|}{C}}H \xrightarrow{NH_3,\ H_2,\ Ni} CH_3CH_2NH_2$$

Acetaldehyde — Ethylamine

Another possibility is reduction of acetamide. This requires an initial oxidation of ethanol to acetic acid.

$$CH_3CH_2OH \xrightarrow[H_2O]{K_2Cr_2O_7,\ H_2SO_4} CH_3CO_2H \xrightarrow[2.\ NH_3]{1.\ SOCl_2} CH_3\overset{\displaystyle O}{\overset{\|}{C}}NH_2 \xrightarrow[2.\ H_2O]{1.\ LiAlH_4} CH_3CH_2NH_2$$

Ethanol — Acetic acid — Acetamide — Ethylamine

(*b*) Acylation of ethylamine with acetyl chloride, prepared in part (*a*), gives the desired amide.

$$\underset{\substack{\text{Acetyl}\\\text{chloride}}}{CH_3\overset{\displaystyle O}{\overset{\|}{C}}Cl} + \underset{\text{Ethylamine}}{2CH_3CH_2NH_2} \longrightarrow \underset{\text{N-Ethylacetamide}}{CH_3\overset{\displaystyle O}{\overset{\|}{C}}NHCH_2CH_3} + \underset{\substack{\text{Ethylammonium}\\\text{chloride}}}{CH_3CH_2\overset{+}{N}H_3 \quad Cl^-}$$

Excess ethylamine can be allowed to react with the hydrogen chloride formed in the acylation reaction. Alternatively, equimolar amounts of acyl chloride and amine can be used in the presence of aqueous hydroxide as the base.

(*c*) Reduction of the *N*-ethylacetamide prepared in part (*b*) yields diethylamine.

$$\underset{\text{N-Ethylacetamide}}{CH_3\overset{\displaystyle O}{\overset{\|}{C}}NHCH_2CH_3} \xrightarrow[\text{2. H}_2\text{O}]{\text{1. LiAlH}_4} \underset{\text{Diethylamine}}{CH_3CH_2NHCH_2CH_3}$$

Diethylamine can also be prepared by reductive amination of acetaldehyde [from part (*a*)] with ethylamine.

$$\underset{\text{Acetaldehyde}}{CH_3\overset{\displaystyle O}{\overset{\|}{C}}H} + \underset{\text{Ethylamine}}{CH_3CH_2NH_2} \xrightarrow[\text{or NaBH}_3\text{CN}]{\text{H}_2,\ \text{Ni}} \underset{\text{Diethylamine}}{CH_3CH_2NHCH_2CH_3}$$

(*d*) The preparation of *N,N*-diethylacetamide is a standard acylation reaction. The necessary acetyl chloride and diethylamine have been prepared in previous parts of this problem.

$$\underset{\text{Acetyl chloride}}{CH_3\overset{\displaystyle O}{\overset{\|}{C}}Cl} + \underset{\text{Diethylamine}}{(CH_3CH_2)_2NH} \xrightarrow{\text{HO}^-} \underset{\text{N,N-Diethylacetamide}}{CH_3\overset{\displaystyle O}{\overset{\|}{C}}N(CH_2CH_3)_2}$$

(*e*) Triethylamine arises by reduction of *N,N*-diethylacetamide or by reductive amination.

$$\underset{\text{N,N-Diethylacetamide}}{CH_3\overset{\displaystyle O}{\overset{\|}{C}}N(CH_2CH_3)_2} \xrightarrow[\text{2. H}_2\text{O}]{\text{1. LiAlH}_4} \underset{\text{Triethylamine}}{(CH_3CH_2)_3N}$$

$$\underset{\text{Acetaldehyde}}{CH_3\overset{\displaystyle O}{\overset{\|}{C}}H} + \underset{\text{Diethylamine}}{(CH_3CH_2)_2NH} \xrightarrow[\substack{\text{or}\\\text{NaBH}_3\text{CN}}]{\text{H}_2,\ \text{Ni}} \underset{\text{Triethylamine}}{(CH_3CH_2)_3N}$$

(*f*) Quaternary ammonium halides are formed by reaction of alkyl halides and tertiary amines.

$$\underset{\text{Ethyl bromide}}{CH_3CH_2Br} + \underset{\text{Triethylamine}}{(CH_3CH_2)_3N} \longrightarrow \underset{\text{Tetraethylammonium bromide}}{(CH_3CH_2)_4\overset{+}{N}\ Br^-}$$

22.30 (*a*) In this problem a primary alkanamine must be prepared with a carbon chain extended by one carbon. This can be accomplished by way of a nitrile.

$$RCH_2NH_2 \implies RCN \implies RBr \implies ROH$$

$$(R— = CH_3CH_2CH_2CH_2—)$$

The desired reaction sequence is therefore:

$$CH_3CH_2CH_2CH_2OH \xrightarrow[\text{or}]{\underset{\text{HBr}}{PBr_3}} CH_3CH_2CH_2CH_2Br \xrightarrow{NaCN} CH_3CH_2CH_2CH_2CN$$

1-Butanol   Butyl bromide   Pentanenitrile

$$\downarrow \begin{array}{l} 1.\ LiAlH_4 \\ 2.\ H_2O \end{array}$$

$$CH_3CH_2CH_2CH_2CH_2NH_2$$

1-Pentanamine

(b) The carbon chain of *tert*-butyl chloride cannot be extended by a nucleophilic substitution reaction; the S_N2 reaction which would be required on the tertiary halide would not work. Therefore the sequence employed in part (a) is not effective in this case. The best route is carboxylation of the Grignard reagent and subsequent conversion of the corresponding amide to the desired primary amine product.

$$(CH_3)_3CCH_2NH_2 \implies (CH_3)_3C\overset{\overset{\displaystyle O}{\|}}{C}NH_2 \implies (CH_3)_3CCO_2H \implies (CH_3)_3CCl$$

The reaction sequence to be used is:

$$(CH_3)_3CCl \xrightarrow[\substack{2.\ CO_2 \\ 3.\ H_3O^+}]{1.\ Mg,\ diethyl\ ether} (CH_3)_3CCO_2H$$

tert-Butyl chloride   2,2-Dimethylpropanoic acid

Once the carboxylic acid has been obtained, it is converted to the desired amine by reduction of the corresponding amide.

$$(CH_3)_3CCO_2H \xrightarrow[\substack{2.\ NH_3}]{1.\ SOCl_2} (CH_3)_3C\overset{\overset{\displaystyle O}{\|}}{C}NH_2 \xrightarrow[\substack{2.\ H_2O}]{1.\ LiAlH_4} (CH_3)_3CCH_2NH_2$$

2,2-Dimethylpropanoic   2,2-Dimethylpropanamide   2,2-Dimethyl-1-propanamine
acid

(c) Oxidation of cyclohexanol to cyclohexanone gives a suitable substrate for reductive amination.

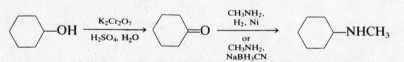

Cyclohexanol   Cyclohexanone   *N*-Methylcyclohexylamine

(d) The desired product is the reduction product of the cyanohydrin of acetone.

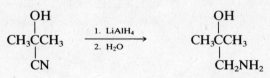

Acetone cyanohydrin   1-Amino-2-methyl-2-propanol

The cyanohydrin is made from acetone in the usual way. Acetone is available by oxidation of isopropyl alcohol.

$$CH_3CHCH_3 \xrightarrow[\text{H}_2\text{O}]{\substack{\text{K}_2\text{Cr}_2\text{O}_7, \\ \text{H}_2\text{SO}_4}} CH_3\overset{\text{O}}{\overset{\|}{\text{C}}}CH_3 \xrightarrow[\text{H}_2\text{SO}_4]{\text{KCN}} CH_3\overset{\text{OH}}{\underset{\text{CN}}{\overset{|}{\underset{|}{\text{C}}}}}CH_3$$
$$\underset{\text{OH}}{|}$$

Isopropyl alcohol   Acetone   Acetone cyanohydrin

(*e*) The target amino alcohol is the product of nucleophilic ring opening of 1,2-epoxypropane by ammonia. Ammonia attacks the less hindered carbon of the epoxide function.

$$CH_3CH\overset{}{—}CH_2 \xrightarrow{\text{NH}_3} CH_3CHCH_2NH_2$$
$$\underset{\text{O}}{\diagdown\diagup} \qquad\qquad \underset{\text{OH}}{|}$$

1,2-Epoxypropane   1-Amino-2-propanol

The necessary epoxide is formed by epoxidation of propene.

$$CH_3CHCH_3 \xrightarrow[\text{heat}]{\text{H}_2\text{SO}_4} CH_3CH{=}CH_2 \xrightarrow{\text{CH}_3\text{CO}_2\text{OH}} CH_3CH\overset{}{—}CH_2$$
$$\underset{\text{OH}}{|} \qquad\qquad\qquad\qquad\qquad \underset{\text{O}}{\diagdown\diagup}$$

Isopropyl alcohol   Propene   1,2-Epoxypropane

(*f*) The reaction sequence is the same as in part (*e*) except that dimethylamine is used as the nucleophile instead of ammonia.

$$CH_3CH\overset{}{—}CH_2 \quad + \quad (CH_3)_2NH \longrightarrow CH_3CHCH_2N(CH_3)_2$$
$$\underset{\text{O}}{\diagdown\diagup} \qquad\qquad\qquad\qquad\qquad\qquad \underset{\text{OH}}{|}$$

1,2-Epoxypropane   Dimethylamine   1-(*N*,*N*-Dimethylamino)-2-
[prepared as in part (*e*)]   propanol

(*g*) The key to performing this synthesis is recognition of the starting material as an acetal of acetophenone. Acetals may be hydrolyzed to carbonyl compounds.

$$\underset{\underset{\text{C}_6\text{H}_5\quad\text{CH}_3}{}}{\overset{\text{O}\quad\text{O}}{\diagup\diagdown}} \xrightarrow{\text{H}_3\text{O}^+} C_6H_5\overset{\text{O}}{\overset{\|}{\text{C}}}CH_3 + HOCH_2CH_2OH$$

2-Methyl-2-phenyl-   Acetophenone   1,2-Ethanediol
1,3-dioxolane

Once acetophenone has been obtained, it may be converted to the required product by reductive amination.

$$C_6H_5\overset{\text{O}}{\overset{\|}{\text{C}}}CH_3 \quad + \quad \underset{\underset{\text{H}}{\text{N}}}{\bigcirc} \xrightarrow[\text{or H}_2, \text{Ni}]{\text{NaBH}_3\text{CN}} \underset{\underset{\underset{\text{C}_6\text{H}_5\text{CHCH}_3}{|}}{\text{N}}}{\bigcirc}$$

Acetophenone   Piperidine   *N*-(1-Phenylethyl)piperidine

22.31 (*a*) The reaction of alkyl halides with *N*-potassiophthalimide (the first step in the Gabriel synthesis of amines) is a nucleophilic substitution reaction. Alkyl bromides are more reactive than alkyl fluorides; i.e., bromide is a better leaving group than fluoride.

$$N\text{-Potassiophthalimide} + FCH_2CH_2Br \longrightarrow 2\text{-Phthalimidoethyl fluoride}$$

N-Potassiophthalimide   1-Bromo-2-fluoroethane   2-Phthalimidoethyl fluoride

(b) In this example one bromide is attached to a primary and the other to a secondary carbon. Phthalimide anion is a good nucleophile and reacts with alkyl halides by the S_N2 mechanism. It attacks the less hindered primary carbon.

$$\longrightarrow NCH_2CH_2CH_2CHCH_3$$

more crowded

less crowded

1,4-Dibromopentane

N-4-Bromopentylphthalimide
(only product, 67% yield)

(c) Both bromines are bonded to primary carbons, but branching at the adjacent carbon hinders nucleophilic attack at one of them.

$$\longrightarrow NCH_2CH_2CCH_2Br$$

more
crowded

less
crowded

1,4-Dibromo-2,2-dimethylbutane

N-4-Bromo-3,3-dimethylphthalimide
(only product, 53% yield)

22.32 (a) Amines are basic and are protonated by hydrogen halides.

$$C_6H_5CH_2NH_2 + HBr \longrightarrow C_6H_5CH_2\overset{+}{N}H_3\ Br^-$$

Benzylamine   Benzylammonium
bromide

(b) Equimolar amounts of benzylamine and sulfuric acid yield benzylammonium hydrogen sulfate as the product.

$$C_6H_5CH_2NH_2 + HOSO_2OH \longrightarrow C_6H_5CH_2\overset{+}{N}H_3\ ^-OSO_2OH$$

Benzylamine   Sulfuric acid   Benzylammonium hydrogen
sulfate

(c) Acetic acid transfers a proton to benzylamine.

$$C_6H_5CH_2NH_2 + CH_3\overset{\overset{O}{\|}}{C}OH \longrightarrow C_6H_5CH_2\overset{+}{N}H_3\ ^-O\overset{\overset{O}{\|}}{C}CH_3$$

Benzylamine   Acetic acid   Benzylammonium acetate

(d) Acetyl chloride reacts with benzylamine by acyl transfer to form an amide.

$$2C_6H_5CH_2NH_2 + CH_3\overset{\overset{\displaystyle O}{\|}}{C}Cl \longrightarrow CH_3\overset{\overset{\displaystyle O}{\|}}{C}NHCH_2C_6H_5 + C_6H_5CH_2\overset{+}{N}H_3 \quad Cl^-$$

Benzylamine   Acetyl   *N*-Benzylacetamide   Benzylammonium
             chloride   chloride

(e) Acetic anhydride also reacts with benzylamine by acyl transfer.

$$2C_6H_5CH_2NH_2 + CH_3\overset{\overset{\displaystyle O}{\|}}{C}O\overset{\overset{\displaystyle O}{\|}}{C}CH_3 \longrightarrow CH_3\overset{\overset{\displaystyle O}{\|}}{C}NHCH_2C_6H_5 + C_6H_5CH_2\overset{+}{N}H_3 \;\; {}^-O\overset{\overset{\displaystyle O}{\|}}{C}CH_3$$

Benzylamine   Acetic anhydride   *N*-Benzylacetamide   Benzylammonium acetate

(f) Primary amines react with ketones to give imines.

$$C_6H_5CH_2NH_2 + CH_3\overset{\overset{\displaystyle O}{\|}}{C}CH_3 \longrightarrow (CH_3)_2C{=}NCH_2C_6H_5$$

Benzylamine   Acetone   *N*-Isopropylidenebenzylamine

(g) These reaction conditions lead to reduction of the imine formed in part (f). The reaction is reductive amination.

$$C_6H_5CH_2NH_2 + CH_3\overset{\overset{\displaystyle O}{\|}}{C}CH_3 \xrightarrow{\;H_2,\,Ni\;} (CH_3)_2CHNHCH_2C_6H_5$$

Benzylamine   Acetone   *N*-Isopropylbenzylamine

(h) Amines are nucleophilic and bring about the opening of epoxide rings.

$$C_6H_5CH_2NH_2 + \underset{\displaystyle O}{CH_2{-}CH_2} \longrightarrow C_6H_5CH_2NHCH_2CH_2OH$$

Benzylamine   Ethylene oxide   2-(*N*-Benzylamino)ethanol

(i) In these nucleophilic ring-opening reactions the amine attacks the less sterically hindered carbon of the ring.

$$C_6H_5CH_2NH_2 + \underset{\displaystyle O}{CH_2{-}CHCH_3} \longrightarrow C_6H_5CH_2NHCH_2\underset{\displaystyle OH}{CHCH_3}$$

Benzylamine   1,2-Epoxypropane   1-(*N*-Benzylamino)-2-propanol

(j) With excess methyl iodide, amines are converted to quaternary ammonium iodides.

$$C_6H_5CH_2NH_2 + \quad 3CH_3I \quad \longrightarrow \quad C_6H_5CH_2\overset{+}{N}(CH_3)_3 \; I^-$$

Benzylamine   Methyl iodide   Benzyltrimethylammonium
                                                 iodide

(k) Nitrous acid forms from sodium nitrite in dilute hydrochloric acid. Nitrosation of benzylamine in water gives benzyl alcohol via a diazonium ion intermediate.

$$C_6H_5CH_2NH_2 \xrightarrow[\;H_2O\;]{NaNO_2,\,HCl} C_6H_5CH_2\overset{+}{N}{\equiv}N \xrightarrow[\;H_2O\;]{-N_2} C_6H_5CH_2OH$$

Benzylamine   Benzyldiazonium ion   Benzyl alcohol

Benzyl chloride will also be formed by attack of chloride on the diazonium ion.

22.33 (a) Aniline is a weak base and yields a salt on reaction with hydrogen bromide.

$$C_6H_5NH_2 + \quad HBr \quad \longrightarrow \quad C_6H_5\overset{+}{N}H_3 \; Br^-$$

Aniline   Hydrogen   Anilinium bromide
        bromide

(b) Aniline acts as a nucleophile toward methyl iodide. With excess methyl iodide, a quaternary ammonium salt is formed.

$$C_6H_5NH_2 + 3CH_3I \longrightarrow C_6H_5\overset{+}{N}(CH_3)_3 \ I^-$$

Aniline   Methyl N,N,N-Trimethylanilinium
iodide   iodide

The reaction is carried out in the presence of a base such as potassium carbonate to neutralize the HI that is formed.

(c) Aniline undergoes nucleophilic addition to aldehydes and ketones to form imines.

$$C_6H_5NH_2 + \overset{\displaystyle O}{\overset{\|}{CH_3CH}} \longrightarrow C_6H_5N{=}CHCH_3 + H_2O$$

Aniline   Acetaldehyde N-Phenylacetaldimine   Water

(d) When an imine is formed in the presence of hydrogen and a suitable catalyst, reductive amination occurs to give an amine.

$$C_6H_5NH_2 + \overset{\displaystyle O}{\overset{\|}{CH_3CH}} \xrightarrow{H_2,\ Ni} C_6H_5NHCH_2CH_3$$

Aniline   Acetaldehyde N-Ethylaniline

(e) Aniline undergoes N-acylation on treatment with carboxylic acid anhydrides.

$$2C_6H_5NH_2 + \overset{\displaystyle O\ \ O}{\overset{\|\ \ \|}{CH_3COCCH_3}} \longrightarrow \overset{\displaystyle O}{\overset{\|}{C_6H_5NHCCH_3}} + C_6H_5\overset{+}{N}H_3 \ \overset{\displaystyle O}{{}^-\overset{\|}{OCCH_3}}$$

Aniline   Acetic anhydride   Acetanilide   Anilinium acetate

(f) Carboxylic acid chlorides bring about N-acylation of arylamines.

$$2C_6H_5NH_2 + \overset{\displaystyle O}{\overset{\|}{C_6H_5CCl}} \longrightarrow \overset{\displaystyle O}{\overset{\|}{C_6H_5NHCC_6H_5}} + C_6H_5\overset{+}{N}H_3 \ Cl^-$$

Aniline   Benzoyl chloride   Benzanilide   Anilinium chloride

(g) Nitrosation of primary arylamines yields aryl diazonium salts.

$$C_6H_5NH_2 \xrightarrow[H_2O,\ 0-5°C]{NaNO_2,\ H_2SO_4} C_6H_5\overset{+}{N}{\equiv}N: \ HSO_4^-$$

Aniline   Benzenediazonium
hydrogen sulfate

The replacement reactions that can be achieved by using diazonium salts are illustrated in parts (h) through (n). In all cases molecular nitrogen is lost from the ring carbon to which it was attached and is replaced by another substituent.

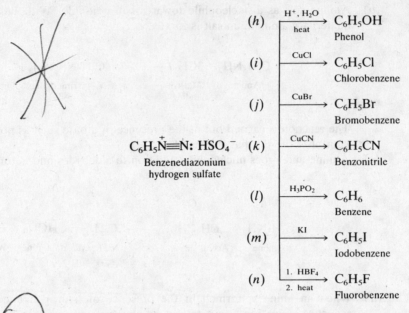

$$C_6H_5\overset{+}{N}\equiv N: \ HSO_4^-$$
Benzenediazonium
hydrogen sulfate

(h) $\xrightarrow[\text{heat}]{H^+, H_2O}$ C$_6$H$_5$OH
Phenol

(i) $\xrightarrow{CuCl}$ C$_6$H$_5$Cl
Chlorobenzene

(j) $\xrightarrow{CuBr}$ C$_6$H$_5$Br
Bromobenzene

(k) $\xrightarrow{CuCN}$ C$_6$H$_5$CN
Benzonitrile

(l) $\xrightarrow{H_3PO_2}$ C$_6$H$_6$
Benzene

(m) $\xrightarrow{KI}$ C$_6$H$_5$I
Iodobenzene

(n) $\xrightarrow[\text{2. heat}]{\text{1. } HBF_4}$ C$_6$H$_5$F
Fluorobenzene

(o) The nitrogens of an aryl diazonium salt are retained on reaction with the electron-rich ring of a phenol. Azo coupling occurs.

$$C_6H_5\overset{+}{N}\equiv N: + C_6H_5OH \longrightarrow C_6H_5N=N-\text{⟨⟩}-OH$$
$$HSO_4^-$$

Benzenediazonium　　　　Phenol　　　　　　p-(Azophenyl)phenol
hydrogen sulfate

(p) Azo coupling occurs when aryl diazonium salts react with *N,N*-dialkylarylamines.

$$C_6H_5\overset{+}{N}\equiv N: + C_6H_5N(CH_3)_2 \longrightarrow C_6H_5N=N-\text{⟨⟩}-N(CH_3)_2$$
$$HSO_4^-$$

Benzenediazonium　　*N,N*-Dimethylaniline　　p-(Azophenyl)-*N,N*-dimethylaniline
hydrogen sulfate

22.34 (a) Amides are reduced to amines by lithium aluminum hydride.

$$C_6H_5NH\overset{O}{\overset{\|}{C}}CH_3 \xrightarrow[\text{2. } H_2O]{\text{1. } LiAlH_4, \text{ diethyl ether}} C_6H_5NHCH_2CH_3$$

Acetanilide　　　　　　　　　　　　　　　　　*N*-Ethylaniline

(b) Acetanilide is a reactive substrate toward electrophilic aromatic substitution. An acetamido group is ortho, para–directing.

$$C_6H_5NH\overset{O}{\overset{\|}{C}}CH_3 \xrightarrow[H_2SO_4]{HNO_3}$$

Acetanilide　　　　　　　*o*-Nitroacetanilide　　*p*-Nitroacetanilide

(c) Sulfonation of the ring occurs.

$$C_6H_5NH\overset{\overset{\displaystyle O}{\|}}{C}CH_3 \xrightarrow[\text{H}_2\text{SO}_4]{\text{SO}_3}$$

NHCCH₃ (with O double bond) on a benzene ring with SO₃H at para position   + ortho isomer

Acetanilide   *p*-Acetamidobenzenesulfonic
                                                  acid

(d) Bromination of the ring takes place.

$$C_6H_5NH\overset{\overset{\displaystyle O}{\|}}{C}CH_3 \xrightarrow[\text{acetic acid}]{\text{Br}_2}$$

NHCCH₃ on benzene ring with Br at para position   + ortho isomer

Acetanilide   *p*-Bromoacetanilide

(e) Acetanilide undergoes Friedel-Crafts alkylation readily.

$$C_6H_5NH\overset{\overset{\displaystyle O}{\|}}{C}CH_3 + (CH_3)_3CCl \xrightarrow{\text{AlCl}_3}$$

NHCCH₃ on benzene ring with C(CH₃)₃ at para position   + ortho isomer

Acetanilide   *tert*-Butyl   *p-tert*-Butylacetanilide
                      chloride

(f) Friedel-Crafts acylation also is easily carried out.

$$C_6H_5NH\overset{\overset{\displaystyle O}{\|}}{C}CH_3 + CH_3\overset{\overset{\displaystyle O}{\|}}{C}Cl \xrightarrow{\text{AlCl}_3}$$

NHCCH₃ on benzene ring with CCH₃ (with O) at para position   + ortho isomer

Acetanilide   Acetyl   *p*-Acetamidoacetophenone
                      chloride

(g) Acetanilide is an amide and can be hydrolyzed when heated with aqueous acid.
     Under acidic conditions the aniline that is formed exists in its protonated form as the
     anilinium cation.

$$C_6H_5NH\overset{\overset{\displaystyle O}{\|}}{C}CH_3 + H_2O + HCl \longrightarrow C_6H_5\overset{+}{N}H_3\ Cl^- + HO\overset{\overset{\displaystyle O}{\|}}{C}CH_3$$

Acetanilide   Water Hydrogen   Anilinium   Acetic acid
                      chloride   chloride

(*h*) Amides are also hydrolyzed in base.

$$C_6H_5NH\overset{O}{\overset{\|}{C}}CH_3 + \ NaOH \xrightarrow{\ H_2O\ } C_6H_5NH_2 + Na^+ \ ^-O\overset{O}{\overset{\|}{C}}CH_3$$

Acetanilide   Sodium   Aniline   Sodium acetate
            hydroxide

22.35 (*a*) The reaction illustrates the preparation of a secondary amine by reductive amination.

Cyclohexanone   Cyclohexylamine   Dicyclohexylamine (70%)

(*b*) Amides are reduced to amines by lithium aluminum hydride.

$$\xrightarrow[\text{2. H}_2\text{O, HO}^-]{\text{1. LiAlH}_4}$$

6-Ethyl-6-azabicyclo[3.2.1]octan-7-one   6-Ethyl-6-azabicyclo[3.2.1]octane

(*c*) Treatment of alcohols with *p*-toluenesulfonyl chloride converts them to *p*-toluenesulfonate esters.

$$C_6H_5CH_2CH_2CH_2OH + CH_3\text{—}\langle\text{—}\rangle\text{—}SO_2Cl$$

3-Phenyl-1-propanol   *p*-Toluenesulfonyl chloride

$$\downarrow \text{pyridine}$$

$$C_6H_5CH_2CH_2CH_2O\overset{O}{\underset{O}{\overset{\|}{\underset{\|}{S}}}}\text{—}\langle\text{—}\rangle\text{—}CH_3$$

3-Phenylpropyl *p*-toluenesulfonate

p-Toluenesulfonate is an excellent leaving group in nucleophilic substitution reactions. Dimethylamine is the nucleophile.

$$C_6H_5CH_2CH_2CH_2OSO_2\text{—}\langle\text{—}\rangle\text{—}CH_3 + (CH_3)_2NH \longrightarrow$$

3-Phenylpropyl
p-toluenesulfonate

$$C_6H_5CH_2CH_2CH_2N(CH_3)_2$$

N,*N*-Dimethyl-3-phenyl-1-propanamine (86%)

(*d*) Amines are sufficiently nucleophilic to react with epoxides. Attack occurs at the less substituted carbon of the epoxide.

2-(2,5-Dimethoxyphenyl)oxirane   Isopropylamine   1-(2,5-Dimethoxyphenyl)-2-
                                                           (isopropylamino)ethanol (67%)

(*e*) α-Halo ketones are reactive substrates in nucleophilic substitution reactions. Dibenzylamine is the nucleophile.

$$(C_6H_5CH_2)_2\overset{..}{N}H + CH_3\overset{O}{\overset{\|}{C}}CH_2\!-\!Cl \longrightarrow CH_3\overset{O}{\overset{\|}{C}}CH_2N(CH_2C_6H_5)_2$$

Dibenzylamine   1-Chloro-2-propanone   1-(Dibenzylamino)-2-
                                              propanone (87%)

Since the reaction liberates hydrogen chloride, it is carried out in the presence of added base—in this case triethylamine—so as to avoid converting the dibenzylamine to its hydrochloride salt.

(*f*) Quaternary ammonium hydroxides undergo a Hofmann elimination reaction when they are heated. A point to be considered here concerns the regioselectivity of Hofmann eliminations: it is the less hindered β proton that is removed by the base, so that usually the less substituted alkene is formed.

trans-1-Isopropenyl-4-
methylcyclohexane (98%)   Trimethylamine

Elimination to give   [CH₃-cyclohexane=C(CH₃)₂]   does not occur.

(*g*) The combination of sodium nitrite and aqueous acid is a nitrosating agent. Secondary alkanamines react with nitrosating agents to give *N*-nitroso amines as the isolated products.

$$(CH_3)_2CHNHCH(CH_3)_2 \xrightarrow[\text{HCl, H}_2\text{O}]{\text{NaNO}_2} (CH_3)_2CHNCH(CH_3)_2$$

Diisopropylamine   *N*-Nitrosodiisopropylamine (91%)

22.36 (*a*) Catalytic hydrogenation reduces nitro groups to amino groups.

1,2-Diethyl-4-nitrobenzene   3,4-Diethylaniline (93–99%)

(*b*) Nitro groups are readily reduced by tin(II) chloride.

$$\text{1,3-Dimethyl-2-nitrobenzene} \xrightarrow[\text{2. HO}^-]{\text{1. SnCl}_2,\ \text{HCl}} \text{2,6-Dimethylaniline}$$

This reaction is the first step in a synthesis of the drug *lidocaine*.

(*c*) The amino group of arylamines is nucleophilic and undergoes acylation on reaction with chloroacetyl chloride.

2,6-Dimethylaniline + ClCH$_2$CCl (Chloroacetyl chloride) ⟶ N-(Chloroacetyl)-2,6-dimethylaniline

Chloroacetyl chloride is a difunctional compound—it is both an acyl chloride and an alkyl chloride. Acyl chlorides react with nucleophiles faster than do alkyl chlorides, so that acylation of the amine nitrogen occurs rather than alkylation.

(*d*) The final step in the synthesis of lidocaine is displacement of the chloride by diethylamine from the α-halo amide formed in part (*c*) in a nucleophilic substitution reaction.

N-(Chloroacetyl)-2,6-dimethylaniline + (CH$_3$CH$_2$)$_2$NH (Diethylamine) ⟶ Lidocaine

The reaction is carried out with excess diethylamine, which acts as a base to neutralize the hydrogen chloride formed.

(*e*) For use as an anesthetic, lidocaine is made available as its hydrochloride salt. Of the two nitrogens in lidocaine, the amine nitrogen is more basic than the amide.

Lidocaine $\xrightarrow{\text{HCl}}$ Lidocaine hydrochloride

(*f*) Lithium aluminum hydride reduction of amides is one of the best methods for the preparation of amines, including arylamines.

$$\text{C}_6\text{H}_5\text{NHCCH}_2\text{CH}_2\text{CH}_3 \xrightarrow[\text{2. H}_2\text{O}]{\text{1. LiAlH}_4} \text{C}_6\text{H}_5\text{NHCH}_2\text{CH}_2\text{CH}_2\text{CH}_3$$

N-Phenylbutanamide → *N*-Butylaniline (92%)

(g) Arylamines react with aldehydes and ketones in the presence of hydrogen and nickel to give the product of reductive amination.

$$C_6H_5NH_2 + CH_3(CH_2)_5\overset{\overset{\displaystyle O}{\|}}{C}H \xrightarrow{H_2,\ Ni} C_6H_5NHCH_2(CH_2)_5CH_3$$

Aniline   Heptanal   *N*-Heptylaniline (65%)

(h) Acetanilide is a reactive substrate toward electrophilic aromatic substitution. On reaction with chloroacetyl chloride, it undergoes Friedel-Crafts acylation, primarily at its para position.

Acetanilide   Chloroacetyl   *p*-Acetamidophenacyl chloride
                  chloride   (79–83%)

Acylation, rather than alkylation, occurs. Acyl chlorides are more reactive than alkyl chlorides toward electrophilic aromatic substitution reactions as a result of the more stable intermediate (acylium ion) formed.

(i) Reduction with iron in hydrochloric acid is one of the most common methods for converting nitroarenes to arylamines.

4-Bromo-4'-nitrobiphenyl   4-Amino-4'-bromobiphenyl (94%)

(j) Primary arylamines are converted to aryl diazonium salts on treatment with sodium nitrite in aqueous acid. When the aqueous acidic solution containing the diazonium salt is heated, a phenol is formed.

4-Amino-4'-bromobiphenyl

4-Bromo-4'-hydroxybiphenyl (85%)

(k) This problem illustrates the conversion of an arylamine to an aryl chloride by the Sandmeyer reaction.

2,6-Dinitroaniline   2-Chloro-1,3-dinitrobenzene
                                                               (71–74%)

(*l*) Diazotization of primary arylamines followed by treatment with copper(I) bromide converts them to aryl bromides.

m-Bromoaniline   *m*-Dibromobenzene
(80–87%)

(*m*) Nitriles are formed when aryl diazonium salts react with copper(I) cyanide.

o-Nitroaniline   *o*-Nitrobenzonitrile
(87%)

(*n*) An aryl diazonium salt is converted to an aryl iodide on reaction with potassium iodide.

2,6-Diiodo-4-nitroaniline   1,2,3-Triiodo-5-nitrobenzene
(94–95%)

(*o*) Aryl diazonium fluoroborates are converted to aryl fluorides when heated. Both diazonium salt functions in the starting material undergo this reaction.

4,4′-Bis(diazonio)biphenylfluoroborate   4,4′-Difluorobiphenyl (82%)

(*p*) Hypophosphorous acid (H_3PO_2) reduces aryl diazonium salts to arenes.

2,4,6-Trinitroaniline   1,3,5-Trinitrobenzene
(60–65%)

(*q*) Ethanol, like hypophosphorous acid, is an effective reagent for the reduction of aryl diazonium salts.

2-Amino-5-iodobenzoic acid   *m*-Iodobenzoic acid
(86–93%)

(r) Diazotization of aniline followed by addition of a phenol yields a bright-red diazo-substituted phenol. The diazonium ion acts as an electrophile toward the activated aromatic ring of the phenol.

$$C_6H_5NH_2 \xrightarrow[\text{H}_2\text{O}]{\text{NaNO}_2,\ \text{H}_2\text{SO}_4} C_6H_5\overset{+}{N}\equiv N: \ HSO_4^-$$

Aniline   Benzenediazonium
                                           hydrogen sulfate

2,3,6-Trimethyl-4-(phenylazo)phenol
(98%)

(s) Nitrosation of N,N-dialkylarylamines takes place on the ring at the position para to the dialkylamino group.

$(CH_3)_2N$ —⟨ring⟩— CH_3 $\xrightarrow[\text{2. HO}^-]{\text{1. NaNO}_2,\ \text{HCl, H}_2\text{O, 5°C}}$ $(CH_3)_2N$ —⟨ring⟩— N=O, CH_3

N,N-Dimethyl-m-toluidine   3-Methyl-4-nitroso-N,N-dimethylaniline
                                                    (83%)

22.37 (a) 4-Methylpiperidine can participate in intermolecular hydrogen bonding in the liquid phase.

These hydrogen bonds must be broken in order for individual 4-methylpiperidine molecules to escape into the gas phase. N-Methylpiperidine lacks a proton bonded to nitrogen and so cannot engage in intermolecular hydrogen bonding. Less energy is required to transfer a molecule of N-methylpiperidine to the gaseous state, and therefore it has a lower boiling point than 4-methylpiperidine.

N-Methylpiperidine; no hydrogen
bonding possible to other
N-methylpiperidine molecules

(b) The two products are diastereomeric quaternary ammonium chlorides that differ in the configuration at the nitrogen atom.

$$\text{4-}tert\text{-Butyl-}N\text{-methylpiperidine} \xrightarrow{\ C_6H_5CH_2Cl\ } \quad + \quad$$

4-*tert*-Butyl-*N*-methylpiperidine

(c) Tetramethylammonium hydroxide cannot undergo Hofmann elimination. The only reaction that can take place is nucleophilic substitution.

$$(CH_3)_3\overset{+}{N}-CH_3 \quad {}^-OH \longrightarrow (CH_3)_3N: \ + \ CH_3OH$$

Tetramethylammonium hydroxide   Trimethylamine   Methanol

(d) The key intermediate in the reaction of an amine with nitrous acid is the corresponding diazonium ion.

$$CH_3CH_2CH_2NH_2 \xrightarrow[\ H_2O\ ]{NaNO_2,\ HCl} CH_3CH_2CH_2-\overset{+}{N}\equiv N:$$

1-Propanamine   Propyldiazonium ion

Loss of nitrogen from this diazonium ion is accompanied by a hydride shift to form a secondary carbocation.

$$CH_3CHCH_2-\overset{+}{N}\equiv N: \longrightarrow CH_3\overset{+}{CHCH_3} + :N\equiv N:$$
$$\quad\ |$$
$$\quad\ H$$

Propyldiazonium ion   Isopropyl cation   Nitrogen

Capture of isopropyl cation by water yields the major product of the reaction, 2-propanol.

$$CH_3\overset{+}{CHCH_3} + H_2O \longrightarrow CH_3CHCH_3 \longrightarrow CH_3CHCH_3 + H^+$$
$$\qquad\qquad\qquad\qquad\qquad\ \ |\qquad\qquad\qquad\ |$$
$$\qquad\qquad\qquad\qquad\quad H\overset{\overset{+}{O}}{\diagdown}H\qquad\qquad OH$$

Isopropyl cation   Water   2-Propanol

22.38 Alcohols are converted to *p*-toluenesulfonate esters by reaction with *p*-toluenesulfonyl chloride. None of the bonds to the stereogenic center are affected in this reaction.

$$CH_3(CH_2)_5 \overset{H}{\underset{CH_3}{\overset{|}{C}}}{-}OH + CH_3 {-}\langle\text{benzene}\rangle{-}SO_2Cl$$

(S)-2-Octanol   *p*-Toluenesulfonyl chloride

↓ pyridine

$$CH_3(CH_2)_5 \overset{H}{\underset{CH_3}{\overset{|}{C}}}{-}OSO_2 {-}\langle\text{benzene}\rangle{-}CH_3$$

(S)-1-Methylheptyl *p*-toluenesulfonate
(compound A)

Displacement of the *p*-toluenesulfonate leaving group by sodium azide proceeds with inversion of configuration.

$$:\overset{-}{N}{=}\overset{+}{N}{=}\overset{-}{N}: \quad \longrightarrow \quad CH_3(CH_2)_5 \overset{H}{\underset{CH_3}{\overset{|}{C}}}{-}OSO_2{-}\langle\text{benzene}\rangle{-}CH_3 \quad \longrightarrow \quad :\overset{-}{N}{=}\overset{+}{N}{=}N{-}\overset{H\ (CH_2)_5CH_3}{\underset{CH_3}{\overset{|}{C}}}$$

(S)-1-Methylheptyl *p*-toluenesulfonate   *(R)*-1-Methylheptyl azide
(compound A)   (compound B)

$$+ \ \overset{-}{O}SO_2{-}\langle\text{benzene}\rangle{-}CH_3$$

Reduction of the azide yields a primary amine. A nitrogen-nitrogen bond is cleaved; all the bonds to the stereogenic center remain intact.

$$:\overset{-}{N}{=}\overset{+}{N}{=}\overset{..}{N}{-}\overset{H\ (CH_2)_5CH_3}{\underset{CH_3}{\overset{|}{C}}} \quad \xrightarrow[\text{2. H}_2\text{O, HO}^-]{\text{1. LiAlH}_4} \quad H_2\overset{..}{N}{-}\overset{H\ (CH_2)_5CH_3}{\underset{CH_3}{\overset{|}{C}}}$$

(R)-1-Methylheptyl azide   *(R)*-2-Octanamine
(compound B)   (compound C)

22.39 *(a)* The overall transformation can be expressed as RBr→RCH$_2$NH$_2$. In many cases this can be carried out via a nitrile, as RBr→RCN→RCH$_2$NH$_2$. In this case, however, the substrate is 1-bromo-2,2-dimethylpropane, an alkyl halide that reacts very slowly in nucleophilic substitution processes. Carbon-carbon bond formation with 1-bromo-2,2-dimethylpropane can be achieved by carboxylation of the corresponding Grignard reagent.

$$(CH_3)_3CCH_2Br \quad \xrightarrow[\substack{\text{2. CO}_2 \\ \text{3. H}_3\text{O}^+}]{\text{1. Mg}} \quad (CH_3)_3CCH_2CO_2H$$

1-Bromo-2,2-dimethylpropane   3,3-Dimethylbutanoic acid (63%)

The carboxylic acid can then be converted to the desired amine by reduction of the derived amide.

$$(CH_3)_3CCH_2CO_2H \quad \xrightarrow[\text{2. NH}_3]{\text{1. SOCl}_2} \quad (CH_3)_3CCH_2\overset{O}{\overset{\|}{C}}NH_2 \quad \xrightarrow[\text{2. H}_2\text{O}]{\text{1. LiAlH}_4} \quad (CH_3)_3CCH_2CH_2NH_2$$

3,3-Dimethylbutanoic acid   3,3-Dimethylbutanamide   3,3-Dimethyl-1-butanamine

(51%)   (57%)

The yields listed in parentheses are those reported in the chemical literature for this synthesis.

(*b*) Consider the starting materials in relation to the desired product:

N-(10-Undecenyl)pyrrolidine   10-Undecenoic acid   Pyrrolidine

The synthetic tasks are to form the necessary carbon-nitrogen bond and to reduce the carbonyl group to a methylene group. This has been accomplished by way of the amide as a key intermediate.

10-Undecenoic acid   N-(10-Undecenoyl)pyrrolidine (75%)

N-(10-Undecenyl)pyrrolidine (66%)

A second approach utilizes reductive amination following conversion of the starting carboxylic acid to an aldehyde.

10-Undecenoic acid   10-Undecen-1-ol   10-Undecenal

The reducing agent in the reductive amination process cannot be hydrogen, because that would result in hydrogenation of the double bond. Sodium cyanoborohydride is required.

10-Undecenal   Pyrrolidine   N-(10-Undecenyl)pyrrolidine

(*c*) It is stereochemistry that determines the choice of which synthetic method to employ in introducing the amine group. The carbon-nitrogen bond must be formed with inversion of configuration at the alcohol carbon. Conversion of the alcohol to its *p*-toluenesulfonate ester ensures that the leaving group is introduced with exactly the same stereochemistry as the alcohol.

$$C_6H_5O \quad OH \quad + CH_3 \!-\!\!\!\bigcirc\!\!\!- SO_2Cl$$

cis-2-Phenoxycyclopentanol   P-Toluenesulfonyl chloride

pyridine ↓

$$C_6H_5O \quad OSO_2 \!-\!\!\!\bigcirc\!\!\!- CH_3$$

cis-2-Phenoxycyclopentyl *p*-toluenesulfonate

Once the leaving group has been introduced with the proper stereochemistry, it can be displaced by a nitrogen nucleophile suitable for subsequent conversion to an amine.

$$C_6H_5O \quad OSO_2 \!-\!\!\!\bigcirc\!\!\!- CH_3 \xrightarrow{NaN_3} C_6H_5O \quad N_3$$

cis-2-Phenoxycyclopentyl *p*-toluenesulfonate   trans-2-Phenoxycyclopentyl azide (90%)

1. LiAlH$_4$
2. H$_2$O ↓

$$C_6H_5O \quad NH_2$$

trans-2-Phenoxycyclopentyl-amine

(As actually reported, the azide was reduced by hydrogenation over a palladium catalyst, and the amine was isolated as its hydrochloride salt in 66 percent yield.)

(*d*) Recognition that the primary amine is derivable from the corresponding nitrile by reduction,

$$C_6H_5CH_2NCH_2CH_2CH_2CH_2NH_2 \;\Longrightarrow\; C_6H_5CH_2NCH_2CH_2CH_2C\!\equiv\!N$$
$$\qquad\qquad | \qquad\qquad\qquad\qquad\qquad\qquad\qquad\quad |$$
$$\qquad\qquad CH_3 \qquad\qquad\qquad\qquad\qquad\qquad\qquad CH_3$$

and that the necessary tertiary amine function can be introduced by a nucleophilic substitution reaction between the two given starting materials suggests the following synthesis:

$$C_6H_5CH_2NH \;+\; BrCH_2CH_2CH_2CN \longrightarrow C_6H_5CH_2NCH_2CH_2CH_2CN$$
$$\qquad\quad | \qquad\qquad\qquad\qquad\qquad\qquad\qquad\qquad\qquad |$$
$$\qquad\quad CH_3 \qquad\qquad\qquad\qquad\qquad\qquad\qquad\qquad CH_3$$

N-Methylbenzylamine   4-Bromobutanenitrile

1. LiAlH$_4$
2. H$_2$O ↓

$$C_6H_5CH_2NCH_2CH_2CH_2CH_2NH_2$$
$$\qquad\qquad |$$
$$\qquad\qquad CH_3$$

N-Benzyl-*N*-methyl-1,4-butanediamine

Alkylation of *N*-methylbenzylamine with 4-bromobutanenitrile has been achieved in 92 percent yield in the presence of potassium carbonate as a weak base to neutralize the hydrogen bromide produced. The nitrile may be reduced with lithium aluminum hydride, as shown in the equation, or by catalytic hydrogenation. Catalytic hydrogenation over platinum gave the desired diamine, isolated as its hydrochloride salt, in 90 percent yield.

(*e*) The overall transformation may be viewed retrosynthetically as follows:

$$ArCH_2N(CH_3)_2 \implies ArCH_2Br \implies ArCH_3$$

$$Ar = NC\!-\!\!\left\langle\!\!\!\bigcirc\!\!\!\right\rangle\!-$$

The sequence which presents itself begins with benzylic bromination with *N*-bromosuccinimide.

p-Cyanotoluene   *p*-Cyanobenzyl bromide

The reaction shown in the equation has been reported in the chemical literature and gave the benzylic bromide in 60 percent yield.

Treatment of this bromide with dimethylamine gives the desired product. (The isolated yield was 83 percent by this method.)

p-Cyanobenzyl bromide   Dimethylamine   *p*-Cyano-*N,N*-dimethylbenzylamine

22.40 (*a*) This problem illustrates the application of the Sandmeyer reaction to the preparation of aryl cyanides. Diazotization of *p*-nitroaniline followed by treatment with copper(I) cyanide converts it to *p*-nitrobenzonitrile.

p-Nitroaniline   *p*-Nitrobenzonitrile

(*b*) An acceptable pathway becomes apparent when it is realized that the amino group in the product is derived from the nitro group of the starting material. Two chlorines are introduced by electrophilic aromatic substitution, the third by a Sandmeyer reaction.

Two of the required chlorine atoms can be introduced by chlorination of the starting material, *p*-nitroaniline.

p-Nitroaniline   2,6-Dichloro-4-nitroaniline

The third chlorine can be introduced via the Sandmeyer reaction. Reduction of the nitro group completes the synthesis of 3,4,5-trichloroaniline.

2,6-Dichloro-4-nitroaniline   1,2,3-Trichloro-5-nitrobenzene   3,4,5-Trichloroaniline

The reduction step has been carried out by hydrogenation with a nickel catalyst in 70 percent yield.

(c) The amino group that is present in the starting material facilitates the introduction of the bromine substituents, and is then removed by reductive deamination.

p-Nitroaniline   2,6-Dibromo-4-nitroaniline   1,3-Dibromo-5-nitrobenzene
    (95%)   (70%)

Hypophosphorous acid has also been used successfully in the reductive deamination step.

(d) Reduction of the nitro group of the 1,3-dibromo-5-nitrobenzene prepared in the preceding part of this problem gives the desired product. The customary reducing agents used for the reduction of nitroarenes would all be suitable.

1,3-Dibromo-5-nitrobenzene   3,5-Dibromoaniline
[prepared from *p*-nitroaniline as in part (c)]   (80%)

(e) The synthetic objective is:

p-Acetamidophenol

This compound, known as *acetaminophen* and used as an aspirin substitute to reduce fever and relieve minor pain, may be prepared from *p*-nitroaniline by way of *p*-nitrophenol.

p-Nitroaniline   *p*-Nitrophenol   *p*-Acetamidophenol

Any of the customary reducing agents suitable for converting aryl nitro groups to arylamines (Fe, HCl; Sn, HCl; H_2, Ni) may be used. Acetylation of *p*-aminophenol may be carried out with acetyl chloride or acetic anhydride. The amino group of *p*-aminophenol is more nucleophilic than the hydroxyl group and is acetylated preferentially.

22.41 (*a*) Replacement of an amino substituent by a bromine is readily achieved by the Sandmeyer reaction.

o-Anisidine   *o*-Bromoanisole (88–93%)

(*b*) This conversion demonstrates the replacement of an amino substituent by fluorine via the Schiemann reaction.

o-Anisidine   *o*-Methoxybenzenediazonium fluoroborate (57%)   *o*-Fluoroanisole (53%)

(*c*) We can use the *o*-fluoroanisole prepared in part (*b*) to prepare 3-fluoro-4-methoxyacetophenone by Friedel-Crafts acylation.

o-Anisidine   *o*-Fluoroanisole   3-Fluoro-4-methoxyacetophenone (70–80%)

Remember from Section 12.16 that it is the more activating substituent that determines the regioselectivity of electrophilic aromatic substitution when an arene bears two different substituents. Methoxy is a strongly activating substituent; fluorine is slightly deactivating. Friedel-Crafts acylation takes place at the position para to the methoxy group.

(*d*) The *o*-fluoroanisole prepared in part (*b*) serves nicely as a precursor to 3-fluoro-4-methoxybenzonitrile via diazonium salt chemistry.

[from part (*b*)]

The desired sequence of reactions to carry out the synthesis is:

o-Anisidine   *o*-Fluoroanisole   2-Fluoro-4-nitroanisole
                                                                       (53%)

3-Fluoro-4-methoxybenzonitrile   4-Amino-2-fluoroanisole
           (46%)   (85%)

Conversion of *o*-fluoroanisole to 4-amino-2-fluoroanisole proceeds in the conventional way by preparation and reduction of a nitro derivative. Once the necessary arylamine is at hand, it is converted to the nitrile by a Sandmeyer reaction.

(*e*) Diazotization followed by hydrolysis of the 4-amino-2-fluoroanisole prepared as an intermediate in part (*d*) yields the desired phenol.

o-Anisidine   4-Amino-2-fluoroanisole   3-Fluoro-4-methoxyphenol
                                                                          (70%)

22.42 (*a*) The carboxyl group of *p*-aminobenzoic acid can be derived from the methyl group of *p*-methylaniline by oxidation. First, however, the nitrogen must be acylated so as to protect the ring from oxidation.

p-Aminobenzoic
    acid   *p*-Methylaniline

The sequence of reactions to be used is:

p-Methylaniline   p-Methylacetanilide   p-Acetamidobenzoic acid

p-Aminobenzoic acid

(b) Attachment of fluoro and propanoyl groups to a benzene ring is required. The fluorine substituent can be introduced by way of the diazonium tetrafluoroborate, the propanoyl group by way of a Friedel-Crafts acylation. Since the fluorine substituent is ortho, para–directing, introducing it first gives the proper orientation of substituents.

Ethyl p-fluorophenyl   Fluorobenzene   Aniline
ketone

Fluorobenzene is prepared from aniline by the Schiemann reaction, shown in Section 22.18. Aniline is, of course, prepared from benzene via nitrobenzene. Friedel-Crafts acylation of fluorobenzene has been carried out with the results shown and gives the required ethyl p-fluorophenyl ketone as the major product.

Fluorobenzene   Propanoyl chloride   Ethyl p-fluorophenyl
ketone (86%)

(c) Our synthetic plan is based on the essential step of forming the fluorine derivative from an amine by way of a diazonium salt.

1-Bromo-2-fluoro-   2,4-Dimethylaniline
3,5-dimethylbenzene

The required substituted aniline is derived from *m*-xylene by a standard synthetic sequence.

m-Xylene → (HNO₃, H₂SO₄) → 1,3-Dimethyl-4-nitrobenzene (98%) → (1. Fe, HCl 2. HO⁻) → 2,4-Dimethylaniline → (Br₂) → 2-Bromo-4,6-dimethylaniline → (1. NaNO₂, HCl, H₂O, 0°C 2. HBF₄ 3. heat) → 1-Bromo-2-fluoro-3,5-dimethylbenzene (60%)

(*d*) In this problem two nitrogen-containing groups of the starting material are each to be replaced by a halogen substituent. The task is sufficiently straightforward that it may be confronted directly.

Replace amino group by bromine:

2-Methyl-4-nitro-1-naphthylamine → (1. NaNO₂, HBr, H₂O 2. CuBr) → 1-Bromo-2-methyl-4-nitronaphthalene (82%)

Reduce nitro group to amine:

1-Bromo-2-methyl-4-nitronaphthalene → (1. Fe, HCl 2. HO⁻) → 4-Bromo-3-methyl-1-naphthylamine

Replace amino group by fluorine:

4-Bromo-3-methyl-1-naphthylamine → (1. NaNO₂, HCl, H₂O, 0–5°C 2. HBF₄ 3. heat) → 1-Bromo-4-fluoro-2-methylnaphthalene (64%)

(e)   Bromination of the starting material will introduce the bromine substituent at the correct position, i.e., ortho to the *tert*-butyl group.

C(CH$_3$)$_3$ Br$_2$, Fe → C(CH$_3$)$_3$ — Br

NO$_2$ NO$_2$

p-tert-Butylnitrobenzene   2-Bromo-1-*tert*-butyl-4-nitrobenzene

The desired product will be obtained if the nitro group can be removed. This is achieved by its conversion to the corresponding amine, followed by reductive deamination.

C(CH$_3$)$_3$ — Br   H$_2$, Ni (or other appropriate reducing agent) → C(CH$_3$)$_3$ — Br   1. NaNO$_2$, H$^+$ 2. H$_3$PO$_2$ → C(CH$_3$)$_3$ — Br

NO$_2$ NH$_2$

2-Bromo-1-*tert*-butyl-4-nitrobenzene   3-Bromo-4-*tert*-butylaniline   *o*-Bromo-*tert*-butylbenzene

(f)   The proper orientation of the chlorine substituent can be achieved only if it is introduced after the nitro group is reduced.

C(CH$_3$)$_3$ — Cl ⇒ C(CH$_3$)$_3$ — Cl (NH$_2$) ⇒ C(CH$_3$)$_3$ (NH$_2$) ⇒ C(CH$_3$)$_3$ (NO$_2$)

The correct sequence of reactions to carry out this synthesis is shown.

C(CH$_3$)$_3$ H$_2$, Ni → C(CH$_3$)$_3$ acetic anhydride → C(CH$_3$)$_3$

NO$_2$ NH$_2$ NHCCH$_3$
                                                              ‖
                                                              O

p-tert-Butylnitrobenzene   *p-tert*-Butylaniline   *p-tert*-Butylacetanilide

                                                                        ↓ Cl$_2$

C(CH$_3$)$_3$ 1. NaNO$_2$, H$^+$ 2. H$_3$PO$_2$ ← C(CH$_3$)$_3$ hydrolysis to remove acetyl group ← C(CH$_3$)$_3$

Cl   Cl   Cl
                            NH$_2$ NHCCH$_3$
                                                          ‖
                                                                 O

m-tert-Butylchlorobenzene   4-*tert*-Butyl-2-chloroaniline   4-*tert*-Butyl-2-chloroacetanilide

(g) The orientation of substituents in the target molecule can be achieved by using an amino group to control the regiochemistry of bromination, then removing it by reductive deamination.

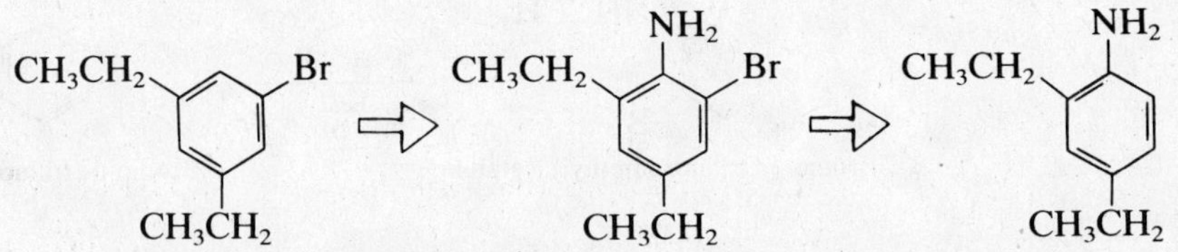

The amino group is introduced in the standard fashion by nitration of an arene followed by reduction.

This analysis leads to the synthesis shown.

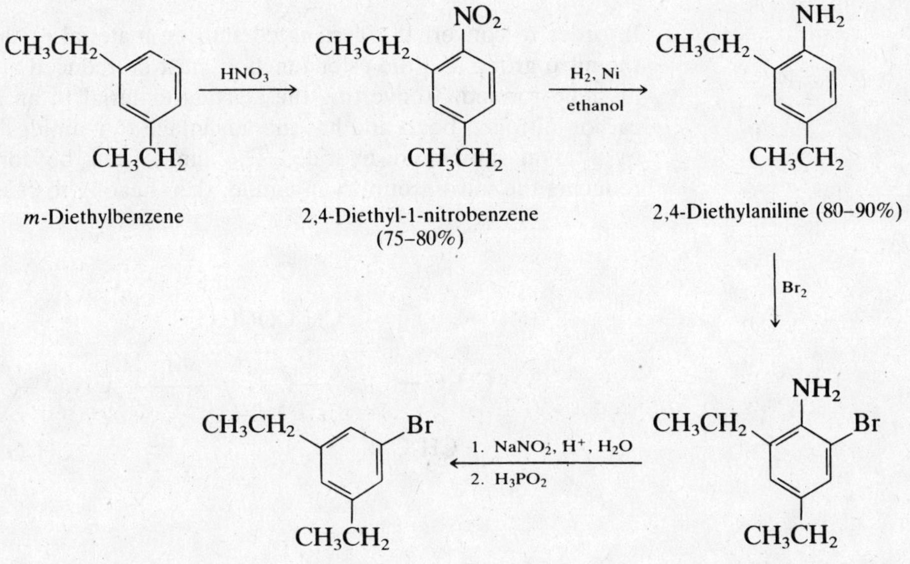

m-Diethylbenzene

2,4-Diethyl-1-nitrobenzene (75–80%)

2,4-Diethylaniline (80–90%)

1-Bromo-3,5-diethylbenzene (70%)

2-Bromo-4,6-diethylaniline (40%)

(h) In this exercise the two nitrogen substituents are differentiated; one is an amino nitrogen, the other an amide nitrogen. By keeping them differentiated they can be manipulated independently. Remove one amino group completely before deprotecting the other.

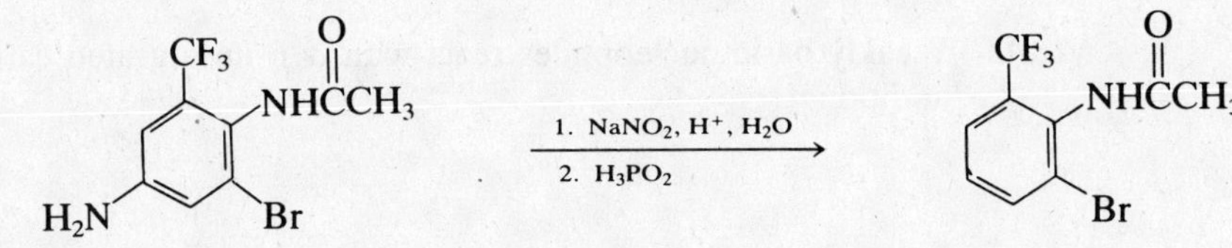

4-Amino-2-bromo-6-(trifluoromethyl)acetanilide

2-Bromo-6-(trifluoromethyl)acetanilide (92%)

Once the acetyl group has been removed by hydrolysis, the molecule is ready for introduction of the iodo substituent by way of a diazonium salt.

2-Bromo-6-(trifluoromethyl)acetanilide

2-Bromo-6-(trifluoromethyl)aniline (69%)

1-Bromo-2-iodo-3-(trifluoromethyl)benzene (87%)

(*i*) In order to convert the designated starting material to the indicated product, both the nitro group and the ester function must be reduced and a carbon-nitrogen bond must be formed. Converting the starting material to an amide gives the necessary carbon-nitrogen bond and has the advantage that amides can be reduced to amines by lithium aluminum hydride. The amide can be formed intramolecularly by reducing the nitro group to an amine, then heating to cause cyclization.

This synthesis is the one described in the chemical literature. Other routes are also possible, but the one shown is short and efficient.

22.43 Weakly basic nucleophiles react with α,β-unsaturated carbonyl compounds by conjugate addition.

Ammonia and its derivatives are very prone to react in this way; thus conjugate addition provides a method for the preparation of β-amino carbonyl compounds.

(*a*)

$$(CH_3)_2C{=}CHCCH_3 \;+\; NH_3 \;\longrightarrow\; (CH_3)_2CCH_2CCH_3$$

4-Methyl-3-penten-2-one   Ammonia   4-Amino-4-methyl-2-pentanone
(63–70%)

(*b*)

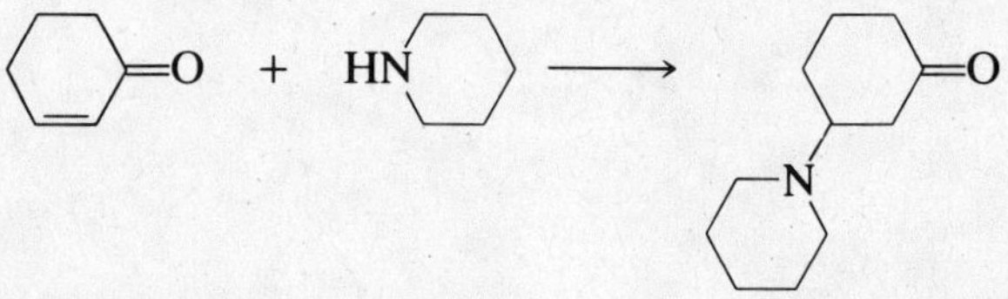

2-Cyclohexen-1-one   Piperidine   3-Piperidinocyclohexanone
(45%)

(*c*)

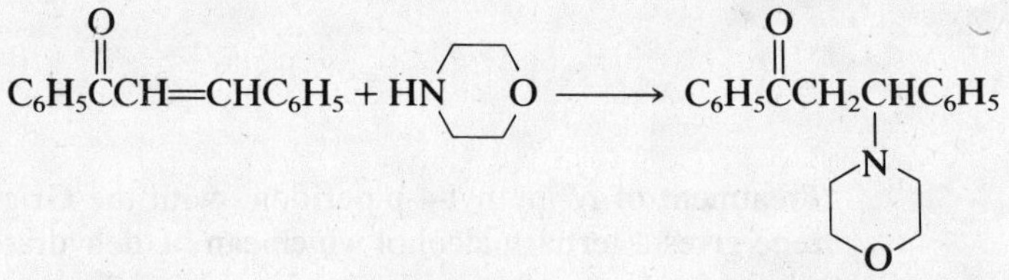

$$C_6H_5CCH{=}CHC_6H_5 + HN\!\!\!\begin{array}{c}\end{array}\!\!\!O \longrightarrow C_6H_5CCH_2CHC_6H_5$$

1,3-Diphenyl-2-
propen-1-one   Morpholine   3-Morpholino-1,3-diphenyl-
1-propanone (91%)

(*d*)   The conjugate addition reaction that takes place in this case is an intramolecular one and occurs in virtually 100 percent yield.

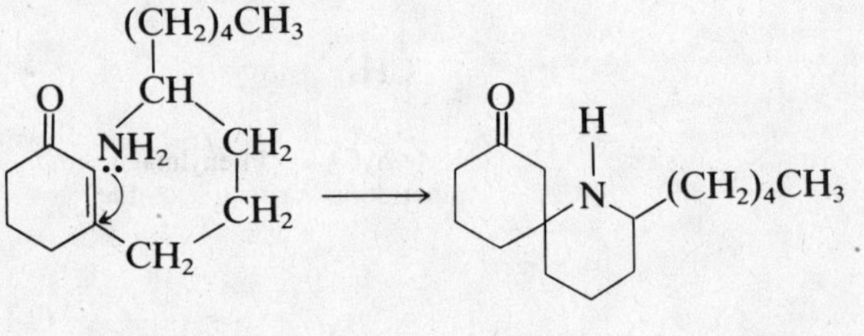

22.44   The first step in the synthesis is the conjugate addition of methylamine to ethyl acrylate. Two sequential Michael addition reactions take place.

$$CH_3NH_2 \;+\; CH_2{=}CHCOCH_2CH_3 \;\longrightarrow\; CH_3NHCH_2CH_2COCH_2CH_3$$

Methylamine   Ethyl acrylate

$$\Big\downarrow {\scriptstyle CH_2=CHCO_2CH_2CH_3}$$

$$CH_3N(CH_2CH_2CO_2CH_2CH_3)_2$$

Conversion of this intermediate to the desired *N*-methyl-4-piperidone requires a Dieckmann cyclization followed by decarboxylation of the resulting β-keto ester.

N-Methyl-4-piperidone

Treatment of *N*-methyl-4-piperidone with the Grignard reagent derived from bromobenzene gives a tertiary alcohol which can be dehydrated to an alkene. Hydrogenation of the alkene completes the synthesis.

N-Methyl-4-phenylpiperidine
(compound D)

22.45 Sodium cyanide reacts with alkyl bromides by the S_N2 mechanism. Reduction of the cyano group with lithium aluminum hydride yields a primary amine. This reveals the structure of mescaline to be 2-(3,4,5-trimethoxyphenyl)ethylamine.

3,4,5-Trimethoxybenzyl
bromide

2-(3,4,5-Trimethoxyphenyl)-
ethanenitrile

2-(3,4,5-Trimethoxyphenyl)ethylamine
(mescaline)

22.46 Reductive amination of a ketone with methylamine yields a secondary amine. Methamphetamine is *N*-methyl-1-phenyl-2-propanamine.

Benzyl methyl   Methylamine
ketone

N-Methyl-1-phenyl-
2-propanamine
(Methamphetamine)

22.47 There is no obvious reason why the dimethylamino group in 4-(*N,N*-dimethylamino)pyridine should be appreciably more basic than it is in *N,N*-dimethylaniline; it is the ring nitrogen of 4-(*N,N*-dimethylamino)pyridine that is more basic. Note that protonation of the ring nitrogen permits delocalization of the dimethylamino lone pair and dispersal of the positive charge.

Most stable protonated form of
4-(*N,N*-dimethylamino)pyridine

22.48 Since there are peaks corresponding to five aromatic protons in the ^{1}H nmr spectrum of each isomer, compounds E and F each contain a monosubstituted benzene ring. Only four compounds of molecular formula $C_8H_{11}N$ meet this requirement.

$C_6H_5CH_2NHCH_3$ $C_6H_5NHCH_2CH_3$ $C_6H_5CHCH_3$ $C_6H_5CH_2CH_2NH_2$
$\qquad\qquad\qquad\qquad\qquad\qquad\qquad\qquad\qquad\qquad\quad$ |
$\qquad\qquad\qquad\qquad\qquad\qquad\qquad\qquad\qquad\qquad\ \ NH_2$

N-Methylbenzylamine   *N*-Ethylaniline   1-Phenylethylamine   2-Phenylethylamine

Neither ^{1}H nmr spectrum is consistent with *N*-methylbenzylamine, which would have two singlets due to the methyl and methylene groups. Likewise, the spectra are not consistent with *N*-ethylaniline, which would exhibit the characteristic triplet-quartet pattern of an ethyl group. While there is a quartet in the spectrum of compound E, it corresponds to only one proton, not the two that an ethyl group requires. The one-proton quartet in compound E arises from an H—C—CH$_3$ unit. Compound E is 1-phenylethylamine.

Compound F has a ^{1}H nmr spectrum that fits 2-phenylethylamine.

overlapping multiplets
centered at δ = 2.7 ppm

While we might expect the signals arising from each set of methylene protons in the CH$_2$CH$_2$ unit to appear as a triplet, the actual spectrum is somewhat more complicated than that. Because the nonequivalent methylene groups of the CH$_2$CH$_2$ unit have similar chemical shifts, the simple first-order splitting rules do not apply.

22.49   Write the structural formulas for the two possible compounds given in the problem and consider how their ^{13}C nmr spectra will differ from each other. Both will exhibit their methyl carbons as quartets at high field, but they differ in the positions of their methylene and quaternary carbons. A carbon bonded to nitrogen is more shielded than one bonded to oxygen, because nitrogen is less electronegative than oxygen.

In one isomer the lowest-field signal is a singlet; in the other it is a triplet. The spectrum shown in Figure 22.11 shows the lowest-field signal as a triplet. Therefore the compound is 2-amino-2-methyl-1-propanol, (CH$_3$)$_2$CCH$_2$OH.
        |
       NH$_2$

This compound *cannot* be prepared by reaction of ammonia with an epoxide, because in basic solution nucleophiles attack epoxides at the less hindered carbon and therefore epoxide ring opening will give 1-amino-2-methyl-2-propanol rather than 2-amino-2-methyl-1-propanol.

2,2-Dimethyloxirane   Ammonia   1-Amino-2-methyl-2-propanol

SELF-TEST

PART A

A-1. Give an acceptable name for each of the following. Identify each compound as a primary, secondary, or tertiary amine.

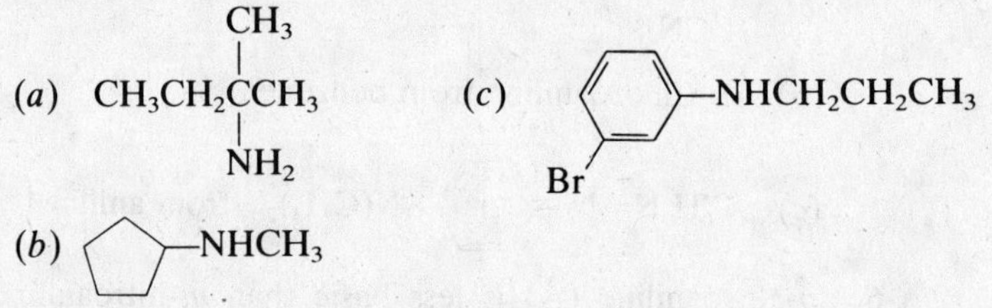

(a)
$$\underset{\underset{NH_2}{|}}{\overset{\overset{CH_3}{|}}{CH_3CH_2CCH_3}}$$

(c)   a benzene ring with Br and $-NHCH_2CH_2CH_3$ substituents

(b)   a cyclopentane ring with $-NHCH_3$

A-2. Provide the correct structure of the reagent omitted from each of the following reactions:

(a) $C_6H_5CH_2Br \xrightarrow[\substack{2.\ LiAlH_4 \\ 3.\ H_2O}]{1.\ ?} C_6H_5CH_2NH_2$

(b) $C_6H_5CH_2Br \xrightarrow[\substack{2.\ LiAlH_4 \\ 3.\ H_2O}]{1.\ ?} C_6H_5CH_2CH_2NH_2$

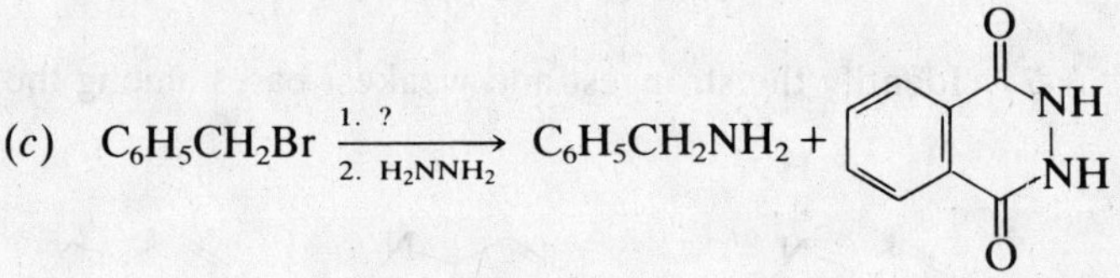

(c) $C_6H_5CH_2Br \xrightarrow[2.\ H_2NNH_2]{1.\ ?} C_6H_5CH_2NH_2 + $ phthalimide structure

A-3. Provide the missing component (reactant, reagent, or product) for each of the following:

(a) CH_3-⟨benzene ring⟩$-NH_2 \xrightarrow[H_2O]{NaNO_2,\ HCl}$?

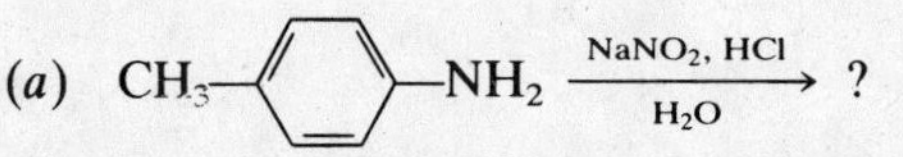

(b)   Product of part (a) $\xrightarrow{CuBr}$?

(c)   Product of part (a) $\xrightarrow{?}$ toluene

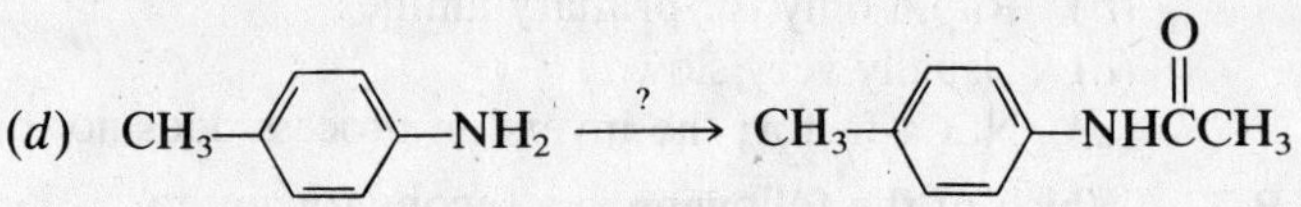

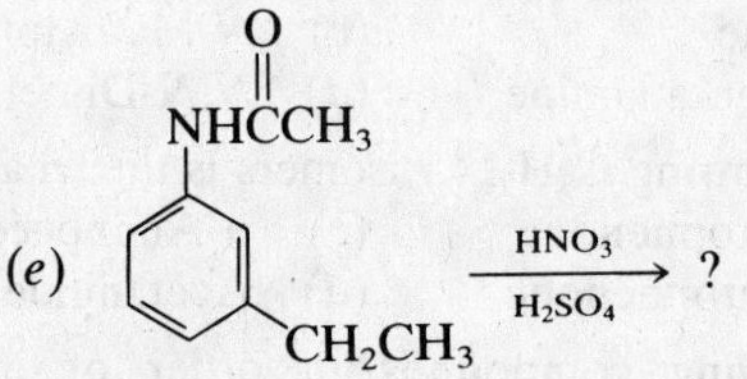

(d) CH_3-⟨benzene ring⟩$-NH_2 \xrightarrow{?} CH_3-$⟨benzene ring⟩$-\overset{\overset{O}{\|}}{NHCCH_3}$

(e)   benzene ring with $\overset{\overset{O}{\|}}{NHCCH_3}$ and CH_2CH_3 substituents $\xrightarrow[H_2SO_4]{HNO_3}$?

A-4. Provide structures for compounds A through E in the following reaction sequences:

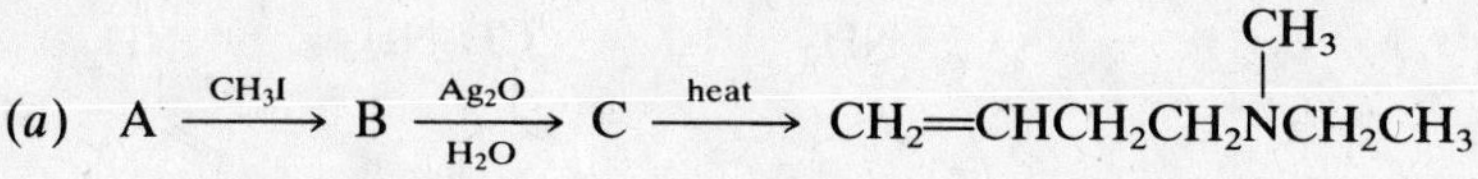

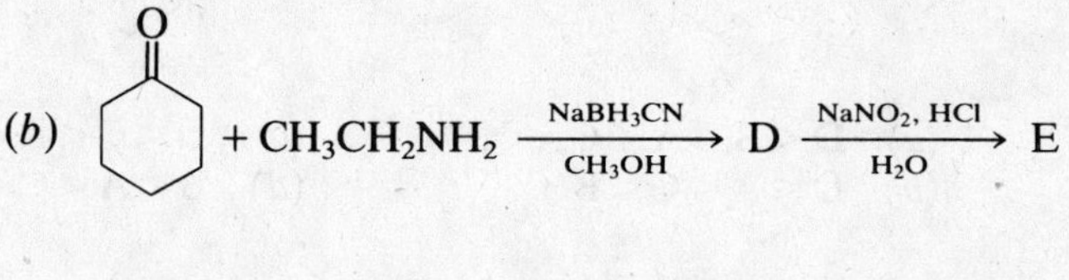

(a) $A \xrightarrow{CH_3I} B \xrightarrow[H_2O]{Ag_2O} C \xrightarrow{heat} \underset{\underset{CH_3}{|}}{CH_2=CHCH_2CH_2NCH_2CH_3}$

(b)   cyclohexanone $+ CH_3CH_2NH_2 \xrightarrow[CH_3OH]{NaBH_3CN} D \xrightarrow[H_2O]{NaNO_2,\ HCl} E$

A-5. Give the series of reaction steps involved in the following synthetic conversions:

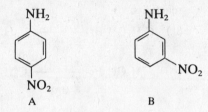

(*a*)   from benzene

(*b*) *m*-Chloroaniline from benzene

(*c*) $C_6H_5N{=}N{-}$⟨⟩$-N(CH_3)_2$ from aniline

A-6. *p*-Nitroaniline (A) is less basic than *m*-nitroaniline (B). Using resonance structures, explain the reason for this difference.

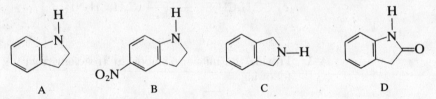

A   B

A-7. Identify the strongest and weakest bases among the following:

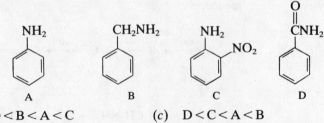

A   B   C   D

PART B

B-1. The rate of inversion of the sp^3 nitrogen in an alkylamine is:
(*a*) Usually very rapid
(*b*) Rapid only for primary amines
(*c*) Usually very slow
(*d*) Not a factor; the inversion process does not occur.

B-2. Which of the following is a secondary amine?
(*a*) 2-Butanamine   (*c*) *N*-Methylpiperidine
(*b*) *N*-Ethyl-2-pentanamine   (*d*) *N,N*-Dimethylcyclohexylamine

B-3. Which of the following C_8H_9NO isomers is the weakest base?
(*a*) *o*-Aminoacetophenone   (*c*) *p*-Aminoacetophenone
(*b*) *m*-Aminoacetophenone   (*d*) Acetanilide

B-4. Rank the following compounds in order of increasing basicity (weakest→ strongest):

A   B   C   D

(*a*) D<B<A<C   (*c*) D<C<A<B
(*b*) D<A<C<B   (*d*) B<A<C<D

B-5.   On the basis of comparison of the carbon-nitrogen bond length (of the amino nitrogen) in compounds A and B, which of the following statements is correct?

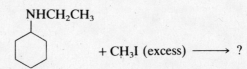

A   B

   (*a*)   The bond in A is shorter.
   (*b*)   The bond in B is shorter.
   (*c*)   The bonds in A and B have the same length.
   (*d*)   No comparison of bond length can be made.

B-6.   Which of the following arylamines will *not* form a diazonium salt on reaction with sodium nitrite in hydrochloric acid?
   (*a*)   *m*-Ethylaniline   (*c*)   *p*-Aminoacetophenone
   (*b*)   4-Chloro-2-nitroaniline   (*d*)   *N*-Ethyl-2-methylaniline

B-7.   The reaction

NHCH$_2$CH$_3$

$+$ CH$_3$I (excess) $\longrightarrow$?

   gives as final product
   (*a*)   A primary amine   (*c*)   A tertiary amine
   (*b*)   A secondary amine   (*d*)   A quaternary ammonium salt

B-8.   A substance is soluble in dilute aqueous HCl and has a single peak in the region 3200 to 3500 cm^{-1} in its infrared spectrum. Which of the following best fits the data?

   (*a*)   —N(CH$_3$)$_2$ (*c*)   —NHCH$_3$

   (*b*)   —NH$_2$ (*d*)   —CO$_2$H

CHAPTER

23

ARYL HALIDES

IMPORTANT TERMS AND CONCEPTS

Nucleophilic Aromatic Substitution Reactions (Secs. 23.5 to 23.8)

Under certain circumstances aryl halides are able to undergo nucleophilic aromatic substitution reactions. One of two conditions must be met: the aromatic ring must bear a strongly electron-withdrawing group, such as nitro, or the nucleophile must be a strong base, such as the amide ion.

When the first of these conditions is met, the reaction proceeds by an *addition-elimination mechanism.* As shown below, the intermediate formed on addition of the nucleophile is a resonance-stabilized carbanion. In order to participate in resonance delocalization of the negative charge, the nitro groups must be situated ortho or para to the halide. Elimination of halide ion restores aromaticity to the system, forming the product.

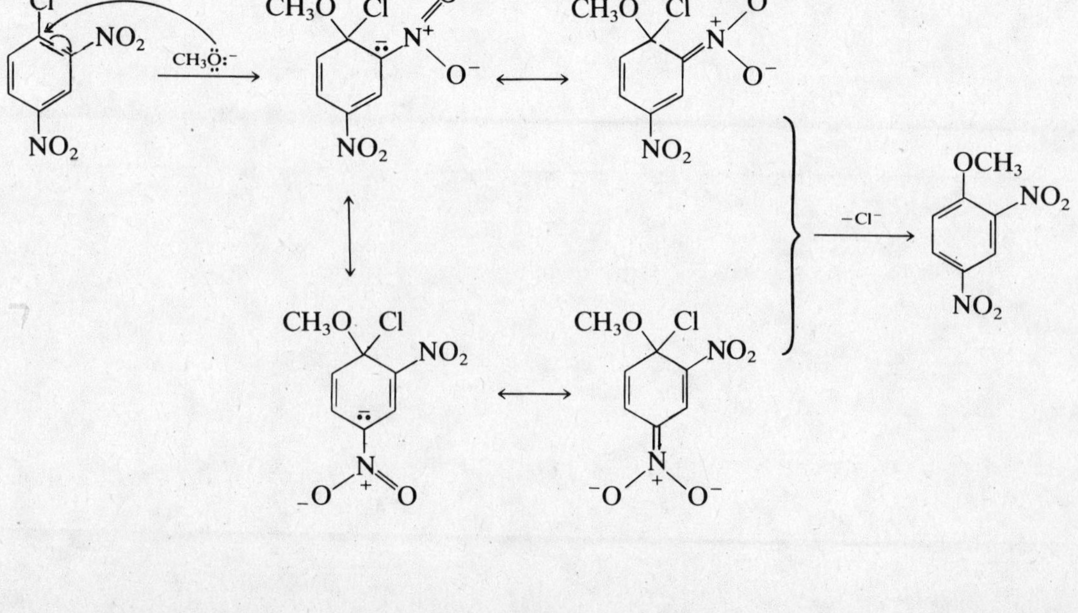

In the presence of a very strong base such as amide ion, aryl halides may undergo a second type of substitution reaction, involving an *elimination-addition mechanism*. The first stage is a base-promoted dehydrohalogenation, forming an intermediate species known as *benzyne*. The second stage involves addition of base (NH_2^-) and protonation of the anion formed.

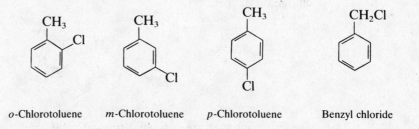

3-Methylbenzyne

SOLUTIONS TO TEXT PROBLEMS

23.1 Among the isomers of C_7H_7Cl are four based on the benzene nucleus, namely, *o, m,* and *p*-chlorotoluene and benzyl chloride.

| *o*-Chlorotoluene | *m*-Chlorotoluene | *p*-Chlorotoluene | Benzyl chloride |

Of this group only benzyl chloride is not an aryl halide; its halogen is not attached to the aromatic ring but to an sp^3 hybridized carbon. Benzyl chloride has the weakest carbon-halogen bond, its measured carbon-chlorine bond dissociation energy being only 293 kJ/mol (70 kcal/mol). Homolytic cleavage of this bond produces a resonance-stabilized benzyl radical.

Benzyl chloride   Benzyl radical   Chlorine atom

23.2 (*b*) The negatively charged sulfur in $C_6H_5CH_2SNa$ is a good nucleophile, which displaces chloride from 1-chloro-2,4-dinitrobenzene.

1-Chloro-2,4-dinitrobenzene   Benzyl 2,4-dinitrophenyl sulfide

(*c*) The nitrogen in ammonia has an unshared electron pair and is nucleophilic; it displaces chloride from 1-chloro-2,4-dinitrobenzene.

1-Chloro-2,4-dinitrobenzene   2,4-Dinitroaniline

(*d*) As with ammonia, methylamine is nucleophilic and displaces chloride.

1-Chloro-2,4-dinitrobenzene   *N*-Methyl-2,4-dinitroaniline

23.3 The most stable resonance structure for the cyclohexadienyl anion formed by reaction of methoxide ion with *o*-fluoronitrobenzene involves the nitro group and has the negative charge on oxygen.

23.4 The positions that are activated toward nucleophilic attack are those that are ortho and para to the nitro group. Among the carbons that bear a bromine leaving group in 1,2,3-tribromo-5-nitrobenzene, only C-2 satisfies this requirement.

1,2,3-Tribromo-5-nitrobenzene   1,3-Dibromo-2-ethoxy-
5-nitrobenzene

23.5 4-Chloropyridine is more reactive toward nucleophiles than 3-chloropyridine because the anionic intermediate formed by reaction of 4-chloropyridine has its charge on nitrogen. Since nitrogen is more electronegative than carbon, the intermediate is more stable.

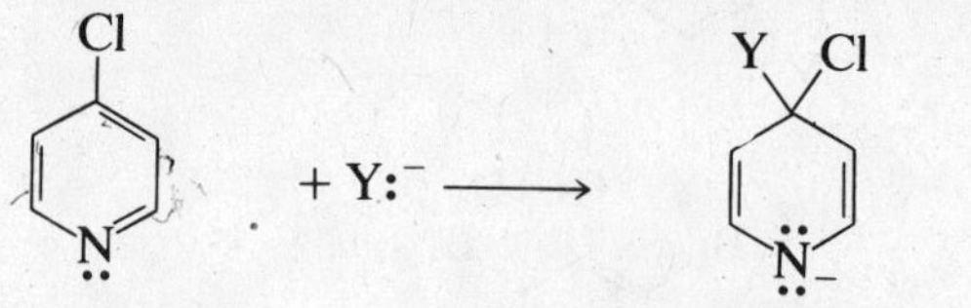

4-Chloropyridine   Anionic intermediate
                                  (more stable)

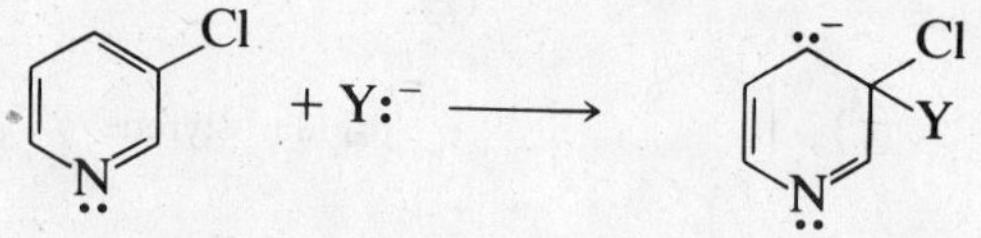

3-Chloropyridine   Anionic intermediate
                                 (less stable)

23.6 The aryl halide is incapable of elimination and so cannot form the benzyne intermediate necessary for substitution by the elimination-addition pathway.

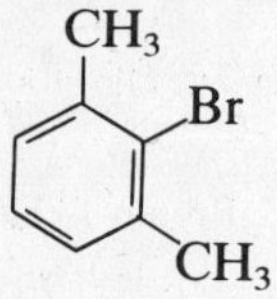

(No protons ortho to bromine; elimination is impossible.)

2-Bromo-1,3-dimethylbenzene

23.7 The aryne intermediate from p-iodotoluene can undergo addition of hydroxide ion at the position meta to the methyl group or para to it. The two isomeric phenols are m- and p-methylphenol.

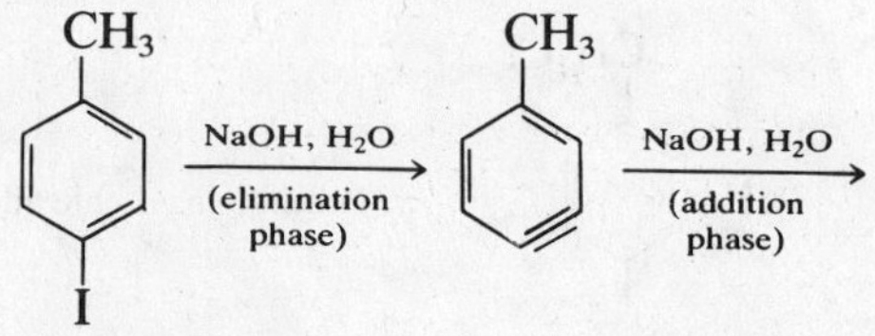

p-Iodotoluene

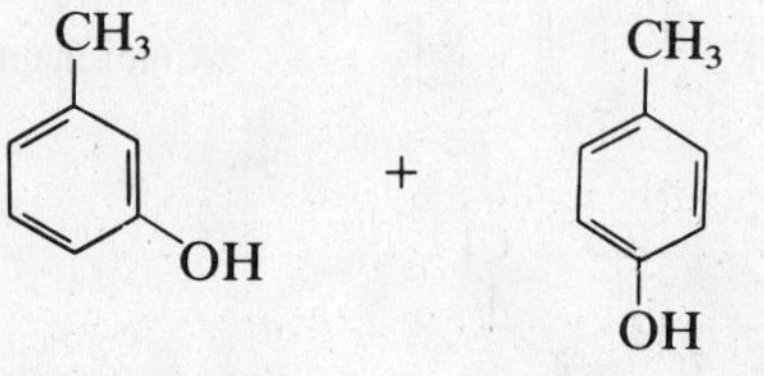

m-Methylphenol p-Methylphenol

23.8 The "triple bond" of benzyne adds to the diene system of furan.

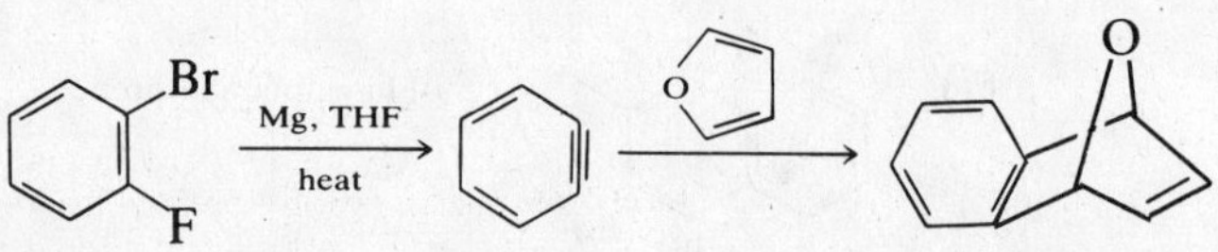

23.9 (a)

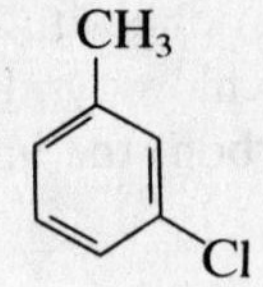

CH₃ (drawn at top)

m-Chlorotoluene

(b)

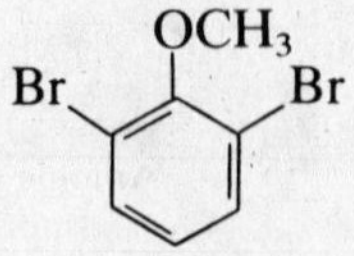

2,6-Dibromoanisole

(c)

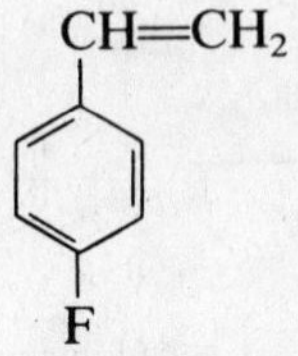

p-Fluorostyrene

(d) I—⟨4-3-2-1—1'-2'-3'-4'⟩—I   4,4'-Diiodobiphenyl

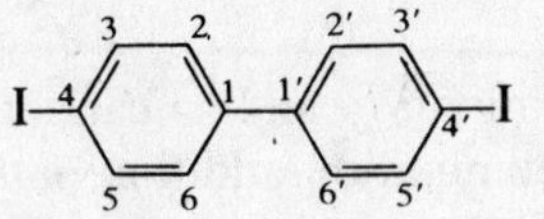

(e)

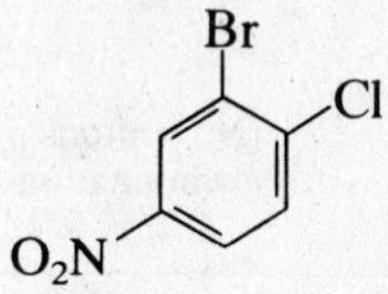

2-Bromo-1-chloro-4-nitrobenzene

(f)

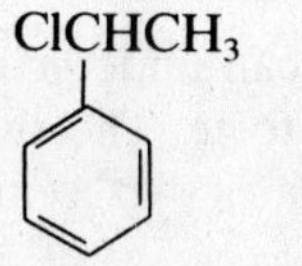

1-Chloro-1-phenylethane
(*Note*: This compound is not an
aryl halide.)

(g)

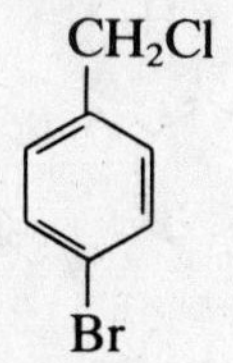

p-Bromobenzyl chloride

(h)

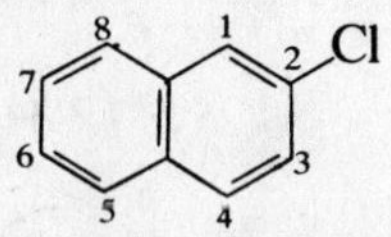

2-Chloronaphthalene

(i)

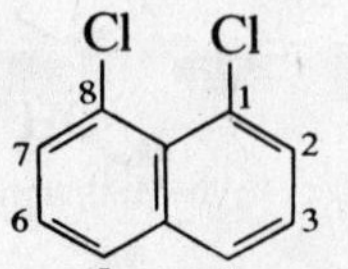

1,8-Dichloronaphthalene

(j)

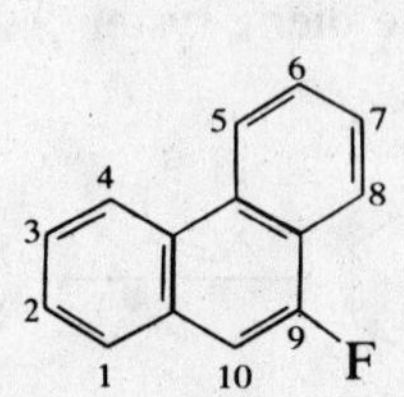

9-Fluorophenanthrene

23.10 (*a*) Chlorine is a weakly deactivating, ortho, para–directing substituent.

Chlorobenzene   Acetyl chloride   *o*-Chloroacetophenone   *p*-Chloroacetophenone

(*b*) Bromobenzene reacts with magnesium to give a Grignard reagent.

$$C_6H_5Br + Mg \xrightarrow[\text{ether}]{\text{diethyl}} C_6H_5MgBr$$

Bromobenzene   Phenylmagnesium bromide

(*c*) Protonation of the Grignard reagent in part (*b*) converts it to benzene.

$$C_6H_5MgBr \xrightarrow[\text{HCl}]{H_2O} C_6H_6$$

Phenylmagnesium bromide   Benzene

(*d*) Aryl halides react with lithium in much the same way that alkyl halides do, to form organolithium reagents.

$$C_6H_5I + 2Li \xrightarrow[\text{ether}]{\text{diethyl}} C_6H_5Li + LiI$$

Iodobenzene   Lithium   Phenyllithium   Lithium iodide

(*e*) With a base as strong as sodium amide, nucleophilic aromatic substitution by the elimination-addition mechanism takes place.

Bromobenzene   Benzyne   Aniline

(*f*) The benzyne intermediate from *p*-bromotoluene gives a mixture of *m*- and *p*-methylaniline.

p-Bromotoluene   4-Methylbenzyne   *m*-Methylaniline   *p*-Methylaniline

(g) Nucleophilic aromatic substitution of bromide by ammonia occurs by the addition-elimination mechanism.

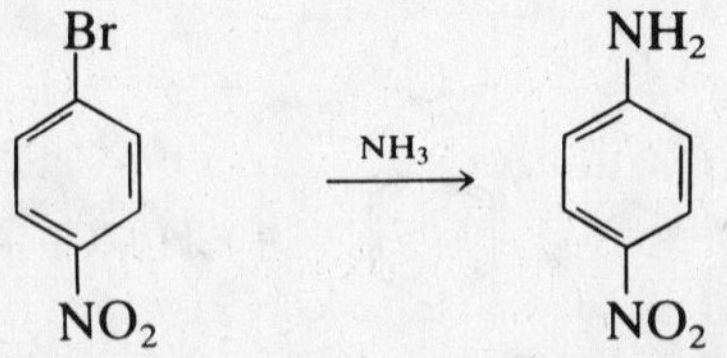

1-Bromo-4-nitrobenzene   *p*-Nitroaniline

(h) The bromine attached to the benzylic carbon is far more reactive than the one on the ring and is the one replaced by the nucleophile.

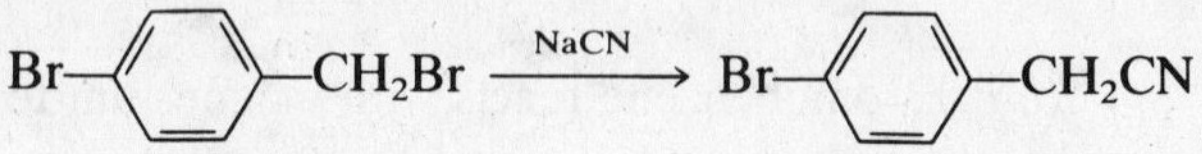

p-Bromobenzyl
bromide

p-Bromobenzyl cyanide

(i) The aromatic ring of *N,N*-dimethylaniline is very reactive and is attacked by *p*-chlorobenzenediazonium ion.

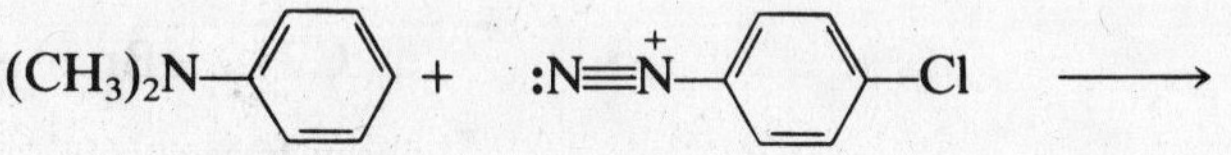

N,N-Dimethylaniline   *p*-Chlorobenzenediazonium ion

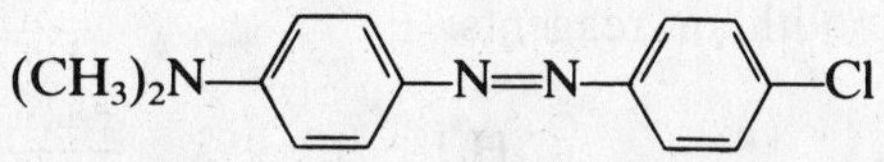

4-(4'-Chlorophenylazo)-*N,N*-dimethylaniline

(j) Hexafluorobenzene undergoes substitution of one of its fluorines on reaction with nucleophiles such as sodium hydrogen sulfide.

Hexafluorobenzene   Sodium hydrogen
                            sulfide

2,3,4,5,6-Pentafluorobenzenethiol

23.11 (a) Since the *tert*-butoxy group replaces fluoride at the position occupied by the leaving group, substitution likely occurs by the addition-elimination mechanism.

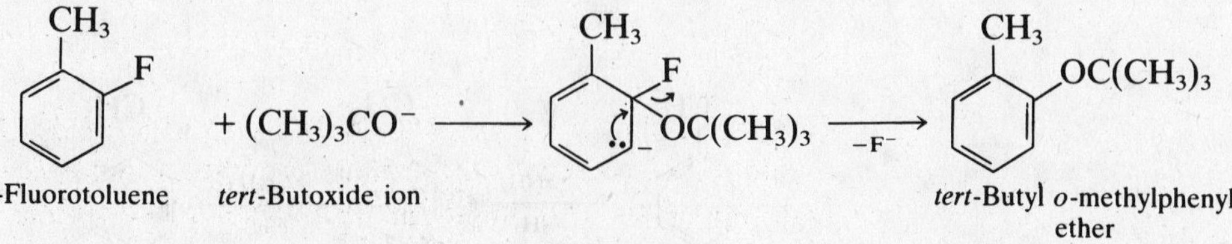

o-Fluorotoluene   *tert*-Butoxide ion

tert-Butyl *o*-methylphenyl
ether

(b) In nucleophilic aromatic substitution reactions that proceed by the addition-elimination mechanism, aryl fluorides react faster than aryl bromides. Since the aryl bromide is more reactive in this case, it must be reacting by a different mechanism, which is most likely elimination-addition.

Bromobenzene   Benzyne   *tert*-Butyl phenyl ether

23.12 (*a*) Two benzyne intermediates are equally likely to be formed. Reaction with amide ion can occur in two different directions with each benzyne, giving three possible products. They are formed in a 1:2:1 ratio.

Ratio: 1   :   2   :   1

(*b*) Only one benzyne intermediate is possible, leading to two products in a 1:1 ratio.

Ratio: 1   :   1

23.13 (*a*) o-Chloronitrobenzene is more reactive than chlorobenzene, because the cyclohexadienyl anion intermediate is stabilized by the nitro group.

Comparing the rate constants for the two aryl halides in this reaction reveals that *o*-chloronitrobenzene is more than 20 billion times more reactive at 50°C.

(*b*) The cyclohexadienyl anion intermediate is more stable, and is formed faster, when the electron-withdrawing nitro group is ortho to chlorine. *o*-Chloronitrobenzene reacts faster than *m*-chloronitrobenzene. The measured difference is a factor of approximately 40,000 at 50°C.

(c) 4-Chloro-3-nitroacetophenone is more reactive, because the ring bears two powerful electron-withdrawing groups in positions where they can stabilize the cyclohexadienyl anion intermediate.

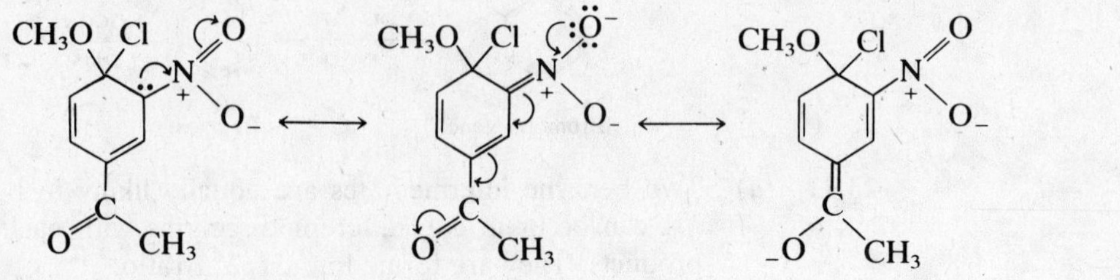

(d) Nitro groups activate aryl halides toward nucleophilic aromatic substitution best when they are ortho or para to the leaving group.

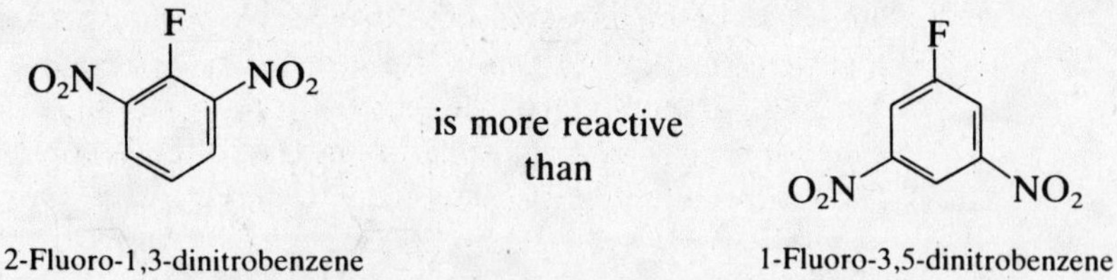

is more reactive than

2-Fluoro-1,3-dinitrobenzene

1-Fluoro-3,5-dinitrobenzene

(e) The aryl halide with nitro groups ortho and para to the bromide leaving group is more reactive than the aryl halide with only one nitro group.

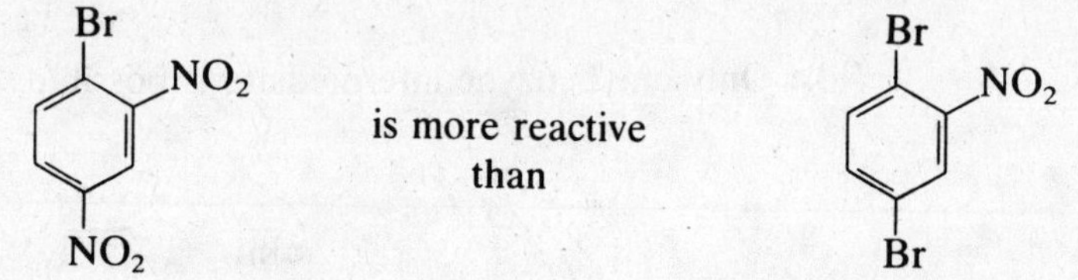

is more reactive than

1-Bromo-2,4-dinitrobenzene

1,4-Dibromo-2-nitrobenzene

23.14 (a) The nucleophile is the lithium salt of pyrrolidine, which reacts with bromobenzene by an elimination-addition mechanism.

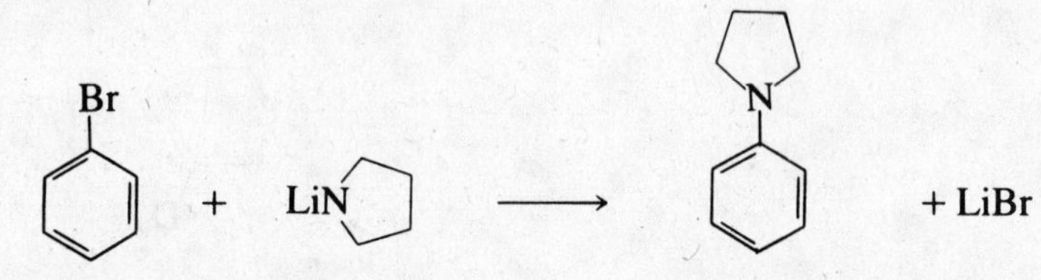

Bromobenzene   Lithium pyrrolidide

N-Phenylpyrrolidine
(observed yield, 84%)

+ LiBr

(b) The nucleophile in this case is piperidine. The substrate, 1-bromo-2,4-dinitrobenzene, is very reactive in nucleophilic aromatic substitution by the addition-elimination mechanism.

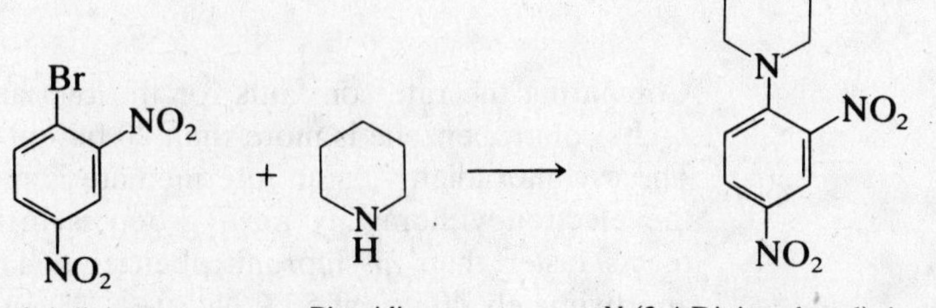

1-Bromo-2,4-dinitrobenzene   Piperidine

N-(2,4-Dinitrophenyl)piperidine

(*c*) Of the two bromine atoms, one is ortho and the other meta to the nitro group. Nitro groups activate positions ortho and para to themselves toward nucleophilic aromatic substitution, and so it will be the bromine ortho to the nitro group that is displaced.

1,4-Dibromo-2-nitrobenzene   Piperidine   N-(4-Bromo-2-nitrophenyl)piperidine

23.15 Since isomeric products are formed by reaction of 1- and 2-bromonaphthalene with piperidine at elevated temperatures, it is reasonable to conclude that these reactions do not involve a common intermediate and hence follow an addition-elimination pathway. Piperidine acts as a nucleophile and substitutes for bromine on the same carbon atom from which bromine is lost.

1-Bromonaphthalene   Piperidine   Compound A

2-Bromonaphthalene   Piperidine   Compound B

When the strong base sodium piperidide is used, reaction occurs by the elimination-addition pathway via a "naphthalyne" intermediate. Only one mode of elimination is possible from 1-bromonaphthalene.

This intermediate can yield both A and B in the addition stage.

Compound A   Compound B

Two modes of elimination are possible from 2-bromonaphthalene:

Compounds A and B   Compound B only

Both naphthalyne intermediates are probably formed from 2-bromonaphthalene, since there is no reason to expect elimination to occur only in one direction.

23.16   Reaction of a nitro-substituted aryl halide with a good nucleophile leads to nucleophilic aromatic substitution. Methoxide will displace fluoride from the ring, preferentially at the positions ortho and para to the nitro group.

1,2,3,4,5-Pentafluoro-   2,3,4,5-Tetrafluoro-   2,3,5,6-Tetrafluoro-
6-nitrobenzene   6-nitroanisole   4-nitroanisole

23.17   (*a*)   This reaction is one of nucleophilic aromatic substitution by the addition-elimination mechanism.

4-Chloro-3-nitrotoluene   4-(Benzylthio)-3-nitrotoluene

The nucleophile, $C_6H_5CH_2S^-$, displaces chloride directly from the aromatic ring. The product in this case was isolated in 57 percent yield.

(*b*)   The nucleophile, hydrazine, will react with 1-chloro-2,4-dinitrobenzene by an addition-elimination mechanism as shown.

1-Chloro-2,4-dinitrobenzene   Hydrazine   2,4-Dinitrophenylhydrazine

The nitrogen atom of hydrazine has an unshared electron pair and is fairly nucleophilic. The product, 2,4-dinitrophenylhydrazine, is formed in quantitative yield.

(*c*)   The problem requires you to track the starting material through two transformations. The first of these is nitration of *m*-dichlorobenzene, an electrophilic aromatic substitution reaction.

m-Dichlorobenzene   2,4-Dichloro-1-nitrobenzene

Since the final product of the sequence has four nitrogen atoms ($C_6H_6N_4O_4$), 2,4-dichloro-1-nitrobenzene is an unlikely starting material for the second transformation. Stepwise nucleophilic aromatic substitution of both chlorines is possible but leads to a compound with the wrong molecular formula ($C_6H_7N_3O_2$).

2,4-Dichloro-1-nitrobenzene   2,4-Diamino-1-nitrobenzene

In order to obtain a final product with the correct molecular formula, the original nitration reaction must lead not to a mononitro but to a dinitro derivative. This is reasonable in view of the fact that this reaction is carried out at elevated temperature (120°C).

m-Dichlorobenzene   1,5-Diamino-2,4-dinitrobenzene
                                                         ($C_6H_6N_4O_4$)

This two-step sequence has been carried out with product yields of 70 to 71 percent in the first step and 88 to 95 percent in the second step.

(*d*)   This problem also involves two transformations, nitration and nucleophilic aromatic substitution. Nitration will take place ortho to chlorine (meta to trifluoromethyl).

1-Chloro-4-(trifluoromethyl)benzene   1-Chloro-2-nitro-4-(trifluoromethyl)benzene

2-Nitro-4-(trifluoromethyl)anisole

(*e*)   The primary alkyl halide is more reactive toward nucleophilic substitution than the aryl halide. A phosphonium salt forms by an S_N2 process.

p-Iodobenzyl bromide   Triphenyl   (p-Iodobenzyl)triphenyl-
                        phosphine   phosphonium bromide (86%)

(*f*) *N*-Bromosuccinimide is a reagent used to substitute benzylic and allylic hydrogens with bromine. The benzylic bromide undergoes S_N2 substitution with the nucleophile, methanethiolate. As in part (*e*), the alkyl halide is more reactive toward substitution than the aryl halide.

2-Bromo-5-methoxy-toluene   (2-Bromo-5-methoxyphenyl)-benzyl bromide   (2-Bromo-5-methoxyphenyl)-methyl sulfide

23.18 The reaction of *p*-bromotoluene with aqueous sodium hydroxide at elevated temperature proceeds by way of a benzyne intermediate.

m-Methylphenol   *p*-Methylphenol

The same benzyne intermediate is formed when *p*-chlorotoluene is the reactant, and so the product ratio must be identical regardless of whether the leaving group is bromide or chloride.

23.19 Dinitration of *p*-chloro(trifluoromethyl)benzene will take place at the ring positions ortho to the chlorine. Compound C is 2-chloro-5-(trifluoromethyl)-1,3-dinitrobenzene. Trifluralin is formed by nucleophilic aromatic substitution of chlorine by dipropylamine. Trifluralin is *N*,*N*-dipropyl-4-(trifluoromethyl)-2,6-dinitroaniline.

p-Chloro(trifluoromethyl)-benzene   2-Chloro-5-(trifluoromethyl)-1,3-dinitrobenzene (compound C)   *N*,*N*-dipropyl-4-(trifluoromethyl)-2,6-dinitroaniline (trifluralin)

23.20 *p*-Chlorobenzenethiolate reacts with *p*-nitrobenzyl chloride by an S_N2 process to give compound D. Reduction of the nitro group yields the aniline derivative, compound E.

Chlorbenside is then formed by a Sandmeyer reaction in which the diazonium ion is replaced by chlorine.

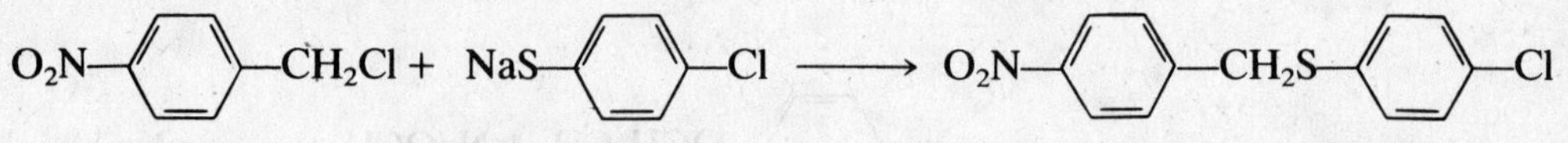

O_2N—〈 〉—CH_2Cl + NaS—〈 〉—Cl ⟶ O_2N—〈 〉—CH_2S—〈 〉—Cl

| *p*-Nitrobenzyl chloride | Sodium *p*-chlorobenzenethiolate | *p*-Chlorophenyl *p*-nitrobenzyl sulfide (compound D) |

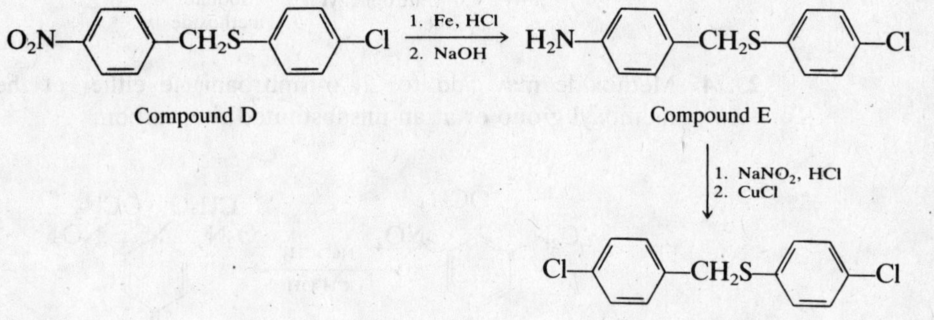

O_2N—〈 〉—CH_2S—〈 〉—Cl $\xrightarrow[\text{2. NaOH}]{\text{1. Fe, HCl}}$ H_2N—〈 〉—CH_2S—〈 〉—Cl

Compound D   Compound E

$\downarrow$ 1. $NaNO_2$, HCl
2. CuCl

Cl—〈 〉—CH_2S—〈 〉—Cl

Chlorbenside

23.21 Benzyne is formed by loss of nitrogen and carbon dioxide.

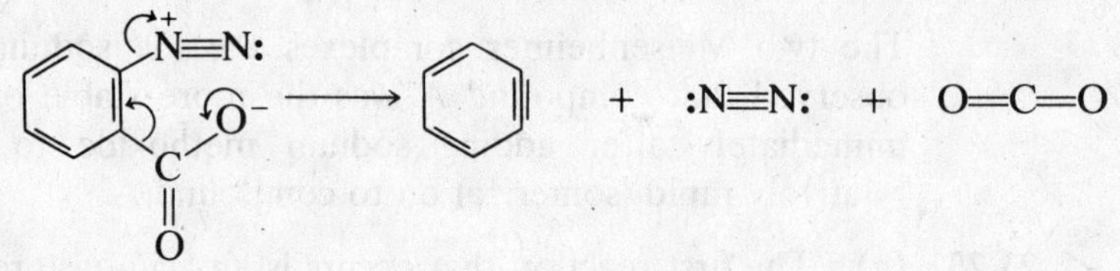

| Benzenediazonium-2-carboxylate | Benzyne | Nitrogen | Carbon dioxide |

23.22 *o*-Bromofluorobenzene yields benzyne on reaction with magnesium (text Sec. 23.9). Triptycene is the Diels-Alder cycloaddition product from the reaction of benzyne with anthracene (compound F). Although anthracene is aromatic, it is able to undergo cycloaddition at the center ring with a dienophile, since the adduct retains the stabilization energy of two benzene rings.

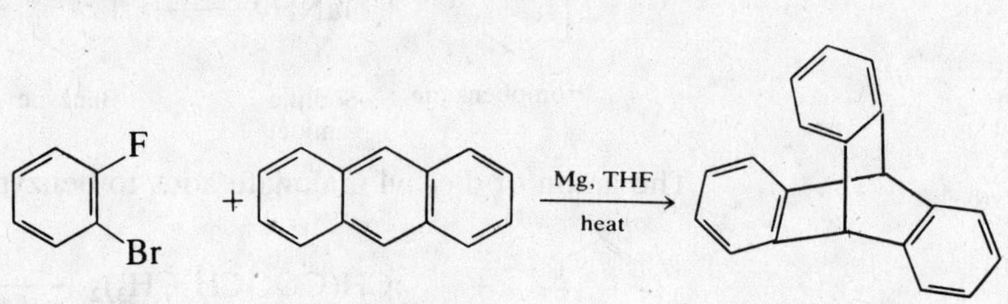

| *o*-Bromofluorobenzene | Anthracene (compound F) | Triptycene |

23.23 (*a*) Ethoxide adds to the aromatic ring to give a cyclohexadienyl anion.

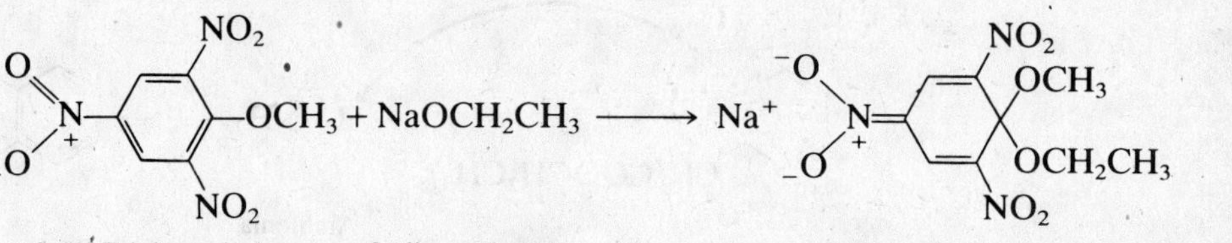

| 2,4,6-Trinitroanisole | Sodium ethoxide | Meisenheimer complex |

(*b*) The same Meisenheimer complex results when ethyl 2,4,6-trinitrophenyl ether reacts with sodium methoxide.

Ethyl 2,4,6-trinitrophenyl ether   Sodium methoxide   Meisenheimer complex

23.24 Methoxide may add to 2,4,6-trinitroanisole either at the ring carbon that bears the methoxyl group or at an unsubstituted ring carbon.

2,4,6-Trinitroanisole   A   B

The two Meisenheimer complexes are the sodium salts of the anions shown. It was observed that compound A was the more stable of the two. Compound B was present immediately after adding sodium methoxide to 2,4,6-trinitroanisole but underwent relatively rapid isomerization to compound A.

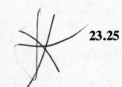

23.25 (*a*) The first reaction that occurs is an acid-base reaction between diethyl malonate and sodium amide.

$$CH_2(COOCH_2CH_3)_2 + NaNH_2 \longrightarrow Na^+:\bar{C}H(COOCH_2CH_3)_2 + NH_3$$

Diethyl malonate   Sodium amide   Diethyl sodiomalonate   Ammonia

A second equivalent of sodium amide converts bromobenzene to benzyne.

Bromobenzene   Sodium amide   Benzyne   Ammonia   Sodium bromide

The anion of diethyl malonate adds to benzyne.

Benzyne   Anion of diethyl malonate

This anion then abstracts a proton from ammonia to give the observed product.

Ammonia   Diethyl phenylmalonate   Amide anion

(*b*) The ester is deprotonated by the strong base sodium amide, after which the ester enolate undergoes an elimination reaction to form a benzyne intermediate. Cyclization to the final product occurs by intramolecular attack of the ester enolate on the reactive triple bond of the aryne.

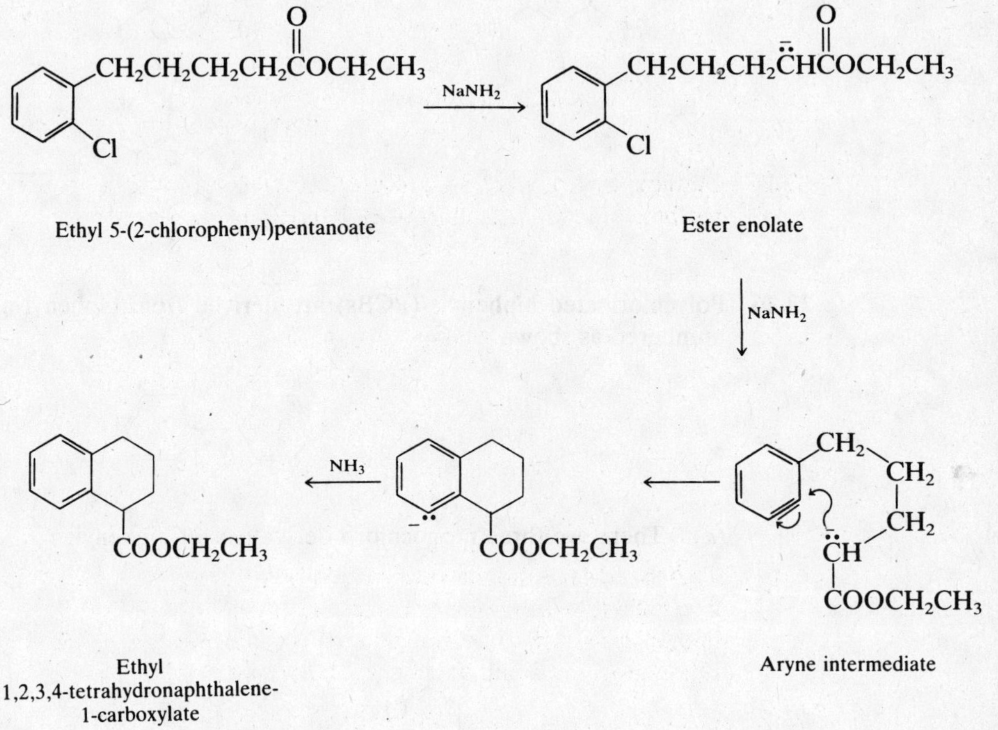
NaNH2
Ethyl 5-(2-chlorophenyl)pentanoate
Ester enolate
NaNH2
NH3
Aryne intermediate
Ethyl 1,2,3,4-tetrahydronaphthalene-1-carboxylate

(*c*) In the presence of very strong bases, aryl halides undergo nucleophilic aromatic substitution by an elimination-addition mechanism. The structure of the product indicates that a nitrogen of the side chain acts as a nucleophile in the addition step.

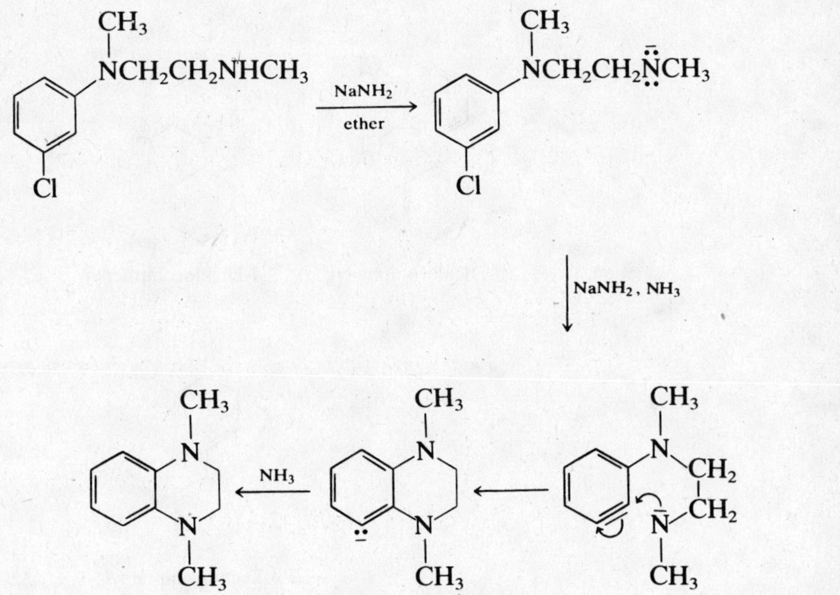
NaNH2
ether
NaNH2, NH3
NH3

(d) On treatment with base, intramolecular nucleophilic aromatic substitution leads to the observed product.

23.26 Polychlorinated biphenyls (PCBs) are derived from biphenyl as the base structure. It is numbered as shown.

(a) There are three monochloro derivatives of biphenyl:

2-Chlorobiphenyl
(o-chlorobiphenyl)

3-Chlorobiphenyl
(m-chlorobiphenyl)

4-Chlorobiphenyl
(p-chlorobiphenyl)

(b) Two chlorine substituents may be in the same ring (six isomers):

2,3-Dichlorobiphenyl   2,4-Dichlorobiphenyl   2,5-Dichlorobiphenyl   2,6-Dichlorobiphenyl

3,4-Dichlorobiphenyl   3,5-Dichlorobiphenyl

The two chlorine substituents may be in different rings (six isomers):

2,2'-Dichlorobiphenyl   2,3'-Dichlorobiphenyl   2,4'-Dichlorobiphenyl

3,3'-Dichlorobiphenyl   3,4'-Dichlorobiphenyl   4,4'-Dichlorobiphenyl

Therefore there are a total of 12 isomeric dichlorobiphenyls.

(c) The number of octachlorobiphenyls will be equal to the number of dichlorobiphenyls (12). In both cases we are dealing with a situation in which 8 of the 10 substituents of the biphenyl system are the same and considering how the remaining two may be arranged. In the dichlorobiphenyls described in part (b), eight substituents are hydrogen and two are chlorine; in the octachlorobiphenyls, eight substituents are chlorine while two are hydrogen.

(d) The number of nonachloro isomers (nine chlorine substituents, one hydrogen substituent) must equal the number of monochloro isomers (one chlorine substituent, nine hydrogen substituents). Therefore, there are three nonachloro derivatives of biphenyl.

23.27 The principal isotopes of chlorine are ^{35}Cl and ^{37}Cl. A cluster of five peaks indicates that DDE contains *four* chlorines.

^{35}Cl	^{35}Cl	^{35}Cl	^{35}Cl
^{35}Cl	^{35}Cl	^{35}Cl	^{37}Cl
^{35}Cl	^{35}Cl	^{37}Cl	^{37}Cl
^{35}Cl	^{37}Cl	^{37}Cl	^{37}Cl
^{37}Cl	^{37}Cl	^{37}Cl	^{37}Cl

The peak at m/z 316 therefore corresponds to a compound $C_{14}H_8Cl_4$ in which all four chlorines are ^{35}Cl. The respective molecular formulas indicate that DDE is the dehydrochlorination product of DDT.

$$C_{14}H_9Cl_5 \xrightarrow{-HCl} C_{14}H_8Cl_4$$

DDT   DDE

The structure of DDT was given in the statement of the problem. This permits the structure of DDE to be assigned:

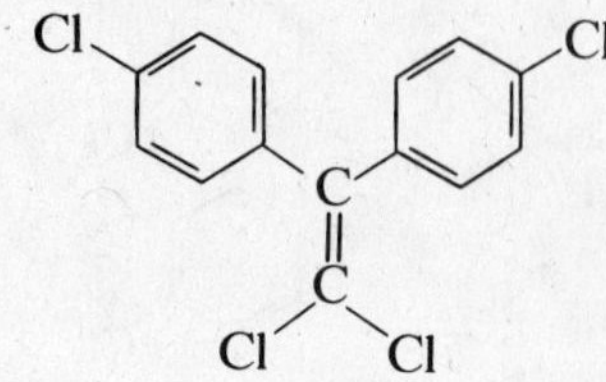

DDE (only reasonable dehydrochlorination product of DDT)

SELF-TEST

PART A

A-1. Give the product(s) obtained from each of the following reactions:

(a)

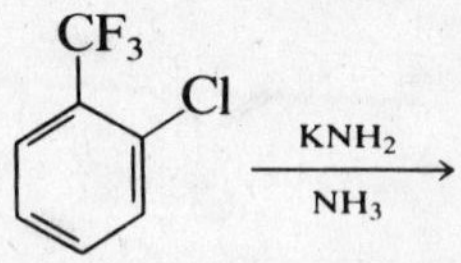

$\xrightarrow[\text{NH}_3]{\text{KNH}_2}$? (two products)

(b)

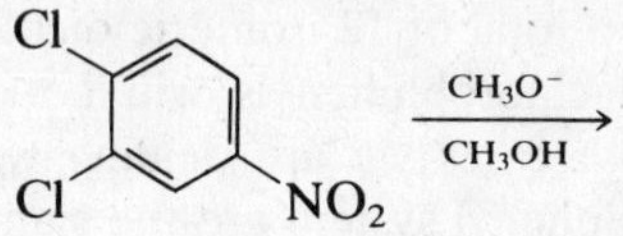

$\xrightarrow[\text{CH}_3\text{OH}]{\text{CH}_3\text{O}^-}$? (monosubstitution)

(c)

$\xrightarrow[\text{NH}_3]{\text{NaNH}_2}$? (two products)

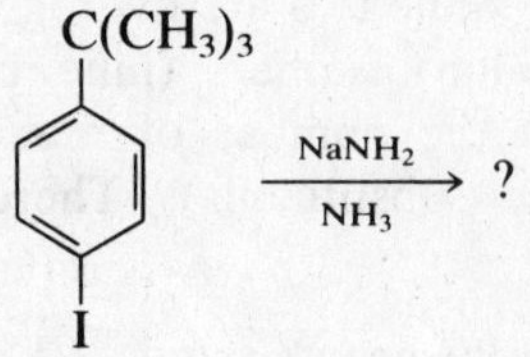

A-2. Suggest synthetic schemes by which chlorobenzene may be converted into:
(a) 2,4-Dinitroanisole (1-methoxy-2,4-dinitrobenzene)
(b) *p*-Isopropylaniline

A-3. Write a mechanism using resonance structures to show how a nitro group directs ortho, para in a nucleophilic substitution reaction.

PART B

B-1. The reaction

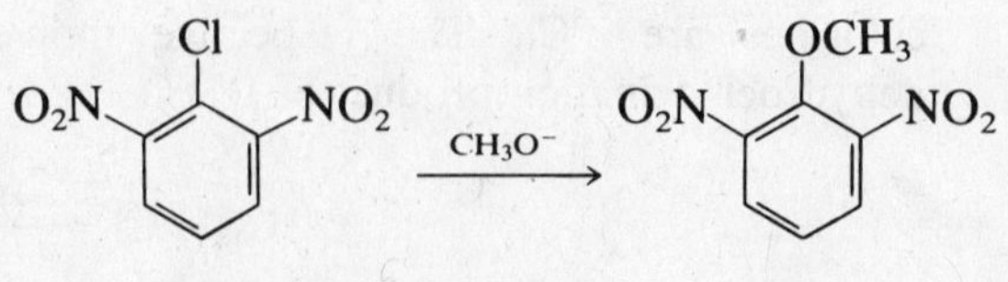

most likely occurs by which of the following mechanisms?
(a) Addition-elimination (c) Both (a) and (b)
(b) Elimination-addition (d) Neither of these

B-2. Rank the following in order of decreasing rate of reaction with ethoxide ion ($CH_3CH_2O^-$) in a nucleophilic aromatic substitution reaction:

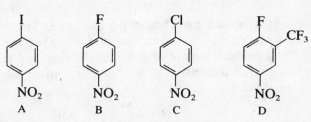

  A   B   C   D

(a) $C > D > A > B$ (c) $C > D > B > A$
(b) $B > A > D > C$ (d) $D > C > B > A$

B-3. The reaction

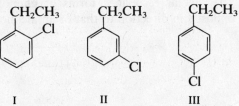

most likely involves which of the following aromatic substitution mechanisms?
(a) Addition-elimination   (c) Elimination-addition
(b) Electrophilic substitution   (d) Both (a) and (c)

B-4. Rank the following aryl halides in order of increasing rate of reaction with CH_3O^- in CH_3OH:

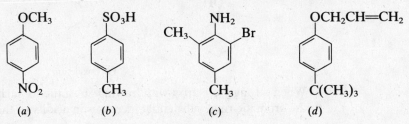

  A   B   C   D

(a) $A < C < D < B$ (c) $B < D < C < A$
(b) $D < B < C < A$ (d) $A < C < B < D$

B-5. Which of the following compounds gives a single benzyne intermediate on reaction with sodium amide?

I   II   III

(a) I only   (c) III only
(b) I and III   (d) I and II

B-6. Which one of the following compounds can be efficiently prepared by a procedure in which nucleophilic aromatic substitution is the last step?

(a)   (b)   (c)   (d)

PHENOLS

IMPORTANT TERMS AND CONCEPTS

Structure and Bonding (Sec. 24.2) The carbon-oxygen bond of phenol is somewhat shorter than that of an alcohol, for two reasons. The oxygen in phenol is bonded to an sp^2 hybridized carbon of the aromatic ring, and the π system of the ring is conjugated with one of the unshared pairs of electrons on the oxygen resulting in partial double bond character.

The dipolar resonance forms of phenol also explain the polarity of phenol, which is opposite in direction to that of an alcohol such as methanol.

Methanol   Phenol

Acidity (Secs. 24.4, 24.5) Phenols are approximately 10^6 to 10^{10} times *more* acidic than alcohols. Stabilization of the phenoxide ion and the resulting acidity of phenol result from delocalization of the negative charge into the aromatic ring.

Phenoxide ion

When strongly electron-withdrawing substituents such as the nitro group are attached to the aromatic ring, substantial increases in acidity of these substituted phenols as against that of phenol itself are noted. For example, *o-* and *p*-nitrophenol are several hundred times more acidic than phenol itself. The largest increase in acidity is noted when the nitro group is directly conjugated with the hydroxyl group, that is, when it is located either ortho or para to the hydroxyl. *m*-Nitrophenol is less acidic than the other nitrophenol isomers; however, it is still more acidic than phenol.

o-Nitrophenol	*p*-Nitrophenol	*m*-Nitrophenol	Phenol
pK_a 7.2	pK_a 7.2	pK_a 8.4	pK_a 10.0

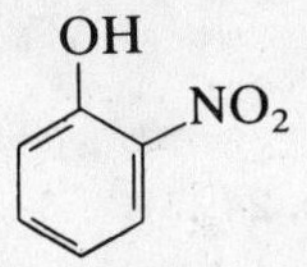

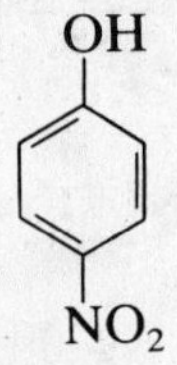

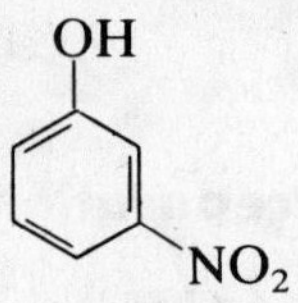

IMPORTANT REACTIONS

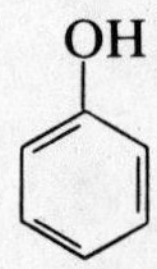

Electrophilic Aromatic Substitution (Sec. 24.8)

General:

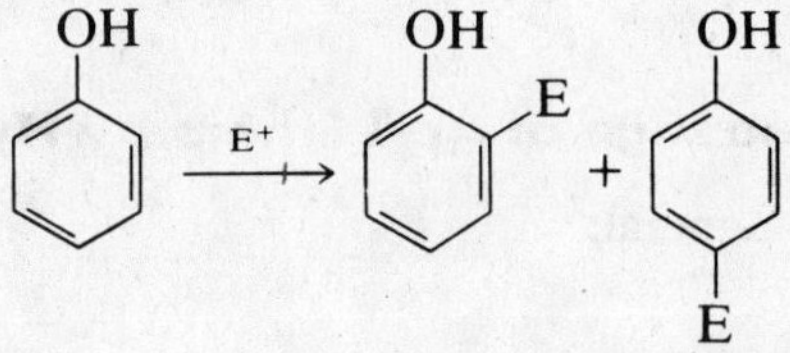

Examples:

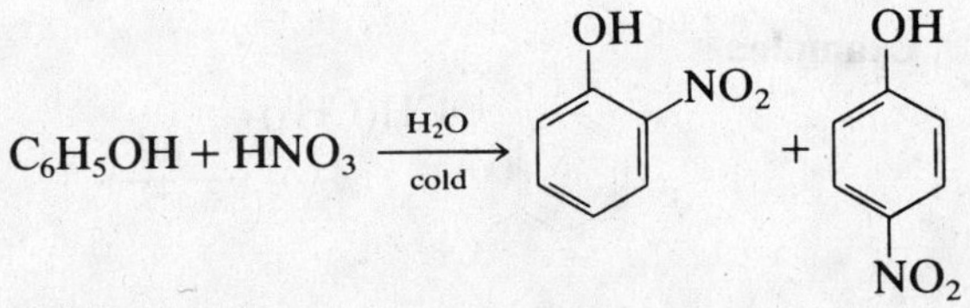

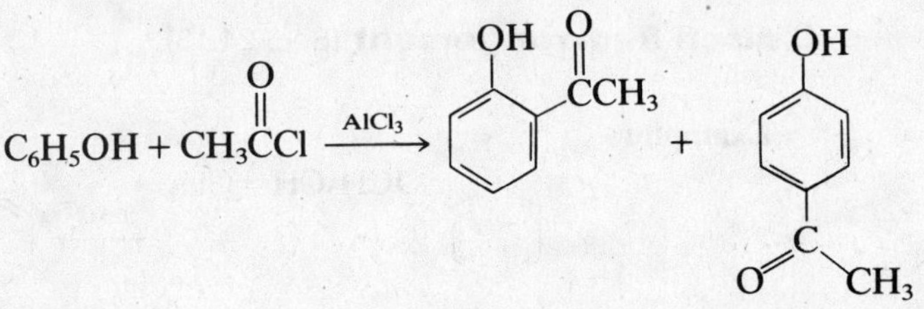

$$C_6H_5OH + HNO_3 \xrightarrow[\text{cold}]{H_2O}$$

$$C_6H_5OH + CH_3CCl \xrightarrow{AlCl_3}$$

Acylation (Sec. 24.9)

General:

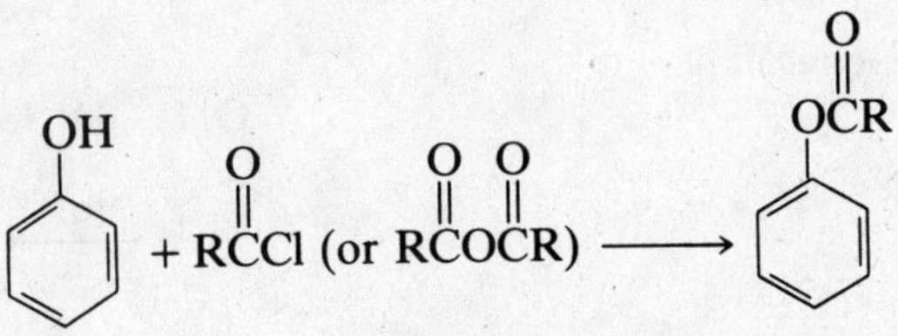

Example:

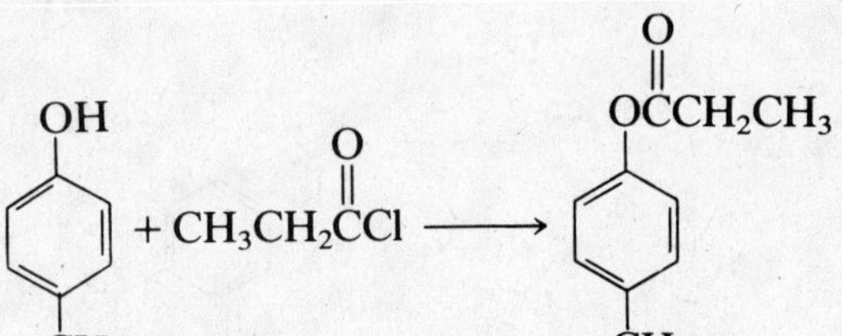

Carboxylation; the Kolbe-Schmitt Reaction (Sec. 24.10)

Example:

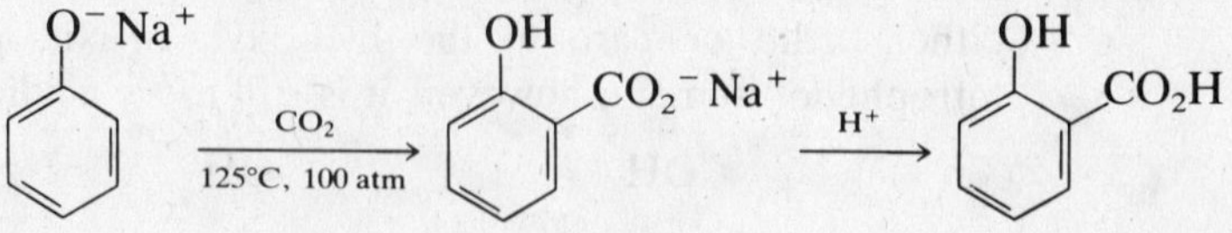

Preparation of Aryl Ethers (Sec. 24.11)

General:

$$ArO^- + RX \longrightarrow ArOR + X^-$$

Example:

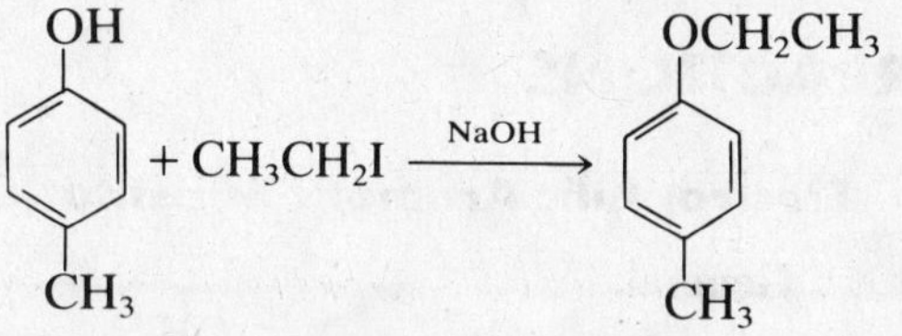

Cleavage of Aryl Ethers by Hydrogen Halides (Sec. 24.12)

General:

$$ArOR + HX \longrightarrow ArOH + RX$$

Example:

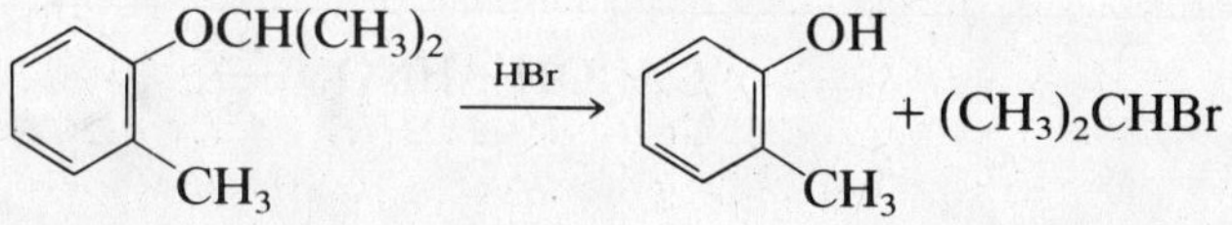

Claisen Rearrangement (Sec. 24.13)

Example:

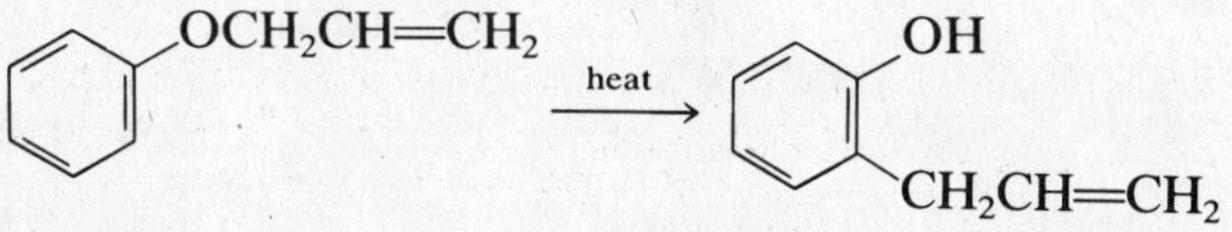

Oxidation of Phenols; Quinones (Sec. 24.14)

General:

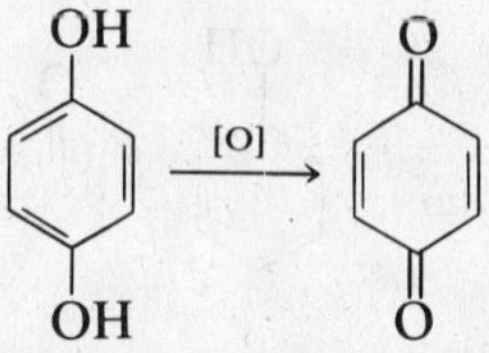

Example:

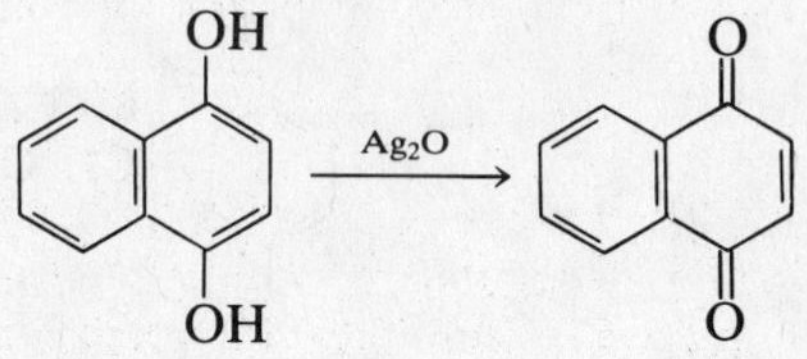

SOLUTIONS TO TEXT PROBLEMS

24.1 (b) A benzyl group ($C_6H_5CH_2$—) is ortho to the phenolic hydroxyl group in *o*-benzylphenol.

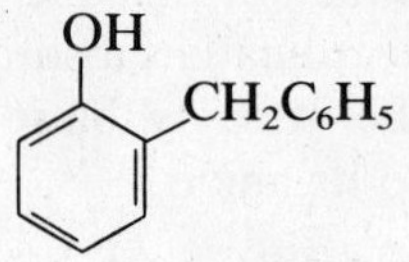

(c) Naphthalene is numbered as shown. 3-Nitro-1-naphthol has a hydroxyl group at C-1 and a nitro group at C-3.

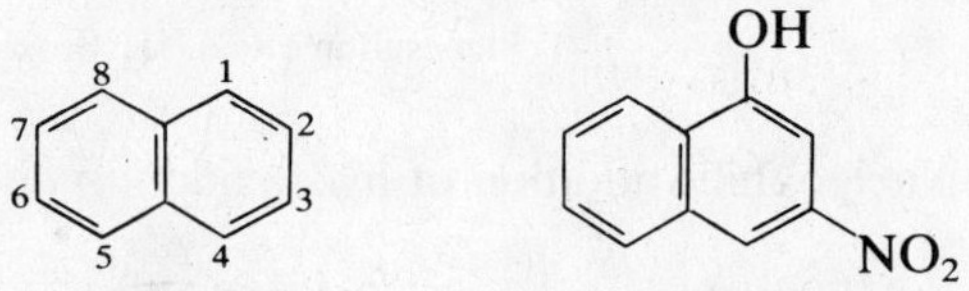

Naphthalene

3-Nitro-1-naphthol

(d) Resorcinol is 1,3-benzenediol. Therefore 4-chlororesorcinol is

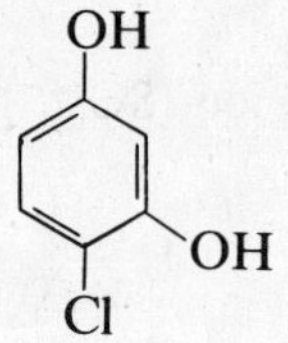

24.2 *p*-Nitrophenol has a larger dipole moment than either phenol or nitrobenzene because the electron-releasing effect of the hydroxyl group and the electron-withdrawing effect of the nitro group reinforce each other. The two groups are directly conjugated; this conjugation can be expressed in resonance terms as shown:

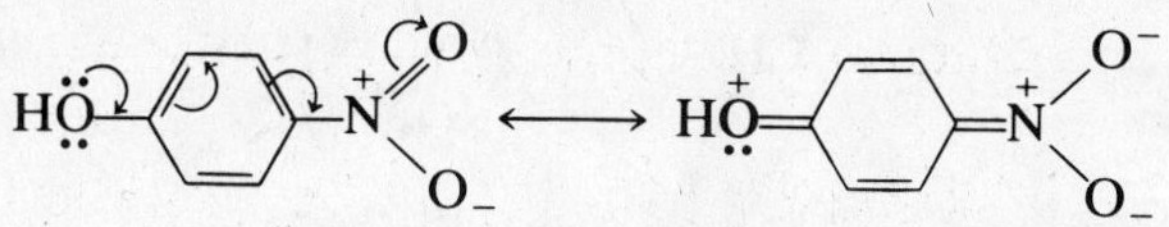

24.3 Intramolecular hydrogen bonding between the hydroxyl group and the ester carbonyl can occur when these groups are ortho to one another.

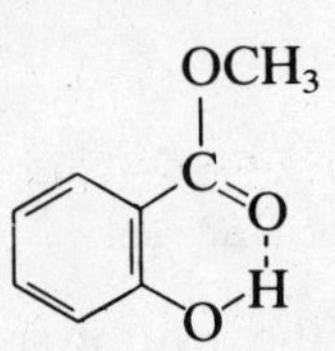

Methyl salicylate

Intramolecular hydrogen bonds form at the expense of intermolecular ones, and intramolecularly hydrogen-bonded species have lower boiling points than intermolecularly hydrogen-bonded ones of the same molecular composition.

24.4 (b) A cyano group withdraws electrons from the ring by resonance. A *p*-cyano substituent is conjugated directly with the negatively charged oxygen and stabilizes the anion more than does an *m*-cyano substituent.

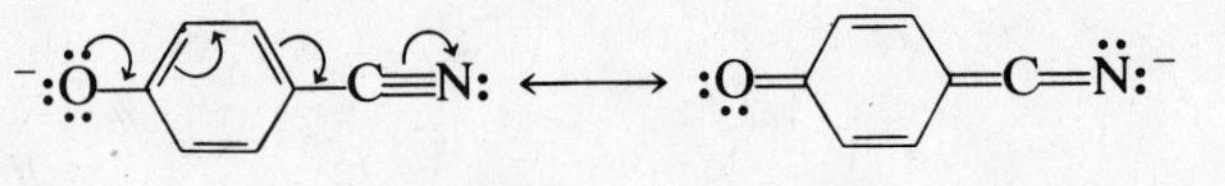

p-Cyanophenol is slightly more acidic than *m*-cyanophenol, the K_a values being 1.0×10^{-8} and 2.8×10^{-9}, respectively.

(c)   The electron-withdrawing inductive effect of the fluorine substituent will be more pronounced at the ortho position than at the para. *o*-Fluorophenol ($K_a = 1.9 \times 10^{-9}$) is a stronger acid than *p*-fluorophenol ($K_a = 1.3 \times 10^{-10}$).

24.5 The text points out that the reaction proceeds by the addition-elimination mechanism of nucleophilic aromatic substitution.

Under the strongly basic conditions of the reaction, *p*-toluenesulfonic acid is first converted to its anion.

$$CH_3 - \text{(ring)} - S(=O)(=O) - O{-}H + {:}\ddot{O}H \longrightarrow CH_3 - \text{(ring)} - S(=O)(=O) - O^- + HOH$$

| *p*-Toluenesulfonic acid | Hydroxide ion | *p*-Toluenesulfonate ion | Water |

Nucleophilic addition of hydroxide ion gives a cyclohexadienyl anion intermediate.

$$CH_3 - \text{(ring)} - SO_3^- + {:}\ddot{O}H \longrightarrow CH_3 - \text{(ring)}(OH)(SO_3^-)$$

| *p*-Toluenesulfonate ion | Hydroxide | Cyclohexadienyl anion |

Loss of sulfite ion (SO_3^{2-}) gives *p*-cresol.

$$CH_3 - \text{(ring)}(OH)(SO_3^-) \longrightarrow CH_3 - \text{(ring)} - OH + SO_3^{2-}$$

| Cyclohexadienyl anion | | *p*-Cresol |

It is also possible that the elimination stage of the reaction proceeds as follows:

$$CH_3 - \text{(ring)}(OH)(SO_3^-) + H_2O \longrightarrow CH_3 - \text{(ring)}(O{-}H)(SO_3^-) + {:}\ddot{O}H$$

Cyclohexadienyl anion intermediate

$$CH_3 - \text{(ring)} - O^- \xleftarrow{HO^-} CH_3 - \text{(ring)}(=O) + SO_3^{2-} + H_2O$$

p-Methylphenoxide ion

24.6 The text states that the hydrolysis of chlorobenzene in base follows an elimination-addition mechanism.

$$\text{(ring)}{-}H \cdots Cl + {:}\ddot{O}H \longrightarrow \text{(benzyne ring)} + H_2O + Cl^-$$

Chlorobenzene   Benzyne

$$\text{(benzyne ring)} + {:}\ddot{O}H \longrightarrow \text{(ring)} - OH \xrightarrow{H_2O} \text{(ring)} - OH$$

Benzyne   Phenol

24.7 (*b*) Proton transfer from sulfuric acid to 2-methylpropene gives *tert*-butyl cation. Since the position para to the hydroxyl substituent already bears a bromine, the *tert*-butyl cation attacks the ring at the position ortho to the hydroxyl.

4-Bromo-2-methylphenol   2-Methylpropene   4-Bromo-2-*tert*-butyl-6-methylphenol
(isolated yield, 70%)

(*c*) Acidification of sodium nitrite produces nitrous acid, which nitrosates the strongly activated aromatic ring of phenols.

2-Isopropyl-5-methylphenol   2-Isopropyl-5-methyl-4-nitrosophenol
(isolated yield, 87%)

(*d*) Friedel-Crafts acylation occurs ortho to the hydroxyl group.

p-Cresol   Propanoyl   1-(2-Hydroxy-5-methylphenyl)-1-propanone
chloride   (isolated yield, 87%)

24.8 (*b*) The hydroxyl group of 2-naphthol is converted to the corresponding acetate ester.

2-Naphthol   Acetic anhydride   2-Naphthyl acetate   Sodium acetate

(*c*) Benzoyl chloride acylates the hydroxyl group of phenol.

Phenol   Benzoyl chloride   Phenyl benzoate   Hydrogen
chloride

24.9 Epoxides are sensitive to nucleophilic ring-opening reactions. Phenoxide ion attacks the less hindered carbon to yield 1-phenoxy-2-propanol.

Phenoxide ion   1,2-Epoxypropane   1-Phenoxy-2-propanol

24.10 The aryl halide must be one that is reactive toward nucleophilic aromatic substitution by the addition-elimination mechanism. *p*-Fluoronitrobenzene is far more reactive than fluorobenzene. The reaction shown yields *p*-nitrophenyl phenyl ether in 92 percent yield.

Potassium   *p*-Fluoronitrobenzene   *p*-Nitrophenyl phenyl ether
phenoxide

24.11 Substituted allyl aryl ethers undergo a Claisen rearrangement similar to the reaction described in text Sec. 24.13 for allyl phenyl ether. 2-Butenyl phenyl ether rearranges on heating to give *o*-(1-methyl-2-propenyl)phenol.

2-Butenyl phenyl ether   *o*-(1-methyl-2-propenyl)phenol

24.12 (*a*)   The parent compound is benzaldehyde. Vanillin bears a methoxy group (CH_3O) at C-3 and a hydroxyl group (HO) at C-4.

Vanillin (4-hydroxy-3-methoxybenzaldehyde)

(*b, c*)   Thymol and carvacrol differ with respect to the position of the hydroxyl group.

Thymol (2-isopropyl-5-methylphenol)   Carvacrol (5-isopropyl-2-methylphenol)

(d) An allyl substituent is —$CH_2CH{=}CH_2$.

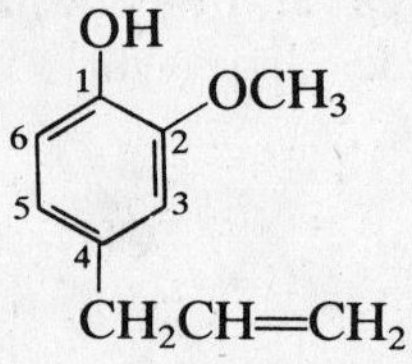

Eugenol (4-allyl-2-methoxyphenol)

(e) Benzoic acid is $C_6H_5CO_2H$. Gallic acid bears three hydroxyl groups, located at C-3, C-4, and C-5.

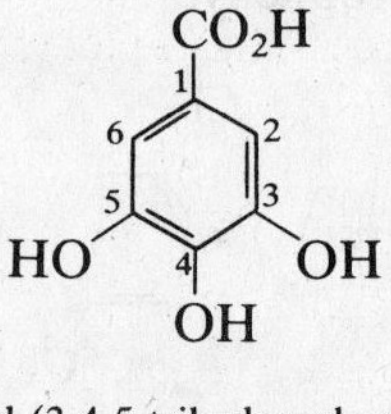

Gallic acid (3,4,5-trihydroxybenzoic acid)

(f) Benzyl alcohol is $C_6H_5CH_2OH$. Salicyl alcohol bears a hydroxyl group at the ortho position.

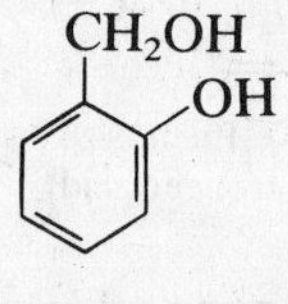

Salicyl alcohol (o-hydroxybenzyl alcohol)

24.13 (a) The compound is named as a derivative of phenol. The substituents (ethyl and nitro) are cited in alphabetical order with numbers assigned in the direction that gives the lowest number at the first point of difference.

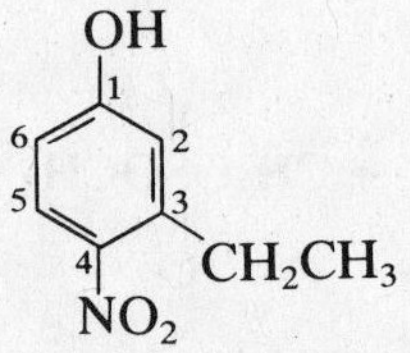

3-Ethyl-4-nitrophenol

(b) An isomer of the compound in part (a) is 4-ethyl-3-nitrophenol.

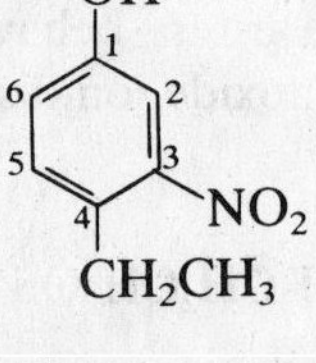

4-Ethyl-3-nitrophenol

(c) The parent compound is phenol. It bears, in alphabetical order, a benzyl group at C-4 and a chlorine at C-2.

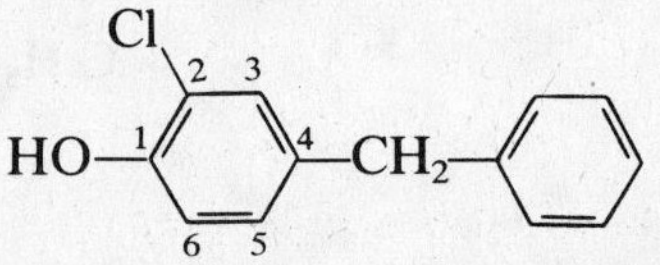

4-Benzyl-2-chlorophenol

(d) This compound is named as a derivative of anisole, $C_6H_5OCH_3$. Since multiplicative prefixes (di, tri-, etc.) are not considered when alphabetizing substituents, isopropyl precedes dimethyl.

4-Isopropyl-2,6-dimethylanisole

(e) The compound is an aryl ester of trichloroacetic acid. The aryl group is 2,5-dichlorophenyl.

2,5-Dichlorophenyl trichloroacetate

24.14 (a) The reaction is an acid-base reaction. Phenol is the acid; sodium hydroxide is the base.

Phenol	Sodium hydroxide	Sodium phenoxide	Water
(stronger acid)	(stronger base)	(weaker base)	(weaker acid)

(b) Sodium phenoxide reacts with ethyl bromide to yield ethyl phenyl ether in a Williamson reaction. Phenoxide ion acts as a nucleophile.

$$C_6H_5ONa + CH_3CH_2Br \longrightarrow C_6H_5OCH_2CH_3 + NaBr$$

Sodium phenoxide   Ethyl bromide   Ethyl phenyl ether   Sodium bromide

(c) Dimethyl sulfate is a methylating agent.

Sodium phenoxide   Dimethyl sulfate   Methyl phenyl ether   Sodium methyl sulfate

The reaction is a nucleophilic substitution (S_N2) at one of the methyl groups of dimethyl sulfate by phenoxide ion acting as the nucleophile.

(d) p-Toluenesulfonate esters behave much like alkyl iodides in nucleophilic substitution reactions. Phenoxide ion displaces p-toluenesulfonate from the primary carbon.

Sodium phenoxide   Butyl p-toluenesulfonate

Butyl phenyl ether   Sodium p-toluenesulfonate

(e) Carboxylic acid anhydrides react with phenoxide anions to yield aryl esters.

$$C_6H_5ONa + CH_3COCCH_3 \longrightarrow C_6H_5OCCH_3 + CH_3CONa$$

Sodium phenoxide　　Acetic anhydride　　　　Phenyl acetate　　Sodium acetate

(f) Acyl chlorides convert phenols to aryl esters.

o-Cresol　　Benzoyl chloride　　　　2-Methylphenyl benzoate　　　Hydrogen chloride

(g) Phenols react as nucleophiles toward epoxides.

m-Cresol　　Ethylene oxide　　　　2-(3-Methylphenoxy)ethanol

The reaction as written conforms to the requirements of the problem that a balanced equation be written. Of course, the reaction will be much faster if catalyzed by acid or base, but the catalysts do not enter into the equation representing the overall process.

(h) Bromination of the aromatic ring of 2,6-dichlorophenol occurs para to the hydroxy group. The more activating group (—OH) determines the orientation of the product.

2,6-Dichlorophenol　　Bromine　　4-Bromo-2,6-dichlorophenol　　Hydrogen bromide

(i) In aqueous solution bromination occurs at all the open positions that are ortho and para to the hydroxyl group.

p-Cresol　　Bromine　　2,6-Dibromo-4-methylphenol　　Hydrogen bromide

(*j*)   Hydrogen bromide cleaves ethers to give an alkyl halide and a phenol.

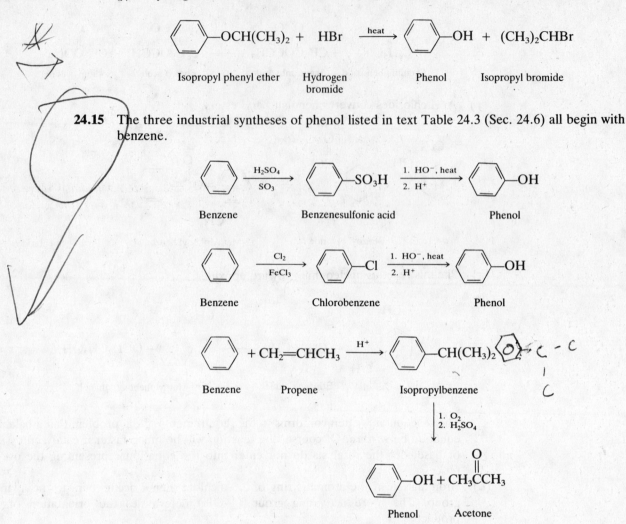

24.15   The three industrial syntheses of phenol listed in text Table 24.3 (Sec. 24.6) all begin with benzene.

The fourth synthesis rests on the fact that a phenolic hydroxyl may be introduced by first preparing aniline and then heating the derived diazonium ion in water.

$$C_6H_5\text{---}OH \Rightarrow C_6H_5\text{---}NH_2 \Rightarrow C_6H_5\text{---}NO_2 \Rightarrow C_6H_6$$

The reaction scheme to be used is:

24.16   (*a*)   Strongly electron-withdrawing groups, particularly those such as —NO$_2$, increase the acidity of phenols by resonance stabilization of the resulting phenoxide anion.

Electron-releasing substituents such as —CH$_3$ exert a very small acid-weakening effect.

2,4,6-Trinitrophenol, more acidic
($K_a = 3.8 \times 10^{-1}$, p$K_a = 0.4$)

2,4,6-Trimethylphenol, less acidic
($K_a = 1.3 \times 10^{-11}$, p$K_a = 10.9$)

Picric acid (2,4,6-trinitrophenol) is a stronger acid by far than 2,4,6-trimethylphenol. All three nitro groups participate in resonance stabilization of the picrate anion.

(*b*)   Stabilization of a phenoxide anion is most effective when electron-withdrawing groups are present at the ortho and para positions, because it is these carbons that bear most of the negative charge in phenoxide anion.

Therefore, 2,6-dichlorophenol is expected to be (and is) a stronger acid than 3,5-dichlorophenol.

2,6-Dichlorophenol, more acidic
($K_a = 1.6 \times 10^{-7}$, p$K_a = 6.8$)

3,5-Dichlorophenol, less acidic
($K_a = 6.5 \times 10^{-9}$, p$K_a = 8.2$)

(c) The same principle is at work here as in part (b). A nitro group para to the phenol oxygen is directly conjugated to it and stabilizes the anion better than one at the meta position.

4-Nitrophenol, stronger acid
$(K_a = 1.0 \times 10^{-8}, pK_a = 7.2)$

3-Nitrophenol, weaker acid
$(K_a = 4.1 \times 10^{-9}, pK_a = 8.4)$

(d) A cyano group is strongly electron-withdrawing, and so 4-cyanophenol is a stronger acid than phenol.

4-Cyanophenol, more acidic
$(K_a = 1.1 \times 10^{-8}, pK_a = 8.0)$

Phenol, less acidic
$(K_a = 1 \times 10^{-10}, pK_a = 10)$

There is resonance stabilization of the 4-cyanophenoxide anion:

(e) The 5-nitro group in 2,5-dinitrophenol is meta to the hydroxyl group and so does not stabilize the resulting anion as much as does an ortho or a para nitro group.

2,6-Dinitrophenol, more acidic
$(K_a = 2.0 \times 10^{-4}, pK_a = 3.7)$

2,5-Dinitrophenol, less acidic
$(K_a = 6.0 \times 10^{-6}, pK_a = 5.2)$

24.17 (a) The rate-determining step in base-catalyzed ester hydrolysis is formation of the tetrahedral intermediate.

Since this intermediate is negatively charged, there will be a small effect favoring its formation when the aryl group bears an electron-withdrawing substituent. Furthermore, this intermediate can either return to starting materials or proceed to products.

The proportion of the tetrahedral intermediate that goes on to products increases as the leaving group ArO^- becomes less basic. This is strongly affected by substituents; electron-withdrawing groups stabilize ArO^-. The prediction is that *m*-nitrophenyl acetate undergoes base-catalyzed hydrolysis faster than phenol. Indeed, this is observed to be the case; *m*-nitrophenyl acetate reacts some 10 times faster than does phenyl acetate at 25°C.

m-Nitrophenyl acetate
(more reactive)

m-Nitrophenoxide anion
(a better leaving group than phenoxide
because it is less basic)

(*b*)   The same principle applies here as in part (*a*). *p*-Nitrophenyl acetate reacts faster than *m*-nitrophenyl acetate (by about 45 percent) largely because *p*-nitrophenoxide is a better, less basic leaving group than *m*-nitrophenoxide.

Resonance in *p*-nitrophenoxide is particularly effective because the *p*-nitro group is directly conjugated to the oxyanion; direct conjugation of these groups is absent in *m*-nitrophenoxide.

(*c*)   The reaction of ethyl bromide with a phenol is an S_N2 reaction in which the oxygen of the phenol is the nucleophile. The reaction is much faster with sodium phenoxide than with phenol, because an anion is more nucleophilic than a corresponding neutral molecule.

Faster reaction:

$$Ar\ddot{O}:^- \quad CH_2{-}Br \longrightarrow ArOCH_2CH_3 + Br^-$$

Slower reaction:

$$Ar\ddot{O}: \quad CH_2{-}Br \longrightarrow Ar\overset{+}{O}CH_2CH_3 + Br^-$$

(*d*)   The answer here also depends on the nucleophilicity of the attacking species, which is a phenoxide anion in both reactions.

$$Ar\ddot{O}:^- \quad CH_2{-}CH_2 \longrightarrow ArOCH_2CH_2O^-$$

The more nucleophilic anion is phenoxide ion, because it is more basic than *p*-nitrophenoxide.

More basic;
better nucleophile

Better delocalization of negative charge makes
this less basic and less nucleophilic.

Rate measurements reveal that sodium phenoxide reacts 17 times faster with ethylene oxide (in ethanol at 70°C) than does its *p*-nitro derivative.

(e) This reaction is electrophilic aromatic substitution. Since a hydroxy substituent is more activating than an acetate group, phenol undergoes bromination faster than does phenyl acetate.

Resonance involving ester group reduces tendency of
oxygen to donate electrons to ring.

24.18 Nucleophilic aromatic substitution by the elimination-addition mechanism is impossible, owing to the absence of any protons that might be abstracted from the substrate. However, the addition-elimination pathway is available.

Hexafluorobenzene   Pentafluorophenol

This pathway is favorable because the cyclohexadienyl anion intermediate formed in the rate-determining step is stabilized by the electron-withdrawing inductive effect of its fluorine substituents.

24.19 (a) Allyl bromide is a reactive alkylating agent and converts the free hydroxyl group of the aryl compound (a natural product known as *guaiacol*) to its corresponding allyl ether.

Guaiacol   Allyl bromide   2-Allyloxyanisole (80–90%)

(b) Dimethyl sulfate is a standard reagent for methylation of phenols.

2,5-Dihydroxyacetophenone   Dimethyl sulfate   2,5-Dimethoxyacetophenone
(71–74%)

(c) Sodium phenoxide acts as a nucleophile in this reaction and is converted to an ether.

Sodium phenoxide   3-Chloro-1,2-propanediol   3-Phenoxy-1,2-
propanediol (61–63%)

(*d*) Orientation in nitration is governed by the most activating substituent, in this case the hydroxyl group.

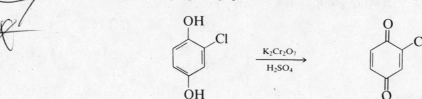

Vanillin

4-Hydroxy-3-methoxy-5-nitrobenzaldehyde (83%)

(*e*) Allyl aryl ethers undergo a Claisen rearrangement on heating. Heating *p*-acetamidophenyl allyl ether gave an 83 percent yield of 4-acetamido-2-allylphenol.

p-Acetamidophenyl allyl ether

4-Acetamido-2-allylphenol

(*f*) The hydroxyl group, as the most activating substituent, controls the orientation of electrophilic aromatic substitution. Bromination takes place ortho to the hydroxyl group.

2-Ethoxy-4-nitrophenol

2-Bromo-6-ethoxy-4-nitrophenol (65%)

(*g*) Oxidation of hydroquinone derivatives (*p*-dihydroxybenzenes) with Cr(VI) reagents is a method for preparing quinones.

2-Chloro-1,4-benzenediol

2-Chloro-1,4-benzoquinone (88%)

(*h*) Silver oxide is a mild oxidizing agent, often used for the preparation of quinones from hydroquinones. It will not affect the allyl groups.

2,3-Diallylhydroquinone

2,3-Diallyl-1,4-benzoquinone (96%)

(*i*)   Aryl esters undergo a reaction known as the *Fries rearrangement* on being treated with aluminum chloride, which converts them to acyl phenols. Acylation takes place para to the hydroxyl in this case.

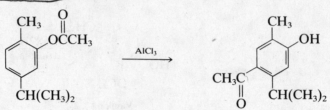

5-Isopropyl-2-methylphenyl acetate   4-Hydroxy-2-isopropyl-5-methylacetophenone
                                                              (90%)

(*j*)   Nucleophilic aromatic substitution takes place to yield a diaryl ether. The nucleophile is the phenoxide ion derived from 2,6-dimethylphenol.

2,6-Dimethylphenol   *p*-Chloronitrobenzene   2,6-Dimethylphenyl *p*-nitrophenyl ether (82%)

(*k*)   Chlorination with excess chlorine occurs at all available positions that are ortho and para to the hydroxyl group.

2,5-Dichlorophenol   Chlorine   2,3,4,6-Tetrachlorophenol   Hydrogen
                                                       (isolated yield, 100%)   chloride

(*l*)   Amines react with esters to give amides. In the case of phenyl ester, phenol is the leaving group.

o-Methylaniline   Phenyl salicylate

N-(*o*-Methylphenyl)salicylamide   Phenol
(isolated yield, 73–77%)

(*m*) Aryl diazonium salts attack electron-rich aromatic rings, such as those of phenols, to give the products of electrophilic aromatic substitution.

| 2,4,5-Trichlorophenol | Benzenediazonium chloride | 2-Benzeneazo-3,4,6-trichlorophenol (80%) |

24.20 In the first step *p*-nitrophenol is alkylated on its phenolic oxygen with diethyl sulfate.

| *p*-Nitrophenol | Diethyl sulfate | Ethyl *p*-nitrophenyl ether |

Reduction of the nitro group gives the corresponding arylamine.

| Ethyl *p*-nitrophenyl ether | *p*-Ethoxyaniline |

Treatment of *p*-ethoxyaniline with acetic anhydride gives the product of *N*-acylation.

| *p*-Ethoxyaniline | Acetic anhydride | *p*-Ethoxyacetanilide (phenacetin) |

24.21 The three parts of this problem make up the series of steps by which *o*-bromophenol is prepared.

(*a*) Since direct bromination of phenol yields both *o*-bromophenol and *p*-bromophenol, it is essential that the para position be blocked prior to the bromination step. In practice, what is done is to disulfonate phenol, which blocks the para and one of the ortho positions.

| Phenol | 4-Hydroxy-1,3-benzenedisulfonic acid (compound A) |

(*b*) Bromination then can be accomplished cleanly at the open position ortho to the hydroxyl group.

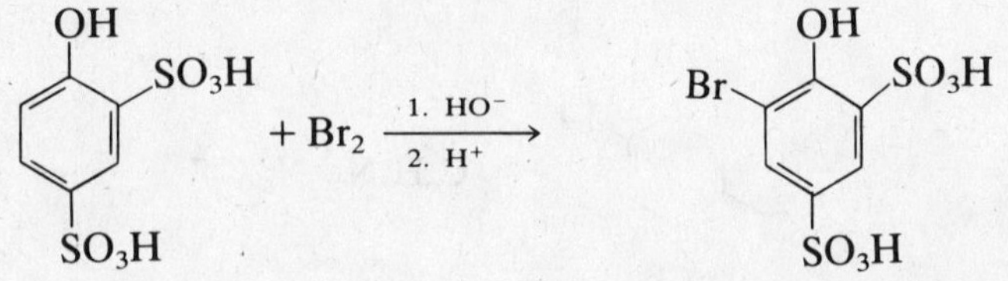

Compound A

5-Bromo-4-hydroxy-1,3-benzenedisulfonic acid (compound B)

(*c*) After bromination the sulfonic acid groups are removed by acid-catalyzed hydrolysis.

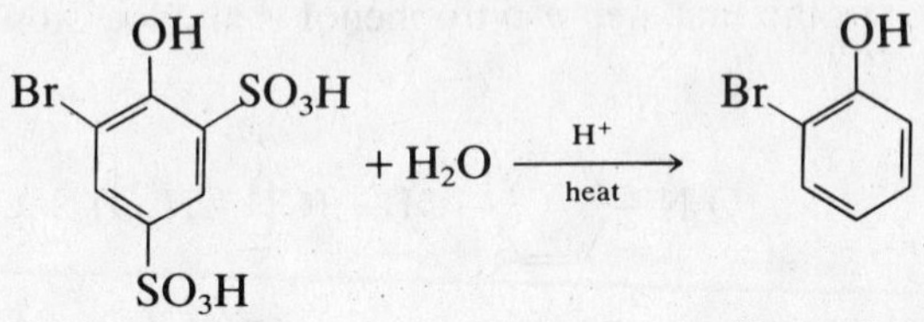

Compound B

o-Bromophenol (compound C)

24.22 Nitration of 3,5-dimethylphenol gives a mixture of the 2-nitro and 4-nitro derivatives.

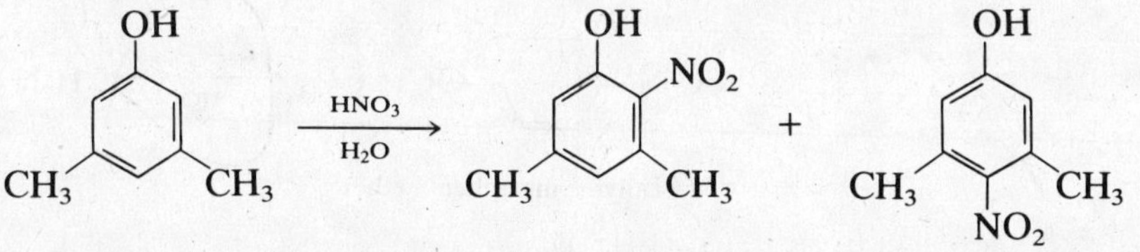

3,5-Dimethylphenol

3,5-Dimethyl-2-nitrophenol

3,5-Dimethyl-4-nitrophenol

The more volatile compound (compound D), isolated by steam distillation, is the 2-nitro derivative. Intramolecular hydrogen bonding is possible between the nitro group and the hydroxyl group.

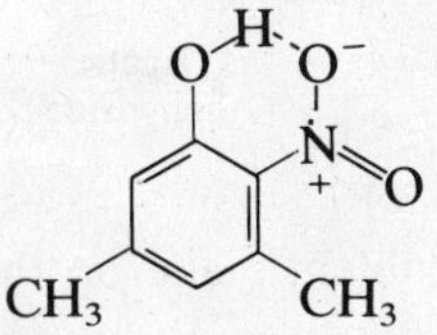

Intramolecular hydrogen bonding in 3,5-dimethyl-2-nitrophenol

The 4-nitro derivative participates in intermolecular hydrogen bonds and has a much higher boiling point; it is compound E.

24.23 The relationship between the target molecule and the starting materials tells us that two processes are required, formation of a diaryl ether linkage and nitration of an aromatic ring. The proper order of carrying out these two separate processes is what needs to be considered.

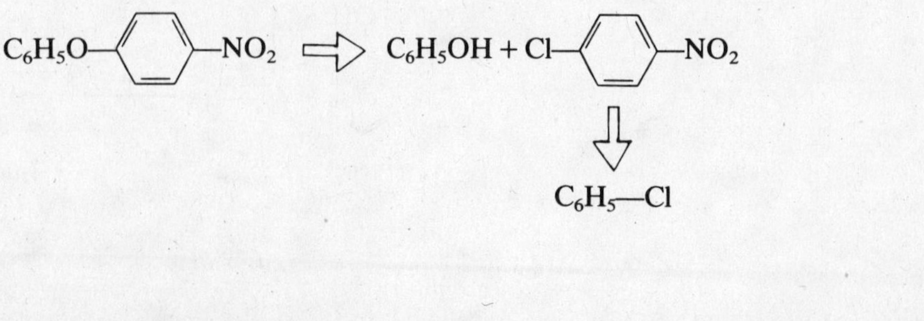

The critical step is ether formation, a step which is feasible for the reactants shown:

Phenol p-Chloronitrobenzene $\xrightarrow[\text{heat}]{\text{KOH}}$

4-Nitrophenyl phenyl ether

The reason this reaction is suitable is that it involves nucleophilic aromatic substitution by the addition-elimination mechanism on a p-nitro-substituted aryl halide. Indeed, this reaction has been carried out and gives an 80 to 82 percent yield. Therefore a reasonable synthesis would begin with the preparation of p-chloronitrobenzene.

Chlorobenzene $\xrightarrow[\text{H}_2\text{SO}_4]{\text{HNO}_3}$ o-Chloronitrobenzene + p-Chloronitrobenzene

Separation of the p-nitro-substituted aryl halide and reaction with phenoxide ion complete the synthesis.

The following alternative route is less satisfactory:

Phenol Chlorobenzene $\xrightarrow{\text{base}}$ Diphenyl ether

Diphenyl ether $\xrightarrow[\text{H}_2\text{SO}_4]{\text{HNO}_3}$ 2-Nitrophenyl phenyl ether + 4-Nitrophenyl phenyl ether

The difficulty with this route concerns the preparation of diphenyl ether. Direct reaction of phenoxide with chlorobenzene is very slow because chlorobenzene is a poor substrate for nucleophilic substitution.

A third route is also unsatisfactory because it, too, requires nucleophilic substitution on chlorobenzene.

Phenol $\xrightarrow{\text{HNO}_3}$ o-Nitrophenol + p-Nitrophenol

p-Nitrophenol + Chlorobenzene $\xrightarrow{\text{base}}$ 4-Nitrophenyl phenyl ether

24.24 The overall transformation that needs to be effected is:

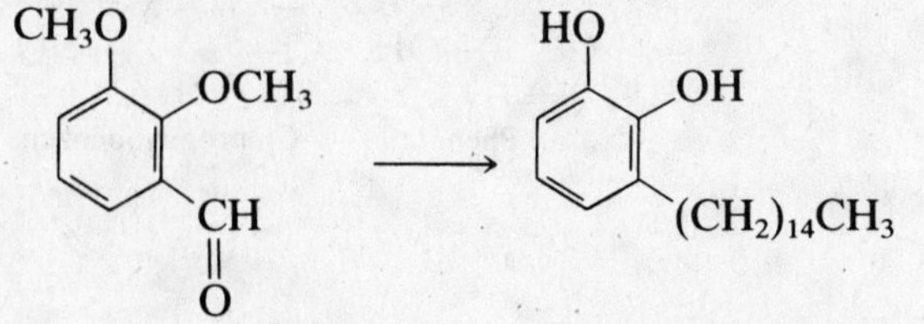

2,3-Dimethoxybenzaldehyde   3-Pentadecylcatechol

A reasonable place to begin is with the attachment of the side chain. The aldehyde function allows for chain extension by a Wittig reaction.

$$Ar{-}CH_2CH_2R \Rightarrow Ar{-}CH{=}CHR \Rightarrow Ar{-}\overset{O}{\overset{\|}{C}}H + (C_6H_5)_3\overset{+}{P}{-}\overset{..}{\overset{-}{C}}HR$$

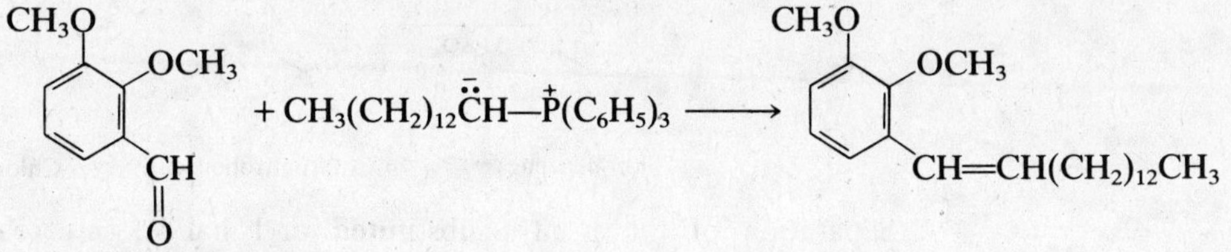

$$+ CH_3(CH_2)_{12}\overset{..}{\overset{-}{C}}H{-}\overset{+}{P}(C_6H_5)_3 \longrightarrow$$

2,3-Dimethoxybenzaldehyde

Hydrogenation of the double bond and hydrogen halide cleavage of the ether functions complete the synthesis.

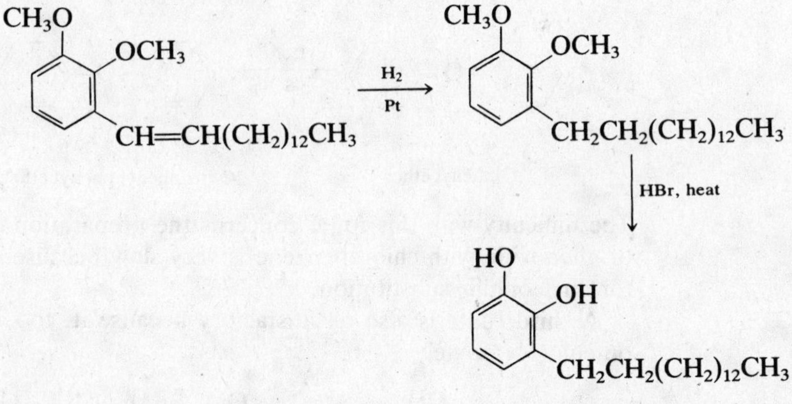

3-Pentadecylcatechol

Other synthetic routes are of course possible. One of the earliest approaches utilized a Grignard reaction to attach the side chain.

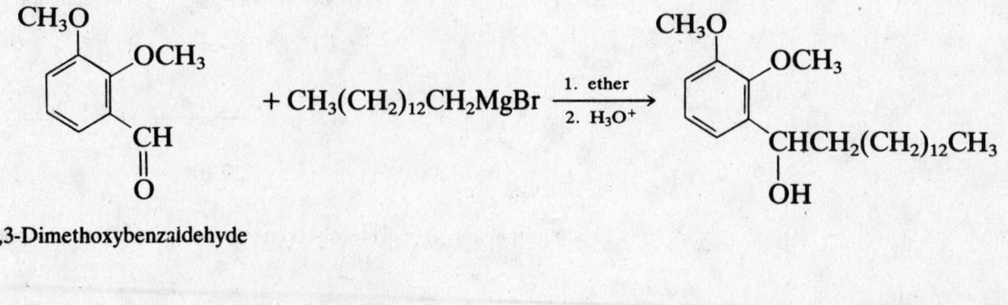

$$+ CH_3(CH_2)_{12}CH_2MgBr \xrightarrow[\text{2. } H_3O^+]{\text{1. ether}}$$

2,3-Dimethoxybenzaldehyde

The resulting secondary alcohol can then be dehydrated to the same alkene intermediate prepared in the preceding synthetic scheme.

$$CH_3O\text{—}\quad OCH_3 \quad \text{—CHCH}_2(CH_2)_{12}CH_3 \quad \xrightarrow{\ H^+\ } \quad CH_3O\text{—}\quad OCH_3 \quad \text{—CH=CH}(CH_2)_{12}CH_3$$
$$\qquad\qquad\qquad\qquad\ |$$
$$\qquad\qquad\qquad\qquad OH$$

Again, hydrogenation of the double bond and ether cleavage leads to the desired 3-pentadecylcatechol.

24.25 Recall that the Claisen rearrangement converts an aryl allyl ether to an ortho-substituted allyl phenol. The presence of an allyl substituent in the product ortho to an aryl ether thus suggests the following retrosynthesis:

$$CH_3O\text{—}\ \overset{OCH_3}{\underset{\underset{OH}{CH_3}}{\bigcirc}}\ \text{—CH}_2\text{CH=CH}_2 \quad \Longrightarrow \quad CH_3O\text{—}\ \overset{OCH_2CH=CH_2}{\underset{\underset{OCCH_3}{\underset{\parallel}{CH_3}\ O}}{\bigcirc}}$$

As reported in the literature synthesis, the starting phenol may be converted to the corresponding allyl ether by reaction with allyl bromide in the presence of base. This step was accomplished in 80 percent yield. Heating the allyl ether yields the *o*-allyl phenol.

$$CH_3O\text{—}\overset{OH}{\underset{\underset{OCCH_3}{\underset{\parallel}{CH_3}\ O}}{\bigcirc}} \xrightarrow[K_2CO_3]{CH_2=CHCH_2Br} CH_3O\text{—}\overset{OCH_2CH=CH_2}{\underset{\underset{OCCH_3}{\underset{\parallel}{CH_3}\ O}}{\bigcirc}}$$

$$\xrightarrow{200°C,\ 3\ h} \quad CH_3O\text{—}\overset{OH}{\underset{\underset{OCCH_3}{\underset{\parallel}{CH_3}\ O}}{\bigcirc}}\text{—CH}_2\text{CH=CH}_2$$

The synthesis is completed by methylation of the phenolic oxygen and saponification of the acetate ester. The final three steps of the synthesis proceeded in an 82 percent overall yield.

24.26 The driving force for this reaction is the stabilization which results from formation of the aromatic ring. A reasonable series of steps is as follows:

Resonance forms of protonated ketone

(Protonated ketone can rearrange by alkyl migration)

(Aromatization of this intermediate occurs by loss of a proton)

24.27 Bromination of *p*-hydroxybenzoic acid takes place in the normal fashion at both positions ortho to the hydroxy group.

p-Hydroxybenzoic acid   3,5-Dibromo-4-hydroxybenzoic acid

A third bromination step, this time at the para position, leads to the intermediate shown.

Aromatization of this intermediate occurs by decarboxylation.

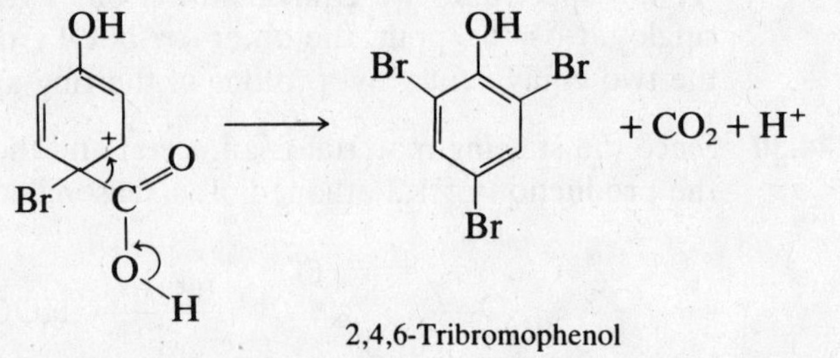

2,4,6-Tribromophenol

24.28 Electrophilic attack of bromine on 2,4,6-tribromophenol leads to a cationic intermediate.

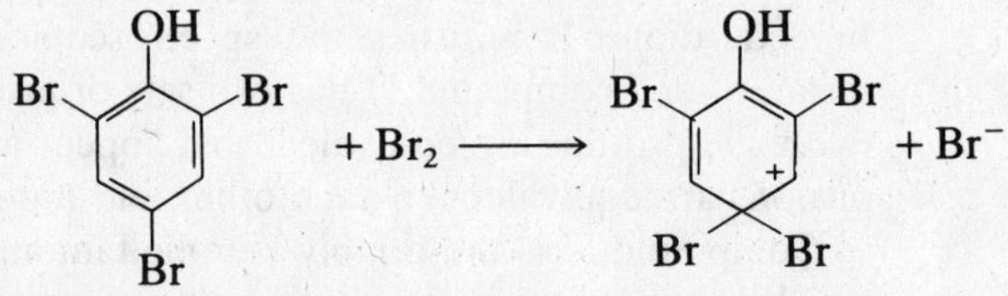

2,4,6-Tribromophenol

Loss of the hydroxyl proton from this intermediate generates the observed product.

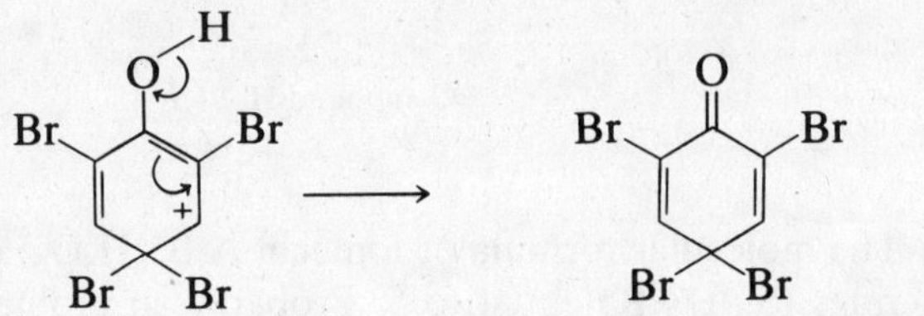

2,4,4,6-Tetrabromocyclohexadienone

24.29 A good way to approach this problem is to assume that bromine attacks the aromatic ring of the phenol in the usual way, that is, para to the hydroxyl group.

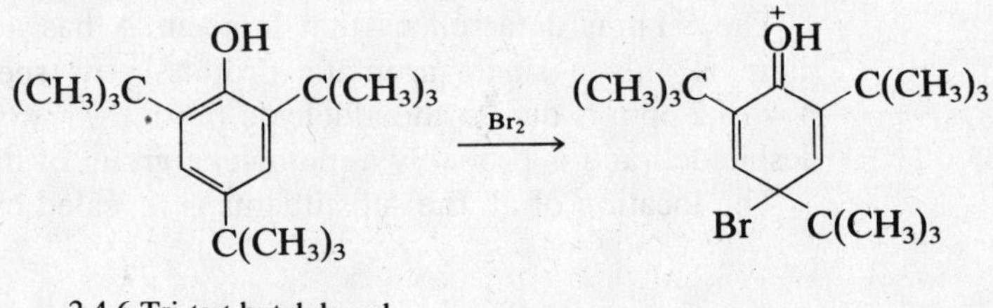

2,4,6-Tri-*tert*-butylphenol

This cation cannot yield the product of electrophilic aromatic substitution by loss of a proton from the ring but can lose a proton from oxygen to give a cyclohexadienone derivative.

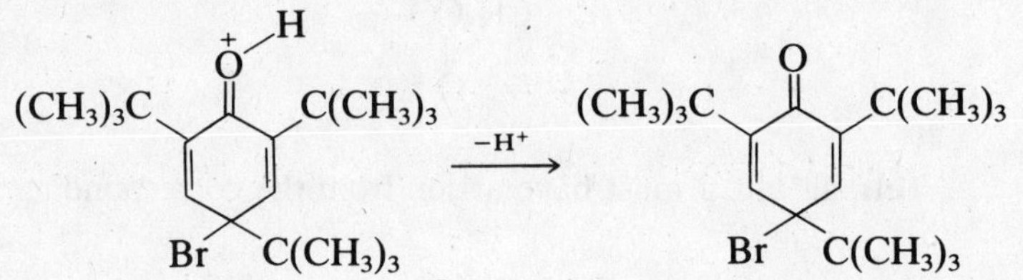

4-Bromo-2,4,6-tri-*tert*-butyl-2,5-cyclohexadienone

This cyclohexadienone is compound F. It has the correct molecular formula ($C_{18}H_{29}BrO$), and the peaks at 1655 and 1630 cm^{-1} in the infrared are consistent with

C=O and C=C stretching vibrations. Its symmetry is consistent with the observed ^{1}H nmr spectrum; two equivalent *tert*-butyl groups at C-2 and C-6 appear as an 18-proton singlet at $\delta = 1.3$ ppm, the other *tert*-butyl group is a 9-proton singlet at $\delta = 1.2$ ppm, and the two equivalent vinyl protons of the ring appear as a singlet at $\delta = 6.9$ ppm.

24.30   Since the starting material is an acetal and the reaction conditions lead to hydrolysis with the production of 1,2-ethanediol, a reasonable reaction course is:

Compound G   1,2-Ethanediol   Compound H

Indeed, dione H satisfies the spectroscopic criteria. Carbonyl bands are seen in the infrared, and compound H has two sets of protons to be seen in its ^{1}H nmr spectrum. The two vinyl protons are equivalent and appear at low field, $\delta = 6.7$ ppm; the four methylene protons are equivalent to each other and are seen at $\delta = 2.9$ ppm.

Compound H is the doubly ketonic tautomeric form of hydroquinone, compound I, to which it isomerizes on standing in water.

Compound H   Compound I
                             (hydroquinone)

24.31   The molecular formula of fomecin A ($C_8H_8O_5$) corresponds to a sum of double bonds and rings (SODAR) equal to 5. Probably an aromatic ring is present, accounting for four of these. Its solubility in dilute sodium hydroxide suggests that fomecin A is a phenol. A carbonyl group is indicated by formation of a 2,4-dinitrophenylhydrazone derivative and is probably an aldehyde function, since fomecin A reduces ammoniacal silver nitrate. Thus, the aromatic ring of a phenol and the carbon-oxygen double bond of an aldehyde account for all the elements of unsaturation implied by the molecular formula.

The ^{1}H nmr data tell us that fomecin A has a pentasubstituted aromatic ring, since there is only a single aromatic proton in its spectrum at $\delta = 6.5$ ppm. The signal at $\delta = 10.2$ ppm is due to the aldehyde proton. A two-proton singlet at $\delta = 4.6$ ppm is fairly deshielded and is probably a methylene group of the type ArCH$_2$O—.

The location of all the substituents is revealed by the isolation of the tetraacetate J.

Compound J

Tetraacetate J must have arisen from the corresponding tetrol:

This tetrol, however, does not have an aldehyde group, while the nmr evidence tells us that one must be present. Notice that the tetrol shown is a cyclic hemiacetal; its noncyclic form must be fomecin A.

Compound J   Fomecin A

24.32   A reasonable first step is protonation of the hydroxyl oxygen.

Cumene hydroperoxide

The weak oxygen-oxygen bond can now be cleaved, with loss of water as the leaving group.

This intermediate bears a positively charged oxygen with only six electrons in its valence shell. Like a carbocation, such a species is highly electrophilic. The electrophilic oxygen attacks the π system of the neighboring aromatic ring to give an unstable intermediate.

Ring opening of this intermediate is assisted by one of the lone pairs of oxygen and restores the aromaticity of the ring.

The cation formed by ring opening is captured by a water molecule to yield the hemiacetal product.

24.33 (*a*) The molecular formula of compound K ($C_9H_{12}O$) tells us that it has a total of four double bonds and rings (SODAR = 4). Peaks in the $\delta = 6.5$ to 7 ppm region of the ^{1}H nmr spectrum indicate an aromatic ring, which accounts for six of the nine carbon atoms of compound K and all its double bonds and rings. Thus any alkyl groups must have only single bonds to all carbons.

The infrared spectrum suggests an alcohol or a phenol, based on a strong absorption at 3300 cm^{-1}. An OH group is also consistent with a broad absorption at $\delta = 5.5$ ppm in the nmr spectrum. A three-proton triplet at high field ($\delta = 0.9$ ppm) must be a CH_3 group attached to a CH_2 unit, which must be different from the CH_2 unit responsible for the two-proton triplet at $\delta = 2.5$ ppm because the triplet nature of the signal at 2.5 ppm means that CH_2 is bonded to another CH_2 and not to CH_3. A propyl group attached to an aromatic ring is a possibility.

$$\underset{\substack{\delta = 2.5\text{ ppm} \\ \text{(benzylic triplet)}}}{Ar—CH_2}\underset{\text{multiplet at }\delta = 1.5\text{ ppm}}{—CH_2—}\overset{\text{triplet at }\delta = 0.9\text{ ppm}}{CH_3}$$

Integration of the spectrum reveals the aromatic ring to be disubstituted. The characteristic pair of doublets with a splitting of about 8 Hz indicates para substitution, permitting the last part of the structure to be assigned. The hydroxyl group can only be located on the aromatic ring and is para to the propyl group. The compound is 4-propylphenol.

$$HO—\langle\text{ring}\rangle—CH_2CH_2CH_3$$

Compound K

(*b*) Compound L has a monosubstituted aromatic ring because there is a five-proton multiplet in the $\delta = 6.5$ to 7.5 ppm region of the ^{1}H nmr spectrum. The lack of absorption in the infrared spectrum around 3300 cm^{-1} means that no hydroxyl group is present. Since the aromatic rings accounts for all the double bonds and rings of the molecule (SODAR = 4), its oxygen atom can only be part of an ether function.

Two-proton signals in the nmr at $\delta = 3.5$ and 4.1 ppm suggest OCH_2 and CH_2Br. Each one is a triplet, indicating that each CH_2 is itself bonded to a CH_2 group. Compound L is 3-bromopropyl phenyl ether.

$$\langle\text{ring}\rangle—O—\underset{\substack{\text{triplet} \\ (\delta = 4.1\text{ ppm})}}{CH_2}—\underset{\substack{\text{quintet} \\ (\delta = 2.3\text{ ppm})}}{CH_2}—\underset{\substack{\text{triplet} \\ (\delta = 3.5\text{ ppm})}}{CH_2}—Br$$

(*c*) Compound M, from its molecular formula, has a total of five double bonds and rings (SODAR = 5). A disubstituted aromatic ring, revealed by two two-proton doublets in the $\delta = 6.5$ to 8 ppm region of the ^{1}H nmr spectrum, accounts for four of those elements of unsaturation. Thus the remaining atoms must include one double bond or one ring. A strong band in the infrared at 1670 cm^{-1} indicates that this remaining element of unsaturation is the double bond of a carbonyl group. We also see an OH group absorption at 3300 cm^{-1} in the infrared spectrum. A one-proton signal at low field ($\delta = 9.1$ ppm) reveals the OH group to be that of a phenol.

Compound M is a disubstituted aromatic ring in which one of the substituents is a hydroxyl group and the other is a four-carbon chain that bears a chlorine substituent and contains a carbonyl group. The substitution pattern on the ring is para, as indicated by the characteristic pair of doublets with a splitting of about 10 Hz. The side chain has three CH_2 groups, as shown by its nmr spectrum.

Compound M must be 3-chloropropyl *p*-hydroxyphenyl ketone.

$$\delta = 9.1\text{ ppm} \quad HO\!-\!\!\!\!\raisebox{-0.5ex}{\text{(ring)}}\!\!\!\!-\!\!\underset{\underset{\text{triplet, }\delta = 3.1\text{ ppm}}{\Big\uparrow}}{\overset{\overset{\text{O}}{\|}}{C}}\!CH_2CH_2CH_2Cl$$

triplet, $\delta = 3.7$ ppm

triplet, $\delta = 3.1$ ppm   multiplet, $\delta = 2.2$ ppm

SELF-TEST

PART A

A-1. Which is the stronger acid, *m*-hydroxybenzaldehyde or *p*-hydroxybenzaldehyde? Explain your answer, using resonance structures.

A-2. The cresols are methyl-substituted phenols. Predict the major products to be obtained from the reactions of *o*-, *m*-, and *p*-cresol with dilute nitric acid.

A-3. Give the structure of the product from the reaction of *p*-cresol with propanoyl chloride, $CH_3CH_2\overset{\overset{\text{O}}{\|}}{C}Cl$, in the presence of $AlCl_3$. What product is obtained in the absence of $AlCl_3$?

A-4. Provide the structure of the reactant, reagent, or product omitted from each of the following:

(*a*) $\bigcirc\!-\!OCH(CH_3)_2 \xrightarrow{\text{HBr}}$? (two products)

(*b*) ? (two compounds) $\longrightarrow$ $\underset{CH_3}{\overset{OCH_2CH(CH_3)_2}{\bigcirc}}$

(*c*) $\underset{}{\overset{O^-Na^+}{\bigcirc}} \xrightarrow[\text{2. }H^+]{\text{1. ?}}$ $\underset{}{\overset{OH}{\bigcirc}}\!CO_2H$

(*d*) $\bigcirc\!\!\!\!<\!\!\!\overset{}{\underset{O}{\rule{1em}{0pt}}} \xrightarrow[\text{heat}]{\text{HBr}}$? (C_8H_9BrO)

A-5. Provide the structures of compounds A and B in the following sequence of reactions:

$$\underset{}{\overset{OH}{\bigcirc}} \xrightarrow[K_2CO_3]{CH_3CH_2CH=CHCH_2Br} A \xrightarrow{\text{heat}} B\ (C_{11}H_{14}O)$$

PART B

B-1. Rank the following in order of decreasing acid strength (most acidic → least acidic):

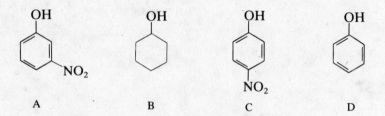

A   B   C   D

   (*a*) B > D > A > C   (*c*) A > C > D > B
   (*b*) C > A > B > D   (*d*) C > A > D > B

B-2. Which of the following choices correctly describes the solubility of benzoic acid (A) and β-naphthol (B) in the aqueous solutions shown?

	Aq NaOH	Aq NaHCO$_3$
(*a*)	A, soluble; B, insoluble	A, insoluble; B, soluble
(*b*)	A, insoluble; B, soluble	A, soluble; B, insoluble
(*c*)	A, soluble; B, soluble	A, soluble; B, insoluble
(*d*)	A, soluble, B, soluble	A, insoluble; B, soluble

B-3. Which of the following phenols has the largest pK_a value (i.e., is least acidic)?

   (*a*) Cl—⟨ ⟩—OH

   (*c*) O$_2$N—⟨ ⟩—OH

   (*b*) CH$_3$—⟨ ⟩—OH

   (*d*) N≡C—⟨ ⟩—OH

B-4. Which of the following reactions is a more effective method for preparing phenyl propyl ether?

   A: $C_6H_5ONa + CH_3CH_2CH_2Br \longrightarrow$
   B: $CH_3CH_2CH_2ONa + C_6H_5Br \longrightarrow$

   (*a*) Reaction A is more effective.
   (*b*) Reaction B is more effective.
   (*c*) Both reactions A and B are effective.
   (*d*) Neither reaction A nor reaction B is effective.

B-5. What reactant gives the product shown on heating with aluminum chloride?

$$CH_3-C_6H_4-\overset{\displaystyle O}{\underset{\displaystyle \|}{C}}-C_6H_4-OH$$

(a) $CH_3-C_6H_4-O-\overset{\displaystyle O}{\underset{\displaystyle \|}{C}}-C_6H_5$

(c) $CH_3-C_6H_4-\overset{\displaystyle O}{\underset{\displaystyle \|}{C}}-O-C_6H_5$

(b) $CH_3-C_6H_3(\overset{\displaystyle O}{\underset{\displaystyle \|}{C}}OH)(O-C_6H_5)$

(d) $C_6H_4(OH)-\overset{\displaystyle O}{\underset{\displaystyle \|}{C}}-C_6H_4-CH_3$

CARBOHYDRATES

IMPORTANT TERMS AND CONCEPTS

Carbohydrate Classification (Secs. 25.1 to 25.4) The basis for carbohydrate classification is the term *saccharide*. *Monosaccharides* are simple carbohydrates, that is, those which are not cleaved on hydrolysis to smaller carbohydrates. *Disaccharides* are cleaved to two monosaccharides on hydrolysis. The terms *oligosaccharides* and *polysaccharides* refer to carbohydrates which on hydrolysis yield 3 to 10 and more than 10 monosaccharide units, respectively.

Monosaccharides that are polyhydroxy aldehydes are known as *aldoses*; those that are polyhydroxy ketones are *ketoses*. A carbohydrate having three, four, five, or six carbon atoms is termed a *triose, tetrose, pentose,* or *hexose,* respectively.

Stereochemical Notation (Sec. 25.2) The stereochemical representation most widely used for carbohydrates is the *Fischer projection,* described in Chapter 7.

Carbohydrates are identified as being either D or L depending on their relationship to dextrorotatory and levorotatory glyceraldehyde, respectively. That is, a D carbohydrate has the same configuration at its highest-numbered stereogenic center as that of (+)-glyceraldehyde.

D-(+)-Glyceraldehyde   D-Lyxose

Cyclic Forms of Carbohydrates (Secs. 25.6 to 25.8) Aldoses and ketoses exist as *cyclic hemiacetals*. Five-membered ring forms are known as *furanose* forms; the six-membered rings are *pyranose* forms. The ring carbon derived from the carbonyl is called the *anomeric* carbon.

Haworth formulas are often used to represent the cyclic hemiacetal forms of carbohydrates, as shown in the following illustrations; while useful for representing the *configurational* relationships of carbohydrates, it should be recognized that Haworth formulas provide no information regarding carbohydrate *conformations*.

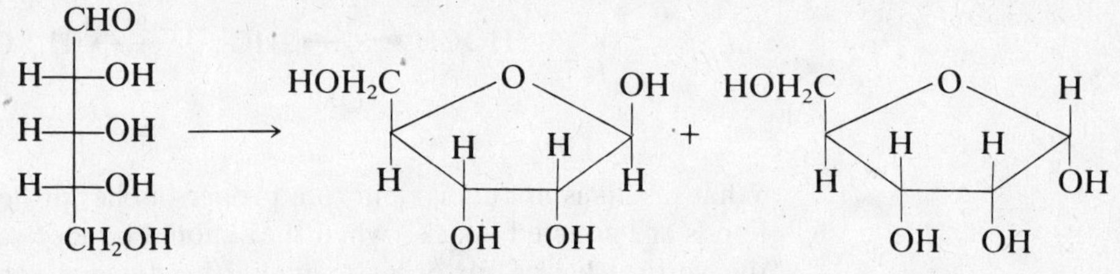

The hydroxyl group on the anomeric carbon may adopt either of two configurations in the cyclic hemiacetal. In the D series the hydroxyl is α when "down" in the Haworth formula and β when "up."

As may be shown by optical rotation measurements, in solution the individual anomers of a carbohydrate rapidly equilibrate with each other by a process known as *mutarotation*.

Glycosides (Secs. 25.13, 25.14) Carbohydrate derivatives formed by the replacement of the anomeric hydroxyl group by some other substituent are known as *glycosides*. Disaccharides are examples of glycosides in which the anomeric hydroxyl has been replaced by a second sugar molecule. Examples include *maltose*, obtained by hydrolysis of starch, and *cellobiose*, obtained by hydrolysis of cellulose.

CARBOHYDRATE REACTIONS

Reduction (Sec. 25.18) Typical procedures for the reduction of carbohydrates employ catalytic hydrogenation or sodium borohydride. The reduction products are known as *alditols*.

Oxidation (Sec. 25.19) Aldoses undergo oxidation of the aldehyde group readily. A solution of copper(II) ion is employed in *Benedict's reagent*; formation of red copper(I) oxide is taken as a positive test. Carbohydrates which give a positive test with Benedict's solution are termed *reducing sugars*. Other reagent solutions used to test for reducing sugars are *Fehling's solution*, a copper(II) tartrate complex; and *Tollens' reagent*, a solution of ammoniacal silver ion.

Carbohydrate derivatives in which the aldehyde group has been oxidized are known as *aldonic acids*. Oxidation with nitric acid leads to formation of *aldaric acids*, in which both the aldehyde and the terminal hydroxyl groups have been oxidized to carboxylic acids.

Osazone Formation (Sec. 25.21) Reaction of aldoses and ketoses with an excess of phenylhydrazine converts two adjacent carbons (the carbonyl and the nearest hydroxyl carbon) to phenylhydrazone units, forming what is known as an *osazone*.

SOLUTIONS TO TEXT PROBLEMS

25.1 (b) Redraw the Fischer projection so as to show the orientation of the groups in three dimensions.

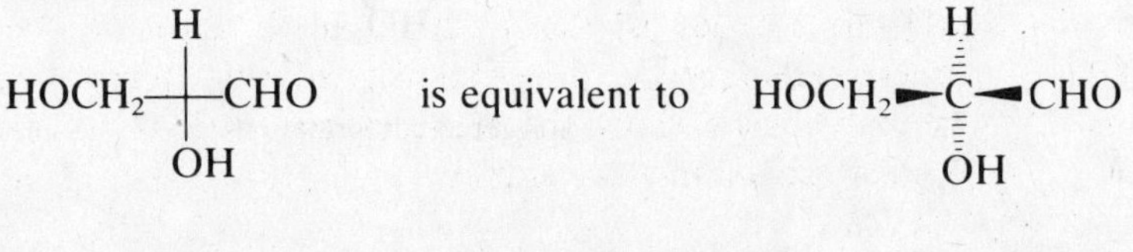

Reorient the three-dimensional representation, putting the aldehyde group at the top and the primary alcohol at the bottom.

$$
\text{HOCH}_2\text{—}\overset{\displaystyle H}{\underset{\displaystyle OH}{C}}\text{—CHO} \xrightarrow{\text{turn }90°} \text{H}\text{—}\overset{\displaystyle CHO}{\underset{\displaystyle CH_2OH}{C}}\text{—OH}
$$

What results is not equivalent to a proper Fischer projection, because the horizontal bonds are directed "back" when they should be "forward." The opposite is true for the vertical bonds. In order to make the drawing correspond to a proper Fischer projection, we need to rotate it 180° around a vertical axis.

$$
\text{H}\text{—}\overset{\displaystyle CHO}{\underset{\displaystyle CH_2OH}{C}}\text{—OH} \longrightarrow \text{HO}\text{—}\overset{\displaystyle CHO}{\underset{\displaystyle CH_2OH}{C}}\text{—H} \quad \text{is equivalent to} \quad \text{HO}\text{—}\overset{\displaystyle CHO}{\underset{\displaystyle CH_2OH}{\vert}}\text{—H}
$$

rotate 180°

Now, having the molecule arranged properly, we see that it is L-glyceraldehyde.

(c) Again proceed by converting the Fischer projection into a three-dimensional representation.

$$
\text{HOCH}_2\text{—}\overset{\displaystyle CHO}{\underset{\displaystyle OH}{\vert}}\text{—H} \quad \text{is equivalent to} \quad \text{HOCH}_2\text{—}\overset{\displaystyle CHO}{\underset{\displaystyle OH}{C}}\text{—H}
$$

Look at the drawing from a perspective that permits you to see the carbon chain oriented vertically with the aldehyde at the top and the CH₂OH at the bottom. Both groups should point away from you. When examined from this perspective, the hydrogen is to the left and the hydroxyl to the right with both pointing toward you.

$$
\text{HOCH}_2\text{—}\overset{\displaystyle CHO}{\underset{\displaystyle OH}{C}}\text{—H} \quad \text{is equivalent to} \quad \text{H}\text{—}\overset{\displaystyle CHO}{\underset{\displaystyle CH_2OH}{C}}\text{—OH}
$$

The molecule is D-glyceraldehyde.

25.2 In order to match the compound with the Fischer projection formulas of the aldotetroses, first redraw it in an eclipsed conformation.

Staggered conformation   Same molecule in eclipsed conformation

The eclipsed conformation shown, when oriented so that the aldehyde carbon is at the top, vertical bonds back, and horizontal bonds pointing outward from their stereogenic centers, is readily transformed into the Fischer projection of L-erythrose.

L-Erythrose

25.3 L-Arabinose is the mirror image of D-arabinose, the structure of which is given in text Figure 25.2. The configuration at *each* stereogenic center of D-arabinose must be reversed to transform it into L-arabinose.

D-(−)-Arabinose   L-(+)-Arabinose

25.4 The configuration at C-5 is opposite to that of D-(+)-glyceraldehyde. Therefore, this particular carbohydrate belongs to the L series. Comparing it with the Fischer projection formulas of the eight D-aldohexoses reveals it to be in the mirror image of D-(+)-talose; it is L-(−)-talose.

25.5 (*b*) The Fischer projection formula of D-arabinose may be found in text Figure 25.2. The Fischer projection and the eclipsed conformation corresponding to it are:

D-Arabinose   Eclipsed conformation of D-arabinose   Conformation suitable for furanose ring formation

Cyclic hemiacetal formation between the carbonyl group and the C-4 hydroxyl yields the α- and β-furanose forms of D-arabinose.

β-D-Arabinofuranose   α-D-Arabinofuranose

(*c*)   The mirror image of D-arabinose [from part (*b*)] is L-arabinose.

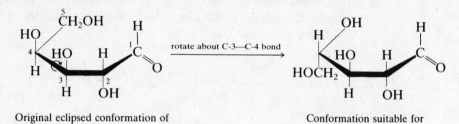

D-Arabinose   L-Arabinose   Eclipsed conformation of L-arabinose

The C-4 atom of the eclipsed conformation of L-arabinose must be rotated 120° in a clockwise sense so as to bring its hydroxyl group into the proper orientation for furanose ring formation.

rotate about C-3—C-4 bond

Original eclipsed conformation of
L-arabinose

Conformation suitable for
furanose ring formation

Cyclization gives the α- and β-furanose forms of L-arabinose.

α-L-Arabinofuranose   β-L-Arabinofuranose

In the L series the anomeric hydroxyl is up in the α isomer and down in the β isomer.

(*d*)   The Fischer projection formula for D-threose is given in text Figure 25.2. Reorientation of that projection into a form that illustrates its potential for cyclization is shown.

is equivalent to

D-Threose

Cyclization yields the two stereoisomeric furanose forms.

β-D-Threofuranose   α-D-Threofuranose

25.6 *(b)*   The Fischer projection and Haworth formula for D-mannose are:

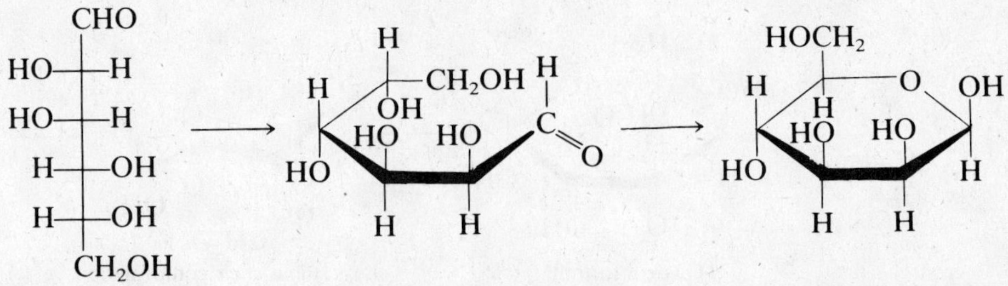

D-Mannose

β-D-Mannopyranose
(Haworth formula)

The Haworth formula is more realistically drawn as the following chair conformation:

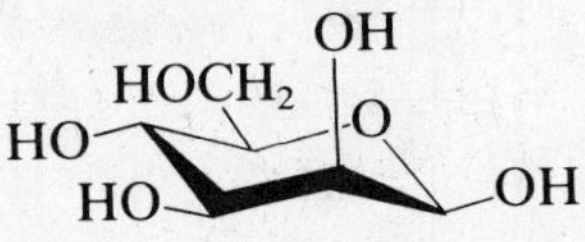

β-D-Mannopyranose

Mannose differs from glucose in configuration at C-2. All hydroxyl groups are equatorial in β-D-glucopyranose; the hydroxyl at C-2 is axial in β-D-mannopyranose.

(c)   The conformational depiction of β-L-mannopyranose begins in the same way as that of β-D-mannopyranose. L-Mannose is the mirror image of D-mannose.

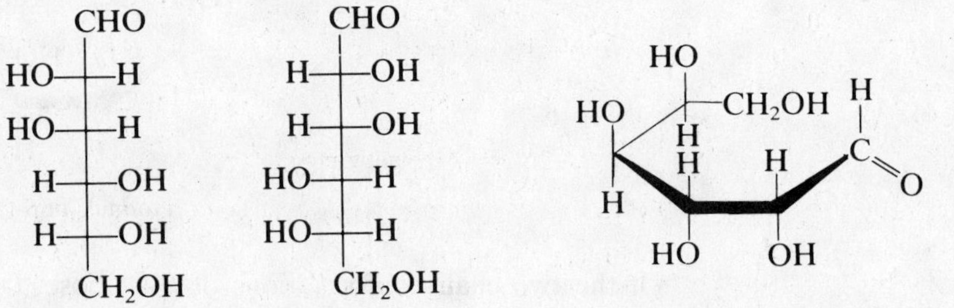

D-Mannose   L-Mannose   Eclipsed conformation of L-mannose

In order to rewrite the eclipsed conformation of L-mannose in a way that permits hemiacetal formation between the carbonyl group and the C-5 hydroxyl, C-5 is rotated 120° in the clockwise sense.

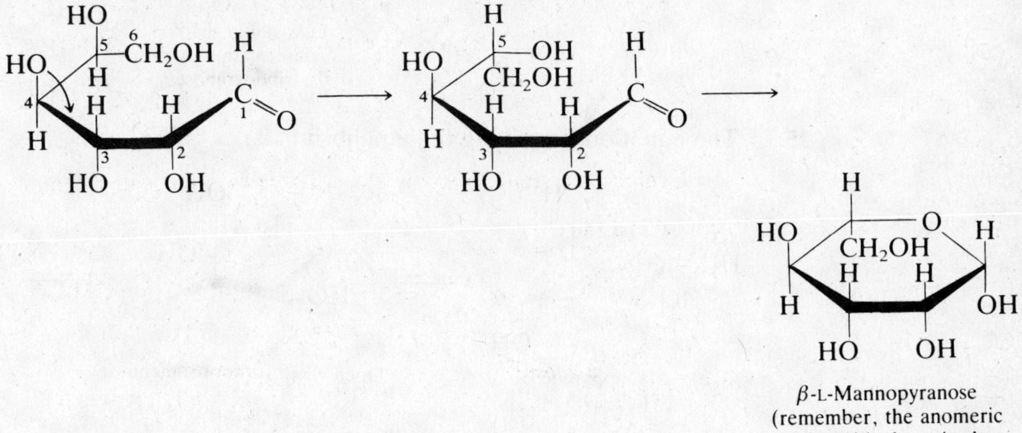

β-L-Mannopyranose
(remember, the anomeric
hydroxyl is down in the
L series)

Translating the Haworth formula into a proper conformational depiction requires that a choice be made between the two chair conformations shown:

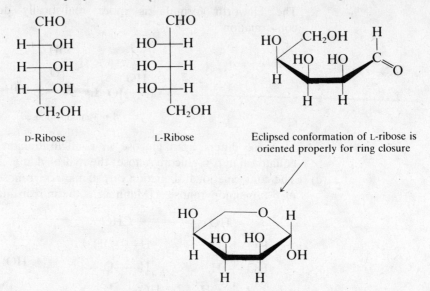

Haworth formula of
β-L-mannopyranose

Less stable chair conformation;
CH₂OH is axial

More stable chair conformation;
CH₂OH is equatorial

(*d*) The Fischer projection formula for L-ribose is the mirror image of that for D-ribose.

D-Ribose

L-Ribose

Eclipsed conformation of L-ribose is
oriented properly for ring closure

Haworth formula of β-L-ribopyranose

Of the two chair conformations of β-L-ribose, the one with the greater number of equatorial substituents is more stable.

Less stable chair
conformation of
β-L-ribopyranose

More stable chair
conformation of
β-L-ribopyranose

25.7 The equation describing the equilibrium is:

α-D-Mannopyranose
$[\alpha]_D^{20}$ +29.3°

Open-chain form of D-mannose

β-D-Mannopyranose
$[\alpha]_D^{20}$ −17.0°

Let A = percent α isomer; $100 - A$ = percent β isomer. Then:

$$A(+29.3°) + (100 - A)(-17.0°) = 100(+14.2°)$$
$$46.3A = 3120$$
$$\text{Percent } \alpha \text{ isomer} = 67\%$$
$$\text{Percent } \beta \text{ isomer} = (100 - A) = 33\%$$

25.8 Review carbohydrate terminology by referring to Table 25.1. A *ketotetrose* is a four-carbon ketose. Writing a Fischer projection formula for a four-carbon ketose reveals that only one stereogenic center is present, and thus there are only two ketotetroses. They are mirror images of each other and are known as D- and L-erythrulose.

D-Erythrulose   L-Erythrulose

25.9 (*b*) Since L-fucose is 6-deoxy-L-galactose, first write the Fischer projection formula of D-galactose, and then transform it to its mirror image, L-galactose. Transform the C-6 CH$_2$OH group to CH$_3$ to produce 6-deoxy-L-galactose.

D-Galactose   L-Galactose   6-Deoxy-L-galactose
(from Figure 25.2)   (L-fucose)

25.10 Acid-catalyzed addition of methanol to the glycal proceeds by regioselective protonation of the double bond in the direction that leads to the more stable carbocation. Here again, the more stable carbocation is the one stabilized by the ring oxygen.

Capture of the carbocation by methanol yields the α and β methyl glycosides.

25.11 The hemiacetal function opens to give an intermediate containing a free aldehyde function. Cyclization of this intermediate can produce either the α or the β configuration at this center.

β configuration of hemiacetal

Key intermediate formed by cleavage of hemiacetal function

α configuration of hemiacetal

Only the configuration of the hemiacetal function is affected in this process. The α configuration of the glycosidic linkage remains unchanged.

25.12 Cellulose is to cellobiose as amylose is to *maltose*. The polysaccharide cellulose and the disaccharide cellobiose contain β(1,4)-linked D-glucose units. Amylose contains α(1,4)-linked D-glucose units, as does the disaccharide *maltose* (Sec. 25.14). As cellulose can be thought of as a collection of cellobiose units, amylose can be thought of as a collection of maltose units.

25.13 Write the chemical equation so that you can clearly relate the product to the starting material.

$$\begin{array}{c} \text{CHO} \\ \text{H}-\!\!\!\!-\text{OH} \\ \text{H}-\!\!\!\!-\text{OH} \\ \text{H}-\!\!\!\!-\text{OH} \\ \text{CH}_2\text{OH} \end{array} \quad \xrightarrow[\text{H}_2\text{O}]{\text{NaBH}_4} \quad \begin{array}{c} \text{CH}_2\text{OH} \\ \text{H}-\!\!\!\!-\text{OH} \\ \text{H}-\!\!\!\!-\text{OH} \\ \text{H}-\!\!\!\!-\text{OH} \\ \text{CH}_2\text{OH} \end{array}$$

D-Ribose → Ribitol — plane of symmetry

Ribitol is a meso form; it is achiral and thus not optically active. A plane of symmetry passing through C-3 bisects the molecule.

25.14 (*b*) Arabinose is a reducing sugar; it will give a positive test with Benedict's reagent, because its open-chain form has a free aldehyde group capable of being oxidized by copper(II) ion.

(*c*) Benedict's reagent reacts with α-hydroxy ketones by way of an isomerization process involving an enediol intermediate.

$$\begin{array}{c} \text{CH}_2\text{OH} \\ \text{C}\!=\!\text{O} \\ \text{CH}_2\text{OH} \end{array} \rightleftharpoons \begin{array}{c} \text{H}\diagdown\text{C}\diagup\text{OH} \\ \text{C}\!-\!\text{OH} \\ \text{CH}_2\text{OH} \end{array} \rightleftharpoons \begin{array}{c} \text{H}\diagdown\text{C}\diagup\text{O} \\ \text{CHOH} \\ \text{CH}_2\text{OH} \end{array} \xrightarrow{\text{Benedict's reagent}} \text{Positive test; Cu}_2\text{O formed}$$

1,3-Dihydroxyacetone   Enediol   Glyceraldehyde

1,3-Dihydroxyacetone gives a positive test with Benedict's reagent.

(*d*)   D-Fructose is an *α*-hydroxy ketone and will give a positive test with Benedict's reagent.

$$
\begin{array}{c}
\text{CH}_2\text{OH} \\
\text{C}=\text{O} \\
\text{HO}\!-\!\!-\!\text{H} \\
\text{H}\!-\!\!-\!\text{OH} \\
\text{H}\!-\!\!-\!\text{OH} \\
\text{CH}_2\text{OH}
\end{array}
\quad \rightleftharpoons \quad
\begin{array}{c}
\text{H}\diagup\!\!\overset{\text{O}}{\underset{}{\text{C}}} \\
\text{CHOH} \\
\text{HO}\!-\!\!-\!\text{H} \\
\text{H}\!-\!\!-\!\text{OH} \\
\text{H}\!-\!\!-\!\text{OH} \\
\text{CH}_2\text{OH}
\end{array}
\quad \xrightarrow[\text{reagent}]{\text{Benedict's}} \quad \text{Positive test; Cu}_2\text{O formed}
$$

D-Fructose

(*e*)   Lactose is a disaccharide and will give a positive test with Benedict's reagent by way of an open-chain isomer of one of the rings. Lactose is a reducing sugar.

$$
\xrightarrow[\text{reagent}]{\text{Benedict's}}
$$

Lactose
(structure presented in Sec. 25.14)

Open-chain form

Positive test; Cu$_2$O formed

(*f*)   Amylose is a polysaccharide. Its glycoside linkages are inert to Benedict's reagent, but the terminal glucose residues at the ends of the chain and its branches are hemiacetals in equilibrium with open-chain structures. A positive test is expected.

25.15   Since the groups at both ends of the carbohydrate chain are oxidized to carboxylic acid functions, two combinations of one CH$_2$OH with one CHO group are possible.

$$
\begin{array}{c}
\text{CHO} \\
\text{H}\!-\!\!-\!\text{OH} \\
\text{HO}\!-\!\!-\!\text{H} \\
\text{H}\!-\!\!-\!\text{OH} \\
\text{H}\!-\!\!-\!\text{OH} \\
\text{CH}_2\text{OH}
\end{array}
\xrightarrow[\text{heat}]{\text{HNO}_3}
\begin{array}{c}
\text{CO}_2\text{H} \\
\text{H}\!-\!\!-\!\text{OH} \\
\text{HO}\!-\!\!-\!\text{H} \\
\text{H}\!-\!\!-\!\text{OH} \\
\text{H}\!-\!\!-\!\text{OH} \\
\text{CO}_2\text{H}
\end{array}
\xleftarrow[\text{heat}]{\text{HNO}_3}
\begin{array}{c}
\text{CH}_2\text{OH} \\
\text{H}\!-\!\!-\!\text{OH} \\
\text{HO}\!-\!\!-\!\text{H} \\
\text{H}\!-\!\!-\!\text{OH} \\
\text{H}\!-\!\!-\!\text{OH} \\
\text{CHO}
\end{array}
\quad
\begin{array}{c}
\text{equivalent} \\
\text{to}
\end{array}
$$

D-Glucose   D-Glucaric acid

$$
\begin{array}{c}
\text{CHO} \\
\text{HO}\!-\!\!-\!\text{H} \\
\text{HO}\!-\!\!-\!\text{H} \\
\text{H}\!-\!\!-\!\text{OH} \\
\text{HO}\!-\!\!-\!\text{H} \\
\text{CH}_2\text{OH}
\end{array}
$$

L-Gulose

L-Gulose yields the same aldaric acid on oxidation as does D-glucose.

25.16 (*b*) The phenylosazone derived from L-erthyrose is the same as the one from L-threose. The two sugars differ in configuration only at C-2.

$$
\begin{array}{ccc}
\text{L-Erythrose} & \text{L-Threose} & \\
\end{array}
$$

CHO		CHO		CH=NNHC$_6$H$_5$
HO——H		H——OH		C=NNHC$_6$H$_5$
HO——H	or	HO——H	$\xrightarrow{\;3C_6H_5NHNH_2\;}$	HO——H
CH$_2$OH		CH$_2$OH		CH$_2$OH
L-Erythrose		L-Threose		

(*c*) The carbohydrate that differs in configuration from D-allose only at C-2 is D-altrose. Both yield the same osazone.

CHO		CHO		CH=NNHC$_6$H$_5$
H——OH		HO——H		C=NNHC$_6$H$_5$
H——OH		H——OH		H——OH
H——OH	or	H——OH	$\xrightarrow{\;3C_6H_5NHNH_2\;}$	H——OH
H——OH		H——OH		H——OH
CH$_2$OH		CH$_2$OH		CH$_2$OH
D-Allose		D-Altrose		

(*d*) First find the structure of D-galactose in Figure 25.2, and then write the structure of its mirror image and convert the terminal CH$_2$OH group to a CH$_3$ group to give the Fischer projection formula of 6-deoxy-L-galactose (L-fucose).

CHO	CHO	CHO
H——OH	HO——H	HO——H
HO——H	H——OH	H——OH
HO——H	H——OH	H——OH
H——OH	HO——H	HO——H
CH$_2$OH	CH$_2$OH	CH$_3$
D-Galactose	L-Galactose	6-Deoxy-L-galactose (L-fucose)

The compound that has the opposite configuration at C-2 to L-fucose and thus will give the same osazone is 6-deoxy-L-talose.

CHO		CHO		CH=NNHC$_6$H$_5$
HO——H		H——OH		C=NNHC$_6$H$_5$
H——OH		H——OH		H——OH
H——OH	or	H——OH	$\xrightarrow{\;3C_6H_5NHNH_2\;}$	H——OH
HO——H		HO——H		HO——H
CH$_3$		CH$_3$		CH$_3$
L-Fucose		6-Deoxy-L-talose		

25.17 In analogy with the D-fructose $\rightleftharpoons$ D-glucose interconversion, dihydroxyacetone phosphate and D-glyceraldehyde 3-phosphate can equilibrate by way of an enediol intermediate.

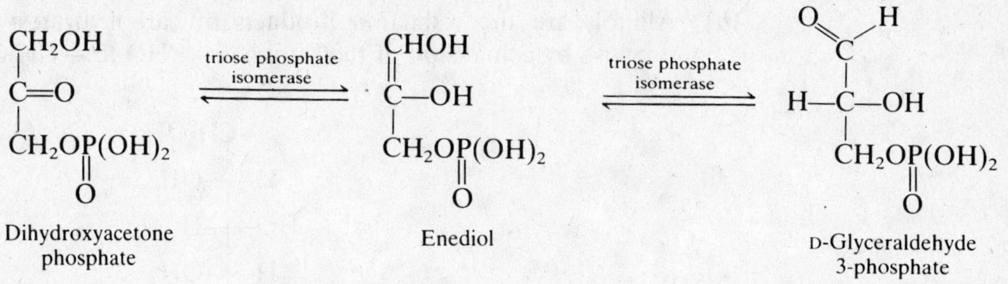

CH₂OH
|
C=O   triose phosphate   CHOH   triose phosphate
| isomerase   ‖   isomerase
CH₂OP(OH)₂   ⇌   C—OH   ⇌
‖ |
O   CH₂OP(OH)₂
                                   ‖
Dihydroxyacetone   O
phosphate   Enediol

O   H
  C
H—C—OH
  |
CH₂OP(OH)₂
‖
O
D-Glyceraldehyde
3-phosphate

25.18 (b) The points of cleavage of D-ribose on treatment with periodic acid are as indicated:

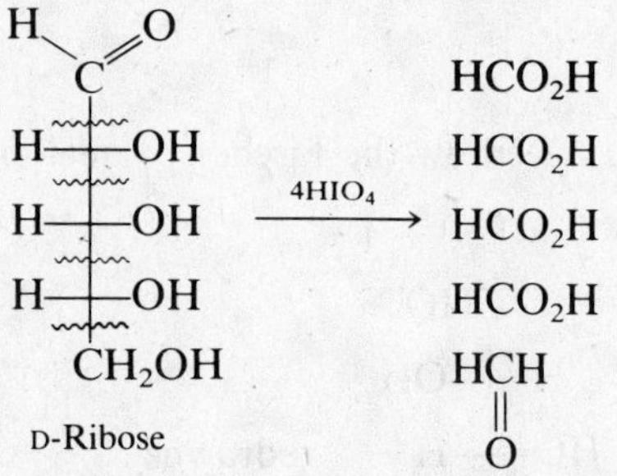

H   O
  C
H——OH
H——OH   4HIO₄
H——OH   ———————→
CH₂OH
D-Ribose

HCO₂H
HCO₂H
HCO₂H
HCO₂H
HCH
‖
O

Four equivalents of periodic acid are required. Four equivalents of formic acid and one of formaldehyde are produced.

(c) Write the structure of methyl β-D-glucopyranoside so as to identify the adjacent alcohol functions.

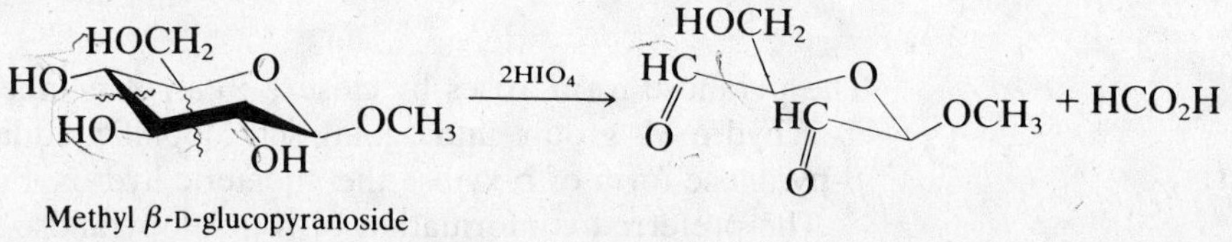

HOCH₂
HO——   O
HO——   OCH₃   2HIO₄
       OH   ———————→

HOCH₂
HC   O
   HC   OCH₃   + HCO₂H
O
‖
O

Methyl β-D-glucopyranoside

Two equivalents of periodic acid are required. One mole of formic acid is produced.

(d) There are two independent vicinal diol functions in this glycoside. Two moles of periodic acid are required per mole of substrate.

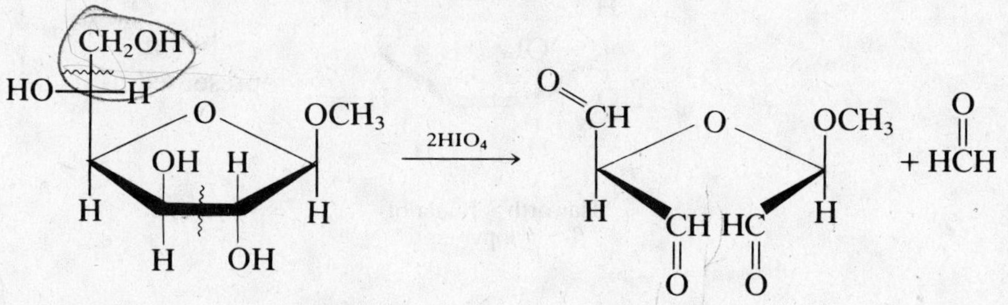

CH₂OH
HO——H
       O   OCH₃   2HIO₄
OH H   ———————→
H   H
H   OH

O
 CH   O   OCH₃   O
              ‖
H   CH HC   H + HCH
    ‖   ‖
    O   O

25.19 (a) The structure shown in Figure 25.2 is D-(+)-xylose; therefore (−)-xylose must be its mirror image and has the L-configuration at C-4.

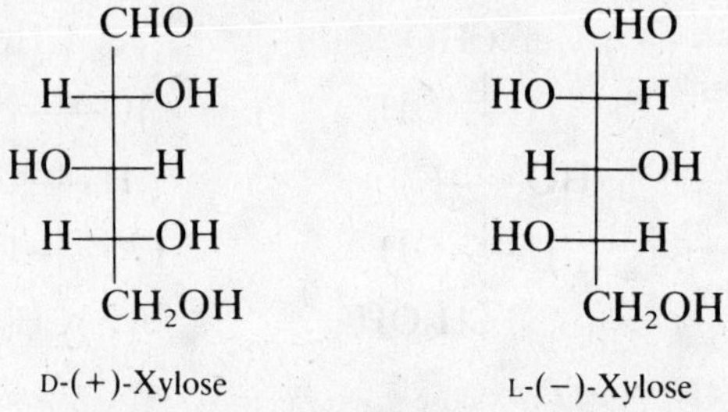

CHO   CHO
H——OH   HO——H
HO——H   H——OH
H——OH   HO——H
CH₂OH   CH₂OH

D-(+)-Xylose   L-(−)-Xylose

(b) Alditols are the reduction products of carbohydrates; D-xylitol is derived from D-xylose by conversion of the terminal —CHO to —CH₂OH.

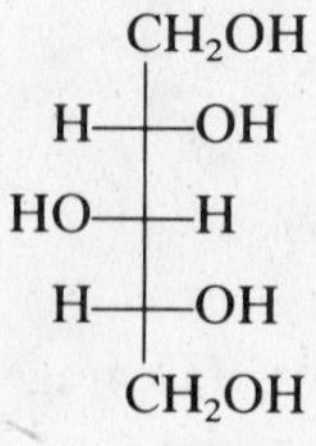

$$
\begin{array}{c}
\mathrm{CH_2OH} \\
\mathrm{H}\!-\!\!-\!\mathrm{OH} \\
\mathrm{HO}\!-\!\!-\!\mathrm{H} \\
\mathrm{H}\!-\!\!-\!\mathrm{OH} \\
\mathrm{CH_2OH}
\end{array}
$$

D-Xylitol

(c) Redraw the Fischer projection of D-xylose in its eclipsed conformation.

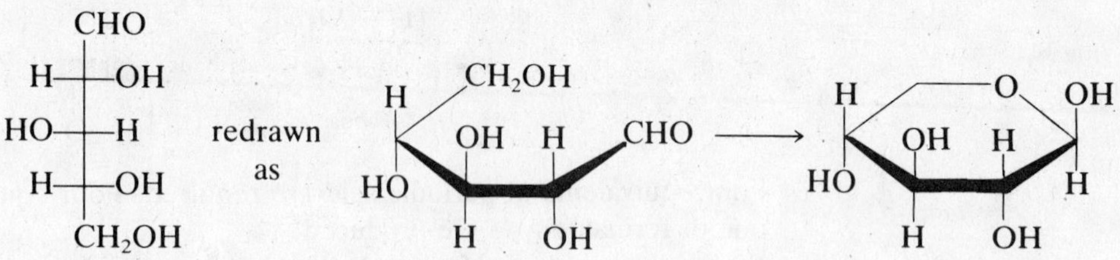

$$
\begin{array}{c}
\mathrm{CHO} \\
\mathrm{H}\!-\!\!-\!\mathrm{OH} \\
\mathrm{HO}\!-\!\!-\!\mathrm{H} \\
\mathrm{H}\!-\!\!-\!\mathrm{OH} \\
\mathrm{CH_2OH}
\end{array}
$$

D-Xylose redrawn as Eclipsed conformation of D-xylose → Haworth formula of β-D-xylopyranose

The pyranose form arises by closure to a six-membered cyclic hemiacetal, with the C-5 hydroxyl group undergoing nucleophilic addition to the carbonyl. In the β-pyranose form of D-xylose the anomeric hydroxyl group is up.

The preferred conformation of β-D-xylopyranose is a chair with all the hydroxyl groups equatorial.

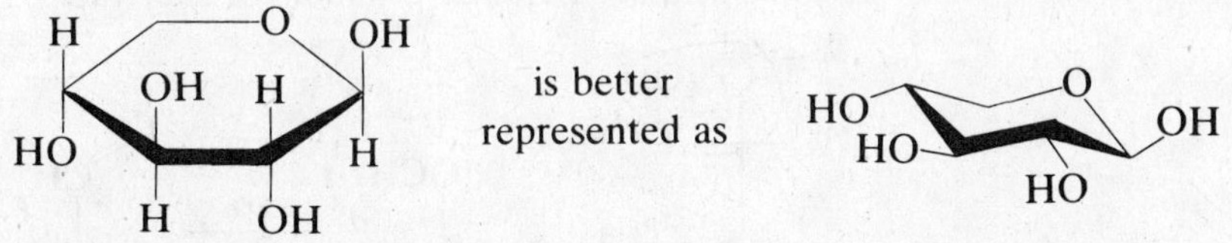

Haworth formula of β-D-xylopyranose is better represented as Chair conformation of β-D-xylopyranose

(d) L-Xylose is the mirror image of D-xylose.

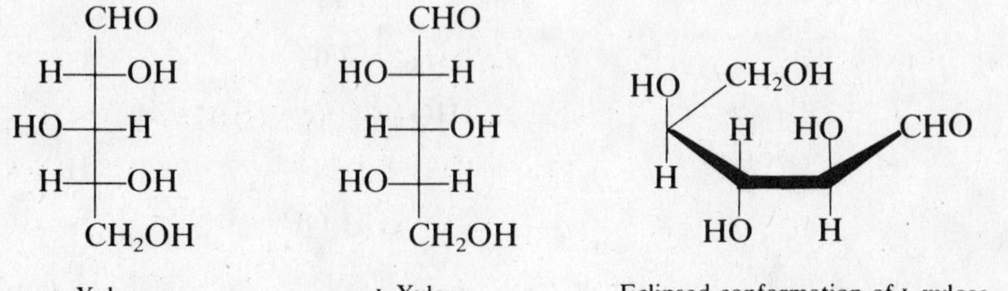

$$
\begin{array}{c}
\mathrm{CHO} \\
\mathrm{H}\!-\!\!-\!\mathrm{OH} \\
\mathrm{HO}\!-\!\!-\!\mathrm{H} \\
\mathrm{H}\!-\!\!-\!\mathrm{OH} \\
\mathrm{CH_2OH}
\end{array}
\qquad
\begin{array}{c}
\mathrm{CHO} \\
\mathrm{HO}\!-\!\!-\!\mathrm{H} \\
\mathrm{H}\!-\!\!-\!\mathrm{OH} \\
\mathrm{HO}\!-\!\!-\!\mathrm{H} \\
\mathrm{CH_2OH}
\end{array}
$$

D-Xylose L-Xylose Eclipsed conformation of L-xylose

In order to construct the furanose form of L-xylose, the hydroxyl at C-4 needs to be brought into the proper orientation to form a five-membered ring.

The anomeric hydroxyl group is up in the L series.

(e) Methyl α-L-xylofuranoside is the methyl glycoside corresponding to the structure just drawn.

(f) Carbohydrates react with many of the compounds that undergo nucleophilic addition to carbonyl groups. Phenylhydrazine reacts with D-xylose to yield the corresponding phenylhydrazone.

(g) With excess phenylhydrazine, D-xylose is converted to its osazone.

(h) Aldonic acids are derived from aldoses by oxidation of the terminal aldehyde to a carboxylic acid.

D-Xylose

D-Xylonic acid

(i) Aldonic acids tend to exist as lactones. A δ-lactone has a six-membered ring.

D-Xylonic acid

redrawn as

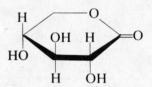

Eclipsed conformation of D-xylonic acid

intramolecular ester formation →

δ-Lactone of D-xylonic acid

(j) A γ-lactone has a five-membered ring.

D-Xylonic acid

rotate about C-3—C-4 bond →

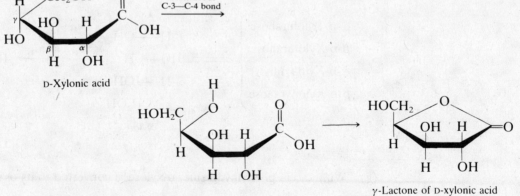

γ-Lactone of D-xylonic acid

(k) Aldaric acids have carboxylic acid groups at both ends of the chain.

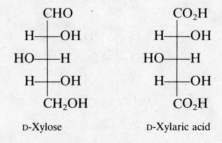

D-Xylose

D-Xylaric acid

25.20 (a) Reduction of aldoses with sodium borohydride yields polyhydroxylic alcohols called *alditols*. Optically inactive alditols are those that have a plane of symmetry, i.e., those that are meso forms. The D-aldohexoses that yield optically inactive alditols are D-allose and D-galactose.

```
   CHO                    CH2OH                 CHO                   CH2OH
 H—OH                   H—OH                  H—OH                  H—OH
 H—OH     NaBH4         H—OH                HO—H       NaBH4       HO—H
 H—OH     ──────>       H—OH                HO—H       ──────>     HO—H
 H—OH                   H—OH                  H—OH                  H—OH
   CH2OH                  CH2OH                 CH2OH                 CH2OH
 D-Allose                 Allitol            D-Galactose            Galactitol
                       (meso compound)                           (meso compound)
```

(b) All the aldonic acids and their lactones obtained on oxidation of the aldohexoses are optically active. The presence of a carboxyl group at one end of the carbon chain and a CH_2OH at the other precludes the existence of meso forms.

(c) Nitric acid oxidation of aldoses converts them to aldaric acids. The same D-aldoses found to yield optically inactive alditols in part (a) yield optically inactive aldaric acids.

```
   CHO                    CO2H                  CHO                   CO2H
 H—OH                   H—OH                  H—OH                  H—OH
 H—OH      HNO3         H—OH                HO—H        HNO3       HO—H
 H—OH      ──────>      H—OH                HO—H        ──────>    HO—H
 H—OH                   H—OH                  H—OH                  H—OH
   CH2OH                  CO2H                  CH2OH                 CO2H
 D-Allose              Allaric acid          D-Galactose          Galactaric acid
                      (meso compound)                            (meso compound)
```

(d) Aldoses that differ in configuration only at C-2 enolize to the same enediol.

```
   CHO                    CHOH                  CHO
 H—OH                      ‖                  HO—H
 H—OH                    C—OH                  H—OH
 H—OH        ⇌         H—OH        ⇌         H—OH
 H—OH                   H—OH                  H—OH
   CH2OH                H—OH                    CH2OH
                          CH2OH
 D-Allose                Enediol              D-Altrose
```

The stereogenic center at C-2 in the D-aldose becomes sp^2 hybridized in the enediol. The other pairs of D-aldohexoses that form the same enediols are:

D-glucose and D-mannose
D-gulose and D-idose
D-galactose and D-talose

(e)　When aldohexoses are converted to osazones, C-2 is converted to $\diagdown C{=}NNHC_6H_5$ irrespective of its configuration. Thus, aldohexoses that differ in configuration only at C-2 yield the same osazone. The answers to this problem are the same as those of part (d).

CHO	CH=NNHC₆H₅	CHO
H——OH	C=NNHC₆H₅	HO——H
H——OH	H——OH	H——OH
H——OH	H——OH	H——OH
H——OH	H——OH	H——OH
CH₂OH	CH₂OH	CH₂OH

$$\text{CHO}\quad \xrightarrow{3C_6H_5NHNH_2}\quad \text{CH=NNHC}_6\text{H}_5 \quad \xleftarrow{3C_6H_5NHNH_2}\quad \text{CHO}$$

D-Allose　　　　　　　　　　　　　　　　　　　　　D-Altrose

The other pairs of D-aldohexoses that form the same osazones are:

D-glucose and D-mannose
D-gulose and D-idose
D-galactose and D-talose

25.21 (a)　To unravel a pyranose form, locate the anomeric carbon and mentally convert the hemiacetal linkage to a carbonyl compound and a hydroxyl function.

disconnect this bond

HOCH₂ ... O ... OH / OH ⟹ HOCH₂ ... open-chain form ... CH=O

Convert the open-chain form to a proper Fischer projection.

rotate about the C-4—C-5 bond

equivalent to

CHO	
HO	H
HO	H
H	OH
HO	H
CH₂OH	

(b) Proceed in the same manner as in part (a) and unravel the furanose sugar by disconnecting the hemiacetal function.

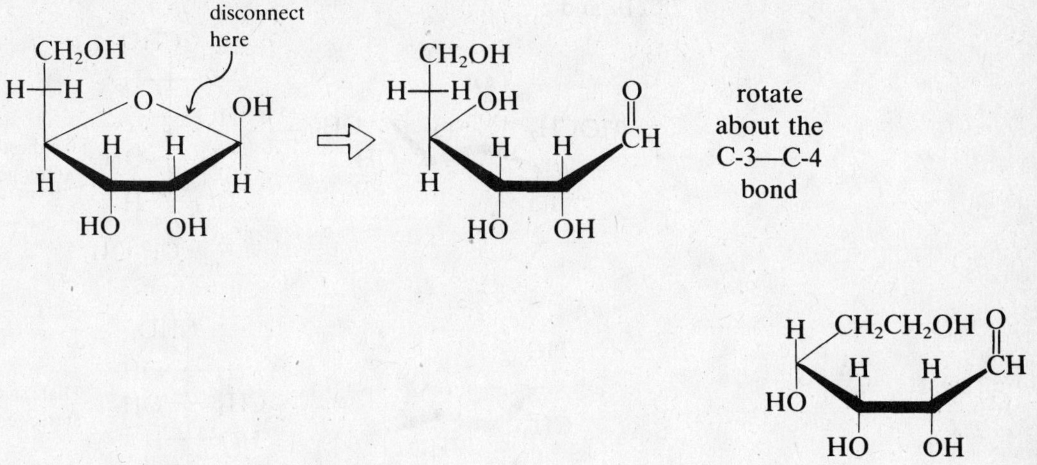

The Fischer projection is:

CHO
H——OH
H——OH
H——OH
H——H
CH₂OH

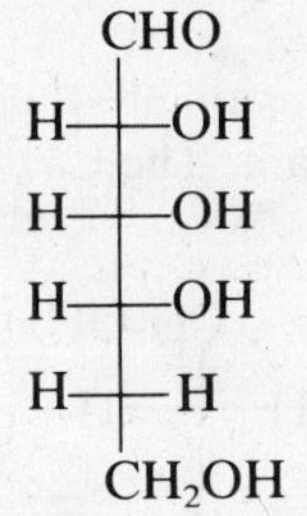

(c) By disconnecting and unraveling as before, the Fischer projection is revealed.

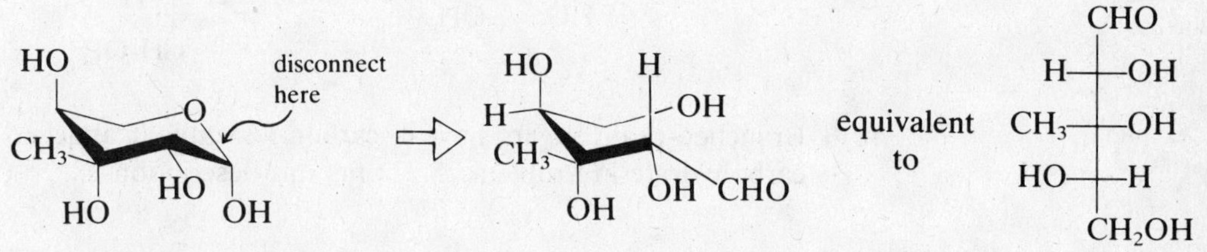

CHO
H——OH
CH₃——OH
HO——H
CH₂OH

equivalent to

(d) Remember in disconnecting cyclic hemiacetals that the anomeric carbon is the one that bears two oxygen substituents.

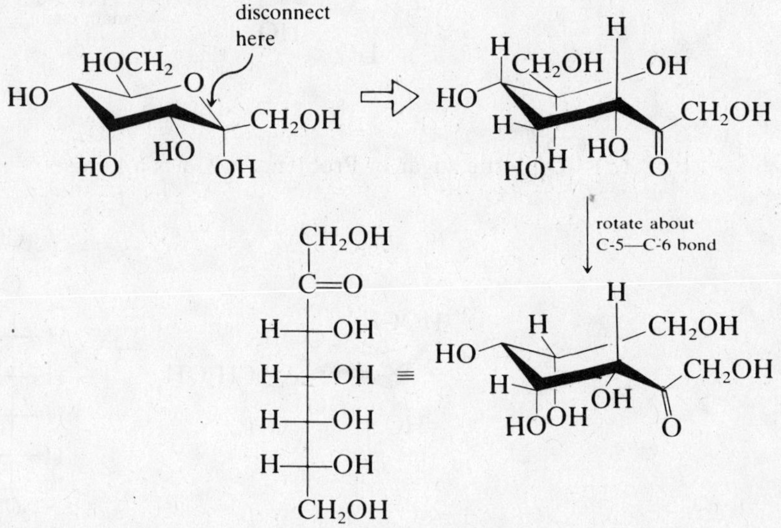

CH₂OH
C=O
H——OH
H——OH
H——OH
H——OH
CH₂OH

≡

25.22 (*a*) The L sugars have the hydroxyl group to the left at the highest-numbered stereogenic center in their Fischer projection. The L sugars are the ones in Problem 25.21*a* and *c*.

Highest-numbered stereogenic center is L.

Highest-numbered stereogenic center is L.

(*b*) Deoxy sugars are those that lack an oxygen substituent on one of the carbons in the main chain. The carbohydrate in Problem 25.21*b* is a deoxy sugar.

no hydroxyl group at C-5

(*c*) Branched-chain sugars have a carbon substituent attached to the main chain; the carbohydrate in Problem 25.21*c* fits this description.

methyl group attached to main chain

(*d*) Only the sugar in Problem 25.21*d* is a ketose.

ketone carbonyl in Fischer projection

(e) A furanose ring is a five-membered cyclic hemiacetal. Only the compound in Problem 25.21*b* is a furanose form.

$$\begin{array}{c}
\text{CH}_2\text{OH} \\
\end{array}$$

H——H   O   OH
   H   H
H   H
   HO   OH

(f) In D sugars, the α configuration corresponds to the condition in which the hydroxyl group at the anomeric carbon is down. The α-D sugar is that in Problem 25.21*d*.

α-Pyranose form of
a D-ketose

HOCH₂   O
HO
                    CH₂OH
      HO
HO   OH

Anomeric hydroxyl is down.

In the α-L series the anomeric hydroxyl is up. Neither of the L sugars—namely, those of Problem 25.21*d* and *c* is α; both are β.

25.23 There are seven possible pentuloses, i.e., five-carbon ketoses. The ketone carbonyl can be located at either C-2 or C-3. When the carbonyl group is at C-2, there are two stereogenic centers, giving rise to four stereoisomers (two pairs of enantiomers).

CH₂OH	CH₂OH	CH₂OH	CH₂OH
C=O	C=O	C=O	C=O
H——OH	HO——H	HO——H	H——OH
H——OH	HO——H	H——OH	HO——H
CH₂OH	CH₂OH	CH₂OH	CH₂OH
D-Erythropentulose (D-ribulose)	L-Erythropentulose (L-ribulose)	D-Threopentulose (D-xylulose)	L-Threopentulose (L-xylulose)

When the carbonyl group is located at C-3, there are only three stereoisomers, because one of them is a meso form and is superposable on its mirror image.

CH₂OH	CH₂OH	CH₂OH
HO——H	H——OH	H——OH
C=O	C=O	C=O
H——OH	HO——H	H——OH
CH₂OH	CH₂OH	CH₂OH

25.24 (a) Carbon-2 is the only stereogenic center in D-apiose.

$$\begin{array}{c}
^1\text{CHO} \\
\text{H}——^2\text{OH} \\
\text{HOCH}_2——^3\text{OH} \\
^4\text{CH}_2\text{OH}
\end{array}$$ D-Apiose

Carbon-3 is not a stereogenic center; it bears two identical CH₂OH substituents.

(b) When D-apiose is converted to an osazone, the stereogenic center at C-2 is transformed to $\text{C}=\text{NNHC}_6\text{H}_5$. The osazone of D-apiose is not optically active.

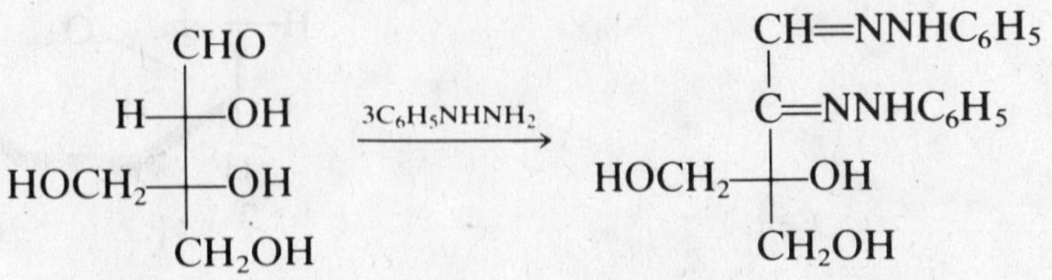

D-Apiose
(optically active)

$\xrightarrow{3C_6H_5NHNH_2}$

Phenylosazone of apiose
(achiral; cannot be
optically active)

(c) The alditol obtained on reduction of D-apiose retains the stereogenic center. It is chiral and optically active.

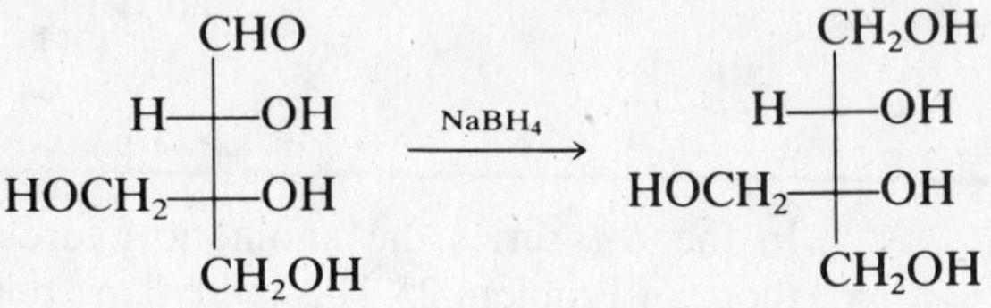

D-Apiose
(optically active)

$\xrightarrow{NaBH_4}$

D-Apitol
(optically active)

(d, e) Cyclic hemiacetal formation in D-apiose involves addition of a CH_2OH hydroxyl group to the aldehyde carbonyl.

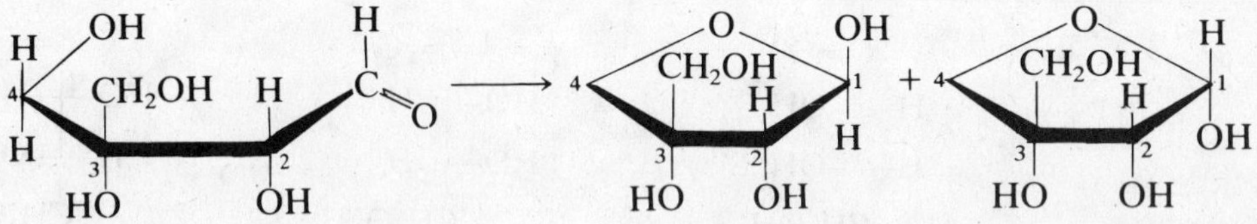

There are three stereogenic centers in the furanose form, namely, the anomeric carbon C-1 and the original stereogenic center C-2, as well as a new stereogenic center at C-3.

In addition to the two furanose forms just shown, two more are possible. Instead of the reaction of the CH_2OH group that was shown to form the cyclic hemiacetal, the other CH_2OH group may add to the aldehyde carbonyl.

$\longrightarrow$ two furanose forms shown previously

rotate C-3
120° about the
C-3—C-4 bond

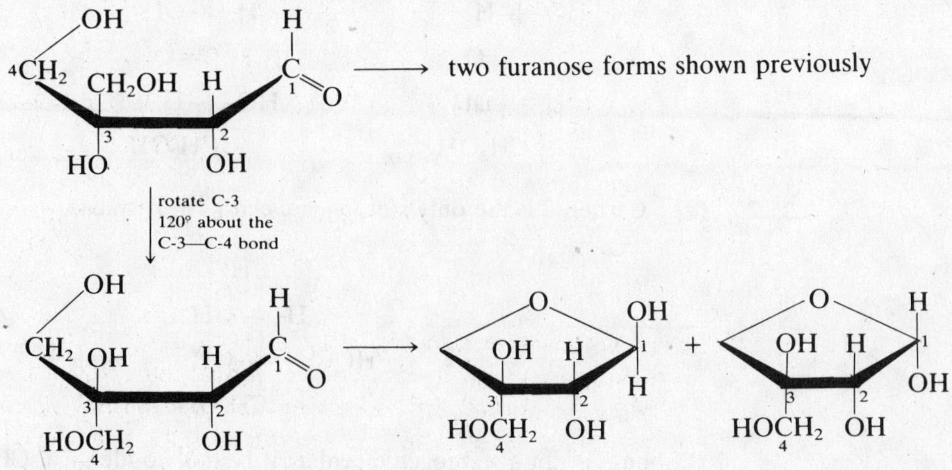

25.25 The most reasonable conclusion is that all four are methyl glycosides. Two are the methyl glycosides of the α- and β-pyranose forms of mannose and two are the methyl glycosides of the α- and β-furanose forms.

D-Mannose w/ Acid catalyst

Methyl
α-D-mannopyranoside

Methyl
β-D-mannopyranoside

Methyl
α-D-mannofuranoside

Methyl
β-D-mannofuranoside

In the case of the methyl glycosides of mannose, comparable amounts of pyranosides and furanosides are formed. The major products are the α isomers.

25.26 (*a*) Disaccharides, by definition, involve an acetal linkage at the anomeric position; thus all the disaccharides must involve C-1. The bond to C-1 can be α or β. The available oxygen atoms in the second D-glucopyranosyl unit are located at C-1, C-2, C-3, C-4, and C-6. Thus, there are 11 possible disaccharides, including maltose and cellobiose, composed of D-glucopyranosyl units.

$\alpha,\alpha(1,1)$	$\alpha,\beta(1,1)$	$\beta,\beta(1,1)$
$\alpha(1,2)$		$\beta(1,2)$
$\alpha(1,3)$		$\beta(1,3)$
$\alpha(1,4)$ (maltose)		$\beta(1,4)$ (cellobiose)
$\alpha(1,6)$		$\beta(1,6)$

(*b*) In order to be a reducing sugar, one of the anomeric positions must be a free hemiacetal. Therefore, all except $\alpha,\alpha(1,1)$, $\alpha,\beta(1,1)$, and $\beta,\beta(1,1)$ are reducing sugars.

25.27 Since gentiobiose undergoes mutarotation, it must have a free hemiacetal group. Formation of two molecules of D-glucose indicates that it is a disaccharide and because that hydrolysis is catalyzed by emulsin, the glycosidic linkage is β. The methylation data, summarized in the following equation, require that the glucose units be present in pyranose forms and be joined by a $\beta(1,6)$ glycoside bond.

R = H: gentiobiose

R = CH₃: gentiobiose octamethyl ether

2,3,4,6-Tetra-*O*-methyl-D-glucose   2,3,4-Tri-*O*-methyl-D-glucose

25.28 Like other glycosides, cyanogenic glycosides are cleaved to a carbohydrate and an alcohol on hydrolysis.

(*a*) In the case of linamarin the alcohol is recognizable as the cyanohydrin of acetone. Once formed, this cyanohydrin dissociates to hydrogen cyanide and acetone.

Linamarin   D-Glucose   Acetone cyanohydrin

$$CH_3CCH_3 + HCN$$
Acetone   Hydrogen cyanide

(*b*) Laetrile undergoes an analogous hydrolytic cleavage to yield the cyanohydrin of benzaldehyde.

Laetrile   D-Glucuronic acid   Benzaldehyde cyanohydrin

$$C_6H_5CH + HCN$$
Benzaldehyde   Hydrogen cyanide

25.29 Comparing D-glucose, D-mannose, and D-galactose, it can be said that the configuration of C-2 has a substantial effect on the relative energies of the α- and β-pyranose forms, but that the configuration of C-4 has virtually no effect. With this observation in mind, write the structures of the pyranose forms of the carbohydrates given in each part.

(*a*) The β-pyranose form of D-gulose is the same as that of D-galactose except for the configuration at C-3.

β-D-Galactopyranose
(64% at equilibrium)

β-D-Gulopyranose

α-D-Gulopyranose
(1, 3 diaxial repulsion between hydroxyl groups)

The axial hydroxyl group at C-3 destabilizes the α-pyranose form more than the β form because of its repulsive interaction with the axially disposed anomeric hydroxyl group. There should be an even higher β/α ratio in D-gulopyranose than in D-galactopyranose. This is so; the observed β/α ratio is 88:12.

(*b*) The β-pyranose form of D-talose is the same as that of D-mannose except for the configuration at C-4.

β-D-Talopyranose

α-D-Talopyranose

α-D-Mannopyranose
(68% at equilibrium)

Since the configuration at C-4 has little effect on the α- to β-pyranose ratio (compare D-glucose and D-galactose), we would expect that talose would behave very much like mannose and that the α-pyranose form would be preferred at equilibrium. This is indeed the case; the α-pyranose form predominates at equilibrium, the observed α/β ratio being 78:22.

(*c*) The pyranose form of D-xylose is just like that of D-glucose except that it lacks a CH₂OH group.

β-D-Glucopyranose
(64% at equilibrium)

β-D-Xylopyranose

α-D-Xylopyranose

We would expect the equilibrium between pyranose forms in D-xylose to be much like that in D-glucose and predict that the β-pyranose form would predominate. It is observed that the β/α ratio in D-xylose is 64:36, exactly the same as in D-glucose.

(*d*) The pyranose form of D-lyxose is like that of D-mannose except that it lacks a CH₂OH group. As in D-mannopyranose, the α form should predominate over the β.

β-D-Lyxopyranose

α-D-Lyxopyranose

α-D-Mannopyranose
(68% at equilibrium)

The observed α/β distribution ratio in D-lyxopyranose is 73:27.

25.30 (*a*) The rate-determining step in glycoside hydrolysis is carbocation formation at the anomeric position. The carbocation formed from methyl α-D-fructofuranoside (compound A) is tertiary and therefore more stable than the one from methyl α-D-glucofuranoside (compound B), which is secondary. The more stable a carbocation is, the more rapidly it will be formed.

Faster:

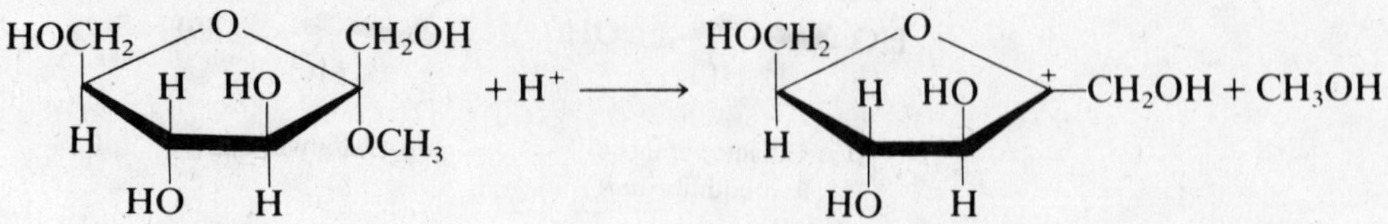

HOCH₂ ... H HO ... CH₂OH ... H ... HO ... OCH₃ + H⁺ ⟶ HOCH₂ ... H HO ... ⁺CH₂OH + CH₃OH

A

Tertiary carbocation

Slower:

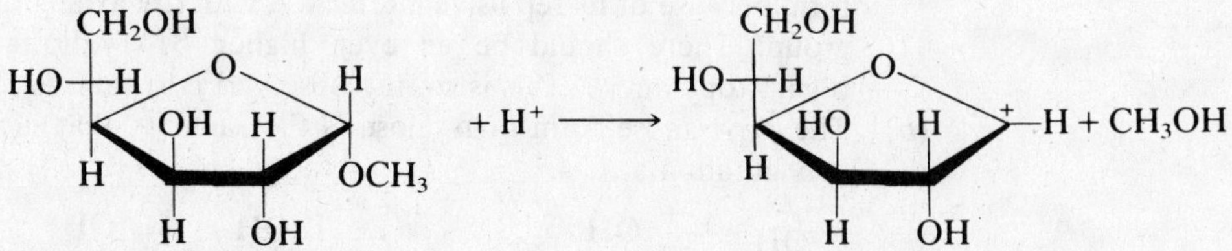

CH₂OH ... HO—H ... OH ... H ... H ... OCH₃ + H⁺ ⟶ CH₂OH ... HO—H ... HO ... H ... ⁺H + CH₃OH

B

Secondary carbocation

(b) The carbocation formed from methyl β-D-glucopyranoside (compound D) is less stable than the one from its 2-deoxy analog (compound C) and is formed more slowly. It is less stable because it is destabilized by the electron-withdrawing inductive effect of the hydroxyl group at C-2.

Faster:

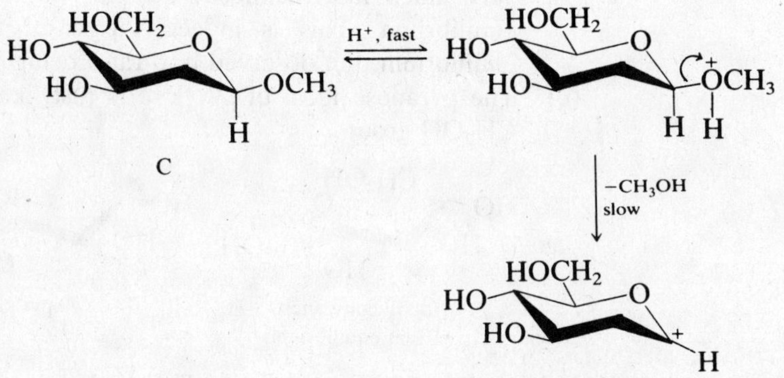

HO ... HOCH₂ ... O ... HO ... OCH₃ ... H $\xrightleftharpoons{H^+, \text{ fast}}$ HO ... HOCH₂ ... O ... HO ... ⁺OCH₃ ... H H

C

$\xrightarrow[\text{slow}]{-CH_3OH}$

HO ... HOCH₂ ... O ... HO ... ⁺ ... H

More stable

Slower:

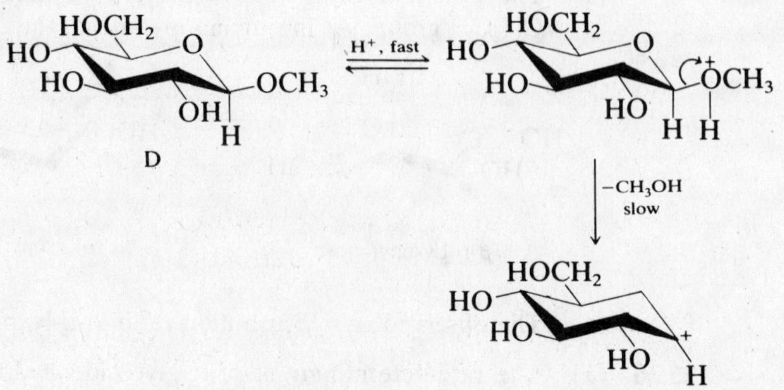

HO ... HOCH₂ ... O ... HO ... OH ... OCH₃ ... H $\xrightleftharpoons{H^+, \text{ fast}}$ HO ... HOCH₂ ... O ... HO ... HO ... ⁺OCH₃ ... H H

D

$\xrightarrow[\text{slow}]{-CH_3OH}$

HO ... HOCH₂ ... O ... HO ... HO ... ⁺ ... H

Less stable

25.31 D-Altrosan is a glycoside. The anomeric carbon—the one with two oxygen substituents— has an alkoxy group attached to it. Hydrolysis of D-altrosan follows the general mechanism for acetal hydrolysis.

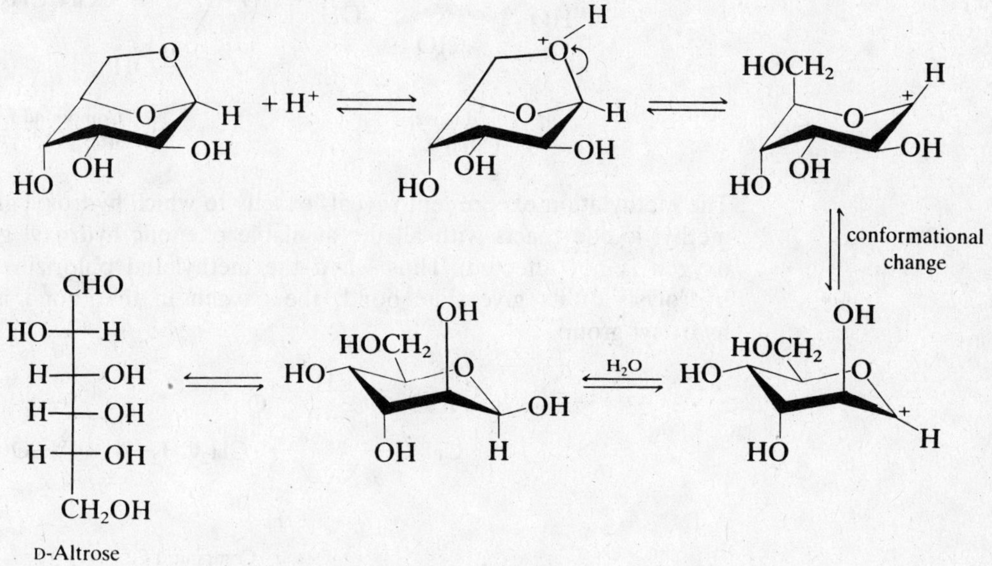

D-Altrose

25.32 Galactose has hydroxyl groups at carbons 2, 3, 4, 5, 6. Therefore, ten trimethyl ethers are possible:

2,3,4	2,4,5	3,4,5	4,5,6
2,3,5	2,4,6	3,4,6	
2,3,6	2,5,6	3,5,6	

To find out which one of these is identical with the degradation product of compound E, carry compound E through the required transformations.

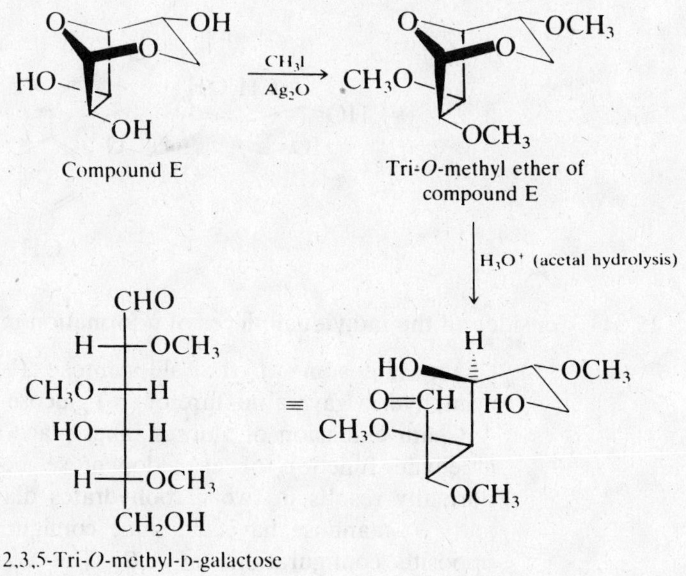

Compound E

Tri-*O*-methyl ether of compound E

2,3,5-Tri-*O*-methyl-D-galactose

25.33 The fact that phlorizin is hydrolyzed to D-glucose and compound F by emulsin indicates that it is a β-glucoside in which D-glucose is attached to one of the phenolic hydroxyls of compound G.

$$C_{21}H_{24}O_{10} + H_2O \xrightarrow{\text{emulsin}}$$

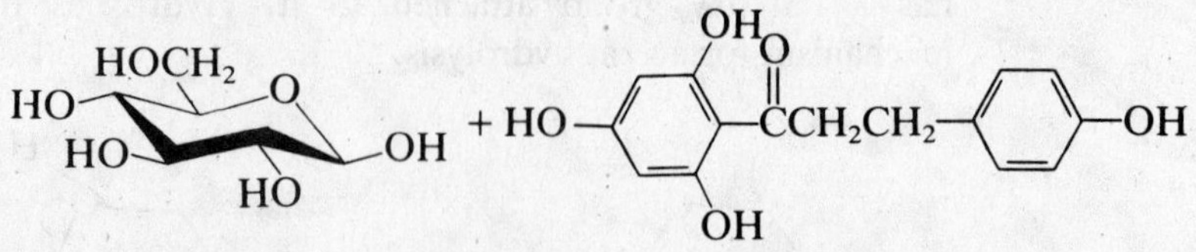

D-Glucose
($C_6H_{12}O_6$)

Compound F
($C_{15}H_{14}O_5$)

The methylation experiment reveals exactly to which hydroxyl glucose is attached. Excess methyl iodide reacts with all the available phenolic hydroxyl groups, but the glycosidic oxygen is not affected. Thus when the methylated phlorizin undergoes acid-catalyzed hydrolysis of its glycosidic bond, the oxygen in that bond is exposed as a phenolic hydroxyl group.

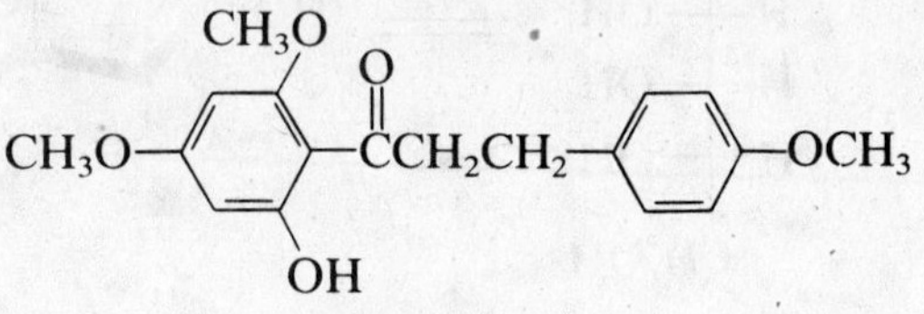

Compound G

This compound must arise by hydrolysis of

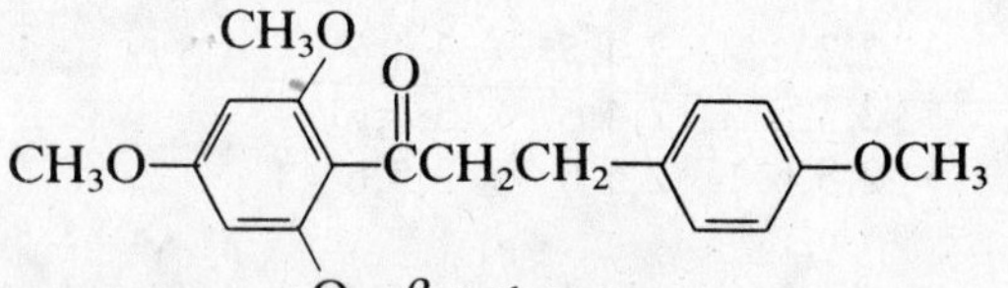

The structure of phlorizin is therefore:

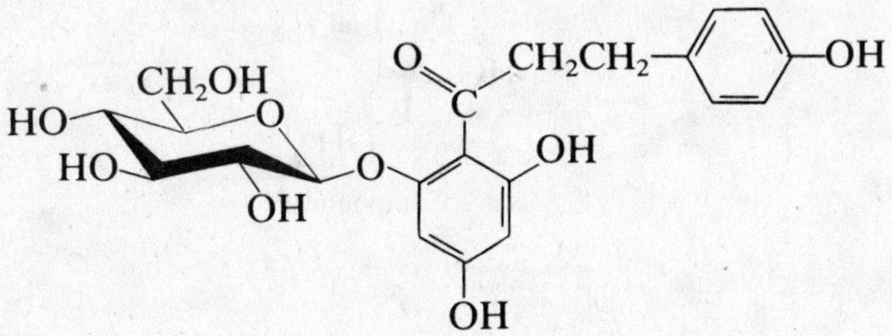

25.34 Consider all the individual pieces of information in the order in which they are presented.

1. Chain extension of the aldopentose (−)-arabinose by way of the derived cyanohydrin gave a mixture of (+)-glucose and (+)-mannose.

   Chain extension of aldoses takes place at the aldehyde end of the chain. The aldehyde function of an aldopentose becomes C-2 of an aldohexose, which normally results in two carbohydrates diastereomeric at C-2. Thus, (+)-glucose and (+)-mannose have the same configuration at C-3, C-4, and C-5; they have opposite configurations at C-2. The configuration at C-2, C-3, and C-4 of (−)-arabinose is the same as that at C-3, C-4, and C-5 of (+)-glucose and (+)-mannose.

2. Oxidation of (−)-arabinose with warm nitric acid gave an optically active aldaric acid.

Since the hydroxyl group at C-4 of (−)-arabinose is at the right in a Fischer projection formula (evidence of step 1), the hydroxyl at C-2 must be to the left in order for the aldaric acid to be optically active.

$$
\begin{array}{ccc}
{}^{1}\text{CHO} & & {}^{1}\text{CO}_2\text{H} \\
\text{HO}-{}^{2}\text{C}-\text{H} & & \text{HO}-{}^{2}\text{C}-\text{H} \\
{}^{3}\text{CHOH} & \xrightarrow{\text{HNO}_3} & {}^{3}\text{CHOH} \\
\text{H}-{}^{4}\text{C}-\text{OH} & & \text{H}-{}^{4}\text{C}-\text{OH} \\
{}^{5}\text{CH}_2\text{OH} & & {}^{5}\text{CO}_2\text{H}
\end{array}
$$

Partial stereostructure of (−)-arabinose   Aldaric acid from (−)-arabinose; optically active irrespective of configuration at C-3

If the C-2 hydroxyl group had been to the right, an optically inactive meso aldaric acid would have been produced.

$$
\begin{array}{ccc}
{}^{1}\text{CHO} & & \text{CO}_2\text{H} \\
\text{H}-{}^{2}\text{C}-\text{OH} & & \text{H}-\text{C}-\text{OH} \\
{}^{3}\text{CHOH} & \xrightarrow{\text{HNO}_3} & \text{CHOH} \\
\text{H}-{}^{4}\text{C}-\text{OH} & & \text{H}-\text{C}-\text{OH} \\
{}^{5}\text{CH}_2\text{OH} & & \text{CO}_2\text{H}
\end{array}
$$

Achiral meso form; cannot be optically active

Therefore we now know the configurations of C-3 and C-5 of (+)-glucose and (+)-mannose and that these two aldohexoses have opposite configurations at C-2, but the same (yet to be determined) configuration at C-4.

$$
\begin{array}{cc}
{}^{1}\text{CHO} & {}^{1}\text{CHO} \\
\text{H}-{}^{2}\text{C}-\text{OH} & \text{HO}-{}^{2}\text{C}-\text{H} \\
\text{HO}-{}^{3}\text{C}-\text{H} & \text{HO}-{}^{3}\text{C}-\text{H} \\
{}^{4}\text{CHOH} & {}^{4}\text{CHOH} \\
\text{H}-{}^{5}\text{C}-\text{OH} & \text{H}-{}^{5}\text{C}-\text{OH} \\
{}^{6}\text{CH}_2\text{OH} & {}^{6}\text{CH}_2\text{OH}
\end{array}
$$

[One of these is (+)-glucose, the other is (+)-mannose.]

3. Both (+)-glucose and (+)-mannose are oxidized to optically active aldaric acids with nitric acid.

Since both (+)-glucose and (+)-mannose yield optically active aldaric acids and both have the same configuration at C-4, the hydroxyl group must lie at the right in the Fischer projection at this carbon.

$$
\begin{array}{ccc}
\text{}^1\text{CHO} & & \text{}^1\text{CHO} \\
\text{H--}^2\text{C--OH} & & \text{HO--}^2\text{C--H} \\
\text{HO--}^3\text{C--H} & & \text{HO--}^3\text{C--H} \\
\text{H--}^4\text{C--OH} & & \text{H--}^4\text{C--OH} \\
\text{H--}^5\text{C--OH} & & \text{H--}^5\text{C--OH} \\
\text{}^6\text{CH}_2\text{OH} & & \text{}^6\text{CH}_2\text{OH}
\end{array}
$$

[One of these is (+)-glucose, the other is (+)-mannose.]

The structures of the corresponding aldaric acids are:

$$
\begin{array}{ccc}
\text{CO}_2\text{H} & & \text{CO}_2\text{H} \\
\text{H--C--OH} & & \text{HO--C--H} \\
\text{HO--C--H} & & \text{HO--C--H} \\
\text{H--C--OH} & & \text{H--C--OH} \\
\text{H--C--OH} & & \text{H--C--OH} \\
\text{CO}_2\text{H} & & \text{CO}_2\text{H}
\end{array}
$$

Both are optically active. Had the C-4 hydroxyl group been to the left, one of the aldaric acids would have been a meso form:

$$
\begin{array}{ccc}
\text{CO}_2\text{H} & & \text{CO}_2\text{H} \\
\text{H--C--OH} & & \text{HO--C--H} \\
\text{HO--C--H} & & \text{HO--C--H} \\
\text{HO--C--H} & & \text{HO--C--H} \\
\text{H--C--OH} & & \text{H--C--OH} \\
\text{CO}_2\text{H} & & \text{CO}_2\text{H}
\end{array}
$$

(This aldaric acid is
optically inactive.)

4. There is another sugar, (+)-gulose, that gives the same aldaric acid on oxidation as does (+)-glucose.

This is the last piece in the puzzle, the one that permits one of the Fischer projections shown in the first part of step 3 to be assigned to (+)-glucose and the other to (+)-mannose. Consider first the structure

$$
\begin{array}{ccc}
\text{CHO} & & \text{CH}_2\text{OH} \\
\text{HO--C--H} & & \text{H--C--OH} \\
\text{HO--C--H} & \text{equivalent to} & \text{HO--C--H} \\
\text{H--C--OH} & & \text{H--C--OH} \\
\text{HO--C--H} & & \text{H--C--OH} \\
\text{CH}_2\text{OH} & & \text{CHO}
\end{array}
$$

Oxidation gives the aldaric acid

$$
\begin{array}{c}
\text{CO}_2\text{H} \\
| \\
\text{H}-\text{C}-\text{OH} \\
| \\
\text{HO}-\text{C}-\text{H} \\
| \\
\text{H}-\text{C}-\text{OH} \\
| \\
\text{H}-\text{C}-\text{OH} \\
| \\
\text{CO}_2\text{H}
\end{array}
$$

This is the same aldaric acid as that provided by one of the structures given as either (+)-glucose or (+)-mannose. Therefore that Fischer projection corresponds to (+)-glucose.

$$
\begin{array}{c}
\text{CHO} \\
| \\
\text{H}-\text{C}-\text{OH} \\
| \\
\text{HO}-\text{C}-\text{H} \\
| \\
\text{H}-\text{C}-\text{OH} \\
| \\
\text{H}-\text{C}-\text{OH} \\
| \\
\text{CH}_2\text{OH}
\end{array}
$$

This must be (+)-glucose.

The structure of (+)-mannose is therefore:

$$
\begin{array}{c}
\text{CHO} \\
| \\
\text{HO}-\text{C}-\text{H} \\
| \\
\text{HO}-\text{C}-\text{H} \\
| \\
\text{H}-\text{C}-\text{OH} \\
| \\
\text{H}-\text{C}-\text{OH} \\
| \\
\text{CH}_2\text{OH}
\end{array}
$$

A sugar that yields the same aldaric acid is:

$$
\begin{array}{c}
\text{CH}_2\text{OH} \\
| \\
\text{HO}-\text{C}-\text{H} \\
| \\
\text{HO}-\text{C}-\text{H} \\
| \\
\text{H}-\text{C}-\text{OH} \\
| \\
\text{H}-\text{C}-\text{OH} \\
| \\
\text{CHO}
\end{array}
$$

This is, in fact, not a different sugar but simply (+)-mannose rotated through an angle of 180°.

SELF-TEST

PART A

A-1. Draw the structures indicated for each of the following:

(*a*) The enantiomer of D-erythrose:

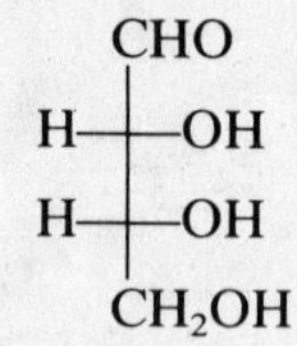

$$
\begin{array}{c}
\text{CHO} \\
\text{H}\!-\!\!-\!\text{OH} \\
\text{H}\!-\!\!-\!\text{OH} \\
\text{CH}_2\text{OH}
\end{array}
$$

(*b*) A diastereomer of D-erythrose

(*c*) The α-furanose form of D-erythrose (use a Haworth formula)

(*d*) The anomer of the structure in part (*c*)

A-2. The structure of D-mannose is

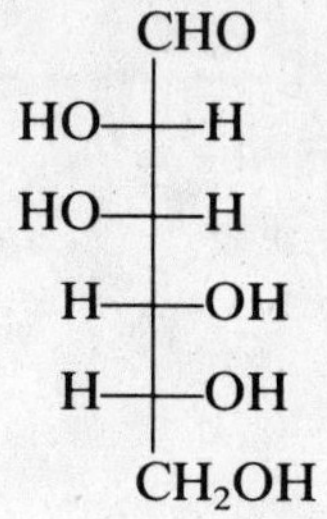

$$
\begin{array}{c}
\text{CHO} \\
\text{HO}\!-\!\!-\!\text{H} \\
\text{HO}\!-\!\!-\!\text{H} \\
\text{H}\!-\!\!-\!\text{OH} \\
\text{H}\!-\!\!-\!\text{OH} \\
\text{CH}_2\text{OH}
\end{array}
\qquad \text{D-Mannose}
$$

Using Fischer projections, draw the product of the reaction of D-mannose with:

(*a*) $NaBH_4$ in H_2O

(*b*) Benedict's reagent

(*c*) Excess phenylhydrazine, $C_6H_5NHNH_2$

(*d*) Excess periodic acid

A-3. Referring to the structure of D-arabinose shown, draw the following:

(*a*) The α-pyranose form of D-arabinose

(*b*) The β-furanose form of D-arabinose

(*c*) The β-pyranose form of L-arabinose

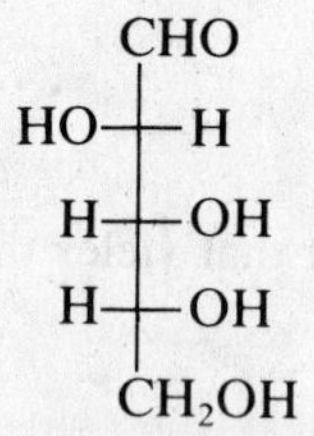

$$
\begin{array}{c}
\text{CHO} \\
\text{HO}\!-\!\!-\!\text{H} \\
\text{H}\!-\!\!-\!\text{OH} \\
\text{H}\!-\!\!-\!\text{OH} \\
\text{CH}_2\text{OH}
\end{array}
$$

D-Arabinose

A-4. Using text Figure 25.2, identify the following carbohydrate:

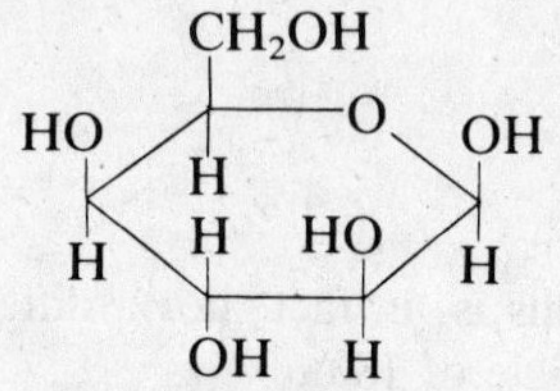

PART B

B-1. Choose the response which provides the best match between the terms given and the structures shown.

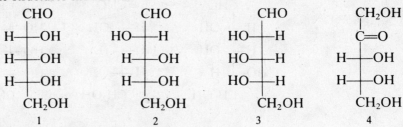

	Diastereomers	Enantiomers
(a)	1, 3, and 4	1 and 3
(b)	1 and 2	1 and 3
(c)	1, 2, and 3	1 and 3
(d)	1 and 4	1 and 2

B-2. A D carbohydrate is
(a) Always dextrorotatory
(b) Always levorotatory
(c) Always the anomer of the corresponding L carbohydrate
(d) None of the above

B-3. Two of the three compounds shown yield the same product on reaction with warm HNO_3. The *exception* is

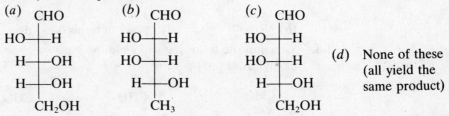

(d) None of these (all yield the same product)

B-4. The optical rotation of the α form of a pyranose is $+150.7°$; that of the β form is $+52.8°$. In solution an equilibrium mixture of the anomers has an optical rotation of $+80.2°$. The percentage of the α form at equilibrium is
(a) 28%   (b) 32%   (c) 68%   (d) 72%

B-5. Which of the following represents the anomer of the compound shown?

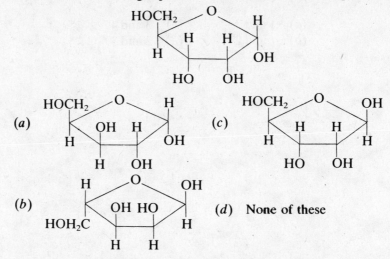

(d) None of these

B-6. Which of the following aldoses yield an optically inactive substance on reaction with sodium borohydride?

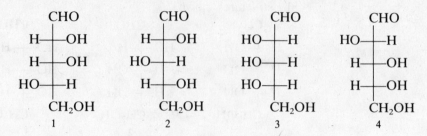

 (*a*) 3 only   (*c*) 2 and 3

 (*b*) 1 and 4   (*d*) All of them (1, 2, 3, and 4)

B-7. Which set of terms correctly identifies the carbohydrate shown?

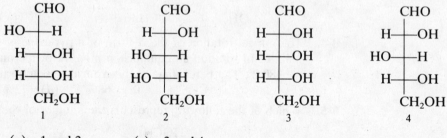

 1. Pentose   **5.** Aldose

 2. Pentulose   **6.** Ketose

 3. Hexulose   **7.** Pyranose

 4. Hexose   **8.** Furanose

 (*a*) 2, 6, 8   (*c*) 1, 5, 8

 (*b*) 2, 6, 7   (*d*) A set of terms other than these

B-8. Which of the following will yield the same osazone on treatment with an excess of phenylhydrazine?

 (*a*) 1 and 3   (*c*) 3 and 4

 (*b*) 1 and 2   (*d*) 2 and 3

ACETATE-DERIVED NATURAL PRODUCTS

IMPORTANT TERMS AND CONCEPTS

Fats and Fatty Acids (Secs. 26.1 to 26.3) Fats are a type of lipid. An operational definition of a *lipid* is a naturally occurring substance that is soluble in nonpolar solvents. Fats are biosynthesized from *acetate*. An important aspect of the biosynthesis of fats is the role of *acetyl coenzyme A*, a thioester which acts as an acyl transfer agent.

Fats and oils are naturally occurring triesters of glycerol, also known as *triacylglycerols*. *Tristearin* is a common component of animal fats.

$$CH_2O\overset{\displaystyle O}{\overset{\|}{C}}(CH_2)_{16}CH_3$$

$$CH_3(CH_2)_{16}\overset{\displaystyle O}{\overset{\|}{C}}OCH$$

$$CH_2O\overset{\displaystyle O}{\overset{\|}{C}}(CH_2)_{16}CH_3$$

Tristearin

The acyl groups have unbranched carbon chains and contain an even number of carbon atoms. Those having 14 to 20 carbon atoms are the most common. Unsaturation, when present, almost always occurs in the form of cis (or *Z*) alkene linkages. The free acids derived from the acyl groups are referred to as *fatty acids*. For example, stearic acid is $CH_3(CH_2)_{16}CO_2H$.

Phospholipids (Sec. 26.4) Phosphorus-containing lipids are known as *phospholipids*. *Phosphatidylcholine*, also known as *lecithin*, is an important phospholipid, which is a principal component of cell membranes.

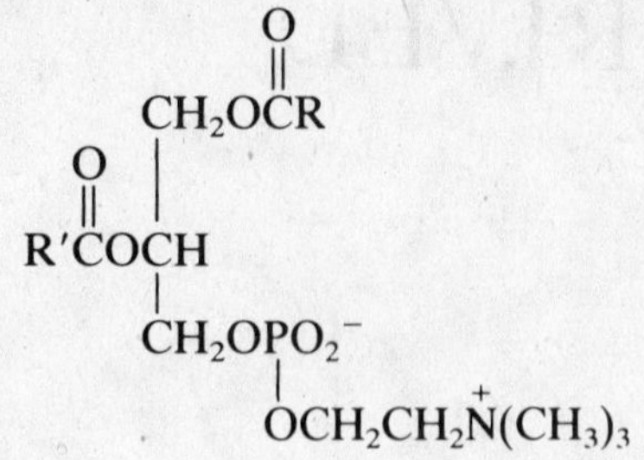

$$
\begin{array}{c}
\overset{\text{O}}{\overset{\|}{\text{CH}_2\text{OCR}}} \\
\overset{\text{O}}{\overset{\|}{\text{R}'\text{COCH}}} \\
\text{CH}_2\text{OPO}_2^- \\
\text{OCH}_2\text{CH}_2\overset{+}{\text{N}}(\text{CH}_3)_3
\end{array}
$$

Phosphatidylcholine
(lecithin)

Cell membranes are thought to be made up of a *lipid bilayer,* as shown in text Figure 26.4.

Terpenes and Terpene Biosynthesis (Secs. 26.7 to 26.10) Terpenes are those naturally occurring substances derived biosynthetically from the *isoprene unit.*

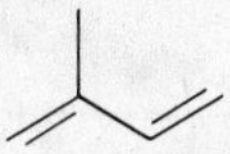

Isoprene

Terpenes are therefore often referred to as *isoprenoid* compounds. They are classified according to the number of carbon atoms they contain, as listed in text Table 26.2.

The biosynthesis of terpenes begins with acetic acid. The *isoprene unit* from which the terpenes are constructed is contained in the key intermediate *isopentenyl pyrophosphate.*

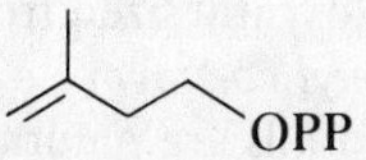

Isopentenyl pyrophosphate

Steroids (Secs. 26.11 to 26.15) Steroids are naturally occurring molecules characterized by the tetracyclic ring skeleton shown in the following:

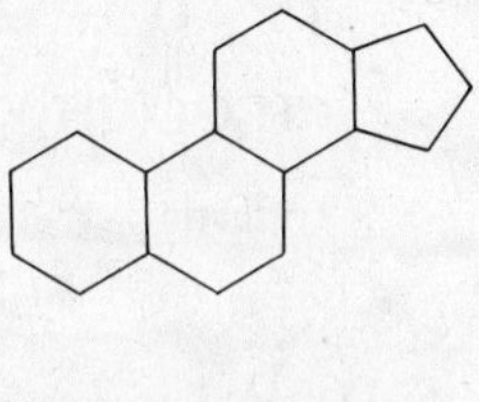

Steroid ring skeleton

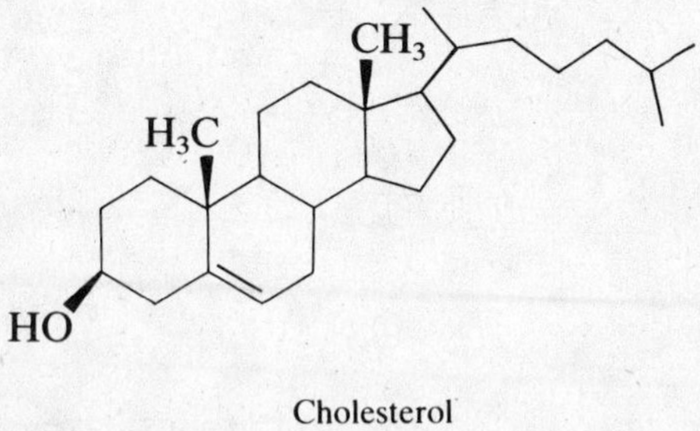

Cholesterol

An important steroid is *cholesterol.* The tetracyclic ring system of cholesterol is biosynthetically derived from the triterpene *squalene,* as shown in text Figure 26.11. A key intermediate, found in significant quantities in lanolin, is *lanosterol.*

Vitamin D and the *bile acids* are two examples of important natural compounds derived biologically from cholesterol. Also, the male sex hormone *testosterone* and the female sex hormones *estrogen* and *progesterone* are steroids formed biosynthetically from cholesterol.

SOLUTIONS TO TEXT PROBLEMS

26.1 The triacylglycerol shown in Figure 26.2*a*, with an oleyl group at C-2 of the glycerol unit and two stearyl groups at C-1 and C-3, yields stearic and oleic acids in a 2:1 molar ratio on hydrolysis. A constitutionally isomeric structure in which the oleyl group is attached to C-1 of glycerol would yield the same hydrolysis products.

$$CH_3(CH_2)_7CH{=}CH(CH_2)_7\overset{\displaystyle O}{\overset{\|}{C}}O\overset{\displaystyle CH_2O\overset{\displaystyle O}{\overset{\|}{C}}(CH_2)_{16}CH_3}{\underset{\displaystyle CH_2O\underset{\displaystyle O}{\overset{\|}{C}}(CH_2)_{16}CH_3}{\overset{|}{C}H}} \qquad \text{or}$$

$$CH_3(CH_2)_{16}\overset{\displaystyle O}{\overset{\|}{C}}O\overset{\displaystyle CH_2O\overset{\displaystyle O}{\overset{\|}{C}}(CH_2)_7CH{=}CH(CH_2)_7CH_3}{\underset{\displaystyle CH_2O\underset{\displaystyle O}{\overset{\|}{C}}(CH_2)_{16}CH_3}{\overset{|}{C}H}}$$

$$\downarrow {\scriptstyle 3H_2O}$$

$$CH_3(CH_2)_7CH{=}CH(CH_2)_7\overset{\displaystyle O}{\overset{\|}{C}}OH + HO\overset{\displaystyle CH_2OH}{\underset{\displaystyle CH_2OH}{\overset{|}{C}H}} + 2HO\overset{\displaystyle O}{\overset{\|}{C}}(CH_2)_{16}CH_3$$

Oleic acid   Glycerol   Stearic acid

26.2 The reaction is an acyl transfer from the sulfur of acetyl coenzyme A to the sulfur of acyl carrier protein.

$$CH_3\overset{\displaystyle O}{\overset{\|}{C}}SCoA + H\overset{..}{\underset{..}{S}}{-}ACP \longrightarrow CH_3\overset{\displaystyle O{-}H}{\underset{\displaystyle S{-}ACP}{\overset{|}{C}{-}SCoA}} \longrightarrow CH_3\overset{\displaystyle O}{\overset{\|}{C}}S{-}ACP + HSCoA$$

| Acetyl coenzyme A | Acyl carrier protein | Tetrahedral intermediate | *S*-Acetyl acyl carrier protein | Coenzyme A |

26.3 Conversion of acyl carrier protein–bound tetradecanoate to hexadecanoate proceeds through the series of intermediates shown.

$$CH_3(CH_2)_{12}\overset{\overset{O}{\|}}{C}S-ACP$$

$$\downarrow \quad HO_2CCH_2\overset{\overset{O}{\|}}{C}S-ACP$$

$$CH_3(CH_2)_{12}\overset{\overset{O}{\|}}{C}CH_2\overset{\overset{O}{\|}}{C}S-ACP$$

$$\downarrow$$

$$CH_3(CH_2)_{12}\overset{\overset{OH}{|}}{C}HCH_2\overset{\overset{O}{\|}}{C}S-ACP$$

$$\downarrow$$

$$CH_3(CH_2)_{12}CH{=}CH\overset{\overset{O}{\|}}{C}S-ACP$$

$$\downarrow$$

$$CH_3(CH_2)_{12}CH_2CH_2\overset{\overset{O}{\|}}{C}S-ACP$$

26.4 The structure of L-glycerol 3-phosphate is shown in a Fischer projection. Translate the Fischer projection formula to a three-dimensional representation:

$$
\begin{array}{ccc}
CH_2OH & CH_2OH & CH_2OH \\
HO{-}\!\!|\!\!{-}H & \equiv \quad HO{-}C{-}H & \text{same as} \quad HO{\cdots}C{-}H \\
CH_2OPO_3H_2 & CH_2OPO_3H_2 & H_2O_3POCH_2 \\
\end{array}
$$

The order of decreasing sequence rule precedence is:

$$HO{-} > H_2O_3POCH_2{-} > HOCH_2{-} > H{-}$$

When the three-dimensional formula is viewed from a perspective in which the lowest-ranked substituent is away from us, we see:

$$
\begin{array}{c}
HOCH_2 \quad OH \\
\curvearrowright C \curvearrowleft \\
| \\
H_2O_3POCH_2
\end{array}
$$

Order of decreasing priority is clockwise, therefore *R*.

The absolute configuration is *R*.

The conversion of L-glycerol 3-phosphate to a phosphatidic acid does not affect any of the bonds to the stereogenic center, nor does it alter the sequence rule ranking of the substituents.

$$
\begin{array}{c}
\overset{\overset{O}{\|}}{R'C}O \quad CH_2O\overset{\overset{O}{\|}}{C}R \\
{\cdots}C{-}H \\
H_2O_3POCH_2
\end{array}
\qquad
R'\overset{\overset{O}{\|}}{C}O{-} > H_2O_3POCH_2{-} > R\overset{\overset{O}{\|}}{C}OCH_2{-} > H{-}
$$

The absolute configuration is *R*.

26.5 Cetyl palmitate (hexadecyl hexadecanoate) is an ester in which both the acyl group and the alkyl group contain 16 carbon atoms.

$$CH_3(CH_2)_{14}\overset{\overset{\displaystyle O}{\|}}{C}O(CH_2)_{15}CH_3$$

Hexadecyl hexadecanoate

26.6 The biosynthetic scheme shown in Figure 26.6 illustrates the conversion of arachidonic acid to PGE_2.

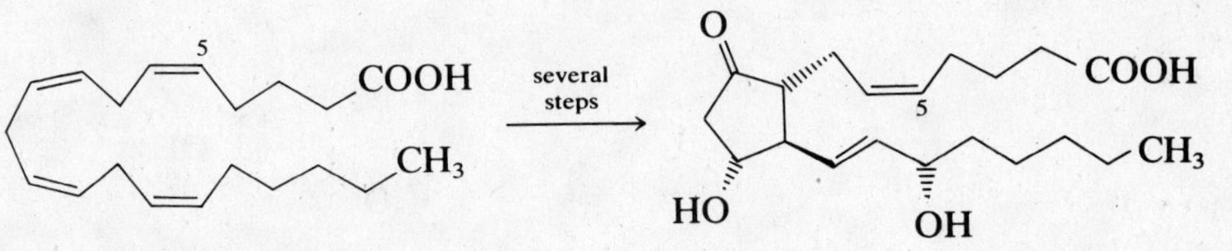

Arachidonic acid

several steps

The structure of PGE_1 is identical to that of PGE_2 except that PGE_1 lacks the double bond at C-5. Therfore the fatty acid that leads to PGE_1 is analogous to arachidonic acid except for the absence of the C-5 double bond.

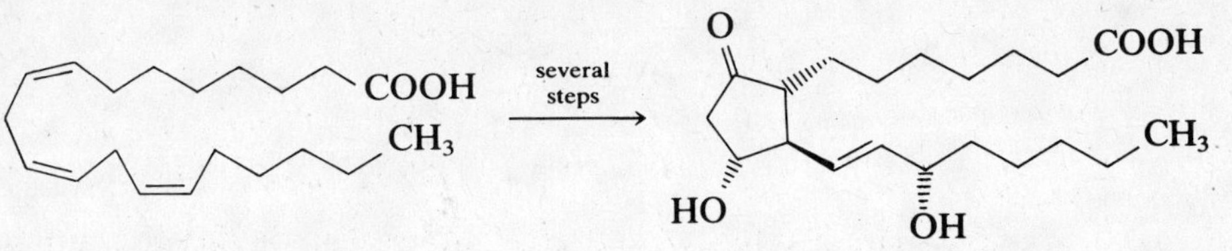

(8Z,11Z,14Z)-8,11,14-Icosatrienoic acid PGE_1

several steps

26.7 Isoprene units are 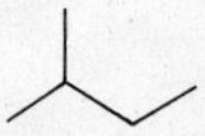 fragments in the carbon skeleton. Functional groups and multiple bonds are ignored when structures are examined for the presence of isoprene units.

α-Phellandrene (two equally correct answers)

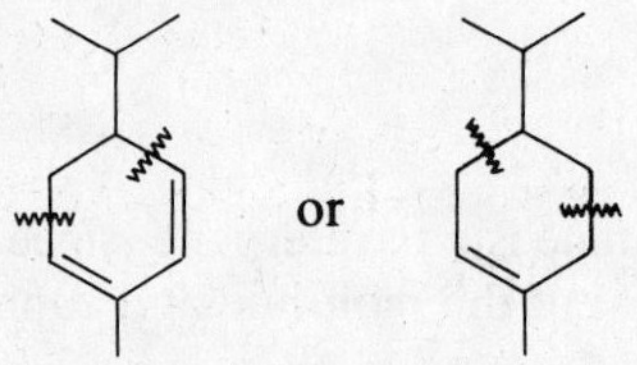

or

Menthol (same carbon skeleton as α-phellandrene but different functionality)

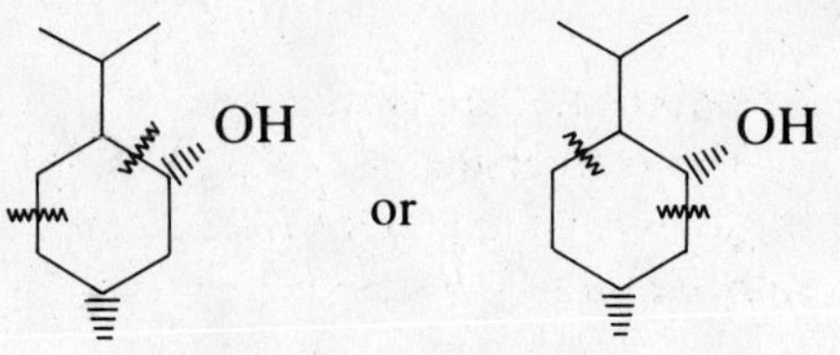

or

Citral

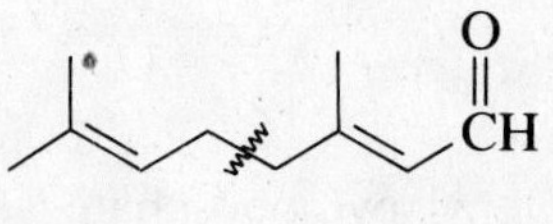

α-Selinene (shown in text Sec. 26.7)

Farnesol

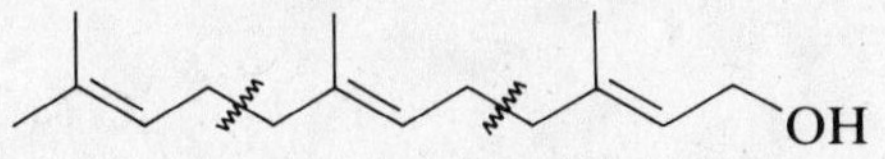

Abscisic acid

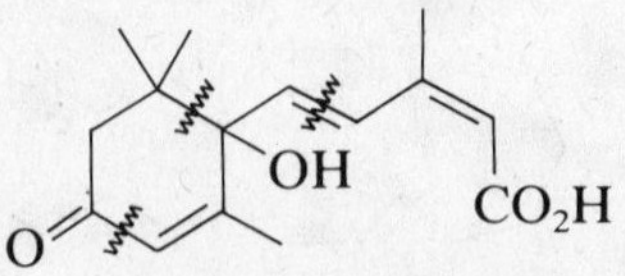

Cembrene (two equally correct answers)

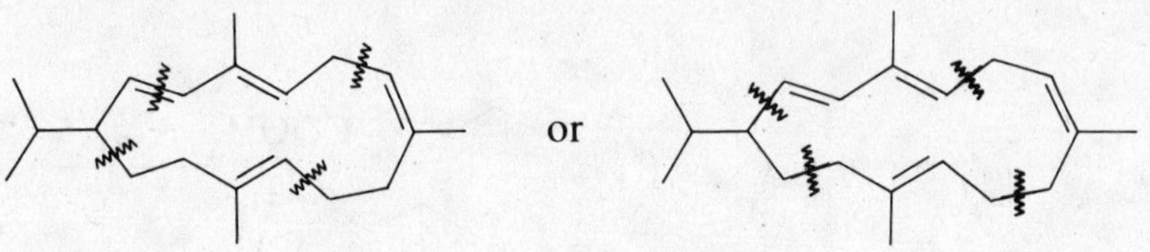

Vitamin A

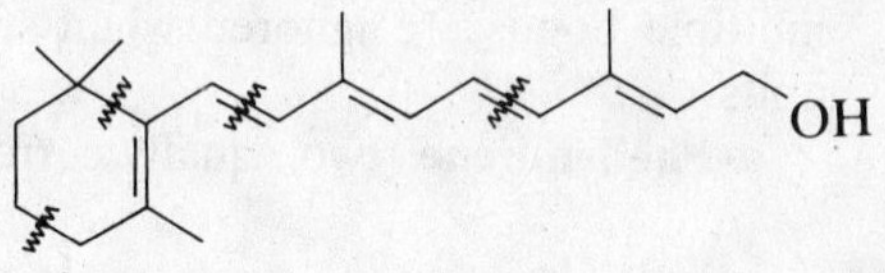

26.8 β-Carotene is a tetraterpene, since it has 40 carbon atoms. The tail-to-tail linkage is at the midpoint of the molecule and connects two 20-carbon fragments.

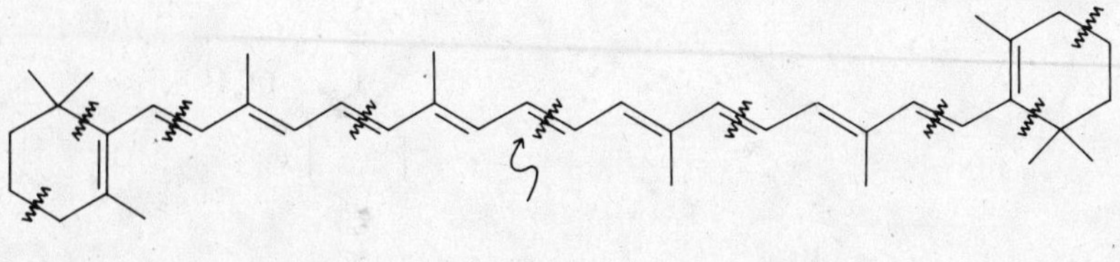

Tail-to-tail link between isoprene units

26.9 Isopentenyl pyrophosphate acts as an alkylating agent toward farnesyl pyrophosphate. Alkylation is followed by loss of a proton from the carbocation intermediate, giving geranylgeranyl pyrophosphate. Hydrolysis of the pyrophosphate yields geranylgeraniol.

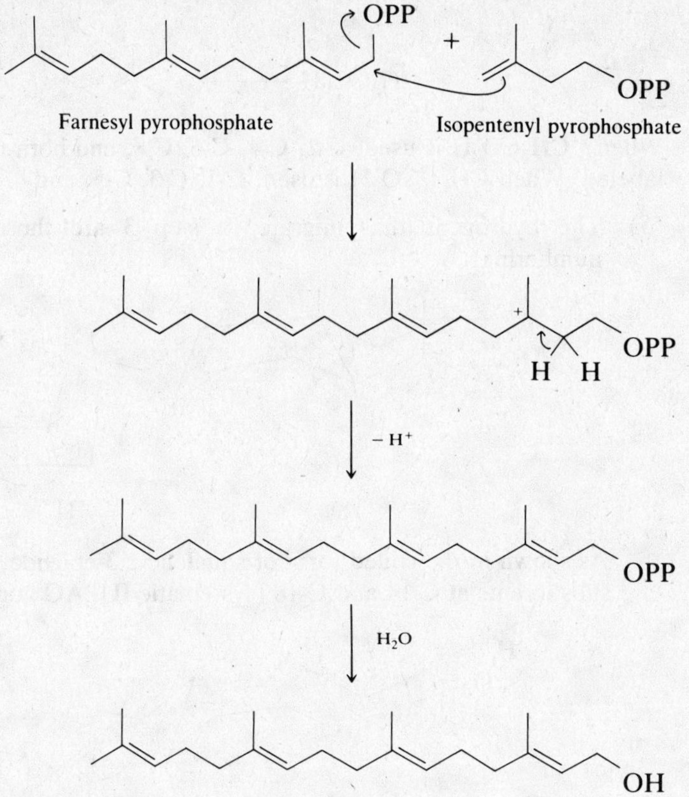

Farnesyl pyrophosphate   Isopentenyl pyrophosphate

$- H^+$

Geranylgeranyl pyrophosphate

H_2O

Geranylgeraniol

26.10 Borneol, the structure of which is given in Figure 26.8, is a secondary alcohol. Oxidation of borneol converts it to the ketone camphor.

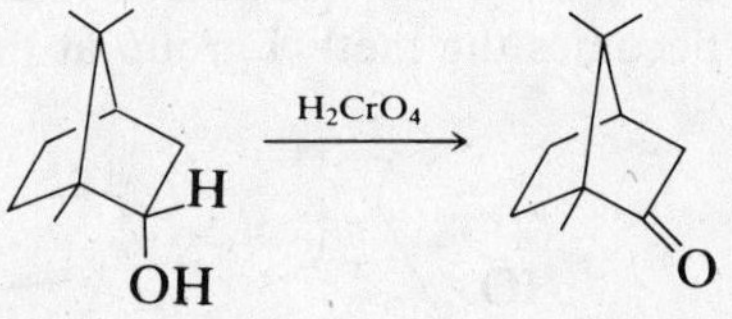

H_2CrO_4

Borneol   Camphor

Reduction of camphor with sodium borohydride gives a mixture of stereoisomeric alcohols, of which one is borneol and the other isoborneol.

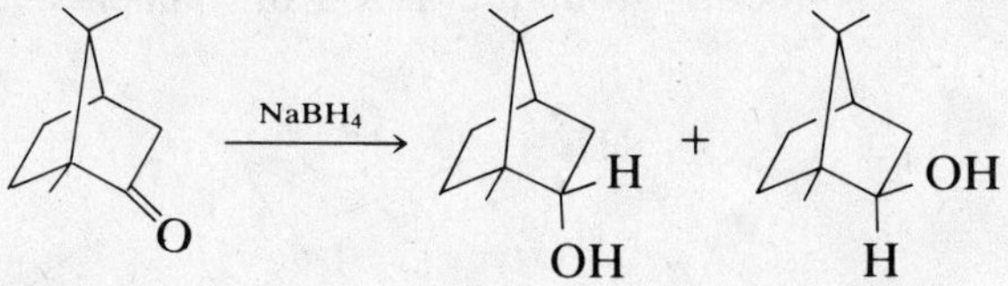

$NaBH_4$

Camphor   Borneol   Isoborneol

26.11 Figure 26.9 in the text describes the distribution of ^{14}C (denoted by *) in citronellal biosynthesized from acetate enriched with ^{14}C in its methyl group.

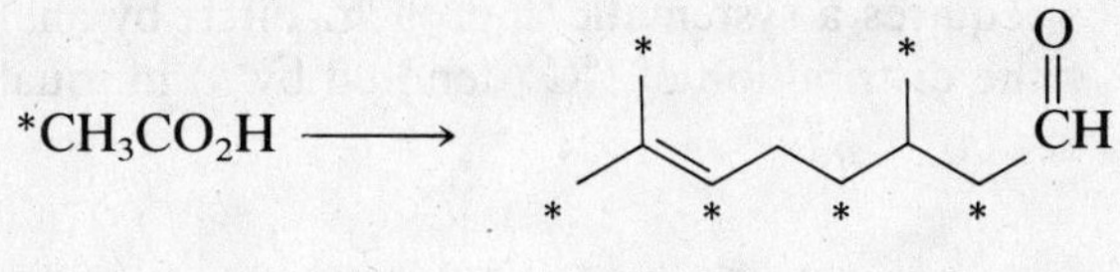

$*CH_3CO_2H \longrightarrow$

If, instead, acetate enriched with ^{14}C at its carbonyl carbon were used, exactly the opposite distribution of the ^{14}C label would be observed.

When $^{14}CH_3CO_2H$ is used, C-2, C-4, C-6, C-8, and both methyl groups of citronellal are labeled. When $CH_3\,^{14}CO_2H$ is used, C-1, C-3, C-5, and C-7 are labeled.

26.12 (*b*) The hydrogens that migrate in step 3 are those at C-13 and C-17 (steroid numbering).

As shown in the coiled form of squalene 2,3-epoxide, these correspond to hydrogen substituents at C-14 and C-18 (systematic IUPAC numbering).

(*c*) The carbon atoms that form the C, D ring junction in cholesterol are C-14 and C-15 of squalene 2,3-epoxide. It is the methyl group at C-15 of squalene 2,3-epoxide that becomes the methyl group at this junction in cholesterol.

(*d*) The methyl groups that are lost are the methyl substituents at C-2 and C-10 plus the methyl group that is C-1 of squalene 2,3-epoxide.

26.13 Tracking the ^{14}C label of $^{14}CH_3CO_2H$ through the complete biosynthesis of cholesterol requires a systematic approach. First, by analogy with Problem 26.11, we can determine the distribution of ^{14}C (denoted by *) in squalene 2,3-epoxide.

Next, follow the path of the ^{14}C-enriched carbons in the cyclization of squalene 2,3-epoxide to lanosterol.

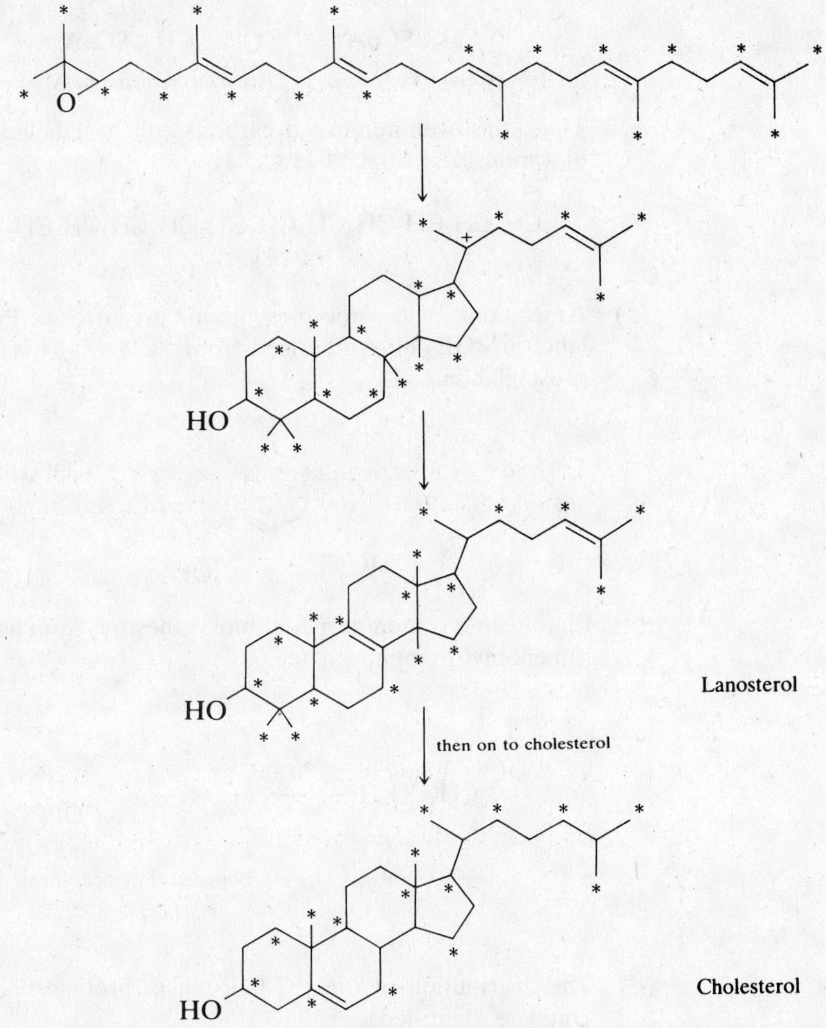

Lanosterol

then on to cholesterol

Cholesterol

26.14 By analogy to the reaction in which 7-dehydrocholesterol is converted to vitamin D$_3$, the structure of vitamin D$_2$ can be deduced from that of ergosterol.

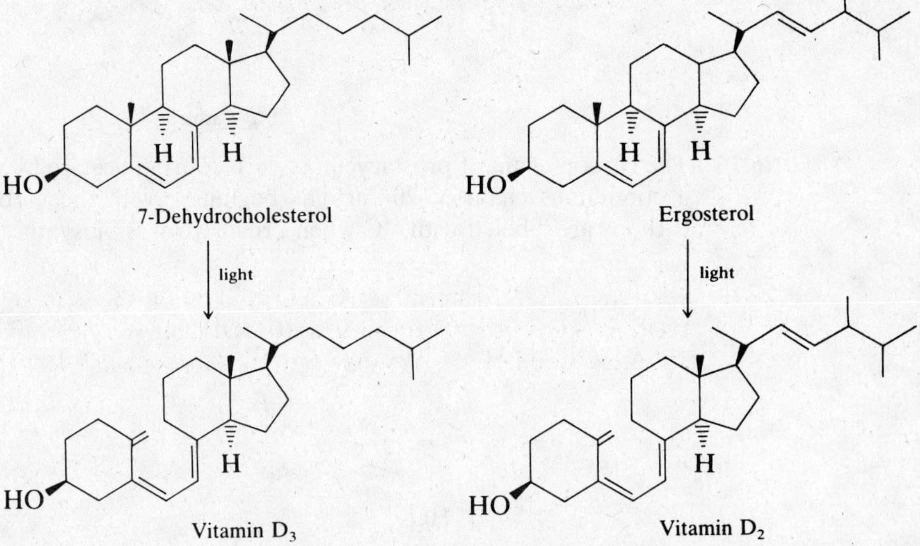

7-Dehydrocholesterol

Ergosterol

light

light

Vitamin D$_3$

Vitamin D$_2$

26.15   (a)   Fatty acid biosynthesis proceeds by the joining of acetate units.

$$CH_3CSCoA \qquad CH_3CCH_2CSCoA \qquad CH_3(CH_2)_{14}CSCoA$$

Acetyl coenzyme A   Acetoacetyl coenzyme A   Palmitoyl coenzyme A

Thus, the even-numbered carbons will be labeled with ^{14}C when palmitic acid is biosynthesized from $^{14}CH_3CO_2H$.

$$CH_3CH_2CH_2CH_2CH_2CH_2CH_2CH_2CH_2CH_2CH_2CH_2CH_2CH_2CH_2CO_2H$$

Palmitic acid

(b)   Arachidonic acid is the biosynthetic precursor of PGE_2. The distribution of the ^{14}C label in PGE_2 biosynthesized from $^{14}CH_3CO_2H$ reflects the fatty acid origin of the prostaglandins.

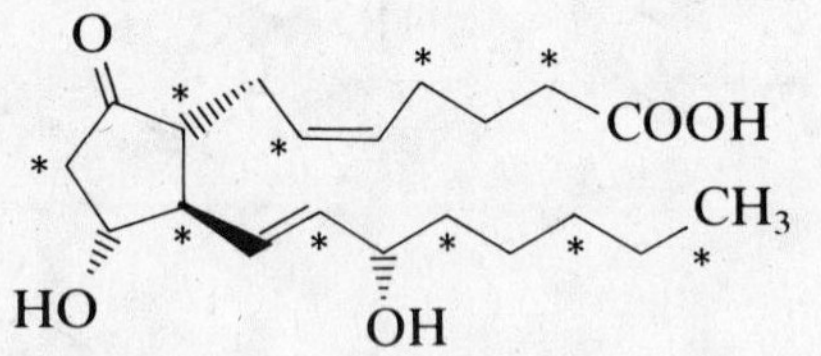

PGE_2

(c)   Limonene is a monoterpene, biosynthesized from acetate by way of mevalonate and isopentenyl pyrophosphate.

$$^{14}CH_3CO_2H \longrightarrow \qquad \longrightarrow$$

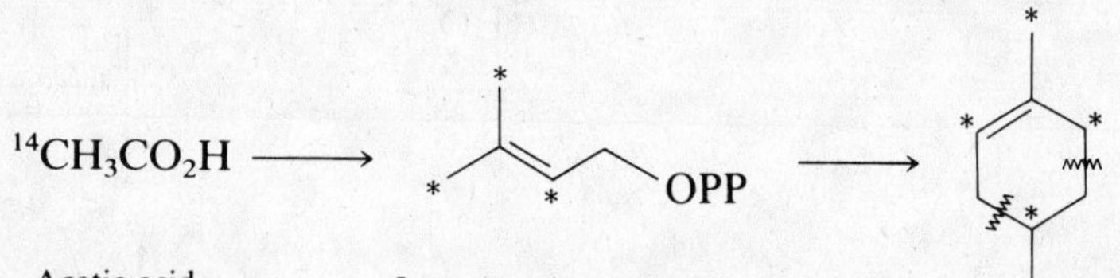

Acetic acid   Isopentenyl pyrophosphate   Limonene

(d)   The distribution of the ^{14}C label in β-carotene becomes evident once its isoprene units are identified.

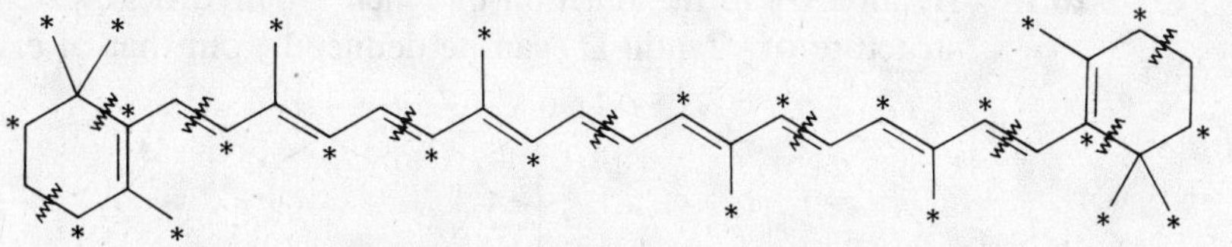

β-Carotene

26.16   The carbon chain of prostacyclin is derived from acetate by way of a C_{20} fatty acid. Trace a continuous chain of 20 carbons beginning with the carboxyl group. Even-numbered carbons are labeled with ^{14}C when prostacyclin is biosynthesized from $^{14}CH_3CO_2H$.

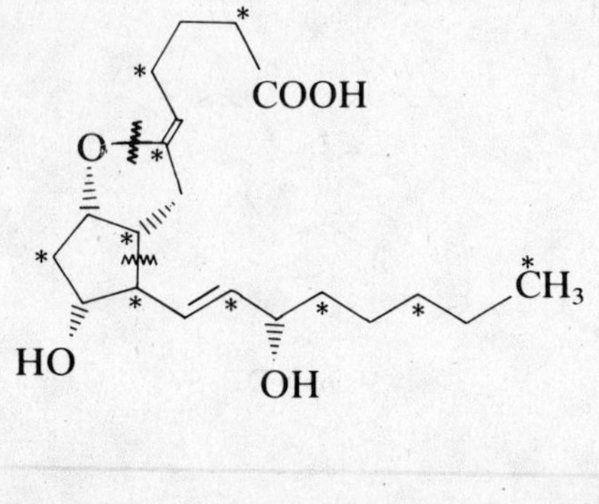

Prostacyclin

26.17 The isoprene units in the designated compounds are shown by disconnections in the structural formulas.

(a) Ascaridole:

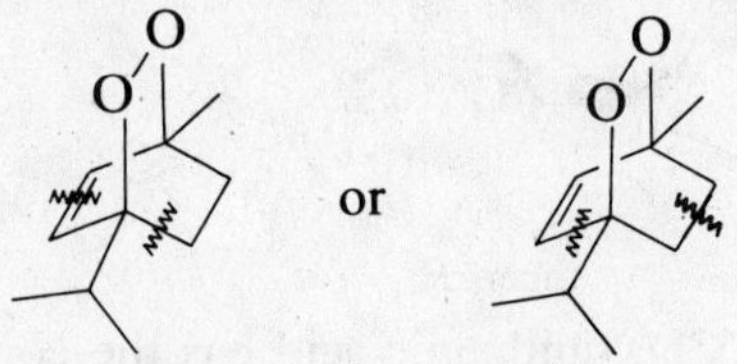

or

(b) Dendrolasin:

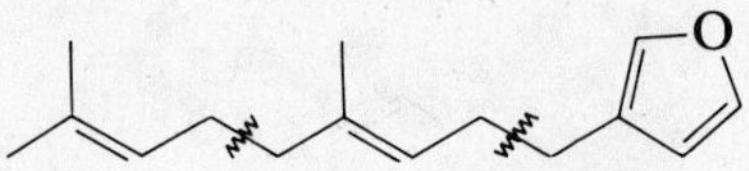

(c) γ-Bisabolene

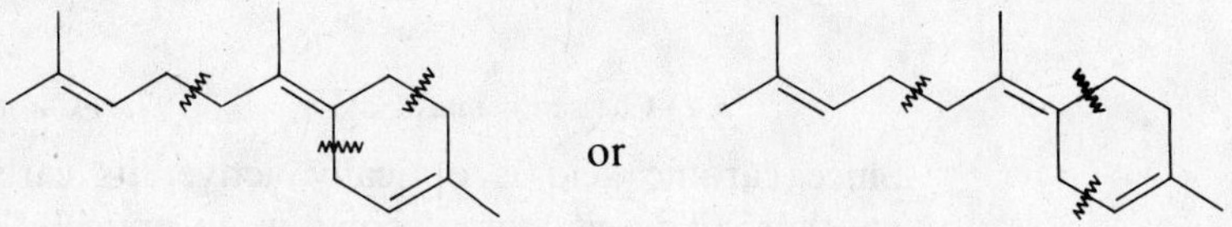

or

(d) α-Santonin

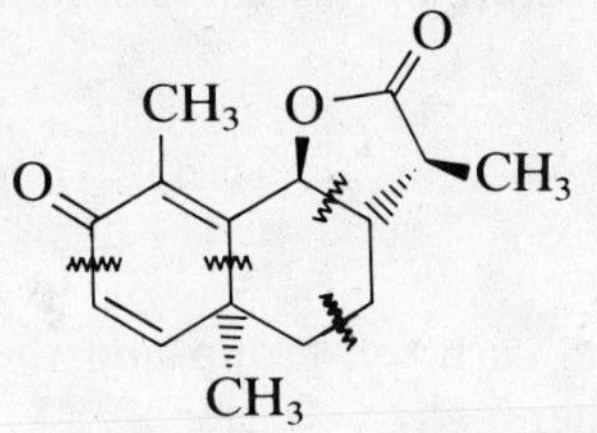

(e) Tetrahymanol

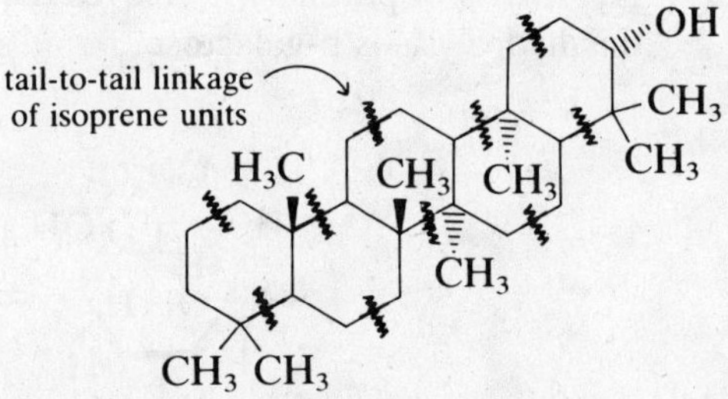

tail-to-tail linkage of isoprene units

26.18 Of the four isoprene units of cubitene, three of them are joined in the usual head-to-tail fashion, but the fourth one is joined in an irregular way.

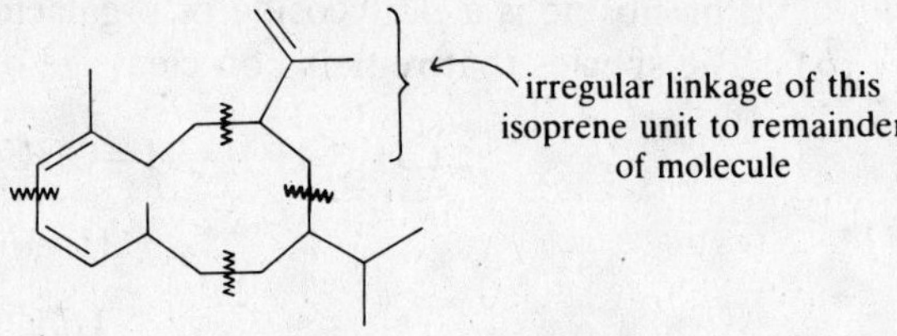

irregular linkage of this isoprene unit to remainder of molecule

26.19 (a) Cinerin I is an ester, the acyl portion of which is composed of two isoprene units, as follows:

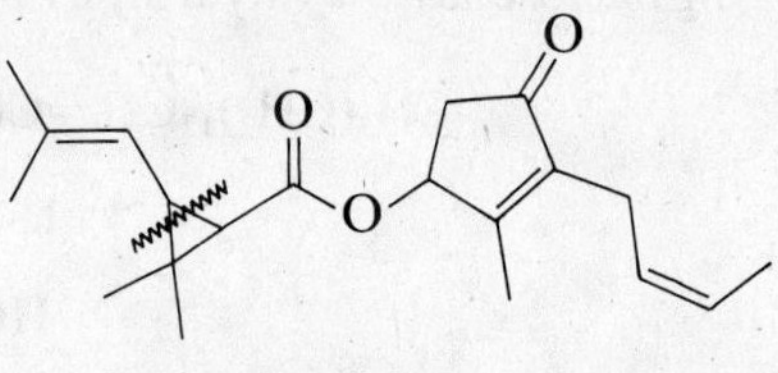

Cinerin I

(*b*) Hydrolysis of cinerin I involves cleavage of the ester unit.

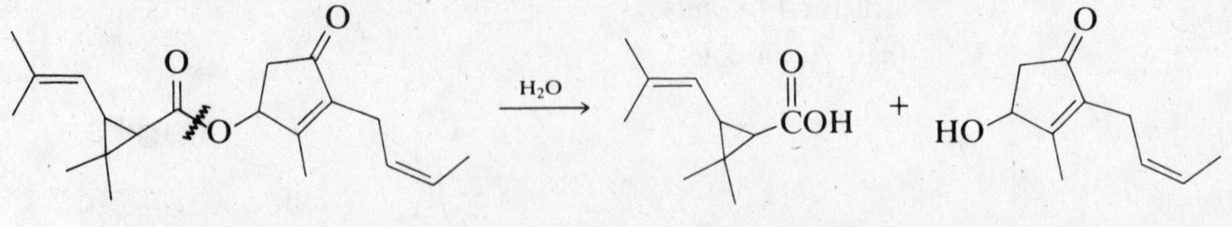
H2O
COH
HO

Cinerin I (+)-Chrysanthemic acid

Chrysanthemic acid has the constitution shown in the equation. Its stereochemistry is revealed by subsequent experiments.

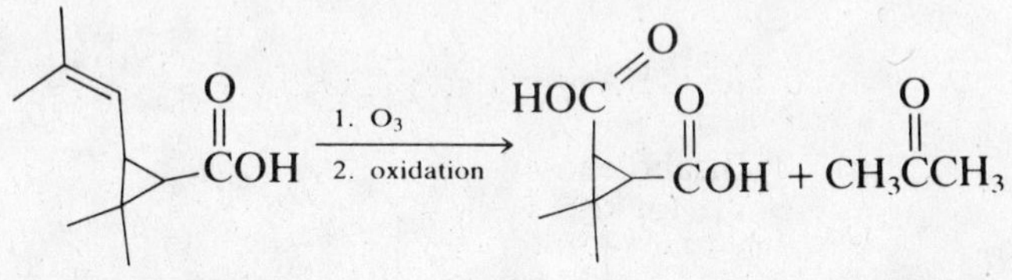
COH
1. O3
2. oxidation
HOC
COH + CH3CCH3

(+)-Chrysanthemic acid (−)-Caronic acid Acetone

Since caronic acid is optically active, its carboxyl groups must be trans to one another. (The cis stereoisomer is an optically inactive meso form.) Therefore the structure of (+)-chrysanthemic acid must be either the following or its mirror image.

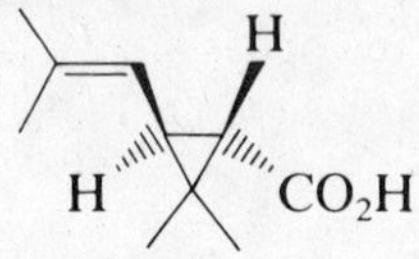
H
H
CO2H

The carboxyl group and the 2-methyl-1-propenyl side chain must be trans to each other.

26.20 (*a*) Hydrolysis of phrenosine cleaves the glycosidic bond. The carbohydrate liberated by this hydrolysis is D-galactose.

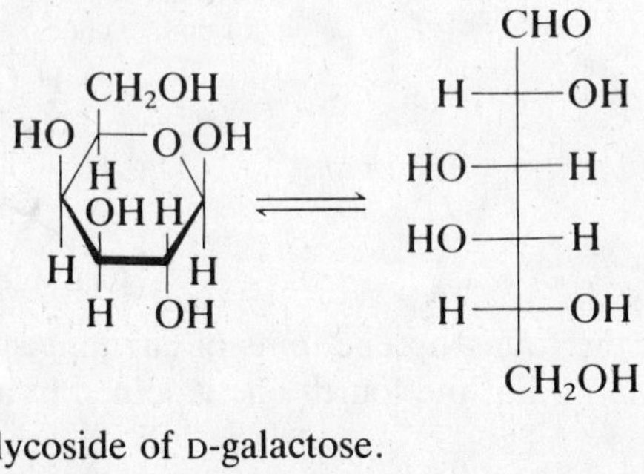
CH2OH
HO
O OH
H
OH H
H
H
H OH
CHO
H
OH
HO
H
HO
H
H
OH
CH2OH

Phrenosine is a β-glycoside of D-galactose.

(*b*) The species that remains on cleavage of the galactose unit has the structure

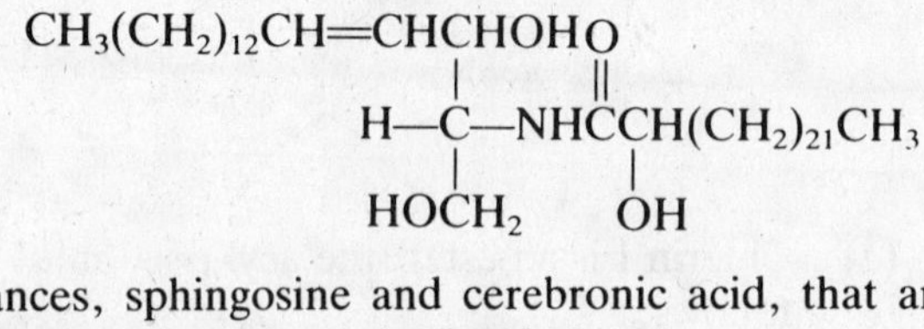
CH3(CH2)12CH=CHCHOHO
H—C—NHCCH(CH2)21CH3
HOCH2
OH

The two substances, sphingosine and cerebronic acid, that are formed along with D-galactose arise by hydrolysis of the amide bond.

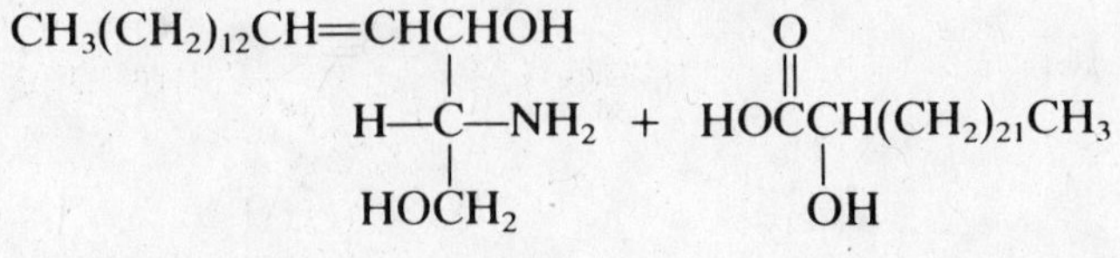
CH3(CH2)12CH=CHCHOH
O
H—C—NH2 + HOCCH(CH2)21CH3
HOCH2
OH

Sphingosine Cerebronic acid

26.21 (*a*)   Catalytic hydrogenation over Lindlar palladium converts alkynes to cis alkenes.

$$CH_3(CH_2)_7C \equiv C(CH_2)_7COOH + H_2 \xrightarrow{\text{Lindlar Pd}}$$

9-Octadecynoic acid
(stearolic acid)

(*Z*)-9-Octadecenoic acid (74%)
(oleic acid)

(*b*)   Carbon-carbon triple bonds are converted to trans alkenes by reduction with lithium and ammonia.

$$CH_3(CH_2)_7C \equiv C(CH_2)_7COOH \xrightarrow[\text{2. } H^+]{\text{1. Li, NH}_3}$$

9-Octadecynoic acid
(stearolic acid)

(*E*)-9-Octadecenoic acid (97%)
(elaidic acid)

(*c*)   The carbon-carbon double bond is hydrogenated readily over a platinum catalyst. Reduction of the ester function does not occur.

$$(Z)\text{-}CH_3(CH_2)_7CH{=}CH(CH_2)_7\overset{O}{\overset{\|}{C}}OCH_2CH_3 \xrightarrow{H_2,\ Pt} CH_3(CH_2)_{16}\overset{O}{\overset{\|}{C}}OCH_2CH_3$$

Ethyl (*Z*)-9-octadecenoate
(ethyl oleate)

Ethyl octadecanoate (91%)
(ethyl stearate)

(*d*)   Lithium aluminum hydride reduces the ester function but leaves the carbon-carbon double bond intact.

$$(Z)\text{-}CH_3(CH_2)_5\underset{\underset{\displaystyle OH}{|}}{C}HCH_2CH{=}CH(CH_2)_7\overset{O}{\overset{\|}{C}}OCH_3$$

Methyl (*Z*)-12-hydroxy-9-octadecenoate
(methyl ricinoleate)

$$\Big\downarrow \begin{array}{l}\text{1. LiAlH}_4\\\text{2. H}_2\text{O}\end{array}$$

$$(Z)\text{-}CH_3(CH_2)_5\underset{\underset{\displaystyle OH}{|}}{C}HCH_2CH{=}CH(CH_2)_7CH_2OH \quad + \quad CH_3OH$$

(*Z*)-9-Octadecen-1,12-diol (52%)   Methanol

(*e*)   Epoxidation of the double bond occurs when an alkene is treated with a peroxy acid. The reaction is stereospecific; substituents that are cis to each other in the alkene remain cis in the epoxide.

$$(Z)\text{-}CH_3(CH_2)_7CH{=}CH(CH_2)_7COOH + C_6H_5CO_2OH$$

Oleic acid   Peroxybenzoic acid

$$\Big\downarrow$$

$$+ C_6H_5\overset{O}{\overset{\|}{C}}OH$$

cis-9,10-Epoxyoctadecanoic acid (62–67%)   Benzoic acid (*f*) Acid-catalyzed hydrolysis of the epoxide yields a diol; its stereochemistry corresponds to net anti hydroxylation of the double bond of the original alkene.

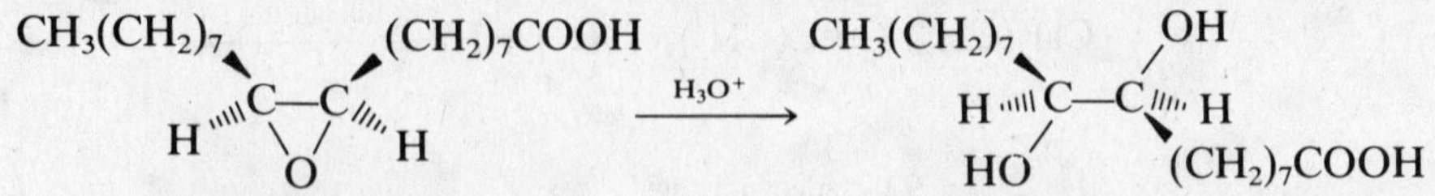
$CH_3(CH_2)_7$
$(CH_2)_7COOH$
H
C
C
H
O
H_3O^+
$CH_3(CH_2)_7$
OH
H
C
C
H
HO
$(CH_2)_7COOH$

cis-9,10-Epoxyoctadecanoic acid 9,10-Dihydroxyoctadecanoic acid

The product is chiral but is formed as a racemic mixture containing equal amounts of the 9*R*,10*R* and 9*S*,10*S* stereoisomers.

(*g*) Hydroxylation of carbon-carbon double bonds with osmium tetraoxide proceeds with syn addition of hydroxyl groups.

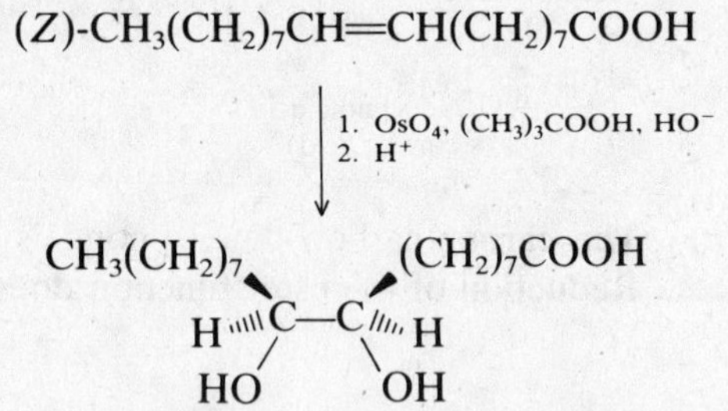
(*Z*)-$CH_3(CH_2)_7CH{=}CH(CH_2)_7COOH$
1. OsO_4, $(CH_3)_3COOH$, HO^-
2. H^+
$CH_3(CH_2)_7$
$(CH_2)_7COOH$
H
C
C
H
HO
OH

9,10-Dihydroxyoctadecanoic acid (70%)

The product is chiral but is formed as a racemic mixture containing equal amounts of the 9*R*,10*S* and 9*S*,10*R* stereoisomers.

(*h*) Hydroboration-oxidation effects the syn hydration of carbon-carbon double bonds with a regioselectivity contrary to Markovnikov's rule. The reagent attacks the less hindered face of the double bond of α-pinene.

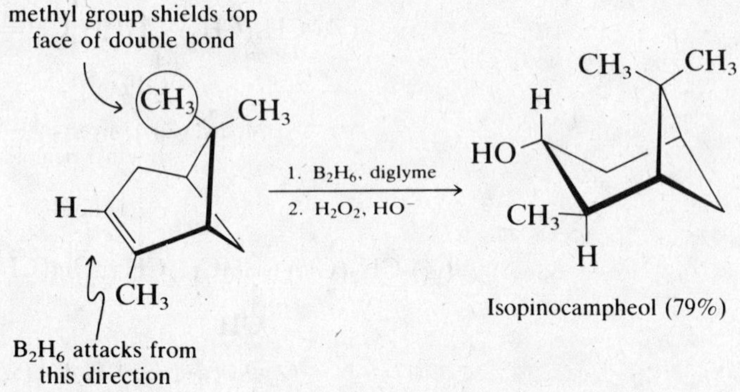
methyl group shields top face of double bond
CH_3
CH_3
H
CH_3
B_2H_6 attacks from this direction
1. B_2H_6, diglyme
2. H_2O_2, HO^-
CH_3
CH_3
H
HO
CH_3
H

Isopinocampheol (79%)

(*i*) The starting alkene in this case is β-pinene. As in the preceding exercise with α-pinene, diborane adds to the bottom face of the double bond.

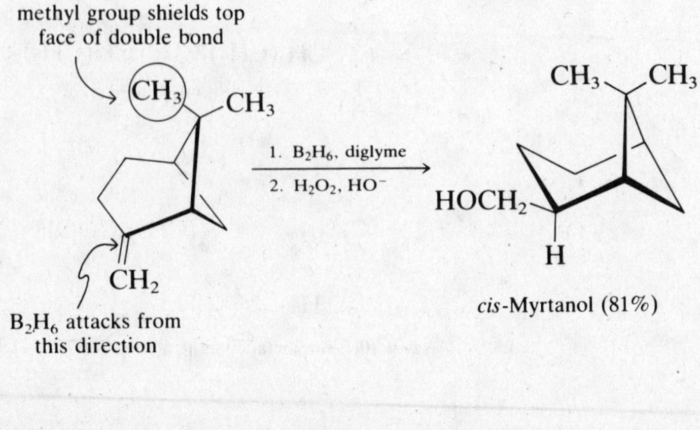
methyl group shields top face of double bond
CH_3
CH_3
CH_2
B_2H_6 attacks from this direction
1. B_2H_6, diglyme
2. H_2O_2, HO^-
CH_3
CH_3
$HOCH_2$
H

cis-Myrtanol (81%)

(*j*) The starting material is an acetal. It undergoes hydrolysis in dilute aqueous acid to give a ketone.

$$\xrightarrow{H_3O^+} 2CH_3OH \; +$$

(95% yield)

(*k*) The conditions described are those of the Reformatskii reaction.

$$\xrightarrow[\text{Zn}]{BrCH_2COOCH_2CH_3}$$

(65% yield)

The zinc enolate adds to the less hindered face of the carbonyl group, that is, trans to the methyl substituent on the adjacent carbon.

26.22 (*a*) There are no direct methods for the reduction of a carboxylic acid to an alkane. A number of indirect methods that may be used, however, involve first converting the carboxylic acid to an alkyl bromide via the corresponding alcohol.

$$CH_3(CH_2)_{16}CO_2H \xrightarrow[\text{2. } H_2O]{\text{1. LiAlH}_4} CH_3(CH_2)_{16}CH_2OH \xrightarrow[\text{or PBr}_3]{\text{HBr, heat}} CH_3(CH_2)_{16}CH_2Br$$

Octadecanoic acid   1-Octadecanol   1-Bromooctadecane

Once the alkyl bromide is in hand, it may be converted to an alkane by conversion to a Grignard reagent followed by addition of water.

$$CH_3(CH_2)_{16}CH_2Br \xrightarrow[\text{diethyl ether}]{\text{Mg}} CH_3(CH_2)_{16}CH_2MgBr \xrightarrow{H_2O} CH_3(CH_2)_{16}CH_3$$

1-Bromooctadecane   Octadecane

Other routes are also possible (for example, E2 elimination from 1-bromooctadecane followed by hydrogenation of the resulting alkene), but require more steps.

(*b*) Retrosynthetic analysis reveals that the 18-carbon chain of the starting material must be attached to a benzene ring.

1-Phenyloctadecane

The desired sequence may be carried out by a Friedel-Crafts acylation reaction, followed by Clemmensen or Wolff-Kishner reduction of the ketone.

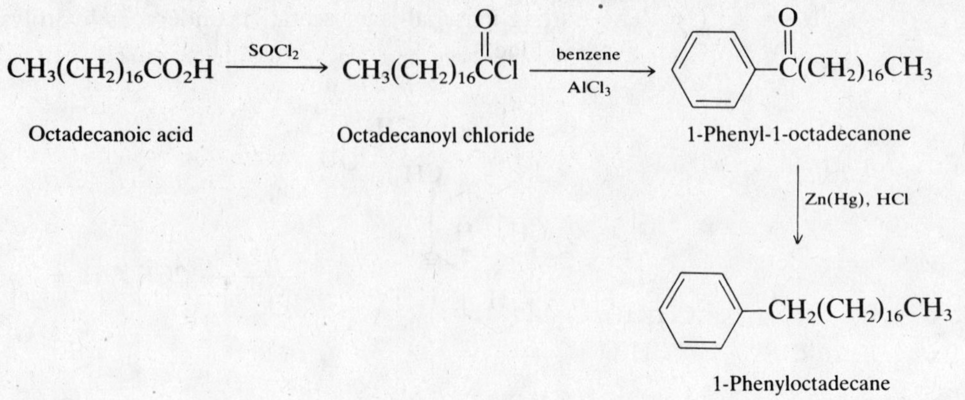

$$CH_3(CH_2)_{16}CO_2H \xrightarrow{\text{SOCl}_2} CH_3(CH_2)_{16}\overset{\displaystyle O}{\overset{\|}{C}}Cl \xrightarrow[\text{AlCl}_3]{\text{benzene}}$$

Octadecanoic acid　　　　　　　　　Octadecanoyl chloride

1-Phenyl-1-octadecanone

$\downarrow$ Zn(Hg), HCl

—CH$_2$(CH$_2$)$_{16}$CH$_3$

1-Phenyloctadecane

Reduction of 1-phenyl-1-octadecanone by the Clemmensen method has been reported to proceed in 77 percent yield.

(c) First examine the structure of the target molecule 3-ethylicosane.

$$CH_3(CH_2)_{16}\underset{\displaystyle \underset{\displaystyle CH_2CH_3}{|}}{CH}CH_2CH_3$$

Retrosynthetic analysis reveals that two ethyl groups have been attached to a C$_{18}$ unit.

$$CH_3(CH_2)_{16}\underset{\displaystyle \underset{\displaystyle CH_2CH_3}{|}}{CH}{-}CH_2CH_3 \implies CH_3(CH_2)_{16}\underset{\displaystyle |}{CH}{-} + 2CH_3CH_2{-}$$

The necessary carbon-carbon bonds can be assembled by the reaction of an ester with two moles of a Grignard reagent.

$$CH_3(CH_2)_{16}\overset{\displaystyle O}{\overset{\|}{C}}OCH_2CH_3 + \quad 2CH_3CH_2MgBr$$

Ethyl octadecanoate　　　　　　Ethylmagnesium bromide
(from octadecanoic acid
and ethanol)

$\downarrow$ 1. diethyl ether
　2. H$_3$O$^+$

$$CH_3(CH_2)_{16}\underset{\displaystyle \underset{\displaystyle CH_2CH_3}{|}}{\overset{\displaystyle \overset{\displaystyle OH}{|}}{C}}CH_2CH_3$$

3-Ethyl-3-icosanol

With the correct carbon skeleton in place, all that is needed is to achieve the correct oxidation state. This can be accomplished by dehydration and reduction.

$$
\begin{array}{c}
\overset{\displaystyle OH}{\underset{\displaystyle CH_2CH_3}{CH_3(CH_2)_{16}\overset{|}{\underset{|}{C}}CH_2CH_3}} \xrightarrow[\text{heat}]{H_2SO_4}
\underset{\displaystyle CH_2CH_3}{CH_3(CH_2)_{16}\overset{|}{C}\!\!=\!\!CHCH_3} + \underset{\displaystyle CH_2CH_3}{CH_3(CH_2)_{15}CH\!\!=\!\!\overset{|}{C}CH_2CH_3}
\end{array}
$$

3-Ethyl-3-icosanol | 3-Ethyl-2-icosene | 3-Ethyl-3-icosene

$$\downarrow \; H_2, Pt$$

$$
\underset{\displaystyle CH_2CH_3}{CH_3(CH_2)_{16}\overset{|}{C}HCH_2CH_3}
$$

3-Ethylicosane

(d) Icosanoic acid contains two more carbon atoms than octadecanoic acid.

$$CH_3(CH_2)_{18}CO_2H \implies CH_3(CH_2)_{16}CH_2Br + \ddot{C}H_2CO_2H$$

Icosanoic acid

A reasonable approach utilizes a malonic ester synthesis as a key step.

$$CH_3(CH_2)_{16}CH_2Br + CH_2(CO_2CH_2CH_3)_2 \xrightarrow{NaOCH_2CH_3} CH_3(CH_2)_{16}CH_2CH(CO_2CH_2CH_3)_2$$

1-Bromooctadecane   Diethyl malonate   Diethyl 2-octadecylmalonate
[prepared as in part (a)]

$$\downarrow \begin{array}{l} 1.\ HO^- \\ 2.\ H^+ \\ 3.\ heat \end{array}$$

$$CH_3(CH_2)_{16}CH_2CH_2CO_2H$$

Icosanoic acid

(e) The carbon chain must be shortened by one carbon atom in this problem. A Hofmann bromoamide rearrangement is indicated.

$$CH_3(CH_2)_{16}CO_2H \xrightarrow[\text{2. NH}_3]{\text{1. SOCl}_2} CH_3(CH_2)_{16}\overset{\displaystyle O}{\overset{\|}{C}}NH_2 \xrightarrow{Br_2,\ HO^-} CH_3(CH_2)_{16}NH_2$$

Octadecanoic acid   Octadecanamide   1-Heptadecanamine

(f) Lithium aluminum hydride reduction of octadecanamide gives the corresponding amine.

$$CH_3(CH_2)_{16}\overset{\displaystyle O}{\overset{\|}{C}}NH_2 \xrightarrow[\text{2. H}_2O]{\text{1. LiAlH}_4} CH_3(CH_2)_{16}CH_2NH_2$$

Octadecanamide   1-Octadecanamine
[from part (e)]

(g) Chain extension can be achieved via cyanide displacement of bromine from 1-bromooctadecane. Reduction of the cyano group completes the synthesis.

$$CH_3(CH_2)_{16}CH_2Br \xrightarrow{KCN} CH_3(CH_2)_{16}CH_2C\!\!\equiv\!\!N$$

1-Bromooctadecane   Nonadecanenitrile
[from part (a)]

$$\downarrow \begin{array}{l} 1.\ LiAlH_4 \\ 2.\ H_2O \end{array}$$

$$CH_3(CH_2)_{16}CH_2CH_2NH_2$$

1-Nonadecanamine

26.23   First acylate the free hydroxyl group with an acyl chloride.

Treatment with aqueous acid brings about hydrolysis of the acetal function.

The two hydroxyl groups of the resulting diol are then esterified with two moles of the second acyl chloride.

26.24   The overall transformation

simply requires conversion of the alcohol function to some suitable leaving group, followed by substitution by an appropriate nucleophile.

3-Methyl-3-buten-1-ol   4-Bromo-2-methyl-   3-Methyl-3-butenyl
                                      1-butene   methyl sulfide

As actually reported in the literature, the alcohol was converted to its corresponding *p*-toluenesulfonate ester and this substance was then used as the substrate in the nucleophilic substitution step to produce the desired sulfide in 76 percent yield.

26.25   The first transformation is an intramolecular aldol condensation. This reaction was carried out under conditions of base catalysis.

6-Methyl-2,5-   (not isolated)   3-Isopropyl-2-
heptanedione   cyclopentenone (71%)

The next step is reduction of a ketone to a secondary alcohol. Lithium aluminum hydride is suitable; it reduces carbonyl groups but leaves the double bond intact.

3-Isopropyl-2-cyclopentenone   3-Isopropyl-2-cyclopenten-1-ol (97%)

Conversion of an alkene to a cyclopropane can be accomplished to using the Simmons-Smith reagent (iodomethylzinc iodide).

3-Isopropyl-2-cyclopenten-1-ol   5-Isopropylbicyclo[3.1.0]hexan-2-ol (66%)

Oxidation of the secondary alcohol to the ketone can be accomplished with any of a number of oxidizing agents. The chemists who reported this synthesis used chromic acid.

5-Isopropylbicyclo[3.1.0]hexan-2-ol   5-Isopropylbicyclo[3.1.0]hexan-2-one (89%)

A Wittig reaction converts the ketone to sabinene.

5-Isopropylbicyclo[3.1.0]hexan-2-one   Sabinene (70%)

26.26 The first step is a 1,4 addition of hydrogen bromide to the diene system of isoprene.

Hydrogen   2-Methyl-   1-Bromo-3-
bromide   1,3-butadiene   methyl-2-butene

This is followed by Markovnikov addition of hydrogen bromide to the remaining double bond.

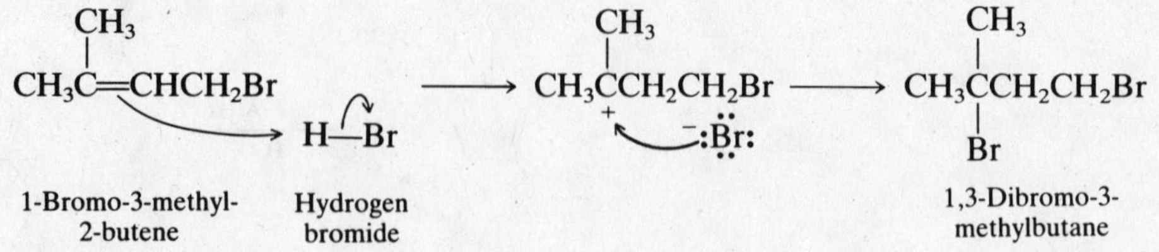

| 1-Bromo-3-methyl-2-butene | Hydrogen bromide | | 1,3-Dibromo-3-methylbutane |

26.27 A reasonable mechanism is protonation of the isolated carbon-carbon double bond, followed by cyclization.

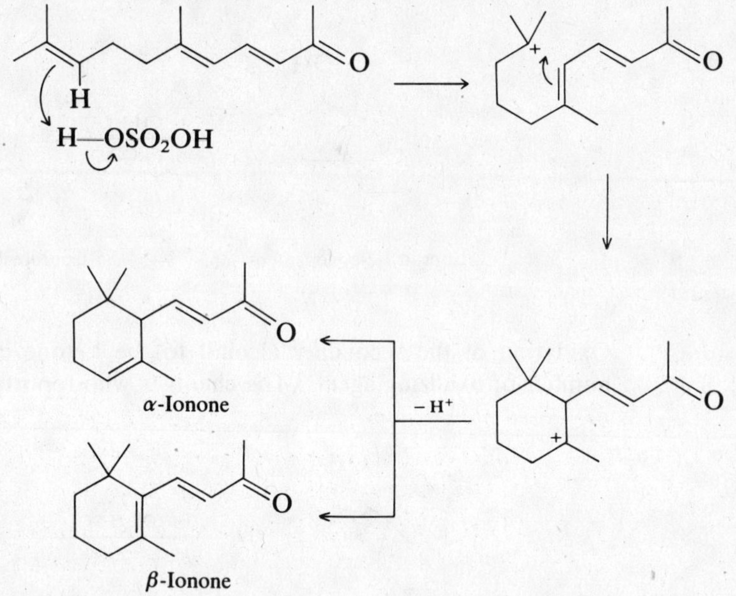

α-Ionone

β-Ionone

26.28 There is a tendency for the double bond to move into conjugation with the carbonyl group. Two mechanisms are more likely than any others under conditions of acid catalysis. One of these simply involves protonation of the double bond followed by loss of a proton from C-4.

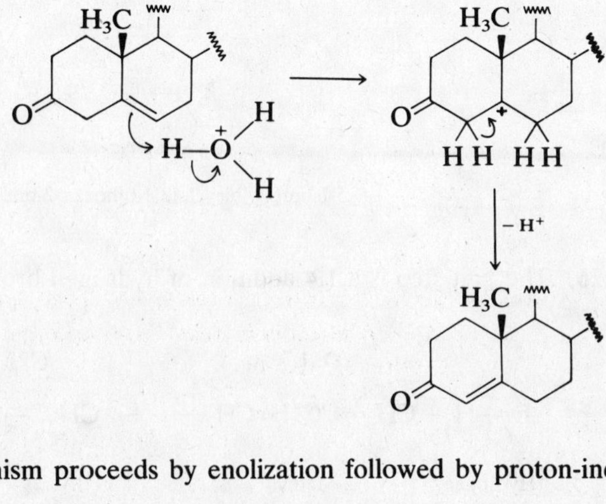

The other mechanism proceeds by enolization followed by proton-induced double bond migration.

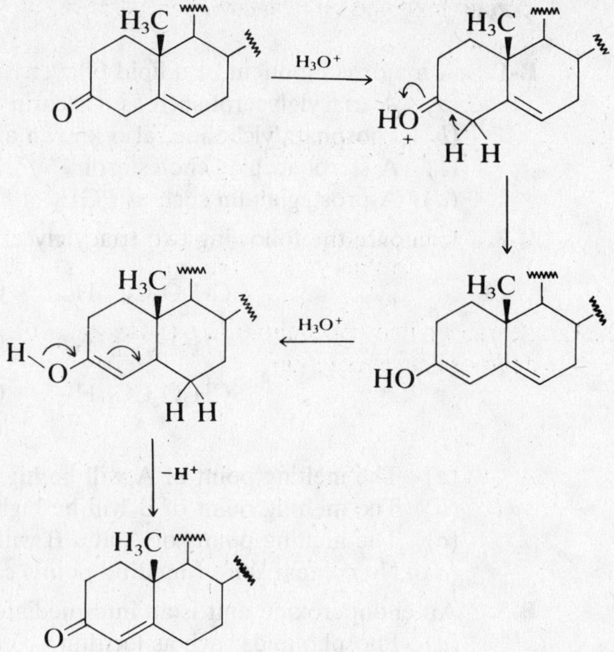

SELF-TEST

PART A

A-1. Write a balanced chemical equation for the saponification of tristearin.

A-2. Both waxes and fats are lipids that contain the ester functional group. In what way do the structures of these lipids differ?

A-3. Classify each of the following isoprenoid compounds as a monoterpene, a diterpene, and so on. Indicate with dashed lines the isoprene units that make up each structure.

(*a*) α-Pinene:

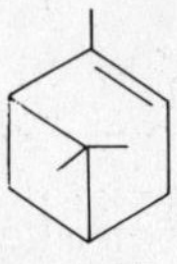

(*b*) Caryophyllene:

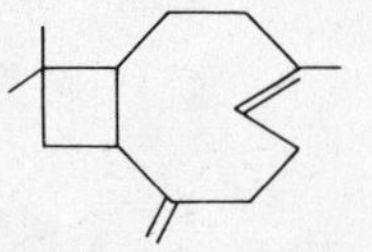

(*c*) Abietic acid:

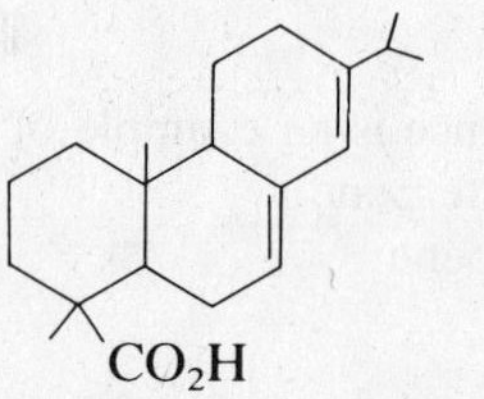

CO₂H

A-4. Propose a series of synthetic steps to carry out the preparation of oleic acid [(*Z*)-9-octadecenoic acid] from compound A. You may use any necessary organic or inorganic reagents.

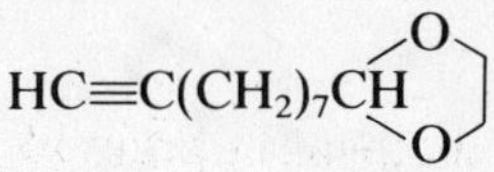

A

PART B

B-1. A major component of a lipid bilayer is:
- (*a*) A triacylglycerol such as tristearin
- (*b*) Phosphatidylcholine, also known as lecithin
- (*c*) A sterol such as cholesterol
- (*d*) A prostaglandin such as PGE_1

B-2. Compare the following two triacylglycerols:

$$CH_2O_2CC_{17}H_{35} \quad CH_2O_2CC_{17}H_{35}$$
$$CHO_2CC_{17}H_{35} \quad CHO_2CC_{17}H_{31}$$
$$CH_2O_2CC_{17}H_{35} \quad CH_2O_2CC_{17}H_{31}$$
$$A \qquad\qquad B$$

- (*a*) The melting point of A will be higher.
- (*b*) The melting point of B will be higher.
- (*c*) The melting points of A and B will be the same.
- (*d*) No comparison of melting points can be made.

B-3. An endoperoxide unit is an intermediate in the biosynthesis of:
- (*a*) Phospholipids such as lecithin
- (*b*) Steroids such as cholesterol
- (*c*) Prostaglandins such as PGE_1
- (*d*) None of these

B-4. Lanosterol, a biosynthetic precursor of cholesterol, exists naturally as a single enantiomer. How many *possible* stereoisomers having the lanosterol skeleton are there?

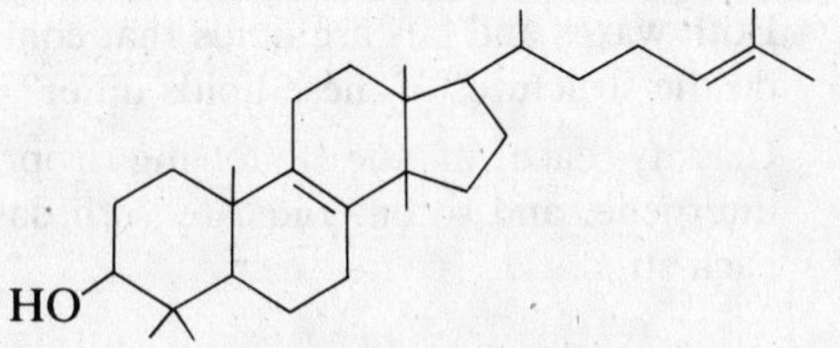

Lanosterol

- (*a*) 7
- (*b*) 64
- (*c*) 128
- (*d*) 256

B-5. The compound whose carbon skeleton is shown, known as selinene, is found in celery.

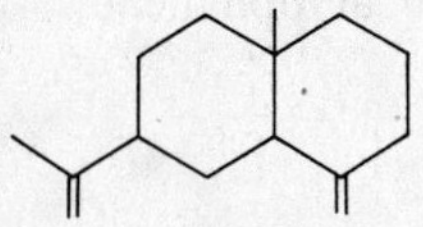

This substance is an example of a
- (*a*) Monoterpene
- (*b*) Diterpene
- (*c*) Sesquiterpene
- (*d*) Triterpene

AMINO ACIDS, PEPTIDES, AND PROTEINS. NUCLEIC ACIDS

IMPORTANT TERMS AND CONCEPTS

Amino Acids (Secs. 27.1 to 27.3) All the amino acids from which proteins are constructed are α-amino acids, and except for proline they contain a primary amino function. The 20 amino acids normally present in proteins are listed in text Table 27.1.

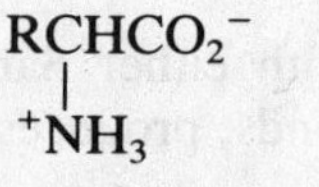

$$RCHCO_2^-$$
$$|$$
$${}^+NH_3$$

Typical α-amino acid

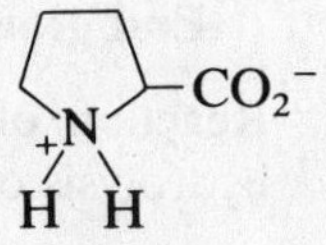

Proline

Except for glycine the amino acids in proteins are chiral and have the L configuration. With the exception of L-cysteine, the chiral amino acids in proteins have the S configuration as specified by the Cahn-Ingold-Prelog method.

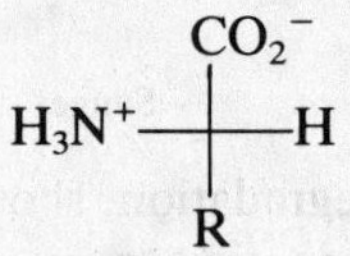

$$CO_2^-$$
$$H_3N^+ \!-\!\!\!-\!\!\!- H$$
$$R$$

L-Amino acid

The physical properties of a typical amino acid suggest that it is a salt. These properties are due to the fact that the amino acid exists as an inner salt, or *zwitterion*. Amino acids are *amphoteric*; that is, they contain both an acidic and a basic functional group. In a strongly acidic medium the predominant species is positively charged; in a strongly basic medium the predominant species is negatively charged.

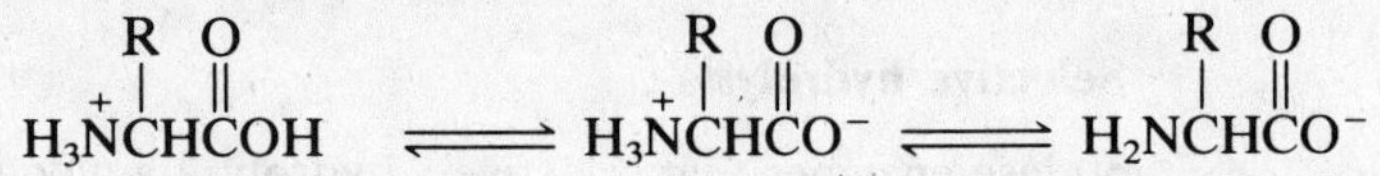

R O	R O	R O
$H_3NCHCOH$	H_3NCHCO^-	H_2NCHCO^-
Cationic species (predominates in strong acid)	Zwitterion (predominates at isoelectric point)	Anionic species (predominates in strong base)

791

The pH at which the zwitterion is the predominant species in solution is the *isoelectric point* of the amino acid. Each amino acid has a characteristic isoelectric point.

Peptides (Sec. 27.7) Peptides are constructed of amino acids linked together by amide bonds. An amide bond between the amino group of one amino acid and the carboxyl of another is known as a *peptide bond*. An example of a dipeptide consisting of two amino acids is glycylalanine. Using the three-letter abbreviations found in Table 27.1, this peptide is Gly-Ala.

$$\overset{+}{H_3N}CH_2\overset{\overset{O}{\|}}{C}-NHCHCO_2^-$$
$$|$$
$$CH_3$$

Gly-Ala

Glycine forms the *N terminus* of the peptide; alanine is the *C terminus*. The *sequence* of amino acids in a peptide is important; Gly-Ala is not the same dipeptide as Ala-Gly.

$$\overset{+}{H_3N}CHC\overset{\overset{O}{\|}}{}-NHCH_2CO_2^-$$
$$|$$
$$CH_3$$

Ala-Gly

Peptide Structure (Secs. 27.8 to 27.12, 27.18, 27.19) The structure of a peptide is described at several levels. The *primary structure* of a peptide is its constitution, that is, the amino acid sequence plus any disulfide links.

The stages of determining the sequence of amino acids in a peptide include end group analysis at both the N and the C terminus and selective hydrolysis.

End group analysis: N terminus

Reaction of the peptide with either Sanger's reagent or dansyl chloride, followed by hydrolysis of the peptide bonds, produces a derivative of the N-terminal amino acid.

Sanger's reagent   Dansyl chloride

The Edman degradation, shown in text Figure 27.10, allows a sequential analysis of the N terminus of a peptide to be undertaken. After each step, the *phenylhydantoin* derivative is isolated and characterized. The peptide chain minus the original N-terminal amino acid remains intact. This "new" peptide is then analyzed and the cycle repeated.

End group analysis: C terminus

Enzymes known as *carboxypeptidases* catalyze the hydrolysis of the C-terminal amino acid, which is identified using an amino acid analyzer.

Selective hydrolysis

Peptidase enzymes that selectively hydrolyze amide bonds between specific pairs of amino acids are used in peptide sequencing. For example, *pepsin* catalyzes the hydrolysis of peptide bonds to the carbonyl groups of methionine and leucine.

The *secondary structure* of a peptide refers to the conformational relationship of nearest-neighbor amino acids with respect to each other. Hydrogen bonding interactions between the N—H group of one amino acid unit and the C=O group of another play an important role in determining the secondary structure of peptides and proteins.

The *tertiary structure* of a peptide or protein refers to the folding of the chain of amino acid units. The chain folding affects both the physical properties and the biological function of a protein.

Nucleic Acids (Secs. 27.23 to 27.29) The biological macromolecules that are involved in the transfer of genetic information and the control of protein biosynthesis are known as *nucleic acids*.

The major kinds of nucleic acids, DNA and RNA, contain heterocycles related to *pyrimidine* and *purine*.

Pyrimidine   Purine

The pyrimidines that occur in RNA are uracil and cytosine; DNA contains thymine and cytosine. Both DNA and RNA contain the purine derivatives adenine and guanine.

A *nucleoside* consists of a purine or pyrimidine *base* attached to a carbohydrate. An example is adenosine.

Adenosine

Nucleotides are phosphoric acid esters of nucleosides.

Nucleic acids such as DNA and RNA are *polynucleotides* in which a phosphate ester links one nucleotide unit with another. Text Figure 27.23 illustrates a generalized view of a polynucleotide chain. DNA exists as a *double helix* in which hydrogen bonds exist between complementary base pairs. Replication of the DNA strands is responsible for the transfer of genetic information.

RNA participates in the biosynthesis of proteins by controlling the sequence of amino acids that are brought together, using triplets of nucleotides known as *codons*.

IMPORTANT REACTIONS

Amino Acid Synthesis (Sec. 27.4)

From α-Halo Carboxylic Acids:

$$\text{RCHCO}_2\text{H} + 2\text{NH}_3 \xrightarrow{\text{H}_2\text{O}} \text{RCHCO}_2^- + \text{NH}_4\text{Br}$$

$$\underset{\text{Br}}{|} \qquad\qquad\qquad\qquad \underset{^+\text{NH}_3}{|}$$

Strecker Synthesis:

$$\underset{O}{\overset{\parallel}{\text{RCH}}} \xrightarrow[\text{NaCN}]{\text{NH}_4\text{Cl}} \underset{\text{NH}_2}{\overset{|}{\text{RCHCN}}} \xrightarrow[\text{2. HO}^-]{\text{1. H}_3\text{O}^+,\ \text{heat}} \underset{^+\text{NH}_3}{\overset{|}{\text{RCHCO}_2^-}}$$

Alkylation of Diethyl Acetamidomalonate:

$$\underset{O}{\overset{\parallel}{\text{CH}_3\text{C}}}\text{NHCH(CO}_2\text{CH}_2\text{CH}_3)_2 \xrightarrow[\text{2. RBr}]{\text{1. NaOCH}_2\text{CH}_3,\ \text{CH}_3\text{CH}_2\text{OH}} \underset{\overset{|}{\text{R}}}{\underset{O}{\overset{\parallel}{\text{CH}_3\text{C}}}\text{NHC(CO}_2\text{CH}_2\text{CH}_3)_2}$$

$$\underset{\overset{|}{\text{R}}}{\underset{O}{\overset{\parallel}{\text{CH}_3\text{C}}}\text{NHC(CO}_2\text{CH}_2\text{CH}_3)_2} \xrightarrow{\text{H}_3\text{O}^+} \underset{\overset{|}{\text{R}}}{\overset{+}{\text{H}_3}\text{NC(CO}_2\text{H})_2} \xrightarrow[-\text{CO}_2]{\text{heat}} \underset{^+\text{NH}_3}{\overset{|}{\text{RCHCO}_2^-}}$$

Amino Acid Reactions (Sec. 27.5)

Amide Formation:

$$\underset{\overset{|}{\text{R}}}{\overset{+}{\text{H}_3}\text{NCHCO}_2^-} + \underset{O\ \ \ O}{\overset{\parallel\ \ \ \parallel}{\text{CH}_3\text{COCCH}_3}} \longrightarrow \underset{\overset{|}{\text{R}}}{\underset{O}{\overset{\parallel}{\text{CH}_3\text{C}}}\text{NHCHCO}_2\text{H}} + \text{CH}_3\text{CO}_2\text{H}$$

Esterification:

$$\underset{^+\text{NH}_3}{\overset{|}{\text{RCHCO}_2^-}} + \text{CH}_3\text{CH}_2\text{OH} \xrightarrow{\text{H}^+} \underset{^+\text{NH}_3}{\overset{|}{\text{RCHCO}_2\text{CH}_2\text{CH}_3}}$$

Peptide Synthesis (Secs. 27.13 to 27.16)

General Strategy:

$$\underset{\overset{|}{\text{R}}}{\underset{O}{\overset{\parallel}{\text{XNHCHC}}}\text{OH}} + \underset{\overset{|}{\text{R}'}}{\text{H}_2\text{N}\underset{O}{\overset{\parallel}{\text{CHC}}}-\text{Y}} \xrightarrow[\text{2. deprotect}]{\text{1. couple}} \underset{\overset{|}{\text{R}}}{\overset{+}{\text{H}_3}\text{N}\underset{O}{\overset{\parallel}{\text{CHC}}}}-\underset{\overset{|}{\text{R}'}}{\text{NHCHCO}_2^-}$$

Amino Group Protection:

Benzyloxycarbonyl group

$$\text{C}_6\text{H}_5\text{CH}_2\text{O}\underset{O}{\overset{\parallel}{\text{C}}}\text{Cl} + \underset{\overset{|}{\text{R}}}{\overset{+}{\text{H}_3}\text{NCHCO}_2^-} \xrightarrow[\text{2. H}^+]{\text{1. NaOH, H}_2\text{O}} \text{C}_6\text{H}_5\text{CH}_2\text{O}\underset{O}{\overset{\parallel}{\text{C}}}\text{NHCHCO}_2\text{H}$$

$$\underset{\overset{|}{\text{R}}}{(\text{Z}-\text{NHCHCO}_2\text{H})}$$

tert-Butoxycarbonyl (Boc) group

$$(\text{CH}_3)_3\text{CO}\underset{O}{\overset{\parallel}{\text{C}}}\text{Cl} + \underset{\overset{|}{\text{R}}}{\overset{+}{\text{H}_3}\text{NCHCO}_2^-} \xrightarrow[\text{2. H}^+]{\text{1. NaOH, H}_2\text{O}} (\text{CH}_3)_3\text{CO}\underset{O}{\overset{\parallel}{\text{C}}}\text{NHCHCO}_2\text{H}$$

$$\underset{\overset{|}{\text{R}}}{(\text{Boc}-\text{NHCHCO}_2\text{H})}$$

Deprotection

$$\text{Z—NHR} \xrightarrow{\text{H}_2,\text{ Pd}} \text{H}_2\text{NR}$$

$$\text{Boc—NHR} \xrightarrow{\text{CF}_3\text{CO}_2\text{H}} \text{H}_3\overset{+}{\text{N}}\text{R}$$

Carboxyl Group Protection:

Esterification

$$\text{H}_3\overset{+}{\text{N}}\text{CHCO}_2^- + \text{C}_6\text{H}_5\text{CH}_2\text{OH} \xrightarrow{\text{H}^+} \text{H}_3\overset{+}{\text{N}}\text{CHCO}_2\text{CH}_2\text{C}_6\text{H}_5$$
$$\underset{\text{R}}{|} \qquad\qquad\qquad\qquad\qquad\qquad\qquad \underset{\text{R}}{|}$$

Deprotection

$$\text{RCOCH}_2\text{C}_6\text{H}_5 \xrightarrow{\text{H}_2,\text{ Pd}} \text{RCO}_2\text{H}$$

Peptide Bond Formation:

DCCI method

$$\underset{\text{R}}{\text{ZNHCHCO}_2\text{H}} + \underset{\text{R}'}{\text{H}_2\text{NCHCO}_2\text{CH}_2\text{C}_6\text{H}_5} \xrightarrow{\text{DCCI}} \underset{\text{R}}{\text{ZNHCHC}}\overset{\text{O}}{\|}\text{—NHCHCO}_2\text{CH}_2\text{C}_6\text{H}_5$$

where DCCI is ◯—N=C=N—◯

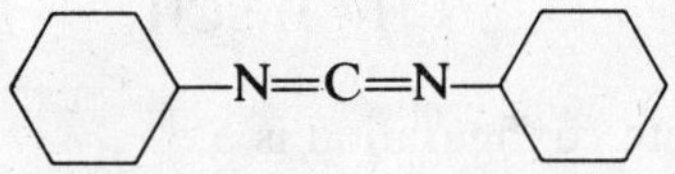

Active ester method

$$\underset{\text{R}}{\text{ZNHCHC}}\overset{\text{O}}{\|}\text{O}\text{—◯—NO}_2 + \underset{\text{R}'}{\text{H}_2\text{NCHCO}_2\text{CH}_2\text{CH}_3} \longrightarrow$$

$$\underset{\text{R}}{\text{ZNHCHC}}\overset{\text{O}}{\|}\text{—NHCHCO}_2\text{CH}_2\text{CH}_3 + \text{HO—◯—NO}_2$$

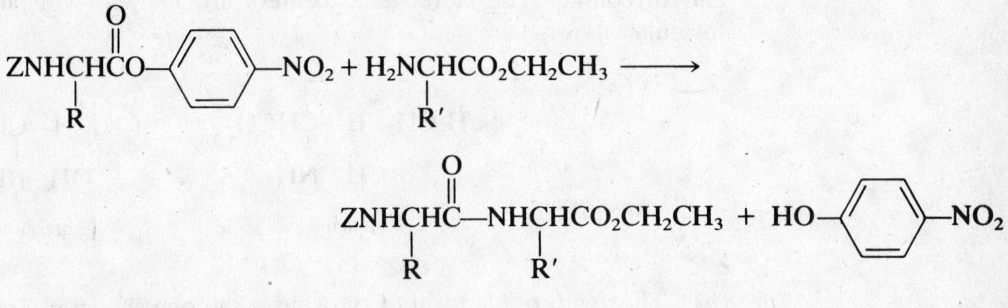

SOLUTIONS TO TEXT PROBLEMS

27.1 (*b*) L-Cysteine is the only amino acid in Table 27.1 that has the *R* configuration at its stereogenic center.

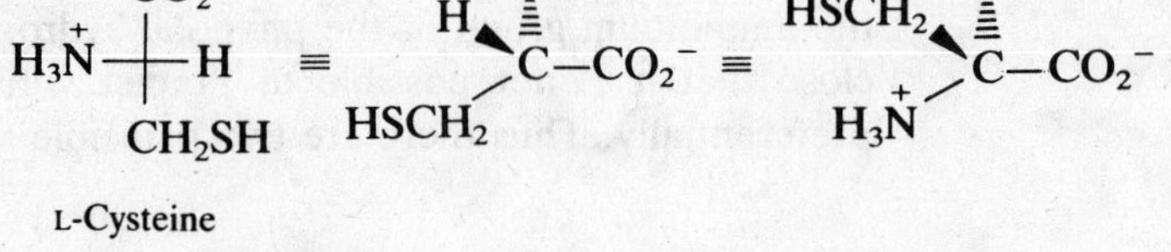

L-Cysteine

The order of decreasing sequence rule precedence is:

$$\text{H}_3\overset{+}{\text{N}}\text{— } > \text{ HSCH}_2\text{— } > \text{ —CO}_2^- > \text{ H—}$$

When the molecule is oriented so that the lowest-ranked substituent (H) is held away from us, the order of decreasing precedence traces a clockwise path.

CH₂SH

H₃N⁺ CO₂⁻   Clockwise; therefore *R*

The reason why L-cysteine has the *R* configuration while all the other L-amino acids have the *S* configuration lies in the fact that the —CH₂SH substituent is the only side chain that outranks —CO₂⁻ according to the sequence rule. Remember, rank order is determined by atomic number at the first point of difference, and —C—S outranks —C—O. In all the other amino acids —CO₂⁻ outranks the substituent at the center. The reversal in the Cahn-Ingold-Prelog descriptor comes not from any change in the spatial arrangement of substituents at the stereogenic center but rather from a reversal in the relative ranks of the carboxylate group and the side chain.

(c) The order of decreasing sequence rule precedence in L-methionine is:

$$H_3\overset{+}{N}— > —CO_2^- > —CH_2CH_2SCH_3 > H—$$

Sulfur is one atom further removed from the chiral center, and so C—O outranks C—C—S.

CO₂⁻

H₃N⁺ —+— H   ≡   CH₃SCH₂CH₂—C—CO₂⁻ (NH₃⁺, H)

CH₂CH₂SCH₃

The absolute configuration is *S*.

27.2 The amino acids in Table 27.1 that have more than one stereogenic center are isoleucine and threonine. The stereogenic centers are marked with an asterisk in the structural formulas shown.

CH₃CH₂CH*—*CHCO₂⁻ (CH₃, NH₃⁺) Isoleucine

CH₃CH*—*CHCO₂⁻ (OH, NH₃⁺) Threonine

27.3 (b) The zwitterionic form of tyrosine is the one shown in Table 27.1.

HO—⟨C₆H₄⟩—CH₂CHCO₂⁻ (NH₃⁺)

(c) As base is added to the zwitterion, a proton is removed from either of two positions, the ammonium group or the phenolic hydroxyl. The acidities of the two sites are so close that it is not possible to predict with certainty which one is deprotonated preferentially. Thus there are two plausible structures for the monoanion:

HO—⟨C₆H₄⟩—CH₂CHCO₂⁻ (NH₂)   and   ⁻O—⟨C₆H₄⟩—CH₂CHCO₂⁻ (NH₃⁺)

In fact, the proton on nitrogen is slightly more acidic than the phenolic hydroxyl, as measured by the pK_a values of the following model compounds:

pK_a 9.75

$$HO-\!\!\!\bigcirc\!\!\!-CH_2CHCO_2^- \qquad CH_3O-\!\!\!\bigcirc\!\!\!-CH_2CHCO_2^-$$

$$\overset{|}{\underset{+}{N(CH_3)_3}} \qquad\qquad\qquad \overset{|}{\underset{+}{NH_3}}$$

pK_a 9.27

(*d*) On further treatment with base, both the monoanions in part (*c*) yield the same dianion.

$$^-O-\!\!\!\bigcirc\!\!\!-CH_2CHCO_2^-$$

$$\overset{|}{NH_2}$$

27.4 At pH 1 the carboxylate oxygen and both nitrogens of lysine are protonated.

$$\overset{+}{H_3}NCH_2CH_2CH_2CH_2CHCO_2H \qquad \text{(Principal form at pH 1)}$$

$$\overset{|}{\underset{+}{NH_3}}$$

As the pH is raised, the carboxyl proton is removed first.

$$\overset{+}{H_3}NCH_2CH_2CH_2CH_2CHCO_2H + HO^- \longrightarrow \overset{+}{H_3}NCH_2CH_2CH_2CH_2CHCO_2^- + H_2O$$

$$\overset{|}{\underset{+}{NH_3}} \qquad\qquad\qquad\qquad\qquad \overset{|}{\underset{+}{NH_3}}$$

The pK_a value for the first ionization of lysine is 2.18 (from Table 27.3), and so this process is virtually complete when the pH is greater than this value.

The second pK_a value for lysine is 8.95. This is a fairly typical value for the second pK_a of amino acids and likely corresponds to proton removal from the nitrogen on the α carbon. The species that results is the predominant one at pH 9:

$$\overset{+}{H_3}NCH_2CH_2CH_2CH_2CHCO_2^- + HO^- \longrightarrow \overset{+}{H_3}NCH_2CH_2CH_2CH_2CHCO_2^- + H_2O$$

$$\overset{|}{\underset{+}{NH_3}} \qquad\qquad\qquad\qquad\qquad \overset{|}{NH_2}$$

$$\text{(Principal form at pH 9)}$$

The pK_a value for the third ionization of lysine is 10.53. This value is fairly high compared with those of most of the amino acids in Tables 27.1 to 27.3 and suggests that this proton is removed from the nitrogen of the side chain. The species that results is the major species present at pH values greater than 10.53.

$$\overset{+}{H_3}NCH_2CH_2CH_2CH_2CHCO_2^- + HO^- \longrightarrow H_2NCH_2CH_2CH_2CH_2CHCO_2^-$$

$$\overset{|}{NH_2} \qquad\qquad\qquad\qquad\qquad \overset{|}{NH_2}$$

$$\text{(Principal form at pH 13)}$$

27.5 In order to convert 3-methylbutanoic acid to valine, a leaving group must be introduced at the α carbon prior to displacement by ammonia. This is best accomplished by bromination under the conditions of the Hell-Volhard-Zelinsky reaction.

$$(CH_3)_2CHCH_2CO_2H \xrightarrow[\text{or } Br_2, PCl_3]{Br_2, P} (CH_3)_2CHCHCO_2H \xrightarrow{NH_3} (CH_3)_2CHCHCO_2^-$$

$$\underset{\text{Br}}{|} \qquad\qquad \underset{\underset{+}{NH_3}}{|}$$

3-Methylbutanoic acid	2-Bromo-3-methylbutanoic acid	Valine

Valine has been prepared by this method. The Hell-Volhard-Zelinsky reaction was carried out in 88 percent yield, but reaction of the α-bromo acid with ammonia was not very efficient, valine being isolated in only 48 percent yield in this step.

27.6 In the Strecker synthesis an aldehyde is treated with ammonia and a source of cyanide ion. The resulting amino nitrile is hydrolyzed to an amino acid.

$$(CH_3)_2CHC\overset{O}{\overset{\|}{H}} \xrightarrow[HCN]{NH_3} (CH_3)_2CHCHC\equiv N \xrightarrow[\text{2. } HO^-]{\text{1. } H_3O^+} (CH_3)_2CHCHCO_2^-$$

$$\underset{NH_2}{|} \qquad\qquad \underset{\underset{+}{NH_3}}{|}$$

2-Methylpropanal	2-Amino-3-methylbutanenitrile	Valine

As actually carried out, the aldehyde was converted to the amino nitrile by treatment with an aqueous solution containing ammonium chloride and potassium cyanide. Hydrolysis was achieved in aqueous hydrochloric acid and gave valine as its hydrochloride salt in 65 percent overall yield.

27.7 The alkyl halide with which the anion of diethyl acetamidomalonate is treated is 2-bromopropane.

$$CH_3\overset{O}{\overset{\|}{C}}NHCH(CO_2CH_2CH_3)_2 + (CH_3)_2CHBr \xrightarrow[CH_3CH_2OH]{NaOCH_2CH_3} CH_3\overset{O}{\overset{\|}{C}}NHC(CO_2CH_2CH_3)_2$$

$$\underset{CH(CH_3)_2}{|}$$

Diethyl acetamidomalonate	2-Bromopropane	Diethyl acetamidoisopropylmalonate

This is the difficult step in the synthesis; it requires a nucleophilic substitution reaction of the S_N2 type involving a secondary alkyl halide. Competition of elimination with substitution results in only a 37 percent observed yield of alkylated diethyl acetamidomalonate.

Hydrolysis and decarboxylation of the alkylated derivative are straightforward and proceed in 85 percent yield to give valine.

$$CH_3\overset{O}{\overset{\|}{C}}NHC(CO_2CH_2CH_3)_2 \xrightarrow[\text{heat}]{HBr, H_2O} H_3\overset{+}{N}C(CO_2H)_2 \xrightarrow[-CO_2]{\text{heat}} H_3\overset{+}{N}CHCO_2^-$$

$$\underset{CH(CH_3)_2}{|} \qquad\qquad \underset{CH(CH_3)_2}{|} \qquad\qquad \underset{CH(CH_3)_2}{|}$$

Diethyl acetamidoisopropylmalonate	2-Aminoisopropylmalonic acid	Valine

27.8 Ninhydrin is the hydrate of a triketone and is in equilibrium with it.

Hydrated form of ninhydrin	Triketo form of ninhydrin

An amino acid reacts with this triketone to form an imine.

$$\text{Triketo form of ninhydrin} \quad + \quad \underset{\underset{+}{NH_3}}{RCHCO_2^-} \quad \xrightarrow{HO^-} \quad \text{Imine}$$

Triketo form of ninhydrin   α-Amino acid   Imine

This imine then undergoes decarboxylation.

$$ \longrightarrow \quad + \quad CO_2$$

The anion that results from the decarboxylation step is then protonated. The product is shown as its diketo form but probably exists as an enol.

$$ + \; H_2O \longrightarrow \quad + \; {}^-OH$$

Hydrolysis of the imine function gives an aldehyde and a compound having a free amino group.

$$ \text{N=CHR} \; + \; H_2O \longrightarrow \quad NH_2 \; + \; RCH$$

This amine then reacts with a second molecule of the triketo form of ninhydrin to give an imine.

$$ NH_2 \; + \; O \longrightarrow $$

Proton abstraction from the neutral imine gives its conjugate base, which is a violet dye.

$$ \longrightarrow \quad + \; H_2O$$

27.9 The carbon that bears the amino group of 4-aminobutanoic acid corresponds to the α carbon of an α-amino acid.

$$CH_2CH_2CH_2CO_2^- \qquad \text{arises by decarboxylation of} \qquad {}^-O_2CCHCH_2CH_2CO_2^-$$

$$|\quad\quad\quad\quad\quad\quad\quad\quad\quad\quad\quad\quad\quad\quad\quad\quad\quad\quad\quad |$$

$$_+NH_3 \qquad\qquad\qquad\qquad\qquad\qquad\qquad\qquad\qquad\qquad\qquad _+NH_3$$

4-Aminobutanoic acid   Glutamic acid

27.10 (*b*) Alanine is the N-terminal amino acid in Ala-Phe. Its carboxyl group is joined to the nitrogen of phenylalanine by a peptide bond.

$$\overset{\displaystyle O}{\overset{\displaystyle \|}{H_3\overset{+}{N}CHC}}\text{—NHCHCO}_2^- \qquad \text{Ala-Phe}$$

$$\quad\quad | \quad\quad\quad\quad\quad\quad |$$

$$\quad\quad CH_3 \quad,\quad\quad CH_2C_6H_5$$

Alanine   Phenylalanine

(*c*) The positions of the amino acids are reversed in Phe-Ala. Phenylalanine is the N terminus and alanine is the C terminus.

$$\overset{\displaystyle O}{\overset{\displaystyle \|}{H_3\overset{+}{N}CHC}}\text{—NHCHCO}_2^- \qquad \text{Phe-Ala}$$

$$C_6H_5CH_2 \quad,\quad\quad CH_3$$

Phenylalanine   Alanine

(*d*) The carboxyl group of glycine is joined by a peptide bond to the amino group of glutamic acid.

$$\overset{\displaystyle O}{\overset{\displaystyle \|}{H_3\overset{+}{N}CH_2C}}\text{—NHCHCO}_2^- \qquad \text{Gly-Glu}$$

$$\quad\quad\quad\quad\quad\quad\quad\quad\quad | $$

$$\quad\quad\quad\quad\quad\quad\quad\quad CH_2CH_2CO_2^-$$

Glycine   Glutamic acid

The dipeptide is written in its anionic form because the carboxyl group of the side chain is ionized at pH 7. Alternatively, it could have been written as a neutral zwitterion with a $CH_2CH_2CO_2H$ side chain.

(*e*) The peptide bond in Lys-Gly is between the carboxyl group of lysine and the amino group of glycine.

$$\overset{\displaystyle O}{\overset{\displaystyle \|}{H_3\overset{+}{N}CHC}}\text{—NHCH}_2CO_2^- \qquad \text{Lys-Gly}$$

$$H_3\overset{+}{N}CH_2CH_2CH_2CH_2$$

Lysine   Glycine

The amino group of the lysine side chain is protonated at pH 7, and so the dipeptide is written here in its cationic form. It could have also been written as a neutral zwitterion with the side chain $H_2NCH_2CH_2CH_2CH_2$.

(*f*) Both amino acids are alanine in D-Ala-D-Ala. The fact that they have the D configuration has no effect on the constitution of the dipeptide.

$$
\overset{+}{H_3N}CH\overset{\displaystyle O}{\overset{\|}{C}}-NHCHCO_2^- \qquad \text{D-Ala-D-Ala}
$$

$$CH_3 \qquad CH_3$$

Alanine   Alanine

27.11 (*b*) When amino acid residues in a dipeptide are indicated without a prefix, it is assumed that the configuration at the α carbon atom is L. Therefore, the stereochemistry of Ala-Phe may be indicated for the zigzag conformation as shown.

The L configuration corresponds to S for each of the stereogenic centers in Ala-Phe.

(*c*) Similarly, Phe-Ala has its substituent at the N-terminal amino acid directed away from us while the C-terminal side chain is pointing toward us, and the L configuration corresponds to S for each stereogenic center.

(*d*) There is only one stereogenic center in Gly-Glu. It has the L (or S) configuration.

(*e*) In order for the N-terminal amino acid in Lys-Gly to have the L (or S) configuration, its side chain must be directed away from us in the conformation indicated.

(*f*) The configuration at both α carbon atoms in D-Ala-D-Ala is exactly the reverse of the configuration of the stereogenic centers in parts (*a*) through (*e*). Both stereogenic centers have the D (or R) configuration.

27.12 Figure 27.6 in the text gives the structure of leucine enkephalin. Methionine enkephalin differs from it only with respect to the C-terminal amino acid. The amino acid sequences of the two pentapeptides are:

Tyr-Gly-Gly-Phe-Leu   Tyr-Gly-Gly-Phe-Met

Leucine enkephalin   Methionine enkephalin

27.13 Unless a peptide is hydrolyzed completely so that all the amide bonds are cleaved, a mixture of smaller fragments is obtained corresponding to all possible cleavage modes.

Val-Phe-Gly-Ala

↓ Hydrolysis

Tripeptides:	Val-Phe-Gly	Phe-Gly-Ala	
Dipeptides:	Val-Phe	Phe-Gly	Gly-Ala
Amino acids:	Val	Phe	Gly	Ala

27.14 The Edman degradation removes the N-terminal amino acid, which is identified as a phenylthiohydantoin derivative. The first Edman degradation of Val-Phe-Gly-Ala gives the phenylthiohydantoin derived from valine; the second gives the phenylthiohydantoin derived from phenylalanine.

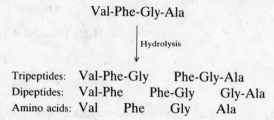

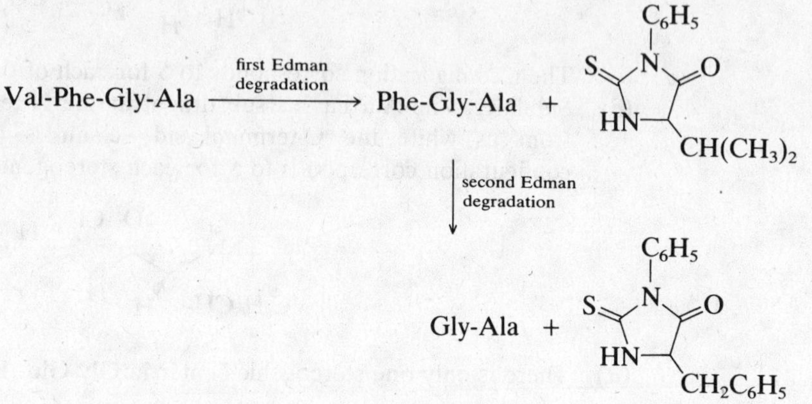

Val-Phe-Gly-Ala $\xrightarrow{\text{first Edman degradation}}$ Phe-Gly-Ala + [PTH derivative with C₆H₅, S, N, O, HN, CH(CH₃)₂]

↓ second Edman degradation

Gly-Ala + [PTH derivative with C₆H₅, S, N, O, HN, CH₂C₆H₅]

(handwritten: PTH Derivative)

27.15 The dipeptide Asp-Phe has the structure shown. Aspartame is its C-terminal methyl ester.

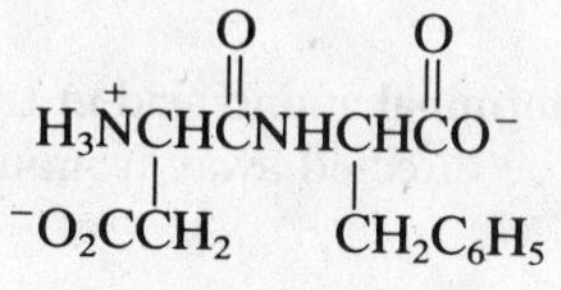

$\overset{+}{H_3}NCHCNHCHCO^-$ ... $\overset{+}{H_3}NCHCNHCHCOCH_3$

(with $^-O_2CCH_2$ and $CH_2C_6H_5$ groups)

Asp-Phe 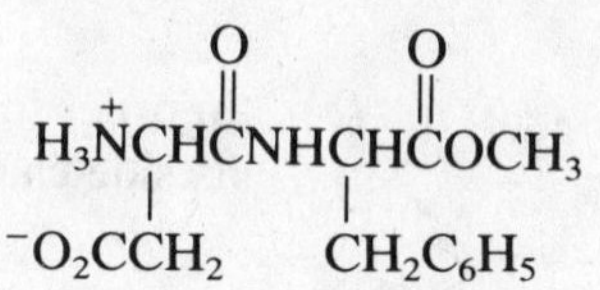   Aspartame

27.16 Lysine has two amino groups. Both amino functions are converted to amides on reaction with benzyloxycarbonyl chloride.

$H_2NCHCO_2^-$ + $2C_6H_5CH_2OCCl$ $\longrightarrow$

$H_2NCH_2CH_2CH_2CH_2$

(handwritten: Benzyloxy carbonyl chloride)

$C_6H_5CH_2OCNHCHCO_2H$

$C_6H_5CH_2OCNHCH_2CH_2CH_2CH_2$

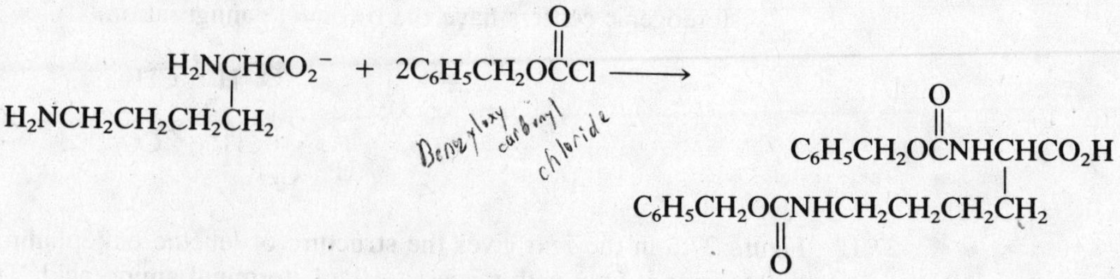

27.17 The peptide bond of Ala-Leu connects the carboxyl group of alanine and the amino group of leucine. Therefore we need to protect the amino group of alanine and the carboxyl group of leucine.

Protect the amino group of alanine as its benzyloxycarbonyl derivative:

$$\underset{\substack{| \\ CH_3 \\ \text{Alanine}}}{H_3\overset{+}{N}CHCO_2^-} + \underset{\text{Benzyloxycarbonyl chloride}}{C_6H_5CH_2O\overset{\displaystyle O}{\overset{\|}{C}}Cl} \longrightarrow \underset{\substack{| \\ CH_3 \\ \text{Z-Protected alanine}}}{C_6H_5CH_2O\overset{\displaystyle O}{\overset{\|}{C}}NHCHCO_2H}$$

Protect the carboxyl group of leucine as its benzyl ester:

$$\underset{\substack{| \\ (CH_3)_2CHCH_2 \\ \text{Leucine}}}{H_3\overset{+}{N}CHCO_2^-} + \underset{\text{Benzyl alcohol}}{C_6H_5CH_2OH} \xrightarrow[\text{2. HO}^-]{\text{1. H}^+\text{, heat}} \underset{\substack{| \\ (CH_3)_2CHCH_2 \\ \text{Leucine benzyl ester}}}{H_2NCHCO_2CH_2C_6H_5}$$

Coupling of the two amino acids is achieved by DCCI-promoted amide bond formation between the free amino group of leucine benzyl ester and the free carboxyl group of Z-protected alanine.

$$\underset{\substack{| \\ CH_3 \\ \text{Z-Protected alanine}}}{C_6H_5CH_2O\overset{\displaystyle O}{\overset{\|}{C}}NHCHCO_2H} + \underset{\substack{| \\ (CH_3)_2CHCH_2 \\ \text{Leucine benzyl ester}}}{H_2NCH\overset{\displaystyle O}{\overset{\|}{C}}OCH_2C_6H_5} \xrightarrow{\text{DCCI}}$$

$$\underset{\substack{\quad| \qquad\qquad | \\ \quad CH_3 \qquad\; CH_2CH(CH_3)_2 \\ \text{Protected dipeptide}}}{C_6H_5CH_2O\overset{\displaystyle O}{\overset{\|}{C}}NHCH\overset{\displaystyle O}{\overset{\|}{C}}NHCH\overset{\displaystyle O}{\overset{\|}{C}}OCH_2C_6H_5}$$

Both the benzyloxycarbonyl protecting group and the benzyl ester protecting group may be removed by hydrogenolysis over palladium. This step completes the synthesis of Ala-Leu.

$$\underset{\substack{\quad| \qquad\quad | \\ \quad CH_3 \quad CH_2CH(CH_3)_2 \\ \text{Protected dipeptide}}}{C_6H_5CH_2O\overset{\displaystyle O}{\overset{\|}{C}}NHCH\overset{\displaystyle O}{\overset{\|}{C}}NHCH\overset{\displaystyle O}{\overset{\|}{C}}OCH_2C_6H_5} \xrightarrow{\text{H}_2\text{, Pd}} \underset{\substack{\quad| \qquad\quad | \\ \quad CH_3 \quad CH_2CH(CH_3)_2 \\ \text{Ala-Leu}}}{H_3\overset{+}{N}CH\overset{\displaystyle O}{\overset{\|}{C}}NHCHCO_2^-}$$

27.18 As in the DCCI-promoted coupling of amino acids, the first step is the addition of the Z-protected amino acid to N,N'-dicyclohexylcarbodiimide (DCCI) to give an O-acylisourea.

$$\underset{\substack{| \\ R \\ \text{Z-Protected} \\ \text{amino acid}}}{ZNHCH\overset{\displaystyle O}{\overset{\|}{C}}OH} + \underset{\text{DCCI}}{C_6H_{11}N{=}C{=}NC_6H_{11}} \longrightarrow \underset{\substack{| \\ R \\ \text{O-Acylisourea}}}{ZNHCH\overset{\displaystyle O}{\overset{\|}{C}}O{-}C\overset{\displaystyle NC_6H_{11}}{\underset{\displaystyle NHC_6H_{11}}{}}}$$

This *O*-acylisourea is attacked by *p*-nitrophenol to give the *p*-nitrophenyl ester of the Z-protected amino acid.

27.19 In order to add a leucine residue to the N terminus of the ethyl ester of Z-Phe-Gly, the benzyloxycarbonyl protecting group must first be removed. This can be accomplished by hydrogenolysis.

Z-Protected ethyl ester of Phe-Gly   Phe-Gly ethyl ester

The reaction shown has been carried out in 100 percent yield. Alternatively, the benzyloxycarbonyl protecting group may be removed by treatment with hydrogen bromide in acetic acid. This latter route has also been reported in the chemical literature and gives the hydrobromide salt of Phe-Gly ethyl ester in 82 percent yield.

Once the protecting group has been removed, the ethyl ester of Phe-Gly is allowed to react with the *p*-nitrophenyl ester of Z-protected leucine to form the protected tripeptide. Hydrogenolysis of the Z-protected tripeptide gives Leu-Phe-Gly as its ethyl ester.

p-Nitrophenyl ester of   Phe-Gly ethyl ester
Z-protected leucine

Z-Protected Leu-Phe-Gly ethyl ester

Leu-Phe-Gly ethyl ester

27.20 Amino acid residues are added by beginning at the C terminus in the Merrifield solid-phase approach to peptide synthesis. Thus the synthesis of Phe-Gly requires glycine to be anchored to the solid support. Begin by protecting glycine as its Boc derivative.

$$(CH_3)_3COCOCl + H_3\overset{+}{N}CH_2CO_2^- \longrightarrow (CH_3)_3COCNHCH_2CO_2H$$

tert-Butoxycarbonyl   Glycine   Boc-Protected glycine
chloride

The protected gylcine is attached via its carboxylate anion to the solid support.

$$(CH_3)_3COCNHCH_2CO_2H \xrightarrow[\text{2. ClCH}_2-\text{resin}]{\text{1. HO}^-} (CH_3)_3COCNHCH_2COCH_2-\text{resin}$$

Boc-Protected glycine

The amino group of glycine is then exposed by removal of the protecting group. Typical conditions for this step involve treatment with hydrogen chloride in acetic acid.

$$(CH_3)_3COCNHCH_2COCH_2-\text{resin} \xrightarrow[\text{acetic acid}]{\text{HCl}} H_2NCH_2COCH_2-\text{resin}$$

Boc-Protected, resin-bound glycine   Resin-bound glycine

In order to attach phenylalanine to resin-bound glycine, we must first protect the amino group of phenylalanine. A Boc-protecting group is appropriate.

$$(CH_3)_3COCOCl + H_3\overset{+}{N}CHCO_2^- \longrightarrow (CH_3)_3COCNHCHCO_2H$$
$$\qquad\qquad\qquad\qquad CH_2C_6H_5 \qquad\qquad\qquad\qquad CH_2C_6H_5$$

tert-Butoxycarbonyl   Phenylalanine   Boc-Protected phenylalanine
chloride

Peptide bond formation occurs when the resin-bound glycine and Boc-protected phenylalanine are combined in the presence of DCCI.

$$(CH_3)_3COCNHCHCO_2H + H_2NCH_2COCH_2-\text{resin} \xrightarrow{\text{DCCI}}$$
$$\qquad CH_2C_6H_5$$

Boc-Protected phenylalanine   Resin-bound glycine

$$(CH_3)_3COCNHCHCNHCH_2COCH_2-\text{resin}$$
$$\qquad\qquad\qquad CH_2C_6H_5$$

Boc-Protected, resin-bound Phe-Gly

Remove the Boc group with HCl and then treat with HBr in trifluoroacetic acid to cleave Phe-Gly from the solid support.

$$(CH_3)_3COCNHCHCNHCH_2COCH_2-\text{resin} \xrightarrow[\text{2. HBr, trifluoroacetic acid}]{\text{1. HCl, acetic acid}} H_3\overset{+}{N}CHCNHCH_2CO_2^-$$
$$\qquad CH_2C_6H_5 \qquad\qquad\qquad\qquad\qquad\qquad\qquad\qquad CH_2C_6H_5$$

Boc-protected, resin-bound Phe-Gly   Phe-Gly

27.21   The numbering of the ring in uracil and its derivatives parallels that in pyrimidine.

Pyrimidine   Uracil   5-Fluorouracil

27.22   (*b*)   Cytidine is present in RNA and so is a nucleoside of D-ribose. The base is cytosine.

(*c*)   Guanosine is present in RNA and so is a guanine nucleoside of D-ribose.

27.23   Table 27.4 in the text lists the messenger RNA codons for the various amino acids. The codons for valine and for glutamic acid are:

Valine:	GUU	GUA	GUC	GUG
Glutamic acid:		GAA		GAG

As can be seen, the codons for glutamic acid (GAA and GAG) are very similar to two of the codons (GUA and GUG) for valine. Replacement of adenine in the glutamic acid codons by uracil causes valine to be incorporated into hemoglobin instead of glutamic acid and is responsible for the sickle cell trait.

27.24   The protonated form of imidazole represented by structure A is stabilized by delocalization of the lone pair of one of the nitrogens. The positive charge is shared by both nitrogens.

A

The positive charge in structure B is localized on a single nitrogen. Resonance stabilization of the type shown in structure A is not possible.

27.25 The following outlines a synthesis of β-alanine in which conjugate addition to acrylonitrile plays a key role:

$$CH_2{=}CHC{\equiv}N \xrightarrow{NH_3} H_2NCH_2CH_2C{\equiv}N \xrightarrow[\text{heat}]{H_2O,\ HO^-} H_3\overset{+}{N}CH_2CH_2CO_2^-$$

Acrylonitrile   3-Aminopropanenitrile   β-Alanine

Addition of ammonia to acrylonitrile has been carried out in modest yield (31 to 33 percent). Hydrolysis of the nitrile group can be accomplished in the presence of either acids or bases. Hydrolysis in the presence of $Ba(OH)_2$ has been reported in the literature to give β-alanine in 85 to 90 percent yield.

27.26 (*a*)   The first step involves alkylation of diethyl malonate by 2-bromobutane.

$$CH_3CH_2\overset{|}{\underset{Br}{C}}HCH_3 + :\bar{C}H(COOCH_2CH_3)_2 \longrightarrow CH_3CH_2\overset{|}{\underset{CH_3}{C}}H{-}CH(COOCH_2CH_3)_2$$

2-Bromobutane   Anion of diethyl malonate   Compound C

In the second step of the synthesis, compound C is subjected to ester saponification. Following acidification, the corresponding diacid (compound D) is isolated.

$$CH_3CH_2\overset{|}{\underset{CH_3}{C}}HCH(COOCH_2CH_3)_2 \xrightarrow[\text{2. HCl}]{\text{1. KOH}} CH_3CH_2\overset{|}{\underset{CH_3}{C}}HCH(COOH)_2$$

Compound C   Compound D ($C_7H_{12}O_4$)

Compound D is readily brominated at its α carbon atom by way of the corresponding enol form.

Compound D   Enol form   Compound E ($C_7H_{11}BrO_4$)

When compound E is heated, it undergoes decarboxylation to give an α-bromo carboxylic acid.

$$CH_3CH_2\overset{|}{\underset{CH_3}{C}}H{-}\overset{|}{\underset{Br}{C}}(COOH)_2 \xrightarrow{\text{heat}} CH_3CH_2\overset{|}{\underset{CH_3}{C}}H{-}\overset{|}{\underset{Br}{C}}HCOOH + CO_2$$

Compound E   Compound F   Carbon dioxide

Treatment of compound F with ammonia converts it to isoleucine by nucleophilic substitution.

$$\underset{\substack{|\\CH_3\ \ Br}}{CH_3CH_2CH-CHCO_2H} + NH_3 \longrightarrow \underset{\substack{|\\CH_3\ \ \overset{+}{N}H_3}}{CH_3CH_2CH-CHCO_2^-}$$

Compound F   Isoleucine (racemic)

(b) The procedure just described can be adapted to the synthesis of other amino acids. The group attached to the α carbon atom is derived from the alkyl halide used to alkylate diethyl malonate. Benzyl bromide (or chloride or iodide) would be appropriate for the preparation of phenylalanine.

$$C_6H_5CH_2Br \longrightarrow \underset{\substack{|\\\overset{+}{N}H_3}}{C_6H_5CH_2CHCO_2^-}$$

Benzyl bromide   Phenylalanine (racemic)

27.27 Acid hydrolysis of compound G converts all its ester functions to free carboxyl groups and cleaves both amide bonds.

Compound G   Water

The hydrolysis product is a substituted derivative of malonic acid and undergoes decarboxylation on being heated. The product of this decarboxylation is aspartic acid (in its protonated form under conditions of acid hydrolysis).

$$\underset{\substack{|\\\overset{+}{H_3N}-C(COOH)_2}}{CH_2COOH} \xrightarrow{\text{heat}} \underset{\substack{|\\\overset{+}{H_3N}-CHCOOH}}{CH_2COOH} + CO_2$$

                                  Aspartic acid   Carbon dioxide

Aspartic acid is chiral, but is formed as a racemic mixture, so that the product of this reaction is not optically active. The starting material (compound G) is achiral and cannot give an optically active product when it reacts with optically inactive reagents.

27.28 The amino acids leucine, phenylalanine, and serine each have one stereogenic center.

$$\underset{\substack{|\\+NH_3}}{RCHCO_2^-}$$

Leucine: $R— = (CH_3)_2CHCH_2—$
Phenylalanine: $R— = C_6H_5CH_2—$
Serine: $R— = HOCH_2—$

When prepared by the Strecker synthesis, each of these amino acids is obtained as a racemic mixture containing 50 percent of the D enantiomer and 50 percent of the L enantiomer.

$$\overset{\displaystyle O}{\overset{\|}{RCH}} + NH_3 + HCN \longrightarrow \underset{\substack{|\\NH_2}}{RCHC\equiv N} \xrightarrow{H_2O} \underset{\substack{|\\+NH_3}}{RCHCO_2^-}$$

Thus, preparation of the tripeptide Leu-Phe-Ser will yield a mixture of eight stereoisomers.

D-Leu-D-Phe-D-Ser   L-Leu-L-Phe-L-Ser
D-Leu-D-Phe-L-Ser   L-Leu-L-Phe-D-Ser
D-Leu-L-Phe-D-Ser   L-Leu-D-Phe-L-Ser
D-Leu-L-Phe-L-Ser   L-Leu-D-Phe-D-Ser

27.29 Bradykinin is a nonapeptide but contains only five different amino acids. Three of the amino acid residues are proline, two are arginine, and two are phenylalanine. There will be five peaks on the strip chart after amino acid analysis of bradykinin.

$$\text{Arg-Pro-Pro-Gly-Phe-Ser-Pro-Phe-Arg} \longrightarrow 2\text{Arg} + 3\text{Pro} + \text{Gly} + 2\text{Phe} + \text{Ser}$$

27.30 Asparagine and glutamine each contain an amide function in their side chain. Under the conditions of peptide bond hydrolysis that characterize amino acid analysis, the side-chain amide is also hydrolyzed, giving ammonia.

27.31 (*a*) 1-Fluoro-2,4-dinitrobenzene reacts with the amino group of the N-terminal amino acid in a nucleophilic aromatic substitution reaction of the addition-elimination type.

DNP-Leu-Gly-Ser

(*b*) Hydrolysis of the product in part (*a*) cleaves the peptide bonds. Leucine is isolated as its 2,4-dinitrophenyl (DNP) derivative, but glycine and serine are isolated as the free amino acids.

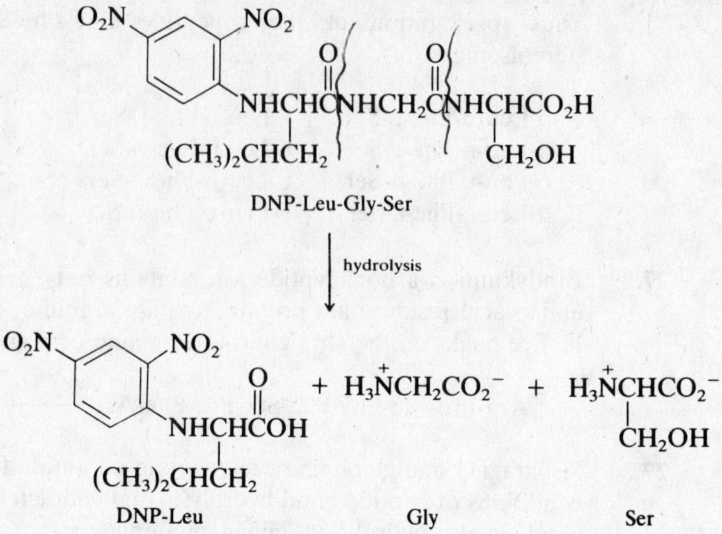

DNP-Leu-Gly-Ser

hydrolysis

DNP-Leu   Gly   Ser

(c) Dansyl chloride is a reagent used for labeling the N-terminal amino acid of a peptide. It is a sulfonyl chloride and reacts with the free amino group to give a sulfonamide.

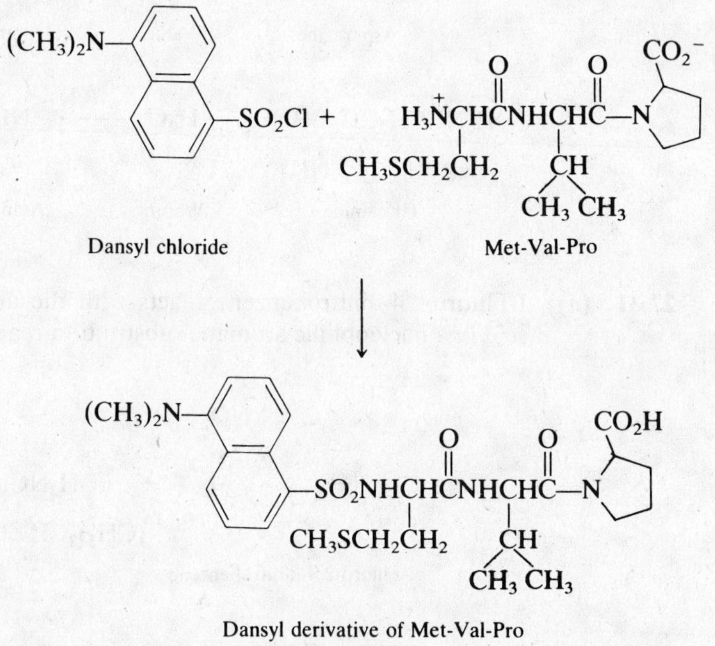

Dansyl chloride   Met-Val-Pro

Dansyl derivative of Met-Val-Pro

Hydrolysis of the dansyl derivative of Met-Val-Pro gives the dansyl derivative of methionine along with the free amino acids valine and proline.

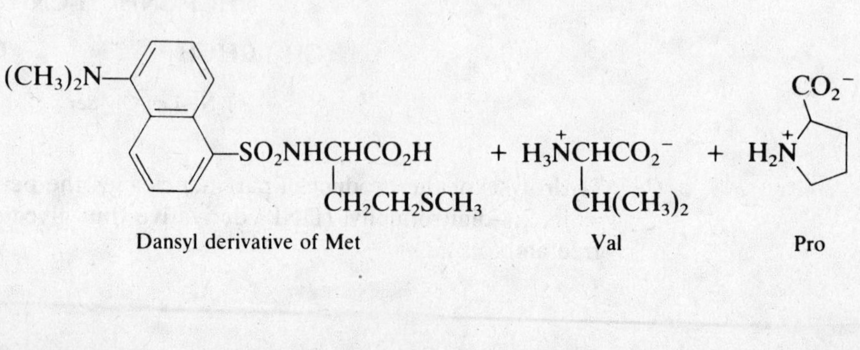

Dansyl derivative of Met   Val   Pro

(d) Phenyl isothiocyanate is a reagent used to identify the N-terminal amino acid of a peptide by the Edman degradation. The N-terminal amino acid is cleaved as a phenylthiohydantoin (PTH) derivative, the remainder of the peptide remaining intact.

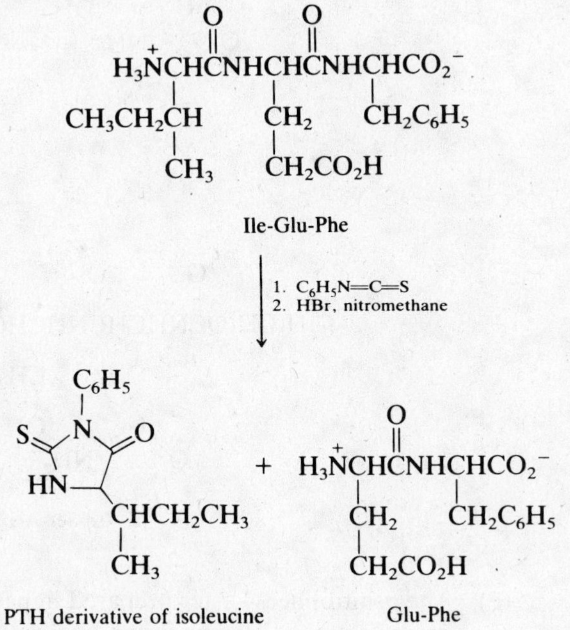

$$H_3\overset{+}{N}CHCNHCHCNHCHCO_2^{-}$$

$$CH_3CH_2CH \qquad CH_2 \qquad CH_2C_6H_5$$

$$CH_3 \qquad CH_2CO_2H$$

Ile-Glu-Phe

1. $C_6H_5N{=}C{=}S$
2. HBr, nitromethane

PTH derivative of isoleucine + Glu-Phe

(e) Benzyloxycarbonyl chloride reacts with amino groups to convert them to amides. The only free amino group in Asn-Ser-Ala is the N terminus. The amide function of asparagine does not react with benzyloxycarbonyl chloride.

$$H_3\overset{+}{N}CHCNHCHCNHCHCO_2^{-} \; + \; C_6H_5CH_2OCCl \longrightarrow$$

Asn-Ser-Ala + Benzyloxycarbonyl chloride

$$C_6H_5CH_2OCNHCHCNHCHCNHCHCO_2H$$

Z-Asn-Ser-Ala

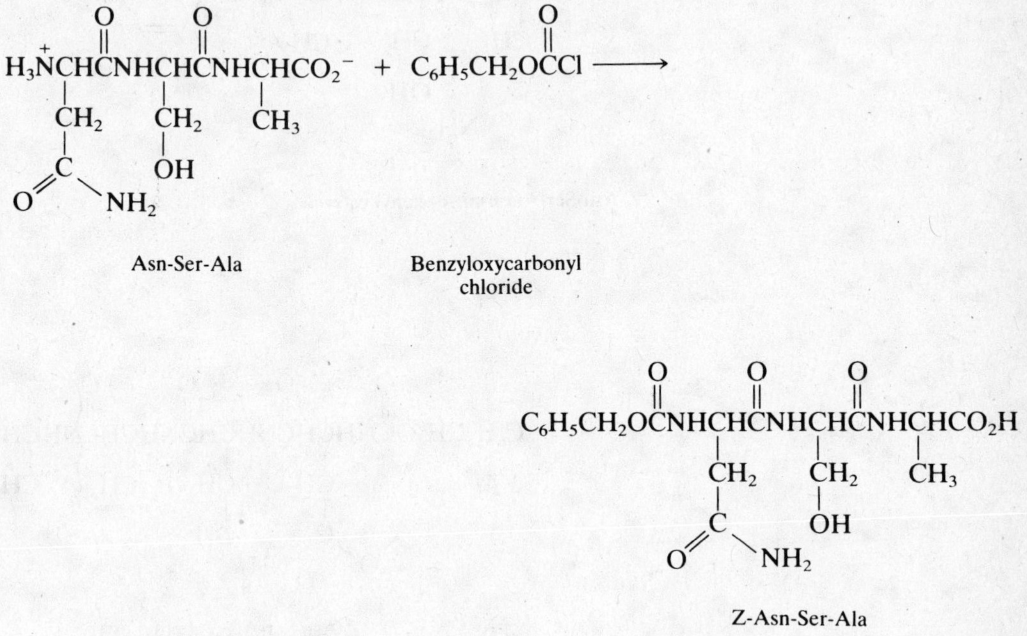

(f) The Z-protected tripeptide formed in part (e) is converted to its C-terminal p-nitrophenyl ester on reaction with p-nitrophenol and N,N'-dicyclohexylcarbodiimide (DCCI).

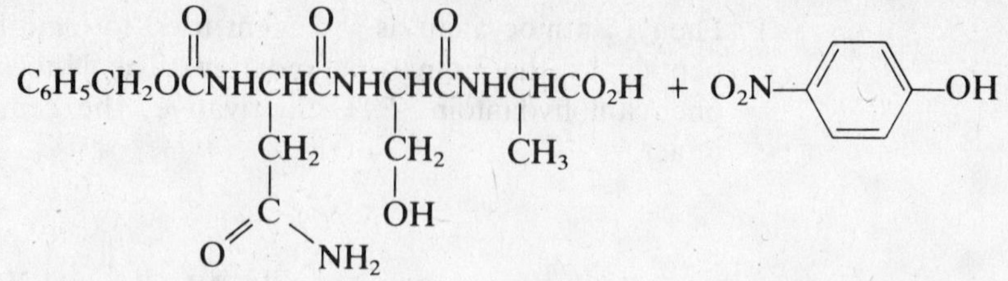

Z-Asn-Ser-Ala   *p*-Nitrophenol

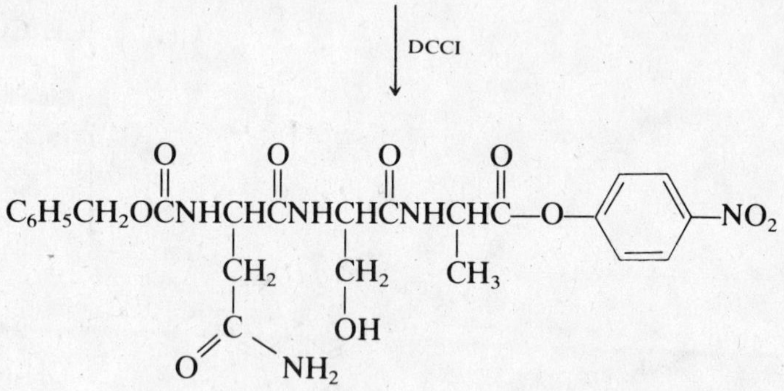

↓ DCCI

Z-Asn-Ser-Ala *p*-nitrophenyl ester

(g) The *p*-nitrophenyl ester prepared in part (*f*) is an "active" ester. The *p*-nitrophenyl group is a good leaving group and can be displaced by the amino nitrogen of valine ethyl ester to form a new peptide bond.

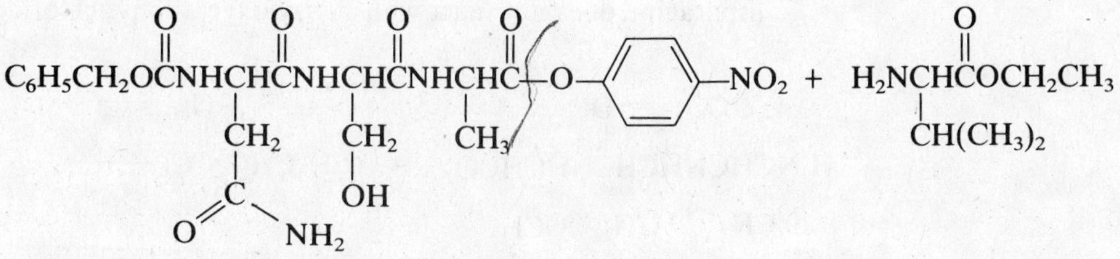

Z-Asn-Ser-Ala *p*-nitrophenyl ester   Valine ethyl ester

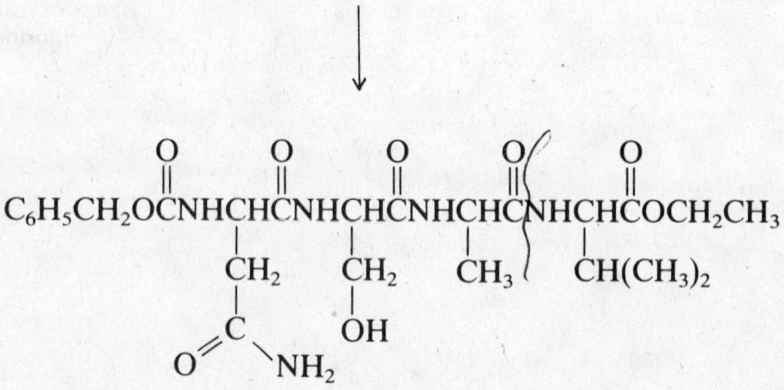

↓

Z-Asn-Ser-Ala-Val ethyl ester

(h) Hydrogenolysis of the Z-protected tetrapeptide ester formed in part (*g*) removes the Z-protecting group.

$$\underset{\substack{| \\ CH_2 \\ | \\ \underset{O}{\overset{\displaystyle \|}{C}}\!\!\!\!-\!\!NH_2}}{}$$

C₆H₅CH₂OCNHCHCNHCHCNHCHCNHCHCOCH₂CH₃ — with carbonyls shown as O above each C:

$$\underset{\substack{CH_2 \\ \| \\ O\!=\!C\!-\!NH_2}}{}\quad \underset{\substack{CH_2 \\ | \\ OH}}{}\quad \underset{CH_3}{}\quad \underset{CH(CH_3)_2}{}$$

Z-Asn-Ser-Ala-Val ethyl ester

↓ H₂, Pd

H₂NCHCNHCHCNHCHCNHCHCOCH₂CH₃

$$\underset{\substack{CH_2 \\ \| \\ O\!=\!C\!-\!NH_2}}{}\quad \underset{\substack{CH_2 \\ | \\ OH}}{}\quad \underset{CH_3}{}\quad \underset{CH(CH_3)_2}{}$$

Asn-Ser-Ala-Val ethyl ester

27.32 Consider, for example, the reaction of hydrazine with a very simple dipeptide such as Gly-Ala. Hydrazine cleaves the peptide by nucleophilic attack on the carbonyl group of glycine.

$$\underset{\text{Gly-Ala}}{\overset{+}{H_3}NCH_2\overset{\displaystyle O}{\overset{\|}{C}}\!\!-\!\!NHCHCO_2^-} + \underset{\text{Hydrazine}}{H_2\ddot{N}NH_2} \longrightarrow \underset{\text{Hydrazide of glycine}}{H_2NCH_2\overset{\displaystyle O}{\overset{\|}{C}}\!\!-\!\!NHNH_2} + \underset{\text{Alanine}}{\overset{+}{H_3}\overset{\displaystyle}{N}CHCO_2^-}$$

with CH₃ substituents on the CH carbons.

It is the C-terminal residue that is cleaved as the free amino acid and identified in the hydrazinolysis of peptides.

27.33 Somatostatin is a tetradecapeptide and so is composed of 14 amino acids. The fact that Edman degradation gave the PTH derivative of alanine identifies this as the N-terminal amino acid. A major piece of information is the amino acid sequence of a hexapeptide obtained by partial hydrolysis:

Ala-Gly-Cys-Lys-Asn-Phe

Using this as a starting point and searching for overlaps with the other hydrolysis products gives the entire sequence.

Ala-Gly-Cys-Lys-Asn-Phe

Asn-Phe-Phe-Trp-Lys

Phe-Trp

Lys-Thr-Phe

Thr-Phe-Thr-Ser-Cys

Thr-Ser-Cys

Ala-Gly-Cys-Lys-Asn-Phe-Phe-Trp-Lys-Thr-Phe-Thr-Ser-Cys

1   2   3   4   5   6   7   8   9   10   11   12   13   14

The disulfide bridge in somatostatin is between cysteine 3 and cysteine 14. Thus the primary structure is:

$$\text{Lys-Asn-Phe-Phe-Trp-Lys}$$
$$\text{Ala-Gly-Cys}$$
$$\text{S-S-Cys-Ser-Thr-Phe-Thr}$$

27.34 It is the C-terminal amino acid that is anchored to the solid support in the preparation of peptides by the Merrifield method. Refer to the structure of oxytocin in Figure 27.7 of the text and note that oxytocin, in fact, has no free carboxyl groups; all the acyl groups of oxytocin appear as amide functions. Thus, the carboxyl terminus of oxytocin has been

modified by conversion to an amide. There are three amide functions of the type $\overset{O}{\overset{\|}{C}}NH_2$, two of which belong to side chains of asparagine and glutamine, respectively. The third

amide belongs to the C-terminal amino acid, glycine, $-NHCH_2\overset{O}{\overset{\|}{C}}OH$, which in oxytocin

has been modified so that it appears as $-NHCH_2\overset{O}{\overset{\|}{C}}NH_2$. Therefore, attach glycine to the solid support in the first step of the Merrifield synthesis. The carboxyl group can be modified to the required amide after all the amino acid residues have been added and the completed peptide is removed from the solid support.

27.35 Purine and its numbering system are as shown:

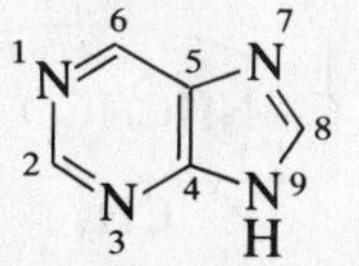

In nebularine, D-ribose in its furanose form is attached to position 9 of purine. The stereochemistry at the anomeric position is β.

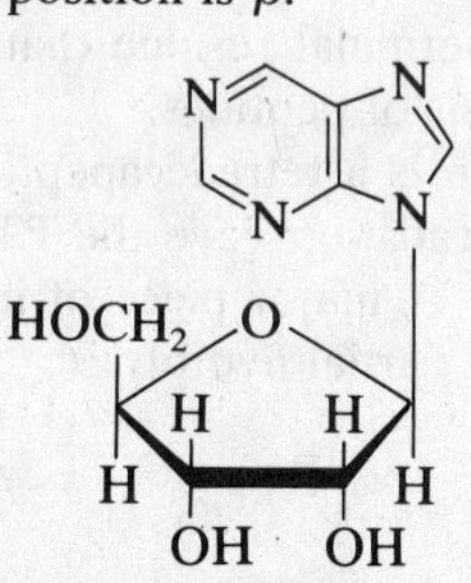

9-β-D-Ribofuranosylpurine (nebularine)

27.36 The problem states that vidarabine is the arabinose analog of adenosine. Arabinose and ribose differ only in their configuration at C-2.

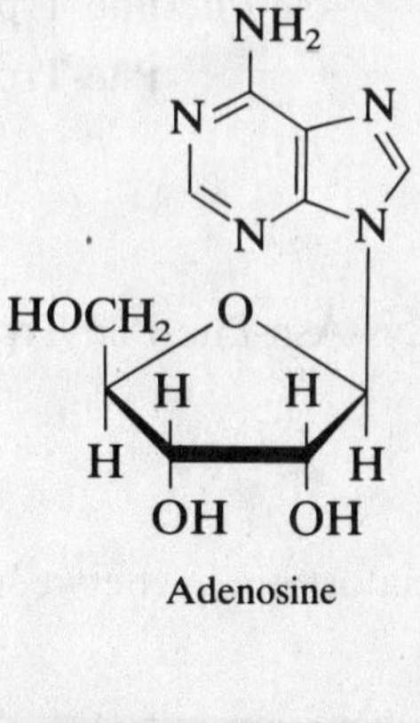

Adenosine 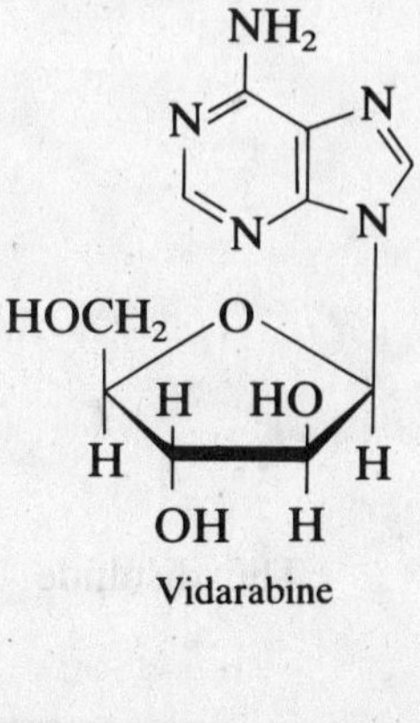   Vidarabine

27.37 Nucleophilic aromatic substitution occurs when 6-chloropurine reacts with hydroxide ion by an addition-elimination pathway.

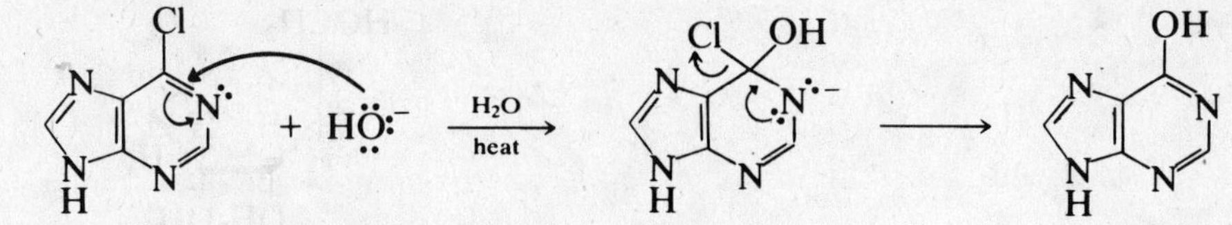

6-Chloropurine

The enol tautomerizes to give hypoxanthine.

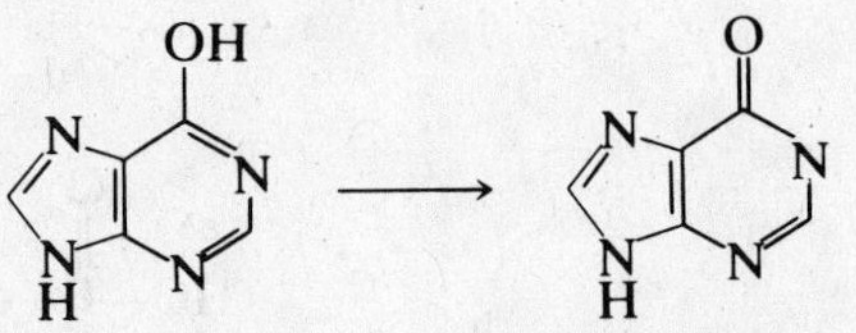

Hypoxanthine

27.38 Nitrous acid reacts with aromatic primary amines to yield diazonium ions.

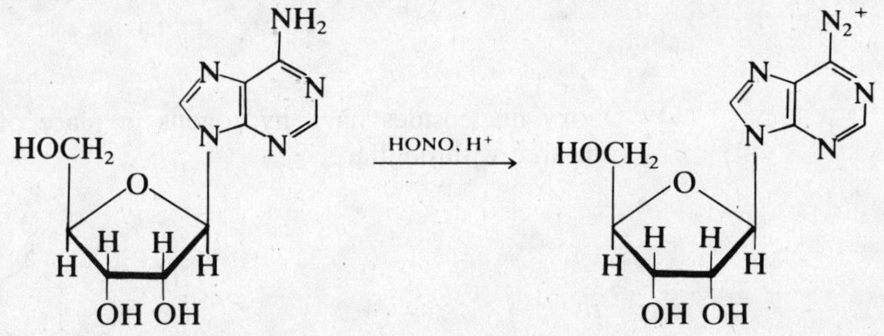

Adenosine

Treatment of the diazonium ion with water yields a phenol. Tautomerization gives inosine.

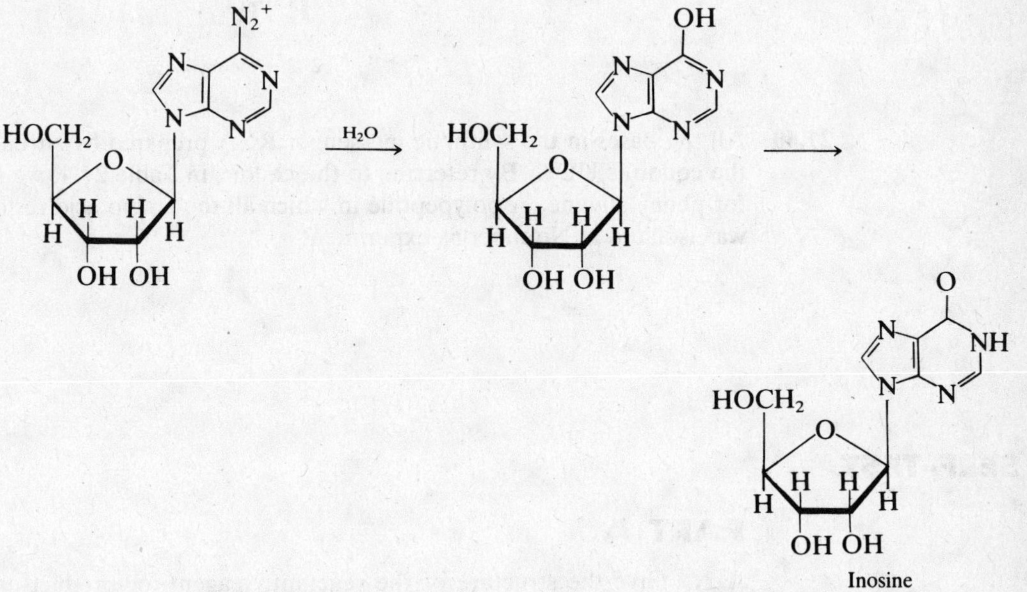

Inosine

27.39 The carbon atoms of the ribose portion of a nucleoside are numbered as follows:

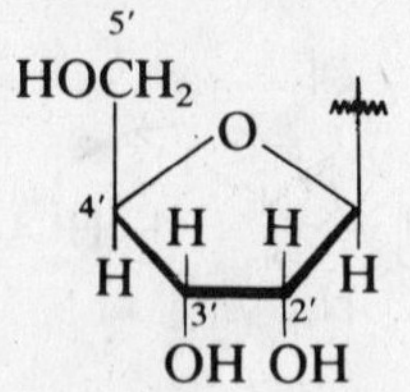

(*a*)   A 5′-nucleotide has a phosphate group attached to the C-5′ hydroxyl.

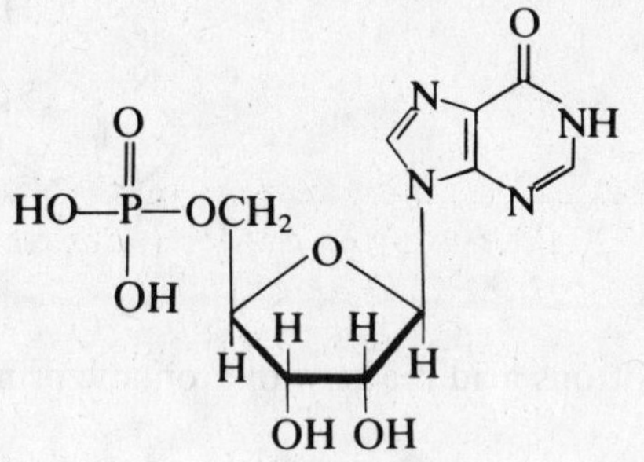

Inosinic acid

(*b*)   Deoxy nucleosides have hydrogens in place of hydroxyl groups at the positions indicated with boldface.

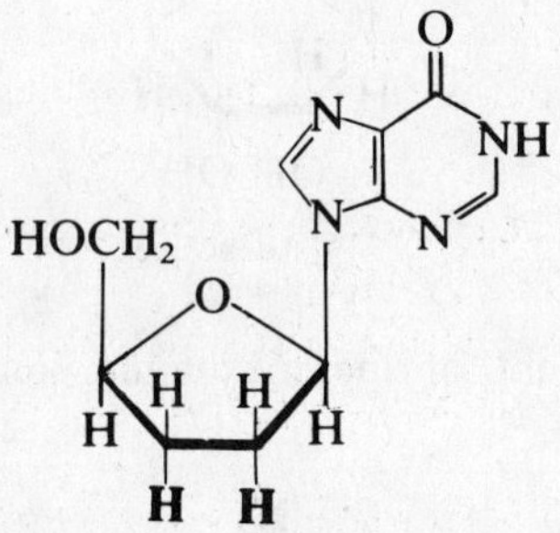

2′,3′-Dideoxyinosine

27.40 All the bases in the synthetic messenger RNA prepared by Nirenberg were U; therefore, the codon is UUU. By referring to the codons in Table 27.4, we see that the UUU codes for phenylalanine. A polypeptide in which all the amino acid residues were phenylalanine was isolated in Nirenberg's experiment.

SELF-TEST

PART A

A-1.   Give the structure of the reactant, reagent, or product omitted from each of the following:

(a) ? $\xrightarrow[\substack{\text{2. } H_3O^+, \text{ heat} \\ \text{3. neutralize}}]{\text{1. } NH_4Cl, NaCN}$ $C_6H_5CH_2\overset{\overset{+}{N}H_3}{\underset{|}{C}}HCO_2^-$

(b) $C_6H_5CH_2O\overset{\overset{O}{\|}}{C}Cl$ + valine $\xrightarrow[\text{2. } H^+]{\text{1. } HO^-, H_2O}$?

(c) Boc-Phe + $H_2NCH_2CO_2CH_2CH_3$ $\xrightarrow{?}$ Boc—$\underset{\underset{CH_2C_6H_5}{|}}{N}HCH\overset{\overset{O}{\|}}{C}NHCH_2CO_2CH_2CH_3$

A-2. Give the structure of the derivative that would be obtained by treatment of Phe-Ala with Sanger's reagent followed by hydrolysis.

A-3. Outline a sequence of steps that would allow the following synthetic conversions to be carried out:

(a) $(CH_3)_2CHCH_2\overset{\overset{+}{N}H_3}{\underset{|}{C}}HCO_2^-$ (leucine) from $CH_3\overset{\overset{O}{\|}}{C}NHCH(CO_2CH_2CH_3)_2$

(b) Leu-Val from leucine and $(CH_3)_2CH\overset{\overset{+}{N}H_3}{\underset{|}{C}}HCO_2^-$ (valine)

A-4. The carboxypeptidase-catalyzed hydrolysis of a pentapeptide yielded phenylalanine (Phe). One cycle of an Edman degradation gave a derivative of leucine (Leu). Partial hydrolysis yielded the fragments Leu-Val-Gly and Gly-Ala among others. Deduce the structure of the peptide.

A-5. Consider the following compound:

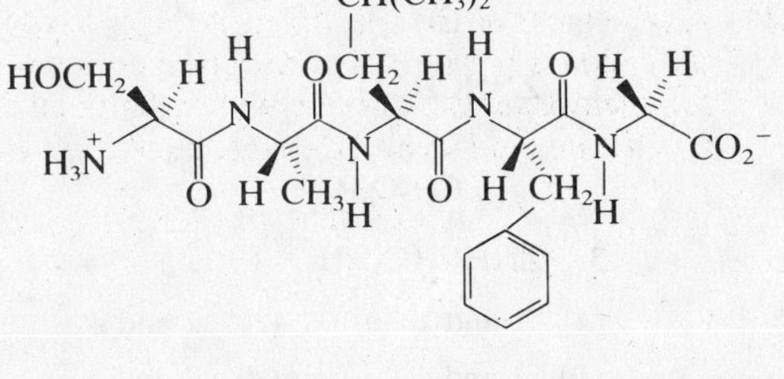

(a) What kind of peptide does this structure represent? (For example, dipeptide)
(b) How many peptide bonds are present?
(c) Give the name for the N-terminal amino acid.
(d) Give the name for the C-terminal amino acid.
(e) Using three-letter abbreviations, write the sequence.

PART B

B-1. Which statement correctly describes the difference in the otherwise similar chemical constituents of DNA and RNA?

(*a*) DNA contains uracil; RNA contains thymine.

(*b*) DNA contains guanine but not adenine; RNA contains both.

(*c*) DNA contains thymine; RNA contains uracil.

(*d*) None of these applies—the chemical constitution is the same.

B-2. Assume that a particular amino acid has an isoelectric point of 6.0. In a solution of pH 1.0, which of the following species will predominate?

(*a*) $\overset{\displaystyle R}{|}$ $H_3\overset{+}{N}CHCO_2H$ (*c*) $\overset{\displaystyle R}{|}$ $H_3\overset{+}{N}CHCO_2^-$

(*b*) $\overset{\displaystyle R}{|}$ H_2NCHCO_2H (*d*) $\overset{\displaystyle R}{|}$ $H_2NCHCO_2^-$

B-3. Choose the response which provides the best match of terms.

	Purine	Pyrimidine
(*a*)	Adenine	Guanine
(*b*)	Thymine	Cytosine
(*c*)	Cytosine	Adenine
(*d*)	Guanine	Cytosine

B-4. Which of the following reagents would be combined in the synthesis of Phe-Ala?

[In phenylalanine (Phe), R in the generalized amino acid formula $\overset{\displaystyle R}{\overset{|}{H_2NCHCO_2H}}$ is $CH_2C_6H_5$, and in alanine (Ala) it is CH_3.]

1. $\overset{\displaystyle CH_3}{\overset{|}{ZNHCHCO_2H}}$ 2. $\overset{\displaystyle CH_3}{\overset{|}{H_2NCHCO_2CH_2C_6H_5}}$

3. $\overset{\displaystyle CH_2C_6H_5}{\overset{|}{ZNHCHCO_2H}}$ 4. $\overset{\displaystyle CH_2C_6H_5}{\overset{|}{H_2NCHCO_2CH_2C_6H_5}}$

(*a*) 1 and 2   (*c*) 2 and 3

(*b*) 1 and 4   (*d*) 3 and 4

B-5. Which amino acid is achiral?

(*a*) $\underset{\overset{|}{\underset{+}{NH_3}}}{CH_2COO^-}$ (*b*) $\underset{\overset{|}{CH_3} \quad \overset{|}{\underset{+}{NH_3}}}{CH_3CHCH_2CHCOO^-}$

(c) CH$_3$CHCHCOO$^-$
       | |
      HO   NH$_3$
             +

(d) CH$_3$CHCOO$^-$
              |
           $_+$NH$_3$

B-6. A nucleoside is a
   (a) Phosphate ester of a nucleotide
   (b) Unit having a sugar bonded to a purine or pyrimidine base
   (c) Chain whose backbone consists of sugar units connected by phosphate groups
   (d) Phosphate salt of a purine or pyrimidine base

APPENDIX A

ANSWERS TO THE SELF-TESTS

CHAPTER 1

A-1. (*a*) P; $1s^22s^22p^63s^23p^3$ (*b*) S^{2-}; $1s^22s^22p^63s^23p^6$

A-2. (*a*) :N̈=C=S̈: Net charge: −1

Formal charge: −1 0 0

(*b*) :N≡C—S̈:

Formal charge: 0 0 −1

A-3. (*a*) 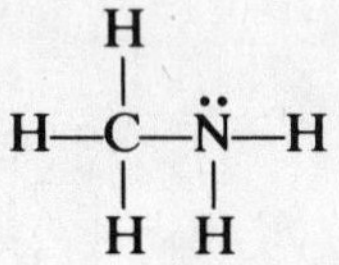 (*b*) 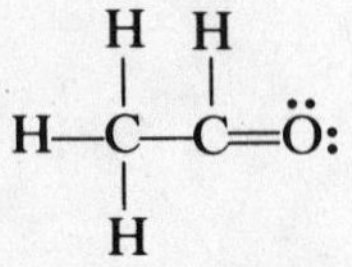

A-4. (*a*) $C_{12}H_{20}O$

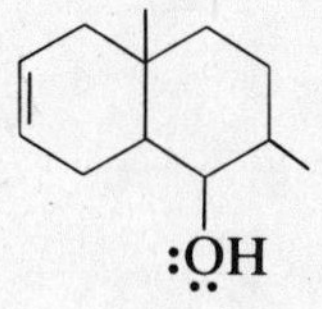

(*c*) $C_{14}H_{24}O$

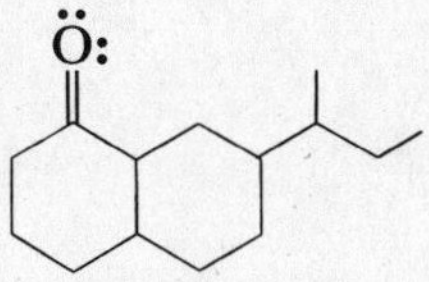

(*b*) $C_{10}H_{22}$

(*d*) C_9H_6BrN

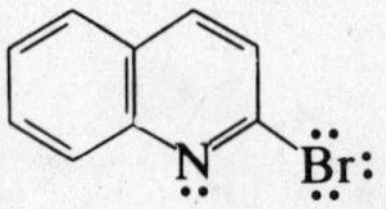

A-5. 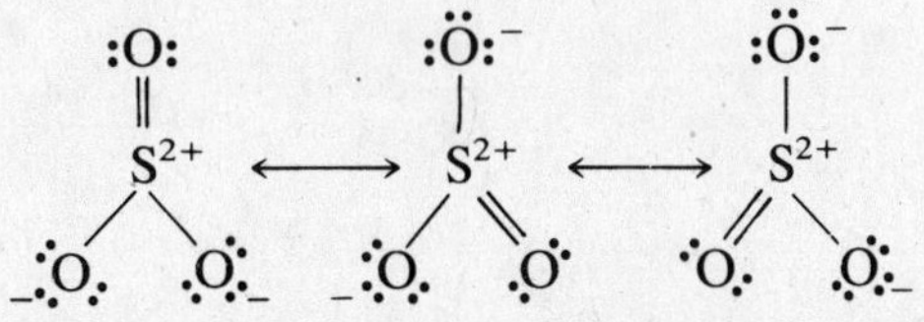

A-6. $:\ddot{O}\!\!-\!\!C\!\!\equiv\!\!N:$ $\longleftrightarrow$ $:\ddot{O}\!\!=\!\!C\!\!=\!\!\ddot{N}:$ Net charge: -1

Formal charge: -1 0   0   0   0 -1

A-7. (*a*)

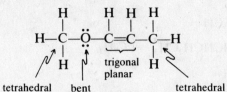

tetrahedral   bent   tetrahedral

(*b*) Pyramidal; $:\underset{\underset{Cl}{|}}{\overset{Cl}{N}}\blacktriangleleft Cl$;   yes, it is polar.

A-8. (*a*) D   (*b*) A, B   (*c*) None (*d*) B   (*e*) None
(*f*) A, D   (*g*) A   (*h*) C

A-9.

A-10. (*a*) $11\,\sigma;\,1\,\pi$ (*b*) $9\,\sigma;\,2\,\pi$

A-11. (*a*) $CH_3\!\!-\!\!CH\!\!=\!\!CH\!\!-\!\!CH_3$ (*b*) $H\!\!-\!\!C\!\!\equiv\!\!C\!\!-\!\!CH_2\!\!-\!\!CH_3$

sp^3 sp^2 sp^3 sp sp^3

B-1. (*b*)   **B-2.** (*c*)   **B-3.** (*d*)   **B-4.** (*a*)
B-5. (*a*)   **B-6.** (*b*)   **B-7.** (*b*)   **B-8.** (*c*)
B-9. (*b*)   **B-10.** (*d*)   **B-11.** (*d*)   **B-12.** (*b*)
B-13. (*d*)

CHAPTER 2

A-1. $CH_3CH_2CH_2CH_2\!\!-$ $CH_3CH_2\underset{\overset{|}{}}{C}HCH_3$

Common: *n*-Butyl   *sec*-Butyl

Systematic: Butyl   1-Methylpropyl

$CH_3\underset{\overset{|}{CH_3}}{C}HCH_2\!\!-$ $CH_3\!\!-\!\!\underset{\underset{CH_3}{|}}{\overset{\overset{CH_3}{|}}{C}}\!\!-\!\!CH_3$

Common: Isobutyl   *tert*-Butyl
Systematic: 2-Methylpropyl   1,1-Dimethylethyl

A-2. (*a*) 28 (8 C—C; 20 C—H)   (*b*) 27 (9 C—C; 18 C—H)

A-3. (*a*) Oxidized   (*b*) Neither   (*c*) Neither   (*d*) Reduced

A-4. (*a*) $CH_3\underset{\overset{|}{CH_3}}{C}H\underset{\overset{|}{CH_3}}{C}H\overset{\overset{CH_3CHCH_3}{|}}{C}HCH_3$ (*b*) Six methyl groups, three isopropyl groups

A-5. 3,4-Dimethylheptane

A-6. Four primary C, three secondary C, two tertiary C

A-7.
$$\text{CH}_3\overset{\overset{\displaystyle \text{CH}_3}{|}}{\text{CH}}\overset{\overset{\displaystyle}{|}}{\underset{\underset{\displaystyle \text{CH}_3}{|}}{\text{CH}}}\text{CH}_2\text{CH}_3 \equiv \text{C}_7\text{H}_{16}$$

$$\text{C}_7\text{H}_{16} + 11\text{O}_2 \longrightarrow 7\text{CO}_2 + 8\text{H}_2\text{O}$$

A-8.

Cyclopentane   Methylcyclobutane   Ethylcyclopropane

1,1-Dimethylcyclopropane   1,2-Dimethylcyclopropane

A-9.
$$\text{CH}_3\overset{\overset{\displaystyle}{|}}{\underset{\underset{\displaystyle \text{CH}_3}{|}}{\text{CH}}}\text{—}\overset{\overset{\displaystyle \text{CH}_2\text{CH}_2\text{CH}_3}{|}}{\underset{\underset{\displaystyle \text{CH}_3}{|}}{\text{C}}}\text{CH}_2\text{CH}_3$$
;   no; 3-ethyl-2,3-dimethylhexane

A-10. $\text{CH}_3\text{CH}_2\text{CH}_2\text{CH}_2\text{CH}_2\text{CH}_2\text{CH}_2\text{CH}_3$

A-11. $x = 11$, $y = 20$, $z = 13$;   less heat is evolved per mole

B-1. (*a*)   **B-2.** (*d*)   **B-3.** (*d*)   **B-4.** (*c*)

B-5. (*b*)   **B-6.** (*a*)   **B-7.** (*c*)   **B-8.** (*c*)

B-9. (*a*)   **B-10.** (*b*)

CHAPTER 3

A-1.

Cl

CH₃

H

H

H

H

(Sawhorse)

CH₃ Cl H

H H H

(Newman)

A-2. (*a*)

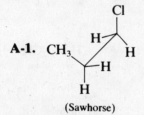

(*b*)

Cl H Cl

Cl Cl

H

Cl Cl Cl

Cl H

H

A-3. (*a*) △   (*b*) ⬠   (*c*) ◻

A-4. $(\text{CH}_3)_2\text{CH}$

H

H

CH₃

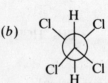

A-5.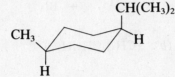

A-6. A is more strained; B is more stable.

A-7. (*a*) C   (*b*) A and B   (*c*) D   (*d*) A

A-8.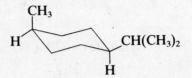

More stable

A-9. *trans*-1-Isopropyl-4-methylcyclohexane has the lower heat of combustion.

A-10. *Angle* strain and *torsional* strain

A-11.

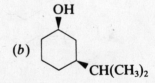

B-1. (*d*)   **B-2.** (*b*)   **B-3.** (*c*)   **B-4.** (*a*)

B-5. (*c*)   **B-6.** (*b*)   **B-7.** (*a*)   **B-8.** (*c*)

B-9. (*e*)   **B-10.** (*c*)

CHAPTER 4

A-1. (*a*) *trans*-1-Bromo-3-methylcyclopentane
  (*b*) 2-Ethyl-4-methyl-1-hexanol

A-2. (*a*) $ICH_2CCH_2CH_2CH_2CH_2CH_3$ (*b*)
        with CH_3 above and Cl below

A-3. Conjugate acid $CH_3\overset{+}{O}H_2$; conjugate base CH_3O^-

A-4. $pK_a = 4.74$

A-5. (*a*) $CH_3CH_2CH_2Cl$ (*b*) $CH_3CH_2\overset{OH}{\underset{}{C}}(CH_3)_2$ (*c*) Na°

A-6. $CH_3CH_2O^- + NH_3 \rightleftharpoons CH_3CH_2OH + NH_2^-$ ($K<1$)

  Conjugate   Conjugate   Stronger   Stronger
  base   acid   acid   base

A-7. (*a*) Three

  (*b, c*) $CH_3\overset{CH_3}{\underset{}{C}}CH_2\overset{CH_3}{\underset{}{C}}HCH_3$ $CH_3\overset{CH_3}{\underset{}{C}}H\overset{CH_3}{\underset{}{C}}HCHCH_3$ $\overset{CH_3}{\underset{}{C}}H_2CH\overset{CH_3}{\underset{}{C}}H_2CHCH_3$

  (Most stable)   (Least stable)

A-8.

$$CH_3\underset{\underset{CH_3}{|}}{\overset{\overset{CH_3}{|}}{C}}CH_3 + Cl_2 \longrightarrow CH_3\underset{\underset{CH_3}{|}}{\overset{\overset{CH_3}{|}}{C}}CH_2Cl + HCl$$

A-9. $Br\cdot + $ $\longrightarrow HBr + $

$+ Br_2 \longrightarrow$ $+ Br\cdot$

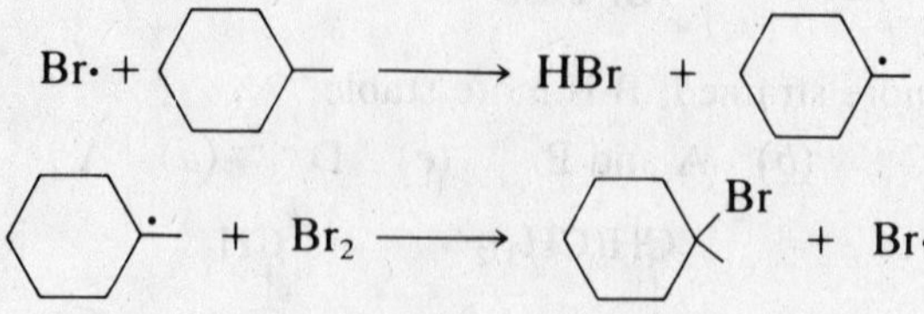

A-10. (a) $(CH_3)_3CCl$ (b) $(CH_3)_2CHCH_2Cl$

A-11.

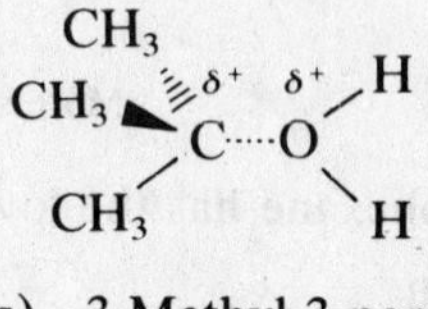

A-12. (a) 3-Methyl-3-pentanol   (b) $KOC(CH_3)_3$ (c) Fluorine (F_2)
        (d) Ethyl radical, $CH_3\dot{C}H_2$ (e) Cl_2

B-1. (c)   **B-2.** (b)   **B-3.** (d)   **B-4.** (c)

B-5. (c)   **B-6.** (d)   **B-7.** (a)   **B-8.** (c)

B-9. (d)   **B-10.** (c)   **B-11.** (a)   **B-12.** (c)

B-13. (c)   **B-14.** (c)

CHAPTER 5

A-1. (a) 2,4,4-Trimethyl-2-pentene   (b) (E)-3,5-Dimethyl-4-octene

A-2. (a) 2,3-Dimethyl-2-pentene

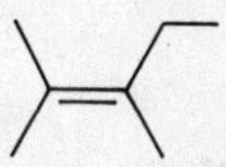

     (b) 1,4-Dimethylcyclohexene

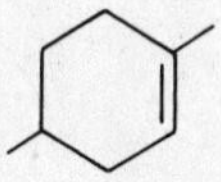

A-3. (a)

       1   2   3   4

     5   6

     (b) Isomer 5   (c) Isomers 1 and 4   (d) Isomers 2 and 3

A-4. Two sp^2 C atoms; four sp^3 C atoms; three sp^2–sp^3 σ bonds

A-5. (a)   (b)

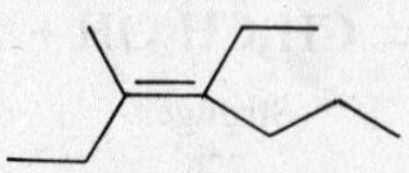

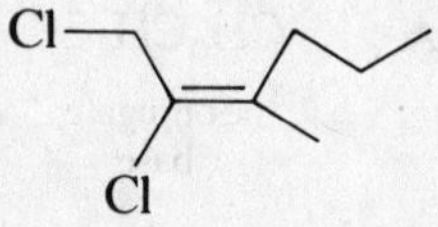

A-6.

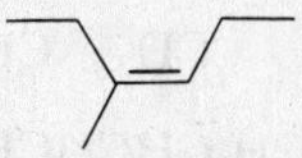

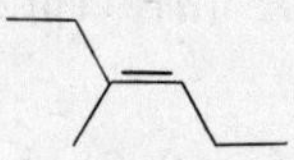

    (Z)-3-Methyl-3-hexene   (E)-3-Methyl-3-hexene

A-7. (a) CH_2=$CCH_2CH_2CH_3$ + $(CH_3)_2C$=$CHCH_2CH_3$ (major)
                    |
                   CH_3

(b)

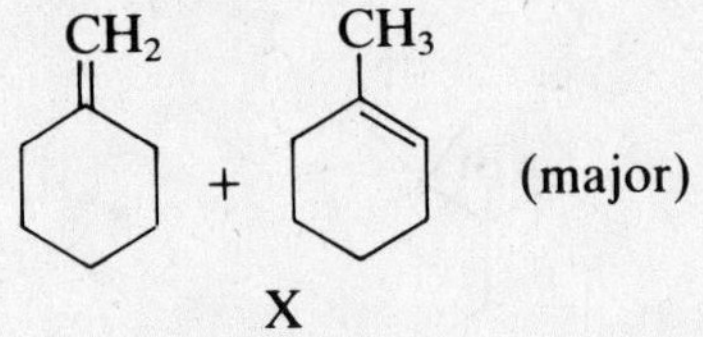

                      X

(c) $(CH_3)_2CHCHCH(CH_3)_2$ (X = Cl, Br, I)
                    |
                   CH_3

(d) CH_2=$CC(CH_3)_3$

A-8.
                   CH_3
                    |
         CH_3CH_2C—CH_2Br
                    |
                   CH_3

A-9. Rearrangement (hydride migration) occurs to form a more stable carbocation:

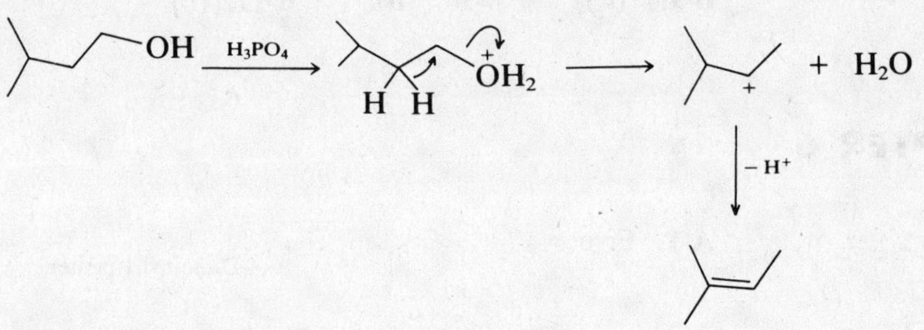

A-10.

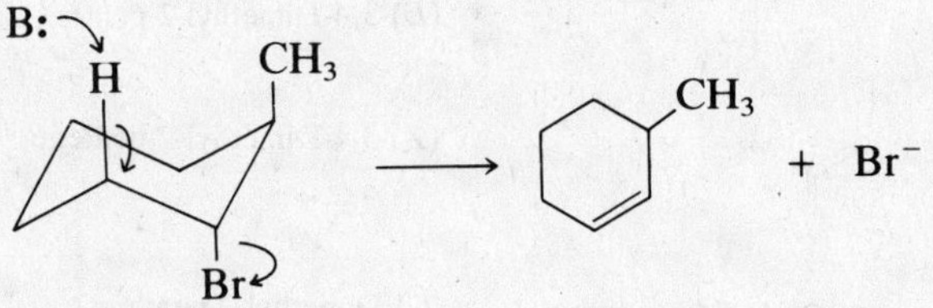

A-11. Cis isomer:

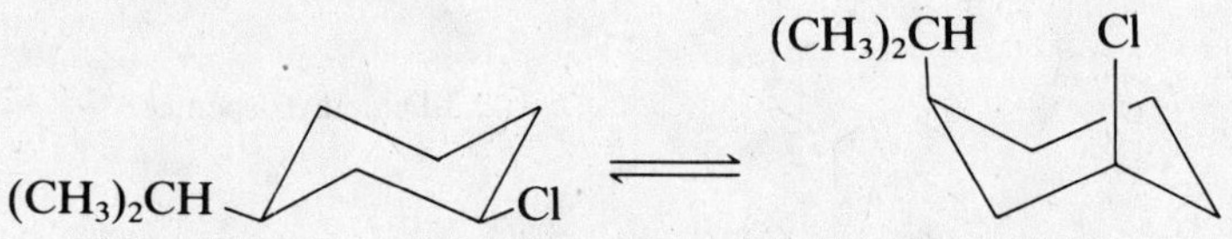

Trans isomer:

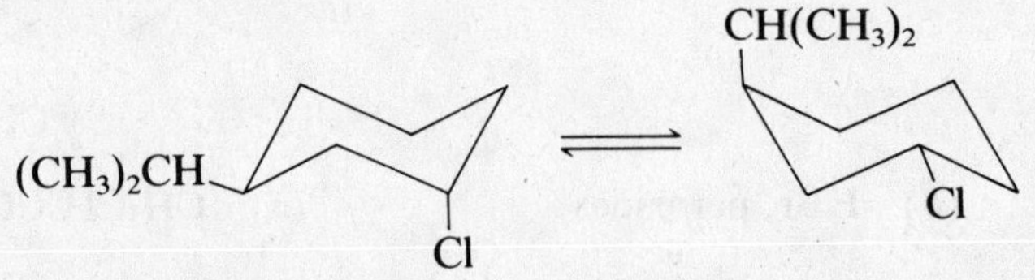

The trans isomer will react faster because its most stable conformation (with the isopropyl group equatorial) has an axial Cl able to undergo E2 elimination.

A-12.

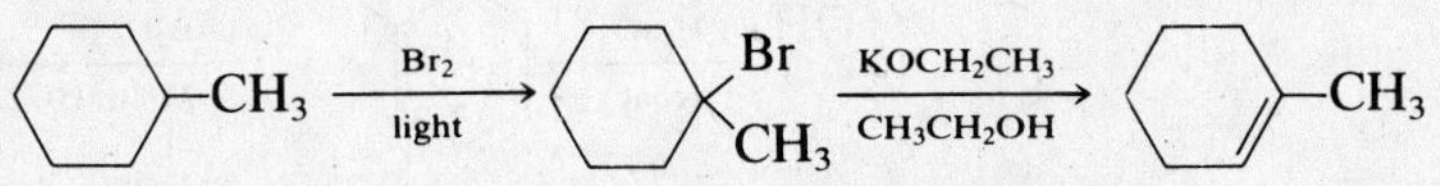

A-13. 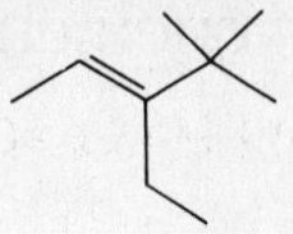   3-Ethyl-4,4-dimethyl-2-pentene

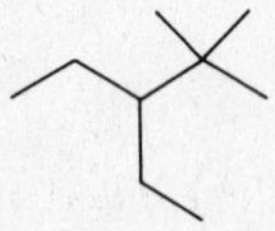

A 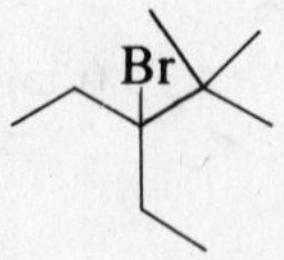   B

A-14.

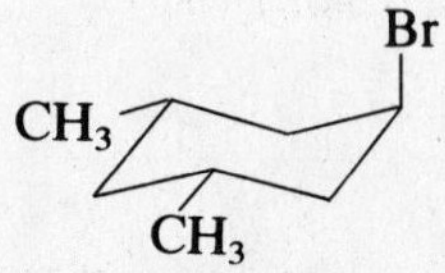

Fastest E2 elimination 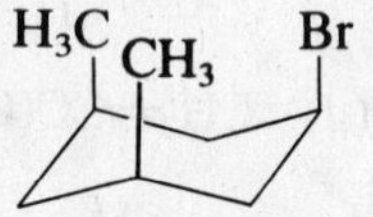   Slowest E2 elimination

B-1.	(c)	**B-2.**	(d)	**B-3.**	(c)	**B-4.**	(a)
B-5.	(a)	**B-6.**	(b)	**B-7.**	(a)	**B-8.**	(b)
B-9.	(d)	**B-10.**	(b)	**B-11.**	(c)		

CHAPTER 6

A-1. Five;

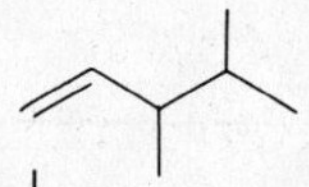

       3,4-Dimethyl-1-pentene

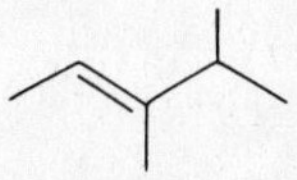

       (E)-3,4-Dimethyl-2-pentene

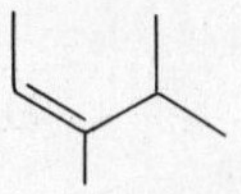

       (Z)-3,4-Dimethyl-2-pentene

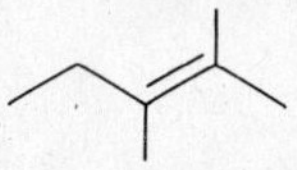

       2,3-Dimethyl-2-pentene

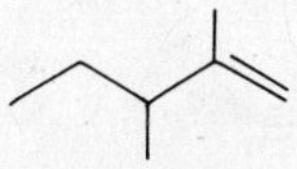

       2,3-Dimethyl-1-pentene

A-2. (a) $(CH_3)_2CCH_2CH_3$ (c)

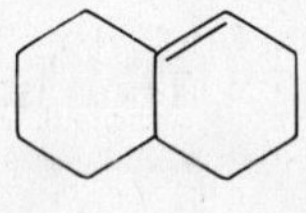

(b)   HBr, peroxides   (d) $CH_3CH_2CCH_2Br$

A-3. (a)

$\xrightarrow[\text{(conc)}]{H_2SO_4}$ $\xrightarrow[\text{2. } H_2O_2, \text{ HO}^-]{\text{1. } B_2H_6}$

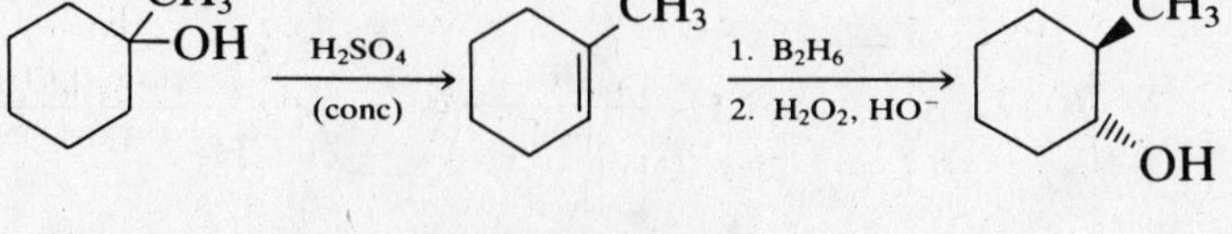

(b)

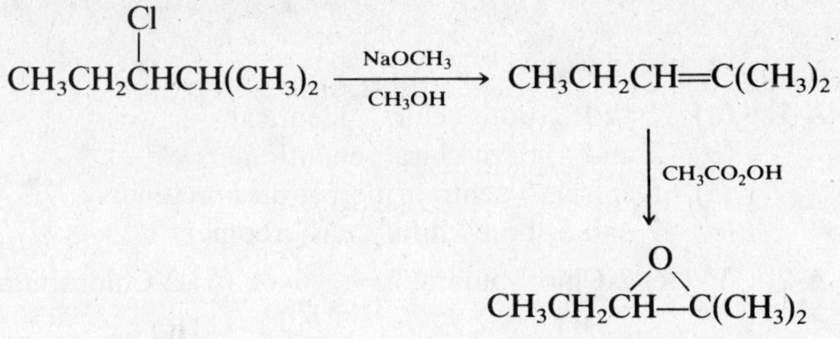

$$CH_3CH_2\overset{\displaystyle Cl}{\underset{\displaystyle |}{C}}HCH(CH_3)_2 \xrightarrow[CH_3OH]{NaOCH_3} CH_3CH_2CH=C(CH_3)_2$$

$$\downarrow CH_3CO_2H$$

$$CH_3CH_2\overset{\displaystyle O}{\overset{\displaystyle \diagup \ \ \diagdown}{CH-C}}(CH_3)_2$$

(c) $(CH_3)_3CCHCH_3 \xrightarrow[CH_3OH]{CH_3O^-Na^+} (CH_3)_3CCH=CH_2 \xrightarrow[Peroxides]{HBr} (CH_3)_3CCH_2CH_2Br$
      |
      Br

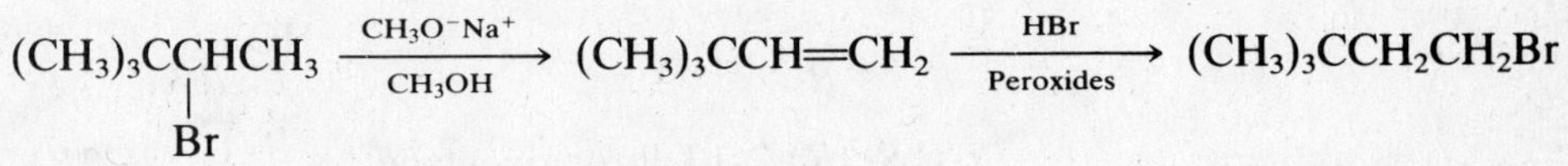

A-4.

$$(CH_3)_2CHC\overset{\displaystyle CH_3}{\overset{\displaystyle |}{=}}CH_2 \xrightarrow{HBr} (CH_3)_2CHC\overset{\displaystyle CH_3}{\underset{\displaystyle |}{\underset{\displaystyle Br}{C}}}CH_3 \xrightarrow[CH_3CH_2OH]{NaOCH_2CH_3} (CH_3)_2C=C(CH_3)_2$$

A   B   C

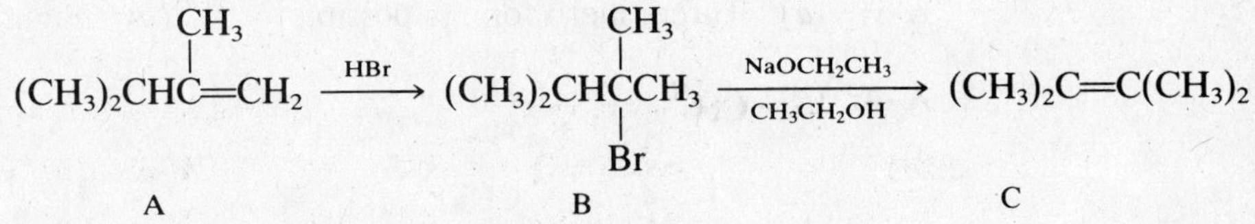

$$C \xrightarrow[\text{2. } H_2O, Zn]{\text{1. } O_3} (CH_3)_2C=O \ (2 \text{ mol})$$

A-5. Initiation:

$$ROOR \xrightarrow[heat]{light \ or} 2RO\cdot$$

$$RO\cdot + HBr \longrightarrow ROH + Br\cdot$$

Propagation: $Br\cdot + CH_3CH_2CH=CH_2 \longrightarrow CH_3CH_2\dot{C}HCH_2Br$

$$CH_3CH_2\dot{C}HCH_2Br + HBr \longrightarrow CH_3CH_2CH_2CH_2Br + Br\cdot$$

A-6.

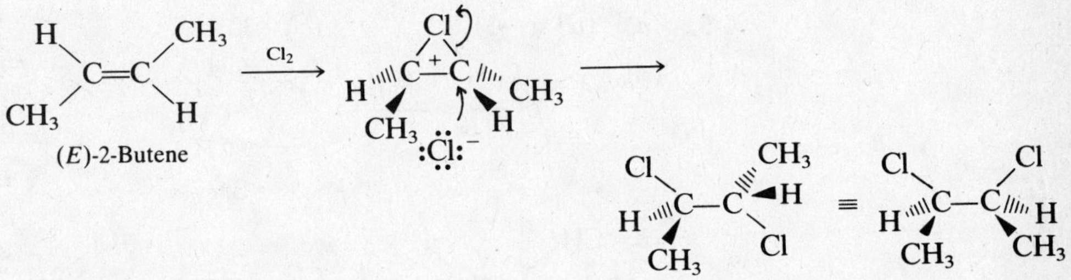

(E)-2-Butene

A-7.

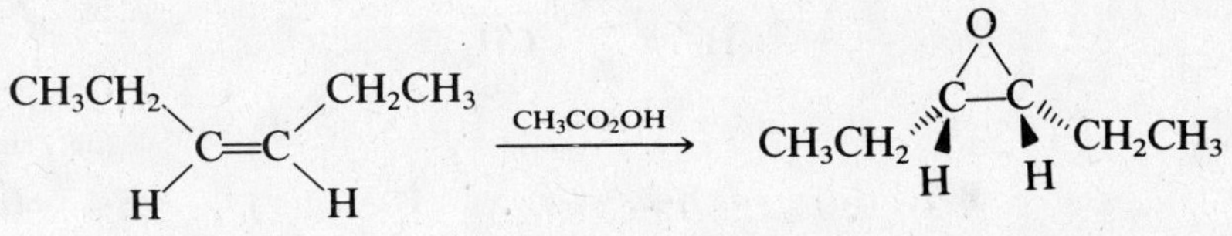

A-8.

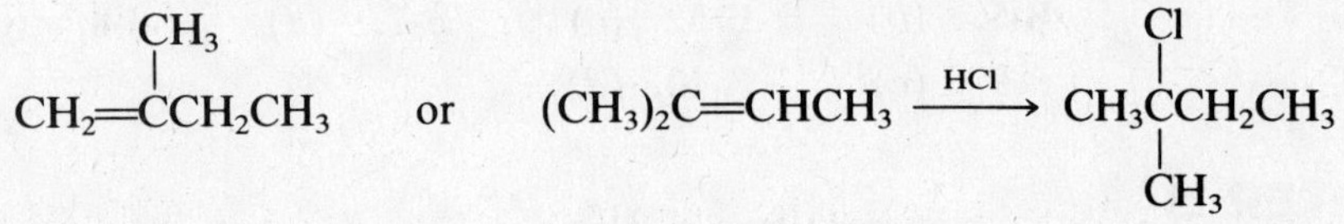

2-Methyl-1-butene   2-Methyl-2-butene   2-Chloro-2-methylbutane

A-9. SODAR = 7; THC contains three rings and four double bonds.

B-1. (b)   **B-2.** (a)   **B-3.** (c)   **B-4.** (d)

B-5. (d)   **B-6.** (b)   **B-7.** (d)   **B-8.** (b)

B-9. (d)   **B-10.** (c)

CHAPTER 7

A-1. (*a*) 1 and 2, both achiral; identical
     (*b*) 3 and 4, both chiral; enantiomers
     (*c*) 5 chiral, 6 achiral (meso); diastereomers
     (*d*) 7 and 8, both chiral; diastereomers

A-2. 3: (*R*)-2-Chlorobutane;   4: (*S*)-2-Chlorobutane

5: 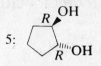   6:

     7: (2*R*,3*R*)-2,3-Dibromopentane;   8: (2*S*,3*R*)-2,3-Dibromopentane

A-3. (*a*) Three; meso form is possible.   (*b*) Eight; no meso form possible.

A–4. (*a*)

CH₃, C–C, Cl, H, OH, H, CH₃

(*b*)

CH₃, H, H, H, Br, CH₃

(*c*)

CH₃
Cl ——— H
H ——— OH
CH₃

CH₃
H ——— H
H ——— Br
CH₃

A-5. (*a*) $[\alpha] = -31.2°$ (*b*) 30% *S*

A-6. (*a*)

CH₃, CH₃, CH₃ →(HBr)→ CH₃, CH₃, CH₃, Br + CH₃, CH₃, Br, CH₃

(*b*)

CH₃, H, H, CH₃ + Cl₂ ⟶ Cl, CH₃, C–C, H, CH₃, Cl, H

Meso form
(only stereoisomer)

B-1. (*c*)   **B-2.** (*b*)   **B-3.** (*b*)   **B-4.** (*d*)
B-5. (*b*)   **B-6.** (*c*)   **B-7.** (*d*)   **B-8.** (*c*)
B-9. (*c*)   **B-10.** (*d*)

CHAPTER 8

A-1. (*a*) CH₃CH₂CH₂CH₂OCH₂CH₃   (*b*)

(cyclopentane with X and CH₃)   (X = OTs, Br, I)

(*c*) CH₃CHCH₂CH₂Cl + NaN₃ ⟶ CH₃CHCH₂CH₂N₃
          |CH₃ |CH₃

(d)

(e)

(f)

A-2. $(CH_3)_2CHS^-Na^+ + CH_3CH_2CH_2Br$

A-3.

$$CH_3CH_2\overset{H}{\underset{CH_3}{\overset{|}{C}}}-OH \xrightarrow[\text{pyridine}]{CH_3-\!\!\!\!\bigcirc\!\!\!\!-SO_2Cl} CH_3CH_2\overset{H}{\underset{CH_3}{\overset{|}{C}}}-OTs \xrightarrow[\text{acetone}]{NaI} I\!-\!\overset{H}{\underset{CH_3}{\overset{|}{C}}}\!\!\!-CH_2CH_3$$

A-4. Step 1. Ionization to form a secondary carbocation:

$$\overset{H_3C}{\underset{CH_3}{\overset{|}{C}}}\!\!\!\overset{Cl}{\underset{}{\overset{|}{C}}}HCH_3 \xrightarrow{H_2O} CH_3\overset{CH_3}{\underset{CH_3}{\overset{|}{C}}}-\overset{+}{C}HCH_3 + Cl^-$$

Step 2. Rearrangement by methyl migration to form a more stable tertiary carbocation:

$$CH_3\overset{CH_3}{\underset{CH_3}{\overset{|}{C}}}-\overset{+}{C}HCH_3 \longrightarrow CH_3\overset{CH_3}{\underset{CH_3}{\overset{|}{\overset{+}{C}}}}-CHCH_3$$

Step 3. Capture of the carbocation by water, followed by deprotonation:

$$CH_3\overset{CH_3}{\underset{CH_3}{\overset{|}{\overset{+}{C}}}}-CHCH_3 \xrightarrow{H_2O} CH_3\overset{\overset{+}{H_2O}\;\;CH_3}{\underset{CH_3}{\overset{|\quad\;|}{C}}}-CHCH_3 \xrightarrow{-H^+} (CH_3)_2\overset{OH}{\underset{}{\overset{|}{C}}}-CH(CH_3)_2$$

A-5. $(CH_3)_3CBr \xrightarrow{\;CH_3OH\;} (CH_3)_3COCH_3$

S_N1, unimolecular substitution; rate $= k[(CH_3)_3CBr]$

A-6. Sodium iodide is soluble in acetone, whereas the by-product of the reaction, sodium bromide, is not. According to Le Châtelier's principle, the reaction will shift in the direction which will replace the component removed from solution, that is, toward product in this case.

A-7.

OH

O⁻K⁺

CH_3CH_2OH

CH_3CH_2Br

A   B   C   D

A-8.

B-1. (*b*)   **B-2.** (*b*)   **B-3.** (*d*)   **B-4.** (*c*)

B-5. (*a*)   **B-6.** (*a*)   **B-7.** (*c*)   **B-8.** (*d*)

B-9. (*c*)   **B-10.** (*a*)

CHAPTER 9

A-1.   (*a*)   4,5-Dimethyl-2-hexyne   (*b*)   4-Ethyl-3-propyl-1-heptyne

A-2.   (*a*) $\underset{\displaystyle \overset{\displaystyle Cl}{\displaystyle |}}{CH_3CH_2CH_2C}=CH_2$ (*d*) $(CH_3)_2CHC\equiv CH$

      (*b*) H_2O, H_2SO_4, $HgSO_4$ (*e*) H_2, Lindlar Pd

      (*c*) $\underset{\displaystyle H}{\overset{\displaystyle CH_3}{}}C=C\underset{\displaystyle H}{\overset{\displaystyle CH_3}{}}$ (*f*) $\underset{\displaystyle Cl}{\overset{\displaystyle CH_3}{}}C=C\underset{\displaystyle CH_2CH_3}{\overset{\displaystyle Cl}{}}$

A-3.   Reaction II is effective; the desired product is formed by an S_N2 reaction:

$$\underset{\displaystyle \overset{\displaystyle CH_3}{\displaystyle |}}{CH_3CH_2CH}C\equiv C:^- Na^+ + CH_3I \longrightarrow \underset{\displaystyle \overset{\displaystyle CH_3}{\displaystyle |}}{CH_3CH_2CH}C\equiv CCH_3 + NaI$$

Reaction I is not effective, owing to E2 elimination from the secondary bromide:

$$\underset{\displaystyle \overset{\displaystyle Br}{\displaystyle |}}{CH_3CH_2CH}CH_3 + CH_3C\equiv C:^- Na^+ \longrightarrow$$

$$CH_3CH=CHCH_3 + CH_3C\equiv CH + NaBr$$

A-4.   (*a*) $CH_2=CHCH_2CH_3 + Br_2 \longrightarrow \underset{\displaystyle \overset{\displaystyle }{\displaystyle Br}}{BrCH_2CHCH_2CH_3} \xrightarrow[NH_3]{NaNH_2}$

$$HC\equiv CCH_2CH_3 \xrightarrow{NaNH_2} Na^{+-}:C\equiv CCH_2CH_3$$

$$\xrightarrow{CH_3CH_2Br} CH_3CH_2C\equiv CCH_2CH_3$$

(*b*) $HC{\equiv}CH \xrightarrow{NaNH_2} HC{\equiv}C{:}^-Na^+ \xrightarrow{(CH_3)_2CHCH_2Br} HC{\equiv}CCH_2CH(CH_3)_2$

$$HC{\equiv}CCH_2CH(CH_3)_2 \xrightarrow[HgSO_4]{H_2O,\ H_2SO_4} CH_3\overset{\overset{\large O}{\|}}{C}CH_2CH(CH_3)_2$$

A-5. 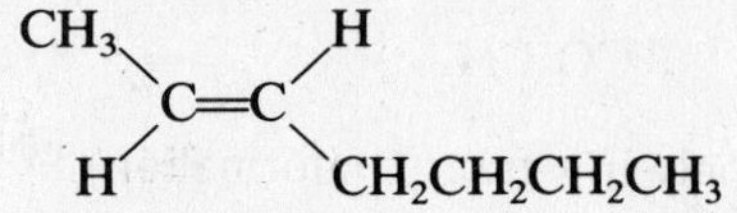 (*E*)-2-Heptene

A-6. $(CH_3)_3CC{\equiv}CH$ (A) $(CH_3)_3CC{\equiv}C{:}^-Na^+$ (B)

$CH_3CH_2CH_2OH$ (C) $CH_3CH_2CH_2Br$ (D)

B-1. (*d*) **B-2.** (*a*) **B-3.** (*d*) **B-4.** (*b*)

B-5. (*b*) **B-6.** (*b*) **B-7.** (*c*)

CHAPTER 10

A-1. $CH_2{=}CHCH_2CH{=}CH_2$ $CH_2{=}CHCH{=}CHCH_3$ (conjugated)

$CH_2{=}CHC(CH_3){=}CH_2$ (Conjugated)

$CH_2{=}C{=}CHCH_2CH_3$, $CH_2{=}C{=}C(CH_3)_2$, $CH_3CH{=}C{=}CHCH_3$ } Allenes

A-2. 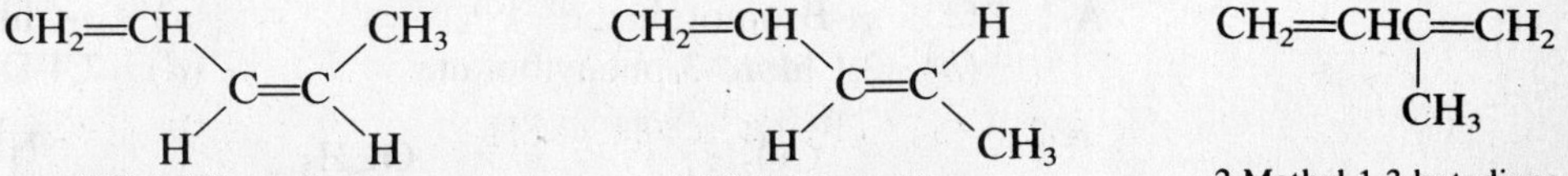

(3*Z*)-1,3-Pentadiene (3*E*)-1,3-Pentadiene 2-Methyl-1,3-butadiene

A-3. 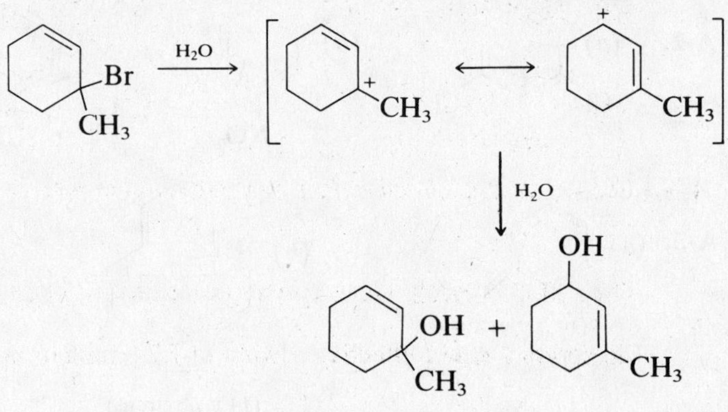

A-4. (*a*) $CH_3CH{=}CHCHBrCHBrCH_3 + CH_3CHBrCH{=}CHCHBrCH_3$

(Direct addition) (Conjugate addition)

(*b*) 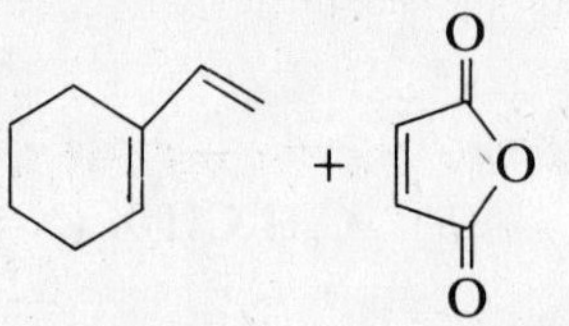 (*c*) 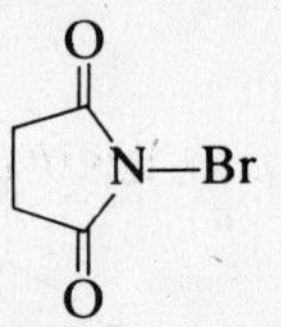 (NBS), heat

(d) [structure: bicyclic diene with CO₂CH₃ groups]

A-5. [cyclopentene structure] (exists only in *s*-trans conformation)

A-6. [reaction scheme: tertiary bromide losing Br⁻ to form carbocation, resonance structures]

B-1. (b)   **B-2.** (c)   **B-3.** (a)   **B-4.** (c)
B-5. (a)   **B-6.** (d)

CHAPTER 11

A-1. (a) *m*-Bromotoluene   (c) *o*-Chloroacetophenone
(b) 2-Chloro-3-phenylbutane   (d) 2,4-Dinitrophenol

A-2. (a) [3,5-dichlorobenzoic acid]   (b) [4-nitroanisole]   (c) [2,4-dimethylaniline]   (d) [3-bromobenzyl chloride]

A-3. (a) [10-membered ring cation] (10 π electrons)   (b) [14-membered ring anion] (14 π electrons)

A-4. Eight π electrons. No, the substance is not aromatic.

A-5. [azulene resonance structures]

A-6. (a) [1-bromotetralin structure]   (b) $C_6H_5CH_2X$ (X = Cl, Br, I, OTs)

(c) $Na_2Cr_2O_7$, H_2SO_4, H_2O, heat   (d)

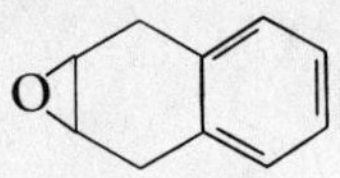

A-7. (I) $C_6H_5CH\!=\!CHCH_3 \xrightarrow{\text{HBr}} C_6H_5\overset{\overset{\displaystyle Br}{|}}{C}HCH_2CH_3$

(II) $C_6H_5CH_2CH_2CH_3 \xrightarrow[\substack{\text{light} \\ \text{(or NBS, heat)}}]{\text{Br}_2} C_6H_5\overset{\overset{\displaystyle Br}{|}}{C}HCH_2CH_3$

B-1. (c)   **B-2.** (c)   **B-3.** (a)   **B-4.** (b)

B-5. (a)   **B-6.** (d)   **B-7.** (b)   **B-8.** (b)

B-9. (a)   **B-10.** (d)

CHAPTER 12

A-1.

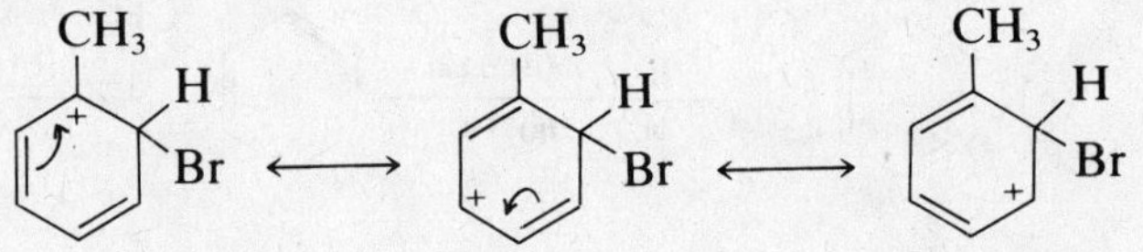

A-2. (a)

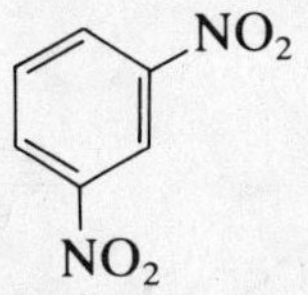

Slower

(b)

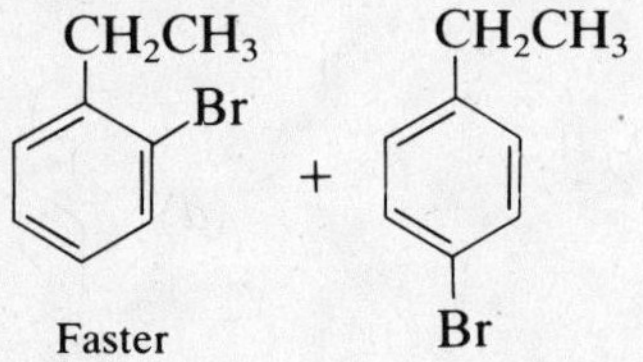

Faster

(c)

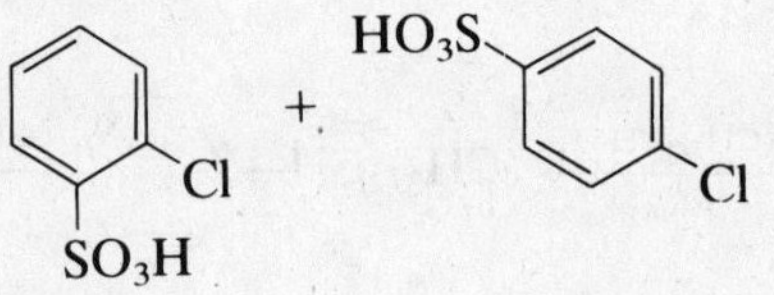

Slower

A-3. (a) NO_2^+ (b) Br—Br---$FeBr_3$ (c) SO_3

A-4. (a)

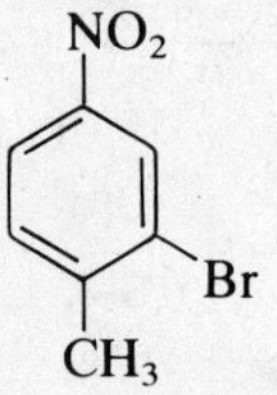

(b)

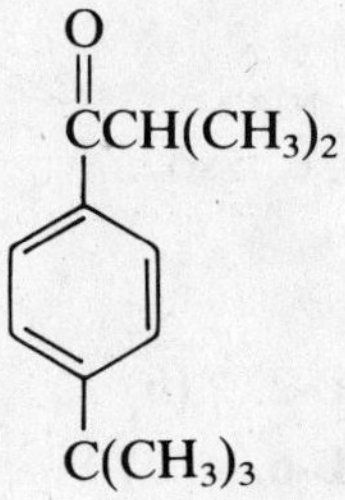

(c)

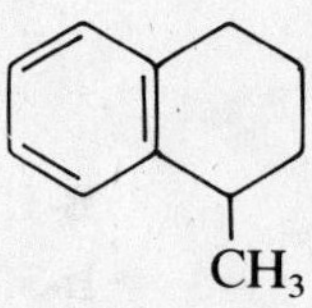

(d) $C_6H_5\overset{\overset{\displaystyle O}{\|}}{C}Cl$, $AlCl_3$ (e)

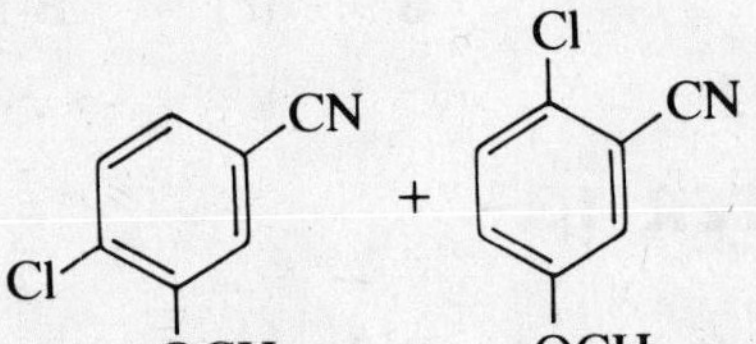

(f)

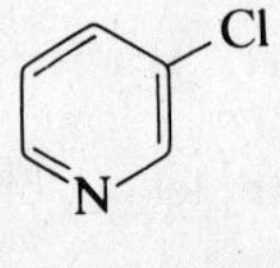

A-5. (*a*)

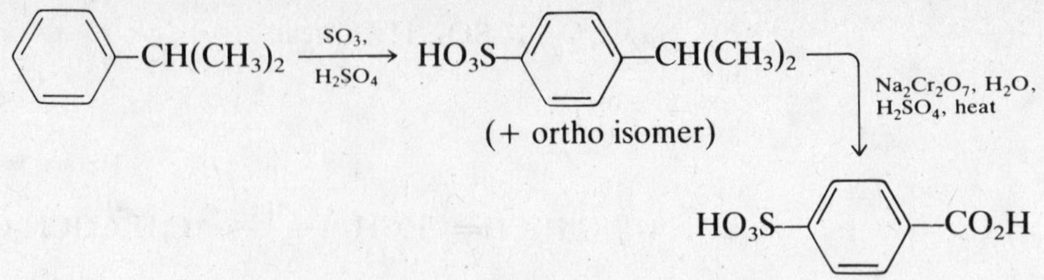

(+ ortho isomer)

(*b*)

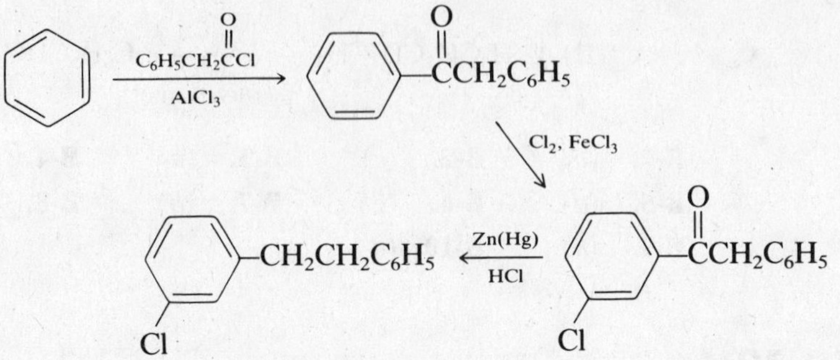

(*c*)

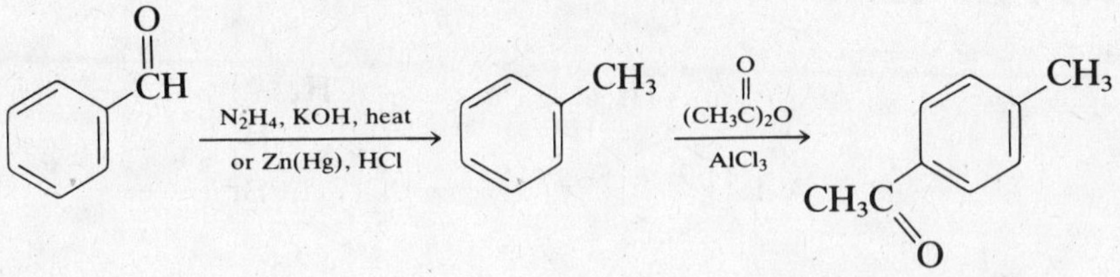

(*d*)

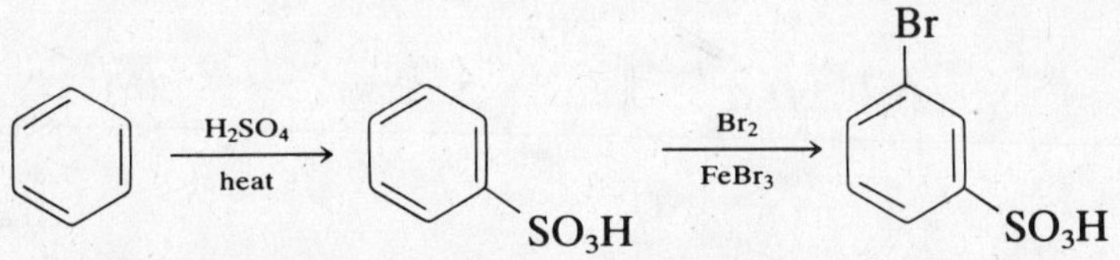

(*e*)

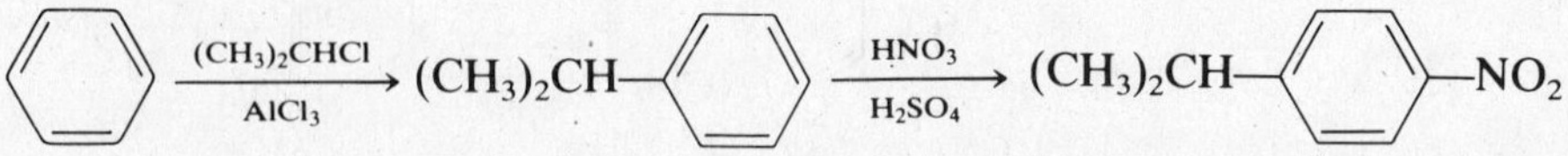

A-6.

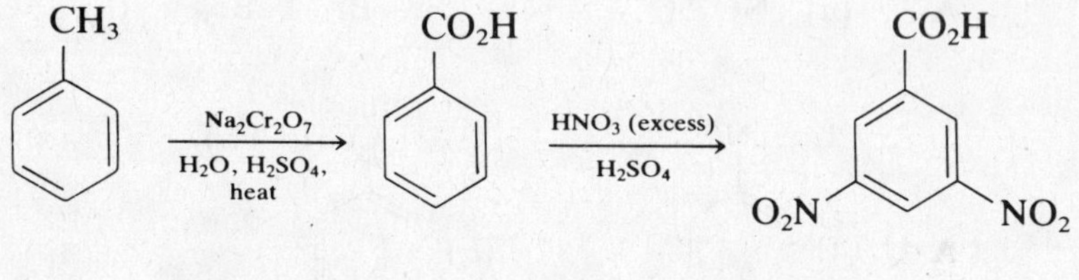

B-1.	(*c*)	**B-2.**	(*b*)	**B-3.**	(*c*)	**B-4.**	(*b*)
B-5.	(*a*)	**B-6.**	(*b*)	**B-7.**	(*c*)	**B-8.**	(*b*)
B-9.	(*c*)	**B-10.**	(*a*)				

CHAPTER 13

A-1.
1:	6.10 ppm	3:	60 MHz
2:	957 Hz	4:	0.00 ppm

A-2. (*a*) Two signals $BrCH_2CH_2CH_2Br$

a   b   a

a: triplet   b: pentet

(*b*) Two signals $\underset{a}{CH_3}\underset{b}{CH_2}\overset{\displaystyle Cl}{\underset{\displaystyle Cl}{C}}\underset{b}{CH_2}\underset{a}{CH_3}$

a: triplet b: quartet

(*c*) Three signals, all singlets

A-3. $CH_3\overset{\displaystyle O}{\overset{\|}{C}}OC(CH_3)_3$ A $CH_3O\overset{\displaystyle O}{\overset{\|}{C}}C(CH_3)_3$ B

A-4. (*a*) ⬡—$CH_2\overset{\displaystyle O}{\overset{\|}{C}}CH_2CH_3$ (*c*) $CH(\overset{\displaystyle O}{\overset{\|}{C}}CH_2CH_3)_3$

(*b*) $(CH_3)_2\overset{\displaystyle HO}{\underset{\displaystyle }{C}}\!-\!\overset{\displaystyle OH}{\underset{\displaystyle }{C}}(CH_3)_2$ (*d*) $(CH_3)_2\overset{\displaystyle OH}{\underset{\displaystyle }{C}}\!-\!C\!\equiv\!N$

A-5. (*a*) Seven signals $\underset{a}{Cl}$—⬡$\overset{b\ c}{}\overset{d}{\underset{e\ f}{}}\overset{\displaystyle O}{\overset{\|}{C}}\underset{g}{CH_2CH_3}$

(*b*) a: singlet   b: doublet   c: doublet   d: singlet
     e: singlet   f: triplet   g: quartet

A-6. $(CH_3)_2CHCH_2CH_2CH(CH_3)_2$

B-1. (*b*)   **B-2.** (*d*)   **B-3.** (*a*)   **B-4.** (*d*)

B-5. (*b*)   **B-6.** (*b*)   **B-7.** (*a*)   **B-8.** (*c*)

B-9. (*a*)   **B-10.** (*a*)

CHAPTER 14

A-1. (*a*) $\overset{\displaystyle X}{⬡}$ (X = Cl, Br, I) $+$ 2Li $\longrightarrow$ $\overset{\displaystyle Li}{⬡}$ $+$ LiX

(*b*) $(CH_3)_3CBr + Mg \longrightarrow (CH_3)_3CMgBr$

(*c*) $2C_6H_5CH_2Li + CuX \longrightarrow (C_6H_5CH_2)_2CuLi + LiX$

(X = Cl, Br, I)

A-2. (*a*) $(C_6H_5)_2\overset{\displaystyle OH}{\underset{\displaystyle }{C}}CH_3$ (*c*) CH_3—⬡—CH_2OH

(*b*) $(CH_3)_2CHCH_2D$ (*d*) $\overset{\displaystyle HO\quad CH_2CH_3}{⬡}$

A-3.

$$CH_2{=}CHCHCH_3 \text{ (with } CH_3 \text{ substituent)} \xrightarrow[\text{2. NaBH}_4]{\text{1. Hg(OAc)}_2, H_2O} CH_3CHCHCH_3 \text{ (with } CH_3 \text{ and } OH \text{ substituents)}$$

$$CH_2{=}CHCHCH_3 \text{ (with } CH_3 \text{ substituent)} \xrightarrow[\text{2. NaBH}_4]{\text{1. Hg(OAc)}_2, CH_3CH_2OH} CH_3CHCHCH_3 \text{ (with } CH_3 \text{ and } OCH_2CH_3 \text{ substituents)}$$

A-4. (*a*) $(CH_3CH_2CH_2)_2CuLi$ (*b*) $(CH_3)_2CHMgX$ (X = Cl, Br, I)
      (*c*) CH_2I_2, Zn(Cu)

A-5. Solvents A, B, and E are suitable; they are all ethers. Solvents C and F have acidic hydrogens and will react with a Grignard reagent. Solvent D is an ester which will react with a Grignard reagent.

A-6. $CH_3(CH_2)_3OH \xrightarrow{PBr_3} CH_3(CH_2)_3Br \xrightarrow{2Li} CH_3(CH_2)_3Li + LiBr$

     $2CH_3(CH_2)_3Li + CuBr \longrightarrow (C_4H_9)_2CuLi \xrightarrow{CH_3(CH_2)_3Br} CH_3(CH_2)_6CH_3$

A-7. (I) $(CH_3)_2CHCCH_3 + CH_3MgBr$ (with $\overset{O}{\overset{\|}{}}$)

     (II) $CH_3CCH_3 + (CH_3)_2CHMgBr$ (with $\overset{O}{\overset{\|}{}}$)

     (III) $(CH_3)_2CHCO_2CH_3 + 2CH_3MgBr$

$$\xrightarrow[\text{2. H}_3O^+]{\text{1. ether}} (CH_3)_2CHC(CH_3)_2 \text{ (with OH substituent)}$$

A-8. $C_6H_5CH_2CH_3 \xrightarrow[\substack{\text{peroxides,}\\ \text{heat}}]{\text{NBS,}} C_6H_5CHCH_3 \text{ (with Br)} \xrightarrow[\text{ether}]{\text{Mg}} C_6H_5CHMgBr \text{ (with } CH_3\text{)}$

$$C_6H_5CHMgBr \text{ (with } CH_3\text{)} \xrightarrow[\text{2. H}_3O^+]{\text{1. } CH_3CH \text{ (with O)}} C_6H_5CHCHCH_3 \text{ (with } CH_3 \text{ and OH)}$$

B-1. (*c*)   **B-2.** (*a*)   **B-3.** (*b*)   **B-4.** (*a*)
B-5. (*b*)   **B-6.** (*c*)   **B-7.** (*b*)

CHAPTER 15

A-1. (*a*) (cyclohexanone structure, $\overset{O}{\overset{\|}{}}$ on ring)

     (*b*) $C_6H_5CO_2CH_2CH_3$

     (*c*) 1. B_2H_6;   2. H_2O_2, HO^-

     (*d*) OsO_4, $(CH_3)_3COOH$, $(CH_3)_3COH$, HO^-

     (*e*) H_2NCNH_2 (with $\overset{S}{\overset{\|}{}}$)

A-2. (a) $C_6H_5CH_2\overset{\displaystyle O}{\overset{\|}{C}}H$ (c) $(C_6H_5CH_2CH_2)_2O$

(b) $CH_3\overset{\displaystyle O}{\overset{\|}{C}}Cl$, pyridine; or   (d) $K_2Cr_2O_7$, H^+, H_2O

$(CH_3\overset{\displaystyle O}{\overset{\|}{C}})_2O$: or CH_3CO_2H, H^+

A-3. (I) $(CH_3)_2CHBr + Mg \longrightarrow (CH_3)_2CHMgBr \xrightarrow[\text{2. }H_3O^+]{\text{1. } \triangle} (CH_3)_2CHCH_2CH_2OH$

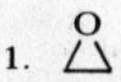

(II) $(CH_3)_2CHCH_2Br + Mg \longrightarrow (CH_3)_2CHCH_2MgBr$

$(CH_3)_2CHCH_2MgBr \xrightarrow[\text{2. }H_3O^+]{\text{1. }H_2C=O} (CH_3)_2CHCH_2CH_2OH$

A-4. (a) [structure] (b) [structure] (c) [structure]

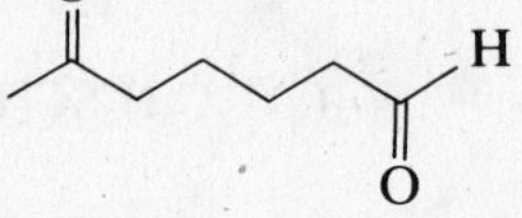

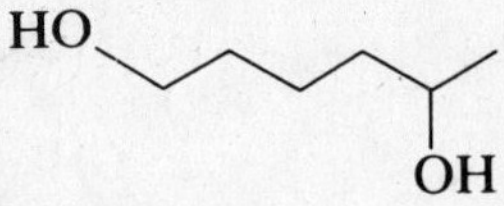

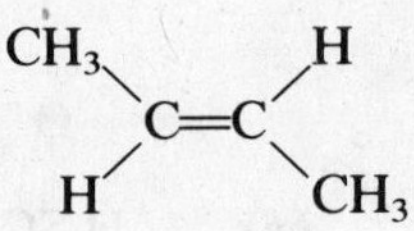

A-5. (a) PCC or PDC or $(C_5H_5N)_2CrO_3$ in CH_2Cl_2
(b) $Na_2Cr_2O_7$, H^+, H_2O
(c) H_2, Pt or Pd
(d) 1. $LiAlH_4$; 2. H_2O
(e) OsO_4, $(CH_3)_3COOH$, $(CH_3)_3COH$, HO^-

A-6. [structures A, B, C]

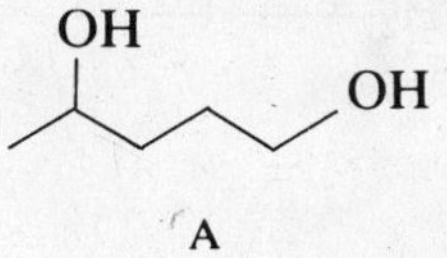

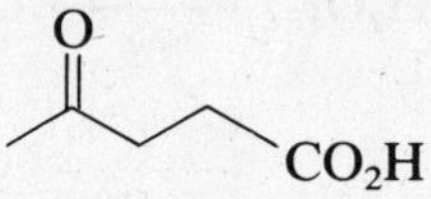

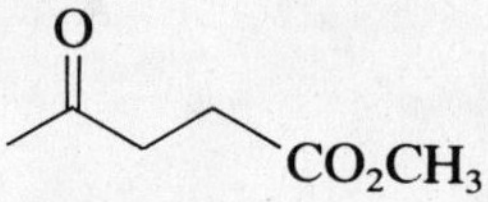

A-7. (a) $(CH_3)_2C=CHCH_3 \xrightarrow[\text{2. }H_2O_2,\ HO^-]{\text{1. }B_2H_6}$ $(CH_3)_2CHCHCH_3 \xrightarrow[CH_2Cl_2]{(C_5H_5N)_2CrO_3} (CH_3)_2CHCCH_3$

(b) [cyclopropyl structures]

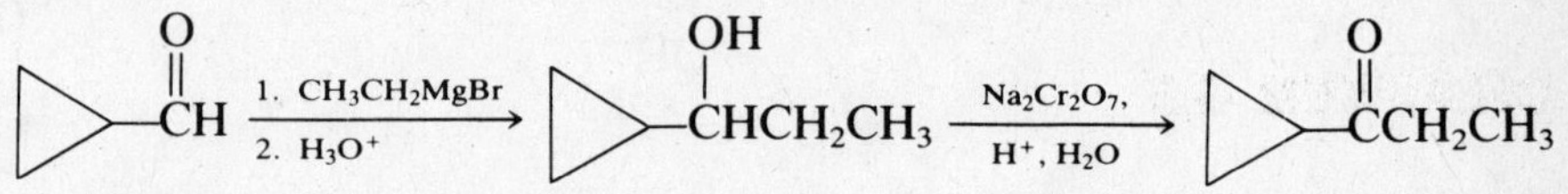

(c) $C_6H_5CH_3 \xrightarrow[\text{peroxides, heat}]{NBS} C_6H_5CH_2Br \xrightarrow{Mg} C_6H_5CH_2MgBr$

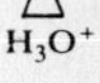

$C_6H_5CH_2CH_2CO_2H \xleftarrow[H^+,\ H_2O]{K_2Cr_2O_7} C_6H_5CH_2CH_2CH_2OH$

$\downarrow CH_3CH_2OH, H^+$

$C_6H_5CH_2CH_2CO_2CH_2CH_3$

B-1. *(b)*   **B-2.** *(d)*   **B-3.** *(c)*   **B-4.** *(c)*

B-5. *(b)*   **B-6.** *(b)*   **B-7.** *(a)*   **B-8.** *(a)*

B-9. *(d)*

CHAPTER 16

A-1. $CH_3OCH_2CH_2CH_3$ Methyl propyl ether

           $CH_3OCH(CH_3)_2$ Isopropyl methyl ether

           $CH_3CH_2OCH_2CH_3$ Diethyl ether

A-2. *(a)*

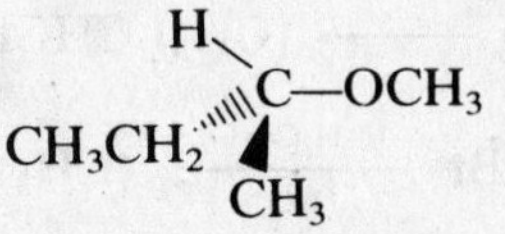

        (d)

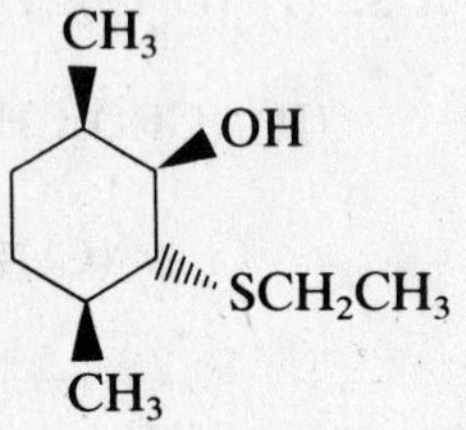

        (b)

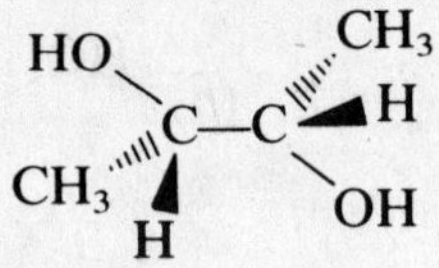

        (e) $C_6H_5SCH_2CH_3$

        (c)

        (f)

A-3. *(a)*

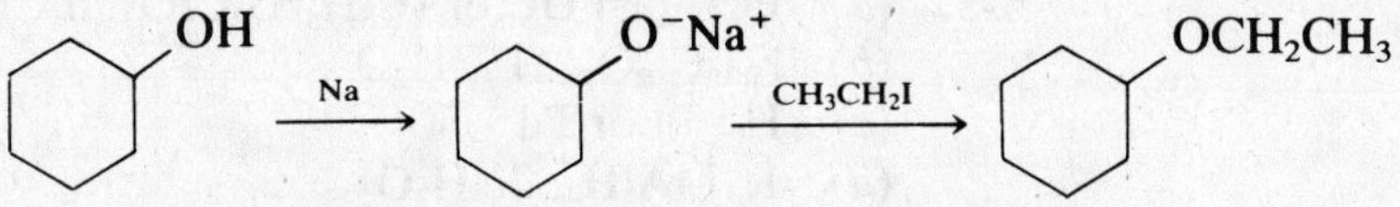

A-4. $CH_3CH_2OH \xrightarrow[\text{heat}]{H_2SO_4} CH_2{=}CH_2 \xrightarrow{CH_3CO_2OH}$

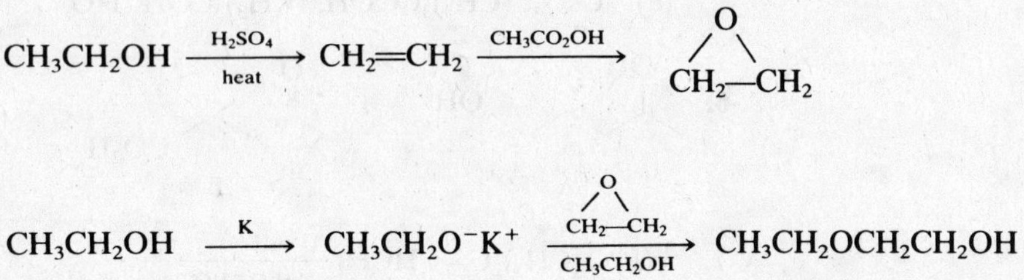

A-5. *(a)*

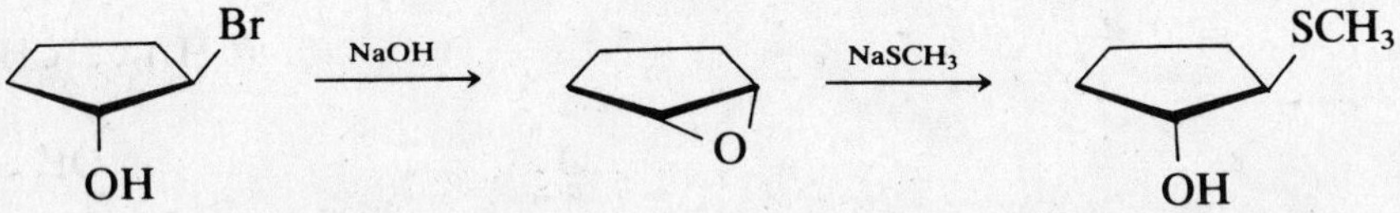

        (b)

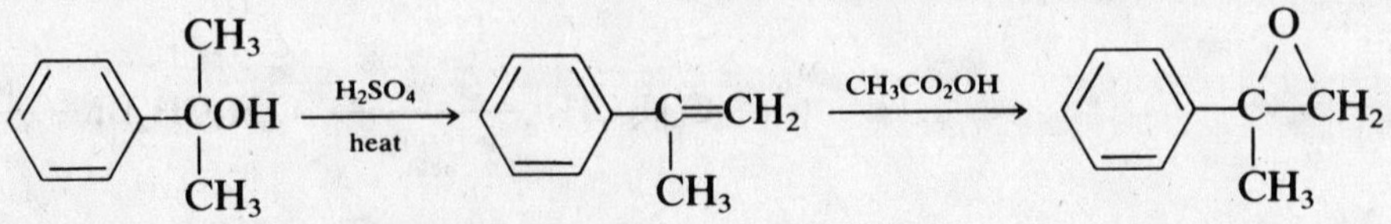

A-6. A:

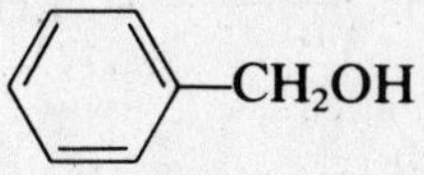

     B:

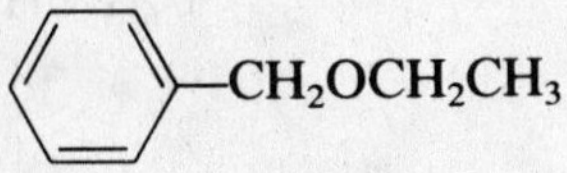

B-1. *(a)*   **B-2.** *(b)*   **B-3.** *(d)*   **B-4.** *(d)*

B-5. *(d)*   **B-6.** *(a)*

CHAPTER 17

A-1. (*a*) 3,4-Dimethylhexanal

(*b*) 2,2,5-Trimethylhexan-3-one

(*c*) *trans*-4-Bromo-2-methylcyclohexanone

A-2. (*a*)

$$\underset{H}{\overset{\overset{\overset{\displaystyle O}{\parallel}}{CH_3C}}{}}C=C\underset{CH_2CH_3}{\overset{H}{}}$$

(*b*) $CH_3\overset{\displaystyle O}{\overset{\parallel}{C}}\overset{\displaystyle O}{\overset{\parallel}{C}}HCCH_3$ (with cyclopropyl)

(*c*) $C_6H_5\overset{CH_3}{\overset{|}{C}}HCH\overset{O}{\overset{\parallel}{CH}}_2CH$ with CH_2CH_3 branch

$$C_6H_5\underset{}{\overset{CH_3}{\overset{|}{CH}}}CH\overset{\displaystyle O}{\overset{\parallel}{CHCH_2CH}}$$
$$\underset{CH_2CH_3}{|}$$

A-3. (*a*) cyclohexane with $\overset{OH}{\underset{CN}{}}$ substituents

(*b*) NH_2OH

(*c*) $(CH_3)_2CH\overset{\displaystyle O}{\overset{\parallel}{CH}}$ + $HOCH_2CH_2CH_2OH$

(*d*) $(C_6H_5)_3\overset{+}{P}—\overset{..}{C}HCH_2CH_3$

(*e*) cyclohexane $=NNHC_6H_5$ with CH_3

(*f*) $CH_3CH_2CH_2CH(OCH_2CH_3)_2$

(*g*) $C_6H_5\overset{\displaystyle O}{\overset{\parallel}{C}}CH_2CH_3$ + $(CH_3)_2NH$

(*h*) $CH_3CH_2CH_2\overset{\displaystyle O}{\overset{\parallel}{C}}OH$

A-4. (*a*) $(C_6H_5)_3\overset{+}{P}—CH_2CH(CH_3)_2$ Br^- $(C_6H_5)_3\overset{+}{P}—\overset{..}{C}HCH(CH_3)_2$
　　　　　　　　　　　A　　　　　　　　　　　　　　　　　B

$$C_6H_5CH=CHCH(CH_3)_2$$
　　　　　　C

(*b*) $C_6H_5CH_2\overset{\displaystyle O}{\overset{\parallel}{C}}CH_2CH_3$ $C_6H_5CH_2O\overset{\displaystyle O}{\overset{\parallel}{C}}CH_2CH_3$
　　　　　　D　　　　　　　　　　　　　　　　E

A-5. (*a*) $CH_3CH_2I + (C_6H_5)_3P \longrightarrow (C_6H_5)_3\overset{+}{P}—CH_2CH_3$ I^-

$$\downarrow C_4H_9Li$$

$$(C_6H_5)_3\overset{+}{P}—\overset{..}{C}HCH_3$$

$$(CH_3)_2C=O + (C_6H_5)_3\overset{+}{P}—\overset{..}{C}HCH_3 \longrightarrow$$

$$(CH_3)_2C=CHCH_3 \xrightarrow{CH_3CO_2OH} (CH_3)_2\overset{\displaystyle O}{\overset{\diagup \diagdown}{C}—CHCH_3}$$

(b) [cyclohexanone structure with CH₃ and OH] $\xrightarrow[\text{H}^+]{\text{OH OH}}$ [ketal structure with O O, CH₃, OH] $\xrightarrow[\text{CH}_2\text{Cl}_2]{(\text{C}_5\text{H}_5\text{N})_2\text{CrO}_3}$ [ketal structure with O O, CH₃, O] $\xrightarrow[\text{2. H}_3\text{O}^+]{\text{1. CH}_3\text{MgI}}$ [cyclohexanone with CH₃, HO CH₃]

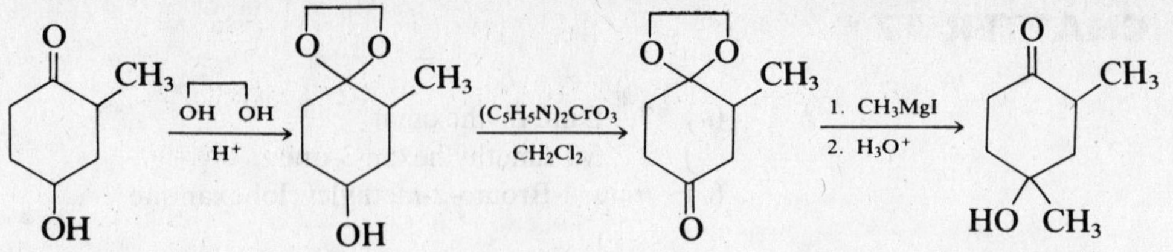

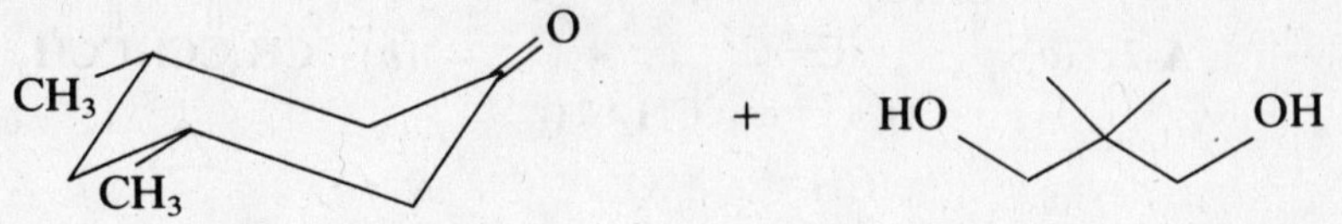

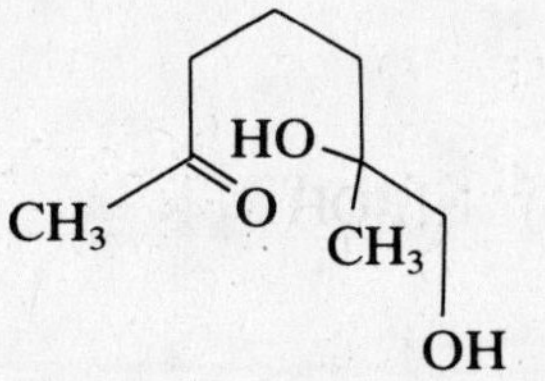

A-6.

[cyclohexanone structure with two CH₃ groups] $+$ HO—[structure]—OH

A-7.

[structure] $\equiv$ $CH_3CCH_2CH_2CH_2CCH_2OH$ with O, OH, CH₃

A-8. (a) (1) CH_3MgI; (2) H_3O^+; (3) H_2SO_4, heat

(b) $(C_6H_5)_3\overset{+}{P}\!-\!\overset{..}{C}H_2$ (from $(C_6H_5)_3P + CH_3I \longrightarrow \xrightarrow{C_4H_9Li}$)

(c) $HOCH_2CH_2OH$, H^+(cat), heat
(d) CH_3CO_2H

B-1. (a)   **B-2.** (c)   **B-3.** (b)   **B-4.** (b)

B-5. (a)   **B-6.** (b)   **B-7.** (a)   **B-8.** (d)

B-9. (b)

CHAPTER 18

A-1. (a) $CH_2\!=\!CCH_2CH_3$ and $CH_3C\!=\!CHCH_3$ (each with OH)

(b) [resonance structures of cyclohexane-1,3-dione enolate] Na^+

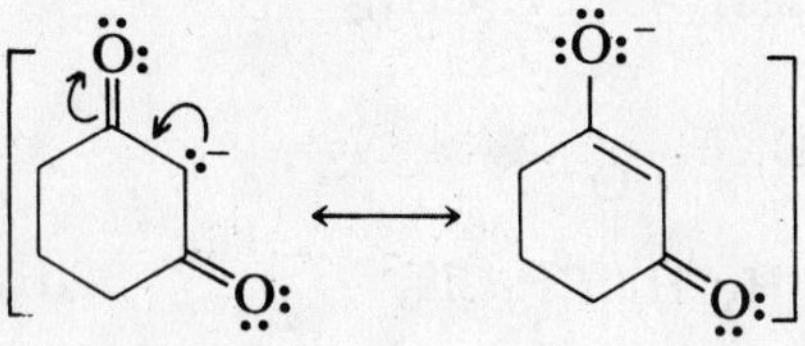

A-2. $C_6H_5CH_2CH\!=\!CCH$ with C_6H_5 (A) $(CH_3CH_2)_2CHCH_2CCH_2CH_3$ with O (B)

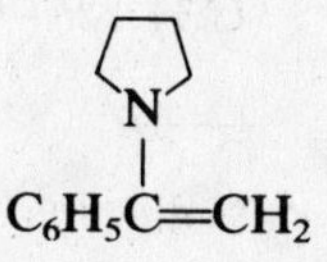

[pyrrolidine enamine structure] $C_6H_5C\!=\!CH_2$ (C) $C_6H_5CCH_2CH_2CH\!=\!CH_2$ with O (D)

A-3.

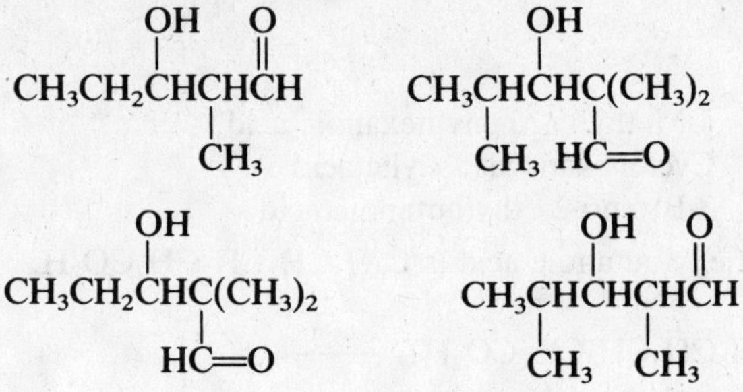

$$CH_3CH_2\overset{OH}{\underset{}{C}}HCH\overset{O}{\underset{CH_3}{C}}H$$

$$CH_3\overset{OH}{\underset{CH_3}{C}}HCHC(CH_3)_2\!\!\!\overset{}{\underset{HC=O}{}}$$

$$CH_3CH_2\overset{OH}{\underset{HC=O}{C}}HC(CH_3)_2$$

$$CH_3\overset{OH}{\underset{CH_3}{C}}HCH\overset{O}{\underset{CH_3}{C}}HCH$$

A-4.

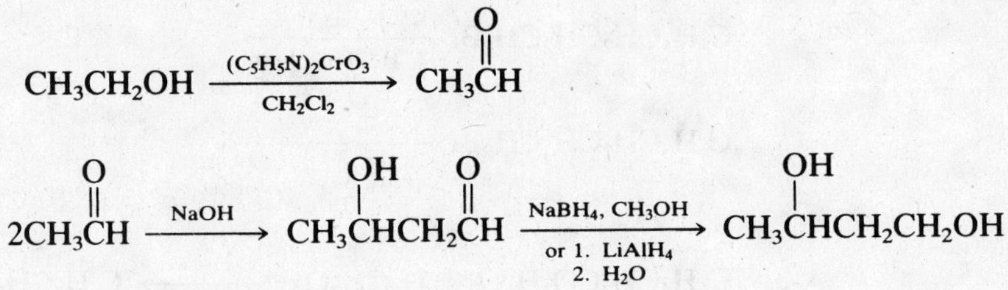

$$CH_3CH_2OH \xrightarrow[CH_2Cl_2]{(C_5H_5N)_2CrO_3} CH_3\overset{O}{C}H$$

$$2CH_3\overset{O}{C}H \xrightarrow{NaOH} CH_3\overset{OH}{C}HCH_2\overset{O}{C}H \xrightarrow[\substack{or\ 1.\ LiAlH_4 \\ 2.\ H_2O}]{NaBH_4,\ CH_3OH} CH_3\overset{OH}{C}HCH_2CH_2OH$$

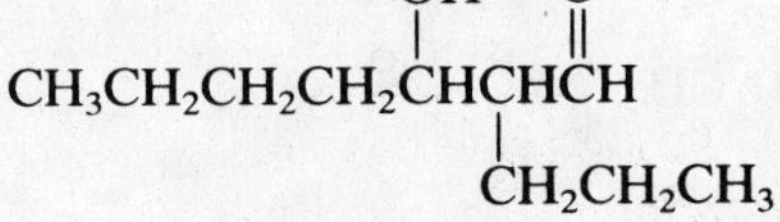

A-5.

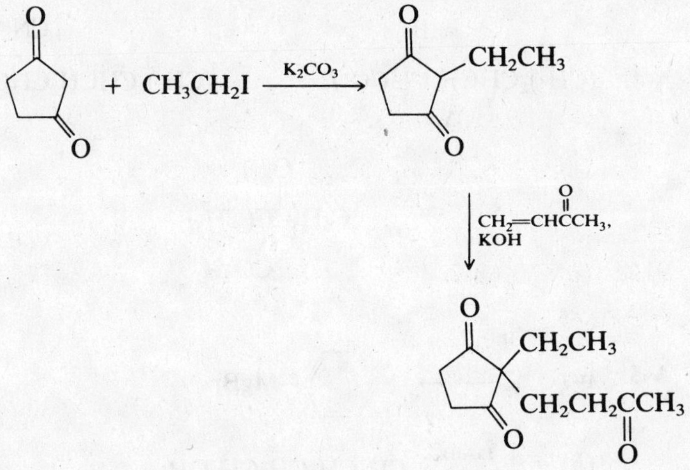

(cyclopentane-1,3-dione) $+ CH_3CH_2I \xrightarrow{K_2CO_3}$ (2-ethyl-cyclopentane-1,3-dione with CH₂CH₃)

$$\xrightarrow[KOH]{CH_2=CHCCH_3,\ \overset{O}{}}$$

(2-ethyl-2-(3-oxobutyl)cyclopentane-1,3-dione with CH₂CH₃ and CH₂CH₂CCH₃)

A-6. (a)

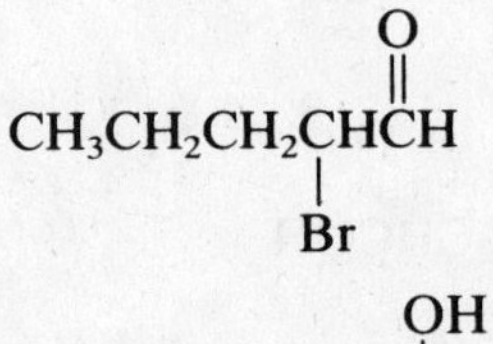

$$CH_3CH_2CH_2\overset{O}{\underset{Br}{C}}HCH$$

(b)

$$CH_3CH_2CH_2CH_2\overset{OH}{\underset{CH_2CH_2CH_3}{C}}HCH\overset{O}{C}H$$

(c)

(phenyl–CH= attached to a cyclopentanone ring with two CH₃ groups)

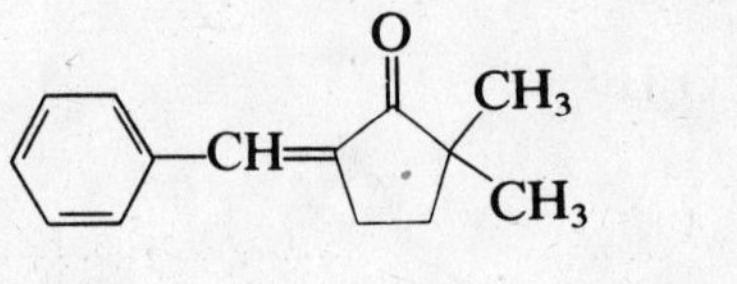

(d)

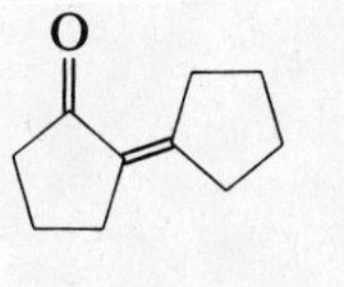

(spiro/cyclopentylidene cyclopentanone)

(e)

(diphenyl structure: C₆H₅–CH(SCH₃)–CH₂–C(=O)–C₆H₅)

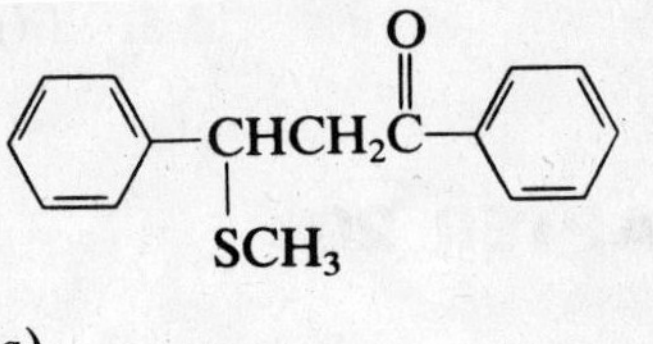

B-1. (a)   **B-2.** (b)   **B-3.** (b)   **B-4.** (a)

B-5. (c)   **B-6.** (c)

CHAPTER 19

A-1. (*a*) 4-Methyl-5-phenylhexanoic acid
(*b*) Cyclohexanecarboxylic acid
(*c*) 3-Bromo-2-ethylbutanoic acid

A-2. 4-Phenylbutanoic acid is $C_6H_5CH_2CH_2CH_2CO_2H$.

$$C_6H_5CH_2CH_2CH(CO_2H)_2 \xrightarrow{\text{heat}}$$

$$C_6H_5CH_2CH_2CH_2Br \xrightarrow[\text{2. H}^+\text{, H}_2\text{O, heat}]{\text{1. CN}^-}$$

$$C_6H_5CH_2CH_2CH_2Br \xrightarrow[\substack{\text{2. CO}_2 \\ \text{3. H}_3\text{O}^+}]{\text{1. Mg}}$$

A-3. $C_6H_5CH_2CO_2H \; + \; CH_3CH_2OH \xrightarrow{\text{H}^+\text{(cat)}} C_6H_5CH_2\overset{\overset{\displaystyle O}{\|}}{C}OCH_2CH_3 \; + \; H_2O$

A-4. $(CH_3)_2CHCH_2\overset{\overset{\displaystyle Br}{|}}{C}HCO_2H$ $(CH_3)_2CHCH_2\overset{\overset{\displaystyle CN}{|}}{C}HCO_2H$
         A   B

$$C_6H_5\overset{\overset{\displaystyle O}{\|}}{C}\underset{\underset{\displaystyle CO_2H}{|}}{C}(CH_3)_2$$
                C

A-5. (*a*) $\xrightarrow{\text{Mg}}$ ⬡—MgBr

     (*b*) $\xrightarrow[\text{2. H}_2\text{O}]{\text{1. LiAlH}_4}$ $CH_3CH_2CH_2CH_2OH$

     (*c*) $\xrightarrow{\text{Br}_2\text{, P}}$ $CH_3CH_2\overset{\overset{\displaystyle Br}{|}}{C}H\overset{\overset{\displaystyle O}{\|}}{C}OH$

     (*d*) $CH_3CH_2\overset{\overset{\displaystyle CN}{|}}{\underset{\underset{\displaystyle OH}{|}}{C}}H \xrightarrow[\text{heat}]{\text{H}^+\text{, H}_2\text{O}}$

A-6. $CH_3CH_2\overset{}{\underset{\underset{\displaystyle Br}{|}}{C}}HCO_2H$

B-1. (*b*)   **B-2.** (*a*)   **B-3.** (*c*)   **B-4.** (*c*)
B-5. (*d*)   **B-6.** (*c*)

CHAPTER 20

A-1. (*a*) Propyl butanoate   (*b*) *N*-Methylbenzamide
(*c*) 4-Methylpentanoyl chloride

A-2. (a) $C_6H_5\overset{O}{C}\overset{O}{C}C_6H_5$ (b) $CH_3\overset{O}{C}NHCHCH_2CH_3$ (c) $C_6H_5\overset{O}{C}OC_6H_5$

                                                               CH_3

A-3. (a) $SOCl_2$ (b) Br_2, $NaOH$, H_2O (c) $C_6H_5\overset{O}{C}OCH_3$

A-4. (a) CH_3CO_2H + ⬡—OH   (b) $C_6H_5\overset{O}{C}O$—⬠

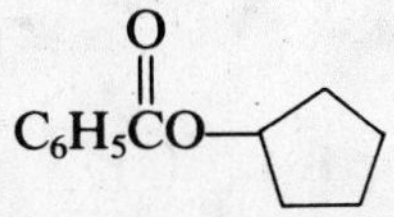

(c)

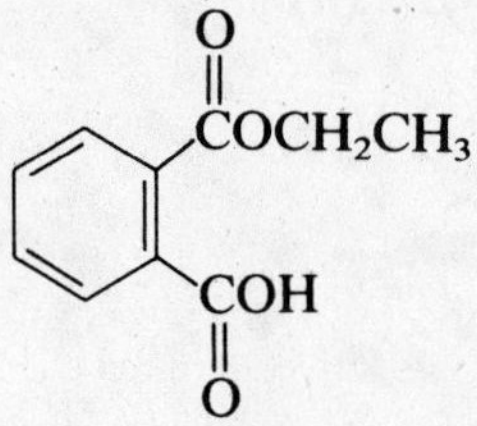

(d) $CH_3CH_2\overset{O}{C}N(CH_3)_2 + CH_3CH_2OH$

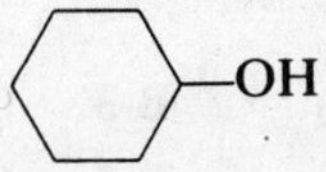

(e) CH_3—⬡—CO_2H + $CH_3\overset{+}{N}H_3$ HSO_4^-

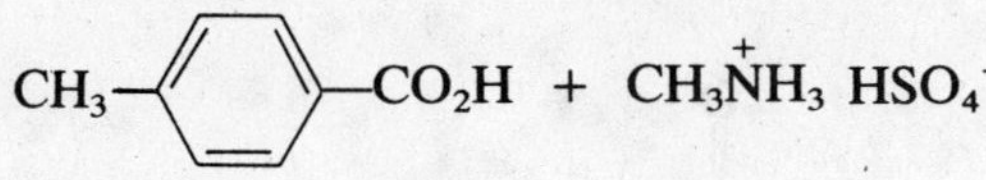

A-5.   from

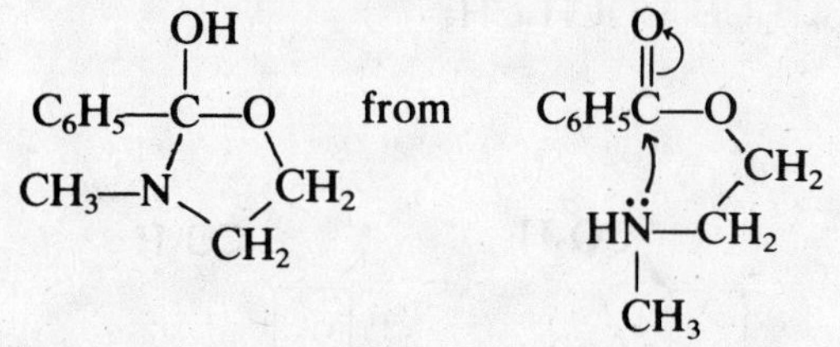

A-6.

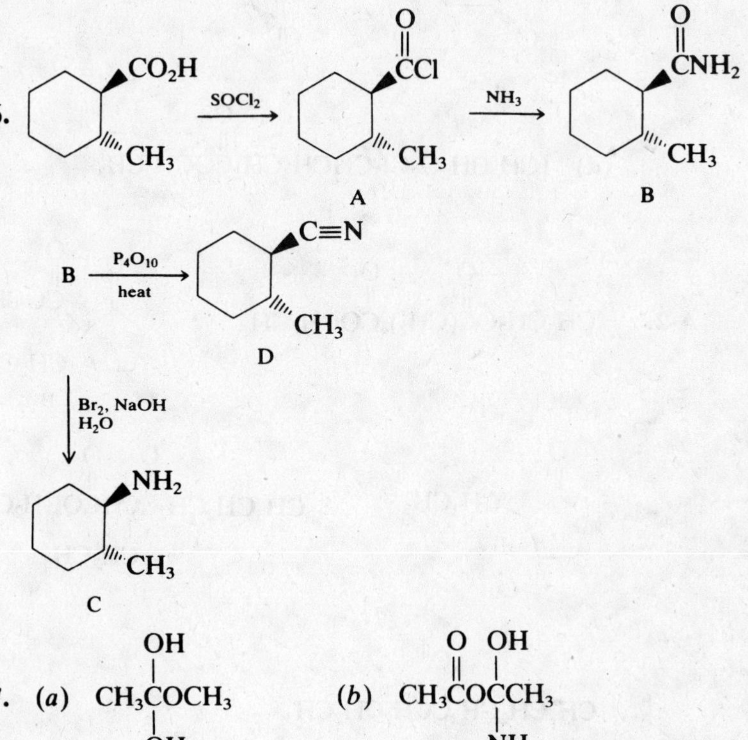

A-7. (a) $CH_3\overset{OH}{\underset{OH}{C}}CH_3$ (b) $CH_3\overset{O}{C}\overset{OH}{\underset{NH_2}{C}}CH_3$

A-8. CH_3—⟨benzene⟩—Br $\xrightarrow{Mg}$ CH_3—⟨benzene⟩—$MgBr$ $\xrightarrow[\text{2. }H_3O^+]{\text{1. }CO_2}$ CH_3—⟨benzene⟩—CO_2H

$\downarrow SOCl_2$

CH_3—⟨benzene⟩—NH_2 $\xleftarrow[H_2O]{Br_2, NaOH}$ CH_3—⟨benzene⟩—$\overset{\overset{O}{\|}}{C}NH_2$ $\xleftarrow{NH_3}$ CH_3—⟨benzene⟩—$\overset{\overset{O}{\|}}{C}Cl$

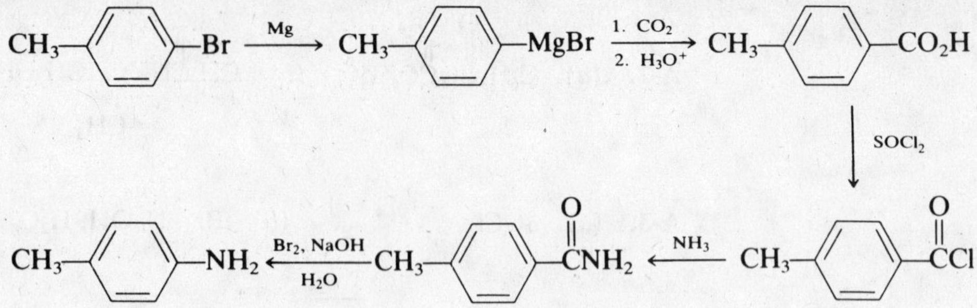

B-1. (a)   **B-2.** (c)   **B-3.** (b)   **B-4.** (c)

B-5. (d)   **B-6.** (c)   **B-7.** (d)

CHAPTER 21

A-1. (a) $CH_3CH_2CH_2\overset{\overset{O}{\|}}{C}\overset{\underset{\underset{CH_2CH_3}{|}}{C}}{C}H\overset{\overset{O}{\|}}{C}OCH_2CH_3$

(e) $CH_3\overset{\overset{O}{\|}}{C}\overset{\underset{\underset{CH_2C_6H_5}{|}}{C}}{C}H\overset{\overset{O}{\|}}{C}OCH_2CH_3$

(b) $C_6H_5CH_2\overset{\overset{O}{\|}}{C}OCH_2CH_3$

(f) 1. HO^-, H_2O
    2. H_3O^+
    3. heat

(c) ⟨cyclobutane with CO_2H and Cl⟩ + ⟨cyclobutane with CO_2H and Cl⟩

(g) $CH_3\overset{\overset{O}{\|}}{C}CH_2CH_2CO_2H$

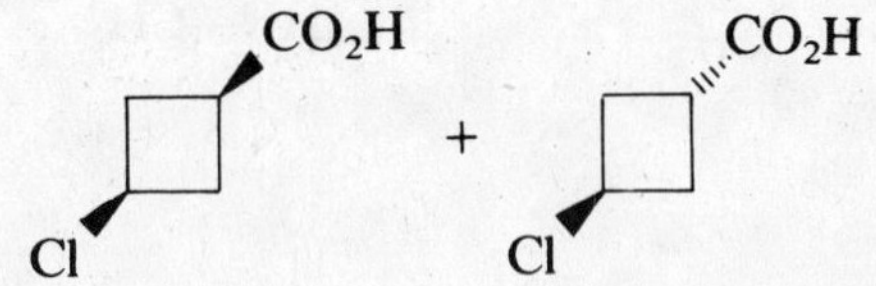

(d) $(CH_3CH_2O_2C)_2CHCH_2CH_2\overset{\overset{O}{\|}}{C}OCH_2CH_3$

A-2. $CH_3CH_2O\overset{\overset{O}{\|}}{C}(CH_2)_4\overset{\overset{O}{\|}}{C}OCH_2CH_3$
A

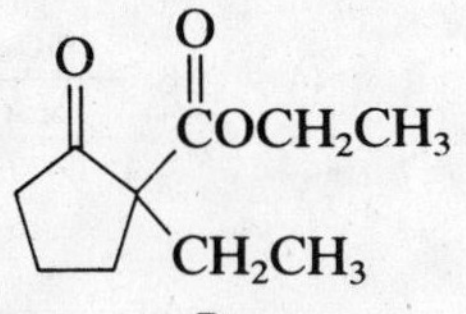

⟨cyclopentanone with $\overset{\overset{O}{\|}}{C}OCH_2CH_3$ and CH_2CH_3⟩
B

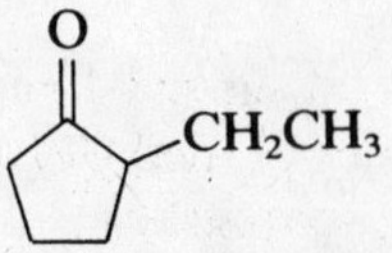

⟨cyclopentanone with CH_2CH_3⟩
C

$CH_3CH_2CH_2\overset{\overset{O}{\|}}{C}\overset{\underset{\underset{CH_2CH_3}{|}}{C}}{C}H\overset{\overset{O}{\|}}{C}OCH_2CH_3$
D

$CH_3CH_2CH_2\overset{\overset{O}{\|}}{C}CH_2CH_2CH_3$
E

A-3. (*a*)

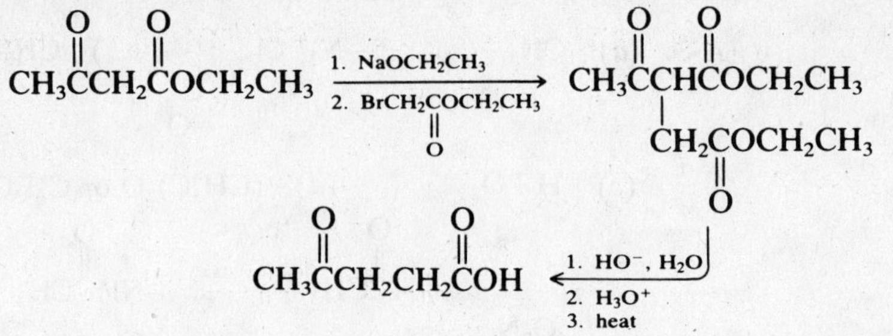

$$CH_3CCH_2COCH_2CH_3 \xrightarrow[\text{2. BrCH}_2\text{COCH}_2\text{CH}_3]{\text{1. NaOCH}_2\text{CH}_3} CH_3CCHCOCH_2CH_3$$

$$CH_2COCH_2CH_3$$

$$CH_3CCH_2CH_2COH \xleftarrow[\substack{\text{2. H}_3\text{O}^+ \\ \text{3. heat}}]{\text{1. HO}^-,\ \text{H}_2\text{O}}$$

(*b*)

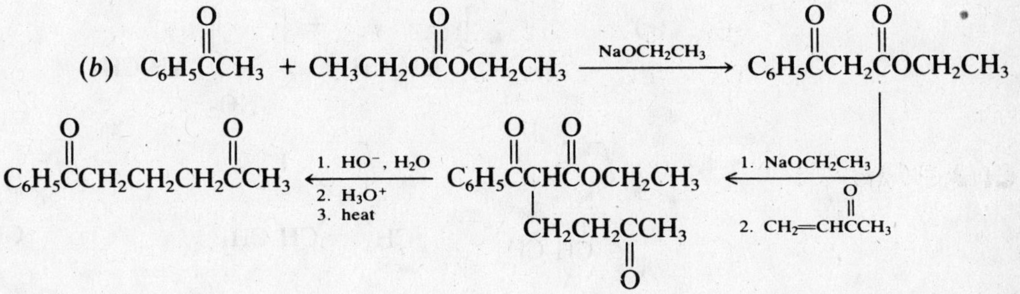

$$C_6H_5CCH_3 + CH_3CH_2OCOCH_2CH_3 \xrightarrow{\text{NaOCH}_2\text{CH}_3} C_6H_5CCH_2COCH_2CH_3$$

$$C_6H_5CCH_2CH_2CH_2CCH_3 \xleftarrow[\substack{\text{2. H}_3\text{O}^+ \\ \text{3. heat}}]{\text{1. HO}^-,\ \text{H}_2\text{O}} C_6H_5CCHCOCH_2CH_3 \xleftarrow[\substack{\text{2. CH}_2=\text{CHCCH}_3}]{\text{1. NaOCH}_2\text{CH}_3}$$

$$CH_2CH_2CCH_3$$

$$O$$

A-4.

$$2CH_3CH_2COCH_2CH_3 \xrightleftharpoons{\text{NaOCH}_2\text{CH}_3}$$

$$CH_3\ddot{C}HCOCH_2CH_3 \longrightarrow CH_3CHCOCH_2CH_3$$

$$CH_3CH_2C\!-\!OCH_2CH_3$$

$$O^-$$

$$+ CH_3CH_2COCH_2CH_3$$

$$-CH_3CH_2O^-$$

$$CH_3CHCOCH_2CH_3$$

$$CH_3\ddot{C}COCH_2CH_3 \xleftarrow{-H^+} CH_3CH_2C=O$$

$$CH_3CH_2C=O$$

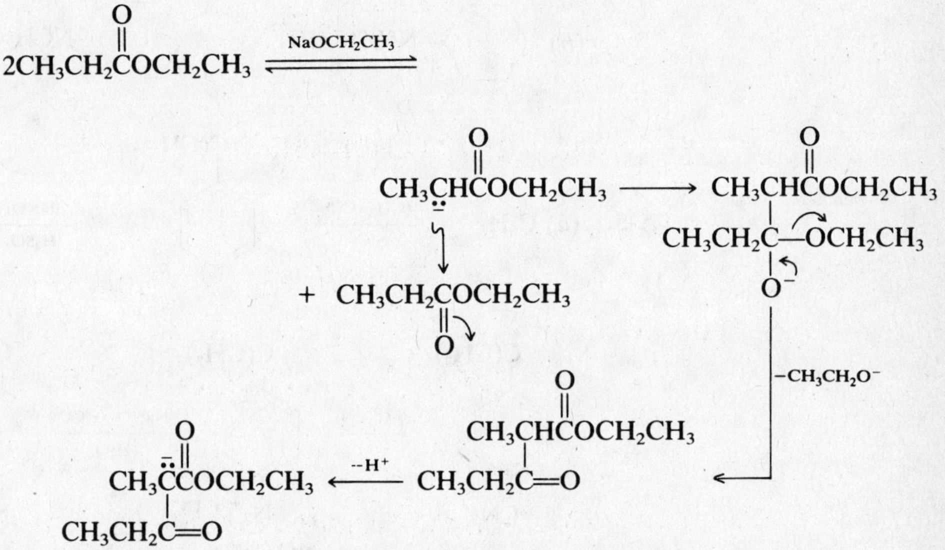

B-1. (*b*) **B-2.** (*d*) **B-3.** (*c*) **B-4.** (*b*)

B-5. (*c*)

CHAPTER 22

A-1. (*a*) 1,1-Dimethylpropylamine or 2-methyl-2-butanamine; primary
     (*b*) *N*-Methylcyclopentylamine or *N*-methylcyclopentanamine; secondary
     (*c*) *m*-Bromo-*N*-propylaniline; secondary

A-2. (*a*) NaN$_3$ (*b*) KCN   (*c*) [phthalimide N$^-$K$^+$ structure]

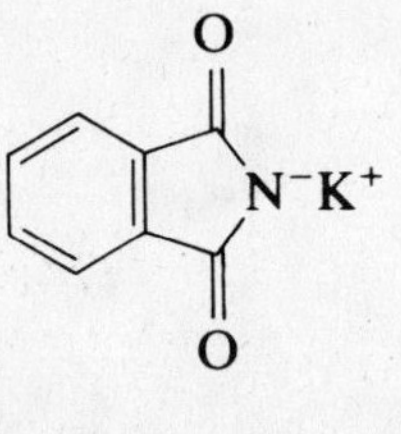

A-3. (a) CH_3—⟨benzene ring⟩—$N_2{}^+ Cl^-$ (b) CH_3—⟨benzene ring⟩—Br

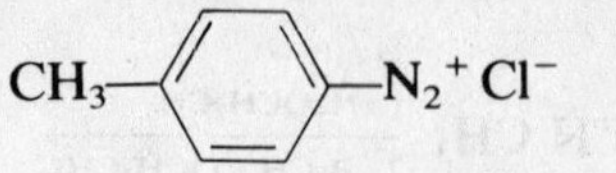

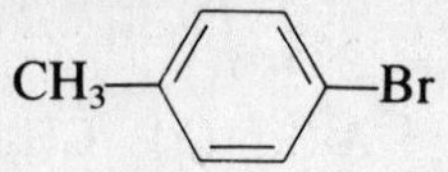

(c) H_3PO_2 (d) $(CH_3\overset{O}{\underset{\|}{C}})_2O$ or $CH_3\overset{O}{\underset{\|}{C}}Cl$

(e)

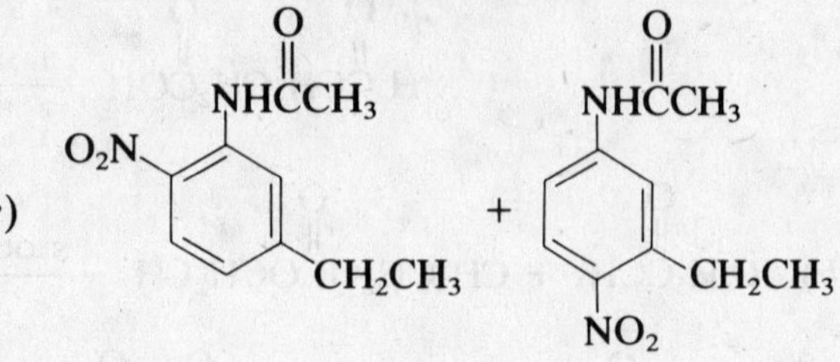

A-4. (a)

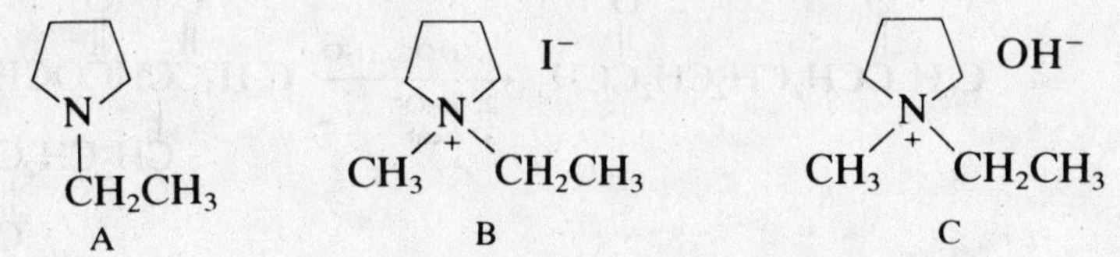

A   B   C

(b)

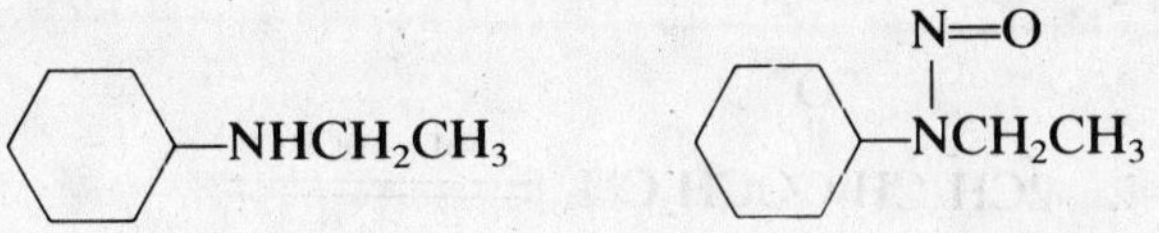

D   E

A-5. (a) $C_6H_6 \xrightarrow[AlCl_3]{(CH_3)_3CCl}$ ⟨t-butylbenzene⟩ $\xrightarrow[H_2SO_4]{HNO_3}$ ⟨p-t-butylnitrobenzene⟩

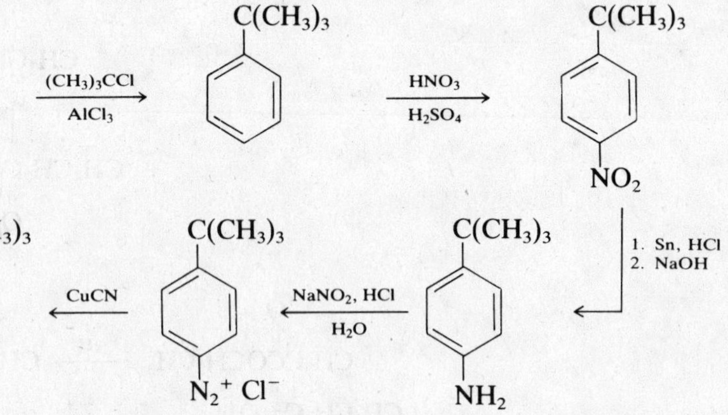

(b) $C_6H_6 \xrightarrow[H_2SO_4]{HNO_3}$ ⟨nitrobenzene⟩ $\xrightarrow{Cl_2, FeCl_3}$ ⟨m-chloronitrobenzene⟩ $\xrightarrow[\text{2. NaOH}]{\text{1. Sn, HCl}}$ ⟨m-chloroaniline⟩

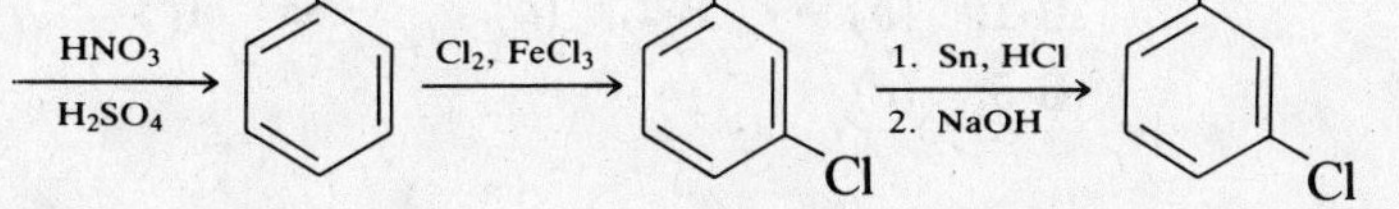

(c) $C_6H_5NH_2 \xrightarrow[H_2O]{NaNO_2, HCl} C_6H_5N_2{}^+ Cl^- \xrightarrow{C_6H_5N(CH_3)_2} C_6H_5N{=}N$—⟨ring⟩—$N(CH_3)_2$

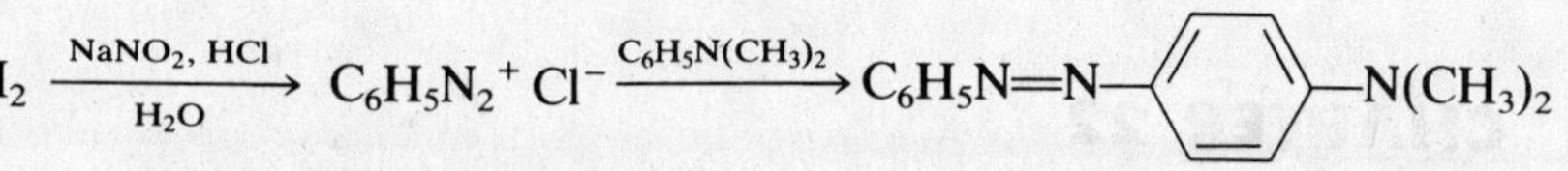

A-6. In the para isomer, resonance delocalization of the electron pair of the amine nitrogen involves the nitro group.

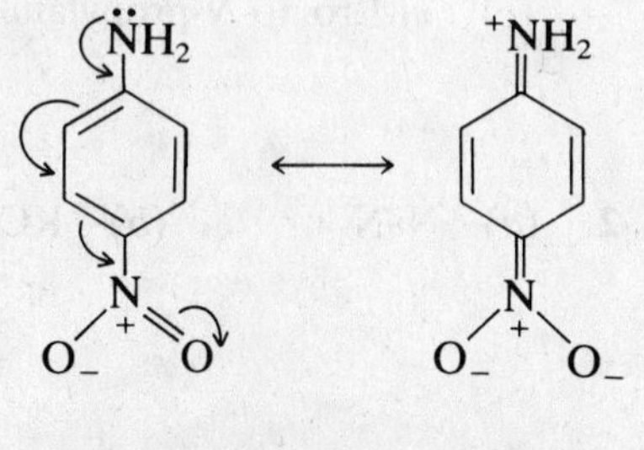

A-7. Strongest base: C, an alkylamine
Weakest base: D, a lactam (cyclic amide)

B-1. (a)	**B-2.** (b)	**B-3.** (d)	**B-4.** (c)
B-5. (a)	**B-6.** (d)	**B-7.** (d)	**B-8.** (c)

CHAPTER 23

A-1. (a)

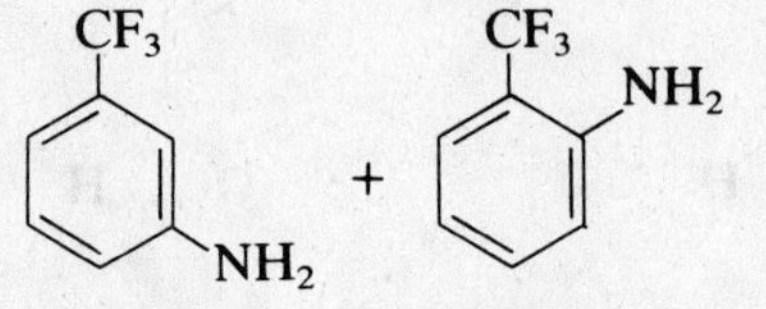

(b)

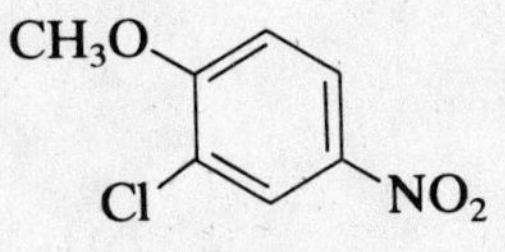

(c)

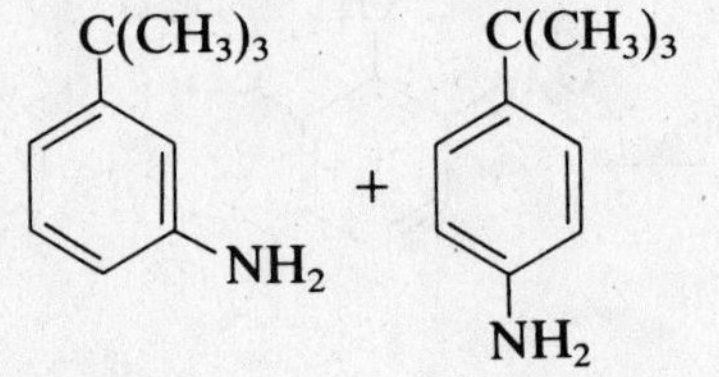

A-2. (a)

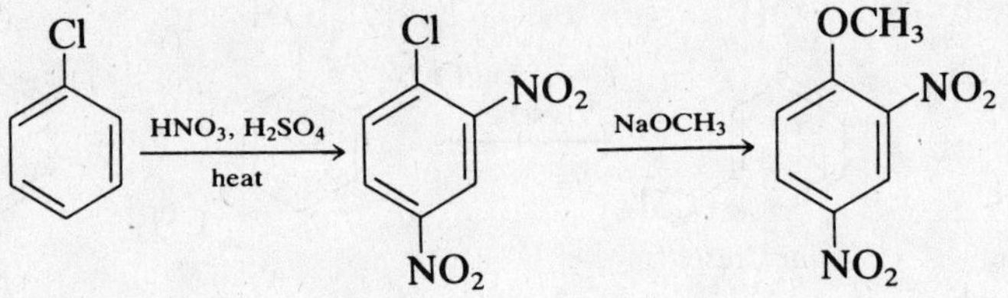

(b)

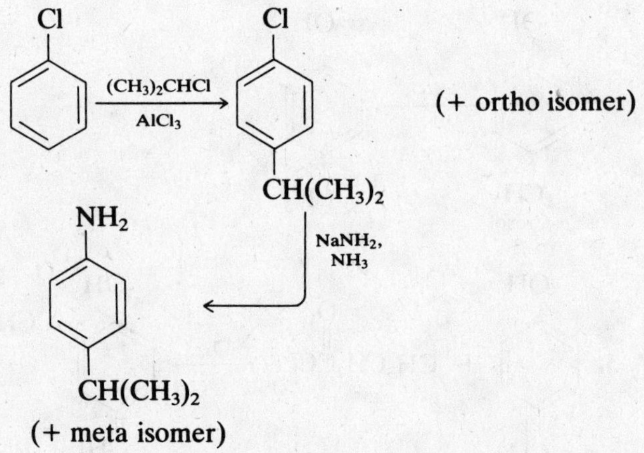

(+ ortho isomer)

(+ meta isomer)

A-3.

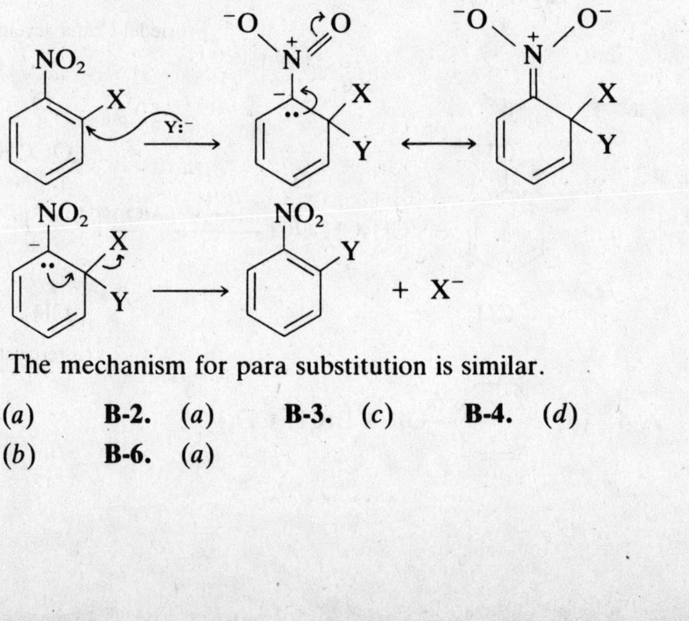

The mechanism for para substitution is similar.

B-1. (a)	**B-2.** (a)	**B-3.** (c)	**B-4.** (d)
B-5. (b)	**B-6.** (a)		

CHAPTER 24

A-1. *p*-Hydroxybenzaldehyde is the stronger acid. The phenoxide anion is stabilized by conjugation with the aldehyde carbonyl.

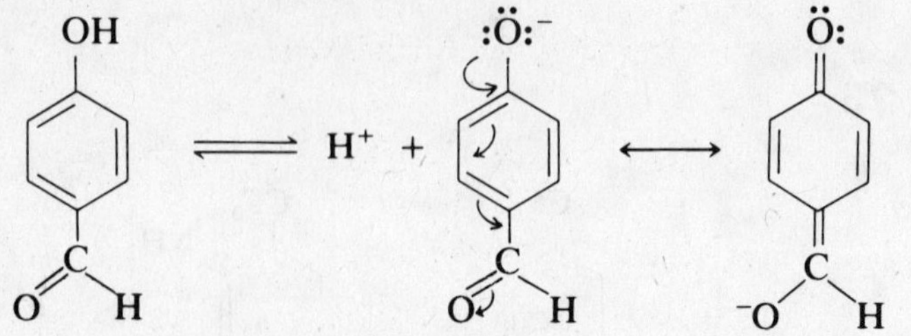

A-2.

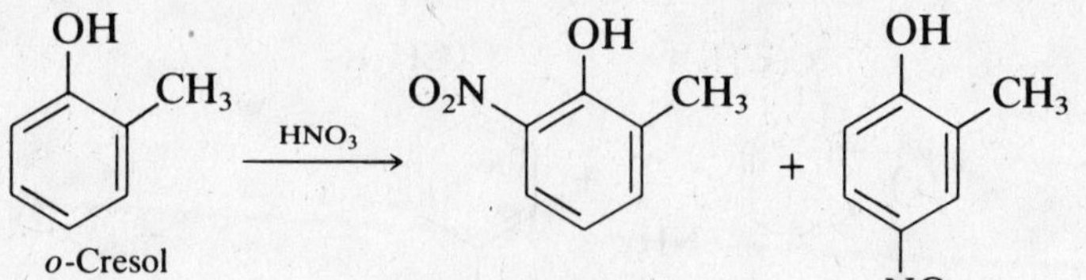

o-Cresol

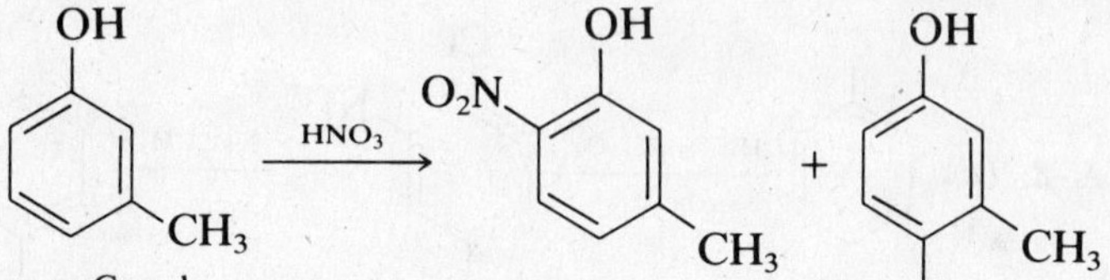

m-Cresol

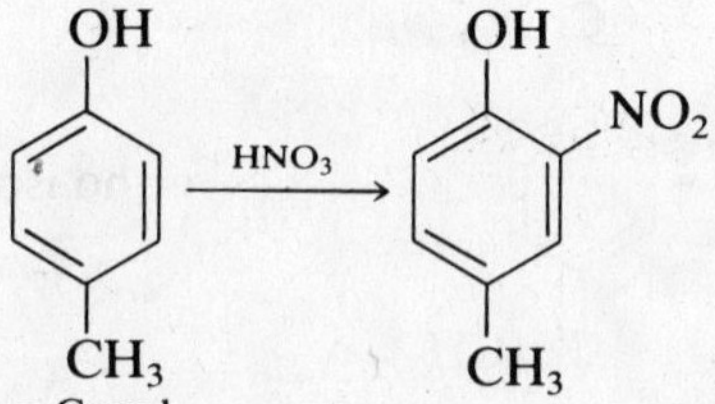

p-Cresol

A-3.

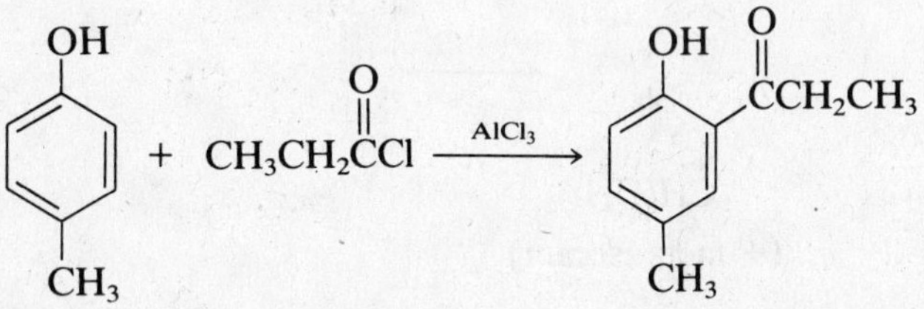

(Friedel-Crafts acylation)

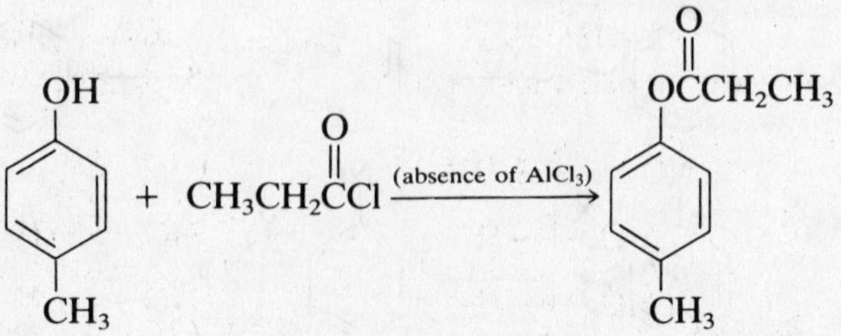

(esterification)

A-4. (*a*) 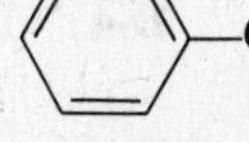 —OH + BrCH(CH₃)₂

(b)

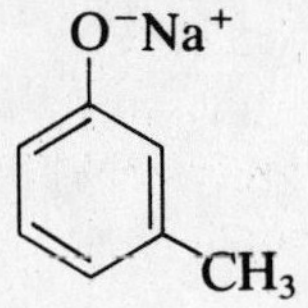

$+ BrCH_2CH(CH_3)_2$

(c) CO_2, 125°C, 100 atm

(d)

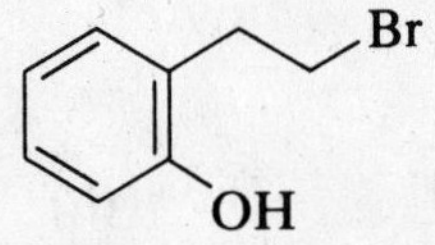

A-5.

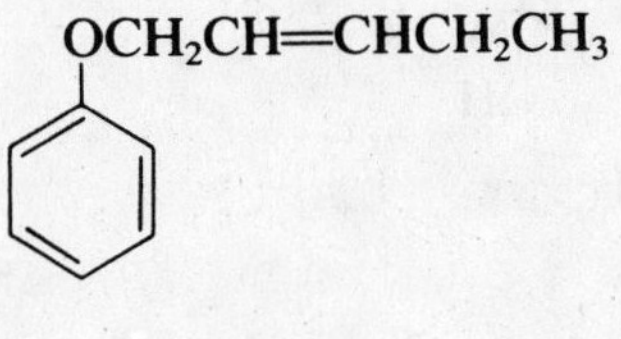

A

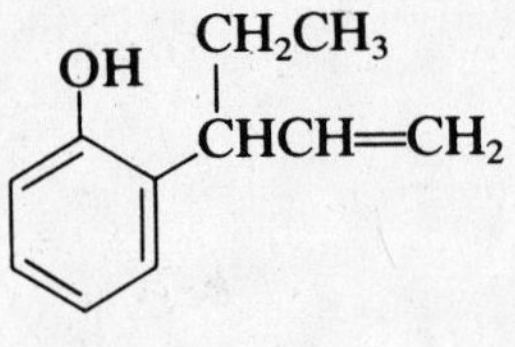

B

B-1. (d)　　**B-2.** (c)　　**B-3.** (b)　　**B-4.** (a)

B-5. (c)

CHAPTER 25

A-1. (a)

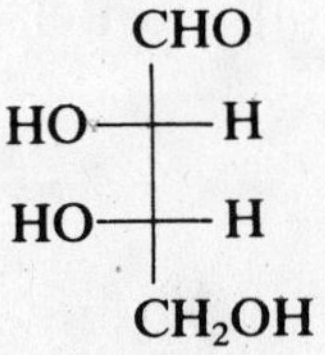

L-Erythrose

(b)

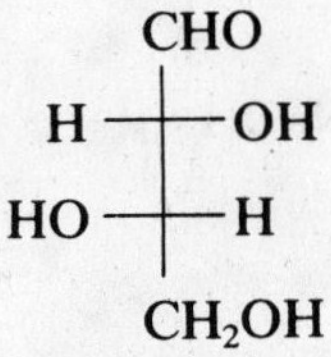

or

L-Threose　　　　　　D-Threose

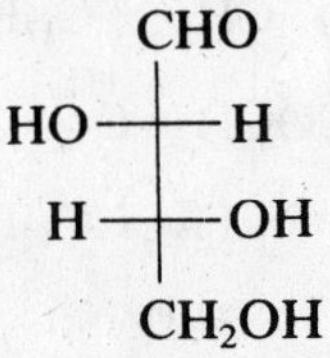

(c)

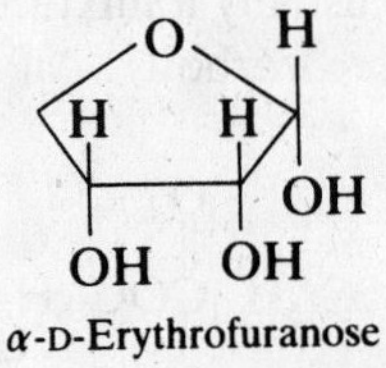

α-D-Erythrofuranose

(d)

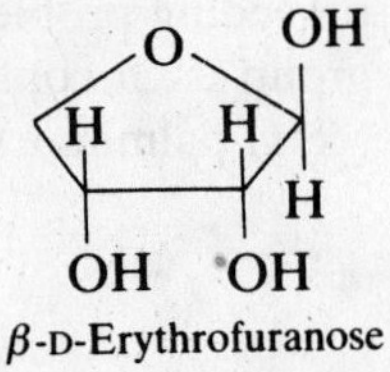

β-D-Erythrofuranose

A-2. (a)

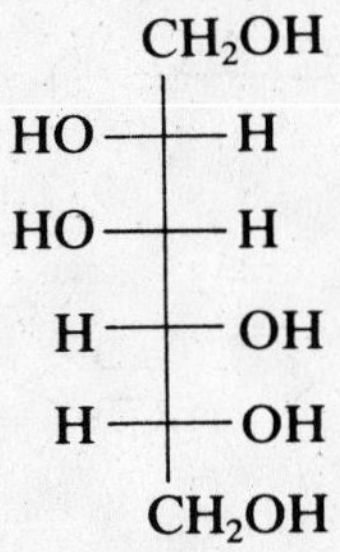

(b)

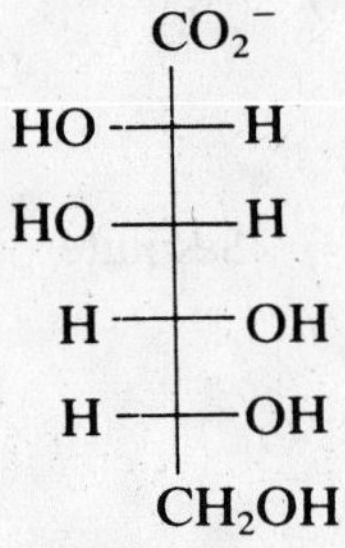

(c)

$$CH{=}NNHC_6H_5$$

$$C{=}NNHC_6H_5$$

HO—+—H

H—+—OH

H—+—OH

$$CH_2OH$$

(d) $$4HCO_2H + H_2C{=}O$$

A-3. (a)

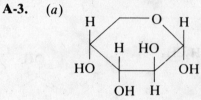

(b)

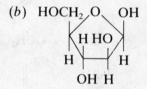

(c) HO

A-4. β-D-Idopyranose (β-pyranose form of D-idose)

B-1. (b)	**B-2.** (d)	**B-3.** (b)	**B-4.** (a)
B-5. (c)	**B-6.** (c)	**B-7.** (a)	**B-8.** (a)

CHAPTER 26

A-1.

$$C_{17}H_{35}\overset{O}{\overset{\|}{C}}O\overset{\begin{array}{c}O\\\|\\CH_2O\overset{O}{\overset{\|}{C}}C_{17}H_{35}\end{array}}{\underset{CH_2O\overset{O}{\overset{\|}{C}}C_{17}H_{35}}{CH}} + 3NaOH \longrightarrow \begin{array}{c}CH_2OH\\CHOH\\CH_2OH\end{array} + 3C_{17}H_{35}CO_2^- Na^+$$

Tristearin

A-2. Fats are triesters of glycerol. A typical example is tristearin, shown in the preceding problem. A wax is usually a mixture of esters in which the alkyl and acyl group each contain 12 or more carbons. An example is hexadecyl hexadecanoate (cetyl palmitate):

$$C_{15}H_{31}\overset{O}{\overset{\|}{C}}OC_{16}H_{33}$$

A-3. (a) Monoterpene;

(b) Sesquiterpene;

(*c*) Diterpene;

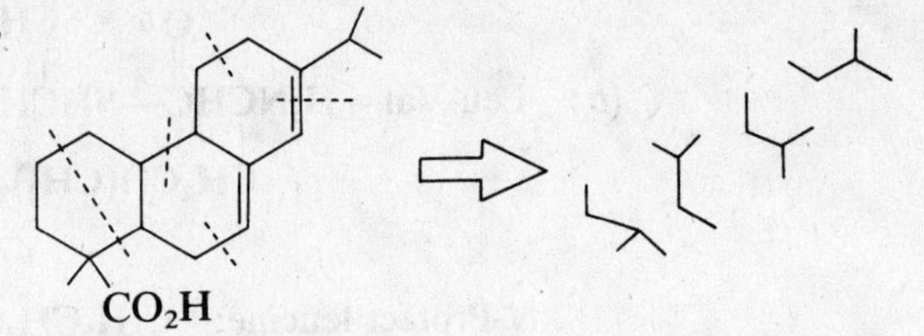

A-4. $CH_3(CH_2)_7$ ... $(CH_2)_7CO_2H$ (C=C with H, H) Oleic acid $\Rightarrow$ $HC\equiv C(CH_2)_7CH$ (dioxolane)

$HC\equiv C(CH_2)_7CH$ (dioxolane) $\xrightarrow{NaNH_2}$ $Na^{+-}:C\equiv C(CH_2)_7CH$ (dioxolane) $\xrightarrow{CH_3(CH_2)_6CH_2Br}$

$CH_3(CH_2)_7C\equiv C(CH_2)_7CH$ (with O double bond) $\xleftarrow{H_3O^+}$ $CH_3(CH_2)_7C\equiv C(CH_2)_7CH$ (dioxolane)

$\downarrow$ 1. $AgNO_3$, HO^-, H_2O
2. H^+

$CH_3(CH_2)_7C\equiv C(CH_2)_7CO_2H$ $\xrightarrow[\text{Lindlar Pd}]{H_2}$ $CH_3(CH_2)_7$ ... $(CH_2)_7CO_2H$ (C=C with H, H)

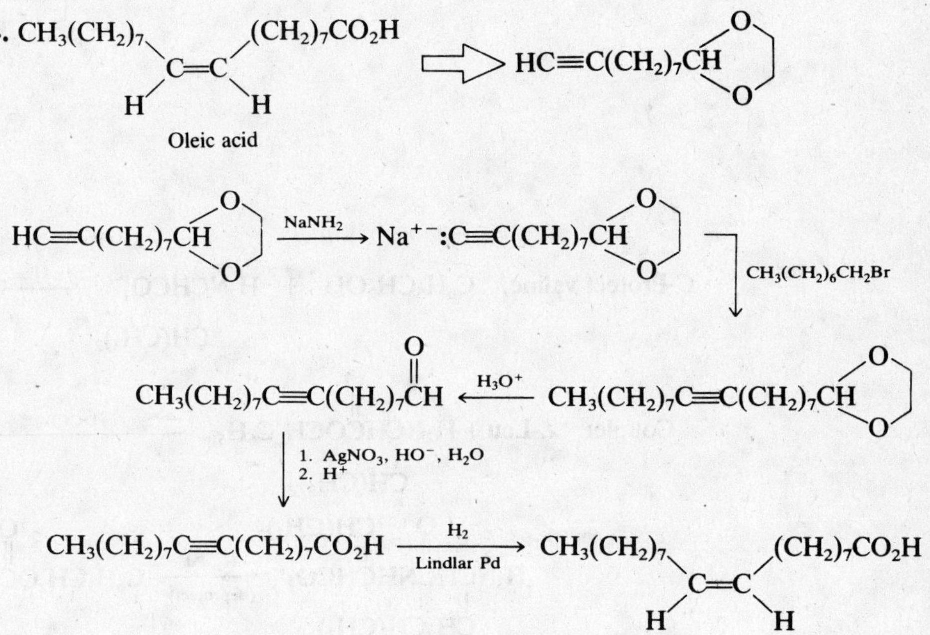

B-1. (*b*)   **B-2.** (*a*)   **B-3.** (*c*)   **B-4.** (*c*)

B-5. (*c*)

CHAPTER 27

A-1. (*a*) $C_6H_5CH_2CH$ (with O)   (*b*) $C_6H_5CH_2OCNHCHCO_2H$ (with O) | $CH(CH_3)_2$

(*c*) DCCI

A-2. O_2N—(benzene ring)—$NHCHCO_2H$ | $CH_2C_6H_5$ with NO_2

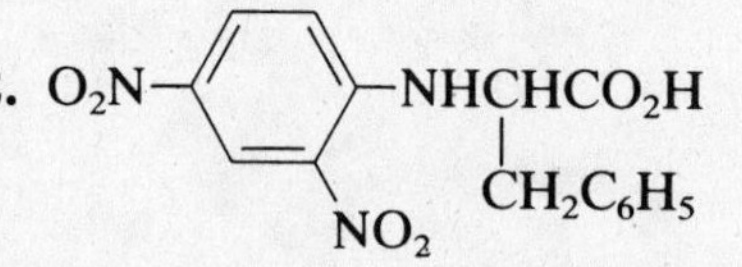

A-3. (*a*) $CH_3CNHCH(CO_2CH_2CH_3)_2$ (with O) $\xrightarrow[\text{ethanol}]{NaOCH_2CH_3}$ $CH_3CNH\ddot{C}(CO_2CH_2CH_3)_2$ (with O)

$\downarrow$ $(CH_3)_2CHCH_2Br$

$H_3\overset{+}{N}CHCO_2^-$ | $CH_2CH(CH_3)_2$ $\xleftarrow[\text{2. heat}]{\text{1. }H_3O^+}$ $CH_3CNHC(CO_2CH_2CH_3)_2$ (with O) | $CH_2CH(CH_3)_2$

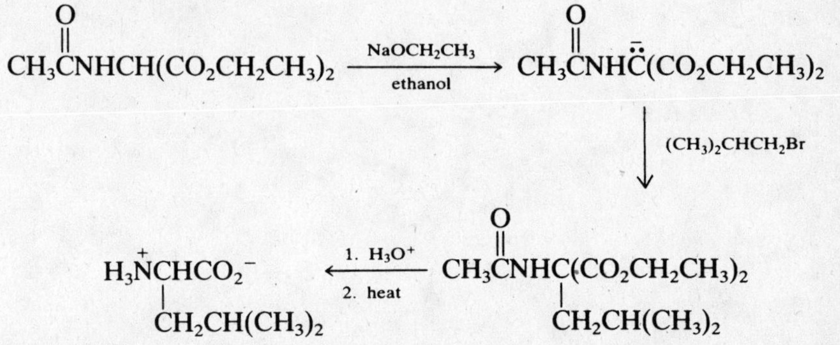

(*b*)   Leu-Val = $\overset{+}{H_3}N\underset{\underset{CH_2CH(CH_3)_2}{|}}{C}H\overset{\overset{O}{\parallel}}{C}-NH\underset{\underset{CH(CH_3)_2}{}}{C}HCO_2^-$

N-Protect leucine: $C_6H_5CH_2O\overset{\overset{O}{\parallel}}{C}Cl + \overset{+}{H_3}N\underset{\underset{CH_2CH(CH_3)_2}{|}}{C}HCO_2^- \quad \xrightarrow[\text{2. } H^+]{\text{1. NaOH, } H_2O}$

$C_6H_5CH_2O\overset{\overset{O}{\parallel}}{C}NH\underset{\underset{CH_2CH(CH_3)_2}{|}}{C}HCO_2H$

(Z-Leu)

C-Protect valine: $C_6H_5CH_2OH + \overset{+}{H_3}N\underset{\underset{CH(CH_3)_2}{|}}{C}HCO_2^- \xrightarrow{H^+} \overset{+}{H_3}N\underset{\underset{CH(CH_3)_2}{|}}{C}H\overset{\overset{O}{\parallel}}{C}OCH_2C_6H_5$

Couple: Z-Leu + $H_2N\underset{\underset{CH(CH_3)_2}{|}}{C}H\overset{\overset{O}{\parallel}}{C}OCH_2C_6H_5 \quad \xrightarrow{\text{DCCl}}$

$\overset{+}{H_3}N\underset{\underset{CH_2CH(CH_3)_2}{|}}{C}H\overset{\overset{O}{\parallel}}{C}NH\underset{\underset{CH(CH_3)_2}{}}{C}HCO_2^- \quad \xleftarrow[\text{(deprotect)}]{H_2,\ Pd} \quad C_6H_5CH_2O\overset{\overset{O}{\parallel}}{C}NH\underset{\underset{CH_2CH(CH_3)_2}{|}}{C}H\overset{\overset{O}{\parallel}}{C}NH\underset{\underset{CH(CH_3)_2}{}}{C}H\overset{\overset{O}{\parallel}}{C}OCH_2C_6H_5$

A-4.   Leu-Val-Gly-Ala-Phe

A-5.   (*a*) Pentapeptide   (*b*) Four   (*c*) Serine
      (*d*) Glycine   (*e*) Ser-Ala-Leu-Phe-Gly

B-1. (*c*)   **B-2.** (*a*)   **B-3.** (*d*)   **B-4.** (*c*)

B-5. (*a*)   **B-6.** (*b*)

TABLES

Table B-1   Bond Dissociation Energies of Some Representative Compounds

Bond	Bond dissociation energy, kJ/mol (kcal/mol)	Bond	Bond dissociation energy, kJ/mol (kcal/mol)
Diatomic molecules			
H—H	435 (104)	H—F	568 (136)
F—F	159 (38)	H—Cl	431 (103)
Cl—Cl	242 (58)	H—Br	366 (87.5)
Br—Br	192 (46)	H—I	297 (71)
I—I	150 (36)		
Alkanes			
CH_3—H	435 (104)	CH_3—CH_3	368 (88)
CH_3CH_2—H	410 (98)	CH_3CH_2—CH_3	355 (85)
$CH_3CH_2CH_2$—H	410 (98)	$(CH_3)_2CH$—CH_3	351 (84)
$(CH_3)_2CH$—H	395 (94.5)	$(CH_3)_3C$—CH_3	334 (80)
$(CH_3)_3C$—H	380 (91)		
Alkyl halides			
CH_3—F	451 (108)	$(CH_3)_2CH$—F	439 (105)
CH_3—Cl	349 (83.5)	$(CH_3)_2CH$—Cl	339 (81)
CH_3—Br	293 (70)	$(CH_3)_2CH$—Br	284 (68)
CH_3—I	234 (56)	$(CH_3)_3C$—Cl	330 (79)
CH_3CH_2—Cl	338 (81)	$(CH_3)_3C$—Br	263 (63)
$CH_3CH_2CH_2$—Cl	343 (82)		
Water and alcohols			
HO—H	497 (119)	CH_3CH_2—OH	380 (91)
CH_3O—H	426 (102)	$(CH_3)_2CH$—OH	385 (92)
CH_3—OH	380 (91)	$(CH_3)_3C$—OH	380 (91)

Note: Bond dissociation energies refer to bonds indicated in structural formula for each substance.

Table B-2   Acid Dissociation Constants

Acid	Formula	Conjugate base	Dissociation constant	pK_a
Hydrogen fluoride	$H—F$	F^-	3.5×10^{-4}	3.5
Acetic acid	$CH_3CO_2—H$	$CH_3CO_2^-$	1.8×10^{-5}	4.7
Hydrogen cyanide	$H—CN$	CN^-	7.2×10^{-10}	9.1
Phenol	$C_6H_5O—H$	$C_6H_5O^-$	1.3×10^{-10}	9.8
Water	$HO—H$	HO^-	1.8×10^{-16}	15.7
Ethanol	$CH_3CH_2O—H$	$CH_3CH_2O^-$	10^{-16}	16
Alkyne (terminal; R = alkyl)	$RC{\equiv}C—H$	$RC{\equiv}C^-$	10^{-26}	26
Ammonia	$NH_2—H$	NH_2^-	10^{-36}	36
Alkene C—H	$RCH{=}CH—H$	$RCH{=}CH^-$	10^{-45}	45
Alkane C—H	$RCH_2CH_2—H$	$RCH_2CH_2^-$	10^{-62}	62

Note: Acid strength decreases from top to bottom of the table; conjugate base strength increases from top to bottom.

Table B-3   Chemical Shifts of Representative Types of Protons

Type of proton	Chemical shift (δ), ppm*	Type of proton	Chemical shift (δ), ppm*
$H—C—R$	0.9–1.8	$H—C—NR$	2.2–2.9
$H—C—C{\equiv}C$	1.6–2.6	$H—C—Cl$	3.1–4.1
$H—C—\overset{\displaystyle O}{C}—$	2.1–2.5	$H—C—Br$	2.7–4.1
$H—C{\equiv}C—$	2.5	$H—C—O$	3.3–3.7
$H—C—Ar$	2.3–2.8	$H—NR$	1–3†
$H—C{=}C{<}$	4.5–6.5	$H—OR$	0.5–5†
$H—Ar$	6.5–8.5	$H—OAr$	6–8†
$H—\overset{\displaystyle O}{C}—$	9–10	$H—O\overset{\displaystyle O}{C}—$	10–13†

* These are approximate values relative to tetramethylsilane; other groups within the molecule can cause a proton signal to appear outside of the range cited.

† The chemical shifts of protons bonded to nitrogen and oxygen are temperature- and concentration-dependent.

Table B-4 Chemical Shifts of Representative Carbons

Type of carbon	Chemical shift (δ), ppm*	Type of carbon	Chemical shift (δ), ppm*
RCH_3	0–35	C=C	100–150
R_2CH_2	15–40		
R_3CH	25–50	(benzene ring)	110–175
RCH_2NH_2	35–50		
RCH_2OH	50–65	C=O	190–220
—C≡C—	65–90		

* Approximate values relative to tetramethylsilane.

Table B-5 Infrared Absorption Frequencies of Some Common Structural Units

Structural unit	Frequency, cm^{-1}	Structural unit	Frequency, cm^{-1}
Stretching vibrations			
Single bonds		*Double bonds*	
—O—H (alcohols)	3200–3600	C=C	1620–1680
—O—H (carboxylic acids)	2500–3600	C=O	
N—H	3350–3500	Aldehydes and ketones	1710–1750
sp C—H	3310–3320	Carboxylic acids	1700–1725
sp^2 C—H	3000–3100	Acid anhydrides	1800–1850 and 1740–1790
sp^3 C—H	2850–2950	Acyl halides	1770–1815
		Esters	1730–1750
sp^2 C—O	1200	Amides	1680–1700
sp^3 C—O	1025–1200		
		Triple bonds	
		—C≡C—	2100–2200
		—C≡N	2240–2280
Bending vibrations of diagnostic value			
Alkenes		*Substituted derivatives of benzene*	
Cis-disubstituted	665–730	Monosubstituted	730–770 and 690–710
Trans-disubstituted	960–980	Ortho-disubstituted	735–770
Trisubstituted	790–840	Meta-disubstituted	750–810 and 680–730
		Para-disubstituted	790–840

ADDITIONAL READINGS

The list of articles that follows is intended to be selective rather than comprehensive. It is limited to short accounts in sources available in practically all college and university libraries. These include the *Journal of Chemical Education* (*JCE*), *Scientific American* (*SAM*), *Chemical and Engineering News* (*CEN*), and *Organic Reactions* (*OR*). More extensive treatments of many of the topics described in the text, along with additional references to the chemical literature, may be found in:

F. A. Carey and R. J. Sundberg, *Advanced Organic Chemistry. Part A: Structure and Mechanisms,* 3d ed., Plenum Press, New York, 1990.

F. A. Carey and R. J. Sundberg, *Advanced Organic Chemistry. Part B: Reactions and Synthesis,* 3d ed., Plenum Press, New York, 1990.

J. March, *Advanced Organic Chemistry. Reactions, Mechanisms, and Structure,* 3d ed., Wiley-Interscience, New York, 1985.

Acids and Bases

D. Kolb, "Acids and Bases," *JCE,* vol. **55,** pp. 459–464, 1978.

G. V. Calder and T. J. Barton, "Actual Effects Controlling the Acidity of Carboxylic Acids," *JCE,* vol. **48,** pp. 338–340, 1971.

Alcohols

R. L. Pruett, "Hydroformylation: An Old Yet New Industrial Route to Alcohols," *JCE,* vol. **63,** pp. 196–198, 1986.

A. W. Ingersoll, "The Resolution of Alcohols," *OR,* vol. **2,** pp. 376–414, 1944.

Aldehydes and Ketones

H. A. Wittcoff, "Acetaldehyde: A Chemical Whose Fortunes Have Changed," *JCE,* vol. **60,** pp. 1044–1047, 1983.

T. Mukaiyama, "The Directed Aldol Reaction," *OR,* vol. **28,** pp. 203–332, 1982.

H. Hart and M. Sasaoka, "Simple Enols: How Rare Are They?" *JCE,* vol. **57,** pp. 685–688, 1980.

E. Vedejs, "Clemmensen Reduction of Ketones in Anhydrous Organic Solvents," *OR,* vol. **22,** pp. 401–422, 1975.

B. P. Mundy, "The Synthesis of Fused Cycloalkenones via Annelation Methods," *JCE,* vol. **50,** pp. 110–113, 1973.

H. Salzman, "Arthur Lapworth: The Genesis of Reaction Mechanism," *JCE,* vol. **49,** pp. 750–752, 1972.

M. J. Jorgenson, "Preparation of Ketones from the Reaction of Organolithium Reagents with Carboxylic Acids," *OR*, vol. **18**, pp. 1–98, 1970.

A. T. Nielsen and W. J. Houlihan, "The Aldol Condensation," *OR*, vol. **16**, pp. 1–438, 1968.

G. Jones, "The Knoevenagel Condensation," *OR*, vol. **15**, pp. 204–600, 1967.

A. Maercker, "The Wittig Reaction," *OR*, vol. **14**, pp. 270–490, 1965.

E. D. Bergmann, D. Ginsburg, and R. Pappo, "The Michael Reaction," *OR*, vol. **10**, pp. 179–556, 1959.

D. A. Shirley, "The Synthesis of Ketones from Acid Halides and Organometallic Compounds of Magnesium, Zinc, and Cadmium," *OR*, vol. **8**, pp. 28–58, 1954.

E. Mosettig, "The Synthesis of Aldehydes from Carboxylic Acids," *OR*, vol. **8**, pp. 218–257, 1954.

D. Todd, "The Wolff-Kishner Reduction," *OR*, vol. **4**, pp. 378–422, 1948.

W. S. Johnson, "The Formation of Cyclic Ketones by Intramolecular Acylation," *OR*, vol. **2**, pp. 114–177, 1944.

E. L. Martin, "The Clemmensen Reduction," *OR*, vol. **1**, pp. 155–209, 1942.

Alkanes

R. E. Davies and P. J. Freyd, "$C_{167}H_{336}$ Is the Smallest Alkane with More Realizable Isomers than the Observed Universe Has 'Particles,'" *JCE*, vol. **66**, pp. 278–281, 1989.

H. R. Henze and C. M. Blair, "The Number of Isomeric Hydrocarbons of the Methane Series," *J. Am. Chem. Soc.*, vol. **53**, pp. 3077–3085, 1931.

Alkenes

R. B. Seymour, "Alkenes and Their Derivatives: The Alchemists' Dream Come True," *JCE*, vol. **66**, pp. 670–672, 1989.

C. C. Wamser and L. T. Scott, "The NBS Reaction: A Simple Explanation for the Predominance of Allylic Substitution over Olefin Addition by Bromine at Low Concentrations," *JCE*, vol. **62**, pp. 650–652, 1985.

J. M. Tedder, "Who is Anti-Markovnikov?" *JCE*, vol. **61**, pp. 237–238, 1984.

W. C. Fernelius, H. Wittcoff, and R. E. Varnerin, "Ethylene: The Organic Chemical Industry's Most Important Building Block," *JCE*, vol. **56**, pp. 385–387, 1979.

M. A. Wilson, "Classification of the Electrophilic Addition Reactions of Olefins and Acetylenes," *JCE*, vol. **52**, pp. 495–498, 1975.

W. R. Dolbier, Jr., "Electrophilic Additions to Alkenes," *JCE*, vol. **46**, pp. 342–344, 1969.

N. Isenberg and M. Grdinic, "A Modern Look at Markovnikov's Rule and the Peroxide Effect," *JCE*, vol. **46**, pp. 601–605, 1969.

E. A. Walters, "Models for the Double Bond," *JCE*, vol. **43**, pp. 134–137, 1966.

J. G. Traynham, "The Bromonium Ion," *JCE*, vol. **40**, pp. 392–395, 1963.

G. Zweifel and H. C. Brown, "Hydration of Olefins, Dienes, and Acetylenes via Hydroboration," *OR*, vol. **13**, pp. 1–54, 1963.

D. Swern, "Epoxidation and Hydroxylation of Ethylenic Compounds with Organic Peracids," *OR*, vol. **7**, pp. 378–434, 1953.

Alkylation Reactions

B. Mundy, "Alkylations in Organic Chemistry," *JCE*, vol. **49**, pp. 91–96, 1972.

C. C. Price, "The Alkylation of Aromatic Compounds by the Friedel-Crafts Method," *OR*, vol. **3**, pp. 1–82, 1946.

Alkyl Halides

J. Correia, "On the Boiling Points of the Alkyl Halides," *JCE*, vol. **65**, pp. 62–64, 1988.

Alkynes

T. L. Jacobs, "The Synthesis of Acetylenes," *OR*, vol. **5**, pp. 1–78, 1949.

Amines

W. G. High, "The Gabriel Synthesis of Benzylamine," *JCE*, vol. **52**, pp. 670–671, 1975.

J. S. Wishnok, "Formation of Nitrosamines in Food and in the Digestive System," *JCE*, vol. **54**, pp. 440–441, 1977.

A. C. Cope and E. R. Trumbull, "Olefins from Amines: The Hofmann Elimination Reaction and Amine Oxide Pyrolysis," *OR*, vol. **11**, pp. 317–494, 1960.

A. Roe, "Preparation of Aromatic Fluorine Compounds from Diazonium Fluoroborates: The Schiemann Reaction," *OR*, vol. **5**, pp. 193–228, 1949.

W. S. Emerson, "The Preparation of Amines by Reductive Alkylation," *OR*, vol. **4**, pp. 174–255, 1948.

E. S. Wallis and J. F. Lane, "The Hofmann Reaction," *OR*, vol. **3**, pp. 267–306, 1946.

N. Kornblum, "Replacement of the Aromatic Primary Amino Group by Hydrogen," *OR*, vol. **2**, pp. 262–340, 1944.

Antibiotics

R. K. Robins, "Synthetic Antiviral Agents," *CEN*, pp. 28–40, January 27, 1986.

E. P. Abraham, "The β-Lactam Antibiotics," *SAM*, pp. 76–86, June 1981.

J. Webb, "Ionophores," *JCE*, vol. **56**, pp. 502–503, 1979.

Aromatic Compounds

R. C. Belioli, "The Misuse of the Circle Notation to Represent Aromatic Rings," *JCE*, vol. **60**, pp. 190–191, 1983.

J. G. Traynham, "Aromatic Substitution Reactions: When You've Said Ortho, Meta, and Para, You Haven't Said It All," *JCE*, vol. **60**, pp. 937–941, 1983.

J. F. Bunnett, "The Remarkable Reactivity of Aryl Halides with Nucleophiles," *JCE*, vol. **51**, 312–315, 1974.

E. Berliner, "The Friedel and Crafts Reaction with Aliphatic Dibasic Acid Anhydrides," *OR*, vol. **5**, pp. 229–289, 1949.

C. M. Suter and A. W. Weston, "Direct Sulfonation of Aromatic Hydrocarbons and Their Halogen Derivatives," *OR*, vol. **3**, pp. 141–197, 1946.

Bioorganic Chemistry

M. Farines, R. Soulier, and J. Soulier, "Analysis of the Triglycerides of Some Vegetable Oils," *JCE*, vol. **65**, pp. 464–466, 1988.

G. M. Bodner, "Metabolism Part I: Glycolysis or the Embden-Meyerhoff Pathway," *JCE*, vol. **63**, pp. 566–570, 1986.

G. M. Bodner, "Metabolism Part II: Tricarboxylic Acid (TCA), Citric Acid, or Krebs Cycle," *JCE*, vol. **63**, pp. 673–677, 1986.

G. M. Bodner, "Metabolism Part III: Lipids," *JCE*, vol. **63**, pp. 772–775, 1986.

M. S. Bretscher, "The Molecules of the Cell Membrane," *SAM*, pp. 100–108, October 1985.

S. H. Snyder, "The Molecular Basis of Communication between Cells," *SAM*, pp. 132–141, October 1985.

M. J. Berridge, "The Molecular Basis of Communication within the Cell," *SAM*, pp. 142–152, October 1985.

K. Folkers, "Perspectives from Research on Vitamins and Hormones," *JCE*, vol. **61**, pp. 747–756, 1984.

J. R. Holum, "The Chemical Composition of the Cell," *JCE*, vol. **61**, pp. 877–881, 1984.

S. Krishnamurthy, "The Intriguing Biological Role of Vitamin E," *JCE*, vol. **60**, pp. 465–467, 1983.

W. F. Wood, "Chemical Ecology: Chemical Communication in Nature," *JCE*, vol. **60**, pp. 531–539, 1983.

L. N. Ferguson, "Bio-Organic Mechanisms II: Chemoreception," *JCE*, vol. **58**, pp. 456–461, 1981.

L. N. Ferguson, "Bioactivity in Organic Chemistry Courses," *JCE*, vol. **57**, pp. 46–48, 1980.

D. A. Labianca, "On the Nature of Cyanide Poisoning," *JCE*, vol. **56**, pp. 385–387, 1979.

Cancer

D. A. Labianca, "The Chimney Sweepers' Cancer: An Interdisciplinary Approach to Carcinogenesis," *JCE*, vol. **59**, pp. 843–846, 1982.

L. S. Alexander and H. M. Goff, "Chemicals, Cancer, and Cytochrome P-450," *JCE*, vol. **59**, pp. 179–182, 1982.

P. Rademacher and H.-G. Hilde, "Chemical Carcinogens," *JCE*, vol. **53**, pp. 757–761, 1976.

L. N. Ferguson, "Cancer: How Can Chemists Help?" *JCE*, vol. **52**, pp. 688–694, 1975.

Carbanions and Enolates

A. D. White, "Alexander Borodin: Full-Time Chemist, Part-Time Musician," *JCE*, vol. **64**, pp. 326–327, 1987.

S. Arseniyadis, K. S. Kyler, and D. S. Watt, "Addition and Substitution Reactions of Nitrile-Stabilized Carbanions," *OR*, vol. **31**, pp. 1–364, 1984.

M. W. Rathke, "The Reformatsky Reaction," *OR*, vol. **22**, pp. 423–460, 1975.

T. M. Harris and C. M. Harris, "The γ-Alkylation and γ-Arylation of Dianions of β-Dicarbonyl Compounds," *OR*, vol. **17**, pp. 155–212, 1969.

J. P. Schaefer and J. J. Bloomfield, "The Dieckmann Condensation," *OR*, vol. **15**, pp. 1–203, 1967.

A. C. Cope, H. L. Holmes, and H. O. House, "The Alkylation of Esters and Nitriles," *OR*, vol. **9**, pp. 107–331, 1957.

C. R. Hauser, F. W. Swamer, and J. T. Adams, "The Acylation of Ketones to Form β-Diketones or β-Keto Aldehydes," *OR*, vol. **8**, pp. 59–196, 1954.

R. L. Shriner, "The Reformatsky Reaction," *OR*, vol. **1**, pp. 1–37, 1942.

C. R. Hauser and B. E. Hudson, Jr., "The Acetoacetic Ester Condensation and Certain Related Reactions," *OR*, vol. **1**, pp. 266–302, 1942.

Carbenes

W. E. Parham and E. E. Schweizer, "Halocyclopropanes from Halocarbenes," *OR*, vol. **13**, pp. 55–90, 1963.

M. Jones, Jr., "Carbenes," *SAM*, pp. 101–113, February 1976.

Carbocations

J. G. Traynham, "Carbonium Ion: Waxing and Waning of a Name," *JCE*, vol. **63**, pp. 930–933, 1986.

Carbohydrates

N. K. Mathur and C. K. Narang, "Chitin and Chitosan, Versatile Polysaccharides from Marine Animals," *JCE*, vol. **67**, pp. 938–942, 1990.

N. Sharon, "Carbohydrates," *SAM*, pp. 90–116, November 1980.

C. S. Hudson, "Emil Fischer's Discovery of the Configuration of Glucose," *JCE*, vol. **18**, pp. 353–357, 1941.

Color

J. Alkema and S. L. Seager, "The Chemical Pigments of Plants," *JCE*, vol. **59**, pp. 183–186, 1982.

M. Sequin-Frey, "The Chemistry of Plant and Animal Dyes," *JCE*, vol. **58**, pp. 301–305, 1981.

Computers

H. W. Orf, "Computer-Assisted Instruction in Organic Synthesis," *JCE*, vol. **52**, pp. 464–467, 1975.

Cyclic Compounds

H. E. Simmons, T. L. Cairns, S. A. Vladuchick, and C. M. Hoiness, "Cyclopropanes from Unsaturated Compounds, Methylene Iodide, and Zinc-Copper Couple," *OR*, vol. **20**, pp. 1–132, 1973.

L. N. Ferguson, "Ring Strain and Reactivity of Alicycles," *JCE*, vol. **47**, pp. 46–53, 1970.

Diels-Alder Reaction

E. Ciganek, "The Intramolecular Diels-Alder Reaction," *OR*, vol. **32**, pp. 1–374, 1984.

L. W. Butz and A. W. Rytina, "The Diels-Alder Reaction: Quinones and Other Cyclenones," *OR*, vol. **5**, pp. 136–192, 1949.

M. C. Kloetzel, "The Diels-Alder Reaction with Maleic Anhydride," *OR*, vol. **4**, pp. 1–59, 1948.

H. L. Holmes, "The Diels-Alder Reaction: Ethylenic and Acetylenic Dienophiles," *OR*, vol. **4**, pp. 60–173, 1948.

Free Radicals

F. R. Mayo, "The Evolution of Free Radical Chemistry at the University of Chicago," *JCE*, vol. **63**, pp. 97–99, 1986.

C. Walling, "The Development of Free Radical Chemistry," *JCE*, vol. **63**, pp. 99–102, 1986.

F. W. Stacey and J. F. Harris, Jr., "Formation of Carbon-Hetero Atom Bonds by Free Radical Chain Additions to Carbon-Carbon Multiple Bonds," *OR*, vol. **13**, pp. 150–376, 1963.

Hydrogenation Reactions

A. J. Birch and D. H. Williamson, "Homogeneous Hydrogenation Catalysts in Organic Synthesis," *OR*, vol. **24**, pp. 1–186, 1976.

H. Adkins, "Catalytic Hydrogenation of Esters to Alcohols," *OR*, vol. **8**, pp. 1–27, 1954.

Insect Chemistry

W. R. Stine, "Chemical Communication by Insects," *JCE*, vol. **63**, pp. 603–605, 1986.

G. D. Prestwich, "The Chemical Defenses of Termites," *SAM*, pp. 78–87, August 1983.

W. F. Wood, "The Sex Pheromones of the Gypsy Moth and the American Cockroach," *JCE*, vol. **59**, pp. 35–36, 1982.

Lactones

H. Zaugg, "β-Lactones," *OR*, vol. **8**, pp. 305–363, 1954.

Mass Spectrometry

M. M. Campbell and O. Runquist, "Fragmentation Mechanisms in Mass Spectrometry," *JCE*, vol. **49**, pp. 104–108, 1972.

Molecular Formulas

V. Pellegrin, "Molecular Formulas of Organic Compounds: The Nitrogen Rule and Degree of Unsaturation," *JCE*, vol. **60**, pp. 626–633, 1983.

Molecular Models

Q. R. Petersen, "Some Reflections on the Use and Abuse of Molecular Models," *JCE*, vol. **47**, pp. 24–29, 1970.

Nuclear Magnetic Resonance

D. A. McQuarrie, "Proton Magnetic Resonance Spectroscopy," *JCE*, vol. **65**, pp. 426–433, 1988.

D. W. Brown, "A Short Set of ^{13}C-NMR Correlation Tables," *JCE*, vol. **62**, pp. 209–212, 1985.

R. S. Macomber, "A Primer on Fourier Transform NMR," *JCE*, vol. **62**, pp. 213–214, 1985.

R. G. Shulman, "NMR Spectroscopy of Living Cells," *SAM*, pp. 86–93, January 1983.

I. L. Pykett, "NMR Imaging in Medicine," *SAM,* pp. 78–89, May 1982.

H. C. Dorn, D. G. I. Kingston, and B. R. Simpers, "Interpretation of a ^{13}C Magnetic Resonance Spectrum," *JCE,* vol. **53,** pp. 584–585, 1976.

Nucleic Acids

J.-S. Taylor, "DNA, Sunlight, and Skin Cancer," *JCE,* vol. **67,** pp. 835–841, 1990.

W. M. Scovell, "Supercoiled DNA," *JCE,* vol. **63,** pp. 562–563, 1986.

G. Felsenfeld, "DNA," *SAM,* pp. 58–67, October 1985.

J. E. Darnell, Jr., "RNA," *SAM,* pp. 68–78, October 1985.

A. T. Wood, "Perspectives in Biochemistry: Methods for DNA Sequencing," *JCE,* vol. **61,** pp. 886–889, 1984.

R. E. Dickerson, "The DNA Helix and How It Is Read," *SAM,* pp. 94–111, December 1983.

Nucleophilic Substitution Reactions

D. J. Raber and J. M. Harris, "Nucleophilic Substitution Reactions at Secondary Carbon Atoms," *JCE,* vol. **49,** pp. 60–64, 1972.

Organometallic Compounds

G. W. Parshall and R. E. Putscher, "Organometallic Chemistry and Catalysis in Industry," *JCE,* vol. **63,** pp. 189–191, 1986.

G. B. Kaufmann, "The Discovery of Ferrocene, the First Sandwich Compound," *JCE,* vol. **60,** pp. 185–186, 1983.

G. H. Posner, "Substitution Reactions Using Organocopper Reagents," *OR,* vol. **22,** pp. 253–400, 1975.

G. H. Posner, "Conjugate Addition Reactions of Organo-copper Reagents," *OR,* vol. **19,** pp. 1–113, 1972.

Peptides and Proteins

A. Veca and J. H. Dreisbach, "Classical Neurotransmitters and Their Significance within the Nervous System," *JCE,* vol. **65,** pp. 108–111, 1988.

R. F. Doolittle, "Proteins," *SAM,* pp. 88–99, October 1985.

N. M. Senozan and R. L. Hunt, "Hemoglobin: Its Occurrence, Structure, and Adaptation," *JCE,* vol. **59,** pp. 173–178, 1982.

F. E. Bloom, "Neuropeptides," *SAM,* pp. 148–168, October 1981.

J. F. Henahan, "R. Bruce Merrifield, Designer of Protein-Making Machine," *CEN,* pp. 22–26, August 2, 1971.

N. F. Albertson, "Synthesis of Peptides with Mixed Anhydrides," *OR,* vol. **12,** pp. 157–355, 1962.

Periodic Acid

E. L. Jackson, "Periodic Acid Oxidation," *OR,* vol. **2,** pp. 341–375, 1944.

Pharmaceutical Industry

N. M. Senozan and S. Nasser-Moaddeli, "Organic Nitrates and Nitrites as Heart Drugs," *JCE,* vol. **61,** pp. 674–677, 1984.

Y. Aharonowitz and G. Cohen, "The Microbiological Production of Pharmaceuticals," *SAM,* pp. 140–152, September 1981.

Phase-Transfer Catalysis

J. M. McIntosh, "Phase-Transfer Catalysis Using Quaternary Onium Salts," *JCE,* vol. **55,** pp. 235–238, 1978.

G. W. Gokel and W. P. Weber, "Phase Transfer Catalysis. Part I: General Principles," *JCE,* vol. **55,** pp. 350–354, 1978.

W. P. Weber and G. W. Gokel, "Phase Transfer Catalysis. Part II: Synthetic Applications," *JCE,* vol. **55,** pp. 429–433, 1978.

Phenols

J. Cason, "Synthesis of Benzoquinones by Oxidation," *OR,* vol. **4,** pp. 305–361, 1948.

A. H. Blatt, "The Fries Reaction," *OR,* vol. **1,** pp. 342–369, 1942.

Polymers

T. M. Lechter and N. S. Lutseke, "A Closer Look at Cotton, Rayon, and Polyester Fibers," *JCE,* vol. **67,** pp. 361–363, 1990.

G. B. Kauffman, "Wallace Hume Carothers and Nylon, The First Completely Synthetic Fiber," *JCE,* vol. **65,** pp. 803–808, 1988.

R. B. Seymour, "Polymers Are Everywhere," *JCE,* vol. **65,** pp. 327–334, 1988.

B. L. Goodall, "The History and Current State of the Art of Propylene Polymerization Catalysts," *JCE,* vol. **63,** 191–195, 1986.

J. K. Smith and D. A. Hounshell, "Wallace H. Carothers and Fundamental Research at DuPont," *Science,* vol. **229,** p. 436, 1985.

C. Cummings, "Neoprene and Nylon Stockings: The Legacy of Wallace Hume Carothers," *JCE,* vol. **61,** pp. 241–242, 1984.

J. J. Eisch, "Karl Ziegler: Master Advocate for the Unity of Pure and Applied Research," *JCE,* vol. **60,** pp. 1009–1014, 1983.

F. W. Harris, "Introduction to Polymer Chemistry," *JCE*, vol. **58,** pp. 837–843, 1981.

J. E. McGrath, "Chain Reaction Polymerization," *JCE*, vol. **58,** pp. 844–861, 1981.

Prostaglandins

N. A. Nelson, R. C. Kelly, and R. A. Johnson, "Prostaglandins and the Arachidonic Acid Cascade," *CEN*, pp. 30–44, August 16, 1982.

Reducing Agents

H. C. Brown, "Hydride Reductions: A 40-Year Revolution in Organic Chemistry," *CEN*, pp. 24–29, March 5, 1979.

W. G. Brown, "Reductions by Lithium Aluminum Hydride," *OR*, vol. **6,** pp. 469–510, 1951.

Saccharin

D. S. Tarbell and A. T. Tarbell, "The Discovery of Saccharin," *JCE*, vol. **55,** pp. 161–162, 1978.

Stereochemistry

D. J. Brand and J. Fisher, "Molecular Structure and Chirality," *JCE*, vol. **64,** pp. 1035–1038, 1987.

J. H. Brewster, "Stereochemistry and the Origins of Life," *JCE*, vol. **63,** pp. 667–670, 1986.

D. H. Rouvray, "Predicting Chemistry from Topology," *SAM*, pp. 40–47, September 1986.

J. E. Huheey, "A Novel Method for Assigning *R, S* Labels to Enantiomers," *JCE*, Vol. **62,** pp. 598–600, 1986.

M. P. Aalund and J. A. Pincock, "A Simple Hand Method for Cahn-Ingold-Prelog Assignment of *R* and *S* Configuration to Chiral Carbons," *JCE*, vol. **63,** pp. 600–601, 1986.

K. N. Carter, "L-Cysteine: The (*R*)-Amino Acid from Protein," *JCE*, vol. **63,** pp. 603–605, 1986.

H. Maehr, "A Proposed New Convention for Graphic Presentation of Molecular Geometry and Topography," *JCE*, vol. **62,** pp. 114–120, 1985.

D. L. Mattern, "Fingertip Assignment of Absolute Configuration," *JCE*, vol. **62,** p. 191, 1985.

P. S. Beauchamp, "'Absolutely' Simple Stereochemistry," *JCE*, vol. **61,** pp. 666–667, 1984.

J. J. McCullough, "Diastereomers, Geometrical Isomers, and Rotation about Bonds," *JCE*, vol. **59,** p. 37, 1982.

E. L. Eliel, 'Stereochemical Non-Equivalence of Ligands and Faces (Heterotopicity)," *JCE*, vol. **57,** pp. 52–55, 1980.

J. K. O'Loane, "Optical Activity in Small Molecules, Nonenantiomorphous Crystals, and Nematic Liquid Crystals," *Chem. Rev.*, vol. **80,** pp. 41–61, 1980.

J. W. Cornforth, "Asymmetry and Enzyme Action," *Science,* vol. **193,** pp. 121–125, 1976.

H. A. M. Snelders, "The Reception of J. H. van't Hoff's Theory of the Asymmetric Carbon Atom," *JCE,* vol. **51,** pp. 2–7, 1974.

W. E. Elias, "The Natural Origin of Optically Active Compounds," *JCE,* vol. **49,** pp. 448–454, 1972.

J. E. Abernethy, "The Concept of Dissymmetric Worlds," *JCE,* vol. **49,** pp. 455–461, 1972.

Structure and Bonding

E. Thall, "Representing Isomeric Structures," *JCE,* vol. **68,** pp. 190–191, 1991.

E. R. Talaty, "Keeping Track of Directions of Atomic Orbitals," *JCE,* vol. **67,** pp. 655–657, 1990.

A. D. Baker and M. D. Baker, "A Geometric Method for Determining the Hückel Molecular Orbital Energy Levels of Open-Chain Fully Conjugated Molecules," *JCE,* vol. **61,** p. 770, 1984.

L. Pauling, "G. N. Lewis and the Chemical Bond," *JCE,* vol. **61,** pp. 201–203, 1984.

L. Salem, "Faithful Couple: The Electron Pair," *JCE,* vol. **55,** pp. 344–348, 1978.

R. J. Gillespie, "The Electron-Pair Repulsion Model for Molecular Geometry," *JCE,* vol. **47,** pp. 18–23, 1970.

E. N. Hiebert, "The Experimental Basis of Kekule's Valence Theory," *JCE,* vol. **36,** pp. 320–327, 1959.

H. M. Leicester, "Contributions of Butlerov to the Development of Structural Theory," *JCE,* vol. **36,** pp. 328–329, 1959.

Substituent Effects

L. M. Stock, "The Origin of the Inductive Effect," *JCE,* vol. **49,** pp. 400–404, 1972.

Sulfur Compounds

E. Block, "The Chemistry of Garlic and Onions," *SAM,* pp. 114–119, March 1985.

K. K. Andersen and D. T. Bernstein, "1-Butanethiol and the Striped Skunk," *JCE,* vol. **55,** pp. 159–160, 1978.

Vision

P. S. Zurer, "The Chemistry of Vision," *CEN,* pp. 24–35, November 28, 1983.